PROBABILITY DISTRIBUTION THEORY AND STATISTICAL INFERENCE

PROBABILITY DISTRIBUTION THEORY AND STATISTICAL INFERENCE

Dr K C Bhuyan
Department of Basic Science
Faculty of Science and Information Technology
American International University, Bangladesh
also Author of *Design of Experiments and Sample Survey*
and Multivariate Analysis

New Central Book Agency (P) Ltd

Co-Published by

Bijay Publications

E-mail : bijaypublications@gmail.com

Exclusively Distributed by

Bijay Pustak Bhandar (P) Ltd

90/6A M. G. Road, Bhawani Dutta Lane, Kolkata-700007

✆ : (033) 4605-7856, 8777898576, 8274916104

E-mail : bijaypustakbhandar@gmail.com

NCBA

REGD OFFICE
8/1 Chintamoni Das Lane, Kolkata 700 009, India
email: ncbakolkata@gmail.com
CORPORATE OFFICE
212 Shahpur Jat, New Delhi 110 049
email: delhincba@gmail.com

BIJAY PUBLICATIONS
90/6A M. G. Road, Bhawani Dutta Lane, Kol-7
E-mail : bijaypublications@gmail.com

BIJAY PUSTAK BHANDAR (P) LTD
90/6A M. G. Road, Bhawani Dutta Lane, Kol-7
E-mail : bijaypustakbhandar@gmail.com

PROBABILITY, DISTRIBUTION THEORY AND STATISTICAL INFERENCE • Dr K C Bhuyan

First Published: 2010
Reprinted: 2015, June 2025

PUBLISHER • TYPESETTER • PRINTER
New Central Book Agency (P) Ltd
8/1 Chintamoni Das Lane, Kolkata 700 009

CO-PUBLISHED BY
Bijay Publications
90/6A M G Road, Bhawani Dutta Lane, Kolkata 700007

TECHNICAL EDITOR
Pradip Chowdhury

PROJECT TEAM
Subir Paul, Prabhat Jas

ISBN: 978 81 7381 628 4

Price ₹ 985.00

Contents

Preface

The objective of statistical analysis is to infer about the characteristics of population. The inference is drawn on the basis of sample observations selected from population and using a function of sample observations. The function of sample observations is known as *statistic.* The researcher uses the *statistic and draws probabilistic conclusions regarding the characteristic of population.* The probabilistic measurement of his confidence regarding conclusion is also provided. The probabilistic conclusion and probabilistic measurement of confidence regarding population parameter are based on the theories of probability and distribution.

The distribution theory originates from probability theory. Again, inference regarding parameter is based on distribution theory. Inference has three aspects, viz. (a) estimation of parameters, (b) test of hypothesis, and (c) confidence interval. All three aspects are based on distribution theory.

The objective of this title is to present the different aspects of probability, random variable, probability distribution, sampling distribution, theory of estimation, test of hypothesis including non-parametric methods and confidence interval. The first four chapters are related to probability, random variable, distribution of random variable and properties of random variable. These four topics cover the syllabus of BSc (Hons) students of statistics. These will also be helpful for the undergraduate students of any field who are studying statistics as one of the major subjects. Chapters V–XXXI are related to distribution theory. Among these chapters, the last four cover sampling distribution. Chapters V–XXV cover the probability distribution and its related information including practical applications of some of the distributions. Theory of estimation, test of hypothesis, confidence interval and non-parametric methods are presented in Chapters XXXII–XXXVI. Sufficient number of examples are given in these five chapters.

The theory of probability or distribution theory or theory of estimation are not new topics. Many researchers published their research findings related to these topics in different books and in renowned journals. In this title all published results are presented with new applications. However, I do acknowledge my enormous intellectual debt to all authors and statisticians (quoted in the title) for using their published works. I am also indebted to Mr Amitabha Sen, Director of NCBA (P) Ltd, for his initiative in publishing this book. I am also thankful to the members of his publishing team.

I feel confident that the book will answer the needs of students for whom it is meant. My effort will be truly rewarded if students are benefitted by using this book. I have tried my best to avoid mistakes. However, any suggestion for the improvement of the book will be gratefully accepted and thankfully acknowledged.

15 April 2010

K C Bhuyan

CHAPTER **1**

Probability

1.1 Short History

The results of experiments in the field of social science, agricultural science, biological science and even in the field of physical and mathematical sciences vary from experiment to experiment when the same experiment is repeated over time or by different investigators. A particular result is not expected in each experiment even if the experiment is repeated under essentially in homogeneous conditions. In such a case the investigator may be interested in the proportion of a particular result if the same experiment is repeated under same conditions. This proportion of experimental result may lead the investigator to interpret the overall performance of the experiment. For eamaple, if an unbiased coin is tossed repeatedly under the same situation, the investigator may be interested in estimating the occurrence of proportion of head (H) or proportion of tail (T). This proportion will not be certain in each experimental condition. The result will be uncertain and it will occur at random. The model by which this random result is explained is known as probability model.

The probability theory has been developed to estimate the proportion of uncertainty of experimental result when experiment is conducted at random. Two French mathematicians B. Pascal and P. Fermat have established the base of probability theory during mid 17th century when they worked to solve the uncertainty in the result of a problem faced by Chevallier de Mere. However, theory on probability has been first developed by an Italian mathematician named Girolamo Cardano (1501-1576). Swiss mathematician J. Bernoulli worked for 20 years on probability theory. The other renowned scientists who have worked on probability theory are T. Bayes, P.S. Laplace, Levy, R.A. Fisher, Chebyshev, A. Marcov, Liaponov, Nevenko, Kolmogoroff. The last five are Russian scientists who have enriched the modern probability theory.

1.2 Terms Related to Probability

RANDOM EXPERIMENT. The act which will be ended showing any of the probable result of the act when it is repeated under homogeneous condition. For example, if an unbiased coin is thrown, the probable results are either head (H) or tail (T) in the upper face of the coin. If this coin is thrown repeatedly under homogeneous condition every time, any of the results H or T will occur. This act of throwing coin is a random experiment. Such random experiments are (i) throwing one or more die/dice, (ii) taking ball(s) from a box at random, (iii) selecting a committee of several members from a group of people, (iv) taking one or more cards from a pack of well-shuffled cards, etc.

OUTCOME. All probable results of a random experiment are known as *outcomes* of the experiment. For example, if two dice are thrown, the outcomes are {11, 12, 13, 14, 15, 16, 21, 22, 23, 24, 25, 26, 31, 32, 33, 34, 35, 36, 41, 42, 43, 44, 45, 46, 51, 52, 53, 54, 55, 56, 61, 62, 63, 64, 65, 66}. The outcomes of tossing two coins are

$$\{HH, HT, TH, TT\}.$$

SAMPLE SPACE. The set formed with all probable results of a random experiment is known as *sample space*. It is denoted by S or Ω. For example, if three coins are tossed, then the sample space is

$$S : \{HHH, HHT, HTT, HTH, TTH, THH, THT, TTT\}.$$

The sample space for the random experiment of throwing a die is

$$S : \{1, 2, 3, 4, 5, 6\}.$$

SAMPLE POINT. Each of the outcomes of the sample space is known as *sample point*. For example, (HH) is a sample point in the sample space $S : \{HH, HT, TH, TT\}$. The sample point is (M_1S_1) in the sample space

$$S : \{M_1M_2, M_1S_1, M_1S_2, M_2S_1, M_2S_2, S_1S_2\}$$

when 2 books are selected at random from a box containing 2 mathematics (M) and 2 statistics (S) books.

EXHAUSTIVE CASES. All probable results of a random experiment are known as *exhaustive cases*. If the birth of a boy (B) and a girl (G) of a couple is assumed as random phenomenon, then the exhaustive cases of birth of 3 children will be

$$BBB, BBG, BGG, BGB, GBB, GGB, GBG, GGG.$$

EQUALLY LIKELY CASES. If the chances of occurrence of all exhaustive cases are same, the cases are termed as *equally likely cases*. For example, if an unbiased coin is tossed, the exhaustive cases are H and T. If the chances of H and T are $\frac{1}{2}$, then H and T are equally likely cases. The equally likely cases are denoted by n. Thus, in tossing one unbiased coin, the equally likely cases are $n = 2$.

NOT EQUALLY LIKELY CASES. If the proportions or chances of occurrence of different exhaustive cases are different, the cases are termed *as not equally likely cases*. For example, if two biased coins are tossed once, the exhaustive cases are (HH, HT, TH, TT). If it is known that proportion of occurrence of H is $\frac{2}{3}$ and the same for the occurrence of T is $\frac{1}{3}$, then the proportions of occurrence of HH, HT, TH and TT are $\frac{4}{9}, \frac{2}{9}, \frac{2}{9}$ and $\frac{1}{9}$ respectively. Here the exhaustive cases are not equally likely.

EVENT. Any statement or description or explanation regarding one or more of the sample points is termed as *event.* The event is denoted by A or B or C or .... For example, if two unbiased dice are thrown, the sample space is

$$S: \left\{ \begin{matrix} 11 & 21 & 31 & 41 & 51 & 61 \\ 12 & 22 & 32 & 42 & 52 & 62 \\ 13 & 23 & 33 & 43 & 53 & 63 \\ 14 & 24 & 34 & 44 & 54 & 64 \\ 15 & 25 & 35 & 45 & 55 & 65 \\ 16 & 26 & 36 & 46 & 56 & 66 \end{matrix} \right\}.$$

We can define events based on this sample space as

A : sum of the upper faces is 8 or more,

B : first die shows number even,

C : dice show same number.

FAVOURABLE CASES. The number of sample points in favour of an event are known as *favourable cases* to that event. It is usually denoted by m. Thus, the favourable cases to the event A mentioned above are

$$(26, 35, 36, 44, 45, 46, 53, 54, 55, 56, 62, 63, 64, 65, 66).$$

Here number of favourable cases are $m = 15$. Number of favourable cases to the event B mentioned above is $m = 18$. The cases are

$$\left\{ \begin{matrix} 21 & 22 & 23 & 24 & 25 & 26 \\ 41 & 42 & 43 & 44 & 45 & 46 \\ 61 & 62 & 63 & 64 & 65 & 66 \end{matrix} \right\}.$$

MUTUALLY EXCLUSIVE CASES. If two or more outcomes of a random experiment cannot occur together, the cases are known as *mutually exclusive cases.* For example, in tossing a coin H and T cannot appear at the same time as upper face. So, H and T are mutually exclusive cases.

MUTUALLY EXCLUSIVE EVENTS. If two or more events defined on the basis of a sample space have no common sample point(s), the events are known as *mutually exclusive events.* For example, let us consider the random experiment of selecting one card from a well-shuffled pack of 52 cards and define the events

A : the selected card is an ace (1)

B : the selected card is a king (K).

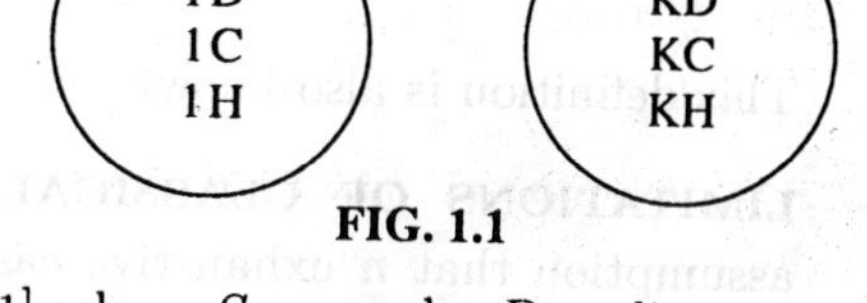

FIG. 1.1

Then the favourable cases of A and B are [FIG. 1.1] where S = spade, D = diamond, C = club, H = heart.

Here A and B have no common sample points; A and B are mutually exclusive events.

NOT MUTUALLY EXCLUSIVE EVENTS. If two or more events defined on a sample space of a random experiment have one or more common sample points, then the events are termed as *not mutually exclusive events.* For example, let

A : the card is an ace, if a card is randomly selected from a well-shuffled pack of 52 cards

and B : the card is a spade.

Then [FIG. 1.2]

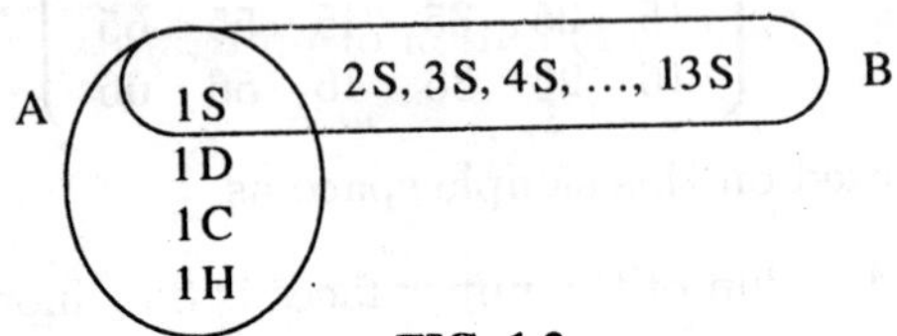

FIG. 1.2

Here A and B have one common point (ace of spade, $1S$). So, A and B are not mutually exclusive events.

VENN DIAGRAM. The diagram which is used to represent the favourable cases to an event is known as *Venn diagram.* For example, let A be the event that indicates same number when two dice are thrown. The favourable cases of A can be shown as follows [FIG. 1.3]:

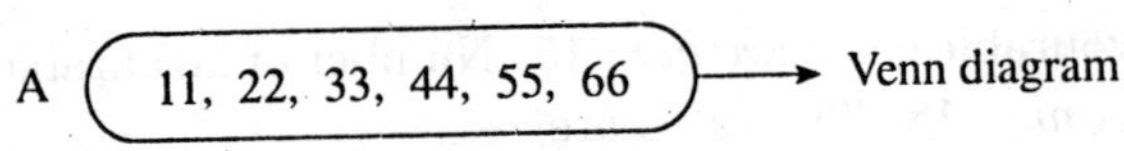

FIG. 1.3

Moreover, the diagram showing the favourable cases of not mutually exclusive events A and B above is also a Venn diagram.

PROBABILITY. Two different definitions are usually used to measure the probability. These are:

(a) Classical definition of probability (Mathematical definition),

(b) Definition based on information (Statistical definition).

CLASSICAL DEFINITION OF PROBABILITY. If an experiment ends in n mutually exclusive, exhaustive and equally likely ways and if $m(\leq n)$ of these ways are in favour of an event A, then the probability of A is given by

$$P(A) = \frac{m}{n}.$$

This definition is also known as *a priori definition of probability.*

LIMITATIONS OF CLASSICAL DEFINITION. The definition is based on the assumption that n exhaustive cases are equally likely also. But all exhaustive cases may not be equally likely. For example, if two biased coins are tossed once, the sample space is

$$S : \{HH, HT, TH, TT\}.$$

Here exhaustive cases are 4, but if it is known that $H : T :: 1 : 3$, then the cases are not equally likely. The different outcomes have different probabilities, viz.

$$P(HH) = \frac{1}{16}, P(HT) = \frac{3}{16}, P(TH) = \frac{3}{16} \text{ and } P(TT) = \frac{9}{16}.$$

In such a case if the event A is defined using outcomes of H and T in two coins, then

$$P(A) = P(HT) + P(TH) = \frac{3}{16} + \frac{3}{16} = \frac{6}{16}.$$

Here we cannot apply the classical definition of probability

$$\left[P(A) = \frac{m}{n}\right].$$

Moreover, this definition cannot be applied if exhaustive cases are infinite and/or unknown.

DEFINITION BASED ON INFORMATION. In explaining the limitations of classical definition of probability, it has already been mentioned that the exhaustive cases may not be equally likely. In that case the probability can be calculated using the information of the experiment. Moreover, the exhaustive cases may not be known. For example, the probability of success and failure in developing a program by a trained personnel may not be equally likely. Usually, we can assume that the probability 'P' of successful development of a program is more. If A is the event that the program will be developed, then $P(A) = P$. Let us consider that the same programmer has developed similar programs under homogeneous conditions and every time the ratio of favourable cases to total cases to the event A tends to P and this ratio will be more nearer to P if the number of trials is increased.

The phenomenon may be explained by another example. Let us consider that an unbiased die is thrown once and A be the event that the upper face of the die is even number. Then

$$P(A) = \frac{1}{2}.$$

Again, consider that the die is thrown 20 times and even number in the upper face is observed 8 times. Then

$$P(A) = \frac{8}{20} = 0.4.$$

This probability is

$$P(A) = \frac{45}{100} = 0.45,$$

if the die is thrown 100 times and even number in the upper face is observed 45 times. The same probability is 0.48 if even number is observed for 240 times when the die is thrown 500 times. It is observed that $P(A)$ tends to the assumed value of $P = 0.5$ when the number of experiments is increased. This fact leads us to explain the statistical definition of probability as follows:

Definition. If an experiment is repeated under essentially homogeneous and identical conditions and if A is any event defined on the basis of outcomes of the repeated trial, the probability of A is the limiting value of the ratio of the number of times the event occurs to the number of trials, where the number of trials is indefinitely large but finite,

$$\text{i.e., } P(A) = \lim_{n\to\infty} \frac{m}{n}.$$

Here,

$$m = \text{number of times event } A \text{ occurs,}$$

$$n = \text{number of repetitions of random experiment.}$$

This definition is also based on relative frequency approach, but relative frequency is calculated objectively by repetitive empirical observations and hence the probability is known as empirical probability.

Limitations of Statistical or Empirical Definition

(a) The relative frequency m/n may not attain a unique value even for a large n.

(b) In case of repetition, the conditions of experimentation may not remain homogeneous or identical.

(c) This probability cannot be calculated in practice.

Besides these two definitions of probability there is another one definition which is known as 'Axiomatic Probability'. The axiomatic definition is due to Russian mathematician A.N. Kolmogorov based on some axioms described below:

Let $S : \{O_1, O_2, O_3, \ldots, O_n\}$ be a sample space such that each of O_i's ($i = 1, 2, \ldots, n$) is an outcome of the experiment. It is also assumed that any of O_i is observed in any of the repetition of the experiment. Let A be an event, then

(i) $P(A) \geq 0$, [$P(A)$ is real and non-negative]

(ii) $P(S) = 1$

(iii) If $A_1, A_2, \ldots, A_k$ are any finite or infinite sequence of disjoint events of S, then

$$P\left(\bigcup_{i=1}^{k} A_i\right) = \sum_{i=1}^{k} P(A_i).$$

Alternatively, let us consider the definition as follows:

Definition. Let us consider the sample S of a random experiment and θ indicates the number of events related to the experiment. Then any set function $P(A)$ for θ is known as probability of the event A if $P(A)$ satisfies the following axioms:

Axiom-1 : $0 \leq P(A) \leq 1$, for any $A \in \theta$

Axiom-2 : $P(S) = 1$

Axiom-3 : If $\{A_k\}$ is the set of disjoint events

$$[A_i \cap A_j = \phi, i \neq j], \text{ then } P\left(\bigcup_{i=1}^{k} = \sum_{i=1}^{k} P(A_i)\right).$$

The subset θ of S is called σ-field if it satisfies the following conditions:

(i) $S \in \theta$.

(ii) If $A \in \theta$, then $\bar{A} \notin \theta$.

(iii) If $A_1, A_2, \ldots \in \theta$, then $\bigcup_{i=1}^{\infty} A_i \in \theta$.

COMPLEMENTARY EVENT. To define mathematical probability it has been mentioned that there are m favourable cases to an event A out of n mutually exclusive, exhaustive and equally likely outcomes of the experiment. The number of outcomes which are not in favour of an event A is $(n - m)$. Then

$$P(\text{Not } A) = \frac{n-m}{n} = 1 - \frac{m}{n} = 1 - P(A).$$

The event not A is denoted by $\bar{A}$ and is known as *complementary event.* It is observed that

$$P(A) + P(\bar{A}) = 1.$$

IMPOSSIBLE EVENT. If any event cannot be defined from a random experiment or if there is no favourable point to an event for an experiment, then the event is called *impossible event.* For an impossible event favourable cases are $m = 0$. If B is an event such as $B : \{\phi\}$, where ϕ is empty set, then

$$P(B) = 0.$$

Here B is called *impossible event.*

Remarks on Mathematical Definition of Probability

It is observed that

$$P(A) = \frac{m}{n} \ (m \leq n).$$

If $m = n$, $P(A) = 1$, then event A is called *sure event.* Again,

$$P \text{ (impossible event)} = 0 \ (m = 0).$$

Thus

(a) $0 \leq P(A) \leq 1$. The exhaustive number of cases and favourable cases are non-negative and hence $P(A) \geq 0$.

(b) Let A be an event defined on a sample space S of a random experiment. If A occurs, it is a success and if A does not occur, it is a failure. In such a case

$$\begin{aligned} P(A) &= P(\text{success}) = p(\text{say}) \text{ and} \\ P(\text{failure}) &= P(\bar{A}) = 1 - p = q \Rightarrow p + q = 1. \end{aligned}$$

►**EXAMPLE 1.1** *In a family there are 3 male and 2 female members. A family work is to be finished by any 2 of them. Find the probability that* (i) *the work will be finished* *by one male and one female member,* (ii) *the work will be finished by two males or by two females.*

Solution.

(i) Let A be the event that the work will be finished by one male and one female members. Since the work is to be finished by any 2 members in the family, it can be finished in $n = {}^5C_2 = 10$ ways. The work can be finished by one male and one female members in $m = {}^3C_1 \times {}^2C_1 = 6$ ways. Therefore,

$$P(A) = \frac{m}{n} = \frac{6}{10} = 0.6.$$

(ii) Since the work is to be finished by 2 of the family members, it can be finished either by one male member and one female member or by two male members or by two female members. The last two statements are complementary to the first statement (A). We need $P(\bar{A})$. Here

$$P(\bar{A}) = 1 - P(A) = 1 - 0.6 = 0.4.$$

►**EXAMPLE 1.2** *Three unbiased coins are tossed once. Find the probability that* (i) *exactly* 2 *heads will appear,* (ii) *there will be no head,* (iii) *at least one head will appear,* (iv) *at best* 2 *heads will appear.*

Solution. The sample space of the experiment is

$$S : \{HHH, HHT, HTT, HTH, TTH, THH, THT, TTT\}.$$

Since the coins are unbiased $[P(H) = 0.5$ and $P(T) = 0.5]$. All cases in sample space are equally likely. Thus, $n = 8$.

(i) Let A be the event that exactly 2 heads will appear. Favourable cases to A are $A : \{HHT, THH, HTH\}, m = 3$.

$$\therefore\ P(A) = \frac{m}{n} = \frac{3}{8}.$$

(ii) Let B be the event that there will be no head. Favourable case to B is $B : \{TTT\}, m = 1$.

$$\therefore\ P(B) = \frac{m}{n} = \frac{1}{8}.$$

(iii) Let C be the event that at least one head will appear. The favourable cases to C are

$$C : \{HHH, HHT, HTT, HTH, TTH, THH, THT\}, m = 7.$$

All cases in the sample space are except the case TTT. Therefore, C is complementary event to B.

$$P(C) = \frac{m}{n} = \frac{7}{8}$$

$$\text{or, } P(C) = P(\bar{B}) = 1 - P(B) = 1 - \frac{1}{8} = \frac{7}{8}.$$

(iv) Let D be the event that at best 2 heads will appear. The favourable cases to D are $D: \{HHT, HTT, HTH, TTH, THH, THT, TTT\}, m = 7$.

$$\therefore\ P(D) = \frac{m}{n} = \frac{7}{8}.$$

Also, find the probabilities of the events A, B, C and D if the coins are biased and

$$H : T :: 2 : 1.$$

Solution. From the given condition $P(H) = \frac{2}{3}, P(T) = \frac{1}{3}$.

(i) $P(A) = P(HHT) + P(THH) + P(HTH)$

$= \frac{2}{3}\cdot\frac{2}{3}\cdot\frac{1}{3} + \frac{1}{3}\cdot\frac{2}{3}\cdot\frac{2}{3} + \frac{2}{3}\cdot\frac{1}{3}\cdot\frac{2}{3} = \frac{12}{27}$.

(ii) $P(B) = P(TTT) = \frac{1}{3}\cdot\frac{1}{3}\cdot\frac{1}{3} = \frac{1}{27}$.

(iii) $P(C) = P(\bar{B}) = 1 - \frac{1}{27} = \frac{26}{27}$.

(iv) Let E be an event that there will be 3 heads.

Then, $P(E) = P(HHH) = \frac{2}{3}\cdot\frac{2}{3}\cdot\frac{2}{3} = \frac{8}{27}$.

Here, $P(D) = P(\bar{E}) = 1 - P(E) = 1 - \frac{8}{27} = \frac{19}{27}$.

►**EXAMPLE 1.3** *Two unbiased dice are thrown once. Find the probability that* (i) *both dice show same number,* (ii) *first die shows even number,* (iii) *both dice show even number,* (iv) *sum of the upper faces of the dice is* 10 *or more,* (v) *sum of the upper faces of the dice is above* 10, (vi) *sum of the upper faces of the dice is less than* 7, (vii) *second die shows number* 5 *or more.*

Solution. The sample space of the experiment is

$$S : \left\{\begin{array}{cccccc} 11 & 21 & 31 & 41 & 51 & 61 \\ 12 & 22 & 32 & 42 & 52 & 62 \\ 13 & 23 & 33 & 43 & 53 & 63 \\ 14 & 24 & 34 & 44 & 54 & 64 \\ 15 & 25 & 35 & 45 & 55 & 65 \\ 16 & 26 & 36 & 46 & 56 & 66 \end{array}\right\}.$$

As the dice are unbiased, all 36 outcomes are equally likely ($n = 36$).

(i) Let A be the event that both dice show same number. The favourable cases to A are $A : \{11, 22, 33, 44, 55, 66\}, m = 6$.

$$\therefore\ P(A) = \frac{m}{n} = \frac{6}{36} = \frac{1}{6}.$$

(ii) Let B be the event that first die shows even number. Favourable cases to B are

$$B : \left\{\begin{array}{cccccc} 21 & 22 & 23 & 24 & 25 & 26 \\ 41 & 42 & 43 & 44 & 45 & 46 \\ 61 & 62 & 63 & 64 & 65 & 66 \end{array}\right\},\ m = 18.$$

$$\therefore\ P(B) = \frac{m}{n} = \frac{18}{36} = \frac{1}{2}.$$

(iii) Let C be the event that both dice show even numbers. Favourable cases to C are

$$C : \begin{Bmatrix} 22 & 42 & 62 \\ 24 & 44 & 64 \\ 26 & 46 & 66 \end{Bmatrix}, \quad m = 9.$$

$$\therefore P(C) = \frac{m}{n} = \frac{9}{36} = \frac{1}{4}.$$

(iv) Let D be the event that sum of the upper faces is 10 or more. Favourable cases to D are $D : \{46, 55, 56, 64, 65, 66\}$, $m = 6$.

$$\therefore P(D) = \frac{m}{n} = \frac{6}{36} = \frac{1}{6}.$$

(v) Let E be the event that sum of the upper faces is above 10. Favourable cases to E are $E : \{56, 65, 66\}$, $m = 3$.

$$\therefore P(E) = \frac{m}{n} = \frac{3}{36} = \frac{1}{12}.$$

(vi) Let F be the event that sum of the upper faces is less than 7. Favourable cases to F are $F : \{11, 12, 13, 14, 15, 21, 22, 23, 24, 31, 32, 33, 41, 42, 51\}$, $m = 15$.

$$\therefore P(F) = \frac{m}{n} = \frac{15}{36} = \frac{5}{12}.$$

(vii) Let G be the event that second die shows number 5 or more. Favourable cases to G are

$$G : \begin{Bmatrix} 15 & 25 & 35 & 45 & 55 & 65 \\ 16 & 26 & 36 & 46 & 56 & 66 \end{Bmatrix}, \quad m = 12.$$

$$\therefore P(G) = \frac{m}{n} = \frac{12}{36} = \frac{1}{3}.$$

►**EXAMPLE 1.4** *An industry supplies sunflower oil in plastic container. It is known from previous experience that* 80% *of the container are found in good condition. Three containers are selected one by one at random from a day's output. Find the probability that* (i) *all containers are good,* (ii) *one container is good,* (iii) *at least two containers are good.*

Solution. Let G : good container; B : bad container.

Given $P(G) = 0.8$, $P(B) = 1 - 0.8 = 0.2$.

The sample space of the experiment is

$$S : \{GGG, GGB, GBB, GBG, BBG, BGG, BGB, BBB\}.$$

(i) Let A be the event that all containers are good.

$$\therefore P(A) = P(GGG) = 0.8 \times 0.8 \times 0.8 = 0.512.$$

(ii) Let B be the event that one container is good.

$$\therefore\ P(B) = P(BBG) + P(BGB) + P(GBB)$$
$$= 0.2 \times 0.2 \times 0.8 + 0.2 \times 0.8 \times 0.2 + 0.8 \times 0.2 \times 0.2 = 0.096.$$

(iii) Let C be the event that at least 2 containers are good.

$$\therefore\ P(C) = P(GGG) + P(GGB) + P(GBG) + P(BGG)$$
$$= 0.8 \times 0.8 \times 0.8 + 0.8 \times 0.8 \times 0.2 + 0.8 \times 0.2 \times 0.8 + 0.2 \times 0.8 \times 0.8$$
$$= 0.896.$$

►**EXAMPLE 1.5** *In a class there are 50 students having roll number 1 to 50. One student is selected at random. Find the probability that the selected student bears the number multiple of 3.*

Solution. One student from 50 students can be selected in $n = {}^{50}C_1 = 50$ ways. Let A be the event that the roll number of the selected student is multiple of 3. Favourable cases to A are $A : \{3, 6, 9, 12, 15, 18, 21, 24, 27, 30, 33, 36, 39, 42, 45, 48\}$, $m = 16$.

$$\therefore\ P(A) = \frac{m}{n} = \frac{16}{50} = \frac{5}{25}.$$

►**EXAMPLE 1.6** *In a town there are 4 hospitals. Four patients have gone to the town for treatment. Find the probability that (i) four patients visit same hospital, (ii) four patients visit 4 different hospitals, (iii) four patients visit any 2 hospitals.*

Solution. Four patients can visit 4 hospitals in $n = 4^4 = 256$ ways.

(i) Let A be the event that four patients visit same hospital. This can be done in $m = 4$ ways.

$$\therefore\ P(A) = \frac{m}{n} = \frac{4}{256} = \frac{1}{64}.$$

(ii) Let B be the event that 4 patients visit 4 hospitals. This can be done in $m = 4! = 24$ ways.

$$\therefore\ P(B) = \frac{m}{n} = \frac{24}{256} = \frac{3}{32}.$$

(iii) Let C be the event that 4 patients visit any 2 hospitals. If the patients visit any 2 hospitals, either one of them can visit one hospital and other 3 can visit any other hospital or 2 patients can visit one hospital and other 2 can visit any other hospital. One or two patients can visit a definite hospital in 30 ways. As there are 4 hospitals, the stated visit can be performed in $m = 30 \times 4 = 120$ ways.

$$\therefore\ P(C) = \frac{m}{n} = \frac{120}{256} = \frac{15}{32}.$$

►**EXAMPLE 1.7** *There are 3 traffic signals on the way from a place P_1 to place P_2 in a city. The green light in the first signal continues for 2 minutes out of 5 minutes,* *the green light is observed for* 1 *minute after every* 3 *minutes in the second signal, the same is observed for* 30 *seconds after every* 2 *minutes. Find the probability that a car in its way from* P_1 *to* P_2 (i) *will not be stopped in any signal,* (ii) *will be stopped at least in one signal,* (iii) *will be stopped in* 3 *signals,* (iv) *will be stopped in* 3*rd signal.*

Solution. The car will move if there is green light (G) in the signal and it will stopped if there is red light (R) in the signal. The sample space of lights in 3 traffic signals is

$$S : \{G_1G_2G_3, G_1G_2R_3, G_1R_2R_3, G_1R_2G_3, R_1R_2G_3, R_1G_2G_3, R_1G_2R_3, R_1R_2R_3\}.$$

Here suffix 1 or 2 or 3 is used to indicate first or second or third, respectively signals. From the given condition

$$P(G_1) = \frac{2}{5}, P(G_2) = \frac{1}{3}, P(G_3) = \frac{1}{4}$$
$$\Rightarrow P(R_1) = \frac{3}{5}, P(R_2) = \frac{2}{3}, P(R_3) = \frac{3}{4}.$$

(i) Let A be the event that the car will not be stopped in any of the signals. Favourable cases to A is $A : \{G_1G_2G_3\}$.

$$\therefore\ P(A) = P(G_1G_2G_3) = \frac{2}{5} \times \frac{1}{3} \times \frac{1}{4} = \frac{1}{30}.$$

(ii) Let B be the event that the car will be stopped at least in one signal. Favourable cases to B are all the sample points except the point $G_1G_2G_3$. This sample point is in favour of A. Thus B is $\bar{A}$.

$$\therefore\ P(B) = 1 - P(A) = 1 - \frac{1}{30} = \frac{29}{30}.$$

(iii) Let C be the event that the car will be stopped in 3 signals. Favourable cases to C is $C : \{R_1R_2R_3\}$.

$$\therefore\ P(C) = P(R_1R_2R_3) = \frac{3}{5} \times \frac{2}{3} \times \frac{3}{4} = \frac{3}{10}.$$

(iv) Let D be the event that the car will be stopped in third signal. Favourable cases to D are $D : \{G_1G_2R_3, G_1R_2R_3, R_1G_2R_3, R_1R_2R_3)\}$.

$$\begin{aligned}\therefore\ P(D) &= P(G_1G_2R_3) + P(G_1R_2R_3) + P(R_1G_2R_3) + P(R_1R_2R_3)\\ &= \frac{2}{5} \times \frac{1}{3} \times \frac{3}{4} + \frac{2}{5} \times \frac{2}{3} \times \frac{3}{4} + \frac{3}{5} \times \frac{1}{3} \times \frac{3}{4} + \frac{3}{5} \times \frac{2}{3} \times \frac{3}{4}\\ &= \frac{3}{4}.\end{aligned}$$

►**EXAMPLE 1.8** *Find the probability that a leap-year has fifty-three Mondays.*

Solution. A leap-year has 366 days. It has 52 full weeks and 2 more days. There will be 53 Mondays if one of the 2 over-days is Monday. The possible outcomes of these two over-days are

(i) Sunday, Monday;	(v) Thursday, Friday;
(ii) Monday, Tuesday;	(vi) Friday, Saturday;
(iii) Tuesday, Wednesday;	(vii) Saturday, Sunday.
(iv) Wednesday, Thursday;	

There are 7 possible outcomes of 2 over-days. Among these 7 outcomes 2 outcomes are in favour of 53 Mondays. Therefore, the required probability is

$$P = \frac{2}{7}.$$

►**EXAMPLE 1.9** *Three letters are to be accommodated in 3 addressed envelopes. If the letters are put at random, what is the probability that these will be put in correct envelopes?*

Solution. Let A be the event that the letters will be put in correct envelopes. The letters can be put in $n = 3! = 6$ ways. But letters can be put correctly in $m = 1$ way.

$$\therefore\ P(A) = \frac{m}{n} = \frac{1}{6}.$$

►**EXAMPLE 1.10** *What is the probability that* (i) *first digit,* (ii) *second digit,* (iii) *both digits will be even if at the first stage one digit from the digits* 1, 2, 3, 4 *and* 5 *is selected at random and at the second stage another one digit from the remaining* 4 *digits is selected at random?*

Solution. According to mentioned conditions of selection, the sample space of the experiment is

$$S : \left\{ \begin{array}{cccc} 12 & 13 & 14 & 15 \\ 21 & 23 & 24 & 25 \\ 31 & 32 & 34 & 35 \\ 41 & 42 & 43 & 45 \\ 51 & 52 & 53 & 54 \end{array} \right\}.$$

The exhaustive cases are 20 and all these cases are equally likely, i.e., $n = 20$.

(i) Let A be the event that the first digit is even. Number of favourable cases to A is $m = 8$.

$$\therefore\ P(A) = \frac{m}{n} = \frac{8}{20} = \frac{2}{5}.$$

(ii) Let B be the event that the second digit is even. Number of favourable cases to B is $m = 8$.

$$\therefore\ P(B) = \frac{m}{n} = \frac{8}{20} = \frac{2}{5}.$$

(iii) Let C be the event that both the digits are even. Number of favourable cases to C is $m = 2$.

$$\therefore\ P(C) = \frac{m}{n} = \frac{2}{20} = \frac{1}{10}.$$

▶**EXAMPLE 1.11** *A company have planned to appoint 2 persons for official work and accordingly an advertisement has been published. Five persons applied for the posts. The applicants are ranked on the basis of their academic performance and experience. The rank is 1 for a candidate who is best in all respects, the rank is 2 for the next best candidate and so on. Two applications have been selected at random. Find the probability that the selected applications will be of* (i) *best two candidates,* (ii) *candidates having rank 1 or 2 and having rank 5,* (iii) *at least one of the best two candidates.*

Solution. The sample space for the experiment is

$$S : \{12, 13, 14, 15, 23, 24, 25, 34, 35, 45\}.$$

(i) Let A be the event that the selected applications are of the best two candidates. Number of favourable case to A is $m = 1$.

$$\therefore\ P(A) = \frac{m}{n} = \frac{1}{10}.$$

(ii) Let B be the event that the application is of a candidate having rank 5 or having rank 1 or 2. Favourable cases to B are $B : \{15, 25\},\ m = 2$.

$$\therefore\ P(B) = \frac{m}{n} = \frac{2}{10} = \frac{1}{5}.$$

(iii) Let C be the event that the applications are of at least one of the best two candidates. Favourable cases to C are

$$C : \{12, 13, 14, 15, 23, 24, 25\},\ m = 7.$$

$$\therefore\ P(C) = \frac{m}{n} = \frac{7}{10}.$$

▶**EXAMPLE 1.12** *Two persons are asked to throw two unbiased dice independently. The condition is that both of them will be rewarded if upper faces of the dice thrown by them show same number. What is the probability of winning the reward by them?*

Solution. If two dice are thrown independently, the possible outcomes are

$$n = 6^2 = 36.$$

These outcomes are equally likely.

Let A be the event that both of them will be rewarded. This is possible if dice shows number $\{11, 22, 33, 44, 55, 66\}$, $m = 6$.

$$\therefore\ P(A) = \frac{m}{n} = \frac{6}{36} = \frac{1}{6}.$$

▶**EXAMPLE 1.13** *Five candidates A, B, C, D and E are planning to contest for M.P. in next national electron. All these 5 candidates have experience of social works for 6, 4, 3, 5 and 2 years, respectively. It is assumed that the chance of any one candidate to be elected is proportional to their years of social work. Find the probability that the candidate will be elected* (i) *who have at least 4 years experience of social work,* (ii) *who have less than 4 years experience.*

Solution.

(i) Let the probabilities of different candidates to be elected are

$$P(A) = P_1, P(B) = P_2, P(C) = P_3, P(D) = P_4 \text{ and } P(E) = P_5.$$

From the assumption, we have

$$P_1 : P_2 : P_3 : P_4 : P_5 :: 6 : 4 : 3 : 5 : 2$$

$$\Rightarrow \frac{P_1}{P_5} = \frac{6}{2}, \frac{P_2}{P_5} = \frac{4}{2}, \frac{P_3}{P_5} = \frac{3}{2}, \frac{P_4}{P_5} = \frac{5}{2}$$

$$\text{or, } P_1 = 3P_5, P_2 = 2P_5, P_3 = \frac{3P_5}{2}, P_4 = \frac{5P_5}{2}$$

The probability of any candidate to be elected is proportional to

$$\left(P_5 + 3P_5 + \frac{4}{2}P_5 + \frac{3}{2}P_5 + \frac{5}{2}P_5\right) = 10P_5.$$

Let H be the event that a candidate has 4 years experience. The candidates A, B and D have such experience. Therefore, the result of proportion in favour of H is proportional to

$$\left(3P_5 + 2P_5 + \frac{15}{2}P_5\right) = \frac{15}{2}P_5.$$

$$\therefore\ P(H) = \frac{\frac{15}{2}P_5}{10P_5} = \frac{15}{20} = \frac{3}{4}.$$

(ii) Let $\bar{H}$ be the event that a candidate has less than 4 years experience.

$$\therefore\ P(\bar{H}) = 1 - P(H) = 1 - \frac{3}{4} = \frac{1}{4}.$$

►**EXAMPLE 1.14** *Three persons are entering a five-storied building. They can go to first floor or second floor or third floor or fifth floor of the building. What is the probability that they will meet in one floor?*

Solution. Let A be the event that all three persons meet in one floor. They can go to any floor in $n = 4^3 = 64$ ways. they can meet in any floor in $m = 4$ ways.

$$\therefore\ P(A) = \frac{m}{n} = \frac{4}{64} = \frac{1}{16}.$$

►**EXAMPLE 1.15** *If the coefficients of the quadratic equation $ax^2 + bx + c = 0$ are determined by throwing an unbiased die, what is the probability of finding real roots of the equation?*

Solution. According to the condition, if

$$b^2 - 4ac \geq 0 \text{ or } b^2 = 4ac,$$

then real roots are available. As the values of a, b and c are observed from the results of die throwing experiment, the values of any one of a, b and c will be either 1 or 2 or 3 or 4 or 5 or 6. The values of a, b and c can occur in $6^3 = 216$ ways. Therefore, $n =$

216 and these numbers are equally likely. The favourable cases to the described event can be observed from the results shown below:

ac	a	c	$4ac$	b	*Number of outcomes* [*under condition* $b^2 \geq 4ac$]
1	1	1	4	2, 3, 4, 5, 6	$5 = 1 \times 5$
2	{1, 2}	{2, 1}	8	3, 4, 5, 6	$8 = 2 \times 4$
3	{1, 3}	{3, 1}	12	4, 5, 6	$6 = 2 \times 3$
4	{1, 4, 2}	{4, 1, 2}	16	4, 5, 6	$9 = 3 \times 3$
5	{1, 5}	{5, 1}	20	5, 6	$4 = 2 \times 2$
6	{1, 6, 3, 2}	{6, 1, 2, 3}	24	5, 6	$8 = 4 \times 2$
7	[ac will not be 7]				
8	{2, 4}	{4, 2}	32	6	$2 = 2 \times 1$
9	3	3	36	6	$1 = 1 \times 1$
Total					**43**

As b can take maximum 6, b^2 will not exceed 36 and hence ac cannot be $10, 11, \ldots$. Therefore, favourable cases to the event mentioned is $m = 43$.

$$\therefore \text{ Required probability} = \frac{m}{n} = \frac{43}{216}.$$

►**EXAMPLE 1.16** *There are 3 white and 4 red balls in an urn. Three persons, one by one, take a ball without replacement. The condition is that one will be rewarded who will take white ball first. What are the probabilities of reward of persons who take ball at first, second and third time, respectively?*

Solution. Let A, B and C be the events that a person will be rewarded first, second and third time, respectively. Since ball is taken by 3 persons one by one without replacement, the sample space is

$$S : \{W, RW, RRW, RRRW, RRRRW\}.$$

Cases in favour of A are $\{W, RRRW\}$.

$$\therefore \ P(A) = P(W) + P(RRRW) = \frac{3}{7} + \frac{4}{7} \cdot \frac{3}{6} \cdot \frac{2}{5} \cdot \frac{3}{4} = \frac{18}{35}.$$

Cases in favour of B are $\{RW, RRRRW\}$.

$$\therefore \ P(B) = P(RW) + P(RRRW) = \frac{4}{7} \cdot \frac{3}{6} + \frac{4}{7} \cdot \frac{3}{6} \cdot \frac{2}{5} \cdot \frac{1}{4} \cdot 1 = \frac{11}{35}.$$

Cases in favour of C are $\{RRW\}$.

$$\therefore\ P(C) = P(RRW) = \frac{4}{7}\cdot\frac{3}{6}\cdot\frac{3}{5} = \frac{6}{35}.$$

►**EXAMPLE 1.17** *Two persons A and B start a game by tossing an unbiased coin. The condition is that he will win the game who can throw head first. If A starts the game, what is the chance of winning by* (i) *A,* (ii) *B?*

Solution. The sample space of the experiment is

$$S : \left\{\begin{array}{cccccc} H & TH & TTH & TTTH & TTTTH & TTTTTH, \ldots \\ A & B & A & B & A & B \end{array}\right\}$$

$$\therefore\ P(A) = P(H) + P(TTH) + P(TTTTH) + \cdots$$

$$= \frac{1}{2} + \frac{1}{2^3} + \frac{1}{2^5} + \cdots = \frac{\frac{1}{2}}{1-\frac{1}{2^2}} = \frac{2}{3}.$$

$$\therefore\ P(B) = 1 - P(A) = 1 - \frac{2}{3} = \frac{1}{3}.$$

►**EXAMPLE 1.18** *There are 3 centres to supply petrol in a petrol pump. In each centre there are 2 pipes. What is the probability that any vehicle can get petrol using any of the pipes?*

Solution. In the petrol pump there are $3 \times 2 = 6$ pipes. Any vehicle can use any of the pipes. Hence, $n = 6$. Let A be the event that a vehicle will get petrol using any of the pipes.

$$\therefore\ P(A) = \frac{1}{6}.$$

►**EXAMPLE 1.19** *If 3 persons are randomly selected, what is the probability that* (i) *the birthday of each of them is Sunday,* (ii) *the birthday of 2 persons is Sunday,* (iii) *the birthday of 2 persons is Sunday and the birthday of one person is Tuesday?*

Solution. The birthday of any person is any day within a week. The permutation of birthday of 3 persons is $n = 7^3$.

(i) Let A be the event that the birthday of 3 persons is Sunday. This can happen in $m = 1$ way.

$$\therefore\ P(A) = \frac{m}{n} = \frac{1}{7^3}.$$

(ii) Let B be the event that the birthday of 2 persons is Sunday. The birthday of 3rd person is any other day except Sunday. This can happen in 6 ways. But Sunday can be birthday of two persons in $3! = 6$ ways. Hence, number of favourable cases to B is $m = 6 \times 6 = 36$.

$$\therefore\ P(B) = \frac{m}{n} = \frac{36}{7^3}.$$

(iii) Let C be the event that the birthday of 2 persons is Sunday and the birthday of another person is Tuesday. Consider that the birthday of first and second person is Sunday and the birthday of third person is Tuesday. This can happen in one way. But this arrangement can occur in 3 ways. Therefore, $m = 3$.

$$\therefore\ P(C) = \frac{m}{n} = \frac{3}{7^3}.$$

►**EXAMPLE 1.20** *In a family there are* 5 *persons. Find the probability that at least* 2 *of them born in the same month.*

Solution. Let us consider the event A that the birth of any two persons does not occur in the same month. Any one can be born in any of the 12 months in a year. Thus, the permutation of birth of 5 persons is $n = 12^5$. If first one born in any month of 12 months, second one can born in the remaining 11 months. Similarly, the birth of 3rd, 4th and 5th persons can take place in 10, 9 and 8 months, respectively. Therefore, the number of favourable cases to A is $m = 12 \times 11 \times 10 \times 9 \times 8 = 95040$.

$$\therefore\ P(A) = \frac{m}{n} = \frac{95040}{12^5}.$$

We need $P(\bar{A})$, where

$$P(\bar{A}) = 1 - P(A) = 1 - \frac{95040}{12^5} = 0.62.$$

►**EXAMPLE 1.21** *In a family there are* 5 *persons. What is the probability that the birthday of at least two of them is same?*

Solution. There are 365 days in a year (if leap-year is not considered). Any day of 365 days can be the birthday of any one of 5 persons. Therefore, the number of outcomes of the birthday is $n = 365^5$. Let A be the event that the birthday of any two of them will not same. Under this condition the birth of first person can take place in any day of 365 days, the birth of second person can take place in any day of the remaining 365 days. Similarly, the birth of third person, fourth person and fifth person can take place in 363, 362 and 361 days, respectively. Therefore, A can occur in $m = 365 \times 364 \times 363 \times 362 \times 361$ ways.

$$\therefore\ P(A) = \frac{m}{n} = \frac{365 \times 364 \times 363 \times 362 \times 361}{365^5}.$$

We need $P(\bar{A})$, where

$$P(\bar{A}) = 1 - P(A) = 1 - \frac{365 \times 364 \times 363 \times 362 \times 361}{365^5}.$$

1.3 Mathematical Preliminaries of Probability

1.3.1 Method of Counting

The mathematical definition of probability is based on counting of exhaustive, mutually exclusive, equally likely outcomes of a random experiment and on favourable outcomes of an event defined on the sample space observed from the experiment. In small-scale random experiment the sample space can be shown and counting of different cases is easier. However, sample space is not easier to show in large-scale experiment and/or in case when outcomes are infinite. Thus, we need some shortcut method of counting exhaustive cases and favourable cases. Two important short cut methods of counting are (i) permutation and (ii) combination. Besides these two methods, there is a fundamental rule of counting which is described below:

Fundamental Rule of Counting

Let us consider that an operation can be done in n_1 ways and another operation can be made in n_2 ways, where the second operation is done on the basis of each of first operation. Thus the two operations together can be performed in $n_1 \times n_2$ ways. For example, let us consider that a statistical analysis needs to be performed. The investigator can develop a program for the analysis or he can use a package program for the purpose. After selection of program, he can use main-frame computer or a PC to do the analysis. In such a case he has the $n = 2 \times 2$ options to perform the analysis. If a person needs to observe the dots in throwing a die, he has the chance of observing either of the dots $1, 2, 3, 4, 5$ and 6. But if 2 dice are thrown once, the investigation has the chance of observing $n = 6^2$ combination of dots of $1, 2, 3, 4, 5$ and 6. The combinations of H and T in tossing n coins are $n = 2^n$, where either H or T $(n = 2)$ is observed in tossing a coin once. A man plans to visit a historical place in a city. First he needs to go to city centre from his home town. He has two options in choosing a transport from his home town to city centre. These are bus and train. After reaching the city centre he can choose three different types of vehicles to reach historical place. These vehicles are bus, taxi, baby taxi. Thus the route of movement of the person can be shown in FIG. 1.4.

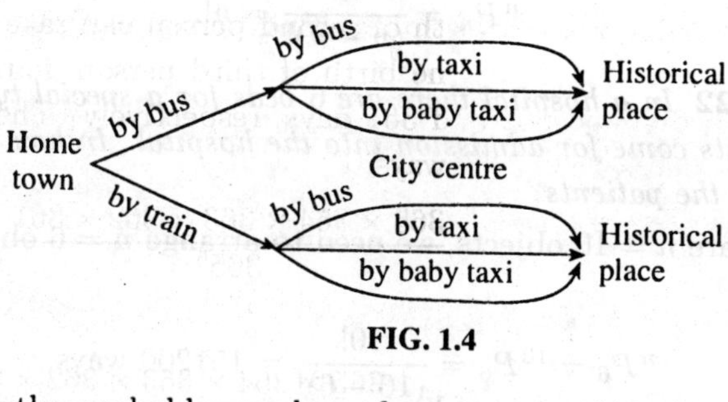

FIG. 1.4

It is observed that the probable number of ways to reach historical place from the residence of the person is $n = 2 \times 3 = 6$.

1.3.2 Factorial

It is a counting rule of occurrence of some sort of action which is specially used in permutation and combination. If we take the product of first n natural numbers, we call it factorial $n(n!)$, where $n! = 1 \times 2 \times 3 \times \cdots \times n$. For example, if $n = 5$, then

$$5! = 1 \times 2 \times 3 \times 4 \times 5 = 120.$$

We have

$$n! = n(n-1)(n-2)(n-3)\cdots 3\times 2\times 1$$
$$= (n-1)!\,n \text{ or, } (n-1)! = \frac{n!}{n}.$$

If $n = 1$, then

$$0! = \frac{1!}{1} = 1.$$

However, rule of factorial is not used for negative integer(s).

1.3.3 Permutation

It is a technique of arrangement of r different objects taken together from n distinguished objects. The total number of such arrangements is denoted by ${}^nP_r$, where

$${}^nP_r = \frac{n!}{(n-r)!}.$$

For example, let there be 3 elements A, B and C. One can arrange 2 of them taken together in ${}^3P_2$ ways.

Here

$${}^3P_2 = \frac{3!}{(3-2)!} = 1\times 2\times 3 = 6.$$

The arrangements are $\{AB, BA, AC, CA, BC, CB\}$. If n objects are arranged all together, the total number of permutations is

$${}^nP_n = \frac{n!}{(n-n)!} = n!.$$

►**EXAMPLE 1.22** *In a hospital there are* 6 *beds for a special type of patients. One day* 10 *such patients come for admission into the hospital. In how many ways the beds can be assigned to the patients?*

Solution. There are $n = 10$ objects, we need to arrange $n = 6$ objects together. This can be done in

$${}^nP_6 = {}^{10}P_6 = \frac{10!}{(10-6)!} = 151200 \text{ ways.}$$

►**EXAMPLE 1.23** *In a box there are* 8 *books. The books are to be arranged in two shelves each of which can contain* 4 *books. Find the number of arrangements of the books.*

Solution. Let us consider that the books are numbered by $1, 2, \ldots, 8$. Out of these books 4 books are to be arranged in a shelf. This can be done in

$${}^8P_4 = \frac{8!}{(8-4)!} = 1680 \text{ ways.}$$

The other rules of permutation are:

RULE 1. The number of different permutations of n different objects taken r at a time with repetition is ${}^nP_r = n^r$.

In particular, ${}^nP_n = n^n$. If $r = 2$, number of arrangements is n^2.

RULE 2. The number of arrangements of n different objects taken all together round a circle is $(n-1)!$.

RULE 3. The number of permutations of n objects taken all at a time, when n_1 objects are alike of one kind, n_2 objects are alike of second kind, ... , n_k objects are alike of k-th kind is

$$\frac{n!}{n_1!\ n_2!\ \cdots\ n_k!}.$$

COROLLARY 1. If the same object is used as many times as needed, then r objects can be chosen from n different objects to fill up r places in n^r ways.

►**EXAMPLE 1.24** *In a show-room there are 10 cars of same model of a company. But 10 cars are of 10 different colours. In how many ways these cars can be arranged in the show-room?*

Solution. Since the cars are of 10 different colours, these can be arranged in

$$10! = 3628800 \text{ ways.}$$

►**EXAMPLE 1.25** *In a departmental store there are 4 sales representatives (M_1, M_2, M_3, M_4) and two cash counters (C_1, C_2). In how many ways the representatives can be assigned in cash counters?*

Solution. Total arrangements is

$$n = 2 \times 3! = 12 \text{ ways.}$$

These arrangements are

$$\begin{array}{lcllllllllllll} C_1 & : & M_1 & M_1 & M_1 & M_2 & M_2 & M_2 & M_3 & M_3 & M_3 & M_4 & M_4 & M_4 \\ C_2 & : & M_2 & M_3 & M_4 & M_1 & M_3 & M_4 & M_1 & M_2 & M_4 & M_1 & M_2 & M_3 \end{array}$$

►**EXAMPLE 1.26** *In a box there are 6 balls numbered by 1, 2, 3, 4, 5 and 6. In how many ways 2 balls can be taken one by one* (i) *with replacement,* (ii) *without replacement?*

Solution.

(i) If balls are taken with replacement, the number of arrangements is $n = 6^2 = 36$. These arrangement are

$$\left\{\begin{array}{cccccc} 11 & 21 & 31 & 41 & 51 & 61 \\ 12 & 22 & 32 & 42 & 52 & 62 \\ 13 & 23 & 33 & 43 & 53 & 63 \\ 14 & 24 & 34 & 44 & 54 & 64 \\ 15 & 25 & 35 & 45 & 55 & 65 \\ 16 & 26 & 36 & 46 & 56 & 66 \end{array}\right\}$$

(ii) If balls are taken without replacement, then the total number of arrangements is

$$^nP_r = {}^6P_2 = \frac{6!}{(6-2)!} = 30.$$

These permutations are

$$\begin{Bmatrix} 12 & 21 & 31 & 41 & 51 & 61 \\ 13 & 23 & 32 & 42 & 52 & 62 \\ 14 & 24 & 34 & 43 & 53 & 63 \\ 15 & 25 & 35 & 45 & 54 & 64 \\ 16 & 26 & 36 & 46 & 56 & 65 \end{Bmatrix}$$

►**EXAMPLE 1.27** *In a box there are 3 marbles identified by 1, 2 and 3. Find the number of arrangements of these marbles taken all together with repetition.*

Solution. This can be done in $n = 3^3 = 27$ ways. The arrangements are

$$\begin{Bmatrix} 111 & 112 & 113 & 121 & 122 & 123 & 131 & 132 & 133 \\ 211 & 212 & 213 & 221 & 222 & 223 & 231 & 232 & 233 \\ 311 & 312 & 313 & 321 & 322 & 323 & 331 & 332 & 333 \end{Bmatrix}$$

►**EXAMPLE 1.28** *In an office there are 5 chairs with handle and 5 others without handle. In how many ways these chairs can be arranged for sitting?*

Solution. Here total objects are $n = n_1+n_2 = 5+5 = 10$. The number of arrangements of these chairs is

$$\frac{n!}{n_1!\, n_2!} = \frac{10!}{5!\, 5!} = 252 \quad \left[\text{or, } \frac{n!}{r!\,(n-r)!} = \frac{10!}{5!\,(10-5)!}\right].$$

►**EXAMPLE 1.29** *In how many ways the cards of a packet of 52 cards can be arranged?*

Solution. In a packet there are 52 cards, out of which 13 are spade, 13 are heart, 13 are diamond, and another 13 are club. Here, $n = 52$, $n_1 = n_2 = n_3 = n_4 = 13$. Total number of arrangements is

$$\frac{n!}{n_1!\, n_2!\, n_3!\, n_4!} = \frac{52!}{13!\, 13!\, 13!\, 13!}.$$

1.3.4 Combination

It is a technique to group n distinct objects taken $r (r \leq n)$ at a time. This grouping can be done in ^nC_r ways, where

$$^nC_r = \frac{n!}{r!\,(n-r)!}.$$

COROLLARY 2. The number of combinations of $(n-r)$ objects from n distinct objects is

$$^nC_{n-r} = \frac{n!}{(n-r)!\, r!} \Rightarrow {}^nC_r = {}^nC_{n-r}.$$

COROLLARY 3. If out of n objects r objects is of one kind and $(n-r)$ objects are of another kind and if all objects are arranged, then number of arrangements is

$$^nP_r = \frac{n!}{r!\,(n-r)!}.$$

Again, if r objects are grouped from n distinct objects, then number of groups is

$$^nC_r = \frac{n!}{r!\,(n-r)!}$$

In such a case, $^nP_r = {}^nC_r$.

COROLLARY 4. $^{(n+1)}C_r = {}^nC_{r-1} + {}^nC_r$.

►**EXAMPLE 1.30** *In how many ways a committee of 3 can be formed from 4 couples, if in the committee* (i) *everyone has equal chance to be included,* (ii) *there are 2 women and one man,* (iii) *there is no particular couple?*

Solution.

(i) There are $n = 8$ persons. We need to form a committee with 3 persons from 8. This can be done in

$$^nC_r = {}^8C_3 = \frac{8!}{3!\,(8-3)!} = 56 \text{ ways.}$$

(ii) From 4 women 2 can be selected in $^4C_2 = 5$ ways and from 4 men one can be selected in $^4C_1 = 4$ ways. After selection of 2 women with every grouping, one man can be accommodated in $^4C_1$ ways. Thus, total number of selection of 3 persons is $^4C_2 \times {}^4C_1 = 24$.

(iii) From 4 couples one couple can be selected in $^4C_1 = 4$ ways. The remaining one person can be selected from the remaining $(8-2) = 6$ persons in $^6C_1 = 6$ ways. Hence, with at least one couple 3 persons can be selected in $^4C_1 \times {}^6C_1 = 24$ ways. Again, from 8 persons 3 persons can be selected in $^8C_3 = 56$ ways. Therefore, the number of groups of 3 in which there is particular couple is $(56 - 24) = 32$.

►**EXAMPLE 1.31** *In a laboratory there are 4 computer scientists, 5 computer engineers and 3 technicians. A committee is to be formed with 3 of them for smooth operation of the laboratory work. In how many ways the committee can be formed* (i) *if there is no restriction on types of persons,* (ii) *with one person of each category?*

Solution.

(i) There are $n = (5+4+3) = 12$ persons. Three persons can be selected from 12 persons in $^{12}C_3$ ways.

$$^nC_r = {}^{12}C_3 = \frac{12!}{3!\,(12-3)!} = 220.$$

(ii) Committee should be formed with 1 computer scientist, 1 computer engineer and 1 technician. This can be done in $^4C_1 \times {}^5C_1 \times {}^3C_1 = 60$ ways.

1.3.5 Set

A set is a collection of objects having given properties and specification according to well-defined rule. For example, the collection of first $n = 5$ natural numbers. The set is denoted by A or B or C or $\cdots$. The set of 5 natural numbers is $A = \{1, 2, 3, 4, 5, 6\}$.

Here, 1 or 2 or 3 or 4 or 5 is the element of set. Each element of the set is occurred once and only once. This means that an element is not repeated in a set and in that sense the elements in a set are independent.

Another examples of sets are

$$\begin{aligned} A &= \{x : \text{even natural numbers up to } 100\}, \\ B &= \{x : \text{all English books in a library}\} \text{ and} \\ C &= \{x : \text{dots shown when a die is thrown}\}. \end{aligned}$$

Here, $C = \{1, 2, 3, 4, 5, 6\}$. But if this experiment of throwing die is repeated for several times, it is laborious to write the set with all possible outcomes. So the outcomes can be expressed describing the elements. For example, let us consider a set of all odd positive numbers. This set can be expressed as

$$D = \{x : x = 2n + 1\}, \ n = 0, 1, 2, \ldots .$$

Here, D is the set of all x, where $x = 2n + 1$, $n = 0, 1, 2, \ldots$. We write $x \in D$, $x = 2n + 1$, $n = 0, 1, 2, \ldots$. In the above set $x \notin D$, if $x = 2n, n = 0, 1, 2, \ldots$.

SUBSET. Let us consider a set $A = \{1, 3, 5, 7, 9\}$ and another set $B = \{5, 7, 9\}$. Here the elements of B belong to the elements of A. Here, B is a subset of A. If every element of a set is the elements of another set, then the former one is called subset of latter one. If $x \in B \Rightarrow x \in A$, then B is the subset of A and it is denoted by $B \subset A$.

By definition of subset, A itself is a subset of A. Two sets A and B are said to be equal or identical if $A \subseteq B$ and $B \subseteq A$, and we write $A = B$ or $B = A$. For example,

$$A = \{5, 6, 7, 8\}, \ B = \{8, 7, 6, 5\}, \text{ then } A = B.$$

Here, A is a subset of B or B is a subset of A.

Again, consider another two sets as follows:

$$\begin{aligned} A &= \{11, 12, 13, 14, 15, 16, 21, 22, 23, 24, 25, 26\} \\ \text{and } B &= \{11, 12\}. \end{aligned}$$

Here, B is a real subset of $A (B \subset A)$. Thus, if all elements of B belong A, then B is called real subset of A.

NULL SET. If there is no element in any set, then the set is called *null set* or *empty set.* It is denoted by ϕ. Let us consider a set of card number 14 if a card is selected from a pack of 52 cards. There is no card with number 14 as there are 13 cards of each suit. Hence, the set with card number 14 is empty set. Again, consider a set

$$A = \{x : x^2 + 1 = 0, \ x \text{ is real}\} = \phi.$$

This set A is empty set as x cannot be real if $x^2 + 1 = 0$. Here all roots of $x^2 + 1 = 0$ will be imaginary.

UNIVERSAL SET. If a set contains all possible elements of a specified property, the set is called *universal set.* Let us consider the set of all outcomes when 3 coins are tossed once. The set is

$$A = \{HHH, HHT, HTT, HTH, TTH, THH, THT, TTT\}.$$

Here, A is called universal set. It is denoted by U or S. Again, let us consider a set B, where

$$B = \{HHH, HHT, THH, HTH\}.$$

Thus, B is a subset of A.

UNION OF SET. Let A and B be two sets. The union of these two sets A and B indicates a new set the elements of which are the elements of A or the elements of B or the elements of both A and B. The union of set A and B is denoted by $A \cup B$. For example, let $A = \{2, 4, 6, 8, 10, 12\}$ and $B = \{4, 8, 12, 16, 20\}$. Then

$$A \cup B = \{2, 4, 6, 8, 12, 16, 20\}.$$

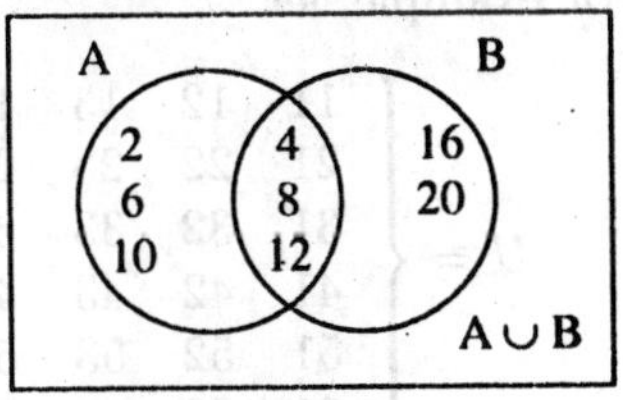

FIG. 1.5

By Venn diagram it is shown in FIG. 1.5.

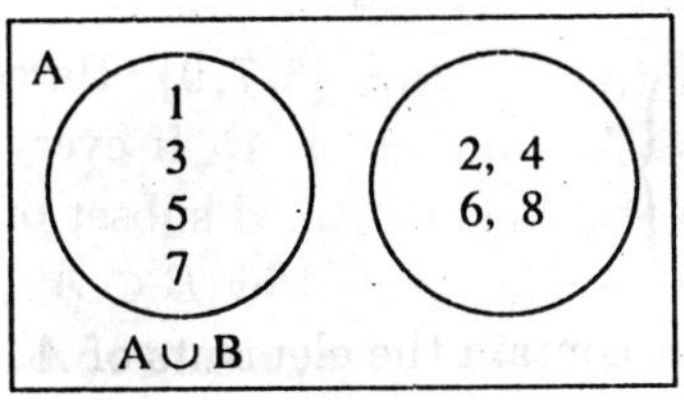

FIG. 1.6

Again, if

$$A = \{1, 3, 5, 7\} \text{ and } B = \{2, 4, 6, 8\},$$

then

$$A \cup B = \{1, 2, 3, 4, 5, 6, 7, 8\} \text{ [FIG. 1.6]}.$$

Here the elements of $A \cup B$ are the elements of both A or B. We usually write

$$A \cup B = \{x : x \in A \text{ or } x \in B\}.$$

If $A_1, A_2, A_3, \ldots, A_n$ are n sets, then union of these n sets is written as

$$\bigcup_{i=1}^{n} A_i = \{x : x \in A_i \text{ for at least one } i = 1, 2, \ldots, n\}.$$

INTERACTION OF SET. Let A and B be two sets such that some elements of A are also the elements of B. The common elements of A and B constitute the intersection of the sets—A and B. The intersection of the set is denoted by $A \cap B$. Thus, if

$$A = \{1, 2, 3, 4, 5, 6\}, \ B = \{2, 4, 6, 8, 10\},$$

then $A \cap B = \{4, 6\}$.

It can be shown by Venn diagram as in FIG. 1.7.

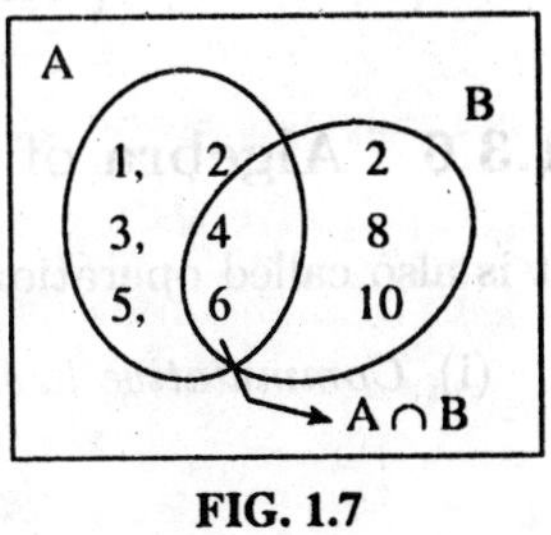

FIG. 1.7

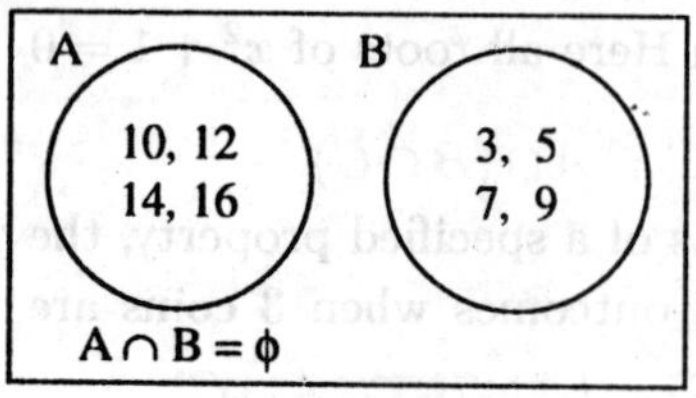

FIG. 1.8

DISJOINT SET. If the sets A and B have no common elements, then the sets are called *disjoint set.* In case of disjoint set the intersection of set is empty. Thus, $A \cap B = \phi$. For example, let $A = \{10, 12, 14, 16\}$ and $B = \{3, 5, 7, 9\}$. Then $A \cap B$ has no element. Here A and B are disjoint sets. The Venn diagram of these two sets is shown in FIG. 1.8.

Here the elements of A do not belong to B or the elements of B do not belong to A.

COMPLEMENTARY SET. Let A be a subset of a universal set U. Then the complementary set of A is a set the elements of which are the elements of U except the elements of A. It is denoted by $\bar{A}$ (Not A). Symbolically,

$$\bar{A} = \{x : x \in U \text{ and } x \notin A\}.$$

For example, let

$$U = \left\{\begin{matrix} 11 & 12 & 13 & 14 & 15 & 16 \\ 21 & 22 & 23 & 24 & 25 & 26 \\ 31 & 32 & 33 & 34 & 35 & 36 \\ 41 & 42 & 43 & 44 & 45 & 46 \\ 51 & 52 & 53 & 54 & 55 & 56 \\ 61 & 62 & 63 & 64 & 65 & 66 \end{matrix}\right\}, A = \left\{\begin{matrix} 11 & 12 & 13 & 14 & 15 & 16 \\ 21 & 22 & 23 & 24 & 25 & 26 \\ 31 & 32 & 33 & 34 & 35 & 36 \end{matrix}\right\}.$$

Then

$$\bar{A} = \left\{\begin{matrix} 41 & 42 & 43 & 44 & 45 & 46 \\ 51 & 52 & 53 & 54 & 55 & 56 \\ 61 & 62 & 63 & 64 & 65 & 66 \end{matrix}\right\}.$$

Here the elements of A are the element of U and $\bar{A}$ does not contain the elements of A but the elements of $\bar{A}$ belong to U.

DIFFERENCE OF SET. Let A and B be two sets. The relative difference of the set A from the set B, denoted by $A - B$ is a set containing those elements of A which do not belong to B. Symbolically,

$$A - B = \{x : x \in A \text{ and } x \notin B\}.$$

For example, let $A = \{HH, HT, TH, TT\}$, $B = \{TT\}$.
Then, $A - B = \{HH, HT, TH\}$.

1.3.6 Algebra of Sets

It is also called operation of sets. Let A and B be two sets. Then

(i) *Commutative law:*

$$A \cup B = B \cup A, A \cap B = B \cap A.$$

(ii) *Associative law:* If A, B and C are three sets, then

$$(A \cap B) \cup C = A \cup (B \cup C); (A \cap B) \cap C = A \cap (B \cap C).$$

(iii) *Distributive law:* For three sets A, B and C

$$A \cap (B \cup C) = (A \cap B) \cup (A \cap C); A \cup (B \cap C) = (A \cup B) \cap (A \cup C).$$

(iv) *Complementary law:*

$$A \cup \bar{A} = S,\ A \cap \bar{A} = \phi,\ A \cup S = S\ (\because\ A \subset S, A \cap S = \phi),$$

where S is an universal set. $A \cup \phi = A, A \cap \phi = \phi$.

(v) *Difference law:*

$$\begin{aligned} A - B &= A \cap \bar{B} \\ A - B &= A - (A \cap B) = (A \cup B) - B \\ A - (B - C) &= (A - B) \cup (A - C) \\ (A \cup B) - C &= (A - C) \cup (B - C) \\ A - (B \cup C) &= (A - B) \cap (A - C) \\ (A \cap B) \cup (A - B) &= A; (A \cap B) \cap (A - B) = \phi \end{aligned}$$

(vi) *De Morgan's law—Dualization law:*

$$\overline{(A \cup B)} = \bar{A} \cap \bar{B},\ \overline{A \cap B} = \bar{A} \cup \bar{B}$$

$$\bigcup_{i=1}^{n} A_i = \bigcap_{i=1}^{n} \bar{A}_i,\ \bigcap_{i=1}^{n} A_i = \bigcup_{i=1}^{n} \bar{A}_i$$

►**EXAMPLE 1.32** *Let A, B and C be three events. Express the following events by appropriate symbols:* (i) *simultaneous occurrence of A, B and C,* (ii) *occurrence of at least one of them,* (iii) *A, B and C are mutually exclusive events,* (iv) *the event B occurs but not A.*

Solution.

(i) $A \cap B \cap C$

(ii) $A \cup B \cup C$

(iii) $A \cap B \cap C = \phi,\ A \cap B = \phi,\ A \cap C = \phi,\ B \cap C = \phi$

(iv) $B \cap \bar{A}$.

1.3.7 Examples on Probability

►**EXAMPLE 1.33** *In a box there are 6 books, three of which are on mathematics and 3 are on statistics. Two books are taken* (i) *one by one with replacement,* (ii) *one by one without replacement, and* (iii) *at random. Find the probability that* (a) *the drawn books are on mathematics,* (b) *one of the drawn books is on statistics,* (c) *at least one book is on mathematics.*

Solution.

(i) Let us denote the mathematics book by M and statistics book by S. Then the sample space of the experiment can be shown as

$$S : \{MM, MS, SM, SS\}.$$

(a) Let A be the event that the drawn books are on mathematics. Favourable case to A is $A : \{MM\}$. There are in total 6 books, one can be selected from box in ${}^6C_1 = 6$ ways. One mathematics book can be selected in ${}^3C_1 = 3$ ways. Therefore, probability of selection of a mathematics book is

$$\frac{{}^3C_1}{{}^6C_1} = \frac{3}{6} = \frac{1}{2}.$$

Since the selection of book is done without replacement, the second mathematics book is also selected with probability $\frac{1}{2}$. Therefore,

$$P(A) = P(MM) = \frac{{}^3C_1}{{}^6C_1} \times \frac{{}^3C_1}{{}^6C_1} = \frac{1}{4}.$$

(b) Let B be the event that one of the drawn books is on statistics. Naturally, the other book is on mathematics. Since number of statistics books is also 3, one statistics book (first one or second one) is selected with probability

$$\frac{{}^3C_1}{{}^6C_1} \left[\begin{array}{l}\because \text{ the selection is done} \\ \text{with replacement}\end{array}\right].$$

The favourable cases to B are $B = \{MS, SM\}$.

$$\therefore\ P(B) = P(MS) + P(SM) = \frac{{}^3C_1}{{}^6C_1} \cdot \frac{{}^3C_1}{{}^6C_1} + \frac{{}^3C_1}{{}^6C_1} \cdot \frac{{}^3C_1}{{}^6C_1} = \frac{1}{2}.$$

(c) Let C be the event that at least one book is on mathematics. The favourable cases to C are $C : \{MM, MS, SM\}$.

$$\begin{aligned}\therefore P(C) &= P(MM) + P(MS) + P(SM) \\ &= \frac{{}^3C_1}{{}^6C_1} \cdot \frac{{}^3C_1}{{}^6C_1} + \frac{{}^3C_1}{{}^6C_1} \cdot \frac{{}^3C_1}{{}^6C_1} + \frac{{}^3C_1}{{}^6C_1} \cdot \frac{{}^3C_1}{{}^6C_1} \\ &= \frac{3}{4}.\end{aligned}$$

Alternatively, let $\bar{C}$ be the event that there will be 2 statistics books. Then

$$P(\bar{C}) = P(SS) = \frac{{}^3C_1}{{}^6C_1} \cdot \frac{{}^3C_1}{{}^6C_1} = \frac{1}{4}.$$

$$\therefore\ P(C) = 1 - P(\bar{C}) = 1 - \frac{1}{4} = \frac{3}{4}.$$

(ii) The sample space is similar to that one given above. Since the drawn is made without replacement, the second draw is made from 5 remaining books. This can be done in ${}^5C_1$ ways. Therefore,

(a) $P(A) = P(MM) = \frac{^3C_1}{^6C_1} \cdot \frac{^2C_1}{^5C_1} = \frac{1}{5}$ $\left[\because \text{ second mathematics book is drawn from 2 mathematics books}\right]$

(b) $P(B) = P(MS) + P(SM) = \frac{^3C_1}{^6C_1} \cdot \frac{^3C_1}{^5C_1} + \frac{^3C_1}{^6C_1} \cdot \frac{^3C_1}{^5C_1} = \frac{3}{5}.$

(c) $P(C) = 1 - P(\bar{C}), P(\bar{C}) = P(SS) = \frac{^3C_1}{^6C_1} \cdot \frac{^2C_1}{^5C_1} = \frac{2}{5}.$

$$\therefore\ P(C) = 1 - \frac{2}{5} = \frac{3}{5}.$$

(iii) Two books from 6 books can be taken in $n = {}^6C_2 = 15$ ways. All 15 cases are exhaustive, mutually exclusive and equally likely.

(a) Two mathematics books can be selected in $m = {}^3C_2 = 3$ ways.

$$\therefore\ P(A) = \frac{m}{n} = \frac{3}{15} = \frac{1}{5}.$$

(b) One mathematics book and one statistics book can be selected in $m = {}^3C_1 \times {}^3C_1 = 9$ ways.

$$\therefore\ P(B) = \frac{m}{n} = \frac{9}{15} = \frac{3}{5}.$$

(c) The event C is that one book will be of mathematics or both will be of mathematics. This can happen in $m = {}^3C_2 + {}^3C_1 \times {}^3C_1 = 12$ ways.

$$\therefore\ P(C) = \frac{m}{n} = \frac{12}{15} = \frac{4}{5}$$

$$\left[\text{or, } P(\bar{C}) = \frac{^3C_2}{^6C_2} = \frac{1}{5},\ P(C) = 1 - P(\bar{C}) = 1 - \frac{1}{5} = \frac{4}{5}\right]$$

►**EXAMPLE 1.34** *A factory of producing fruit juice has three machines, viz. A, B and C. The machines produces 500, 800 and 700 packets of juice per hour, respectively. The packets produced per hour are under investigation.*

(i) *What is the probability that one selected packet is a product of machine A?*

(ii) *If two packets are selected at random, what is the probability that* (a) *both packets are the products of machine A and* (b) *one packet is produced by B and one packet is produced by C?*

(iii) *It is known that 2% packets of A, 1% of B and 3% of C are usually found defective. If two packets are selected* (a) *without replacement,* (b) *at random, what is the probability that one will be defective and one will be non-defective?*

(iv) *If two packets are selected at random, what is the probability that one defective will be of A and another defective will be of B?*

Solution. There are $500 + 800 + 700 = 2000$ packets.

(i) One packet can be selected from 2000 packets in $n = {}^{2000}C_1 = 2000$ ways. Let A be the event that the selected packet is produced by machine A. The product of A can be selected in $m = {}^{500}C_1 = 500$ ways.

$$\therefore\ P(A) = \frac{m}{n} = \frac{500}{2000} = \frac{1}{4}.$$

(ii) Two packets can be selected in $n = {}^{2000}C_2 = 1999000$ ways.

(a) Let B be the event that both selected packets are the products of A. This can be done in $m = {}^{500}C_2 = 124750$ ways.

$$\therefore P(B) = \frac{m}{n} = \frac{124750}{1999000} = 0.0624.$$

(b) Let C be the event that one packet is produced by machine B and another one packet is produced by machine C. These selection can be done in $m = {}^{800}C_1 \times {}^{700}C_1 = 560000$ ways.

$$\therefore P(C) = \frac{m}{n} = \frac{560000}{1999000} = 0.2801.$$

(iii) Total defective packets are $10 + 8 + 21 = 39$.

(a) The sample space of selection of packets is

$S: \{DD, DG, GD, GG\}$, where $D =$ defective, $G =$ good.

Let A be the event that one defective and one non-defective packets are selected. Favourable cases to A are $A: \{DG, GD\}$.

$$\therefore P(A) = P(DG) + P(GD)$$
$$= \frac{{}^{39}C_1}{{}^{2000}C_1} \cdot \frac{{}^{1961}C_1}{{}^{1999}C_1} + \frac{{}^{1961}C_1}{{}^{2000}C_1} \cdot \frac{{}^{39}C_1}{{}^{1961}C_1} = 0.03826.$$

(b) If two packets are selected at random, then equally likely cases are $n = {}^{2000}C_2 = 1999000$.

The event A can occur in $m = {}^{39}C_1 \times {}^{1961}C_1 = 76479$.

$$\therefore P(A) = \frac{m}{n} = \frac{76479}{1999000} = 0.03826.$$

(iv) Let E be the event that one defective will be of machine A and another one will be of machine B. This can happen in $m = {}^{10}C_1 \times {}^{8}C_1 = 80$ ways.

$$\therefore P(E) = \frac{m}{n} = \frac{80}{1999000} = 0.00004.$$

►**EXAMPLE 1.35** *A company has announced an advertisement to appoint 2 managers. In response to the advertisement 5 male and 3 female candidates have submitted their candidature. Find the probability that* (i) *two male candidates,* (ii) *one male and one female candidates will be selected.*

Solution. Total applicants are $5 + 3 = 8$. Two will be selected from 8 in $n = {}^{8}C_2 = 28$ ways.

(i) Let A be the event that 2 male candidates will be selected. This will occur in $m = {}^{5}C_2 = 10$ ways.

$$\therefore P(A) = \frac{m}{n} = \frac{10}{28} = \frac{5}{14}.$$

(ii) Let B be the event that one male candidate and one female candidate will be selected. This will occur in $m = {}^5C_1 \times {}^3C_1 = 15$ ways.

$$\therefore\ P(B) = \frac{m}{n} = \frac{15}{28}.$$

►**EXAMPLE 1.36** *In an urn there are 12 bananas and 8 apples. Two fruits are distributed among some persons. Find the probability that* (i) *a person will get* 2 *bananas,* (ii) *a person will get one banana and one apple.*

Solution. There are $12 + 8 = 20$ fruits. Two fruits can be distributed in $n = {}^{20}C_2 = 190$ ways.

(i) Let A be the event that a person gets 2 bananas. He can get it in $m = {}^{12}C_2 = 66$ ways.

$$\therefore\ P(A) = \frac{m}{n} = \frac{66}{190} = \frac{33}{95}.$$

(ii) Let B be the event that a person gets one banana and one apple. He can get it in $m = {}^{12}C_1 \times {}^8C_1 = 96$ ways.

$$\therefore\ P(B) = \frac{m}{n} = \frac{96}{190} = \frac{48}{95}.$$

►**EXAMPLE 1.37** *Five engineers,* 10 *technicians and* 5 *technical assistants are working in a factory. The owner of the factory has planned to form a committee of* 4 *with these expert people so that machines of the factory are well-maintained. Find the probability that, the committee will be formed with* (i) *one engineer,* 2 *technicians and* 1 *technical assistant,* (ii) *one engineer and* 3 *technical assistants,* (iii) *no engineer.*

Solution. There are $5 + 10 + 5 = 20$ experts. A committee of 4 can be formed in $n = {}^{20}C_4 = 4845$ ways.

(i) Let A be the event that the committee will be formed with one engineer, two technicians and one technical assistant. This can be done in $m = {}^5C_1 \times {}^{10}C_2 \times {}^5C_1 = 1125$ ways.

$$\therefore\ P(A) = \frac{m}{n} = \frac{1125}{4845} = \frac{75}{323}.$$

(ii) Let B be the event that the committee will be formed with 1 engineer and 3 technical assistants. This can happen in $m = {}^5C_1 \times {}^5C_3 = 50$ ways.

$$\therefore\ P(B) = \frac{m}{n} = \frac{50}{4845} = \frac{10}{969}.$$

(iii) Let C be the event that there will be no engineer in the committee. It can occur if committee is formed with (a) 4 technicians, (b) 3 technicians and 1 technical assistant, (c) 2 technicians and 2 technical assistants, (d) 1 technician and 3 technical assistants, (e) 4 technical assistants. Therefore, total number of favourable cases to C is

$$m = {}^{10}C_4 + {}^{10}C_3 \times {}^5C_1 + {}^{10}C_2 \times {}^5C_2 + {}^{10}C_1 \times {}^5C_3 + {}^5C_4 = 1365.$$

$$\therefore\ P(C) = \frac{m}{n} = \frac{1365}{4845} = \frac{91}{323}.$$

►EXAMPLE 1.38 *A farmer has collected 10 pineapples and 5 jackfruits from his garden. Two pineapples and 2 jackfruits are defectives. If a man buys one fruit, what is the probability that he will buy non-defective fruit?*

Solution. Total fruits are $10 + 5 = 15$. One can be bought in $n = {}^{15}C_1 = 15$ ways. The defective fruits are $2 + 2 = 4$. One defective fruit can be bought in ${}^4C_1 = 4$ ways. Let A be the event that one defective fruit can be bought. We need $P(\bar{A})$. Here

$$P(A) = \frac{m}{n} = \frac{4}{15}.$$

$$\therefore\ P(\bar{A}) = 1 - P(A) = 1 - \frac{4}{15} = \frac{11}{15}.$$

►EXAMPLE 1.39 *In a small garden there are 5 rose plants, 4 merrygold plants and 3 daffodil plants. One day it is observed that 2 plants are destroyed. Find the probability that the destroyed plants are* (i) *of rose plants,* (ii) *one rose plant and one merrygold plant,* (iii) *not daffodil plants.*

Solution. There are $5 + 4 + 3 = 12$ plants. Two destroyed plants can occur in $n = {}^{12}C_2 = 66$ ways.

(i) Let A be the event that 2 rose plants are destroyed. This can occur in $m = {}^5C_2 = 10$ ways.

$$\therefore\ P(A) = \frac{m}{n} = \frac{10}{66} = \frac{5}{33}.$$

(ii) Let B be the event that one plant of rose and one plant of merrygold are destroyed. This can occur in $m = {}^5C_1 \times {}^4C_1 = 20$ ways.

$$\therefore\ P(B) = \frac{m}{n} = \frac{20}{66} = \frac{10}{33}.$$

(iii) Let C be the event that no plant of daffodil is destroyed. This can occur if destroyed plants are (a) of rose, (b) of merrygold, (c) one of rose and one of merrygold. This can occur in $m = {}^5C_2 + {}^4C_2 + {}^5C_1 \times {}^4C_1 = 36$ ways. We need $P(\bar{C})$. Here

$$P(C) = \frac{m}{n} = \frac{36}{66}.$$

$$\therefore\ P(\bar{C}) = 1 - P(C) = 1 - \frac{36}{66} = \frac{5}{11}.$$

►EXAMPLE 1.40 *In a room there are 3 sockets to put electric bulb and 5 electric bulbs are available in the room. But, out of 5 bulbs 3 bulbs are defective. What is the probability that there will be light if bulbs are put in the sockets?*

Solution. Let us consider that there will be no light in the room and this is event A. There are 5 bulbs and 3 sockets in the room. Three bulbs can be put in 3 sockets in $n = {}^5C_3 = 10$ ways. The event A will occur if all 3 defective bulbs are put in the socket. This can happen in $m = {}^3C_3 = 1$ way.

$$\therefore\ P(A) = \frac{m}{n} = \frac{1}{10}.$$

We need $P(\bar{A})$. Here

$$P(\bar{A}) = 1 - P(A) = 1 - \frac{1}{10} = \frac{9}{10}.$$

►**EXAMPLE 1.41** *What is the probability that in selecting* 3 *candidates from a group of* 20 *candidates* (i) *one particular candidate is selected,* (ii) *one particular candidate is not selected?*

Solution. Three candidates can be selected in $n = {}^{20}C_3 = 1140$ ways.

(i) Let A be the event that 3 will be selected including a particular candidate. If particular candidate is selected, other 2 will be selected from the remaining 19 candidates. This can occur in $m = {}^{19}C_2 = 171$ ways.

$$\therefore\ P(A) = \frac{m}{n} = \frac{171}{1140} = \frac{3}{20}.$$

(ii) If A be the event that one particular candidate is included in the selection, then $\bar{A}$ is the event that a particular candidate is not selected.

$$\therefore\ P(\bar{A}) = 1 - P(A) = 1 - \frac{3}{20} = \frac{17}{20}.$$

►**EXAMPLE 1.42** *There are* 8 *bulbs in an urn. If two bulbs are selected at random from the urn, what is the probability that the selected bulbs are* (i) *of special type,* (ii) *no one of the special type,* (iii) *except both of special type?*

Solution. Two bulbs can be selected in $n = {}^8C_2 = 28$ ways.

(i) Let A be the event that 2 bulbs will be of special type. This can happen in $m = 1$ way.

$$\therefore\ P(A) = \frac{m}{n} = \frac{1}{28}.$$

(ii) Let B be the event that no one of the special type. Hence 2 bulbs are to be selected from the remaining $(8 - 2) = 6$ bulbs. This can happen in $m = {}^6C_2 = 15$ ways.

$$\therefore\ P(B) = \frac{m}{n} = \frac{15}{28}.$$

(iii) If the selection is made except both of special type, then the event is $\bar{A}$.

$$\therefore\ P(\bar{A}) = 1 - P(A) = 1 - \frac{1}{28} = \frac{27}{28}.$$

►**EXAMPLE 1.43** *Twelve balls are to be distributed in* 3 *boxes. What is the probability that* 4 *balls will go to one box,* 5 *balls will go to another box and the remaining* 3 *balls will go to a third box?*

Solution. Twelve balls can be distributed in three boxes in $n = 3^{12}$ ways.

Let B_1, B_2 and B_3 be the boxes. Then B_1 will get 5 balls, B_2 will get 4 balls and B_3 will get 3 balls. This can happen in

$$\frac{12!}{5!\ 4!\ 3!} \text{ ways.}$$

But the distribution of balls in boxes B_1, B_2 and B_3 can again be performed in 3 ways. Hence, total number of favourable cases is $m = \frac{3\times 12!}{5!\ 4!\ 3!}$.

$$\therefore\ P(A) = \frac{m}{n} = \frac{3 \times 12!}{5!\ 4!\ 3!\ (3^{12})}.$$

Here A be the event that balls are distributed in 3 boxes.

►**EXAMPLE 1.44** *There are 5 children in a family. Let the children be A, B, C, D and E. It is decided that any three of them will be trained as athletes. What is the probability that among 3 children C and D will be included?*

Solution. Three children can be selected from 5 children in $n = {}^5C_3 = 10$ ways. The sample space of the outcomes is

$$S : \{ABC, ABD, ABE, ACD, ACE, ADE, BCD, BCE, BDE, CDE\}.$$

Let A be the events that among 3 children C and D are included. Favourable cases to A are $A : \{ACD, BCD, CDE\}$, $m = 3$.

$$\therefore\ P(A) = \frac{m}{n} = \frac{3}{10}.$$

►**EXAMPLE 1.45** *Five unbiased coins are tossed. Find the probability of observing* (i) 5 *heads,* (ii) *at best one head,* (iii) *at least 2 heads,* (iv) *no heads.*

Solution. The exhaustive, mutually exclusive and equally outcomes of the experiment of tossing 5 coins is $n = 2^5 = 32$.

(i) Let A be the event that there will be 5 heads. Out of 5 coins 5 heads can occur in $m = {}^5C_5 = 1$ way.

$$\therefore\ P(A) = \frac{m}{n} = \frac{1}{32}.$$

(ii) Let B be the event that there will be at best one head. Number of heads will be zero or 1. This can occur in $m = {}^5C_0 + {}^5C_1 = 6$ ways.

$$\therefore\ P(B) = \frac{m}{n} = \frac{6}{32}.$$

(iii) Let C be the event that there will be at least 2 heads. Number of heads will be 2 or 3 or 5 (0 head or 1 head will not be observed). Hence, C is the complementary event of B.

$$\therefore\ P(C) = P(\bar{B}) = 1 - P(B) = 1 - \frac{6}{32} = \frac{26}{32}.$$

(iv) Let D be the events that there will be no heads (implies that all will be tails). This can occur in $m = {}^5C_0 = 1$ way.

$$\therefore\ P(D) = \frac{m}{n} = \frac{1}{32}.$$

►**EXAMPLE 1.46** *Three unbiased dice are thrown. Find the probability that sum of the dots of upper faces of dice is* 15.

Solution. The exhaustive, mutually exclusive and equally likely cases of the experiment is $n = 6^3 = 216$.

Let A be the event that sum of the dots in the upper faces of dice is 15. Favourable cases to A are $A : \{366, 465, 456, 555, 564, 546, 663, 636\}$, $m = 8$.

$$\therefore P(A) = \frac{m}{n} = \frac{8}{216}.$$

►**EXAMPLE 1.47** *Six faces of a die are marked with* $-2, -1, 0, 1, 2, 3$. *If this die is thrown thrice, what is the probability that sum of the upper faces shown is* 6*?*

Solution. If the die is thrown thrice, the equally likely, exhaustive and mutually exclusive cases are $n = 6^3 = 216$.

Let A be the event that sum of the upper faces shown is 6. This can occur if the die shows 1, 2 and 3. These 3 digits can be arranged among themselves in $m = 3! = 6$ ways.

$$\therefore \ P(A) = \frac{m}{n} = \frac{6}{216} = \frac{1}{36}.$$

►**EXAMPLE 1.48** *Let b balls be distributed in N boxes. Find the probability that a particular box receives $r (< b)$ balls.*

Solution. The total number of ways in which b balls can be distributed among N boxes is $n = N^b$.

Let A be the event that a particular box receives r balls. These r balls can be distributed to any particular box in ${}^bC_r$ ways. Then the remaining $(b - r)$ balls can be distributed in $(N - 1)$ boxes in $(N-1)^{(b-r)}$ ways. Therefore, favourable cases to A are $m = {}^bC_r \ (N - 1)^{(b-r)}$.

$$\therefore \ P(A) = \frac{m}{n} = \frac{{}^bC_r(N-1)^{(b-r)}}{N^b}.$$

►**EXAMPLE 1.49** *In a shop there are* 9 *computers. These computers are to be arranged in* 3 *shelves. What is the probability that in one shelf there will be* 4 *computers, in another shelf there will be* 2 *computers and in third shelf there will be* 3 *computers?*

Solution. The computers can be arranged in 3 shelves in $n = 3^9$ ways.

Let the shelves be A, B and C. Consider that A will contain 4, B will contain 2 and C will contain 3 computers. This can be done in $\frac{9!}{4!\ 2!\ 3!}$ ways.

But 4, 2 and 3 computers can be accommodated among 3 shelves in 3 ways. So total number of cases by which computers can be arranged in 3 shelves is $m = \frac{3 \times 9!}{4!\ 2!\ 3!}$.

$$\therefore \text{ required probability} = \frac{m}{n} = \frac{3 \times 9!}{4!\ 2!\ 3! \times 3^9}.$$

►**EXAMPLE 1.50** *In a club there are* 20 *male and* 15 *female members. Four tickets will be distributed among* 4 *members to enjoy a drama. What is the probability that, the tickets will be given to* (i) 4 *female members,* (ii) 2 *male and* 2 *female members?*

Solution. Total members are $20 + 15 = 35$. From these 35 members 4 members can be selected in $n = {}^{35}C_4$ ways.

(i) Let A be the event that tickets will be given to 4 female members. This can be done in $m = {}^{15}C_4$ ways.

$$\therefore\ P(A) = \frac{{}^{15}C_4}{{}^{35}C_4} = \frac{39}{1496}.$$

(ii) Let B be the events that tickets will be given to 2 male and 2 female members. This can be done in $m = {}^{20}C_2 \times {}^{15}C_2$ ways.

$$\therefore\ P(B) = \frac{m}{n} = \frac{{}^{20}C_2 \times {}^{15}C_2}{{}^{35}C_4} = \frac{285}{748}.$$

►**EXAMPLE 1.51** *A company has appointed* 10 *employees in* 5 *days at a rate of* 2 *employees per day. The company has a plan to form a committee of* 4 *of these* 10 *employees. What is the probability that in the committee* (i) *there will be no two employees appointed in one day,* (ii) *there will employees appointed only on same day?*

Solution. The committee can be formed in $n = {}^{10}C_4 = 210$ ways.

(i) Let A be the event that there will be no two employees appointed in one day. Let the employees be $E_{11}, E_{12}, E_{21}, E_{22}, E_{31}, E_{32}, E_{41}, E_{42}, E_{51}, E_{52}$. The event A can occur in the following ways:

(a) 4 can be selected from $E_{11}, E_{21}, E_{31}, E_{41}, E_{51}$
(b) 4 can be selected from $E_{12}, E_{22}, E_{32}, E_{42}, E_{52}$
(c) three of (a) and one of (b) not related to (a)
(d) two from (a) and two from (b) not related to (a)
(e) one from (a) and 3 from (b) not related to (a)

Therefore, total number of ways for the cases (a), (b), (c), (d) and (e) is

$${}^5C_4 + {}^5C_4 + {}^5C_3 \times {}^2C_1 + {}^5C_2 \times {}^3C_2 + {}^5C_1 \times {}^4C_3 = 80 = m.$$
$$\therefore\ P(A) = \frac{m}{n} = \frac{80}{210} = \frac{8}{21}.$$

(ii) Let B be the event that the committee will be formed with employees appointed in the same day. There are 5 groups of employees. From these 5 groups, 2 groups can be selected in $m = {}^5C_2 = 10$ ways.

$$\therefore\ P(B) = \frac{m}{n} = \frac{10}{210} = \frac{1}{21}.$$

►**EXAMPLE 1.52** *What is the probability of obtaining even values as sum of two digits selected from the numbers* $1, 2, \ldots, 10$?

Solution. Let A be the event that sum of two digits are even. Sum can be even if the selected digits are $\{13, 15, 17, 19, 24, 26, 28, 210, 35, 37, 39, 46, 48, 410, 57, 59, 68, 610, 79, 810\}, m = 20$.

Two digits can be selected in $n = {}^{10}C_2 = 45$.

$$\therefore\ P(A) = \frac{m}{n} = \frac{20}{45} = \frac{4}{9}.$$

►EXAMPLE 1.53 *Two digits are selected one by one without replacement from the digits* 1, 2, 3, 4, ... , 9. *Find the probability that* (i) *sum of the digits will be odd,* (ii) *sum of the digits will be more than* 10, (iii) *sum of the digits will be less than or equal to* 10.

Solution. Two digits from 9 digits can be selected one by one without replacement in $n = 9^2$ ways.

(i) Let A be the event that sum of the digits are odd. Favourable cases to A are

$$A : \{12, 14, 16, 18, 21, 23, 25, 27, 29, 32, 34, 36, 38, 41, 43, 45, 47, 49, 52, 54, 56, 58, 61, 63, 65, 67, 69, 72, 74, 76, 78, 81, 83, 85, 87, 89, 92, 94, 96, 98\}, \ m = 40.$$

$$\therefore \ P(A) = \frac{40}{81}.$$

(ii) Let B be the event that sum of the digits will be more than 10. Favourable cases to B are

$$B : \{29, 92, 38, 83, 39, 93, 47, 74, 48, 84, 49, 94, 56, 65, 57, 75, 58, 85, 59, 95, 67, 76, 68, 86, 69, 96, 78, 87, 79, 97, 89, 98, 66, 77, 88, 99\}, \ m = 36$$

$$\therefore \ P(B) = \frac{m}{n} = \frac{36}{81}$$

(iii) Let C be the event that the sum of the digits will be less than or equal to 10. In 36 cases sum of the digits is more than 10. In 9 cases $\{19, 91, 28, 82, 37, 73, 46, 64, 55\}$. Sum of the digits is 10 or more. Therefore, m for C is $(81 - 45) = 36$.

$$\therefore \ P(C) = \frac{m}{n} = \frac{36}{81}.$$

►EXAMPLE 1.54 *In an urn there are* 25 *tickets numbered* 1, 2, ... , 25. *Four tickets are randomly selected from the urn. What is the probability that the second positional ticket will be* 20 *if the selected tickets arranged in ascending order of number?*

Solution. Four tickets from 25 tickets can be selected in $n = {}^{25}C_4 = 12650$ ways.

Let A be the event that the ticket of second-positioned one, when arranged in array, is 20. Ticket of the first position may be any one of 1, 2, ... , 19. This can happen in ${}^{19}C_1$ ways. The number of ticket in 3rd and 4th positions must be any two of the following numbers: 21, 22, 23, 24, 25. These tickets can be grouped in ${}^{5}C_2 = 10$ ways. Therefore, total number of favourable cases to A is $m = 19 \times 10 = 190$.

$$\therefore \ P(A) = \frac{m}{n} = \frac{190}{12650} = \frac{19}{1265}.$$

►EXAMPLE 1.55 *From a pack of well-shuffled* 52 *cards* 4 *cards are selected at random. Find the probability that* (i) *all* 4 *cards are spades,* (ii) *two are spades and* 2 *are hearts,* (iii) *two are black cards and* 2 *are red cards,* (iv) *all* 4 *are aces,* (v) *the cards are of same colour,* (vi) *the cards are of same number,* (vii) *the cards will consist of a knave, a queen, a king and an ace,* (viii) *all* 4 *are honours cards of same suit.*

Solution. Four cards from 52 cards can be selected in

$$n = {}^{52}C_4 = 270725 \text{ ways.}$$

(i) Let A be the event that the cards are spades. There are 13 spades. Four spades can be selected from 13 spades in $m = {}^{13}C_4 = 715$ ways.

$$\therefore P(A) = \frac{m}{n} = \frac{715}{270725} = \frac{143}{54145}.$$

(ii) Let B be the event that 2 cards are spades and 2 are hearts. There are 13 hearts and 2 hearts can be taken in ${}^{13}C_2 = 78$ ways and each of these ways is associated with each of selection of 2 spades. Two spades can be taken in ${}^{13}C_2 = 78$ ways. Therefore, favourable cases to B are $m = 78 \times 78 = 6084$.

$$\therefore P(B) = \frac{m}{n} = \frac{6084}{270725} = 0.02247.$$

(iii) Let C be the event that 2 are black cards and 2 are red cards. There are 26 red and 26 black cards in a packet. Favourable cases to C are $m = {}^{26}C_2 \times {}^{26}C_2 =$ 105625.

$$\therefore P(C) = \frac{m}{n} = \frac{105625}{270725} = \frac{4225}{10829} = 0.39016.$$

(iv) Let D be the event that all 4 cards are aces. There are 4 aces. Hence, D can occur in $m = {}^4C_4 = 1$ way.

$$\therefore P(D) = \frac{m}{n} = \frac{1}{270725}.$$

(v) Let E be the event that the cards are of same colour. All 4 cards may be black or they may be red. This can occur in $m = {}^{26}C_4 + {}^{26}C_4 = 29900$ ways.

$$\therefore P(E) = \frac{m}{n} = \frac{29900}{270725} = \frac{1196}{10829}.$$

(vi) Let F be the event that the cards are of same number. There are 4 suits in a packet and the cards in each suit are numbered from $1, 2, \ldots, 13$. There are 4 same number. Hence, same number can be selected in ${}^4C_4 = 1$ way. But this is true for any number 1 or 2 or ... or 13. Hence, total favourable cases to F are 13×1.

$$\therefore P(F) = \frac{m}{n} = \frac{13}{270725} = \frac{1}{20825}.$$

(vii) Let G be the event that the cards are knave, king, queen and ace. There are 4 knaves, 4 kings, 4 queens and 4 aces. A knave, a queen, a king or an ace can be selected in ${}^4C_1 = 4$ ways. But any of the drawing is associated with other. Hence, favourable cases to G are $m = 4 \times 4 \times 4 \times 4 = 256$.

$$\therefore P(G) = \frac{m}{n} = \frac{256}{270725}.$$

(viii) Let H be the event that the cards are honours cards of same suit. In a suit there are 5 honours cards, 4 of these can be selected in ${}^5C_4 = 5$ ways. But this is true for all 4 suits. Hence, favourable cases to H are $m = 4 \times {}^5C_4 = 20$.

$$\therefore\ P(H) = \frac{20}{270725}.$$

►**EXAMPLE 1.56** *Find the probability that at a game of bridge one hand will get 9 cards of same suit.*

Solution. The cards are dealt among 4 persons, where each hand gets 13 cards. This can be done in $n = {}^{52}C_{13}$ ways.

Let A be the event that a hand gets 9 cards of same suit. He will get 4 cards of the remaining suits. Thus 4 cards will be selected from the remaining 39 cards in ${}^{39}C_4$ ways. Again, 9 cards of a suit can be selected in ${}^{13}C_9$ ways and each of this selection is associated with the selection of 4 cards from other suits. So total number of ways a person can get 9 cards of one suit and 4 cards of other suits is ${}^{39}C_4 \times {}^{13}C_9$. If one hand get 9 cards of one suit, other 3 hands cannot get the same distribution of cards. Hence, favourable cases to A is $m = 3 \times {}^{39}C_4 \times {}^{13}C_9$.

$$\therefore\ P(A) = \frac{m}{n} = \frac{3 \times {}^{39}C_4 \times {}^{13}C_9}{{}^{52}C_{13}}.$$

►**EXAMPLE 1.57** *From a well-shuffled pack of* 52 *cards two cards are selected at random. Find the probability that* (i) *one card will be an ace and other will be non-ace,* (ii) *two cards will be of consecutive numbers.*

Solution. Two cards from 52 cards can be selected in $n = {}^{52}C_2 = 1326$ ways.

(i) Let A be the event that one of the cards is an ace and other card is non-ace. This can be done in $m = {}^4C_1 \times {}^{48}C_1 = 192$ ways.

$$\therefore\ P(A) = \frac{m}{n} = \frac{192}{1326}.$$

(ii) Let B be the event that two cards will be of consecutive numbers. The number of the cards of each suit are $1 =$ ace, $2, 3, 4, \ldots, 10, J = 11, K = 12, Q = 13$ (say). These 13 numbers can be arranged by consecutive numbers taking two at a time in 12 ways [12, 23, 34, 45, 56, 67, 78, 89, 910, 1011, 1112, 1213]. These types of arrangements is true for all 4 suits of cards or for any two types of cards. If the case arises for a particular suit, it can be happened in $4 \times 12 = 48$ ways. Again, if the arrangement is made with 2 different suits, this can be done using 4 suits in ${}^4P_2 = 12$ ways. Among these 12 arrangements each arrangement can have consecutive number in 12 ways. Therefore, using any two suits of cards two consecutive number can be obtained in $12 \times 12 = 144$ ways. Therefore, favourable cases to B are $m = 48 + 144 = 192$.

$$\therefore\ P(B) = \frac{m}{n} = \frac{192}{1326}.$$

►**EXAMPLE 1.58** *If the cards of a well-shuffled packet is distributed among* 4 *persons, what is the probability that a person will get* 4 *cards of same number?*

Solution. The cards can be dealt among 4 persons in $n = {}^{52}C_{13}$ ways.

Let A be the event that a hand will get 4 cards of same number. There are 13 cards of one suit and there are 4 suits. So 4 cards of same number can occur in 13 ways. The remaining 9 cards will be distributed from the remaining 48 cards and this can be done in ${}^{48}C_9$ ways. This arrangement of cards is true for every 4 persons. So favourable cases to A are $m = 13 \times 4 \times {}^{48}C_9$ ways.

$$\therefore\ P(A) = \frac{m}{n} = \frac{13 \times 4 \times {}^{48}C_9}{{}^{52}C_{13}}.$$

►EXAMPLE 1.59 *From a well-shuffled packet of* 52 *cards, cards are distributed one by one and observe the type of card. The distribution is continued until an ace is observed. Find the probability that* (i) *the* 4*th card is an ace,* (ii) *the* 13*th card is a king.*

Solution. There are 52 cards among which 4 are aces and 48 are non-aces. Let the card be denoted by N if it is not an ace and by A if it is an ace. Then the arrangement of card is nothing but the arrangement of $48N$ and $4A$'s. This can be done in

$$n = \frac{52!}{48!\ 4!} = {}^{52}C_4 \text{ ways.}$$

(i) Let A be the event that the 4th card is an ace. The favourable cases to A are $A : \{NNNA\}$. This combination is associated with the arrangement of the remaining $45N$'s and $3A$'s which can be done in

$$\frac{48!}{45!\ 3!} = {}^{48}C_3 = m.$$

All these cases are favourable to the event A.

$$\therefore\ P(A) = \frac{m}{n} = \frac{{}^{48}C_3}{{}^{52}C_4}.$$

(ii) Let B be the event that the 13th card is a king. There are 4 kings and 48 non-kings, where king can be denoted by K and non-king by N. total arrangement can be made in

$$n = \frac{52!}{48!\ 4!} = {}^{52}C_4 \text{ ways.}$$

The 13th card will be a king if the drawn cards are arranged as follows:

$$\{NNNNNNNNNNNNK\}$$

This arrangement can occur in

$$m = \frac{39!}{36!\ 3!} = {}^{39}C_3 \text{ ways.}$$

$$\therefore\ P(B) = \frac{m}{n} = \frac{{}^{39}C_3}{{}^{52}C_4}.$$

►EXAMPLE 1.60 *Four letters are written and there are* 4 *envelopes with correct address written on the envelope. The letters are put into the envelopes in such a way* *that each envelope gets one letter. Find the probability that* (i) *at least one letter is not put in correct envelope,* (ii) *no letter is put in correct envelope.*

Solution. Four letters can be put in 4 envelopes in $n = 4! = 24$ ways.

(i) Let A be the event that letters are put in correct envelopes; then $\bar{A}$ indicates that at least one letter is not put in the correct envelope. The letters can be put in correct envelopes in $m = 1$ way. So

$$P(A) = \frac{m}{n} = \frac{1}{24}$$

$$\text{and } P(\bar{A}) = 1 - P(A) = 1 - \frac{1}{24} = \frac{23}{24}.$$

(ii) Let B be the event that no letter is put in correct envelope. Consider that the letters are 1, 2, 3 and 4, and the corresponding correct envelopes are also 1, 2, 3 and 4. No letter is put in the correct envelope if the letters are accommodated as follows:

Envelopes :	1	2	3	4
Letters :	2	1	4	3
	2	4	1	3
	2	3	4	1
	3	1	4	2
	3	4	2	1
	3	4	1	2
	4	1	2	3
	4	3	1	2
	4	3	2	1

$$\therefore \quad m = 9, \; P(B) = \frac{m}{n} = \frac{9}{24}.$$

►EXAMPLE 1.61 *In a box there are* 30 *tickets numbered* 1, 2, 3, ... , 30. *Five tickets are drawn at random from the box and arranged in ascending order of magnitude that* $x_1 < x_2 < x_3 < x_4 < x_5$, *where* x_i*'s indicate the number of tickets. Find the probability that the third position of ticket is occupied by the ticket numbered* 20.

Solution. Five tickets out of 30 tickets can be selected in $n = {}^{30}C_5$ ways.

Let A be the event that the number of tickets in third position when arranged in ascending order is 20. Therefore, the first two positions will be occupied by the tickets numbered 1 to 19 and the last two positions will be occupied by the tickets numbered 21 to 30. The first two positions and last two positions can be occupied by ${}^{19}C_2 \times {}^{10}C_2$ ways. These cases are in favour of A. Hence, $m = 7695$.

$$\therefore \; P(A) = \frac{7695}{{}^{30}C_5}.$$

►EXAMPLE 1.62 *Two persons* x_1 *and* x_2 *stand in a queue with other* 10 *persons at random. Find the probability that there will be* 3 *persons between* x_1 *and* x_2.

Solution. Total persons in queue are $(10 + 2) = 12$. They can stand in a queue in $n = 12!$ ways.

Let A be the event that there will be 3 persons between x_1 and x_2. These 3 persons can stand between x_1 and x_2 as follows:

(i) x_1 — — — x_2 — — — — — — —
(ii) — x_1 — — — x_2 — — — — — —
(iii) — — x_1 — — — x_2 — — — — —
(iv) — — — x_1 — — — x_2 — — — —
(v) — — — — x_1 — — — x_2 — — —
(vi) — — — — — x_1 — — — x_2 — —
(vii) — — — — — — x_1 — — — x_2 —
(viii) — — — — — — — x_1 — — — x_2

But x_1 and x_2 can change their positions in 2 ways. Therefore, total number of ways in which there will be 3 persons between x_1 and x_2 is $2 \times 8 = 16$. The remaining 10 positions can be occupied by the remaining 10 persons in 10! ways. Hence, total number of ways in which 3 persons can be accommodated between x_1 and x_2 is $16 \times 10!$.

$$\therefore\ P(A) = \frac{m}{n} = \frac{16 \times 10!}{12!}.$$

►**EXAMPLE 1.63** *Twelve persons amongst whom are x_1 and x_2 sit at random at a round table. Find the probability that there are two persons between x_1 and x_2.*

Solution. Twelve persons can sit around a table in $n = (12 - 1)!$ ways.

Let A be the event that there will be two persons between x_1 and x_2. These two persons can sit between x_1 and x_2 as follows:

(i) x_1 — — x_2 — — — — — — — —
(ii) — x_1 — — x_2 — — — — — — —
(iii) — — x_1 — — x_2 — — — — — —
(iv) — — — x_1 — — x_2 — — — — —
(v) — — — — x_1 — — x_2 — — — —
(vi) — — — — — x_1 — — x_2 — — —
(vii) — — — — — — x_1 — — x_2 — —
(viii) — — — — — — — x_1 — — x_2 —
(ix) — — — — — — — — x_1 — — x_2

There are 9 positions in which 2 persons can sit between x_1 and x_2. The remaining 10 positions around round table can be filled up in $(10 - 1)! = 9!$ ways. Therefore, favourable cases to A are $m = 9 \times 9!$.

$$\therefore\ P(A) = \frac{m}{n} = \frac{9 \times 9!}{11!} = \frac{9}{110}.$$

►**EXAMPLE 1.64** *Find the probability that in a random arrangement of the letters of the word* 'UNIVERSITY' *the two I's do not come together.*

Solution. There are 10 letters in the word UNIVERSITY, out of which 2 words are same. These words can be arranged in $n = \frac{10!}{2!} = 181400$ ways.

Let A be the event that two I's can occur together. This can happen in the following ways:

(i) I I — — — — — — — —

(ii) — I I — — — — — — —

(iii) — — I I — — — — — —

(iv) — — — I I — — — — —

(v) — — — — I I — — — —

(vi) — — — — — I I — — —

(vii) — — — — — — I I — —

(viii) — — — — — — — I I —

(ix) — — — — — — — — I I

Two I's can take place together in 9 ways. The other 8 places can be filled up in 8! ways. Therefore, favourable cases to A are $m = 9 \times 8!$.

$$\therefore\ P(A) = \frac{m}{n} = \frac{9 \times 8!}{181400} = \frac{1}{5}.$$

The event $\bar{A}$ indicates that two I's cannot come together.

$$\therefore\ P(\bar{A}) = 1 - P(A) = 1 - \frac{1}{5} = \frac{4}{5}.$$

►**EXAMPLE 1.65** *If the letters of the word* 'STATISTICS' *are arranged at random, what is the probability that* 3 *S's do not occupy consecutive places?*

Solution. There are 10 letters in the word. The number of different letters are S = 3, T = 3, I = 2, A = 1 and C = 1. These letters can be arranged in

$$n = \frac{10!}{3!\ 3!\ 2!\ 1!\ 1!} \text{ ways.}$$

Let A be the event that 3 S's can take consecutive places. This can happen in the following ways:

(i) S S S — — — — — — —

(ii) — S S S — — — — — —

(iii) — — S S S — — — — —

(iv) — — — S S S — — — —

(v) — — — — S S S — — —

(vi) — — — — — S S S — —

(vii) — — — — — — S S S —

(viii) — — — — — — — S S S

In each arrangement of the 8 cases, there are 7 other places which can be filled up with the remaining 7 letters. This can be done in

$$\frac{7!}{3!\ 2!\ 1!\ 1!}$$

Total number of cases by which A can occur is

$$m = \frac{8 \times 7!}{3!\ 2!\ 1!\ 1!}.$$

$$\therefore\ P(A) = \frac{m}{n} = \frac{8 \times 7!}{3!\ 2!\ 1!\ 1!} \frac{3!\ 3!\ 2!\ 1!\ 1!}{10!} = \frac{1}{15}.$$

Let $\bar{A}$ be the event that 3 S's cannot occur together. Thus

$$P(\bar{A}) = 1 - P(A) = 1 - \frac{1}{15} = \frac{14}{15}.$$

►**EXAMPLE 1.66** *In a box there are 6 letters A, H, M, O, S, T. From the box one letter is drawn at a time and recorded. Find the probability that the letters in the order drawn spell* 'THOMAS'.

Solution. There are 6 letters. These can be drawn in $n = 6!$ ways to form a word.

Let A be the event the letters in the order drawn spell the word 'THOMAS'. This event can occur only in $m = 1$ way.

$$\therefore\ P(A) = \frac{m}{n} = \frac{1}{6!}.$$

►**EXAMPLE 1.67** *Find the probability that a hand of* 5 *cards in poker contains four cards of one kind.*

Solution. In a packet of cards there are 52 cards. Five cards from 52 cards can be distributed in $n = {}^{52}C_5 = 2598960$ ways.

Let A be the event that the poker hand contains 4 cards of one kind. There are 13 kinds of card of a suit and 4 cards of one kind for 4 suits. One kind of 13 cards can be selected in ${}^{13}C_1$ ways and 4 cards of same kind can be selected in ${}^4C_4$ ways. The fifth card is selected from the remaining 48 cards in ${}^{48}C_1$ ways. Therefore, favourable cases to A are $m = {}^{13}C_1 \times {}^4C_4 \times {}^{48}C_1$.

$$\therefore\ P(A) = \frac{m}{n} = \frac{13 \times 1 \times 48}{2598960}.$$

►**EXAMPLE 1.68** *Find the probability that a poker hand contains a full house, that is, three of one kind and two of another kind.*

Solution. From 52 cards a poker hand contains 5 cards in $n = {}^{52}C_5 = 2598960$ ways.

Let A be the event that a poker hand contains a full house. There are 13 kinds of card in 4 different suits. Three cards of one kind and 2 cards of another kind can be distributed in ${}^4C_3 \times {}^4C_2$ ways. But each kind of 2 cards be selected in ${}^{13}C_2$ ways. Therefore, favourable cases to B are $m = {}^{13}C_2 \times {}^4C_3 \times {}^4C_2$.

$$\therefore\ P(A) = \frac{m}{n} = \frac{78 \times 4 \times 6}{2598960}.$$

►**EXAMPLE 1.69** *Find the probability that a poker hand contains at least one ace.*

Solution. The poker hand can get the cards in $n = {}^{52}C_5$ ways. Let A be the event that the poker hand does not contain an ace. Without ace there are 48 cards. The poker hand can get 5 cards from 48 cards in ${}^{48}C_5$ ways.

$$P(A) = \frac{{}^{48}C_5}{{}^{52}C_5}.$$

But $\bar{A}$ is the event that the poker hand contains at least an ace.

$$\therefore\ P(\bar{A}) = 1 - P(A) = 1 - \frac{{}^{48}C_5}{{}^{52}C_5}.$$

►EXAMPLE 1.70 *What is the probability that a poker hand contains a royal flush (that 5 honours card of same suit)?*

Solution. Five cards from 52 cards can be distributed in

$$n = {}^{52}C_5 = 2598960 \text{ ways.}$$

Let A be the event that a poker hand contains a royal flush, that is, the cards $(10, J, Q, K, A)$ of same suit. As there are 4 suits of cards A can occur in 4 ways.

$$\therefore\ P(A) = \frac{m}{n} = \frac{4}{2598960}.$$

►EXAMPLE 1.71 *What is the probability that a poker hand does not contain a flush?*

Solution. Five cards from 52 cards can be distributed in

$$n = {}^{52}C_5 = 2598960 \text{ ways.}$$

Let A be the event that the poker hand contains a flush, that is, five cards of the same suit. There are 13 cards in each suit and there are 4 suits. Favourable cases to A are $m = 4 \times {}^{13}C_5$.

$$\therefore\ P(A) = \frac{m}{n} = \frac{4 \times {}^{13}C_5}{{}^{52}C_5}.$$

►EXAMPLE 1.72 *What is the probability that a poker hand contains a straight?*

Solution. Five cards can be distributed to a poker hand in

$$n = {}^{52}C_5 = 2598960 \text{ ways.}$$

Let A be the event that the poker hand contains a straight, that is, five cards in sequence, where ace is the lowest card or the highest card [i.e., the sequences are $1, 2, 3, 4, 5$; $2, 3, 4, 5, 6$; ... or $9, 10, J, Q, K$ and $10, J, Q, K, A$]. The sequence can occur in 10 ways. There are 13 kinds of cards and 4 suits and hence there are 4 cards of same kind. Therefore, the straight can be formed in $m = 10 \times 4^5$ ways.

$$\therefore\ P(A) = \frac{m}{n} = \frac{10 \times 4^5}{2598960}.$$

►EXAMPLE 1.73 *What is the probability that x, y and z with the first, second and third prizes, respectively, in a drawing if 200 people enter a contest and* (i) *no one can win more than one prize,* (ii) *winning more than one prize is allowed?*

Solution.

(i) As there are 3 prizes, the prizes can be distributed in ${}^{200}C_3$ ways. But the prizes can be arranged in $3! = 6$ ways. Hence, total number of ways in which the prizes can be distributed is $n = 6 \times {}^{200}C_3 = 7880400$. Let A be the event that x, y and z win the first, second and third prize, respectively. This can occur in $m = 1$ way.

$$\therefore\ P(A) = \frac{m}{n} = \frac{1}{7880400}.$$

(ii) Since any one can win more than one prize, the prizes can be distributed in $n = 200^3 = 8000000$ ways. Let B be the event that x, y and z win the first, second and third prize, respectively. This can occur in $m = 1$ way.

$$\therefore\ P(B) = \frac{m}{n} = \frac{1}{8000000}.$$

1.4 Theorems on Probability of Events

❐ **THEOREM 1.1** *Probability of impossible event is zero.*

Proof. Let S be the sample space of a random experiment and A be the impossible event defined on the sample. Favourable points to $A = \phi$ and hence $S \cup \phi = S$.

$$\begin{aligned} \therefore\ P(S \cup \phi) &= P(S) \\ P(S) + P(\phi) &= P(S) \\ \Rightarrow\ P(\phi) &= 0. \end{aligned}$$

❐ **THEOREM 1.2** *Probability of the complementary event $\bar{A}$ is given by $P(\bar{A}) = 1 - P(A)$, where A is the event.*

Proof. Let S be the sample space of a random experiment and A be the event defined on the sample space S. Then $\bar{A}$ is the complementary event, where A and $\bar{A}$ are mutually exclusive (disjoint) events.

Moreover,

$$\begin{aligned} A \cup \bar{A} &= S \\ P(A \cup \bar{A}) &= P(S) \\ P(A) + P(\bar{A}) &= 1\ \Rightarrow\ P(\bar{A}) = 1 - P(A). \end{aligned}$$

❐ **THEOREM 1.3** *For any two events A and B, $P(\bar{A} \cap B) = P(B) - P(A \cap B)$.*

Proof. Let S be the sample space of a random experiment. Consider two events A and B defined on the sample space. The two events on the sample space can be shown by Venn diagram as follows [FIG. 1.9]:

It is observed from the Venn diagram that the events $A \cap B$ and $\bar{A} \cap B$ are disjoints. We see that

$$(A \cap B) \cup (\bar{A} \cap B) = B$$
$$\therefore\ P(A \cap B) \cup P(\bar{A} \cap B) = P(B)$$
$$\Rightarrow P(A \cap B) + P(\bar{A} \cap B) = P(B)$$
$$\therefore\ P(\bar{A} \cap B) = P(B) - P(A \cap B).$$

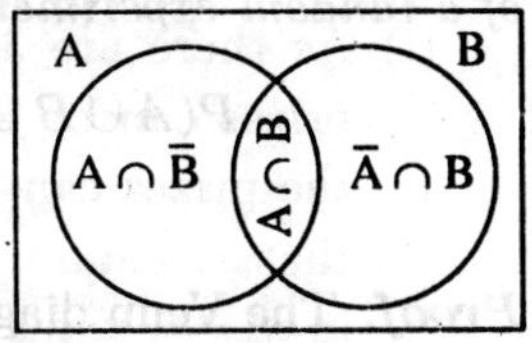

FIG. 1.9

❑ THEOREM 1.4 (Addition Law of Probability)

Let A and B be two events defined on the same sample space S of an experiment. Then

$$P(A \cup B) = P(A) + P(B) - P(A \cap B)$$

if A and B are not mutually exclusive events

$$= P(A) + P(B)$$

if A and B are mutually exclusive events.

Proof. The Venn diagram of the events is shown in FIG. 1.10.

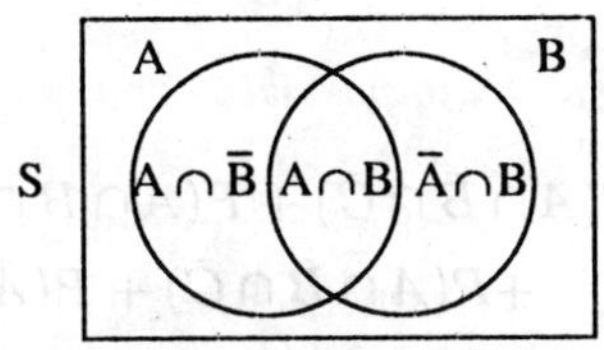

FIG. 1.10

It is observed that $A \cap \bar{B}$ and $A \cap B$ are mutually exclusive events, again $A \cap B$ and $\bar{A} \cap B$ are mutually exclusive events.

Hence,

$$\begin{aligned} A \cup B &= A \cap \bar{B} + A \cap B + \bar{A} \cap B \\ &= A + \bar{A} \cap B \\ &= A + B - A \cap B \end{aligned}$$

$$\therefore\ P(A \cup B) = P(A) + P(B) - P(A \cap B).$$

But if A and B are mutually exclusive events, then

$$A \cap B = \phi \text{ (empty set) and } P(A \cap B) = 0.$$

Therefore, $P(A \cup B) = P(A) + P(B)$.

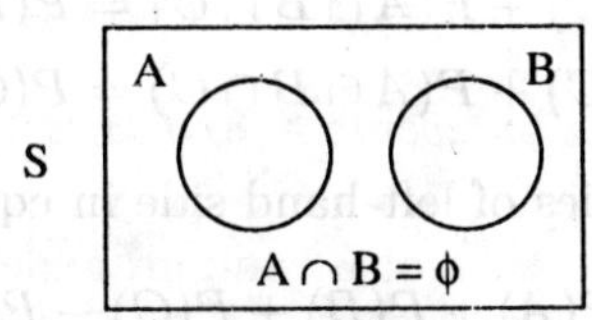

FIG. 1.11

❑ THEOREM 1.5 *If A, B and C are any three events defined on a sample space S of a random experiment, then*

$$\begin{aligned}P(A \cup B \cup C) &= P(A)+P(B)+P(C)-P(A \cap B) \\ &\quad -P(A \cap C)-P(B \cap C)+P(A \cap B \cap C).\end{aligned}$$

Proof. The Venn diagram of three events can be shown in FIG. 1.12.

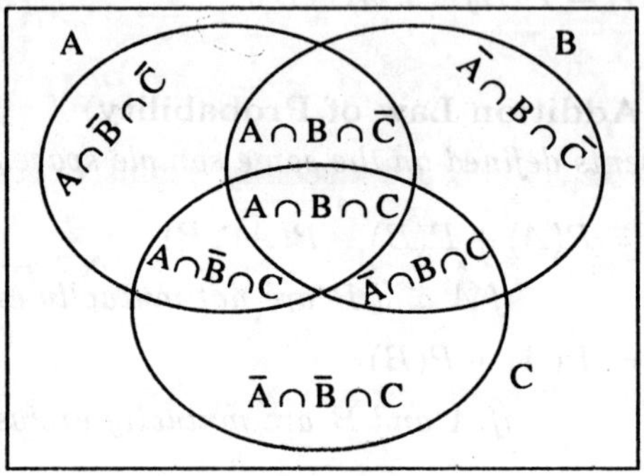

FIG. 1.12

Now,

$$\begin{aligned}P(A \cup B \cup C) &= P(A \cap \bar{B} \cap \bar{C})+P(A \cap B \cap \bar{C})+P(A \cap B \cap C) \\ &\quad +P(A \cap \bar{B} \cap \bar{C})+P(\bar{A} \cap B \cap \bar{C}) \\ &\quad +P(\bar{A} \cap B \cap C)+P(\bar{A} \cap \bar{B} \cap C) \qquad \textbf{(a)}\end{aligned}$$

[$\because$ all the events are disjoint].

We have,

$$\begin{aligned}P(A) &= P(A \cap \bar{B} \cap \bar{C})+P(A \cap B \cap \bar{C})+P(A \cap B \cap C)+P(A \cap \bar{B} \cap C) \\ P(B) &= P(\bar{A} \cap B \cap \bar{C})+P(A \cap B \cap \bar{C})+P(A \cap B \cap C)+P(\bar{A} \cap B \cap C) \\ P(C) &= P(A \cap \bar{B} \cap C)+P(A \cap B \cap C)+P(\bar{A} \cap B \cap C)+P(\bar{A} \cap \bar{B} \cap C).\end{aligned}$$

Again,

$$\begin{aligned}P(A) &= P(A \cap \bar{B} \cap \bar{C})+P(A \cap B)+P(A \cap \bar{B} \cap C) \\ P(B) &= P(\bar{A} \cap B \cap \bar{C})+P(A \cap B \cap \bar{C})+P(B \cap C) \\ P(C) &= P(\bar{A} \cap \bar{B} \cap C)+P(A \cap C)+P(\bar{A} \cap B \cap C) \\ P(A \cap \bar{B} \cap \bar{C})+P(A \cap \bar{B} \cap C) &= P(A)-P(A \cap B) \\ P(\bar{A} \cap B \cap \bar{C})+P(A \cap B \cap \bar{C}) &= P(B)-P(B \cap C) \\ P(\bar{A} \cap \bar{B} \cap C)+P(\bar{A} \cap B \cap C) &= P(C)-P(A \cap C).\end{aligned}$$

Replacing these probabilities of left-hand side in equation (a), we have

$$\begin{aligned}P(A \cup B \cup C) &= P(A)+P(B)+P(C)-P(A \cap B)-P(B \cap C) \\ &\quad -P(A \cap C)+P(A \cap B \cap C).\end{aligned}$$

COROLLARY 5. Let θ be any subset of the sample space, then if $A_1, A_2, \ldots, A_k \in \theta$, then

$$P(A_1 \cup A_2 \cup \cdots \cup A_k) = \sum_{i=1}^{k} P(A_i) - \sum_{\substack{i \neq j=1 \\ i<j=2}}^{k} P(A_i \cap A_j)$$

$$+ \sum_{i<j<1<3}^{k} P(A_i \cap A_j \cap A_l) - \cdots$$

$$+(-1)^n P(A_i \cap A_2 \cap \cdots \cap A_k).$$

However, if $A_1, A_2, \ldots, A_k$ are mutually exclusive events, then

$$P(A_1 \cup A_2 \cup \cdots \cup A_k) = \sum_{i=1}^{k} P(A_i).$$

This can be proved in a similar way as it is done for Theorem 1.5.

COROLLARY 6. If A, B and C are three mutually exclusive events, then

$$P(A \cup B \cup C) = P(A) + P(B) + P(C).$$

COROLLARY 7. If A and $B \in \theta$ and θ is a subset of a sample space of a random experiment, then

$$P(A \cup B) \leq P(A) + P(B).$$

The sign of equality remains valid if $P(A \cap B) = 0$. In a similar way, we can write

$$P\left(\bigcup_{i=1}^{k} A_i\right) \leq \sum_{i=1}^{k} P(A_i).$$

COROLLARY 8. If A and $B \in \theta$, where θ is a subset of a sample space of a random experiment and if $A \subseteq B$, then

$$P(A) \leq P(B).$$

Here as $A \subseteq B, A \cap B = A$. Hence,

$$P(B - A) = P(B) - P(A \cap B) = P(B) - P(A) \geq 0.$$

$$\therefore P(A) \leq P(B).$$

COROLLARY 9. If A and $B \in \theta$, where θ is a subset of a sample space S of a random experiment, then

$$P(A \cap B) \geq 1 - P(\bar{A}) - P(\bar{B}).$$

If A and B are not mutually exclusive events, then

$$\overline{A \cap B} = \bar{A} \cup \bar{B}.$$

Again, from the definition of complementary event, we can write

$$\begin{aligned} P(A\cap B) &= 1-P(\overline{A\cap B})\\ &= 1-P(\bar{A}\cup\bar{B})\\ &= 1-\{P(\bar{A})+P(\bar{B})-P(\bar{A}\cap\bar{B})\}\\ &= 1-P(\bar{A})-P(\bar{B})+P(\bar{A}\cap\bar{B}).\\ \therefore\; P(A\cap B) &\geq 1-P(\bar{A})-P(\bar{B}). \end{aligned}$$

Similarly, if $A_1, A_2, \ldots, A_k \in \theta$, then

$$P\left(\bigcap_{i=1}^{k} A_i\right) = 1-P\left(\bigcup_{i=1}^{k} \bar{A}_i\right) \geq 1-\sum_{i=1}^{k} P(\bar{A}_i).$$

❒ THEOREM 1.6 (Boole's Inequality)

Let $A_1, A_2, \ldots, A_n$ be n events defined on a sample space S of a random experiment. Then

(i) $P\left(\bigcap_{i=1}^{n} A_i\right) \geq \sum_{i=1}^{n} P(A_i)-(n-1)$

(ii) $P\left(\bigcup_{i=1}^{n} A_i\right) \leq \sum_{i=1}^{n} P(A_i)$

Proof.

(i) We know that, if A_1 and A_2 are two not mutually exclusive events, then

$$\begin{aligned} P(A_1\cup A_2) &= P(A_1)+P(A_2)-P(A_1\cap A_2)\leq 1\\ \therefore\; P(A_1\cap A_2) &\geq P(A_1)+P(A_2)-1 \text{ [Bonferroni's inequality]} \qquad \text{(a)} \end{aligned}$$

This theorem is true for $n=2$ events. Thus, if $n=r$ events, then

$$P(A_1\cap A_2\cap\cdots\cap A_r) \geq P(A_1)+P(A_2)+\cdots+P(A_r)-(r-1). \qquad \text{(b)}$$

Again,

$$\begin{aligned} P\left(\bigcap_{i=1}^{r+1} A_i\right) &= P\left(\bigcap_{i=1}^{r} A_i\cap A_{r+1}\right)\\ &\geq P\left(\bigcap_{i=1}^{r} A_i\right)+P(A_{r+1})-1 \text{ [according to (a)]}\\ &\geq \sum_{i=1}^{r} P(A_i)-(r-1)+P(A_{r+1})-1 \text{ [according to (b)].}\\ \therefore\; P\left(\bigcap_{i=1}^{r+1} A_i\right) &\geq \sum_{i=1}^{r+1} P(A_i)-r. \end{aligned}$$

Thus, the theorem is true for $n=r+1$ events. Therefore, by mathematical induction the theorem is true for any value of n.

(ii) From theorem (i) above, we have

$$\begin{aligned} P(\bar{A}_1 \cap \bar{A}_2 \cap \cdots \cap \bar{A}_n) &\geq P(\bar{A}_1) + P(\bar{A}_2) + \cdots + P(\bar{A}_n) - (n-1) \\ &= 1 - P(A_1) + 1 - P(A_2) + \cdots \\ &\quad +1 - P(A_n) - (n-1) \\ &= 1 - P(A_1) - P(A_2) - \cdots - P(A_n). \\ \therefore\ P(A_1) + P(A_2) + \cdots + P(A_n) &\geq 1 - P(\bar{A}_1 \cap \bar{A}_2 \cap \cdots \cap \bar{A}_n) \\ &= 1 - P(\overline{A_1 \cap A_2 \cap \cdots \cap A_n}) \\ &= P(A_1 \cup A_2 \cup \cdots \cup A_n). \\ \therefore\ P\left(\bigcup_{i=1}^{n} A_i\right) &\leq \sum_{i=1}^{n} P(A_i). \end{aligned}$$

❑ **THEOREM 1.7** *Let $A_1, A_2, \ldots, A_n$ be n events defined on a sample space S of a random experiment. Then*

$$P\left(\bigcup_{i=1}^{n} A_i\right) \geq \sum_{i=1}^{n} P(A_i) - \sum_{i<j}^{n}\sum P(A_i \cap A_j).$$

Proof. For three events A_1, A_2 and A_3, we have

$$\begin{aligned} P(A_1 \cup A_2 \cup A_3) &= P(A_1) + P(A_2) + P(A_3) - P(A_1 \cap A_2) - P(A_1 \cap A_3) \\ &\quad - P(A_2 \cap A_3) + P(A_1 \cap A_2 \cap A_3) \\ \Rightarrow P(A_1 \cup A_2 \cup A_3) &\geq \sum_{i=1}^{3} P(A_i) - \sum_{i<j=1}^{3}\sum P(A_i \cap A_j). \end{aligned}$$

The theorem is true for $n = 3$ events. If $n = r$, then

$$P\left(\bigcup_{i=1}^{r} A_i\right) \geq \sum_{i=1}^{r} P(A_i) - \sum_{i<j=1}^{r}\sum P(A_i \cap A_j). \tag{a}$$

Now,

$$\begin{aligned} P\left(\bigcup_{i=1}^{r+1} A_i\right) &= P\left(\bigcup_{i=1}^{r} A_i \cup A_{r+1}\right) \\ &= P\left(\bigcup_{i=1}^{r} A_i\right) + P(A_{r+1}) - P\left[\left(\bigcup_{i=1}^{r} A_i\right) \cap A_{r+1}\right] \\ &= P\left(\bigcup_{i=1}^{r} A_i\right) + P(A_{r+1}) - P\left[\bigcup_{i=1}^{r} A_i \cap A_{r+1}\right] \\ &\geq \sum_{i=1}^{r} P(A_i) - \sum_{i<j=1}^{r}\sum P(A_i \cap A_j) \text{ [from (a)]} \\ &\quad + P(A_{r+1}) - P\left[\bigcup_{i=1}^{r} (A_i \cap A_{r+1})\right] \end{aligned} \tag{b}$$

Again, from Boole's inequality, we have

$$P\left[\bigcup_{i=1}^{r}(A_i \cap A_{r+1})\right] \le \sum_{i=1}^{r} P(A_i \cap A_{r+1})$$

$$\Rightarrow -P\left[\bigcup_{i=1}^{r}(A_i \cap A_{r+1})\right] \ge -\sum_{i=1}^{r} P(A_i \cap A_{r+1}).$$

Therefore, from (b), we have

$$P\left(\bigcup_{i=1}^{r+1} A_i\right) \ge \sum_{i=1}^{r+1} P(A_i) - \sum\sum_{i<j=1}^{r} P(A_i \cap A_j) - \sum_{j=1}^{r} P(A_i \cap A_{r+1})$$

$$\Rightarrow P\left(\bigcup_{i=1}^{r+1} A_i\right) \ge \sum_{i=1}^{r+1} P(A_i) - \sum\sum_{i<j=1}^{r} P(A_i \cap A_j).$$

The theorem is true for $n = r+1$ events. Also, it is true for $n = 3$ events. Hence, by mathematical induction, we can say that the result is true for any value of n.

►**EXAMPLE 1.74** *From a pack of* 52 *cards one card is taken at random. Find the probability that the card* (i) *is either an ace or a spade,* (ii) *is either a spade or a heart.*

Solution. From a pack of 52 cards one card can be taken in

$$n = {}^{52}C_1 = 52 \text{ ways.}$$

(i) Let A be the event that the card is an ace and B be the event that the card is a spade.

Here the cards are numbered by $1, 2, \ldots, 13$ and S = spade, H = heart, C = club and D = diamond [FIG. 1.13].

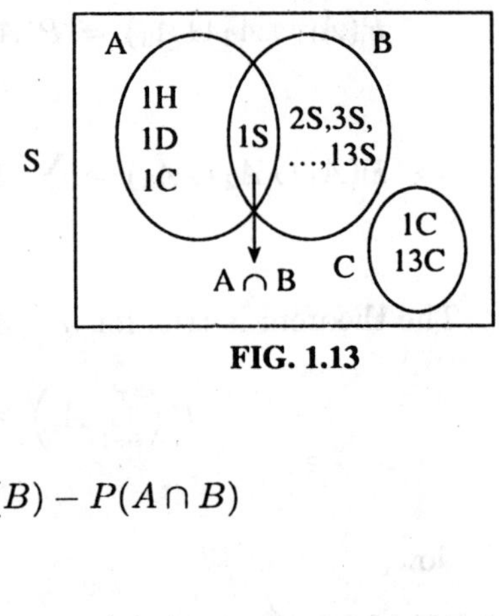

FIG. 1.13

We need $P(A \text{ or } B)$.

$$\begin{aligned} P(A \text{ or } B) &= P(A \cup B) = P(A) + P(B) - P(A \cap B) \\ &= \frac{4}{52} + \frac{13}{52} - \frac{1}{15} = \frac{4}{13}. \end{aligned}$$

(ii) Let C be the event that the card is a heart. We need $P(B \text{ or } C)$

$$\begin{aligned} P(B \text{ or } C) &= P(B \cup C) \\ &= P(B) + P(C) \quad [\because B \text{ and } C \text{ are mutually exclusive events}] \\ &= \frac{13}{52} + \frac{13}{52} = \frac{1}{2}. \end{aligned}$$

►**EXAMPLE 1.75** *Two unbiased dice are thrown once. Find the probability that* (i) *sum of the upper faces of the dice is* 10 *or more or first die shows number* 6, (ii) *second die shows* 6 *or sum of the upper faces of the dice is less than* 7.

Solution. The sample space of the experiment is

$$S : \left\{ \begin{array}{cccccc} 11 & 21 & 31 & 41 & 51 & 61 \\ 12 & 22 & 32 & 42 & 52 & 62 \\ 13 & 23 & 33 & 43 & 53 & 63 \\ 14 & 24 & 34 & 44 & 54 & 64 \\ 15 & 25 & 35 & 45 & 55 & 65 \\ 16 & 26 & 36 & 46 & 56 & 66 \end{array} \right\}, \quad n = 36.$$

(i) Let A be the event that sum of the upper faces of the dice is 10 or more and B be the event that first die shows number 6. Favourable cases to A are

$$A : \{46, 55, 56, 64, 65, 66\}$$

and favourable cases to B are $B : \{61, 62, 63, 64, 65, 66\}$.

$$\therefore \; A \cap B : \{64, 65, 66\}.$$

We need

$$\begin{aligned} P(A \cup B) &= P(A) + P(B) - P(A \cap B) \\ &= \frac{6}{36} + \frac{6}{36} - \frac{3}{36} = \frac{1}{4}. \end{aligned}$$

(ii) Let A be the event that second die shows 6 and B be the event that sum of the upper faces of the dice is less than 7. Favourable cases to A are

$$A : \{16, 26, 36, 46, 56, 66\}$$

and favourable cases to B are

$$B : \{11, 12, 13, 14, 15, 21, 22, 23, 24, 31, 32, 33, 41, 42, 51\}.$$

We need

$$\begin{aligned} P(A \cup B) &= P(A) + P(B) \; [\because A \text{ and } B \text{ are mutually exclusive events}] \\ &= \frac{6}{36} + \frac{15}{36} \\ &= \frac{7}{12}. \end{aligned}$$

►**EXAMPLE 1.76** *A company produces computer spare parts using two machines A and B. In a day machines A and B produces 400 and 300 spare parts, respectively. It is known that 10% spare parts of A and 20% of B are usually found defective. From a day's output a spare part is selected at random. What is the probability that it is the product of A or it is a defective spare part?* ●

Solution. Let A be the event that the product is produced by A and B be the event that the product is defective. There are 40 (10% of 400) +60 (20% of 300) = 100 defective spare parts. One spare part can be selected from (400 + 300) = 700 spare parts in $n = {}^{700}C_1 = 700$ ways.

We need $P(A \cup B)$.

$$\begin{aligned} P(A \cup B) &= P(A) + P(B) - P(A \cap B) \\ &= \frac{400}{700} + \frac{100}{700} - \frac{40}{700} \\ &= \frac{23}{35}. \end{aligned}$$

►**EXAMPLE 1.77** *From a group of 4 persons P_1, P_2, P_3 and P_4 a committee of two persons are to be formed. What is the probability that the committee consists of either 2 persons from the group of P_1, P_2 and P_3 or from the group of P_2, P_3 and P_4?*

Solution. Two persons from 4 persons can be selected in $n = {}^4C_2 = 6$ ways.

Let A be the event that 2 persons are selected from P_1, P_2 and P_3, and B be the event that 2 persons are selected from P_2, P_3 and P_4. We need $P(A \cup B)$. Here the favourable cases to A and B are, respectively

$$A : \{P_1P_3, P_1P_2, P_2P_3\} \text{ and } B : \{P_2P_3, P_2P_4, P_3P_4\}.$$

$$\therefore \quad A \cap B : \{P_2P_3\}$$

$$P(A \cup B) = P(A) + P(B) - P(A \cap B) = \frac{3}{6} + \frac{3}{6} - \frac{1}{6} = \frac{5}{6}.$$

►**EXAMPLE 1.78** *In an examination 20% students failed in statistics, 30% failed in mathematics and 10% students failed in both the courses. One student is selected at random. Find the probability that he has failed in a course.*

Solution. Let A be the event that the student failed in statistics and B be the event that the student failed in mathematics. We need $P(A \cup B)$.

$$\begin{aligned} P(A \cup B) &= P(A) + P(B) - P(A \cap B) \\ &= 0.20 + 0.30 - 0.10 = 0.40. \end{aligned}$$

►**EXAMPLE 1.79** *A student is expected to sit for the examination in three courses C_1, C_2 and C_3. From his class performance it is observed that he has the chances to pass* 0.80, 0.70 *and* 0.75, *respectively in three courses. What is the least probability that the student will pass in all three courses?*

Solution. Given $P(C_1) = 0.80, P(C_2) = 0.70$ and $P(C_3) = 0.75$. Therefore, $P(\bar{C}_1) = 0.20, P(\bar{C}_2) = 0.30$ and $P(\bar{C}_3) = 0.25$.

We need $P(C_1 \cap C_2 \cap C_3)$. We know

$$\begin{aligned} P(C_1 \cap C_2 \cap C_3) &\geq 1 - P(\bar{C}_1) - P(\bar{C}_2) - P(\bar{C}_3) \\ &= 1 - 0.20 - 0.30 - 0.25 = 0.25. \end{aligned}$$

$$\therefore \quad P(C_1 \cap C_2 \cap C_3) \geq 0.45.$$

►**EXAMPLE 1.80** *The chance of a student to pass the subject Computer Science is* 0.75 *and the chance for English is* 0.80. *He has the chance to pass at least one of the subjects is* 0.85. *What is the probability that the student will pass* (i) *in both subjects,* (ii) *in one subject,* (iii) *in no subject?*

Solution. Let A be the event that the student passes in Computer Science and B be the event that the student passes in English. Given

$$P(A) = 0.75,\ P(B) = 0.80 \text{ and } P(A \cup B) = 0.85.$$

(i) We need $P(A \cap B)$.

$$\begin{aligned} P(A \cup B) &= P(A) + P(B) - P(A \cap B) \\ \therefore P(A \cap B) &= P(A) + P(B) - P(A \cup B) \\ &= 0.75 + 0.80 - 0.85 = 0.70. \end{aligned}$$

(ii) The student passes one subject. The cases in favour of passing a subject is $\bar{A} \cap B$ or $A \cap \bar{B}$.

We need $P(\bar{A} \cap B) + P(A \cap \bar{B})$.

$$\begin{aligned} P(A \cap \bar{B}) &= P(A) - P(A \cap B) = 0.75 - 0.70 = 0.05 \\ P(\bar{A} \cap B) &= P(B) - P(A \cap B) = 0.80 - 0.70 = 0.10 \\ \therefore P(\bar{A} \cap B) + P(A \cap \bar{B}) &= 0.05 + 0.10 = 0.15. \end{aligned}$$

(iii) We need $1 - P(A \cup B)$.

$$\therefore 1 - P(A \cup B) = 1 - 0.85 = 0.15.$$

►**EXAMPLE 1.81** *What is the probability that a positive integer not exceeding* 50 *selected at random is divisible by* 3 *or* 5*?*

Solution. One positive integer can be selected in

$$n = {}^{50}C_1 = 50 \text{ ways.}$$

Let A be the event that the selected integers are divisible by 3 and B be the event that the selected integers are divisible by 5. Favourable cases to A and B are, respectively

$$\begin{aligned} A &: \{3, 6, 9, 12, 15, 18, 21, 24, 27, 30, 33, 36, 39, 42, 45, 48\} \\ \text{and } B &: \{5, 10, 15, 20, 25, 30, 35, 40, 45, 50\}. \\ \therefore\ A \cap B &: \{15, 30, 45\}. \end{aligned}$$

We need $P(A \cup B)$.

$$\begin{aligned} P(A \cup B) &= P(A) + P(B) - P(A \cap B) \\ &= \frac{16}{50} + \frac{10}{50} - \frac{3}{50} \\ &= \frac{23}{50}. \end{aligned}$$

►**EXAMPLE 1.82** *What is the probability that a positive integer from integers* $1, 2, \ldots, 20$ *selected at random is divisible by* 3 *or it is a prime number?*

Solution. One positive integer can be selected in

$$n = {}^{20}C_1 = 20 \text{ ways.}$$

Let A be the event that the selected integers are divisible by 3 and B be the event that the selected integers are prime number. Favourable cases to A and B are, respectively

$$A : \{3, 6, 9, 12, 15, 18\}$$
$$\text{and } B : \{1, 3, 5, 7, 11, 13, 17, 19\}.$$

We need $P(A \cup B)$.

$$\begin{aligned} P(A \cup B) &= P(A) + P(B) - P(A \cap B) \\ &= \frac{6}{20} + \frac{8}{20} - \frac{1}{20} \\ &= \frac{13}{20} \ [\because A \cap B : \{3\}]. \end{aligned}$$

►**EXAMPLE 1.83** *A survey is conducted over* 100 *persons living in a city centre and observed that, among these* 100 *persons* 65 *persons read the newspaper A,* 40 *persons read the paper B,* 70 *persons read the paper C,* 30 *persons read both A and B,* 40 *persons read both A and C,* 25 *persons read both B and C and* 20 *persons read all three papers. A person is selected at random. What is the probability that he reads* (i) *only A,* (ii) *only A or B,* (iii) *none of three papers,* (iv) *at least one paper,* (v) *exactly* 2 *papers?*

Solution. The Venn diagram of the results is in FIG. 1.14.

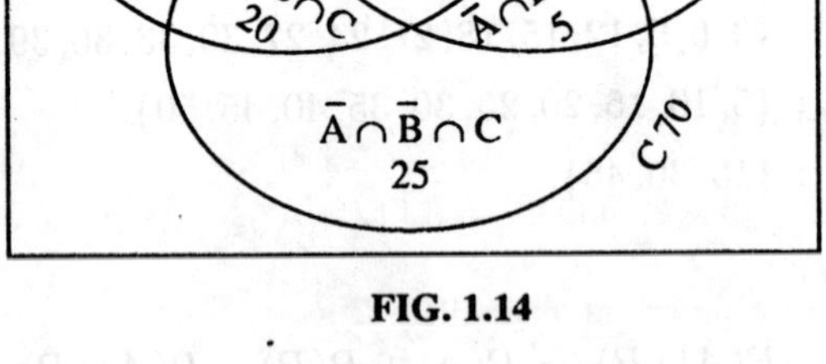

FIG. 1.14

(i) The event to read only A is $A \cap \bar{B} \cap \bar{C}$.

$$\begin{aligned} P(A \cup \bar{B} \cap \bar{C}) &= P(A) - P(A \cap B) - P(A \cap C) + P(A \cap B \cap C) \\ &= \frac{65}{100} - \frac{30}{100} - \frac{40}{100} + \frac{20}{100} \\ &= 0.15. \end{aligned}$$

(ii) The favourable events to read only A or B are $A \cap \bar{B} \cap \bar{C}$, $A \cap B \cap \bar{C}$ and

$\bar{A} \cap B \cap \bar{C}$.

$$A \cap B \cap \bar{C} = (A \cap B) - (A \cap B \cap C).$$

$$\therefore\ P(A \cap B \cap \bar{C}) = P(A \cap B) - P(A \cap B \cap C)$$

$$= \frac{30}{100} - \frac{20}{100}$$

$$= \frac{10}{100}.$$

$$\therefore\ P(\text{only reading } A \text{ or } B) = P(A \cap \bar{B} \cap \bar{C}) + P(A \cap B \cap \bar{C}) + P(\bar{A} \cap B \cap \bar{C})$$

$$= \frac{15}{100} + \frac{10}{100} + \frac{5}{100}$$

$$= \frac{30}{100} = 0.3.$$

(iii) The event in favour of reading either of 3 is $A \cup B \cup C$.

We need $1 - P(A \cup B \cup C)$.

$$\therefore\ P(A \cup B \cup C) = P(A) + P(B) + P(C) - P(A \cap B) - P(A \cap C)$$

$$P(B \cap C) + P(A \cap B \cap C)$$

$$= \frac{65}{100} + \frac{40}{100} + \frac{70}{100} - \frac{30}{100} - \frac{40}{100} - \frac{25}{100} + \frac{20}{100}$$

$$= 1.0.$$

$$\therefore\ 1 - P(A \cup B \cup C) = 1 - 1.0 = 0.0.$$

(iv) The event in favour of reading at least one paper is $A \cup B \cup C$.

$$\therefore\ P(A \cup B \cup C) = P(A) + P(B) + P(C) - P(A \cap B) - P(A \cap C)$$

$$-P(B \cap C) + P(A \cap B \cap C)$$

$$= 1.0 \ [\text{(iii)}]$$

(v) The events in favour of reading exactly two papers are

$$A \cap B \cap \bar{C}, \bar{A} \cap B \cap C \text{ and } A \cap \bar{B} \cap C.$$

From Venn diagram, we have

$$A \cap B \cap \bar{C} = A \cap B - A \cap B \cap C.$$

Favourable cases to this event are $(30 - 20) = 10$

$$A \cap \bar{B} \cap \bar{C} = A \cap C - A \cap B \cap C$$

Favourable cases to this event are $(40 - 20) = 20$.

$$\therefore\ P(\text{reading exactly two papers})$$

$$= P(A \cap B \cap \bar{C}) + P(\bar{A} \cap B \cap C) + P(A \cap \bar{B} \cap C)$$

$$= \frac{10}{100} + \frac{5}{100} + \frac{20}{100} = 0.35.$$

►**EXAMPLE 1.84** *Let A and B be events such that $P(A) = 0.7$ and $P(B) = 0.5$. Show that $P(A \cap B) \geq 0.2$.*

Solution. We know,

$$P(A \cap B) \geq P(A) + P(B) - 1$$
$$= 0.7 + 0.5 - 1 = 0.2.$$
$$\therefore\ P(A \cap B) \geq 0.2.$$

1.5 Conditional Probability and Related Theorems

Definition. Let A and B be any two events defined on a sample space. Then the conditional probability of A given that B has already been occurred is given by

$$P(A/B) = \frac{P(A \cap B)}{P(B)},\ \text{given that } P(B) \neq 0.$$

Here $P(A/B)$ is called *conditional probability* of A under the condition that B has already been occurred. Similarly, the conditional probability of B when A has already been occurred is given by

$$P(B/A) = \frac{P(A \cap B)}{P(A)},\ \text{given that } P(A) \neq 0.$$

Proof. Consider that the sample space contains n equally likely, exhaustive and mutually exclusive cases. Let us define an event A with n_A outcomes ($n_A \leq n$) of the sample space. Also, define another event B such that it has n_B outcomes ($n_B \leq n$) of the sample space. Let n_{AB} be the favourable cases to the event $A \cap B$. Then,

$$P(A) = \frac{n_A}{n},\ P(B) = \frac{n_B}{n},\ P(A \cap B) = \frac{n_{AB}}{n}.$$

But,

$$P(A \cap B) = \frac{n_{AB}}{n_A} \cdot \frac{n_A}{n}.$$

Here n_{AB} is the proportion of observations out of n_A observations which are in favour of A. This proportion is in favour of B when A has already been occurred. We can write

$$P(B/A) = \frac{n_{AB}}{n_A}.$$
$$\therefore P(A \cap B) = P(B/A)P(A)$$
$$\text{or, } P(B/A) = \frac{P(A \cap B)}{P(A)}.$$

Here,

$$P(A \cap B) = P(B/A)P(A)$$
$$\text{or, } P(A \cap B) = P(A/B)P(B)$$

is known as *multiplication rule* of probability. The rule can be generalised for k events.

❑ **THEOREM 1.8** *Let there be k events $A_1, A_2, \dots, A_k$ each one is defined on a sample space S. Then,*

$$P(A_1 \cap A_2 \cap \cdots \cap A_k) = P(A_1)P(A_2/A_1)P(A_3/A_1 \cap A_2) \cdots P(A_k/A_1 \cap A_2 \cap \cdots \cap A_{k-1}).$$

Here $P(A_i/A_j \cap A_l \cap \cdots \cap A_s)(j \neq l \neq \cdots \neq s)$ is called conditional probability of the event A_i given that $A_j, A_l, \dots, A_s$ have already happened.

Proof. Let us prove the theorem for 3 events A_1, A_2 and A_3. We can write

$$\begin{aligned} P(A_1 \cap A_2 \cap A_3) &= P(A_1)P(A_2 \cap A_3/A_1) \\ &= P(A_1)P(A_2/A_1)P(A_3/A_1 \cap A_2). \end{aligned}$$

The rule of multiplication is true for $n = 2$. Let us suppose that the rule is true for $n = r$. Then,

$$P(A_1 \cap A_2 \cap \cdots \cap A_r) = P(A_1)P(A_2/A_1)P(A_3/A_1 \cap A_2) \cdots P(A_r/A_1 \cap A_2 \cap \cdots \cap A_{r-1}).$$

Now,

$$P(A_1 \cap A_2 \cap \cdots \cap A_{r+1}) = P(A_1)P(A_2/A_1)P(A_3/A_1 \cap A_2) \cdots P(A_{r+1}/A_1 \cap A_2 \cap \cdots \cap A_r)$$

The rule is true for $n = r + 1$ events. Therefore, by mathematical induction it can be said that the rule is true for any value of n.

Independent Events

Two events A and B are said to be independent, if and only if $P(A \cap B) = P(A)P(B)$.

Pairwise Independent Events

A set of events $A_1, A_2, \dots, A_k$ are said to be pairwise independent, if

$$P(A_i \cap A_j) = P(A_i)P(A_j), i \neq j = 1, 2, \dots, k.$$

The events $A_1, A_2, \dots, A_k$ are said to be independent, if

$$P(A_1 \cap A_2 \cap \cdots \cap A_k) = P(A_1)P(A_2) \cdots P(A_k).$$

These k events are mutually independent, if

(i) $P(A_i \cap A_j) = P(A_i)P(A_j), i \neq j = 1, 2, \dots, k$

(ii) $P(A_i \cap A_j \cap A_l) = P(A_i)P(A_j)P(A_l), i \neq j \neq l = 1, 2, \dots, k$

(iii) $P(A_i \cap A_j \cap A_l \cap A_m) = P(A_i)P(A_j)P(A_l)P(A_m) \cdots, i \neq j \neq l \neq m = 1, 2, \dots, k$

(iv) $P(A_1 \cap A_2 \cap A_3 \cap \cdots \cap A_k) = P(A_1)P(A_2)P(A_3) \cdots P(A_k)$.

Three events A_1, A_2 and A_3 are said to be mutually independent, if

(i) $P(A_1 \cap A_2) = P(A_1)P(A_2)$

(ii) $P(A_1 \cap A_3) = P(A_1)P(A_3)$

(iii) $P(A_2 \cap A_3) = P(A_2)P(A_3)$

(iv) $P(A_1 \cap A_2 \cap A_3) = P(A_1)P(A_2)P(A_3)$.

❑ **THEOREM 1.9** *Let A, B and C be 3 events defined on a sample S. Then,*

$$P(A \cup B/C) = P(A/C) + P(B/C) - P(A \cap B/C).$$

Proof. We have

$$P(A \cup B) = P(A) + P(B) - P(A \cap B) \text{ [Addition law of probability]}$$

Also, we can write

$$P(A \cap C) \cup (B \cap C)] = P(A \cap C) + P(B \cap C) - P(A \cap B \cap C).$$

Dividing both sides by $P(C)$ $[P(C) \neq 0]$, we have

$$\frac{P(A \cap C) \cup (B \cap C)}{P(C)} = \frac{P(A \cap C)}{P(C)} + \frac{P(B \cap C)}{P(C)} - \frac{P(A \cap B \cap C)}{P(C)}$$

$$\Rightarrow \frac{P[(A \cup B) \cap C)]}{P(C)} = P(A/C) + P(B/C) - P(A \cap B/C)$$

$$\text{or, } P[(A \cup B)/C] = P(A/C) + P(B/C) - P(A \cap B/C).$$

❑ **THEOREM 1.10** *If A, B and C are 3 events defined on a sample space S, then $P(A \cap \bar{B}/C) + P(A \cap B/C) = P(A/C)$.*

Proof.

$$P(A \cap \bar{B}/C) + P(A \cap B/C) = \frac{P(A \cap \bar{B} \cap C)}{P(C)} + \frac{P(A \cap B \cap C)}{P(C)}$$

$$= \frac{P(A \cap \bar{B} \cap C) + P(A \cap B \cap C)}{P(C)}$$

$$= \frac{P(A \cap C)}{P(C)} = P(A/C).$$

❑ **THEOREM 1.11** *Let A, B and C be 3 events defined on a sample space S; then $P(B/A) \leq P(C/A)$ if $B \subset C$ and $P(A) > 0$.*

Proof.

$$P(C/A) = \frac{P(A \cap C)}{P(A)} = \frac{P[(B \cap C \cap A) \cup (\bar{B} \cap C \cap A)]}{P(A)}$$

$$= \frac{P[(B \cap C \cap A)]}{P(A)} + \frac{P[(\bar{B} \cap C \cap A]}{P(A)}$$

$$= P(B \cap C/A) + P(\bar{B} \cap C/A).$$

Now, since $B \subset C, B \cap C = B$,

$$\therefore \; P(C/A) = P(B/A) + P(\bar{B} \cap C/A) \Rightarrow \; P(B/A) \leq P(C/A).$$

❏ **THEOREM 1.12** *If A and B are two independent events, then* (i) *A and $\bar{B}$ are independent,* (ii) *$\bar{A}$ and $\bar{B}$ are independent.*

Proof.

(i) Let us write

B \ A	B	$\bar{B}$	*Total*
A	$A \cap B$	$A \cap \bar{B}$	$A \cap B + A \cap \bar{B}$
$\bar{A}$	$\bar{A} \cap B$	$\bar{A} \cap \bar{B}$	$\bar{A} \cap B + \bar{A} \cap \bar{B}$
Total	$A \cap B + \bar{A} \cap B$	$A \cap \bar{B} + \bar{A} \cap \bar{B}$	S

It is observed that

$$\begin{aligned} P(A \cap \bar{B}) &= P(A) - P(A \cap B) \\ &= P(A) - P(A)P(B) \; [\because \; A \text{ and } B \text{ are independent events}] \\ &= P(A)[1 - P(B)] = P(A)P(\bar{B}). \end{aligned}$$

Hence, A and $\bar{B}$ are independent events.

(ii) Since A and B are independent events, $P(A \cap B) = P(A)P(B)$.

Again,

$$\begin{aligned} P(\bar{A} \cap \bar{B}) &= P(\overline{A \cup B}) = 1 - P(A \cup B) \\ &= 1 - P(A) - P(B) + P(A \cap B) \\ &= P(\bar{A}) - P(B) + P(A)P(B) \\ &= P(\bar{A}) - P(B)[1 - P(A)] = P(\bar{A}) - P(B)P(\bar{A}) \\ &= P(\bar{A})[1 - P(B)] = P(\bar{A})P(\bar{B}) \end{aligned}$$

Hence, $\bar{A}$ and $\bar{B}$ are independent events.

❏ **THEOREM 1.13** *If A, B and C are* 3 *mutually independent events, then $A \cup B$ and C are also independent.*

Proof. We need to prove that $P[(A \cup B) \cap C] = P(A \cap B)P(C)$.

$$\begin{aligned} \text{L.H.S.} &= P[(A \cap B) \cup (B \cap C)] \; [\text{Distributive law}] \\ &= P(A \cap C) + P(B \cap C) - P(A \cap B \cap C) \\ &= P(A)P(C) + P(B)P(C) - P(A)P(B)P(C) \\ &\qquad [\because \; A, B \text{ and } C \text{ are mutually independent}] \\ &= P(C)[P(A) + P(B) - P(A)P(B)] \\ &= P(C)P(A \cup B) = \text{R.H.S.} \end{aligned}$$

Hence, $(A \cup B)$ and C are independent.

❑ **THEOREM 1.14** *Let A, B and C be 3 events defined on a sample S. If A, B and C are pairwise independent and if A and $(B \cup C)$ are also independent, then A, B and C are mutually independent.*

Proof. Since A, B and C are pairwise independent

$$\begin{aligned} P(A \cap B) &= P(A)P(B) \\ P(A \cap C) &= P(A)P(C) \\ P(B \cap C) &= P(B)P(C). \end{aligned}$$

Also, $P[A \cap (B \cup C)] = P(A)P(B \cup C)$.

But,

$$\begin{aligned} P[A \cap (B \cup C)] &= P[(A \cap B) \cup (A \cap C)] \\ &= P(A \cap B) + P(A \cap C) - P(A \cap B \cap C) \\ &= P(A)P(B) + P(A)P(C) - P(A \cap B \cap C) \qquad \text{(a)} \end{aligned}$$

Again,

$$\begin{aligned} P(A)P(B \cup C) &= P(A)[P(B) + P(C) - P(B \cap C)] \\ &= P(A)P(B) + P(A)P(C) - P(A)P(B \cap C) \qquad \text{(b)} \end{aligned}$$

From (a) and (b), we have

$$P(A \cap B \cap C) = P(A)P(B \cap C) = P(A)P(B)P(C).$$

Hence, A, B and C are mutually independent.

❑ **THEOREM 1.15** *Let A and B be two events defined on a sample space S. Prove that $P(A \cap B) \leq P(A) \leq P(A \cup B) \leq P(A) + P(B)$.*

Proof. We have

$$\begin{aligned} A &= (A \cap B) \cup (A \cap \bar{B}) \\ P(A) &= P[(A \cap B) \cup (A \cap \bar{B})] \\ &= P(A \cap B) + P(A \cap \bar{B}) \\ \Rightarrow\ P(A) &\geq P(A \cap B) \text{ as } P(A \cap \bar{B}) \geq 0. \\ \therefore\ P(A \cap B) &\leq P(A). \end{aligned}$$

Similarly, $P(B) \geq P(A \cap B)$.

$$\therefore\ P(B) - P(A \cap B) \geq 0.$$

Again,

$$\begin{aligned} P(A \cup B) &= P(A) + P(B) - P(A \cap B) \\ \Rightarrow\ P(A \cup B) &\geq P(A), \quad \because\ P(B) - P(A \cap B) \geq 0 \\ \therefore\ P(A) &\leq P(A \cup B). \end{aligned}$$

Also,

$$P(A \cup B) \leq P(A) + P(B), \quad \because \ P(A \cap B) \geq 0$$

$$\therefore \ P(A \cap B) \leq P(A) \leq P(A \cup B) \leq P(A) + P(B).$$

►**EXAMPLE 1.85** *Three students A, B and C are asked to develop a computer program. Previous records certify that A, B and C become successful in 60%, 70% and 50% cases, respectively. If they work done independently, what is the probability that the program will be developed?*

Solution. The program will be developed if it is developed either by A or by B or by C. So we need $P(A \cup B \cup C)$.

$$P(A \cup B \cup C) = P(A) + P(B) + P(C) - P(A \cap B) - P(A \cap C) - P(B \cap C) + P(A \cap B \cap C).$$

From the given information $P(A) = 0.6$, $P(B) = 0.7$ and $P(C) = 0.5$.

$$\therefore \ P(A \cap B) = P(A)P(B) = 0.6 \times 0.7 = 0.42 \ [\because \ A \text{ and } B \text{ work independently}].$$

Similarly,

$$P(A \cap C) = P(A)P(C) = 0.6 \times 0.5 = 0.30$$

$$P(B \cap C) = P(B)P(C) = 0.7 \times 0.5 = 0.35$$

$$P(A \cap B \cap C) = P(A)P(B)P(C) = 0.6 \times 0.7 \times 0.5 = 0.21.$$

$$\therefore \ P(A \cup B \cup C) = 0.6 + 0.7 + 0.5 - 0.42 - 0.30 - 0.35 + 0.21 = 0.94.$$

►**EXAMPLE 1.86** *In a lot there are 2000 electric bulbs. Among these bulbs 1800 are good and 500 are bought from company B. Out of the good bulbs 300 are bought from company B. A bulb is randomly selected and found that it is bought from company B. What is the probability that it is a good bulb?*

Solution. Let A be the event that the bulb is good and B be the event that the bulb is bought from company B. We need $P(A/B)$ [FIG. 1.15].

$$P(B) = \frac{500}{2000}$$

$$P(A \cap B) = \frac{300}{2000},$$

$$P(A/B) = \frac{P(A \cap B)}{P(B)} = \frac{300/2000}{500/2000} = \frac{3}{5}.$$

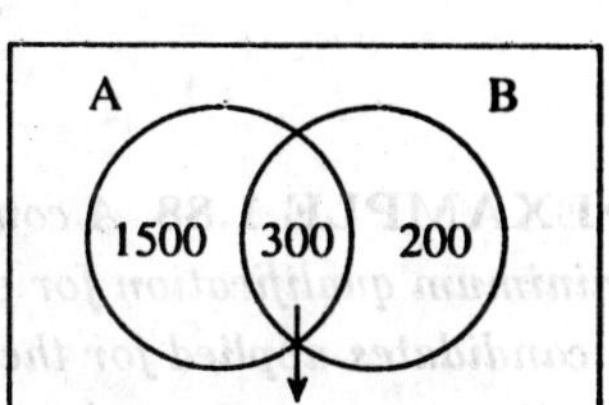

FIG. 1.15

►**EXAMPLE 1.87** *A graduate of computer science applies for a post in two industries X and Y. The probability of his being selected in industry X is 0.6 and that in Y is 0.5. The probability of at least one of his applications being rejected is 0.4. Find the probability that he will be selected in one of the industries.*

Solution. Let A be the event that the graduate will be selected in industry X and B be the event that the graduate will be selected in industry Y.

Given,

$$P(A) = 0.6, P(B) = 0.5 \text{ and } P(\bar{A} \cup \bar{B}) = 0.4.$$

We need $P(A \cup B)$. We know

$$\begin{aligned} P(A \cap B) &= 1 - P(\bar{A} \cup \bar{B}) \\ &= 1 - 0.4 = 0.6. \\ \therefore\ P(A \cup B) &= P(A) + P(B) - P(A \cap B) \\ &= 0.6 + 0.5 - 0.6 = 0.5. \end{aligned}$$

❐ **THEOREM 1.16** *Let A and B be two events defind on a sample space S. Given that* $P(B) \neq 1$. *Then,*

(i) $P(A/\bar{B}) = \dfrac{P(A) - P(A \cap B)}{1 - P(B)}$;

(ii) $P(A \cap B) \geq P(A) + P(B) - 1$.

Proof.

(i) We know [FIG. 1.16],

$$\begin{aligned} P(A/\bar{B}) &= \frac{P(A \cap \bar{B})}{P(\bar{B})} \\ &= \frac{P(A) - P(A \cap B)}{1 - P(B)} \\ &\quad [\because A \cap \bar{B} = A - A \cap B] \end{aligned}$$

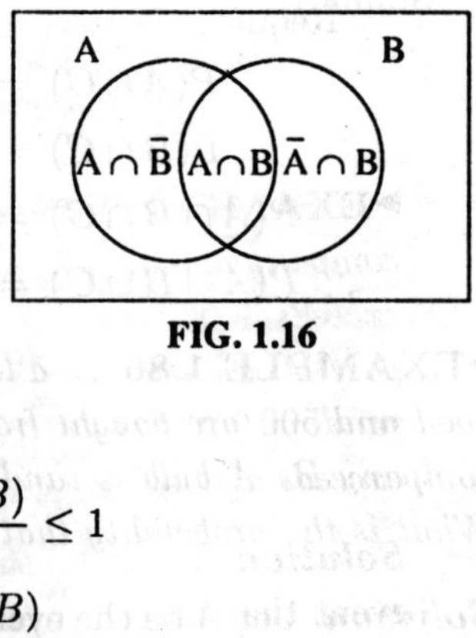

FIG. 1.16

(ii) Since $P(A/\bar{B}) \leq 1$, we have

$$\begin{aligned} P(A/\bar{B}) &= \frac{P(A) - P(A \cap B)}{1 - P(B)} \leq 1 \\ \text{or,}\quad P(A) - P(A \cap B) &\leq 1 - P(B) \\ \text{or,}\quad P(A) + P(B) - 1 &\leq P(A \cap B). \\ \therefore\quad P(A \cap B) &\geq P(A) + P(B) - 1. \end{aligned}$$

►**EXAMPLE 1.88** *A company announced an advertisement for* 3 *separate posts. The minimum qualification for each post is graduate. For the first post* 4 *candidates applied,* 5 *candidates applied for the second post and* 8 *candidates applied for the third post. All applicants are independent. If a candidate is randomly selected, what is the probability that he will be selected for either of a post?*

Solution. Let A be the event that a candidate will get first post, B be the event that the candidate will get second post and C be the event that a candidate will get third post. We need $P(A \cup B \cup C)$. Since the candidates are independent, selection for a post is also independent.

Hence,

$$\begin{aligned}
P(A\cap B) &= P(A)P(B) = \frac{1}{4}\cdot\frac{1}{5} = \frac{1}{20}\\
P(A\cap C) &= P(A)P(C) = \frac{1}{4}\cdot\frac{1}{8} = \frac{1}{32}\\
P(B\cap C) &= P(B)P(C) = \frac{1}{5}\cdot\frac{1}{8} = \frac{1}{40}\\
P(A\cap B\cap C) &= P(A)P(B)P(C) = \frac{1}{4}\cdot\frac{1}{5}\cdot\frac{1}{8} = \frac{1}{160}.\\
\therefore\ P(A\cup B\cup C) &= P(A)+P(B)+P(C)-P(A\cap B)-P(A\cap C)\\
&\quad -P(B\cap C)+P(A\cap B\cap C)\\
&= \frac{1}{4}+\frac{1}{5}+\frac{1}{8}-\frac{1}{20}-\frac{1}{32}-\frac{1}{40}+\frac{1}{160}\\
&= \frac{76}{160} = \frac{19}{40}.
\end{aligned}$$

Alternative. The candidate will not be selected for any post.

This is the event $\bar{A}\cap\bar{B}\cap\bar{C}$.

$$P(\bar{A}\cap\bar{B}\cap\bar{C} = P(\bar{A})P(\bar{B})P(\bar{C}) = \frac{3}{4}\times\frac{4}{5}\times\frac{7}{8} = \frac{84}{160}.$$

Required probability is

$$1-P(\bar{A}\cap\bar{B}\cap\bar{C}) = 1-\frac{84}{160} = \frac{19}{40}.$$

►**EXAMPLE 1.89** *The probability that an employee will continue to his post in a company is* 0.6. *The probability that the monthly salary of an employee of the company exceeds* 40000.00 *is* 0.05. *The probability that either an employee will continue to his post or an employee's salary exceeds* 40000.00 *is* 0.68. *If an employee is selected at random, what is the probability that he will continue under the condition that his salary exceeds* 40000.00*?*

Solution. Let A be the event that an employee will continue to his post and B be the event that an employee's salary exceeds 40000.00. We need $P(A/B)$. Given

$$\begin{aligned}
P(A) &= 0.6\\
P(B) &= 0.05\\
P(A\cup B) &= 0.68\\
P(A\cup B) &= P(A)+P(B)-P(A\cap B)\\
\therefore\ P(A\cap B) &= P(A\cup B)-P(A)-P(B)\\
&= 0.68-0.6-0.05 = 0.03.\\
\therefore\ P(A/B) &= \frac{P(A\cap B)}{P(B)} = \frac{0.03}{0.05} = \frac{3}{5}.
\end{aligned}$$

►**EXAMPLE 1.90** *There are 3 persons—A, B and C who are contesting for a post. The probabilities of appointment of these persons are* 0.30, 0.25 *and* 0.40, *respectively. The probabilities that development work will be promoted by these persons if they are* *appointed are* 0.4, 0.5 *and* 0.6, *respectively. What is the probability that the development will be promoted by the newly appointed persons?*

Solution. Given

$$P(A) = 0.30, P(B) = 0.25, P(C) = 0.40.$$

Let D be the event that the development work will be promoted by them if they are appointed. Then

$$P(D/A) = 0.4, P(D/B) = 0.5, P(D/C) = 0.6.$$

We need

$$P(A \cap D) + P(B \cap D) + P(C \cap D).$$

We have

$$P(A \cap D) = P(D/A)P(A) = 0.4 \times 0.30 = 0.12$$
$$P(B \cap D) = P(D/B)P(B) = 0.5 \times 0.25 = 0.125$$
$$P(C \cap D) = P(D/C)P(C) = 0.6 \times 0.40 = 0.24.$$
$$\therefore P(A \cap D) + P(B \cap D) + P(C \cap D) = 0.12 + 0.125 + 0.24 = 0.485.$$

►**EXAMPLE 1.91** *A company made an announcement that a manager would be appointed. One hundred candidates applied for the post. They were classified according to their qualification and experience as follows:*

	Qualification		*Experience (in yrs.)*	*Total*
	M.B.A.	*B.B.A.*		
	20	40	0	60
	2	13	5	15
	13	12	10	25
Total	35	65	—	100

What is the probability that (i) *one will be selected having* 5 *years experience under the condition that he is an M.B.A.,* (ii) *one will be selected having B.B.A. degree under the condition that he has* 10 *years experience?*

Solution.

(i) Let A be the event that a candidate has M.B.A. degree and B be the event that the candidate has 5 years experience. We need $P(B/A)$.

$$P(B/A) = \frac{P(A \cap B)}{P(A)} = \frac{5/100}{35/100} = \frac{1}{7}.$$

(ii) Let A be the event that the candidate has B.B.A. degree and B be the event that the candidate has 10 years experience. We need $P(A/B)$.

$$P(A/B) = \frac{P(A \cap B)}{P(B)} = \frac{12/100}{25/100} = \frac{12}{25}.$$

►**EXAMPLE 1.92** *Seventy male and 30 female workers are working in an industry. Ten percent male and 50% female workers are working for less than one year. From this latter group one worker is selected at random. What is the probability that the worker is a male?*

Solution. The distribution of employees according to sex and experience is as follows:

Experience	*Sex*		*Total*
	Male	*Female*	
Less than 1 year	7	15	22
Other	63	15	78
Total	70	30	100

Let A be the event that a worker has less than one year experience and B be the event that the worker is a male. We need $P(B/A)$. From the given information

$$P(A) = \frac{22}{100}$$
$$P(A \cap B) = \frac{7}{100}$$
$$P(B/A) = \frac{P(A \cap B)}{P(A)} = \frac{7/100}{22/100} = \frac{7}{22}.$$

►**EXAMPLE 1.93** *A representative has been selected for marketing the items A and B produced in an industry. The representative is successful to market the items A and B in 80% and 20% cases, respectively. A and B are marketed independently. If the representative succeeds in marketing at least one of the items, what is the probability that the item is A?*

Solution. Let A be the event that the marketed item is A and B be the event that the marketed item is B. Since A and B are marketed independently,

$$P(A \cap B) = P(A)P(B) = 0.8 \times 0.2 = 0.16.$$

Here, $P(A) = 0.8, P(B) = 0.2$. The probability of marketing at least one item is

$$P(A \cup B) = P(A) + P(B) - P(A \cap B) = 0.8 + 0.2 - 0.16 = 0.84.$$

We need $P(A/A \cup B)$.

$$P(A/A \cup B) = \frac{P[A \cap (A \cap B)]}{P(A \cap B)} = \frac{P(A)}{P(A \cap B)}$$
$$= \frac{0.80}{0.84} = \frac{20}{21}.$$

►**EXAMPLE 1.94** *A machine has 3 components—A, B and C. The machine runs well if all three components run well. However, the probabilities that, in any day, during work the machines A, B and C fail to work are 0.20, 0.25 and 0.30, respectively. Find the probability that in a day the machine will not work.*

Solution. The event $A \cap B \cap C$ indicates that the machine will run.

We need $1 - P(A \cap B \cap C)$. Given

$$P(\bar{A}) = 0.20, P(\bar{B}) = 0.25 \text{ and } P(\bar{C}) = 0.30.$$
$$\therefore\ P(A) = 0.80, P(B) = 0.75 \text{ and } P(C) = 0.70.$$

Since the components are independent

$$P(A \cap B \cap C) = P(A)P(B)P(C) = 0.80 \times 0.75 \times 0.70 = 0.42.$$
$$\therefore\ 1 - P(A \cap B \cap C) = 1 - 0.42 = 0.58.$$

►**EXAMPLE 1.95** *An article manufactured by a company consists of two parts A and B. In the process of manufacture of part A,* 10% *are found defective. The same percentage for B is* 12. *Find the probability that the assembled part is not defective.*

Solution. It will not be defective if both A and B are not defective. The parts A and B are independent. Given

$$P(\bar{A}) = 0.10 \text{ and } P(\bar{B}) = 0.12.$$
$$\therefore\ P(A) = 0.90 \text{ and } P(B) = 0.88.$$
$$\therefore\ P(A \cap B) = P(A)P(B) = 0.90 \times 0.88 = 0.792.$$

►**EXAMPLE 1.96** *The upper part, lower part and the heel of a shoe are manufactured independently. In manufacturing these parts* 15%, 10% *and* 12%, *respectively, are found defective. What is the probability that a randomly selected pair of shoes will not be defective?*

Solution. Let A be the event that indicates upper part of shoe, B be the event that indicates lower part of shoe and C be the event that indicates heel of shoe. The probabilities of non-defective parts are, respectively

$$P(A) = 1 - P(\bar{A}) = 1 - 0.15 = 0.85$$
$$P(B) = 1 - P(\bar{B}) = 1 - 0.10 = 0.90$$
$$P(C) = 1 - P(\bar{C}) = 1 - 0.12 = 0.88.$$

The probability of non-defective is

$$P(A \cap B \cap C) = 0.85 \times 0.90 \times 0.88 = 0.6732.$$

We need to find the probability that a pair of shoes is non-defective.

$$\begin{aligned} P[\text{non-defective pair}] &= P[\text{non-defective shoe of left leg}] \\ &\quad \times P[\text{non-defective shoe of right leg}] \\ &= 0.6732 \times 0.6732 = 0.45319824. \end{aligned}$$

►**EXAMPLE 1.97** *Find the probability of observing exactly* 4 *heads under the condition that first time head appears when a fair coin is tossed* 5 *times.*

Solution. If a fair coin is tossed 5 times, equally likely cases are $n = 2^5$.

Let A be the event that 4 heads appear and B be the event that first coin shows head. Favourable cases to B are 2^4 (if first coin shows head, the remaining results can appear in 2^4 ways). Among these 2^4 cases 3 heads will be observed in ${}^4C_3 = 4$ ways. This 4 cases are in favour of $A \cap B$.

$$P(B) = \frac{2^4}{2^5} = \frac{1}{2}.$$

We need $P(A/B)$.

$$P(A/B) = \frac{P(A \cap B)}{P(B)} = \frac{4/2^5}{1/2} = \frac{1}{4}.$$

►**EXAMPLE 1.98** *Two fair dice are thrown once. Find the probability that sum of the upper faces is* 10 *or above provided first die shows even number.*

Solution. The equally likely cases are $n = 6^2 = 36$. Let A be the event that sum of the upper faces is 10 or above and B be event that first die shows even number.

Favourable cases to A are $A : \{46, 55, 56, 64, 65, 66\}$. There are 18 cases in favour of B. Here, $A \cap B : \{46, 64, 66\}$.

We need $P(A/B)$.

$$P(A/B) = \frac{P(A \cap B)}{P(B)} = \frac{3/36}{18/36} = \frac{1}{6}.$$

►**EXAMPLE 1.99** *In a packet there are* 4 *bulbs of* 40 *watt each and* 3 *bulbs of* 60 *watts each. Two bulbs are taken one by one without replacement. Find the probability that* (i) *both are* 40 *watts bulbs,* (ii) *one is of* 40 *watts and another is of* 60 *watts.*

Solution.

(i) Let A be the event that first bulb is of 40 watts and B be the event that second bulb is also of 40 watts. We need $P(A \cap B)$.

$$P(B/A) = \frac{P(A \cap B)}{P(A)} \Rightarrow P(A \cap B) = P(A)P(B/A).$$

Here,

$$P(B/A) = \frac{3}{6} \text{ and } P(A) = \frac{4}{7}.$$

$$\therefore\ P(A \cap B) = \frac{4}{7} \cdot \frac{3}{6} = \frac{2}{7}.$$

(ii) Let A be the event that the first one is of 40 watts and B be the event that the second one is of 60 watts. We need $P(A \cap B)$.

$$P(A \cap B) = P(A)P(B/A) = \frac{4}{7} \cdot \frac{3}{6} = \frac{2}{7}.$$

But A can occur in second drawing and B can occur in first drawing. In that case

$$P(A \cap B) = P(B)P(B/A) = \frac{3}{7} \cdot \frac{4}{6} = \frac{2}{7}.$$

Hence, the required probability is

$$P(1 \text{ of } 40 \text{ watts and } 1 \text{ of } 60 \text{ watts}) = P(A \cap B) + P(A \cap B)$$
$$= \frac{2}{7} + \frac{2}{7} = \frac{4}{7}.$$

►**EXAMPLE 1.100** *In an urn there are 10 mobile phone sets. Out of 10 sets 4 are defective. Each of the sets is inspected one by one at random. What is the probability that the 7th inspected set is the last defective one?*

Solution. Let A be the event that 3 defective sets are observed in 6 inspection and B be the event that 7th inspected set is defective. We need $P(A \cap B)$. We have

$$P(A) = \frac{{}^4C_3 \times {}^6C_3}{{}^{10}C_6} = \frac{8}{21}.$$

After the inspection there are 4 sets, out of which 1 is defective. Hence,

$$P(B/A) = \frac{1}{4}.$$

$$\therefore\ P(A \cap B) = P(A)P(B/A) = \frac{8}{21} \times \frac{1}{4} = \frac{2}{21}.$$

►**EXAMPLE 1.101** *In an examination* 30% *students failed in Statistics,* 25% *students failed in English and* 10% *students failed in both the subjects. Find the probability that, a randomly selected student* (i) *failed in Statistics if it is known that he failed in English,* (ii) *failed in Statistics but passed in English.*

Solution.

(i) Let A be the event that the student failed in Statistics and B be the event that the student failed in English. We need $P(B/A)$.
Given $P(A) = 0.30$, $P(A \cap B) = 0.10$, $P(B) = 0.25$.

$$P(A/B) = \frac{P(A \cap B)}{P(B)} = \frac{0.10}{0.25} = \frac{2}{5}.$$

(ii) Let A be the event that the student failed in Statistics and B be the event that the student passed in English. We need $P(A \cap B)$.

$$P(B/A) = \frac{20}{30}$$
$$P(A \cap B) = P(B/A)P(A) = \frac{20}{30} \times \frac{30}{100} = 0.2.$$

►**EXAMPLE 1.102** *A student is expected to answer a set of* 10 *questions by giving 'yes' or 'no' sign. He does not know the correct answer and decides to answer the question as 'yes' if H is observed in throwing an unbiased coin, otherwise the anwser is 'no'. What is the probability of answering at least* 5 *correct questions in this process?*

Solution. Let H be the event that correct answer is made and T be the event that incorrect is made. We have

$$P(H) = \frac{1}{2},\ P(T) = \frac{1}{2}.$$

Since the coin is tossed indepedently, the probability of no correct answer is

$$P(TTTTTTTTTT) = \frac{1}{2^{10}}.$$

This probability is true for any type of answer. Now, the probability of at least 5 correct answers is

$$1 - \frac{1}{2^{10}}\left[{}^{10}C_0 + {}^{10}C_1 + {}^{10}C_2 + {}^{10}C_3 + {}^{10}C_4 + {}^{10}C_5\right] = \frac{319}{512}.$$

►**EXAMPLE 1.103** *A person aged x years will die within a year with probability P. Let $A_1, A_2, \ldots, A_n$ be the n persons. If all of them are in x years, what is the probability that A_1 will die within a year and only A_1 will die first?*

Solution. Let $H_i (i = 1, 2, \ldots, n)$ be the event that A_i will die within a year. Given $P(H_i) = P$. Then,

$$P(\bar{H}_i) = 1 - P(H_i) = 1 - P.$$

None of $A_1, A_2, \ldots, A_n$ will die within a year with probability

$$P(\bar{H}_1 \cap \bar{H}_2 \cap \cdots \cap \bar{H}_n) = (1 - P)^n.$$

Then the probability that at least one will die is

$$1 - P(\bar{H}_1 \cap \bar{H}_2 \cap \cdots \cap \bar{H}_n) = 1 - (1 - P)^n.$$

Now, out of n persons A_1 will die first with probability $\frac{1}{n}$. Therefore, the required probability is

$$\frac{1}{n}\left[1 - (1 - P)^n\right].$$

►**EXAMPLE 1.104** *Two computer scientists A and B have participated in a contest of developing a computer programming. Both of them will try together. One will win who will finish the work first. The probability that A will win is $\frac{3}{5}$. If A tries 4 times, what is the probability that* (i) *A will loss the contest every time,* (ii) *A will win at least once?*

Solution. Let A be the event that he will win in ith attempt ($i = 1, 2, 3, 4$). Given

$$P(A_i) = \frac{3}{5}, \; P(\bar{A}_i) = \frac{2}{5}.$$

(i) We need

$$P(\bar{A}_1 \cap \bar{A}_2 \cap \bar{A}_3 \cap \bar{A}_4) = \frac{2}{5}\,\frac{2}{5}\,\frac{2}{5}\,\frac{2}{5} = \frac{16}{625}.$$

(ii) We need

$$1 - P(\bar{A}_1 \cap \bar{A}_2 \cap \bar{A}_3 \cap \bar{A}_4) = 1 - \frac{16}{625} = \frac{609}{625}.$$

1.6 Marginal Probability

Let us consider that a random experiment will be ended showing n mutually exclusive, exhaustive and equally likely outcomes. These n outcomes constitute a sample space and n sample points are favourable to k mutually exclusive events $A_1, A_2, \ldots, A_k$ such that $N_{1.}$ outcomes are in favour of the event A_1, $N_{2.}$ outcomes are in favour of the event A_2 and, ..., $N_{k.}$ outcomes are in favour of the event A_k.

Again, consider that $N_{.j}$ outcomes are in favour of another event B_j $(j = 1, 2, \ldots, m)$, where $B_1, B_2, \ldots, B_m$ are mutually exclusive events such that n_{ij} outcomes are in favour of an event

$$A_i \cap B_j (i = 1, 2, \ldots, k;\ j = 1, 2, \ldots, m) \text{ and } \sum_i \sum_j n_{ij} = n.$$

The outcomes of the event $A_i \cap B_j$ are shown below:

A \ B	B_1	B_2	$\cdots$	B_m	Total $N_{i.}$
A_1	n_{11}	n_{12}	$\cdots$	n_{1m}	$N_{1.}$
A_2	n_{21}	n_{22}	$\cdots$	n_{2m}	$N_{2.}$
$\vdots$	$\cdots$	$\cdots$	$\cdots$	$\cdots$	$\vdots$
A_k	n_{k1}	n_{k2}	$\cdots$	n_{km}	$N_{k.}$
Total $N_{.j}$	$N_{.1}$	$N_{.2}$	$\cdots$	$N_{.m}$	n

Here,

$$P(A_1) = \frac{N_{1.}}{n}, P(A_2) = \frac{N_{2.}}{n}, \ldots, P(A_i) = \frac{N_{i.}}{n}, \ldots, P(A_k) = \frac{N_{k.}}{N}$$
$$P(B_1) = \frac{N_{.1}}{n}, P(B_2) = \frac{N_{.2}}{n}, \ldots, P(B_j) = \frac{N_{.j}}{n}, \ldots, P(B_m) = \frac{N_{.m}}{n}$$

The probabilities $P(A_i)$ and $P(B_j)$ $(i = 1, 2, \ldots, k;\ j = 1, 2, \ldots, m)$ are known as *marginal probabilities* as these probabilities depend on the marginal total outcomes in the sample space.

Definition. The marginal probability is the ratio of n mutually exclusive, exhaustive and equally likely outcomes of an experiment to the favourable outcomes of an event when the favourable cases are splitted into several mutually exclusive cases according to some other mutually exclusive events.

In other words, if A and B are defined on a sample space of a random experiment and if A has k levels $A_1, A_2, \ldots, A_k$ such that the outcomes of A_i $(i = 1, 2, \ldots, k)$ are favourable to that of the mutually exclusive events $B_1, B_2, \ldots, B_m$, then

$$P(A_i) = \frac{N_{i.}}{n} \text{ and } P(B_j) = \frac{N_{.j}}{n}$$

are called marginal probabilities of i-th level of A and j-th level of B, respectively,

where

$$N_{i\cdot} = \sum_j n_{ij} \text{ and } N_{\cdot j} = \sum_i n_{ij}$$

and n_{ij} are favourable cases to the event $A_i \cap B_j$.

Here, we have

$$P(A_i \cap B_j) = \frac{n_{ij}}{n}.$$

This probability is called the *probability of joint event* of the intersection

$$A_i \cap B_j \ (i = 1, 2, \ldots, k; \ j = 1, 2, \ldots, m).$$

We can also define conditional probability of A_i, given that B_j has already occurred. Thus,

$$P(A_i/B_j) = \frac{P(A_i \cap B_j)}{P(B_j)},$$

$$P(B_j) = \frac{N_{\cdot j}}{n} \neq 0.$$

Similarly,

$$P(B_j/A_i) = \frac{P(A_i \cap B_j)}{P(A_i)},$$

$$P(A_i) = \frac{N_{i\cdot}}{n} \neq 0.$$

►**EXAMPLE 1.105** *An advertisement has been announced to promote two persons of an organisation, where a couple is working. The probabilities of husband and wife to be promoted are* $\frac{1}{5}$ *and* $\frac{1}{7}$ *respectively. Find the probability that* (i) *both of them will be promoted,* (ii) *either of the couple will be promoted,* (iii) *none will be promoted,* (iv) *husband will be promoted under the condition that wife will not be promoted,* (v) *any one of the couple will be promoted.*

Solution. Let A be the event that husband will be promoted and B be the event that wife will be appointed. The probabilities of the events are shown below:

A \ B	$P(B)$	$P(\bar{B})$	*Total*
A	$P(A \cap B)$	$P(A \cap \bar{B})$	1/5
$\bar{A}$	$P(\bar{A} \cap B)$	$P(\bar{A} \cap \bar{B})$	4/5
Total	1/7	6/7	1

(i) The promotion of each of husband and wife is independent. Hence,

$$P(A \cap B) = P(A)P(B) = \frac{1}{5} \times \frac{1}{7} = \frac{1}{35}.$$

Here the event $A \cap B$ indicates that both husband and wife are promoted.

(ii) The event either of the couple will be promoted is $P(A \cap \bar{B}) + P(\bar{A} \cap B)$. Here,

$$P(A \cap \bar{B}) = P(A) - P(A \cap B) = \frac{1}{5} - \frac{1}{35} = \frac{6}{35}$$

$$P(\bar{A} \cap B) = P(B) - P(A \cap B) = \frac{1}{7} - \frac{1}{35} = \frac{4}{35}.$$

$$\therefore\ P(A \cap \bar{B}) + P(\bar{A} \cap B) = \frac{6}{35} + \frac{4}{35} = \frac{2}{7}.$$

(iii) The event $\bar{A} \cap \bar{B}$ indicates that none of the couple is promoted.

$$P(\bar{A} \cap \bar{B}) = P(\bar{A}) - P(\bar{A} \cap B) = \frac{4}{5} - \frac{4}{35} = \frac{24}{35}.$$

(iv) We need $P(A/\bar{B})$, where

$$P(A/\bar{B}) = \frac{P(A \cap \bar{B})}{P(\bar{B})} = \frac{6/35}{6/7} = \frac{1}{5}.$$

(v) We need $P(A \cup B)$.

$$P(A \cup B) = P(A) + P(B) - P(A \cap B) = \frac{1}{5} + \frac{1}{7} - \frac{1}{35} = \frac{11}{35}.$$

►**EXAMPLE 1.106** *A laboratory has 100 computers out of which 60 are Philips computer and 40 are Samsung computer. It is known that 10% computers of Philips and 15% of Samsung create problem during work. One day a computer is selected at random. Find the probability that* (i) *the selected one is a defective computer of Philips,* (ii) *the selected one is of Samsung, given that it is defective,* (iii) *the selected one is defective.* (iv) *Are the events a defective computer and a Philips computer independent?*

Solution. Let A be the event that the computer is of Philips and B be the event that the computer is defective. The events are shown below:

B \ A	A	$\bar{A}$	*Total*
B	$A \cap B$	$\bar{A} \cap B$	12
$\bar{B}$	$A \cap \bar{B}$	$\bar{A} \cap \bar{B}$	88
Total	60	40	100

$\bar{A}$ = Samsung computer, $\quad \bar{B}$ = defective computer.

Favourable cases to $A \cap B$ are 6 and hence favourable cases to $\bar{A} \cap B = 12 - 6 = 6$. Similarly, favourable cases to $A \cap \bar{B}$ and $\bar{A} \cap \bar{B}$ are

$$A \cap \bar{B} = 60 - 6 = 54$$

$$\text{and } \bar{A} \cap \bar{B} = 40 - 6 = 34.$$

$$\therefore\ P(A \cap B) = \frac{6}{100},\ P(\bar{A} \cap B) = \frac{6}{100},$$

$$P(A \cap \bar{B}) = \frac{54}{100},\ P(\bar{A} \cap \bar{B}) = \frac{34}{100}.$$

(i) We need $P(A \cap B)$, $P(A \cap B) = \frac{6}{100}$.

(ii) We need $P(\bar{A}/B)$, $P(\bar{A}/B) = \frac{P(\bar{A} \cap B)}{P(B)} = \frac{6/100}{12/100} = \frac{1}{2}$.

(iii) We need $P(B)$, $P(B) = \frac{12}{100} = \frac{3}{25}$.

(iv) $P(A)P(B) = \frac{60}{100} \times \frac{12}{100} = \frac{9}{125} \neq P(A \cap B)$.

Hence, A and B are not independent.

►**EXAMPLE 1.107** ***The probability that a* 50*-year old man will be alive at* 60 *is* 0.60 *and that a* 40*-year old woman will be alive at* 50 *is* 0.75*. Find the probability that a man who is* 50 *and his wife who is* 40 (i) *both will be alive* 10 *years hence,* (ii) *none will be alive* 10 *years hence,* (iii) *one wil be alive,* (iv) *any one of them will be alive.***

Solution. Let A be the event that the man will be alive 10 years hence and B be the event that the woman will be alive 10 years hence. The probabilities of the events and complementary events are shown below:

B \ A	A	$\bar{A}$	*Total*
B	$P(A \cap B)$	$P(\bar{A} \cap B)$	0.75
$\bar{B}$	$P(A \cap \bar{B})$	$P(\bar{A} \cap \bar{B})$	0.25
Total	0.60	0.40	1.00

(i) We need $P(A \cap B)$. Here A and B are independent events.

$$\therefore \; P(A \cap B) = P(A) \cdot P(B) = 0.60 \times 0.75 = 0.45.$$

(ii) We need $P(\bar{A} \cap \bar{B})$,

$$\begin{aligned} P(\bar{A} \cap \bar{B}) &= P(\bar{A}) - P(\bar{A} \cap B) \\ &= 0.40 - 0.30 = 0.10 \\ P(\bar{A} \cap B) &= P(B) - P(A \cap B) \\ &= 0.75 - 0.45 = 0.30. \end{aligned}$$

(iii) We need $P(\bar{A} \cap B) + P(A \cap \bar{B})$,

$$\begin{aligned} P(\bar{A} \cap B) + P(A \cap \bar{B}) &= 0.30 + 0.15 = 0.45 \\ P(A \cap \bar{B}) &= P(A) - P(A \cap B) \\ &= 0.60 - 0.45 = 0.15. \end{aligned}$$

(iv) We need $P(A \cup B)$,

$$\begin{aligned} P(A \cup B) &= P(A) + P(B) - P(A \cap B) \\ &= 0.60 + 0.75 - 0.45 = 0.90. \end{aligned}$$

►**EXAMPLE 1.108** *A social investigation on 1000 families provides the classified results of the families according to three different characters of the families. The classified results are shown below:*

Possession of T.V.	*Family income (in Rs.) per year*				*Total*
	125000.00		125000.00$^+$		
	Possession of telephone				
	Yes	*No*	*Yes*	*No*	
Yes	340	107	228	44	719
No	210	18	37	16	281
Total	550	125	265	60	1000

A family is randomly selected. Find the probability that

(i) *the family possesses television (T.V.),*

(ii) *it has telephone and has income* 125000.00$^+$, *under this condition it has T.V.,*

(iii) *it has telephone under the condition that it has T.V.*

Solution.

(i) Let A be the event that the family possesses T.V. and B be the event that the income of the family is 125000.00$^+$.
We need $P(A)$.

$$P(A) = \frac{719}{1000} = 0.719.$$

(ii) Let A be the event that the family has telephone. B be the event that it has income 125000.00$^+$ and C be the event that it has television (T.V). We need $P(C/A \cap B)$.
Here,

$$P(A \cap B) = \frac{265}{1000}$$

$$P(C/A \cap B) = \frac{P(A \cap B \cap C)}{P(A \cap B)} = \frac{228/1000}{265/1000}$$

$$= \frac{228}{265} = 0.8604.$$

(iii) Let A be the event that the family possesses telephone and B be the event that it has T.V. We need $P(A/B)$.

$$P(A/B) = \frac{P(A \cap B)}{P(B)} = \frac{568/1000}{719/1000} = \frac{568}{719} = 0.78999.$$

►**EXAMPLE 1.109** *If a machine is operated properly, it produces 90% non-defective items, otherwise produces 40% non-defective items. The machine is operated properly in 80% cases. Find the probability that the machine produces non-defective items under the condition that it operates properly.*

Solution. Let A be the event that the machine is operated properly, B be the event that the machine is not operated properly and E be the event that the machine produces non-defective items. We need $P(A/E)$. Given,

$$\begin{aligned} P(A) &= 0.8,\ P(B) = 0.2, \\ P(E/A) &= 0.9,\ P(E/B) = 0.4. \\ P(E) &= P(A)P(E/A) + P(B)P(E/B) \\ &= 0.8 \times 0.9 + 0.2 \times 0.4 = 0.8. \\ \therefore\ P(A/E) &= \frac{P(A)P(E/A)}{P(E)} = \frac{0.72}{0.8} = 0.9. \end{aligned}$$

►**EXAMPLE 1.110** *Two different groups of employees of a company are interested to join the management of the company. The probabilities of joining are* 0.6 *and* 0.4. *If first group can join, the company will be able to produce new items with probability* 0.5. *The same probability for the second group is* 0.7. *Find the probability that the company will be able to produce new items.*

Solution. Let A be the event that the first group will join the management, B be the event that the second group will join the management and E be the event that the company will produce new items. We need $P(E)$. Given,

$$\begin{aligned} P(A) &= 0.6,\ P(B) = 0.4, \\ P(E/A) &= 0.5, P(E/B) = 0.7. \\ P(A \cap E) &= P(A)P(E/A) = 0.6 \times 0.5 = 0.30. \\ P(B \cap E) &= P(B)P(E/B) = 0.4 \times 0.7 = 0.28. \\ \therefore\ P(E) &= P(A \cap E) + P(B \cap E) \\ &= 0.30 + 0.28 = 0.58. \end{aligned}$$

►**EXAMPLE 1.111** *A student has selected* 3 *courses A, B and C to study in a semester. His chances of pass in these courses are* 0.8, 0.6, 0.9, *respectively. Find the probability that he will pass in at least one course.*

Solution. The event $\bar{A} \cap \bar{B} \cap \bar{C}$ indicates that the students will not pass in any course. We need $1 - P(\bar{A} \cap \bar{B} \cap \bar{C})$. As passing each course is independent,

$$\begin{aligned} P(\bar{A} \cap \bar{B} \cap \bar{C}) &= P(\bar{A})P(\bar{B})P(\bar{C}) \\ &= 0.2 \times 0.4 \times 0.1 = 0.008. \\ \therefore\ \text{required probability} &= 1 - P(\bar{A} \cap \bar{B} \cap \bar{C}) \\ &= 1 - 0.008 = 0.992. \end{aligned}$$

1.7 Bayes Theorem

Let $H_1, H_2, \dots, H_n$ be n mutually exclusive events defined on a sample space S such that $P(H_i) \neq 0$ $(i = 1, 2, \dots, n)$. Consider an arbitrary event E, where E is defined using the subset of $\overset{n}{\underset{i=1}{\cup}} H_i$ such that $P(E) \neq 0$. Then Bayes theorem states that

$$P(H_i/E) = \frac{P(H_i \cap E)}{P(E)}, \; i = 1, 2, \ldots, n$$
$$= \frac{P(H_i)P(E/H_i)}{\sum_{i=1}^{n} P(H_i)P(E/H_i)}.$$

Proof. The events can be shown by the following Venn diagram [FIG. 1.17]:

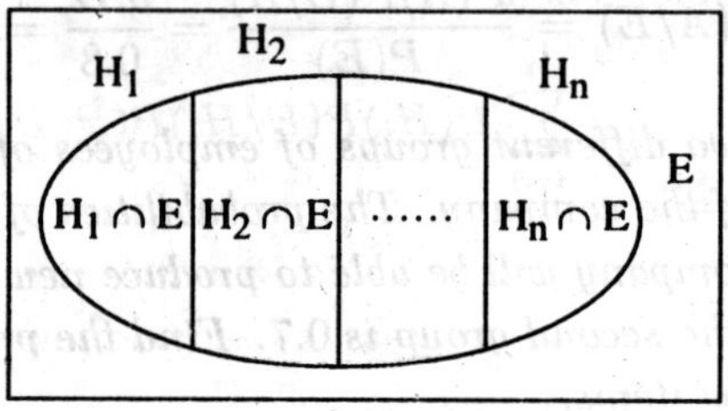

FIG. 1.17

Since $E \subset \overset{n}{\underset{i=1}{\cup}} H_i$, we have

$$E = E \cap \left(\overset{n}{\underset{i=1}{\cup}} H_i\right) = \overset{n}{\underset{i=1}{\cup}} H_i \cap E$$
$$= H_1 \cap E + H_2 \cap E + \cdots + H_n \cap E.$$

Since $H_i \cap E \subset H_i$ $(i = 1, 2, \ldots, n)$ and are mutually exclusive events, we have by addition law of probability

$$P(E) = P\left(\overset{n}{\underset{i=1}{\cup}} H_i \cap E\right)$$
$$= P(H_1 \cap E) + P(H_2 \cap E) + \cdots + P(H_n \cap E).$$

Using the definition of conditional probability, we can write

$$P(H_i \cap E) = P(H_i)P(E/H_i), \; i = 1, 2, \ldots, n.$$

$$\therefore P(H_i/E) = \frac{P(H_i \cap E)}{P(E)}$$
$$= \frac{P(H_i)P(E/H_i)}{\sum_{i=1}^{n} P(H_i)P(E/H_i)}$$
$$= \frac{P(H_i \cap E)}{P(H_1 \cap E) + P(H_2 \cap E) + \cdots + P(H_n \cap E)}.$$

REMARK 1. Here $P(H_i)$ $(i = 1, 2, \ldots, n)$ are called *a priori probabilities* as these exist before we get any information from the experiment itself.

REMARK 2. The probabilities $P(H_i/E)$ $(i = 1, 2, \ldots, n)$ are called *posterior probabilities* because these are determined after the result of the experiment is recorded.

REMARK 3. The probabilities $P(E/H_i)$ $(i = 1, 2, \ldots, n)$ are called ***likelihoods*** as these indicate how likely the event E under consideration is to occur, given each and every a priori probability.

Bayes Theorem for Future Event

Let us consider another event B which is occurred under the condition that the events $H_i \cap E$ $(i = 1, 2, \ldots, n)$ are occurred and $P(H_i \cap E) \neq 0$. Also, $P(B/H_i \cap E)$ $(i = 1, 2, \ldots, n)$ are available. Then,

$$P(B/E) = \frac{\sum_{i=1}^{n} P(H_i)P(E/H_i)P(B/H_i \cap E)}{\sum_{i=1}^{n} P(H_i)P(E/H_i)}$$

$$= \frac{\sum P(H_i \cap E)P(B/H_i \cap E)}{P(E)}.$$

Proof. Since the occurrence of the event E implies the occurrence of one and only one of the events $H_1, H_2, \ldots, H_n$, the event B (given that E has occurred) can occur in the following mutually exclusive ways:

$$H_1 \cap B, H_2 \cap B, \ldots, H_n \cap B.$$

$$\therefore\ B = H_1 \cap B + H_2 \cap B + \cdots + H_n \cap B.$$

Hence,

$$(B/E) = [(H_1 \cap B)/E] \cup [(H_2 \cap B)/E] \cup \cdots \cup [(H_n \cap B)/E].$$

$$\therefore\ P(B/E) = P[(H_1 \cap B)/E + P(H_2 \cap B)/E + \cdots + P(H_n \cap B)/E]$$

$$= \sum_{i=1}^{n} P[(H_i \cap B)/E]$$

$$= \sum_{i=1}^{n} P(H_i/E)P[B/(H_i \cap E)].$$

But

$$P(H_i/E) = \frac{P(H_i)P(E/H_i)}{P(E)}$$

$$= \frac{P(H_i)P(E/H_i)}{\sum_{i=1}^{n} P(H_i)P(E/H_i)}.$$

$$\therefore\ P(B/E) = \frac{\sum_{i=1}^{n} P(H_i)P(E/H_i)P(B/H_i \cap E)}{\sum_{i=1}^{n} P(H_i)P(E/H_i)}.$$

In some cases H_i $(i = 1, 2, \ldots, n)$ events are occurred in such a way that B and E are independent events. Then,

$$P(B/H_i \cap E) = P(B/H_i)$$

$$\text{and } P(B/E) = \frac{\sum_{i=1}^{n} P(H_i)P(E/H_i)P(B/H_i)}{\sum_{i=1}^{n} P(H_i)P(E/H_i)}.$$

►**EXAMPLE 1.112** *The labourers of an industry are supposed to get bonus on an occasion. Among the labourers 40% are skilled and 60% are unskilled. The bonus is given to 80% skilled and 50% unskilled labourers. A labourer is selected for bonus. Find the probability that the selected one is an unskilled labourer. Another one is selected for future bonus plan. What is the probability that he is skilled labourer. It is given that 50% labourers who are eligible for bonus may have the bonus.*

Solution. Let H_1 be the event that a labourer is skilled, H_2 be the event that a labourer is unskilled and E be the event that a labourer gets bonus. Given,

$$P(H_1) = 0.40,\ P(H_2) = 0.60,$$
$$P(E/H_1) = 0.80,\ P(E/H_2) = 0.50.$$

We need $P(H_2/E)$.

$$P(H_1 \cap E) = P(H_1)P(E/H_1) = 0.4 \times 0.6 = 0.24$$
$$P(H_2 \cap E) = P(H_2)P(E/H_2) = 0.6 \times 0.5 = 0.30$$
$$P(E) = P(H_1 \cap E) + P(H_2 \cap E) = 0.24 + 0.30 = 0.54$$
$$P(H_2/E) = \frac{P(H_2 \cap E)}{P(E)} = \frac{0.30}{0.54} = \frac{5}{9}.$$

Again, let A be the event that a skilled labourer will get bonus. Given that, $P(A/H_1 \cap E) = 0.50$, $P(A/H_2 \cap E) = 0.00$. We need $P(A/E)$.

$$P(A/E) = \frac{P(H_1 \cap E)P(A/H_1 \cap E) + P(H_2 \cap E)P(A/H_2 \cap E)}{P(E)}$$
$$= \frac{0.24 \times 0.50 + 0.30 \times 0.00}{0.54} = \frac{0.12}{0.54} = \frac{2}{9}.$$

►**EXAMPLE 1.113** *In a laboratory there are* 100 *computers,* 60 *of which are Philips,* 30 *are Samsung and* 10 *are ICL. The companies of the computers assure that within one year any computer may create problem with probability* 0.2, 0.3 *and* 0.4, *respectively. During the year of service, a computer is found defective. What is the probability that* (i) *it is a Philips computer,* (ii) *it is either Samsung or ICL?*

Solution. Let H_1 be the event that the computer is Philips, H_2 be the event that the computer is Samsung, C be the event that the computer is ICL and B be the event that the computer is defective.

Given,

$$\begin{aligned}
P(H_1) &= \frac{60}{100},\ P(H_2) = \frac{30}{100},\ P(H_3) = \frac{10}{100},\\
P(E/H_1) &= 0.2,\ P(E/H_2) = 0.3 \text{ and } P(E/H_3) = 0.4.\\
P(H_1 \cap E) &= P(H_1)P(E/H_1) = 0.6 \times 0.2 = 0.12\\
P(H_2 \cap E) &= P(H_2)P(E/H_2) = 0.3 \times 0.3 = 0.09\\
P(H_3 \cap E) &= P(H_3)P(E/H_3) = 0.1 \times 0.4 = 0.04\\
P(E) &= P(H_1 \cap E) + P(H_2 \cap E) + P(H_3 \cap E)\\
&= 0.12 + 0.09 + 0.04 = 0.25.
\end{aligned}$$

(i) $P(H_1/E) = \dfrac{P(H_1 \cap E)}{P(E)} = \dfrac{0.12}{0.25} = 0.48.$

(ii) $P(H_2/E) + P(H_3/E) = \dfrac{P(H_2 \cap E)}{P(E)} + \dfrac{P(H_3 \cap E)}{P(E)}$

$$= \frac{0.09}{0.25} + \frac{0.04}{0.25} = 0.52.$$

►**EXAMPLE 1.114** *Sixty per cent members of a club are male. Fifty per cent of the male members and 80% of the female members are non-smokers. One non-smoking member will be elected as President of the club. Find the probability that the female member will be elected.*

Solution. Let H_1 be the event that a member is male, H_2 be the event that a member is female and E be the event that a member is non-smoker.

Given,

$$\begin{aligned}
P(H_1) &= 0.6,\ P(H_2) = 0.4,\\
P(E/H_1) &= 0.5,\ P(E/H_2) = 0.8.\\
P(H_1 \cap E) &= P(H_1)P(E/H_1) = 0.6 \times 0.5 = 0.30\\
P(H_2 \cap E) &= P(H_2)P(E/H_2) = 0.4 \times 0.8 = 0.32.
\end{aligned}$$

We need $P(H_2/E)$.

$$\begin{aligned}
P(E) &= P(H_1 \cap E) + P(H_2 \cap E) = 0.30 + 0.32 = 0.62.\\
\therefore\ P(H_2/E) &= \frac{P(H_2 \cap E)}{P(E)} = \frac{0.32}{0.62} = \frac{16}{31}.
\end{aligned}$$

►**EXAMPLE 1.115** *A factory produces computer spare parts using three machines A, B and C. Machine A produces 125 parts per hour. The similar production of B and C are 150 and 225, respectively. Machines A, B and C produce 10, 15 and 15 defective parts per hour. A product of an hour is randomly selected and found that it is defective. What is the probability that the selected product is of machine B?*

Solution. Let H_1 be the event that machine A is used, H_2 be the event that machine B is used, H_3 be the event that machine C is used and E be the event that a product is defective.

Given,

$$P(H_1) = \frac{125}{500}, P(H_2) = \frac{150}{500}, P(H_3) = \frac{225}{500}.$$
$$P(E/H_1) = \frac{10}{125}, P(E/H_2) = \frac{15}{150}, P(E/H_3) = \frac{15}{225}.$$

We need $P(H_2/E)$.

$$P(H_1 \cap E) = P(H_1)P(E/H_1) = \frac{125}{500} \times \frac{10}{125} = 0.02$$
$$P(H_2 \cap E) = P(H_2)P(E/H_2) = \frac{150}{500} \times \frac{15}{150} = 0.03$$
$$P(H_3 \cap E) = P(H_3)P(E/H_3) = \frac{225}{500} \times \frac{15}{225} = 0.03$$
$$P(E) = P(H_1 \cap E) + P(H_2 \cap E) + P(H_3 \cap E)$$
$$= 0.02 + 0.03 + 0.03 = 0.08$$
$$P(H_2/E) = \frac{P(H_2 \cap E)}{P(E)} = \frac{0.03}{0.08} = \frac{3}{8}.$$

►**EXAMPLE 1.116** *In a departmental store three sale girls A, B and C are working and they are equally expert in serving the customer. But their behaviours are not similar with the customer. A is well-behaved for 90% of the time, B and C are well-behaved for 70% and 60% times, respectively. One day a customer has been ill-behaved. Find the probability that he has been ill-behaved by B.*

Solution. Let H_1 be the event that the sale girl is A, H_2 be the event that the sale girl is B, H_3 be the event that the sale girl is C and E be the event that the sale girl is ill-behaved.

Given,

$$P(H_1) = \frac{1}{3}, P(H_2) = \frac{1}{3}, P(H_3) = \frac{1}{3},$$
$$P(E/H_1) = 0.1, P(E/H_2) = 0.3, P(E/H_3) = 0.4.$$

We need $P(H_2/E)$.

$$P(H_1 \cap E) = P(H_1)P(E/H_1) = \frac{1}{3} \times 0.1$$
$$P(H_2 \cap E) = P(H_2)P(E/H_2) = \frac{1}{3} \times 0.3$$
$$P(H_3 \cap E) = P(H_3)P(E/H_3) = \frac{1}{3} \times 0.4$$
$$P(E) = P(H_1 \cap E) + P(H_2 \cap E) + P(H_3 \cap E)$$
$$= \frac{1}{3}[0.1 + 0.3 + 0.4] = \frac{1}{3} \times 0.8$$
$$P(H_2/E) = \frac{P(H_2 \cap E)}{P(E)} = \frac{\frac{1}{3} \times 0.3}{\frac{1}{3} \times 0.8}$$
$$= \frac{3}{8}.$$

►**EXAMPLE 1.117** *There are 3 urns of same type and in each urn different amounts of bulbs are kept. These are shown below:*

Urn-1 : 3 bulbs of 40 watts and 2 bulbs of 60 watts,

Urn-2 : 4 bulbs of 40 watts and 6 bulbs of 60 watts,

Urn-3 : 5 bulbs of 40 watts and 5 bulbs of 60 watts.

An urn is selected at random and a bulb from that selected urn is selected at random and found that it is of 60 watts. What is the probability that it is selected from urn-2? If the selected bulb (i) is replaced, (ii) is not replaced and a second bulb is selected, what is the probability of selection of a 60-watt bulb at the second selection?

Solution. Let H_1 be the event that urn-1 is selected, H_2 be the event that urn-2 is selected, H_3 is the event that urn-3 is selected and E be the event that a 60-watt bulb is selected.

We need $P(H_2/E)$. Given,

$$P(H_1) = P(H_2) = P(H_3) = \frac{1}{3},$$

$$P(E/H_1) = \frac{2}{5},\ P(E/H_2) = \frac{6}{10},\ P(E/H_3) = \frac{5}{10}.$$

$$P(H_1 \cap E) = P(H_1)P(E/H_1) = \frac{1}{3} \times \frac{2}{5} = \frac{2}{15}$$

$$P(H_2 \cap E) = P(H_2)P(E/H_2) = \frac{1}{3} \times \frac{6}{10} = \frac{6}{30}$$

$$P(H_3 \cap E) = P(H_3)P(E/H_3) = \frac{1}{3} \times \frac{5}{10} = \frac{5}{30}$$

$$P(E) = P(H_1 \cap E) + P(H_2 \cap E) + P(H_3 \cap E)$$

$$= \frac{2}{15} + \frac{6}{30} + \frac{5}{30} = \frac{15}{30}.$$

$$\therefore\ P(H_2/E) = \frac{P(H_2 \cap E)}{P(E)} = \frac{6/30}{15/30} = \frac{2}{5}.$$

(i) Let A be the event that second selected bulb is of 60 watts. Then,

$$P(A/H_1 \cap E) = \frac{2}{5}$$

$$P(A/H_2 \cap E) = \frac{6}{10}$$

$$P(A/H_3 \cap E) = \frac{5}{10}.$$

We need $P(A/E)$, where

$$P(A/E) = \frac{\sum_{i=1}^{3} P(H_i \cap E)P(A/H_i \cap E)}{P(E)}$$

$$= \frac{\frac{2}{15} \times \frac{2}{5} + \frac{6}{30} \times \frac{6}{10} + \frac{5}{30} \times \frac{5}{10}}{\frac{15}{30}} = \frac{\frac{77}{300}}{\frac{15}{30}} = \frac{77}{150}$$

(ii) $P(A/H_1 \cap E) = \frac{1}{4}, P(A/H_2 \cap E) = \frac{5}{9}, P(A/H_3 \cap E) = \frac{4}{9}.$

$$\therefore\ P(A/E) = \frac{\sum_{i=1}^{3} P(H_i \cap E)P(A/H_i \cap E)}{P(E)}$$

$$= \frac{\frac{2}{15} \times \frac{1}{4} + \frac{6}{30} \times \frac{5}{9} + \frac{5}{30} \times \frac{4}{9}}{\frac{15}{30}} = \frac{59}{135}.$$

►**EXAMPLE 1.118** *A company has two plants A and B to manufacture iron rod. Plant A produces 70% of the total output. At plant A 90% rods are manufactured according to a given specification and at plant B 80% rods are manufactured according to the given specification.*

(i) *Find the probability that a customer obtains rods according to given specification if he buys the rods from the above-mentioned company.*

(ii) *Find the probability that the customer buys the rods of plant A if it is known that the rods are of given specification.*

Solution. Let H_1 be the event that the rods are produced by plant A, H_2 be the event that the rods are produced by plant B and E be the event that the rods are of given specification. Given,

$$P(H_1) = 0.70,\ P(H_2) = 0.30,\ P(E/H_1) = 0.90, P(E/H_2) = 0.80.$$

(i) We need $P(E)$.

$$\begin{aligned} P(H_1 \cap E) &= P(H_1)P(E/H_1) = 0.7 \times 0.9 = 0.63 \\ P(H_2 \cap E) &= P(H_2)P(E/H_2) = 0.3 \times 0.8 = 0.24 \\ P(E) &= P(H_1 \cap E) + P(H_2 \cap E) = 0.63 + 0.24 = 0.87. \end{aligned}$$

(ii) We need $P(H_1/E)$.

$$P(H_1/E) = \frac{P(H_1 \cap E)}{P(E)} = \frac{0.63}{0.87} = \frac{21}{29}.$$

►**EXAMPLE 1.119** *A and B are two direct independent witnesses of an incidence. They can speak truth about the incidence with probabilities* 0.8 *and* 0.7, *respectively. Both of them comment same about the incidence. What is the probability that* (i) *they speak the truth,* (ii) *they speak false?*

Solution. Let H_1 be the event of speaking truth by A, H_2 be the event of speaking truth by B. The events $H_1 \cap H_2$ and $\bar{H}_1 \cap \bar{H}_2$ are in favour of same comment by both of them. Here,

$$\begin{aligned} P(\bar{H}_1) &= 1 - P(H_1) = 1 - 0.8 = 0.2 \\ P(\bar{H}_2) &= 1 - P(H_2) = 1 - 0.7 = 0.3. \end{aligned}$$

Both of them speak independently. Hence,

$$P(H_1 \cap H_2) = P(H_1)P(H_2) = 0.7 \times 0.8 = 0.56$$
$$P(\bar{H}_1 \cap \bar{H}_2) = P(\bar{H}_1)P(\bar{H}_2) = 0.2 \times 0.3 = 0.06.$$

(i) We need the probability of speaking truth under the condition of same comment. The required probability is

$$\frac{P(H_1 \cap H_2)}{P(H_1 \cap H_2) + P(\bar{H}_1 \cap \bar{H}_2)} = \frac{0.56}{0.62} = \frac{28}{31}.$$

(ii) We need the probability of speaking false under the condition that they make same comment. The required probability is

$$\frac{P(\bar{H}_1 \cap \bar{H}_2)}{P(H_1 \cap H_2) + P(\bar{H}_1 \cap \bar{H}_2)} = \frac{0.06}{0.62} = \frac{3}{31}.$$

►**EXAMPLE 1.120** *An insurance company categorized the drivers in 3 classes when they want to insure the vehicles. The first category drivers are those who make accidents in 5% days of the year. The percentages of days of accidents for second and third category drivers are 10 and 25. The percentages of drivers of these 3 categories are 20, 35 and 45, respectively. A driver made a contact for insurance of his vehicles and he made an accident within first month of insurance. What is the probability that the driver is of third category?*

Solution. Let H_1 be the event that the driver is of first category, H_2 be the event that the driver is of second category, H_3 be the event that the driver is of third category and E be the event that a driver faces the accident. Given,

$$P(H_1) = 0.20,\ P(H_2) = 0.35,\ P(H_3) = 0.45,$$
$$P(E/H_1) = 0.05,\ P(E/H_2) = 0.10,\ P(E/H_3) = 0.25.$$

We need $P(H_3/E)$.

$$P(H_1 \cap E) = P(H_1)P(E/H_1) = 0.20 \times 0.05 = 0.01$$
$$P(H_2 \cap E) = P(H_2)P(E/H_2) = 0.35 \times 0.10 = 0.035$$
$$P(H_3 \cap E) = P(H_3)P(E/H_3) = 0.45 \times 0.25 = 0.1125$$
$$P(E) = P(H_1 \cap E) + P(H_2 \cap E) + P(H_3 \cap E)$$
$$= 0.01 + 0.035 + 0.1125 = 0.1575.$$
$$\therefore\ P(H_3/E) = \frac{P(H_3 \cap E)}{P(E)} = \frac{0.1125}{0.1575} = 0.7143.$$

►**EXAMPLE 1.121** *In a coffee-house two persons A and B supply coffee to the customer. A serves 70% customers and B serves 30% customers of the house. A serves in due time in 80% cases and B serves in due time in 50% cases. One day a person is not served in due time. What is the probability that he is served by B?*

Solution. Let H_1 be the event that the server is A, H_2 be the event that the server is B and E be the event that the service is not made in due time.

Given,

$$P(H_1) = 0.7,\ P(H_2) = 0.3,$$
$$P(E/H_1) = 0.2,\ P(E/H_2) = 0.5.$$

We need $P(H_2/E)$.

$$\begin{aligned} P(H_1 \cap E) &= P(H_1)P(E/H_1) = 0.7 \times 0.2 = 0.14 \\ P(H_2 \cap E) &= P(H_2)P(E/H_2) = 0.3 \times 0.5 = 0.15 \\ P(E) &= P(H_1 \cap E) + P(H_2 \cap E) = 0.14 + 0.15 = 0.29 \\ P(H_2/E) &= \frac{P(H_2 \cap E)}{P(E)} = \frac{0.15}{0.29} = \frac{15}{29}. \end{aligned}$$

►**EXAMPLE 1.122** *In an office there are two computers which are used for official work without any preference of using a particular computer. But one computer creates problem 20% time when is in use. The same percentage for another computer is 30. One day during work the computer creates problem. What is the probability that first-mentioned computer is in use?*

Solution. Let H_1 be the event that the computer is first-mentioned one, H_2 be the event that the computer is second-mentioned one and E be the event that the computer creates problem. We need $P(H_1/E)$. Given,

$$\begin{aligned} P(H_1) &= 0.5,\ P(H_2) = 0.5, \\ P(E/H_1) &= 0.2,\ P(E/H_2) = 0.3. \\ P(H_1 \cap E) &= P(H_1)P(E/H_1) = 0.5 \times 0.2 = 0.10 \\ P(H_2 \cap E) &= P(H_2)P(E/H_2) = 0.5 \times 0.3 = 0.15 \\ P(E) &= P(H_1 \cap E) + P(H_2 \cap E) \\ &= 0.10 + 0.15 = 0.25 \\ P(H_1/E) &= \frac{P(H_1 \cap E)}{P(E)} = \frac{0.10}{0.25} = \frac{2}{5}. \end{aligned}$$

1.8 More Problems on Probability

►**EXAMPLE 1.123** *What is the probability of selecting 2 pairs of same card if 5 cards are taken from a pack of 52 well-shuffled cards?*

Solution. Five cards from the pack can be taken in $n = {}^{52}C_5$ ways. There are 13 cards of same suit. Two cards can be taken from 13 cards in ${}^{13}C_2$ ways. There are 4 suits, a pair of each suit can occur in ${}^4C_2$ ways. Then 2 pairs of same suit can occur in ${}^{13}C_2 \times {}^4C_2 \times {}^4C_2$ ways. Remaining one card can be taken from $(52 - 8) = 44$ cards in ${}^{44}C_1$ ways. Therefore, two pairs of card can occur in $m = {}^{13}C_2 \times {}^4C_2 \times {}^4C_2 \times {}^{44}C_1$ ways.

$$\therefore \text{Required problem} = \frac{m}{n} = \frac{{}^{13}C_2 \times {}^4C_2 \times {}^4C_2 \times {}^{44}C_1}{{}^{52}C_5}.$$

►**EXAMPLE 1.124** *A random experiment is conducted* 10 *times to observe the digits* 0 *or* 1. *What is the probability of observing* 0 *in at least one trial?*

Solution. Let A be the event that at least one 0 is observed. The experiment will show $n = 2^{10}$ results $= 1024$. Here $\bar{A}$ is the event that all show 1. Then, $P(\bar{A}) = 1 - P(A)$. All show 1 in one way. Hence,

$$P(\bar{A}) = \frac{1}{1024}$$

$$P(A) = 1 - P(\bar{A}) = 1 - \frac{1}{1024} = \frac{1023}{1024}.$$

►**EXAMPLE 1.125** *What is the probability of obtaining exactly* 4 *heads under the condition that first one is head when an unbiased coin is tossed* 5 *times?*

Solution. Let A be the event that head is shown first time and B be the event that exactly 4 heads are observed. We need $P(B/A)$. The coin shows $n = 2^5$ mutually exclusive, exhaustive and equally likely outcomes. First time H is shown in $2^{5-1} = 16$ ways. Therefore, $P(A) = \frac{16}{32}$.

Except the head of first trial 3 heads can be observed from the 4 trials in ${}^4C_3$ ways. This event of observing three heads in presence of first head is $A \cap B$.

$$\therefore \; P(A \cap B) = \frac{{}^4C_3}{2^5} = \frac{4}{32}$$

$$P(B/A) = \frac{P(A \cap B)}{P(A)} = \frac{4/32}{16/32} = \frac{1}{4}.$$

►**EXAMPLE 1.126** *In a box there are* 60 *books of* 5 *titles which are sent to a seller by the publisher. There are* 12 *books of each title and one book of each title is complimentary copy. From the box one book is selected at random. Find the probability that the selected book is*

(i) *of first title or a complimentary copy,*

(ii) *a complimentary copy when it is known that the copy is of first title. Are the events of occurrence of complimentary copy and of first title independent?*

Solution. Let A be the event that the selected book is of first title and B be the event that the selected book is a complimentary copy.

(i) We need $P(A \cap B)$.

One book from 60 books can be taken in $n = {}^{60}C_1 = 60$ ways. Favourable cases to A are ${}^{12}C_1 = 12$, favourable cases to B are ${}^5C_1 = 5$. But complimentary copy of first title is common with both A and B.

$$P(A \cup B) = P(A) + P(B) - P(A \cap B)$$

$$= \frac{12}{60} + \frac{5}{60} - \frac{1}{60} = \frac{4}{15}.$$

(ii) We need $P(B/A)$.

$$P(B/A) = \frac{P(A \cap B)}{P(A)} = \frac{1/60}{12/60} = \frac{1}{12}.$$

$$P(A)P(B) = \frac{12}{60} \times \frac{5}{60} = \frac{1}{60} = P(A \cap B).$$

Hence, A and B are independent events.

►**EXAMPLE 1.127** *Two persons A and B hit a target by throwing bullets. When A throws 9 bullets, B throws 10 bullets in the same time. Eight bullets out of 10 of A and 7 bullets out of 10 of B hit the target. A bullet is found on the target. What is the probability that it is thrown by B?*

Solution. Let H_1 be the event that the bullet is thrown by A, H_2 be the event that the bullet is thrown by B and E be the event that the bullets hit the target. We need $P(H_2/E)$. Given,

$$P(H_1) = \frac{9}{19},\ P(H_2) = \frac{10}{19},\ P(E/H_1) = \frac{8}{10},\ P(E/H_2) = \frac{7}{10}$$

$$P(H_1 \cap E) = P(H_1)P(E/H_1) = \frac{9}{19} \times \frac{8}{10} = \frac{72}{190}$$

$$P(H_2 \cap E) = P(H_2)P(E/H_2) = \frac{10}{19} \times \frac{7}{10} = \frac{70}{190}$$

$$P(E) = P(H_1 \cap E) + P(H_2 \cap E) = \frac{72}{190} + \frac{70}{190} = \frac{142}{190}$$

$$\therefore\ P(H_2/E) = \frac{P(H_2 \cap E)}{P(E)} = \frac{70/190}{142/190} = \frac{35}{71}.$$

►**EXAMPLE 1.128** *A computer scientist received interview cards for 3 different jobs. For one job, there are 5 candidates, for second job the candidates are 8 in number and for third job the number of candidates are 10. What is the probability that the scientist will get at least a job?*

Solution. Let A be the event of selection for 1st job, B be the event of selection for 2nd job and C be the event of selection for 3rd job. We need $P(A \cup B \cup C)$.

$$P(A) = \frac{1}{5}, P(B) = \frac{1}{8}, P(C) = \frac{1}{10}.$$

$$P(A \cap B) = P(A)P(B) = \frac{1}{5} \times \frac{1}{8} = \frac{1}{40} \quad \left[\because \begin{array}{c} A, B \text{ and } C \text{ are} \\ \text{independent events} \end{array}\right]$$

$$P(A \cap C) = P(A)P(C) = \frac{1}{5} \times \frac{1}{10} = \frac{1}{50}$$

$$P(B \cap C) = P(B)P(C) = \frac{1}{8} \times \frac{1}{10} = \frac{1}{80}$$

$$P(A \cap B \cap C) = P(A)P(B)P(C) = \frac{1}{5} \times \frac{1}{8} \times \frac{1}{10} = \frac{1}{400}$$

$$\begin{aligned} P(A \cup B \cup C) &= P(A) + P(B) + P(C) - P(A \cap B) - P(A \cap C) \\ &\quad - P(B \cap C) + P(\cap B \cap C) \\ &= \frac{1}{5} + \frac{1}{8} + \frac{1}{10} - \frac{1}{40} - \frac{1}{50} - \frac{1}{80} + \frac{1}{400} = 0.37. \end{aligned}$$

►**EXAMPLE 1.129** *In a research laboratory* 10 *male and* 6 *female researchers are working. Fifty percent of both male and female have Ph.D. degree. One of the researchers is selected at random. Find the probability that the selected one is*

(i) *either a female or a Ph.D. degree-holder,*

(ii) *male under the condition that he has Ph.D. degree.*

Solution. Let A be the event that a researcher has Ph.D. degree and B be the event that the researcher is a female.

(i) We need $P(A \cap B)$ [FIG. 1.18].

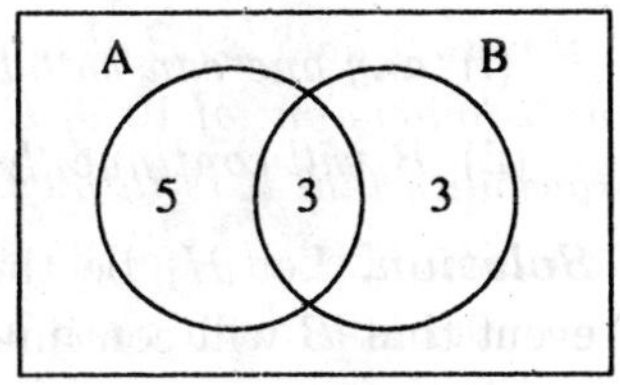

FIG. 1.18

$$P(A) = \frac{8}{16}, \ P(B) = \frac{6}{16}, \ P(A \cap B) = \frac{3}{16}.$$

$$\begin{aligned} P(A \cup B) &= P(A) + P(B) - P(A \cap B) \\ &= \frac{8}{16} + \frac{6}{16} - \frac{3}{16} = \frac{11}{16}. \end{aligned}$$

(ii) Let C be the event that the researcher is a male. We need $P(C/A)$ [FIG. 1.19].

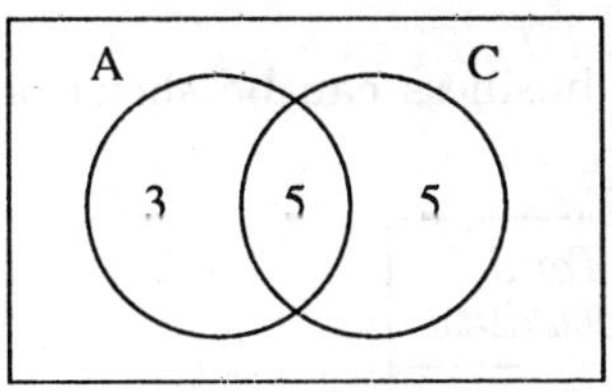

FIG. 1.19

$$\begin{aligned} P(A \cap C) &= \frac{5}{16} \\ P(C/A) &= \frac{P(A \cap C)}{P(A)} = \frac{5/16}{8/16} = \frac{5}{8}. \end{aligned}$$

►**EXAMPLE 1.130** *The product of a company reaches to the consumer after three steps. The company supplies it to the distributor, distributor supplies it to the whole-saler and whole-saler supplies to the retailer. Then consumer can buy the product. The company supplies the product rapidly with probability* 0.8, *the similar probabilities for other two suppliers are* 0.6 *and* 0.9, *respectively. Find the probability that*

(i) *the product will go to the consumer rapidly,*

(ii) *the product will reach rapidly in at least* 2 *levels,*

(iii) *the product will reach rapidly in at best* 2 *levels.*

Solution. Let S_i $(i = 1, 2, 3)$ be the event that at i-th stage the product reaches rapidly and N_i $(i = 1, 2, 3)$ be the event that at i-th stage it is not reached. Given,

$$\begin{aligned} P(S_1) &= 0.8, P(S_2) = 0.6, P(S_3) = 0.9 \\ P(N_1) &= 0.2, P(N_2) = 0.4, P(N_3) = 0.1 \end{aligned}$$

The levels of supply can be arranged as follows:

$$\begin{aligned} S : \{&N_1N_2N_3, N_1N_2S_3, N_1S_2S_3, N_1S_2N_3, \\ &S_1S_2S_3, S_1S_2N_1, S_1N_2N_3, S_1N_2S_3\} \end{aligned}$$

(i) We need $P(S_1S_2S_3) = 0.8 \times 0.6 \times 0.9 = 0.432$.

(ii) We need $P(S_1S_2S_3) + P(N_1S_2S_3) + P(S_1S_2N_3) + P(S_1N_2S_3) = 0.8 \times 0.6 \times 0.9 + 0.2 \times 0.6 \times 0.9 + 0.8 \times 0.6 \times 0.1 + 0.8 \times 0.4 \times 0.9 = 0.876$.

(iii) The event $S_1S_2S_3$ indicates that the product will reach rapidly in all levels. We need $1 - P(S_1S_2S_3)$.

$$1 - P(S_1S_2S_3) = 1 - 0.8 \times 0.6 \times 0.9 = 0.568.$$

►**EXAMPLE 1.131** *Two persons A and B have started a business on first January on the condition that any one can withdraw his share independently. A will continue for the whole year with probability* 0.7 *and B will continue for the whole year with probability* 0.8. *Find the probability that*

(i) *any one can withdraw his share before one year,*

(ii) *B will continue the business under the condition that A will withdraw his share.*

Solution. Let H_1 be the event that A will continue for the whole year, H_2 be the event that B will continue for the whole year. Given,

$$P(H_1) = 0.7, P(\bar{H}_1) = 0.3, P(H_2) = 0.8, P(\bar{H}_2) = 0.2.$$

The events continuation of business or withdraw from business can be shown as follows:

H_2 \ H_1	H_2	$\bar{H}_2$	*Total probability*
H_1	$H_1 \cap H_2$	$H_1 \cap \bar{H}_2$	0.7
$\bar{H}_1$	$\bar{H}_1 \cap H_2$	$\bar{H}_1 \cap \bar{H}_2$	0.3
Total probability	0.8	0.2	1.00

Here $H_1 \cap H_2$ indicates that both A and B will continue the business. As they take independent decision,

$$P(H_1 \cap H_2) = P(H_1)P(H_2) = 0.7 \times 0.8 = 0.56.$$

(i) We need $P(H_1 \cap \bar{H}_2) + P(\bar{H}_1 \cap H_2)$.

$$\begin{aligned} &P(H_1 \cap \bar{H}_2) + P(\bar{H}_1 \cap H_2) \\ &\quad = [P(H_1) - P(H_1 \cap H_2)] + [P(H_2) - P(H_1 \cap H_2)] \\ &\quad = (0.70 - 0.56) + (0.8 - 0.56) = 0.38. \end{aligned}$$

(ii) We need $P(H_2/\bar{H}_1)$.

$$P(H_2/\bar{H}_1) = \frac{P(H_1 \cap \bar{H}_2)}{P(\bar{H}_1)} = \frac{0.14}{0.30} = \frac{7}{15}.$$

►**EXAMPLE 1.132** *There are 2 drawers in each of 3 similar boxes. There are one red ball and one white ball in one drawer of first box. In each drawer of second box there is one red ball and in each drawer of third box there is one white ball. A box is randomly selected and its drawer is also randomly selected and found a red ball. What is the probability that in other drawer of the selected box there is a red ball?*

Solution. As the second box contains one red ball in each drawer, second ball will be red if second box is selected. Otherwise second ball cannot be red under the condition of first red ball. Therefore, the required probability is the probability of selection of second box. This probability is

$$\text{Required probability} = \frac{1}{3}.$$

EXERCISE

1. (a) What do you mean by probability? Give examples of equally likely cases, exhaustive cases, mutually exclusive cases, event, mutually exclusive events.

(b) An industry produces computer spare parts using three machines, A, B and C, where the machines produces 500, 800 and 700 parts respectively per hour. From a day's production a part is randomly selected. Find the probability that the selected one is the production of machine A.

(c) If two parts are randomly selected, what is the probability that (i) both are the products of machine A, (ii) one is the product of machine B and another is the product of machine C?

(d) It is known that 2% products of machine A, 1% of machine B and 3% of machine C are usually found defectives. If two parts are selected at random one by one with replacement, find the probability that (i) both will be defective, (ii) one will be defective. If two parts are selected at random one by one without replacement, what will be probability that (iii) both will be non-defective, (iv) one will be defective?

(e) If two are selected at random, what is the probability that one is the defective product of machine A and another is that of machine B?

[**Ans.** (b) 0.25; (c) (i) 0.0624, (ii) 0.2801; (d) (i) 0.00038, (ii) 0.0382, (iii) 0.9614, (iv) 0.0383; (e) 0.00004.]

2. (a) Define probability of an event. State and prove the addition law of probability for three events. Also prove that $P(\bar{A}/B) = 1 - P(A/B)$.

(b) Three workers are to be appointed from 3 groups of workers. In first group there are 3 males and 1 female workers, second and third group have 2 males and 2 females workers and 1 male and 3 female workers, respectively. If one worker is selected from each group, what is the probability that one male and two female workers are selected? **[Ans. 0.40625]**

(c) The probability that a person aged X years will die within one year is P. If all of n persons $A_1, A_2, \ldots, A_n$ are of age X years, what is the probability that A_1 will die within one year and A_1 is the first one to die? [**Ans.** $\frac{1}{n}[1 - (1 - P)^n]$]

3. (a) Define probability with example. Explain the limitations of classical definition of probability.

(b) For any three events A, B and C, prove that

$$P(A \cup B/C) = P(A/C) + P(B/C) - P(A \cap B/C).$$

(c) Two computer scientists A and B have joined a competition of developing program. He will win the competition who will finish earlier. If the probability of winning by A is 0.6, then what is the probability that (i) A will be defeated every time, (ii) A will win at least once? **[Ans.** (i) 0.0265, (ii) 0.9735]

4. (a) Write down some examples of random experiment, mutually exclusive events, equally likely cases and exhaustive cases. If A and B are two mutually exclusive events, show that $P(A/A \cup B) = P(A)/[P(A) + P(B)]$, provided $P(A \cup B) \neq 0$.

(b) Prove that, if A and B are any two events, then $P(A \cup B) = 1 - P(\bar{A} \cap \bar{B})$.

(c) A show has three separate parts, viz. lower part, upper part and middle part. The probabilities that these parts will be defective are 0.1, 0.05 and 0.15, respectively. If a person buy a pair of shoes, what is the probability that the shoes will non-defective? **[Ans.** 0.72675]

5. (a) State addition law of probability for three events.

(b) Let $A_1, A_2, \ldots, A_n$ be events defined on a sample space S. Then show that

$$P(\cap A_i) \geq \sum_{i=1}^{n} P(A_i) - (n-1)$$

Also prove that

$$P\left(\bigcup_{i=1}^{n} A_i\right) \geq \sum_{i=1}^{n} P(A_i) > \sum_{i \neq j}^{n} P(A_i \cap A_j)$$

(c) An organization conducting social research has investigated randomly selected 1250 families. The families are classified according to monthly income (in Rs.), availability of telephone and television as follow:

Availability of telephone	Monthly income (in Rs.) 40000.00		Above 40000.00		Total
	Availability of television				
	Yes	No	Yes	No	
Yes	270	200	180	100	750
No	180	100	120	100	500
Total	450	300	300	200	1250

A family is selected at random. Find the probability that (i) the family has a television, (ii) the income of the family is above 40000.00 and has telephone, (iii) the family has television under the condition that it has telephone.

[Ans. (i) 0.06, (ii) 0.06, (iii) 7/15.]

6. (a) Define marginal probability, conditional probability and independent events with examples. Write down the conditions for n events to be complete independent.

(b) For three events A, B and C prove that

$$P(A \cup B/C) = P(A/B) + P(B/C), \text{ if } A \cap B = \phi.$$

(c) A bag contains 3 red and 2 white balls and another bag contains 4 red and 3 white balls. A ball is drawn at random from first bag and transferred to second bag without noting its colour. Find the probability of drawing a red ball from second bag after transformation of a ball from first bag. **[Ans.** 23/40]

7. (a) Define conditional probability, independent events and complementary event.

(b) If A and B are two independent events, then show that (i) A and $\bar{B}$, (ii) $\bar{A}$ and $\bar{B}$ are independent events. Also, for any two events A and B, prove that

$$P(A \cap B) \leq P(A) \leq P(A \cup B) \leq P(A) + P(B).$$

(c) A machine has three components A, B and C. The machine runs well if all three components are in good condition. The probabilities of creating problem by the machines in any working day are 0.12, 0.15, 0.20, respectively. What is the probability that one day the machine does not work. **[Ans. 0.4016]**

8. (a) Define conditional probability, independent events and marginal probability. Let A, B and C be three events defined on a sample space S. If A, B and C are pair wise independent and if A and $(B \cup C)$ are also independent, then show that A, B and C are mutually independent.

(b) Two products A and B of an industry are expected to be marketed by a representative. Previous experience provides information that the representative becomes successful in doing so in 80% and 20% cases, respectively. The products A and B are marketed independently. What is the probability that the product will be A if the representative becomes successful in marketing at least one product? **[Ans. 20/21]**

9. (a) Let A, B and C be three events defined on a sample space S. Then $P(B/A) \leq P(C/A)$ if $B \subset C$ and $P(A) > 0$.

(b) A computer graduate applies for a post in two industries X and Y. The probability of his being selected in industry X is 0.6 and the probability of being selected in industry Y is 0.5. The probability of at least one of his application being rejected is 0.4. Find the probability that he will be selected in one of the industry. **[Ans. 0.5]**

(c) In an examination 20% students have failed in Statistics course, 30% have failed in Mathematics course and 10% have failed in both courses. A student is randomly selected, find the probability that he has failed in a course. **[Ans. 0.4]**

10. (a) For any two events A and B, show that $P(A \cap B) \geq P(A) + P(B) - 1$.

(b) What is the probability that a positive integer not exceeding 100 selected at random is divisible by 5 or 7? **[Ans. 0.32]**

(c) What is the probability that exactly 4 heads appear when a fair coin is tossed 5 times, given that the first toss show a head. **[Ans. 0.25]**

11. (a) Define conditional probability and independence of events. Write down the limitations of classical definition of probability.

(b) Sixty-five percent students of a college hostel read English newspaper, 25% students read Bengali newspaper and 10% students read both English and Bengali newspaper. A student is randomly selected, find the probability that he reads any one of the paper. **[Ans. 0.85]**

(c) Three persons A, B and C fire from gun at a target. The probability of hitting the target by A, B and C are, respectively 0.3, 0.25 and 0.1. One bullet is found out at the target. Find the probability that the bullet is from the gun of (i) A, (ii) A or B. **[Ans. (i) 0.2024, (ii) 0.36]**

12. (a) Write down some examples of random experiment, not equally likely cases, not mutually exclusive events, and independent events.

(b) State and prove Bayes theorem. Also, explain the theorem for future events.

(c) Two friends A and B have started a business on first of January on the understanding that they can take decision independently to withdraw their share at any time. The probability that A will continue the business for one complete year is 0.7. The same probability for B is 0.9. Find the probability that (i) any one of them can withdraw the share before completion of the year, (ii) both of them will stop the business. [**Ans.** (i) 0.32, (ii) 0.04]

13. (a) State and prove addition law of probability. Show that, for any two events A and B, $P(A) \leq P(B)$, if $A \subset B$.

(b) In a game of poker 5 cards are dealt to each player after thoroughly shuffling the 52 cards. It is known as poker hand. Find the probability that a poker hand gets 3 aces and two queens. [**Ans.** 0.0000092]

(c) In a box there are 10 mobile phone sets, out of which 4 are defectives. The sets are inspected one by one at random. What is the probability that at seventh inspection the last defective is found out. [**Ans.** 0.0952]

14. (a) Define a priori and a posteriori probabilities. State and prove Bayes theorem.

(b) In a departmental store three girls A, B and C work as sales girl. All three are equally experienced. But A behaves well 90% time with the customers. The same percentages for B and C are, respectively 75 and 60. One day a customer is not well-behaved. What is the probability that he is not well-behaved by C? [**Ans.** 0.1333]

(c) There are three similar boxes. The boxes contain bulbs of different numbers as follows:

Box-1 : 3 bulbs of 40 watts, 2 bulbs of 60 watts

Box-2 : 2 bulbs of 40 watts, 3 bulbs of 60 watts

Box-3 : 4 bulbs of 40 watts, 4 bulbs of 60 watts.

A box is randomly selected and then a bulb is selected at random from the selected box. If the bulb is of 40 watts, (i) what is the probability that it is selected from Box-1? If a second bulb is selected from the selected box (ii) by replacing the first selected bulb, (iii) without replacing the first selected bulb, what is the probability that second selected bulb is of 40 watts? [**Ans.** (i) 0.4, (ii) 0.5133, (iii) 0.4095]

15. (a) Define the terms event, complementary event, mutually exclusive events and probability of an event. State and prove addition law of probability.

(b) Three students A, B and C of a computer science class are asked to develop a computer program. Previous experience provides information that they become successful in 70%, 60% and 80% cases, respectively. Find the probability that the program is developed. [**Ans.** 0.976]

(c) Sixty percent workers of a computer center are females. Five females and 15 males are trained. One worker is selected at random. Find the probability that the selected one is trained one under the condition that she is female. [**Ans.** 1/12]

16. (a) Define conditional probability and independent events with examples. State and prove Bayes theorem.

(b) A husband at the age of 60 years has the chance of 0.45 to reach the next marriage day. The same chance for wife at the age of 54 is 0.52. Find the probability that both of them will die before the next marriage day. **[Ans.** 0.264]

(c) A factory produces computer spare parts using machines A, B and C. The machines produces daily 1,000, 1200 and 1800 parts, respectively. It is known that 10% products of Machine A, 12% products of Machine B and 18% products of Machine C are usually defectives. One day a defective spare part is identified. What is the probability that it is the product of Machine C? **[Ans. 0.57]**

17. (a) Explain the axiomatic definition of probability. State and prove multiplication rule of probability.

(b) A, B and C are independent witnesses of an incidence which is known to have occurred. A speaks the truth 3 times out of 4, B speaks the truth 4 times out of 5 and the same case of C is 5 out of 6. What is the probability that the occurrence will be reported truthfully by majority of 3 witnesses. **[Ans.** 31/60]

(c) Forty per cent officers of a government office have cars. Six officers are randomly selected from the office with replacement. Find the probability that out of 6 officers (i) four officers have car, (ii) at least one officer has car.

[Ans. (i) 0.13824, (ii) 0.953344]

18. (a) Define a priori and a posteriori probability with example. State and prove Bayes theorem.

(b) A machine produces 90% non-defective items if it is run properly, otherwise it produces 40% non-defective items. It is observed that the machine is run properly in 80% cases. One day a non-defective item is found out. Find the probability that it is the product when machine is run properly. **[Ans.** 0.9]

(c) Two unbiased dice are thrown once. Find the probability that the sum of the upper faces of the dice is not 9 or 10. **[Ans.** 5/6]

19. (a) Write down examples of random experiment, not equally likely cases, exhaustive cases, event, sample space, mutually exclusive cases.

(b) In an industry there are 10 machines made of Korea and 5 machines made of China. Five machines made of Korea and 2 machines made of China do not work properly. One day a machine is identified which does not work properly. Find the probablity that it is made of Korea. **[Ans.** 1/3]

(c) A family has 2 children. The birth of a boy and a girl are equally likely. Find the probability that (i) both children are boys under the condition that first child is a boy, (ii) at least one of the child is a boy. **[Ans.** (i) 0.5, (ii) 0.75]

20. (a) Define marginal probability and conditional probability. State and prove Bayes theorem.

(b) Two persons A and B fire a target by gun. When A fires 9 bullets, B fires 10 bullets at that time. Eighty per cent bullets of A can hit the target, while 70% bullets of B can hit the target. The target is hit by a bullet. Find the probability that it is the bullet fired by B. **[Ans.** 35/71]

(c) If 4 digits are selected by a machine and if the selected digits are in proper order, then a person wins a lottery. Find the probability of winning the lottery. What is the probability of winning the lottery if 3 digits out of 4 are in proper order.

[Ans. 0.0001, 0.0036]

21. (a) Define marginal probability, conditional probability and independent events. State and prove addition law of probability for n events.

(b) What is the probability that a poker hand does not contain the queen or heart? **[Ans.** 47/72]

(c) Find the probability that a family of 5 children does not have a boy, if the probability of a boy and a girl are equally likely and if the sexes of children are independent. **[Ans.** 1/32]

22. (a) There are three sockets to use electric bulbs in a room. There are 5 bulbs in the house, out of which 3 are defectives. Find the probability that there will be light in the room. **[Ans.** 0.9]

(b) If two numbers are selected one by one with replacement from numbers 1, 2, 3, . . . , 10, what is the probability that (i) both numbers are even, (ii) sum of the numbers are more than 10. **[Ans.** (i) 0.5, (ii) 0.55]

(c) The cards of a pack of 52 cards are distributed one by one and record is maintained about the type of card. The distribution goes on until an ace is observed. Find the probability that (i) the fourth card, (ii) the thirteenth card, is an ace. **[Ans.** (i) 0.0639, (ii) 0.337]

23. (a) If A is an event defined on a finite sample space S, where the favourable sample points of A are nil such that $P(A) = 0$. But, if favourable points of A is the union of several primary events. Prove.

(b) The owner of a garments industry exports 70% of the produced garments. Thirty per cent of the products are marketed in the local market. Thirty per cent of the products sold in the local market are defective. Five per cent garments of export quality are also defective. A defective product is identified. Find the probability that it is produced for export. **[Ans.** 0.28]

(c) There are three computers A, B and C in a computer center which are used for typing purpose. It is known that these computers do not work properly in 10%, 20% and 30% working times. One day a person starts work but find that computer does not work properly. What is the probability that it is computer C? **[Ans.** 0.5]

24. (a) Define event, sample space, not equally likely cases, not mutually exclusive events with example. State and prove multiplication rule of probability for n events.

(b) In a class there are 30 students bearing roll number 1 to 30. A student is randomly selected. Find the probability that the roll number of the selected student is multiple of 3 or 4. **[Ans.** 14/30]

(c) Twelve balls are distributed in 4 boxes. Find the probability that the first box will contain 4 balls. **[Ans.** 0.1936]

25. (a) Define marginal probability and conditional probability. Also, define complete independence of events.

(b) A problem of mathematics is given to four students A, B, C and D to solve. Previous knowledge provides information that their capacity in terms of probabilities to solve the problem are 0.5, 0.33, 0.25 and 0.20 respectively. Find the probability that the problem is solved. **[Ans.** 0.8]

(c) Three unbiased dice are thrown once. Find the probability that (i) sum of the upper faces of dice is 10, (ii) sum of the upper faces of dice is 10 given that first die shows 2. **[Ans.** (i) 1/8, (ii) 5/36]

CHAPTER **2**

Random Variable and Its Distribution

2.1 Introduction

The outcomes of a random experiment, if these are finite in number, are expressed in the sample space and some sample points are counted which are in favour of an event defined on the sample space. But sample space may not always be shown, specially if the outcomes are large in number or infinite in number. In such case the count of favourable cases to an event is also not done easily. Moreover, outcomes of some experiment may be shown by qualitative character. For example, if 10 coins are tossed, we expect 2^{10} outcomes which will be shown in combination of 'head' (H) and 'tail' (T). Here writing the sample space is not easy. However, the results can be shown in set notation using a variable X, where in each sample point a real number can be assigned.

Let S be the sample space of an experiment, the sample points of which are $s_1, s_2, \ldots, s_n$ (say). The sample points can be expressed by the variable X; where the values of X are $x_1, x_2, \ldots, x_k (k \leq n)$ and each one is real number. The assignment of real number to the sample points of sample space is the subject matter of random variable.

2.2 Definition of Random Variable

Let S be the sample space of a random experiment containing sample points s_1, $s_2, \ldots, s_n$ (say) such that $s \in S$. The real-valued function $X(s) = x$ of X assigned to each sample point is called *random variable.* The space of X is constituted with a set of real values and is denoted by

$$R = \{X : x = X(s), s \in S\}.$$

For example, let us consider the sample space of random experiment of throwing two coins once as follows:

$$S : \{HH, HT, TH, TT\}$$

Consider that X is the number of heads in throwing two coins once. Then $X(s) = 0$, if sample point is TT, $X(s) = 1$, if sample points are HT and TH and $X(s) = 2$, if sample point is HH. Thus, X is a real-valued function, where real values are assigned to all sample points in S. The space of X is written as

$$R = \{X : x = 0, 1, 2\}.$$

Again, consider an event A on the sample space S, where A is constituted with subset of R such that $P(X \in A)$ is the probability of the event A which can be measured. This means that the variable X is a function of real values, where each real value is associated with probability. This X is called *random variable* which generated from random experiment and it takes different values with associated probability.

In throwing two unbiased coins once, we have

$$P(X = 0) = \frac{1}{4}, P(X = 1) = \frac{2}{4}, P(X = 2) = \frac{1}{4}$$

where X is the number of heads. Here

$$R : \{2, 1, 1, 0\}.$$

The probabilities and values of X can be arranged as follows:

X :	0	1	2
$P(X = x)$:	1/4	1/2	1/4

The random variable defined above is associated to a simple event constituted with one or more sample points in the sample space. This type of random variable is called *one-dimensional random variable.* If two events can be defined with one or more sample points of a sample space of the random experiment, we may have two real-valued functions, where each pair of values is associated with probability. Then the random variable is known as *two-dimensional random variable.* For example, consider the random experiment of throwing two unbiased dice once and from this experiment let us define one random variable X, where X indicates sum of the upper faces even. Consider another random variable Y which indicates that first die shows even numbers. Here the range space of X is

$$R_1 : \{0, 2, 4, 6, 8, 10, 12\}$$

and the range space of Y is

$$R_2 : \{0, 2, 4, 6\}.$$

The favourable points to 0 in R_1 are

$$\left\{\begin{array}{l} 12, 14, 16, 21, 23, 25, 32, 34, 36, \\ 41, 43, 45, 52, 54, 56, 61, 63, 65. \end{array}\right\}$$

Some of the points favourable to 0 in R_1 are also favourable to 0 in R_2. These points are

$$\{12, 14, 16, 32, 34, 36, 52, 54, 56\}.$$

These points are jointly in favour of $X = 0, Y = 0$. The probability of these points is

$$P(X = 0, Y = 0) = \frac{9}{36}.$$

Here X and Y take real values with associated probabilities and hence X and Y are called two-dimensional random variable. The other sample points in favour of different values of X and Y are shown below:

Y \ X	0	2	4	6	8	10	12
0	12, 14, 16, 32, 34, 36, 52, 54, 56	11	13, 31	51, 15, 33	35, 53	55	—
2	21, 23, 25	—	22	24	26	—	—
4	41, 43, 45	—	—	42	44	46	—
6	61, 63, 65	—	—	—	62	64	66

The probabilities associated with different joint values of X and Y are shown below:

$$P(X = x, Y = y)$$

Y \ X	0	2	4	6	8	10	12	Total
0	9/36	1/36	2/36	3/36	2/36	1/36	—	18/36
2	3/36	—	1/36	1/36	1/36	—	—	6/36
4	3/36	—	—	1/36	1/36	1/36	—	6/36
6	3/36	—			1/36	1/36	1/36	6/36
Total	18/36	1/36	3/36	5/36	5/36	3/36	1/36	1.00

Here different real values for X and Y are assigned to different sample points in the sample space S and each pair of real values is associated with probability. Hence X and Y are called *two-dimensional random variable.* In a similar way, n-dimensional random variable can be defined based on sample points of a sample space.

2.3 Types of Random Variable

The random variable mentioned above takes only integral values (X = sum of upper faces even when two unbiased dice are thrown). But variable can take integral as well as fractional values and hence the random variable is of two types, viz., (i) discrete random variable, (ii) continuous random variable.

DISCRETE RANDOM VARIABLE. If a random variable takes only integral values in the limit 0 to countable infinity, then the random variable is known as discrete. For example, number of heads in coin-tossing experiment, sum of upper faces in throwing dice, number of computers of special type when selected from a lot of different types of computers, number of mis-call in a telephone per day, number of printing mistakes per page, etc.

CONTINUOUS RANDOM VARIABLE. The random variable which takes different values, integral as well as fractional, in the limit $-\infty$ to ∞ is called *continuous random variable.* For example, let us consider an experiment of selection of some families at random to observe the reporting error of income of those families. The variable reporting error of income may take $-$ve as well as $+$ve integral or fractional values. Usually, age, height, weight, income, production, etc., are continuous random variable if these variables are observed from any random experiment.

2.4 Probability Mass Function, Probability Density Function and Probability Distribution

Let X be a random variable defined on the basis of simple event on a sample space S. The range space of X can be written as

$$R = \{X : X(s), s \in S\}.$$

Let A be any event depending on the subset of R. Then $P(A) = P(X \in A)$ is used to denote the probability of the event A defined on the subset of R. This function $P(A)$ is used to calculate probabilities for different values of X [if X is discrete] or is used to calculate probabilities of an interval x to $x + \Delta x$ [if X is continuous random variable], where Δx is a small change in the values of X. Thus, $P(A) = P(X \in A)$ is called a *probability function* of X.

Definition. Let $x_1, x_2, \ldots, x_n$ be n values of a discrete random variable X such that $P(X = x_1), P(X = x_2), \ldots, P(X = x_n)$ be the corresponding probabilities. This $P(X = x_i; i = 1, 2, \ldots, n)$ is called a *probability mass function* (probability function) of the random variable X if the following two conditions are satisfied:

(i) $P(X = x_i) \geq 0$,

(ii) $\sum_{i=1}^{n} P(X = x_i) = 1$.

PROBABILITY DENSITY FUNCTION (p.d.f.). If X is continuous random variable and $f(x)$ be the probability function of X, then $f(x)$ is called probability density function of X if the following two conditions are satisfied:

(i) $f(x) \geq 0$,

(ii) $\int_{-\infty}^{\infty} f(x)dx = 1$.

PROBABILITY DISTRIBUTION. The set of paired values of $\{x, P(X = x)\}$ for discrete random variable X and paired values $\{x, f(x)\}$ for continuous random variable X are known as probability distribution.

►**EXAMPLE 2.1** *In an urn there are 4 white and 6 red balls. Two balls are taken at random from the urn. Let X be the number of white balls. Find probability function of X and write down the probability distribution.*

Solution. The sample space of the experiment can be shown as $S : \{RR, RW, WR, WW\}$, where $R =$ red ball, $W =$ white ball.

$$\therefore \ R : \{X : x = 0, 1, 2\}.$$

Here $X(WW) = 2, X(RW, WR) = 1, X(RR) = 0$.
Now, the probability distribution can be written as

X	0	1	2	Total
$P(X = x)$	15/45	24/45	6/45	1.00

Here $P(X = x) > 0$ and $\sum P(X = x) = 1$ and hence $P(X = x)$ is a probability function and $\{x, P(X = x)\}$ is the probability distribution.

►**EXAMPLE 2.2** *Let X be a random variable whose range space is $R : \{X : 0 \leq x \leq 2\}$ and $f(x) = kx$. Show that $f(x)$ is a p.d.f.*

Solution. $f(x)$ will be a p.d.f. if $\int_0^2 f(x)dx = 1$. We have

$$\int_0^2 f(x)dx = 1 \text{ or, } k\int_0^1 xdx = 1 \text{ or, } k\frac{x^2}{2}\Big|_0^2 = 1$$

$$\text{or, } \frac{k}{2}[4 - 0] = 1, \ \therefore \ k = \frac{1}{2}$$

$$\therefore \ f(x) = \frac{x}{2}, 0 \leq x \leq 2.$$

Here $f(x) \geq 0$ for any value of X and $\int f(x)dx = \frac{1}{2}\int_0^2 xdx = 1$.
$\therefore \ f(x)$ is a p.d.f.

►**EXAMPLE 2.3** *From a well-shuffled pack of* 52 *cards* 3 *cards are taken at random. Find the probability of getting* (i) *at least* 2 *red cards,* (ii) *at best* 2 *red cards,* (iii) *no red card.*

Solution. Let X be the random variable indicating red cards. The range space of X is

$$R : \{X : x = 0, 1, 2, 3\}.$$

The probability distribution of X is

X	0	1	2	3	Total
$P(X = x)$	2600/22100	8450/22100	8450/22100	2600/22100	1.00

Here $P(X = x) > 0$ and $\sum P(X = x) = 1$. So $P(X = x)$ is a probability function and $\{x, P(X = x)\}$ is the probability distribution.

(i) We need $P(X \geq 2)$

$$P(X \geq 2) = P(X = 2) + P(X = 3) = \frac{8450}{22100} + \frac{2600}{22100} = 0.5.$$

(ii) We need $P(X \leq 2)$

$$P(X \leq 2) = P(X=0)+P(X=1)+P(X=2)$$
$$= \frac{2600}{22100}+\frac{8450}{22100}+\frac{8450}{22100} = 0.88235.$$

(iii) We need $P(X=0)$

$$P(X=0) = \frac{2600}{22100} = 0.11765.$$

►**EXAMPLE 2.4** *Given a function of continuous random variable X as follows*

$$f(x) = kx(1-x),$$

with range space $R : \{X : 0 \leq x \leq 1\}$.

Is $f(x)$ *a density function? If so, find* $P(A_1)$, *where* $A_1 = \{X : 0 \leq x \leq 1/3\}$ *and* $P(A_2)$ *where* $A_2 = \{X : x \geq 1/2\}$.

Solution. $f(x)$ will be a p.d.f. if $\int_0^1 f(x)dx = 1$. We have

$$k\int_0^1 x(1-x)dx = 1 \text{ or, } k\left[\frac{x^2}{2}\Big|_0^1 - \frac{x^3}{3}\Big|_0^1\right] = 1 \text{ or, } k = 6$$
$$\therefore\ f(x) = 6x(1-x), 0 \leq x \leq 1.$$

This $f(x)$ is a p.d.f. as $f(x) \geq 0, \int f(x)dx = 1$.

(i) We need $P(A_1)$, where $A_1 = \{X : 0 \leq x \leq 1/3\}$

$$P(A) = P(0 \leq x \leq 1/3) = \int_0^{1/3} 6x(1-x)dx$$
$$= 6\left[\frac{x^2}{2}\Big|_0^{1/3} - \frac{x^3}{3}\Big|_0^{1/3}\right] = 6\left[\frac{1}{18}-\frac{1}{81}\right] = \frac{7}{27}.$$

(ii) We need $P(A_2)$, where $A_2 = \{X : x \geq 1/2\}$

$$P(A) = P\left(x \geq \frac{1}{2}\right) = 6\int_{1/2}^1 x(1-x)dx$$
$$= 6\left[\frac{x^2}{2}\Big|_{1/2}^1 - \frac{x^3}{3}\Big|_{1/2}^1\right] = \frac{1}{2}.$$

2.4.1 Joint Probability Density (Mass) Function

Let X and Y be two random variables defined simultaneously on the same sample space S. The variables X and Y are called *two-dimensional random variables.* The probability $P(X = x, Y = y)$ for each pair of values of X and Y, if these are discrete random variables, is known as joint probability mass function if the following two conditions are satisfied:

(i) $P(X = x, Y = y) \geq 0$,

(ii) $\sum_x \sum_y P(X = x, Y = y) = 1$.

If X and Y are continuous random variables and $f(x, y)$ is a function of X and Y, then $f(x, y)$ is called *joint probability density function* of X and Y if the following two conditions are satisfied:

(i) $f(x, y) \geq 0$ for all values of X and Y,

(ii) $\int_x \int_y f(xy)dxdy = 1.$

2.4.2 Marginal Probability Density Function

Let X and Y be two-dimensional random variables defined on a sample space S. The probability function $P(X = x, Y = y)$ of pairs of values $(X = x, Y = y)$ is known as *joint probability function.* The probability function of X for all values of Y is given by

$$g(x) = \sum_y P(X = x, Y = y).$$

This probability function $g(x)$ is known as marginal probability function of X. Similarly,

$$h(y) = \sum_x P(X = x, Y = y)$$

is called *marginal probability function* of Y.

If two-dimensional random variables X and Y are continuous, then the marginal probability density functions of X and Y are given, respectively, by

$$g(x) = \int_y f(x, y)dy \text{ and } h(y) = \int_x f(x, y)dx.$$

If R is the range space of two-dimensional random variables X and Y, and if A be any event defined on the range space, then

$$P(A) = P[(X, Y) \in A] = \sum_A \sum_A P(X = x, Y = y),$$

where X and Y are discrete random variables. If X and Y are continuous random variables, then

$$P(A) = P[(X, Y) \in A] = \int_A \int_A f(x, y)dxdy.$$

For n-dimensional random variables $X_1, X_2, \ldots, X_n$ these two probabilities are given, respectively, by

$$\begin{aligned} P(A) &= P[(X_1, X_2, \ldots, X_n) \in A] \\ &= \sum\sum \cdots \sum_A P(X_1 = x_1, X_2 = x_2, \ldots, X_n = x_n) \\ \text{and } P(A) &= P[(X_1, X_2, \ldots, X_n) \in A] \\ &= \iint_A \cdots \int f(x_1, x_2, \ldots, x_n)dx_1 \cdots dx_n. \end{aligned}$$

2.4.3 Conditional Probability Density Function

Let X and Y be two-dimensional random variables defined simultaneously on a sample space S. The joint probability function of X and Y is $P(X = x, Y = y)$. Again, the marginal probability functions of X and Y are $g(x)$ and $h(y)$, where $g(x) > 0$ and $h(y) > 0$. Then the conditional probability function of X given $Y = y$ is given by

$$g(x/Y = y) = \frac{P(X = x, Y = y)}{h(y)}, \quad h(y) > 0.$$

Similarly, the conditional probability function of Y given $X = x$ is given by

$$h(y/X = x) = \frac{P(X = x, Y = y)}{g(x)}, \quad g(x) > 0.$$

If X and Y are two-dimensional continuous random variables, then

$$\begin{aligned} g(x/Y = y) &= \frac{f(x,y)}{h(y)}, \quad h(y) > 0 \\ \text{and } h(y/X = x) &= \frac{f(x,y)}{g(x)}, \quad g(x) > 0. \end{aligned}$$

►**EXAMPLE 2.5** *An unbiased die is thrown once and a biased coin is tossed twice in an attempt. It is known that $H : T :: 1 : 3$ occurs when coin is tossed.*

Let X be the event that the die shows odd numbers and Y be the event that the coins show head H.

Find (i) *the joint probability function of X and Y,* (ii) *marginal probability functions of X and Y, and* (iii) *conditional probability function of X for all values of Y, and conditional probability function of Y for all values of X.* (iv) *Find the probability that the die shows even numbers and the coins show 2 heads.*

Solution. The sample space of the experiment is

$$S : \left\{ \begin{array}{l} 1\ HH,\ HT,\ TH,\ TT \\ 2\ HH,\ HT,\ TH,\ TT \\ 3\ HH,\ HT,\ TH,\ TT \\ 4\ HH,\ HT,\ TH,\ TT \\ 5\ HH,\ HT,\ TH,\ TT \\ 6\ HH,\ HT,\ TH,\ TT \end{array} \right\}$$

The range space of X is $R_1 : \{X : x = 0, 1, 3, 5\}$ and the range space of Y is $R_2 : \{Y : y = 0, 1, 2\}$. There are 24 sample points in the sample space. All 24 cases are not equally likely as

$$P(H) = \frac{1}{4} \text{ and } P(T) = \frac{3}{4}.$$

The sample points in favour of different values of X and Y are shown below:

Y \ X	0	1	3	5
0	$2TT, 4TT, 6TT$	$1TT$	$3TT$	$5TT$
1	$2TH, 2HT, 4TH,$ $4HT, 6TH, 6HT$	$1TH, 1HT$	$3TH, 3HT$	$5TH, 5HT$
2	$2HH, 4HH, 6HH$	$1HH$	$3HH$	$5HH$

(i), (ii) The joint probability function and marginal probability functions of X and Y are shown below:

Y \ X	0	1	3	5	Total marginal probability function, $h(y)$
0	27/96	9/96	9/96	9/96	54/96
1	18/96	6/96	6/96	6/96	36/96
2	3/96	1/96	1/96	1/96	6/96
Total, $g(x)$	48/96	16/96	16/96	16/96	1.00

(iii) The conditional probability function of X for different values of Y is

$$g(x/Y = y)$$

Y \ X	0	1	3	5	Total
0	27/54	9/54	9/54	9/54	1.00
1	18/36	6/36	6/36	6/36	1.00
2	3/6	1/6	1/6	1/6	1.00

The conditional probability function of Y for different values of X is

$$h(y/X = x)$$

Y \ X	0	1	3	5
0	27/48	9/16	9/16	9/16
1	18/48	6/16	6/16	6/16
2	3/48	1/16	1/16	1/16
Total	1.00	1.00	1.00	1.00

(iv) Let A be the event that the die shows even numbers and the coins show heads.

Here range space of A is

$$R : \{X, Y : (x, y) = (0, 2)\}$$

$$P(A) = P(X = 0, Y = 2) = \frac{3}{96} = \frac{1}{32}.$$

If B be the events that the die shows odd numbers and the coin shows heads. Then the range space of B is

$$R : \{X, Y : (x, y) : (1, 0), (3, 0), (5, 0), (1, 1), (3, 1), (5, 1), (1, 2), (3, 2), (5, 2)\}$$

$$\therefore\ P(B) = \sum_{x=1,3,5;y=0,1,2}\sum P(X = x, Y = y)$$

$$= \frac{9}{96} + \frac{6}{96} + \frac{1}{96} + \frac{9}{96} + \frac{6}{96} + \frac{1}{96} + \frac{9}{96} + \frac{6}{96} + \frac{1}{96} = \frac{1}{2}.$$

►**EXAMPLE 2.6** *The two-dimensional random variables X and Y have the joint probability density function*

$$f(x, y) = \begin{cases} 21x^2y^3, & 0 < x < y < 1 \\ 0, & \textit{otherwise.} \end{cases}$$

Find marginal and conditional distributions of X and Y. Also, find the probability of the event A, where $A : \{Y : y \geq 1/2\}$.

Solution. The marginal density function of X is

$$g(x) = \int_y f(x, y)dy = \int_x^1 21x^2y^3dy = 21x^2\int_x^1 y^3dy$$

$$= 21x^2\left(\frac{1}{4} - \frac{x^4}{4}\right) = 21x^2\left(\frac{1 - x^4}{4}\right)$$

Marginal distribution of X is

$$g(x) = \begin{cases} \frac{21}{4}x^2(1 - x^4), & 0 < x < 1 \\ 0, & \text{otherwise.} \end{cases}$$

The marginal density function of Y is

$$h(y) = \int_x f(x, y)dx = \int_0^y 21x^2y^3dx = 21y^3\int_0^y x^2dx$$

$$= 21y^3 \cdot \frac{y^3}{3} = 7y^6.$$

The marginal distribution of Y is

$$h(y) = \begin{cases} 7y^6, & 0 < y < 1 \\ 0, & \text{otherwise.} \end{cases}$$

The conditional density function of X for given Y is

$$g(x/Y=y)=\frac{f(xy)}{h(y)}=\frac{21x^2y^3}{7y^6}=\begin{cases}3\frac{x^2}{y^3}; & 0<x<y, 0<y<1\\ 0, & \text{otherwise.}\end{cases}$$

The conditional density function of Y for given X is

$$h(y/X=x)=\frac{f(xy)}{g(x)}=\frac{21x^2y^3}{\frac{21x^2(1-x^4)}{4}}=\begin{cases}\frac{4y^3}{1-x^4}; & 0<x<y, 0<y<1\\ 0, & \text{otherwise.}\end{cases}$$

$$P(A)=P\left(y\geq\frac{1}{2}\right)=\int_{1/2}^{1}h(y)dy=7\int_{1/2}^{1}y^6dy$$
$$=\left(1-\frac{1}{2^7}\right)=\frac{127}{128}.$$

►**EXAMPLE 2.7** *The joint probability density function of two-dimensional random variables X and Y is given by*

$$f(x,y)=4x(1-y),\ 0<x<1, 0<y<1.$$

Find marginal density functions of X and Y. Conditional density functions of X and Y. Also, find the probability of the event A, where

$$A:\left\{X,Y:(x,y):x>\frac{1}{2},y<\frac{1}{2}\right\}.$$

Solution. The marginal density function of X is

$$g(x)=\int_y f(x,y)dy=\int_0^1 4x(1-y)dy$$
$$=4x\left(1-\frac{1}{2}\right)=2x,\ 0<x<1.$$

The marginal density function of Y is

$$h(y)=\int_x f(x,y)dy=\int_0^1 4x(1-y)dx$$
$$=4(1-y)\frac{1}{2}=2(1-y),\ 0<y<1.$$

The conditional density function of X given Y is

$$g(x/y)=\frac{f(xy)}{h(y)}=\frac{4x(1-y)}{2(1-y)}$$
$$=2x,\ 0<x<1,\ 0<y<1.$$

The conditional density function of Y for given X is

$$\begin{aligned}
h(y/x) &= \frac{f(xy)}{g(x)} = \frac{4x(1-y)}{2x} = 2(1-y),\ 0 < x < 1, 0 < y < 1. \\
P(A) &= \int_{1/2}^{1}\int_{0}^{1/2} f(xy)dxdy \\
&= \int_{1/2}^{1}\int_{0}^{1/2} 4x(1-y)dxdy \\
&= 4\int_{1/2}^{1} xdx \int_{0}^{1/2}(1-y)dy \\
&= 4\left[\left(\frac{1}{2}-\frac{1}{8}\right)\left(\frac{1}{2}-0\right)-\left(\frac{1}{8}-0\right)\right] = \frac{9}{16}.
\end{aligned}$$

2.4.4 Independent Probability Density Function (Independent Random Variables)

Let X and Y be two-dimensional random variables defined on a sample space S and let $P(X = x, Y = y)$ be the joint probability function of the discrete random variables X and Y. The marginal probability function of X and Y are, respectively

$$\begin{aligned}
g(x) &= \sum_y P(X = x, Y = y) \\
\text{and } h(y) &= \sum_x P(X = x, Y = y).
\end{aligned}$$

The conditional probability function of X given $Y = y$ is

$$\begin{aligned}
g(x/Y = y) &= \frac{P(X = x, Y = y)}{h(y)} \\
P(X = x, Y = y) &= h(y)g(x/Y = y) \\
&= g(x/Y = y) = \sum_x P(X = x, Y = y) \\
\sum_y P(X = x, Y = y) &= \sum_y g(x/Y = y)\sum_x P(X = x, Y = y) \\
g(x) &= g(x/Y = y)\sum\sum P(X = x, Y = y) \\
&= g(x/Y = y).
\end{aligned}$$

This is possible, if X and Y are independent. Hence, we can write

$$P(X = x,\ Y = y) = g(X = x)h(Y = y).$$

Here $g(X = x)$ and $h(Y = y)$ are two indepenent probability density functions. The variables X and Y are stochastic independent. For continuous random variables X and Y, we can write

$$f(x, y) = g(x/y)h(y).$$

If X does not **depend on Y, the marginal probability** density function of X is given by

$$\begin{aligned} g(x) &= \int_y f(x,y)dy = \int_y f(x/y)h(y)dy \\ &= g(x/y)\int_y h(y)dy \\ &= g(x/y) \\ \therefore\ f(x,y) &= g(x)h(y). \end{aligned}$$

Here X and Y are stochastic independent random variables.

Definition. Two random variables X and Y are said to be independent, if and only if the product of their marginal probability density functions is equal to their joint probability density functions, i.e.,

$$f(x,y) = g(x)h(y) \text{ or, } P(X = x, Y = y) = g(X = x)h(Y = y).$$

Here $g(x)$ and $h(y)$ are independent probability density functions.

MUTUALLY STOCHASTIC INDEPENDENCE. Let $X_1, X_2, \ldots, X_n$ be n-dimensional random variables defined on the same sample space S and let $f(x_1, x_2, \ldots, x_n)$ be the joint probability density function of $X_1, X_2, \ldots, X_n$. The marginal probability density functions of these random variables are, respectively $f_1(x_1), f_2(x_2), \ldots, f_n(x_n)$. The random variables $X_1, X_2, \ldots, X_n$ are mutually stochastic independent, if and only if

$$\begin{aligned} f(x_1, x_2, \ldots, x_n) &= f_1(x_1)f_2(x_2)\cdots f_n(x_n) \\ \text{and } f_{ij}(x_i, x_j) &= f_i(x_i)f_j(x_j),\ i \neq j = 1, 2, \ldots, n. \end{aligned}$$

This definition implies that if X_i and X_j are independent, all X_i's will not be stochastic independent, if

$$f(x_1, x_2, \ldots, x_n) \neq f_1(x_1)f_2(x_2)\cdots f_n(x_n).$$

►**EXAMPLE 2.8** *The joint probability function of two-dimensional discrete random variables is given below:*

Y \ X	1	2	3	4	5	6	*Total* $h(Y = y)$
0	1/24	1/24	1/24	1/24	1/24	1/24	6/24
1	3/24	3/24	3/24	3/24	3/24	3/24	18/24
Total $g(X = x)$	4/24	4/24	4/24	4/24	4/24	4/24	1.00

Find conditional distribution of X for given Y. Are X and Y independent? Let A is defined in the range space of X and Y as

$$A : \{(X, Y) : x \geq 3,\ y = 1\}.$$

Find probability of A.

Solution. The marginal probability function of X and Y are shown in last row and last column, respectively. The conditional probability function of X given Y is

$$g(x/Y = y) = \frac{P(X = x, Y = y)}{h(y)}.$$

The different values of $g(x/Y = y)$ are shown below:

$g(x/Y = y)$

Y \ X	1	2	3	4	5	6	Total
0	1/6	1/6	1/6	1/6	1/6	1/6	1.00
1	1/6	1/6	1/6	1/6	1/6	1/6	1.00

We have

$$P(X = 1, Y = 0) = \frac{1}{24}.$$

Also

$$g(X = 1)h(Y = 0) = \frac{4}{24} \times \frac{6}{24} = \frac{1}{24}$$

$$P(X = 1, Y = 1) = \frac{3}{24}$$

$$g(X = 1)h(Y = 1) = \frac{4}{24} \times \frac{18}{24} = \frac{3}{24} = P(X = 1, Y = 1)$$

$$P(X = 2, Y = 0) = \frac{1}{24}$$

$$g(X = 2)h(Y = 0) = \frac{4}{24} \times \frac{6}{24} = P(X = 2, Y = 0)$$

In a similar way, we can show that

$$P(X = x_i, Y = y_j) = g(X = x_i)h(Y = y_j) \text{ for all } i \text{ and } j.$$

Hence X and Y are independent random variables

$$P(A) = \sum_{x=3}^{6}\sum_{y=1}^{1} P(X = x, Y = y) = \frac{3}{24} + \frac{3}{24} + \frac{3}{24} + \frac{3}{24} = \frac{1}{2}.$$

►**EXAMPLE 2.9** *Let X and Y be two two-dimensional random variables defined on the same sample S. The joint probability function of X and Y is given by*

$$P(X = x, Y = y) = \frac{1}{15}(2x + y); \; x = 0, 1, 2; \; y = 0, 1.$$

Find the conditional probability function of X given Y. Are X and Y independent?

Solution. The joint probability function can be written as

Y \ X	0	1	2	Total $h(Y = y)$
0	0	2/15	4/15	6/15
1	1/15	3/15	5/15	9/15
Total $g(X = x)$	1/15	5/15	9/15	1.00

Here marginal probability functions are shown in last row and last column, where

$$g(x) = \sum_y P(X = x, Y = y) \text{ and } h(y) = \sum_x P(X = x, Y = y).$$

We need $g(x/Y = y)$. The probabilities are shown below:

Y \ X	0	1	2
0	0	2/6	4/6
1	1/9	3/9	5/9

$$g(x/Y = y) = \frac{P(X = x, Y = y)}{h(y)}$$

We have

$$P(X = 0, Y = 0) = 0$$

$$g(X = 0)h(Y = 0) = \frac{6}{15} \times \frac{1}{15} \neq P(X = 0, Y = 0)$$

Hence X and Y are not independent random variables.

►**EXAMPLE 2.10** *The two-dimensional continuous random variables X and Y have the joint probability density function*

$$f(x, y) = 8xy, \ 0 < x < y < 1.$$

Find marginal probability density functions and conditional probability density functions of X and Y. Are X and Y independent random variables? Also calculate probability of the event A, where $A : \{(X, Y) : x < \frac{1}{2}, y > \frac{1}{2}\}$.

Solution. The marginal probability density functions of X and Y are, respectively

$$\begin{aligned} g(x) &= \int_y f(x, y)dy = \int_x^1 8xydy = 8x \int_x^1 ydy \\ &= \begin{cases} 4x(1 - x^2), & 0 < x < 1 \\ 0, & \text{otherwise} \end{cases} \end{aligned}$$

$$\begin{aligned} \text{and } h(y) &= \int_0^y f(x, y)dx = 8y \int_0^y xdx = 4y(y^2 - 0) \\ &= \begin{cases} 4y^3, & 0 < y < 1 \\ 0, & \text{otherwise.} \end{cases} \end{aligned}$$

The conditional probability density functions of X given Y and of Y given X are, respectively

$$g(x/y) = \frac{f(x,y)}{h(y)} = \frac{8xy}{4y^3} = \begin{cases} 2xy^{-2}, & 0 < x < 1 \\ 0, & \text{otherwise} \end{cases}$$

$$\text{and } h(y/x) = \frac{f(xy)}{g(x)} = \frac{8xy}{4x(1-x^2)} = \begin{cases} \frac{2y}{1-x^2}, & 0 < y < 1 \\ 0, & \text{otherwise.} \end{cases}$$

Here $g(x)h(y) = 4x(1-x^2)4y^3 = 16x(1-x^2)y^3 \neq f(x,y)$.
Hence X and Y are not independent random variables

$$P(A) = \int_0^{1/2}\int_{1/2}^{1} 8xydxdy = 8\int_0^{1/2} xdx \int_{1/2}^{1} ydy$$
$$= 8\left[\frac{1}{8} - 0\right]\left[\frac{1}{2} - \frac{1}{8}\right] = \frac{3}{8}.$$

►**EXAMPLE 2.11** *The two-dimensional continuous random variables X and Y have the joint probability density function*

$$f(xy) = \begin{cases} \frac{3}{2}y^2; & 0 \le x \le 2, 0 \le y \le 1 \\ 0, & \textit{otherwise.} \end{cases}$$

Find conditional probability density functions of both the variables. Are the variables independent? Find probability of the event A, where $A: \{X : \frac{1}{2} \le x \le 1\}$.

Solution. The marginal probability density functions of X and Y are, respectively

$$g(x) = \int_y f(x,y)dy = \frac{3}{2}\int_0^1 y^2 dy = \frac{1}{2}(1-0) = \begin{cases} \frac{1}{2}, & 0 \le x \le 2 \\ 0, & \text{otherwise} \end{cases}$$

$$\text{and } h(y) = \int_x f(x,y)dx = \frac{3}{2}y^2\int_0^2 dx = \begin{cases} 3y^2, & 0 \le y \le 1 \\ 0, & \text{otherwise} \end{cases}$$

The conditional probability density functions of X and Y are, respectively

$$g(x/y) = \frac{f(x,y)}{h(y)} = \frac{\frac{3}{2}y^2}{3y^2} = \begin{cases} \frac{1}{2}, & 0 \le x \le 2 \\ 0, & \text{otherwise} \end{cases}$$

$$\text{and } h(y/x) = \frac{f(x,y)}{g(x)} = \frac{\frac{3}{2}y^2}{\frac{1}{2}} = 3y^2,\ 0 \le y \le 1.$$

Here

$$g(x)h(y) = \frac{1}{2} \times 3y^2 = \frac{3}{2}y^2 = f(x,y).$$

Hence X and Y are independent random variables.

$$P(A) = \int_A g(x)dx = \frac{1}{2}\int_{1/2}^{1} dx = \frac{1}{2}\left(1 - \frac{1}{2}\right) = \frac{1}{4}.$$

►**EXAMPLE 2.12** *The two-dimensional random variables X and Y have the joint probability density function*

$$f(x, y) = \begin{cases} kx(x - y); \; 0 \le x \le 2, -x < y < x \\ 0, \quad \textit{otherwise.} \end{cases}$$

Find conditional probability density function of Y. Also, find probability of the event $A : \{X : x \ge 1\}$.

Solution. First we need to find the value of k, where k is available solving the equation

$$\iint f(x, y)dxdy = 1$$

$$\text{or, } k\int_0^2\int_{-x}^{x} x(x - y)dxdy = 1$$

$$\text{or, } k\left[\int_0^2\int_{-x}^{x} x^2 dxdy - \int_0^2 xdx \int_{-x}^{x} ydy\right] = 1$$

$$\text{or, } k\left[\int_0^2 x^2(x + x)dy - \frac{1}{2}(4 - 0)\frac{1}{2}(x^2 - x^2)\right] = 1$$

$$\text{or, } 2k\int_0^2 x^3 dx = 1 \text{ or, } k = \frac{1}{8}$$

$$\therefore \; f(x, y) = \frac{1}{8}x(x - y), 0 \le x \le 2, -x < y < x.$$

The marginal probability density functions of X is

$$\begin{aligned} g(x) &= \int_{-x}^{x} f(x, y)dy = \frac{1}{8}\left[\int_{-x}^{x} x^2 dy - x\int_{-x}^{x} ydy\right] \\ &= \frac{1}{8}\left[x^2(x + x) - \frac{x}{2}(x^2 - x^2)\right] \\ &= \frac{x^3}{4}, 0 \le x \le 2. \end{aligned}$$

The conditional density function of Y for given $X = x$ is

$$\begin{aligned} h(y/x) &= \frac{f(x, y)}{g(x)} = \frac{\frac{1}{8}x(x - y)}{\frac{x^3}{4}} \\ &= \frac{1}{2}\frac{x - y}{x^2}, -x < y < x, 0 \le x \le 2. \end{aligned}$$

Also,

$$P(A) = \int_1^2 g(x)dx = \frac{1}{4}\int_1^2 x^3 dx = \frac{1}{16}x^4\Big|_1^2 = \frac{15}{16}.$$

►EXAMPLE 2.13 *Two-dimensional random variables X and Y have the joint probability density function*

$$f(x, y) = 2, 0 < x < 1, 0 < y < x.$$

Find conditional probability density function of Y. Also, find $P\left(X \geq \frac{1}{2}\right)$.

Solution. The marginal probability density function of x is

$$g(x) = \int_y f(x, y)dy = 2\int_0^x dy = 2x, 0 < x < 1.$$

The conditional probability density function of Y is

$$h(y/x) = \frac{f(x, y)}{g(x)} = \frac{2}{2x} = \frac{1}{x}, 0 < y < x$$

$$P\left(x \geq \frac{1}{2}\right) = \int_{1/2}^1 g(x)dx = 2\int_{1/2}^1 x = 1 - \frac{1}{4} = \frac{3}{4}.$$

►EXAMPLE 2.14 *Two-dimensional random variables X and Y have the joint probability function*

$$f(x, y) = 2(2 - x - y), 0 \leq x \leq y \leq 1.$$

Find conditional probability function of X and Y. Are X and Y independent? Also, find $P\left(x \leq \frac{1}{2}\right)$.

Solution. The marginal probability functions of X and Y are, respectively

$$\begin{aligned} g(x) &= \int_y f(x, y)dy = 2\int_x^1 (2 - x - y)dy \\ &= 2\left[2(1 - x) - x(1 - x) - \frac{1}{2}(1 - x^2)\right] \\ &= \begin{cases} 3x^2 - 6x + 3, \ 0 < x < 1 \\ 0, \quad \text{otherwise} \end{cases} \end{aligned}$$

$$\begin{aligned} \text{and } h(y) &= \int_x f(x, y)dx = 2\int_0^y (2 - x - y)dx \\ &= 2\left[2(y - 0) - \frac{1}{2}(y^2 - 0) - y(y - 0)\right] \\ &= \begin{cases} 4y - 3y^2, \ 0 < y < 1 \\ 0, \quad \text{otherwise.} \end{cases} \end{aligned}$$

The conditional probability distribution of X given $Y = y$ is

$$g(x/y) = \frac{f(x, y)}{h(y)} = \begin{cases} \frac{2(2-x-y)}{4y-3y^2}, \ 0 < x < y \\ 0, \quad \text{otherwise} \end{cases}$$

The conditional probability distribution of Y for given x is

$$h(y/x) = \frac{f(x,y)}{g(x)} = \frac{2(2-x-y)}{3x^2-6x+3}$$

$$= \begin{cases} \frac{2(2-x-y)}{3(x-1)^2}, & x < y < 1 \\ 0, & \text{otherwise} \end{cases}$$

$$g(x)h(y) = 3(x-1)^2 \times y(4-3y) \neq f(x,y).$$

Hence X and Y are not independent random variables.

$$P\left(x \leq \frac{1}{2}\right) = \int_{1/2}^{1} g(x)dx = \int_{1/2}^{1} (3x^2 - 6x + 3)dx$$

$$= 3\frac{x^3}{3}\bigg|_{1/2}^{1} - 6\frac{x^2}{2}\bigg|_{1/2}^{1} + 3x\bigg|_{1/2}^{1}$$

$$= \left(1 - \frac{1}{8}\right) - 3\left(1 - \frac{1}{4}\right) + 3\left(1 - \frac{1}{2}\right)$$

$$= \frac{1}{8}.$$

►**EXAMPLE 2.15** *The two-dimensional random variables X and Y have the joint probability density function*

$$f(x,y) = x + y; 0 < x < 1, 0 < y < 1.$$

Find conditional probability distribution of X and Y. Also, find $P\left(x > \frac{1}{2}, y > \frac{1}{2}\right)$. Are X and Y independent?

Solution. The marginal probability distribution of X and Y are, respectively

$$g(x) = \int_y f(x,y)dy = \int_0^1 (x+y)dy$$

$$= x\int_0^1 dy + \int_0^1 ydy = x(1-0) + \frac{1}{2}(1-0)$$

$$= \begin{cases} \frac{2x+1}{2}, & 0 < x < 1 \\ 0, & \text{otherwise} \end{cases}$$

$$\text{and } h(y) = \int_x f(x,y)dx = \int_0^1 (x+y)dx$$

$$= \int_0^1 xdx + y\int_0^1 dx = \left(\frac{1}{2} - 0\right) + y(1-0)$$

$$= \begin{cases} \frac{2y+1}{2}, & 0 < y < 1 \\ 0, & \text{otherwise.} \end{cases}$$

Here

$$g(x)h(y) = \frac{2x+1}{2} \times \frac{2y+1}{2} \neq f(xy).$$

Hence X and Y are not independent.

$$
\begin{aligned}
P\left(x \geq \frac{1}{2}, y \geq \frac{1}{2}\right) &= \int_{1/2}^{1}\int_{1/2}^{1}(x+y)dxdy \\
&= \int_{1/2}^{1}\int_{1/2}^{1} xdxdy + \int_{1/2}^{1}\int_{1/2}^{1} ydxdy \\
&= \int \left[x \ \ y\Big|_{1/2}^{1}\right] dx + \int \left[y \ \ x\Big|_{1/2}^{1}\right] dy \\
&= \int_{1/2}^{1} \frac{x}{2} dx + \int_{1/2}^{1} \frac{y}{2} dy \\
&= \frac{1}{4}\left(1-\frac{1}{4}\right) + \frac{1}{4}\left(1-\frac{1}{4}\right) = \frac{3}{8}
\end{aligned}
$$

The conditional distribution of X and Y are, respectively

$$
g(x/y) = \frac{f(x,y)}{h(y)} = \frac{x+y}{\frac{2y+1}{2}} = \begin{cases} \frac{2(x+y)}{2y+1}; & 0 < y < 1, 0 < x < 1 \\ 0, & \text{otherwise} \end{cases}
$$

$$
\text{and } h(y/x) = \frac{f(x,y)}{g(x)} = \frac{x+y}{\frac{2x+1}{2}} = \begin{cases} \frac{2(x+y)}{2x+1}; & 0 < x < 1, 0 < y < 1 \\ 0, & \text{otherwise.} \end{cases}
$$

►**EXAMPLE 2.16** *Two-dimensional discrete random variables X and Y have the joint probability function*

$$
P(X = x, Y = y) = \frac{1}{18}(x+2y); x = 1, 2; y = 1, 2.
$$

Find conditional probability function of X and Y.

Solution. The marginal probability functions of X and Y are, respectively

$$
\begin{aligned}
g(x) &= \sum_{y} P(X = x, Y = y) \\
&= \frac{x+2\cdot 1}{18} + \frac{x + 2\cdot 2}{18} = \frac{2x+6}{18}; \ x = 1, 2 \\
\text{and } h(y) &= \sum_{x} P(X = x, Y = y) \\
&= \frac{1+2y}{18} + \frac{2+2y}{18} = \frac{3+4y}{18}; \ y = 1, 2.
\end{aligned}
$$

The conditional probability functions of X and Y are, respectively

$$
g(x/y) = \frac{P(X = x, Y = y)}{h(Y = y)} = \frac{x+2y}{18} \times \frac{18}{3+4y} = \frac{x+2y}{3+4y}; \ \begin{matrix} x = 1, 2 \\ y = 1 \text{ or}, 2 \end{matrix}
$$

$$
\text{and } h(y/x) = \frac{P(X = x, Y = y)}{g(X = x)} = \frac{x+2y}{18} \times \frac{18}{2x+6} = \frac{x+2y}{2x+6}; \ \begin{matrix} y = 1, 2 \\ x = 1 \text{ or}, 2. \end{matrix}
$$

►EXAMPLE 2.17 *The two-dimensional random variables X and Y have the joint probability function*

$$f(x,y) = \frac{9(1+x+y)}{2(1+x)^4(1+y)^4};\ 0 < x < \infty,\ 0 < y < \infty.$$

Find conditional probability density function of X and Y. Are X and Y stochastically independent?

Solution. The marginal distributions of X and Y are, respectively

$$\begin{aligned} g(x) &= \int_0^\infty f(x,y)dy \\ &= \frac{9}{2(1+x)^4}\left[\int_0^\infty \frac{1+y}{(1+y)^4} + x\int_0^\infty \frac{1}{(1+y)^4}dy\right] \\ &= \frac{9}{2(1+x)^4}\left[\int_0^\infty (1+y)^{-3}dy + x\int_0^\infty (1+y)^{-4}dy\right]. \end{aligned}$$

Let $z = 1+y, dy = dz, 0 < z < \infty \Rightarrow 1 \le z \le \infty$

$$\begin{aligned} g(x) &= \frac{9}{2(1+x)^4}\left[\int_1^\infty z^{-3}dz + x\int_1^\infty z^{-4}dz\right] \\ &= \frac{9}{2(1+x)^4}\left[\frac{z^{-2}}{-2}\Big|_1^\infty + x\frac{z^{-3}}{-3}\Big|_1^\infty\right] \\ &= \frac{9}{2(1+x)^4}\left[-\frac{1}{2}(0-1) - \frac{x}{3}(0-1)\right] \\ &= \frac{9}{2(1+x)^4}\left[\frac{1}{2}+\frac{x}{3}\right] = \begin{cases} \frac{3(3+2x)}{4(1+x)^4}, & 0 < x \\ 0, & \text{otherwise.} \end{cases} \end{aligned}$$

Similarly,

$$h(y) = \int_x f(x,y)dx = \frac{3(3+2y)}{4(1+y)^4},\ 0 < y < \infty.$$

The conditional probability distribution of X given Y is

$$\begin{aligned} g(x/y) &= \frac{f(x,y)}{h(y)} = \frac{9(1+x+y)}{2(1+x)^4(1+y)^4} \times \frac{4(1+y)^4}{3(3+2y)} \\ &= \frac{6(1+x+y)}{(1+x)^4(3+2y)},\ 0 < x < \infty. \end{aligned}$$

The conditional probability distribution of Y given X is

$$\begin{aligned} h(y/x) &= \frac{f(x,y)}{g(x)} = \frac{9(1+x+y)}{2(1+x)^4(1+y)^4} \times \frac{4(1+x)^4}{3(3+2x)} \\ &= \frac{6(1+x+y)}{(1+y)^4(3+2x)},\ 0 < y < \infty \\ g(x)h(y) &= \frac{3(3+2x)}{4(1+x)^4} \times \frac{3(3+2y)}{4(1+y)^4} \neq f(x,y). \end{aligned}$$

Hence X and Y are not stochastically independent.

►EXAMPLE 2.18 *The joint probability density function of three-dimensional random variables X, Y and Z is*

$$f(x,y,z) = \exp\{-(x+y+z)\},\ 0 \le x,y,z \le \infty$$

Are X, Y and Z mutually stochastically independent? Find conditional density function of X given Y and Z.

Solution. The marginal density function of X is

$$g(x) = \int_y \int_z e^{-(x+y+z)} dydz = e^{-x} \int_0^\infty e^{-y} dy \int_0^\infty e^{-z} dz = e^{-x},\ 0 \le x \le \infty.$$

Similarly,

$$h(y) = e^{-y},\ 0 \le y \le \infty \text{ and } k(z) = e^{-z},\ 0 \le z \le \infty$$

$$\begin{aligned} f_1(x,y) &= \int_z f(x,y,z)dz = \int_0^\infty e^{-(x+y+z)} dz \\ &= e^{-(x+y)} \int_0^\infty e^{-z} dz = e^{-(x+y)},\ 0 \le x,\ y \le \infty \end{aligned}$$

Similarly,

$$\begin{aligned} f_2(x,z) &= e^{-(x+z)},\ 0 \le x,\ z \le \infty \\ f_3(y,z) &= e^{-(y+z)},\ 0 \le y,\ z \le \infty \\ g(x)h(y) &= e^{-x} \times e^{-y} = e^{-(x+y)} = f_1(x,y) \\ g(x)k(z) &= e^{-x} \times e^{-z} = e^{-(x+z)} = f_2(x,z) \\ h(y)k(z) &= e^{-y} \times e^{-z} = e^{-(y+z)} = f_3(y,z) \end{aligned}$$

Again,

$$\begin{aligned} g(x)h(y)k(z) &= e^{-x} \times e^{-y} \times e^{-z} = e^{-(x+y+z)} \\ &= = f(x,y,z),\ 0 \le x,y,z \le \infty. \end{aligned}$$

Hence X, Y and Z are mutually stochastically independent.
The conditional density function of X given Y and Z is given by

$$g(x/Y=y, Z=z) = \frac{f(x,y,z)}{f_3(y,z)} = \frac{e^{-(x+y+z)}}{e^{-(y+z)}} = e^{-x}$$

$$\therefore\ g(x/Y=y, Z=z) = \begin{cases} e^{-x}, & 0 \le x \le \infty,\ 0 \le y \le \infty,\ z \le z \le 0 \\ 0, & \text{otherwise.} \end{cases}$$

►EXAMPLE 2.19 *The joint probability function of two-dimensional discrete random variable is*

$$P(x,y) = \begin{cases} \frac{1}{16};\ x = 1,2,3,4; y = 1,2,3,4 \\ 0,\ \textit{otherwise.} \end{cases}$$

Find conditional probability function of X when $Y = 2$ and conditional probability function of Y when $X = 1$. Also, calculate $P(x > 2, y > 2)$. Are X and Y independent.

Solution. The marginal probability function of X is

$$g(x) = \sum_y P(x, y) = \sum_{y=1}^{4} \frac{1}{16} = \frac{1}{4}, \; x = 1, 2, 3, 4.$$

Similarly,

$$h(y) = \sum_x P(x, y) = \sum_{x=1}^{4} \frac{1}{16} = \frac{1}{4}, \; y = 1, 2, 3, 4.$$

We have

$$g(x)h(y) = \frac{1}{4} \times \frac{1}{4} = \frac{1}{16} = P(x, y).$$

So X and Y are independent random variables. The conditional probability function of X given $Y = 2$ is

$$g(x/Y = 2) = \frac{P(x, y)}{h(Y = 2)} = \frac{1}{16} \times \frac{4}{1} = \frac{1}{4}; \; x = 1, 2, 3, 4; \; y = 2$$

Similarly,

$$h(y/X = 1) = \frac{P(x, y)}{g(X = 1)} = \frac{1}{16} \times \frac{4}{1} = \frac{1}{4}, \; y = 1, 2, 3, 4; \; x = 1$$

$$P(x > 2, y > 2) = \sum_{x=3}^{4} \sum_{y=3}^{4} P(x, y) = \frac{1}{16} + \frac{1}{16} + \frac{1}{16} + \frac{1}{16} = \frac{1}{4}.$$

❐ **THEOREM 2.1** *If X and Y are two stochastically independent random variables, and if $g(x)$ and $h(y)$ are the marginal probability density functions of X and Y respectively, then*

$$P(a < x < b, c < y < d) = P(a < x < b)P(c < y < d).$$

Proof. As X and Y are independent random variables, the joint probability function of X and Y can be written as $f(x, y) = g(x)h(y)$, where $f(x, y)$ is the joint probability density function of X and Y.

$$\begin{aligned} P(a < x < b, c < y < d) &= \int_x \int_y f(x, y)dxdy = \int_x g(x)dx \int_y h(y)dy \\ &= \int_a^b g(x)dx \int_c^d h(y)dy = P(a < x < b)P(c < y < d). \end{aligned}$$

If X and Y are discrete random variables, and if $P(x, y)$ is the joint probability function, then $P(x, y) = g(x)h(y)$.

We can write

$$\begin{aligned} P(a < x < b, c < y < d) &= \sum_{a<x<b} g(x) \sum_{c<y<d} h(y) \\ &= P(a < x < b, \; c < y < d). \end{aligned}$$

►**EXAMPLE 2.20** *The two-dimensional continuous random variables X and Y have the joint probability function*

$$f(x,y) = \frac{3}{2}y^2;\ 0 < x < 2;\ 0 < y < 1.$$

Find $P\left(x \geq 1, y \leq \frac{1}{2}\right)$.

Solution. We have

$$g(x) = \int_0^1 f(x,y)dy = \frac{3}{2}\int_0^1 y^2 dy = \frac{1}{2}y^3\Big|_0^1 = \frac{1}{2};\ 0 < x < 2$$

$$h(y) = \int_0^2 f(x,y)dx = \frac{3}{2}y^2\int_0^2 dx = \frac{3y^2}{2}x\Big|_0^2 = 3y^2;\ 0 < y < 1.$$

Here

$$g(x)h(y) = \frac{1}{2} \times 3y^2 = \frac{3}{2}y^2 = f(x,y).$$

Hence X and Y are independent.

Now,

$$\begin{aligned}
P\left(x \geq 1, y \leq \frac{1}{2}\right) &= P(x \geq 1)P\left(y \leq \frac{1}{2}\right) \\
&= \int_1^2 g(x)dx \int_0^{1/2} h(y)dy \\
&= \frac{1}{2}\int_1^2 dx \cdot 3\int_0^{1/2} y^2 dy \\
&= \frac{1}{2}x\Big|_1^2 \times 3\frac{y^3}{3}\Big|_0^{1/2} \\
&= \frac{1}{2} \times \left(\frac{1}{8} - 0\right) \\
&= \frac{1}{16}.
\end{aligned}$$

2.4.5 Uniform Probability Distribution

Let X be a random variable discrete or continuous. If the probability density functions for all values of X are uniform, then the distribution is known as *uniform probability distribution.* For example, for continuous random variable

$$\begin{aligned}
f(x) &= 1,\ 0 < x < 1 \\
f(x) &= \frac{1}{b-a},\ a < x < b \\
f(x) &= \frac{1}{2},\ 0 < x < 2.
\end{aligned}$$

For discrete random variable

$$p(x) = \frac{1}{4},\ x = 1, 2, 3, 4.$$

2.5 Distribution Function

Let X be a discrete random variable having probability function

$$P(X = x)(x = 0,\ 1, 2, \ldots, \infty)$$

where x is a particular value of X. Then distribution function of X is given by

$$F(x) = P(X \leq x) = \sum_{i=0}^{x} P(X = i).$$

The function $F(x)$ is also known as *cumulative probability function* or *cumulative distribution function* (c.d.f.). The c.d.f. of continuous random variable $X(-\infty \leq x \leq \infty)$ is given by

$$F(x) = P(X \leq x) = \int_{-\infty}^{x} f(x)dx.$$

The function indicates that it is a positive and real one-valued number.

►**EXAMPLE 2.21** *The discrete random variable X has the probability function*

$$P(X = x) = \frac{x}{15};\ x = 1, 2, 3, 4, 5.$$

Find distribution function of X.

Solution. We have

$$F(x) = P(X \leq x) = \sum_{i=0}^{x} P(X = i).$$

Now,

$$F(x) = \begin{cases} 0, & x < 1 \\ \frac{1}{15}, & 1 \leq x < 2 \\ \frac{3}{15}, & 2 \leq x < 3 \\ \frac{6}{15}, & 3 \leq x < 4 \\ \frac{10}{15}, & 4 \leq x < 5 \\ 1, & 5 \leq x \end{cases}$$

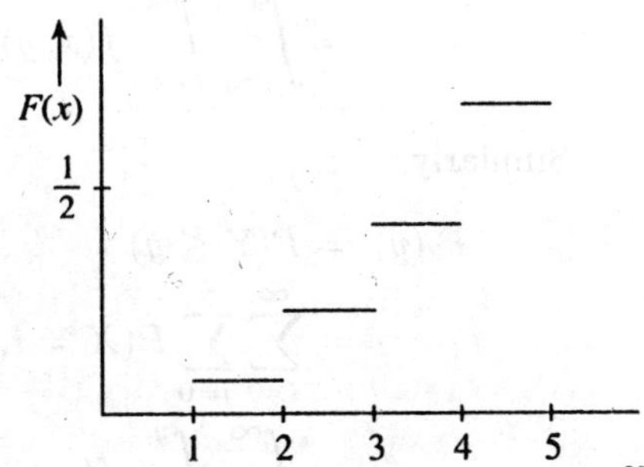

FIG. 2.1

The graphical representation is shown in FIG. 2.1.

►**EXAMPLE 2.22** *The continuous random variable X has the p.d.f.*

$$f(x) = \begin{cases} \frac{x}{8}, & 0 < x < 4 \\ 0, & \textit{otherwise}. \end{cases}$$

Find distribution function of X.

Solution. According to definition of c.d.f., we can write

$$F(x) = P(X \le x) = \int_{-\infty}^{x} f(x)dx = \begin{cases} 0, & x < 4 \\ \frac{1}{16}, & 0 < x \le 1 \\ \frac{3}{16}, & 1 < x \le 2 \\ \frac{5}{16}, & 2 < x \le 3 \\ \frac{7}{16}, & 3 < x < 4 \end{cases}$$

DISTRIBUTION FUNCTION OF TWO-DIMENSIONAL RANDOM VARIABLES. Let X and Y be the two-dimensional random variables defined on the same sample space and $P(X = x, Y = y)$ be the probability function if X and Y are discrete random variables. Then the joint distribution of X and Y is given by

$$F(x, y) = P(X \le x, Y \le y) = \sum_{i=0}^{x}\sum_{j=0}^{y} P(X = i, Y = j).$$

If X and Y are continuous random variables having joint p.d.f. $f(x, y)$, then

$$F(x, y) = P(X \le x, Y \le y) = \int_{-\infty}^{x}\int_{-\infty}^{y} f(x, y)dxdy.$$

The marginal distribution functions of X and Y are also be obtained as follows:

$$\begin{aligned} F_1(x) &= P(X \le x) \\ &= \sum_{i=0}^{x}\sum_{j=0}^{\infty} P(X = i, Y = j), \text{ if } X \text{ is discrete random variable} \\ &= \int_{-\infty}^{x}\int_{-\infty}^{\infty} f(x, y)dxdy, \text{ if } X \text{ is continuous random variable.} \end{aligned}$$

Similarly,

$$\begin{aligned} F_2(y) &= P(Y \le y) \\ &= \sum_{i=0}^{\infty}\sum_{j=0}^{y} P(X = i, Y = j), \text{ if } Y \text{ is discrete random variable} \\ &= \int_{-\infty}^{\infty}\int_{-\infty}^{y} f(x, y)dxdy, \text{ if } Y \text{ is continuous random variable.} \end{aligned}$$

If $X_1, X_2, \ldots, X_n$ are n-dimensional random variables, then

$$\begin{aligned} F(x_1, x_2, \ldots, x_n) &= \sum_{0}^{x_1}\sum_{0}^{x_2}\cdots\sum_{0}^{x_n} P(X_1 = x_1, X_2 = x_2, \ldots, X_n = x_n), \\ &\qquad \text{if variables are discrete} \\ &= \int_{-\infty}^{x_1}\int_{-\infty}^{x_2}\cdots\int_{-\infty}^{x_n} f(x_1, x_2, \ldots, x_n)dx_1dx_2\cdots dx_n, \\ &\qquad \text{if variables are continuous.} \end{aligned}$$

►**EXAMPLE 2.23** *The joint probability density function of continuous random variables X and Y is*

$$f(x,y) = \begin{cases} \frac{1}{8}(6-x-y); & 0 < x < 2,\ 2 < y < 4 \\ 0, & \textit{otherwise.} \end{cases}$$

Find joint distribution function of X and Y. Hence or otherwise, find (i) $P(X < 1, Y < 3)$, (ii) $P(X < 1)$, (iii) $P(Y \leq 3)$.

Solution. According to definition of joint distribution function, we can write

$$\begin{aligned} F(x,y) &= P(X \leq x, Y \leq y) \\ &= \int_{-\infty}^{x}\int_{-\infty}^{y} f(x,y)dxdy = \frac{1}{8}\int_0^x\int_2^y (6-u-v)dudv \\ &= \frac{6}{8}\int_0^x\int_2^y dudv - \frac{1}{8}\int_0^x\int_2^y ududv - \frac{1}{8}\int_0^x\int_2^y vdudv \\ &= \frac{6}{8}u\Big|_0^x v\Big|_2^y - \frac{1}{8}\frac{u^2}{2}\Big|_0^x v\Big|_2^y - \frac{1}{8}u\Big|_0^x \frac{v^2}{2}\Big|_2^y \\ &= \frac{6x(y-2)}{8} - \frac{1}{16}x^2(y-2) - \frac{1}{16}x(y^2-4). \end{aligned}$$

Now,

$$P(X < 1, Y < 3) = \frac{6\cdot 1(3-2)}{8} - \frac{1}{16}\cdot 1^2(3-2) - \frac{1}{16}\cdot 1(9-4) = \frac{3}{8}.$$

Alternatively, we have

$$\begin{aligned} F(X=1, Y=3) &= \int_0^1\int_2^3 \frac{1}{8}(6-x-y)dxdy \\ &= \frac{6}{8}\int_0^1\int_2^3 dxdy - \frac{1}{8}\int_0^1\int_2^3 xdxdy - \frac{1}{8}\int_0^1\int_2^3 ydxdy \\ &= \frac{6}{8}(1-0)(3-2) - \frac{1}{16}(1-0)(3-2) - \frac{1}{16}(1-0)(9-4) = \frac{3}{8}. \end{aligned}$$

Again,

$$\begin{aligned} F_1(x) &= P(X \leq x) \\ &= \int_0^x\int_2^4 f(x,y)dxdy = \frac{1}{8}\int_0^x\int_2^4 (6-u-v)dudv \\ &= \frac{6}{8}\int_0^x\int_2^4 dudv - \frac{1}{8}\int_0^x\int_2^4 ududv - \frac{1}{8}\int_0^x\int_2^4 vdudv \\ &= \frac{6}{8}(x-0)(4-2) - \frac{1}{16}(x^2-0)(4-2) - \frac{1}{16}(x-0)(16-4) \\ &= \frac{12x}{8} - \frac{x^2}{8} - \frac{6x}{8} = \frac{1}{8}(6x - x^2),\ 0 < x < 2. \end{aligned}$$

$$\therefore\ F(X=1) = P(X \leq 1) = \frac{6\cdot 1 - 1^2}{8} = \frac{5}{8}.$$

Again,

$$
\begin{aligned}
P(Y \leq 3) &= \int_0^2 \int_2^3 f(x,y)dxdy \\
&= \frac{1}{8}\int_0^2 \int_2^3 (6-x-y)dxdy \\
&= \frac{6}{8}\int_0^2 \int_2^3 dxdy - \frac{1}{8}\int_0^2 \int_2^3 xdxdy - \frac{1}{8}\int_0^2 \int_2^3 ydxdy \\
&= \frac{6}{8}(2-0)(3-2) - \frac{1}{16}(4-0)(3-2) - \frac{1}{16}(2-0)(9-4) \\
&= \frac{5}{8}.
\end{aligned}
$$

►**EXAMPLE 2.24** *The joint probability function of two-dimensional discrete random variables X and Y is*

X \ Y	0	1	2	3	$g(y)$
0	1/12	2/12	—	2/12	5/12
1	—	1/12	1/12	1/12	3/12
2	2/12	—	—	2/12	4/12
$h(y)$	3/12	3/12	1/12	5/12	1.00

Find joint distribution function of X and Y. Also, find distribution function of X.

Solution. The joint distribution function of X and Y is shown below:

X \ Y	$y < 0$	$0 \leq y < 1$	$1 \leq y < 2$	$2 \leq y < 3$	$y \leq 3$
$x < 0$	0	0	0	0	0
$0 \leq x < 1$	0	1/12	3/12	3/12	5/12
$1 \leq x < 2$	0	1/12	4/12	5/12	8/12
$x \leq 2$	0	3/12	6/12	7/12	1.00

The probability function of X and its distribution function are shown below:

X	0	1	2	$x < 0$
$g(x)$	5/12	3/12	4/12	0
$F(x) = P(X \leq x)$	5/12	8/12	1.00	

►**EXAMPLE 2.25** *The p.d.f. of continuous random variable X is*

$$f(x) = \frac{1}{b-a};\ a < x < b.$$

Find distribution function of X.

Solution. We can write

$$F(x) = P(X \leq x) = \frac{1}{b-a}\int_a^x f(x)dx = \frac{x-a}{b-a}$$

$$F(x) = P(X \leq x) = \begin{cases} 0, & x < a, x < b \\ \frac{-a}{b-a}, & a < x < 0 \\ \frac{b}{b-a}, & 0 < x < b \end{cases}$$

These are true for any real values of a and b.

►**EXAMPLE 2.26** ***The joint probability density function of two-dimensional continuous random variables X and Y is***

$$f(x,y) = \begin{cases} 4x(1-x^2); \ 0 < x < 1; \ 0 < y < 1 \\ 0, \quad \textit{otherwise.} \end{cases}$$

Find joint distribution function of X and Y. Also, find distribution function of X.

Solution. The joint distribution function of X and Y is shown below:

$$\begin{aligned} F(x,y) &= P(X \leq x, Y \leq y) \\ &= \int_0^x \int_0^y f(x,y)dxdy \\ &= \int_0^x \int_0^y 4u(1-u^2)dudv \\ &= 4\left[\int_0^x \int_0^y ududv - \int_0^x \int_0^y u^3 dudv\right] \\ &= 4\left[\left(\frac{x^2}{2} - 0\right)(y-0) - \left(\frac{x^4}{4} - 0\right)(y-0)\right] \\ &= 2x^2y - x^4y. \end{aligned}$$

Also, we have

$$\begin{aligned} F_1(x) &= P(X \leq x) = \int_0^x \int_0^1 f(x,y)dxdy \\ &= 4\left[\int_0^x \int_0^1 ududv - \int_0^x \int_0^1 u^3 dudv\right] \\ &= 4\left[\left(\frac{x^2}{2} - 0\right)(1-0) - \left(\frac{x^4}{4} - 0\right)(1-0)\right] \\ &= 2x^2 - x^4 = x^2(2-x^2), \ 0 < x < 1. \end{aligned}$$

►**EXAMPLE 2.27** ***The continuous random variable X has the p.d.f.***

$$f(x) = \frac{1}{x^2}, \ 1 < x < \infty.$$

Find distribution function of X.

Solution. The distribution function is given by

$$F(x) = \begin{cases} 0, & x < 1 \\ \displaystyle\int_1^x \frac{1}{u^2}du = \left(1 - \frac{1}{x}\right), & 1 \le x \end{cases}$$

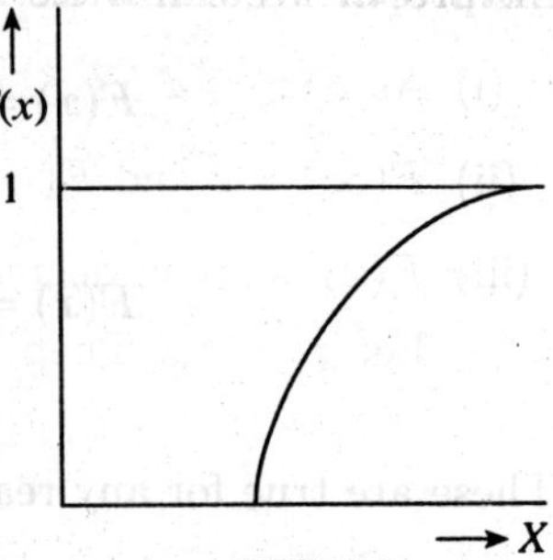

FIG. 2.2

This $F(x)$ is continuous for all real values of X. The value of $F(x)$ is continuous at every point in the right-hand side. This can be shown graphically as in FIG. 2.2.

►**EXAMPLE 2.28** *The joint probability density function of three-dimensional random variables X, Y and Z is given by*

$$f(x, y, z) = \begin{cases} e^{-(x+y+z)}; & 0 < x,\ y,\ z < \infty \\ 0, & otherwise. \end{cases}$$

Find joint distribution function of X, Y and Z. Also, find distribution function of X.

Solution. We have

$$\begin{aligned} F(x, y, z) &= \int_0^x \int_0^y \int_0^z f(x, y, z)dxdydz \\ &= \int_0^x \int_0^y \int_0^z e^{-(u+v+z)}dudvdz \\ &= \begin{cases} (1 - e^{-x})(1 - e^{-y})(1 - e^{-z}); & 0 < x, y, z < \infty \\ 0, & \text{otherwise.} \end{cases} \end{aligned}$$

Again,

$$\begin{aligned} F(x) = P(X \le x) &= \int_0^x \int_0^\infty \int_0^\infty f(x, y, z)dxdydz \\ &= \int_0^x e^{-u}du \int_0^\infty e^{-y}dy \int_0^\infty e^{-z}dz \\ &= (1 - e^{-x}) \left[-e^{-y}\Big|_0^\infty\right] \left[-e^{-z}\Big|_0^\infty\right] \\ &= \begin{cases} 1 - e^{-x}, & 0 < x < \infty \\ 0, & \text{otherwise.} \end{cases} \end{aligned}$$

2.5.1 Properties of Distribution Function

Let X be a continuous random variable having p.d.f. $f(x)$ and $F(x)$ be the distribution function of X, such that

$$\lim_{x\to\infty} F(x) = F(\infty) \quad \text{and} \quad \lim_{x\to\infty} F(x) = F(-\infty)$$

The properties of this distribution function are discussed below:

(i) As $F(x) = P(X \leq x)$ and $0 \leq P(X \leq x) \leq 1, 0 \leq F(x) \leq 1$.

(ii) $F(\infty) = 1$ and $F(-\infty) = 0$.

(iii) $F(x)$ is a non-decreasing function.

Let $x_1 < x_2$. Then $\{X : x \leq x_2\} = \{X : x \leq x_1\} \cup \{X : x_1 < X < x_2\}$.

$$\begin{aligned} \therefore\ P(X \leq x_2) &= P(X \leq x_1) + P(x_1 < X < x_2) \\ F(x_2) &= F(x_1) + P(x_1 < X < x_2) \\ \therefore\ P(x_1 < X < x_2) &= F(x_2) - F(x_1) \geq 0 \end{aligned}$$

Hence $F(x)$ is a non-decreasing function.

(iv) If $a < b$, then $P(a < X \leq b) = F(b) - F(a)$.

$$P(a < X \leq b) = \int_a^b f(x)dx = \int_a^0 f(x)dx + \int_0^b f(x)dx$$

We have $F(b) = P(X \leq b)$ and $F(a) = P(X \leq a)$.

$$P(a < X \leq b) = \int_0^b f(x)dx - \int_0^a f(x)dx = F(b) - F(a).$$

(v) If X is a continuous random variable and $F(x)$ is the distribution function, then the p.d.f. of X is given by

$$f(x) = \frac{\partial F(x)}{\partial x} \text{ (if differentiation is possible).}$$

As

$$F(x) = \int_{-\infty}^{x} f(x)dx, \ f(x) = \frac{\partial F(x)}{\partial x}.$$

But, if X is a discrete random variable, the probability function of $X = x_i$ is given by

$$P(X = x_i) = F(x_i) - F(x_{i-1}).$$

For two-dimensional random variables X and Y, $f(x, y)$ is given by

$$f(x, y) = \frac{\partial F(x, y)}{\partial x \partial y}, \text{ if differentiation is possible.}$$

(vi) If X and Y are independent random variables, then $F(x, y) = F(x)F(y)$, where

$$F(x) = P(X \leq x) \text{ and } F(y) = P(Y \leq y).$$

Similarly, for n-dimensional independent random variables $X_1, X_2, \ldots, X_n$, we have

$$F(x_1, x_2, \ldots, x_n) = F(X_1 \leq x_1)F(X_2 \leq x_2) \cdots F(X_n \leq x_n).$$

►**EXAMPLE 2.29** *The distribution function of continuous random variable X is given by*

$$F(x) = \begin{cases} 0, \ x < 0 \\ \frac{x}{2}, \ 0 \le x < 1 \\ \frac{1}{2}, \ 1 \le x < 2 \\ \frac{x}{4}, \ 2 \le x < 4 \\ 1, \ x \ge 4. \end{cases}$$

Find probability density function of X.

Solution. Here for $x = 0, x = 1, x = 2$ and $x = 4$, the value of distribution function remains somewhere; then size of $F(x)$ is for $x < 0$ and $0 \le x < 1; 0 \le x < 1$ and $1 \le x < 2; 1 \le x < 2$ and $2 \le x < 4$ and $x \ge 4$. Thus, the distribution is continuous. Hence the probability density function is

$$f(x) = \frac{\partial F(x)}{\partial x}.$$

$$\therefore \ f(x) = \begin{cases} 0, \ x < 0 \\ \frac{1}{2}, \ 0 \le x < 1 \\ 0, \ 1 \le x < 2 \\ \frac{1}{4}, \ 2 \le x < 4 \\ 0, \ x \ge 4. \end{cases}$$

►**EXAMPLE 2.30** *The distribution function of a discrete random variable X is given by*

$$F(x) = \begin{cases} 0, \ x < 1 \\ \frac{3}{8}, \ 1 \le x < 2 \\ \frac{4}{8}, \ 2 \le x < 3 \\ 1, \ 3 \le x. \end{cases}$$

Find the probability function of X.

Solution. We know $P(X = x_i) = F(x_i) - F(x_{i-1})$. Therefore,

$$F(x = 0) = 0, F(x \le 1) = \frac{3}{8}, F(x \le 2) = \frac{4}{8}, F(x \le 3) = 1$$

We have

$$\begin{aligned} P(X = 1) &= F(x \le 1) - F(x \le 0) = \frac{3}{8} - 0 = \frac{3}{8} \\ P(X = 2) &= F(x \le 2) - F(x \le 1) = \frac{4}{8} - \frac{3}{8} = \frac{1}{8} \\ P(X = 3) &= F(x \le 3) - F(x \le 2) = 1 - \frac{4}{8} = \frac{4}{8}. \end{aligned}$$

Hence the probability distribution of X is given by

X	1	2	3
$P(X=x)$	3/8	1/8	4/8

►EXAMPLE 2.31 *The distribution function of continuous random variable X is given by*

$$F(x) = \begin{cases} 0, & x < 0 \\ 3x^2 - 2x^3, & 0 \le x < 1 \\ 1, & 1 \le x \end{cases}$$

Find p.d.f. of X.

Solution. The p.d.f. of X is

$$f(x) = \frac{\partial F(x)}{\partial x} = \frac{\partial(3x^2 - 2x^3)}{\partial x} = 6x(1-x), 0 < x < 1.$$

Again,

$$f(x) = 0, x < 0 \text{ and } x > 1.$$

$$\therefore \ f(x) = \begin{cases} 6x(1-x), & 0 < x < 1 \\ 0, & \text{otherwise.} \end{cases}$$

►EXAMPLE 2.32 *The distribution function of a continuous random variable X is given by*

$$F(x) = \begin{cases} 1 - \frac{1}{x^2}, & 1 \le x \\ 0, & x < 1 \end{cases}$$

Find p.d.f. of X.

Solution. The p.d.f. of X is

$$f(x) = \frac{\partial F(x)}{\partial x} = \frac{\partial(1 - 1/x^2)}{\partial x} = \frac{2}{x^3}, \ 1 < x < \infty.$$

►EXAMPLE 2.33 *The joint distribution function of two-dimensional continuous random variables X and Y is*

$$F(x, y) = \begin{cases} 0; \ x < 0, \ y < 0 \\ xy; \ 0 \le x < 1, \ 0 \le y < 1 \\ y; \ 1 \le x, \ 0 \le y < 1 \\ x; \ 1 \le y, \ 0 \le x < 1 \\ 1; \ x \ge 1, \ y \ge 1 \end{cases}$$

Find joint p.d.f. of X and Y.

Solution. We have

$$f(x,y) = \frac{\partial F(x,y)}{\partial x \partial y}.$$

$$\therefore f(x,y) = \begin{cases} 1;\ 0 < x < 1,\ 0 < y < 1 \\ 0,\ \text{otherwise.} \end{cases}$$

►**EXAMPLE 2.34** *The joint distribution function of two-dimensional discrete random variables X and Y is shown below:*

Y \ X	$x < 1$	$1 \le x < 2$	$2 \le x < 3$	$3 \le x < 4$	$4 \le x < 5$	$5 \le x < 6$	$6 \le x$
$y < 0$	0	0	0	0	0	0	0
$0 \le y < 1$	0	1/12	2/12	3/12	4/12	5/12	6/12
$1 \le y$	0	2/12	4/12	6/12	8/12	10/12	1

Are X and Y independent?

Solution. It is observed that

$$F(X < 1, Y < 0) = 0 \text{ and } F(X \le 1, Y < 0) = \frac{1}{12}.$$

$$\therefore P(X = 1, Y = 0) = F(X \le 1, Y \le 0) - F(X \le 0, Y < 0)$$

$$= \frac{1}{12} - 0 = \frac{1}{12}$$

Again,

$$F(X \le 1, Y \le 1) = \frac{2}{12}.$$

$$\therefore P(X = 1, Y = 1) = F(X \le 1, Y \le 1) - F(X \le 1, Y < 0)$$

$$= \frac{2}{12} - \frac{1}{12} = \frac{1}{12}$$

Similarly, $P(X = x, Y = y)$, where $x = 1, 2, 3, 4, 5, 6$ and $y = 0, 1$ are shown below:

Y \ X	1	2	3	4	5	6	$h(y)$	$F(y)$
0	1/12	1/12	1/12	1/12	1/12	1/12	1/2	1/2
1	1/12	1/12	1/12	1/12	1/12	1/12	1/2	1
$g(x)$	2/12	2/12	2/12	2/12	2/12	2/12	1.00	
$F(x)$	2/12	3/12	6/12	8/12	10/12	1		

$$F(x) = 0,\ x < 1$$

$$F(y) = 0,\ y < 0$$

$$F(y = 0) = \begin{cases} \frac{1}{2},\ 0 \le y < 1 \\ 1,\ 1 \le y \end{cases}$$

Now,

$$F(x=1)F(y=0) = \frac{2}{12}\times\frac{1}{2} = \frac{1}{12} = F(x=1, y=0)$$
$$F(x=1)F(y=1) = \frac{2}{12}\times 1 = \frac{2}{12} = F(x=1, y=1)$$

In a similar way, we can show that

$$F(x=6)F(y=1) = 1\times 1 = 1 = F(x=6, y=1).$$

Thus, X and Y are independent random variables.

2.6 Functions of Random Variables and Their Distribution

Let X be a random variable, continuous or discrete and $Y_1 = X + a$, where a is any constant. Here Y_1 is a function of random variable X and Y_1 is also a random variable. Similarly, if X_1 and X_2 are two-dimensional random variables, then $Y_2 = CX_1$, $Y_3 = X_1 + X_2$, $Y_4 = CX_1 + DX_2$, $Y_5 = X_1X_2$, $Y_6 = X_1/X_2$, $Y_7 = \min(X_1, X_2)$, $Y_8 = \max(X_1X_2)$ are functions of random variables. Similar functions of random variables can also be defined for k-dimensional random variables $X_1, X_2, \ldots, X_k$.

In the present section, we shall discuss different functions of random variables and their distributions.

2.6.1 Methods of Deriving the Distribution of Functions of Random Variables

Let $X_1, X_2, \ldots, X_k$ be k-dimensional random variables and $f(x_1, x_2, \ldots, x_k)$ be the probability density function of these random variables. Let $Y_i = h_i(X_1, X_2, \ldots, X_k)$; $i = 1, 2, \ldots, n$ be the i-th function of the k-dimensional random variables $X_1, X_2, \ldots, X_k$. We need to find the probability density function of Y_i. The different methods of deriving the probability density functions of $Y_i (i = 1, 2, \ldots, n)$ are

(i) Method of distribution function,

(ii) Transformation of variable technique,

(iii) Analytical method,

(iv) Method of moment generating function,

(v) Method of characteristic function.

Some of the methods are discussed with examples in the following sub-sections.

2.6.2 Application of Method of Distribution Function

Let X_1, X_2, ..., X_k be k-dimensional random variables defined on the same sample space S. Let the joint probability density function of X_1, X_2, ..., X_k be

$f(x_1, x_2, \ldots, x_k)$. The joint distribution function of $X_1, X_2, \ldots, X_k$ is $F(x_1, x_2, \ldots, x_k)$, where

$$F(x_1, x_2, \ldots, x_k) = \int_{-\infty}^{x_1} \int_{-\infty}^{x_2} \cdots \int_{-\infty}^{x_k} f(u_1, u_2, \ldots, u_k) du_1 du_2 \cdots du_k.$$

Let us consider that $Y_j = h_j(X_1, X_2, \ldots, X_k)$ is a function of $X_1, X_2, \ldots, X_k$ $(j = 1, 2, \ldots, n)$ and the joint probability function of Y_j exists. Then the distribution function of $Y_1, Y_2, \ldots, Y_n$ is

$$\begin{aligned} F(y_1, y_2, \ldots, y_n) &= P(Y_1 \leq y_1, Y_2 \leq y_2, \ldots, Y_n \leq y_n) \\ &= P[h_1(x_1, x_2, \ldots, x_k) \leq y_1, \ldots, h_n(x_1, x_2, \ldots, x_n) \leq y_n] \\ &= \iint \cdots \int f(x_1, x_2, \ldots, x_k) dx_1, dx_2, \ldots, dx_k \\ &\qquad h_1(x_1, x_2, \ldots, x_k) \leq y_1, \ldots, h_n(x_1, x_2, \ldots, x_k) \leq y_n. \end{aligned}$$

Then the joint probability density function of $Y_1, Y_2, \ldots, Y_n$ is

$$f(y_1, y_2, \ldots, y_n) = F'(y_1, y_2, \ldots, y_n) = \frac{\partial^n F(y_1, y_2, \ldots, y_n)}{\partial y_1 \partial y_2 \cdots \partial y_n}.$$

This is possible if differentiation exists and if the variables are continuous random variables.

For discrete random variables $X_1, X_2, \ldots, X_k$ the joint distribution function is given by

$$\begin{aligned} F(y_1, y_2, \ldots, y_n) &= \sum \cdots \sum P(X_1 \leq x_1, X_2 \leq x_2, \ldots, X_k \leq x_k) \\ &\qquad h_1(x_1, x_2, \ldots, x_k) \leq y_1, \ldots, h_n(x_1, x_2, \ldots, x_k) \leq y_n \\ \text{or, } F(Y) = P(Y \leq y) &= \sum \cdots \sum P(X_1 \leq x_1, X_2 \leq x_2, \ldots, X_k = x_k) \\ &\qquad h_1(x_1, x_2, \ldots, x_k) \cdots h_n(x_1, x_2, \ldots, x_k) \leq y \end{aligned}$$

Then $P(Y = y_i) = F(y_i) - F(y_{i-1})$.

►**EXAMPLE 2.35** *The distribution function of the discrete random variable X is given by*

X	0	1	2	3	4
$F(x)$	1/16	5/16	11/16	15/16	1

Find the probability function of x.

Solution. Since X is discrete random variable

$$P(X_i = x_i) = F(x_i) - F(x_{i-1})$$

Hence the probability function of X is given by

X	0	1	2	3	4
$P(X = x_i)$	1/16	4/16	6/16	4/16	1/16

►EXAMPLE 2.36 *The continuous random variable X has the distribution function*

$$F(x) = \begin{cases} 0, & x < 2 \\ \frac{x^2}{12}, & 2 \le x < 4 \\ 1, & x \ge 4. \end{cases}$$

Find the probability density function of X.

Solution. The probability density function of X is given by

$$f(x) = F'(x) = \frac{\partial F(x)}{\partial x} = \begin{cases} \frac{x}{6}, & 2 \le x \le 4 \\ 0, & \text{otherwise.} \end{cases}$$

►EXAMPLE 2.37 *The continuous random variable X has the p.d.f.*

$$f(x) = \frac{1}{\sqrt{2\pi}} e^{-x^2/2}; \ -\infty < x < \infty.$$

Find the p.d.f. of $Y = X^2$.

Solution. We have

$$\begin{aligned} F(y) &= P(Y \le y) = 0, \text{ if } y \le 0, \text{ and} \\ &= P(X^2 \le y) = P\left[-\sqrt{y} \le X \le \sqrt{y}\right], \text{ if } y \ge 0 \\ &= 2\int_0^{\sqrt{y}} \frac{1}{\sqrt{2\pi}} e^{-\frac{1}{2}u^2} du = \frac{1}{\Gamma\frac{1}{2}2^{\frac{1}{2}}} \int_0^z \frac{1}{\sqrt{z}} e^{-\frac{1}{2}z} dz \\ &= \frac{1}{\sqrt{2z}\Gamma\frac{1}{2}} \int_0^z \frac{1}{\sqrt{z}} e^{-\frac{1}{2}z} dz, \ z > 0. \end{aligned}$$

$$\therefore \ f(y) = \begin{cases} \frac{1}{\Gamma\frac{1}{2} \cdot 2^{1/2}} e^{-\frac{1}{2}y} y^{\frac{1}{2}-1}, & y > 0 \\ 0, & \text{otherwise.} \end{cases}$$

►EXAMPLE 2.38 *The probability density function of a continuous random variable X is given by*

$$f(x) = \begin{cases} \frac{1}{\lambda} e^{-x/\lambda}; & x > 0, \lambda > 0 \\ 0, & \textit{otherwise.} \end{cases}$$

Find the p.d.f. of $Y = \sqrt{X}$.

Solution. $F(y) = P(Y \le y) = P\left(X^{1/2} \le y\right) = P(X \le y^2), y > 0$ and $P(Y \le y) = 0, y \le 0$.

$$\therefore \ f(y) = \frac{\partial F(y)}{\partial y}$$

$$F(y) = \int_0^{y^2} f(x)dx = \int_0^{y^2} \frac{1}{\lambda} e^{-x/\lambda} dx = 1 - e^{-y^2/\lambda}$$

$$\therefore \ f(y) = \frac{2y}{\lambda} e^{-y^2/\lambda}, \ y > 0.$$

2.6.3 Distribution of Sum and Difference of Two Random Variables

Let X and Y be two-dimensional random variables defined on a sample space S. Assume that the joint p.d.f. of X and Y is $f(x, y)$. Let us define new variables U and V, where $U = X + Y$ and $V = X - Y$. The problem is to find out the distribution of U and Y.

The distribution function of U is written as

$$\begin{aligned} F_u(u) &= P[U \le u] = P[X+Y \le u] = \iint\limits_{x+y\le u} f(x,y)dxdy \\ &= \int_{-\infty}^{\infty}\left[\int_{-\infty}^{u-x} f(x,y)dxdy\right] = \int_{-\infty}^{\infty}\left[\int_{-\infty}^{u} f(x,z-x)dx\right] \end{aligned}$$

where $y = z - x \Rightarrow dy = dz$.

Now,

$$\begin{aligned} f_1(u) &= \frac{\partial F_u(u)}{\partial u} = \frac{d}{du}\left[\int_{-\infty}^{u}\left\{\int_{-\infty}^{\infty} f(x,z-x)dx\right\}dz\right] \\ &= \int_{-\infty}^{\infty} f(x,u-x)dx \end{aligned}$$

However, if X and Y are independent, then

$$f_1(u) = \int_{-\infty}^{\infty} f_y(u-x)f_x(x)dx = \int_{-\infty}^{\infty} f_x(u-y)f_y(y)dy.$$

As

$$\begin{aligned} P[U \le u] &= P[X+Y \le u] = \int_{-\infty}^{\infty} P[X+Y \le u|X=x]f_x(x)dx \\ &= \int_{-\infty}^{\infty} P[X+Y \le u]f_x(x)dx = \int_{-\infty}^{\infty} F_y(u-x)f_x(x)dx \end{aligned}$$

Therefore,

$$\begin{aligned} f_1(u) &= \frac{\partial F_u(u)}{\partial u} = \frac{d}{du}\left[\int_{-\infty}^{\infty} F_y(u-x)f_x(x)dx\right] \\ &= \int_{-\infty}^{\infty} \frac{dF_y(u-x)}{du} f_x(x)dx = \int_{-\infty}^{\infty} f_y(u-x)f_x(x)dx. \end{aligned}$$

This formula of deriving $f_1(u)$ is known as *convolution rule.* In a similar way the density function of $V = X - Y$ can be derived.

►**EXAMPLE 2.39** *The joint probability density function of two-dimensional random variables X and Y is given by*

$$f(x,y) = \begin{cases} 2, & 0 \le x \le y \le 1 \\ 0, & \textit{otherwise.} \end{cases}$$

Find the distribution of $U = X + Y$.

Solution. Since the limit of the variables is $0 \le x \le y \le 1$, u cannot be less than or equal to 0. The maximum value of u may be 2. Hence, if $u < 0$, $P(U \le u) = 0$ and if $U < 2$, $P(U \le u) = 1$. However, for $0 \le u \le 2$,

$$\begin{aligned} F(u) &= P(U \le x) = P(X + Y \le u) \\ &= \int_{0 \le x \le y = 1} \cdots \int_{x+y \le u} f(x, y)dxdy. \end{aligned}$$

Here also, we observed two events. These are $0 \le u \le 1$ and $1 \le u \le 2$. Let $0 \le u \le 1$. Then,

$$\begin{aligned} F(u) &= P(U \le u) \\ &= \int_{x=0}^{u/2} \left[\int_{y=x}^{u-x} 2dy\right] \\ &= 2\int_0^{u/2} (u - 2x)dx = \frac{u^2}{2}. \\ \therefore\ f(u) &= \frac{dF(u)}{du} = u,\ 0 \le u \le 1. \end{aligned}$$

Again, let $1 \le u \le 2$, then

$$\begin{aligned} F(u) &= P(U \le u) = 1 - P(U > u) \\ &= 1 - \int_{y=\frac{u}{2}}^{1} \left[\int_{x=u-y}^{y} 2dx\right] dy \\ &= 1 - 2\int_{u/2}^{1} (2y - u)dy = 1 - \frac{(u-2)^2}{2}. \\ \therefore\ f(u) &= \frac{dF(u)}{du} = -(u - 2) = 2 - u,\ 1 \le u \le 2. \end{aligned}$$

Therefore, $U = X + Y$ has the p.d.f.

$$f(u) = \begin{cases} u, & 0 \le u \le 1 \\ 2 - u, & 1 \le u \le 2 \\ 0, & \text{otherwise.} \end{cases}$$

►**EXAMPLE 2.40** *The joint probability density function of two-dimensional random variables is given by*

$$f(x, y) = 1,\ 0 < x < 1,\ 0 < y < 1.$$

Find the p.d.f. of $U = X - Y$.

Solution. We have

$$g(x) = \int_0^1 f(x, y)dy = \int_0^1 dy = 1,\ 0 < x < 1$$

$$\text{and } h(y) = \int_0^1 f(x, y)dx = \int_0^1 dx = 1,\ 0 < y < 1.$$

It is observed that X and Y are independent random variables. Thus,

$$F(u) = P(U \leq u) = \int_{x-y \leq u} \int f(x, y) dx dy.$$

But $P(U \leq u) = 0$, if $U \leq -1$ and $P(U \leq u) = 1$, if $u \geq 1$.
Now, if $|u| < 1$, there may be two events which can be shown in FIG. 2.3.

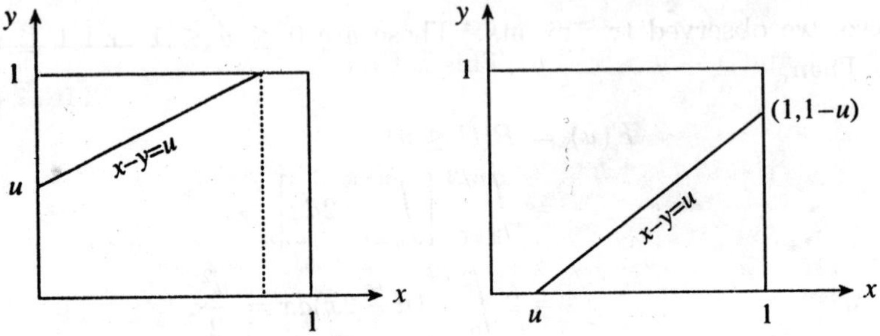

FIG. 2.3

Now,

$$\{(x, y) : -1 < x - y \leq u \leq 0\}, \; \{(x, y) : 0 < x - y \leq y \leq 1\}.$$

Therefore, if $-1 < u < 0$, then

$$\begin{aligned} P[U \leq u] &= \int_{x=0}^{u+1} \left[\int_{y=x-u}^{1} 1 dy \right] dx \\ &= \int_{0}^{u+1} (1 - x + u) dx = \frac{(u+1)^2}{2}. \end{aligned}$$

If $0 < u < 1$, then

$$\begin{aligned} P[U \leq u] &= 1 - P[U \leq u] \\ &= 1 - \int_{x=u}^{1} \left[\int_{y=0}^{x-u} 1 dy \right] dx \\ &= 1 - \int_{u}^{1} (x - u) dx = 1 - \frac{(u-1)^2}{2}. \end{aligned}$$

$$\therefore \; f(u) = \begin{cases} 1 + u, & -1 < u \leq 0 \\ 1 - u, & 0 < u < 1 \\ 0, & \text{otherwise} \end{cases} = \begin{cases} 1 - |u|, & |u| < 1 \\ 0, & \text{otherwise.} \end{cases}$$

►**EXAMPLE 2.41** *The joint p.d.f. of continuous random variables X and Y is given by*

$$f(x, y) = \begin{cases} e^{-x}, & 0 \leq y \leq x \leq \infty \\ 0, & \text{otherwise.} \end{cases}$$

Find the p.d.f. of $U = X + Y$ and $V = X - Y$.

Solution. The joint distribution function of U and V is

$$F(u,v) = P[U \le u,\ V = v]$$
$$= \iint_R f(x,y)\,dx\,dy$$

where $R = \{(x,y) : x+y \le u,\ x-y \le v,\ 0 \le y \le x \le \infty\}$ and $x+y > x-y$. This can be shown in FIG. 2.4.

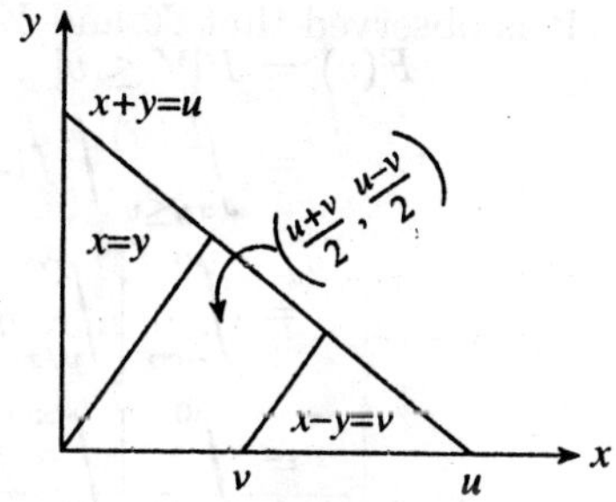

FIG. 2.4

$$\{(x,y) : x+y \le u,\ x-y \le v,\ 0 \le y \le x \le \infty\}$$

Now,

$$P[U \le u, V \le v] = \int_{y=0}^{(u-v)/2} \left[\int_{x=y}^{y+u} e^{-x}dx\right] dy + \int_{y=\frac{u-v}{2}}^{u/2} \left[\int_{x=y}^{u-v} e^{-x}dx\right] dy$$
$$= \int_0^{(u-v)/2} e^{-y}(1-e^{-y})dy + \int_{(u-v)/2}^{u/2} \left(e^{-y} - e^{-u+y}\right) dy$$
$$= (1-e^{-y})\left(1-e^{-(u-v)/2}\right) + \left(e^{-(u-v)/2} - e^{-u/2}\right)$$
$$-e^{-u}\left(e^{u/2} - e^{(u-v)/2}\right)$$
$$= 1 - e^{-v} - 2e^{-u/2} + 2e^{-(u+v)/2}.$$

Therefore,

$$f(u,v) = \begin{cases} \frac{1}{2}e^{-(u+v)/2}, & 0 \le u \le v \le \infty \\ 0, & \text{otherwise.} \end{cases}$$

Now,

$$f_1(u) = \begin{cases} e^{-u}, & u \ge 0 \\ 0, & \text{otherwise} \end{cases}$$

$$\text{and } f_2(v) = \begin{cases} e^{-v/2}\left(1-e^{-v/2}\right), & v \ge 0 \\ 0, & \text{otherwise.} \end{cases}$$

2.6.4 Distribution of Product and Quotient of Two Random Variables

Let X and Y be two-dimensional continuous random variables having joint p.d.f. $f(x,y)$. Let $V = XY$ and $U = X/Y$. We need the density function of U and V.

We have

$$\begin{aligned}
F(v) &= P[V \le v] \\
&= \int_{xy\le v}\int f(x,y)dxdy \\
&= \int_{-\infty}^{0}\left[\int_{v/x}^{\infty} f(x,y)dx\right] + \int_{0}^{\infty}\left[\int_{-\infty}^{v/x} f(x,y)dy\right] dx \\
&= \int_{-\infty}^{0}\left[\int_{v}^{\infty} f\left(x,\frac{z}{x}\right)\frac{dz}{x}\right] dx + \int_{0}^{\infty}\left[\int_{-\infty}^{v} f\left(x,\frac{z}{x}\right)\frac{dz}{x}\right] dx \\
&= \int_{-\infty}^{0}\left[\int_{-\infty}^{0} \frac{1}{-x} f\left(x,\frac{z}{x}\right) dx\right] dz + \int_{-\infty}^{v}\left[\int_{0}^{v} \frac{1}{x} f\left(x,\frac{z}{x}\right) dx\right] dz \\
&= \int_{-\infty}^{v}\left[\int_{-\infty}^{\infty} \frac{1}{|x|} f\left(x,\frac{z}{x}\right) dx\right] dz. \\
\therefore\ f(v) &= \frac{dF(v)}{dv} = \int_{-\infty}^{\infty}\frac{1}{|x|} f\left(x,\frac{v}{x}\right) dx.
\end{aligned}$$

Similarly, we can write

$$f(u) = \int_{-\infty}^{\infty} |y| f(uy,y)dy.$$

►**EXAMPLE 2.42** *The joint probability density function of continuous random variables X and Y is given by*

$$f(x,y) = 1,\ 0 < x < 1,\ 0 < y < 1.$$

Find the probability density function of $V = XY$.

Solution. As $V = XY, v \le 0, v \ge 1$ and $0 < v < 1$. However, for $v \le 0, F(v) = 0$ and for $v \ge 1$, $F(v) = 1$. In such a situation we need $F(v)$ for $0 < v < 1$.

Now,

$$\begin{aligned}
F(v) &= P[V \le v] = P[XY \le v] \\
&= \int_{xy\le v}\int f(x,y)dxdy = 1 - \int_{xy>v}\int f(x,y)dxdy \\
&= 1 - \int_{x=v}^{1}\left[\int_{y=v/x}^{1} 1dy\right] dx = 1 - \int_{v}^{1}\left(1-\frac{v}{x}\right)dx = v(1-\ln v). \\
\therefore\ f(v) &= \frac{dF(v)}{dv} = -\ln v \\
\therefore\ f(v) &= \begin{cases} -\ln v, & 0 \le v \le 1 \\ 0, & \text{otherwise.} \end{cases}
\end{aligned}$$

2.6.5 Transformation of Variable Technique

Let X be a random variable and $Y = g(x)$ be a function of X. We need the probability function/probability density function of Y.

Consider that X is a discrete random variable and it has n values $x_1, x_2, \ldots, x_n$. The probability function of $X = x_i (i = 1, 2, \ldots, n)$ is $P(X = x_i)$. Then probability function of $Y = y_i (i = 1, 2, \ldots, n)$ is $P(Y = y_i)$, where $P(X = x_i) = P(Y = y_i)$. Here the probable values of $Y = g(x)$ are obtained from the function of X. If Y is the sum of $x_1, x_2, \ldots, x_k$ (say), then

$$P(Y = y_i) = \sum_{i=1}^{k} P(X = x_i).$$

Again, let us consider that X is a continuous random variable having p.d.f. $f(x)$. Let $Y = k(x)$ be a function of X. Then the p.d.f. of Y is given by

$$h(Y = y) = f(x)\left|\frac{dx}{dy}\right|.$$

Here $f(x)$ is expressed in terms of y. The function $h(Y = y)$ is available only if $Y = k(x)$ is a monotone increasing or decreasing function and $k(x)$ can be differentiated. Also, $k(x)$ is continuous for all values of X.

The phenomenon can be proved as follows:

Proof. Let $Y = k(x)$ be perfectly monotone increasing function; then $\frac{dy}{dx} > 0$ and $F_y(y) = P(Y \leq y) = P[k(x) \leq y] = P[X \leq k^{-1}(y)]$. As $Y = k(x)$ is a monotone increasing function, $k^{-1}(y)$ exists. Hence

$$F_y(y) = F_x\left[k^{-1}(y)\right] = F_x(x).$$

Now, differentiating $F_y(y)$, we have

$$h(Y = y) = \frac{d}{dy}[F_x(x)] = \frac{d}{dx}F_x(x)\frac{dx}{dy} = f(x)\frac{dx}{dy}. \tag{a}$$

Again, consider that X is perfectly monotone decreasing function and hence $\frac{dy}{dx} < 0$. Then

$$\begin{aligned} F_y(y) &= P[Y \leq y] = P[k(x) \leq y] = P\left[X \geq k^{-1}(y)\right] \\ &= 1 - P\left[X \leq k^{-1}(y)\right] = 1 - F_x\left[k^{-1}(y)\right] \\ &= 1 - F_x(x). \end{aligned}$$

At this stage also $x = k^{-1}(y)$ exists and

$$\begin{aligned} h(y) &= \frac{d}{dy}[1 - F_x(x)] = \frac{d}{dx}[1 - F_x(x)]\frac{dx}{dy} \\ &= -f(x)\frac{dx}{dy} = f(x)\left(-\frac{dx}{dy}\right). \end{aligned} \tag{b}$$

Here the negative sign in $h(y)$ is correct as $\frac{dy}{dx} < 0$. Therefore, from (a) and (b), we have

$$h(y) = f(x)\left|\frac{dx}{dy}\right|.$$

Let us consider that some values of Y are increasing for some values of X and for other values of X the values of Y are decreasing. In such a situation

$$h(y) = \sum f(x) \left|\frac{dx}{dy}\right|,$$

where Σ depends on the existence of $X = k^{-1}(Y)$.

►**EXAMPLE 2.43** *The probability distribution of the random variable X is given as follows:*

X	0	1	2	3	4
$P(X = x)$	1/16	4/16	6/16	4/16	1/16

Find the probability distribution of $Y = k(x) = 3x + 1$.

Solution. Here the values of Y for different values of X are shown below:

X	0	1	2	3	4
Y	1	4	7	10	13

Therefore, the probability distribution of Y will be as follows:

Y	1	4	7	10	13
$P(Y = y)$	1/16	4/16	6/16	4/16	1/16

►**EXAMPLE 2.44** *The probability distribution of a discrete random variable X is given by*

X	0	1	2	3	4
$P(X = x)$	0.2	0.3	0.2	0.1	0.2

Find the probability distribution of $Y = k(x) = (x - 1)^2$.

Solution. The different values of Y for given values of X are shown below:

X	0	1	2	3	4
Y	1	0	1	4	9

Hence the probability distribution of Y is as follows:

Y	0	1	4	9
$P(Y = y)$	0.3	0.4	0.1	0.2

►**EXAMPLE 2.45** *The probability distribution of a discrete random variable X is given by*

$$P(X = x) = {}^nC_x p^x q^{n-x}, \quad x = 0, 1, 2, \ldots, n.$$

Find the probability distribution of $Y = k(x) = ax + b$, where a and b are two positive constants.

Solution. We have

$$P(Y = y) = P(ax + b = y) = P\left[X = \frac{y - b}{a}\right].$$

$$\therefore\ P(Y = y) = \binom{n}{\frac{y-b}{a}} p^{\frac{y-b}{a}} q^{n-\frac{y-b}{a}}$$

where $y = b, (a + b), (2a + b), \ldots, (na + b)$.

►**EXAMPLE 2.46** *The p.d.f. of continuous random variable X is given by*

$$f(x) = \begin{cases} 1, & 0 < x < 1 \\ 0, & otherwise. \end{cases}$$

Find the p.d.f. of $Y = k(x) = X^2$.

Solution. Since $0 < x < 1$, the transformation is one-to-one [as there is no negative value of X]. Therefore,

$$x = \sqrt{y} \text{ and } \frac{dy}{dx} - \frac{1}{2}y^{-\frac{1}{2}}.$$

Hence

$$h(y) = f(x)\left|\frac{dx}{dy}\right| = \begin{cases} \frac{1}{2\sqrt{y}}, & 0 < y < 1 \\ 0, & \text{otherwise.} \end{cases}$$

►**EXAMPLE 2.47** *The p.d.f. of the continuous random variable X is given by*

$$f(x) = \begin{cases} e^x, & x < 0 \\ 0, & otherwise. \end{cases}$$

Find the p.d.f. of $Y = g(x) = X^2$.

Solution. The limit of X is $-\infty < x < 0$. Hence $y = x^2$ is a perfectly decreasing function of X, where $x = -\sqrt{y}$.

Thus

$$\frac{dx}{dy} = -\frac{1}{2\sqrt{y}}.$$

We have

$$f(y) = f(x)\left|\frac{dx}{dy}\right| = e^{-\sqrt{y}}\frac{1}{2\sqrt{y}}, \quad y > 0.$$

►**EXAMPLE 2.48** *The p.d.f. of continuous random variable X is given by*

$$f(x) = \frac{1}{\sqrt{2\pi}}e^{-\frac{1}{2}x^2}, \quad -\infty < x < \infty.$$

Find the p.d.f. of $Y = k(x) = X^2$.

Solution. Here $Y = X^2$ is perfectly decreasing function of X for $-\infty < x < 0$ and $Y = X^2$ is perfectly increasing function of X for $0 < x < \infty$. Thus, we have two solutions from $Y = X^2$ and the solutions are

$$x = -\sqrt{y} \text{ for } -\infty < x < 0$$
$$\text{and } x = \sqrt{y} \text{ for } 0 < x < \infty.$$

Therefore,

$$h(y) = f(x_1)\left|\frac{dx}{dy}\right| + f(x_2)\left|\frac{dx}{dy}\right|, \; y > 0$$
$$= f\left(\sqrt{-y}\right)\frac{1}{2\sqrt{y}} + f\left(\sqrt{y}\right)\frac{1}{2\sqrt{y}} = \begin{cases} \frac{1}{2\sqrt{\pi y}}e^{-\frac{y}{2}}, & y > 0 \\ 0, & \text{otherwise.} \end{cases}$$

►**EXAMPLE 2.49** *The p.d.f. of continuous random variable X is given by*

$$f(x) = \begin{cases} \frac{x}{8}, & 0 < x < 4 \\ 0, & \textit{otherwise.} \end{cases}$$

Find the p.d.f. of $Y = k(x) = X^2$.

Solution. As X has no negative value, the transformation of X to Y is one-to-one increasing.

Thus,

$$x = \sqrt{y} \Rightarrow \frac{dx}{dy} = \frac{1}{2\sqrt{y}}.$$

Hence,

$$h(y) = f(x)\left|\frac{dx}{dy}\right| = \frac{\sqrt{y}}{8}\frac{1}{2\sqrt{y}} = \begin{cases} \frac{1}{16}, & 0 < y < 16 \\ 0, & \text{otherwise.} \end{cases}$$

2.6.6 Transformation of Variable Technique in Two-dimensional Random Variables

Let us consider the two-dimensional continuous random variables X and Y having joint p.d.f. $f(x, y)$. Consider two functions of X and Y as $U = g(X, Y)$ and $V = h(X, Y)$, where both the functions are continuous and are differentiable. Then the Jacobian of the transformation is

$$|J| = \left|\frac{\partial(x, y)}{\partial(u, v)}\right| = \begin{vmatrix} \frac{\partial x}{\partial u} & \frac{\partial y}{\partial u} \\ \frac{\partial x}{\partial v} & \frac{\partial y}{\partial v} \end{vmatrix}.$$

The transformation of U and V is done in such a way that $X = g^{-1}(U, V)$ and $Y = h^{-1}(U, V)$ exist. In such a case J may be positive or may be negative. Then

$$k(u, v) = f(x, y)|J|,$$

where $f(x, y)$ is expressed in terms of u and v.

Let us consider a particular function $U = X + Y$. Then

$$h(u) = \int_{-\infty}^{\infty} f(v, u - v)dv, \text{ where } V = X.$$

Proof. Let $u = x + y,\ v = x$. Then $x = v$ and $y = u - v$ and Jacobian of the transformation is

$$|J| = \left|\frac{\partial(x, y)}{\partial(u, v)}\right| = \begin{vmatrix} \frac{\partial x}{\partial u} & \frac{\partial y}{\partial u} \\ \frac{\partial x}{\partial v} & \frac{\partial y}{\partial v} \end{vmatrix} = \begin{vmatrix} 0 & 1 \\ 1 & -1 \end{vmatrix} = |-1| = 1.$$

Then the joint p.d.f. of U and V is

$$g(u, v) = f(x, y)|J| = f(v, u - v)|J|.$$

Now, the p.d.f. of U is

$$h(u) = \int_{-\infty}^{\infty} g(u, v)dv = \int_{-\infty}^{\infty} f(v, u - v)|J|dv.$$

However, if X and Y are independent,

$$\begin{aligned} f(x, y) &= f_x(x)f_y(y) \\ g(u, v) &= f_x(x)f_y(y)|J| = f_x(u)f_y(u - v)|J|. \end{aligned}$$

Hence,

$$h(u) = \int_{-\infty}^{\infty} f_x(u)f_y(u - v)|J|dv.$$

Here $h(\cdot)$ is called *convolution of* $f_x(\cdot)$ *and* $f_y(\cdot)$.

In a similar way it can be shown that

$$g(u, v) = f(x, y)|J| = f(u + v, v)|J|, \text{ where } U = Y - X, V = X.$$

►EXAMPLE 2.50 *Two-dimensional continuous random variables X and Y have the joint probability density function*

$$f(x, y) = \begin{cases} 4xye^{-(x^2+y^2)}; & x \geq 0,\ y \geq 0 \\ 0, & \textit{otherwise.} \end{cases}$$

Find the p.d.f. of $U = \sqrt{X^2 + Y^2}$.

Solution. Let $u = \sqrt{x^2 + y^2}$ and $v = x$. $\therefore\ v \geq 0,\ u \geq 0$.

Also, $u \geq 0$ and $0 \leq v \leq u$.

The Jacobian of the transformation is

$$\frac{1}{J} = \frac{\partial(u, v)}{\partial(x, y)} = \begin{vmatrix} \frac{\partial u}{\partial x} & \frac{\partial v}{\partial x} \\ \frac{\partial u}{\partial y} & \frac{\partial v}{\partial y} \end{vmatrix} = \frac{y}{\sqrt{x^2 + y^2}}$$

Thus the joint p.d.f. of U and V is

$$\begin{aligned} g(u,v) &= f(x,y)|J| = 4xye^{-(x^2+y^2)}|J| \\ &= 4x\sqrt{x^2+y^2}e^{-(x^2+y^2)} \\ &= \begin{cases} 4uve^{-u^2}; & u \geq 0,\ 0 \leq v \leq u \\ 0, & \text{otherwise.} \end{cases} \end{aligned}$$

Hence the density function of $U = \sqrt{X^2+Y^2}$ is

$$h(u) = \int_0^u g(u,v)dv = 4ue^{-u^2}\int_0^v vdv = 2u^3e^{-u^2},\ u \geq 0.$$

$$\therefore\ h(u) = \begin{cases} 2u^3e^{-u^2}, & u \geq 0 \\ 0, & \text{otherwise.} \end{cases}$$

►**EXAMPLE 2.51** *Let x_1 and x_2 be two observations of a random sample of size 2 from a population having density function*

$$f(x) = \begin{cases} e^{-x}, & x > 0 \\ 0, & \textit{otherwise.} \end{cases}$$

Find the p.d.f. of $U = X_1 + X_2$ and $V = \frac{X_1}{X_1+X_2}$.

Solution. Let $u = x_1 + x_2$ and $v = \frac{x_1}{x_1+x_2}$. Since $x > 0, u > 0$ and $0 < v < 1$. Here $x_1 = uv, x_2 = u(1-v)$.

The Jacobian of the transformation is

$$J = \frac{\partial(x_1,x_2)}{\partial(u,v)} = \begin{vmatrix} \frac{\partial x_1}{\partial u} & \frac{\partial x_2}{\partial u} \\ \frac{\partial x_1}{\partial v} & \frac{\partial x_2}{\partial v} \end{vmatrix} = \begin{vmatrix} v & 1-v \\ u & -u \end{vmatrix} = -u.$$

We have

$$f(x_1,x_2) = e^{-(x_1+x_2)}.$$

$$\therefore\ g(u,v) = f(x_1,x_2)|J| = ue^{-u}.$$

Now,

$$g(u) = \int_0^1 g(u,v)dv = ue^{-u}\int_0^1 dv = ue^{-u},\ u > 0$$

$$\text{and } h(v) = \int_0^\infty ue^{-u}du = 1,\ 0 < v < 1.$$

2.6.7 Method of Moment Generating Function

Let $X_1, X_2, \ldots, X_k$ be k-dimensional continuous random variables having joint p.d.f. $f(x_1, x_2, \ldots, x_k)$. Let us consider n functions of $X_1, X_2, \ldots, X_k$, where the functions are $Y_1 = g_1(x_1, x_2, \ldots, x_k)$, $Y_2 = g_2(x_1, x_2, \ldots, x_k), \ldots, Y_n = g_n(x_1, x_2, \ldots, x_k)$. The problem is to derive the joint p.d.f. of $Y_1, Y_2, \ldots, Y_n$ and hence or otherwise the marginal density function of $Y_j (j = 1, 2, \ldots, n)$. Using the definition of moment generating function, we can write

$$\begin{aligned} &M_{y_1, y_2, \ldots, y_n}(t_1, t_2, \ldots, t_n) \\ &= E\left[\exp(t_1 Y_1 + t_2 Y_2 + \cdots + t_n Y_n\right] \\ &= \int_{-\infty}^{\infty}\int_{-\infty}^{\infty}\cdots\int_{-\infty}^{\infty}[\exp t_1 g_1(x_1, x_2, \ldots, x_k) \\ &\quad + \exp t_2 g_2(x_1, x_2, \ldots, x_k) + \cdots \\ &\quad + \exp t_n g_n(x_1, x_2, \ldots, x_k)] dx_1 dx_2 \cdots dx_k. \end{aligned}$$

If there exists integration shown above, we can find joint moment generating function of $Y_1, Y_2, \ldots, Y_n$ and if this moment generating function is the moment generating function of a known distribution, then we can write the p.d.f. of $Y_1, Y_2, \ldots, Y_n$. Also, by integrating we can find the marginal density function of any of the Y_j. However, the application of this technique is limited, specially if $n > 1$. Since for $n > 1$, the joint p.d.f. of the random variables and hence their moment generating function are seldom known to us.

►**EXAMPLE 2.52** *The p.d.f. of the random variable X is given by*

$$f(x) = \frac{1}{\sigma\sqrt{2\pi}} e^{-\frac{1}{2}\left(\frac{X-\mu}{\sigma}\right)^2}, \; -\infty < x < \infty.$$

Find the p.d.f. of $Y = \frac{1}{2}\left(\frac{X-\mu}{\sigma}\right)^2$.

Solution. The moment generating function of Y is

$$\begin{aligned} M_y(t) &= E[e^{tY}] = E\left[e^{-\frac{1}{2}\left(\frac{X-\mu}{\sigma}\right)^2 t}\right] \\ &= \frac{1}{\sigma\sqrt{2\pi}}\int_{-\infty}^{\infty} e^{-\frac{1}{2}\left(\frac{X-\mu}{\sigma}\right)^2(1-t)} dx. \end{aligned}$$

Let

$$\begin{aligned} z &= \frac{x-\mu}{\sigma} \Rightarrow dx = \sigma dz, \; -\infty < z < \infty. \\ M_y(t) &= \frac{1}{\sqrt{2\pi}}\int_{-\infty}^{\infty} e^{-\frac{1}{2}z^2(1-t)} dz \\ &= \frac{(1-t)^{-1/2}}{(1-t)^{-1/2}\sqrt{2\pi}}\int_{-\infty}^{\infty} e^{-\frac{1}{2}z^2(1-t)} dz \\ &= (1-t)^{-1/2}. \end{aligned}$$

It is observed that $M_y(t)$ is the value of moment generating function of Gamma distribution with parameter 1/2. Therefore,

$$g(y) = \begin{cases} \frac{1}{\Gamma\frac{1}{2}} e^{-y} y^{\frac{1}{2}-1}, & y > 0 \\ 0, & \text{otherwise.} \end{cases}$$

2.6.8 Method of Characteristic Function

Let $X_1, X_2, \ldots, X_n$ be the n-dimensional random variables having joint p.d.f. $f(x_1, x_2, \ldots, x_n)$. Let us assume that $X_1, X_2, \ldots, X_n$ be independent and the characteristic function of $X_j (j = 1, 2, \ldots, n)$ be $\psi_j(t_j)$. Then the characteristic function of $Z = x_1 + x_2 + \cdots + x_n$ is

$$\psi_z(t) = \prod_{j=1}^{n} \psi_j(t).$$

Now, if $\psi_z(t)$ is the characteristic function of any known distribution, then we can write the p.d.f. of Z. Also, using inversion theorem or Levy's theorem we can find the p.d.f. of any function of $X_1, X_2, \ldots, X_n$. However, this technique has limited application since we are not very much familiar with characteristic function of joint distribution when $n > 1$ and when $X_1, X_2, \ldots, X_n$ are not independent.

►**EXAMPLE 2.53** *Let $X_1, X_2, \ldots, X_k$ be k-independent random variables such that*

$$P(X_j = x_j) = \binom{n_j}{x_j} p^{x_j} q^{n_j - x}, \quad x_j = 0, 1, 2, \ldots, n_j$$
$$j = 1, 2, \ldots, k.$$

Find the p.d.f. of $Z = X_1 + X_2 + \cdots + X_k$.

Solution. According to $P(X_j = x_j)$ each X_j is binomial variable. The characteristic function of binomial distribution is

$$\psi_j(t) = \left(q + pe^{it}\right)^{n_j}.$$

Now, since $X_1, X_2, \ldots, X_k$ are independent, we can write

$$\begin{aligned}\psi_z(t) &= \prod_{j=1}^{k} \psi_j(t) = \prod_{j=1}^{k} \left(q + pe^{it}\right)^{n_j} \\ &= \left(q + pe^{it}\right)^{n_1 + n_2 + \cdots + n_k}\end{aligned}$$

It is observed that $\psi_z(t)$ is the characteristic function of binomial variable having parameters p and $n = n_1 + n_2 + \cdots + n_k$. Thus,

$$P(Z = z) = \binom{n}{z} p^z q^{n-z}, \quad z = 0, 1, 2, \ldots, n.$$

►**EXAMPLE 2.54** *Let $X_1, X_2, \ldots, X_k$ be k-independent Gamma variate such that*

$$f(x_j) = \frac{1}{\Gamma r} \lambda^r e^{-\lambda x_j} x_j^{r-1}, \quad x_j \geq 0.$$

Find the p.d.f. of $\bar{X} = \frac{1}{k}(X_1 + X_2 + \cdots + X_k)$.

Solution. The characteristic function of Gamma distribution is

$$\psi_{x_j}(t) = \left(1 - \frac{it}{\lambda}\right)^{-r}.$$

Hence,

$$\psi_{\bar{x}}(t) = \prod^{k} \left(1 - \frac{it}{\lambda}\right)^{-r} = \left(1 - \frac{it}{\lambda}\right)^{-rk}$$

which is a characteristic function of Gamma variate. Therefore,

$$h(\bar{x}) = \begin{cases} \dfrac{(h\lambda)^{kr} e^{-k\lambda\bar{x}} (\bar{x})^{kr-1}}{\overline{|kr}}, & 0 < \bar{x} < \infty \\ 0, & \text{otherwise.} \end{cases}$$

2.7 More Examples on Distribution of Functions of Random Variables

►**EXAMPLE 2.55** *Let X_1 and X_2 be $N(0,1)$. Find the distributions of* (i) $Y = X_1 + X_2$, (ii) $Y_1 = X_1 - X_2$, *and* (iii) $Y_2 = X_1 X_2$ *and* $Y_3 = X_1/X_2$.

Solution.

(i) The density function of Y is given by

$$\begin{aligned} h(y) &= \int_{-\infty}^{\infty} f(y - x_2) f(x_2) dx_2 \\ &= \frac{1}{2\pi} \int_{-\infty}^{\infty} e^{-\frac{1}{2}(y-x_2)^2 - \frac{1}{2}x_2^2} dx_2 = \frac{1}{2\pi} \int_{-\infty}^{\infty} e^{-\frac{1}{2}(2x_2^2 - 2yx_2 + y^2)} dx_2 \\ &= \frac{1}{\sqrt{2}\sqrt{2\pi}} \int_{-\infty}^{\infty} \frac{1}{\sqrt{2\pi}} e^{-\frac{1}{2}(w-y)^2} dw e^{-y^2/4}, \text{ where } w = x_2\sqrt{2}. \\ \therefore\ h(y) &= \frac{1}{\sqrt{2}} \frac{1}{\sqrt{2\pi}} e^{-y^2/4} \end{aligned}$$

Thus Y is $N(0,2)$ variate.

(ii) Let $Y_1 = X_1 - X_2$; then

$$\begin{aligned} H(y) &= P(x_1 - x_2 \le y_1) = P(x_1 \le y_1 + x_2) \\ &= \int_{-\infty}^{\infty} \left[\int_{-\infty}^{y_1 + x_2} f(x_1, x_2) dx_1 \right] dx_2 \end{aligned}$$

If X_1 and X_2 are independent,

$$\begin{aligned} H(y) &= \int_{-\infty}^{\infty} F(y_1 + x_2) f(x_2) dx_2 \\ \text{and } h(y) &= \int_{-\infty}^{\infty} f(y_1 + x_2) f(x_2) dx_2 = \frac{1}{2\pi} \int_{-\infty}^{\infty} e^{-\frac{1}{2}(y_1+x_2)^2 - \frac{1}{2}x_2^2} dx_2 \\ &= \frac{1}{2\pi} \int_{-\infty}^{\infty} e^{-\frac{1}{2}(y_1^2 + 2y_1x_2 + x_2^2) - \frac{1}{2}x_2^2} dx_2 = \frac{1}{2\pi} \int_{-\infty}^{\infty} e^{-\frac{1}{2}(y_1^2 + 2y_1x_2 + 2x_2^2)} dx_2 \\ &= \frac{1}{\sqrt{2}} \frac{1}{\sqrt{2\pi}} \int \frac{1}{\sqrt{2\pi}} e^{-\frac{1}{2}(w+y_1)^2} e^{-\frac{1}{2}\frac{y_1^2}{4}} dx_2 = \frac{1}{\sqrt{2}} \frac{1}{\sqrt{2\pi}} e^{-\frac{1}{2}\frac{y_1^2}{4}} \end{aligned}$$

Thus Y_1 is also $N(0,2)$.

(iii) Similarly, one can find the distributions of $Y_2 = X_1 X_2$ and $Y_3 = X_1/X_2$. Here, for independent X_1 and X_2, we have

$$H(y) = \int_{-\infty}^{\infty} F(y_1 + x_2) f(x_2) dx_2$$

$$\text{and } h(y) = \int_{-\infty}^{\infty} f(y_1 + x_2) f(x_2) dx_2, \text{ when } Y_1 = X_1 - X_2.$$

If $Y = X_1 X_2$, then

$$H(y_2) = P(x_1 x_2 \leq y_2) = P\left(x_1 \leq \frac{y_2}{x_2}\right).$$

For a fixed x_2

$$H(y_2) = \int_{-\infty}^{y_2/x_2} f(x_1, x_2) dx_1$$

Unconditional

$$H(y) = \int_{-\infty}^{\infty} \int_{-\infty}^{y_2/x_2} f(x_1, x_2) dx_1 dx_2.$$

But, if X_1 and X_2 are independent,

$$\begin{aligned} H(y) &= \int_{-\infty}^{\infty} F(y_2/x_2) f(x_2) dx_2 \\ \text{and } h(y) &= \int_{-\infty}^{\infty} \frac{1}{|x_2|} f(y_2/x_2) f(x_2) dx_2 \\ &= \int_{0}^{\infty} \frac{1}{x_2} f(y_2/x_2) f(x_2) dx_2 \\ &\quad - \int_{-\infty}^{0} \frac{1}{x_2} f(y_2/x_2) f(x_2) dx_2. \end{aligned}$$

If $Y_3 = X_1/X_2$, then

$$\begin{aligned} H(y_3) &= P\left[\frac{x_1}{x_2} \leq y_3\right] \\ &= P[x_1 \leq y_3 x_2]. \end{aligned}$$

Now, for fixed x_2

$$H(y_3) = \int_{-\infty}^{y_3 x_2} f(x_1, x_2) dx_1$$

Unconditional

$$H(y) = \int_{-\infty}^{\infty} \left[\int_{-\infty}^{y_3 x_2} f(x_1, x_2) dx_1\right] dx_2.$$

If X_1 and X_2 are independent, then

$$H(y) = \int_{-\infty}^{\infty} F(y_3x_2)f(x_2)dx_2.$$

$$\begin{aligned}\therefore\ h(y_3) &= \int_{-\infty}^{\infty} |x_2| f(y_3x_2)f(x_2)dx_2 \\ &= \int_{0}^{\infty} x_2(y_3x_2)f(x_2)dx_2 - \int_{-\infty}^{0} x_2 f(y_3x_2)f(x_2)dx_2 \\ &= \frac{1}{2\pi}\frac{1}{(1+y_3)^2}\int_{-\infty}^{\infty} |y| \exp\left[e^{-\frac{1}{2}}\frac{(1+y_3^2)y^2}{(1+y_3)^2}dy\right], \text{ where } y = x_1 + x_2.\end{aligned}$$

Let

$$\begin{aligned}u &= \frac{1}{2}\frac{(1+y_3^2)y^2}{(1+y_3)^2} \\ \Rightarrow du &= \frac{(1+y_3^2)}{(1+y_3)^2}ydy. \\ \therefore\ h(y_3) &= \frac{1}{2\pi}\frac{1}{(1+y_3)^2}\frac{(1+y_3)^2}{1+y_3^2}\cdot 2\int_0^{\infty} e^{-u}du \\ &= \frac{1}{\pi}\frac{1}{1+y_3^2}.\end{aligned}$$

►**EXAMPLE 2.56** *Let $X_1, X_2, \ldots, X_n$ be n variables which form a random sample of n observations from a population having p.d.f.*

$$f(x) = 1,\ 0 < x < 1.$$

Let $Y_1 = \min(X_1, X_2, \ldots, X_n)$ and $Y_n = \max(X_1, X_2, \ldots, X_n)$. Find the distribution of $W = Y_n - Y_1$.

Solution. The joint density function of Y_1 and Y_n is given by

$$g(y_1, y_n) = n(n-1)\left[F(y_n) - F(y_1)\right]^{n-2} f(y_1)f(y_n).$$

Let $Z = Y_1$, then the transformation from Y_1 and Y_n to W and Z will be one-to-one linear transformation. Thus

$$y_1 = g_1^{-1}(w, z) = z \text{ and } y_n = g_2^{-1}(w, z) = w + z.$$

Here $|J| = 1$. Now, the joint distribution of W and Z is

$$h(w, z) = n(n-1)[F(w+z) - F(z)]^{n-2} f(z)f(w+z).$$

Here $F(x) = x$ for $0 < x < 1$. Therefore,

$$h(w, z) = 0 \text{ unless } 0 < w < 1 \text{ and } 0 < z < 1 - w.$$

Under these conditions

$$h(w, z) = n(n-1)w^{n-2}.$$

Now, for $0 < w < 1$, the p.d.f. of w is

$$h_1(w) = \int_0^{1-w} n(n-1)w^{n-2}dz = n(n-1)w^{n-2}(1-w)$$

and $h_1(w) = 0$, otherwise.

►**EXAMPLE 2.57** *Let X be a random variable having p.d.f.*

$$f(x) = \begin{cases} 3x^2, \ 0 < x < 1 \\ 0, \quad \text{otherwise.} \end{cases}$$

Find the distribution of $Y = 1 - X^2$.

Solution. As $0 < x < 1$, Y also varies in the interval $(0, 1)$. For $0 < y < 1$, the inverse function is

$$x = g_1(y) = (1-y)^{1/2} \text{ and } \frac{dx}{dy} = -\frac{1}{2(1-y)^{1/2}}.$$

$$\therefore \ h(y) = 3(1-y)\frac{1}{2(1-y)^{1/2}} = \frac{3}{2}(1-y)^{1/2}, \ 0 < y < 1.$$

►**EXAMPLE 2.58** *Let X_1 and X_2 be two independent random variables having the joint probability density function*

$$f(x_1, x_2) = \begin{cases} 4x_1x_2; \ 0 < x_1 < 1, \ 0 < x_2 < 1 \\ 0, \quad \text{otherwise.} \end{cases}$$

Find the distribution of $Y_1 = X_1/X_2$ and $Y_2 = X_1X_2$.

Solution. We have $X_1 = \sqrt{Y_1Y_2}$ and $X_2 = (Y_2/Y_1)^{1/2}$. Here $Y_1 > 0, Y_2 > 0, Y_1Y_2 < 1$ and $Y_2/Y_1 < 1$. As the values (x_1, x_2) of X_1 and X_2 vary over the set $\mathcal{A}$ (shown below), the values (y_1, y_2) of Y_1 and Y_2 will vary over the set $\mathcal{B}$ (shown below) containing all points such that $y_1 > 0, y_2 > 0, (y_1y_2)^{1/2} < 1$ and $(y_2/y_1)^{1/2} < 1$. The sets $\mathcal{A}$ and $\mathcal{B}$ are sketched below FIG. 2.5:

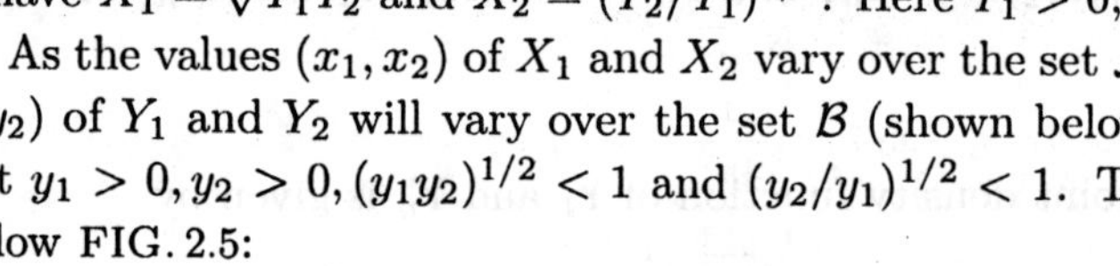

FIG. 2.5

Thus there is one-to-one relation between the points in $\mathcal{A}$ and the points in $\mathcal{B}$. For $(y_1, y_2) \in \mathcal{B}$, this transformation is given by the relation

$$x_1 = g_1^{-1}(y_1, y_2) = (y_1y_2)^{1/2}$$

$$\text{and } x_2 = g_2^{-1}(y_1y_2)) = \left(\frac{y_2}{y_1}\right)^{1/2}.$$

The Jacobian of the transformation is

$$J = \begin{vmatrix} \dfrac{\partial x_1}{\partial y_1} & \dfrac{\partial x_1}{\partial y_2} \\ \dfrac{\partial x_2}{\partial y_1} & \dfrac{\partial x_2}{\partial y_2} \end{vmatrix} = \begin{vmatrix} \dfrac{1}{2}\left(\dfrac{y_2}{y_1}\right)^{1/2} & \dfrac{1}{2}\left(\dfrac{y_1}{y_2}\right)^{1/2} \\ -\dfrac{1}{2}\left(\dfrac{y_2}{y_1^3}\right)^{1/2} & \dfrac{1}{2}\left(\dfrac{1}{y_1 y_2}\right)^{1/2} \end{vmatrix} = \frac{1}{2y_1}.$$

The joint p.d.f. of Y_1 and Y_2 is

$$h(y_1, y_2) = \begin{cases} 4(y_1y_2)^{1/2}(y_2/y_1)^{1/2}|J|; & y_1 > 0,\ y_2 > 0 \\ 0, & \text{otherwise.} \end{cases}$$

$$= \begin{cases} 2y_2/y_1; & \text{for } (y_1, y_2) \in \mathcal{B} \\ 0, & \text{otherwise.} \end{cases}$$

►**EXAMPLE 2.59** *Let X_1 and X_2 be two random variables having the joint p.d.f.*

$$f(x_1, x_2) = \begin{cases} \dfrac{1}{4a^2}\left[1 + x_1x_2(x_1^2 - x_2^2)\right]; & |x_1|, |x_2| \le a,\ a > 0 \\ 0, & \textit{otherwise.} \end{cases}$$

Find the distribution of $Y = X_1 + X_2$.

Solution. Let $x_1 = x_1, y - x_1 = x_2$.

We have $|x_1| \le a,\ x_1 - a \le y \le x_1 + a$ or, $y - a \le x_1 \le y + a$.

Thus, $y - a \le -a,\ y + a \le a,$ for $y \le 0$

$y - a > -a,\ y + a > -a,$ for $y > 0$.

Now,

$$f(x_1, x_2) = \frac{1}{4a^2}\left[1 + x_1x_2\left(x_1^2 - x_2^2\right)\right].$$

$$\therefore\ f(x_1, y - x_1) = \frac{1}{4a^2}\left[1 + x_1(y - x_1)\left\{x_1^2 - (y - x_1)^2\right\}\right]$$

$$= \frac{1}{4a^2}\left[1 + 3y^2x_1^2 - 2yx_1^3 - x_1y^3\right].$$

The density function of Y is

$$h(y) = \int \frac{1}{4a^2}\left[1 + 3y^2x_1^2 - 2yx_1^3 - x_1y^3\right]dx_1.$$

Integration will be done from $-a$ to $y + a$ when $x_1 \le 0$ and $y - a$ to a when $x_1 > 0$. Thus,

$$h(y) = \begin{cases} \displaystyle\int_{-a}^{y+a} \frac{1}{4a^2}(1 + 3x_1^2y^2 - 2x_1^3y - x_1y^3)dx_1, & -2a \le y \le 0 \\ \displaystyle\int_{y-a}^{a} \frac{1}{4a^2}(1 + 3x_1^2y^2 - 2x_1^3y - x_1y^3)dx_1, & 0 < y \le 2a \end{cases}$$

$$= \begin{cases} \frac{2a+y}{4a^2}, & -2a \le y \le 0 \\ \frac{2a-y}{4a^2}, & 0 < y \le 2a. \end{cases}$$

►EXAMPLE 2.60 *The joint p.d.f. of X_1 and X_2 is given by*

$$f(x_1, x_2) = \begin{cases} x_1 + x_2; \ 0 < x_1 < 1, \ 0 < x_2 < 1 \\ 0, \quad \textit{otherwise.} \end{cases}$$

Find the distribution of $Y_1 = X_1/X_2$.

Solution. Let $y_1 = x_1/x_2$ and $y_2 = x_1 x_2$.

The Jacobian of the transformation is

$$J = \frac{1}{2y_1} \text{ [Example 2.58].}$$

$$\therefore \ h(y_1, y_2) = \begin{cases} \left\{(y_1 y_2)^{1/2} + \left(\dfrac{y_2}{y_1}\right)^{1/2}\right\} \dfrac{1}{2y_1}; \text{ for } (y_1, y_2) \in \mathcal{B} \\ 0, \quad \text{otherwise.} \end{cases}$$

Here $0 < y_2 < y_1 < \infty$. Again, $0 < y_1 \leq 1$ and $1 < y_1 < \infty$.

Therefore,

$$h(y_1) = \int_0^{y_1} \frac{\sqrt{y_2}(1 + y_1)}{2y_1^{3/2}} dy_2, \ 0 < y_1 \leq 1$$

$$h(y_1) = \frac{1}{3}(1 + y_1), \ 0 < y_1 \leq 1.$$

Again,

$$h(y_1) = \begin{cases} \dfrac{1}{3y_1^3}(1 + y_1), \ y_1 > 0 \\ 0, \quad y_1 \leq 0. \end{cases}$$

►EXAMPLE 2.61 *Let X_1 and X_2 be two random sample observations from a population having p.d.f.*

$$f(x) = \frac{1}{2}, \ 0 < x < 2.$$

Find the distribution of $Y = X_1 + X_2$.

Solution. Let us consider the transformation $y = x_1 + x_2$ and $y_1 = x_1 - x_2$.

Thus, we have

$$x_1 = \frac{1}{2}(y_1 + y_2)$$

$$\text{and } x_2 = \frac{1}{2}(y_1 - y_2)$$

$$J = \begin{vmatrix} \dfrac{\partial x_1}{\partial y_1} & \dfrac{\partial x_1}{\partial y_2} \\ \dfrac{\partial x_2}{\partial y_1} & \dfrac{\partial x_2}{\partial y_2} \end{vmatrix} = \begin{vmatrix} \dfrac{1}{2} & \dfrac{1}{2} \\ \dfrac{1}{2} & -\dfrac{1}{2} \end{vmatrix} = -\frac{1}{2}$$

The shape of the space $\mathcal{A}$ and $\mathcal{B}$ will be as follows FIG. 2.6:

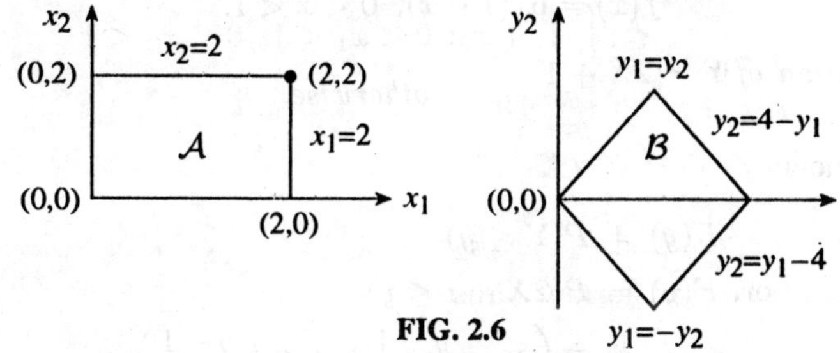

FIG. 2.6

From the above transformation, we have

$$-y_1 < y_2 < y_1, \qquad y_1 - 4 < y_2 < 4 - y_1$$
$$-y_2 < y_1 < y_2 + 2, \; y_2 < y_1 < 4 - y_2$$

The joint p.d.f. of Y_1 and Y_2 is

$$g(y_1, y_2) = f(x_1, x_2)|J| = \frac{1}{8}.$$

$$\therefore \; h(y_1) = \begin{cases} \dfrac{1}{8}\displaystyle\int_{-y_1}^{y_1} dy_2, & 0 < y_1 \le 2 \\ \dfrac{1}{8}\displaystyle\int_{y_1-4}^{4-y_1} dy_2, & 2 < y_1 \le 4 \end{cases}$$

$$= \begin{cases} \dfrac{y_1}{4}, & 0 < y_1 \le 2 \\ \dfrac{4-y_1}{4}, & 2 < y_1 \le 4 \\ 0, & \text{otherwise.} \end{cases}$$

►**EXAMPLE 2.62** *Let X be a random variable having p.d.f.*

$$f(x) = \begin{cases} 1, \; 0 < x < 1 \\ 0, \; \textit{otherwise.} \end{cases}$$

Find the distribution of $Y = 2 \ln X$.

Solution. Let $y = -2 \ln y$. It gives $x = e^{-y/2}$,

$$J = \frac{dx}{dy} = -\frac{1}{2}e^{-y/2}.$$

Here

$$\mathcal{A} = \{x, \; 0 < x < 1\}$$
$$\mathcal{B} = \{y, \; 0 < y < \infty\}.$$
$$\therefore \; h(y) = f(x)|J| = \frac{1}{2}e^{-y/2}, \; 0 < y < \infty.$$

►EXAMPLE 2.63 *Let X be a random variable having density function*

$$f(x) = 6x(1-x),\ 0 < x < 1.$$

Find the distribution of $Y = 2X + 1$.

Solution. We know

$$\begin{aligned} F(y) &= P(Y \le y) \\ \text{or, } F(y) &= P(2X+1 \le y) \\ &= P\left(X \le \frac{y-1}{2}\right) = F\left(\frac{y-1}{2}\right). \end{aligned}$$

But

$$\begin{aligned} F_x(t) &= \int_0^t 6x(1-x)dx = 3t^2 - 2t^3,\ 0 < t < 1. \\ \therefore\ F_y(y) &= F_x\left(\frac{y-1}{2}\right) = 3\left(\frac{y-1}{2}\right)^2 - 2\left(\frac{y-1}{2}\right)^3 \\ &= -\frac{y^3}{4} + \frac{3y^2}{2} - \frac{9y}{4} + 1,\ 1 < y < 3. \\ \therefore\ h(y) &= -\frac{3y^2}{4} + 3y - \frac{9}{4},\ 1 < y < 3. \end{aligned}$$

►EXAMPLE 2.64 *Let* X_1 *and* X_2 *be two random observations selected from a population having p.d.f.*

$$f(x) = \frac{1}{2}e^{-\frac{x}{2}},\ x > 0.$$

Find the distribution of $Y = \frac{1}{2}(X_1 - X_2)$.

Solution. Let us consider the transformations

$$y_1 = \frac{1}{2}(x_1 - x_2) \text{ and } y_2 = x_2 \Rightarrow x_1 = 2y_1 + y_2 \text{ and } x_2 = y_2,$$

where for $0 < x_1, -2y_1 < y_2$ and $0 < y_2$.

$$\therefore\ -\infty < y_1 < \infty;\ y_2 > -2y_1 \text{ and } y_2 > 0.$$

$$J = \frac{\partial(x_1, x_2)}{\partial(y_1, y_2)} = \begin{vmatrix} \frac{\partial x_1}{\partial y_1} & \frac{\partial x_1}{\partial y_2} \\ \frac{\partial x_2}{\partial y_1} & \frac{\partial x_2}{\partial y_2} \end{vmatrix} = \begin{vmatrix} 2 & 1 \\ 0 & 1 \end{vmatrix} = 2$$

We have

$$\begin{aligned} f(x_1, x_2) &= \frac{1}{4}e^{-\frac{1}{2}(x_1+x_2)}. \\ \therefore\ g(y_1, y_2) &= \frac{1}{4}e^{-\frac{1}{2}(2y_1+y_2+y_2)} \cdot 2 = \frac{1}{2}e^{-(y_1+y_2)}. \end{aligned}$$

Now,

$$h(y_1) = \begin{cases} \frac{1}{2}e^{-y_1}\int_{-2y_1}^{\infty} e^{-y_2}dy_2, & -\infty < y_1 < 0 \\ \frac{1}{2}e^{-y_1}\int_{0}^{\infty} e^{-y_2}dy_2, & 0 < y_1 < \infty \end{cases}$$

$$= \begin{cases} \frac{1}{2}e^{y_1}, & -\infty < y_1 < 0 \\ \frac{1}{2}e^{-y_1}, & 0 < y_1 < \infty \end{cases}$$

$$\therefore \; h(y_1) = \frac{1}{2}e^{-|y_1|}, \; -\infty < y_1 < \infty.$$

►**EXAMPLE 2.65** *Two random variables X_1 and X_2 follow the probability distribution having p.d.f. $f(x) = \lambda e^{-\lambda x}$; $x > 0$, $\lambda > 0$. Find the joint p.d.f. of $Y_1 = X_1 + X_2$ and $Y_2 = X_1/X_2$.*

Solution. Let $Y_1 = X_1 + X_2$ and $Y_2 = X_1/X_2$. The Jacobian of the transformation is

$$\frac{1}{J} = \left|\frac{\partial(y_1, y_2)}{\partial(x_1, x_2)}\right| = \begin{vmatrix} \frac{\partial y_1}{\partial x_1} & \frac{\partial y_1}{\partial x_2} \\ \frac{\partial y_2}{\partial x_1} & \frac{\partial y_2}{\partial x_2} \end{vmatrix} = \begin{vmatrix} 1 & 1 \\ \frac{1}{x_2^2} & -\frac{x_1}{x_2} \end{vmatrix}$$

$$= -\frac{x_1}{x_2^2} - \frac{1}{x_2}; y_1 > 0, y_2 > 0.$$

$$\therefore \; g(y_1, y_2) = f(x_1, x_2)\left|\frac{1}{J}\right| = \lambda^2 e^{-\lambda y_1}\frac{y_1^2/(1+y_2)^2}{y_1}$$

$$= \lambda^2 y_1 e^{-\lambda y_1}\frac{1}{(1+y_2)^2}, y_1 > 0, y_2 > 0.$$

►**EXAMPLE 2.66** *The joint density function of the random variables X_1, X_2, X_3 and X_4 is given by*

$$f(x_1, x_2, x_3, x_4) = \frac{1}{(2\pi)^2}e^{-\frac{1}{2}(x_1^2+x_2^2+x_3^2+x_4^2)};$$
$$-\infty < x_1, x_2, x_3, x_4 < \infty.$$

Find the distribution of $Y = X_1X_2 - X_3X_4$.

Solution. The characteristic function of Y is given by

$$\psi_y(t) = \frac{1}{(2\pi)^2}\int_{-\infty}^{\infty}\int_{-\infty}^{\infty}\int_{-\infty}^{\infty}\int_{-\infty}^{\infty} e^{it(x_1x_2 - x_3x_4)}e^{-\frac{1}{2}\sum^4 x_i^2}dx_1\cdots dx_4$$

$$= \frac{1}{2\pi}\int_{-\infty}^{\infty}\int_{-\infty}^{\infty} e^{-\frac{1}{2}(x_1^2 - 2itx_1x_2 + x_2^2)}dx_1dx_2$$

$$\times\frac{1}{2\pi}\int_{-\infty}^{\infty}\int_{-\infty}^{\infty} e^{-\frac{1}{2}(x_3^2 + 2itx_3x_4 + x_4^2)}dx_3dx_4$$

$$= \frac{1}{\sqrt{1+t^2}}\cdot\frac{1}{\sqrt{1+t^2}} = \frac{1}{1+t^2}.$$

Using the result $\frac{1}{\pi}\int_{-\infty}^{\infty}\int_{-\infty}^{\infty} e^{-(ax^2+2hxy+by^2)}dxdy = \frac{1}{\sqrt{ab-h^2}}$, where a, b and h are all positive.

$$\therefore\ h(y) = \frac{1}{2\pi}\int_{-\infty}^{\infty}\frac{e^{-ity}}{1+t^2}dt = \frac{1}{2\pi}\int_{-\infty}^{\infty}\frac{\cos ty}{1+t^2}dt = \frac{1}{2}e^{-|y|}.$$

►**EXAMPLE 2.67** *Let X_1, X_2 and X_3 be three random variables which follow $N(0,1)$. Let us find the joint density function of*

$$Y_1 = X_1, Y_2 = \frac{1}{2}(X_1+X_2) \text{ and } Y_3 = \frac{1}{3}(X_1+X_2+X_3).$$

Also, find the marginal density function of Y_3.

Solution. Let

$$y_1 = x_1,\ y_2 = \frac{1}{2}(x_1+x_2)$$

$$\text{and } y_3 = \frac{1}{3}(x_1+x_2+x_3)$$

$$\Rightarrow x_1 = y_1,\ x_2 = 2y_2 - y_1, x_3 = 3y_3 - 2y_2$$

$$J = \begin{vmatrix} \frac{\partial x_1}{\partial y_1} & \frac{\partial x_1}{\partial y_2} & \frac{\partial x_1}{\partial y_3} \\ \frac{\partial x_2}{\partial y_1} & \frac{\partial x_2}{\partial y_2} & \frac{\partial x_2}{\partial y_3} \\ \frac{\partial x_3}{\partial y_1} & \frac{\partial x_3}{\partial y_2} & \frac{\partial x_3}{\partial y_3} \end{vmatrix} = \begin{vmatrix} 1 & 0 & 0 \\ -1 & 2 & 0 \\ 0 & -2 & 3 \end{vmatrix} = 6$$

$$\begin{aligned}\therefore\ g(y_1,y_2,y_3) &= f(x_1,x_2,x_3)|J| \\ &= \frac{6}{(\sqrt{2\pi})^3}\exp\left[-\frac{1}{2}\left\{y_1^2+(2y_2-y_1)^2+(3y_3-2y_2)^2\right\}\right] \\ &= \frac{6}{(\sqrt{2\pi})^3}\exp\left[-\frac{1}{2}\left(2y_1^2-4y_1y_2+8y_2^2-12y_2y_3+9y_3^2\right)\right]\end{aligned}$$

The marginal density function of Y_3 is

$$\begin{aligned}h(y_3) &= \int_{-\infty}^{\infty}\int_{-\infty}^{\infty} g(y_1,y_2,y_3)dy_1dy_2 \\ &= \frac{6}{(\sqrt{2\pi})^3}\int_{-\infty}^{\infty}\exp\left[-\frac{1}{2}(6y_2^2-12y_2y_3+9y_3^2)dy_1\right] \\ &\quad\times\int_{-\infty}^{\infty}\exp\left[-\frac{1}{2}(2y_1^2-4y_1y_2+2y_2^2)dy_2\right] \\ &= \frac{6}{(\sqrt{2\pi})^2}\frac{1}{\sqrt{2}}\int_{-\infty}^{\infty}\exp\left[-\frac{1}{2}(6y_2^2-12y_2y_3+6y_3^2)\right] \\ &\qquad\exp\left[-\frac{1}{2}(3y_3^2)\right]dy_2 \\ &= \frac{\sqrt{3}}{\sqrt{2\pi}}\exp\left[-\frac{1}{2}(3y_3^2)\right],\ -\infty < y_3 < \infty.\end{aligned}$$

►EXAMPLE **2.68** *Let X_1 and X_2 be two sample observations from $N(0,1)$. Find the distribution of*

$$Y_1 = \frac{1}{2}(X_1 + X_2) \textit{ and } Y_2 = \frac{1}{2}(X_1 - X_2)^2.$$

Solution. Consider the transformation

$$y_1 = \frac{1}{2}(x_1 + x_2) \text{ and } y_2 = \frac{1}{2}(x_1 - x_2)^2.$$

The transformation gives two inverse functions

$$\begin{aligned} x_1 &= y_1 - \sqrt{y_2/2}, \quad x_1 = y_1 + \sqrt{y_2/2} \\ x_2 &= y_1 + \sqrt{y_2/2}, \ x_2 = y_1 - \sqrt{y_2/2}. \end{aligned}$$

Thus, the transformation is not one-to-one. Here

$$\mathcal{A} = \{(x_1, x_2); -\infty < x_1 < \infty, \ -\infty < x_2 < \infty\}$$

maps onto

$$\mathcal{B} = \{(y_1, y_2); -\infty < y_1 < \infty, 0 \le y_2 \le \infty\}.$$

To each point in $\mathcal{B}$, exclusive of points where $y_2 = 0$, there correspond two points in $\mathcal{A}$. The problem is due to those points of $\mathcal{A}$ that lie on the line whose equation is $x_2 = x_1$ and at each of these points $y_2 = 0$. Let us define, without altering the distribution of probability, $f(x_1, x_2) = 0$ at each point where $x_1 = x_2$. Thus, we have

$$\mathcal{A} = \{(x_1, x_2); -\infty < x_1 < \infty, -\infty < x_2 < \infty\}$$

but $x_1 \neq x_2$. This gives two distinct sets

$$\mathcal{A}_1 = \{(x_1, x_2); x_2 > x_1\} \text{ and } \mathcal{A}_2 = \{(x_1, x_2); x_2 < x_1\}.$$

Now the transformation maps onto

$$\mathcal{B} = \{(y_1, y_2); -\infty < y_1 < \infty, \ 0 < y_2 < \infty\}.$$

From two groups of inverse functions, we have $|J_1| = |J_2| = 1/\sqrt{2y_2}$. Therefore,

$$\begin{aligned} g(y_1, y_2) &= \frac{1}{2\pi} \exp\left[-\frac{\left(y_1 - \sqrt{y_2/2}\right)^2}{2} - \frac{\left(y_1 + \sqrt{y_2/2}\right)^2}{2}\right] \frac{1}{\sqrt{2y_2}} \\ &\quad + \frac{1}{2\pi} \exp\left[-\frac{\left(y_1 + \sqrt{y_2/2}\right)^2}{2} - \frac{\left(y_1 - \sqrt{y_2/2}\right)^2}{2}\right] \frac{1}{\sqrt{2y_2}} \\ &= \sqrt{\frac{2}{2\pi}} e^{-y_1^2} \frac{1}{\sqrt{2}\Gamma\frac{1}{2}} y_2^{\frac{1}{2}-1} e^{-y_2/2}; \ -\infty < y_1 < \infty, \ 0 < y_2 < \infty. \end{aligned}$$

►EXAMPLE 2.69 *Let X_1 and X_2 be two random variables having joint density function*

$$f(x_1, x_2) = 1;\ 0 < x_1 < 1,\ 0 < x_2 < 1.$$

Find the distribution of $Y = X_1X_2$.

Solution. Let us consider the transformation $y = x_1x_2$. The shape of the space due to this transformation is as follows FIG. 2.7:

Here $0 < x < y,\ 0 < x < 1$

$$F_y(y) = \int_0^y \int_0^1 dydx + \int_y^1 \int_0^{y/x} dydx$$

$$= \int_0^y dx + \int_y^1 \frac{y}{x} dx = y - y\log y,\ 0 < y < 1.$$

$$\therefore\ h(y) = -\log y,\ 0 < y < 1.$$

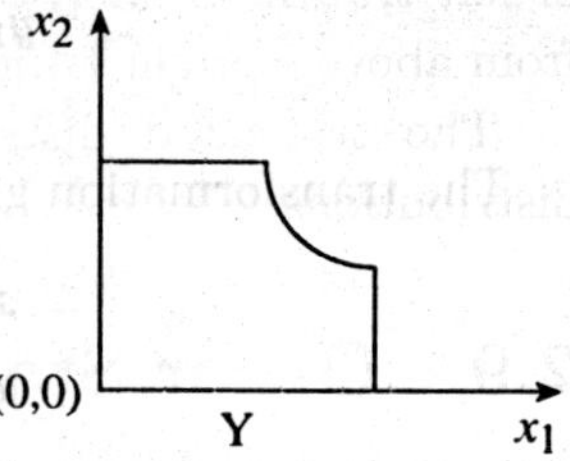

FIG. 2.7

►EXAMPLE 2.70 *Let X_1 and X_2 be two random variables having the joint p.d.f.*

$$f(x_1, x_2) = \frac{1}{4};\ -1 < x_1 < 1,\ -1 < x_2 < 1.$$

Find the p.d.f. of $Y = X_2/(1 + X_1^2)$.

Solution. Let $y_1 = x_1$ and $y_2 = x_2/(1 + x_1^2)$. Thus, we have

$$x_1 = y_1 \text{ and } x_2 = (1 + y_1^2)y_2$$

$$J = \begin{vmatrix} \dfrac{\partial x_1}{\partial y_1} & \dfrac{\partial x_1}{\partial y_2} \\ \dfrac{\partial x_2}{\partial y_1} & \dfrac{\partial x_2}{\partial y_2} \end{vmatrix} = \begin{vmatrix} 1 & 0 \\ 2y_2y_1 & 1 + y_1^2 \end{vmatrix} = 1 + y_1^2.$$

$$\therefore\ g(y_1, y_2) = \frac{1 + y_1^2}{4};\ -1 < y_1 < 1,\ -\frac{1}{1 + y_1^2} < y_2 < \frac{1}{1 + y_1^2}$$

Hence,

$$h(y) = \begin{cases} \dfrac{2}{3}, & |y_2| < \dfrac{1}{2} \\ \dfrac{1}{6}\left[\dfrac{2|y_2| + 1}{|y_2|}\sqrt{\dfrac{1}{|y_2|} - 1}\right], & \dfrac{1}{2} < |y_2| < 1. \end{cases}$$

2.8 Truncated Distribution

Let X be a random variable with different values of a given region R. Let us take $\bar{R}$ as the complement of R and assume that $x_1, x_2, \ldots, x_n$ are in $\bar{R}$, where $\bar{R}$ is a finite or infinite interval. The distribution of X excluding the values of $x_1, x_2, \ldots, x_n$ is called *truncated distribution.* If $\bar{R}$ is a finite interval with end points A and B, the distribution is doubly truncated. Here A and B are truncation points.

Let X' be a variable having a distribution formed by doubly truncating the distribution of a continuous random variable X, then the p.d.f. of X' in terms of the probability density function and cumulative distribution function is given by

$$P_{x'}(x') = P_x(x')/[F_x(B) - F_x(A)], \ A \le x' \le B.$$

Here $\bar{R}$ consists of all values greater than A, the distribution is truncated from below or left truncated. If $\bar{R}$ consists of all values less than B, the distribution is truncated from above or right truncated.

The truncated distribution will be discussed along with different probability distributions.

2.9 Order Statistics

Let $y_1 < y_2 < \cdots < y_n$ be the values of $x_1, x_2, \ldots, x_n$ in ascending order of magnitude. The joint distribution of $y_1, y_2, \ldots, y_n$ is given by

$$g(y_1, y_2, \ldots, y_n) = (n!) f(y_1) f(y_2) \cdots f(y_n),$$

where the density function of x is $f(x)$. Here $y_i (i = 1, 2, \ldots, n)$ is called the *i-th order statistic.* The distribution of order statistics has been discussed in Chapter 31.

2.10 Censoring

Let us consider the random sample of observations $x_1, x_2, \ldots, x_n$. Let

$$y_1 < y_2 < \cdots < y_n$$

be the values of $x_1, x_2, \ldots, x_n$ in ascending order of magnitude. The joint distribution of $y_1, y_2, \ldots, y_n$ is

$$f(y_1, y_2, \ldots, y_n) = (n!) f(y_1) f(y_2) \cdots f(y_n),$$

such that

$$a < y_1 < y_2 < \cdots < y_n < b \text{ and } a < x < b.$$

Sometimes a predetermined number of order statistics is omitted in studying the distribution of order statistics. The omission of order statistics is called *censoring.* If the smallest n_1 values $y_1, y_2, \ldots, y_{n_1}$ from n values ($n_1 < n$) are omitted, it is called *censoring from below*; if the largest $n_2 (n_2 < n)$ values $y_{n-n_2+1}, y_{n-n_2+2}, \ldots, y_n$ of n values are omitted, it is called *censoring from above.* Again, if smallest n_1 and largest $n_2 (n_1 < n, n_2 < n)$ are omitted, it is the case of *double censoring.* Thus, there is an analogy between truncation and censoring. In the former case, it modifies the distribution directly and in the latter case, it modifies the selection of the values of random variable. By censoring we ignore some observed values which are larger (or smaller) than other observed values, while in truncation we omit values outside predetermined limits.

2.11 Mixture Distributions

Let $\{F_i(x_1, x_2, \ldots, x_n);\ i = \ldots -1, 0, 1, 2, \ldots\}$ represent different proper cumulative distribution functions and $\sum_{i=-\infty}^{\infty} a_i = 1$ for every $a_i \geq 0$. Then

$$F(x_1, x_2, \ldots, x_n) = \sum_{i=-\infty}^{\infty} a_i F_i(x_1, x_2, \ldots, x_n)$$

is called a *mixture of distributions* $\{F_i\}$. Here $F(x_1, x_2, \ldots, x_n)$ is also a proper cumulative distribution function. The functions $F_i(x_1, x_2, \ldots, x_n)$ can be written in the form

$$F(x_1, x_2, \ldots, x_n; \theta_{1i}, \theta_{2i}, \ldots, \theta_{si}).$$

In such a case, a_i's are the probabilities in a (discrete) joint distribution of the θ's. If θ's have a joint continuous distribution with p.d.f. $P(\theta_1, \theta_2, \ldots, \theta_s)$, then the cumulative distribution function of the mixture is given by

$$\begin{aligned} F(x_1, x_2, \ldots, x_n) = & \int_{-\infty}^{\infty}\int_{-\infty}^{\infty}\cdots\int_{-\infty}^{\infty} P(\theta_1, \theta_2, \ldots, \theta_s) \\ & \times F(x_1, x_2, \ldots, x_n; \theta_1, \theta_2, \ldots, \theta_s) d\theta_1 d\theta_2 \cdots d\theta_s. \end{aligned}$$

This can be written as

$$F(x_1, x_2, \ldots, x_n) = E_{\theta_1, \theta_2, \ldots, \theta_s}[F(x_1, x_2, \ldots, x_n;\ \theta_1, \theta_2, \ldots, \theta_s],$$

where E (expectation) is defined for the variables $\theta_1, \theta_2, \ldots, \theta_s$.

To find the cumulative distribution function of the mixture the integration is taken in respect of the variable $\theta_1, \theta_2, \ldots, \theta_s$ and hence in the mixture distribution the parameters $\theta_1, \theta_2, \ldots, \theta_s$ do not appear. However, F_i's may have their parameters which are not eliminated by integration or summation.

2.12 Distribution of Cumulative Density Function (c.d.f.)

Let X be a random variable having a continuous density function $f(x, \theta)$. The c.d.f. of X is $F(x)$. Let us define a new random variable $Y = F(x)$. This Y has rectangular distribution having interval $[0, 1]$. Here the characteristic function of Y is

$$\psi_y(t) = E\left[e^{ity}\right] = E\left[e^{itF(x)}\right] = \int_{-\infty}^{\infty} e^{itF(x)} dF(x)$$

$$\text{or, } \psi_y(t) = \left.\frac{e^{itF(x)}}{it}\right|_0^1 = \frac{e^{it} - 1}{it}, \text{ where } 0 < F(x) < 1.$$

Also, by inversion theorem, it can be shown that Y has rectangular distribution. It can also easily be shown that $Y_1 = 1 - F(x)$ has rectangular distribution. This result shows that any continuous variable can be transformed to a uniform variable. As $Y = F(x)$

or $Y_1 = 1 - F(x)$ has rectangular distribution, the new random variables $Z_1 = -\ln Y$ and $Z_2 = -2\ln Y$ have gamma distribution.

We know $y = 1 - F(x)$ has rectangular distribution. Let

$$z_1 = -\ln y \Rightarrow y = e^{-z_1}, \quad \left|\frac{dy}{dz}\right| = e^{-z_1}.$$

$$\therefore \; f(z_1) = e^{-z_1} = \frac{1}{\Gamma 1} e^{-z_1} z_1^{1-1}, \; 0 < z_1 < \infty$$

which is the p.d.f. of a gamma variate.

Again, let

$$z_2 = -2\ln y \Rightarrow \; y = e^{-z_2/2}, \quad \left|\frac{dy}{dx}\right| = \frac{1}{2}e^{-z_2/2}.$$

$$\therefore \; f(z_2) = \frac{1}{2}e^{-z_2/2}.$$

►**EXAMPLE 2.71** *Given*

$$f(x) = 2(1-x), \; 0 < x < 1.$$

Find the distribution of $Y = (1 - X)^2$.

Solution. We have

$$F(x) = 2\int_0^1 (1-x)dx = 2x - x^2 = 1 - (1-x)^2.$$

Let $y = 1 - F(x) = (1-x)^2$. But Y has uniform distribution having the interval $[0, 1]$. Again, let

$$z = -\ln y = -\ln(1-x)^2 \Rightarrow (1-x)^2 = e^{-z},$$

where $0 < z < \infty$. Thus, $(1-x)^2$ is a gamma variate.

2.13 Sampling Distribution

Let $x_1, x_2, \ldots, x_n$ be the sample observations selected randomly from a population having density function $f(x, \theta)$, where θ is the parameter of the distribution. Let us define a real-valued function $T = t(x_1, x_2, \ldots, x_n)$ of the sample observations $x_1, x_2, \ldots, x_n$. The function T is called the *statistic*. As T is the function of random variable, it itself is a random variable. The distribution of T can, in principle, be derived from the joint distribution $f(x_1, x_2, \ldots, x_n;\, \theta)$. This distribution of T is called *sampling distribution* as it is originated from the distribution of sample statistic. For example, let us consider that in a population there are N population units. One can select a simple random sample of size $n (\leq N)$ without replacement. The possible number of random samples are ${}^NC_n$. From different samples the sample means are, say, $\bar{x}_1, \bar{x}_2, \ldots, \bar{x}_{{}^NC_n}$ and the sample variances are $s_1^2, s_2^2, \ldots, s_{{}^NC_n}^2$. Each of $\bar{x}_i$ and s_i^2 $(i = 1, 2, \ldots, {}^NC_n)$ is a function of sample observations (statistic). The distribution of $\bar{x}_i$ or any function of $\bar{x}_i$ is known as sampling distribution. Similarly, the distribution of s_i^2 or any function of s_i^2 is known as *sampling distribution.*

The objective of the sample selection is to estimate the parameter θ in $f(x, \theta)$ and θ is estimated by statistic. However, one needs a good estimator. The criterion of a good estimator is studied using the probability distribution of the estimator. One of the criteria of a good estimator is that the estimator will not differ from parameter θ by more than a specified value so that the mean square error of the estimator becomes minimum. It is then necessary to determine the sampling distribution of the estimator. If θ is the population mean, then for large sample one can make probabilistic inference about θ using law of large number or central limit theorem (Chapter 3). But such inference can be drawn if the sample size is smaller. In that case sampling distribution of the statistics is required.

The sampling distribution (distribution of function of random variables) can be derived when the form of the population density function is known. Important sampling distributions are presented in Chapter 26, Chapter 27 and Chapter 28.

EXERCISE

1. Let X_1 and X_2 be two random variables having joint p.d.f.

$$f(x_1, x_2) = \begin{cases} 2(x_1 + x_2), & 0 \le x_1 \le x_2 \le 1 \\ 0, & \text{otherwise.} \end{cases}$$

Find the p.d.f. of $Y = X_1 + X_2$.

2. The density function of a continuous random variable is given by

$$f(x) = 2xe^{-x^2}, \; 0 < x < \infty.$$

Find the distribution of $Y = X^2$.

3. The p.d.f. of a continuous random variable X is

$$f(x) = \begin{cases} 3x^2, & 0 < x < 1 \\ 0, & \text{otherwise.} \end{cases}$$

Find the distribution of $Y = 1 - X^2$.

4. Let X be a random variable having p.d.f.

$$f(x) = \begin{cases} e^{-x}, & x > 0 \\ 0, & x \le 0. \end{cases}$$

Determine the p.d.f. of $Y_1 = \sqrt{X}$, $Y_2 = 3X + 5$, $Y_3 = X^2$, $Y_4 = X_1 + X_2$, $Y_5 = X_1/(X_1 + X_2)$.

5. Given

$$f(x) = \begin{cases} 2x, & 0 < x < 1 \\ 0, & \text{otherwise} \end{cases}$$

of a random variable X. Find p.d.f. of $Y = 8X^3$.

6. Let X be a random variable with probability function

$$p(x) = \begin{cases} \left(\frac{1}{2}\right)^x; & x = 1, 2, 3, \ldots \\ 0, & \text{otherwise.} \end{cases}$$

Find p.d.f. of $Y = X^3$.

7. If $X_1, X_2, \ldots, X_n$ are n-independent and identically distributed random variables and if each of which has p.d.f.

$$f(x) = \frac{1}{\theta} e^{-x/\theta}; \; x > 0, \; \theta > 0$$

then find the p.d.f. of $Y = \sum_{i=1}^{n} X_i$.

8. Let X_1 and X_2 be two sample observations from a standard normal distribution. Find the distribution of $Y = \frac{1}{2}(X_2 - X_1)^2$.

9. The joint probability function of two random variables X_1 and X_2 is

$$P(x_1, x_2) = \begin{cases} \frac{x_1 x_2}{36}; & x_1 = 1, 2, 3; \; x_2 = 1, 2, 3 \\ 0, & \text{otherwise.} \end{cases}$$

Find the distribution of $Y_1 = X_1 X_2$ and $Y_2 = X_2$.

10. The joint density function of three random variables X_1, X_2 and X_3 is given by

$$f(x_1, x_2, x_3) = \begin{cases} 8x_1 x_2 x_3; & 0 < x_i < 1, \; i = 1, 2, 3 \\ 0, & \text{otherwise.} \end{cases}$$

Find the joint distribution of $Y_1 = X_1, Y_2 = X_1 X_2$ and $Y_3 = X_1 X_2 X_3$.

11. The characteristic function of a random variable X is given by

$$\psi_x(t) = e^{-|t|}, \; t > 0.$$

Find the density function of X.

12. The joint probability function of two random variables X_1 and X_2 is given by

$$P(x_1, x_2) = \frac{2!}{x_1! x_2! (2 - x_1 - x_2)!} \left(\frac{3}{6}\right)^{x_1} \left(\frac{2}{6}\right)^{x_2} \left(\frac{1}{6}\right)^{2 - x_1 - x_2}$$
$$x_1 = 0, 1, 2, ; \; x_2 = 0, 1, 2.$$

Find the probability function of $Y = X_1 + X_2$.

13. Given

$$f(x) = \frac{1}{2}, \; -1 < x < 1$$

for a random variable X. Find the density function of (i) $Y_1 = X^2$, (ii) $Y_2 = X_1 + X_2$, (iii) $Y_3 = X_1 - X_2$.

14. Let $x_1, x_2, \dots, x_n$ be the sample observations from Bernoulli population having probability function

$$p(x) = p^x (1-p)^{1-x}; \; x = 0, 1.$$

Find the distribution of Y.

15. Let X_1 and X_2 be two independent random variables having common density function

$$f(x) = \frac{1}{\beta^2} e^{-x/\beta}; \; 0 < x < \infty, \; \beta > 0.$$

Find the p.d.f. of (i) $Y_1 = X_1 + X_2$, (ii) $Y_2 = \frac{X_1}{X_1+X_2}$.

16. If X_1 and X_2 are two random variables having common p.d.f.

$$f(x) = \frac{1}{2} e^{-x/2}, \; x > 0.$$

Find the density function of $Y = \frac{1}{2}(X_1 - X_2)$.

17. Let X_1 and X_2 be two random variables having joint p.d.f.

$$f(x_1, x_2) = \frac{1}{4}; \; -1 < x_1 < 1, \; -1 < x_2 < 1.$$

Find the distribution of $Y_1 = X_1$ and $Y_2 = \frac{X_2}{(1+X_1^2)}$.

18. If X_1, X_2 and X_3 are three observations from a population having p.d.f.

$$f(x) = e^{-x}, \; 0 < x < \infty.$$

Find the p.d.f. of (i) $Y = X_1 + X_2 + X_3$, (ii) $Y = 1/X_1$.

19. Let X be a random variable with p.d.f.

$$f(x) = \frac{2x}{R^2}, \; 0 < x < R.$$

Find the distribution of (i) $Y_1 = X^2$, (ii) $Y_2 = X/R$.

20. Let X_1 and X_2 be two random variables having joint p.d.f.

$$f(x_1, x_2) = \frac{1}{8}(6 - x_1 - x_2); \; 0 < x_1 < 2, \; 2 < x_2 < 4.$$

Determine the p.d.f. of $Y = X_1 + X_2$.

CHAPTER 3

Expectation, Variance and Other Moments of Random Variable

3.1 Introduction

The study of random variable and its distribution helps us in formulating a mathematical model for the results of random experiment. The model is used to find the probabilities of different events defined on the range space of the random variable. Except the calculation of probabilities of different events, we may be interested in different characteristics, viz., mean, variance, moments, skewness and kurtosis of the distribution of random variable. Mean is calculated to obtain a measure of central tendency, where such a measure is the first raw moment or first general moment about zero. As a measure of dispersion of the distribution, second central moment/variance is calculated. In this chapter we shall discuss the different moments and their properties of the distribution of random variable.

3.2 Expected Value of a Random Variable

Let X be a discrete random variable having probability function $p(x)$. Consider a function of X as $Y = g(x)$, where Y is also a random variable. Then expected value of Y is defined by

$$\begin{aligned} E(Y) &= E[g(x)] \\ &= \sum_{i=0}^{\infty} g(x_i)P(X = x_i), \text{ if} \sum_{i=0}^{\infty} |g(x_i)|P(X = x_i) < \infty \text{ [convergent]}. \end{aligned}$$

Here $E[Y]$ is the mean of the random variable Y. If X is a continuous random variable having p.d.f. $f(x)$, then $E(Y)$ is given by

$$\begin{aligned} E[Y] &= E[g(x)] \\ &= \int_{-\infty}^{\infty} g(x)f(x)dx, \text{ if } \int_{-\infty}^{\infty} |g(x)|f(x)dx < \infty \text{ [convergent]}. \end{aligned}$$

Again, let $X_1, X_2, \dots, X_n$ be n random variables and

$$Y = g(x_1, x_2, \dots, x_n)$$

be a function of X's. Then expected value of Y is given by

$$E(Y) = \sum_{x_1}\sum_{x_2}\cdots\sum_{x_n} g(x_1, x_2, \dots, x_n)P(x_1, x_2, \dots, x_n),$$

if $X_1, X_2, \dots, X_n$ are discrete random variables and if

$$\sum_{x_1}\sum_{x_2}\cdots\sum_{x_n} g(x_1, x_2, \dots, x_n)P(x_1, x_2, \dots, x_n) \text{ exists.}$$

For n-dimensional continuous random variables

$$E(Y) = \int_{-\infty}^{\infty}\int_{-\infty}^{\infty}\cdots\int_{-\infty}^{\infty} g(x_1, x_2, \dots, x_n)f(x_1, x_2, \dots, x_n)dx_1, dx_2, \dots, dx_n,$$

if

$$\int_{-\infty}^{\infty}\int_{-\infty}^{\infty}\cdots\int_{-\infty}^{\infty} g(x_1, x_2, \dots, x_n)f(x_1, x_2, \dots, x_n)dx_1, dx_2, \dots, dx_n$$

exists.

If the function of X is $Y = g(x) = X$, then

$$\begin{aligned} E(Y) &= E(X) \\ &= \sum_{x=0}^{\infty} xp(x), \text{ if } X \text{ is discrete random variable and if the summation is convergent} \\ &= \int_{-\infty}^{\infty} xf(x), \text{ if } X \text{ is continuous random variable and if the integration is convergent.} \end{aligned}$$

►**EXAMPLE 3.1** *Two unbiased dice are thrown. Let X be the random variable indicating sum of the upper faces of the dice. Find the expected value of the sum of the upper faces of the dice.*

Solution. Let X be the sum of the upper faces of the dice. Then the probability distribution of X will be as follows:

X	2	3	4	5	6	7	8	9	10	11	12
$p(x)$	1/36	2/36	3/36	4/36	5/36	6/36	5/36	4/36	3/36	2/36	1/36

$$\begin{aligned} E(X) &= \sum xp(x) \\ &= 2 \times \frac{1}{36} + 3 \times \frac{2}{36} + 4 \times \frac{3}{36} + 5 \times \frac{4}{36} + 6 \times \frac{5}{36} + 7 \times \frac{6}{36} \\ &\quad + 8 \times \frac{5}{36} + 9 \times \frac{4}{36} + 10 \times \frac{3}{36} + 11 \times \frac{2}{36} + 12 \times \frac{1}{36} \\ &= 7. \end{aligned}$$

►**EXAMPLE 3.2** *Given a probability distribution as follows:*

$$f(x) = \frac{x}{18}, \; 0 < x < 6.$$

Find $E(X)$.

Solution.

$$E(X) = \int_{-\infty}^{\infty} xf(x)dx = \frac{1}{18}\int_0^6 x \cdot x \cdot dx = \frac{1}{18}\frac{x^3}{3}\bigg|_0^6 = 4.$$

3.3 Properties of Expected Value

Let X be a random variable, discrete or continuous. Its expected value is given by

$$E(X) = \sum_{x=0}^{\infty} xp(x), \text{ if } X \text{ is discrete random variable}$$

$$= \int_{-\infty}^{\infty} xf(x)dx, \text{ if } X \text{ is continuous random variable.}$$

Let us now discuss the properties of this expected value which are presented as the expected value of functions of X.

(i) Let C be any constant and X be a random variable having p.d.f. $f(x)$. Then $E(C) = C$.

Proof.

$$E[C] = \int_{-\infty}^{\infty} Cf(x)dx = C\int_{-\infty}^{\infty} f(x)dx = C.$$

Similarly, it can be shown if X is a discrete random variable.

(ii) Let C be any constant and X be a random variable. Then

$$E[CX + B] = CE(X) + B,$$

where B is any other constant.

Proof. By definition, we can write

$$E[CX + B] = \int_{-\infty}^{\infty} (CX + B)f(x)dx,$$

assuming X as continuous random variable

$$= C\int_{-\infty}^{\infty} xf(x)dx + B\int_{-\infty}^{\infty} f(x)dx$$

$$= CE(X) + B$$

This implies that $E(CX) = CE(X)$.

(iii) Let X and Y be two-dimensional random variables having joint p.d.f. $f(x, y)$. Then $E[AX \pm BY] = AE(X) \pm BE(Y)$, where A and B are any two constants.

Proof. By definition, we can write

$$\begin{aligned}
E[AX \pm BY] &= \int_{-\infty}^{\infty}\int_{-\infty}^{\infty}(Ax \pm By)f(xy)dxdy\\
&= \int_{-\infty}^{\infty} Ax\int_{-\infty}^{\infty} f(xy)dydx \pm \int_{-\infty}^{\infty} Bydy\int_{-\infty}^{\infty} f(xy)dx\\
&= A\int_{-\infty}^{\infty} xg(x) \pm B\int_{-\infty}^{\infty} yh(y)dy\\
&\quad\left[\because \int_{-\infty}^{\infty} f(xy)dx = h(y) \text{ and } \int_{-\infty}^{\infty} f(xy)dy = g(x)\right]\\
&= AE(X) \pm BE(Y).
\end{aligned}$$

If $A = 1$ and $B = 1$, then $E[AX \pm BY] = E(X) \pm E(Y)$.

COROLLARY 1. If $X_1, X_2, \ldots, X_k$ are k-dimensional random variables and $C_1, C_2, \ldots, C_k$ are any k constants, then

$$E\left[\sum_{i=1}^{k} C_i X_i\right] = \sum_{i=1}^{k} C_i E(X_i).$$

If $C_1 = C_2 = \cdots = C_k = 1$, then

$$\begin{aligned}
E\left[\sum_{i=1}^{k} C_i X_i\right] &= E[X_1 + X_2 + \cdots + X_k]\\
&= E(X_1) + E(X_2) + \cdots + E(X_k).
\end{aligned}$$

(iv) If X and Y be two-dimensional independent random variables having joint p.d.f. $f(xy)$, then $E(XY) = E(X)E(Y)$.

Proof. We have

$$\begin{aligned}
E(XY) &= \int_{-\infty}^{\infty}\int_{-\infty}^{\infty} xyf(xy)dxdy\\
&= \int_{-\infty}^{\infty} xg(x)\int_{-\infty}^{\infty} yh(y)dy \quad [\because X \text{ and } Y \text{ are independent}]\\
&= E(X)E(Y).
\end{aligned}$$

COROLLARY 2. If $X_1, X_2, \ldots, X_n$ are n independent random variables, then

$$E(X_1, X_2, \ldots, X_n) = E(X_1)E(X_2)\cdots E(X_n).$$

All the properties discussed above can also be proved for discrete random variable(s).

(v) If $m \leq X \leq M$, then $m \leq E(X) \leq M$.

(vi) If $X_1 \leq X_2$, then $E(X_1) \leq E(X_2)$.

(vii) For a random variable $X, |E(X)| \leq E|X|$.

►EXAMPLE 3.3 *The probability function of two-dimensional random variables X and Y is shown below:*

Y \ X	0	1	2	3	$h(y)$
1	1/18	3/18	1/18	1/18	6/18
2	2/18	2/18	1/18	2/18	7/18
3	1/18	1/18	2/18	1/18	5/18
$g(x)$	4/18	6/18	4/18	4/18	1.00

Find (i) $E(3X-2Y)$, (ii) $E(XY)$, (iii) $E(X-3)$, (iv) $E(Y+1)$.

Solution.

(i) We know $E[3X-2Y]=3E(X)-2E(Y)$. Now

$$E(X) = \sum_{x=0}^{3} xg(x) = 0\times\frac{4}{18}+1\times\frac{6}{18}+2\times\frac{4}{18}+3\times\frac{4}{18}=\frac{26}{18}$$

$$E(Y) = \sum_{y=1}^{3} yh(y) = 1\times\frac{6}{18}+2\times\frac{7}{18}+3\times\frac{5}{18}=\frac{35}{18}$$

$$\therefore\ E[3X-2Y] = 3E(X)-2E(Y)=3\times\frac{26}{18}-2\times\frac{35}{18}=\frac{4}{9}.$$

(ii) $$E(XY) = \sum_{x=0}^{3}\sum_{y=1}^{3} xyP(x,y)$$

$$= 0\cdot1\cdot\frac{1}{18}+0\cdot2\cdot\frac{2}{18}+0\cdot3\cdot\frac{1}{18}+1\cdot1\cdot\frac{3}{18}+1\cdot2\cdot\frac{2}{18}+1\cdot3\cdot\frac{1}{18}$$
$$+2\cdot1\cdot\frac{1}{18}+2\cdot2\cdot\frac{1}{18}+2\cdot3\cdot\frac{1}{18}+3\cdot1\cdot\frac{1}{18}+3\cdot2\cdot\frac{2}{18}+3\cdot3\cdot\frac{1}{18}$$
$$=\frac{23}{9}$$

(iii) $E(X-3)=E(X)-3=\dfrac{26}{18}-3=-\dfrac{14}{9}.$

(iv) $E(Y+1)=E(Y)+1=\dfrac{35}{18}+1=\dfrac{53}{18}.$

►EXAMPLE 3.4 *The joint probability density function of two-dimensional random variables X and Y is shown below:*

$$f(x,y)=\begin{cases}2, & 0\le x\le y\le 1\\ 0, & \textit{otherwise.}\end{cases}$$

Find (i) $E(X-Y)$, (ii) $E(XY)$, (iii) $E(2X+3)$, (iv) $E(3Y-1)$.

Solution. We have

$$g(x) = \int_y f(x,y)dy = \int_x^1 2dy = \begin{cases} 2(1-x), \; 0 \le x \le 1 \\ 0, \quad \text{otherwise} \end{cases}$$

$$\text{and } h(y) = \int_x f(xy)dx = \int_0^y 2dy = \begin{cases} 2y, \; 0 \le y \le 1 \\ 0, \; \text{otherwise.} \end{cases}$$

Now,

$$E(X) = \int_0^1 xg(x)dx$$

$$= 2\int_0^1 x(1-x)dx = 2\left[\frac{x^2}{2}\bigg|_0^1 - \frac{x^3}{3}\bigg|_0^1\right] = \frac{1}{3}$$

$$E(Y) = \int_0^1 yh(y)dy = 2\int_0^1 y^2dy = \frac{2}{3}.$$

(i) $E(X-Y) = E(X) - E(Y) = \frac{1}{3} - \frac{2}{3} = -\frac{1}{3}.$

(ii) $E(XY) = \iint xyf(xy)dxdy = 2\iint xydxdy$

$$= 2\int_0^1 xdx \int_x^1 ydy = 2\int_0^1 xdx\frac{1}{2}(1-x^2)$$

$$= \int_0^1 (x-x^3)dx = \frac{x^2}{2}\bigg|_0^1 - \frac{x^4}{4}\bigg|_0^1 = \frac{1}{2}.$$

(iii) $E[2X+3] = 2E(X) + 3 = 2 \times \frac{1}{3} + 3 = \frac{11}{3}.$

(iv) $E[3Y-1] = 3E(Y) - 1 = 3 \times \frac{2}{3} - 1 = 1.$

►**EXAMPLE 3.5** *The probability function of two-dimensional discrete random variables X and Y is shown below:*

Y \ X	1	2	3	4	5	6	$h(y)$
0	1/12	1/12	1/12	1/12	1/12	1/12	6/12
1	1/12	1/12	1/12	1/12	1/11	1/12	6/12
$g(x)$	2/12	2/12	2/12	2/12	2/12	2/12	1.00

Find (i) $E[3X+4Y]$, (ii) $E(XY)$, (iii) $E[2X-5]$, (iv) $E(3Y-4)$.

Solution. Here the marginal probability functions of X and Y are $g(x)$ and $h(y)$. It is observed that

$$P(X=1, Y=0) = \frac{1}{12}.$$

Again

$$g(X=1)h(Y=0) = \frac{2}{12} \times \frac{6}{12} = \frac{1}{12} = P(X=1, Y=0)$$

$$P(X=2, Y=0) = \frac{1}{12}.$$

Again

$$g(X=2)h(Y=0) = \frac{2}{12} \times \frac{6}{12} = \frac{1}{12} = P(X=2, Y=0).$$

In a similar way, it is observed that

$$P(X=6, Y=1) = \frac{1}{12} = g(X=6)h(Y=1).$$

Therefore, X and Y are independent random variables. Now

$$E(X) = \sum_{x=1}^{6} xg(x) = \frac{2}{12}(1+2+3+4+5+6) = \frac{42}{12}$$

$$\text{and } E(Y) = \sum_{y=0}^{1} yh(y) = \frac{6}{12}(0+1) = \frac{6}{12}.$$

(i) $E[3X+4Y] = 3E(X) + 4E(Y) = 3 \times \frac{42}{12} + 4 \times \frac{6}{12} = 12.5.$

(ii) Here X and Y are independent. Hence

$$E(XY) = E(X)E(Y) = \frac{42}{12} \times \frac{6}{12} = 1.75.$$

(iii) $E[2X-5] = 2E(X) - 5 = 2 \times \frac{42}{12} - 5 = 2.$

(iv) $E[3Y-4] = 3E(Y) - 4 = 3 \times \frac{6}{12} - 4 = -2.5.$

►**EXAMPLE 3.6** *The joint p.d.f. of two-dimensional random variables X and Y is shown below:*

$$f(x,y) = \begin{cases} \frac{1}{2}xe^{-y}; \ 0 < x < 2, \ y > 0 \\ 0, \quad \textit{otherwise.} \end{cases}$$

Find (i) $E(X-2Y)$, (ii) $E(2X+Y)$, (iii) $E(XY)$, (iv) $E(X-1)$, (v) $E(3Y-2)$.

Solution. We have

$$g(x) = \int_y f(xy)dy = \frac{1}{2}x\int_0^\infty e^{-y}dy = \frac{1}{2}x, \ 0 < x < 2$$

$$\text{and } h(y) = \int_x f(xy)dx = e^{-y}\frac{1}{2}\int_0^2 xdx = e^{-y}, \ y > 0$$

$$g(x)h(y) = \frac{1}{2}x \times e^{-y} = f(x,y).$$

Here X and Y are independent

$$E(X) = \int_0^2 xg(x)dx = \frac{1}{2}\int_0^2 x^2 dx = \frac{1}{2}\frac{x^3}{3}\Big|_0^2 = \frac{8}{6}$$

$$E(Y) = \int_0^\infty yh(y)dy = \int_0^\infty ye^{-y}dy = 1.$$

(i) $E[X-2Y] = E(X) - 2E(Y) = \frac{8}{6} - 2\times 1 = -\frac{2}{3}.$

(ii) $E[2X+Y] = 2E(X) + E(Y) = 2\times\frac{8}{6} + 1 = \frac{11}{3}.$

(iii) $E(XY) = E(X)E(Y)$ [$\because$ X and Y are independent]

$$= \frac{8}{6}\times 1 = \frac{4}{3}.$$

(iv) $E(X-1) = E(X) - 1 = \frac{8}{6} - 1 = \frac{1}{3}.$

(v) $E[3Y-2] = 3E(Y) - 2 = 3\times 1 - 2 = 1.$

►**EXAMPLE 3.7** *The continuous random variable X has the p.d.f.*

$$f(x) = \begin{cases} \frac{2}{x^3}, & 1 < x < \infty \\ 0, & otherwise. \end{cases}$$

Find $E(2X-1)$.

Solution. Here

$$E(X) = \int_1^\infty xf(x)dx = 2\int_1^\infty \frac{1}{x^2}dx = 2$$

$$\therefore\ E(2X-1) = 2E(X) - 1 = 2\times 2 - 1 = 3.$$

3.4 Conditional Expectation

Let X and Y be two-dimensional random variables having joint p.d.f. $f(x,y)$. Let $u = u(XY)$ be a function of X and Y. The conditional expectation of U under the condition $X = x$ is defined by

$$E[u(XY)/X = x] = \int_{-\infty}^{\infty} u(x,y)f(y/x)dy.$$

If X and Y are discrete random variables having joint probability function

$$P(X = x, Y = y),$$

then

$$E[u(XY)/X = x] = \sum_y u(x,y)P(y/X = x).$$

Let $u(x,y) = y$, then $E(Y/X = x) = E(Y/X)$. Here $E(Y/X)$ and $E[u(X,Y)/X]$ are functions of X. Let this function be $v(X)$ such that $v(X) = E[u(Y)/X]$. The expected value of $v(X)$ is $E[v(X)] = E[E\{u(Y)/X\}]$. Therefore,

$$
\begin{aligned}
E[E\{u(Y)/X\}] = E[v(X)] &= \int_{-\infty}^{\infty} v(x)g(x)dx \\
&= \int_{-\infty}^{\infty} E[u(Y/X)]g(x)dx \\
&= \int_{-\infty}^{\infty}\left[\int_{-\infty}^{\infty} u(y)f(y/x)\right] g(x)dx \\
&= \int_{-\infty}^{\infty}\int_{-\infty}^{\infty} u(y)f(y/x)g(x)dxdy \\
&= \int_{-\infty}^{\infty}\int_{-\infty}^{\infty} u(y)f(x,y)dxdy = E[u(Y)]
\end{aligned}
$$

Thus, $E[u(Y)] = E[E\{u(Y)/X\}]$ and $E(Y) = E[E\{Y/X\}]$.

►**EXAMPLE 3.8** *The joint probability function of two-dimensional random variables X and Y is given below:*

X \ Y	0	1	2	$g(x)$
1	1/12	3/12	1/12	5/12
2	2/12	1/12	1/12	4/12
3	1/12	1/12	1/12	3/12
$h(y)$	4/12	5/12	3/12	1.00

Find (i) $E(X/Y)$, (ii) $E(X^2/Y)$, (iii) $E(Y/X)$, (iv) $E(Y)$.

Solution.

(i) The conditional probability function of X/Y is shown below:

Probability function	Values of X		
	1	2	3
$P(x/y=0)$	1/4	2/4	1/4
$P(x/y=1)$	3/5	1/5	1/5
$P(x/y=2)$	1/3	1/3	1/3

Hence

$$
\begin{aligned}
E(X/Y=0) &= \sum xP(x/y=0) = 1\times\frac{1}{4}+2\times\frac{2}{4}+3\times\frac{1}{4} = 2 \\
E(X/Y=1) &= \sum xP(x/y=1) = 1\times\frac{3}{5}+2\times\frac{1}{5}+3\times\frac{1}{5} = \frac{8}{5} \\
E(X/Y=2) &= \sum xP(x/y=2) = 1\times\frac{1}{3}+2\times\frac{1}{3}+3\times\frac{1}{3} = 2
\end{aligned}
$$

These expected values are functions of Y and let us consider this function as $U(Y)$.

(ii) $E(X^2/Y)$

$$\begin{aligned} E(X^2/Y=0) &= \sum x^2 P(x/y=0) = 1^2 \times \frac{1}{4} + 2^2 \times \frac{2}{4} + 3^2 \times \frac{1}{4} = \frac{9}{2} \\ E(X^2/Y=1) &= \sum x^2 P(x/y=1) = 1^2 \times \frac{3}{5} + 2^2 \times \frac{1}{5} + 3^2 \times \frac{1}{5} = \frac{16}{5} \\ E(X^2/Y=2) &= \sum x^2 P(x/y=2) = 1^2 \times \frac{1}{3} + 2^2 \times \frac{1}{3} + 3^2 \times \frac{1}{3} = \frac{14}{3} \end{aligned}$$

These expected values are also functions of Y. Let the function be $U_1(Y)$.

(iii) The conditional probability function of $Y/X = x$ is shown below:

Probability function	Values of Y		
	0	1	2
$P(y/x=1)$	1/5	3/5	1/5
$P(y/x=2)$	2/4	1/4	1/4
$P(y/x=3)$	1/3	1/3	1/3

Hence

$$\begin{aligned} E(Y/X=1) &= \sum yP(y/x=1) = 0 \times \frac{1}{5} + 1 \times \frac{3}{5} + 2 \times \frac{1}{5} = 1 \\ E(Y/X=2) &= \sum yP(y/x=2) = 0 \times \frac{2}{4} + 1 \times \frac{1}{4} + 2 \times \frac{1}{4} = \frac{3}{4} \\ E(Y/X=3) &= \sum yP(y/x=3) = 0 \times \frac{1}{3} + 1 \times \frac{1}{3} + 2 \times \frac{1}{3} = 1 \end{aligned}$$

These expected values are functions of X. Let this function be $V(X)$.

(iv)
$$\begin{aligned} E(Y) &= E[E(Y/X=x)] \\ &= E[V(X)] \\ &= \sum V(x)g(x) \\ &= 1 \times \frac{5}{12} + \frac{3}{4} \times \frac{4}{12} + 1 \times \frac{3}{12} \\ &= \frac{11}{12}. \end{aligned}$$

►**EXAMPLE 3.9** *The joint p.d.f. of the two-dimensional random variables X and Y is given below:*

$$f(x,y) = \begin{cases} k(6-x-y); & 0 < x < 2,\ 0 < y < 2 \\ 0, & \textit{otherwise}. \end{cases}$$

Find (i) $E(Y/X)$, (ii) $E(X^2/Y)$, (iii) $E(Y)$.

Solution. Here

$$\iint f(x,y)dxdy = 1 \text{ or, } k\int_0^2\int_0^2 (6-x-y)dxdy = 1$$

$$\text{or, } k\left[\int_0^2\int_0^2 6dxdy - \int_0^2\int_0^2 xdxdy - \int_0^2\int_0^2 ydxdy\right] = 1$$

$$\Rightarrow k = \frac{1}{16}$$

$$\therefore f(x,y) = \frac{1}{16}[6-x-y];\ 0<x<2,\ 0<y<2.$$

Now

$$g(x) = \int_y f(x,y)dy = \frac{1}{16}\int_0^2 (6-x-y)dy$$

$$= \frac{1}{16}[6(2-0)-x(2-0)-(2-0)] = \frac{1}{16}(10-2x),\ 0<x<2.$$

Similarly,

$$h(y) = \frac{1}{16}(10-2y),\ 0<y<2$$

$$\therefore f_1(y/x) = \frac{f(x,y)}{g(x)} = \frac{\frac{1}{16}[6-x-y]}{\frac{1}{16}(10-2x)}$$

$$= \frac{6-x-y}{10-2x}, 0<x<2,\ 0<y<2.$$

(i) $$E[Y/X] = \int_y yf_1(y/x)dy = \int_0^2 \frac{y(6-x-y)}{10-2x}dy$$

$$= \frac{1}{10-2x}\left[6\int_0^2 ydy - x\int_0^2 ydy - \int_0^2 y^2dy\right]$$

$$= \frac{28-6x}{3(10-2x)} = V(X).$$

(ii) $$g_1(x/y) = \frac{f(x,y)}{h(y)} = \frac{6-x-y}{10-2y};\ 0<x<2,\ 0<y<2$$

$$E(X^2/Y) = \int_x x^2 g_1(x/y)dx = \frac{1}{10-2y}\int_0^2 x^2(6-x-y)dx$$

$$= \frac{1}{10-2y}\left[6\int_0^2 x^2dx - \int_0^2 x^3dx - y\int_0^2 x^2dx\right]$$

$$= \frac{36-8y}{2(10-2y)}.$$

(iii) $$E(Y) = E[E(Y/X)] = E[V(X)] = \int_0^2 v(x)g(x)dx$$

$$= \int \frac{28-6x}{3(10-2x)}\frac{1}{16}(10-2x)dx = \frac{11}{12}.$$

3.5 Variance and Covariance of Random Variables

Let X be a random variable and $E(X)$ be the expected value of X. Let us consider a function of X such that $U = [X - E(X)]^2$. Then variance of X is the mean of U, i.e., $V(X) = E[X - E(X)]^2$, where $V(X) = \sigma^2$ = variance of X.

On simplification, we can write

$$V(X) = E(X^2) - [E(X)]^2.$$

Here

$$E(X^2) = \int_{-\infty}^{\infty} x^2 f(x)dx, \text{ if } X \text{ is a continuous random variable}$$

$$= \sum_{x=0}^{\infty} x^2 p(x), \text{ if } X \text{ is a discrete random variable.}$$

It is noted that U is a function of X and $E(U)$ is the variance of X. The expected value of X is $E(X)$. This expected value of X is a measure of central tendency, but $E[U] = V(X)$ is a measure of dispersion of the random variable X, where dispersion is the deviation of observations from central value $[E(X)]$. However, as a measure of dispersion $E[X - E(X)]$ is not used as $E[X - E(X)] = 0$.

COVARIANCE. Let X and Y be two-dimensional random variables. We need a measure of dispersion of the two-dimensional random variables X and Y. The simultaneous variation of two-dimensional random variables X and Y is known as covariance, where covariance of X and Y is given by

$$\text{cov}(XY) = E[X - E(X)][Y - E(Y)] = E(XY) - E(X)E(Y)$$

$$= 0, \text{ if } X \text{ and } Y \text{ are independent random variables.}$$

Here

$$E(XY) = \int_{-\infty}^{\infty}\int_{-\infty}^{\infty} xyf(x,y)dxdy, \text{ if both } X \text{ and } Y \text{ are continuous random variables}$$

$$= \sum_x \sum_y xyP(x,y), \text{ if both } X \text{ and } Y \text{ are discrete random variables.}$$

►**EXAMPLE 3.10** *The joint probability function of two-dimensional random variables X and Y is given below:*

X \ Y	0	1	2	3	$g(x)$
0	2/8	1/8	1/8	1/8	5/8
1	1/8	1/8	1/8	—	3/8
$h(y)$	3/8	2/8	2/8	1/8	1.00

Find (i) $V(X)$, (ii) $V(Y)$, (iii) cov(XY).

Solution. Here $g(x)$ and $h(y)$ are the marginal probability functions of X and Y, respectively. We have

$$\text{(i)}\quad E(X) = \sum_{x=0}^{1} xg(x) = 0 \times \frac{5}{8} + 1 \times \frac{3}{8} = \frac{3}{8}$$

$$E(X^2) = \sum_{x=0}^{1} x^2 g(x) = 0^2 \times \frac{5}{8} + 1^2 \times \frac{3}{8} = \frac{3}{8}$$

$$V(X) = E(X^2) - [E(X)]^2 = \frac{3}{8} - \left(\frac{3}{8}\right)^2 = \frac{15}{64}.$$

$$\text{(ii)}\quad E(Y) = \sum_{y=0}^{3} yh(y) = 0 \times \frac{3}{8} + 1 \times \frac{2}{8} + 2 \times \frac{2}{8} + 3 \times \frac{1}{8} = \frac{9}{8}$$

$$E(Y^2) = \sum_{y=0}^{3} y^2 h(y) = 0^2 \times \frac{3}{8} + 1^2 \times \frac{2}{8} + 2^2 \times \frac{2}{8} + 3^2 \times \frac{1}{8} = \frac{19}{8}$$

$$V(Y) = E(Y^2) - [E(Y)]^2 = \frac{19}{8} - \left(\frac{9}{8}\right)^2 = \frac{71}{64}.$$

$$\text{(iii)}\quad E(XY) = \sum_{x=0}^{1}\sum_{y=0}^{3} xyP(x,y)$$

$$= 0 \times 0 \times \frac{2}{8} + 0 \times 1 \times \frac{1}{8} + 0 \times 2 \times \frac{1}{8} + 0 \times 3 \times \frac{1}{8}$$

$$+1 \times 0 \times \frac{1}{8} + 1 \times 1 \times \frac{1}{8} + 1 \times 2 \times \frac{1}{8} + 1 \times 3 \times 0 = \frac{3}{8}.$$

$$\text{cov}(XY) = E(XY) - E(X)E(Y) = \frac{3}{8} - \frac{3}{8} \times \frac{3}{8} = \frac{15}{64}.$$

►EXAMPLE 3.11 *The joint p.d.f. of two-dimensional random variables X and Y is given below:*

$$f(x,y) = \begin{cases} 21x^2y^3, & 0 < x < y < 1 \\ 0, & \textit{otherwise.} \end{cases}$$

Find (i) $V(X)$, (ii) $V(Y)$, (iii) $\text{cov}(XY)$.

Solution. The marginal p.d.f. of X and Y are shown below:

$$g(x) = \int_y f(xy)dy = 21x^2 \int_x^1 y^3 dy = \frac{21}{4}x^2(1 - x^4),\ 0 < x < 1$$

$$h(y) = \int_x f(xy)dx = 21y^3 \int_0^y x^2 dx = 7y^6,\ 0 < y < 1.$$

$$\text{(i)}\quad E(X) = \int_0^1 xg(x)dx = \frac{21}{4}\int_0^1 x^3(1 - x^4)dx = \frac{21}{4}\left[\frac{1}{4} - \frac{1}{8}\right] = \frac{21}{32}$$

$$E(X^2) = \int_0^1 x^2 g(x)dx = \frac{21}{4}\int x^4(1 - x^4)dx = \frac{21}{4}\left[\frac{1}{5} - \frac{1}{9}\right] = \frac{7}{15}$$

$$V(X) = E(X^2) - [E(X)]^2 = \frac{7}{15} - \left(\frac{21}{32}\right)^2 = 0.036.$$

(ii) $$E(Y) = \int_0^1 yh(y)dy = 7\int_0^1 y^7\,dy = \frac{7}{8}$$

$$E(Y^2) = \int_0^1 y^2h(y)dy = 7\int y^8\,dy = \frac{7}{9}$$

$$V(Y) = E(Y^2) - [E(Y)]^2 = \frac{7}{9} - \left(\frac{7}{8}\right)^2 = \frac{7}{576}.$$

(iii) $$E(XY) = \int_x\int_y xyf(x,y)dxdy$$

$$= 21\int_0^1\int_x^1 x^3y^4dxdy = \frac{21}{5}\int_0^1 x^3(1-x^5)dx$$

$$= \frac{21}{5}\left(\frac{1}{4}-\frac{1}{9}\right) = \frac{7}{12}$$

$$\text{cov}(XY) = E(XY) - E(X)E(Y) = \frac{7}{12} - \frac{21}{32}\times\frac{7}{8} = \frac{7}{768}.$$

3.6 Properties of Variance

1. Let C be any constant. Then $V(C) = 0$.

Proof. By definition, we have $V(C) = E[C^2] - [E(C)]^2 = C^2 - C^2 = 0$.

2. Let A and B be any two constants, and X is a random variable.

Then $V(AX \pm B) = A^2V(X)$.

Proof. By definition, we have

$$\begin{aligned} V(AX \pm B) &= E(AX \pm B)^2 - [E(AX \pm B)]^2 \\ &= E[AX \pm B - AE(X) \mp B]^2 \\ &= A^2E[X - E(X)]^2 = A^2V(X). \end{aligned}$$

COROLLARY 3. If $A = 1$, then $V(X \pm B) = V(X)$.

3. If X and Y are two random variables, and A and B are any two constants, then

$$\begin{aligned} V(AX \pm BY) &= A^2V(X) + B^2V(Y) \pm 2AB\,\text{cov}(XY) \\ &= A^2V(X) + B^2V(Y), \quad \text{if } X \text{ and } Y \text{ are independent random variables.} \end{aligned}$$

Proof. By definition, we have

$$\begin{aligned} V(AX \pm BY) &= E[AX \pm BY - E(AX \pm BY)]^2 \\ &= E[AX \pm BY - AE(X) \mp BE(Y)]^2 \\ &= E[\{AX - AE(X)\} \pm \{BY - BE(Y)\}]^2 \\ &= A^2E[X - E(X)]^2 + B^2E[Y - E(Y)]^2 \\ &\quad \pm 2ABE[X - E(X)][Y - E(Y)] \\ &= A^2V(X) + B^2V(Y) \pm 2AB\,\text{cov}(XY). \end{aligned}$$

But, if X and Y are independent, $\text{cov}(XY) = 0$. Therefore

$$V(AX \pm BY) = A^2V(X) + B^2V(Y),$$

if X and Y are independent.

COROLLARY 4. If $A = 1, B = 1$,

$$V(X \pm Y) = V(X) + V(Y) \pm 2\,\text{cov}(XY) = V(X) + V(Y),$$

if X and Y are independent.

COROLLARY 5. If $X_1, X_2, \ldots, X_k$ are k random variables, then

$$\begin{aligned} V(X_1 \pm X_2 \pm \cdots \pm X_k) &= \sum_{i=1}^{k} V(X_i) \pm 2 \sum_{i<j=1}^{k} \text{cov}(X_i, X_j) \\ &= \sum_{i=1}^{k} V(X_i) \pm \sum_{i \neq j=1}^{k} \sum \text{cov}(X_i, X_j) \\ &= \sum_{i=1}^{k} V(X_i), \text{ if } X_i\text{'s are independent.} \end{aligned}$$

4. Let X be a random variable such that $E(X) = \mu$ and $V(X) = \sigma^2$. Then, if $E[X^2] < \infty$

$$E(Z) = 0, V(Z) = 1, \text{ where } Z = \frac{X - \mu}{\sigma}.$$

Here

$$E(Z) = E\left[\frac{X-\mu}{\sigma}\right] = \frac{1}{\sigma}[E(X) - \mu] = 0$$

$$\text{and } V(Z) = V\left[\frac{X-\mu}{\sigma}\right] = \frac{1}{\sigma^2}V(X - \mu)$$

$$= \frac{1}{\sigma^2}V(X) = \frac{\sigma^2}{\sigma^2} = 1.$$

This Z is called *standardized random variable.*

5. If $X_1, X_2, \ldots, X_k$ are k-dimensional random variables such that $E[X_i] = \mu_i$, $\text{cov}(X_i, X_j) = \sigma_{ij} (i \neq j = 1, 2, \ldots, k)$ and $V(X_i) = \sigma_i^2 = \sigma_{ii}$. The variance-covariance matrix of X_i's is $\sum = [\sigma_{ij}]$.

Let

$$H = \sum_{i=1}^{k} C_i X_i$$

be a linear function of X_i's, where $C_1, C_2, \ldots, C_k$ are constants. Then the mean and variance of H is

$$E[H] = E\left[\sum_{i=1}^{k} C_i X_i\right] = \sum_{i=1}^{k} C_i E(X_i) = \sum_{i=1}^{k} C_i \mu_i$$

$$\text{and } V[H] = V\left(\sum_{i=1}^{k} C_i X_i\right) = \sum C_i^2 \sigma_i^2 + \sum\sum_{i \neq j} C_i C_j \sigma_{ij}$$

However, if $X_1, X_2, \ldots, X_k$ are uncorrelated, then

$$V[H] = \sum C_i^2 \sigma_i^2 \; [\because \; \text{cov}(X_i, X_j) = 0]$$

The necessary and sufficient condition for perfect linear dependence of $X_1, X_2, \ldots, X_k$ is

$$\text{Rank}\left(\sum\right) = k - 1$$

where $\sum = [\sigma_{ij}]$ is a matrix of order $k \times k$. However, the necessary and sufficient condition for $\sum$ to be positive definite matrix is that there is no linear dependence among $X_1, X_2, \ldots, X_k$.

6. Let $X_1, X_2, \ldots, X_k$ be k-dimensional random variables and

$$H = \sum_{i=1}^{k} C_i X_i$$

be a linear function of $X_1, X_2, \ldots, X_k$, where $C_1, C_2, \ldots, C_k$ are constants. Then, if

$$V(H) = \sum_i \sum_j C_i C_j \sigma_{ij}$$

is positive definite matrix, $\sum\sum C_i C_j \sigma^{ij}$ is also positive definite, where σ^{ij} are the elements of $\sum^{-1}$ matrix.

7. Let $X_1, X_2, \ldots, X_k$ and $Y_1, Y_2, \ldots, Y_m$ be two sets of random variables. Consider two sets of constants $a_1, a_2, \ldots, a_k$ and $b_1, b_2, \ldots, b_m$. Then

$$\text{cov}\left[\sum_{i=1}^{k} C_i X_i, \sum_{j=1}^{m} b_j Y_j\right] = \sum_{i=1}^{k}\sum_{j=1}^{m} a_i b_j \, \text{cov}(X_i, Y_j).$$

8. Let X_1 and X_2 be bi-variate random variables such that

$$E(X_1) = \mu_1 \text{ and } E(X_2) = \mu_2.$$

Then

$$E\left[\frac{X_1}{X_2}\right] = \frac{\mu_1}{\mu_2} - \frac{1}{\mu_2^2}\text{cov}(X_1, X_2) + \frac{\mu_1}{\mu_2^3} V(X_2)$$

$$\text{and } V\left[\frac{X_1}{X_2}\right] = \left(\frac{\mu_1}{\mu_2}\right)^2 \left[\frac{V(X_1)}{\mu_1^2} + \frac{V(X_2)}{\mu_2^2} - \frac{2\,\text{cov}(X_1, X_2)}{\mu_1 \mu_2}\right].$$

9. Let X_1, X_2 be bi-variate random variables such that

$$E(X_1) = \mu_1 \text{ and } E(X_2) = \mu_2.$$

Then

$$\begin{aligned} E[X_1X_2] &= \mu_1\mu_2 + \text{cov}(X_1, X_2) \\ \text{and } V[X_1X_2] &= \mu_2^2V(X_1) + \mu_1^2V(X_2) + 2\mu_1\mu_2\,\text{cov}(X_1, X_2) \\ &\quad -[\text{cov}(X_1, X_2)]^2 + E[(X_1 - \mu_1)^2(X_2 - \mu_2)^2] \\ &\quad +2\mu_2E[(X_1 - \mu_1)^2(X_2 - \mu_2)^2 + 2\mu_1E[(X_1 - \mu_1)^2(X_2 - \mu_2)^2] \end{aligned}$$

But, if X_1 and X_2 are independent random variables, then

$$V[X_1X_2] = \mu_2^2V(X_1) + \mu_1^2V(X_2) + V(X_1)V(X_2).$$

10. Let $X_1, X_2, \ldots, X_n$ be a set of identically distributed random variables. Consider mean and variance of the random variables are μ_x and σ_x^2, respectively. Also, consider a discrete random variable N, when

$$S_N = \sum_{i=1}^{N} x_i = Nx_i \ (i = 1, 2, \ldots, n).$$

Then $E[S_N] = E[E(S_N/N)]$. But $E[S_N/N = a] = E[X_1 + X_2 + \cdots + X_n] = n\mu_x$. Therefore,

$$\begin{aligned} E[S_N/N] &= N\mu_x \text{ and } E[E(S_N/N)] = E(N\mu_x) = \mu_x N, \text{ where } E[N] = \mu_N \\ \text{and } V(S_N) &= E[V(S_N/N)] + V[E(S_N/N)] \\ &= E(N\sigma_x^2) + V(N\mu_x) = \sigma_x^2E(N) + \mu_x^2V(N), \text{ where } V(N) = \sigma_N^2. \end{aligned}$$

3.7 Conditional Variance

Let X and Y be two-dimensional random variables having joint p.d.f. $f(x, y)$. The conditional density function of Y given $X = x$ is $f(y/X = x)$. The variance of Y based on this conditional distribution of Y is known as conditional variance of Y and is given by

$$V(Y/X = x) = E[Y^2/X = x] - [E(Y/X = x)]^2.$$

❐ **THEOREM 3.1** $V(Y) = E[V(Y/X] + V[E(Y/X)]$.

Proof.

$$\begin{aligned} E[V(Y/X)] &= E[E(Y^2/X)] - E[\{E(Y/X)\}^2] \\ &= E[Y^2] - [E(Y)]^2 - E[\{E(Y/X\}^2] + \{E(Y)\}^2 \\ &= V(Y) - E[\{E(Y/X)\}]^2 + \{E(Y/X)\}^2 \\ &= V(Y) - V[E(Y/X)] \\ \therefore\ V(Y) &= E[V(Y/X)] + V[E(Y/X)]. \end{aligned}$$

Definition. If g_1 and g_2 are two functions of the bi-variate random variables X and Y, then

(i) $E[g_1(Y) + g_2(Y/X = x) = E[g_1(Y/X = x)] + E[g_2(Y/X = x)]$

(ii) $E[g_1(Y)g_2(Y/X = x)] = g_2(Y)E[g_1(Y/X = x)]$.

►**EXAMPLE 3.12** *The joint probability function of two-dimensional random variables X and Y is shown below:*

Y / X	P(x, y)						
X	1	2	3	4	5	6	$g(x)$
0	1/12	1/12	1/12	1/12	1/12	1/12	1/2
1	1/12	1/12	1/12	1/12	1/12	1/12	1/2
$h(y)$	1/6	1/6	1/6	1/6	1/6	1/6	1.00

Find (i) $V(2X - 3Y)$, (ii) $V(Y/X = x)$ *and hence* $V(Y)$.

Solution. We have

$$E(X) = \sum xg(x) = 0 \times \frac{1}{2} + 1 \times \frac{1}{2} = \frac{1}{2}$$

$$E(X^2) = \sum x^2 g(x) = 0^2 \times \frac{1}{2} + 1^2 \times \frac{1}{2} = \frac{1}{2}$$

$$V(X) = E(X^2) - [E(X)]^2 = \frac{1}{2} - \left(\frac{1}{2}\right)^2 = \frac{1}{4}$$

$$E(Y) = \sum yh(y) = 1 \times \frac{1}{6} + 2 \times \frac{1}{6} + 3 \times \frac{1}{6} + 4 \times \frac{1}{6} + 5 \times \frac{1}{6} + 6 \times \frac{1}{6} = \frac{7}{2}$$

$$E(Y^2) = \sum y^2 h(y) = 1^2 \times \frac{1}{6} + 2^2 \times \frac{1}{2} + 3^2 \times \frac{1}{6} + 4^2 \times \frac{1}{6} + 5^2 \times \frac{1}{6} + 6^2 \times \frac{1}{6} = \frac{91}{6}$$

$$V(Y) = E(Y^2) - [E(Y)]^2 = \frac{91}{6} - \left(\frac{7}{2}\right)^2 = \frac{35}{12}$$

$$E(XY) = \sum\sum xyP(x, y)$$

$$= 0 \times 1 \times \frac{1}{12} + 0 \times 2 \times \frac{1}{12} + 0 \times 3 \times \frac{1}{12} + 0 \times 4 \times \frac{1}{12}$$

$$+ 0 \times 5 \times \frac{1}{12} + 0 \times 6 \times \frac{1}{12}$$

$$= 1 \times 1 \times \frac{1}{12} + 1 \times 2 \times \frac{1}{12} + 1 \times 3 \times \frac{1}{12} + 1 \times 4 \times \frac{1}{12}$$

$$+ 1 \times 5 \times \frac{1}{12} + 1 \times 6 \times \frac{1}{12} = \frac{7}{4}$$

$$\text{cov}(XY) = E(XY) - E(X)E(Y) = \frac{7}{4} - \frac{1}{2} \times \frac{7}{2} = 0.$$

(i) $$V(2X - 3Y) = 2^2 V(X) + 3^2 V(Y) - 2 \cdot 2 \cdot 3\,\text{cov}(XY)$$

$$= 4 \times \frac{1}{4} + 9 \times \frac{35}{12} = \frac{317}{12}.$$

(ii) The conditional distribution of Y given $X = x$ is shown below.

We have $h(y/X = x) = \dfrac{P(x, y)}{g(x)}$.

Probability function	Values of Y					
	1	2	3	4	5	6
$h(y/x = 0)$	1/6	1/6	1/6	1/6	1/6	1/6
$h(y/x = 1)$	1/6	1/6	1/6	1/6	1/6	1/6

$$E(Y/X = 0) = \sum yh(y/x = 0)$$

$$= 1 \times \frac{1}{6} + 2 \times \frac{1}{6} + 3 \times \frac{1}{6} + 4 \times \frac{1}{6} + 5 \times \frac{1}{6} + 6 \times \frac{1}{6} = \frac{7}{2}$$

$$E(Y/X = 1) = \sum yh(y/x = 1)$$

$$= 1 \times \frac{1}{6} + 2 \times \frac{1}{6} + 3 \times \frac{1}{6} + 4 \times \frac{1}{6} + 5 \times \frac{1}{6} + 6 \times \frac{1}{6} = \frac{7}{2}$$

$$E(Y^2/X - 0) - \sum y^2 h(y/x = 0)$$

$$= 1^2 \times \frac{1}{6} + 2^2 \times \frac{1}{6} + 3^3 \times \frac{1}{6} + 4^2 \times \frac{1}{6} + 5^2 \times \frac{1}{6} + 6^2 \times \frac{1}{6} = \frac{91}{6}$$

Similarly,

$$E(Y^2/X = 1) = \frac{91}{6}.$$

$$\therefore\ V(Y/X = 0) = E(Y^2/X = 0) - [E(Y/X = 0)]^2 = \frac{91}{6} - \left(\frac{7}{2}\right)^2 = \frac{35}{12}.$$

Similarly,

$$V(Y/X = 1) = \frac{35}{12}.$$

Now

$$E[V(Y/X)] = \sum V(Y/X)g(x) = \frac{35}{12} \times \frac{1}{2} + \frac{35}{12} \times \frac{1}{2} = \frac{35}{12}$$

$$V[E(Y/X)] = E[E(Y/X)]^2 - \{E[E(Y/X)]\}^2$$

$$E[E(Y/X)]^2 = \sum [E(Y/X)]^2 g(x) = \left(\frac{7}{2}\right)^2 \times \frac{1}{2} + \left(\frac{7}{2}\right)^2 \times \frac{1}{2} = \left(\frac{7}{2}\right)^2 = \frac{49}{4}$$

$$E[E(Y/X)] = \sum E(Y/X)g(x) = \frac{7}{2} \times \frac{1}{2} + \frac{7}{2} \times \frac{1}{2} = \frac{7}{2}$$

$$\therefore\ V[E(Y/X)] = E[E(Y/X)]^2 - \{E[E(Y/X)]\}^2 = \frac{49}{4} - \left(\frac{7}{2}\right)^2 = 0$$

Hence

$$V(Y) = E[V(Y/X)] + V[E(Y/X)] = \frac{35}{12} + 0 = \frac{35}{12}.$$

EXAMPLE 3.13 *The joint p.d.f. of two-dimensional random variables X and Y is given by*

$$f(x,y) = 21x^2y^3,\ 0 < x < y < 1.$$

Find (i) $V(X-2Y)$, (ii) $V(X/Y)$ *and* $V(X)$.

Solution. We have

$$h(y) = \int_0^y f(x,y)dx = 7y^6,\ 0 < y < 1$$

$$g(x) = \int_x^1 f(x,y)dy = \frac{21}{4}x^2(1-x^4),\ 0 < x < 1$$

$$E[X] = \int_0^1 xg(x)dx = \frac{21}{4}\int_0^1 x^3(1-x^4)dx = \frac{21}{32}$$

$$E[X^2] = \int_0^1 x^2g(x)dx = \frac{21}{4}\int x^4(1-x^4)dx = \frac{21}{45}$$

$$E[Y] = \int_0^1 yh(y)dy = 7\int_0^1 y^7dy = \frac{7}{8}$$

$$E[Y^2] = \int_0^1 y^2h(y)dy = 7\int_0^1 y^8dy = \frac{7}{9}$$

$$\therefore\ V(X) = E(X^2) - E(X)]^2 = \frac{21}{45} - \left(\frac{21}{32}\right)^2 = 0.036$$

$$V(Y) = E(Y^2) - [E(Y)]^2 = \frac{7}{9} - \left(\frac{7}{8}\right)^2 = 0.012$$

$$E(XY) = \int_0^1\int_x^1 xyf(xy)dxdy = \int_0^1\int_x^1 21xyx^2y^3dxdy$$

$$= 21\int_0^1 x^3dx\int_x^1 y^4dy = \frac{21}{5}\int_0^1 x^3(1-x^5)dx = \frac{7}{12}$$

$$\text{cov}(XY) = E(XY) - E(X)E(Y) = \frac{7}{12} - \frac{21}{32}\times\frac{7}{8} = 0.0091.$$

(i) $V(X-2Y) = V(X) + 2^2V(Y) - 2\cdot 1\cdot 2\,\text{cov}(XY)$

$= 0.036 + 4\times 0.012 - 4\times 0.0091 = 0.0476.$

(ii) $$f_1(x/y) = \frac{f(x,y)}{h(y)} = \frac{21x^2y^3}{7y^6} = 3y^{-3}x^2,\ 0 < x < y < 1$$

$$E(X/Y) = \int xf_1(x/y)dx = 3y^{-3}\int_0^y x^3dx = \frac{3}{4}y$$

$$E(X^2/Y) = \int x^2f_1(x/y)dx = 3y^{-3}\int_0^y x^4dx = \frac{3}{5}y^2$$

$$V(X/Y) = E(X^2/Y) - [E(X/Y)]^2 = \frac{3}{5}y^2 - \left(\frac{3}{4}y\right)^2 = \frac{3y^2}{80}$$

$$V(X) = E[V(X/Y)] + V[E(X/Y)]$$

$$E[V(X/Y)] = \int V(X/Y)h(y)dy = \frac{21}{80}\int_0^1 y^8 dy = \frac{7}{240}$$

$$V[E(X/Y)] = E[E(X/Y)]^2 - \{E[E(X/Y)]\}^2$$

$$E[E(X/Y)]^2 = \int [E(X/Y)]^2 h(y)dy = \frac{63}{16}\int_0^1 y^8 dy = \frac{7}{16}$$

$$E(X) = E[E(X/Y)] = \int E(X/Y)h(y)dy = \frac{21}{4}\int_0^1 y^7 dy = \frac{21}{32}$$

$$V[E(X/Y)] = E(E(X/Y))^2 - \{E[E(X/Y)]\}^2 = \frac{7}{16} - \left(\frac{21}{32}\right)^2 = \frac{7}{1024}$$

$$\therefore\ V(X) = E[V(X/Y)] + V[E(X/Y)] = \frac{7}{240} + \frac{7}{1024} = 0.036.$$

►**EXAMPLE 3.14** *The joint p.d.f. of two-dimensional random variables X and Y is given below:*

$$f(x, y) = \frac{1}{8}(6 - x - y);\ 0 < x < 2,\ 2 < y < 4.$$

Find (i) $E(Y/X)$, (ii) $V(Y/X)$, (iii) $V(3X - 4Y)$.

Solution. We have

$$g(x) = \int_2^4 f(x, y)dy = \frac{1}{8}\int_2^4 (6 - x - y)dy = \frac{1}{8}(6 - 2x),\ 0 < x < 2$$

$$h(y) = \int_0^2 f(x, y)dx = \frac{1}{8}\int_0^2 (6 - x - y)dx = \frac{1}{8}(10 - 2y),\ 2 < y < 4$$

$$f_1(y/x) = \frac{f(x, y)}{g(x)} = \frac{\frac{1}{8}(6 - x - y)}{\frac{1}{8}(6 - 2x)} = \frac{6 - x - y}{6 - 2x};\ 0 < x < 2,\ 2 < y < 4.$$

(i) $$E(Y/X) = \int_2^4 y f_1(y/x)dy = \frac{1}{6 - 2x}\int_2^4 (6 - x - y)y dy$$

$$= \frac{1}{6 - 2x}\left[\int_2^4 6y dy - x\int_2^4 y dy - \int_2^4 y^2 dy\right] = \frac{52 - 18x}{3(6 - 2x)} = v(x).$$

(ii) $$E(Y^2/X) = \int y^2 f_1(y/x)dy = \int_2^4 \frac{6 - x - y}{6 - 2x} y^2 dy$$

$$= \frac{1}{6 - 2x}\left[\int_2^4 6y^2 dy - x\int_2^4 y^2 dy - \int_2^4 y^3 dy\right]$$

$$= \frac{156 - 56x}{3(6 - 2x)} = v_1(x)$$

$$V[Y/X] = E[Y^2/X] - [E(Y/X)]^2$$

$$= \frac{156 - 56x}{3(6 - 2x)} - \left[\frac{52 - 18x}{3(6 - 2x)}\right]^2 = \frac{2(6x^2 - 36x + 52)}{9(6 - 2x)^2}.$$

(iii) $E(Y) = E[E(Y/X)] = E\left[\frac{52-18x}{3(6-2x)}\right]$

$$= \int_0^2 \frac{52-18x}{3(6-2x)} g(x)dx = \int_0^2 \frac{52-18x}{3(6-2x)} \frac{1}{8}(6-2x)dx$$

$$= \frac{1}{12}\int_0^2 (26-9x)dx = \frac{17}{6}$$

Again

$$E(Y) = \int_2^4 yh(y)dy = \frac{1}{4}\int_2^4 (5-y)ydy = \frac{17}{6}$$

$$E(Y^2) = \int_2^4 y^2 h(y)dy = \frac{1}{4}\int_2^4 (5-y)y^2 dy = \frac{25}{3}.$$

$$\therefore V(Y) = E(Y^2) - [E(Y)]^2 = \frac{25}{3} - \left(\frac{17}{6}\right)^2 = \frac{11}{36}$$

$$E(X) = \int_0^2 xg(x)dx = \frac{1}{8}\int_0^2 (6-2x)xdx = \frac{5}{6}$$

$$E(X^2) = \int_0^2 x^2 g(x)dx = \frac{1}{8}\int_0^2 (6-2x)x^2 dx = 1$$

$$V(X) = E(X^2) - [E(X)]^2 = 1 - \left(\frac{5}{6}\right)^2 = \frac{11}{36}$$

$$E(XY) = \int_0^2\int_2^4 xyf(x,y)dxdy = \frac{1}{8}\iint xy(6-x-y)dxdy = \frac{7}{3}$$

$$\text{cov}(XY) = E(XY) - E(X)E(Y) = \frac{7}{3} - \frac{5}{6}\times\frac{17}{6} = \frac{-1}{36}.$$

$$\therefore V(3X-4Y) = 3^2V(X) + 4^2V(Y) - 2\cdot 3\cdot 4\cdot \text{cov}(XY)$$

$$= 9\times\frac{11}{36} + 16\times\frac{11}{36} + 24\times\frac{1}{36} = \frac{299}{36}.$$

►**EXAMPLE 3.15** *The joint p.d.f. of two-dimensional random variables X and Y is given below:*

$$f(x,y) = 2;\ 0 < x < y,\ 0 < y < 1.$$

Find cov(XY).

Solution.

$$E(XY) = \iint xyf(x,y)dxdy = 2\int_0^y xdx\int_0^1 ydy = \frac{1}{3}$$

$$E(X) = 2\int_0^y\int_x^1 xdxdy = \frac{1}{3}$$

$$E(Y) = 2\int_0^y dx\int_x^1 ydy = \frac{2}{3}$$

$$\text{cov}(XY) = E(XY) - E(X)E(Y) = \frac{1}{3} - \frac{1}{3}\times\frac{2}{3} = \frac{1}{9}.$$

3.8 Theorems on Mean and Variance

3.8.1 Cauchy-Schwartz Inequality

Let X and Y be two random variables having finite variances σ_x^2 and σ_y^2, respectively. Then

$$[E(XY)]^2 \leq E(X^2)E(Y^2).$$

Proof. Since X and Y have finite variances, $E(X^2)$ and $E(Y^2)$ exist. Also, $E(X), E(Y)$ and $E(XY)$ exist. Let us define a real-valued function of the real-value 't' such that

$$h(t) = E[tX - Y]^2 \geq 0 = t^2E(X^2) + E(Y^2) - 2tE(XY) \geq 0.$$

Here $h(t)$ is obviously a quadratic function of 't' and $h(t) \geq 0$. But, if $h(t) > 0$, its roots are not real. Therefore, for a quadratic function $ax^2 + bx + c = 0$, $b^2 - 4ac \leq 0$.

We can write

$$4[E(XY)]^2 - 4E(X^2)E(Y^2) \leq 0 \text{ or, } [E(XY)]^2 \leq E(X^2)E(Y^2).$$

The equality sign holds if $h(t) = 0$. This implies that

$$E(Xt - Y)^2 = 0.$$

This is true, if for any value of t, say t_0

$$P[t_0X = Y] = 1.$$

3.8.2 Chebyshev's Inequality

Let X be a random variable having probability density function $f(x)$, $E(X) = \mu$ and $V(X) = \sigma^2$, where σ^2 is a finite quantity. Then for all $k > 0$

$$P[|X - \mu| \geq k\sigma] \leq \frac{1}{k^2}.$$

Proof. Let us assume that for random variable X having p.d.f. $f(x)$, $E[u(X)]$ exists, where $u(X)$ is a function of X.

$$P[u(X) \geq C] \leq \frac{E[u(X)]}{C}$$

where C is any constant. Also, consider that $u(X) = (X - \mu)^2$ and $C = k^2\sigma^2$.

Then

$$P[(X - \mu)^2 \geq k^2\sigma^2] \leq \frac{E(X - \mu)^2}{k^2\sigma^2}$$

$$\text{or, } P[(X - \mu)^2 \geq k^2\sigma^2] \leq \frac{\sigma^2}{k^2\sigma^2}$$

$$\text{or, } P[(X - \mu)^2 \geq k^2\sigma^2] \leq \frac{1}{k^2}$$

$$\text{or, } P[|X - \mu| \geq k\sigma] \leq \frac{1}{k^2}.$$

This inequality is also available in the following form:

$$P[|X-\mu| \leq k\sigma] \geq 1-\frac{1}{k^2} \text{ where } k \geq 1.$$

Thus, we have

$$P[\mu - k\sigma < X < \mu + k\sigma] \geq 1-\frac{1}{k^2}.$$

REMARK 1. $\frac{1}{k^2}$ is the upper limit of $P[|X-\mu| \geq k\sigma]$. Thus, if k increases, the probability $P[|X-\mu| \geq k\sigma]$ decreases. For example, let $k=5$ and 10. Then

$$P[|X-\mu| \geq 5\sigma] \leq \frac{1}{25} = 0.04$$

$$\text{and } P[|X-\mu| \geq 10\sigma] \leq \frac{1}{100} = 0.01.$$

REMARK 2. Since

$$P[|X-\mu| \leq k\sigma] \geq 1-\frac{1}{k^2},\ k \geq 1,$$

we can write

$$P[\mu - k\sigma < x < \mu + k\sigma] \geq 1-\frac{1}{k^2}.$$

Here μ, σ and k are parameters. If the values of μ, σ and k are given, then the probability of $X \in (\mu - k\sigma, \mu + k\sigma)$ is at least $(1-1/k^2)$. But in practice μ and σ^2 are not known. We need to estimate confidence bounds for μ and σ in such a way that μ and σ fall in that bounds with probability $1-\alpha$, where $0<\alpha<1$.

The confidence bounds are derived as follows:

Let μ be unknown and σ is known. Consider that μ falls within the limit $[a(X), b(X)]$ with probability $1-\alpha$ or more, i.e.,

$$P[a(X) < \mu < b(X)] \geq 1-\alpha.$$

Let $1-\alpha = 1-(1/k^2)$. Then $k = 1/\sqrt{\alpha} > 1$. Thus

$$P\left[\mu - \frac{\sigma}{\sqrt{\alpha}} < X < \mu + \frac{\sigma}{\sqrt{\alpha}}\right] \geq 1-\alpha$$

$$\text{or, } P\left[X - \frac{\sigma}{\sqrt{\alpha}} < \mu < X + \frac{\sigma}{\sqrt{\alpha}}\right] \geq 1-\alpha,$$

where

$$a(X) = X - \frac{\sigma}{\sqrt{\alpha}} \text{ and } b(X) = X + \frac{\sigma}{\sqrt{\alpha}}.$$

Similarly, if μ is known, we can write

$$P\left[0 < \sigma < \left|X - \frac{\mu}{\sqrt{\alpha}}\right|\right] \leq \alpha.$$

The above discussed Chebyshev's inequality is based on the following theorem:

❐ **THEOREM 3.2** *Let $f(x)$ be the p.d.f. of the random variable X. Let us define a non-negative function $u(X)$ such that $E[u(X)]$ exists. Then for any constant C*

$$P[u(X) \geq C] \leq \frac{E[u(X)]}{C}.$$

Proof. Consider that for random variable X, $A = \{x : u(x) \geq c\}$. Then

$$E[u(X)] = \int_{-\infty}^{\infty} u(x)f(x)dx = \int_A u(x)f(x)dx + \int_{A^*} u(x)f(x)dx.$$

Here both the terms in the right-hand side are positive. Thus

$$E[u(X)] \geq \int_A u(X)f(x)dx \text{ or, } \int_{A^*} u(x)f(x)dx.$$

But, as $x \in A$, $u(x) \geq C$ and if $u(x)$ is replaced by C, then

$$E[u(X)] \geq C \int_A f(x)dx.$$

Because $\int_A f(x)dx = P(X \in A) = P[u(X) \geq C]$

$$\therefore\ E[u(X)] \geq CP[u(X) \geq C]$$

Therefore, $P[u(X) \geq C] \leq \dfrac{E[u(X)]}{C}$.

PROPOSITION 1. Let $X_1, X_2, \ldots, X_n$ be independent and identically distributed random variables such that $E(X) = \mu$ and $V(X) = \sigma^2$. Then (i) $E(\bar{X}) = \mu$, (ii) $V(\bar{X}) = \frac{\sigma^2}{n}$, (iii) $\text{cov}(\bar{X}, X_i - \bar{X}) = 0$.

Proof.

(i) We have

$$\bar{X} = \frac{1}{n}\sum X_i$$

$$E(\bar{X}) = E\frac{1}{n}\sum X_i = \frac{1}{n}\sum E(X_i) = \mu.$$

(ii)
$$V(\bar{X}) = V\left(\frac{1}{n}\sum X_i\right) = \frac{1}{n^2}\sum V(X_i)$$

$$= \frac{n\sigma^2}{n^2} = \frac{\sigma^2}{n}\ [\because\ \text{cov}(X_i, X_j) = 0,\ i \neq j = 1, 2, \ldots, n].$$

(iii)
$$\text{cov}(\bar{X}, X_i - \bar{X}) = \text{cov}(\bar{X}, X_i) - \text{cov}(\bar{X}, \bar{X})$$

$$= \text{cov}\left[\frac{1}{n}\sum X_i, X_i\right] - V(\bar{X}).$$

$$= \frac{1}{n}V(X_i) - \frac{\sigma^2}{n} = \frac{\sigma^2}{n} - \frac{\sigma^2}{n}$$

$$= 0\ [\because\ \text{cov}(X_i, X_j) = 0,\ i \neq j].$$

3.8.3 Markov's Inequality

If X is a random variable such that X takes only non-negative values, then for any value $a > 0$

$$P[X \geq a] \leq \frac{E(X)}{a}.$$

Proof. We know

$$\begin{aligned} E(X) &= \int_0^\infty x f(x)dx \\ &= \int_0^a x f(x)dx + \int_a^\infty x f(x)dx \geq \int_a^\infty x f(x)dx \end{aligned}$$

$$\text{or, } E(X) \geq a \int_a^\infty f(x)dx = aP(X \geq a)$$

$$\therefore \frac{E(X)}{a} \geq P(X \geq a) \Rightarrow P(X \geq a) \leq \frac{E(X)}{a}.$$

►**EXAMPLE 3.16** *An unbiased die is thrown* 4 *times to observe the dots* 6. *Let p be the proportion of times in which die shows* 6. *Show that the relative frequency of observing* 6 *tends to probability of success, where observing* 6 *is considered as success.*

Solution. Let X be the number of success (be the number of times in which 6 is observed). The die is thrown $n = 4$ times. Here

$$P\,(\text{success}) = p \text{ and } P\,(\text{failure}) = 1 - p = q.$$

The probability function of X is shown below:

X	0	1	2	3	4
$p(x)$	625/1296	500/1296	150/1296	20/1296	1/1296

Now,

$$\begin{aligned} E(X) &= \sum xp(x) \\ &= 0 \times \frac{625}{1296} + 1 \times \frac{500}{1296} + 2 \times \frac{150}{1296} + 3 \times \frac{20}{1296} + \frac{4}{1296} \\ &= \frac{864}{1296} = \frac{2}{3} = np \\ E(X^2) &= \sum x^2 p(x) \\ &= 0^2 \times \frac{625}{1296} + 1^2 \times \frac{500}{1296} + 2^2 \times \frac{150}{1296} + 3^2 \times \frac{20}{1296} + 4^2 \frac{1}{1296} = 1 \\ V(X) &= E(X^2) - [E(X)]^2 \\ &= 1 - \frac{4}{9} = \frac{5}{9} = npq. \end{aligned}$$

Here $n = 4,\ p = 1/6,\ q = 5/6$.

The relative frequency of observing $6X$ times is X/n. Consider that ϵ is a number, where $\epsilon > 0$. Then

$$P\left[\left|\frac{X}{n} - p\right| \geq \epsilon\right] = P[|X - np| \geq \epsilon n]$$
$$= P\left[|X - \mu| \geq \epsilon\sqrt{\frac{n}{p(1-p)}\sigma^2}\right].$$

Let $k = \epsilon\sqrt{n/p(1-p)}$. Then applying Chebyshev's inequality, we can write

$$P\left[|X - \mu| \geq \epsilon\sqrt{\frac{n}{p(1-p)}\sigma^2}\right] \leq \frac{p(1-p)}{n\epsilon^2} \quad [\mu = E(X) = np]$$

$$\text{or, } P\left[\left|\frac{X}{n} - p\right| \leq \epsilon\right] \leq \frac{p(1-p)}{n\epsilon^2}.$$

It is observed that for large value of n and for any fixed value of $\epsilon(\epsilon > 0)$ the right-hand side of the inequality will be zero. That is

$$\lim_{n\to\infty} P\left[\left|\frac{X}{n} - p\right| \geq \epsilon\right] = 0 \to \lim_{n\to\infty} P\left[\left|\frac{X}{n} - p\right| < \epsilon\right] = 1.$$

This last result is true for any value of $\epsilon > 0$. Thus, we can say that the relative frequency of success tends to probability of success.

►**EXAMPLE 3.17** *The p.d.f. of the random variable X is given below:*

$$f(x) = \begin{cases} 2x, & 0 < x < 1 \\ 0, & \textit{otherwise.} \end{cases}$$

Show that, if k increases $P[(|x - \mu|) \geq k\sigma]$ decreases.

Solution. Here

$$E(X) = \int_0^1 xf(x)dx = 2\int_0^1 x^2 dx = \frac{2}{3}$$

$$E(X^2) = \int_0^1 x^2 f(x)dx = 2\int_0^1 x^3 dx = \frac{1}{2}$$

$$V(X) = E(X^2) - [E(X)]^2 = \frac{1}{2} - \left(\frac{2}{3}\right)^2 = \frac{1}{18}.$$

Now, using Chebyshev's inequality, we can write

$$P\left[\left|X - \frac{2}{3}\right| \geq k\sqrt{\frac{1}{18}}\right] \leq \frac{1}{k^2}.$$

Let $k = 5$, then

$$P\left[\left|X - \frac{2}{3}\right| \geq \frac{5}{\sqrt{18}}\right] \leq \frac{1}{25} \text{ or, } P\left[X \geq \frac{2}{3} - \frac{5}{\sqrt{18}}\right] \leq 0.04$$

$$\text{and } P\left[X \geq \frac{2}{3} + \frac{5}{\sqrt{18}}\right] \leq 0.04.$$

Again, let $k = 10$, then

$$P\left[\left|X - \frac{2}{3}\right| \geq \frac{1}{\sqrt{18}}\right] \leq \frac{1}{100} \text{ or, } P\left[X \geq \frac{2}{3} - \frac{10}{\sqrt{18}}\right] \leq 0.01$$

$$\text{and } P\left[X \geq \frac{2}{3} + \frac{10}{\sqrt{18}}\right] \leq 0.01.$$

3.9 Moments of Random Variable

For a random variable X we have considered $E(Y_1)$, where $Y_1 = g_1(X)$ is a function of X. If $g_1(X) = X$, then $E(Y_1) = E[g_1(X)] = E(X)$ and if $g_2(X) = X^2$, then $E(Y_2) = E[g_2(X)] = E(X^2)$. Also, we have defined $V(X) = \sigma^2 = E(X^2) - [E(X)]^2$. Here $E(X)$ and $V(X)$ are mean and variance, respectively, of random variable X.

The expected value of other functions of X can also be calculated. Let us consider such three functions of X as follows:

$$Y_3 = g_3(X) = X^r, \; Y_4 = g_4(X) = (X - A)^r \text{ and } Y_5 = g_5(X) = (X - \mu)^r$$

where $E(X) = \mu = \text{Mean } (X)$. Here

$$E(Y_3) = E[g_3(X)] = E(X^r) = \mu'_r = r\text{-th raw moment of } X.$$

Similarly, r-th moment about any arbitrary value A is defined by

$$E[Y_4] = E[g_4(X)] = E[X - A]^r = r\text{-th general moment.}$$

This moment can also be written as μ'_r if $A = 0$. Here $E(X - A)^r$ is the mean deviation of r-th order of X when deviation is taken from an arbitrary value A. If the mean deviation is taken from central value mean $[E(X)]$, then mean deviation of r-th order is known as r-th central moment and it is given by

$$\mu_r = E[X - \mu]^r = E[X - E(X)]^r.$$

Here

$$\mu'_r = E(X^r) = \sum_{x=0}^{\infty} x^r p(x), \text{ if } X \text{ is a discrete random variable}$$

$$= \int_{-\infty}^{\infty} x^r f(x), \text{ if } X \text{ is a continuous random variable.}$$

When $r = 0$, we have

$$\mu'_0 = E[X^0] = \int x^0 f(x)dx = 1.$$

For $r = 1, 2, 3$ and 4, we have $\mu'_1 = E(X)$, $\mu'_2 = E(X^2)$, $\mu'_3 = E(X^3)$ and $\mu'_4 = E(X^4)$. Again,

$$E[X - E(X)]^0 = E[X - \mu]^0 = \int (x - \mu)^0 f(x)dx = 1 = \mu_0.$$

Similarly,

$$\begin{aligned}\mu_1 &= E[X-\mu] = \int (x-\mu)f(x)dx \\ &= \int xf(x)dx - \mu\int f(x)dx = \mu - \mu = 0 \\ \mu_2 &= E[X-\mu]^2 = \int (x-\mu)^2 f(x)dx \\ &= \int x^2 f(x)dx + \mu^2 \int f(x)dx - 2\mu\int xf(x)dx = \mu_2' - \mu_1'^2.\end{aligned}$$

Similarly, it can be shown that

$$\begin{aligned}\mu_3 &= \mu_3' - 3\mu_2'\mu_1' + 2\mu_1'^3 \\ \mu_4 &= \mu_4' - 4\mu_3'\mu_1' + 6\mu_2'\mu_1'^2 - 3\mu_1'^4.\end{aligned}$$

Besides these general (raw) moments and central moments, we have another set of moments which are known as factorial moment. The r-th factorial moment is defined by

$$\mu_{[r]} = E[X(X-1)(X-2)\cdots(X-r+1)].$$

Thus, when $r = 1$, we have

$$\mu_{[1]} = E(X) = \mu_1'.$$

When $r = 2$, we have

$$\mu_{[2]} = E[X(X-1)] = E(X^2) - E(X) = \mu_2' - \mu_1'.$$

Similarly, for $r = 3$ and $r = 4$, we have

$$\begin{aligned}\mu_{[3]} &= \mu_3' - 3\mu_2' + 2\mu_1' \\ \mu_{[4]} &= \mu_4' - 6\mu_3' + 11\mu_2' - 6\mu_1'.\end{aligned}$$

The central moments are used to measure the coefficient of skewness and coefficient of kurtosis, where these coefficients are, respectively

$$\beta_1 = \frac{\mu_3^2}{\mu_2^3} \text{ and } \beta_2 = \frac{\mu_4}{\mu_2^2}.$$

3.10 Moment Generating Function

Let X be a random variable having p.d.f. $f(x)$. The moment generating function of X is defined by

$$\begin{aligned}M_x(t) = E[e^{tx}] &= \sum_{x=0}^{\infty} e^{tx}p(x), \text{ if } X \text{ is discrete random variable} \\ &= \int_{-\infty}^{\infty} e^{tx} f(x)dx, \text{ if } X \text{ is continuous random variable}\end{aligned}$$

provided $E[e^{tx}]$ for all values in the limit $-h < t < h,\ h > 0$.

The function $M_x(t)$ is

$$M_x(t) = E\left[1 + tX + \frac{t^2}{2!}X^2 + \frac{t^3}{3!}X^3 + \frac{t^4}{4!}X^4 + \cdots\right]$$
$$= 1 + tE(X) + \frac{t^2}{2!}E(X^2) + \frac{t^3}{3!}E(X^3) + \frac{t^4}{4!}E(X^4) + \cdots$$
$$= 1 + t\mu_1' + \frac{t^2}{2!}\mu_2' + \frac{t^3}{3!}\mu_3' + \frac{t^4}{4!}\mu_4' + \cdots.$$

It is observed that

$$\mu_r' = \text{coefficient of } \frac{t^r}{r!} \text{ in } M_x(t).$$

Also, we have

$$M_x(t) = \int_{-\infty}^{\infty} e^{tx} f(x)dx.$$

Differentiating both sides w.r.t. 't' and putting $t = 0$, we get

$$\left.\frac{dM_x(t)}{dt}\right|_{t=0} = \left.\int_{-\infty}^{\infty} xe^{tx} f(x)dx\right|_{t=0} = \int_{-\infty}^{\infty} xf(x)dx = \mu_1'$$
$$\left.\frac{d^2M_x(t)}{dt^2}\right|_{t=0} = \left.\int_{-\infty}^{\infty} x^2e^{tx} f(x)dx\right|_{t=0} = \int_{-\infty}^{\infty} x^2f(x)dx = \mu_2'.$$

In a similar way we can find μ_r', where

$$\mu_r' = \left.\frac{d^rM_x(t)}{dt^r}\right|_{t=0}.$$

3.10.1 Factorial Moment Generating Function

Let X be a random variable and $f(x)$ is the p.d.f. of X (if X is continuous random variable). Then factorial moment generating function of X is defined by

$$M_{[x]}(t) = E\left[t^X\right] = \int_{-\infty}^{\infty} t^x f(x)dx.$$

If X is discrete random variable having probability function $p(x)$, then

$$M_{[x]}(t) = \sum_{x=0}^{\infty} t^x p(x).$$

Also, we have

$$\mu_{[r]} = \left.\frac{d^rM_{[x]}(t)}{dt^r}\right|_{t=0}.$$

►**EXAMPLE 3.18** *The probability function of the random variable X is given by*

$$p(x) = \frac{e^{-\lambda}\lambda^x}{x!}, \quad x = 0, 1, 2, \ldots; \ \lambda > 0.$$

Find first four moments and factorial moments.

Solution. The moment generating function is

$$M_x(t) = E\left[e^{tX}\right] = \sum_{x=0}^{\infty} e^{tx} p(x) = \sum_{x=0}^{\infty} e^{-\lambda} \frac{(\lambda e^t)^x}{x!} = e^{-\lambda} e^{\lambda e^t} = e^{\lambda[e^t - 1]}$$

$$= \left[1 + \lambda(e^t - 1) + \frac{\lambda^2(e^t - 1)^2}{2!} + \frac{\lambda^3(e^t - 1)^3}{3!} + \frac{\lambda^4(e^t - 1)^4}{4!} + \cdots\right]$$

$$= \left[1 + \lambda\left(t + \frac{t^2}{2!} + \frac{t^3}{3!} + \frac{t^4}{4!} + \cdots\right) + \frac{\lambda^2}{2!}\left(t + \frac{t^2}{2!} + \cdots\right)^2\right.$$

$$\left. + \frac{\lambda^3}{3!}\left(t + \frac{t^2}{2!} + \cdots\right)^3 + \frac{\lambda^4}{4!}\left(t + \frac{t^2}{2!} + \cdots\right)^4 + \cdots\right]$$

Here

$$\mu_1' = \text{coefficient of } t \text{ in } M_x(t) = \lambda$$

$$\mu_2' = \text{coefficient of } \frac{t^2}{2!} \text{ in } M_x(t) = \lambda^2 + \lambda$$

$$\mu_3' = \text{coefficient of } \frac{t^3}{3!} \text{ in } M_x(t) = \lambda + 3\lambda^2 + \lambda^3$$

$$\mu_4' = \lambda + 3\lambda^2 + 4\lambda^2 + 6\lambda^3 + \lambda^4 = \lambda^4 + 6\lambda^3 + 7\lambda^2 + \lambda.$$

Hence

$$\mu_2 = \mu_2' - \mu_1'^2 = \lambda$$

$$\mu_3 = \mu_3' - 3\mu_2'\mu_1' + 2\mu_1'^3 = \lambda$$

$$\mu_4 = \mu_4' - 4\mu_3'\mu_1' + 6\mu_2'\mu_1'^2 - 3\mu_1'^4 = 3\lambda^2 + \lambda.$$

Also, we have

$$\beta_1 = \frac{\mu_3^2}{\mu_2^3} = \frac{1}{\lambda} \text{ and } \beta_2 = \frac{\mu_4}{\mu_2^2} = 3 + \frac{1}{\lambda} > 3.$$

The distribution is positively skewed and leptokurtic. The factorial moment generating function is

$$M_{[x]}(t) = E\left[t^X\right] = \sum_{x=0}^{\infty} \frac{t^x e^{-\lambda} \lambda^x}{x!} = e^{-\lambda}\left[1 + \frac{\lambda t}{1!} + \frac{(\lambda t)^2}{2!} + \cdots\right]$$

$$= e^{-\lambda} e^{\lambda t} = e^{\lambda(e^t - 1)}.$$

Now,

$$\mu_{[1]} = \left.\frac{dM_{[x]}(t)}{dt}\right|_{t=1} = \left.\lambda e^{\lambda(e^t-1)}\right|_{t=1} = \lambda = \mu_1'$$

$$\mu_{[2]} = \left.\frac{d^2 M_{[x]}(t)}{dt^2}\right|_{t=1} = \left.\lambda^2 e^{\lambda(e^t-1)}\right|_{t=1} = \lambda^2 = \mu_2' - \mu_1'$$

$$\mu_{[3]} = \left.\frac{d^3 M_{[x]}(t)}{dt^3}\right|_{t=1} = \left.\lambda^3 e^{\lambda(e^t-1)}\right|_{t=1} = \lambda^3 = \mu_3' - 3\mu_2' + 2\mu_1'$$

$$\mu_{[4]} = \left.\frac{d^4 M_{[x]}(t)}{dt^4}\right|_{t=1} = \left.\lambda^4 e^{\lambda(e^t-1)}\right|_{t=1} = \lambda^4 = \mu_4' - 6\mu_2' + 11\mu_2' - 6\mu_1'.$$

The central moments can also be found out using above factorial moments, where

$$\begin{aligned} \mu_1' &= \lambda,\ \mu_2' = \mu_{[2]} + \mu_1' = \lambda^2 + \lambda \\ \mu_3' &= \mu_{[3]} + 3\mu_{[2]} + \mu_{[1]} = \lambda^3 + 3\lambda^2 + \lambda \\ \mu_4' &= \mu_{[4]} + 6\mu_{[3]} + 7\mu_{[2]} + \mu_{[1]} = \lambda^4 + 6\lambda^3 + 7\lambda^2 + \lambda \end{aligned}$$

and hence $\mu_2 = \lambda,\ \mu_3 = \lambda,\ \mu_4 = 3\lambda^2 + \lambda$.

►**EXAMPLE 3.19** *The p.d.f. of continuous random variable X is given by*

$$f(x) = e^{-x},\ x > 0.$$

Find moment generating function of X and hence comment on the shape of the distribution of X.

Solution. The moment generating function of X is

$$\begin{aligned} M_x(t) &= E\left[e^{tX}\right] \\ &= \int_0^\infty e^{tx}e^{-x}dx = \int_0^\infty e^{-x(1-t)}dx = (1-t)^{-1} \\ &= 1 + t + t^2 + t^3 + t^4 + \cdots . \end{aligned}$$

Here μ_r' = coefficient of $\frac{t^r}{r!}$ in $M_x(t)$. Thus

$$\begin{aligned} \mu_1' &= 1,\ \mu_2' = 2,\ \mu_3' = 6,\ \mu_4' = 24 \\ \mu_2 &= \mu_2' - \mu_1'^2 = 2 - (1)^2 = 1 \\ \mu_3 &= \mu_3' - 3\mu_2'\mu_1' + 2\mu_1'^3 = 6 - 3\cdot 2\cdot 1 + 2(1)^3 = 2 \\ \mu_4 &= \mu_4' - 4\mu_3'\mu_1' + 6\mu_2'\mu_1'^2 - 3\mu_1'^3 \\ &= 24 - 4\cdot 6\cdot 1 + 6\cdot 2\cdot (1)^2 - 3(1)^2 = 9. \\ \therefore\ \beta_1 &= \frac{\mu_3^2}{\mu_2^3} = \frac{(2)^2}{(1)^3} = 4, \\ \beta_2 &= \frac{\mu_4}{\mu_2^2} = \frac{9}{(1)^2} = 9 > 3. \end{aligned}$$

Here $\mu_3 > 0$. The distribution is positively skewed and leptokurtic.

►**EXAMPLE 3.20** *The probability function of the discrete random variable X is given by*

$$p(x) = {}^nC_x p^x q^{n-x},\ x = 0, 1, 2, \ldots, n.$$

Find moment generating function of X and hence find first two moments.

Solution. The m.g.f. of X is

$$M_x(t) = E\left[e^{tX}\right] = \sum_{x=0}^{n} e^{tx}p(x) = \sum_{x=0}^{n} {}^nC_x(pe^t)^x q^{n-x} = (q + pe^t)^n.$$

Now,

$$\mu_1' = \left.\frac{dM_x(t)}{dt}\right|_{t=0} = n(q+pe^t)^{n-1}pe^t\Big|_{t=0} = np$$

$$\mu_2' = \left.\frac{d^2M_x(t)}{dt^2}\right|_{t=0} = \left[n(n-1)p(q+pe^t)^{n-2}pe^t + npe^t(q+pe^t)^{n-1}\right]\Big|_{t=0}$$

$$= n(n-1)p^2 + np.$$

$$\therefore\ \mu_2 = \mu_2' - \mu_1'^2 = (n(n-1)p^2 + np - (np)^2 = npq.$$

►**EXAMPLE 3.21** *The probability function of a discrete random variable X is given below:*

X	0	1	2	3
$p(x)$	0.4	0.2	0.3	0.1

Find first four moments of the distribution and hence comment on its shape characteristics.

Solution. We have

$$\begin{aligned}
\mu_1' &= E(X) = \sum xp(x) = 0\times 0.4 + 1\times 0.2 + 2\times 0.3 + 3\times 0.1 = 1.1\\
\mu_2' &= E(X^2) = \sum x^2p(x) = 0^2\times 0.4 + 1^2\times 0.2 + 2^2\times 0.3 + 3^2\times 0.1 = 2.3\\
\mu_3' &= E(X^3) = \sum x^3p(x) = 0^3\times 0.4 + 1^3\times 0.2 + 2^3\times 0.3 + 3^3\times 0.1 = 5.3\\
\mu_4' &= E(X^4) = \sum x^4p(x) = 0^4\times 0.4 + 1^4\times 0.2 + 2^4\times 0.3 + 3^4\times 0.1 = 13.1\\
\mu_2 &= \mu_2' - \mu_1'^2 = 2.3 - (1.1)^2 = 1.09\\
\mu_3 &= \mu_3' - 3\mu_2'\mu_1' + 2\mu_1'^3 = 5.3 - 3\times 2.3\times 1.1 + 2(1.1)^3 = -0.09\\
\mu_4 &= \mu_4' - 4\mu_3'\mu_1' + 6\mu_2'\mu_1'^2 - 3\mu_1'^4\\
&= 13.1 - 4\times 5.3\times 1.1 + 6\times 2.3(1.1)^2 - 3(1.1)^4 = 2.0857\\
\beta_1 &= \frac{\mu_3^2}{\mu_2^3} = \frac{(-0.09)^2}{(1.09)^3} = 0.00625\\
\gamma_1 &= \sqrt{\beta_1} = -0.079\ (\because\ \mu_3 < 0)\\
\beta_2 &= \frac{\mu_4}{\mu_2^2} = \frac{2.0857}{(1.09)^2} = 1.76 < 3.
\end{aligned}$$

The distribution is negatively skewed and platykurtic.

►**EXAMPLE 3.22** *The p.d.f. of a continuous random variable X is given below:*

$$f(x) = \begin{cases} 2x,\ 0 < x < 1\\ 0,\ \ \textit{otherwise.}\end{cases}$$

Find first four moments of X and hence comment on the shape characteristics of the distribution of X.

Solution. We have

$$\mu_1' = E(X) = \int x f(x)dx = 2\int_0^1 x \cdot x dx = \frac{2}{3}$$

$$\mu_2' = E(X^2) = \int x^2 f(x)dx = 2\int_0^1 x^2 \cdot x dx = \frac{1}{2}$$

$$\mu_3' = E(X^3) = \int x^3 f(x)dx = 2\int_0^1 x^3 \cdot x dx = \frac{2}{5}$$

$$\mu_4' = E(X^4) = \int x^4 f(x)dx = 2\int_0^1 x^4 \cdot x dx = \frac{1}{3}$$

$$\mu_2 = \mu_2' - \mu_1'^2 = \frac{1}{2} - \left(\frac{2}{3}\right)^2 = \frac{5}{18}$$

$$\mu_3 = \mu_3' - 3\mu_2'\mu_1' + 2\mu_1'^3 = \frac{2}{5} - 3\cdot\frac{1}{2}\cdot\frac{2}{3} + 2\left(\frac{2}{3}\right)^3 = \frac{-1}{135}$$

$$\mu_4 = \mu_4' - 4\mu_3'\mu_1' + 6\mu_2'\mu_1'^2 - 3\mu_1'^4$$

$$= \frac{1}{3} - 4\times\frac{2}{5}\times\frac{2}{3} + 6\times\frac{1}{2}\left(\frac{2}{3}\right)^2 - 3\times\left(\frac{2}{3}\right)^4 = \frac{1}{135}$$

$$\beta_1 = \frac{\mu_3^2}{\mu_2^3} = \frac{(-1/135)^2}{(5/18)^3} = 0.00256$$

$$\gamma_1 = \sqrt{\beta_1} = -0.05 \ (\because \ \mu_3 < 0)$$

$$\beta_2 = \frac{\mu_4}{\mu_2^2} = \frac{(1/135)}{(5/18)^2} = 0.096 < 3.$$

The distribution is negatively skewed and platykurtic.

3.11 Joint Moment Generating Function

Let $X_1, X_2, \ldots, X_k$ be k-dimensional random variables having joint p.d.f. $f(x_1, x_2, \ldots, x_k)$. Then the joint moment generating function of $X_1, X_2, \ldots, X_k$ is given by

$$M_{x_1,x_2,\ldots,x_k}(t_1, t_2, \ldots, t_k)$$
$$= E\left[e^{t_1X_1}e^{t_2X_2}\cdots e^{t_kX_k}\right]$$
$$= \int_{-\infty}^{\infty}\int_{-\infty}^{\infty}\cdots\int_{-\infty}^{\infty}\exp\left(\sum_{j=1}^{k} x_j t_j\right) f(x_1 x_2 \cdots x_k)dx_1\cdots dx_k.$$

However, if X_j's are discrete random variables with joint probability function $p(x_1, x_2, \ldots, x_k)$, then

$$M_{x_1,x_2,\ldots,x_k}(t_1, t_2, \ldots, t_k) = E\left[\exp\sum_{j=1}^{k} X_j t_j\right]$$

$$= \sum_{x_1}\sum_{x_2}\cdots\sum_{x_k}\exp\left[\sum_{j=1}^{k} x_j t_j\right] p(x_1, x_2, \ldots, x_k)$$

Here also the moments are found out by differentiating joint moment generating function with respect to t_j's and putting $t_j = 0$.

►**EXAMPLE 3.23** *The joint p.d.f. of three-dimensional random variables X, Y and Z is given by*

$$f(x, y, z) = e^{-(x+y+z)};\ x > 0,\ y > 0,\ z > 0.$$

Find joint m.g.f. of X, Y and Z, and hence find moments of X, Y and Z.

Solution. The joint m.g.f. of X, Y and Z is given by

$$\begin{aligned} M_{x,y,z}(t_1, t_2, t_3) &= E\left[e^{(t_1x+t_2y+t_3z)}\right] \\ &= \int_0^\infty\int_0^\infty\int_0^\infty \exp(t_1x + t_2y + t_3z) f(x, y, z)dxdydz \\ &= \int_0^\infty\int_0^\infty\int_0^\infty e^{t_1x}e^{-x}dxe^{t_2y}e^{-y}dye^{t_3z}e^{-z}dxdydz \\ &= \int_0^\infty\int_0^\infty\int_0^\infty e^{-x(1-t_1)}e^{-y(1-t_2}e^{-z(1-t_3)}dxdydz \\ &= (1-t_1)^{-1}(1-t_2)^{-1}(1-t_3)^{-1} \end{aligned}$$

Now,

$$\begin{aligned} \mu_1'(x) &= \left.\frac{dM_{x,y,z}(t_1, t_2, t_3)}{dt_1}\right|_{t_1=0,t_2=0,t_3=0} = (1-t_1)^{-2}\Big|_{t_1=0,t_2=0,t_3=0} = 1 \\ \mu_2'(x) &= \left.\frac{d^2M_{x,y,z}(t_1, t_2, t_3)}{dt_1^2}\right|_{t_1=0,t_2=0,t_3=0} = -2(1-t_1)^{-3}\Big|_{t_1=0,t_2=0,t_3=0} = 2 \\ \therefore\ \mu_{2(x)} &= \mu'_{2(x)} - \mu'^2_{1(x)} = 2 - (1)^2 = 1 \\ E(XY) &= \left.\frac{dM_{x,y,z}(t_1, t_2, t_3)}{dt_1dt_2}\right|_{t_1=0,t_2=0,t_3=0} = (1-t_1)^{-2}(1-t_2)^{-2}\Big|_{t_1=0,t_2=0} \\ &= 1\cdot 1 = 1 \\ \mu'_{1(y)} &= \left.\frac{dM_{x,y,z}(t_1, t_2, t_3)}{dt_2}\right|_{t_1=0,t_2=0,t_3=0} = (1-t_2)^{-2}\Big|_{t_1=0,t_2=0,t_3=0} = 1 \\ \therefore\ \text{cov}(XY) &= E(XY) - E(X)E(Y) = 1 - 1\cdot 1 = 0 \\ \mu'_{2(y)} &= \left.\frac{d^2M_{x,y,z}(t_1, t_2, t_3)}{dt_2^2}\right|_{t_1=0,t_2=0,t_3=0} = 2(1-t_2)^{-3}\Big|_{t_1=0,t_2=0,t_3=0} = 2 \\ \mu_{2(y)} &= \mu'_{2(y)} - \mu'^2_{1(y)} = 2 - (1)^2 = 1 \\ \mu'_{1(z)} &= \left.\frac{dM_{x,y,z}(t_1, t_2, t_3)}{dt_3}\right|_{t_1=0,t_2=0,t_3=0} = (1-t_3)^{-2}\Big|_{t_1=0,t_2=0,t_3=0} = 1 \end{aligned}$$

$$\mu'_{2(z)} = \left.\frac{d^2M_{x,y,z}(t_1,t_2,t_3)}{dt_3^2}\right|_{t_1=0,t_2=0,t_3=0} = 2(1-t_3)^{-2}\Big|_{t_1=0,t_2=0,t_3=0} = 2$$

$$\mu_{2(z)} = \mu'_{2(z)} - \mu'^2_{1(z)} = 2 - (1)^2 = 1$$

$$E(XZ) = \left.\frac{dM_{x,y,z}(t_1,t_2,t_3)}{dt_1, dt_2}\right|_{t_1=0,t_2=0,t_3=0} = (1-t_1)^{-2}(1-t_3)^{-2}\Big|_{t_1=0,t_2=0,t_3=0}$$

$$= 1\cdot 1 = 1$$

$$\text{cov}(XZ) = E(XZ) - E(X)E(Z) = 1 - 1\cdot 1 = 0$$

$$E(YZ) = \left.\frac{dM_{x,y,z}(t_1,t_2,t_3)}{dt_2dt_3}\right|_{t_1=0,t_2=0,t_3=0} = 1\cdot 1 = 1$$

$$\text{cov}(YZ) = E(YZ) - E(Y)E(Z) = 1 - 1\cdot 1 = 0.$$

Therefore, X, Y and Z are mutually independent.

3.12 Characteristics of Moment Generating Function

1. Let X be a random variable and A is any constant, then

$$M_{ax}(t) = M_x(At).$$

We have

$$M_{ax}(t) = E\left[e^{AtX}\right] = M_x(At).$$

2. Let $X_1, X_2, \ldots, X_k$ be k-independent random variables and

$$Y = g_1(x_1, x_2, \ldots, x_k) = \sum_{j=1}^{k} X_j$$

be a function of X_j's. Then m.g.f. of Y is given by

$$M_y(t) = M_{x_1}(t)M_{x_2}(t)\cdots M_{x_k}(t).$$

We have

$$\begin{aligned} M_y(t) &= E\left[e^{tY}\right] = E\left[e^{t(X_1+X_2+\cdots+X_k)}\right] \\ &= E\left[e^{tX_1}\right]E\left[e^{tX_2}\right]\cdots E\left[e^{tX_k}\right] \quad [\because X_j\text{'s are independent}] \\ &= M_{x_1}(t)M_{x_2}(t)\cdots M_{x_k}(t). \end{aligned}$$

3. Let X be a random variable and

$$Y = g(X) = \frac{x-a}{h}$$

be a function of X. Then $M_y(t) = e^{-at/h}M_x(t/h)$.

We have

$$M_y(t) = E\left[e^{tY}\right] = E\left[e^{t\left(\frac{X-a}{h}\right)}\right] = E\left[e^{tX/h}e^{-at/h}\right]$$
$$= e^{-at/h}E\left[e^{tX/h}\right] = e^{-at/h}M_x(t/h).$$

4. The moment generating functions of some random variables exist but moments do not exist.

For example, let us consider the discrete random variable X having probability function

$$p(x) = \begin{cases} \frac{1}{x(x+1)}; \ x = 1, 2, \ldots \\ 0, \quad \text{otherwise.} \end{cases}$$

Then

$$E(X) = \sum_{x=1}^{\infty} xp(x) = \sum_{x=1}^{\infty} \frac{1}{x+1}$$
$$= \frac{1}{2} + \frac{1}{3} + \frac{1}{4} + \cdots = \sum_{x=1}^{\infty} \frac{1}{x} - 1.$$

But $\sum_{x=1}^{\infty} \frac{1}{x}$ is a divergent series and $E(X)$ does not exist. As a result no other moments are available. However, moment generating function of X exists. Here

$$M_x(t) = E\left[e^{tX}\right] = \sum_{x=1}^{\infty} \frac{e^{tx}}{x(1+x)}.$$

Let $z = e^t$, then

$$M_x(t) = \sum_{x=1}^{\infty} \frac{z^x}{x(1+x)} = \frac{z}{1 \cdot 2} + \frac{z^2}{2 \cdot 3} + \frac{z^3}{3 \cdot 4} + \cdots$$
$$= z\left(1 - \frac{1}{2}\right) + z^2\left(\frac{1}{2} - \frac{1}{3}\right) + z^3\left(\frac{1}{3} - \frac{1}{4}\right) + \cdots$$
$$= \left(z + \frac{z^2}{2} + \frac{z^3}{3} + \cdots\right) - \left(\frac{z}{2} + \frac{z^2}{3} + \frac{z^3}{4} + \cdots\right)$$
$$= -\log(1-z) - \frac{1}{z}\left(\frac{z^2}{2} + \frac{z^3}{3} + \frac{z^4}{4} + \cdots\right)$$
$$= -\log(1-z) + 1 + \frac{1}{z}\log(1-z), \ |z| < 1$$
$$= 1 + \log(1-z)\left(\frac{1}{z} - 1\right), \ |z| < 1$$
$$= 1 + (e^{-t} - 1)\log(1 - e^t) \ [\because \ |z| < 1] = |e^t| < 1.$$

But, if $t = 0$, then $M_x(t) = 1$.

However, if $t > 0$, $M_x(t)$ does not exist.

5. The moments and moment generating function of a random variable may exist. But moments are not found out from moment generating function.

Let us consider the random variable whose p.d.f. is

$$f(x) = \begin{cases} 2x, \; 0 < x < 1 \\ 0, \text{ otherwise.} \end{cases}$$

We have

$$\mu_r' = E(X^r) = \int_0^1 x^r f(x)dx = 2\int_0^1 x^{r+1}dx = \frac{2}{r+2}.$$

Here μ_r' is available for all real values of r.

The moment generating function of X is

$$M_x(t) = E\left[e^{tX}\right] = 2\int_0^1 xe^{tx}dx = e^t\frac{t-1}{t^2}.$$

The moments are not available from this moment generating function.

Again

$$P(X = \pm 2^x) = \begin{cases} \frac{e^{-1}}{2x!}; \; x = 0, 1, 2, \ldots \\ 0, \; \text{ otherwise.} \end{cases}$$

The even order moments of this random variable X is

$$\mu_{2r}' = E\left[X^{2r}\right] = \sum_{x=0}^{\infty}[\pm 2^x]^{2r}\frac{e^{-1}}{2x!} = \frac{1}{2e}\sum_{x=0}^{\infty}\frac{(2r)^x}{x!}$$

$$= \frac{1}{2e}\exp e^{2r} = \exp\left[2^{2r} - 1\right].$$

The distribution is symmetric with respect to $X = 0$. Thus $\mu_{2r+1} = 0$.

The moment generating function of X is

$$M_x(t) = \sum_{x=0}^{0}\left(e^{t2^x} + e^{-t2^x}\right)\frac{e^{-1}}{2x!} = e^{-1}\sum_{x=0}^{\infty}\frac{\cosh(t2^x)}{x!}.$$

The series is convergent at $t = 0$. Thus $M_x(t)$ exists at $t = 0$. This means that some moments of a random variable are available but moment generating function does not exist.

6. The moment generating function is used to derive the probability function of a random variable.

Let

$$X \sim N(\mu, \sigma^2) \text{ and } Z = \frac{1}{2}\left(\frac{X-\mu}{\sigma}\right)^2.$$

We need to find the density function of Z. The m.g.f. of Z is

$$\begin{aligned} M_z(t) &= E\left[e^{tZ}\right] \\ &= E\left[\exp t\frac{1}{2}\left(\frac{X-\mu}{\sigma}\right)^2\right] \\ &= \frac{1}{\sigma\sqrt{2\pi}}\int_{-\infty}^{\infty}\exp\frac{1}{2}\left(\frac{X-\mu}{\sigma}\right)^2 t\exp-\frac{1}{2}\left(\frac{x-\mu}{\sigma}\right)^2 dx \end{aligned}$$

Let

$$y = \frac{x-\mu}{\sigma} \Rightarrow x = \mu + \sigma y,\ dx = \sigma dy,\ -\infty < y < \infty.$$

$$\therefore\ M_z(t) = \frac{1}{\sqrt{2\pi}}\int_{-\infty}^{\infty} e^{-\frac{1}{2}y^2(1-t)}dy.$$

Again, let

$$\begin{aligned} y\sqrt{1-t} &= u \Rightarrow y = \frac{u}{\sqrt{1-t}} \\ dy &= \frac{du}{\sqrt{1-t}},\ -\infty < u < \infty. \\ \therefore\ M_z(t) &= \frac{1}{\sqrt{2\pi}}\int_{-\infty}^{\infty} e^{-\frac{1}{2}u^2}\frac{du}{\sqrt{1-t}} = (1-t)^{-1/2} \end{aligned}$$

This $M_z(t)$ is the m.g.f. of gamma distribution with parameter $\frac{1}{2}$. Hence

$$g(z) = \frac{1}{\Gamma\frac{1}{2}}e^{-z}z^{\frac{1}{2}-1},\ z > 0.$$

3.13 Cumulant and Cumulant Generating Function

Let X be a random variable and $M_x(t)$ be the m.g.f. of X. Then the cumulant generating function (c.g.f.) of X is defined by

$$k_x(t) = \ln M_x(t) = \ln E\left[e^{tX}\right]$$

provided $E\left[e^{tX}\right]$ exists, i.e., $\ln M_x(t)$ is to be expanded in powers of t and the series is convergent. Here

$$\begin{aligned} k_x(t) &= \ln M_x(t) \\ &= \ln\left[1 + t\mu_1' + \frac{t^2}{2!}\mu_2' + \frac{t^3}{3!}\mu_3' + \frac{t^4}{4!}\mu_4' + \cdots\right] \\ &= \left(t\mu_1' + \frac{t^2}{2!}\mu_2' + \frac{t^3}{3!}\mu_3' + \frac{t^4}{4!}\mu_4' + \cdots\right) - \frac{1}{2}\left(t\mu_1' + \frac{t^2}{2!}\mu_2' + \cdots\right)^2 \\ &\quad + \frac{1}{3}\left(t\mu_1' + \frac{t^2}{2}\mu_2' + \cdots\right)^3 - \frac{1}{4}\left(t\mu_1' + \frac{t^2}{2!}\mu_2' + \cdots\right)^4 + \cdots \\ &= k_1 t + k_2\frac{t^2}{2!} + k_3\frac{t^3}{3!} + k^4\frac{t^4}{4!} + \cdots \end{aligned}$$

$$
\begin{aligned}
\text{where} \quad k_1 &= \text{coefficient of } t = \mu_1' \\
k_2 &= \text{coefficient of } t^2/2! = \mu_2' - \mu_1'^2 = \mu_2 \\
k_3 &= \text{coefficient of } t^3/3! = \mu_3' - 3\mu_2'\mu_1' + 2\mu_1'^3 = \mu_3 \\
k_4 &= \text{coefficient of } t^4/4! = \mu_4' - 4\mu_3'\mu_1' + 6\mu_2'\mu_1'^2 - 3\mu_1'^4 - 3(\mu_2' - \mu_1'^2)^2 \\
&= \mu_4 - 3\mu_2^2 \\
\therefore \; \mu_4 &= k_4 + 3\mu_2^2
\end{aligned}
$$

Thus k_r = coefficient of $t^r/r!$ in $k_x(t)$.

Here k_r = r-th cumulant of X. The value of k_r is also found out from r-th derivative of $k_x(t)$ and putting $t = 0$. That is

$$k_r = \left.\frac{d^r k_x(t)}{dt^r}\right|_{t=0}.$$

►**EXAMPLE 3.24** *The p.d.f. of X is given by*

$$f(x) = \begin{cases} \frac{1}{\lambda} e^{-x/\lambda}; & x > 0, \; \lambda > 0 \\ 0, & \textit{otherwise.} \end{cases}$$

Find β_1 and β_2 of the distribution using cumulants and comment on the shape characteristics of the distribution.

Solution. The m.g.f. of X is

$$
\begin{aligned}
M_x(t) &= E\left[e^{tX}\right] \\
&= \int_0^\infty e^{tx} f(x)dx = \frac{1}{\lambda}\int_0^\infty e^{-x/\lambda} e^{tx}\,dx \\
&= \frac{1}{\lambda}\int_0^\infty e^{-x\left(\frac{1-\lambda t}{\lambda}\right)}dx = (1-\lambda t)^{-1}.
\end{aligned}
$$

The c.g.f. of X is

$$
\begin{aligned}
k_x(t) &= \ln M_x(t) = \ln(1-\lambda t)^{-1} = -\ln(1-\lambda t) \\
&= \lambda t + \frac{\lambda^2 t^2}{2} + \frac{\lambda^3 t^3}{3} + \frac{\lambda^4 t^4}{4} + \cdots.
\end{aligned}
$$

We know

$$k_r = \text{coefficient of } \frac{t^r}{r!} \text{ in } k_x(t).$$

Thus

$$
\begin{aligned}
k_1 &= \text{coefficient of } t = \lambda = \mu_1' \\
k_2 &= \text{coefficient of } t^2/2! = \lambda^2 = \mu_2 \\
k_3 &= \text{coefficient of } t^3/3! = 2\lambda^3 = \mu_3 \\
k_4 &= \text{coefficient of } t^4/4! = 6\lambda^4 = \mu_4 - 3\mu_2^2 \\
\therefore \; \mu_4 &= k_4 + 3\mu_2^2 = 6\lambda^4 + 3(\lambda^2)^2 = 9\lambda^4
\end{aligned}
$$

Hence

$$\beta_1 = \frac{\mu_3^2}{\mu_2^3} = \frac{4\lambda^6}{\lambda^6} = 4, \ \gamma_1 = \sqrt{\beta_1} = +2 \ [\because \ \mu_3 > 0]$$

$$\beta_2 = \frac{\mu_4}{\mu_2^2} = \frac{9\lambda^4}{\lambda^4} = 9 > 3.$$

Therefore, the distribution is positively skewed and leptokurtic.

►**EXAMPLE 3.25** *The p.d.f. of a random variable X is given by*

$$f(x) = \frac{1}{\sigma\sqrt{2\pi}} e^{-\frac{1}{2}\left(\frac{x-\mu}{\sigma}\right)^2}, \ -\infty < x < \infty.$$

Find β_1 and β_2 of the distribution using cumulants, and comment on the shape characteristics of the distribution.

Solution. The m.g.f. of X is

$$M_x(t) = E\left[e^{tX}\right] = \int_{-\infty}^{\infty} e^{tx} f(x)dx = \frac{1}{\sigma\sqrt{2\pi}} \int_{-\infty}^{\infty} e^{tx} e^{-\frac{1}{2}\left(\frac{x-\mu}{\sigma}\right)^2} dx,$$

Let

$$z = \frac{x-\mu}{\sigma} \Rightarrow x = \mu + \sigma z, \ dx = \sigma dz, \ -\infty < z < \infty.$$

$$\begin{aligned}
\therefore \ M_x(t) &= \frac{1}{\sqrt{2\pi}} \int_{-\infty}^{\infty} e^{t(\mu+\sigma z)} e^{-\frac{1}{2}z^2} dz \\
&= \frac{e^{\mu t + \frac{1}{2}t^2\sigma^2}}{\sqrt{2\pi}} \int_{-\infty}^{\infty} e^{-\frac{1}{2}(z^2 - 2\sigma z + t^2\sigma^2)} dz = \frac{e^{\mu t + \frac{1}{2}t^2\sigma^2}}{\sqrt{2\pi}} \int_{-\infty}^{\infty} e^{-\frac{1}{2}(z-t\sigma)^2} dz.
\end{aligned}$$

Let $z - t\sigma = u \Rightarrow dz = du, \ -\infty < u < \infty.$

$$\therefore \ M_x(t) = \frac{e^{\mu t + \frac{1}{2}\sigma^2 t^2}}{\sqrt{2\pi}} \int_{-\infty}^{\infty} e^{-\frac{1}{2}u^2} du = e^{\mu t + \frac{1}{2}\sigma^2 t^2}.$$

The c.g.f. of X is $k_x(t) = \ln M_x(t) = \ln e^{\mu t + \frac{1}{2}t^2\sigma^2} = \mu t + \frac{1}{2}t^2\sigma^2$.
Therefore,

$$\begin{aligned}
k_1 &= \text{coefficient of } t \text{ in } k_x(t) = \mu = \mu_1' \\
k_2 &= \text{coefficient of } t^2/2! \text{ in } k_x(t) = \sigma^2 = \mu_2 \\
k_3 &= \text{coefficient of } t^3/3! \text{ in } k_x(t) = 0 = \mu_3.
\end{aligned}$$

Also,

$$k_4 = 0 \Rightarrow \mu_4 = k_4 + 3\mu_2^2 = 3\sigma^4.$$

Now,

$$\beta_1 = \frac{\mu_3^2}{\mu_2^3} = 0, \ \gamma_1 = \sqrt{\beta_1} = 0 \ [\because \ \mu_3 = 0]$$

$$\beta_2 = \frac{\mu_4}{\mu_2^2} = \frac{3\sigma^4}{\sigma^4} = 3.$$

Therefore, the distribution is symmetric and mesokurtic.

►**EXAMPLE 3.26** *The r-th raw moment (moment about zero) of a random variable X is given below:*

$$\mu'_r = (r+1)!\ 2^r.$$

Find cumulants of the variable X and comment on the shape characteristics of the distribution of X.

Solution. The m.g.f. of X is given by

$$M_x(t) = \sum_{r=0}^{\infty} \frac{t^r}{r!}\mu'_r$$

Thus, we have

$$\begin{aligned} M_x(t) &= \sum_{r=0}^{\infty} \frac{t^r}{r!}(r+1)!2^r = \sum_{r=0}^{\infty}(2t)^r(r+1) \\ &= 1 + 2\cdot(2t) + 3\cdot(2t)^2 + 4\cdot(2t)^3 + 5\cdot(2t)^4 + \cdots \\ &= (1-2t)^{-2} \end{aligned}$$

Hence the cumulant generating function of X is

$$\begin{aligned} k_x(t) &= \ln(1-2t)^{-2} = -2\ln(1-2t) \\ &= 2\left[2t + \frac{(2t)^2}{2} + \frac{(2t)^3}{3} + \frac{(2t)^4}{4} + \cdots\right] \end{aligned}$$

Here

$$\begin{aligned} k_1 &= \text{coefficient of } t \text{ in } k_x(t) = 4 = \mu'_1 \\ k_2 &= \text{coefficient of } t^2/2! \text{ in } k_x(t) = 8 = \mu_2 \\ k_3 &= \text{coefficient of } t^3/3! \text{ in } k_x(t) = 32 = \mu_3 \\ k_4 &= \text{coefficient of } t^4/4! \text{ in } k_x(t) = 192 \Rightarrow \mu_4 = k_4 + 3\mu_2^2 = 384 \end{aligned}$$

Hence

$$\begin{aligned} \beta_1 &= \frac{\mu_3^2}{\mu_2^3} = \frac{(32)^2}{(8)^3} = 2 \\ \gamma_1 &= \sqrt{2} = 1.41 > 0\ [\because\ \mu_3 > 0] \\ \beta_2 &= \frac{\mu_4}{\mu_2^2} = \frac{384}{(8)^2} = 6 > 3. \end{aligned}$$

The distribution is positively skewed and leptokurtic.

3.14 Properties of Cumulant Generating Function

1. Let $X_1, X_2, \ldots, X_k$ be k-independent random variables and

$$Y = g(X) = \sum_{j=1}^{k} X_j$$

be a function of X_j's. Then

$$k_y(t) = \sum_{j=1}^{k} \ln M_{x_j}(t_j).$$

Since $M_y(t) = M_{(x_1+x_2+\cdots+x_k)}(t) = M_{x_1}(t)M_{x_2}(t)\cdots M_{x_k}(t)$, then

$$\begin{aligned} k_y(t) = \ln M_y(t) &= \ln\left[M_{x_1}(t), M_{x_2}(t)\ldots M_{x_k}(t)\right] \\ &= \sum_{j=1}^{k} \ln M_{x_j}(t). \end{aligned}$$

2. Let X be a random variable and $Y = g(X)$ be a function of X such that

$$Y = g(X) = \frac{X-a}{h}$$

where a and h are any two constants. Then

$$k_y(t) = \frac{k_x(t)-a}{h} \text{ and } k_r(y) = \frac{k_r(x)}{h^r},\ r = 2, 3, \ldots .$$

Here

$$Y = \frac{X-a}{h} \text{ and } M_y(t) = e^{-at/h}M_x(t/h).$$

$$\therefore\ k_y(t) = \ln M_y(t) = -\frac{ah}{h} + \ln M_x\left(\frac{t}{h}\right).$$

Expanding both sides, we have

$$\begin{aligned} & k_1'(t) + \frac{t^2}{2!}k_2' + \frac{t^3}{3!}k_3' + \frac{t^4}{4!}k_4' + \cdots \\ &= -\frac{at}{h} + k_1\left(\frac{t}{h}\right) + k_2\left(\frac{t}{h}\right)^2\frac{1}{2!} + k_3\left(\frac{t}{h}\right)^3\frac{1}{3!} + k_4\left(\frac{t}{h}\right)^4\frac{1}{4!} + \cdots . \end{aligned}$$

Here k_r' and k_r are the r-th cumulants of Y and X, respectively. Equating the coefficients of $\frac{t^r}{r!}$ from both sides, we have

$$k_1' = \frac{k_1 - a}{h},\ k_r' = \frac{k_r}{h^r},\ r = 2, 3, \ldots .$$

►**EXAMPLE 3.27** *The joint p.d.f. of X and Y is given by*

$$f(x,y) = e^{-(x+y)};\ x > 0,\ y > 0.$$

Find cumulants of $U = X + Y$, and hence β_1 and β_2 of U.

Solution. Here $f(x, y) = e^{-x}e^{-y}$; $x > 0$, $y > 0$. Therefore, X and Y are independent random variables.

We have

$$k_x(t) = \ln M_x(t) = -\ln(1-t) \text{ [Example 3.19]}.$$

Similarly,

$$\begin{aligned}
k_y(t) &= \ln M_y(t) = -\ln(1-t).\\
\therefore\ k_u(t) &= k_x(t) + k_y(t) = -2\ln(1-t) = 2\left(t + \frac{t^2}{2} + \frac{t^3}{3} + \frac{t^4}{4} + \cdots\right)\\
\therefore\ k_1 &= \mu_1' = 2,\ k_2 = \mu_2 = 2,\ k_3 = \mu_3 = 4,\\
k_4 &= 12 \Rightarrow \mu_4 = k_4 + 3\mu_2^2 = 12 + 12 = 24\\
\beta_1 &= \frac{\mu_3^2}{\mu_2^3} = \frac{4^2}{2^3} = 2\\
\beta_2 &= \frac{\mu_4}{\mu_2^2} = \frac{24}{4} = 6.
\end{aligned}$$

3.15 More Information on Expectation

CONTINUOUS CONVEX FUNCTION. Let X be a random variable and $g(x)$ be a function of X. The function $g(x)$ is called *continuous convex function* if for any interval I and for any value of x_1 and x_2, $(x_1 + x_2)/2 \in I$.

Thus we can write

$$g\left(\frac{x_1 + x_2}{2}\right) \leq \frac{1}{2}g(x_1) + \frac{1}{2}g(x_2).$$

This inequality can also be written in other form. Let $g(x)$ be measurable and $0 \leq \lambda \leq 1$. Then

$$g\left[\lambda x_1 + (1-\lambda)x_2\right] \leq \lambda g(x_1) + (1-\lambda)g(x_2).$$

It is noted that, if $\lambda = \frac{1}{2}$, then the latter inequality takes the form of former one. However, if $g(x)$ is not measurable, the latter inequality does not exist. It is also noted that if $x \in$ Interior of I, i.e., if $g''(x) \leq 0$, $g(x)$ is called *convex function* in the mid-point of interval. Again, if x_0 is any point in the interval I, then $y = a + bx$ passes through the interval $[x_0, g(x_0)]$ and for all $x \in I$, $g(x) \geq ax + b$.

CONTINUOUS CONCAVE FUNCTION. Any function $g(x)$ is called *continuous concave function* on any interval I if $[-g(x)]$ is convex.

Let us now discuss an important theorem based on convex function.

JENSEN INEQUALITY. Let us consider that $E(X)$ of a random variable X exists and $E(X)$ falls in the interval I. Also, consider that $g(X)$ is a convex function; then

$$E[g(X)] \geq g[E(X)].$$

Proof. The probable interval of X may be $I = [-\infty, \infty]$, $I = [a, \infty]$, $I = [-a, \infty]$, $I = [-\infty, b]$, $I = [a, b]$, etc. We need to show that $E(X) \in I$.

If $E(X)$ exists, then $-\infty < E(X) < \infty$. Again, if $P[X \geq a] = 1$, $E(X) \geq a$ and if $P[X \leq b] = 1$, then $E(X) \leq b$. Hence $E(X) \in I$ and $E(X)$ will be the end point of left side or right side of I or mid-point of I. Let us consider that the end point of left side of I is a, then if $X \geq a$ and $E(X) = a$,

$$\begin{aligned} P[X - a \geq 0] &= 1 \Rightarrow P(X = a) = 1 \text{ or, } P[X - a = 0] = 1 \\ \therefore\; E[g(X)] &= E[g(a)] \; [\because\; P[g(X) = g(a)] = 1 \\ &= g(a) \; [\because\; g(a) \text{ is a constant}] \\ &= g[P(X)]. \end{aligned}$$

If the end of right-hand side of I is b and $E(X) = b$, similar result is observed. It has already been mentioned that if x_0 is a point in the mid-point of interval I and if $y = a + bx$ passes through the interval $[x_0, g(x_0)]$ and if for all values of $x \in I$, $g(x) \geq ax + b$. Therefore,

$$\begin{aligned} E[g(X)] \geq E[(ax + b)] &= aE(X) + b = ax_0 + b \\ &= g(x_0) = g[E(X)] \Rightarrow E[g(X)] \geq g[E(X)] \end{aligned}$$

The equality sign holds good, if and only if $P[g(X) = aX + b] = 1$.

COROLLARY 6. Let X be a random variable such that the probability of falling all values of X in an interval I is 1. In that case, if $g(X)$ is a concave function over I, then

$$E[g(X)] \leq g[E(X)].$$

This is true if $E(X)$ exists.

REMARK 3. Let X be a random variable and $g(X)$ be a function of X such that $g(X) = X^2$. This implies that $g''(X) = 2 > 0$. Then $g(X)$ is a convex function. Now, if $E(X^2)$ exists, then

$$E(X^2) \geq [E(X)]^2.$$

REMARK 4. Let the values of X be positive $[X > 0]$. Consider that $E(X)$ and $E[1/X]$ exist. Then

$$E[1/X] \geq 1/E(X).$$

This is true, since $g(X) = 1/X$ implies that $g''(X) = \frac{2}{X^2} > 0$. Therefore, $g(X)$ is a convex function.

REMARK 5. Let the values of random variable X be positive $[X > 0]$ and $g(X) = \sqrt{X}$ such that $g''(X) = -\frac{1}{4}X^{-3/2} < 0$ $[\because\; X > 0]$. Hence $g(X)$ is a concave function. Thus

$$E\left[\sqrt{X}\right] = \sqrt{E(X)}.$$

REMARK 6. Let the values of random variable X be positive $[X > 0]$ and $g(X) = \ln X$. Since $X > 0, \ln X$ is a concave function. If $E[X]$ exists,

$$E[\ln X] \leq \ln[E(X)].$$

REMARK 7. Consider $g(X) = E\left[e^{tX}\right]$ for a random variable X and is true for all values of X and t. Then, if $E\left[e^{tX}\right]$ and $E(X)$ exist,

$$E\left[e^{tX}\right] \geq e^{tE(X)}.$$

In other words, $M_x(t) \geq e^{tE(X)}$. If $E(X) = 0, M_x(t) \geq 1 \; \forall \; t$. This implies that, if $M_x(t)$ exists, its minimum value will be 1. In such a case $E(X) = 0$. The lower limit of $M_x(t)$ is available at a point $t = 0$.

❑ **THEOREM 3.3** *Let X be a random variable, $f(X)$ and $g(X)$ be the monotone functions over a subset of any real line and the range of X falls within this subset. Then if $E[f(X)]$ and $E[g(X)]$ exist,*

$$E[f(X) \cdot g(X)] \geq E[f(X)] \; E[g(X)]$$

if $f(X)$ and $g(X)$ are the monotone functions in the same direction. If these functions are in opposite directions,

$$E[f(X) \cdot g(X)] \leq E[f(X)] \; E[g(X)].$$

Proof. Let us consider that $f(X)$ and $g(X)$ be the monotone functions in the same direction. Also, consider that x and y fall on the domain of $f(X)$ and $g(X)$, respectively. If $f(X)$ and $g(X)$ are monotone increasing function, then

$$\begin{aligned} & y \geq x \Rightarrow f(y) \geq f(x) \text{ and } g(y) \geq g(x) \\ \text{or, } & f(y) - f(x) \geq 0 \text{ and } g(y) - g(x) \geq 0 \\ \text{or, } & [f(y) - f(x)], \; [g(y) - g(x)] \geq 0 \end{aligned}$$

Again, if $f(X)$ and $g(X)$ are the monotone decreasing functions for $y \geq x$, then

$$\begin{aligned} & f(y) \leq f(x) \text{ and } g(y) \leq g(x) \\ \text{or, } & f(y) - f(x) \leq 0 \text{ and } g(y) - g(x) \leq 0 \\ \text{or, } & f(x) - f(y) \geq 0 \text{ and } g(x) - g(y) \geq 0 \\ \text{or, } & [f(y) - f(x)], \; [g(y) - g(x)] \geq 0 \end{aligned}$$

It is noted that the similar results are observed for both monotone increasing and decreasing functions of $f(X)$ and $g(X)$.

Let us consider that X and Y be independent and identically distributed random variables. Then

$$\begin{aligned} & E[\{f(Y) - f(X)\}\{g(Y) - g(X)\}] \geq 0 \\ \text{or, } & E[f(Y) \; g(Y)] - E[f(Y) \; g(X)] - E[f(X) \; g(Y)] + E[f(X) \; g(X)] \geq 0 \\ \text{or, } & E[f(Y) \; g(Y)] - E[f(Y) \; g(X)] - E[f(X) \; g(Y)] - E[f(X) \; g(X)] \geq 0 \qquad \text{(a)} \end{aligned}$$

Again, as X and Y are independent and identically distributed random variables,

$$\begin{aligned} E[f(Y) \; g(Y)] &= E[f(X) \; g(X)] \\ E[f(X) \; g(Y)] &= E[f(X)] \; E[g(X)] = E[f(X)] \; E[g(X)] \end{aligned}$$

Putting these values in (a), we get

$$2E[f(X)\ g(X)] - 2E[f(X)]\ E[g(X)] \geq 0$$
$$E[f(X)\ g(X)] \geq E[f(X)]\ E[g(X)]$$

The theorem can be proved in a similar way if $f(X)$ and $g(X)$ are monotone functions in opposite direction.

Let us now note the expected values based on the above theorem for special values of $f(X)$ and $g(X)$.

(i) Let X be a random variable such that $X > 0$. If $E(X)$ exists, then for $\alpha > 0$, $\beta > 0$, $f(X) = X^{\alpha}$ and $g(X) = X^{\beta}$

$$E\left[X^{\alpha}, X^{\beta}\right] \geq E\left[X^{\alpha}\right] E\left[X^{\beta}\right]$$
$$\text{or, } E\left[X^{\alpha+\beta}\right] \geq E\left[X^{\alpha}\right] E\left[X^{\beta}\right].$$

However, if $\alpha = \beta = 1$, then

$$E(X^2) \geq \{E(X)\}^2.$$

(ii) Consider that the random variable X takes only positive values $[X > 0]$. Also, consider that $\alpha > 0$, $E[X^{\alpha}]$ and $E[X^{-1}]$ exist. Then

$$E\left[X^{\alpha} \cdot X^{-1}\right] \leq E\left[X^{\alpha}\right] E\left[X^{-1}\right]$$
$$\text{or, } E\left[X^{\alpha}\right] E\left[X^{-1}\right] \geq E\left[X^{\alpha} \cdot X^{-1}\right]$$

If $\alpha = 1$, then

$$E[X]E[X^{-1}] \geq E[X^0] = 1.$$

If $\alpha = 2$, then

$$E\left[X^2\right] E\left[X^{-1}\right] \geq E\left[X^2 \cdot X^{-1}\right] = E[X]$$
$$\text{or, } E[X^2] \geq \frac{E[X]}{E[X^{-1}]} \geq \frac{1}{\{E\left[X^{-1}\right]\}^2}.$$

(iii) Consider that the moment generating function of the random variable X exists for all values of t, where the moment generating function is

$$M_x(t) = E\left[e^{tX}\right].$$

Then

$$M_x(u+v) = E\left[e^{(u+v)X}\right] = E\left[e^{uX}\right]\left[e^{vX}\right].$$

But

$$E\left\{\left[e^{uX}\right]\left[e^{vX}\right]\right\} \geq E\left[e^{uX}\right] E\left[e^{vX}\right].$$

Here as $M_x(u) = E\left[e^{uX}\right]$ and $M_x(v) = E\left[e^{vX}\right]$,

$$M_x(u+v) \geq M_x(u)M_x(v), \text{ where } u, v \geq 0.$$

3.16 Generalised Form of Bienayme-Chebyshev's Inequality

Let X be a random variable and $g(X)$ be a non-negative function of X. Then for every value of $k > 0$

$$P[g(X) \geq k] \leq \frac{E[g(x)]}{k}.$$

Proof. Consider a set of values of X such that $g(X) \geq k$. Let this set be S,

$$[S = \{x : g(x) \geq k\}].$$

Then

$$\int_g dF(x) = P[X \in S] = P[g(X) \geq k]$$

where $F(x)$ is the distribution function of X. Thus, we have

$$E[g(X)] = \int_{-\infty}^{\infty} g(x)dF(x) = \int_S g(x)dF(x) + \int_{\bar{S}} g(x)dF(x).$$

Here $P(\bar{S}) = 1 - P(S)$.

$$\therefore \ E[g(X)] \geq \int_S g(x)dF(x).$$

This is true since every integral value of right-hand side is positive.

$$\begin{aligned}\therefore \ E[g(X)] &\geq k\int_S dF(x) \ [\because \ g(X) \geq k] \\ &= kP[g(X) \geq k] \\ \therefore \ P[g(X) \geq k] &\leq \frac{E[g(X)]}{k}.\end{aligned}$$

From this inequality one can derive easily the Chebyshev's inequality. Let

$$g(X) = [X - E(X)]^2 = (X - \mu)^2 \ [\because \ E(X) = \mu].$$

Let us replace k of the above inequality by $k\sigma^2$. Then, we have

$$P\left[|X - \mu|^2 \geq k^2\sigma^2\right] \leq \frac{E(X - \mu)^2}{k^2\sigma^2} = \frac{\sigma^2}{k^2\sigma^2} = \frac{1}{k^2}.$$

$$\therefore \ P[|X - \mu| \geq k\sigma] \leq \frac{1}{k^2}.$$

3.17 Characteristic Function

It has already been observed that moments of a distribution are observed from moment generating function. Moment generating function is also used to derive the density function of some functions of random variables. It has also been observed that all moments of certain distributions may not be found out from moment generating function. Again, for certain distributions moment generating functions do not exist. To avoid such problems we need some functions which may be used to find the moments of the distribution and to derive the probability function of the distribution. Two such functions are characteristic function and probability generating function. In this section we shall study characteristic function. Probability generating function is presented in Chapter 4.

Definition. Let X be a random variable and its p.d.f. be $f(x)$ [assuming X as continuous random variable]. Then the characteristic function of X is defined by

$$\psi_x(t) = E\left[e^{itX}\right] = \int_{-\infty}^{\infty} e^{itx} f(x)dx$$

where $-h < t < h$ and $h > 0$.

However, if X is a discrete random variable having probability function $p(x)$, the characteristic function is given by

$$\psi_x(t) = E\left[e^{itX}\right] = \sum_{x=0}^{\infty} e^{itx} p(x).$$

If $F(x)$ is the distribution function of continuous random variable X, then

$$\psi_x(t) = E\left[e^{itX}\right] = \int_{-\infty}^{\infty} e^{itx} dF(x).$$

It is observed that $\psi_x(t)$ is a complex function of real value t and

$$|\psi_x(t)| = \left|\int_{-\infty}^{\infty} e^{itx} f(x)\right| \leq \int_{-\infty}^{\infty} \left|e^{itx}\right| f(x)dx.$$

If $\left|e^{itx}\right| = |\cos tx + i \sin tx|^{1/2} = \left|\cos^2 tx + \sin^2 tx\right|^{1/2} = 1$, then

$$|\psi_x(t)| \leq \int_{-\infty}^{\infty} f(x)dx = 1.$$

It is observed that $\psi_x(t) \leq 1$ and hence it can be found out in every case. Here $i = \sqrt{1}$ and t is real.

Let us consider that $(2s)$-th moment of a distribution exists. This $(2s)$-th moment can be observed from $\psi_x(t)$, where

$$\psi_x(t) = \int_{-\infty}^{\infty} \left[1 + \sum_{r=1}^{2s-1} \frac{(itx)^r}{r!} + \frac{(itx)^{2s}}{(2s)!} (\cos t'x + \sin t''x)\right]$$

where t' and t'' are two values in the interval $[0, t]$.

Let us assume that X is non-degenerate, then $\mu'_{2s} \neq 0$. In such a case $\psi_x(t)$ can be written as

$$\psi_x(t) = 1 + \sum_{r=1}^{2s-1} \left[\frac{(it)^r}{r!}\mu'_r + \frac{(it)^{2s}}{(2s)!}\mu'_{2s} g_1(t, s)\right].$$

Here $g_1(t, s)$ is a complex function and its real and imaginary components are, respectively

$$E\left[x^{2s}\cos t'x/\mu'_{2s}\right] \text{ and } E\left[x^{2s}\sin t''x/\mu'_{2s}\right].$$

Here $|g_1(t, s)| \leq 1$ and t' and t'' fall in the interval $[0, t]$.

$$\lim_{t\to 0} g_1(t, s) = 1.$$

Similarly, if $(2s-1)$-th moment exists, then $\psi_x(t)$ can be written as

$$\psi_x(t) = 1 + \sum_{r=1}^{2s-2}\frac{(it)^r}{r!}\mu'_r + \frac{(it)^{2s-1}}{(2s-1)!}\gamma'_{2s-1}g_2(t, s)$$

where $g_2(t, s)$ is a complex function, and its real and imaginary values are, respectively

$$E\left[x^{2s-1}\cos t_1 x/\gamma'_{2s-1}\right] \text{ and } E\left[x^{2s-1}\sin t_2 x/\gamma'_{2s-1}\right].$$

Here t_1 and t_2 are two values which fall in the interval $[0, t]$ and

$$\gamma'_{2s-1} = E\left[|X|^{2s-1}\right].$$

This $\gamma'_{2s-1} \neq 0$ if X is non-degenerate and $|g_2(t, s)| \leq 1$.

Moreover,

$$\lim_{t\to 0} g_2(t, s) = \frac{\mu'_{2s-1}}{\gamma'_{2s-1}}.$$

The above characteristic function is used to derive the probability function of the function of random variable. It is also used to find out the moments of the distribution, where

$$\mu'_r = (-i)^r \frac{d^r\psi_x(t)}{dt^r}\bigg|_{t=0}.$$

The series expansion of $\psi_x(t)$ nearest to $t = 0$ is

$$\psi_x(t) = 1 + \sum_{h=1}^{r}\frac{(it)^h}{h!}\mu'_r + O(t^r).$$

Here

$$\lim_{t\to 0}\frac{O(t^r)}{t^r} = 0.$$

This implies that the coefficient of $(it)^r/r!$ is μ'_r when $\psi_x(t)$ is expanded. The logarithmic series expansion of $\psi_x(t)$ gives

$$\log\psi_x(t) = \sum_{h=1}^{r}\frac{(it)^h k_h}{h!} + O(t^r).$$

Therefore, k_r = coefficient of $\dfrac{(it)^r}{r!}$ in $\log\psi_x(t)$.

►EXAMPLE 3.28 *The continuous random variable X has the probability density function*

$$f(x) = \theta e^{\theta x};\ x > 0,\ \theta > 0.$$

Find moments of the distribution and comment on its shape characteristics.

Solution. The characteristic function of X is

$$\begin{aligned}\psi_x(t) &= E\left[e^{itX}\right] = \int_{-\infty}^{\infty} \theta e^{itx} e^{-\theta x} dx = \theta \int_{0}^{\infty} e^{-x(\theta - it)} dx = \theta(\theta - it)^{-1} \\ &= \left(1 - \frac{it}{\theta}\right)^{-1} = \frac{it}{\theta} + \frac{(it)^2}{\theta^2} + \frac{(it)^3}{\theta^3} + \frac{(it)^4}{\theta^4} + \cdots.\end{aligned}$$

Thus

$$\begin{aligned}\mu_1' &= \text{coefficient of } it \text{ in } \psi_x(t) = \frac{1}{\theta} \\ \mu_2' &= \text{coefficient of } \frac{(it)^2}{2!} \text{ in } \psi_x(t) = \frac{2}{\theta^2} \\ \mu_3' &= \text{coefficient of } \frac{(it)^3}{3!} \text{ in } \psi_x(t) = \frac{6}{\theta^3} \\ \mu_4' &= \text{coefficient of } \frac{(it)^4}{4!} \text{ in } \psi_x(t) = \frac{24}{\theta^4} \\ \therefore\ \mu_2 &= \mu_2' - \mu_1'^2 = \frac{2}{\theta^2} - \frac{1}{\theta^2} = \frac{1}{\theta^2} \\ \mu_3 &= \mu_3' - 3\mu_2'\mu_1' + 2\mu_1'^3 = \frac{6}{\theta^3} - 3 \cdot \frac{2}{\theta^2} \cdot \frac{1}{\theta} + 2\left(\frac{1}{\theta}\right)^3 = \frac{2}{\theta^3} \\ \mu_4 &= \mu_4' - 4\mu_3'\mu_1' + 6\mu_2'\mu_1'^2 - 3\mu_1'^4 \\ &= \frac{24}{\theta^4} - 4 \cdot \frac{6}{\theta^3} \cdot \frac{1}{\theta} + 6 \cdot \frac{2}{\theta^2} \cdot \frac{1}{\theta^2} - 3\left(\frac{1}{\theta}\right)^4 = \frac{9}{\theta^4}.\end{aligned}$$

Again

$$\begin{aligned}\log \psi_x(t) &= \log\left(1 - \frac{it}{\theta}\right)^{-1} = -\log\left(1 - \frac{it}{\theta}\right) \\ &= \frac{it}{\theta} + \frac{(it)^2}{2\theta^2} + \frac{(it)^3}{3\theta^3} + \frac{(it)^4}{4\theta^4} + \cdots.\end{aligned}$$

Here

$$\begin{aligned}k_1 &= \mu_1' = \text{coefficient of } it \text{ in } \log \psi_x(t) = \frac{1}{\theta} \\ k_2 &= \mu_2 = \text{coefficient of } \frac{(it)^2}{2!} \text{ in } \log \psi_x(t) = \frac{1}{\theta^2} \\ k_3 &= \mu_3 = \text{coefficient of } \frac{(it)^3}{3!} \text{ in } \log \psi_x(t) = \frac{2}{\theta^3} \\ k_4 &= \text{coefficient of } \frac{(it)^4}{4!} \text{ in } \log \psi_x(t) = \frac{6}{\theta^4} \\ \therefore\ \mu_4 &= k_4 + 3\mu_2^2 = \frac{6}{\theta^4} + \frac{3}{\theta^4} = \frac{9}{\theta^4}\end{aligned}$$

Hence

$$\beta_1 = \frac{\mu_3^2}{\mu_2^3} = \frac{4/\theta^6}{1/\theta^6} = 4$$

$$\gamma_1 = \sqrt{\beta_1} = +2 \ [\because \ \mu_3 > 0]$$

$$\beta_2 = \frac{\mu_4}{\mu_2^2} = \frac{9/\theta^4}{1/\theta^4} = 9 > 3.$$

The distribution is positively skewed and leptokurtic.

►EXAMPLE 3.29 *The continuous random variable X has the p.d.f.*

$$f(x) = \frac{1}{\Gamma n} e^{-x} x^{n-1}, \ x > 0.$$

Find its moments using characteristic function and comment on shape characteristics of the distribution.

Solution. The characteristic function of X is

$$\psi_x(t) = E\left[e^{itX}\right] = \int e^{itx} f(x) dx$$

$$= \frac{1}{\Gamma n} \int_0^\infty x^{n-1} e^{-x(1-it)} dx = (1-it)^{-n}.$$

Then

$$\log \psi_x(t) = \log(1-it)^{-n} = -n \log(1-it)$$

$$= n\left[it + \frac{(it)^2}{2} + \frac{(it)^3}{3} + \frac{(it)^4}{4} + \cdots\right].$$

Now

$$k_1 = \text{coefficient of } it \text{ in } \log \psi_x(t) = n = \mu_1'$$

$$k_2 = \text{coefficient of } \frac{(it)^2}{2!} \text{ in } \log \psi_x(t) = n = \mu_2$$

$$k_3 = \text{coefficient of } \frac{(it)^3}{3!} \text{ in } \log \psi_x(t) = 2n = \mu_3$$

$$k_4 = \text{coefficient of } \frac{(it)^4}{4!} \text{ in } \log \psi_x(t) = 6n$$

$$\therefore \ \mu_4 = k_4 + 3\mu_2^2 = 6n + 3n^2.$$

Also, we have

$$\beta_1 = \frac{\mu_3^2}{\mu_2^3} = \frac{4n^2}{n^3} = \frac{4}{n}$$

$$\gamma_1 = \sqrt{\beta_1} = \frac{2}{\sqrt{n}} > 0 \ [\because \ \mu_3 > 0]$$

$$\beta_2 = \frac{\mu_4}{\mu_2^2} = \frac{3n^2 + 6n}{n^2} = 3 + \frac{6}{n} > 3.$$

The distribution is positively skewed and leptokurtic.

LEVY'S THEOREM. Let X be a random variable having characteristic function $\psi_x(t)$ and distribution function $F(x)$. If $F(x)$ is continuous at $x = x' \pm \delta$, $\delta > 0$, then

$$F(x'+\delta) - F(x'-\delta) = \lim_{A\to\infty} \frac{1}{\pi}\int_{-A}^{A} \frac{\sin\delta t}{t} e^{itx'} \psi_x(t)dt. \tag{a}$$

Again, if

$$\int_{-\infty}^{\infty} |\psi_x(t)|dt < +\infty, \tag{b}$$

then at the point $x = x'$ a probability density function exists, where the density function is

$$f(x') = \frac{1}{2\pi}\int_{-\infty}^{\infty} e^{itx'} \psi_x(t)dt.$$

Proof. The equation (a) above can be written as

$$G(x', A, \delta) = \frac{1}{\pi}\int_{-A}^{A} \frac{\sin\delta t}{t} e^{itx'} \int_{-\infty}^{\infty} e^{itx} dF(x).$$

Again, as

$$\left|\frac{\sin\delta t}{t} e^{-it(x'-x)}\right| < d$$

we can write

$$G(x', A, \delta) = \int_{-\infty}^{\infty} m(x, x', A, \delta)dF(x)$$

where

$$\begin{aligned} m(x, x', A, \delta) &= \frac{1}{\pi}\int_{-A}^{A} \frac{\sin\delta t}{t} e^{it(x-x')} dt \\ &= \frac{2}{\pi}\int_{0}^{A} \frac{\sin\delta t}{t} \cos[t(x-x')]dt \\ &= \frac{1}{\pi}\int_{0}^{A} \frac{\sin(x-x'+\delta)t}{t} dt - \frac{1}{\pi}\int_{0}^{A} \frac{\sin(x-x'-\delta)t}{t} dt. \end{aligned}$$

Now

$$\lim_{A\to\infty} G(x', A, \delta) = \int_{-\infty}^{\infty} \lim_{A\to\infty} m(x, x', A, \delta)dF(x).$$

But

$$\lim_{A\to\infty} \int_{0}^{A} \frac{\sin u}{u} du = \frac{\pi}{2} \text{ and hence}$$

$$\lim_{A\to\infty} m(x, x', A, \delta) = \begin{cases} 0, & x < x' - \delta,\ x > x' + \delta \\ \frac{1}{2}, & x = x' - \delta,\ x = x' + \delta \\ 1, & x' - \delta < x < x' + \delta \end{cases}$$

Now, as $F(x)$ is continuous at the point $x = x' \pm \delta$, we have

$$\lim_{A\to\infty} G(x', A, \delta) = \int_{x'-\delta}^{x'+\delta} dF(x) = F(x'+\delta) - F(x'-\delta).$$

Again

$$\frac{F(x'+\delta) - F(x'-\delta)}{2\delta} = \frac{1}{2\pi}\int_{-\infty}^{\infty} \frac{\sin \delta t}{\delta t} e^{-itx'} \psi_x(t)dt$$

and as

$$\left|\frac{\sin \delta t}{\delta t} e^{-itx'} \psi_x(t)\right| < \psi_x(t)$$

we get $f(x')$ by differentiating $F(x)$ at the point $x = x'$ and if

$$\int_{-\infty}^{\infty} |\psi_x(t)|dt < \infty$$

then

$$\lim_{\delta\to 0}\left[\frac{F(x'+\delta) - F(x'-\delta)}{2\delta}\right] = \frac{1}{2\pi}\int_{-\infty}^{\infty} \lim_{\delta\to 0} \frac{\sin \delta t}{\delta t}\left[e^{-itx}\right] \psi_x(t)dt.$$

But, as $\lim_{\delta\to 0} \frac{\sin \delta t}{\delta t} = 1$, we get

$$f(x') = \frac{1}{2\pi}\int_{-\infty}^{\infty} e^{-itx'} \psi_x(t)dt.$$

Using this theorem we can derive the density function. This can be observed from the following example.

►**EXAMPLE 3.30** *Let $X_1, X_2, \ldots, X_k$ be k-independent random variables having the joint p.d.f.*

$$f(x_1, x_2, \ldots, x_k) = \begin{cases} \exp\left(-\sum_{i=1}^{k} x_i\right); & x_i > 0,\ i = 1, 2, \ldots, k \\ 0, & \textit{otherwise.} \end{cases}$$

Find the density function of $U = \sum_{i=1}^{k} X_i$.

Solution. The characteristic function of U is

$$\begin{aligned} \psi(t) &= \int_0^{\infty}\int_0^{\infty}\cdots\int_0^{\infty} e^{-\Sigma x_i + it\Sigma x_i} dx_1 dx_2 \cdots dx_k \\ &= \int_0^{\infty}\int_0^{\infty}\cdots\int_0^{\infty} e^{-\Sigma x_i(1-it)} dx_1 dx_2 \cdots dx_k \\ &= \int_0^{\infty} e^{-x_1(1-it)} dx_1 \int_0^{\infty} e^{-x_2(1-it)} dx_2 \cdots \int_0^{\infty} e^{-x_k(1-it)} dx_k \\ &= (1-it)^{-k}. \end{aligned}$$

According to the Levy's theorem, the density function of U is

$$f(u) = \frac{1}{2\pi}\int_{-\infty}^{\infty} e^{-itu}(1-it)^{-k}dt.$$

Let $z = -u(1-it)$, then we have $u^{k-1}e^{-u}H$, where

$$H = \frac{1}{2\pi}\int_{-L+i\infty}^{-L-i\infty} e^{-z}(-z)^{-k}dz.$$

Here H is called *Hankel integral* and $H = 1/\Gamma k$. Therefore,

$$f(u) = \begin{cases} \frac{1}{\Gamma k}u^{k-1}e^{-u}, & u > 0 \\ 0, & u < 0. \end{cases}$$

3.18 Characteristic Function of k-dimensional Random Variables

Let $X_1, X_2, \ldots, X_k$ be k-dimensional random variables having joint p.d.f. $f(x_1, x_2, \ldots, x_k)$ and joint distribution function $F(x_1, x_2, \ldots, x_k)$. Then the characteristic function of $X_1, X_2, \ldots, X_k$ is

$$\begin{aligned} \psi(t_1, t_2, \ldots, t_k) &= E\left[\exp\left\{i\sum_{j=1}^{k} t_j X_j\right\}\right] \\ &= \int_{-\infty}^{\infty}\int_{-\infty}^{\infty}\cdots\int_{-\infty}^{\infty} \exp\left\{i\sum_{j=1}^{k} t_j X_j\right\} f(x_1, x_2, \ldots, x_k)dx_1\cdots dx_k \\ &= \int_{-\infty}^{\infty}\int_{-\infty}^{\infty}\cdots\int_{-\infty}^{\infty} \exp\left\{i\sum_{j=1}^{k} t_j X_j\right\} dF(x_1, x_2, \cdots, x_k). \end{aligned}$$

However, if $X_1, X_2, \ldots, X_k$ are discrete random variables, then

$$\psi(t_1, t_2, \ldots, t_k) = \sum_{x_1}\sum_{x_2}\cdots\sum_{x_k} \exp\left\{i\sum_{j=1}^{k} t_j X_j\right\} p(x_1, x_2, \ldots, x_k)$$

where $p(x_1, x_2, \ldots, x_k)$ is the joint probability function of $X_1, X_2, \ldots, X_k$.

This characteristic function also provides moments, where

$$\mu'_{r_1, r_2, \ldots, r_k} = (-i)^r \frac{\delta^r \psi(t_1, t_2, \ldots, t_k)}{\delta t_1^r \delta t_2^r \cdots \delta t_k^r}\bigg|_{t_1=0, t_2=0, \ldots, t_k=0}$$

Also, we can have the r-th moment of j-th variable as follows:

$$\mu'_{r_j} = (-i)^r \frac{\delta^r \psi(t_1, t_2, \ldots, t_k)}{\delta t_j^r}\bigg|_{t_1=0, t_2=0, \ldots, t_k=0}$$

We can also apply Levy's theorem to find out p.d.f. using this joint characteristic function. In this case

$$f(x_1', x_2', \ldots, x_k') = \left(\frac{1}{2\pi}\right)^k \int_{-\infty}^{\infty}\int_{-\infty}^{\infty}\cdots\int_{-\infty}^{\infty} \exp\left(-i\sum_{j=1}^{k} t_j x_j\right)\psi(t_1, t_2, \ldots, t_k)dt_1\cdots dt_k.$$

►**EXAMPLE 3.31** *The joint p.d.f. of two-dimensional random variables X_1 and X_2 is given by*

$$f(x_1, x_2) = e^{-(x_1+x_2)};\ x_1 > 0,\ x_2 > 0.$$

Find characteristic function of the random variables and hence moments.

Solution. The joint characteristic function of X_1 and X_2 is

$$\begin{aligned}\psi(t_1, t_2) &= E\left[e^{it_1X_1+it_2X_2}\right]\\ &= \int_0^\infty\int_0^\infty e^{it_1x_1+it_2x_2}e^{-(x_1+x_2)}dx_1dx_2\\ &= \int_0^\infty e^{-x_1(1-it_1)}dx_1\int_0^\infty e^{-x_2(1-it_2)}dx\ [\because X_1 \text{ and } X_2 \text{ are independent}]\\ &= (1-it_1)^{-1}(1-it_2)^{-1}.\end{aligned}$$

Hence

$$\begin{aligned}\mu'_{1(x_1)} &= (-i)^1\frac{d\psi(t_1,t_2)}{dt_1}\bigg|_{t_1=0,t_2=0}\\ &= -(-i)(-i)(1-it_1)^{-2}(1-it_2)^{-1}\bigg|_{t_1=0,t_2=0} = 1\\ \mu'_{2(x_1)} &= (-i)^2\frac{d^2\psi(t_1t_2)}{dt_1}\bigg|_{t_1=0,t_2=0}\\ &= (-i)^2 2(-i)^2(1-it_1)^{-3}(1-it_2)^{-1}\bigg|_{t_1=0,t_2=0} = 2\\ \therefore\ \mu_2 &= \mu_2' - \mu_1'^2 = 2-(1)^2 = 1.\end{aligned}$$

Other moments are also available in a similar way.

3.19 Properties of Characteristic Function

(i) $\psi(0) = \int_{-\infty}^{\infty} dF(x) = 1$

(ii) $|\psi(t)| \le 1 = \psi(0)$

(iii) $\psi(t)$ is a continuous function of t within the interval $[-\infty, \infty]$.

(iv) The necessary and sufficient conditions of a two-dimensional random variable to be independent is $\psi(t_1, t_2) = \psi(t_1)\psi(t_2)$, where $\psi(t_1), \psi(t_2)$ and $\psi(t_1, t_2)$ are characteristic functions of X_1, X_2 and joint characteristic functions of X_1 and X_2, respectively.

Similarly, the necessary and sufficient conditions of k-dimensional random variables to be independent is

$$\psi(t_1, t_2, \ldots, t_k) = \psi(t_1)\psi(t_2)\cdots\psi(t_k).$$

(v) Let $X_1, X_2, \ldots, X_k$ be k-independent random variables such that the characteristic function of j-th variable is $\psi_j(t_j)$. Consider

$$U = \sum_{j=1}^{k} C_j X_j$$

is a function of X_j's. Then the characteristic function of U is

$$\psi_u(t) = \prod_{i=1}^{k} \psi_j(t_j).$$

(vi) Let X be a random variable and $Y = CX$ be a function of X. Then characteristic function of Y is $\psi_{cx}(t) = \psi_x(ct)$.

(vii) Let X_1 and X_2 be two random variables having distribution functions $F_1(x_1)$ and $F_2(x_2)$, respectively. The characteristic functions of X_1 and X_2 are $\psi_1(t_1)$ and $\psi_2(t_2)$, respectively. Then the necessary and sufficient conditions for $F_1(x_1) \equiv F_2(x_2)$ is $\psi_1(t_1) \equiv \psi_2(t_2)$.

(viii) Let the distribution function of a random variable be symmetric at near to zero. Then $1 - F(x) = F(-x)$. The characteristic function of such a random variable X is a real-valued function and it is an even function of t.

(ix) Let X be a random variable having characteristic function $\psi(t)$. Let us consider a function of X such that $U = \frac{X-a}{h}$. Then $\psi_u(t) = e^{-iat/h}\psi_x(t/h)$.

(x) The necessary and sufficient condition for a characteristic function $\psi(t)$ of X is $\psi(0) = 1$.

It has been mentioned that the p.d.f. of a random variable or its function can be found out uniquely from characteristic function. We have discussed the characteristic function of a random variable. But in sampling distribution the convergence in probability of a sequence of random variables is important. This convergence in probability can be easily discussed with the help of characteristic function of sequence of random variables. Let us now discuss the sequence of random variables and its related characteristics.

3.20 Sequences of Random Variables

3.20.1 Stochastic Process

Let $X_1, X_2, \ldots, X_n$ be n-dimensional random variables and $g_n(X_1, X_2, \ldots, X_n)$ be a function of these random variables. Here $g_n(\cdot)$ is also a random variable. There are many components of this random variable and a random variable which has such many components is called a *stochastic process*.

If a stochastic process has countably infinite component, it is called a *sequence of random variable.* The main task of study of stochastic process or a sequence of random variables is to derive a limiting distribution of $g_n(x_1, x_2, \cdots, x_n)$ when $n \to \infty$ or to derive a distribution function of it, if it exists.

From the definition of stochastic process it can be said that the stochastic process is a family of several random variables and it is denoted by $\{x_\alpha : \alpha \in A\}$, where range A is an interval of real axis or a sequence of several points in real axis. Let

$$F_{\alpha_1, \alpha_2, \dots, \alpha_n}(x_{\alpha_1}, x_{\alpha_2}, \cdots, x_{\alpha_n})$$

be a distribution function, if it exists, for a finite set of components $[x_{\alpha_1}, x_{\alpha_2}, \cdots, x_{\alpha_n}]$, where this set exists in the range space. This distribution is similar to that of each of marginal distribution function.

3.20.2 Probability Measure for a Stochastic Process

Let us consider that in the stochastic process $[x_1, x_2, \dots]$ there are countable infinite components. Also consider that the range spaces for the random variables $X_1, X_2, \dots, X_n$ are

$$R_1^{(1)}, R_2^{(2)}, \dots, R_n^{(n)},$$

respectively. Then the range space R for the stochastic process is $R_1^{(1)} \times R_2^{(2)} \times \cdots$. In such a case the sequence $[b_1, b_2, \dots]$ of countable infinite real numbers is a point in R. Here R may be called a *sample space* of the stochastic process. Let us consider a finite coordinate of $(b_{\alpha_1}, b_{\alpha_2}, \cdots, b_{\alpha_m})$ of b. This coordinate is a point in m-dimensional Euclidean space

$$R_m^{(\alpha_1, \alpha_2, \dots, \alpha_m)} = R_1^{(\alpha_1)} \times R_1^{(\alpha_2)} \times R_1^{(\alpha_3)} \times \cdots \times R_1^{(\alpha_m)}.$$

Let this point be b'. Here b' is the projection of b in R.

$$b' = h_{\alpha_1, \alpha_2, \dots, \alpha_m}(b) \text{ or, } b = h^{-1}_{\alpha_1, \alpha_2, \dots, \alpha_m}(b).$$

If E is the set of all points in R and if E projects E' in $R_m^{(\alpha_1, \alpha_2, \dots, \alpha_m)}$, then E is called *cylinder set* and

$$E = h^{-1}_{\alpha_1, \alpha_2, \dots, \alpha_m}(E).$$

Now, if E' is a Borel set in $R_m^{(\alpha_1, \alpha_2, \dots, \alpha_m)}$, cylinder set E in R with respect to E' is known as *Borel cylinder set.* Let E' is a Borel set in sample space $R_m^{(\alpha_1, \alpha_2, \cdots, \alpha_m)}$ and E is the preimage in R. Then

$$P[E] = P_{\alpha_1, \alpha_2, \dots, \alpha_m}(E')$$

where $P_{\alpha_1, \alpha_2, \dots, \alpha_m}(E')$ is the probability of E' derived from the distribution function of $(x_{\alpha_1, \alpha_2, \dots, \alpha_m})$ and $(x_{\alpha_1, \alpha_2, \dots, \alpha_m})$ is the identified component from the stochastic process $(x_1, x_2, \dots)$.

3.20.3 Characteristic Function of Sequence of Random Variables

Let $\psi_1(t), \psi_2(t), \ldots$ be a sequence of characteristic function of the sequence of random variables $(x_1, x_2, \ldots)$, respectively. Let us consider that $\psi(t)$ is continuous at $t = 0$. Then the necessary and sufficient condition for the sequence $(x_1, x_2, \ldots)$ to be convergence in distribution function $F(x)$ is that $\psi_1(t), \psi_2(t), \ldots$ are convergence in the limit $\psi(t)$. Under this condition $\psi(t)$ is similar to the characteristic function of $F(x)$ [*Levy* (1937), *Cramer* (1937)]. Thus the characteristic function of sequence $(x_1, x_2, \ldots)$ is the characteristic function of $F(x)$.

Let $\{x_1, x_2, \ldots\}$ be a sequence of random variable. The respective characteristic functions of the variables are $\psi_1(t), \psi_2(t), \ldots$. This sequence of characteristic function converges in probability to a constant c under the necessary and sufficient condition

$$\lim_{n\to\infty} \psi_n(t) = e^{ict}.$$

3.21 Convergence in Probability

Let $(x, x_1, x_2, \ldots)$ be a stochastic process. Consider that, for any arbitrary value $\epsilon > 0$

$$\lim_{n\to\infty} P[|x_n - x| \geq \epsilon] = 0.$$

Then the stochastic process $(x, x_1, x_2, \ldots)$ will be convergence in probability at random variable x. This convergence in probability condition can be written as

$$P \lim_{n\to\infty} x_n = x.$$

If X is degenerate random variable, then the stochastic process $(x, x_1, x_2, \ldots)$ is convergence in probability at a constant x_0, i.e.,

$$P(X = x_0) = 1.$$

The simplest criteria of convergence in probability can be described as follows:

Let $(x, x_1, x_2, \ldots)$ be a stochastic process. Then $(x_1, x_2, \ldots)$ will be convergence in probability at a random variable x if

$$\lim_{n\to\infty} E[x_n - x]^2 = 0.$$

It is true as we can write from Chebyshev's inequality that

$$P[|x_n - x| \geq \epsilon] \leq \frac{E[x_n - x]^2}{\sigma^2}.$$

Again, as

$$\lim_{n\to\infty} E[x_n - x]^2 = 0$$

$$\lim_{n\to\infty} P[|x_n - x| \geq \epsilon] \leq 0.$$

Thus $(x_1, x_2, \dots)$ is convergence in probability at x. The simplest and important example of convergence in probability of a stochastic process at a constant is the weak law of large number.

3.22 Weak Law of Large Number

Let $x_1, x_2, \dots, x_n$ be a sequence of random variables having mean of each zero and variances $\sigma_1^2, \sigma_2^2, \dots, \sigma_m^2$, respectively. Consider that

$$C_n^2 = \sigma_1^2 + \sigma_2^2 + \cdots + \sigma_n^2 \text{ and } \lim_{n\to\infty} \frac{C_n^2}{n^2} = 0.$$

Then the sequence $(\bar{x}_1, \bar{x}_2, \dots)$ is convergence in probability at zero, where

$$\bar{x}_i = \frac{1}{n}\sum_{i=1}^{n} x_i.$$

Proof. From the given condition

$$E[\bar{x}_n] = 0 \text{ and } E[\bar{x}_n^2] = C_n^2/n^2.$$

Then from Chebyshev's inequality, we can write

$$P[|\bar{x}_n| \leq \epsilon] \geq 1 - \frac{C_n^2}{n^2\epsilon^2}$$

where $\epsilon > 0$ is an arbitrary value. Thus as $n \to \infty$ and $C_n^2/n^2 \to 0$

$$P[|\bar{x}_n| < \epsilon] = 1.$$

This is true if $\text{cov}(x_i, x_j) = 0,\ i \neq j$.

From the convergence in probability, it can be said that if $\bar{x}_n$ is a sample mean from n observations it will be convergence in probability at μ, where μ is the population mean of a population from which sample is selected. The population variance is σ^2.

Since $E[\bar{x}_n] = \mu$ and $V(\bar{x}) = \sigma^2/n$, then for an arbitrary value $\epsilon > 0$, we can write

$$P[|\bar{x}_n - \mu| \geq \epsilon] = P\left[|\bar{x} - \mu| \geq \frac{k\sigma}{\sqrt{n}}\right]$$

where $k = \epsilon\sqrt{n}/\sigma$. But according to Chebyshev's inequality

$$P[|\bar{x}_n - \mu|] \leq \frac{1}{k^2} = \frac{\sigma^2}{n\epsilon^2}.$$

Therefore, for any arbitrary value $\epsilon > 0$

$$\lim_{n\to\infty} P[|\bar{x}_n - \mu| \geq \epsilon] \leq \lim_{n\to\infty} \frac{\sigma^2}{n\epsilon^2} = 0.$$

The weak law of large number states that the sequence $(\bar{x}_1, \bar{x}_2, \dots)$ is convergence in probability at zero. But this sequence will be convergence in probability at any point which is far away from zero. This convergence is known as *strong law of large numbers.*

3.23 Strong Law of Large Numbers

Let $(x_1, x_2, \dots)$ be a sequence of independent random variables such that $E(x_j) = 0$ and $V(x_j) = \sigma_j^2$; $j = 1, 2, \dots$. Then, if

$$\sum_{j=1}^{\infty} \sigma_j^2 / j^2 < +\infty$$

$(\bar{x}_1, \bar{x}_2, \dots)$ is convergence in probability at nearest to zero, where

$$\bar{x}_n = \frac{1}{n} \sum_{i=1}^{n} x_i.$$

That is, if $\delta > 0, \epsilon > 0$ are arbitrary values and $N_{\delta,\epsilon}$ is any number, then

$$P(|\bar{x}_n| < \epsilon, n = N, N+1, \dots, N+k) \geq 1 - \delta.$$

This is true for all $N > N_{\delta,\epsilon}$ and for all k.

❑**THEOREM 3.4** *Let $(x, x_1, x_2, \dots)$ be a stochastic process and the distribution functions of the elements of this process be $F(x), F(x_1), \dots$, respectively. If $(x_1, x_2, \dots)$ is convergence in probability at X, then the sequence $F(x), F(x_1), \dots$ is convergence in probability at every continuous point of $F(x)$.*

Proof. Let us consider the stochastic process $(x, x_1, x_2, \dots)$ the sample space of which be R. Consider that $F(x)$ is continuous at $x = x_0$ and x' is a constant such that $x' < x_0$. Now, consider 3 events $x \leq x', x_n \leq x_0$ and $|x_n - x| > (x_0 - x')$. These events indicate that the first event exists in the combination of second and third events. Then

$$F(x') = P(x \leq x') \leq P(x_n \leq x_0) + P[|x_n - x| > x_0 - x'].$$

But, as the sequence $(x_1, x_2, \dots)$ is convergence in probability at X

$$\lim_{n \to \infty} P[|x_n - x| > (x_0 - x')] = 0.$$

Therefore,

$$F(x') \leq \lim_{n \to \infty} \inf F_n(x_0).$$

Similarly, if we consider $x'' > x_0$, then

$$F(x'') \geq \lim_{n \to \infty} \sup F_n(x_0).$$

That is

$$F(x') \leq \lim_{n \to \infty} \inf F_n(x_0) \leq \lim_{n \to \infty} \sup F_n(x_0) \leq F(x'').$$

But, as $F(x)$ is continuous at $x = x_0$

$$\lim_{x' \to x_0} F(x') = \lim_{x'' \to x_0} F(x'') = F(x_0).$$

Therefore, $\lim_{n \to \infty} F_n(x_0) = F(x_0)$.

❑ **THEOREM 3.5** *Let $(x, x', x_1, x_2, \ldots)$ be a stochastic process. If $(x_1, x_2, \ldots)$ is such that every element of the process is convergence in probability at every point of x and x', then x and x' are similar random variables.*

Proof. It is to be noted that

$$P\left[|x - x'| > \frac{1}{N}\right] = P\left[|(x_n - x') - (x_n - x)| > \frac{1}{N}\right].$$

Let us define three events E, E_1 and E_2 on the sample space R such that the favourable cases to the events are

$$|(x_n - x') - (x_n - x)| > \frac{1}{N}, \; |x_n - x| > \frac{1}{2N} \text{ and } |(x_n - x')| > \frac{1}{2N}$$

respectively. Then $E \subset (E_1 \cup E_2)$ and hence

$$P(E) \leq P(E_1) + P(E_2)$$

$$\text{or, } P\left[|x_n - x'| > \frac{1}{N}\right] \leq P\left[|x_n - x| > \frac{1}{2N}\right] + P\left[|x_n - x'| > \frac{1}{2N}\right].$$

But, if $n \to \infty$,

$$P\left[|x_n - x'| > \frac{1}{N}\right] > 0.$$

Let us now consider

$$|x - x'| > \frac{1}{N} = G_N.$$

Then

$$G_1 \subset G_2 \subset \cdots \text{ and } \lim_{n\to\infty} G_N = G.$$

Here G indicates all the sample points of R under the condition $x \neq x'$. Hence

$$P(x \neq x') = P[G_1 \cup G_2 \cup \cdots] \leq P(G_1) + P(G_2) + \cdots.$$

But as $P(G_1) = P(G_2) = \cdots = 0, P(x \neq x') = 0$ and thus x and x' are identical random variables.

❑ **THEOREM 3.6** *Let $(x, x_1, x_2, \ldots)$ be a stochastic process. The sequence $(x_1, x_2, \ldots)$ is such that each element of it is convergence in probability at X. Now, if $g(x)$ is a continuous function of X, then*

$$P \lim_{n\to\infty} g(x_n) = g(x).$$

Proof. Let us consider that $g(x)$ is identically continuous in an interval $[-M, M]$. Now, as X is a random variable

$$P[|x| > M] < \frac{\epsilon}{2} \tag{a}$$

where $\epsilon > 0$ is an arbitrary value.

Since ϵ and M are selected as above, a number $\delta(\epsilon, M)$ exists such that

(i) $|x| \leq M$

(ii) $|x_n - x| < \delta(\epsilon, M)$

(iii) $|g(x_n) - g(x)| < \epsilon$.

Now, if for sample space (i), (ii) and (iii) the events E_1, E_2 and E_3, respectively are considered, then $\bar{E}_3 \subset (\bar{E}_1 \cup \bar{E}_2)$. Hence

$$P(\bar{E}_3) \leq P(\bar{E}_1) + P(\bar{E}_2).$$

This implies that

$$P[|g(x_n) - g(x)| \geq \epsilon] \leq P[|x_n - x| \geq \delta(\epsilon, M)] + P(|x| > M).$$

But a function $n(\epsilon, \delta, M)$ exists in such a way that $n > n(\epsilon, \delta, M)$. Therefore

$$P[|x_n - x| \geq \delta(\epsilon, M)] < \frac{\epsilon}{2}.$$

But M is selected in such a way that (a) is true. Thus $P[|g(x_n) - g(x)|] \geq \epsilon < \epsilon$.

$$\therefore \ P \lim_{n\to\infty} g(x_n) = g(x).$$

❑**THEOREM 3.7** *Let $(x, x_1, x_2, \dots)$ be a stochastic process, where $(x_1, x_2, \dots)$ are almost certainly convergence in probability at X. Then the sequence $(x_1, x_2, \dots)$ is convergence in probability at X.*

Proof. Let us consider two events E_n and F_n in the sample space R such that

$$[|x_{n+j} - x| < \epsilon, \ j = 0, 1, 2, \dots] \text{ and } [|x_n - x| < \epsilon]$$

respectively. Then $E_n \subset F_n$. If E exists in the convergent set $(x_1, x_2, \dots)$, then for an arbitrary value $\epsilon > 0$ and for n

$$P[|x_n - x| < \epsilon] \geq P[|x_{n+j} - x| < \epsilon], \ j = 0, 1, 2, \dots.$$

Now, if $n \to \infty$,

$$\lim_{n\to\infty} P[|x_n - x| < \epsilon] \geq \lim_{n\to\infty} P[|x_{n+j} - x| < \epsilon]; \ j = 0, 1, 2, \dots$$
$$= P\left(\lim_{n\to\infty} E_n\right) = P(E) = 1.$$

Therefore,

$$\lim_{n\to\infty} P[|x_n - x| < \epsilon] = 1.$$

ALMOST CERTAIN CONVERGENCE. Let $(x_1, x_2, \dots)$ be a stochastic process the range space of which is R and P be the probability. Also consider that E is the set of all sequence $(b_1, b_2, \dots)$ and is convergence.

Now, let $P(E) = 1$ and X be a random variable. The value of the variable is $\lim_{n\to\infty} b_n$, if the sequence $(b_1, b_2, \dots)$ exists in E and is 0 if the sequence $(b_1, b_2, \dots)$ does not exist in E. In such a case the sequence is almost convergence at X. •

3.24 Central Limit Theorem

Let $(X_1, X_2, \ldots, X_n)$ be a sequence of independent random variables such that

$$E(X_j) = \mu_j, V(X_j) = \sigma_j^2$$

and the distribution function $F_j(x_j)$ $(j = 1, 2, \ldots, n)$. Then

$$\lim_{n\to\infty} P\left[\frac{\sum_{j=1}^{n}(x_j - \mu_j)}{\sqrt{\sum_j \sigma_j^2}} \le y\right] = \frac{1}{\sqrt{2\pi}}\int_{-\infty}^{y} e^{-\frac{1}{2}u^2}du.$$

This central limit theorem is also applicable for k-dimensional random variables. Let $x_{1j}, x_{2j}, \ldots, x_{kj}$; $j = 1, 2, \ldots, n$ be the sample observations of size n from k-dimensional random variables. Let us consider that the mean vector of the k-dimensional population be $\boldsymbol{\mu} = [\mu_1, \mu_2, \ldots, \mu_k]'$. The variance-covariance matrix of this population is

$$\sum = [\sigma_{ij}]_{k\times k}.$$

The sample mean vector and the sample variance-covariance matrix are $\bar{X} = [\bar{x}_1, \bar{x}_2, \ldots, \bar{x}_k]'$ and $S^2 = [s_{ij}]$ respectively. The variance-covariance matrix of $\bar{X}$ is

$$n^{-1}\sum = n^{-1}[\sigma_{ij}]_{k\times k}.$$

The asymptotic distribution of $\bar{X}$ is $N_k[\boldsymbol{\mu}, n^{-1}\sum]$, i.e.,

$$\lim_{n\to\infty} P\left[\sqrt{n}\,\sum^{-1/2}(\bar{X} - \boldsymbol{\mu}) \le Y\right]$$

$$= \frac{\sqrt{|\sigma^{ij}|}}{(2\pi)^{k/2}}\int_{-\infty}^{y_1}\int_{-\infty}^{y_2}\cdots\int_{-\infty}^{y_k}\exp\left(-\frac{1}{2}\sum\sigma^{ij}u_iu_j\right)d_{u_1}\cdots d_{u_k}$$

where

$$Y = [y_1, y_2, \ldots, y_k]' \text{ and } U = \sqrt{n}\,\sum^{-1/2}(\bar{X} - \boldsymbol{\mu}).$$

REMARK 8. Let $\{x_n\}$ be a sequence of random variables such that $V(x_j) = \sigma_j^2$ $(j = 1, 2, \ldots, n)$ and this variance is finite. Then for $n \to \infty$ and for any arbitrary value ϵ

$$P\left[\left|\sum_{j=1}^{n}\frac{x_j - E(x_j)}{n}\right| > \epsilon\right] \to 0.$$

The result is obvious from Chebyshev's inequality, where

$$\lim_{n\to\infty} P\left[\left|\sum_{j=1}^{n}\frac{x_j - E(x_j)}{n}\right| > \epsilon\right] \le \frac{V\left[\sum_{j=1}^{n}x_j\right]}{n^2\epsilon^2} \to 0.$$

3.25 Kolmogorov's Inequality

Let $x_1, x_2, \ldots, x_n$ be a set of random variables such that $E(x_j) = 0$ and $V(x_j) = \sigma_j^2$ $(j = 1, 2, \ldots, n)$. If $C_n^2 = \sigma_1^2 + \sigma_2^2 + \cdots + \sigma_n^2$, then

$$P[x_1 + x_2 + \cdots + x_\alpha < \lambda C_n] \geq 1 - \frac{1}{\lambda^2}$$

where $\alpha = 1, 2, \ldots, n$.

Proof. Let us define an event F_α in the sample R in which $|x_1 + x_2 + \cdots + x_\alpha| \geq \lambda C_n$ $(\alpha = 1, 2, \ldots, n)$ exists. Also, consider

$$\begin{aligned} E_1 &= F_1 \\ E_2 &= \bar{F}_1 \cap F_2 \\ E_3 &= \bar{F}_1 \cap \bar{F}_2 \cap F_3, \cdots \\ E_n &= \bar{F}_1 \cap \bar{F}_2 \cap \cdots \cap \bar{F}_{n-1} \cap F_n \text{ and } E_1 \cup E_2 \cup \cdots \cup E_n. \end{aligned}$$

Consider a complementary event G_n of $E_1 \cup E_2 \cup \cdots \cup E_n$, where G_n is a set of R and these are true for $|x_1 + x_2 + \cdots + x_\alpha| < \lambda C_n$. Here $E_1, E_2, \ldots, G_n$ are mutually exclusive events and their combination is the whole sample space R. Therefore,

$$P(G_n) = 1 - [P(E_1) + P(E_2) + \cdots + P(E_n)].$$

Again, consider a random variable Z_α $(\alpha = 1, 2, \ldots, n)$ the value of which is 1 at every point of E_α and zero at every point of $\bar{E}_\alpha$. Also, consider that a similar random variable Z_{n+1} can be defined based on G_n. Then

$$\begin{aligned} E[x_1 + x_2 + \cdots + x_n]^2 &= E[Z_1(x_1 + x_2 + \cdots + x_n)^2 + \cdots \\ &+ E[Z_n(x_1 + x_2 + \cdots + x_n)^2 + E[Z_{n+1}(x_1 + x_2 + \cdots + x_n)^2]. \end{aligned} \tag{a}$$

But as any function of $(x_1, x_2, \ldots, x_\alpha)$ is independent of the remaining x's,

$$E[g_\alpha, X_\beta] = E[g_\alpha]E(X_\beta)] = 0, \ \beta = \alpha + 1, \ldots, n.$$

Here g_α is the function of $(x_1, x_2, \ldots, x_\alpha)$ and X_β is the function of the remaining x's. Hence

$$\begin{aligned} E[Z_\alpha(x_1, x_2 + \cdots + x_n)^2] &= E[Z_\alpha(x_1 + x_2 + \cdots + x_\alpha)^2] \\ &+ E[Z_\alpha(x_1 + x_2 + \cdots + x_n)^2] \geq E[Z_\alpha(x_1 + x_2 + \cdots + x_\alpha)^2]. \end{aligned}$$

But as $(x_1 + x_2 + \cdots + x_\alpha)^2 \geq \lambda^2 C_n^2$ at every point of $E_\alpha(\alpha = 1, 2, \ldots, n)$,

$$E[Z_\alpha(x_1 + x_2 + \cdots + x_n)^2 \geq \lambda^2 C_n^2 P(E_\alpha). \tag{b}$$

As $E[x_1 + x_2 + \cdots + x_n]^2 = C_n^2$ we can apply (b) at the right-hand side of (a) and deleting $E[Z_{n+1}(x_1 + x_2 + \cdots + x_n)^2$, we get

$$C_n^2 \geq \lambda^2 C_n [P(E_1) + P(E_2) + \cdots + P(E_n)].$$

Hence $P(C_n) \geq 1 - 1/\lambda^2$.

3.26 Khintchine Theorem

Let $(x_1, x_2, \ldots, x_n)$ be a set of sample observation drawn from a population having distribution function $F(x)$, where $E(x) = \mu$ and μ is finite. Then, if $n \to \infty$, $\bar{x}$ converges in probability at μ.

Proof. The characteristic function of $f(x)$, where $E(x) = \mu$ can be written as

$$\psi(t) = 1 + i\mu t + O(t).$$

Again, from the characteristic function of $\bar{x}$, we can write

$$[\psi(t/n)]^n = \left[1 + \frac{i\mu t}{n} + O\left(\frac{t}{n}\right)\right]^n.$$

Now, if $n \to \infty$, then for any value of t, $O(t/n) \to 0$, i.e.,

$$\lim_{n\to\infty}\left[\psi\left(\frac{t}{n}\right)\right]^n = e^{i\mu t}.$$

This function is the characteristic function of degenerate function $E(x-\mu)$. Therefore, if $n \to \infty$, $\bar{x}$ converges at $E(x-\mu)$. In other words, if $n \to \infty$, $\bar{x}$ converges in probability at μ. Here the proof is discussed for finite value of μ. But the similar result can be derived if μ and σ^2 are available. In that case applying Chebyshev's inequality it can be shown that $\bar{x}$ converges in probability at μ. Since $E(\bar{x}) = \mu$, $V(\bar{x}) = \sigma^2/n$, we can write for any arbitrary value $\epsilon > 0$

$$P[|\bar{x} - \mu| \geq \epsilon] = P\left[|\bar{x} - \mu| \geq \frac{\epsilon}{\sigma/\sqrt{n}} \cdot \sigma\sqrt{n}\right] \leq \frac{\sigma^2}{n\epsilon^2}.$$

$$\therefore \lim_{n\to\infty} P(|\bar{x} - \mu| \leq \epsilon) = 0.$$

Hence $\bar{x}$ is convergence in probability at μ.

The above result can be presented by a basic theorem of Lineberg. The theorem is presented below:

LINEBERG THEOREM. Let $(x_1, x_2, \ldots, x_n)$ be a set of sample observations from a population having $E(x) = \mu$ and $V(x) = \sigma^2$, where σ^2 is finite. Now, if Z and $\bar{x}$ are the sample total and sample mean, respectively, then

$$\lim_{n\to\infty} P\left[\frac{Z - n\mu}{\sigma\sqrt{n}} \leq y\right] = \lim_{n\to\infty} P\left[\frac{\sqrt{n}(\bar{x} - \mu)}{\sigma} \leq y\right] = \frac{1}{\sqrt{2\pi}}\int_{-\infty}^{y} e^{-\frac{1}{2}y^2}\,dy$$

where $y \sim N(0, 1)$.

►**EXAMPLE 3.32** *The p.d.f. of a random variable X is given by*

$$f(x) = \frac{1}{\sqrt{2\pi}}e^{-\frac{1}{2}x^2}, \quad -\infty < x < \infty.$$

Find characteristic function of X and hence using Lineberg theorem show that $(Z - n\mu)/\sigma\sqrt{n} \sim N(0, 1)$, where $Z = \sum x_j$.

Solution. The characteristic function is

$$\begin{aligned}\psi(t) &= E\left[e^{itX}\right]\\ &= \frac{1}{\sqrt{2\pi}}\int_{-\infty}^{\infty} e^{itx}e^{-\frac{1}{2}x^2}\,dx = \frac{1}{\sqrt{2\pi}}e^{\frac{1}{2}i^2t^2}\int_{-\infty}^{\infty} e^{-\frac{1}{2}(x-it)^2}\,dx\\ &= \frac{1}{\sqrt{2\pi}}\int_{-\infty}^{\infty} e^{\frac{1}{2}i^2t^2}e^{-\frac{1}{2}u^2}\,du,\ \text{where } u = x - it, dx = du, -\infty < u < \infty\\ &= \frac{e^{\frac{1}{2}i^2t^2}}{\sqrt{2\pi}}\int_{-\infty}^{\infty} e^{-\frac{1}{2}u^2}\,du = e^{\frac{1}{2}i^2t^2}\end{aligned}$$

Thus

$$\mu_1' = E(X) = 0,\ \mu_2 = \text{coefficient of } \frac{i^2t^2}{2} \text{ in } \psi(t) = 1.$$

Let $v = (Z - n\mu)/\sigma\sqrt{n}$ and its characteristic function be $\psi_n(t)$. Then

$$\psi_n(t) = E\left[e^{it\left(\frac{Z-n\mu}{\sigma\sqrt{n}}\right)}\right] = E\left[e^{it\sum_{j=1}^{n}(x_j-\mu)/\sigma\sqrt{n}}\right] = [\psi(t)]^n$$

where $\psi(t)$ is the characteristic function of $(x-\mu)/\sigma\sqrt{n}$.

Again,

$$\psi(t) = 1 + iE\left[\frac{x-\mu}{\sigma\sqrt{n}}\right]t - \frac{1}{2}E\left[\frac{x-\mu}{\sigma\sqrt{n}}\right]^2 t + O\left(\frac{t^2}{n}\right).$$

As $n \to \infty$,

$$O\left(\frac{t^2}{n}\right) \to 0.$$

This is true for any value of t and

$$O\left(\frac{t^2}{n}\right) = 0 \text{ if } t = 0.$$

Therefore, for any value of t

$$\lim_{n\to\infty}\psi_n(t) = \lim_{n\to\infty}\left[1 + \frac{t^2}{2n} + O\left(\frac{t^2}{n}\right)\right]^n = e^{-\frac{1}{2}t^2}.$$

This is the characteristic function of $N(0,1)$ and hence $(Z - n\mu)/\sigma\sqrt{n} \to N(0,1)$ when $n \to \infty$.

EXERCISE

1. (a) Define random variable and its expectation. Show that for the random variables X_1 and X_2

$$E(X_1 + X_2) = E(X_1) + E(X_2).$$

(b) The probability distribution of number of failures of a computer system during work in a weak is shown below:

No. of failures (x)	0	1	2	3	4	5	6
$P(X = x)$	0.02	0.18	0.32	0.14	0.16	0.14	0.08

(i) Calculate expected number of failures in a week.
(ii) Show that $E(2X + 1) = 7.12$.

2. (a) If X is a random variable such that $h(x) < g(x)$, then show that

$$E[h(x)] < E[g(x)].$$

(b) For two dimensional discrete random variables X and Y show that

$$E(X + Y) = E(X) + E(Y).$$

(c) The probability distribution of two dimensional random variables X and Y is given below:

X \ Y	0	1	2	3
0	0.1	0.0	0.2	0.1
1	0.0	0.2	0.1	0.0
2	0.2	0.0	0.0	0.1

Find the value of $E(2X - 3Y)$.

3. (a) Define expectation of function of random variables. If $X_1, X_2, \ldots, X_k$ are k-independent random variables, show that

$$E(X_1, X_2, \ldots, X_k) = E(X_1)E(X_2)\cdots E(X_k).$$

What is the value of $E\left(\sum_{i=1}^{k} X_i\right)$?

(b) Given the joint p.d.f. of two continuous random variables X and Y as follows:

$$f(xy) = \begin{cases} \frac{1}{16}(6 - x - y); & 0 < x < 2,\ 0 < y < 2 \\ 0, & \text{otherwise} \end{cases}$$

Are X and Y independent random variables? If not, calculate $E(Y/X)$ and $E(X/Y)$.

4. (a) What do you mean by conditional expectation and conditional variance?

(b) For two random variables X and Y, show that

$$V(Y) = E[V(Y/X)] + V[E(Y/X)]$$

(c) Given

$$f(xy) = \frac{1}{8}(6 - x - y);\ 0 < x < 2,\ 2 < y < 4$$

Find (i) $E(X/Y)$, (ii) $V(X/y)$, (iii) $E(X^2/Y)$.

5. (a) What do you mean by expectation and variance of a random variable. State and prove the properties of expectation of a random variable.

(b) The joint probability function of two discrete random variables X and Y is shown below:

X \ Y	0	1	2
0	$\frac{1}{8}$	—	$\frac{2}{8}$
1	—	$\frac{2}{8}$	$\frac{1}{8}$
2	$\frac{1}{8}$	$\frac{1}{8}$	—

Find (i) $E(Y/X)$, (ii) $V(Y/X)$, (iii) $V(2X - 3Y)$.

(c) Let $X_1, X_2, \ldots, X_n$ be mutually and identically distributed random variables with mean μ and variance σ^2. Show that

$$E\left[\sum_{k=1}^{n}(X_k - \bar{X})^2\right] = (n-1)\sigma^2, \text{ where } \bar{X} = \frac{1}{n}\sum_{k=1}^{n} X_i.$$

6. (a) Distinguish between indpendent random variables and orthogonal random variables. If X and Y are uncorrelated, determine the conditions under which they are orthogonal. Also, show that

$$\text{cov}^2(XY) \le V(X)V(Y).$$

(b) The joint p.d.f. of two continuous random variables X and Y is given by

$$f(x, y) = \begin{cases} 21x^2y^3, & 0 < x < y < 1 \\ 0, & \text{otherwise} \end{cases}$$

Find (i) $E(X/Y)$, (ii) $V(X/Y)$, (iii) $V(X)$, (iv) $\text{cov}(XY)$.

7. (a) Define conditional variance. Let X and Y be two-diemnsional random variables. Then, show that

$$V(Y/X) = E[Y^2/X] - [E(Y/X)]^2.$$

(b) Two unbiased dice are thrown once. Let X be the random variable indicating sum of the upper faces of dice and Y be the random variable showing even numbers in first die. Find (i) $E(X - 2Y)$, (ii) $V(Y/X)$, (iii) $V(X)$.

(c) The joint p.d.f. of two-dimensional random variables X and Y is given by

$$f(xy) = \begin{cases} 2; & 0 < x < y,\ 0 < y < 1 \\ 0, & \text{otherwise} \end{cases}$$

Calculate (i) $V(Y/X)$, (ii) $V(X)$.

8. (a) State and prove Cauchy-Schwartz inequality.

   (b) Let $f(x)$ be the p.d.f. of the random variable X. Let us define a non-negative function $u(X)$ such that $E[u(X)]$ exists. Then, show that

$$P[u(X) \geq c] = \frac{E[u(X)]}{c}.$$

   (c) Given

$$f(x) = \begin{cases} 2x, & 0 < x < 1 \\ 0, & \text{otherwise} \end{cases}$$

   Find (i) $E(2X-3)$, (ii) $V(3X-2)$, (iii) $E(X^2+X+1)$, (iv) $E(5X)$.

9. (a) State and prove Cauchy-Schwartz inequality and Markov's inequality.

   (b) Let $X_1, X_2, \ldots, X_n$ be independent and identically distributed random variables such that $E(X) = \mu$ and $V(X) = \sigma^2$. Then find (i) $E(\bar{X})$, (ii) $V(\bar{X})$, (iii) $\text{cov}(\bar{X}, X_i - \bar{X})$.

10. (a) Define moment of a random variable. What are the different types of moment. Deduce the relation between first four central moments and general moments.

    (b) Find first four central moments of the distribution of discrete random variable X, where

$$p(X = x) = \binom{n}{x} p^x q^{n-x}; \ x = 0, 1, 2, \ldots, n.$$

11. (a) Define variance and covariance of random variables. State and prove the properties of variance of random variables.

    (b) The joint p.d.f. of two random variables X and Y is given by

$$f(x, y) = \begin{cases} 2(x + y - 3xy^2); & 0 < x < 1, \ 0 < y < 1 \\ 0, & \text{otherwise} \end{cases}$$

    Find mean and variance of the random variables.

12. (a) Define moment and moment generating function of a random variable. State and prove the properties of moment generating function.

    (b) Find first 4 central moments of the distribution of random variable X, where

$$p(X = x) = \frac{\bar{e}^{\lambda} \lambda^{\lambda}}{x!}, \ x = 0, 1, 2, \ldots.$$

    Comment on the shape characteristic of the distribution of X.

13. (a) State and prove the weak law of large number. Deduce as a corollary Bernoulli theorem and comment on its application.

    (b) The probability function of the discrete random variable X is given by

$$p(X = x) = q^{x-1} p; \ x = 1, 2, \ldots,; \ p + q = 1$$

    Find moment generating function of X and hence find mean and variance of X.

(c) The p.d.f. of the continuous random variable X is given by

$$f(x) = e^{-x}, \; x > 0.$$

Find m.g.f. of X and hence find β_1 and β_2 of the distribution of X.

14. (a) Define cumulant generating function and characteristic function of a random variable. Deduce the relation among cumulants and central moments. State and prove the properties of cumulant generating function.

(b) The p.d.f. of the continuous random variable X is given by

$$f(x) = \begin{cases} \theta\, \bar{e}^{\theta x}; & \theta > 0, \; x > 0 \\ 0, & \text{otherwise} \end{cases}$$

Find cumulant generating function of X and hence find β_1 and β_2 of the random variable X.

(c) Given

$$f(x) = \begin{cases} 2x, & 0 < x < 1 \\ 0, & \text{otherwise} \end{cases}$$

Find β_1 and β_2 of the random variable X and hence comment on the shape characteristic of the distribution of X.

15. (a) State and prove Jensen inequality.

(b) Let X be a random variable, $f(X)$ and $g(X)$ be the monotone functions of X over a subset of any real line, where the range of X falls within this subset. Then, if $E[f(X)]$ and $E[g(X)]$ exist

$$E[f(X), G(X)] \geq E[f(X)]E[g(X)].$$

(c) If $g(X)$ is the non-negative function of X, then for every value of $k > 0$

$$P[g(X) \geq k] \leq \frac{E[g(X)]}{k}.$$

16. The continuous random variable X has the p.d.f.

$$f(x) = \begin{cases} \frac{1}{\Gamma n} e^{-x} x^{n-1}, & x > 0 \\ 0, & \text{otherwise} \end{cases}$$

Find characteristic function of X and hence find first four central moments of the distribution.

17. (a) What do you mean by convergence in probability? State and prove weak law of large numbers. What is strong law of large numbers?

(b) Find characteristic function of X, where $X \sim N(\mu, \sigma^2)$.

(c) State and prove Khintchine theorem.

18. (a) Find characteristic function of X, where $X \sim N(\mu, \sigma^2)$ and hence using the Lineberg theorem, show that

$$(Z - n\mu)/\sigma\sqrt{n} \sim N(0, 1), \text{ where } Z = \sum_{i=1}^{n} x_i.$$

(b) State and prove Kolmogorov's inequality.

19. (a) Define characteristic function and write down the properties of characteristic function.

(b) The p.d.f. of the continuous random variable X is given by

$$f(x) = \frac{1}{\lambda} e^{-x/\lambda}; \ \lambda > 0, \ x > 0.$$

Find characteristic function of the distribution of X and hence find first four central moments of X.

20. (a) Let $X_1, X_2, \ldots, X_n$ be n-independent random variables having joint p.d.f.

$$f(x_1, x_2, \ldots, x_n) = \exp\left(-\sum^{n} x_i\right); \ x_i > 0, \ i = 1, 2, \ldots, n$$

Find the p.d.f. of $U = \sum^{n} x_i$ using Levy's theorem.

(b) State and prove the generalized form of Bienayme-Chebyshev's inequality.

CHAPTER 4

Probability Generating Function

4.1 Definition

Let $(a_0, a_1, a_2, \ldots, a_n)$ be a sequence of real numbers. Then the generating function of this sequence is given by

$$A(S) = \sum_{k=0}^{\infty} a_k S^k.$$

The right-hand side of this function converges at zero for all values of S, where

$$|S| < h.$$

If the sequence $\{a_n\}$ is bounded, then by comparing with geometric series it can be said that $A(S)$ converges for at least $|S| < 1$.

Let us consider that for all values of n, $a_n = 1$. Then

$$A(S) = 1 + S + S^2 + \cdots = 1/(1 - S).$$

Again, for a sequence $\{0, 0, 1, 1, \ldots)$

$$A(S) = S^2 + S^3 + \cdots = S^2/(1 - S).$$

For all $a_n = 1/n$, $A(S) = e^S$. In this way for other sequence the generating function $A(S)$ can be found out.

If the values of a sequence $\{a_n\}, n = 0, 1, \ldots$ are the positive integral n values of a random variable X, i.e., if $a_k = P(X = k), k = 0, 1, \ldots,$ then

$$\begin{aligned} P(S) &= E[S^X] \\ &= \sum_{k=0}^{\infty} a_k S^k = \sum_{k=0}^{\infty} S^k P(X = k). \end{aligned}$$

If this $P(S)$ can be defined for at least $|S| < 1$, then it is called ***probability generating function*** of X. Here the coefficient of S^k is $P(X = k)$. This implies that if $P(S)$ can be defined, probability function of X can be found out.

Again, let $S = e^t$, then

$$P(S = e^t) = E\left[e^{tX}\right] = M_x(t).$$

Thus moment generating function and hence moments can be derived if probability generating function exists.

4.2 Properties of Probability Generating Function

1. Let $P(S)$ be the probability generating function of a random variable X. Then for $|S| < 1$, $P(S)$ can be differentiated innumerable times, where the r-th differentiated result is

$$P^{(r)}(S) = \sum_{n=r}^{\infty} n(n-1)\cdots(n-r+1)P(X=n)S^{n-r}.$$

A particular value of this differentiated result is

$$P^{(r)}(0) = r!P(X = r).$$

Proof. The value of $P^{(r)}(S)$ can be written as

$$P^{(r)}(S) = \sum_{i=0}^{\infty}(r+1)(r+1-1)\cdots(l+1)P(X = r+l)S^l \ [n = r+l].$$

Again, from $P^r(0) = r!\ P(X = r)$, we can write

$$P(X = r) = (r!)^{-1}P^{(r)}(0).$$

Therefore, $P(X = r), r = 0, 1, \ldots$ is obtained from r-th differentiated result of $P(S)$ at the point zero.

2. If $P(S)$ is differentiated r times and put $S = 1$, then the r-th factorial moment can be found out. That is

$$\mu_{[r]} = E[X(X-1)\cdots(X-r+1)] = \left.\frac{d^r P(S)}{dS^r}\right|_{S=1}.$$

Proof. We have

$$P(S) = E[S^X] = \sum_{k=0}^{\infty} S^k P(X = k).$$

Then

$$\frac{d^r P(S)}{dS^r} = \sum_{x \in S} x(x-1)(x-2)\cdots(x-r+1)S^{X-r}P(X = k).$$

Let us put $S = 1$. Then

$$\left.\frac{d^r P(S)}{dS^r}\right|_{S=1} = \sum_{x \in S} x(x-1)(x-2)\cdots(x-r+1)P(X = k) = \mu_{[r]}.$$

3. If $E(X)$ of a random variable exists, then

$$E(X) = \left.\frac{dP(S)}{dS}\right|_{S=1}.$$

Again, if $E(X^2)$ exists, then

$$V(X) = \left.\frac{d^2P(S)}{dS^2}\right|_{S=1} + \left.\frac{dP(S)}{dS}\right|_{S=1} - \left[\left.\frac{dP(S)}{dS}\right|_{S=1}\right]^2.$$

Proof. We know

$$V(X) = E(X^2) - [E(X)]^2 = E[X(X-1)] + E(X) - [E(X)]^2.$$

But

$$\left.\frac{dP(S)}{dS}\right|_{S=1} = E(X) \text{ and } \left.\frac{d^2P(S)}{dS^2}\right|_{S=1} = E[X(X-1)].$$

Therefore,

$$V(X) = \left.\frac{d^2P(S)}{dS^2}\right|_{S=1} + \left.\frac{dP(S)}{dS}\right|_{S=1} - \left[\left.\frac{dP(S)}{dS}\right|_{S=1}\right]^2.$$

4. Let $X_1, X_2, \ldots, X_n$ be n-independent random variables. Then for $|S| \leq 1$,

$$\sum_{i=1}^{P_n} X_i(S) = \prod_{i=1}^{n} P_{x_i}(S) = [P_{x_i}(S)]^n$$

where $\sum_{i=1}^{P_n} X_i(S)$ is the probability generating function of $\sum_{i=1}^{n} X_i$ and $P_{x_i}(S)$ is the probability generating function of X_i.

Proof. Since X_i's are independent, S^{X_i} $(i = 1, 2, \ldots, n)$ are also independent. Then

$$\sum_{i=1}^{P_n} X_i(S) = E\left[S^{\sum_i X_i}\right] = E\left[\prod_{i=1}^{n} S_i^{X_i}\right]$$

$$= \prod_{i=1}^{n} E\left[S_i^{X_i}\right] = \prod_{i=1}^{n} P_{x_i}(S) = [P_{x_i}(S)]^n.$$

5. The distribution function can be derived uniquely from probability generating function.

Proof. Let X and Y be two random variables having probability generating functions $P(S)$ and $Q(S)$, respectively. That is

$$P(S) = \sum_{k=0}^{\infty} S^k P(X = k)$$

$$\text{and } Q(S) = \sum_{k=0}^{\infty} S^k P(Y = k), \text{ where } |S| < h,\ h > 0.$$

But

$$P(X = k) = \frac{1}{k!}\frac{d^k P(S)}{dS^k}\bigg|_{S=0} \quad \text{and } P(Y = k) = \frac{1}{k!}\frac{d^k Q(S)}{dS^k}\bigg|_{S=0}.$$

But the right-hand sides of both the equations are same. Therefore,

$$P(X = k) = P(Y = k), \ k = 0, 1, 2, \ldots.$$

Thus X and Y have the same distribution.

6. Let the moment generating function of random variable X be $M(S)$ such that $M(S)$ is finite and $|S| < S_0, \ S_0 > 0$. Then

(a) moments of all order of X exist,

(b) Taylor series expansion toward zero of $M(S)$ is possible.

Here the r-th moment of X is

$$\mu'_r = \frac{d^r M(S)}{dS^r}\bigg|_{S=0}.$$

Proof. To prove this let us consider the following aspect:

Let X be a random variable having distribution function F. Consider that $E\left[e^{SX}\right]$ is finite for an open interval including $S = 0$. Then the moment generating function of X is

$$\begin{aligned} M(S) &= E\left[e^{SX}\right] \\ &= \sum_{j=1}^{\infty} e^{Sx_j} P(X = x_j), \text{ if } X \text{ is discrete random variable} \\ &= \int_{-\infty}^{\infty} e^{Sx} f(x)dx, \text{ if } X \text{ is continuous random variable.} \end{aligned}$$

But, if X is a random variable of non-negative values having probability generating function $P(S)$, then

$$M(S) = E\left[e^{SX}\right] = P\left[e^{S}\right].$$

This implies that, if $S = 0$,

$$M(S) = P[e^0] = 1.$$

Therefore,

$$\mu'_r = \frac{d^r M(S)}{dS^r}\bigg|_{S=0}.$$

Now, let us consider the proof of the property (6) given above

$$\exp[|SX|] \leq \exp(SX) + \exp(-SX).$$

Since $M(S)$ is finite for $|S| < S_0, \ S_0 > 0$, we can write

$$E\left[e^{|SX|}\right] < \infty, \ S < |S_0|.$$

Let $S \in (0, S_0)$. Then for every value of $n \geq 0$,

$$\exp[|SX|] = \exp[S|X|] > \frac{S^n|X|^n}{n!}.$$

Hence

$$E[|X|^n] < n; \; S^{-n}E[|SX|] < \infty.$$

Again, for $|S| < S_0$ and for any $n \geq 0$,

$$\left|\sum_{k=0}^{n} \frac{S^k}{k!} X^k\right| \leq \sum_{k=0}^{\infty} \frac{|SX|^k}{k!} = \exp[|SX|].$$

Hence

$$\left|\sum_{k=0}^{\infty} \frac{S^k}{k!} E[X^k]\right| < \infty$$

$$M(S) = E\left[e^{SX}\right] = E\left[\sum_{k=0}^{\infty} \frac{(Xk)^k}{k!}\right] = \sum_{k=0}^{\infty} \frac{S^k E[X^k]}{k!}.$$

Therefore, Taylor series expansion of $M(S)$ is possible. Again, as $M(S)$ is power series, it can be differentiated and

$$E[X^k] = \mu_r' = \left.\frac{d^r M(S)}{dS^r}\right|_{S=0}, \; r \geq 1.$$

It can also be decided that, if $M(S)$ is a moment generation function, then

$$\psi(S) = \ln M(S)$$

exists and it can be differentiated. Thus

$$E(X) = \left.\frac{d\psi(S)}{dS}\right|_{S=0} \text{ and } V(X) = \left.\frac{d^2\psi(S)}{dS^2}\right|_{S=0}.$$

Because

$$\psi'(S) = \frac{d\psi(S)}{dS} = \frac{d\ln M(S)}{dS} = \frac{dM(S)}{M(S)} = \frac{M'(S)}{M(S)}.$$

Now, if $S = 0$,

$$\begin{aligned}\psi'(0) &= \left.\frac{d\psi(S)}{dS}\right|_{S=0} = \frac{M'(0)}{M(0)} = M'(0) \; [\because \; M(0) = 1] \\ &= \left.\frac{dM(S)}{dS}\right|_{S=0} = E(X).\end{aligned}$$

Again,

$$\psi''(S) = \frac{M''(S)M(S) - [M'(S)]^2}{[M(S)]^2}.$$

Therefore, for $S = 0$

$$\psi''(0) = M''(0) - [M''(0)]^2 = E(X^2) - [E(X)]^2 = V(X).$$

This moment generating function also provides probability function.

7. Let $X_1, X_2, \ldots, X_n$ be n-independent random variables. The moment generating functions of these random variables are $M_1(S), M_2(S), \ldots, M_n(S)$. These moment generating functions are finite over $(-S_0, S_0)$ for $S_0 > 0$. Then the moment generating function of $S_n = \sum_{i=1}^{n} X_i$ is

$$M(S) = \prod_{i=1}^{n} M_i(S), \; S \in (-S_0, S_0).$$

If X_i's are identically distributed random variables, then

$$M(S) = [M_i(S)]^n, \; i = 1, 2, \ldots, n.$$

►**EXAMPLE 4.1** *The probability function of the discrete random variable X is given below:*

$$P(x) = {}^nC_x p^x q^{n-x}, \; x = 0, 1, 2, \ldots, n.$$

Find probability generating function of X and hence or otherwise find the moments of the distribution.

Solution. We know

$$\begin{aligned} P(S) = E[S^X] &= \sum_{x=0}^{n} S^x p(x) = \sum_{x=0}^{n} S^x \, {}^nC_x p^x q^{n-x} \\ &= \sum_{x=0}^{n} {}^nC_x (pS)^x q^{n-x} = (q + pS)^n. \end{aligned}$$

Now

$$\begin{aligned} \mu_1' = E(X) &= \left.\frac{dM(S)}{dS}\right|_{S=1} = n(q+pS)^{n-1}p\Big|_{S=1} = np \\ \mu_2 = V(X) &= \left.\frac{d^2P(S)}{dS^2}\right|_{S=1} + \left.\frac{dP(S)}{dS}\right|_{S=1} - \left[\left.\frac{dP(S)}{dS}\right|_{S=1}\right]^2 \\ &= n(n-1)p^2(q+pS)^{n-2}\Big|_{S=1} + np - n^2p^2 \\ &= npq. \end{aligned}$$

Also, we know the moment generating function of X is

$$M(S) = E\left[e^{SX}\right] = \sum_{x=0}^{n} {}^nC_x \left(pe^S\right)^x q^{n-x} = \left(q + pe^S\right)^n$$

$$\begin{aligned} \mu_3' = \left.\frac{d^3M(S)}{dS^3}\right|_{S=0} &= \Big[n(n-1)(n-2)\left(q+pe^S\right)^{n-3} p^2 e^{2S} \\ &\quad + n(n-1)\left(q+pe^S\right)^{n-2} pe^S pe^S + np\left(q+pe^S\right)^{n-1} pe^S \\ &\quad + 2n(n-1)p^2e^{2S}(q+pe^S)^{n-2}\Big]\Big|_{S=0} \\ &= np + n(n-1)p^2[3 + (n-2)p]. \end{aligned}$$

Similarly,

$$\mu_4' = \left.\frac{d^4P(S)}{dS^4}\right|_{S=0} = np[1+7(n-1)p] + n(n-1)(n-2)p^3\,[6+(n-3)p]\,.$$

►**EXAMPLE 4.2** *The probability generating function of a random variable X is given by*

$$P(S) = \left(\frac{1+S}{2}\right)^n.$$

Find probability function of X.

Solution. We have

$$\begin{aligned} P(S) &= \left(\frac{1}{2}\right)^n (1+S)^n \\ &= \left(\frac{1}{2}\right)^n \left[1 + {}^nC_1 S + {}^nC_2 S^2 + \cdots + S^n\right]. \end{aligned}$$

Here $P(X=x)$ is the coefficient of S^n. Therefore,

$$\begin{aligned} P(X=x) &= \left(\frac{1}{2}\right)^n {}^nC_x \\ &= {}^nC_x \left(\frac{1}{2}\right)^x \left(\frac{1}{2}\right)^{n-x}, \quad x = 0,1,2,\ldots,n. \end{aligned}$$

►**EXAMPLE 4.3** *The probability generating function of a random variable X is given by*

$$P(S) = \left(\frac{1+3S}{4}\right)^n.$$

Find probability function of X.

Solution. We have

$$\begin{aligned} P(S) &= \left(\frac{1}{4} + \frac{3}{4}S\right)^n \\ &= \left(\frac{1}{4}\right)^n + {}^nC_1\left(\frac{1}{4}\right)^{n-1}\left(\frac{3}{4}S\right) + {}^nC_2\left(\frac{1}{4}\right)^{n-2}\left(\frac{3}{4}\right)^2 + \cdots \\ &\quad + {}^nC_x\left(\frac{1}{4}\right)^x\left(\frac{3}{4}\right)^{n-x} + \cdots + \left(\frac{3}{4}S\right)^n. \end{aligned}$$

Here $P(X=x)$ is the coefficient of S^n. Therefore,

$$P(X=x) = {}^nC_x\left(\frac{1}{4}\right)^x\left(\frac{3}{4}\right)^{n-x}, \quad x = 0,1,2,\ldots,n.$$

►**EXAMPLE 4.4** *The probability function of a random variable X is given by*

$$P(x) = \frac{1}{N}, \quad x = 1,2,\ldots,N.$$

Find probability generating function of X and hence find its mean and variance.

Solution. We know

$$P(S) = E\left[S^X\right] = \sum_{x=1}^{N} S^x \frac{1}{N} = \frac{1}{N}\sum_{x=1}^{N} S^x$$

$$= \frac{1}{N}\left[S + S^2 + S^3 + \cdots + S^n\right] = \frac{S\left(1 - S^N\right)}{1 - S}, |S| < 1.$$

Now

$$\text{Mean } (X) = E(X) = \mu_1' = \left.\frac{dP(S)}{dS}\right|_{S=1}$$

$$= \frac{1}{N}\frac{dP(S)}{dS}\left.\left[S + S^2 + \cdots + S^N\right]\right|_{S=1}$$

$$= \frac{1}{N}[1 + 2 + 3 + \cdots + N] = \frac{N+1}{2}$$

$$V(X) = \mu_2 = \left.\frac{d^2P(S)}{dS^2}\right|_{S=1} + \left.\frac{dP(S)}{dS}\right|_{S=1} - \left[\left.\frac{dP(S)}{dS}\right|_{S=1}\right]^2$$

$$= \frac{1}{N}\left.\left[2 + 6S + 12S^2 + \cdots + N(N-1)S^{N-2}\right]\right|_{S=1} + \frac{N+1}{2} - \frac{(N+1)^2}{4}$$

$$= \frac{N^2-1}{3} + \frac{N+1}{2} - \frac{(N+1)^2}{4} = \frac{N^2-1}{12}.$$

►**EXAMPLE 4.5** *The probability generating function of a random variable X is given by*

$$P(S) = \ln(1 - SP)/\ln(1 - P),\ |S| < \frac{1}{P}.$$

Find probability function of X.

Solution. We can write

$$P(S) = \frac{-1}{\ln(1-P)}\left[SP + \frac{S^2P^2}{2} + \frac{S^3P^3}{3} + \cdots + \frac{S^xP^x}{x} + \cdots\right].$$

Here the coefficient of S^x is $P(X = x)$. Thus

$$P(X = x) = \frac{-P^x}{x\ln(1-P)} = \frac{P^x}{\ln(1-P)^{-x}},\ x = 1, 2, \ldots .$$

►**EXAMPLE 4.6** *The probability function of the random variable X is given by*

$$P(x) = \frac{e^{-\lambda}}{1 - e^{-\lambda}}\frac{\lambda^x}{x!},\ x = 1, 2, \ldots ;\ \lambda > 0.$$

Find probability generating function of X.

Solution. We can write

$$P(S) = E\left[S^X\right] = \frac{e^{-\lambda}}{1 - e^{-\lambda}}\sum_{x=1}^{\infty}\frac{(\lambda S)^x}{x!}$$

$$= \frac{e^{-\lambda\left(e^{\lambda S} - 1\right)}}{1 - e^{-\lambda}} = \frac{e^{-\lambda}}{1 - e^{-\lambda}}\left(e^{\lambda S} - 1\right).$$

►EXAMPLE 4.7 *The moment generating function of a random variable X is*

$$M(S) = (1 - S)^{-n}.$$

Find moments of the random variable and comment.

Solution. We know

$$\mu_r' = \left.\frac{d^r M(S)}{dS^r}\right|_{S=0}.$$

Hence

$$\begin{aligned}
\mu_1' &= \left.\frac{dM(S)}{dS}\right|_{S=0} = -n(1-S)^{-n-1}(-1)\Big|_{S=0} = n\\
\mu_2' &= \left.\frac{d^2M(S)}{dS^2}\right|_{S=0} = -(n+1)n(1-S)^{-n-2}(-1)\Big|_{S=0} = n(n+1)\\
\mu_3' &= \left.\frac{d^3M(S)}{dS^3}\right|_{S=0} = -n(n+1)(n+2)(1-S)^{-n-3}(-1)\Big|_{S=0}\\
&= n(n+1)(n+2)\\
\mu_4' &= \left.\frac{d^4M(S)}{dS^4}\right|_{S=0} = -n(n+1)(n+2)(n+3)(1-S)^{-n-4}(-1)\Big|_{S=0}\\
&= n(n+1)(n+2)(n+3)\\
\mu_2 &= \mu_2' - \mu_1'^2 = n(n+1) - n^2 = n\\
\mu_3 &= \mu_3' - 3\mu_2'\mu_1' + 2\mu_1'^3 = n(n+1)(n+2) - 3n^2(n+1) + 2n^3 = 2n\\
\mu_4 &= \mu_4' - 4\mu_3'\mu_1' + 6\mu_2'\mu_1'^2 - 3\mu_1'^4\\
&= n(n+1)(n+2)(n+3) - 4n^2(n+1)(n+2) + 6n^3(n+1) - 3n^4\\
&= 3n(n+2)\\
\therefore\ \beta_1 &= \frac{\mu_3^2}{\mu_2^3} = \frac{4n^2}{n^3} = \frac{4}{n},\quad \mu_3 = 2n > 0\\
\beta_2 &= \frac{\mu_4}{\mu_2^2} = \frac{3n(n+2)}{n^2} = 3 + \frac{6}{n} > 3.
\end{aligned}$$

The distribution is positively skewed and leptokurtic.

►EXAMPLE 4.8 *Let $X_1, X_2, \ldots, X_k$ be k-independent random variables. The probability function of X_i $(i = 1, 2, \ldots, k)$ is*

$$P(x_i) = p(1-p)^x,\ x = 1, 2, \ldots.$$

Find probability generating function of $\sum_{i=1}^{n} X_i$ and hence its probability function.

Solution. We know

$$\begin{aligned}
P(S) = E\left[S^X\right] &= \sum_{x=1}^{\infty} S^x p(1-p)^x\\
&= \frac{pS}{1 - S(1-p)},\ |S| \le (1-p)^{-1}.
\end{aligned}$$

Since $X_1, X_2, \ldots, X_k$ are independent,

$$P_{\Sigma x_i}(S) = \prod_{i=1}^{k} P_{x_i}(S) = \left[\frac{pS}{1-S(1-p)}\right]^k$$

$$\therefore\ p(Y = \Sigma X_j) = \binom{y-1}{k-1} p^k (1-p)^{y-k};\ y = k, k+1, k+2, \ldots.$$

►**EXAMPLE 4.9** *The probability function of a discrete random variable X is given by*

$$P(x) = \frac{1}{2^x},\ x = 1, 2, \ldots.$$

Find probability generating function of X. Also, find mean and variance of X.

Solution. The probability function of X is

$$\begin{aligned} P(S) &= E\left[S^X\right] \\ &= \sum_{x=1}^{\infty} S^x P(x) = \sum_{x=1}^{\infty} \frac{S^x}{2^x} = \sum_{x=1}^{\infty} \left(\frac{S}{2}\right)^x = \frac{\frac{S}{2}}{1-\frac{S}{2}},\ |S| < 1. \end{aligned}$$

Also, we have

$$\begin{aligned} E(X) &= \mu_1' = \left.\frac{dP(S)}{dS}\right|_{S=1} \\ &= \left.\left[\left(1-\frac{S}{2}\right)^{-1}\frac{1}{2} + \frac{S}{2}\left(1-\frac{S}{2}\right)^{-2}\frac{1}{2}\right]\right|_{S=1} = 2 \\ V(X) &= \mu_2 = \left.\frac{d^2P(S)}{dS^2}\right|_{S=1} + \left.\frac{dP(S)}{dS}\right|_{S=1} - \left[\left.\frac{dP(S)}{dS}\right|_{S=1}\right]^2 \\ &= \left.\left[\frac{1}{4}\left(1-\frac{S}{2}\right)^{-2} + \frac{1}{4}\left(1-\frac{S}{2}\right)^{-2} + \frac{S}{4}\times 2\left(1-\frac{S}{2}\right)^{-3}\cdot\frac{1}{2}\right]\right|_{S=1} + 2 - 2^2 \\ &= 4 + 2 - 4 = 2. \end{aligned}$$

CHAPTER 5

General Discrete Distributions

5.1 System of Discrete Distributions

Carver (1919) and Katz (1948) considered the system of discrete distributions using the difference equation

$$\Delta f_{r-1} = \frac{(a-r)f_{r-1}}{b_0 + b_1 r + b_2 r(r-1)}. \tag{5.1}$$

This system is analogous to the system of differential equation

$$\frac{df}{dx} = \frac{(a-x)f}{b_0 + b_1 x + b_2 x^2} \tag{5.2}$$

to define the Pearson system of continuous distributions. In system (5.1), if $b_2 = 0$, a special case of the power series distributions discussed by Tweedie (1947) and Noack (1950) are obtained. Ord (1967) discussed the range, density function and critical values of distributions contained in the system.

Moments Relations

Multiplying (5.1) throughout by

$$r^{(j)} = r(r-1), \ldots, (r-j+1)$$

and summing over $[u, v]$, the range of r, the factorial moments about the origin, $\mu'_{(j)}$, satisfy the relation

$$\begin{aligned} &\{(j+2)b_2 - 1\}\mu'_{(j+1)} + \{(j+1)(b_1 + 2jb_2) + a - 2j - 1\}\mu'_{(j)} \\ &+\{b_0 + b_{1j} + b_{2j}(j-1) + a - j\}j\mu'_{(j-1)} + E_{j+1} = 0 \text{ for } j = 0, 1, 2, \ldots \end{aligned}$$

with $\mu'_{(-1)} = 0$, $\mu'_{(0)} = 1$ and

$$\begin{aligned} E_{j+1} &= \{b_0 + b_1 u + b_2 u(u-1)\}u^{(j)} f_u \\ &\quad -\{a + b_0 + (v+1)(b_1 - 1 + vb_2)\}(v+1)^{(j)} f_v. \end{aligned}$$

The lower and upper limits (u, v) of variation of j for which $f_j \neq 0$ are taken ($-\infty$ to $+\infty$) or $(0, N)$, where N may be a positive integer or infinite. Ord considered the two cases where the range is

(i) doubly infinite; or

(ii) has $b_0 = 0$, lower terminal zero and has when N is infinite

$$a + (b_1 + b_2 N - 1)(N + 1) = 0$$

then $E_{j+1} = 0$ for all j.

The second case covers all the complete distributions based on hypergeometric series. Ord showed the expressions for the parameters, in terms of moments which are shown in Table-1.

Table-1: Values of the parameters in terms of the first three (or four) moments

Parameter	*Discrete with range* $(-\infty$ *to* $\infty)$	*Discrete with range* $[0, N]$
a	$\frac{1}{2} - (\mu_3/\mu_2)(\beta_2 + 3 - 1/\mu_2)D_F$	$(1 - 2b_2)(1 + \mu) - b_1$
b_0	$\mu_2(4\beta_2 - 3\beta_1 - 1/\mu_2)/D_F$	0
b_1	$1 - a$	$\{\mu_2 - b_2(3\mu_2^2 + \mu^2 - \mu - 2)\}$
b_2	$(2\beta_2 - 3\beta_1 - 6 + 1/\mu)/D_F$	$\{\mu(\mu_3 + \mu_2) - 2\mu_2^2\}/D_G$
	$D_F = 2(5\beta_2 - 6\beta_1 - 9 + 1/\mu_2)$	$D_G = 4\mu\mu_3 + 2\mu_2(2 + \mu^2 - 3\mu_2)$

Ord mentioned that the form of the density function will depend on the behaviour of the roots of the denominator in equation (5.1). He defined the criterion

$$k = (b_1 - b_2 - 1)^2/\{4b_2(b_0 + a)\}$$

corresponding to the alternative form of (5.1)

$$\Delta f_{r-1} = \frac{(a - r)f_r}{(a + b_0) + (b_1 - 1)r + b_2 r(r - 1)}$$

This equation has real roots in the denominator unless $0 < k < 1$. The types of distributions derived by Ord are presented in Table-2.

Table-2: Distributions contained in system (5.1) with range, density function and critical values

Type (d)	*Name*	*Density*	*Criteria*	*Range*	*Comments*
I(a)	Hypergeometric	$\binom{N_p}{r}\binom{N_q}{n-r}/\binom{N}{n}$	$I < 1$, $K > 1$	$[0, m]$ $m = \min$ $[n, N_p]$	J- or Bell-shaped
I(b)	Negative hyper-geometric or beta-binomial	$\binom{k+r-1}{r}\binom{N-k-r}{N_p-r}/\binom{N}{N_p}$	$K > 0$	$[0, N_p]$	J- or Bell-shaped

Type (d)	*Name*	*Density*	*Criteria*	*Range*	*Comments*
I(e)	—	$\binom{A}{r}\binom{C}{B-r}/\binom{A+C}{B}$	$K > 1$	$[0, \infty)$	A, C non-integer, but have the same integral part
I(u)	—	$\alpha\left\{\binom{A}{C+r}\binom{B}{D-r}\right\}^{-1}$	$K > 1$	$[0, n]$	U-shaped
VI	Beta-Pascal	$\left(\frac{A}{k+A}\right)\binom{k-r-1}{r}$ $\binom{A+B-1}{A}/$ $\binom{K+A+B+r-1}{k+A}$	$I > 1$, $K > 1$	$[0, \infty)$	J- or Bell-shaped
IV	—	$\alpha Q(r, a, d)/$ $Q(r, k+a, b), r > 0$	$0 < K < 1$	$(-\infty, \infty)$	k a positive integer, Bell-shaped
II(a) II(b) II(u)	As for type I $(\cdot)$	Similar expression for $r < 0$	$I < 1$ $K = 1$ $K = 0$ $K = 1$	As for type I $(\cdot)$	Symmetric forms of type I $(\cdot)$
V	—	As type IV, but $b = 0$	$K = 0$	$[0, \infty)$ or $[-\infty, \infty]$	Limiting form of IV
III(B)	Binomial	$\binom{n}{r}p^r(1-p)^{n-r}$	$I < 1$, $K \to \infty$	$[0, n]$	Limiting form of I(a), I(b)
III(N)	Negative binomial or Pascal	$\binom{k+r-1}{r}p^k(1-p)^r$	$I > 1$, $K \to \infty$	$[0, \infty)$	Limiting form of I(b), VI
III(P)	Poisson	$e^{-m}m^r/r!$	$I = 1$, $K \to \infty$	$[0, \infty)$	Limiting form of III(B), III(N)
VII	Discrete Student's t	$\alpha\left[\prod_{j=1}^{k}\{(j+r+a)^2+b^2\}\right]^{-1}$	$0 < K < 1$	$[-\infty, \infty]$	Nearly symmetric form of IV

NOTES:

(1) $Q(r, a, d) = (a^2 + d^2)\{(a+1)^2 + d^2\} \cdots \{(a+r)^2 + d^2\}$

(2) Except for random variable r, all symbols are parameters and $p = 1 - q$.

(3) α is a constant such that the total measure is 1.

(4) The type VIId distribution is also discussed by Ord; all its finite odd order moments about mean is zero.

To distinguish the different distributions Ord discussed the (β_1, β_2)-plane. This plane determines the regions for hypergeometric series distributions. The plane is shown in FIG. 5.1.

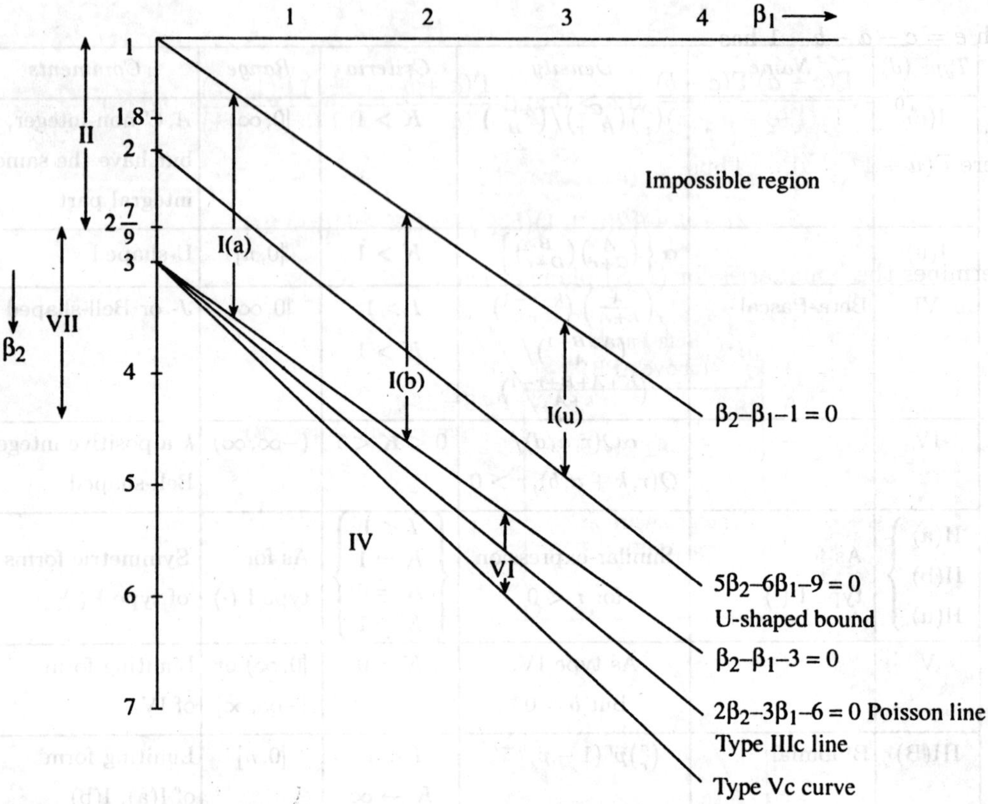

FIG. 5.1 : Plot of β_1 against β_2 for the discrete system

NOTES:

(1) The binomial is described by the family of straight lines $\beta_2 - \beta_1 - 3 + \frac{2}{n} = 0$ and the negative binomial by $\beta_2 - \beta_1 - 3 - \frac{2}{k} = 0$.

(2) The type IIId distributions may be shown to satisfy $2\beta_2 - 3\beta_1 - 6 + \frac{1}{\mu_2} = 0$

(3) By analogy with the Pearson system, it is conjectured that the U-shaped distributions lie above the line $5\beta_2 - 6\beta_1 - 9 = 0$.

(4) Type IId covers the β_2-axis over the interval $[0, 3]$, while type VIId covers this axis over $\left[2\frac{7}{4}, \infty\right]$.

(5) The boundaries of the type IVd and type Vd regions are uncertain, designated by lines without arrowheads, dotted by the type Vc curve.

There is another way of distinguishing the distributions defined over $[0, N]$, without any overlap of regions. This is according to Ord is (I, S) plane, where $I = \mu_2/\mu_1'$ and $S = \mu_3/\mu_2$. The (I, S) plane was used for hypergeometric series distributions, where the hypergeometric series is

$$F(a, b; c : 1) = f_0 \left\{ 1 + \frac{ab}{CI} + \frac{ab(a+1)(b+1)}{C(c+1)2!} + \dots \right\}$$

with $e = c - a - b - 1$ has

$$f_0 = \frac{\Gamma(c-a)\,\Gamma(c-b)}{\Gamma c\,\Gamma(c-a-b)} \text{ if } e > 0 \text{ and } \frac{\Gamma(a+b-c+1)\,\Gamma(1-c)}{\Gamma(a-c+1)\,\Gamma(b-c+1)} \text{ if } e < 0$$

where $\Gamma(u+1) = u\Gamma u$. Thus

$$S = 2I - I + \{2(a+1)(b+1)\}/\{(e-1)(e-2)\}$$

determines the boundaries for (I, S) plane as shown in FIG. 5.2.

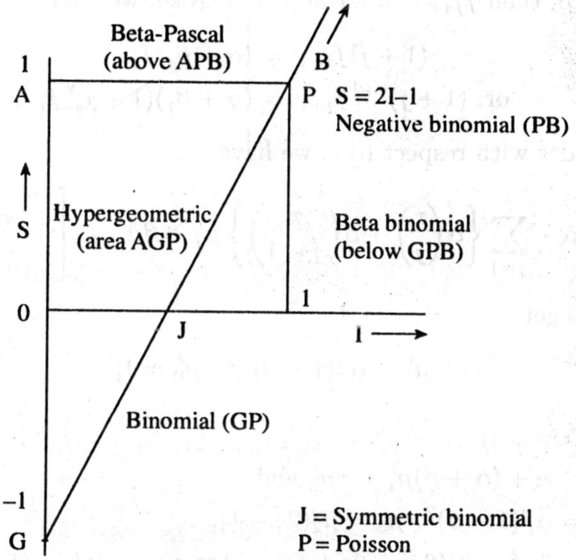

FIG. 5.2 : *I*, *S* diagram for the real roots distributions

The type I(e) is excluded from the diagram as it covers the entire region $I > 0, S >$
Now, for a given mean value it can be shown that

$$I(H) < I(B) < \left\{ \begin{array}{c} I(Be - B) \\ I(P) = 1 \end{array} \right\} < I(Be - N)$$

Here H, B, Be, P and N are used for hypergeometric, binomial, beta, Poisson and negative binomial respectively. The above ordering of the distributions requires knowledge of the parameters. When the moments involve the terminal frequencies, through single or double truncation or otherwise, a graphical method for choosing the form of distribution has been discussed by Ord (1967) plotting $u_r = rf_r/f_{r-1}$ against r, where

$$af_{r-1} - \{b_0 + b_1 r + b_2 r(r-1)\}\Delta f_{r-1} = rf_{r-1}.$$

The graph gives straight line for the binomial, negative binomial, Poisson and rectangular distributions and their truncated forms, including the logarithmic. For the U-shaped distributions, the graph is plotted for $U_r = rf_{r-1}/f_r$ against r, which yields a straight line for type III(u). From observed data I and S are estimated from sample moments which provides an objective means of selecting a model. Katz (1963) discussed the simpler systems of distributions for which

$$\frac{f_{j+1}}{f_j} = \frac{\alpha+\beta_j}{1+j};\ j=0,1,2,\ldots;\ \alpha>0;\ \beta<1.$$

This is based on his earlier work on systems of distributions for which

$$f_{j+1}/f_j = (\text{linear function of } j)/(\text{quadratic function of } j).$$

Now, if $\alpha+\beta_j<0$, then $f_{j+r}=0$ for all $r>0$. Also, we have

$$\begin{aligned}(1+j)f_{j+1} &= (\alpha+\beta_j)f_j\\ \text{or, } (1+j)^{r+1}f_{j+1} &= (\alpha+\beta_j)(1+j)^r f_j.\end{aligned}$$

Summing both sides with respect to r, we have

$$\mu'_{r+1} = \alpha + \sum_{j=1}^{r}\left\{\alpha\binom{r}{j}+\beta\binom{r}{j-1}\right\}\mu'_j + \beta\mu'_{r+1}\quad\left[\because\ \sum_{j=0}^{\infty} f_j = 1\right].$$

Putting $r=0$, we get

$$\mu'_1 = \alpha/(1-\beta)\ [\because\ \mu'_0 = 1].$$

Similarly,

$$\begin{aligned}\mu'_2 &= \alpha+(\alpha+\beta)\mu'_1+\beta\mu'_2 \text{ and}\\ \mu_2 &= \alpha/(1-\beta)^2,\ \mu_3=\mu_2(2c-1),\\ \mu_4 &= 3\mu_2^2+\mu_2(6c^2-6c+1),\ \text{where } c=\mu_2/\mu'_1=(1-\beta)^{-1}.\end{aligned}$$

The moment ratios are

$$\begin{aligned}\alpha_3 &= \sqrt{\beta_1} = \left(2-c^{-1}\right)\alpha^{-1/2} = (1+\beta)\alpha^{-1/2}\\ \alpha_4 &= \beta_2 = 3+\left(6-6c^{-1}+c^{-2}\right)\alpha^{-1} = 3+\left(1+4\beta+\beta^2\right)\alpha^{-1}\end{aligned}$$

From Katz's system, we have

$$(j+1)f_{j+1} = (\alpha+\beta_j)f_j.$$

Summing both sides over all values of j not less than $[\mu'_1]$, we get

$$\sum_{j=[\mu'_1]}^{\infty}(j+1)f_{j+1} = \alpha\sum_{j=[\mu'_1]}^{\infty} f_j + \beta\left[f_{[\mu'_1]} + \sum_{j=[\mu'_1]}^{\infty}(j+1)f_{j+1}\right].$$

Noting that $\mu'_1=\alpha(1-\beta)^{-1}$, we get

$$(1-\beta)\sum_{j=[\mu'_1]}^{\infty}(j+1-\mu'_1)f_{j+1} = (\alpha+\beta[\mu'_1])\,f_{[\mu'_1]}.$$

Hence, the mean deviation for this class of distribution is

$$\text{mean deviation} = 2f_{[\mu_1']\mu_2}.$$

This provides an approximate result. But, if μ_1' is an integer, or if $\beta = 0$, this formula is exact for calculating mean deviation.

The probability generating function derived from Katz's system of equations satisfies the equation

$$g'(t) = \alpha g(t) + \beta t g'(t), \text{ i.e., } \frac{d \log g(t)}{dt} = \frac{\alpha}{1-\beta t}.$$

Hence, if $\beta \neq 0$

$$g(t) = [(1-\beta t)/(1-\beta]^{-\alpha/\beta}.$$

As $g(1) = 1$, the factorial moment generating function is given by

$$[(1-\beta-\beta t)/(1-\beta)]^{-\alpha/\beta}$$

and the r-th factorial moment is

$$\mu_{(r)} = (-\beta/(1-\beta)^r(-\alpha/\beta)^{(r)}.$$

When $\beta = 0$, the Katz's system gives $f_{j+1} = \alpha f_j/(j+1)$

$$\therefore\ f_j = (\alpha^j/J!)\, f_0.$$

Corresponding to this value the distribution is Poisson with

$$f_j = (\alpha^j/J!)e^{-\alpha}.$$

A similar analysis shows that if $0 < \beta < 1$, then

$$f_j = \binom{\frac{\alpha}{\beta}+j-1}{j}\beta^j(1-\beta)^{\alpha/\beta},\ j = 0, 1, 2, \ldots$$

which is a negative binomial distribution with parameters α/β and $\beta/(1-\beta)$.

If $\beta < 0$ and $(-\alpha/\beta)$ is an integer, then $0 < -\beta/(j-\beta) < 1$ and

$$f_j = \binom{-\alpha/\beta}{j}\left(\frac{-\beta}{1-\beta}\right)^j\left(1-\frac{-\beta}{1-\beta}\right)^{-\frac{\alpha}{\beta}-j}.$$

This is a binomial distribution with parameters $-\alpha/\beta$ and $-\beta/(1-\beta)$.

To decide about the kind of distribution to use for a set of data Guldberg (1931) made use of the identity

$$\frac{f_{j+1}}{f_j}(j+1) + \frac{\mu_1'-\mu_2}{\mu_2}j = \alpha + \beta_j + \frac{(1-\beta)^{-1}-(1-\beta)^{-2}}{(1-\beta)^{-2}}$$
$$= \alpha = \mu_1'^2/\mu_2.$$

Hence the quantities

$$\left[T_j + \frac{\mu_1' - \mu_2}{\mu_2} j\right] \frac{\mu_2}{\mu_1'^2}$$

are constant and equal to one. Thus to satisfy

$$\frac{f_{j+1}}{f_j} = \frac{\alpha + \beta_j}{1 + j}, \; j = 0, 1, 2, \ldots; \; \alpha > 0; \; \beta < 1$$

the sample values

$$\left[t_j + \frac{\bar{x} - s^2}{s^2} j\right] \frac{s^2}{\bar{X}^2} = t_j \frac{s^2}{\bar{X}^2} + \frac{\bar{X} - s^2}{\bar{X}^2} j$$

should be equal to 1. Here

$$T_j = (j+1) f_{j+1}/f_j, \bar{x} = n^{-1} \sum_{j=1}^{n} x_j, s^2 = (n-1)^{-1} \sum_{j=1}^{n} (x_j - \bar{x})^2.$$

5.2 Power Series Distributions

The distributions for which the intervals between values of any one random variable with non-zero probabilities are all integral multiples of one quantity are called *lattice distributions.* Again, for most of the discrete distributions the values of the random variables cannot be negative. These non-negative lattice distributions can be classified in many ways. It can be classified on the form of the probability f_j which is a function of j, ($j = 0, 1, 2, \ldots$). Noack (1950) defined such a function as

$$f_j = a_j \theta^j / P(\theta) \text{ for } j = 0, 1, 2, \ldots$$

where $a_j \theta^j \geq 0$, a_j is a function of j or constant, and

$$P(\theta) = \sum_{j=0}^{\infty} a_j \theta^j$$

is convergent for $|\theta| \leq r$. This distribution f_j is a power-series distribution. The r-th factorial moment of this distribution is defined by

$$EX(X-h)(X-2h)\cdots(X-rh+h)$$

i.e. the coefficient of $z^r/r!$ in the expansion of $E(1+Zh)^{X/h}$. Hence the factorial-moment generating function is

$$F(Z) = E[1 + Zh]^{X/h}$$

when $h \to 0$, $F(Z) \to \psi(Z) = E(e^{ZX})$, which is the moment generating function. The relations between the moment generating function and factorial moment generating function are

$$F(Z) = \psi\left\{\frac{1}{h}\log(1+Zh)\right\} \text{ and } \psi(Z) = F\{(e^{Zh} - 1)/h\}.$$

Hence the relation between factorial cumulants and the cumulants can be established. Let $K_{[r]}$ is the r-th factorial cumulant and K_r is the r-th cumulant, then $K_{[r]} \to K_r$, if $h \to 0$. These relationships were established by Khatri (1959). According to him the j-th factorial cumulant, $K_{[j]}$ of the power series distribution is given by

$$K_{[j]} = \theta^{jh}\left(\theta^{-h+1}\frac{d}{d\theta}\right)^j \log P(\theta)$$

$$\text{and } K_{[j+1]} = \theta\frac{dK_{[j]}}{d\theta} - jhK_{[j]}, \text{ where}$$

$$\left(\theta^{-h+1}\frac{d}{d\theta}\right)^j = \left(\theta^{-h+1}\frac{d}{d\theta}\right)\left(\theta^{-h+1}\frac{d}{d\theta}\right)\cdots j \text{ times.}$$

When $h = 0$, it gives

$$K_j = \left(\theta\frac{d}{d\theta}\right)^j \log P(\theta) = \theta\frac{dK_{j-1}}{d\theta}$$

and for $h = 1$

$$K_{[j]} = \theta^r\frac{d^r}{d\theta^r}\log P(\theta) = \theta\frac{dK_{[j-1]}}{d\theta} - (j-1)K_{[j-1]}.$$

From K_j, we have

$$K_1 = \theta\frac{d}{d\theta}\log P(\theta) = \frac{d}{y}\log f(y)\bigg/\frac{d\log\theta}{dy}$$

where $f(y) = f(g(y))$

$$\text{and } K_2 = \theta\frac{dK_1}{d\theta} = \frac{dK_1}{dy}\bigg/\frac{d\log\theta}{dy}.$$

Therefore,

$$\frac{d\log\theta}{dy} = \frac{dK_1}{dy}\bigg/K_2 = \frac{d}{dy}\log f(y)/K_1.$$

Hence

$$\log\theta = \int\left(\frac{dK_1}{dy}\bigg/K_2\right)dy + c_3 = \log f_1(y) + \log c_1, \text{ say,}$$

$\theta = c_1 f_1(y)$

$$\text{and } \log f(y) = \int\left(K_1\frac{dK_1}{dy}\bigg/K_2\right)dy + c_4 = \log f_2(y) + \log c_2, \text{ say,}$$

i.e., $f(y) = c_2 f_2(y)$, where c_1, c_2, c_3 and c_4 are constants.

Let a_j be the coefficient of $\{f_1(y)\}^j$ in the expansion of $f_2(y)$, then $c_2c_1^{-j}a_j$ will be the coefficient of θ^x in the expansion of $f(y)$. Hence, the probability that the random variable X takes the value j is

$$c_2c_1^{-j}a_j\theta^j/f(y) = a_j\{f_1(y)\}^j/f_2(y).$$

Hence, without loss of generality, it can be assumed that in the functions of θ and $f(y)$,

$c_1 = c_2 = 1$. Thus

$$\theta = \exp\left\{\int\left(\frac{dK_1}{df}\Big/K_2\right)df\right\}$$

$$\text{and } P(\theta) = \exp\left\{\int\left(K_1\frac{dK_1}{df}\Big/K_2\right)df\right\}.$$

From these two expressions it can be calculated that for any two distributions having K_1 and K_2 equal, the two distributions are equal.

Let us consider that

$$P(\theta) = (1-\theta)^{-n} \text{ for } 0 \le \theta \le 1,\ n > 0, \theta = p/(1+p) = p/q,\ p \ge 0.$$

Then

$$f_j = \frac{\Gamma(n+j)}{j!\,\Gamma n}p^j q^{-n-j},\ j = 0, 1, 2, \ldots$$

which gives

$$K_r = \theta\frac{dK_{r-1}}{d\theta} = pq\frac{dK_{r-1}}{dp}.$$

From this cumulant relation, we have

$$K_1 = np,\ K_2 = npq,\ K_3 = npq(q+p),\ K_4 = npq(1+6pq).$$

These are the cumulants for negative binomial distributions. Again, let

$$P(\theta) = (1+\theta)^n,\ \theta \ge 0,$$

n is a positive integer,

$$\theta = p/(1-p),\ 0 \le p \le 1.$$

Then

$$f_j = \binom{n}{j}p^j(1-p)^{n-j},\ j = 0, 1, 2, \ldots.$$

Then

$$K_r = \theta\frac{dK_{r-1}}{d\theta} = p(1-p)\frac{dK_{r-1}}{dp}.$$

This gives $K_1 = np$ and $K_2 = np(1-p)$. These are the cumulants for binomial distribution.

5.3 Generalized Discrete Distributions

Let $g_1(t)$ is the probability generating function of a distribution f_1 and $g_2(t)$ is the probability generating function of another distribution f_2. If the argument t of $g_1(t)$ is replaced by $g_2(t)$, then the resulting function $g_1(g_2(t))$ is also a probability generating function. This function $g_1(g_2(t))$ is a polynomial function of t with non-negative coefficients. If $t = 1$, then $g_2(1) = 1$ and $g_1(g_2(1)) = g_1(1) = 1$. Thus there is a probability distribution corresponding to $g_1(g_2(t))$. This distribution is called a *'generalized' f_1 distribution.*

Let f_1 be the binomial distribution, where

$$f_1 = \binom{n}{x} p^x q^{n-x}, \; x = 0, 1, 2, \ldots, n$$

and f_2 be the Poission distribution with probability function

$$f_2 = \frac{e^{-\theta}\theta^x}{x!}, \; x = 0, 1, 2, \ldots.$$

Then

$$g_2(t) = E(t^x) = \sum_{x=0}^{\infty} e^{-\theta}\frac{(\theta t)^x}{x!} = e^{-\theta}e^{\theta t} = e^{-\theta(1-t)}.$$

Now

$$\begin{aligned} g_1(g_2(t)) &= E\left[e^{-\theta(1-t)}\right]^x = \sum_{x=0}^{n} \left\{e^{-\theta(1-t)}\right\}^x \binom{n}{x} p^x q^{n-x} \\ &= \sum_{x=0}^{n} \binom{n}{x} \left\{e^{-\theta(1-t)}\right\}^x q^{n-x} = \left[q + pe^{-\theta(1-t)}\right]^n \end{aligned}$$

which is the probability generating function of the generalized f_1 distribution.

Similarly, if

$$\begin{aligned} f_1 &= \frac{e^{-\theta}\theta^x}{x!}, \; x = 0, 1, \ldots \\ \text{and } f_2 &= -\frac{1}{\log(1-\lambda)}\frac{\lambda^x}{x}, \; x = 1, 2, \ldots \end{aligned}$$

then the probability generating function of generalized f_1 is given by

$$\begin{aligned} \exp\left[\theta\left\{\frac{\log(1-t\lambda)}{\log(1-\lambda)} - 1\right\}\right] &= e^{-\theta}(1-t\lambda)^{\theta/\log(1-\lambda)} \\ &= \left[(1-\lambda)^{-1} - \lambda(1-\lambda)^{-1}t\right]^{\theta/\log(1-\lambda)} \end{aligned}$$

which is the probability generating function of negative binomial distribution with parameters

$$\theta[-\log(1-\lambda)]^{-1}, \; \lambda(1-\lambda)^{-1}.$$

Again, if

$$\begin{aligned} f_1 &= \binom{n}{x} p^x q^{n-x}, \; x = 0, 1, 2, \ldots, n \\ \text{and } f_2 &= -\frac{1}{\log(1-\lambda)}\frac{\lambda^x}{x}, \; x = 1, 2, \ldots \end{aligned}$$

then the probability generating function of the generalized f_1 distribution is given by

$$\left[q+p\frac{\log(1-t\lambda)}{\log(1-\lambda)}\right]^n.$$

The mean and variance of this distribution are

$$\mu_1' = Np\lambda(1-\lambda)^{-1}[-\log(1-\lambda)]^{-1}$$
$$\mu_2 = Np\lambda(1-\lambda)^{-2}[-\log(1-\lambda)]^{-1}\{1-p\lambda[-\log(1-\lambda)]^{-1}\}.$$

If

$$f_1 = \binom{n+x-1}{n-1}\left(\frac{P}{Q}\right)^x\left(1-\frac{P}{Q}\right)^n, \quad x=0,1,2,\ldots$$
$$\text{and } f_2 = -\frac{1}{\log(1-\lambda)}\frac{\lambda^x}{x}, \quad x=1,2,\ldots$$

then the probability generating function of the generalized f_1 distribution is given by

$$\left[Q-P\frac{\log(1-\lambda t)}{\log(1-\lambda)}\right]^{-n} = \left[1+\{-\log(1-\lambda)\}^{-1}\left\{1+\log\left(\frac{-1}{1-\lambda}-\frac{\lambda t}{1-\lambda}\right)\right\}\right]^{-n}.$$

The mean and variance of this distribution are

$$\mu_1' = NP\lambda(1-\lambda)^{-1}[-\log(1-\lambda)]^{-1}$$
$$\mu_2 = NP\lambda(1-\lambda)^{-2}[-\log(1-\lambda)]^{-1}\left[1+P\lambda\{-\log(1-\lambda)\}^{-1}\right].$$

EXERCISE

1. Explain Pearsonian system of distributions. If the differential equation

$$\frac{df}{dx} = \frac{(a-x)f}{b_0-b_1x+b_2x^2}$$

is used to define the Pearson system of equations, then find the moment relations of different system of distributions.

2. Define power series distribution and find the r-th factorial moment of such distribution. Also, deduce the relation between moment generating function and factorial moment generating function of power series distribution.

3. Define probability generating function of a random variable X. Let $g_1(t)$ and $g_2(2)$ be the probability generating functions of the distributions f_1 and f_2, respectively. What is the distribution of $g_1(g_2(t))$?

If f_1 is the binomial distribution and f_2 is the Poisson distribution, then find the probability generating function of $g_1(g_2(t))$. Also, find the mean and variance of this distribution.

CHAPTER 6

Bernoulli Distribution

6.1 Bernoulli Distribution

Random variable is defined on the basis of sample point(s) in a sample space of a random experiment. Let us consider that a random variable takes two values 0 and 1, where 0 is assigned to some of the sample points and 1 is assigned to the remaining sample points in the sample space of a random experiment. Such random experiment is termed as *Bernoulli experiment.* For example, a human being is selected at random. The sample space of this experiment is $S : \{M, F;\ M =$ Male, $F =$ Female$\}$. Here outcomes of the experiment are either M or/and F and these outcomes are mutually exclusive. Let us define a random variable X, where X is the number of male. Here value of $X = 0$ when sample point is F and $X = 1$ when sample point is M. Thus the experiment is Bernoulli experiment and the random variable X is Bernoulli variable. The other examples of Bernoulli random variable are, number of non-defective computers in selecting one computer from a lot of defective and non-defective computers, number of good product of an industry in selecting one product from a lot of defective and non-defective products of a day, number of head in tossing a coin once, etc.

Let p be the probability in selecting a good computer from a lot of computers containing good and defective computers. Then the probability of selecting defective computer is $(1-p) = q$. Let us consider that X be the number of good computer, where X takes values 0 or 1. Then

$$P(X = 0) = 1 - p = q \text{ and } P(X = 1) = p,\ 0 < p < 1.$$

This X is Bernoulli variate and $P(X = x)$ is the probability function of X. The value $\{x, P(X = x)\}$ is the Bernoulli distribution.

6.2 Moments of Bernoulli Distribution

The moment generating function of Bernoulli distribution is

$$M_x(t) = E\left[e^{tX}\right] = \sum e^{tx} P(X = x) = (1 - p) + pe^t.$$

The r-th general moment about origin is obtained from the equation

$$\mu'_r = \left.\frac{d^r M_x(t)}{dt^r}\right|_{t=0}.$$

Thus, when $r = 1$, $\mu'_1 = p$. Similarly, $\mu'_2 = p, \mu'_3 = p, \mu'_4 = p$.

The central moments are

$$\begin{aligned}
\mu_2 &= \mu'_2 - \mu'^2_1 = p - p^2 = pq \\
\mu_3 &= \mu'_3 - 3\mu'_2\mu'_1 + 2\mu'^3_1 = p - 3p^2 + 2p^3 = pq(1-2p) \\
\mu_4 &= \mu'_4 - 4\mu'_3\mu'_1 + 6\mu'_2\mu'^2_1 - 3\mu'^4_1 = p - 4p^2 + 6p^3 - 3p^4 \\
&= p(1-p^3) - 2p^2q(1+q).
\end{aligned}$$

Using these moments, we have

$$\beta_1 = \frac{\mu_3^2}{\mu_2^3} = \frac{(1-2p)^2}{pq} \text{ and } \beta_2 = \frac{\mu_4}{\mu_2^2} = \frac{(1-p^3) - 2pq(1+q)}{pq^2}.$$

These two results indicate that, when

$p = 0.5$, the distribution is symmetric and platykurtic,

$p > 0.5$, the distribution is negatively skewed,

$p < 0.5$, the distribution is positively skewed.

When $p = 0.5$, $\beta_2 = 1$ and β_2 is always less than 3 for any value of $p(0 < p < 1)$.

6.3 Applications of Bernoulli Distribution

►**EXAMPLE 6.1** *A biased coin is tossed once. It is known that head and tail come in the ratio 1 : 3. Find the probability distribution of number of head. Find mean and variance of number of head.*

Solution. Let X be the random variable showing head, where $X = 0$ or 1 as sample points are T or H, respectively. From the given condition

$$P(X=0) = \frac{3}{4} \text{ and } P(X=1) = \frac{1}{4}.$$

Then the Bernoulli distribution is

X	0	1
$P(X=x)$	3/4	1/4

Again, we have

$$\begin{aligned}
E(X) &= \Sigma xp(x) = 0 \times \frac{3}{4} + 1 \times \frac{1}{4} = \frac{1}{4} \\
E(X^2) &= \Sigma x^2p(x) = 0^2 \times \frac{3}{4} + 1^2 \times \frac{1}{4} = \frac{1}{4} \\
\therefore\ \mu_2 &= E(X^2) - [E(X)]^2 = \frac{1}{4} - \frac{1}{16} = \frac{3}{16}.
\end{aligned}$$

►**EXAMPLE 6.2** *An unbiased die is thrown once. Let X be the random variable indicating even dots in the upper face of the die. Find mean and variance of X.*

Solution. Here

$$X = \begin{cases} 0, \text{ if the die shows odd numbers} \\ 1, \text{ if the die shows even numbers.} \end{cases}$$

Then the Bernoulli probability distribution is

X	0	1
$P(X = x)$	1/2	1/2

Hence

$$E(X) = \mu_1' = p = \frac{1}{2},$$

where

$$P(X = 1) = p = \frac{1}{2}$$

$$\text{and } V(X) = \mu_2 = pq = \frac{1}{2} \cdot \frac{1}{2} = \frac{1}{4}.$$

►**EXAMPLE 6.3** *In an office there are 10 Samsung and 15 Philips computers. One computer is selected at random. Find the probability distribution of selecting a Philips computer. What is the expected number of selection of Philips computer?*

Solution. Let X be the random variable indicating number of selected Philips computer, where $X = 0$ or 1. Then the probability distribution of X is

X	0	1
$P(X = x)$	10/25	15/25

Here

$$X = \begin{cases} 1, \text{ when a selected computer is Philips computer} \\ 0, \text{ otherwise.} \end{cases}$$

The expected number of selected Philips computer is

$$E(X) = \sum xP(X = x) = 0 \times \frac{10}{25} + 1 \times \frac{15}{25} = \frac{3}{5}.$$

►**EXAMPLE 6.4** *From previous experience it is known that out of 500 blood samples one blood sample shows HIV positive. One day one blood sample is selected at random. Find the probability that the selected blood sample shows HIV positive. Find the variance of the blood samples containing HIV positive.*

Solution. Let X be the number of blood sample containing HIV positive, where $X = 0$ or 1 ($X = 1$, when the sample shows HIV positive; $X = 0$, otherwise). Therefore, the probability distribution of X is

X	0	1
$P(X = x)$	0.998	0.002

$$\therefore\ P(X=1) = 0.002$$
$$V(X) = E(X^2) - [E(X)]^2$$
$$E(X) = \sum xp(x) = 0 \times 0.998 + 1 \times 0.002 = 0.002$$
$$E(X^2) = \sum x^2 p(x) = 0^2 \times 0.998 + 1^2 \times 0.002 = 0.002$$
$$\therefore\ V(X) = 0.002 - (0.002)^2 = 0.001996.$$

►**EXAMPLE 6.5** *Twenty percent vehicles running in a highway are operated by CNG. One day one vehicle is selected at random. Find the probability distribution and variance of the distribution of the number of vehicle operated by CNG.*

Solution. Let X be the number of vehicle operated by CNG, where $X = 0$, when the vehicle is not operated by CNG and $X = 1$, when the vehicle is operated by CNG. The probability distribution of X is

X	0	1
$P(X = x)$	0.8	0.2

$$V(X) = E(X^2) - [E(X)]^2$$
$$E(X) = \sum xp(x) = 0 \times 0.8 + 1 \times 0.2 = 0.2$$
$$E(X^2) = \sum x^2 p(x) = 0^2 \times 0.8 + 1^2 \times 0.2 = 0.2$$
$$V(X) = 0.2 - (0.2)^2 = 0.16.$$

►**EXAMPLE 6.6** *Ninety percent computers of a laboratory operate perfectly everyday. One day one computer is selected at random. Find the probability distribution of computer which is not operated perfectly. What is the expected number of computers which are not operated properly.*

Solution. Let X be the number of computer which is not operated perfectly, where $X = 0$, if computer operates perfectly and $X = 1$, otherwise. Then the probability distribution of X is

X	0	1
$P(X = x)$	0.9	0.1

$$E(X) = \sum xp(x) = 0 \times 0.9 + 1 \times 0.1 = 0.1.$$

►**EXAMPLE 6.7** *There are* 10 *books on computer science,* 5 *books on mathematics and* 10 *books on statistics in a box. One book is selected at random. Let X be the number of books of computer science. Find the expected number of computer science book. Also, find $V(X)$.*

Solution. Let

$$X = \begin{cases} 1, \text{ if the selected book is of computer science} \\ 0, \text{ otherwise.} \end{cases}$$

Then the probability distribution of X is

X	0	1
$P(X = x)$	15/25	10/25

Expected number of computer science book is

$$E(X) = \sum xp(x) = 0 \times \frac{15}{25} + 1 \times \frac{10}{25} = \frac{2}{5}$$

Again,

$$E(X^2) = \sum x^2 p(x) = 0^2 \times \frac{15}{25} + 1^2 \times \frac{10}{25} = \frac{2}{5}$$

$$\therefore\ V(X) = E(X^2) - [E(X)]^2 = \frac{2}{5} - \frac{4}{25} = \frac{6}{25}.$$

►**EXAMPLE 6.8** *Eighty percent patients of a diabetic hospital are suffering from hypertension. One patient is randomly selected. Find the probability distribution of the patients who are suffering from hypertension. Also, find the mean and variance of the distribution.*

Solution. Let X be the number of patient suffering from hypertension, where

$$X = \begin{cases} 1, \text{ if the patient suffers from hypertension} \\ 0, \text{ otherwise.} \end{cases}$$

The probability distribution of X is

X	0	1
$P(X = x)$	0.2	0.8

The mean of the distribution is

$$E(X) = \sum xp(x) = 0 \times 0.2 + 1 \times 0.8 = 0.8.$$

The variance of the distribution is

$$V(X) = E(X^2) - [E(X)]^2, \text{ where}$$

$$E(X^2) = \sum x^2 p(x) = 0^2 \times 0.2 + 1^2 \times 0.8 = 0.8$$

$$\therefore\ V(X) = 0.8 - (0.8)^2 = 0.16.$$

►**EXAMPLE 6.9** *Ten percent products of a company are usually found defective everyday. Everyday one product is selected at random and the experiment is conducted uniformly for* 300 *days. Let* X *be the number of defective product. The distribution of*

X over 300 days is shown below:

X	0	1	*Total*
No. of days, O_i	260	40	$300 = N$

Fit a probability model of the above data, test the goodness of fit.

Solution. From the given condition, the probability distribution of X is (assuming Bernoulli distribution for X)

X	0	1
$P(X = x)$	0.9	0.1

The expected number of days of observing defective product is

$$E_i = NP(X = x_i) : 270,\ 30.$$

Here E_i is observed in fitting the Bernoulli model for the data. To study the goodness of fit, we need to calculate χ^2-statistic, where

$$\chi^2 = \sum \frac{O_i^2}{E_i} - N = 303.70 - 300 = 3.70.$$

Since $\chi^2 < \chi^2_{0.05,1} = 3.84$, the fitted distribution is satisfactory for the data.

►**EXAMPLE 6.10** *In a slum area 40% people are doing some small-scale business and 60% are involved in different activities. One people is selected at ramdom in a day and the process is continued uniformly over 100 days. Let X be the number of people engaged in business. The distribution of X over 100 days is as follows:*

X	0	1	*Total*
No. of days, O_i	65	35	100

Fit Bernoulli distribution for the data and comment.

Solution. From the given condition the Bernoulli distribution of X is

X	0	1
$P(X = x)$	0.6	0.4

where

$$X = \begin{cases} 1, \text{ if the people is engaged in business} \\ 0, \text{ otherwise.} \end{cases}$$

The expected number of days to observe businessman is

$$E_i = NP(X = x_i) : 60,\ 40, \text{ where } N = 100.$$

This E_i is observed according to Bernoulli probability law. Now,

$$\chi^2 = \sum \frac{O_i^2}{E_i} - N = 101.04 - 100 = 1.04.$$

Since $\chi^2 < \chi^2_{0.05,1} = 3.84$, the fitted distribution is satisfactory for the data.

EXERCISE

1. (a) What do you mean by Bernoulli experiment? Write down some examples of Bernoulli experiment and Bernoulli variable.

(b) Deduce the condition under which a Bernoulli distribution is negatively skewed.

(c) An unbiased die is tossed once. Find the mean and variance of observing one dot in the upper face of die. What is the shape characteristic of the distribution of one dot?

2. (a) Explain with example the Bernoulli trial and Bernoulli variable. Derive moment generating function of Bernoulli distribution and hence find the values of β_1 and β_2 of the distribution.

(b) The hardware of computers of a laboratory fail usually 20% of the time of work. One day a computer is randomly selected. Find mean and variance of the number of failure of hardware of the selected computer.

3. (a) Write down some experiments, where experimental results can be expressed as Bernoulli variable. Show that the Bernoulli distribution is positively skewed and leptokurtic.

(b) One ship out of 10 is sunk during war. Find the probability distribution of a randomly selected ship which come back safely after a voyage. In a random experiment 20 ships out of 25 are found to return safely. Fit a suitable distribution for the number of ships which are sunk.

CHAPTER **7**

Binomial Distribution

7.1 Definition and Origin

It has already been observed that the Bernoulli variate takes only two values 0 and 1. The results of the Bernoulli trial can be classified as 'Success—S' or 'Failure—F', where

$$P(S) = P(X = 1) = p \text{ and } P(F) = P(X = 0) = 1 - p = q.$$

This Bernoulli trial can be repeated under the following conditions:

(i) Bernoulli trial is repeated n times,

(ii) the results of each trial are classified into two classes, viz., 'Success—S' and 'Failure—F'. Success and Failure are mutually exclusive,

(iii) for every trial $P(S) = p$, and

(iv) each trial is independent of other.

If a Bernoulli trial is repeated according to the above conditions, the trial is known as *binomial experiment* and the distribution of 'Success' or 'Failure' of this experiment is known as *binomial distribution.* The number of 'Success' or 'Failure' as is required is the binomial variable.

The binomial distribution can be defined in terms of the binomial expansion of $(q + p)^n$, where $q + p = 1, p > 0, q > 0$ and n is a positive integer. The expansion is

$$(q + p)^n = \sum_{x=0}^{n} \binom{n}{x} p^x q^{n-x}.$$

In the expansion, the $(x + 1)$th term can be written as a function of x which is

$$p(x) = \binom{n}{x} p^x q^{n-x}, \quad x = 0, 1, 2, \ldots, n$$

where n is a positive integer and $0 < p < 1$. Under this condition it is clear that $p(x) \geq 0$ and

$$\sum p(x) = \sum_{x=0}^{n} \binom{n}{x} p^x q^{n-x} = (p + q)^n = 1.$$

Therefore, $p(x)$ is a probability function. This function $p(x)$ is the distribution c the random variable X and it is called *binomial distribution* with parameters n and p This distribution was derived by J. Bernoulli in his treatise *Ars Conjectanc* published in 1713. The distribution is derived for the sum of n Bernoulli trials. the random variables $X_1, X_2, \ldots$ are independent and identically distributed, and each random variable X_i has a Bernoulli distribution with parameter p, then it i said that the random variables $X_1, X_2, \ldots, X_n$ form an infinite sequence of Bernoul trials with parameter p. In such a case if the random variables $X_1, X_2, \ldots, X_n$ ar independent and each has a Bernoulli distribution with parameter p, then it is sai that the variables $X_1, X_2, \ldots, X_n$ form n Bernoulli trials with parameter p and

$$X = X_1 + X_2 + \cdots + X_n$$

has a binomial distribution with parameters n and p.

The binomial distribution serves as an excellent mathematical model in a numbe of experimental situations. Suppose that the outcome of a random experiment ca be classified in one of two mutually exclusive and exhaustive ways, say, success c failure. For example, in tossing an unbiased coin the outcome 'Head' or 'Tail'; i a production process the production of 'Defective' or 'Non-defective' items; in a li testing experiment the incidence of 'Alive' or 'Death', etc. Let the random experimer be repeated n independent times. Let us call the occurrence of an event as 'succes and its non-occurrence a 'failure'. Let p be the probability of success and q be th probability of failure in a single trial, so that $p + q = 1$. Assume that the probability is same in every trial. The number of success in n trials may be $0, 1, 2, \ldots, x, \ldots, n$ a it is a random variable. Consider that out of n trials the first x trials are successes an the last $(n - x)$ trials are failures. The sequence of success and failure can be writte as

$$\underbrace{SSS \cdots S}_{x}\ \underbrace{FFFF \cdots F}_{n-x}$$

where S indicates success and F indicates failure.

Now, by the theorem of compound probability, the probability that the first x tria are successes and the last $(n - x)$ trials are failures is $p^x q^{n-x}$. But x trials of succe can occur in $\binom{n}{x}$ mutually exclusive ways. Hence, the probability of x successes in series of n independent trials is given by

$$p(x) = \binom{n}{x} p^x p^{n-x}, \quad x = 0, 1, 2, \ldots, n.$$

7.2 First Four Moments of the Binomial Distribution

The r-th moment about origin is defined by

$$\mu'_r = E(x^r) = \sum_{x=0}^{n} \binom{n}{x} x^r p^x q^{n-x}.$$

Putting $r = 1$, we have

$$\begin{aligned}
\mu_1' &= \sum_{x=0}^{n} \binom{n}{x} x p^x q^{n-x} \\
&= 0 \cdot q^n + \binom{n}{1} pq^{n-1} + \binom{n}{2} 2p^2 q^{n-2} + \cdots + \binom{n}{n} np^n \\
&= np \left[q^{n-1} + \binom{n}{1} pq^{n-2} + \cdots + \binom{n-1}{n-1} p^{n-1} \right] \\
&= np(q+p)^{n-1} \\
&= np.
\end{aligned}$$

Therefore, the mean of the binomial distribution is np.

When $r = 2$, we get

$$\begin{aligned}
\mu_2' &= \sum_{x=0}^{n} \binom{n}{x} x^2 p^x q^{n-x} \\
&= \sum_{x=0}^{n} \{x + x(x-1)\} \binom{n}{x} p^x q^{n-x} \\
&= \sum_{x=0}^{n} x \binom{n}{x} p^x q^{n-x} + \sum_{x=0}^{n} x(x-1) \binom{n}{x} p^x q^{n-x} \\
&= np + n(n-1)p^2 \sum \binom{n-2}{x-2} p^{x-2} q^{n-x} \\
&= np + n(n-1)p^2 (q+p)^{n-2} \\
&= np + [1 + np - p] \\
&= np(q + np)
\end{aligned}$$

Thus, the variance of the binomial distribution is obtained by

$$\mu_2 = \mu_2' - \mu_1'^2 = np(q+np) - n^2p^2 = npq.$$

As $q < 1$, the variance of binomial distribution is less than its mean.

When $r = 3$, we have

$$\begin{aligned}
\mu_3' &= \sum_{x=0}^{n} x^3 \binom{n}{x} p^x q^{n-x} \\
&= \sum \{x(x-1)(x-2) + 3x(x-1) + x\} \binom{n}{x} p^x q^{n-x} \\
&= \sum x \binom{n}{x} p^x q^{n-x} + 3 \sum x(x-1) \binom{n}{x} p^x q^{n-x} \\
&\quad + \sum x(x-1)(x-2) \binom{n}{x} p^x q^{n-x} \\
&= np + 3n(n-1)p^2 + n(n-1)(n-2)p^3 (q+p)^{n-3} \\
&= np + 3n(n-1)p^2 + n(n-1)(n-2)p^3.
\end{aligned}$$

When $r = 4$, we get

$$\begin{aligned}
\mu_4' &= \sum_{x=0}^{n} x^4 \binom{n}{x} p^x q^{n-x} \\
&= \sum \{x + 7x(x-1) + 6x(x-1)(x-2) \\
&\quad + x(x-1)(x-2)(x-3)\} \binom{n}{x} p^x q^{n-x} \\
&= \sum x \binom{n}{x} p^x q^{n-x} + 7 \sum x(x-1) \binom{n}{x} p^x q^{n-x} \\
&\quad + 6 \sum x(x-1)(x-2) \binom{n}{x} p^x q^{n-x} \\
&\quad + \sum x(x-1)(x-2)(x-3) \binom{n}{x} p^x q^{n-x} \\
&= np + 7n(n-1)p^2 + 6n(n-1)(n-2)p^3 \\
&\quad + p^4 n(n-1)(n-2)(n-3)(q+p)^{n-4} \\
&= np + 7n(n-1)p^2 + 6n(n-1)(n-2)p^3 + n(n-1)(n-2)(n-3)p^4.
\end{aligned}$$

We know

$$\begin{aligned}
\mu_3 &= \mu_3' - 3\mu_2'\mu_1' + 2\mu_1'^3 \\
&= np + 3n(n-1)p^2 + n(n-1)(n-2)p^3 - 3n^2p^2(q+np) + 2n^3p^3 \\
&= npq(q-p).
\end{aligned}$$

Again $\mu_4 = \mu_4' - 4\mu_3'\mu_1' + 6\mu_2'\mu_1'^2 - 3\mu_1'^4$.

Putting the values of μ_4', μ_3', μ_2' and μ_1' and on simplification, we get

$$\mu_4 = npq[1 + 3(n-2)pq].$$

The moment ratios are

$$\alpha_3 = \sqrt{\beta_1} = \sqrt{\frac{\mu_3^2}{\mu_2^3}} = \sqrt{\left[\frac{\{npq(q-p)\}^2}{(npq)^3}\right]} = (q-p)/\sqrt{npq}$$

$$\text{and } \alpha_4 = \beta_2 = \frac{\mu_4}{\mu_2^2} = \frac{npq[1+3(n-2)pq}{(npq)^2} = 3 + \frac{1}{npq}(1-6pq).$$

7.3 Generating Functions of the Binomial Distribution

The moment generating function about origin is given by

$$\begin{aligned}
M_x(t) = E(e^{tx}) &= \sum_{x=0}^{n} \binom{n}{x} e^{tx} p^x q^{n-x} \\
&= \sum_{x=0}^{n} \binom{n}{x} q^{n-x} \left(pe^t\right)^x = \left(pe^t + q\right)^n.
\end{aligned}$$

The r-th moment about origin is obtained from $M_x(t)$ by the relation

$$\left.\frac{d^r M_x(t)}{dt^r}\right|_{t=0}$$

Thus

$$\left.\frac{dM_x(t)}{dt}\right|_{t=0} = \cdots \left.\frac{d(pe^t+q)^n}{dt}\right|_{t=0} \text{ gives } \mu_1' = np.$$

Similarly, for $r = 2$, we have

$$\mu_2' = npq + n^2p^2.$$

Also

$$\mu_2 = \mu_2' - \mu_1'^2 = npq + n^2p^2 = npq.$$

The cumulant generating function is given by

$$\begin{aligned}
K_x(t) &= \log M_x(t) = \log\left(pe^t+q\right)^n = n\log(pe^t+q) \\
&= n\log\left[p\left(1+t+\frac{t^2}{2!}+\frac{t^3}{3!}+\frac{t^4}{4!}+\cdots\right)+q\right] \\
&= n\log\left[1+pt+p\frac{t^2}{2!}+\frac{pt^3}{3!}+p\frac{t^4}{4!}+\cdots\right] \\
&= n\left[pt+p\frac{t^2}{2!}+p\frac{t^3}{3!}+p\frac{t^4}{4!}+\cdots\right]-\frac{n}{2}\left[pt+p\frac{t^2}{2!}+\cdots\right]^2 \\
&\quad +\frac{n}{3}\left[pt+p\frac{t^2}{2!}+\cdots\right]^3-\cdots.
\end{aligned}$$

In $K(t)$ the coefficient of $t^r/r!$ is the K_r. Thus

$$\begin{aligned}
K_1 &= np = \mu_1',\ K_2 = npq = \mu_2, \\
K_3 &= npq(q-p) = \mu_3 \text{ and } K_4 = npq(1-6pq).
\end{aligned}$$

The factorial moment generating function of binomial distribution is given by

$$\begin{aligned}
M_{(x)}(t) = E(1+t)^x &= \sum_{x=0}^{n}\binom{n}{x}p^x q^{n-x}(1+t)^x \\
&= \sum_{x=0}^{n}\binom{n}{x}\{p(1+t)\}^x q^{n-x} = \{q+p(1+t)\}^n = (1+pt)^n.
\end{aligned}$$

The r-th factorial moment is the coefficient of $t^r/r!$ in the expansion of $M_{(x)}(t)$. Thus

$$\mu_{(r)} = n^{(r)}p^r.$$

In particular,

$$\begin{aligned}
\mu_{(1)} &= np,\ \mu_{(2)} = n(n-1)p^2,\ \mu_{(3)} = n(n-1)(n-2)p^3, \\
\mu_{(4)} &= n(n-1)(n-2)(n-3)p^4.
\end{aligned}$$

From these factorial moments, moments about origin can be found out, where

$$\begin{aligned}
\mu_1' &= \mu_{(1)} = np \\
\mu_2' &= \mu_{(2)} + \mu_{(1)} = np + n(n-1)p^2 \\
\mu_3' &= \mu_{(3)} + 3\mu_{(2)} + \mu_{(1)} \\
&= n(n-1)(n-2)p^3 + 3n(n-1)p^2 + np \\
\mu_4' &= \mu_{(4)} + 6\mu_{(3)} + 7\mu_{(2)} + \mu_{(1)} \\
&= n(n-1)(n-2)(n-3)p^4 + 6n(n-1)(n-2)p^3 \\
&\quad +7n(n-1)p^2 + np.
\end{aligned}$$

The r-th factorial moments can also be obtained by differentiating $(q+p)^n r$ times partially with respect to p and multiplying by p^r. Thus

$$n^{(r)}(q+p)^{n-r}p^r = \sum_{x=0}^{n} x^{(r)}\binom{n}{x}p^x q^{n-x} = \mu_{(r)}$$

$$\therefore\ \mu_{(r)} = n^{(r)}p^r.$$

Romanovsky (1923) derived a recursion formula for the central moments which is given by

$$\mu_{r+1} = pq\left(nr\mu_{r-1} + \frac{d\mu_r}{dp}\right).$$

We know

$$\mu_r = \sum_{x=0}^{n}\binom{n}{x}(x-np)^r p^x q^{n-x}.$$

Now

$$\begin{aligned}
\frac{d\mu_r}{dp} &= \sum_{x=0}^{n}\binom{n}{x}xp^{x-1}q^{n-x}(x-np)^r \\
&\quad -\sum_{x=0}^{n}\binom{n}{x}(n-x)q^{n-x-1}(x-np)^r p^x \\
&\quad -\sum_{x=0}^{n}\binom{n}{x}nrp^x q^{n-x}(x-np)^{r-1} \\
&= \sum_{x=0}^{n}\binom{n}{x}p^{x-1}q^{n-x-1}(x-np)^r[x-np] - nr\mu_{r-1} \\
pq\frac{d\mu_r}{dp} &= \sum_{x=0}^{n}\binom{n}{x}p^x q^{n-x}(x-np)^{r+1} - npqr\mu_{r-1} \\
&= \mu_{r+1} - npqr\mu_{r-1}. \\
\therefore\ \mu_{r+1} &= pq\left(\frac{d\mu_r}{dp} + nr\mu_{r-1}\right).
\end{aligned}$$

The probability generating function of binomial distribution is defined by

$$\begin{aligned} g(t) = E(t^x) &= \sum_{x=0}^{n} \binom{n}{x} p^x q^{n-x} t^x \\ &= \sum_{x=0}^{n} \binom{n}{x} (pt)^x q^{n-x} = (q+pt)^n. \end{aligned}$$

The characteristic function of binomial distribution is given by

$$\begin{aligned} \psi(t) = E(e^{itx}) &= \sum_{x=0}^{n} \binom{n}{x} p^x q^{n-x} e^{itx} \\ &= \sum_{x=0}^{n} \binom{n}{x} (pe^{it})^x q^{n-x} = (q+pe^{it})^n. \end{aligned}$$

The recursion formula for cumulants of binomial distribution is given by

$$K_{r+1} = pq\frac{dK_r}{dp}, \quad r > 1.$$

We know $M_x(t) = (q+pe^t)^n$ and the cumulant generating function is

$$K(t) = \log M_x(t) = n\log(q+pe^t).$$

Also

$$\begin{aligned} K_r &= \frac{d^r}{dt^r}\left[n\log(q+pe^t)\right]_{t=0}. \\ \therefore \quad \frac{dK_r}{dp} &= n\left[\frac{d^r}{dt^r}\left(\frac{-1+e^t}{q+pe^t}\right)\right]_{t=0}. \end{aligned}$$

Again

$$\begin{aligned} K_{r+1} &= \left[\frac{d^{r+1}}{dt^{r+1}} n\log(q+pe^t)\right]_{t=0} = np\left[\frac{d^r}{dt^r}\frac{e^t}{q+pe^t}\right]_{t=0} \\ &= n\left[\frac{d^r}{dt^r}\left(1-\frac{q}{q+pe^t}\right)\right]_{t=0} = -nq\left[\frac{d^r}{dt^r}\left(\frac{1}{q+pe^t}\right)\right]_{t=0}. \end{aligned}$$

Now

$$\begin{aligned} K_{r+1} - pq\frac{dK_r}{dp} &= -nq\left[\frac{d^r}{dt^r}\left(\frac{1}{q+pe^t}\right)\right]_{t=0} - npq\left[\frac{d^r}{dt^r}\left(\frac{e^t-1}{q+pe^t}\right)\right]_{t=0} \\ &= -nq\left[\frac{d^r}{dt^r}\frac{1+pe^t-p}{q+pe^t}\right]_{t=0} = -nq\frac{d^r}{dt^r}(1) = 0. \\ \therefore \quad K_{r+1} &= pq\frac{dK_r}{dp}. \end{aligned}$$

It may be concluded that, when $p = q = \frac{1}{2}$, the distribution transforms to symmetric one since $\beta_1 = 0$. But, if $p \neq q$, the distribution becomes skewed. Particularly, if $p < q$, let $p = 0.1$ and $p = 0.2$, then $\gamma_1 > 0$ ($\mu_3 > 0$), the distribution becomes positively skewed. On the other hand, if $p > q$, let $p = 0.8, p = 0.9$, then

$\gamma_1 < 0$ ($\mu_3 < 0$), the distribution becomes negatively skewed. Moreover, it is observed that, if $n \to \infty, \beta_1 \to 0$ irrespective of the value of p. In such a case the distribution becomes symmetric. Further, if $p = q = \frac{1}{2}$, $\gamma_2 = \beta_2 - 3 < 0$, the distribution becomes platykurtic. But, when $p < q$, let $p = 0.1$ or $p = 0.2$, then $\gamma_2 > 0$ and the distribution becomes leptokurtic. Here also, it is observed that, if $n \to \infty$, $\beta_2 \to 3$ and the distribution tends to mesokurtic.

7.4 Applications of Binomial Distribution

It has already been mentioned that the experimental results of random experiment which can be classified into two classes, viz., success (S) and failure (F), are expressed by binomial probability law. The distribution of successes or failures is binomial distribution. Therefore, the results of the following random experiments can be expressed by binomial probability law:

(i) the selection of persons classified as male or female,

(ii) selection of children and classified as boys or girls,

(iii) selection of industrial product to classify as defective or non-defective,

(iv) identification of present or absent employees in a factory,

(v) identification of skilled or unskilled labourers in an industry,

(vi) identification of good or bad machines in an industry,

(vii) to hit the target successfully by gun or bomb,

(viii) to study the successful implementation of a project,

(ix) successful operation of a machine in a factory,

(x) to conduct the opinion survey supporting or opposing an issue,

(xi) to study the effectiveness of a medicine or a chemical,

(xii) to study the successful landing of planes in an airport,

(xiii) to study the successful implementation of family planning program in a society,

(xiv) to observe the number of dead patients admitted in a hospital,

(xv) to observe the successful delivery of newborn baby in a maternity centre.

►**EXAMPLE 7.1** *Six biased coins are tossed once. From previous experience it is known that $H : T :: 2 : 3$. Find the probability that* (i) *there will be 4 heads,* (ii) *there will be at best 3 heads,* (iii) *there will be at least 2 heads,* (iv) *there will be no head. If the coins are unbiased, what would be the probabilities of the above events?*

Solution. Let X be the number of heads (H). From the given condition

$$P(H) = p = \frac{2}{5} \text{ and } P(T) = q = \frac{3}{5}.$$

Thus

$$P(X = x) = \binom{6}{x}\left(\frac{2}{5}\right)^x\left(\frac{3}{5}\right)^{6-x}, \quad x = 0, 1, \ldots, 6.$$

(i) We need $P(X=4)$

$$P(X=4)=\binom{6}{4}\left(\frac{2}{5}\right)^4\left(\frac{3}{5}\right)^2=\frac{5184}{15625}.$$

(ii) We need $P(X \le 3)$

$$\begin{aligned}P(X\le 3) &= P(X=0)+P(X=1)+P(X=2)+P(X=3)\\ &= \binom{6}{0}(0.6)^6+\binom{6}{1}(0.4)(0.6)^5+\binom{6}{2}(0.4)^2(0.6)^4\\ &\quad+\binom{6}{3}(0.4)^3(0.6)^3=0.8208.\end{aligned}$$

(iii) We need $P(X\ge 2)$

$$\begin{aligned}P(X\ge 2) &= 1-P(X<2)=1-P(X=0)-P(X=1)\\ &= 1-\binom{6}{0}(0.6)^6-\binom{6}{1}(0.4)(0.6)^5=0.76672.\end{aligned}$$

(iv) We need $P(X=0)$

$$P(X=0)=\binom{6}{0}(0.6)^6=0.046656.$$

If the coins are unbiased, then

$$\begin{aligned}P(X=x) &= \binom{6}{x}\left(\frac{1}{2}\right)^6,\ x=0,1,2,\ldots,6\\ P(X=4) &= \left(\frac{1}{2}\right)^6 {}^6C_4=\frac{15}{64}\\ P(X\le 3) &= P(X=0)+P(X=1)+P(X=2)+P(X=3)\\ &= \left(\frac{1}{2}\right)^6[{}^6C_0+{}^6C_1+{}^6C_2+{}^6C_3]=\frac{21}{32}\\ P(X\ge 2) &= 1-P(X<2)=1-P(X=0)-P(X=1)\\ &= 1-\left(\frac{1}{2}\right)^6[{}^6C_0+{}^6C_1]=\frac{57}{64}\\ P(X=0) &= \left(\frac{1}{2}\right)^6 {}^6C_0=\frac{1}{64}.\end{aligned}$$

►**EXAMPLE 7.2** *Five unbiased dice are thrown once. Find the probability that, on the upper faces* (i) *there will be no even numbers,* (ii) *there will be even numbers on 3 dice,* (iii) *at best 2 dice show even numbers,* (iv) *at least 3 dice show even numbers.*

Solution. Let X be the random variable indicating even numbers on the upper face of the dice. As the dice are unbiased,

$$P\text{ (even number)} = P\text{ (odd number)} = \frac{1}{2}.$$

Therefore,

$$P(X = x) = \binom{n}{x} p^x q^{n-x}, \ x = 0, 1, 2, \ldots, n.$$

Here $n = 5$, $p = q = \frac{1}{2}$.

$$\therefore \ P(X = x) = \binom{5}{x}\left(\frac{1}{2}\right)^5, \ x = 0, 1, \ldots, 5.$$

(i) We need $P(X = 0)$

$$P(X = 0) = \binom{5}{0}\left(\frac{1}{2}\right)^5 = \frac{1}{32}.$$

(ii) We need $P(X = 3)$

$$P(X = 3) = \binom{5}{3}\left(\frac{1}{2}\right)^5 = \frac{5}{16}.$$

(iii) We need $P(X \leq 2)$

$$\begin{aligned} P(X \leq 2) &= P(X = 0) + P(X = 1) + P(X = 2) \\ &= \left(\frac{1}{2}\right)^5 \left[\binom{5}{0} + \binom{5}{1} + \binom{5}{2}\right] = \frac{1}{2} \end{aligned}$$

(iv) We need $P(X \geq 3)$

$$\begin{aligned} P(X \geq 3) &= 1 - P(X < 3) = 1 - P(X \leq 2) \\ &= 1 - \frac{1}{2} = \frac{1}{2}. \end{aligned}$$

►**EXAMPLE 7.3** *A departmental store takes the responsibility of supply of their commodities in the customer's house. Usually, in* 95% *cases the store does its duty perfectly. Find the probability that, in a randomly selected week* (i) 2 *days the store fails to do the job perfectly,* (ii) *at best* 2 *days the store fails to do the job perfectly.* (iii) *If the job performance is observed for* 52 *weeks, then in how many weeks the store fails* 2 *days per week to do the job perfectly?*

Solution. Let X be the number of days in which the store fails to do the job perfectly. Then

$$P(X = x) = \binom{n}{x} p^x q^{n-x}, \ x = 0, 1, 2, \ldots, n.$$

Given $n = 7, p = 0.05, q = 0.95$

$$\therefore \ P(X = x) = \binom{7}{2}(0.05)^x(0.95)^{7-x}, \ x = 0, 1, 2, \ldots, 7.$$

(i) We need $P(X = 2)$

$$P(X = 2) = \binom{7}{2}(0.05)^2(0.95)^5 \approx 0.04062$$

(ii) We need $P(X \leq 2)$

$$\begin{aligned} P(X \leq 2) &= P(X=0) + P(X=1) + P(X=2) \\ &= \binom{7}{0}(0.95)^7 + \binom{7}{1}(0.05)(0.95)^6 + \binom{7}{2}(0.05)^2(0.95)^5 \\ &\approx 0.99624. \end{aligned}$$

(iii) Number of weeks in which 2 days per week the store fails to do the job perfectly is given by

$$NP(X=2) = 52 \times 0.04062 \approx 2.$$

►**EXAMPLE 7.4** *Three planes out of* 5 *return safely when the planes go to attack a city. One day* 6 *planes went to attack the city. Find the probability that, out of the* 6 *planes* (i) 3 *will return safely,* (ii) *at best* 2 *will return safely,* (iii) *at least* 3 *will return safely,* (iv) *no one will return safely.*

Solution. Let X be the number of planes which return safely. Then

$$P(X=x) = \binom{n}{x} p^x q^{n-x}, \quad x = 0, 1, 2, \ldots, n.$$

Given $n = 6, p = \frac{3}{5} = 0.6,\ q = 1 - p = 0.4.$

$$\therefore\ P(X=x) = \binom{6}{x}(0.6)^x(0.4)^{6-x}, \quad x = 0, 1, 2, \ldots, 6.$$

(i) We need $P(X=3)$

$$P(X=3) = \binom{6}{3}(0.6)^3(0.4)^3 \approx 0.27648.$$

(ii) We need $P(X \leq 2)$

$$\begin{aligned} P(X \leq 2) &= P(X=0) + P(X=1) + P(X=2) \\ &= \binom{6}{0}(0.4)^6 + \binom{6}{1}(0.6)(0.4)^5 + \binom{6}{2}(0.6)^2(0.4)^2 \approx 0.1792. \end{aligned}$$

(iii) We need $P(X \geq 3)$

$$P(X \geq 3) = 1 - P(X < 3) = 1 - P(X \leq 2) = 1 - 0.1792 \approx 0.8208.$$

(iv) We need $P(X=0)$

$$P(X=0) = \binom{6}{0}(0.4)^6 \approx 0.004096.$$

►**EXAMPLE 7.5** *The probability of facing trouble by an engine of a plane on movement is* $1-p$. *The probability is true independently for all planes. The plane successfully moves if* 50% *engines are in well-working condition. If a plane has* 4 *engines and if it is needed to transform to a plane of* 2 *engines, what would be the value of* p?

Solution. It is assumed that all engines are run or stopped independently. Let X be the number of engines which are in working condition. Then for any plane of 4 engines

$$P(X = x) = \binom{4}{x} p^x (1-p)^{4-x}, \; x = 0, 1, 2, 3, 4.$$

As 50% good engines are sufficient for successful movement of the plane, the probability of successful movement of a plane is

$$\binom{4}{2} p^2(1-p)^2 + \binom{4}{3} p^3(1-p) + \binom{4}{4} p^4$$
$$= 6p^2(1-p)^2 + 4p^3(1-p) + p^4.$$

Again, if the plane contains 2 engines, its probability for successful movement is

$$\binom{2}{1} p(1-p) + \binom{2}{2} p^2 = 2p(1-p) + p^2.$$

Now, if the plane of 4 engines is transformed into a plane of 2 engines, then the following condition should be satisfied:

$$6p^2(1-p)^2 + 4p^3(1-p) + p^4 \geq 2p(1-p) + p^2.$$

On simplification $p \geq 2/3$.

►**EXAMPLE 7.6** *In an office there are* 10 *computers. The probability of successful operation by any computer is* 0.8. *Four computers are selected at random in a day. Find the probability that, out of* 4 *computers* (i) *all are found in non-working condition,* (ii) *one will be out-of-order,* (iii) *at best one will be out-of-order.*

Solution. Let X be the number of out-of-order computers. Then

$$P(X = x) = \binom{n}{x} p^x q^{n-x}, \; x = 0, 1, 2, \ldots, n.$$

Given $n = 4, \; p = 0.2, \; q = 0.8$.

$$\therefore \; P(X = x) = \binom{4}{x} (0.2)^x (0.8)^{4-x}, \; x = 0, 1, 2, 3, 4.$$

(i) We need $P(X = 4)$

$$P(X = 4) = \binom{4}{4} (0.2)^4 = 0.0016.$$

(ii) We need $P(X = 1)$

$$P(X = 1) = \binom{4}{1} (0.2)(0.8)^3 = 0.4096.$$

(iii) We need $P(X \leq 1)$

$$P(X \leq 1) = P(X = 0) + P(X = 1)$$
$$= \binom{4}{0} (0.8)^4 + \binom{4}{1} (0.2)(0.8)^3 = 0.8192.$$

►**EXAMPLE 7.7** *A computer-producing company ensures the customers that 10% of their products need repair work within one year of purchase. A customer purchased 10 computers from the company and observed that 2 computers needed repair within one year of purchase. Is the information conformable to the assurance?*

Solution. Let X be the number of computers which need repair. Then

$$P(X = x) = \binom{10}{x}(0.1)^x(0.9)^{10-x}, \ x = 0, 1, 2, \dots, 10.$$

The probability that at least two computers need repair is

$$P(X \geq 2) = 1 - P(X < 2) = 1 - P(X = 0) - P(X = 1)$$
$$= 1 - \binom{10}{0}(0.9)^{10} - \binom{10}{1}(0.1)(0.9)^9 \approx 0.2639.$$

As the probability of 2 or more defective computers is less, the company's assurance is not meaningful.

►**EXAMPLE 7.8** *The chance of winning a lottery from the lottery tickets of a company is 20%. A man purchased 8 tickets. Out of 8 tickets, what is the probability of winning* (i) *from 2 lottery tickets,* (ii) *no chance of winning?*

Solution. Let X be the number of winning tickets. Given that

$$P(X = x) = \binom{8}{x}(0.2)^x(0.8)^{8-x}, \ x = 0, 1, 2, \dots, 8.$$

(i) We need $P(X = 2)$

$$P(X = 2) = \binom{8}{2}(0.2)^2(0.8)^6 \approx 0.2936.$$

(ii) We need $P(X = 0)$

$$P(X = 0) = \binom{8}{0}(0.8)^8 \approx 0.16777.$$

►**EXAMPLE 7.9** *In a poultry hatchery 50% eggs are transformed to new birds. If 10 eggs are used in the hatchery,* (i) *what is the probability that at best 5 eggs are transformed to new birds,* (ii) *what is the probability that at least 2 eggs will be spoiled?*

Solution.

(i) Let X be the number of eggs which are transformed to new birds. Then

$$P(X = x) = \binom{n}{x}p^x q^{n-x}, \ x = 0, 1, 2, \dots, n.$$

Given $n = 10$, $p = q = \frac{1}{2}$.

$$\therefore \ P(X = x) = \binom{10}{x}\left(\frac{1}{2}\right)^{10}, \ x = 0, 1, 2, \dots, 10.$$

We need $P(X \leq 5)$

$$\begin{aligned} P(X \leq 5) &= P(X=0)+P(X=1)+P(X=2) \\ &\quad +P(X=3)+P(X=4)+P(X=5) \\ &= \left(\frac{1}{2}\right)^{10}\left[\binom{10}{0}+\binom{10}{1}+\binom{10}{2}+\binom{10}{3}+\binom{10}{4}+\binom{10}{5}\right] \\ &\approx 0.6230. \end{aligned}$$

(ii) Let Y be the number of spoiled eggs. Here

$$P(Y=y)=\binom{10}{y}\left(\frac{1}{2}\right)^{10}, \quad y=0,1,2,\ldots,10.$$

We need $P(Y \geq 2)$

$$\begin{aligned} P(Y \geq 2) &= 1-P(Y<2)=1-P(Y=0)-P(Y=1) \\ &= 1-\left(\frac{1}{2}\right)^{10}\left[\binom{10}{0}+\binom{10}{1}\right] \approx 0.98926. \end{aligned}$$

►**EXAMPLE 7.10** *In a questionnaire there are* 6 *questions. Each question has* 5 *different answers, one of which is correct. Find the probability that, in answering the questions one examinee will answer correctly* (i) *to questions* 1 *and* 2, (ii) *any two questions,* (iii) *at least* 2 *questions.*

Solution. Let X be the number of questions for which answer is correct. From the given condition

$$P(X=x)=\binom{6}{x}(0.2)^x(0.8)^{6-x}, \quad x=0,1,2,\ldots,6.$$

(i) We need the answer $CCWWWW$, where C is for correct answer and W is for wrong answer. Then

$$P(CCWWWW)=(0.2)^2(0.8)^4 \approx 0.0164.$$

(ii) We need $P(X=2)$

$$P(X=2)=\binom{6}{2}(0.2)^6(0.8)^4 \approx 0.24576.$$

(iii) We need $P(X \geq 2)$

$$\begin{aligned} P(X \geq 2) &= 1-P(X<2)=1-P(X=0)-P(X=1) \\ &= 1-\binom{6}{0}(0.8)^6-\binom{6}{1}(0.2)(0.8)^7 \approx 0.34464. \end{aligned}$$

►**EXAMPLE 7.11** *The probability of death of people from heart attack is* 0.3. *Ten people are selected from an area where death rate is alike. Find the probability that, out of* 10 *people* 4 *will die due to heart attack.*

Solution. Let X be the number of dead people due to heart attack. Given

$$P(X=x)=\binom{10}{x}(0.3)^x(0.7)^{10-x},\ x=0,1,2,\ldots,n.$$

We need $P(X=4)$

$$P(X=4)=\binom{10}{4}(0.3)^4(0.7)^6\approx 0.2001.$$

►**EXAMPLE 7.12** *Out of a fleet of* 10 *warships* 7 *can return safely after attacking the enemy. One day a fleet of* 10 *warships went to attack the enemy. Find the probability that, out of* 10 *warships* (i) *one failed to return,* (ii) *at least* 2 *ships failed to return safely,* (iii) 8 *warships came back safely,* (iv) *at best* 4 *warships returned safely.*

Solution. Let X be the random variable indicating the number of warships which fail to return safely and Y be the random variable indicating number of ships which return safely. From the given condition

$$\begin{aligned} P(X=x) &= \binom{10}{x}(0.3)^x(0.7)^{10-x},\ x=0,1,2,\ldots,10 \\ \text{and } P(Y=y) &= \binom{10}{y}(0.7)^y(0.3)^{10-y},\ y=0,1,2,\ldots,10. \end{aligned}$$

(i) We need $P(X=1)$

$$P(X=1)=\binom{10}{1}(0.3)(0.7)^9\approx 0.1211.$$

(ii) We need $P(X\geq 2)$

$$\begin{aligned} P(X\geq 2) &= 1-P(X<2) \\ &= 1-P(X=0)-P(X=1) \\ &= 1-\binom{10}{0}(0.7)^{10}-\binom{10}{1}(0.3)(0.7)^9 \\ &\approx 0.8506. \end{aligned}$$

(iii) We need $P(Y=8)$

$$P(Y=8)=\binom{10}{8}(0.7)^8(0.3)^2=0.2335.$$

(iv) We need $P(Y\leq 4)$

$$\begin{aligned} P(Y\leq 4) &= P(Y=0)+P(Y=1)+P(Y=2)+P(Y=3)+P(Y=4) \\ &= \binom{10}{0}(0.3)^{10}+\binom{10}{1}(0.7)(0.3)^9+\binom{10}{2}(0.3)^2(0.7)^8 \\ &\quad +\binom{10}{3}(0.3)^3(0.7)^7+\binom{10}{4}(0.3)^4(0.7)^6 \\ &\approx 0.047349. \end{aligned}$$

►**EXAMPLE 7.13** *Ten percent labourers of a factory are usually found absent daily. A list of* 10 *labourers is prepared randomly and one day their presence is varified. Find the probability that, out of* 10 *labourers one day* (i) *two will be absent,* (ii) *at least two will be absent,* (iii) *at best one will be absent.* (iv) *If the observation is made for* 100 *days, in how many days at best one will be absent?*

Solution. Let X be the number of absent labourers in a day. From the given condition, we have

$$P(X = x) = \binom{10}{x}(0.1)^x(0.9)^{10-x}, \ x = 0, 1, 2, \ldots, 10.$$

(i) We need $P(X = 2)$

$$P(X = 2) = \binom{10}{2}(0.1)^2(0.9)^8 \approx 0.19371.$$

(ii) We need $P(X \geq 2)$

$$\begin{aligned} P(X \geq 2) &= 1 - P(X < 2) \\ &= 1 - P(X = 0) - P(X = 1) \\ &= 1 - \binom{10}{0}(0.9)^{10} - \binom{10}{1}(0.1)(0.9)^9 \approx 0.2639. \end{aligned}$$

(iii) We need $P(X \leq 1)$

$$\begin{aligned} P(X \leq 1) &= P(X = 0) + P(X = 1) \\ &= \binom{10}{0}(0.9)^{10} + \binom{10}{1}(0.1)(0.9)^9 \approx 0.7361. \end{aligned}$$

(iv) At best one labourer will be absent in

$$NP(X \leq 1) = 100 \times 0.7361 \approx 74 \text{ days}.$$

►**EXAMPLE 7.14** *The health authority of an area claims that* 60% *eligible couples adopted family planning. To verify the fact,* 5 *couples were selected at random. Find the probability that out of* 5 *couples* (i) *one couple did not adopt family planning,* (ii) *at least one couple did not adopt family planning.* (iii) *Is the claim of the authority correct?*

Solution. Let X be the number of couples who did not adopt family planning. From the given condition, we have

$$P(X = x) = \binom{5}{x}(0.4)^x(0.6)^{5-x}, \ x = 0, 1, \ldots, 5.$$

(i) We need $P(X = 1)$

$$P(X = 1) = \binom{5}{1}(0.4)(0.6)^4 \approx 0.2592.$$

(ii) We need $P(X \geq 1)$

$$\begin{aligned} P(X \geq 1) &= 1 - P(X < 1) \\ &= 1 - P(X = 0) \\ &= 1 - \binom{5}{0}(0.6)^5 = 0.92224. \end{aligned}$$

(iii) If claim of the authority is correct, then out of 5 couples at best 2 couples do not adopt the family planning. Let us calculate $P(X \leq 2)$.

$$\begin{aligned} P(X \leq 2) &= P(X = 0) + P(X = 1) + P(X = 2) \\ &= \binom{5}{0}(0.6)^5 + \binom{5}{1}(0.4)(0.6)^4 + \binom{5}{2}(0.4)^2(0.6)^3 \\ &= 0.68256. \end{aligned}$$

Since this probability is larger (> 0.5), the claim of the authority is considered true.

►**EXAMPLE 7.15** *A plane lands 4 times in a day in different airports. The successful landing rate of the plane is 90%. Find the probability that, in a randomly selected day* (i) *the plane lands successfully all the 4 times,* (ii) *at least 3 times it lands successfully,* (iii) *at best 3 times it lands successfully,* (iv) *all the times it fails to land successfully.*

Solution. Let X be the random variable indicating the number of successful landing of the plane. From the given condition, we have

$$P(X = x) = \binom{4}{x}(0.9)^x(0.1)^{4-x}, \; x = 0, 1, 2, 3, 4.$$

(i) We need $P(X = 4)$

$$P(X = 4) = \binom{4}{4}(0.9)^4 = 0.6561.$$

(ii) We need $P(X \geq 3)$

$$\begin{aligned} P(X \geq 3) &= P(X = 3) + P(X = 4) \\ &= \binom{4}{3}(0.9)^3(0.1) + \binom{4}{4}(0.9)^4 \\ &= 0.9477. \end{aligned}$$

(iii) We need $P(X \leq 3)$

$$P(X \leq 3) = 1 - P(X = 4) = 1 - 0.6561 = 0.3439.$$

(iv) We need $P(X = 0)$

$$P(X = 0) = \binom{4}{0}(0.1)^4 = 0.0001.$$

►EXAMPLE 7.16 *In an airport 100 planes land everyday. The average successful landing rate of the planes is 80%. Find the mean and variance of the number of planes which land successfully in the airport. A particular plane land 5 times a day in that airport. Throughout the year how many days this plane lands at least 4 times successfully?*

Solution. Let X be the number of planes which land successfully in an airport. Given that $n = 100,\ p = 0.8$, so $q = 0.2$. Therefore, mean and variance of number of planes which land successfully are

$$\begin{aligned} E(X) &= np = 100 \times 0.8 = 80 \\ \text{and } V(X) &= npq = 100 \times 0.8 \times 0.2 = 16, \text{ respectively.} \end{aligned}$$

For a particular plane

$$P(X = x) = \binom{5}{x}(0.8)^x(0.2)^{5-x},\ x = 0, 1, 2, \ldots, 5.$$

The probability of landing successfully at least 4 times is

$$\begin{aligned} P(X \geq 4) &= P(X = 4) + P(X = 5) \\ &= \binom{5}{4}(0.8)^4(0.2) + \binom{5}{5}(0.8)^5 = 0.73728. \end{aligned}$$

Throughout the year this plane lands in $NP(X \geq 4) = 365 \times 0.73728 \approx 269$ days successfully at least 4 times a day.

►EXAMPLE 7.17 *A plane having 4 engines when fly is experienced of 2% failure of any of its engines. Find the probability that, for successful fly of the plane,* (i) *at best 3 engines will be on start,* (ii) *at least 3 engines need to be operated.* (iii) *If 100 such planes are observed, in how many cases the plane successfully flies with at least 3 engines on start?*

Solution. Let X be the number of engines on start for successful fly of the plane. From the given condition

$$P(X = x) = \binom{4}{x}(0.98)^x(0.02)^{4-x},\ x = 0, 1, 2, 3, 4.$$

(i) We need $P(X \leq 3)$

$$P(X \leq 3) = 1 - P(X = 4) = 1 - \binom{4}{4}(0.98)^4 = 0.07763184.$$

(ii) We need $P(X \geq 3)$

$$\begin{aligned} P(X \geq 3) &= P(X = 3) + P(X = 4) \\ &= \binom{4}{3}(0.98)^3(0.02) + \binom{4}{4}(0.98)^4 \approx 0.997663. \end{aligned}$$

(iii) Out of 100 planes $NP(X \geq 3) = 100 \times 0.997663 \approx 100$ planes need at least 3 engines on operation for successful fly.

►**EXAMPLE 7.18** *Ten observations are selected randomly from a binomial population having mean* = 4 *and variance* = 2.4. *Find the probability that, out of* 10 *observations at least one observation is not selected properly.*

Solution. Let X be the binomial variable. Given

$$E(X) = 4,\ V(X) = 2.4,\ n = 10 \Rightarrow\ p = 0.4 \text{ and } q = 0.6.$$

$$\therefore\ P(X = x) = \binom{10}{x}(0.4)^x(0.6)^{10-x},\ x = 0, 1, 2, \ldots, 10.$$

Let S be the case of successful selection of an observation from binomial population. Then

$$P(S) = 0.4.$$

If an observation is not selected successfully, then $P(S) \neq 0.4$. Let $P(S) = 0.6$. Then

$$P(Y = y) = \binom{10}{y}(0.6)^y(0.4)^{10-y},\ y = 0, 1, \ldots, 10.$$

Here Y is the random variable the observation of which is selected with probability 0.6. We need $P(Y \geq 1)$

$$P(Y \geq 1) = 1 - P(Y < 1) = 1 - P(Y = 0)$$
$$= 1 - \binom{10}{0}(0.4)^{10} \approx 0.999895.$$

►**EXAMPLE 7.19** *The mean and variance of a binomial distribution are, respectively* $E(X) = 5, V(X) = 4$. *Find the probability that an observation of this distribution exceeds* 2.

Solution. Let X be the binomial random variable. Given

$$E(X) = np = 5 \text{ and } V(X) = npq = 4$$
$$\Rightarrow\ q = 0.8,\ p = 0.2,\ n = 25.$$
$$\therefore\ P(X = x) = \binom{25}{x}(0.2)^x(0.8)^{25-x},\ x = 0, 1, 2, \ldots, 25.$$

We need $P(X > 2)$

$$P(X > 2) = 1 - P(X \leq 2) = 1 - P(X = 0) - P(X = 1) - P(X = 2)$$
$$= 1 - \binom{25}{0}(0.8)^{25} - \binom{25}{1}(0.2)(0.8)^{24} - \binom{25}{2}(0.2)^2(0.8)^{23}$$
$$\approx 0.90177.$$

►**EXAMPLE 7.20** *A random sample of* 5 *observation is selected from a binomial distribution. The mean of these* 5 *observations is found* 2. *Another random sample of* 15 *observations is selected, independently of first sample, from another binomial population. The mean of these* 15 *observations is found* 6. *Find the probability that the sum of the values of any two observations of two samples will be* 2 *or more.*

Solution. Let X and Y be the first and second binomial random variable, respectively. Observed that $E(X) = 2$, $E(Y) = 6$. The sample sizes are $n_1 = 5, n_2 = 15$.

Therefore, $n_1p_1 = 2 \Rightarrow p_1 = 0.4$ and $n_2p_2 = 6 \Rightarrow p_2 = 0.4$.

$$\therefore\ q_1 = 0.6 = q_2.$$

We have to find $P(X+Y \geq 2)$. We know that

$$M_x(t) = (q_1 + P_1e^t)^{n_1},\ M_y(t) = (q_2 + p_2e^t)^{n_2}.$$

Since $p = p_1 = p_2,\ q_1 = q_2 = q$

$$M_{x+y}(t) = (q+pe^t)^{n_1+n_2} = (0.6+0.4e^t)^{5+15} = M_z(t)$$

where $X+Y = Z$,

$$\begin{aligned} P(X+Y \geq 2) &= P(Z \geq 2) = 1 - P(Z < 2) = 1 - P(Z=0) - p(Z=1) \\ &= 1 - \binom{20}{0}(0.6)^{20} - \binom{20}{1}(0.4)(0.6)^{19} \approx 0.99947. \end{aligned}$$

►**EXAMPLE 7.21** *Five percent products of a factory are usually found defective. Five products of a day are selected randomly and the experiment is repeated on two consecutive days. Find the probability that, in the samples* (i) *there will be 2 defective products,* (ii) *there will be at least 2 defective products.* (iii) *If the experiment is repeated 100 times, in how many cases the number of defective products will be at least 2?*

Solution. Let X and Y be the number of defective products in first and second sample, respectively. From the given condition

$$\begin{aligned} P(X=x) &= \binom{5}{x}(0.05)^x(0.95)^{5-x},\ x = 0,1,\ldots,5 \\ P(Y=y) &= \binom{5}{y}(0.05)^y(0.95)^{5-y},\ y = 0,1,\ldots,5. \end{aligned}$$

Since two samples are selected independently and since in each case

$$\begin{aligned} P\ (\text{defective}) &= p = 0.05 \\ P(X+Y=z) &= \binom{10}{z}(0.05)^z(0.95)^{10-z},\ z = 0,1,2,\ldots,10. \end{aligned}$$

(i) We need $P(Z=2)$

$$P(Z=2) = \binom{10}{2}(0.05)^2(0.95)^8 \approx 0.07463.$$

(ii) We need $P(Z \geq 2)$

$$\begin{aligned} P(Z \geq 2) &= 1 - P(Z<2) = 1 - P(Z=0) - P(Z=1) \\ &= 1 - \binom{10}{0}(0.95)^{10} - \binom{10}{1}(0.05)(0.95)^9 \approx 0.086138. \end{aligned}$$

(iii) Out of $N = 100$ repetitions $NP(Z \geq 2) = 100 \times 0.086138 \approx 9$ cases will show that there will be at least 2 defective products.

▸EXAMPLE 7.22 *An unbiased die is tossed 6 times. If the die shows even numbers, the experimental results are termed as success. Find percentage of success of the experiment.*

Solution. Let X be the random variable indicating even numbers on the upper face of the die. Then

$$P(X=x)=\left(\frac{1}{2}\right)^6\binom{6}{x},\quad x=0,1,2,\ldots,6.$$

The experiment will be ended in success if the die shows 2, 4, 6. The number of successes is $x=3$. Therefore,

$$P(X=3)=\left(\frac{1}{2}\right)^6\binom{6}{3}=0.3125.$$

Hence in 31.25% cases the experiment will be ended in success.

▸EXAMPLE 7.23 *The telephone lines are not connected in 10% cases if the attempt is made during 10.00 AM to 11.00 AM. One day 12 telephone numbers are selected at random and tried to connect them during 10.00 AM to 11.00 AM. Find the probability that, out of 12 telephones (i) all are connected properly, (ii) two are not connected, (iii) at least 2 are not connected. (iv) If the experiment is repeated by 25 persons, in how many cases connection is made in all telephones?*

Solution. Let X be the number of telephones which are not connected properly. Given

$$P(X=x)=\binom{12}{x}(0.1)^x(0.9)^{12-x},\quad x=0,1,2,\ldots,12.$$

(i) We need $P(X=0)$

$$P(X=0)=\binom{12}{0}(0.9)^{12}\approx 0.2824.$$

(ii) We need $P(X=2)$

$$P(X=2)=\binom{12}{2}(0.1)^2(0.9)^{10}\approx 0.23013.$$

(iii) We need $P(X\geq 2)$

$$\begin{aligned}P(X\geq 2)&=1-P(X<2)\\&=1-P(X=0)-P(X=1)\\&=1-\binom{12}{0}(0.9)^{12}-\binom{12}{1}(0.1)(0.9)^{11}\approx 0.341.\end{aligned}$$

(iv) The probability that the connection is made in all telephones is

$$P(X=0)=0.2824.$$

Therefore, out of $25=N$ persons $NP(X=0)=25\times 0.2824\approx 7$ persons connected the telephone property.

►**EXAMPLE 7.24** *Thirty percent industrial workers of an industry usually face accident. Ten workers of that industry are selected randomly for a survey. Find the probability that, among* 10 *workers,* (i) *one faced an accident,* (ii) *at least one faced an accident,* (iii) *at best one faced an accident.* (iv) *If such a survey is conducted in* 100 *similar industries, in how many industries no worker will be found who faces accident?*

Solution. Let X be the random variable indicating number of workers who face accident. From the given condition

$$P(X = x) = \binom{10}{x}(0.3)^x(0.7)^{10-x}, \quad x = 0, 1, \ldots, 10.$$

(i) We need $P(X = 1)$

$$P(X = 1) = \binom{10}{1}(0.3)(0.7)^9 \approx 0.1211.$$

(ii) We need $P(X \geq 1)$

$$\begin{aligned} P(X \geq 1) &= 1 - P(X < 1) = 1 - P(X = 0) \\ &= 1 - \binom{10}{0}(0.7)^{10} \approx 0.9717. \end{aligned}$$

(iii) We need $P(X \leq 1)$

$$\begin{aligned} P(X \leq 1) &= P(X = 0) + P(X = 1) \\ &= \binom{10}{0}(0.7)^{10} + \binom{10}{1}(0.3)(0.7)^9 \approx 0.1494. \end{aligned}$$

(iv) The probability that no one will face accident is

$$P(X = 0) = \binom{10}{0}(0.7)^{10} \approx 0.0283.$$

Therefore, out of 100 industries in $NP(X = 0) = 100 \times 0.0283 \approx 3$ industries, no worker will face accident.

►**EXAMPLE 7.25** *Eighty percent students of a private university are from urban area. Ten students of that university are randomly selected to conduct a survey. Find the probability that, out of* 10 *students,* (i) *everyone will be from urban area,* (ii) *at least one will not be from urban area,* (iii) *at best two will not be from urban area.* (iv) *In how many private universities all students will be from urban area, if there are* 50 *universities?*

Solution. Let X be the random variable indicating the number of urban students. Then from the given condition, we have

$$P(X = x) = \binom{10}{x}(0.8)^x(0.2)^{10-x}, \quad x = 0, 1, 2, \ldots, 10.$$

(i) We need $P(X = 10)$

$$P(X = 10) = \binom{10}{10}(0.8)^{10} \approx 0.10737.$$

(ii) We need $P(Y \geq 1)$, where

$$P(Y = y) = \binom{10}{y}(0.2)^y(0.8)^{10-y}, \; y = 0, 1, \ldots, 10$$

$$P(Y \geq 1) = 1 - P(Y < 1) = 1 - P(Y = 0)$$

$$= 1 - \binom{10}{0}(0.8)^{10} \approx 0.89263.$$

(iii) We need $P(Y \leq 2)$

$$P(Y \leq 2) = P(Y = 0) + P(Y = 1) + P(Y = 2)$$

$$= \binom{10}{0}(0.8)^{10} + \binom{10}{1}(0.2)(0.8)^9 + \binom{10}{2}(0.2)^2(0.8)^2$$

$$\approx 0.6778.$$

Out of 50 universities $NP(X = 0) = 50 \times 0.10737 \approx 5$ universities have their all students from urban area.

►**EXAMPLE 7.26** *There are* 6 *children in a family living in a rural area. But the distribution of boys and girls is not known. It is assumed that boys and girls are equally probable. Find the probability that* (i) *all are girls,* (ii) *at least* 3 *will be boys,* (iii) *at best* 2 *will be boys.* (iv) *If such* 50 *families live in the area, how many of them will have all girls?*

Solution. Let X be the number of boys. We have

$$P(X = x) = \binom{6}{x}\left(\frac{1}{2}\right)^6, \; x = 0, 1, 2, \ldots, 6.$$

(i) All will be girls if $X = 0$

$$P(X = 0) = \binom{6}{0}\left(\frac{1}{2}\right)^6 = \frac{1}{64}.$$

(ii) We need $P(X \geq 3)$

$$P(X \geq 3) = 1 - P(X < 3)$$

$$= 1 - P(X = 0) - P(X = 1) - P(X = 2)$$

$$= 1 - \left(\frac{1}{2}\right)^6\left[\binom{6}{0} + \binom{6}{1} + \binom{6}{2}\right] = \frac{21}{32}.$$

(iii) We need $P(X \leq 2)$

$$P(X \leq 2) = P(X = 0) + P(X = 1) + P(X = 2)$$

$$= \left(\frac{1}{2}\right)^6\left[\binom{6}{0} + \binom{6}{1} + \binom{6}{2}\right] = \frac{11}{32}.$$

(iv) From the given condition approximately in 1 family $[NP(X = 0) = \frac{50}{64}]$ all will be girls.

►**EXAMPLE 7.27** *Inoculation against Hepatitis-B has been done in an area. Previous experience says that the inoculation is effective in* 80% *cases. After* 6 *months* 5 *of the inoculated persons are randomly selected. Find the probability that, out of* 5 *persons* (i) *inoculation is not found effective for one person,* (ii) *inoculation is not found effective for at least* 1 *person.* (iii) *If the survey on* 5 *persons is conducted by* 25 *research workers, in how many cases inoculation will be found uneffective for one person?*

Solution. Let X be the random variable indicating number of persons for whom inoculation is not effective. From the given condition

$$P(X = x) = \binom{5}{x}(0.2)^x(0.8)^{5-x}, \quad x = 0, 1, 2, \ldots, 5.$$

(i) We need $P(X = 1)$

$$P(X = 1) = \binom{5}{1}(0.2)(0.8)^4 \approx 0.16777.$$

(ii) We need $P(X \geq 1)$

$$\begin{aligned} P(X \geq 1) &= 1 - P(X < 1) = 1 - P(X = 0) \\ &= 1 - \binom{5}{0}(0.8)^5 \approx 0.67232. \end{aligned}$$

(iii) Since probability of a person for whom inoculation is not effective is

$$P(X = 1) = 0.16777$$

in one survey, therefore, in $NP(X = 1) = 25 \times 0.16777 \approx 4$ cases similar result is expected to be observed.

►**EXAMPLE 7.28** *In a garments industry* 80% *female workers are paid less than* Rs.1500.00 *per month. A random sample of* 10 *workers are selected at random. Find the probability that, out of* 10 *female workers,* (i) *all of them get less than* Rs. 1500.00 *per month,* (ii) *at least* 4 *workers get less than* Rs. 1500.00, (iii) *at best* 2 *workers get more than* Rs. 1500.00 *per month.* (iv) *There are* 500 *female workers in the industry. What is the expected number of female workers who get salary less than* Rs. 1500.00 *per month.*

Solution. Let X be the random variable indicating numbers of female workers who get salary less than Rs.1500.00. From the given condition

$$P(X = x) = \binom{10}{x}(0.8)^x(0.2)^{10-x}, \quad x = 0, 1, 2, \ldots, 10.$$

(i) We need $P(X = 10)$

$$P(X = 10) = \binom{10}{10}(0.8)^{10} \approx 0.10737.$$

(ii) We need $P(X \geq 4)$

$$\begin{aligned} P(X \geq 4) &= 1 - P(X < 4) \\ &= 1 - P(X=0) - P(X=1) - P(X=2) - P(X=3) \\ &= 1 - \binom{10}{0}(0.2)^{10} - \binom{10}{1}(0.8)(0.2)^9 \\ &\quad - \binom{10}{2}(0.2)^2(0.8)^8 - \binom{10}{3}(0.2)^3(0.8)^7 \\ &\approx 0.999135. \end{aligned}$$

(iii) Let Y be the number of female workers who get more than Rs.1500.00 per month. Then

$$P(Y=y) = \binom{10}{y}(0.2)^y(0.8)^{10-y}, \; y = 0, 1, \ldots, 10.$$

We need $P(Y \leq 2)$

$$\begin{aligned} P(Y \leq 2) &= P(Y=0) + P(Y=1) + P(Y=2) \\ &= \binom{10}{0}(0.8)^{10} + \binom{10}{1}(0.2)(0.8)^9 + \binom{10}{2}(0.2)^2(0.8)^8 \\ &\approx 0.6778. \end{aligned}$$

(iv) Given $n = 500$ and $p = 0.8$. Therefore, expected number of female workers who get salary less than Rs. 1500.00 is $E(X) = np = 500 \times 0.8 = 400$.

►**EXAMPLE 7.29** *The probability of hitting a target by a person is* 0.75. *If this person shoots* 5 *times, what is the probability that* (i) *he will hit the target at least once,* (ii) *at best* 2*?* (iii) *If* 100 *persons of the same experience try to hit the target, then how many of them will fail to hit the target?* (iv) *How many shoots are needed for a person to hit the target at least once with probability more than* 0.7*?*

Solution. Let X be the random variable indicating the number of hitting the target. From the given condition, we have

$$P(X=x) = \binom{5}{x}(0.75)^x(0.25)^{5-x}, \; x = 0, 1, 2, \ldots, 5.$$

(i) We need $P(X \geq 1)$

$$\begin{aligned} P(X \geq 1) &= 1 - P(X < 1) = 1 - P(X=0) \\ &= 1 - \binom{5}{0}(0.25)^5 \approx 0.99902. \end{aligned}$$

(ii) We need $P(X \leq 2)$

$$\begin{aligned} P(X \leq 2) &= P(X=0) + P(X=1) + P(X=2) \\ &= \binom{5}{0}(0.25)^5 + \binom{5}{1}(0.75)(0.25)^4 + \binom{5}{2}(0.75)^2(0.25)^3 \\ &\approx 0.1035. \end{aligned}$$

(iii) If any person fails to hit the target, then $X = 0$.

$$P(X = 0) = \binom{5}{0}(0.25)^5 \approx 0.000976.$$

If $N = 100$ such persons try to hit the target, then

$$NP(X = 0) = 100 \times 0.000976 = 0.0976.$$

Thus no one will fail to hit the target.

(iv) Let n be the number of shoots needed to hit the target at least once with probability $p = 0.7$, i.e.,

$$\begin{aligned} P(X \geq 1) &= 1 - P(X < 1) = 1 - P(X = 0) \\ &= 1 - \binom{n}{0}(0.75)^0(0.25)^n = 1 - (0.25)^n. \end{aligned}$$

The condition is that $1 - (0.25)^n > 0.7$ or, $(0.25)^n < 0.3 \Rightarrow n = 1$.

►**EXAMPLE 7.30** *An industrialist ensures that at best 10% of his product be defective. To investigate his assurance it has been decided to conduct a survey on the understanding that, if at best 2 products out of 20 products are found defective, then the assurance is to be considered true. In practice, if the probability of producing a defective item is 0.08, what is the probability of accepting the claim of the industrialist?*

Solution. Let X be the random variable indicating the number of defective items produced in the industry. Then

$$P(X = x) = \binom{20}{x}(0.08)^x(0.92)^{20-x}, \; x = 0, 1, 2, \ldots, 20.$$

If the assurance of the producer is true, then $X \leq 2$.

$$\begin{aligned} P(X \leq 2) &= P(X = 0) + P(X = 1) + P(X = 2) \\ &= \binom{20}{0}(0.92)^{20} + \binom{20}{1}(0.08)(0.92)^{19} + \binom{20}{2}(0.08)^2(0.92)^{18} \\ &\approx 0.7879. \end{aligned}$$

►**EXAMPLE 7.31** *Five rooms in a hotel are air-conditioned. The owner of the hotel has decided to rent a television in each room if the customer demands it. The owner has bought 4 televisions on the understanding that 80% customers may demand it. All five rooms are occupied by the customers always. What would be the demand in percentage of days of the television? What is the probability that there will be a demand of the number of televisions more than that the owner possesses?*

Solution. Let X be the random variable indicating the number of demanded television sets. Then

$$P(X = x) = \binom{5}{x}(0.8)^x(0.2)^{5-x}, \; x = 0, 1, \ldots, 5.$$

We need $P(X = 4)$

$$P(X = 4) = \binom{5}{4}(0.8)^4(0.2) = 0.4096.$$

Therefore, there may be demand for all television sets in 41% days. The probability of more demand than the number of available sets is $P(X > 4)$. Here

$$P(X > 4) = P(X = 5) = \binom{5}{5}(0.8)^5 = 0.32768.$$

►**EXAMPLE 7.32** *The government has taken steps to improve the law and order situation in the country. But 80% people of the country do not support the steps. However, to verify the fact a survey is planned and 20 people are selected at random for the purpose. Find the probability that, out of 20 people (i) at least two persons will not agree with the government steps, (ii) 4 persons agree with the government steps.*

Solution. Let X be the number of persons who do not agree with the government steps. Then

$$P(X = x) = \binom{20}{x}(0.8)^x(0.2)^{20-x}, \; x = 0, 1, 2, \ldots, 20.$$

(i) We need $P(X \geq 2)$

$$\begin{aligned} P(X \geq 2) &= 1 - P(X < 2) \\ &= 1 - P(X = 0) - P(X = 1) \\ &= 1 - \binom{20}{0}(0.2)^{20} - \binom{20}{1}(0.8)(0.2)^{19} \approx 1.00. \end{aligned}$$

(ii) Let Y be the random variable indicating the number of people who support the government actions. Then

$$P(Y = y) = \binom{20}{y}(0.2)^y(0.8)^{20-y}, \; y = 0, 1, \ldots, 20.$$

We need $P(Y = 4)$

$$P(Y = 4) = \binom{20}{4}(0.2)^4(0.8)^{16} \approx 0.2182.$$

►**EXAMPLE 7.33** *A pair of genes is identified to decide the eye colour of human beings. One of the genes is of father and another one is of mother. It is known that some of the genes are dominating (d) and some are recessive (r) to change the body formation of human. Any person has dd, rd, rr pairs of genes. Here 'dd' is perfect dominance, 'rr' is perfect recessive and 'rd' is hybrid pairs of genes. The pairs dd and rr look alike. Let there are 4 children of a pair of hybrid parents. What is the probability that, out of 4 children 3 children have dd pair of genes?*

Solution. Lex X be the random variable indicating the number of children having *dd* pair of genes. It is assumed that for a children one gene is of father and another one is of mother. Hence

$$P(dd) = \frac{1}{4}, P(rr) = \frac{1}{4}, P(rd) = \frac{2}{4}.$$

Therefore, the probability that the off-spring has dominance gene is given by

$$P(dd) + P(rd) = \frac{1}{4} + \frac{2}{4} = \frac{3}{4}.$$

Hence

$$P(X = x) = \binom{4}{x}\left(\frac{3}{4}\right)^x \left(\frac{1}{4}\right)^{4-x}, \quad x = 0, 1, 2, 3, 4.$$

We need $P(X = 3)$

$$P(X = 3) = \binom{4}{3}\left(\frac{3}{4}\right)^3 \frac{1}{4} = \frac{27}{64}.$$

►**EXAMPLE 7.34** *In a box there are* 3 *white and* 4 *black balls. One ball is taken at random from the box and the process is continued* 4 *times with replacement. Find the probability that, out of first* 4 *balls* (i) *one ball is white,* (ii) *at best one ball is white.* (iii) *If the experiment is repeated by* 100 *persons, how many of them get one white ball?* (iv) *Find the mean and variance of number of white balls.*

Solution. Since one ball is taken from the box with replacement,

$$P(1W) = \frac{3}{7} \text{ and } P(1B) = \frac{4}{7} \text{ [}W\text{: white ball, }B\text{: black ball].}$$

These probabilities are true at every draw of the ball. Now, let X be the random variable indicating the number of white balls. Then

$$P(X = x) = \binom{4}{x}\left(\frac{3}{7}\right)^x \left(\frac{4}{7}\right)^{4-x}, \quad x = 0, 1, 2, 3, 4.$$

(i) We need $P(X = 1)$

$$P(X = 1) = \binom{4}{1}\left(\frac{3}{7}\right)\left(\frac{4}{7}\right)^3 \approx 0.31987.$$

(ii) We need $P(X \le 1)$

$$\begin{aligned} P(X \le 1) &= P(X = 0) + P(X = 1) \\ &= \binom{4}{0}\left(\frac{4}{7}\right)^4 + \binom{4}{1}\left(\frac{3}{7}\right)\left(\frac{4}{7}\right)^3 \approx 0.42644. \end{aligned}$$

(iii) If the experiment is repeated by 100 persons, then $NP(X = 1) = 100 \times 0.31987$ ≈ 32 persons will get one white ball.

(iv) The mean and variance of number of white balls are

$$E(X) = np = 4 \times \frac{3}{7} = \frac{12}{7}$$

$$\text{and } V(X) = npq = 4 \times \frac{3}{7} \times \frac{4}{7} = \frac{48}{49} \text{ respectively.}$$

7.5 Fitting of Binomial Distribution

We know

$$p(x) = \binom{n}{x} p^x q^{n-x}$$

$$\text{and } P(x+1) = \binom{n}{x+1} p^{x+1} q^{n-x-1}, \; x = 0, 1, 2, \ldots, n.$$

$$\therefore \frac{p(x+1)}{p(x)} = \frac{n-x}{x+1}\frac{p}{q}.$$

Thus the recurrence relation for binomial probability is given by

$$p(x+1) = \frac{n-x}{x+1}\frac{p}{q} p(x).$$

Now, the binomial distribution can be fitted to the data of a frequency distribution which arises from a random experiment showing results as successes and failures. From such an experiment the value of p can be calculated from $\mu_1' = np$, where $p = \mu_1'/n$. Using this value of p one can calculate $p(x = 0)$. Then for different values of x, $p(x+1)$ can be calculated which will give expected frequency of binomial distribution, where

$$\text{expected frequency} = N \times p(x).$$

Here N = total frequency of the observed data. The recurrence relation shows that $P(X = x)$ increases with x so long as $x < np - q = (n+1)p - 1$, and decreases with x if $x > np - q$. Hence as x increases, the values of the $P(X = x)$ increase to a maximum value at the integer x satisfying

$$(n+1)p - 1 < x \leq (n+1)p$$

and thereafter they decrease. If $(n+1)p$ is an integer, then

$$P(X = (n+1)p - 1) = P(X = (n+1)p).$$

The common value is the maximum among the values of $P(X = x)$. This result helps us to find the most probable number of successes in a series of n independent trials. Thus, if $p = \frac{1}{3}$ and $n = 11$, then $(n+1)p - 1 = 3$ so that 3 or 4 are the most probable number of successes. Again, if $np = k_1 + f$, where k_1 is an integer and f is a proper fraction, $p(x)$ is maximum when $x = (np + p)$. Thus, if $p = \frac{3}{4}$ and $n = 10$, then $x = (10 \times \frac{3}{4} + \frac{3}{4}) = 8$ is the most probable number of successes.

►**EXAMPLE 7.35** *A group of doctors started operation of eye through 'Visiting Eye Clinic'. Per visit they operated* 6 *patients in one place and in a year they visited* 100 *places. Below is given the number of successful operations in different places:*

Number of successful operations, x	0	1	2	3	4	5	6	*Total*
Number of places, $f = O_i$	0	3	8	10	30	40	9	100

Solution. Here

$$n = 6,\ N = 100,\ \bar{x} = \frac{\Sigma fx}{N} = \frac{423}{100} = np$$

$$\therefore\ p = \frac{423}{600} = 0.705 \text{ and } q = 1 - p = 0.295.$$

Thus

$$P(x=0) = \binom{6}{0} p^0 q^{n-0} = (0.295)^6 = 0.00066.$$

Here the probability function for the binomial distribution to be fitted is

$$p(x) = \binom{6}{x}(0.705)^x(0.295)^{6-x},\ x = 0, 1, 2, \ldots, 6.$$

Calculation for finding probabilities and expected frequencies

x	$\frac{n-x}{x+1} \cdot \frac{p}{q}$	$p(x)$ $\left[p(x+1) = \frac{n-x}{x+1}\frac{p}{q}p(x)\right]$	Expected frequency $N \times p(x) = E_i$
0	14.34	0.00065	0
1	5.97	0.00946	1
2	3.19	0.05652	6
3	1.79	0.18030	18
4	0.96	0.32274	32
5	0.40	0.30983	31
6	0	0.12393	12

Now, it can be verified whether the observed frequencies are consistent with the theoretical frequencies of binomial distribution. This can be done by calculating χ^2-statistic, where

$$\chi^2 = \sum_{i=1}^{7} \frac{O_i^2}{E_i} - N = 11.71.$$

Here $\chi^2_{0.05,6} = 12.59 > \chi^2$. Hence it can be concluded that the observed frequencies are consistent with the theoretical frequencies of binomial distribution. Here the most probable number of success is $(np + p) = 5$.

7.6 Truncated Binomial Distribution

Let X be a random variable with probability function

$$P(X = x) = \binom{n}{x} p^x q^{n-x},\ x = 0, 1, 2, \ldots, n.$$

Let us consider another random variable X_1 such that the values of X_1 are

$$x_1 = r_1, \ldots, n - r_2,$$

where $0 \leq x < r_1$ and $n - r_2 < x < n$. Thus

$$0 \leq r_1 \leq n - r_2 \leq n.$$

Here the values of X_1 are the values of X except the values at two end points. Thus the total probability of x except the two end points is

$$P(r_1 \leq X \leq n - r_2) = \sum_{x=r_1}^{n-r_2} \binom{n}{x} p^x q^{n-x}.$$

Now, the probability function of X_1 is defined by

$$P(X_1 = x_1) = \frac{\binom{n}{x_1} p^{x_1} q^{n-x_1}}{\sum\limits_{x=r_1}^{n-r_2} \binom{n}{x} p^x q^{n-x}}, \quad x_1 = r_1, \ldots, n - r_2.$$

Here the distribution of X_1 is called *doubly truncated binomial distribution.* If only the values $0, 1, \ldots, (r_1 - 1)(r_1 \geq 1)$ or the values $n - r_2 + 1, \ldots, n(r_2 \geq 1)$ are omitted, then the distribution is called *singly truncated binomial distribution.* The truncated binomial distribution by omitting value 0 only, is given by

$$P(X_1 = x_1) = \frac{\binom{n}{x_1} p^{x_1} q^{n-x_1}}{1 - q^n}, \quad x_1 = 1, 2, \ldots, n.$$

The r-th moment about origin of this distribution is given by

$$\mu'_{r(x_1)} = E(X_1^r) = \frac{1}{1 - q^n} \sum_{x_1=1}^{n} \binom{n}{x_1} p^{x_1} q^{n-x_1} x_1^r = \frac{1}{1 - q^n} \mu'_{r(x)}.$$

In particular, when $r = 1$ and $r = 2$

$$\begin{aligned}
\mu'_{1(x_1)} &= \frac{1}{1 - q^n} \sum_{x_1=1}^{n} \binom{n}{x_1} p^{x_1} q^{n-x_1} x_1 = \frac{np}{1 - q^n} \\
\text{and } \mu'_{2(x_1)} &= \frac{1}{1 - q^n} \sum_{x_1=1}^{n} \binom{n}{x_1} p^{x_1} q^{n-x_1} x_1^2 \\
&= \frac{1}{1 - q^n} \left[\sum_{x_1=1}^{n} \binom{n}{x_1} p^{x_1} q^{n-x_1} x_1 + \sum_{x_1=1}^{n} \binom{n}{x_1} p^{x_1} q^{n-x_1} x(x-1) \right] \\
&= \frac{np}{1 - q^n} + \frac{n(n-1)p^2}{1 - q^n}. \\
\therefore \mu_{2(x_1)} &= \mu'_{2(x_1)} - \mu'^2_{1(x_1)} \\
&= \frac{npq}{1 - q^n} - \frac{n^2 p^2 q^n}{(1 - q^n)^2}.
\end{aligned}$$

If $r = -1$, Grab and Savage (1954) note the approximation formula

$$E(X_1^{-1}) = (np - q)^{-1}.$$

The first three moments about zero of doubly truncated distribution are given by

$$\begin{aligned}\mu_1' &= C_1 - C_2 + np,\\ \mu_2' &= r_1C_1 - (n - r_2 + 1)C_2 + \mu_1'(n-1)p + np\\ \text{and } \mu_3' &= r_1^2C_1 - (n - r_2 + 1)^2C_2 + \mu_2'(n-2)p + \mu_1'(2n-1)p + np\end{aligned}$$

where

$$C_1 = r_1 q\binom{n}{r_1}p^{r_1}q^{n-r_1}\left[\sum_{x_1=r_1}^{n-r_2}\binom{n}{x_1}p^{x_1}q^{n-x_1}\right]^{-1}$$

$$\text{and } C_2 = (n - r_2 + 1)q\binom{n}{n-r_2+1}p^{n-r_2+1}q^{r_2-1}\left[\sum_{x_1=r_1}^{n-r_2}\binom{n}{x_1}p^{x_1}q^{n-x_1}\right]^{-1}.$$

7.7 Mixtures of Binomial Distributions

The mixture of binomial distribution is commonly called *compound distribution.* This distribution arises due to the assumption that the parameter of the original distribution follows a probability distribution. For example, suppose that the production of items in an industry can be classified in one of two mutually exclusive and exhaustive cases, say, defective or non-defective. Let us assume that the probability of producing non-defective items is $P \cdot (0 < P < 1)$. The number of non-defective items may vary from day-to-day production. This means that the value of P varies within the range 0 to 1. Again, if an investigation is made to study the number of non-defective items on randomly selected days, then n, the number of trials, will follow a probability distribution. Assume that the probability of non-defective items (X) in a day's production is given by

$$P(X = x) = \binom{n}{x}p^x q^{n-x}, \; x = 0, 1, 2, \ldots, n.$$

Also, suppose that the probability of production of items (y) in randomly selected days is given by

$$P(y) = \frac{e^{-\lambda}\lambda^y}{y!}, \; y = 0, 1, 2, \ldots.$$

Then the probability of total of x non-defective items in selected days is given by

$$\begin{aligned}P_x &= P(X = x)\\ &= e^{-\lambda}\sum_{y\geq \frac{x}{n}}\binom{ny}{x}p^x q^{ny-x}(\lambda^y/y!)\\ &= \frac{e^{-\lambda}(p/q)^x}{x!}\sum_{y\geq\frac{x}{n}}(ny)^{(x)}[(\lambda q^n)^y/y!].\end{aligned}$$

This distribution is one of the compound binomial distribution which is known as the Poisson binomial distribution. The probability generating function of this distribution is

$$g(t) = E[(q+pt)^{ny}] = \exp[\lambda\{(q+pt)^n - 1\}]$$

where y is a Poisson variate with $E(y) = \lambda$.

Now, putting $t = 0$, we get

$$\begin{aligned}
P_0 &= \exp[-\lambda(1-q^n)] \\
P_1 &= \left.\frac{dg(t)}{dt}\right|_{t=0} = n\lambda pq^{n-1}\exp[-\lambda(1-q^n)] \\
P_2 &= \left.\frac{d^2g(t)}{dt^2}\right|_{t=0} = \frac{1}{2}n\lambda p^2[n-1+n\lambda q^n]q^{n-2}\exp[-\lambda(1-q^n)]
\end{aligned}$$

It is seen that if $n\lambda pq^{n-1} < 1$ but $n-1+n\lambda q^n > 2q/p$, then $P_0 > P_1 < P_2$. Therefore, there is a local minimum at $P_1(x=1)$.

The successive P_x's can be calculated from the recurrence relation

$$P_x = \frac{n\lambda p}{x+1}\sum_{y=0}^{x}\binom{n-1}{y}p^y q^{n-y-1}P_{x-y}.$$

Shumway and Gurland (1960) proposed another formula which is given by

$$P_x = \exp[-\lambda(1-q^n)]\left(\frac{p}{q}\right)^x (x!)^{-1}\mu'_{(x)}$$

where $\mu'_{(x)}$ is the x-th factorial moment of xy. This gives

$$\frac{P_{x+1}}{P_x} = \frac{p}{(x+1)q}R_{[x]}, \text{ where } R_{[x]} = \frac{\mu'_{(x+1)}}{\mu'_{(x)}}.$$

The mean and the first three central moments of the Poisson binomial distribution are

$$\begin{aligned}
\mu'_1 &= n\lambda p \\
\mu_2 &= n^2p^2\lambda + npq\lambda \\
\mu_3 &= n^3p^3\lambda + 3n^2p^2q\lambda + npq(q-p)\lambda \\
\mu_4 &= 3(n^2p^2\lambda + npq\lambda)^2 + 6n^3p^3q\lambda + n^2p^2q(7-11p)\lambda + npq(1-6pq)\lambda.
\end{aligned}$$

The mixture of binomial distributions also arises from component distributions having common value of n, but different values of p. This type of mixture distribution has probabilities of the form

$$P(X=x) = \binom{n}{x}\sum_{j=1}^{\infty} w_j p_j^x(1-p_j)^{n-x};\ w_j \geq 0, \sum_{j=1}^{\infty} w_j = 1 \qquad (7.1)$$

when p has a discrete distribution, and

$$P(X=x) = \binom{n}{x}\int_0^1 f(p)p^x(1-p)^{n-x}dp \qquad (7.2)$$

with $f(p) \geq 0$,

$$\int_0^1 f(p)dp = 1 \text{ when } p \text{ has a continuous distribution.}$$

If $x_1, x_2, \ldots, x_N$ each have mixture of binomial distribution given by (7.1) with the summation from $j = 1$ to $j = R$, and $n \geq 2R-1$, then the first $2R-1$ reduced factorial moments

$$F_S = \frac{1}{N}\sum_{j=1}^{N} \frac{x_j(x_j-1)\cdots(x_j-S+1)}{n(n-1)\cdots(n-S+1)}, \quad S = 1, 2, \ldots, 2R-1$$

are calculable from $x_1, x_2, \ldots, x_N$, and

$$E[F_S] = \sum_{j=1}^{R} w_j p_j^S = E(X^S)\mu'_S.$$

When $R = 2$, it corresponds to a mixture of two binomials with probability function

$$P(X = x) = \alpha\binom{n_1}{x}p_1^x q_1^{n-x} + (1-\alpha)\binom{n_2}{x}p_2^x q_2^{n-x}$$

with $0 < \alpha < 1$, $0 < q_j = 1 - p_j < 1$ $(j = 1, 2)$, $n_1 = n_2 = n$.

The mixtures of binomial distributions with common n were studied by Lexis (1875) which are called *Lexian distributions.* Lexis obtained the general forms of mean and variances of such mixtures distributions. These are given by

$$\begin{aligned} E(X) &= nE(p) \\ V(X) &= nE(p)[1 - E(p)] + n(n-1)(p). \end{aligned}$$

The formula for variance indicates that it is always greater than the value for a binomial distribution with parameters $n, E(p)$.

Binomial-Binomial Distribution

Let X be a random variable with probability function

$$P(X = x) = \binom{n}{x}p^x q^{n-x}, \quad x = 0, 1, 2, \ldots, n.$$

Let us assume that np is also distributed with probability function

$$P(n) = \binom{n_1}{j}p_1^j q_1^{n_1-j}, \quad j = 0, 1, 2, \ldots, n_1.$$

Then the mixture distribution of these two binomial distributions is called *binomial-binomial distribution* and its probability function is given by

$$P_x = P(X = x) = \sum_{j \geq \frac{x}{n}} \binom{n_1}{j}p_1^j q_1^{n_1-j}\binom{n_j}{x}p^x q^{n_j-x}$$

$$(x = 0, 1, 2, \ldots, nn_1; \; q = 1 - p; \; q_1 = 1 - p_1.$$

The probability generating function of this distribution is given by

$$g(t) = [q_1 + p_1(q + pt)^n]^{n_1}.$$

If $n = 1$, then

$$g(t) = [q_1 + p_1 q + p_1 pt]^{n_1}$$

which is the probability generating function of binomial distribution with parameters n_1, pp_1. This latter probability generating function arises from a two-stage experiment with n_1 repetitions, where p be the probability of success at the first stage and p_1 be the probability of success at the second stage and at both stages the experiment is performed independently. This is a binomial-binomial distribution with varying n.

The compound binomial distribution with varying n also follows Poisson distribution. This can be shown as follows:

Let X be a random variable with

$$P(X = x) = \binom{n}{x} p^x q^{n-x}, \quad x - 0, 1, 2, \ldots, n.$$

Here x is regarded as number of successes in n independent trials with constant probability of success for each trial.

Suppose that the number of trials n, instead of being considered as a fixed constant, is also a random variable with probability function

$$P(n = k) = \frac{e^{-\lambda}\lambda^k}{k!}, \quad k = 0, 1, 2, \ldots.$$

The joint probability function of X and n is given by

$$P(X = x \cap n = k) = P(n = k)P(X = x|n = k) = \frac{e^{-\lambda}\lambda^k}{k!}\binom{k}{x} p^x q^{k-x}$$

since $P(X = x|n = k)$ is the probability of x successes in k trials. Obviously, $x \leq k$. Therefore, the marginal distribution of X is given by

$$\begin{aligned}
P(X = x) &= \sum_{k=x}^{\infty} P(X = x \cap n = k) = e^{-\lambda} p^x \sum_{k=x}^{\infty} \binom{k}{x} \frac{\lambda^k q^{k-x}}{k!} \\
&= \frac{e^{-\lambda}(\lambda p)^x}{x!} \sum_{k=x}^{\infty} \frac{(\lambda q)^{k-x}}{(k-x)!} = \frac{e^{-\lambda}(\lambda p)^x}{x!} \sum_{j=0}^{\infty} \frac{(\lambda q)^j}{j!} \quad (j = k - x) \\
&= \frac{e^{-\lambda}(\lambda p)^x}{x!} e^{\lambda q} = \frac{e^{-\lambda p}(\lambda p)^x}{x!}.
\end{aligned}$$

The probability generating function of this distribution is

$$g(t) = E(t^x) = \sum_{x=0}^{\infty} \frac{e^{-\lambda p}(\lambda p)^x t^x}{x!} = e^{-\lambda p} \sum_{x=0}^{\infty} \frac{(\lambda pt)^x}{x!} = e^{-\lambda p(1-t)}.$$

So far we have discussed the discrete mixtures of binomial distributions with common n as well as varying n. There are more mixtures of binomial distributions where

n or p has discrete distributions. Some of these are binomial-negative binomial distribution, binomial-hypergeometric distribution, etc. For details readers need to go through Johnson and Kotz (1969). Let us now consider the continuous mixtures of binomial distributions with common n and varying p. The first one is the binomial-beta distribution, where p has the density function

$$f(p) = \frac{1}{B(a,b)} p^{a-1} q^{b-1};\ a > 0; b > 0;\ 0 < p < 1$$

with $f(p) \geq 0$ and

$$\int_0^1 f(p)dp = 1.$$

Then the mixture distribution is given by

$$\begin{aligned} P(X = x) &= \binom{n}{x} \int_0^1 f(p) p^x (1-p)^{n-x} dp \\ &= \frac{1}{B(a,b)} \binom{n}{x} \int_0^1 p^{x+a-1} (1-p)^{n-x+b-1} dp \\ &= \frac{\Gamma n+1}{\Gamma x+1\, \Gamma n-x+1} \frac{\Gamma x+a\, \Gamma n-x+b\, \Gamma a+b}{\Gamma n+a+b\, \Gamma a\, \Gamma b} \\ x &= 0, 1, 2, \ldots, n. \end{aligned} \tag{7.3}$$

The distribution given by (7.3) is called *Pólya-Eggenberger distribution.* The mean and variance of this distribution are respectively

$$\begin{aligned} E(x) &= na/(a+b) \\ \text{and } V(x) &= nab(a+b)^{-2} + n(n-1)ab(a+b)^{-2}(a+b+1)^{-1}. \end{aligned}$$

The Pólya-Eggenberger distribution can be used to study the probability of defective items found in routine sampling inspection plan where the number of defective items vary. This distribution can also be used to find the probability of drawing coloured balls from an urn. Suppose that there are a white balls and b black balls in an urn. One ball is drawn at random, and then replaced, together with s balls of same colour. Thus every time, if the experiment is repeated, the probability of drawing a particular coloured ball is changing. Suppose that the drawing is repeated n times, and X represents the total number of times a white ball is drawn. Then

$$P(X = x) = \binom{n}{x} \frac{[a(a+s)\cdots(a+\overline{k-1}\cdot s)\ b(b+s)\cdots(b+\overline{n-x-1}\cdot s)]}{(a+b)(a+b+s)\cdots(a+b+\overline{n-1}\cdot s)}$$

is called the *Pólya-Eggenberger distribution* of X with parameters n, a, b, s.

The recurrence relation among the moments about zero of this distribution is given by

$$\mu'_{r+1} = (a+b+rs)^{-1} \sum_{j=0}^{r} \left[an\binom{r}{j} - (a-sn)\binom{r}{j+1} - s\binom{r}{j+2} \right] \mu'_{r-j}.$$

The r-th factorial moment is

$$\mu_{(r)} = E\left[x^{(r)}\right]$$

$$= n^{(r)} \prod_{j=0}^{r-1} \left[\frac{P+j\alpha}{1+j\alpha}\right] \sum_{x-r=0}^{n-r} \binom{n-r}{x-r} \frac{\prod_{j=0}^{x-r-1} (P_1+j\alpha_1) \prod_{j=0}^{n-x-1} (Q_1+j\alpha_1)}{\prod_{j=0}^{n-x-1} (1+j\alpha_1)}.$$

Here

$$\begin{aligned} P_1 &= (P+r\alpha)(1+r\alpha)^{-1}; \\ Q_1 &= Q(1+r\alpha)^{-1}; \ \alpha_1 = \alpha(1+r\alpha)^{-1}; \\ P &= a/(a+b); \\ Q &= 1-P = b/(a+b) \text{ and } \alpha = s/(a+b). \end{aligned}$$

In particular,

$$\begin{aligned} \mu_1' &= nP \\ \mu_2 &= n(n-1)P(P+\alpha)(1-\alpha)^{-1} + nP - (nP)^2 \\ &= nPQ(1+n\alpha)(1+\alpha)^{-1} \\ \mu_3 &= nPQ(Q-P)(1+n\alpha)(1+2n\alpha)(1+\alpha)^{-1}(1+2\alpha)^{-1} \\ \mu_4 &= \frac{nPQ(1+n\alpha)[(1+2n\alpha)(1+3n\alpha)(1-3PQ) + (n-1)\{\alpha+3PQ(1+n\alpha)\}]}{(1+\alpha)(1+2\alpha)(1+3\alpha)}. \end{aligned}$$

From these moments we have index of skewness

$$\alpha_3 = \sqrt{\beta_1} = \frac{(Q-P)(1+2n\alpha)\sqrt{(1+\alpha)}}{(1+2\alpha)[nPQ(1+n\alpha)]^{1/2}}$$

and index of kurtosis

$$\alpha_4 = \beta_2 = \frac{3(n-1)(1+\alpha)}{n(1+2\alpha)(1+3\alpha)} + \frac{(1+\alpha)}{nPQ(1+2\alpha)(1+3\alpha)(1+n\alpha)} \{(1+2n\alpha)(1+3n\alpha)(1+3PQ)+(n-1)\alpha\}.$$

Binomial-Triangular Distribution

This distribution arises from a binomial variable X with parameters n and p, where p varies from set to set of trials. Hald (1968) has obtained this compound binomial distribution assuming $w(p)$ as the density function of p. Thus

$$\begin{aligned} P(X=x) &= \binom{n}{x} \int_0^1 p^x q^{n-x} w(p)dp \\ &= n^{-1} w(x/n) \left[1 + b_1(x/n)n^{-1} + b_2(x/n)n^{-2} + O\left(n^{-3}\right)\right] \end{aligned}$$

where

$$b_1(h) = [w(h)]^{-1}\left[-w(h)+(1-2h)w'(h)+\frac{1}{2}h(1-h)w^{(2)}(h)\right]$$

$$\text{and } b_2(h) = [w(h)]^{-1}\Big[w(h)-3(1-2h)w'(h)+\{1-6h(1-h)\}w^{(2)}(h)$$
$$+\frac{5}{6}h(1-h)(1-2h)w^{(3)}(h)+\frac{1}{8}h^2(1-h)^2w^{(4)}(h)\Big].$$

7.8 Properties of Binomial Distribution

Mode of the Binomial Distribution

Mode is the value of x for which $p(x)$ is maximum. We have already discussed that $p(x)$ increases to a maximum value at the integer x satisfying

$$(n+1)p-1 < x < (n+1)p.$$

This also can be shown from the relation

$$\begin{aligned}\frac{p(x)}{p(x-1)} &= \binom{n}{x}p^x q^{n-x}\Big/\binom{n}{x-1}p^{x-1}q^{n-x+1}\\ &= \frac{(n-x+1)p}{xq}\\ &= \frac{xq+(n-x+1)p-xq}{xq}\\ &= 1+\frac{(n+1)p-x(p+q)}{xq}\\ &= 1+\frac{(n+1)p-x}{xq}.\end{aligned}$$

Now, this $(n+1)p$ may be an integer or it may be a fraction. Let us consider that $(n+1)p$ is an integer and $(n+1)p = k$. Then

$$\frac{p(x)}{p(x-1)} = 1+\frac{k-x}{xq}.$$

From this relation it is obvious that

$$\frac{p(x)}{p(x-1)}\left.\begin{matrix} >1, & \text{if } x = 1,2,\ldots,k-1\\ =1, & \text{if } x = k\\ <1, & \text{if } x = k+1,k+2,\ldots,n.\end{matrix}\right.$$

From these relationships we have

$$p(0) < p(1) < \cdots < p(k-1) = p(k) > p(k+1) > \cdots > p(n).$$

Thus the binomial distribution is bimodal and modal values are k and $k-1$. Also consider that $(n+1)p$ is not an integer and $(n+1)p = k+f$, where k is an integer and

f is a fractional part such that $0 < f < 1$. Therefore, we have

$$\frac{p(x)}{p(x-1)} = 1 + \frac{(k+f)-x}{xq}$$

$$\text{and} \quad \frac{p(x)}{p(x-1)} \left.\begin{matrix} > 1, & \text{if } x = 0,1,2,\dots,k \\ < 1, & \text{if } x = k+1, k+2, \dots, n. \end{matrix}\right.$$

Thus

$$\frac{p(1)}{p(0)} > 1, \ \frac{p(2)}{p(1)} > 1, \cdots, \frac{p(k)}{p(k-1)} > 1$$

$$\text{and} \quad \frac{p(k+1)}{p(k)} < 1, \ \frac{p(k+2)}{p(k+1)} < 1, \ \cdots, \ \frac{p(n)}{p(n-1)} < 1.$$

$$\therefore \ p(0) < p(1) < \cdots < p(k-1) < p(k) > p(k+1) > \cdots > p(n).$$

This indicates that binomial distribution is unimodal and the modal value is k, the integral part of $(n+1)p$.

Mean Deviation About Mean of Binomial Distribution

The mean deviation γ_1 about the mean np of the binomial distribution is given by

$$\begin{aligned}
\gamma_1 &= \sum_{x=0}^{n} |x - np| \binom{n}{x} p^x q^{n-x} \\
&= \sum_{x=0}^{np} -(x-np) \binom{n}{x} p^x q^{n-x} + \sum_{x=np}^{n} (x-np) \binom{n}{x} p^x q^{n-x} \\
&= 2 \sum_{x>np}^{n} (x-np) \binom{n}{x} p^x q^{n-x} = 2 \sum_{x=\mu}^{n} (x-np) \binom{n}{x} p^x q^{n-x}
\end{aligned}$$

where μ is the greatest integer contained in $(np+1)$

$$\begin{aligned}
&= 2 \sum_{\mu}^{n} \left[\{xq - (n-x)p\} \binom{n}{x} p^x q^{n-x} \right] \\
&= 2 \sum_{\mu}^{n} \left[\frac{n!}{(x-1)!(n-x)!} p^x q^{n-x+1} - \frac{n!}{x!(n-x-1)!} p^{x+1} q^{n-x} \right] \\
&= 2 \sum_{\mu}^{n} [t_{x-1} - t_x], \text{ where } t_x = \frac{n!}{x!(n-x-1)!} p^{x+1} q^{n-x} \\
&= 2 [t_{\mu-1} - t_n] = 2t_{\mu-1} \ [\because t_n = 0] \\
\therefore \ \gamma_1 &= 2t_{\mu-1} = 2 \frac{n!}{(\mu-1)!(n-\mu)!} p^{\mu} q^{n-\mu+1} \\
&= 2npq \binom{n-1}{\mu-1} p^{\mu-1} q^{n-\mu}.
\end{aligned}$$

Using Stirling's approximation for $n!$, where

$$n! \doteq \sqrt{2\pi}\, e^{-n} n^{n+1/2} \text{ when } n \to \infty$$

we have

$$\gamma_1 \doteq \sqrt{\frac{2}{\pi}npq} \text{ as } n \to \infty.$$

Thus the ratio of the mean deviation to the standard deviation approaches the limiting value $\sqrt{2/\pi} \doteq 0.798$ as n tends to infinity.

Additive Property of Binomial Distribution

Let $X_1 \sim b(n_1, p)$ and $X_2 \sim b(n_2, p)$, where $b(n, p)$ indicates binomial variate with parameters n and p. Then $Y = X_1 + X_2$ is also a binomial variate with parameters

$$n = n_1 + n_2 \text{ and } p.$$

We know, the moment generating function of binomial distribution is

$$M_x(t) = \left(q + pe^t\right)^n.$$

Thus

$$M_{x_1}(t) = \left(q + pe^t\right)^{n_1} \text{ and } M_{x_2}(t) = \left(q + pe^t\right)^{n_2}.$$

Also

$$\begin{aligned} M_{x_1+x_2}(t) &= M_{x_1}(t) \cdot M_{x_2}(t) \text{ if } X_1 \text{ and } X_2 \text{ are independent} \\ M_y(t) &= \left(q + pe^t\right)^{n_1+n_2} \end{aligned}$$

which is the m.g.f. of binomial variate with parameters $n = n_1 + n_2$ and p.

In general, if $X_1, X_2, \ldots, X_k$ are k independent binomial variates with parameters $(n_1, p), (n_2, p), \ldots, (n_k, p)$ respectively, then $Y = X_1 + X_2 + \cdots + X_k$ is also a binomial variate with parameters $n = \sum_{i=1}^{k} n_i$ and p. But, if $X_i \sim b(n_i, p_i)$, then $Y = \sum^{k} X_i$ is not a binomial variate, since

$$M_y(t) = \left(q_1 + p_1e^t\right)^{n_1}\left(q_2 + p_2e^t\right)^{n_2} \cdots \left(q_k + p_ke^t\right)^{n_k}$$

cannot be expressed in the form of $(q+pe^t)^n$. Thus, in general, the binomial distribution does not possess the additive or reproductive property. If X_1 and X_2 are independent binomial variates with parameters (n_1, p) and (n_2, p) respectively, then the conditional distribution $X_1/X_1 + X_2 = X$ is given by

$$P(X_1 = k|X) = \frac{\binom{n_1}{k}p^kq^{n_1-k}\binom{n_2}{x-k}p^{x-k}q^{n_2-(x-k)}}{\binom{n_1+n_2}{x}p^xq^{n_1+n_2-x}} = \frac{\binom{n_1}{k}\binom{n_2}{x-k}}{\binom{n_1+n_2}{x}}$$

where $\max(0, x - n_2) \le k \le \min(n_1, x)$. Here x is the value of the random variable X. The above conditional distribution is hypergeometric distribution. The distribution of the difference of $X_1 - X_2$ is given by

$$P(X_1 - X_2 = k) = \sum_{x_1}\binom{n_1}{x_1}\binom{n_2}{x_1 - k}p^{2x_1-k}q^{n_1+n_2-(2x_1-k)}$$

where the summation is between the limits

$$\max(0, k) \leq x_1 \leq \min(n_1, n_2 + k).$$

When $p = q = 1/2$,

$$P(X_1 - X_2 = k) = \binom{n_1 + n_2}{n_2 + k} \Big/ 2^{n_1+n_2}, \quad -n_2 \leq k \leq n_1.$$

Standardized Binomial Variate

Let X be a binomial variate with parameters n and p. Then

$$Z = (x - np)/\sqrt{npq}$$

is called the *standardized binomial variate.* The m.g.f. of Z is given by

$$
\begin{aligned}
M_z(t) &= E\left[e^{tz}\right] = E\left[e^{t(x-\mu)/\sigma}\right], \text{ where } \mu = np \text{ and } \sigma = \sqrt{npq}. \\
\therefore\ M_z(t) &= e^{-\mu t/\sigma} M_x(t/\sigma) = e^{-npt/\sqrt{npq}}\left(q + pe^{t/\sqrt{npq}}\right)^n \\
&= \left[e^{-pt/\sqrt{npq}}\left(q + pe^{t/\sqrt{npq}}\right)\right]^n = \left[qe^{-pt/\sqrt{npq}} + pe^{qt/\sqrt{npq}}\right]^n \\
&= \left[q\left\{1 - \frac{pt}{\sqrt{npq}} + \frac{p^2t^2}{2npq} + O\left(n^{-3/2}\right)\right\}\right. \\
&\qquad \left. +p\left\{1 + \frac{qt}{\sqrt{npq}} + \frac{q^2t^2}{2npq} + O\left(n^{-3/2}\right)\right\}\right]^n \\
\therefore\ M_z(t) &= \left[(q+p) + \frac{t^2pq}{2npq}(p+q) + O\left(n^{-3/2}\right)\right]^n \\
&= \left[1 + \frac{t^2}{2n} + O\left(n^{-3/2}\right)\right]^n \\
K_z(t) &= \log M_z(t) = n \log\left[1 + \frac{t^2}{2n} + O\left(n^{-3/2}\right)\right] \\
&= n\left[\left\{\frac{t^2}{2n} + O\left(n^{-3/2}\right)\right\} - \frac{1}{2}\left\{\frac{t^2}{2n} + O\left(n^{-3/2}\right)\right\}\right] \\
&= \frac{t^2}{2} + O\left(n^{-1/2}\right).
\end{aligned}
$$

Now

$$\underset{\lim n \to \infty}{K_z(t)} = \frac{t^2}{2} \Rightarrow \lim_{n \to \infty} M_z(t) = e^{t^2/2}$$

which is the m.g.f. of standard normal variate. Hence by the uniqueness theorem of moment generating functions, standard binomial variate tends to standard normal variate as n tends to infinity. That is, for any real numbers $\alpha, \beta (\alpha < \beta)$

$$\lim_{n \to \infty} P[\alpha < z < \beta] = \frac{1}{\sqrt{2\pi}} \int_\alpha^\beta e^{-\frac{1}{2}z^2}\, dz.$$

The result is known as *Laplace's theorem.*

Using Gram-Charlier expansion a natural modification of the normal approximation is done. In this modification one term in addition to the leading normal term is taken. The approximation is given by

$$P(X \leq k) \doteq \frac{1}{\sqrt{2\pi}} \int_{-\infty}^{(k+\frac{1}{2}-np)/\sqrt{npq}} e^{-\frac{1}{2}z^2} dz$$
$$-\frac{1}{6}\frac{q-p}{\sqrt{npq}}\left[\frac{\left(k+\frac{1}{2}-np\right)^2}{npq}-1\right] \times \frac{1}{\sqrt{2\pi}} \exp\left[-\frac{\left(k+\frac{1}{2}-np\right)^2}{2npq}\right].$$

This approximation includes continuity correction. For individual binomial probabilities, the normal approximation with continuity correction gives

$$P(X = k) \doteq \frac{1}{\sqrt{2\pi}} \int_{(k-\frac{1}{2}-np)/\sqrt{npq}}^{(k+\frac{1}{2}-np)/\sqrt{npq}} e^{-\frac{1}{2}z^2} dz$$

A nearly equivalent approximation is

$$P(X = k) \doteq \frac{1}{\sqrt{npq}} \frac{1}{\sqrt{2\pi}} \exp\left[-\frac{1}{2}\frac{(k-np)^2}{npq}\right].$$

The approximations of binomial probabilities are applied to the incomplete beta function ratio $I_p(k, n-k+1)$. This approximation is based on

$$\begin{aligned} P(X \geq k) &= \sum_{x=k}^{n} \binom{n}{x} p^x q^{n-x} \\ &= B(k, n-k+1)^{-1} \int_0^p t^{k-1}(1-t)^{n-k} dt \\ &= I_p(k, n-k+1). \end{aligned}$$

The 'arcsine' transformation

$$y = \sin^{-1} \sqrt{\left(x+\frac{3}{8}\right) \Big/ \left(n+\frac{3}{4}\right)}$$

to binomial probabilities produces a random variable y which is approximately normally distributed with

$$E(y) = \sin^{-1}\sqrt{p} \text{ and } V(y) = \frac{1}{4n}.$$

This approximation is given by

$$P(X \leq k) \doteq \frac{1}{\sqrt{2\pi}} \int_{-\infty}^{Y} e^{-\frac{1}{2}z^2} dz,$$

where

$$Y = 2\sqrt{n}\left[\sin^{-1}\sqrt{\left(k+\frac{3}{8}\right) \Big/ \left(n+\frac{3}{4}\right)} - \sin^{-1}\sqrt{p}\right].$$

❑ **THEOREM 7.1** *If $X \sim b(n,p)$, then*

$$\sum_{x=0}^{k} b(n,p) = (n-k)\binom{n}{k}\int_0^q t^{n-k-1}(1-t)^k dt.$$

Proof. We have

$$P(X=x) = \binom{n}{x}p^x q^{n-x},\ x = 0,1,2,\ldots,n$$

$$\begin{aligned}
\sum_{x=0}^{k} b(n,p) &= \sum_{x=0}^{k} \frac{n!}{x!(n-x)!}p^x q^{n-x} \\
&= \sum_{x=0}^{k} \frac{n!}{k!(n-k-1)!}\frac{k!}{x!(k-x)!}\frac{(k-x)!(n-k-1)!}{(n-x)!}p^x \times q^{n-k}q^{k-x} \\
&= \sum_{x=0}^{k} \frac{n!}{k!(n-k-1)!}q^{n-k}\frac{k!}{x!(k-x)!}p^x q^{k-x}\frac{(n-k-1)!(k-x)!}{(n-x)!} \\
&= \frac{n!}{k!(n-k-1)!}q^{n-k}\sum_{x=0}^{k}\binom{k}{x}p^x q^{k-x}\frac{\lfloor n-k \lfloor k-x+1}{\lfloor n-x+1} \\
&= \frac{n!}{k!(n-k-1)!}q^{n-k}\sum_{x=0}^{k}\binom{k}{x}p^x q^{k-x}\beta(n-k,k-x+1) \\
&= \frac{n!}{k!(n-k-1)!}q^{n-k}\sum_{x=0}^{k}\left[\binom{k}{x}p^x q^{k-x}\int_0^1 y^{k-x}(1-y)^{n-k-1}dy\right] \\
&= \frac{n!}{k!(n-k-1)!}q^{n-k}\sum_{x=0}^{k}\binom{k}{x}p^x\int_0^1 (ky)^{k-x}(1-y)^{n-k-1}dy \\
&= \frac{n!}{k!(n-k-1)!}q^{n-k}\int_0^1 (1-y)^{n-k-1}(p+qy)^k dy.
\end{aligned}$$

Let

$$y = 1 - \frac{t}{q} \Rightarrow dy = \frac{dt}{q}.$$

Thus we have

$$\begin{aligned}
\sum_{x=0}^{k} b(n,p) &= \frac{n!}{k!(n-k-1)!}\int_0^q t^{n-k-1}(1-t)^k dt \\
&= \frac{n!}{k!(n-k)!}\frac{(n-k)!}{(n-k-1)!}\int_0^q t^{n-k-1}(1-t)^k dt \\
&= \binom{n}{k}(n-k)\int_0^q t^{n-k-1}(1-t)^k dt.
\end{aligned}$$

This result enables us to represent the cumulative binomial probabilities in terms of Incomplete Beta Functions.

7.9 Two-Dimensional Binomial Distribution

Let the results of a random experiment be of three types when the experiment is repeated n times. Let E_1, E_2 and E_3 be three events based on n experimental results. These events are mutually exclusive. Let us assume that $P(E_1) = p_1$, $P(E_2) = p_2$ and $P(E_3) = p_3$ such that $p_1 + p_2 + p_3 = 1$. Assume that the sample points of the event E_1 are expressed by a variable X_1 and the sample points in favour of X_2 are expressed by a variable X_2 such that $x_1 + x_2 \le n$, where x_1 and x_2 are the particular values of X_1 and X_2, respectively. For example, consider that in a laboratory there are Samsung and Philips computers, and n of the computers are randomly selected. Among the selected computers n_1 computers are Samsung and n_2 computers are Philips such that $n_1 + n_2 = n$. Let X_1 be the random variable indicating number of Samsung computers and X_2 be the event indicating number of Philips computers. If p_1 is the probability of selecting Samsung computers and p_2 is the probability of Philips computers, then

$$P(X_1 = x_1, X_2 = x_2) = \frac{n!}{x_1!x_2!} \prod_{j=1}^{2} p_j^{x_j}, \text{ where } 0 \le x_j \le n,$$

$$\sum_{j=1}^{2} x_j = n, \ \sum_{j=1}^{2} p_j = 1.$$

Here X_1 and X_2 are two-dimensional binomial variables and $P(X_1 = x_1, X_2 = x_2)$ is the joint probability function of two-dimensional binomial variables.

The probability function can also be written as

$$\begin{aligned} P(X_1 = x_1) &= \frac{n!}{x_1!(n-x_1)!} p_1^{x_1}(1-p_1)^{n-x_1}, \ x_1 = 0, 1, 2, \ldots, n_1 \\ &= \frac{n!}{x_1!(n-x_1)!} p_1^{x} p_3^{n-x_1}, \text{ where } p_3 = 1 - p_1. \end{aligned}$$

The probability function can also be written as

$$P(X_1 = x_1, X_2 = x_2) = \frac{n!}{x_1!x_2!(n-x_1-x_2)!} p_1^{x_1} p_2^{x_2} p_3^{n-x_1-x_2}$$

This latter function is also known as trinomial probability function.

7.10 Moment Generating Function of Bivariate Binomial Distribution

The moment generating function is given by

$$\begin{aligned} M_{x_1,x_2}(t_1, t_2) &= E[\exp(X_1 t_1 + X_2 t_2)] \\ &= \sum_{x_1=0}^{n} \sum_{x_2=0}^{n-x_1} \frac{n!}{x_1!x_2!(n-x_1-x_2)!} \left(p_1 e^{t_1}\right)^{x_1} \left(p_2 e^{t_2}\right)^{x_2} p_3^{n-x_1-x_2} \\ &= \sum_{x_1=0}^{n} \sum_{x_2=0}^{n-x_1} \frac{n!}{x_1!(n-x_1)!} \left(p_1 e^{t_1}\right)^{x_1} \times \frac{(n-x_1)!}{x_2!(n-x_1-x_2)!} \left(p_2 e^{t_2}\right)^{x_2} p_3^{n-x_1-x_2} \\ &= \sum_{x_1=0}^{n} \frac{n!}{(n-x_1)!x_1!} \left(p_1 e^{t_1}\right)^{x_1} \left(p_1 + p_2 e^{t_2}\right)^{n-x_1} \\ &= \left(p_3 + p_1 e^{t} + p_2 e^{t_2}\right)^{n}, \text{ where } t_1 \text{ and } t_2 \text{ are real numbers.} \end{aligned}$$

Now

$$E(X_1) = \mu_1' = \left.\frac{dM_{x_1,x_2}(t_1,t_2)}{dt_1}\right|_{t_1=0,t_2=0}$$

$$= \left.\left[np_1\left(p_3 + p_1 e^{t} + p_2 e^{t_2}\right)^{n-1} e^{t_1}\right]\right|_{t_1=0,\ t_2=0}$$

$$= np_1(p_3 + p_1 + p_2) = np_1$$

Similarly,

$$E(X_2) = \left.\frac{dM_{x_1,x_2}(t_1,t_2)}{dt_2}\right|_{t_1=0,t_2=0}$$

$$= np_2\left(p_3 + p_1 e^{t_1} + p_2 e^{t_2}\right)^{n-1} \left. e^{t_2}\right|_{t_1=0,\ t_2=0} = np_2$$

$$E(X_1^2) = \left.\frac{d^2M_{x_1,x_2}(t_1,t_2)}{dt_1^2}\right|_{t_1=0,\ t_2=0}$$

$$= \left.\left[n(n-1)p_1 e^{t_1}\left(p_3 + p_1 e^{t_1} + p_2 e^{t_2}\right)^{n-2} p_1 e^{t_1} + np_1 e^{t_1}\left(p_3 + p_1 e^{t_1} + p_2 e^{t_2}\right)^{n-1}\right]\right|_{t_1=0,\ t_2=0}$$

$$= n(n-1)p_1^2 + np_1$$

$$V(X_1) = E(X_1^2) - [E(X_1)]^2$$

$$= n(n-1)p_1^2 + np_1 - (np_1)^2 = np_1(1-p_1)$$

$$E(X_1X_2) = \left.\frac{d^2M_{x_1,x_2}(t_1,t_2)}{dt_1dt_2}\right|_{t_1=0,\ t_2=0}$$

$$= \left.\left[n(n-1)p_1 e^{t_1}\left(p_3 + p_1 e^{t_1} + t_2 e^{t_2}\right)^{n-2} p_2 e^{t_2}\right]\right|_{t_1=0,\ t_2=0}$$

$$= n(n-1)p_1p_2(p_3 + p_1 + p_2)^{n-2} = n(n-1)p_1p_2.$$

$$\therefore\ \operatorname{cov}(X_1, X_2) = E(X_1X_2) - E(X_1)E(X_2)$$

$$= n(n-1)p_1p_2 - (np_1)(np_2) = -np_1p_2.$$

Again,

$$M_{x_1,x_2}(t_1, 0) = E\left[e^{t_1X_1}\right]$$

$$= \sum_{x_1=0}^{n}\sum_{x_2=0}^{n-x_1} \frac{n!\,(p_1e^{t_1})^{x_1}}{x_1!x_2!(n-x_1-x_2)!} p_2^{x_2} p_3^{n-x_1-x_2}$$

$$= \sum_{x_1=0}^{n}\sum_{x_2=0}^{n-x_1} \frac{n!}{(n-x_1)!x_1!}\left(p_1e^{t_1}\right)^{x_1} \frac{(n-x_1)!}{x_2!(n-x_1-x_2)!} p_2^{x_2} p_3^{n-x_1-x_2}$$

$$= \sum_{x_1=0}^{n} \frac{n!}{(n-x_1)!x_1!}\left(p_1e^{t_1}\right)^{x_1}(p_2+p_3)^{n-x_1}$$

$$= \left(p_2 + p_3 + p_1e^{t_1}\right)^n = \left(1 - p_1 + p_1e^{t_1}\right)^n\ [p_1 + p_2 + p_3 = 1].$$

Similarly, it can be shown that $M_{x_1,x_2}(0,t_2) = (1-p_2+p_2e^{t_2})^n$. It is noted that

$$M_{x_1,x_2}(t_1,t_2) \neq M_{x_1,x_2}(t,0)M_{x_1,x_2}(0,t_2).$$

Thus X_1 and X_2 are not stochastically independent. But

$$X_1 \sim b(n,p_1) \text{ and } X_2 \sim b(n,p_2).$$

7.11 Conditional Distribution in Case of Bivariate Binomial Distribution

It has already been observed that

$$\begin{aligned} P(X_1 = x_1) &= \frac{n!}{x_1!(n-x_1)!}p_1^{x_1}(1-p_1)^{n-x_1} \\ \text{and } P(X_2 = x_2) &= \frac{n!}{x_2!(n-x_2)!}p_2^{x_2}(1-p_2)^{n-x_2}, \text{ where } p_1+p_2+p_3 = 1. \end{aligned}$$

Again,

$$P(X_1 = x_1, X_2 = x_2) = \frac{n!}{x_1!x_2!(n-x_1-x_2)!}p_1^{x_1}p_2^{x_2}p_3^{n-x_1-x_2}.$$

Now, the conditional probability function of X_2 given $X_1 = x_1$ is

$$\begin{aligned} P(X_2/X_1 = x_1) &= \frac{P(X_1 = x_1, X_2 = x_2)}{P(X_1 = x_1)} \\ &= \frac{\frac{n!}{x_1!x_2!(n-x_1-x_2)!}p_1^{x_1}p_2^{x_2}p_3^{n-x_1-x_2}}{\frac{n!}{x_1!(n-x_1)!}p_1^{x_1}(1-p_1)^{n-x_1}} \\ &= \frac{(n-x_1)!}{x_2!(n-x_1-x_2)!}\left(\frac{p_2}{1-p_1}\right)^{x_2}\left(\frac{p_3}{1-p_1}\right)^{n-x_1-x_2} \\ x_2 &= 0,1,2,\ldots,n-x_1. \end{aligned}$$

It is observed that

$$X_2/X_1 = x_1 \sim b(n-x_1, p_2/1-p_1).$$

Hence the mean and variance of $X_2/X_1 = x_1$ are

$$(n-x_1)\frac{p_2}{1-p_1} \text{ and } (n-x_1)\frac{p_2}{1-p_1}\left(1-\frac{p_2}{1-p_1}\right), \text{ respectively.}$$

Similarly, it can be shown that

$$\begin{aligned} X_1/X_2 &= x_2 \sim b(n-x_2, p_1/1-p_2) \\ \text{and } E(X_1/X_2 = x_2) &= (n-x_2)\frac{p_1}{1-p_2} \\ V(X_1/X_2 = x) &= (n-x_2)\frac{p_1}{1-p_2}\left(1-\frac{p_1}{1-p_2}\right). \end{aligned}$$

7.12 More Examples on Binomial Distribution

►**EXAMPLE 7.36** *The plane of a company has* 100 *seats. It is recorded that* 5% *passengers of the plane do not avail the plane in each flight. to avoid the problem of unutilized seats, the company sells* 105 *tickets for each flight. Find the probability that all passengers will be able to board on every flight.*

Solution. Let X be the random variable indicating the number of passengers who fail to board on. Then

$$P(X=x)=\binom{105}{x}(0.05)^x(0.95)^{105-x},\ x=0,1,2,\dots,105.$$

If all passengers can board on, the number of passengers who fail to do so are 5 or more. Thus we need $P(X\geq 5)$.

$$\begin{aligned}P(X\geq 5)&=1-P(X<5)\\&=1-P(X=0)-P(X=1)-P(X=2)\\&\quad -P(X=3)-P(X=4)\\&=1-\binom{105}{0}(0.95)^{105}-\binom{105}{1}(0.05)(0.95)^{104}\\&\quad -\binom{105}{2}(0.05)^2(0.95)^{103}-\binom{105}{3}(0.05)^3(0.95)^{102}\\&\quad -\binom{105}{4}(0.05)^4(0.95)^{101}\\&\approx 0.609714.\end{aligned}$$

►**EXAMPLE 7.37** *Five unbiased dice are independently thrown* 200 *times. The success of the experiment is to observe* 1 *or* 3 *on the upper faces of the dice. The number of dice showing* 1 *or* 3 *on the upper faces along with frequencies are shown below:*

Show 1 *or* 3	0	1	2	3	4	5	*Total*
Frequency, O_i	50	41	38	42	20	9	$200=N$

Fit a suitable distribution for the above data.

Solution. Let X be the random variable showing 1 or 3 on the upper faces of the dice. Since the dice are unbiased, for any die $P(1 \text{ or } 3)=\frac{1}{3}$ and $P(2,4,5,6)=\frac{2}{3}$. Hence

$$P(X=x)=\binom{5}{x}\left(\frac{1}{3}\right)^x\left(\frac{2}{3}\right)^{5-x},\ x=0,1,2,3,4,5.$$

Therefore, a binomial distribution can be fitted for the data. Using the binomial probability law, we have

$$\frac{p(x+1)}{p(x)}=\frac{n-x}{x+1}\frac{p}{q}\ \text{or},\ p(x+1)=\frac{n-x}{n+1}\frac{p}{q}p(x)$$

where $p=1/3$ and $q=2/3$.

We have,

$$P(x=0) = \binom{5}{0}\left(\frac{1}{3}\right)^0\left(\frac{2}{3}\right)^5 = 0.13169.$$

The probabilities for other values of X along with the expected frequency

$$[E_i = Np(x_i)]$$

are shown below:

x	$\frac{n-x}{x+1}$	$\frac{n-x}{x+1}\frac{p}{q}$	$p(x)$	O_i	$E_i = Np(x_i)$
0	5	2.5	0.13169	50	$26.3 \approx 26$
1	2	1	0.13292	41	$65.8 \approx 66$
2	1	0.5	0.3292	38	$65.8 \approx 66$
3	0.5	0.25	0.1646	42	$32.9 \approx 33$
4	0.2	0.10	0.04115	20	$8.2 \approx 8$
5	—	—	0.00416	9	1
Total	—	—	1.00	200	200

Since $E_5 < 5$, it is added to E_4 and hence we have 5 classes.

The goodness of fit test can be judged calculating the chi-square statistic, where

$$\chi^2 = \sum_{i=0}^{4} \frac{O_i^2}{E_i} - N = 290.40 - 200 = 90.40.$$

As $\chi^2 > \chi^2_{0.05,4} = 9.488$, the binomial distribution is not suitable for the given data.

►**EXAMPLE 7.38** *Five percent products of a factory are usually found defective. Ten products of each day throughout the month of January are randomly selected and inspected. The number of defective products of different days are shown below:*

Defective products	0	1	2	3	4	5	6	7	8	9	10	*Total*
No. of days, O_i	13	12	4	1	1	0	0	0	0	0	0	31

Fit a suitable probability distribution for the distribution of defective products.

Solution. Let X be the random variable indicating number of defective products. Since products are either good or defective, the binomial probability law is assumed to be suitable for the distribution of defective products. Thus

$$P(X=x) = \binom{10}{x}(0.05)^x(0.95)^{10-x}, \ x = 0, 1, \ldots, 10.$$

We have $P(x=0) = (0.95)^{10} \approx 0.59874$. The probabilities of other values of x and the expected number of days $[E_i = Np(x_i)]$ of producing different amounts of defective products are shown below:

x	$\frac{n-x}{x+1}$	$\frac{n-x}{x+1}\frac{p}{q}$	$p(x)$	O_i	$E_i = Np(x_i)$
0	10	0.5263	0.59874	13	$18.5 \approx 18$
1	4.5	0.2368	0.31512	12	$9.8 \approx 10$
2	2.67	0.1405	0.07463	4	$2.3 \approx 2$
3	1.75	0.0921	0.01048	1	0.32
4	1.20	0.0631	0.00096	1	0.03
5	0.83	0.0437	0.00006	0	0.0018
6	0.57	0.03	0.000002	0	0.00008
7	0.375	0.0197	0.00000	0	0.00000
8	0.22	0.0116	0.00000	0	0.00000
9	0.10	0.0052	0.00000	0	0.00000
10	—	—	0.000068	0	0.00000
Total			1.00	31	31

Since $n \geq 10, k > 3, \frac{N^2}{k} > 10$ and E_5 to $E_{11} < 0.25$, all are added to E_4, where modified $E_4 = 2$ [Kochler and Larntz (1980)]. Therefore, the chi-square statistic to judge the suitability of the binomial distribution of the given data is

$$\chi^2 = \sum_{i=1}^{4} \frac{O_i^2}{E_i} - N = \frac{13^2}{18} + \frac{12^2}{10} + \frac{4^2}{2} + \frac{2^2}{1} - 31 = 4.79.$$

Here $\chi^2 < \chi^2_{0.05,3} = 7.81$ and hence it may be concluded that binomial distribution is suitable for the given data.

►**EXAMPLE 7.39** *The assumed adoption rate in an area is* 60%. *To verify the fact, it is decided to investigate* 10 *randomly selected couples. The investigation is repeated by* 500 *investigators. The number of adopter couples in each set and the number of investigators are shown below:*

No. of adopter couples	0	1	2	3	4	5	6	7	8	9	10	*Total*
No. of investigators, O_i	2	1	7	25	62	105	110	95	65	25	3	500

Fit a suitable distribution for the above data.

Solution. Since the couples may or may not adopt family planning, the distribution of adopter couples is expected to follow binomial distribution. Let X be the random variable indicating number of adopter couples. Then

$$P(X = x) = \binom{10}{x}(0.6)^x(0.4)^{10-x}, \quad x = 0, 1, 2, \ldots, 10.$$

We have

$$P(x=0)=\binom{10}{0}(0.4)^{10}\approx 0.0001048.$$

The probabilities of other values of x and the expected number of investigators $[E_i = Np(x_i)]$ are shown below:

x	$\frac{n-x}{x+1}$	$\frac{n-x}{x+1}\frac{p}{q}$	$p(x)$	O_i	$E_i = Np(x_i)$
0	10	15	0.0001048	2	$0.052 \approx 0$
1	4.5	6.75	0.001572	1	$0.786 \approx 1$
2	2.67	4.005	0.010611	7	$5.30 \approx 5$
3	1.75	2.625	0.042497	25	$21.25 \approx 21$
4	1.20	1.8	0.111554	62	$55.77 \approx 56$
5	0.83	1.245	0.200797	105	$100.39 \approx 100$
6	0.57	0.855	0.249992	110	$124.99 \approx 125$
7	0.375	0.5625	0.213743	95	$106.87 \approx 107$
8	0.22	0.33	0.12023	65	$60.11 \approx 60$
9	0.10	0.15	0.039676	25	$19.83 \approx 20$
10	—	—	0.0092232	3	5
Total			1.00	500	500

Here $n > 10, k > 3, \frac{N^2}{k} > 10$. But $E_1 = 0$ and hence E_1 is added to E_2. However, adjusted $E_2 = 1(> 0.25)$. Now,

$$\chi^2 = \sum_{i=2}^{11} \frac{O_i^2}{E_i} - N = 504.07 - 500 = 4.07.$$

Here $\chi^2 < \chi^2_{0.05,9} = 16.919$ and hence it may be concluded that binomial distribution is suitable for the given data.

►**EXAMPLE 7.40** *The male and female population in an area are assumed in the ratio* 50 : 50. *To verify the ratio it is decided to select a group of* 8 *persons and count the number of males. The experiment is repeated by* 100 *investigators. The result of different investigators are shown below:*

No. of males	0	1	2	3	4	5	6	7	8	*Total*
No. of investigators, O_i	3	2	7	24	30	20	8	3	3	100

Do you think that the assumption of population composition is correct?

Solution. Let X be the random variable indicating number of males in a sample of 8 people. If the assumption is correct, then

$$P(X=x)=\binom{8}{x}\left(\frac{1}{2}\right)^8, \quad x=0,1,2,\ldots,8.$$

The probability distribution of X gives the probabilities of different values of X and the expected number of investigators $[E_i = Np(x_i)]$ who observed different number of males are shown below:

x	0	1	2	3	4	5	6	7	8
$p(x)$	$(1/2)^8$	$8/2^8$	$28/2^8$	$56/2^8$	$70/2^8$	$56/2^8$	$28/2^8$	$8/2^8$	$1/2^8$
E_i	0.4	3.1	10.9	21.9	27.3	21.9	10.9	3.1	0.4

Since $n < 10$ and E_1, E_2 as well as E_8, E_9 are less than 5, we need to add E_1, E_2 and E_3 and E_7, E_8 and E_9. The other E_i's > 5. The adjusted E_i's and O_i's are

O_i	12	24	30	20	14	Total
E_i	14	22	27	22	15	100

The χ^2-value of the above data is

$$\chi^2 = \sum_{i=1}^{5} \frac{O_i^2}{E_i} - N = 101.05 - 100 = 1.05.$$

Since $\chi_0^2 < \chi_{0.05,4}^2 = 9.488$, it may be concluded that the data fit well to binomial distribution. Hence the assumption of $M : F :: 50 : 50$ is true.

►**EXAMPLE 7.41** *Ten singers joined in a song competition. Each singer is evaluated by a panel of judge and each is ranked on the basis of average judgement. The ranks of different singers are shown below:*

Singer	1	2	3	4	5	6	7	8	9	10
Rank (x)	5	4	3	3	7	10	8	9	9	3

Do you think that the average result of judgement has a median 5*?*

Solution. Let X be the random variable indicating the result of judgement. If the median of the distribution of X is 5, some values of X will be less than 5 and some values will be more than 5. Hence with respect to median the distribution of results of judgement is expected to be binomial. We have

$$x\text{-median} : 0 \quad - \quad - \quad - \quad + \quad + \quad + \quad + \quad + \quad -$$

The distribution of signs should be binomial. Let k be the number of signs which is fewer. Then

$$p(k) = \binom{n}{k} p^k q^{n-k}, \; k = 0, 1, 2, \ldots, n.$$

Here $p = q = 0.5$. The probability that k takes a specific value k_0 or less is given by

$$P(k \le k_0) = \sum_{k=0}^{k_0} P(k) = \sum_{k=0}^{k_0} \binom{n}{k} (0.5)^k.$$

If this probability is less than or equal to $\alpha/2$, then it is to be decided with $100\alpha\%$ confidence that the median of the distribution is not the given one or the data do not follow binomial distribution as assumed above. We have $k_0 = 4$ and $n_1 = n - 1 = 9$ ($\because$ one value of x-median $= 0$). Then

$$P(k \leq 4) = (0.5)^9 \sum_{k=0}^{5} \binom{n}{k}$$

$$= (0.5)^9 \left[\binom{9}{0} + \binom{9}{1} + \binom{9}{2} + \binom{9}{3} + \binom{9}{4}\right] = 0.5.$$

Here $P(k \leq 4) = 0.5 > 0.025$. Therefore, the distribution of ranks has median 5. the conclusion is made with 5% level of significance.

►**EXAMPLE 7.42** *Ten labourers of a factory are selected at random and their monthly salaries are recorded. The data are given below:*

Monthly salary (x in Rs.) : 2200.00, 2300.00, 1800.00, 1800.00,
2100.00, 2300.00, 1700.00, 1900.00,
2500.00, 2400.00.

Do you think that the median of the distribution of monthly salaries is Rs. 2000.00*?*

Solution. Let X be the random variable indicating monthly salary of the labourers. If median of the distribution of X is 2000.00, half of the values of X are expected to be less than 2000.00 and another half are expected to be more than 2000.00. We have

$$x\text{-median}: \; + \; + \; - \; - \; + \; + \; - \; - \; + \; +$$

Therefore, the distribution of signs of (x-median) should follow binomial distribution.
Let k be the number of fewer signs. Then

$$P(k \leq k_0) = \sum_{k=0}^{k_0} P(k) = \sum_{k=0}^{k_0} \binom{10}{k} (0.5)^{10}.$$

Here $k_0 = 4$ (no. of negative signs).

$$\therefore \; P(k \leq 4) = (0.5)^{10} \left[\binom{10}{0} + \binom{10}{1} + \binom{10}{2} + \binom{10}{3} + \binom{10}{4}\right]$$

$$\approx 0.3769.$$

It is observed that $P(k \leq 4) = 0.3769 > 0.025$. Therefore, at 5% level of significance it may be concluded that the median of the distribution of salary is Rs. 2000.00.

EXERCISE

1. (a) Distinguish between Bernoulli variable and binomial variable. Write down some examples of binomial variable. Deduce the binomial probability function of x successes in a random experiment of n trials.

   (b) Show that the mean of binomial distribution is greater than its variance.

(c) For a discrete random variable X the mean and variance are 6 and 4, respectively. If 10 observations of X are randomly selected, what is the probability that the value of X is at least 2?

2. (a) Write down some examples of binomial variable. Eighty per cent computers of a laboratory run well every time. What is the shape characteristics of the distribution of well-run computers of the laboratory?

(b) If $X_1 \sim b(n_1, p)$ and $X_2 \sim b(n_2, p)$, then show that the distribution of $X_1 + X_2$ is also binomial.

(c) In an experiment of throw of biased die 10 times the expectation of observing even number 5 times is double of the expectation of observing even number 4 times. Find the number of times in which there is no even number if the experiment of 10 throws of a die is repeated 10000 times.

3. (a) Under what condition a Bernoulli variable transforms to binomial variable? Write down some examples of binomial variable.

(b) Find moment generating function of binomial distribution. Hence or otherwise find first four central moments of binomial distribution.

(c) Eight unbiased coins are tossed once. Find the probability of obtaining (i) 4 heads, (ii) at least 2 heads, (iii) at best 3 heads.

4. (a) Mention some experiments, where experimental results can be expressed by binomial probability law. Deduce the binomial probability function of a random variable X.

(b) For binomial distribution, prove that

$$k_{r+1} = pq\frac{dk_r}{dp},$$

where k_r = r-th cumulant of binomial distribution.

(c) A company produces electric bulbs. Ten percent of the produced bulbs are usually found defective. One day 10 bulbs are selected at random. Find the probability that out of 10 bulbs (i) at least 2 are defectives, (ii) 3 are defectives, (iii) at best 2 are defectives.

What are the means and variance of non-defective bulbs?

5. (a) What is the difference between Bernoulli distribution and binomial distribution? Write down some examples of binomial variable.

Find the characteristic function of binomial distribution.

(b) One ship out of 10 is sunk during war in making a voyage. Find the probability that out of 8 ships (i) a convey of 4 ships will arrive safely, (ii) at least 4 ships will arrive safely.

Find mean and variance of number of ships which arrive safely.

6. (a) What is binomial variable? Derive binomial distribution of x successes in n trials. Find mode of binomial distribution.

(b) The probability of hitting a target by a person is 0.3. If the person fires 7 times, what is the probability of hitting the target (i) twice, (ii) at least twice?

(c) Derive the recursion formula for the central moments of binomial distribution.

7. (a) What do you mean by truncated binomial distribution? Find out the probability function of doubly truncated binomial distribution.

(b) Let

$$x \sim b(n,p) \text{ and } p(n) = \binom{n_1}{j} p_1^j q^{n_1-j} q_1, j = 0, 1, 2, \ldots, n_1.$$

Write down the probability function of this mixture of binomial distribution. If p is also a random variable with p.d.f.

$$f(p) = \frac{1}{B(a,b)} p^{a-1} q^{b-1}; \; a > 0, b > 0; \; 0 < p < 1$$

then find the mixture of binomial-beta distribution.

8. (a) **Write down some examples of binomial variable. Derive the probability function of binomial variable.**

(b) **Find mode of the binomial distribution.**

(c) **Under what condition a binomial distribution becomes negatively skewed distribution?**

9. (a) **Write down the difference between Bernoulli variable and binomial variable. If $X \sim b(n,p)$, then show that**

$$\sum_{x=0}^{k} b(n,p) = (n-k)\binom{n}{k} \int_0^q t^{n-k-1}(1-t)^k dt$$

(b) **If $X \sim b(n,p)$, then show that**

$$z = \frac{X - np}{\sqrt{npq}} \sim N(0,1).$$

(c) **Twenty per cent workers of an industry are injured during work. One day 10 workers are randomly selected. Find the probability that, out of 10 workers (i) 3 are injured (ii) at best one is injured, (iii) at least 2 are injured.**
Find mean and variance of the number of workers who are not injured.

CHAPTER 8

Poisson Distribution

8.1 Definition and Applications

The series

$$1+\theta+\frac{\theta^2}{2!}+\frac{\theta^3}{3!}+\cdots=\sum_{x=0}^{\infty}\frac{\theta^x}{x!}$$

converges to e^{θ} for all values of θ. Now, we can define a function

$$P(X=x)=\begin{cases}\dfrac{\theta^x e^{-\theta}}{x!}, & x=0,1,2,\ldots\\ 0, & \text{elsewhere}\end{cases}$$

where $\theta > 0$. As $\theta > 0$, the function $P(X=x) \geq 0$ and

$$\sum_{x=0}^{\infty}P(x)=\sum_{x=0}^{\infty}\frac{e^{-\theta}\theta^x}{x!}=e^{-\theta}\sum_{x=0}^{\infty}\frac{\theta^x}{x!}=1.$$

Thus $P(x)$ satisfies the condition of being a p.d.f. of a random variable X. Any random variable X having the p.d.f. of the form of $P(x)$ is said to have a Poisson distribution, and $P(x)$ is called the *Poisson p.d.f.*.

This distribution was discovered by the French mathematician and physicist Simeon Denis Poisson (1781-1840) who published it in 1837. He approached the distribution by considering limiting forms of the binomial distribution.

Let,

(i) for binomial distribution, the number of parameter (trial) n is indefinitely large,

$$\text{i.e., } n \to \infty.$$

(ii) p, the probability of success for each trial is constant and indefinitely small,

$$\text{i.e., } p \to 0.$$

(iii) $np = \theta$ (say) is finite. Thus $p=\theta/n$, $q=1-\theta/n$ where θ is a positive real number.

We know that the binomial probability is given by

$$b(x;n,p) = \binom{n}{x} p^x q^{n-x} = \binom{n}{x}\left(\frac{p}{1-p}\right)^x (1-p)^n$$

$$= \frac{n(n-1)(n-2)\cdots(n-x+1)}{x!} \frac{\left(\frac{\theta}{n}\right)^x}{\left(1-\frac{\theta}{n}\right)^x}\left(1-\frac{\theta}{n}\right)^n$$

$$= \frac{\left(1-\frac{1}{n}\right)\left(1-\frac{2}{n}\right)\cdots\left(1-\frac{x-1}{n}\right)\theta^x}{x!\left(1-\frac{\theta}{n}\right)^x}\left(1-\frac{\theta}{n}\right)^n.$$

Thus

$$\underset{n\to\infty}{b(x;n,p)} = \frac{e^{-\theta}\theta^x}{x!}, x = 0, 1, 2, \ldots$$

since

$$\lim_{n\to\infty}\left(1-\frac{\theta}{n}\right)^n = e^{-\theta}, \text{ and } \lim_{n\to\infty}\left(1-\frac{\theta}{n}\right)^a = 1,$$

where a is not a function of n. Thus Poisson distribution is a limiting case of binomial distribution. The Poisson distribution may also be obtained independently without considering it as a limiting form of the binomial distribution.

Let us consider that X_t be the number of telephone calls received in time interval 't' on a telephone switch-board. Now, consider that

(i) the probability of getting a call in small time interval is θdt, where θ is a positive constant and dt denotes a small increment in time 't';

(ii) the probability of getting more than one call in small time interval is very small, i.e., is of the order $(dt)^2$ such that

$$\lim_{dt\to 0}\frac{O(dt)^2}{dt} = 0;$$

(iii) the probability of any particular call in the time interval $(t, t+dt)$ is independent of the actual time t and also of all previous calls.

Under the above conditions it can be shown that the probability of getting x calls in time 't' say $P_x(t)$ is given by

$$P_x(t) = \frac{e^{-\theta t}(\theta t)^x}{x!}, \quad x = 0, 1, 2, \ldots.$$

Let P [of at least one call during $(t, t+dt)$] $= \theta dt + O(dt)^2$ and P [of more than one call during $(t, t+dt)$] $= O(dt)^2$. The event of getting exactly x calls in time $t+dt$ can occur in the following two mutually exclusive ways:

(i) x calls in $(0, t)$ and none during $(t, t+dt)$ and the probability of this event is $P_x(t)\{1-(\theta dt + O(dt)^2\}$.

(ii) Exactly $(x-1)$ calls during $(0, t)$ and one call in $(t, t+dt)$ and the probability of this event is $P_{x-1}(t)(\theta dt)$.

Now, from the addition law of probability, we get

$$\begin{aligned} P_x(t+dt) &= P_x(t)\{1-\theta dt - O(dt)^2 + P_{x-1}(t)\theta dt \\ &= P_x(t)(1-\theta dt) + P_{x-1}(t)\theta dt + O(dt)^2 \end{aligned} \tag{8.1}$$

From (8.1) we can write

$$\frac{P_x(t+dt)-P_x(t)}{dt} = -\theta P_x(t) + \theta P_{x-1}(t) + \frac{O(dt)^2}{dt}$$

$$\text{and } \lim_{dt\to 0} \frac{P_x(t+dt)-P_x(t)}{dt} = -\theta P_x(t) + \theta P_{x-1}(t).$$

$$\therefore\ P'_x(t) = -\theta P_x(t) + \theta P_{x-1}(t) \tag{8.2}$$

where $P'_x(t)$ is the differentiation $P_x(t)$ with respect to 't'. But when $x = 0$,

$$P_{x-1}(t) = P_{-1}(t) = P\{(-1) \text{ calls in time '}t\text{'}\} \equiv 0.$$

Thus from (8.1), we get

$$P_0(t+dt) = P_0(t)\{1-\theta dt\} + O(dt)^2$$

which on taking the limit $dt \to 0$, gives

$$P'_0(t) = -\theta P_0(t) \text{ or, } \frac{P'_0(t)}{P_0(t)} = -\theta. \tag{8.3}$$

Integrating (8.3) with respect to 't', we get

$$\log P_0(t) = -\theta t + C$$

where C is an arbitrary constant to be determined from the condition

$$P_0(0) = 1.$$

Hence, $\log 1 = C$, so that $C = 0$.

$$\therefore\ \log P_0(t) = -\theta t \text{ or, } P_0(t) = e^{-\theta t}.$$

Putting this value of $P_0(t)$ in (8.2), we get, with $x = 1$

$$P'_1(t) = -\theta P_1(t) + \theta e^{-\theta t} \text{ or, } P'_1(t) + \theta P_1(t) = \theta e^{-\theta t}.$$

From this differential equation with integrating factor $e^{\theta t}$, we get the solution

$$e^{\theta t} P_1(t) = \theta \int e^{\theta t} e^{-\theta t} dt + C_1 = \theta t + C_1$$

where C_1 is an arbitrary constant to be determined from $P_1(0) = 0$, which gives

$$C_1 = 0.$$

$$\therefore\ P_1(t) = e^{-\theta t}\theta t.$$

Substituting this value of $P_1(t)$ in (8.2) with $x = 2$, we get

$$P'_2(t) + \theta P_2(t) = \theta e^{-\theta t}\theta t.$$

The solution of this differential equation is given by

$$P_2(t)e^{\theta t} = \theta^2 \int te^{-\theta t}e^{\theta t}dt + C_2 = \frac{\theta^2 t^2}{2} + C_2$$

where C_2 is an arbitrary constant to be determined from $P_2(0) = 0$ which gives $C_2 = 0$. Hence

$$P_2(t) = e^{-\theta t}\frac{(\theta t)^2}{2}.$$

Proceeding step by step in similar way, we get

$$P_x(t) = \frac{e^{-\theta t}(\theta t)^x}{x!}, x = 0, 1, 2, \ldots.$$

Here telephone calls are coming at random points of time and space. There are many other instances which can be thought of as a process that generates a number of changes in a fixed interval of time or space. If such a process leads to a Poisson distribution, that process is called *Poisson process.*

W.S. Gosset, in 1907, used Poisson distribution to represent the number of particles falling in small area when a large number of such areas are spread at random over a surface. The Poisson distribution has been used in many other cases. Following are some instances where it can be applied successfully:

(i) The number of defective items in a packet manufactured by a company.

(ii) Number of air accidents in some unit of time.

(iii) Number of printing mistakes at each page of a book.

(iv) Number of eggs transform to birds/fishes in poultry or hatchery.

(v) Number of cars passing a crossing per minute during the busy hours of a day.

(vi) Number of suicides reported in a city.

(vii) Number of deaths from a disease such as heart attack or cancer.

(viii) The emission of radioactive (alpha) articles.

(ix) Number of telephone calls received at a particular time in a telephone exchange.

8.2 First Four Moments of the Poisson Distribution

The r-th moment about origin is given by

$$\mu'_r = E(x^r) = \sum_{x=0}^{\infty} x^r \frac{e^{-\theta}\theta^x}{x!}.$$

When $r = 1$, we get

$$\mu'_1 = \sum_{x=0}^{\infty} x\frac{e^{-\theta}\theta^x}{x!} = e^{-\theta}\theta\sum_{x=1}^{\infty}\frac{\theta^{x-1}}{(x-1)!}$$

$$= \theta e^{-\theta}\left[1 + \theta + \frac{\theta^2}{2!} + \frac{\theta^3}{3!} + \cdots\right] = \theta e^{-\theta}e^{\theta} = \theta.$$

Hence mean of the Poisson distribution is θ.

When $r = 2$, we get

$$\begin{aligned}\mu_2' &= \sum_{x=0}^{\infty} x^2 \frac{e^{-\theta}\theta^x}{x!} = \sum_{x=0}^{\infty} \{x(x-1)+x\} \frac{e^{-\theta}\theta^x}{x!} \\ &= \sum_{x=0}^{\infty} x \frac{e^{-\theta}\theta^x}{x!} + \sum_{x=0}^{\infty} x(x-1) \frac{e^{-\theta}\theta^x}{x!} \\ &= \theta + e^{-\theta}\theta^2 \sum_{x=0}^{\infty} \frac{\theta^{x-2}}{(x-2)!} = \theta + \theta^2 e^{-\theta} e^{\theta} = \theta^2 + \theta.\end{aligned}$$

Hence the variance of Poisson distribution is

$$\mu_2 = \mu_2' - \mu_1'^2 = \theta^2 + \theta - \theta^2 = \theta.$$

Thus the mean and the variance of Poisson distribution are same.

When $r = 3$, we get

$$\begin{aligned}\mu_3' &= \sum_{x=0}^{\infty} x^3 \frac{e^{-\theta}\theta^x}{x!} = \sum_{x=0}^{\infty} \{x(x-1)(x-2) + 3x(x-1) + x\} \frac{e^{-\theta}\theta^x}{x!} \\ &= \sum_{x=0}^{\infty} x \frac{e^{-\theta}\theta^x}{x!} + 3\sum_{x=0}^{\infty} x(x-1) \frac{e^{-\theta}\theta^x}{x!} + \sum_{x=0}^{\infty} x(x-1)(x-2) \frac{e^{-\theta}\theta^x}{x!} \\ &= \theta + 3\theta^2 + \theta^3 \sum_{x=3}^{\infty} \frac{e^{-\theta}\theta^{x-3}}{(x-3)!} \\ &= \theta + 3\theta^2 + \theta^3 e^{-\theta} e^{\theta} = \theta^3 + 3\theta^2 + \theta.\end{aligned}$$

Hence the third central moment of Poisson distribution is

$$\mu_3 = \mu_3' - 3\mu_2'\mu_1' + 2\mu_1'^3 = \theta^3 + 3\theta^2 + \theta - 3(\theta^2 + \theta)\theta + 2\theta^3 = \theta.$$

When $r = 4$, we get

$$\begin{aligned}\mu_4' &= \sum_{x=0}^{\infty} x^4 \frac{e^{-\theta}\theta^x}{x!} \\ &= \sum_{x=0}^{\infty} \{x(x-1)(x-2)(x-3) + 6x(x-1)(x-2) + 7x(x-1) + x\} \frac{e^{-\theta}\theta^x}{x!} \\ &= \sum_{x=0}^{\infty} x \frac{e^{-\theta}\theta^x}{x!} + 7\sum_{x=0}^{\infty} x(x-1) \frac{e^{-\theta}\theta^x}{x!} \\ &\quad + 6\sum_{x=0}^{\infty} x(x-1)(x-2) \frac{e^{-\theta}\theta^x}{x!} + \sum_{x=0}^{\infty} x(x-1)(x-2)(x-3) \frac{e^{-\theta}\theta^x}{x!} \\ &= \theta + 7\theta^2 + 6\theta^3 + e^{-\theta}\theta^4 \sum_{x=4}^{\infty} \frac{\theta^{x-4}}{(x-4)!} \\ &= \theta + 7\theta^2 + 6\theta^3 + e^{-\theta}\theta^4 e^{\theta} = \theta^4 + 6\theta^3 + 7\theta^2 + \theta.\end{aligned}$$

Hence the fourth central moment is given by

$$\begin{aligned}\mu_4 &= \mu_4' - 4\mu_3'\mu_1' + 6\mu_2'\mu_1'^2 - 3\mu_1'^4 \\ &= \theta^4 + 6\theta^3 + 7\theta^2 + \theta - 4(\theta^3 + 3\theta^2 + \theta)\theta + 6(\theta^2 + \theta)\theta^2 - 3\theta^4 = 3\theta^2 + \theta.\end{aligned}$$

The indices of skewness and kurtosis of Poisson distribution are given, respectively, by

$$\begin{aligned}\alpha_3 &= \sqrt{\beta_1} = \left(\frac{\mu_3^2}{\mu_2^3}\right)^{1/2} = \left(\frac{\theta^2}{\theta^3}\right)^{1/2} = \frac{1}{\sqrt{\theta}} \\ \text{and } \alpha_4 &= \beta_4 = \frac{\mu_4}{\mu_2^2} = \frac{3\theta^2 + \theta}{\theta^2} = 3 + \frac{1}{\theta}.\end{aligned}$$

From α_3 it can be concluded that the Poisson distribution is always a positively skewed distribution since $\theta > 0$. The distribution tends to symmetric one if $\theta \to \infty$. The index of kurtosis, α_4, indicates that $\beta_2 > 3$ and the distribution is leptokurtic. However, $\beta_2 \to 3$ if $\theta \to \infty$.

8.3 Generating Functions of the Poisson Distribution

The moment generating function of the distribution is given by

$$\begin{aligned}M_x(t) = E\left(e^{tx}\right) &= \sum_{x=0}^{\infty} e^{tx} e^{-\theta} \frac{\theta^x}{x!} = e^{-\theta} \sum_{x=0}^{\infty} \frac{(e^t\theta)^x}{x!} \\ &= e^{-\theta}\left[1 + \theta e^t + \frac{\theta^2 e^{2t}}{2!} + \frac{\theta^3 e^{3t}}{3!} + \cdots\right] \\ &= e^{-\theta} e^{\theta e^t} = e^{\theta(e^t - 1)} = e^{\theta\left[1 + t + \frac{t^2}{2!} + \frac{t^3}{3!} + \frac{t^4}{4!} + \cdots - 1\right]} \\ &= \theta\left\{t + \frac{t^2}{2!} + \frac{t^3}{3!} + \frac{t^4}{4!} + \cdots\right\} + \frac{\theta^2}{2!}\left\{t + \frac{t^2}{2!} + \cdots\right\}^2 \\ &\quad + \frac{\theta^3}{3!}\left\{t + \frac{t^2}{2!} + \cdots\right\}^3 + \frac{\theta^4}{4!}\left\{t + \frac{t^2}{2!} + \cdots\right\}^4 + \cdots\end{aligned}$$

Here μ_r' is the coefficient of $\frac{t^r}{r!}$ in the expansion of $M_x(t)$. Thus

$$\begin{aligned}\mu_1' &= \theta \\ \mu_2' &= \theta^2 + \theta \\ \mu_3' &= \theta^3 + 3\theta^2 + \theta \\ \mu_4' &= \theta^4 + 6\theta^3 + 7\theta^2 + \theta.\end{aligned}$$

The cumulant generating function of the Poisson distribution is given by

$$\begin{aligned}K_x(t) &= \log M_x(t) = \log e^{\theta(e^t - 1)} = \theta(e^t - 1) \\ &= \theta\left[1 + t + \frac{t^2}{2!} + \frac{t^3}{3!} + \frac{t^4}{4!} + \cdots - 1\right].\end{aligned}$$

The r-th cumulant of the distribution is the coefficient of $\frac{t^r}{r!}$ in the expansion of $K_x(t)$. Thus

$$K_1 = \theta, K_2 = \theta, K_3 = \theta, K_4 = \theta.$$

But

$$\begin{aligned} K_1 &= \mu_1' = \theta,\ K_2 = \mu_2 = \theta, \\ K_3 &= \mu_3 = \theta,\ K_4 = \mu_4 - 3\mu_2^2. \\ \therefore\ \mu_4 &= \theta + 3\theta^2. \end{aligned}$$

The characteristic function of the Poisson distribution is given by

$$\begin{aligned} \psi_x(t) = E\left[e^{itx}\right] &= \sum_{x=0}^{\infty} e^{itx}\frac{e^{-\theta}\theta^x}{x!} = e^{-\theta}\sum_{x=0}^{\infty}\frac{(\theta e^{it})^x}{x!} \\ &= e^{-\theta}e^{\theta e^{it}} = e^{\theta(e^{it}-1)}. \end{aligned}$$

The factorial moment generating function of the Poisson distribution is given by

$$\begin{aligned} M_{(x)}(t) &= E(1+t)^x = \sum_{x=0}^{\infty}(1+t)^x\frac{e^{-\theta}\theta^x}{x!} \\ &= e^{-\theta}\sum_{x=0}^{\infty}\frac{\{\theta(1+t)\}^x}{x!} = e^{-\theta}e^{\theta(1+t)} = e^{\theta t} \\ &= 1 + \theta t + \frac{\theta^2t^2}{2!} + \frac{\theta^3t^3}{3!} + \frac{\theta^4t^4}{4!} + \cdots. \end{aligned}$$

The r-th factorial moment is the coefficient of $t^r/r!$ in the expansion of $M_{(x)}(t)$. Thus

$$\mu_{(r)} = \theta^r.$$

In particular

$$\mu_{(1)} = \theta = \mu_1',\ \mu_{(2)} = \theta^2,\ \mu_{(3)} = \theta^3,\ \mu_{(4)} = \theta^4.$$

From these factorial moments, moments about origin can be found out, where

$$\begin{aligned} \mu_1' &= \mu_{(1)} = \theta \\ \mu_2' &= \mu_{(2)} + \mu_{(1)} = \theta^2 + \theta \\ \mu_3' &= \mu_{(3)} + 3\mu_{(2)} + \mu_{(1)} = \theta^3 + 3\theta^2 + \theta \\ \mu_4' &= \mu_{(4)} + 6\mu_{(3)} + 7\mu_{(2)} + \mu_{(1)} = \theta^4 + 6\theta^3 + 7\theta^2 + \theta. \end{aligned}$$

The recurrence relation for the moments of the Poisson distribution is given by as follows:

The r-th central moment is defined by

$$\mu_4 = E[x - E(x)]^r = \sum_{x=0}^{\infty}(x-\theta)^r\frac{e^{-\theta}\theta^x}{x!}.$$

Differentiating with respect to θ, we get

$$\begin{aligned}
\frac{d\mu_r}{d\theta} &= \sum_{x=0}^{\infty} \frac{e^{-\theta}\theta^x}{x!}(-1)r(x-\theta)^{r-1} + \sum_{x=0}^{\infty} \frac{(x-\theta)^r}{x!}\{xe^{-\theta}\theta^{x-1} - \theta^x e^{-\theta}\} \\
&= -r\sum_{x=0}^{\infty}(x-\theta)^{r-1}\frac{e^{-\theta}\theta^x}{x!} + \sum_{x=0}^{\infty}\frac{(x-\theta)^r}{x!}\theta^{x-1}e^{-\theta}(x-\theta) \\
&= -r\sum_{x=0}^{\infty}(x-\theta)^{r-1}\frac{e^{-\theta}\theta^x}{x!} + \frac{1}{\theta}\sum_{x=0}^{\infty} e^{-\theta}(x-\theta)^{r+1}\frac{\theta^x}{x!} \\
&= -r\mu_{r-1} + \frac{1}{\theta}\mu_{r+1}.
\end{aligned}$$

$$\therefore\ \mu_{r+1} = r\theta\mu_{r-1} + \theta\frac{d\mu_r}{d\theta}.$$

Putting $r = 1, 2$ and 3 successively, we get

$$\begin{aligned}
\mu_2 &= \theta\mu_0 + \theta\frac{d\mu_1}{d\theta} = \theta\ (\because\ \mu_0 = 1 \text{ and } \mu_1 = 0) \\
\mu_3 &= 2\theta\mu_1 + \theta\frac{d\mu_2}{d\theta} = \theta \\
\mu_4 &= 3\theta\mu_2 + \theta\frac{d\mu_3}{d\theta} = 3\theta^2 + \theta.
\end{aligned}$$

The probability generating function of the Poisson distribution is defined by

$$\begin{aligned}
g(t) = E(t^x) &= \sum_{x=0}^{\infty} t^x e^{-\theta}\frac{\theta^x}{x!} = \sum_{x=0}^{\infty} e^{-\theta}\frac{(t\theta)^x}{x!} \\
&= e^{-\theta}e^{t\theta} = e^{-\theta(1-t)}.
\end{aligned}$$

The recursion formula for cumulants of Poisson distribution is given by

$$K_{r+1} = \frac{\theta dK_r}{d\theta}$$

We know, $M_x(t) = e^{\theta(e^t-1)}$.

The cumulant generating function is $K_x(t) = \theta(e^t - 1)$.

Hence

$$\begin{aligned}
K_r &= -\frac{d^r}{dt^r}\theta(e^t-1)\Big|_{t=0} \\
\therefore\ \frac{dK_r}{d\theta} &= \frac{d^r}{dt^r}(e^t-1)\Big|_{t=0} \\
\frac{\theta dK_r}{d\theta} &= \frac{d^r}{dt^r}\theta(e^t-1)\Big|_{t=0}
\end{aligned}$$

Again

$$K_{r+1} = \frac{d^{r+1}}{dt^{r+1}}\theta(e^t-1)\Big|_{t=0} = \frac{d^r}{dt^r}\theta e^t\Big|_{t=0}.$$

Now

$$K_{r+1} - \frac{\theta dK_r}{d\theta} = \frac{d^r}{dt^r}\left[\theta e^t - \theta e^t + \theta\right]_{t=0} = \frac{d^r}{dt^r}\theta = 0.$$

$$\therefore\ K_{r+1} = \frac{\theta dK_r}{d\theta}.$$

8.4 Fitting of Poisson Distribution

We know

$$P(x) = \frac{e^{-\theta}\theta^x}{x!},\ x = 0, 1, 2, \ldots$$

$$\text{and } P(x+1) = \frac{e^{-\theta}\theta^{x+1}}{(x+1)!},\ x = 0, 1, 2, \ldots$$

$$\frac{P(x+1)}{P(x)} = \frac{\theta}{x+1} \Rightarrow P(x+1) = \frac{\theta}{x+1}P(x).$$

This is the recurrence relation for Poisson probability for different values of x. Here θ is the mean of the distribution which can be replaced by the sample mean of a frequency distribution for which Poisson distributed is to be fitted. Let

$$\theta = \frac{1}{N}\Sigma fx = \bar{x},\ \text{then } P(0) = e^{-\theta}$$

and the successive values of probability will be calculated by the recurrence relation. The expected frequency is given by $NP(x)$, $N = \Sigma f$.

►**EXAMPLE 8.1** *Four hundred samples of each of* 1 *c.c. of a liquid were investigated to study the number of bacteria present in each sample. The maximum number of bacteria was found* 8 *in the samples. Below is given the number of bacteria* (x) *in different samples* (f)*:*

Number of bacteria, x	0	1	2	3	4	5	6	7	8	*Total*
Number of samples, f	107	142	92	45	8	4	2	0	0	400

Fit a Poisson distribution using the above data.

Solution. The mean of this distribution is

$$\bar{X} = \frac{\Sigma fx}{N} = \frac{525}{400} = 1.31.$$

The variance is

$$\sigma^2 = \frac{1}{N}\left[\Sigma fx^2 - \frac{(\Sigma fx)^2}{N}\right] = \frac{1}{400}\left[1215 - \frac{(525)^2}{400}\right] = 1.31.$$

As mean and variance of the distribution is same, we can fit Poisson distribution to the given data. Here

$$P(0) = e^{-\theta} = e^{-1.31} = 0.26982.$$

The probability function to be fitted is

$$P(x) = \frac{e^{-1.31}(1.31)^x}{x!}, \; x = 0, 1, 2, \ldots, 8.$$

Calculation for finding probabilities and expected frequencies

x	$\frac{\theta}{x+1}$	$P(x)$ $\left[P(x+1) = \frac{\theta}{x+1}P(x)\right]$	Expected frequency $N \times P(x)$
0	1.31	0.26982	108
1	0.655	0.35346	141
2	0.437	0.23152	93
3	0.327	0.10117	40
4	0.262	0.03308	13
5	0.218	0.00867	3
6	0.187	0.00189	1
7	0.164	0.00035	1
8	0.145	0.00004	0

From the fitted distribution, we have

$$\chi^2 = \sum \frac{O_i^2}{E_i} - N = 4.91.$$

As $\chi^2_{0.05,8} = 15.5 > \chi^2$, it can be concluded that the observed frequencies are consistent with the theoretical frequencies of Poisson distribution. In this distribution the maximum probability is $P(X = 1) = 0.35346$ and hence the most probable number of bacteria in the sample is one.

8.5 Truncated Poisson Distribution

Let X be a random variable with probability function

$$P(x) = \frac{e^{-\theta}\theta^x}{x!}, x = 0, 1, 2, \ldots.$$

Let us define another random variable X_1 such that the values of x_1 are $x_1 = 1, 2, \ldots$. This X_1 is the truncated variable X where the form of truncation is the omission of the zero class. The probability function of X_1 is given by

$$P(X_1 = x) = \frac{e^{-\theta}\theta^{x_1}}{x!(1 - e^{-\theta})} = \frac{1}{(e^\theta - 1)}\frac{\theta^{x_1}}{x!}, \; x = 1, 2, \ldots.$$

This distribution is usually called *positive Poisson distribution.* The r-th factorial moment of this distribution is given by

$$\mu_{(r)} = E\left[x^{(r)}\right] = \sum_{x=1}^{\infty} x^{(r)} \left(e^\theta - 1\right)^{-1} \frac{\theta^x}{x!}$$

$$= \frac{\theta^r}{(e^\theta - 1)} \sum_{x=r}^{\infty} \frac{\theta^{x-r}}{(x-r)!} = \frac{e^\theta \theta^r}{(e^\theta - 1)} = \frac{\theta^r}{(1 - e^{-\theta})}.$$

In particular,

$$\mu_{(1)} = \frac{\theta}{1-e^{-\theta}}, \ \mu_{(2)} = \frac{\theta^2}{(1-e^{-\theta})}.$$

Hence

$$\mu_1' = \mu_{(1)} = \theta\left(1-e^{-\theta}\right)^{-1}$$

$$\mu_2' = \mu_{(2)} + \mu_{(1)} = \left(1-e^{-\theta}\right)^{-1}\theta(1+\theta).$$

$$\therefore \ \mu_2 = \mu_2' - \mu_1'^2 = \left(1-e^{-\theta}\right)^{-1}\theta(1+\theta) - \frac{\theta^2}{(1-e^{-\theta})^2}$$

$$= \left(1-e^{-\theta}\right)^{-1}\theta\left[1+\theta-\theta\left(1-e^{-\theta}\right)^{-1}\right].$$

If the first r_1 values of x are omitted $[0, 1, \ldots, (r_1-1)]$, then the left truncated Poisson distribution is given by

$$P(x) = \frac{e^{-\theta}\theta^x}{x!} \Big/ \left[1 - \sum_{x=0}^{r_1-1} e^{-\theta}\frac{\theta^x}{x!}\right], \ x = r_1, r_1+1, \ldots .$$

The right truncated Poisson distribution is obtained when the values of x greater than r_2 are omitted. In that case the probability function is given by

$$P(x) = \frac{e^{-\theta}\theta^x}{x!} \Big/ \sum_{x=0}^{r_2} e^{-\theta}\frac{\theta^x}{x!}, \ x = 0, 1, 2, \ldots, r_2$$

$$= \frac{\theta^x}{x!} \Big/ \sum_{x=0}^{r_2} \frac{\theta^x}{x!}.$$

The doubly truncated distribution of a random variable X is one for which the first r_1 values $[0, 1, 2, \ldots, (r_1-1)]$ and the values greater than r_2 of x are omitted. The probability function of this doubly truncated Poisson distribution is given by

$$P(x) = \frac{\theta^x}{x!} \Big/ \sum_{x=r_1}^{r_2} \frac{\theta^x}{x!}, \quad x = r_1, r_1+1, \ldots, r_2, \quad r_2 \geq r_1 > 0.$$

8.6 Mixtures of Poisson Distribution

We have considered the random variable X to be distributed as Poisson distribution if the probability function of X is given by

$$P(x) = e^{-\theta}\frac{\theta^x}{x!}, \ x = 0, 1, 2, \ldots$$

where θ is a parameter. If in this probability function θ is considered as random variable with p.d.f. $f(\theta)$, then a new distribution

$$P(x) = \frac{1}{x!}E\left(e^{-\theta}\theta^x\right), \ x = 0, 1, 2, \ldots$$

is obtained. This new distribution is usually called *compound Poisson distribution.*

Poisson-Rectangular Distribution

This is a compound Poisson distribution derived from Poisson distribution with parameter θ and assuming θ as a random variable with p.d.f.

$$f(\theta) = \frac{1}{b-a}, \; a \leq \theta \leq b.$$

Thus the compound Poisson distribution is given by

$$P(x) = \frac{1}{x!}\frac{1}{(b-a)}\int_a^b e^{-\theta}\theta^x d\theta, \; x = 0, 1, 2, \ldots.$$

We know that if

$$f(y) = \frac{1}{\Gamma\alpha}e^{-y}y^{\alpha-1}, \; 0 < y < \infty$$

with $\alpha > 0$, then

$$\begin{aligned}
P(y > Y) &= \frac{1}{\Gamma\alpha}\int_Y^\infty e^{-y}y^{\alpha-1}dy \\
&= \frac{1}{\Gamma\alpha}Y^{\alpha-1}e^{-Y} + \int_Y^\infty \frac{1}{\Gamma\alpha-1}y^{\alpha-2}e^{-y}dy \\
&= \frac{1}{\Gamma\alpha}Y^{\alpha-1}e^{-Y} + \frac{1}{\Gamma\alpha-1}Y^{\alpha-2}e^{-Y} + \cdots + Ye^{-Y} + e^{-Y} \\
&= \frac{e^{-Y}Y^{\alpha-1}}{(\alpha-1)!} + \frac{e^{-Y}Y^{\alpha-2}}{(\alpha-2)!} + \cdots + \frac{e^{-Y}Y^{2-1}}{\Gamma 2} + \frac{e^{-Y}Y^{1-1}}{\Gamma 1} \\
&= \text{probability that a random variable having a Poisson} \\
&\quad \text{distribution with expected value } Y \text{ is less than } \alpha. \qquad (8.4)
\end{aligned}$$

Using (8.4) it can be written that

$$\begin{aligned}
P(x) &= \frac{1}{(b-a)}\left[e^{-a}\left(1 + \frac{a}{1!} + \frac{a^2}{2!} + \cdots + \frac{a^x}{x!}\right)\right. \\
&\quad \left. - e^{-b}\left(1 + \frac{b}{1!} + \frac{b^2}{2!} + \cdots + \frac{b^x}{x!}\right)\right].
\end{aligned}$$

The mean and the first three central moments are given by

$$\begin{aligned}
\mu_1' &= \frac{1}{2}(a+b) \\
\mu_2 &= \frac{1}{2}(a+b) + \frac{1}{12}(b-a)^2 \\
\mu_3 &= \frac{1}{2}(a+b) + \frac{1}{4}(b-a)^2 \\
\mu_4 &= \frac{1}{2}(a+b) + \frac{1}{3}(4b^2 + ab + 4a^2) + \frac{1}{4}(a+b)(b-a)^2 + \frac{1}{80}(b-a)^4
\end{aligned}$$

From this compound Poisson distribution it can be shown that

$$P(X = x+1) = P(X = x) + (b-a)^{-1}\left(e^{-a}a^{x+1} - e^{-b}b^{x+1}\right)/(x+1)!.$$

Thus the distribution is unimodal, with mode at the greatest integer x for which

$$x \leq (b-a)[\log(b/a)]^{-1}.$$

However, there are two equal modal values if $(b-a)[\log(b/a)]^{-1}$ is an integer.

Poisson-Gamma Distribution

Let X be a random variable with probability function

$$P(x) = \frac{e^{-\theta}\theta^x}{x!}, \; x = 0, 1, 2, \ldots$$

Again, θ is a random variable with p.d.f.

$$f(\theta) = \frac{1}{\Gamma\alpha\beta^\alpha} e^{-\theta/\beta}\theta^{\alpha-1}, \; 0 < \theta < \infty, \; (\alpha, \beta) > 0.$$

Now, the compound Poisson distribution is given by

$$\begin{aligned} P(x) &= \frac{1}{\Gamma\alpha\beta^\alpha x!}\int_0^\infty e^{-\theta}e^{-\theta/\beta}\theta^{x+\alpha-1}d\theta \\ &= \frac{1}{\Gamma\alpha\beta^\alpha x!}\int_0^\infty e^{-\theta(\frac{1+\beta}{\beta})}\theta^{\alpha+x-1}d\theta \\ &= \frac{\Gamma\alpha+x}{\Gamma\alpha\,\Gamma(x+1)}\left(\frac{\beta}{1+\beta}\right)^x\left(\frac{1}{1+\beta}\right)^\alpha \\ &= \binom{\alpha+x-1}{\alpha-1}\left(\frac{\beta}{1+\beta}\right)^x\left(\frac{1}{1+\beta}\right)^\alpha, \; x = 0, 1, 2, \ldots \end{aligned}$$

which is a negative binomial distribution with parameters (α, β) [see Chapter 5].

Poisson-Truncated Normal Distribution

Let X be a random variable with probability function

$$P(x) = \frac{e^{-\theta}\theta^x}{x!}, \; x = 0, 1, 2, \ldots$$

and θ follows truncated normal distribution with parameters μ and $\sigma^2[N(\mu, \sigma^2)]$ where the point of truncation is in the left and is usually taken at zero, ensuring θ has zero probability of being negative. Thus the p.d.f. of this compound distribution is given by

$$\begin{aligned} P(x) &= \frac{\frac{1}{\sqrt{2\pi}\sigma}\frac{1}{x!}\int_0^\infty e^{-\theta}\theta^x e^{-\frac{1}{2\sigma^2}(\theta-\mu)^2}d\theta}{\frac{1}{\sqrt{2\pi}\sigma}\int_0^\infty e^{-\frac{1}{2\sigma^2}(\theta-\mu)^2}d\theta} \\ &= \frac{\exp\left[\frac{1}{2}\sigma^2-\mu\right]\int_0^\infty \theta^x \exp\left[-\frac{1}{2}\left(\frac{\theta-\mu+\sigma^2}{\sigma}\right)^2\right]d\theta}{x!\int_0^\infty \exp\left[-\frac{1}{2}\left(\frac{\theta-\mu}{\sigma}\right)^2 d\theta\right]}. \end{aligned}$$

Poisson-Binomial Distribution

Let X be a random variable with probability function

$$P(x) = \frac{e^{-\theta j}(\theta j)^x}{x!}, \; x = 0, 1, 2, \ldots$$

and j is a random variable with probability function

$$P(j) = \binom{N}{j} p^j q^{N-j}, \; q = 1 - p.$$

Then the compound distribution of X is given by

$$P(x) = \frac{\theta^x}{x!} \sum_{j=0}^{N} e^{-\theta j} j^x \binom{N}{j} p^j q^{N-j}, \; x = 0, 1, 2, \ldots .$$

On simplification it gives

$$P(x) = \frac{\theta^x}{x!} \sum_{j=0}^{N} (\Delta^j O^x / j!) N^{(j)} \left(pe^{-\theta}\right)^j \left(q + pe^{\theta}\right)^{N-j}$$

since

$$\sum_{j=0}^{N} e^{-\theta j} j^{(x)} \binom{N}{j} p^j q^{n-j} = N^{(x)} (pe^{-\theta})^x (q + pe^{-\theta})^{N-x}.$$

The factorial moment of this compound distribution is given by

$$\mu_{(r)} = E\left[x^{(r)}\right] = \theta^r E(j^r).$$

In particular, when $r = 1$ and $r_2 = 2$, we have

$$\begin{aligned}
\mu_1' &= \mu_{(1)} = Np\theta, \; \mu_{(2)} = Np\theta^2(q + Np) \\
\mu_2' &= \mu_{(2)} + \mu_{(1)} = Np\theta^2(q + Np) + Np\theta. \\
\therefore \; \mu_2 &= \mu_2' - \mu_1'^2 = Np\theta^2(q + Np) + Np\theta - N^2p^2\theta^2 \\
&= Np\theta + Npq\theta^2.
\end{aligned}$$

Similarly,

$$\begin{aligned}
\mu_3 &= Np\theta + 3Npq\theta^2 + Npq(q - p)\theta^3 \\
\mu_4 &= 3(Np\theta + Npq\theta^2)^2 + Np\theta + 7Npq\theta^2 \\
&\quad + 6Npq(q - p)\theta^3 + Npq(1 - 6pq)\theta^4.
\end{aligned}$$

Poisson-Negative Binomial Distribution

Let X be a random variable with probability function

$$P(x) = \frac{e^{-\theta j}(\theta j)^x}{x!}, \; x = 0, 1, 2, \ldots$$

and j is a random variable with probability function

$$P(j) = \binom{N+j-1}{N-1} Q^{-N}(P/Q)^j,\ j = 0, 1, 2, \dots, \text{ where } Q = 1 + P.$$

Then the compound Poisson distribution is given by

$$P(x) = \frac{\theta^x}{x!} \sum_{j=0}^{\infty} e^{-\theta j} j^x \binom{N+j-1}{N-1} Q^{-N}(P/Q)^j.$$

The probability generating function of this distribution is given by

$$E[\exp\{j\theta(t-1)\}] = [Q - P\exp\{\theta(t-1)\}]^{-N}.$$

The moments of this distribution are

$$\begin{aligned}
\mu_1' &= NP\theta \\
\mu_2 &= NP\theta + NPQ\theta^2 \\
\mu_3 &= NP\theta + 3NPQ\theta^2 + NPQ(Q+P)\theta^3 \\
\mu_4 &= 3(NP\theta + NPQ\theta^2)^2 + NP\theta + 7NPQ\theta^2 \\
&\quad + 6NPQ(Q+P)\theta^3 + NPQ(1+6PQ)\theta^4.
\end{aligned}$$

Poisson-Hypergeometric Distribution

Let X be a random variable with probability function

$$P(x) = \frac{e^{-\theta j}(\theta j)^x}{x!},\ x = 0, 1, 2, \dots$$

and j is a random variable with probability function

$$P(j) = \binom{X}{j}\binom{N-X}{n-j} \Big/ \binom{N}{n}$$

for which $\max(0, n-N+X) \le x \le \min(n, X)$. Then the compound Poisson distribution is given by

$$P(x) = \frac{1}{\binom{N}{n}} \frac{\theta^x}{x!} \sum_j e^{-\theta j} j^x \binom{X}{j}\binom{N-X}{n-j}.$$

The summation is taken over the range for which $0 \le j \le X$ and $0 \le n - j \le N - X$. The probability generating function of this distribution is given by

$$\frac{(N-n)!(N-X)!}{N!(N-n-X)!} F\left(-n, -X;\ N - X - n + 1;\ e^{\theta(t-1)}\right).$$

Poisson-Logseries Distribution

Let X be a random variable with probability function

$$P(x) = \frac{e^{-\theta j}(\theta j)^x}{x!},\ x = 0, 1, 2, \dots$$

and j is also a random variable with probability function

$$P(j) = -\frac{1}{\log(1-\lambda)}\frac{\lambda^j}{j},\ j = 1, 2, \ldots.$$

Then the compound Poisson distribution is given by

$$\begin{aligned} P(x) &= -\frac{1}{\log(1-\lambda)}\sum_{j=1}^{\infty}\frac{\lambda^j}{j}\frac{e^{-\theta j}(\theta j)^x}{x!} \\ &= -\frac{1}{\log(1-\lambda)}\frac{1}{x!}\theta^x\sum_{j=1}^{\infty} j^{x-1}\left(\lambda e^{-\theta}\right)^j \end{aligned}$$

$$\text{for } x = 0, 1, 2, \ldots;\ \lambda < 1.$$

We know

$$\sum_{j=1}^{\infty} j^{(r)}\left(\lambda e^{-\theta}\right)^j = r!\left(\lambda e^{-\theta}\right)^r\left(1-\lambda e^{-\theta}\right)^{-(r+1)}.$$

Thus

$$P(x) = -\frac{1}{\log(1-\lambda)}\frac{\theta^x}{x!}\sum_{j=0}^{x-1}\left(\Delta^j O^{x-1}\right)\left(\lambda e^{-\theta}\right)^j\left(1-\lambda e^{-\theta}\right)^{-(j+1)},$$

$$\text{for } x > 0$$

$$\begin{aligned} \text{and } P(x=0) &= \log\left(1-\lambda e^{-\theta}\right)/\log(1-\lambda), \text{ for } x = 0 \\ P(x=1) &= -\frac{1}{\log(1-\lambda)}\lambda\theta e^{-\theta}\left(1-\lambda e^{-\theta}\right)^{-1}, \text{ for } x = 1. \end{aligned}$$

In general, it can be written

$$P(x=k) = \frac{\theta^k}{k!}\left[\log\left(1-\lambda e^{-\theta}\right)/\log(1-\lambda)\right]\mu_k', \text{ for } x = k$$

where μ_k' is the k-th moment about zero of a logseries distribution with parameter $\lambda e^{-\theta}$. The probability generating function of this distribution is given by

$$\log\left\{1-\lambda e^{\theta(t-1)}\right\}/[\log(1-\lambda)].$$

The mean and variance of this distribution are given, respectively, by

$$\begin{aligned} \mu_1' &= -\frac{1}{\log(1-\lambda)}\frac{\theta\lambda}{1-\lambda} \\ \text{and } \mu_2 &= -\frac{1}{\log(1-\lambda)}\frac{\theta\lambda}{1-\lambda} \\ &\quad +[-\log(1-\lambda)]^{-1}(1-\lambda)^{-2}\theta^2\lambda\left[1-\lambda\{-\log(1-\lambda)\}^{-1}\right]. \end{aligned}$$

Katti and Rao (1965) modified this Poisson-logseries distribution by adding extra probability to the event $(x = 0)$. According to them, the log-zero-Poisson distribution is represented by

$$P(x=0) = w + (1-w)\left[\log\left(1-\lambda e^{-\theta}\right)\right][\log(1-\lambda)]^{-1}$$

$$\text{and } P(x=k) = (1-w)[-\log(1-\lambda)]^{-1}\frac{\theta^k}{k!}\sum_{j=1}^{\infty} j^{k-1}\left(\lambda e^{-\theta}\right)^j$$

$$(k \geq 1;\ 0 < w < 1)$$

The probability generating function of this log-zero-Poisson distribution is given by

$$w + (1-w)\left[\log\left(1-\lambda e^{\theta(t-1)}\right)\right][\log(1-\lambda)]^{-1}.$$

Poisson-Poisson Distribution

This is known by Neyman's Type A distribution and is obtained by compounding a Poisson distribution by a Poisson distribution. Let X be a random variable with probability function

$$P(x) = \frac{e^{-0j}(\theta j)^x}{x!},\ x = 0, 1, 2, \ldots$$

and j is another random variable with probability function

$$P(j) = \frac{e^{-\lambda}\lambda^j}{j!},\ j = 0, 1, 2, \ldots.$$

Then the compound Poisson distribution for $x > 0$ is given by

$$P(x) = \sum_{j=1}^{\infty}\frac{e^{-\lambda}\lambda^j}{j!}e^{-j\theta}\frac{(j\theta)^x}{x!} \tag{8.5}$$

and when $x = 0$

$$P(x=0) = e^{-\lambda} + \sum_{j=1}^{\infty}\frac{e^{-\lambda}\lambda^j}{j!}e^{-j\theta}.$$

Equation (8.5) can be written as

$$\begin{aligned} P(x) &= \frac{\theta^x}{x!}\sum_{j=1}^{\infty} e^{-\lambda}\frac{\lambda^j}{j!}e^{-j\theta}\sum_{r=1}^{x} j^{(r)}\left[\Delta^r O^x/r!\right] \\ &= \frac{\theta^x}{x!}e^{-\lambda}\sum_{r=1}^{x}\frac{\Delta^r O^x}{r!}\sum_{j=r}^{\infty}\frac{\left(\lambda e^{-\theta}\right)^j}{(j-r)!} \\ &= \frac{\theta^x}{x!}e^{-\lambda}\sum_{r=1}^{x}\left(\lambda e^{-\theta}\right)^r\left[\Delta^r O^x/r!\right]\exp(\lambda e^{-\theta}). \end{aligned}$$

In particular,

$$P(x=1) = \lambda\theta e^{-\theta}\exp[-\lambda(1-e^{-\theta})]$$

$$\text{and } P(x=2) = \frac{1}{2}\lambda\theta^2 e^{-\theta}(1+\lambda e^{-\theta})\exp[-\lambda(1-e^{-\theta}).$$

It can be shown that $P(x=1) < P(x=0)$ if $\lambda\theta e^{-\theta} < 1$ and $P(x=1) < P(x=2)$ if $\frac{1}{2}\theta(1+\lambda e^{-\theta}) > 1$. This inequality indicates that the distribution has a local minimum at $x = 1$. To hold this inequality the values of λ and θ are to be chosen for a given value of $\lambda\theta e^{-\theta}$. For further study of the modality of the Type A distribution, Barton (1957) and Shenton and Bowman (1967) can be consulted. The probability generating function of this distribution is

$$g(t) = \exp\left[-\lambda\left(1 - e^{-\theta(1-t)}\right)\right].$$

The r-th factorial moment of this distribution is

$$\mu_{(r)} = E\left[\theta^r j^r\right] = \theta^r \sum_{j=1}^{r} \frac{\Delta^j O^r}{j!}\lambda^j.$$

Using the value of $\mu_{(r)}$, we have

$$\begin{aligned}
\mu_1' &= \lambda\theta \\
\mu_2 &= \lambda\theta(1+\theta) \\
\mu_3 &= \lambda\theta(1+3\theta+\theta^2) \\
\mu_4 &= 3\lambda^2\theta^2(1+\theta)^2 + \lambda\theta(1+7\theta+6\theta^2+\theta^3).
\end{aligned}$$

The indices of skewness and kurtosis are given by

$$\begin{aligned}
\alpha_3 &= \sqrt{\beta_1} = \frac{\mu_3}{\mu_2^{3/2}} = (1+3\theta+\theta^2)(1+\theta)^{-3/2}(\lambda\theta)^{-1/2} \\
\beta_2 &= \alpha_4 = \frac{\mu_4}{\mu_2^2} = 3 + (1+7\theta+6\theta^2+\theta^3)(1+\theta)^{-2}(\lambda\theta)^{-1}.
\end{aligned}$$

Shenton (1949) has established the recurrence relations among the cumulants. These relationships are

$$\begin{aligned}
K_{r+1} &= \theta\left\{\sum_{j=0}^{r-1}\binom{r}{j}K_{r-j} + \lambda\right\} \\
K_{r+1} &= \theta\left(K_r + \frac{\partial k_r}{\partial\lambda}\right).
\end{aligned}$$

8.7 Properties of Poisson Distribution

We have already observed that

$$\frac{P(x+1)}{P(x)} = \frac{\theta}{x+1}.$$

From this relation it can be concluded that $P(x)$ increases with x to a maximum at $x = \theta$ and thereafter decreases as x increases. This relation helps to study the mode of the Poisson distribution.

Mode of Poisson Distribution

We have

$$\frac{P(x)}{P(x-1)} = \frac{\theta}{x}.$$

Suppose that θ is not an integer and let S is the integral part of θ. Then

$$\frac{P(1)}{P(0)} > 1, \ldots, \frac{P(S-1)}{P(S-2)} > 1, \frac{P(S)}{P(S-1)} > 1$$

$$\text{and } \frac{P(S+1)}{P(S)} < 1, \frac{P(S+2)}{P(S+1)} < 1, \ldots.$$

Combining the above expressions, we have

$$\begin{aligned} & P(0) < P(1) < P(2) < \cdots < P(S-2) < P(S-1) \\ & < P(S) > P(S+1)P(S+2)\cdots. \end{aligned}$$

Thus $P(S)$ is the maximum value and the distribution is unimodal. The integral part of θ is the modal value. Again, let $\theta = k$ (say) is an integer. Then

$$\frac{P(1)}{P(0)} > 1, \frac{P(2)}{P(1)} > 1, \ldots, \frac{P(k-1)}{P(k-2)} > 1$$

$$\text{and } \frac{P(k)}{P(k-1)} = 1, \frac{P(k+1)}{P(k)} < 1, \frac{P(k+2)}{P(k+1)} < 1, \ldots.$$

$$\begin{aligned} \therefore\ & P(0) < P(1) < P(2) < \cdots < P(k-2) < P(k-1) \\ & = P(k) > P(k+1)\cdots. \end{aligned}$$

Hence two maximum values are $P(k-1)$ and $P(k)$, and the distribution is bimodal. The modal values are $(\theta - 1)$ and θ.

Additive Property of Poisson Distribution

Let X_1 and X_2 be two Poisson variates with means θ_1 and θ_2 respectively. The moment generating functions of these two variables are

$$M_{x_1}(t) = e^{\theta_1(e^t-1)} \text{ and } M_{x_2}(t) = e^{\theta_2(e^t-1)}.$$

Let $Y = X_1 + X_2$ and assume that X_1 and X_2 are independent. Then

$$\begin{aligned} M_y(t) &= M_{x_1+x_2}(t) \\ &= M_{x_1}(t) \cdot M_{x_2}(t) \\ &= e^{\theta_1(e^t-1)} \cdot e^{\theta_2(e^t-1)} \\ &= e^{(\theta_1+\theta_2)(e^t-1)} \end{aligned}$$

which is the m.g.f. of a Poisson variate with mean $\theta_1 + \theta_2$. Thus Y is a Poisson variate with mean $\theta_1 + \theta_2$. In general, if $X_1, X_2, \ldots, X_k$ are Poisson variates with means $\theta_1, \theta_2, \ldots, \theta_k$ respectively, then their sum, $\sum_{i=1}^{k} X_i$, is also a Poisson variate with mean $\sum_{i=1}^{k} \theta_i$.

However, the difference of two independent Poisson variates is not a Poisson variate, since

$$M_{x_1-x_2}(t) = M_{x_1+(-x_2)}(t) = M_{x_1}(t) \cdot M_{x_2}(-t).$$

We have

$$\begin{aligned} M_{x_1}(t) &= e^{\theta_1(e^t-1)} \\ \text{and } M_{x_2}(-t) &= e^{\theta_2(e^{-t}-1)}. \\ \therefore\ M_{x_1-x_2}(t) &= e^{\theta_1(e^t-1)} \cdot e^{\theta_2(e^{-t}-1)} \end{aligned}$$

which cannot be written in the form of m.g.f. of Poisson variate. Hence $X_1 - X_2$ is not a Poisson variate.

The distribution of $X_1 - X_2$ has been studied by Irwin (1937) assuming

$$E(X_1) = E(X_2) = \theta.$$

Let $Y = X_1 - X_2$, then

$$\begin{aligned} P(X_1 - X_2 = y) &= e^{-2\theta} \sum_{j=y}^{\infty} \theta^{j+(j-y)} [j!(j-y)!]^{-1} \\ &= e^{-2\theta} \theta^{\frac{y}{2}} \sum_{j=0}^{\infty} \frac{(\theta^2)^j}{j!\,\Gamma\left(\frac{y}{2}+j+1\right)} \\ &= e^{-2\theta} I_{\frac{y}{2}}(2\theta) \end{aligned}$$

where $I_{\frac{y}{2}}(2\theta)$ is a modified Bessel function.

Skellam (1946) and de Castro (1952) discussed the distribution of $X_1 - X_2$ when

$$E(x_1) = \theta_1 \neq E(x_2) = \theta_2.$$

In that case

$$\begin{aligned} P(X_1 - X_2 = y) &= e^{-(\theta_1+\theta_2)} \sum_{j=y}^{\infty} \theta_1^j \theta_2^{j-y} [j!(j-y)!]^{-1} \\ &= e^{-(\theta_1+\theta_2)} \left(\frac{\theta_1}{\theta_2}\right)^{y/2} I_{\frac{1}{2}y}\left(2\sqrt{\theta_1\theta_2}\right). \end{aligned}$$

The cumulants of the distribution of $X_1 - X_2$ are

$$\begin{aligned} K_r(X_1 - X_2) &= K_r(X_1) + (-1)^r K_r(X_2) \\ &= \theta_1 + (-1)^r \theta_2. \end{aligned}$$

❑ **THEOREM 8.1** *Let X_1 and X_2 be two independent Poisson variates with parameters θ_1 and θ_2 respectively. Then the conditional distribution of X_1 given $X_1 + X_2 = n$ is a binomial distribution.*

Proof. As X_1 and X_2 are Poisson variates, $X_1 + X_2$ is also a Poisson variate with parameter $\theta_1 + \theta_2$.

Now

$$\begin{aligned} P(X_1 = r|X_1 + X_2 = n) &= \frac{P(X_1 = r \cap X_1 + X_2 = n)}{P(X_1 + X_2 = n)} \\ &= \frac{P(X_1 = r \cap X_2 = n - r)}{P(X_1 + X_2 = n)} \\ &= \frac{P(X_1 = r)P(X_2 = n - r)}{P(X_1 + X_2 = n)} \\ &= \frac{\left[e^{-\theta_1}\theta_1^r/r!\right]\left[e^{-\theta_2}\theta_2^{n-r}/(n-r)!\right]}{e^{-\theta_1-\theta_2}\frac{(\theta_1+\theta_2)^n}{n!}} \\ &= \frac{n!}{r!(n-r)!}\left(\frac{\theta_1}{\theta_1+\theta_2}\right)^r\left(\frac{\theta_2}{\theta_1+\theta_2}\right)^{n-r} \\ &= \binom{n}{r}p^r(1-p)^{n-r}, \text{ where } p = \frac{\theta_1}{\theta_1+\theta_2}. \end{aligned}$$

❐ **THEOREM 8.2** *If X is a Poisson variate with parameter $\theta = 1$, then mean deviation about mean is $2/e$.*

Proof. Here

$$\begin{aligned} P(x) &= \frac{e^{-\theta}\theta^x}{x!}, \; x = 0, 1, 2, \ldots \\ &= e^{-1}/x! \; [\because \; \theta = 1]. \end{aligned}$$

Mean deviation about mean is given by

$$\begin{aligned} E|x - \theta| &= E|x - 1| = \sum_{x=0}^{\infty} |x - 1|\frac{e^{-1}}{x!} \\ &= e^{-1}\left[1 + \frac{1}{2} + \frac{2}{3!} + \frac{3}{4!} + \cdots\right] \\ &= e^{-1}\left[1 + \left(1 - \frac{1}{2!}\right) + \left(\frac{1}{2!} - \frac{1}{3!}\right) + \left(\frac{1}{3!} - \frac{1}{4!}\right) + \cdots\right] \end{aligned}$$

As

$$\frac{n}{(n+1)!} = \frac{(n+1) - 1}{(n+1)!} = \frac{1}{n!} - \frac{1}{(n+1)!}.$$

$$\therefore \; E|x - 1| = e^{-1}(1 + 1) = \frac{2}{e}.$$

From this it may be said that when $\theta = 1$, the mean deviation about mean is $(2/e)$ times the standard deviation $[V(X) = \theta = 1$, standard deviation $= 1]$.

❐ **THEOREM 8.3** *Let X be a Poisson variate with mean θ and μ_r be the r-th central moment of X. Then*

$$\mu_{r+1} = \theta\left[\binom{r}{1}\mu_{r-1} + \binom{r}{2}\mu_{r-2} + \cdots + \binom{r}{r}\mu_0\right].$$

Proof. We know

$$P(x) = \frac{e^{-\theta}\theta^x}{x!},\ x = 0, 1, 2, \ldots$$

$$\begin{aligned}
\mu_{r+1} &= E(x-\theta)^{r+1} = \sum_{x=0}^{\infty}(x-\theta)^{r+1}\frac{e^{-\theta}\theta^x}{x!} = \sum_{x=0}^{\infty}(x-\theta)^r(x-\theta)\frac{e^{-\theta}\theta^x}{x!} \\
&= \sum_{x=0}^{\infty}x(x-\theta)^r\frac{e^{-\theta}\theta^x}{x!} - \theta\sum_{x=0}^{\infty}(x-\theta)^r\frac{e^{-\theta}\theta^x}{x!} = \sum_{x=1}^{\infty}(x-\theta)^r\frac{e^{-\theta}\theta^x}{(x-1)!} - \theta\mu_r \\
&= \sum_{x=1}^{\infty}(x-\theta+1-1)^r\frac{e^{-\theta}\theta^x}{(x-1)!} - \theta\mu_r = \theta\sum_{x=0}^{\infty}(x-\theta+1)^r\frac{e^{-\theta}\theta^x}{x!} - \theta\mu_r \\
&= \theta\left[\sum_{x=0}^{\infty}(x-\theta)^r\frac{e^{-\theta}\theta^x}{x!} + \binom{r}{1}\sum_{x=0}^{\infty}(x-\theta)^{r-1}\frac{e^{-\theta}\theta^x}{x!} + \cdots\right. \\
&\qquad \left. + \binom{r}{r-1}\sum_{x=0}^{\infty}(x-\theta)\frac{e^{-\theta}\theta^x}{x!} + 1\right] - \theta\mu_r \\
&= \theta\left[\mu_r + \binom{r}{1}\mu_{r-1} + \binom{r}{2}\mu_{r-2} + \cdots + \binom{r}{r}\mu_0\right] - \theta\mu_r \\
&= \theta\left[\binom{r}{1}\mu_{r-1} + \binom{r}{2}\mu_{r-2} + \cdots + \binom{r}{r}\mu_0\right].
\end{aligned}$$

Standardized Poisson Variate

Let X be a Poisson variate with mean θ. Then

$$Z = \frac{X-\theta}{\sqrt{\theta}}$$

is a standardized Poisson variate. Now

$$\begin{aligned}
E(Z) &= \frac{E(X)-\theta}{\sqrt{\theta}} = 0 \\
V(Z) &= E\left(\frac{X-\theta}{\sqrt{\theta}}\right)^2 = \frac{1}{\theta}E(X-\theta)^2 = \theta = 1.
\end{aligned}$$

Thus Z has mean zero and standard deviation 1. The m.g.f. of Z is given by

$$\begin{aligned}
M_z(t) &= E\left[e^{tz}\right] = E\left[e^{t\frac{X-\theta}{\sqrt{\theta}}}\right] = e^{-t\sqrt{\theta}}E\left[e^{tX/\sqrt{\theta}}\right] \\
&= e^{-t\sqrt{\theta}}\sum_{x=0}^{\infty}e^{tx/\sqrt{\theta}}\frac{e^{-\theta}\theta^x}{x!} = e^{-t\sqrt{\theta}}\sum_{x=0}^{\infty}e^{-\theta}\frac{\left(\theta e^{t/\sqrt{\theta}}\right)}{x!} \\
&= e^{-t\sqrt{\theta}} \cdot e^{-\theta} \cdot e^{\theta e^{t/\sqrt{\theta}}} = e^{-\theta-t\sqrt{\theta}+\theta e^{t/\sqrt{\theta}}} \\
&= 1 + \left(\frac{t^2}{2}+\cdots\right) + \frac{1}{2!}\left(\frac{t^2}{2}+\cdots\right)^2 + \cdots \\
&= e^{\frac{1}{2}t^2} \text{ if } \theta \to \infty.
\end{aligned}$$

$$\therefore \lim_{\theta\to\infty} M_z(t) = e^{\frac{1}{2}t^2}$$

which is the m.g.f. of a standard normal variate. Thus standard Poisson variate tends to standard normal variate as θ tends to infinity. That is

$$\lim_{\theta\to\infty} P[\alpha < z < \beta] = \frac{1}{\sqrt{2\pi}} \int_{\alpha}^{\beta} e^{-\frac{1}{2}z^2} dz.$$

Individual Poisson probabilities may be approximated by the formula

$$P(X = k) = \frac{1}{\sqrt{2\pi}} \int_{\alpha_1}^{\alpha_2} e^{-\frac{1}{2}z^2} dz$$

where

$$\alpha_1 = \left(k - \theta + \frac{1}{2}\right) \theta^{-\frac{1}{2}}$$

$$\text{and } \alpha_2 = \left(k - \theta - \frac{1}{2}\right) \theta^{-\frac{1}{2}}.$$

A modified form of this probability is given by

$$P(X = k) \doteq \frac{1}{\sqrt{2\pi\theta}} e^{-\frac{1}{2}u^2} \left[1 + \frac{1}{6}(3u - u^3)\theta^{-1/2}\right] + R$$

where

$$u = (k - \theta)\theta^{-1/2} \text{ and } |R| < 0.0748\theta^{-3/2} - 0.0055\theta^{-2} + 0.3724\theta^{-5/2}$$
$$+0.5595\theta^{-3} + \theta^{-1/4}\left(1 + \frac{1}{2}\theta^{-1/2}\right) \exp\left[-2\sqrt{\theta}\right].$$

This approximation has been done by Makabe and Morimura (1955). They also have suggested the following approximation to sums of individual probabilities

$$\sum_{x=k_1}^{k_2} e^{-\theta}\theta^x/x! = \frac{1}{\sqrt{2\pi}} \int_{x_1}^{x_2} \exp(-z^2/2)dz + \frac{1}{6}\frac{1}{\sqrt{2\pi\theta}}$$
$$[(1 - x_2^2)] \exp(-x_2^2/2) - (1 - x_1^2) \exp(-x_1^2/2)] + R'$$

where

$$x_1 = \left(k_1 - \theta - \frac{1}{2}\right) \theta^{-1/2}, \; x_2 = \left(k_2 - \theta + \frac{1}{2}\right) \theta^{-1/2}$$
$$|R'| \le 0.0544\theta^{-1} + 0.0108\theta^{-3/2} + 0.2743\theta^{-2} + 0.0065\theta^{-5/2}$$
$$+ \left(1 + \frac{1}{2}\theta^{-1/2}\right) \exp\left(-2\sqrt{\theta}\right).$$

For further approximations to the Poisson distribution Anscombe (1948) and Freeman and Tukey (1950) may be consulted.

❒ **THEOREM 8.4** *Let X be a Poisson variate with parameter θ. Then the distribution fucntion of X is given by*

$$F(x) = \frac{1}{\Gamma x + 1} \int_{\theta}^{\infty} e^{-t} t^x dt; \; x = 0, 1, 2, \ldots.$$

Proof. We have

$$P(x) = \frac{e^{-\theta}\theta^x}{x!}, \; x = 0, 1, 2, \ldots.$$

Consider the incomplete gamma integral

$$\begin{aligned}
I_x &= \frac{1}{x!}\int_0^\infty e^{-t}t^x dt, \text{ (x is positive integer)} \\
&= -\frac{e^{-t}t^x}{x!}\bigg|_\theta^\infty + \frac{1}{(x-1)!}\int_\theta^\infty e^{-t}t^{x-1}dt = \frac{e^{-\theta}\theta^x}{x!} + I_{x-1}
\end{aligned}$$

which is a reduction formula for I_x.

Repeated applications of the value of I_x gives

$$I_x = \frac{e^{-\theta}\theta^x}{x!} + \frac{e^{-\theta}\theta^{x-1}}{(x-1)!} + \cdots + \frac{e^{-\theta}\theta}{1!} + I_0.$$

But

$$I_0 = \int_\theta^\infty e^{-t}dt = -e^{-t}\bigg|_\theta^\infty = e^{-\theta}.$$

$$\begin{aligned}
\therefore\ I_x &= e^{-\theta} + \theta e^{-\theta} + \frac{\theta^2 e^{-\theta}}{2!} + \cdots + \frac{\theta^x e^{-\theta}}{x!} \\
&= P(X=0) + P(X=1) + \cdots + P(X=x) = P(X \le x) = F(x). \\
\therefore\ F(x) &= \frac{1}{x!}\int_\theta^\infty e^{-t}t^x dt = \frac{1}{\Gamma x+1}\int_\theta^\infty e^{-t}t^x dt.
\end{aligned}$$

Thus cumulative Poisson probability can be represented in terms of incomplete gamma integral.

Hyper-Poisson Distribution

In Section (8.1) we have discussed the experimental results where Poisson distribution can be applied. Along with other application of this distribution we can mention that Poisson distribution can be used when the events occur randomly in time. An important type of this event is the Poisson processes in queueing theory. A Poisson process is one in which the intervals between successive events have independent identical exponential distributions and, consequently, the number of events in a specified time interval has a Poisson distribution. Morlat (1952) generalized the Poisson law by assuming that the common distribution of the times (t) is a general gamma distribution with origin at zero, where

$$P(t) = \frac{1}{\Gamma\alpha}t^{\alpha-1}e^{-t},\ t \ge 0.$$

Then the probability of k events in time t is

$$P(k) = \frac{1}{\Gamma k\alpha}\int_0^T [1 - y^\alpha\Gamma(k\alpha)/\Gamma\{(k+1)\alpha\}]\, y^{k\alpha-1}e^{-y}dy.$$

If α is an integer, we have

$$P(k) = \sum_{j=k\alpha}^{(k+1)\alpha-1}\frac{e^{-T}T^j}{j!}.$$

A generalization of the Poisson distribution has been suggested by Bardwell and Crow (1964). This generalization has been done using the following definition of the Poisson distribution

$$P(X=k) = C_\theta \frac{\theta^k}{1(1+1)\cdots(1+\overline{k-1})} = \frac{C_\theta \theta^k}{1^{[K]}}$$

where $C_\theta = e^{-\theta}, k = 0, 1, \ldots ;\ \theta > 0$.

The corresponding generalization is of the form

$$P(X=k) = C_{\theta,\lambda} \frac{\theta^k}{\lambda^{[k]}},\ k = 0, 1, \ldots ;\ \theta > 0;\ \lambda > 0.$$

According to Bardwell and Crow this family of distributions is called *hyper-Poisson distribution.* These are again classified into sub-Poisson ($\lambda < 1$) and super-Poisson ($\lambda > 1$) distributions. Here $C_{\theta,\lambda}$ is the reciprocal of a confluent hyper-geometric function and is given by

$$C_{\theta,\lambda} = [M(1;\lambda;\theta)]^{-1}.$$

For this hyper-Poisson distribution the recurrence relation for probabilities is given by

$$(\lambda + k)P(X = k+1) = \theta P(X = k),\ k = 0, 1, \ldots .$$

Summing the probabilities obtained by this recurrence relation and writing

$$\lambda + k = \lambda - 1 + (k+1)$$

we have $(\lambda - 1)[1 - P(X=0)] + E(X) = \theta$

$$\therefore\ E(X) = \theta + (1-\lambda)[1 - P(X=0)] = \theta \frac{\partial}{\partial \theta} \log M(1;\lambda;\theta).$$

Here

$$\left\{1 - \theta(\lambda-1)^{-1}\right\} M(1;\lambda;\theta) + \theta(\lambda-1)^{-1} \frac{\partial M(1;\lambda;\theta)}{\partial \theta} = M(0;\lambda;\theta) = 1.$$

By multiplying the recurrence relation of probabilities by $(k+1)^r$ and summing, we get the relation among the moments of the hyper-Poisson distribution as follows:

$$\lambda\mu_r' + (\mu_{r+1}' - \mu_r') = \theta\left[\mu_r' + \binom{r}{1}\mu_{r-1}' + \cdots + \binom{r}{r-1}\mu_1' + 1\right]$$

$$\text{or, } \mu_{r+1}' = (\theta - \lambda + 1)\mu_r' + \theta\left[\binom{r}{1}\mu_{r-1}' + \binom{r}{2}\mu_{r-2}' + \cdots + \binom{r}{r-1}\mu_1' + 1\right],\ r = 1, 2, \ldots .$$

From this relation of moments it can be said that all moments of this distribution are finite. The variance is given by

$$\begin{aligned} V(X) &= (\theta - \lambda + 1)\mu_1' + \theta - \mu_1'^2 \\ &= \theta + (1-\lambda)[\theta - \lambda + 1 - (1-\lambda)P(X=0)]P(X=0). \end{aligned}$$

The probability generating function is given by

$$g(t) = M(1; \lambda; \theta t)/M(1; \lambda; \theta).$$

Here

$$\begin{aligned} M(1; \lambda; \theta) &= 1 + \frac{1}{\lambda}\frac{\theta}{1!} + \frac{1(1+1)}{\lambda(\lambda+1)}\frac{\theta^2}{2!} + \cdots \\ &= \sum_{j=0}^{\infty} \frac{1^{[j]}}{\lambda^{[j]}}\frac{\theta^j}{j!}, \ \lambda > 0. \end{aligned}$$

❑ **THEOREM 8.5** *If X is a Poisson variate with parameter θ and Y is another discrete variate whose conditional distribution for a given X is given by*

$$P(Y = r|X = x) = \binom{x}{r} p^r (1-p)^{x-r}$$
$$0 < p < 1; \ r = 0, 1, 2, \ldots, x;$$

then the unconditional distribution of Y is a Poisson distribution with parameter θp.

Proof. We know that

$$\begin{aligned} P(x) &= \frac{e^{-\theta}\theta^x}{x!}; \ x = 0, 1, 2, \ldots \\ \text{and } P(Y = r|X = x) &= \binom{x}{r} p^r (1-p)^{x-r}; \ r = 0, 1, 2, \ldots, x. \\ \therefore \ P(X = x \cap Y = r) &= P(X = x)P(Y = r|X = x) \\ &= \frac{e^{-\theta}\theta^x}{x!}\binom{x}{r} p^r (1-p)^{x-r}. \end{aligned}$$

The unconditional distribution of Y is given by

$$\begin{aligned} P(Y = r) &= \sum_{x=r}^{\infty} \frac{e^{-\theta}\theta^x}{x!}\binom{x}{r} p^r (1-p)^{x-r} \\ &= e^{-\theta}\sum_{x=r}^{\infty}\binom{x}{r}\frac{\theta^x p^r (1-p)^{x-r}}{x!} \\ &= e^{-\theta}\sum_{x=r}^{\infty}\frac{1}{r!(x-r)!}\theta^x p^r (1-p)^{x-r} \\ &= \frac{e^{-\theta}}{r!}\sum_{x=r}^{\infty}\frac{(\theta p)^r}{(x-r)!}\theta^{x-r}(1-p)^{x-r} \\ &= \frac{e^{-\theta}(\theta p)^r}{r!}\sum_{x=r}^{\infty}\frac{\{\theta(1-p)\}^{x-r}}{(x-r)!} \\ &= \frac{e^{-\theta}(\theta p)^r}{r!}e^{\theta(1-p)} \\ &= \frac{e^{-\theta p}(\theta p)^r}{r!}; \ r = 0, 1, 2, \ldots. \end{aligned}$$

8.8 Multivariate Poisson Distribution

Let $n_1, n_2, \ldots, n_k$ be the numbers of occurrences of the k mutually exclusive events $E_1, E_2, \ldots, E_k$ respectively in a series of N independent trials $\left(\text{with } \sum_{j=1}^{k} n_j = N\right)$. Then the joint distribution of the random variables $n_1, n_2, \ldots, n_k$ is

$$P(n_1, n_2, \ldots, n_k) = N! \prod_{j=1}^{k} (p_j^{n_j}/n_j!) \tag{8.6}$$

with $0 \le n_j$; $\sum_{j=1}^{k} n_j = N$. Here p_j is the probability of occurrence of E_j. In such a case, if it is assumed that $p_1, p_2, \ldots, p_{k-1}$ tend to zero so that $p_k \to 1$ and N tends to infinity in such a way that $N_{p_j} = \theta_j$ $(j = 1, 2, \ldots, k-1)$, then the joint density of the random variables $n_1, n_2, \ldots, n_{k-1}$ under $\lim_{N\to\infty}$ is defined by

$$\lim_{\substack{N\to\infty \\ P_j\to 0}} P(n_1, n_2, \ldots, n_k) = \prod_{j=1}^{k-1} \left[e^{-\theta_j}\theta_j^{n_j}/n_j!\right].$$

This distribution is called *multivariate Poisson distribution.* It is simply the joint distribution of $(k-1)$ independent Poisson variates.

Holgate (1964) derived the bivariate Poisson distribution. For this, let us consider u, v_1 and v_2 be the three independent Poisson variates with parameters λ, θ_1 and θ_2 respectively. Also define

$$X_1 = u + v_1 \text{ and } X_2 = u + v_2.$$

Here X_1 and X_2 are Poisson variates with parameters $\lambda + \theta_1$ and $\lambda + \theta_2$ respectively. Now the joint distribution of X_1 and X_2 is defined by

$$P(X_1 = x_1,\ X_2 = x_2) = e^{-(\lambda+\theta_1+\theta_2)} \sum_{j=0}^{\min(x_1,x_2)} \frac{\lambda^j}{j!} \frac{\theta_1^{x_1-j}}{(x_1-j)!} \frac{\theta_2^{x_2-j}}{(x_2-j)!}.$$

The moment generating function of this distribution is

$$\begin{aligned} E\left[e^{t_1x_1+t_2x_2}\right] &= E\left[e^{(t_1+t_2)u+t_1v_1+t_2v_2}\right] \\ &= \exp\left[-\lambda(1-e^{t_1+t_2}) - \theta_1(1-e^{t_1}) - \theta_2(1-e^{t_2})\right]. \end{aligned}$$

The cumulant generating function is

$$\lambda(e^{t_1+t_2}-1) + \theta_1(e^{t_1}-1) + \theta_2(e^{t_2}-1).$$

The probabilities of $P(x_1, x_2)$ can be computed using the following recurrence relations:

$$\begin{aligned} x_1P(x_1,x_2) &= \theta_1P(x_1-1,x_2) + \lambda P(x_1-1,x_2-1) \\ x_2P(x_1,x_2) &= \theta_2P(x_1,x_2-1) + \lambda P(x_1-1,x_2-1). \end{aligned}$$

8.9 Examples on Poisson Distribution

►**EXAMPLE 8.2** *Two blades out of* 1000 *blades produced in an industry are usually found defective. In an hour the industry produces* 5000 *blades per plant. Find the probability that, in a randomly selected plant,* (i) 4 *defective blades are produced,* (ii) *at least* 2 *defective blades are produced,* (iii) *at best* 2 *defective blades are produced, in a randomly selected hour.* (iv) *If the production is observed in* 5 *randomly selected plants, what is the probability that in no plant* 4 *defective blades are produced?*

Solution. Let X be the number of defective blades produced in an hour in a plant. The average defective blades is $\lambda = np = 5000 \times 0.002$ (p = probability of defective blades = 0.002) = 10. Therefore, the probability distribution of X is

$$P(X = x) = \frac{e^{-\lambda}\lambda^x}{x!} = \frac{e^{-10}10^x}{x!}, \quad x = 0, 1, 2, \ldots .$$

(i) We need $P(X = 4)$

$$P(X = 4) = \frac{e^{-10}10^4}{4!} = 0.01892.$$

(ii) We need $P(X \geq 2)$

$$\begin{aligned} P(X \geq 2) &= 1 - P(X < 2) \\ &= 1 - [P(X = 0) + P(X = 1)] \\ &= 1 - e^{-10}\left[\frac{10^0}{0!} + \frac{10^1}{1!}\right] = 0.9995. \end{aligned}$$

(iii) We need $P(X \leq 2)$

$$\begin{aligned} P(X \leq 2) &= P(X = 0) + P(X = 1) + P(X = 2) \\ &= e^{-10}\left[\frac{10^0}{0!} + \frac{10^1}{1!} + \frac{10^2}{2!}\right] = 0.00277. \end{aligned}$$

(iv) Let Y be the number of plants in which 4 defective blades are produced in an hour. Given $n = 5$, $p = P(X = 4) = 0.01892$, $q = 1 - p = 0.98108$. Therefore

$$P(Y = y) = {}^5C_y(0.01892)^y(0.98108)^{5-y}, \; y = 0, 1, 2, 3, 4, 5.$$

We need $P(Y = 0)$

$$P(Y = 0) = {}^5C_0(0.01892)^0(0.98108)^{5-0} = (0.98108)^5.$$

►**EXAMPLE 8.3** *The average number of road accidents in a highway is* 4 *per day. Find the probability that, in a randomly selected day,* (i) *there are* 4 *accidents,* (ii) *there are not more than* 4 *accidents,* (iii) *there are at least* 4 *accidents.* (iv) *If the number of accidents are observed for a week, what is the probability that in* 2 *days there are at least* 4 *accidents?*

Solution. Let X be the number of road accidents per day in a highway. Average number of accidents per day is $\lambda = 4$. Therefore, the probability distribution of X is

$$P(X = x) = \frac{e^{-\lambda}\lambda^x}{x!} = \frac{e^{-4}4^x}{x!}, \quad x = 0, 1, 2, \ldots .$$

(i) We need $P(X = 4)$

$$P(X = 4) = \frac{e^{-4}4^4}{4!} = 0.1954.$$

(ii) We need $P(X \leq 4)$

$$\begin{aligned} P(X \leq 4) &= P(X = 0) + P(X = 1) + P(X = 2) + P(X = 3) + P(X = 4) \\ &= e^{-4}\left[\frac{4^0}{0!} + \frac{4^1}{1!} + \frac{4^2}{2!} + \frac{4^3}{3!} + \frac{4^4}{4!}\right] = 0.6288. \end{aligned}$$

(iii) We need $P(X \geq 4)$

$$\begin{aligned} P(X \geq 4) &= 1 - P(X < 4) \\ &= 1 - [P(X = 0) + P(X = 1) + P(X = 2) + P(X = 3)] \\ &= 1 - e^{-4}\left[\frac{4^0}{0!} + \frac{4^1}{1!} + \frac{4^2}{2!} + \frac{4^3}{3!}\right] = 0.5665. \end{aligned}$$

(iv) Let Y be the number of days in which there are at least 4 accidents.

Given $n = 7$, $p = 0.5665$ and $q = 1 - p = 0.4335$. Therefore,

$$P(Y = y) = {}^7C_y(0.5665)^y(0.4335)^{7-y}, \quad y = 0, 1, \ldots, 7.$$

We need $P(Y = 2)$

$$P(Y = 2) = {}^7C_2(0.5665)^2(0.4335)^{7-2} = 0.1032.$$

►**EXAMPLE 8.4** *The average printing mistakes per page is* 1. *Find the probability that, in a randomly selected page,* (i) *there are* 5 *mistakes,* (ii) *there are at best* 2 *mistakes,* (iii) *there are at least* 3 *mistakes.* (iv) *If* 5 *pages are randomly observed, what is the probability that in* 3 *pages there are* 5 *mistakes?*

Solution. Let X be the number of mistakes per page. Given $E(X) = \lambda = 1$. Therefore,

$$P(X = x) = \frac{e^{-\lambda}\lambda^x}{x!} = \frac{e^{-1}1^x}{x!} = \frac{e^{-1}}{x!}, \quad x = 0, 1, 2, \ldots .$$

(i) We need $P(X = 5)$

$$P(X = 5) = \frac{e^{-1}}{5!} = 0.0031.$$

(ii) We need $P(X \leq 2)$

$$\begin{aligned} P(X \leq 2) &= P(X = 0) + P(X = 1) + P(X = 2) \\ &= e^{-1}\left[\frac{1}{0!} + \frac{1}{1!} + \frac{1}{2!}\right] = 0.9197. \end{aligned}$$

(iii) We need $P(X \geq 3)$

$$P(X \geq 3) = 1 - P(X < 3) = 1 - P(X \leq 2) = 1 - 0.9197 = 0.0803.$$

(iv) Let Y be the number of pages in which there are 5 mistakes.
Given $n = 5$, $p = P(X = 5) = 0.0031$, $q = 1 - p = 0.9969$. Therefore,

$$P(Y = y) = {}^nC_y p^y q^{n-y} = {}^5C_y (0.0031)^y (0.9969)^{5-y},\ y = 0, 1, \ldots, 5.$$

We need $P(Y = 3)$

$$P(Y = 3) = {}^5C_3 (0.0031)^3 (0.9969)^{5-3} = 0.000000296.$$

►**EXAMPLE 8.5** *The mean and variance of a discrete random variable is 3. An observation of this variable is selected at random. Find the probability that the selected observation is* (i) 3, (ii) *more than* 3, (iii) *less than* 4. (iv) *If* 10 *such observations are selected at random, what is the probability that the value of* 2 *observations is* 3?

Solution. Let X be the random variable. Given $E(X) = V(X) = 3$. Since X is discrete random variable and $E(X) = V(X)$, X is assumed to follow Poisson distribution having $E(X) = \lambda = 3$. Therefore,

$$P(X = x) = \frac{e^{-\lambda}\lambda^x}{x!} = \frac{e^{-3}3^x}{x!},\ x = 0, 1, 2, \ldots.$$

(i) We need $P(X = 3)$

$$P(X = 3) = \frac{e^{-3}3^3}{3!} = 0.22404.$$

(ii) We need $P(X > 3)$

$$\begin{aligned} P(X > 3) &= 1 - P(X \leq 3) \\ &= 1 - [P(X = 0) + P(X = 1) + P(X = 2) + P(X = 3)] \\ &= 1 - e^{-3}\left[\frac{3^0}{0!} + \frac{3^1}{1!} + \frac{3^2}{2!} + \frac{3^3}{3!}\right] = 0.3528. \end{aligned}$$

(iii) We need $P(X < 4)$

$$P(X < 4) = P(X = 0) + P(X = 1) + P(X = 2) + P(X = 3) = 0.6472.$$

(iv) Let Y be the number of observations the value of which is 3. Given $n = 10$, $P(X = 3) = p = 0.22404$, $q = 1 - p = 0.77596$. Therefore,

$$P(Y = y) = {}^nC_y p^y q^{n-y} = {}^{10}C_y (0.22404)^y (0.77596)^{10-y}, y = 0, 1, \ldots, 10.$$

We need $P(Y = 2)$

$$P(Y = 2) = {}^{10}C_2 (0.22404)^2 (0.77596)^{10-2} = 0.2969.$$

►**EXAMPLE 8.6** *Two percent fuses produced by a company are usually found defective. What is the probability that, in a randomly selected box of* 300 *fuses, there* *will be* (i) *no defective fuse,* (ii) *at best 2 defective fuses,* (iii) *at least 3 defective fuses?* (iv) *If 5 boxes are randomly selected, what is the probability that in 2 boxes there will be 2 defective fuses?*

Solution. Let X be the number of defective fuses. The probability of defective fuses is $p = 0.02$. Given $n = 300$. Hence $E(X) = np = 300 \times 0.02 = 6$. Therefore,

$$P(X = x) = \frac{e^{-\lambda}\lambda^x}{x!} = \frac{e^{-6}6^x}{x!},\ x = 0, 1, 2, \ldots .$$

(i) We need $P(X = 0)$

$$P(X = 0) = \frac{e^{-6}6^0}{0!} = 0.00248.$$

(ii) We need $P(X \leq 2)$

$$\begin{aligned} P(X \leq 2) &= P(X = 0) + P(X = 1) + P(X = 2) \\ &= e^{-6}\left[\frac{6^0}{0!} + \frac{6^1}{1!} + \frac{6^2}{2!}\right] = 0.06197. \end{aligned}$$

(iii) We need $P(X \geq 3)$

$$P(X \geq 3) = 1 - P(X < 3) = 1 - P(X \leq 2) = 1 - 0.06197 = 0.93803.$$

(iv) Let Y be the number of boxes in which there are 2 defective fuses. Given $n = 5$, $p = P(X = 2) = \frac{e^{-6}6^2}{2!} = 0.0446$, $q = 1 - p = 0.9554$. Therefore,

$$P(Y = y) = {}^nC_y p^y q^{n-y} = {}^5C_y(0.0446)^y(0.9554)^{5-y},\ y = 0, 1, \ldots, 5.$$

We need $P(Y = 2)$

$$P(Y = 2) = {}^5C_2(0.0446)^2(0.9554)^{5-2} = 0.0173.$$

►**EXAMPLE 8.7** *The probability of no air accident in a randomly selected day is* e^{-3}. *Find the probability that in a day there will be 2 air accidents. If the number of accidents is observed for the month of January, what is the percentage of days in which there are 2 air accidents?*

Solution. Let X be the number of air accidents in a day. Given

$$\begin{aligned} P(X = 0) &= e^{-3} = \frac{e^{-3}3^0}{0!} \Rightarrow \lambda = 3. \\ \therefore\ P(X = x) &= \frac{e^{-3}3^x}{x!},\ x = 0, 1, 2, \ldots . \end{aligned}$$

We need $P(X = 2)$

$$P(X = 2) = \frac{e^{-3}3^2}{2!} = 0.22404.$$

The number of accidents in the month of January is

$$NP(X = 2) = 31 \times 0.22404 = 6.9 \approx 7.$$

Therefore, percentage of days in which there are 2 air accidents is

$$\frac{7}{31} \times 100 = 22.58\%.$$

►**EXAMPLE 8.8** *A distributor of electric bulbs knows that* 5% *bulbs are usually found defective. The distributor sells the bulbs in packets of* 200 *bulbs and guarantees* 95% *non-defective bulbs. Find the probability that a particular packet will violate the guarantee.*

Solution. Let X be the number of defective bulbs and p be the probability of defective bulbs. Given $p = 0.05$, $n = 200$. Hence $E(X) = \lambda = np = 200 \times 0.05 = 10$. Therefore,

$$P(X = x) = \frac{e^{-\lambda}\lambda^x}{x!} = \frac{e^{-10}10^x}{x!}, \; x = 0, 1, 2, \ldots .$$

The guarantee will be violated if a packet contains more than $5\% \times 200 = 10$ defective bulbs. We need $P(X > 10)$.

$$\begin{aligned} P(X > 10) &= 1 - P(X \leq 10) = 1 - \sum_{i=0}^{10} P(X = i) \\ &= 1 - e^{-10}\sum_{x=0}^{10}\frac{10^x}{x!} = 0.4170. \end{aligned}$$

►**EXAMPLE 8.9** *On an average* 4 *persons die every year due to snakebite. Find the probability that, in a randomly selected year,* (i) 2 *persons die,* (ii) *number of deaths does not exceed* 4, (iii) *number of deaths will not be less than* 2.

Solution. Let X be the number of deaths due to snakebite. Given $E(X) = \lambda = 4$. Therefore,

$$P(X = x) = \frac{e^{-4}4^x}{x!}, \; x = 0, 1, 2, \ldots .$$

(i) We need $P(X = 2)$

$$P(X = 2) = \frac{e^{-4}4^2}{2!} = 0.1465.$$

(ii) We need $P(X \leq 4)$

$$\begin{aligned} P(X \leq 4) &= P(X = 0) + P(X = 1) + P(X = 2) + P(X = 3) + P(X = 4) \\ &= e^{-4}\left[\frac{4^0}{0!} + \frac{4^1}{1!} + \frac{4^2}{2!} + \frac{4^3}{3!} + \frac{4^4}{4!}\right] = 0.6288. \end{aligned}$$

(iii) We need $P(X \geq 2)$

$$\begin{aligned} P(X \geq 2) &= 1 - P(X < 2) = 1 - [P(X = 0) + P(X = 1)] \\ &= 1 - e^{-4}\left[\frac{4^0}{0!} + \frac{4^0}{1!}\right] = 0.9084. \end{aligned}$$

EXERCISE

1. (a) Write down some examples of Poisson variable. Derive the Poisson distribution as a limiting case of binomial distribution. For Poisson distribution, prove that

$$\mu_{r+1} = r\lambda\mu_{r-1} + \lambda\frac{\delta\mu_r}{\partial\lambda}, \text{ where}$$
$$\mu_i = i\text{-th central moment of Poisson distribution having mean } \lambda.$$

(b) Let x be a discrete random variable having $E(x) = V(x) = 3$. Six observations of X are randomly selected, find the probability that the value of X is at best 2.

2. (a) Under what conditions a binomial distribution transforms to Poisson distribution? Write down some examples of Poisson variable. Derive Poisson probability function without considering it as a limiting form of binomial distribution.

(b) Derive the moment generating function of Poisson distribution and hence comment on shape characteristics of Poisson distribution.

(c) The average number of defective bulbs produced per hour in an industry is 2. Find the probability that in a randomly selected hour the number of defective bulbs are (i) 2, (ii) at least 2, (iii) at best 2.

If the production of bulbs are observed for 5 randomly selected hours, what is the probability that in 2 hours the number of defective bulbs are at best 2?

3. (a) What do you mean by truncation? Find the probability function of truncated Poisson distribution, where truncation is done at point zero. Find the factorial moment of this truncated distribution.

(b) A company supplies blades in packets, The average defective packets in a lot are usually 4. A lot is selected at random. Find the probability that (i) there are 4 packets in the lot which are defective, (iii) there are at least 3 defective packets, (iii) there is no defective packet.

(c) The percentage of printing mistakes of a page is 0.1. If the page contains 400 words, what is the probability that in a randomly selected page there will be at best one printing mistake.

4. (a) What do you mean by mixture distribution? Derive the probability function of Poisson-binomial mixture distribution. Also, find the moments of this compound distribution.

(b) State and prove properties of Poisson distribution.

(c) If two per cent fuses produced by a company are usually found defective, what is the probability that in a box of 200 fuses there will be (i) no defective fuse, (ii) 2 defective fuses, (iii) at least one defective fuse?

5. (a) If X_1 and X_2 are two independent Poisson variates with parameters λ_1 and λ_2, respectively, find the cnditional probability distribution of X_1 under the condition that $X_1 + X_2 = n$. What is the distribution of standardized Poisson variate?

(b) The probability of no accident in a highway in a day is e^{-2}. Find the probability that in a randomly selected day there will be (i) two accidents, (ii) at least 2 accidents, (iii) at best 2 accidents.

If accidents are observed for a week, what is the probability that in 3 days there will be at best 2 accidents?

6. (a) Under what conditions a binomial distribution transforms to Poisson distribution? Write down some examples of Poisson variable.

(b) If X is a Poisson variate with parameter λ and γ is another discrete random variable whose conditional probability function for a given X is given by

$$p(Y = y/X = x) = \binom{x}{r} p^r (1-p)^{x-r}.\ r = 0, 1, \ldots, x$$

then find the unconditional distribution of Y.

(c) A distributor of seeds knows that 3% seeds do not germinate. The distributor sells seeds in packets of 150 seeds and guarantees 90% germination. Find the probability that a particular packet will violate the guaranteee.

7. (a) What do you mean by compound distribution? Derive the probability function of Poisson-gamma compound distribution.

(b) Prove that for a Poisson distribution with mean 1 the mean deviation about mean is $2/e$.

(c) Show that for a Poisson distribution with mean λ

$$\mu_{r+1} = \lambda \left[\binom{r}{1} \mu_{r-1} + \binom{r}{2} \mu_{r-2} + \cdots + \binom{r}{r} \mu_0 \right]$$

where μ_i = i-th central moment of Poisson distribution.

8. (a) Show that, if $X_1, X_2, \ldots, X_k$ are k independent Poisson variates with parameters $\lambda_1, \lambda_2, \ldots, \lambda_k$, respectively, then the distribution of $\sum_{i=1}^{k} X_i$ is also a Poisson variate. Is the distribution of $X_1 - X_2$ Poisson?

(b) Find the mode of Poisson distribution.

(c) Show that $K_{r+1} = \lambda \left[\dfrac{\partial k_r}{\partial \lambda}\right]$, where λ is the mean of Poisson distribution and K_i is the i-th cumulant of the distribution.

CHAPTER 9

Negative Binomial Distribution

9.1 Definition and Applications

The negative binomial distribution is defined as the distribution of a random variable X for which

$$P(x) = \binom{N+x-1}{N-1}\left(\frac{P}{Q}\right)^x\left(1-\frac{P}{Q}\right)^N;\ x = 0, 1, 2, \ldots.$$

Here N and P are the parameters of the distribution. This probability of X is defined in terms of the expansion of the negative binomial expression $(Q-P)^{-N}$, where $Q-P=1$, $P>0$. The $(x+1)$-th term in the expansion of $(Q-P)^{-N}$ is given by

$$Q^{-N}\binom{N+x-1}{N-1}\left(\frac{P}{Q}\right)^x = \binom{N+x-1}{N-1}\left(\frac{P}{Q}\right)^x\left(1-\frac{P}{Q}\right)^N. \tag{9.1}$$

An alternative form of the probability function of X can be derived from the consideration of binomial probability situation. Suppose we have a succession of n Bernoulli trials. Let us assume that (i) the trials are independent, (ii) the probability of success 'p' in a trial remains constant from trial to trial.

Let $P(x)$ be the probability that there are x failures preceding the r-th success in $(x+r)$ trials.

Now, the last trial must be a success, whose probability is p. In the remaining $(x+r-1)$ trials there must be $(r-1)$ successes whose probability is given by

$$\binom{x+r-1}{r-1}p^{r-1}q^x.$$

Now, by the compound probability law $P(x)$ is given by

$$\begin{aligned} P(x) &= \binom{x+r-1}{r-1}p^{r-1}q^x \cdot p \\ &= \binom{x+r-1}{r-1}p^r q^x, x = 0, 1, \ldots. \end{aligned}$$

However, if we put $r = N$, $P = 1/Q$ and $q = P/Q$ so that $Q - P = 1$, then

$$\begin{aligned} P(x) &= \binom{N+x-1}{N-1} Q^{-N} \left(\frac{P}{Q}\right)^x \\ &= \binom{N+x-1}{N-1} \left(\frac{P}{Q}\right)^x \left(1 - \frac{P}{Q}\right)^N \end{aligned}$$

which is the same form as it is in (9.1).

Again, we know

$$\begin{aligned} \binom{x+r-1}{r-1} &= \binom{x+r-1}{x} \left[\because \binom{n}{r} = \binom{n}{n-r}\right] \\ &= \frac{(x+r-1)(x+r-2)\cdots(r+1)r}{x!} \\ &= \frac{(-1)^x(-r)(-r-1)\cdots(-r-x+2)(-r-x+1)}{x!} \\ &= (-1)^x \binom{-r}{x} \end{aligned}$$

$$\therefore\ P(x) = \binom{-r}{x} p^r (-q)^x;\ x = 0, 1, 2, \ldots \tag{9.2}$$

which is the $(x+1)$-th term in the expansion of $p^r(1-q)^{-r}$. This expansion is a binomial one with negative index. Hence the distribution is called *negative binomial distribution.* When $p = 1/Q$ and $q = P/Q$, we have

$$P(x) = \binom{-r}{x} Q^{-r} \left(-\frac{P}{Q}\right)^x;\ x = 0, 1, 2, \ldots.$$

From (9.2), we have

$$\begin{aligned} \sum_{x=0}^{\infty} P(x) &= p^r \sum_{x=0}^{\infty} \binom{-r}{x} (-q)^x \\ &= p^r (1-q)^{-r} = 1. \end{aligned}$$

Hence (9.2) is a probability function.

The negative binomial distribution provides an excellent model for a frequency distribution where the variance of the distribution is larger than the mean. It is very often a first choice as alternative when it is felt that a Poisson distribution might be inadequate. The commonest examples of such behaviour are the frequency distributions of plant density obtained by quadrant sampling where plants are clustered. Bacterial clustering, e.g., deaths of insects, number of insect bites leads to the negative binomial distribution. This distribution also arises in inverse sampling from a binomial distribution or as a weighted average of Poisson distribution. Greenwood and Yule (1920) obtained the distribution as a consequence of certain simple assumptions in accident proneness models. The probability function of such model can be derived from a mixture of Poisson distribution.

Let X be a random variable with probability function

$$P(x) = \frac{e^{-\theta}\theta^x}{x!}, \quad x = 0, 1, 2, \ldots$$

and θ be the random variable with probability function

$$f(\theta) = \frac{1}{\Gamma\alpha\beta^\alpha} e^{-\theta/\beta}\theta^{\alpha-1}; \; \theta > 0; \; \alpha > 0; \; \beta > 0.$$

Then the mixture of Poisson distribution is given by

$$\begin{aligned} P(x) &= \frac{1}{\Gamma\alpha\beta^\alpha}\int_0^\infty \theta^{\alpha-1}e^{-\theta/\beta}\frac{\theta^x e^{-\theta}}{x!}d\theta \\ &= \frac{1}{\Gamma\alpha\beta^\alpha}\int_0^\infty \theta^{\alpha+x-1}e^{-\theta\left(1+\frac{1}{\beta}\right)}d\theta \\ &= \binom{\alpha+x-1}{\alpha-1}\left(\frac{\beta}{1+\beta}\right)^x\left(\frac{1}{1+\beta}\right)^\alpha \end{aligned}$$

which is the probability function of negative binomial distribution with parameters (α, β). Here θ represents the expected number of accidents for an individual.

The negative binomial distribution can be used generally when mixtures of random sources are to be represented. It also arises as the distribution of the sum of n independent variables each having the same logarithmic series distribution, when n has a Poisson distribution.

9.2 First Four Moments of the Negative Binomial Distribution

The moment generating function of the negative binomial distribution is given by

$$\begin{aligned} M_x(t) = E\left[e^{tx}\right] &= \sum_{x=0}^{\infty} e^{tx}\binom{-r}{x}Q^{-r}\left(-\frac{P}{Q}\right)^x \\ &= \sum_{x=0}^{\infty}\binom{-r}{x}Q^{-r}\left(-\frac{Pe^t}{Q}\right)^x = (Q - Pe^t)^{-r}. \end{aligned}$$

Now

$$\begin{aligned} \mu_1' &= \left.\frac{\partial M_x(t)}{\partial t}\right|_{t=0} \\ &= \left[-r\left(Q - Pe^t\right)^{-r-1}\left(-Pe^t\right)\right]_{t=0} = rP. \end{aligned}$$

$\therefore$ The mean of the negative binomial distribution is rP and it is NP when $r = N$.

$$\begin{aligned} \mu_2' &= \left.\frac{\partial^2 M_x(t)}{\partial t^2}\right|_{t=0} \\ &= \left[rPe^t(Q - Pe^t)^{-r-1} + (-r-1)rPe^t(Q - Pe^t)^{-r-2}(-Pe^t)\right]_{t=0} \\ &= rP + r(r+1)P^2. \end{aligned}$$

$\therefore$ The variance of the negative binomial distribution is

$$\begin{aligned}\mu_2 &= \mu_2' - \mu_1'^2 \\ &= rP + r(r+1)P^2 - (rP)^2 \\ &= rP(P+1) = rPQ = NPQ, \text{ when } r = N.\end{aligned}$$

We have $Q - P = 1 \Rightarrow Q > 1$. Thus $rP < rPQ$ which implies that mean < variance.

$$\begin{aligned}\mu_3' &= \left.\frac{\partial^3 M_x(t)}{\partial t^3}\right|_{t=0} \\ &= \Big[rPe^t(Q-Pe^t)^{-r-1} + (-r-1)rPe^t(Q-Pe^t)^{-r-2}(-Pe^t) \\ &\quad +2r(r+1)P^2e^{2t}(Q-Pe^t)^{-r-2} \\ &\quad +(-r-2)r(r+1)P^2e^{2t}(Q-Pe^t)^{-r-3}(-Pe^t)\Big]_{t=0} \\ &= rP + r(r+1)P^2 + 2r(r+1)P^2 + r(r+1)(r+2)P^3 \\ &= rP + 3r(r+1)P^2 + r(r+1)(r+2)P^3 \\ \mu_3 &= \mu_3' - 3\mu_2'\mu_1' + 2\mu_1'^3 \\ &= rP + r(r+1)P^2 + 2r(r+1)P^2 + r(r+1)(r+2)P^3 \\ &\quad -3\left[rP + r(r+1)P^2\right]rP + 2(rP)^3 \\ &= rP\left[1 + 3P + 2P^2\right] \\ &= rP(1+P)(1+2P) \\ &= rPQ(Q+P) = NPQ(Q+P), \text{ when } r = N \\ \mu_4' &= \left.\frac{\partial^4 M_x(t)}{\partial t^4}\right|_{t=0} \\ &= \Big[rPe^t(Q-Pe^t)^{-r-1} + (-r-1)rPe^t(Q-Pe^t)^{-r-2}(-Pe^t) \\ &\quad +6r(r+1)P^2e^{2t}(Q-Pe^t)^{r-2} \\ &\quad +3r(r+1)(-r-2)P^2e^{2t}(Q-Pe^t)^{-r-3}(-Pe^t) \\ &\quad +3r(r+1)(r+2)P^3e^{3t}(Q-Pe^t)^{-r-3} \\ &\quad +r(r+1)(r+2)(-r-3)P^3e^{3t}(Q-Pe^t)^{-r-4}(-Pe^t)\Big]_{t=0} \\ &= rP + r(r+1)P^2 + 6r(r+1)P^2 + 6r(r+1)(r+2)P^3 \\ &\quad +r(r+1)(r+2)(r+3)P^4 \\ &= rP\left[1 + 7(r+1)P + 6(r+1)(r+2)P^2 + (r+1)(r+2)(r+3)P^3\right] \\ \mu_4 &= \mu_4' - 4\mu_3'\mu_1' + 6\mu_2'\mu_1'^2 - 3\mu_1'^4 \\ &= rP\{1 + 7(r+1)P + 6(r+1)(r+2)P^2 + (r+1)(r+2)(r+3)P^3\} \\ &\quad -4r^2P^2\{1 + 3(r+1)P + (r+1)(r+2)P^2\} \\ &\quad +6r^3P^3\{1 + (r+1)P\} - 3r^4P^4 \\ &= rPQ[1 + 3PQ(r+2)] \\ &= NPQ[1 + 3PQ(N+2)], \text{ when } r = N.\end{aligned}$$

The moment ratios are

$$\alpha_3 = \sqrt{\beta_1} = \sqrt{\frac{\mu_3^2}{\mu_2^3}} = \sqrt{\frac{\{NPQ(Q+P)\}^2}{(NPQ)^3}} = \frac{Q+P}{\sqrt{NPQ}}.$$

But $p = 1/Q$ and $q = P/Q$, i.e., $Q = 1/p$ and $P = q_\theta$.

$$\begin{aligned}
\therefore\ \alpha_3 &= \frac{1+q}{\sqrt{Nq}} = \frac{1+q}{\sqrt{rq}},\ \text{when } N = r\\
\alpha_4 &= \beta_4 = \frac{\mu_4}{\mu_2^2} = \frac{rPQ[1+3PQ(r+2)]}{(rPQ)^2} = \frac{1+3PQ(r+2)}{rPQ}\\
&= 3 + \frac{1+6PQ}{rPQ} = \frac{p^2+3q(r+2)}{rq},\ \text{when } P = \frac{q}{p}, Q = \frac{1}{p}.\\
\therefore\ \gamma_2 &= \beta_2 - 3 = \frac{p^2+6q}{rq}
\end{aligned}$$

Also we have

$$\gamma_2 = \beta_2 - 3 = \frac{1+6PQ}{rPQ} = \frac{1+6PQ}{NPQ},\ \text{when } r = N.$$

Thus β_2 is greater than 3 and the distribution is leptokurtic.

9.3 Generating Functions of the Negative Binomial Distribution

In the previous section we have discussed the m.g.f. of negative binomial distribution. The cumulant generating function is given by

$$\begin{aligned}
K_x(t) &= \log M_x(t)\\
&= -r\log(Q - Pe^t)\\
&= -r\log\left[Q - P\left\{1 + t + \frac{t^2}{2!} + \frac{t^3}{3!} + \frac{t^4}{4!} + \cdots\right\}\right]\\
&= -r\log\left[1 - P\left\{1 + t + \frac{t^2}{2!} + \cdots\right\}\right]\\
&= r\left\{P\left(1 + t + \frac{t^2}{2!} + \frac{t^3}{3!} + \frac{t^4}{4!} + \cdots\right) + \frac{P^2}{2}\left(1 + t + \frac{t^2}{2!} + \cdots\right)^2\right.\\
&\quad \left. + \frac{P^3}{3}(1 + t + \cdots)^3 + \frac{P^4}{4}(1 + t + \cdots)^4 + \cdots\right\}.
\end{aligned}$$

Thus

$$\begin{aligned}
K_1 &= \mu_1' = rP\\
K_2 &= \mu_2 = rPQ\\
K_3 &= \mu_3 = r\{P + 3P^2 + 2P^3\} = rPQ(Q+P)\\
K_4 &= r\{P + 7P^2 + 12P^3 + 6P^4\} = rPQ(1+6PQ)
\end{aligned}$$

The probability generating function of negative binomial distribution is given by

$$\begin{aligned} g(t) = E(t^x) &= \sum_{x=0}^{\infty} \binom{-r}{x} p^r (-q)^x t^x \\ &= p^r \sum_{x=0}^{\infty} \binom{-r}{x} (-qt)^x = p^r (1-qt)^{-r} \\ &= \left(\frac{p}{1-qt}\right)^r = (Q-Pt)^{-r} \quad \left[\because\ p = \frac{1}{Q}, q = \frac{P}{Q}\right] \\ &= (Q-Pt)^{-N}, \text{ when } r = N. \end{aligned}$$

The factorial moment generating function of negative binomial distribution is given by

$$\begin{aligned} M_{(x)}(t) = E(1+t)^x &= \sum_{x=0}^{\infty} \binom{-r}{x} p^r (-q)^x (1+t)^x \\ &= p^r \sum_{x=0}^{\infty} \binom{-r}{x} \{-q(1+t)\}^x = p^r \{1 - q(1+t)\}^{-r} \\ &= \left[\frac{p}{1-q(1+t)}\right]^r = \{Q - P(1+t)\}^{-r} \\ &= \{Q - P(1+t)\}^{-N}, \text{ when } r = N. \end{aligned}$$

The j-th factorial moment is the coefficient of $t^j/j!$ in the expansion of $M_{(x)}(t)$. Thus

$$\mu_{(j)} = (N+j-1)^{(j)} P^j.$$

In particular,

$$\begin{aligned} \mu_1' &= \mu_{(1)} = NP \\ \mu_2' &= \mu_{(2)} + \mu_{(1)} = N(N+1)P^2 + NP. \end{aligned}$$

The j-th factorial moment can also be defined by

$$\begin{aligned} \mu_{(j)} &= E\left[x^{(j)}\right] = \sum_{x=j}^{\infty} x^{(j)} \binom{N+x-1}{N-1} \left(\frac{P}{Q}\right)^x \left(1 - \frac{P}{Q}\right)^N \\ &= (N+j-1)^{(j)} \left(\frac{P}{Q}\right)^j Q^{-N} \sum_{x=j}^{\infty} \binom{N+j+x-j-1}{N+j-1} \left(\frac{P}{Q}\right)^{x-j} \\ &= (N+j-1)^{(j)} P^j. \end{aligned}$$

The characteristic function of the negative binomial distribution is given by

$$\begin{aligned} \psi(t) &= E\left[e^{itx}\right] = \sum_{x=0}^{\infty} \binom{-r}{x} p^r (-q)^x e^{itx} = p^r \sum_{x=0}^{\infty} \binom{-r}{x} (-qe^{it})^x \\ &= p^r \left\{1 - qe^{it}\right\}^{-r} = \left[\frac{p}{1-qe^{it}}\right]^r = \left(Q - Pe^{it}\right)^{-r}. \end{aligned}$$

The recurrence relation for central moments of the negative binomial distribution is given by

$$\mu_{r+1} = q\left[\frac{\partial \mu_r}{\partial q} + \frac{rk}{p^2}\mu_{r-1}\right]$$

where μ_{r+1} is the $(r+1)$-th central moment of the negative binomial distribution with probability function

$$P(x) = \binom{k+x-1}{x} q^x p^k, \; x = 0, 1, 2, \ldots .$$

This is probability function of a negative binomial distribution with parameters k and p. Hence the mean of this distribution is kq/p. Thus

$$\mu_r = \sum_{x=0}^{\infty}\left(x - k\frac{q}{p}\right)^r \binom{k+x-1}{x} q^x p^k.$$

Differentiating partially w.r.t. q, we have

$$\frac{\partial \mu_r}{\partial q} = \sum_x\left[r\left(x - \frac{kq}{p}\right)^{r-1} \times \frac{\partial}{\partial q}\left(x - \frac{kq}{p}\right)\binom{k+x-1}{x} q^x p^k\right]$$
$$+\sum_x\left[\left(x - \frac{kq}{p}\right)^r \binom{k+x-1}{x}\left\{xq^{x-1}p^k + q^x k p^{k-1}\frac{\partial p}{\partial q}\right\}\right].$$

But

$$\frac{\partial p}{\partial q} = \frac{\partial}{\partial q}(1-q) = -1$$

$$\text{and } \frac{\partial}{\partial q}\left(x - \frac{kq}{p}\right) = \frac{\partial}{\partial q}\left[x - k\left(\frac{1}{p} - 1\right)\right] = \frac{k}{p^2}\frac{\partial p}{\partial q} = -\frac{k}{p^2}.$$

$$\therefore \; \frac{\partial \mu_r}{\partial q} = -\frac{rk}{p^2}\sum_x\left(x - \frac{kq}{p}\right)^{r-1}\binom{k+x-1}{x} p^k q^x$$
$$+\sum_x\left(x - \frac{kq}{p}\right)^r \binom{k+x-1}{x} q^{x-1}p^k\left(x - \frac{kq}{p}\right)$$
$$= -\frac{rk}{p^2}\mu_{r-1} + \frac{1}{q}\sum_x\left(x - \frac{kq}{p}\right)^{r+1}\binom{k+x-1}{x} p^k q^x$$
$$= -\frac{rk}{p^2}\mu_{r-1} + \frac{1}{q}\mu_{r+1}.$$

$$\therefore \; \mu_{r+1} = q\left[\frac{\partial \mu_r}{\partial q} + \frac{rk}{p^2}\mu_{r-1}\right]$$

The recurrence relation among the cumulants of the negative binomial distribution is given by

$$K_{r+1} = PQ\frac{\partial K_r}{\partial Q}, \; r \geq 1.$$

We have $K_x(t) = \log M_x(t) = -r\log(Q - Pe^t)$.

Also

$$K_r = \frac{\partial^r}{\partial t^r}\left[-r\log(Q-Pe^t)\right]_{t=0}.$$

$$\therefore \frac{\partial K_r}{\partial Q} = -r\left[\frac{\partial^r}{\partial t^r}\cdot\frac{1-e^t}{Q-Pe^t}\right]_{t=0}.$$

Again

$$\begin{aligned}
K_{r+1} &= \left[\frac{\partial^{r+1}}{\partial t^{r+1}}\left(-r\log(Q-Pe^t)\right)\right]_{t=0}\\
&= -r\left[\frac{\partial^r}{\partial t^r}\frac{-Pe^t}{Q-Pe^t}\right]_{t=0} = rP\left[\frac{\partial^r}{\partial t^r}\frac{e^t}{Q-Pe^t}\right]_{t=0}\\
&= -r\left[\frac{\partial^r}{\partial t^r}\left(1-\frac{Q}{Q-Pe^t}\right)\right]_{t=0}\\
&= rQ\left[\frac{\partial^r}{\partial t^r}\frac{1}{Q-Pe^t}\right]_{t=0}
\end{aligned}$$

Now

$$\begin{aligned}
K_{r+1} - PQ\frac{\partial K_r}{\partial Q} &= rQ\left[\frac{\partial^r}{\partial t^r}\left\{\frac{1}{Q-Pe^t}+\frac{P(1-e^t)}{Q-Pe^t}\right\}\right]_{t=0}\\
&= rQ\left.\frac{\partial^r}{\partial t^r}(1)\right|_{t=0}\\
&= 0.\\
\therefore K_{r+1} &= PQ\frac{\partial K_r}{\partial Q}.
\end{aligned}$$

9.4 Fitting of Negative Binomial Distribution

We have

$$P(x) = \binom{x+r-1}{r-1}p^r q^x;\; x = 0,1,2,\ldots$$

$$\text{and } P(x+1) = \binom{x+r}{r-1}p^r q^x.$$

$$\begin{aligned}
\therefore \frac{P(x+1)}{P(x)} &= \frac{(x+r)!(r-1)!x!}{(r-1)!(x+1)!(x+r-1)!}q\\
&= \frac{x+r}{x+1}q
\end{aligned}$$

Thus

$$P(x+1) = \frac{x+r}{x+1}qP(x)$$

which is the recurrence relation for negative binomial probability, where q and r are to be found out from the mean and variance of the frequency distribution to which negative binomial distribution is to be fitted. Also $P(X=0)$ can be calculated as

$$P(X=0) = p^r.$$

►**EXAMPLE 9.1** *Fit a negative binomial distribution to the following frequency distribution:*

Values of x	0	1	2	3	4	5	6	*Total*
Frequency of x, f	150	88	22	10	4	2	1	277

Solution. We have

$$\mu_1' = \frac{\Sigma fx}{N} = \frac{194}{277} = 0.70036$$

$$V(x) = \frac{1}{N}\left[\Sigma fx^2 - \frac{(\Sigma fx)^2}{N}\right] = \frac{1}{277}\left[416 - \frac{(194)^2}{277}\right] = 1.0113.$$

Here, mean < variance. So negative binomial distribution can be fitted. We have

$$\mu_1' = r \cdot \frac{q}{p} \text{ and } \mu_2 = \frac{rq}{p^2}.$$

Thus

$$r \cdot \frac{q}{p} = 0.70036 \text{ and } \frac{rq}{p^2} = 1.0113.$$

From these two values we have

$$p = \frac{0.70036}{1.0113} = 0.69253 \text{ and hence}$$

$$q = 1 - p = 0.30747.$$

$$\therefore\ r = \frac{\mu_1' \times p}{q} = \frac{0.70036 \times 0.69253}{0.30747} = 1.57746$$

$$\therefore\ P(X = 0) = p^r = (0.69253)^{1.57746} = 0.56014$$

Calculation for finding probabilities and expected frequencies

x	$\frac{x+r}{x+1}q$	$P(x)$ $\left[P(x+1) = \frac{x+r}{x+1}qP(x)\right]$	Expected frequency $= NP(x)$
0	0.48502	0.56014	155
1	0.39625	0.27168	75
2	0.36665	0.10765	30
3	0.35186	0.03947	11
4	0.34298	0.01389	4
5	0.33706	0.00476	1
6	0.33283	0.00160	1

From the fitted distribution, we have

$$\chi^2 = \sum_{i=1}^{7} \frac{O_i^2}{E_i} - N = 5.64.$$

As $\chi^2 < \chi^2_{0.05,6} = 12.59$, it can be concluded that the observed frequencies are consistent with the theoretical frequencies of negative binomial distribution.

9.5 Properties of Negative Binomial Distribution

Mode of the Distribution

We have

$$\begin{aligned}\frac{P(x+1)}{P(x)} &= \frac{x+r}{x+1}q \\ &= \frac{N+x}{x+1}\frac{P}{Q}\left[\because r=N, q=\frac{P}{Q}\right]\end{aligned}$$

From this it can be seen that

$$P(x+1) \gtrless P(x) \text{ according as } x \lesseqgtr NP-Q.$$

Therefore, if $NP > Q$, there is a mode at the least integer not less than $NP-Q$ and if $NP < Q$, the mode is at $x=0$. If $NP=Q$, then there are two equal modes.

Poisson Distribution as a Limiting Case of the Negative Binomial Distribution

As r tends to infinity and P tends to zero with rP remaining fixed, say θ (finite), the negative binomial distribution tends to Poisson distribution. Thus

$$\begin{aligned}\lim_{r\to\infty} P(x) &= \lim_{r\to\infty}\binom{x+r-1}{r-1}p^r q^x \\ &= \lim_{r\to\infty}\binom{x+r-1}{r-1}Q^{-r}\left(\frac{P}{Q}\right)^x \\ &= \lim_{r\to\infty}\frac{(x+r-1)(x+r-2)\cdots(r+1)r}{x!}(1+P)^{-r}\left(\frac{P}{1+P}\right)^x \\ &= \lim_{r\to\infty}\left[\frac{1}{x!}\left(1+\frac{x-1}{r}\right)\left(1+\frac{x-2}{r}\right)\cdots\right. \\ &\qquad \left.\left(1+\frac{1}{r}\right)\cdot 1\cdot r^x(1+P)^{-r}\left(\frac{P}{1+P}\right)^x\right] \\ &= \frac{1}{x!}\lim_{r\to\infty}\left[(1+P)^{-r}\left(\frac{rP}{1+P}\right)^x\right] \\ &= \frac{\theta^x}{x!}\lim_{r\to\infty}\left[\left(1+\frac{\theta}{r}\right)\right]^{-r}\lim_{r\to\infty}\left(1+\frac{\theta}{r}\right)^{-x} \\ &= \frac{\theta^x}{x!}e^{-\theta}\cdot 1 \\ &= e^{-\theta}\frac{\theta^x}{x!}\end{aligned}$$

which is the probability function of Poisson variate with parameter θ.

Here negative binomial distribution is approximated to Poisson distribution. The other approximations of this distribution can be similarly done as it is done in binomial distribution [Johnson and Kotz (1969)].

Standardized Negative Binomial Distribution

If X is a negative binomial variate with mean NP and variance NPQ, then

$$Z = \frac{x - NP}{\sqrt{NPQ}}$$

is called a ***standardized negative binomial variate.*** This Z is similar to the standardized binomial variate. Hence, it can be easily shown that Z tends to standard normal variate as N tends to infinity.

The negative binomial variate can be approximated to standard normal variate by the transformation

$$Z_1 = \sqrt{N} \sinh^{-1} \sqrt{x/N}$$

where x is the negative binomial variate and Z_1 is approximately distributed as a standard normal variate.

Mean Deviation about Mean of Negative Binomial Distribution

The mean deviation is defined by

$$\gamma_1 = \sum_{x=0}^{\infty} |x - NP| \binom{N+x-1}{N-1} \left(\frac{P}{Q}\right)^x \left(1 - \frac{P}{Q}\right)^N$$

which on simplification gives

$$\gamma_1 = 2m \binom{N+m-1}{N-1} \left(\frac{P}{Q}\right)^m Q^{-(N-1)}$$

where m is the greatest integer not greater than NP.

9.6 Truncated Negative Binomial Distribution

If X is a negative binomial variate, then

$$P(x) = \binom{N+x-1}{N-1} \left(\frac{P}{Q}\right)^x \left(1 - \frac{P}{Q}\right)^N, \quad x = 0, 1, 2, \ldots.$$

The new random variable X_1 can be defined, where $x_1 = 1, 2, \ldots,$ where X_1 is the random variable X with no record of zero. The $P(X = 0) = Q^{-N}$. Now, the probability function of X_1 is defined by

$$P(X_1 = x) = \frac{1}{(1 - Q^{-N})} \binom{N+x-1}{N-1} \left(\frac{P}{Q}\right)^x \left(1 - \frac{P}{Q}\right)^N, \quad (x = 1, 2, \ldots)$$

which is the probability function of the truncated negative binomial distribution.

The factorial moment generating function of this truncated distribution is given by

$$\mu_{(x)}(t) = \left(1 - Q^{-N}\right)^{-1} \{Q - P(1+t)\}^{-N}.$$

The j-th factorial moment is

$$\mu_{(j)} = \left(1 - Q^{-N}\right)^{-1} (N + j - 1)^{(j)} P^j.$$

In particular,

$$\mu_1' = NP\left(1 - Q^{-N}\right)^{-1}$$

$$\mu_2' = \frac{1}{(1 - Q^{-N})}\left[NPQ + (NP)^2\right]$$

$$\mu_2 = \frac{NPQ + (NP)^2}{1 - Q^{-N}} - \frac{(NP)^2}{(1 - Q^{-N})^2}$$

$$= \frac{(1 - Q^{-N})\left\{NPQ + (NP)^2\right\} - (NP)^2}{(1 - Q^{-N})^2}.$$

9.7 Mixtures of Negative Binomial Distribution

Let X be a negative binomial variate with probability function

$$P(x) = \binom{N + x - 1}{N - 1} p^N q^x, \ x = 0, 1, 2, \ldots$$

$$= \binom{N + x - 1}{N - 1}\left(\frac{P}{Q}\right)^x \left(1 - \frac{P}{Q}\right)^N.$$

Let p is also a random variable with p.d.f.

$$f(p) = \frac{1}{B(\alpha, \beta)} p^{\alpha - 1} (1 - p)^{\beta - 1}; \ 0 < p < 1.$$

Then the compound negative binomial distribution is given by

$$P(x) = \binom{N + x - 1}{N - 1} \frac{1}{B(\alpha, \beta)} \int_0^1 p^N q^x p^{\alpha - 1} (1 - p)^{\beta - 1} dp$$

$$= \binom{N + x - 1}{N - 1} \frac{1}{B(\alpha, \beta)} \int_0^1 p^{N + \alpha - 1} (1 - p)^{\beta + x - 1} dp$$

$$= \binom{N + x - 1}{N - 1} \frac{B(N + \alpha, \beta + x)}{B(\alpha, \beta)}$$

$$= \frac{N}{N + x} \frac{\binom{N + \alpha - 1}{\alpha - 1}\binom{x + \beta - 1}{\beta - 1}}{\binom{N + x + \alpha + \beta - 1}{\alpha + \beta - 1}}.$$

This compound negative binomial distribution is a generalized hypergeometric distribution.

Negative Binomial-Gamma Distribution

Let X be a random variable with probability function

$$P(x) = \binom{N + x - 1}{N - 1}\left(\frac{P}{Q}\right)^x \left(1 - \frac{P}{Q}\right)^N; \ x = 0, 1, 2, \ldots.$$

Also consider that N is a random variable with density function

$$f(N) = \frac{1}{\Gamma\alpha\beta^{\alpha}} N^{\alpha-1} e^{-N/\beta};\ N \geq 0.$$

Then the compound negative binomial distribution is defined by

$$P(x) = \frac{1}{\Gamma\alpha\beta^{\alpha}} \left(\frac{P}{Q}\right)^{x} \int_{0}^{\infty} \binom{N+x-1}{N-1}$$

$$N^{\alpha-1} e^{-N/\beta} \left(1 - \frac{P}{Q}\right)^{N} dN;\ x = 0, 1, 2, \ldots$$

The probability generating function of this distribution is given by

$$\begin{aligned} g(t) &= E\left[t^{x}\right] \\ &= \frac{1}{\Gamma\alpha\beta^{\alpha}} \int_{0}^{\infty} N^{\alpha-1} e^{-N/\beta} \sum_{x=0}^{\infty} \binom{N+x-1}{N-1} \left(\frac{Pt}{Q}\right)^{x} \left(1 - \frac{P}{Q}\right)^{N} dN \\ &= \frac{1}{\Gamma\alpha\beta^{\alpha}} \int_{0}^{\infty} N^{\alpha-1} e^{-N/\beta} (Q - Pt)^{-N} dN \\ &= \left[1 + \beta \log(Q - Pt)\right]^{-\alpha}. \end{aligned}$$

Negative Binomial-Poisson Distribution

Let X be a random variable with probability function

$$P(x) = \binom{-Kj}{x} Q^{-Kj} \left(-\frac{P}{Q}\right)^{x};\ x = 0, 1, 2, \ldots .$$

Also assume that j is a random variable with probability function

$$P(j) = \frac{e^{-\theta}\theta^{j}}{j!};\ j = 0, 1, 2, \ldots .$$

Then the compound negative binomial distribution is given by

$$\begin{aligned} P(x) &= e^{-\theta} \left(-\frac{P}{Q}\right)^{x} \sum_{j=0}^{\infty} \binom{-Kj}{x} \left(\theta Q^{-K}\right)^{j} / j! \\ &= \frac{e^{-\theta}}{x!} \left(\frac{P}{Q}\right)^{x} \sum_{j=1}^{\infty} (Kj)^{(x)} \left(\theta Q^{-K}\right)^{j} / j!;\ x = 1, 2, \ldots \\ &= \left[\left(\frac{P}{Q}\right)^{x} \Big/ x!\right] \mu_{[x]} \exp\left[-\theta\left(1 - Q^{-K}\right)\right] \end{aligned}$$

where $\mu_{[x]} = E\left[(Kj)^{[x]}\right]$ with j distributed as a Poisson variate the mean value of which is θQ^{-K}.

The probability generating function of this distribution is given by

$$g(t) = E\left[(Q - Pt)^{-jK}\right] = \exp\left[\theta\left\{(Q - Pt)^{-K} - 1\right\}\right]$$

where j has a Poisson distribution with parameter θ.

If the negative binomial distribution is defined by

$$P(X=k)=\binom{k-1}{N-1}Q^{-k}P^{k-N},\ k=N,N+1,\ldots$$

then the negative binomial-Poisson distribution is given by

$$P(X=k)=e^{-\theta}\left(\frac{P}{Q}\right)^{k}\sum_{j\leq k/K}\binom{k-1}{Kj-1}\left[\left(\theta P^{-K}\right)^{j}/j!\right],\ (k\geq K)$$

with individual probability $P(X=0)=e^{-\theta}$.

This form of $P(X=k)$ is called the *generalized Pólya-Aeppli distribution* and when $K=1$ it corresponds to *Pólya-Aeppli* distribution.

The *Pólya-Aeppli* distribution arises from a model where objects occur in clusters. The number of clusters follows Poisson distribution while the number of objects per cluster has the geometric distribution [Chapter 6].

Let Y be the number of objects per cluster, where

$$P(Y=k)=qp^{k-1};\ k=1,2,\ldots;\ q=1-p;\ p=P/Q;\ q=Q^{-1}.$$

Then the distribution is defined by

$$\begin{aligned}P(X=k)&=e^{-\theta}p^{k}\sum_{j=1}^{k}\binom{k-1}{j-1}\left[(\theta q/p)^{j}/j!\right],\ k\geq 1\\ \text{and } P(X=0)&=e^{-\theta}.\end{aligned}$$

The factorial moment generating function of this distribution is given by

$$E\left[q^{j}(1+t)^{j}\left(q-p^{t}\right)^{-j}\right]=\exp\left[\theta t\left(q-p^{t}\right)^{-1}\right]$$

where j has Poisson distribution with parameter θ. The factorial cumulant generating function is

$$\theta t(q-p^{t})^{-1}$$

and the r-th factorial cumulant is

$$K_{(r)}=r!\theta q^{-1}(p/q)^{r-1};\ r=1,2,\ldots.$$

Now, the ordinary cumulants are

$$\begin{aligned}K_1&=\mu_1'=\theta/q\\ K_2&=\mu_2=\frac{\theta}{q}(1+p/q)=\theta/q^{2}\\ K_3&=\mu_3=\frac{\theta}{q}(1+3p/q+p^{2}/q^{2})=\frac{\theta}{q^{3}}(1+2pq)\\ K_4&=\frac{\theta}{q^{4}}[1+(4-p)pq].\end{aligned}$$

Hence

$$\alpha_3 = \sqrt{\beta_1} = (1+2pq)/\sqrt{\theta}$$
$$\text{and } \alpha_4 = \beta_2 = 3 + [1+(4-p)pq]\theta^{-1}.$$

From this geometric Poisson distribution we also have

$$P(X=1) = e^{-\theta}p(\theta q/p) = \theta q e^{-\theta}$$
$$\text{and } P(X=2) = e^{-\theta}p^2\left[(\theta q/p) + \frac{1}{2}(\theta q/p^2)\right] = \theta q\left(p+\frac{1}{2}\theta q\right)^{-\theta}.$$

Thus the distribution has a local minimum at $k=1$ if $\theta q < 1$ and $p + \frac{1}{2}\theta q > 1$.

Negative Binomial-Binomial Distribution

Let X be a random variable with probability function

$$P(x) = \binom{-n_j}{x} Q^{-n_j}\left(-\frac{P}{Q}\right)^x;\ x = 0,1,2,\ldots.$$

Again, j is a random variable with probability function

$$P(j) = \binom{N_1}{j} p^j q^{N_1-j};\ j = 0,1,2,\ldots,N_1.$$

Then the compound negative binomial distribution is given by

$$P(x) = \left(-\frac{P}{Q}\right)^x \sum_{j=0}^{N_1} \binom{-n_j}{x}\binom{N_1}{j}(pQ^{-n})^j q^{N_1-j}.$$

The probability generating function of this distribution is given by

$$g(t) = E\left[(Q-Pt)^{-nj}\right] = \left[q + p(Q-Pt)^{-n}\right]^{N_1}.$$

Negative Binomial-Logarithmic Series Distribution

Let X be a random variable with probability function

$$P(x) = \binom{-n_j}{x} Q^{-n_j}\left(-\frac{P}{Q}\right)^x;\ x = 0,1,2\ldots.$$

Also, j is a random variable with probability function

$$P(j) = -\frac{1}{\log(1-\theta)}\frac{\theta^j}{j!};\ j = 0,1,2,\ldots.$$

Then the compound negative binomial distribution is given by

$$P(x) = -\frac{1}{\log(1-\theta)}\left(-\frac{P}{Q}\right)^x \sum_{j=0}^{\infty}\binom{-n_j}{x}\frac{(\theta Q^{-n})^j}{j!}$$
$$= -\frac{1}{\log(1-\theta)}\left(\frac{P}{Q}\right)^x \frac{1}{x!}\sum_{j=1}^{\infty}(n_j)^{(x)}\left(\theta Q^{-n}\right)^j/j!,\ x = 1,2,\ldots.$$

The probability generating function of this distribution is

$$\begin{aligned} g(t) &= E\left[(Q-Pt)^{-n_j}\right] \\ &= \left[\log\{1-\theta(Q-Pt)^{-n}\}\right]/[\log(1-\theta)] \end{aligned}$$

where j has a logarithmic series distribution with parameter θ. The probability generating function stands as

$$\frac{\log(Q-Pt-\theta)-\log(Q-Pt)}{\log(1-\theta)}$$

when $n=1$. In that case the r-th factorial moment is

$$\mu_{(r)} = [-\log(1-\theta)]^{-1} r! \left[(Q-\theta)^{-r} - Q^{-r}\right].$$

9.8 Multivariate Negative Binomial Distribution

The probability generating function of negative binomial distribution is $(Q-Pt)^{-N}$. Analogous to this probability generating function the probability generating function of k-variate negative binomial distribution is defined by

$$\left(Q - \sum_{i=1}^{k} P_i t_i\right)^{-N} \tag{9.3}$$

with $P_i > 0$ for all i, $N > 0$ and

$$Q = 1 + \sum_{i=1}^{k} P_i.$$

From (9.3) the distribution function can be written as

$$P(n_1, n_2, \ldots, n_k) = \frac{\Gamma\left(N + \sum_{i=1}^{k} n_i\right)}{\left(\prod_{i=1}^{k} n_i!\right)\Gamma N} Q^{-N} \prod_{i=1}^{k} (P_i/Q)^{n_i}. \tag{9.4}$$

This is the multivariate negative binomial distribution with parameters $N, P_1, P_2, \ldots, P_k$. The distribution can also be represented by

$$P(n_1, n_2, \ldots, n_k) = (-N)^{\Sigma n_i \theta - N - \Sigma n_i} \prod_{i=1}^{k} (\theta_i^{n_i}/n_i!)$$

with $i = 1, 2, \ldots, k$; $\theta = Q^{-1}, \theta_i = -P_i/Q$. Here N can be fractional.

The joint factorial moment of this distribution is defined by

$$\mu_{(r_1, r_2, \ldots, r_k)} = E\left[\prod_{i=1}^{k} n_i^{(r_i)}\right] = N^{\left(\sum_{i=1}^{k} r_i\right)} \prod_{i=1}^{k} P_i^{r_i}.$$

The moment generating function of this distribution is

$$\left(Q-\sum_{i=1}^{k} P_i e^{t_i}\right)^{-N}.$$

Thus the moments of individual n's are those of negative binomial variates with parameters N, P_j. The cumulant generating function is given by

$$-N\log\left(Q-\sum_{i=1}^{k} P_i e^{t_i}\right)$$

and the factorial moment generating function is

$$\left(1-\sum_{i=1}^{k} P_i t_i\right)^{-N}.$$

The factorial cumulant generating function is

$$-N\log\left(1-\sum_{i=1}^{k} P_i t_i\right).$$

Wishart (1949) has given formulas for several mixed cumulants. Some of these are

$$\begin{aligned}
K_{\cdots 21\cdot} &= NP_iP_j(1+2P_i)\\
K_{\cdot 111\cdot} &= 2NP_iP_jP_k\\
K_{\cdot\cdot 31\cdot} &= NP_iP_j[1+6P_i(1+P_i)]\\
K_{\cdot\cdot 22\cdot} &= NP_iP_j[(1+2P_i)(1+2P_j)+2P_iP_j]\\
K_{\cdot 211\cdot} &= 2NP_iP_jP_k(1+3P_i)\\
K_{\cdot 1111\cdot} &= 6NP_iP_jP_kP_l
\end{aligned}$$

The correlation coefficient between n_i and n_j $(i \neq j)$ is

$$\rho_{ij} = \sqrt{P_iP_j(1+P_i)^{-1}(1+P_j)^{-1}}.$$

EXERCISE

1. (a) Explain the experiment, where experimental results can be expressed by negative probability law. Find moment generating function of negative binomial distribution and hence find the values of β_1 and β_2 of the distribution.

   (b) Two per cent bulbs produced by a company are usually found defective. Find the probability that 6 bulbs are to be investigated to find 2 defective bulbs in the production of a day.

2. (a) How would you derive the probability function of negative binomial distribution from the results of Bernoulli experiment? Mention some examples of negative binomial variable.

(b) Show that negative binomial distribution is leptokurtic.

(c) Two cricket teams A and B have decided to play a series of 5 tests under the condition that the team will win the series if it wins first 3 tests. If the probability of winning the game is 0.5 for both teams, what is the probability that the team A will win the series?

3. (a) The blood sugar levels of some diabetic patients have been recorded before the treatment of the patients starts. It is assumed that the blood sugar level of 70% patients will be reduced under treatment. After a predetermined period of treatment the blood sugar level of patient is investigated one by one. Find the probability that 5 patients after investigation of 15 patients will be identified whose blood sugar levels are reduced. Also, find the probability that 16 patients need to be investigated to identify 5 patients whose blood sugar levels are reduced.

(b) Show that the probability function of negative binomial distribution can be derived from the mixture of Poisson and gamma distribution.

4. (a) Write down some examples of negative binomial variables. Find first four central moments of negative binomial distribution.

(b) Show that, for a negative binomial distribution

$$\mu_{r+1} = q\left[\frac{\partial \mu_r}{\partial q} + \frac{rk}{p^2}\mu_{r-1}\right]$$

where p is the probability of success of Bernoulli variable.

5. (a) Show that Poisson distribution is a limiting case of negative binomial distribution.

(b) State and prove the properties of negative binomial distribution.

(c) Three per cent vehicles running in a highway face accidents per day. Find the probability that 6 vehicles need to be investigated to observe one vehicle which has faced accident. Also, find the probability of observing at least 6 vehicles for the same purpose.

6. (a) Find the probability generating function of negative binomial distribution. Hence or otherwise find the first four factorial moments of the distribution.

(b) Derive the truncated negative binomial distribution, where truncation is done at point $x = 0$.

(c) Deduce the relation

$$k_{r+1} = PQ\frac{\partial k_r}{\partial Q},$$

where k_r and k_{r+1} are the r-th and $(r+1)$ the cumulant of negative binomial distribution.

CHAPTER **10**

Geometric Distribution

10.1 Definition

The probability function of negative binomial distribution is given by

$$P(x) = \binom{N+x-1}{N-1}\left(\frac{P}{Q}\right)^x\left(1-\frac{P}{Q}\right)^N \quad ; \; x = 0, 1, 2, \ldots$$

The probability function takes the form

$$\begin{aligned} P(x) &= Q^{-1}\left(\frac{P}{Q}\right)^x, \text{ when } N = 1 \\ &= pq^x \quad \left[\because\; P = Q^{-1} \text{ and } q = \frac{P}{Q}\right] \end{aligned}$$

The probabilities of the random variable X for $x = 0, 1, 2, \ldots$ are in geometric progression with common ratio q. Thus a random variable X is said to have a geometric distribution if it assumes only the non-negative values and its probability function is given by

$$\begin{aligned} P(x) &= pq^x; \; x = 0, 1, 2, \ldots \; ; \; 0 < p < 1 \\ &= 0, \text{ otherwise.} \end{aligned}$$

Here p is the parameter of geometric distribution.

This distribution also arises from Bernoulli trials in which the outcome of any trial is either a success or a failure and the probability of success of any trial is p. Let us consider that there is an infinite sequence of trials. If we assume that the number of failures that occur before the first success is X, then X will have a geometric distribution with parameter p and the probability of $X = x$ is given by

$$P(x) = pq^x; \; x = 0, 1, 2, \ldots .$$

The geometric distribution also arises from *lack of memory*. Let us consider that an event E can occur at one of the times $t = 0, 1, 2, \ldots$ and the occurrence time X has a geometric distribution with

$$P(X = t) = pq^t; \; t = 0, 1, 2, \ldots .$$

Suppose the event E has not occurred before k and $Y = X - k$ is the amount of additional time needed for E to occur. Then it can be easily shown that

$$P(Y = t|X \geq k) = P(X = t) = pq^t.$$

This implies that the additional time to wait has the same distribution as the distribution of initial time to wait.

10.2 Moments of Geometric Distribution

The r-th moment about origin is defined by

$$\mu_r' = \sum_{x=0}^{\infty} x^r pq^x.$$

In particular, when

$$r = 1,\ \mu_1' = \sum_{x=0}^{\infty} xpq^x = pq\sum_{x=1}^{\infty} xq^{x-1} = pq(1-q)^{-2} = \frac{q}{p}$$

$$r = 2,\ \mu_2' = \sum_{x=0}^{\infty} x^2pq^x = \sum_{x=0}^{\infty} xpq^x + \sum_{x=0}^{\infty} x(x-1)pq^x$$

$$= \frac{q}{p} + pq^2\sum_{x=2}^{\infty} x(x-1)q^{x-2} = \frac{q}{p} + 2pq^2\sum \frac{x(x-1)}{2\cdot 1}q^{x-2}$$

$$= \frac{q}{p} + 2pq^2(1-q)^{-3} = \frac{q}{p} + \frac{2q^2}{p^2}$$

$$\mu_2 = \mu_2' - \mu_1'^2 = \frac{q}{p} + \frac{2q^2}{p^2} - \frac{q^2}{p^2} = \frac{q}{p} + \frac{q^2}{p^2} = \frac{q}{p^2}.$$

The moments can also be obtained from moment generating function, where the m.g.f. is defined by

$$M_x(t) = E\left[e^{tx}\right] = \sum_{x=0}^{\infty} e^{tx}pq^x = p\sum_{x=0}^{\infty}(qe^t)^x = p(1-qe^t)^{-1}.$$

Now

$$\mu_3' = \left.\frac{\partial^3 M_x(t)}{\partial t^3}\right|_{t=0}$$

$$= pq\left[2q\left\{\frac{3qe^{3t}}{(1-qe^t)^4} + \frac{2e^{2t}}{(1-qe^t)^3}\right\} + \left\{\frac{2qe^{2t}}{(1-qe^t)^3} + \frac{e^t}{(1-qe^t)^2}\right\}\right]_{t=0}$$

$$= pq\left[\frac{2q(3q+2p)}{p^4} + \frac{2q+p}{p^3}\right]$$

$$= pq\left[\frac{2q(q+2)}{p^4} + \frac{q+1}{p^3}\right]$$

$$= \frac{q}{p^2}\left[\frac{6q^2}{p^2} + \frac{6q}{p} + 1\right]$$

$$\mu_4' = \left.\frac{\partial^4 M_x(t)}{\partial t^4}\right|_{t=0}$$

$$= pq\left[2q\left\{3q\left(\frac{4qe^{4t}}{(1-qe^t)^5}+\frac{3e^{3t}}{(1-qe^t)^4}\right)+\frac{6qe^{3t}}{(1-qe^t)^4}+\frac{4e^{2t}}{(1-qe^t)^3}\right\}\right.$$

$$\left.+2q\left\{\frac{3qe^{3t}}{(1-qe^t)^4}+\frac{2e^{2t}}{(1-qe^t)^3}\right\}+\frac{2qe^{2t}}{(1-qe^t)^3}+\frac{e^t}{(1-qe^t)^2}\right]_{t=0}$$

$$= pq\left[2q\left\{\frac{12q^2}{p^5}+\frac{9q}{p^4}+\frac{6q}{p^4}+\frac{4q}{p^3}\right\}+\frac{6q^2}{p^4}+\frac{4q}{p^3}+\frac{2q}{p^3}+\frac{1}{p^2}\right]$$

$$= \frac{q}{p}\left[\frac{24q^3}{p^3}+\frac{36q^2}{p^2}+\frac{14q}{p}+1\right].$$

$$\therefore\ \mu_3 = \mu_3' - 3\mu_2'\mu_1' + 2\mu_1'^3$$

$$= \frac{q}{p}\left[\frac{6q^2}{p^2}+\frac{6q}{p}+1-\frac{3q}{p}\left(1+\frac{2q}{p}\right)+\frac{2q^2}{p^2}\right]$$

$$= \frac{q}{p}\left[\frac{1+q}{p^2}\right]$$

$$\mu_4 = \mu_4' - 4\mu_3'\mu_1' + 6\mu_2'\mu_1'^2 - 3\mu_1'^4$$

$$= \frac{q}{p}\left[\frac{24q^3}{p^3}+\frac{36q^2}{p^2}+\frac{14q}{p}+1-4\frac{q}{p}\left(\frac{6q^2}{p^2}+\frac{6q}{p}+1\right)\right.$$

$$\left.+6\frac{q}{p}\left(\frac{q}{p}+\frac{2q^2}{p^2}\right)-3\frac{q^3}{p^3}\right]$$

$$= \frac{q}{p^2}\left[1+\frac{9q}{p^2}\right]$$

$$\alpha_3 = \sqrt{\beta_1} = \frac{\frac{q}{p}\left[\frac{1+q}{p^2}\right]}{\frac{q^{3/2}}{p^3}} = \frac{1+q}{\sqrt{q}}$$

$$\alpha_4 = \beta_2 = \frac{\frac{q}{p^2}\left[1+\frac{9q}{p^2}\right]}{\frac{q^2}{p^4}} = \frac{p^2}{q}\left[1+\frac{9q}{p^2}\right] = \frac{p^2}{q}+9.$$

$$\therefore\ \gamma_2 = \beta_2 - 3 = \frac{p^2}{q}+6.$$

This distribution is also leptokurtic and it has a positive tail. Also, $\alpha_3 \to \infty$ as $q \to 0$.

10.3 Generating Functions of Geometric Distribution

The m.g.f. of geometric distribution is

$$M_x(t) = p(1-qe^t)^{-1}.$$

The cumulant generating function is defined by

$$K_x(t) = \log p(1-qe^t)^{-1}.$$

Again, the factorial moment generating function of this distribution is defined by

$$\begin{aligned} M_{(x)}(t) &= \sum_{x=0}^{\infty}(1+t)^x pq^x \\ &= p(1+t)\{1-q(1+t)\}^{-1} = p(1+t)(p-qt)^{-1}. \end{aligned}$$

The probability generating function is given by

$$g(t) = p\sum_{x=0}^{\infty}(tq)^x = p(1-qt)^{-1}.$$

10.4 Properties of Geometric Distribution

❑ **THEOREM 10.1** *If X has the geometric density with parameter p, then*

$$P(X \geq i+j|X \geq i) = P(X \geq j), \text{ for } i,j = 0,1,2,\ldots.$$

Proof.

$$\begin{aligned} P(X \geq i+j|X \geq i) &= \frac{P(X \geq i+j)}{P(X \geq i)} = \frac{\sum\limits_{x=i+j}^{\infty} p(1-p)^x}{\sum\limits_{x=i}^{\infty} p(1-p)^x} \\ &= \frac{(1-p)^{i+j}}{(1-p)^i} = (1-p)^j = P(X \geq j). \end{aligned}$$

❑ **THEOREM 10.2** *If X_1 and X_2 are two independent random variables having the same geometric distribution, then the conditional distribution of $X_1|X_1+X_2=n$ is uniform.*

Proof. Given that $P(X_1 = k) = P(X_2 = k) = pq^k;\ k = 0,1,2,\ldots.$

$$\begin{aligned} P[X_1 = r|(X_1+X_2) = n] &= \frac{P[X_1 = r \cap (X_1+X_2) = n]}{P(X_1+X_2 = n)} \\ &= \frac{P[X_1 = r \cap X_2 = n-r]}{P(X_1+X_2=n)} \\ &= \frac{P(X_1 = r)P(X_2 = n-r)}{\sum\limits_{s=0}^{n} P(X_1 = s \cap X_2 = n-s)} \\ &= \frac{P(X_1 = r)P(X_2 = n-r)}{\sum\limits_{s=0}^{n} P(X_1 = s)P(X_2 = n-s)} \\ &= \frac{pq^r pq^{n-r}}{\sum\limits_{s=0}^{n}[pq^s pq^{n-s}]} = \frac{p^2q^n}{\sum\limits_{s=0}^{n} p^2q^n} \\ &= \frac{p^2q^n}{(n+1)p^2q^n} = \frac{1}{n+1};\ r = 0,1,2,\ldots,n. \end{aligned}$$

❒ THEOREM 10.3 *If X_1 and X_2 are two independent random variables having the same geometric distribution, then the distribution of $Z = \max(X_1, X_2)$ is given by*

$$P(Z = l) = pq^l[2 - q^l - q^{l+1}]; \; l = 0, 1, 2, \ldots.$$

Proof. We have $P(X_1 = k) = P(X_2 = k) = pq^k; \; k = 0, 1, 2, \ldots.$

$$\begin{aligned} P(Z = l) &= P(Z \leq l) - P(Z \leq l - 1) \\ &= P[\max(X_1, X_2) \leq l] - P[\max(X_1, X_2) \leq l - 1] \\ &= P(X_1 \leq l \cap X_2 \leq l) - P(X_1 \leq l - 1 \cap X_2 \leq l - 1) \\ &= P(X_1 \leq l)P(X_2 \leq l) - P(X_1 \leq l - 1)P(X_2 \leq l - 1) \\ &= \sum_{k=0}^{l} \left(pq^k\right)^2 - \sum_{k=0}^{l-1} \left(pq^k\right)^2 \\ &= \left[\frac{p\left(1 - q^{l+1}\right)}{1 - q}\right]^2 - \left[\frac{p\left(1 - q^l\right)}{1 - q}\right]^2 = pq^l\left[2 - q^l - q^{l+1}\right]. \end{aligned}$$

The recurrence relation for the probabilities of geometric distribution is given by

$$P(x + 1) = qp(x).$$

❒ THEOREM 10.4 *If X_1 and X_2 are two independent random variables having the same geometric distribution, then $P(X_1 = X_2) = \frac{p}{1+q}$.*

Proof. We have $P(X_1 = r) = P(X_2 = r) = pq^r; \; r = 0, 1, 2, \ldots$

Now

$$\begin{aligned} P(X_1 = X_2) &= \sum_{r=0}^{\infty} P(X_1 = r \cap X_2 = r) = \sum_{r=0}^{\infty} P(X_1 = r)P(X_2 = r) \\ &= \sum_{r=0}^{\infty} pq^r pq^r = p^2 \sum_{r=0}^{\infty} q^{2r} = p^2(1 + q^2 + q^4 + \cdots) \\ &= \frac{p^2}{1 - q^2} = \frac{p}{1 + q}. \end{aligned}$$

❒ THEOREM 10.5 *If X_1 and X_2 are two independent random variables having the same geometric distribution, then*

$$P(X_1 + X_2 = n) = (n + 1)p^2q^n; \; n = 0, 1, 2, \ldots.$$

Proof. We have $P(X_1 = k) = P(X_2 = k) = pq^k; \; k = 0, 1, 2, \ldots.$

$$\begin{aligned} P(X_1 + X_2 = n) &= \sum_{s=0}^{n} P(X_1 = s \cap X_2 = n - s) \\ &= \sum_{s=0}^{n} P(X_1 = s)P(X_2 = n - s) \\ &= \sum_{s=0}^{n} pq^s pq^{n-s} = p^2 \sum_{s=0}^{n} q^n = (n + 1)p^2q^n; \; n = 0, 1, 2, \ldots. \end{aligned}$$

❒THEOREM 10.6 *The mixture of Poisson exponential is a geometric distribution.*

Proof. Let X be a random variable with probability function

$$P(x) = \frac{e^{-\theta}\theta^x}{x!};\ x = 0, 1, 2, \ldots.$$

Again, θ is a random variable with density function

$$f(\theta) = e^{-\theta};\ \theta \geq 0.$$

Then the mixture distribution of X is given by

$$P(x) = \frac{1}{x!}\int_0^\infty e^{-\theta}\theta^x e^{-\theta}\,d\theta = \frac{1}{x!}\int_0^\infty e^{-2\theta}\theta^x\,d\theta$$

$$= \frac{x!}{x!2^{x+1}} = \frac{1}{2^{x+1}} = \frac{1}{2}\left(\frac{1}{2}\right)^x = pq^x, \text{ with } p = \frac{1}{2}, q = \frac{1}{2}.$$

EXERCISE

1. (a) How geometric distribution is related to Bernoulli distribution and negative binomial distribution? Explain. Find the probability function of geometric distribution.
   (b) Find moment generating function of geometric distribution. Hence or otherwise find first four central moments of the distribution.
   (c) A couple has decided to give birth of child until a son is born. What is the probability that their third child is the first son? Assume that the probability of birth of a son or a daughter is 0.5.
2. (a) Under what condition a negative binomial distribution transforms to geometric distribution?
   Show that, if X has the geometric distribution with parameter p, then
   $$P(X \geq i + j / X \geq i) = (1 - p)^j$$
   (b) Discuss the characteristic of geometric distribution.
   (c) If HIV is observed in the blood of one person out of 1000, what is the probability that HIV will be positive in the blood of a person after 9th person?
3. (a) Write down some examples of the variable which follows geometric distribution. If X_1 and X_2 are two random variables having the same geometric distribution, then the conditional distribution of $X_1/(X_1 + X_2)$ follows uniform distribution. Prove.
   (b) Show that the mixture of Poisson exponential distribution is a geometric distribution.
   (c) A quality control team of an industry inspects the product until a defective item is found out. What is the probability that the fifth inspected item is the first defective one. It is known that 10% products of the industry are usually found defective.

CHAPTER **11**

Hypergeometric Distribution

11.1 Definition and Applications

A discrete random variable X is said to follow the hypergeometric distribution if it takes only non-negative values and its probability function is defined by

$$P(X = k) = \frac{\binom{M}{k}\binom{N-M}{n-k}}{\binom{N}{n}}; \; k = 0, 1, 2, \dots, \min(n, M).$$

The terms in the right-hand side of this probability function are the successive terms in the expansion of

$$\frac{(N-n)!(N-M)!}{N!(N-M-n)!}F(-n, -M;\; N-M-n+1;\; 1),$$

where

$$F(\alpha; \beta; \gamma; z) = 1 + \frac{\alpha\beta}{\gamma} \cdot \frac{z}{1!} + \frac{\alpha(\alpha+1)\beta(\beta+1)}{\gamma(\gamma+1)} \frac{z^2}{2!} + \cdots$$

is a hypergeometric function. Thus the probability function is called *hypergeometric distribution.*

The distribution arises in an experiment where units are drawn without replacement from a finite population. The sampling units are drawn randomly but the units are not independently drawn. For example, let us consider an urn which contains N balls, M of which are white and $(N - M)$ are black. Let us draw a sample of n balls at random without replacement. Then the probability of getting k white balls out of n balls $(k < n)$ is

$$P(X = k) = \frac{\binom{M}{k}\binom{N-M}{n-k}}{\binom{N}{n}}.$$

In this distribution N, M and n are the three parameters. The value of k will be in the limits

$$\max(0, n - N + M) \le k \le \min(n, M).$$

This distribution frequently replaces a binomial distribution when it is desirable to make allowance for finiteness of sample size. The distribution is applied to estimate the size of animal population from *capture-recapture* data. It is also applied to estimate the number of children attacked by an infectious disease when a fixed number (N) are exposed to it.

11.2 Mean and Variance of Hypergeometric Distribution

The mean μ_1' of this distribution is given by

$$\begin{aligned}
\mu_1' &= E(x) \\
&= \sum_{k=0}^{n} kP(X=k) \\
&= \sum_{k=0}^{n} k\left\{\frac{\binom{M}{k}\binom{N-M}{n-k}}{\binom{N}{n}}\right\} \\
&= \frac{M}{\binom{N}{n}}\sum_{k=1}^{n}\binom{M-1}{k-1}\binom{N-M}{n-k} \\
&= \frac{M}{\binom{N}{n}}\sum_{x=0}^{m}\binom{A}{x}\binom{N-A-1}{m-x}, \\
&\qquad \text{where } x = k-1, m = n-1, M-1 = A \\
&= \frac{M}{\binom{N}{n}}\binom{N-1}{m} = \frac{M}{\binom{N}{n}}\binom{N-1}{n-1} = \frac{nM}{N}.
\end{aligned}$$

Again

$$\begin{aligned}
\mu_2' = E(x^2) &= E[x(x-1)+x] = E(x-1)x + E(x) \\
&= \sum_{k=1}^{n}(k-1)k\left\{\binom{M}{k}\binom{N-M}{n-k}\Big/\binom{N}{n}\right\} + \frac{nM}{N} \\
&= \frac{M(M-1)}{\binom{N}{n}}\sum_{k=2}^{n}\binom{M-2}{k-2}\binom{N-M}{n-k} + \frac{nM}{N} \\
&= \frac{M(M-1)}{\binom{N}{n}}\binom{N-2}{n-2} + \frac{nM}{N} \\
&= \frac{M(M-1)n(n-1)}{N(N-1)} + \frac{nM}{N}.
\end{aligned}$$

$\therefore$ The variance of the distribution is given by

$$\begin{aligned}
\mu_2 &= \mu_2' - \mu_1'^2 \\
&= \frac{M(M-1)n(n-1)}{N(N-1)} + \frac{nM}{N} - \frac{(nM)^2}{N^2} \\
&= \frac{nM(N-M)(N-n)}{N^2(N-1)}.
\end{aligned}$$

11.3 Generating Functions of Hypergeometric Distribution

The r-th factorial moment of this distribution is given by

$$\begin{aligned}
\mu_{(r)} &= E\left[x^{(r)}\right] = \sum_{k=r}^{n} k^{(r)} P(X=k) = \sum_{k=r}^{n} k^{(r)} \binom{M}{k}\binom{N-M}{n-k} \Big/ \binom{N}{n} \\
&= \sum_{k=r}^{n} \frac{k(k-1)(k-2)\cdots(k-r+1)M!}{(M-k)!k!} \binom{N-M}{n-k} \Big/ \binom{N}{n} \\
&= \sum_{k=r}^{n} \frac{M(M-1)(M-2)\cdots(M-r+1)}{(k-r)!(M-k)!} \binom{N-M}{n-k} \Big/ \binom{N}{n} \\
&= \sum_{k=r}^{n} M^{(r)} \binom{M-r}{k-r}\binom{N-M}{n-k} \Big/ \binom{N}{n} \\
&= M^{(r)} \sum_{j=0}^{n-r} \binom{M-r}{j}\binom{(N-r)-(M-r)}{(n-r)-j} \Big/ \binom{N}{n}, \text{ where } j = k-r \\
&= \frac{M^{(r)} n^{(r)}}{N^{(r)}} \sum_{j=0}^{n-r} \binom{M-r}{j}\binom{(N-r)-(M-r)}{(n-r)-j} \Big/ \binom{N-r}{n-r} \left[\because n^{(r)}\tbinom{N}{n} = N^{(r)}\tbinom{N-r}{n-r}\right] \\
&= \frac{(M^{(r)} n^{(r)}}{N^{(r)}} \sum_{j=0}^{n-r} h(j; N-r, M-r, n-r) = \frac{M^{(r)} n^{(r)}}{N^{(r)}}.
\end{aligned}$$

In particular, when $r = 1$ and $r = 2$, we have

$$\begin{aligned}
\mu_1' &= \frac{nM}{N}, \ \mu_{(2)} = \frac{M^{(2)} n^{(2)}}{N^{(2)}} = \frac{M(M-1)n(n-1)}{N(N-1)} \\
\mu_2' &= \mu_{(2)} + \mu_{(1)} = \frac{M(M-1)n(n-1)}{N(N-1)} + \frac{nM}{N}.
\end{aligned}$$

Similarly, we have

$$\begin{aligned}
\mu_{(3)} &= \frac{M(M-1)(M-2)n(n-1)(n-2)}{N(N-1)(N-2)} \\
\mu_3' &= \mu_{(3)} + 3\mu_{(2)} + \mu_{(1)} \\
&= \frac{M(M-1)(M-2)n(n-1)(n-2)}{N(N-1)(N-2)} + \frac{3M(M-1)n(n-1)}{N(N-1)} + \frac{nM}{N} \\
&= \frac{nM}{N}\left[\frac{(M-1)(n-1)}{(N-1)}\left\{\frac{(M-2)(n-2)}{N-2} + 3\right\} + 1\right] \\
\mu_{(4)} &= \frac{M(M-1)(M-2)(M-3)(n-1)(n-2)(n-3)n}{N(N-1)(N-2)(N-3)} \\
\mu_4' &= \mu_{(4)} + 6\mu_{(3)} + 7\mu_{(2)} + \mu_{(1)} \\
&= \frac{nM}{N}\left[\frac{(M-1)(M-2)(n-1)(n-2)}{(N-1)(N-2)}\left\{\frac{(M-3)(n-3)}{(N-3)} + 6\right\}\right. \\
&\qquad \left. + \frac{7(M-1)(n-1)}{N-1} + 1\right].
\end{aligned}$$

Using these results we can find the values of μ_3 and μ_4. Also two moment ratios can be found out, where these are

$$\alpha_3 = \sqrt{\beta_1} = \frac{\left(1-\frac{M}{N}\right)-M/N}{\left[\left(\frac{N-n}{N-1}\right)\frac{nM}{N}\left(1-\frac{M}{N}\right)\right]^{1/2}}\frac{N-2n}{N-2}$$

$$\alpha_4 = \beta_2 = \frac{3(N-1)(N+6)}{(N-2)(N-3)} + \frac{N(N-1)(N+1)}{(N-n)(N-2)(N-3)}\left[1-\frac{6N}{N+1}\left(pq+\frac{n(N-n)}{N^2}\right)\right]\frac{1}{npq},$$

where $p = M/N$, $q = 1-p$.

The moment generating function of this distribution is given by

$$M_x(t) = E\left(e^{tx}\right) = \frac{(N-n)!(N-M)!}{N!}F\left(-n,-M;\ N-M-n+1;\ e^t\right).$$

The function $M_x(t)$ satisfies the differential equation

$$\left(1-e^t\right)\left[\frac{d^2M_x(t)}{dt^2}-(n+M)\frac{dM_x(t)}{dt}+nMM_x(t)\right] -nMM_x(t)+\frac{NdM_x(t)}{dt}=0.$$

There is a finite difference relation among the central moments of hypergeometric distribution which is given by

$$N\mu_{r+1} = \{(1+E)^r - E^r\}[\mu_2+\alpha\mu_1+\beta\mu_0],$$

where E is the displacement operator ($E^p\mu_s \equiv \mu_{s+p}$);

$$\alpha = -(M+n-2Mn/N);$$
$$\beta = \frac{M}{N}(1-M/N)N(N-n);\ \mu_1 = 0 \text{ and } \mu_0 = 1.$$

11.4 Properties of Hypergeometric Distribution

The recurrence relation of the distribution

$$P(X=k) = \binom{M}{k}\binom{N-M}{n-k}\Big/\binom{N}{n}$$

is given by

$$\frac{P(X=k+1)}{P(X=k)} = \frac{\binom{M}{k+1}\binom{N-M}{n-k-1}}{\binom{M}{k}\binom{N-M}{n-k}} = \frac{(n-k)(M-k)}{(k+1)(N-M-n+k+1)}.$$

$$\therefore\ P(X=k+1) = \frac{(n-k)(M-k)}{(k+1)(N-M-n+k+1)}P(X=k).$$

This relationship holds good for the parameters n, $N$and M, where the distribution is

$$P(X=k|n,N,M).$$

Also we have

$$P(X=k|M+1,n,N) = \frac{(M+1)(N-M-n+k)}{(N-M)(M+1-k)}P(X=k|M,n,N)$$
$$P(X=k|M,n+1,N) = \frac{(N-M-n+k)(n+1)}{(n+1-k)(N-n)}P(X=k|M,n,N)$$
$$P(X=k|M,n,N+1) = \frac{(N+1-n)(N+1-M)}{(N+1-n-M+k)(N+1)}P(X=k|M,n,N).$$

It can be seen that $P(X=k+1|M,n,N) \gtrless P(X=k|M,n,N)$ according as

$$k \lessgtr \frac{(n+1)(M+1)}{N+2} - 1.$$

Hence $P(X=k|M,n,N)$ increases with k, and reaches a maximum value at the greatest integer which does not exceed $(n+1)(M+1)/(N+2)$. There are two equal maximum values at

$$k = \frac{(n+1)(M+1)}{N+2} - 1 \text{ and } k = \frac{(n+1)(M+1)}{N+2}$$

when $(n+1)(M+1)/(N+2)$ is an integer. The maximum value of k is approximately nM/N if N and M are large.

Mean Deviation

The mean deviation of hypergeometric distribution is given by

$$\gamma_1 = E\left|x - \frac{nM}{N}\right| = \frac{2m}{N}(N-M-n+m)\binom{M}{m}\binom{N-M}{n-m}\bigg/\binom{N}{n}$$

where m is the greatest integer not exceeding $1 + nM/N$.

Approximation to Binomial Distribution

Hypergeometric distribution tends to binomial distribution as $N \to \infty$ and $M/N = p$. We have

$$P(X=k|M,n,N) = \frac{\binom{M}{k}\binom{N-M}{n-k}}{\binom{N}{n}} = \frac{M!}{k!(M-k)!}\frac{(N-M)!}{(n-k)!(N-M-n+k)!}\frac{n!(N-n)!}{N!}$$
$$= \frac{(M-1)((M-2)\cdots(M-k+1)}{k!}$$
$$\times\frac{(N-M)(N-M-1)\cdots(N-M-n+k+1)}{(n-k)!}$$
$$\times\frac{n!}{N(N-1)(N-2)\cdots(N-n+1)}$$
$$= \frac{n!}{k!(n-k)!}\frac{M}{N}\left(\frac{M}{N}-\frac{1}{N}\right)\left(\frac{M}{N}-\frac{2}{N}\right)\cdots\left(\frac{M}{N}-\frac{k-1}{N}\right)$$
$$\times\frac{\left(1-\frac{M}{N}\right)\left(1-\frac{M}{N}-\frac{1}{N}\right)\cdots\left(1-\frac{M}{N}-\frac{n-k-1}{N}\right)}{\left(1-\frac{1}{N}\right)\left(1-\frac{2}{N}\right)\cdots\left(1-\frac{n-1}{N}\right)}.$$

Now limit $N \to \infty$ and putting $M/N = p$, we have

$$\lim_{N\to\infty} P(X = k|M, n, N) = \frac{n!}{k!\,(n-k)!} \underbrace{pp\cdots p}_{k \text{ times}} \underbrace{(1-p)(1-p)\cdots(1-p)}_{n-k \text{ times}}$$

$$= \binom{n}{k} p^k q^{n-k}$$

which is the probability function of a binomial variable.

As binomial distribution can be approximated by Poisson or normal, there is a corresponding Poisson or normal approximation to the hypergeometric distribution. The Poisson approximation is given by

$$P(X = k|M, n, N) = \exp[-nM/N] \left(\frac{nM}{N}\right) \frac{k}{k!}$$

when M/N is small but n is large. If n is large, but M/N not small, then the cumulative probability using continuity correction can be written as

$$P(X \le K) = \sum_{k \le K} P(X = k|M, n, N).$$

This probability can be approximated by

$$\frac{1}{\sqrt{2\pi}} \int_{-\infty}^{Y} \exp\left(\frac{1}{2}u^2\right) du \text{ with}$$

$$Y = \left(K - \frac{nM}{N} + \frac{1}{2}\right) / \{(N-n)(N-1)^{-1} n(M/N)(1 - M/N)\}^{1/2}.$$

Pearson (1906) has approximated hypergeometric distributions by continuous Pearson type distributions. Davies (1933, 1934) continued this type of approximation and showed that for the hypergeometric distribution $\beta_2 - \beta_1 > 3$, while

$$\beta_1(\beta_2 + 3)^2 > 4(2\beta_2 - 3\beta_1 - 6)(4\beta_2 - 3\beta_1).$$

In most cases Pearson curves of Types VI or III fit to the hypergeometric distribution.

►**EXAMPLE 11.1** *The estimate of fish in a lake is given by* $N = \frac{rs}{x}$ *where 'r' and 's' are the numbers of catch of fish in two instances, and x is the value of the random variable X indicating marked fish.*

Solution. Let there be N fish in a lake. To estimate the value of N, let us assume that a catch of r fish at a time is made and all of these fish are returned alive into the lake after marking each with some spot. A short time later with an assumption of random dispersion of spotted fish a second catch of s fish is made [the duration between two catches should not be long enough so that natural changes can affect the population size]. Let us assume that 'r' and 's' be regarded as fixed predetermined constants. Let there be X spotted fish among the 's' caught fish.

Then $P(x) = \binom{r}{x}\binom{N-r}{s-x} \Big/ \binom{N}{s}$ where x is an integer with

$$\max(0, s - N + r) \le x \le \min(r, s).$$

From the recurrence relation, we have

$$\frac{P(x)}{P(x-1)} = \frac{(N-r)(N-s)}{N(N-r-s+x)}.$$

Here

$$\frac{P(x)}{P(x-1)} > 1 \text{ if and only if } N > \frac{rs}{x}.$$

Thus

$$P(x) > P(x-1) \text{ when } N > \frac{rs}{x} \quad (1)$$

$$P(x) < P(x-1) \text{ when } N < \frac{rs}{x} \quad (2)$$

From (1) and (2) it can be said that $P(x)$ will reach maximum when N is approximately equal to rs/x.

$$\therefore\ N = rs/x.$$

11.5 Mixtures of Hypergeometric Distributions

Hypergeometric-Binomial Distribution

Let X be a random variable with probability function

$$P(X=k|n,M,N) = \frac{\binom{M}{k}\binom{N-M}{n-k}}{\binom{N}{n}}, \quad k = 0,1,2,\ldots,n$$

and M be a random variable with probability function

$$P(M) = \binom{N}{M} p^M q^{N-M}; \quad M = 0,1,2,\ldots,N.$$

Then the compound hypergeometric distribution is given by

$$\begin{aligned} P(X=k) &= \frac{1}{\binom{N}{n}} \sum_{M=0}^{N} \binom{M}{k}\binom{N-M}{n-k}\binom{N}{M} p^M q^{N-M} \\ &= \frac{n!}{k!(n-k)!} p^x q^{n-k} \sum_{M-k=0}^{N-n} \binom{N-n}{M-k} p^{M-k} q^{N-n-M+k} \\ &= \binom{n}{k} p^k q^{n-k}; \quad k = 0,1,\ldots,n \end{aligned}$$

which is the probability function of binomial distribution.

Hypergeometric-Hypergeometric Distribution

Let X be a random variable with probability function

$$P(X=k) = \frac{\binom{M}{k}\binom{N-M}{n-k}}{\binom{N}{n}}, \quad k = 0,1,2,\ldots,n$$

and M is a random variable with probability function

$$P(M = k) = \frac{\binom{M_1}{k}\binom{N_1-M_1}{N-k}}{\binom{N_1}{N}}, \; k = 0, 1, 2, \ldots, N.$$

Then the compound distribution is given by

$$\begin{aligned} P(X = k) &= \frac{1}{\binom{N}{n}} \sum_{k=0}^{N} \binom{M}{k}\binom{N-M}{n-k}\binom{M_1}{k}\binom{N_1-M_1}{N-k}\frac{1}{\binom{N_1}{N}} \\ &= \frac{\binom{M_1}{k}\binom{N_1-M_1}{n-k}}{\binom{N_1}{n}}, \; k = 0, 1, 2, \ldots, n. \end{aligned}$$

Hypergeometric-Mixed Binomial Distribution

If X is a random variable which follows hypergeometric distribution with parameters (n, M, N) and if M follows mixed binomial distribution with probability function

$$P(M = m) = \sum_{i=1}^{n_1} w_i \binom{N}{m} p_i^m q_i^{N-m}$$

with $0 < p_i < 1, 0 < w_i < 1, q_i = 1 - p_i, \sum_{i=1}^{n_1} w_i = 1$. Then the compound distribution of X will take the form similar to the compound distribution of hypergeometric-binomial distribution.

More details about compound hypergeometric distributions are discussed by Hald (1960). For a hypergeometric distribution with parameters (n, M, N) if M is supposed to have a compounding distribution, then according to Hald the r-th factorial moment of a compound hypergeometric distribution is

$$\frac{n^{(r)}}{N^{(r)}} \times (r\text{-th factorial moment of } M).$$

11.6 More about Hypergeometric Distribution

Positive Hypergeometric Distribution

The hypergeometric distribution has the parameters (n, M, N). If $n > N - M$, there is no zero observed value of the random variable X. The distribution formed by omitting the zero observed value is positive hypergeometric distribution. Again, if $n \le N - M$, then the distribution is given by

$$P(X = k) = \frac{\binom{M}{k}\binom{N-M}{n-k}}{\binom{N}{n} - \binom{N-M}{n}}, \; k = 1, 2, \ldots, \min(n, M).$$

Negative Hypergeometric Distribution

The hypergeometric distribution arises in finding the probability of white balls out of n balls drawn at random without replacement from an urn containing N balls, of which M are white and $(N-M)$ are red. If sampling is continued until 'a' white balls $(0 < a \leq M)$ or, alternatively 'b' red balls $(0 < b \leq N-M)$ are obtained, then the distribution of the number of draws needed is a negative hypergeometric distribution. The probability function of this distribution is given by

$$P(X = k) = \frac{\binom{M}{a-1}\binom{N-M}{k-a}}{\binom{N}{k-1}} \frac{M-a+1}{N-k+1} \text{ with } a \leq k \leq N-M+a.$$

The r-th ascending factorial moment of this distribution is given by

$$\mu_{[r]} = E\left[x^{[r]}\right] = E[x(x+1)\cdots(x+r-1)] = a^{[r]}(N+1)^{[r]}/(M+1)^{[r]}.$$

In particular

$$\mu_1' = \frac{a(N+1)}{M+1}$$

$$\text{and } \mu_2 = \frac{a(N+1)(N-M)(M+1-a)}{(M+1)^2(M+2)}.$$

As N and M tend to infinity with M/N a fixed constant (between 0 and 1), this distribution tends to a negative binomial distribution.

Noncentral Hypergeometric Distribution

The distribution arises in an experimental situation where balls are drawn from an urn without replacement. The assumption is that the probability of drawing a white ball, given that there are M_1 white and $(N_1 - M_1)$ red balls is not M_1/N_1. This probability is $M_1[M_1 + \theta(N_1 - M_1)]^{-1}$ with $\theta \neq 1$. The probability function of this distribution, as suggested by Wallenius (1963), is

$$P(X = k) = \binom{M}{k}\binom{N-M}{n-k}\int_0^1 (1-t^c)^k(1-t^{\theta c})^{n-k}dt$$

with $c = [M - k + \theta(N - M - n + k]^{-1}$. For n small compared to M and $N - M$, the variable X is approximately distributed binomially with parameters n,

$$[1 + \theta M/(N-M)]^{-1}.$$

Extended Hypergeometric Distribution

If X_1 and X_2 are two independent binomial variables with parameters (n_1, p_1) and (n_2, p_2) respectively, then the distribution of $X_1|X_1 + X_2$ or $X_2|X_1 + X_2$ [Section 9.7] is called *extended hypergeometric distribution* [Harkness (1965)]. The r-th factorial moment of this distribution is given by

$$\mu_{(r)} = \frac{F(r-n_1, r-m; r-m+n_2+1; p_1q_1/p_2q_1)}{F(-n_1, -m; -m-n_2+1; p_1q_2/p_2q_1)}$$

with $X_1 + X_2 = m$ and

$$F(\alpha;\beta;\gamma;x) = \frac{\Gamma\gamma}{\Gamma\beta\Gamma(\gamma-\beta)}\int_0^1 t^{\beta-1}(1-t)^{\gamma-\beta-1}(1-tx)^{-\alpha}dt, \ (\beta > 0).$$

11.7 Multivariate Hypergeometric Distribution

In Section 1 we defined hypergeometric distribution. This distribution can be generalized to k random variables. Suppose the N individuals in the population can be classified into k classes such that the j-th class has N_j individuals with

$$\sum_{j=1}^{k} N_j = N.$$

Now, suppose that a sample of n individuals is drawn without replacement from N individuals. Let the number of individual of j-th class in the sample be n_j ($j = 1, 2, \ldots, k$). Then the joint distribution of the random variables $n_1, n_2, \ldots, n_k$ is defined by

$$P(n_1, n_2, \ldots, n_k) = \prod_{j=1}^{k} \binom{N_j}{n_j} \Big/ \binom{N}{n} \text{ with } \sum_{j=1}^{k} n_j = n,\ 0 \le n_j \le N_j.$$

This distribution is called $(k-1)$ variate (multivariate) hypergeometric distribution since there are $(k-1)$ distinct variables as

$$n_k = n - \sum_{j=1}^{k-1} n_j.$$

The parameters of this distribution are $n, N_1, N_2, \ldots, N_k$. When $k = 2$, it can be written as

$$P(n_1, n_2) = \frac{\binom{N_1}{n_1}\binom{N-N_1}{n-n_1}}{\binom{N}{n}};\ n_2 = n - n_1$$

which is an ordinary hypergeometric distribution. Again, as $N \to \infty$ with $N_j/N = p_j$ constant, then the multivariate hypergeometric distribution tends to multinomial distribution (Chapter 8) with parameters $n, p_1, p_2, \ldots, p_k$.

The marginal distribution of n_j is hypergeometric with probability function

$$P(n_j) = \frac{\binom{N_j}{n_j}\binom{N-N_j}{n-n_j}}{\binom{N}{n}};\ n_j = 0, 1, 2, \ldots, n.$$

We have already observed that the r-th factorial moment of hypergeometric distribution is

$$\mu_{(r)} = \frac{M^{(r)} n^{(r)}}{N^{(r)}} = \frac{M^{(r)}\binom{N-r}{n-r}}{\binom{N}{n}}.$$

By straight-forward extension of this factorial moment the joint r-th factorial moment of multivariate hypergeometric distribution can be written as

$$\mu_{[r_1][r_2]\cdots[r_k]} = E\left[\prod_{j=1}^{k} n_j^{[r_j]}\right] = \frac{N_1^{[r_1]} N_2^{[r_2]} \cdots N_k^{[r_k]} \binom{N-r_1-r_2-\cdots-r_k}{n-r_1-r_2-\cdots-r_k}}{\binom{N}{n}}$$

where $0 \le r_j \le N_j$; $j = 1, 2, \ldots, k$ and $r_1 + r_2 + \cdots + r_k \le n$. Now, for j-th variable if $r_j = 1$, we have

$$\mu_{[1]}(n_j) = \mu'_1(n_j) = \frac{nN_j}{N} = np_j \text{ with } \frac{N_j}{N} = p_j.$$

Also we have

$$\begin{aligned}\mu_2(n_j) &= np_j(1-p_j)\left(\frac{N-n}{N-1}\right)\\ \text{and } \operatorname{cov}(n_j, n_{j'}) &= -np_jp_{j'}\left(\frac{N-n}{N-1}\right);\ j \ne j'.\end{aligned}$$

EXERCISE

1. (a) Write down some experimental results which can be expressed by hypergeometric probability law. Find factorial moments of hypergeometric distribution and hence or otherwise find mean and variance of the distribution.

(b) In a laboratory there are 8 non-defective and 2 defective computers. A random sample of 5 computers is selected. Find the probability that the sample contains (i) one defective computer, (ii) at least one defective computer.

2. (a) Discuss the properties of hypergeometric distribution. Show that hypergeometric distribution can be approximated by binomia distribution. Write down condition that a hypergeometric distribution becomes positive hypergeometric distribution. In a survey to estimate the number of tigers in a forest 5 tigers are catched and allowed to leave after giving a special mark to each of them. It is assumed that in the forest there are 50 tigers. After some time another 5 tigers are catched. Find the probability that second time (i) 3 of the marked tigers are catched, (ii) all 5 are marked tigers.

3. (a) Discuss the appliceation of hypergeometric distribution and multivariate hypergeometric distribution. Show that the mixture of hypergeometric-hypergeometric distribution is also a hypergeometric distribution.

(b) A computer centre is run with 5 programmers and 5 computer engineers. A committee of 4 is to be formed form these 10 persons for a special job. What is the probability that the committee will be formed with (i) all computer engineers, (ii) at least one programmer?

CHAPTER **12**

Multinomial Distribution

12.1 Definition

In binomial distribution the experimental results of n independent trials are dichotomised as success or failure, where success and failure are two mutually exclusive events. There are situations where the outcomes of a trial, out of a series of N independent trials, can be classified as k mutually exclusive events $E_1, E_2, \ldots, E_k$ with probability of occurrence p_j of an event E_j $(j = 1, 2, \ldots, k)$. As an example, we can mention the analysis of multiway contingency table in which the cell frequencies of the joint incidence of two-, three- or k-factors each at a number of different levels represent the favourable points of mutually exclusive events.

Let $n_1, n_2, \ldots, n_k$ represent the numbers of occurrences of the events $E_1, E_2, \ldots, E_k$ respectively, in N trials

$$\left(\sum_{j=1}^{k} n_j = N\right).$$

Then the joint distribution of the random variables $n_1, n_2, \ldots, n_k$ is defined by

$$P(n_1, n_2, \ldots, n_k) = N! \prod_{j=1}^{k} (p_j^{n_j}/n_j!) \tag{12.1}$$

$$0 \leq n_j; \ \sum_{j=1}^{k} n_j = N.$$

The probability function (12.1) is termed as multinomial distribution with parameters $N, p_1, p_2, \ldots, p_k$.

The probability function (12.1) can be regarded as the coefficient of $\prod_{j=1}^{k} t_j^{n_j}$ in the multinomial expansion of

$$(t_1p_1 + t_2p_2 + \cdots + t_kp_k)^N$$

and hence (12.1) is called *multinomial distribution*. The marginal distribution of any

n_j is defined by

$$P(n_j) = \binom{N}{n_j} p_j^{n_j} (1-p_j)^{N-n_j}\ ;\ n_j = 0, 1, \ldots, N.$$

Also the distribution (12.1) reduces to binomial distribution when $k = 2$.

12.2 Moments of Multinomial Distribution

The moment generating function is given by

$$\begin{aligned}
M(t) &= M_{n_1, n_2, \ldots, n_k}(t_1, t_2, \ldots, t_k) = E\left[e^{\sum_{j=1}^{k} t_j n_j}\right] \\
&= \sum_j \frac{N!}{n_1!\, n_2! \ldots n_k!} p_1^{n_1} p_2^{n_2} \cdots p_k^{n_k} e^{\sum_{j=1}^{k} t_j n_j} \\
&= \sum_j \frac{N!}{\prod_{j=1}^{k} n_j!} \left(p_1 e^{t}\right)^{n_1} \left(p_2 e^{t_2}\right)^{n_2} \cdots \left(p_k e^{t_k}\right)^{n_k} \\
&= \left(p_1 e^{t_1} + p_2 e^{t_2} + p_3 e^{t_3} + \cdots + p_k e^{t_k}\right)^N,
\end{aligned}$$

where $t = (t_1, t_2, \ldots, t_k)$.

Now

$$\begin{aligned}
\mu_1'(n_j) &= \left.\frac{\partial M(t)}{\partial t_j}\right|_{t=0} \\
&= N p_j e^{t_j} \left(p_1 e^{t_1} + p_2 e^{t_2} + \cdots + p_k e^{t_k}\right)^{N-1}\Big|_{t=0} = N p_j. \\
\mu_2'(n_j) &= \left.\frac{\partial^2 M(t)}{\partial t_j^2}\right|_{t=0} \\
&= N p_j \Big[e^{t_j}\left(p_1 e^{t_1} + p_2 e^{t_2} + \cdots + p_k e^{t_k}\right)^{N-1} \\
&\quad + (N-1) e^{2t_j} p_j \left(p_1 e^{t_1} + p_2 e^{t_2} + \cdots + p_k e^{t_k}\right)^{N-2}\Big]_{t=0} \\
&= N p_j \left[1 + (N-1) p_j\right] \\
\therefore\ \mu_2(n_j) &= \mu_2'(n_j) - \mu_1'^2(n_j) \\
&= N p_j \left[1 + (N-1) p_j\right] - (N p_j)^2 = N p_j q_j \\
\underset{i \neq j}{E(n_i, n_j)} &= \left.\frac{\partial^2 M(t)}{\partial t_i \partial t_j}\right|_{t=0} \\
&= N p_j e^{t_j} (N-1) \left[p_1 e^{t_1} + p_2 e^{t_2} + \cdots + p_k e^{t_k}\right]^{N-2} p_i e^{t_i}\Big|_{t=0} \\
&= N(N-1) p_i p_j \\
\therefore\ \operatorname{cov}(n_i, n_j) &= E(n_i, n_j) - E(n_i) E(n_j) \\
&= N(N-1) p_i p_j - N^2 p_i p_j = -N p_i p_j.
\end{aligned}$$

Thus the correlation coefficient of n_i, n_j is

$$r(n_i, n_j) = \frac{-Np_ip_j}{N(p_jq_jp_iq_i)^{1/2}} = -\sqrt{\frac{p_ip_j}{q_iq_j}}.$$

We have already mentioned that the distribution of n_j is binomial and hence the other moments of n_j will be same as those of binomial distribution with parameters N, p_j.

The mixed factorial moment of multinomial distribution is given by

$$\mu_{(r_1,r_2,\ldots,r_k)} = E\left[n_1^{(r_1)} n_2^{(r_2)} \cdots n_k^{(r_k)}\right] = N^{\left(\sum_{j=1}^{k} r_j\right)} p_1^{r_1} p_2^{r_2} \cdots p_k^{r_k}.$$

The cumulant generating function is given by

$$K(t) = N \log\left(\sum_{j=1}^{k} p_j e^{t_j}\right).$$

The recurrence relations of cumulants as suggested by Guldburg (1935) can be derived from this formula of c.g.f., where the relation is

$$K_{r_1,r_2,\ldots,r_i+1,r_i+2,\ldots,r_k} = (p_i/p_j)\frac{\partial}{\partial(p_i/p_j)}[K_{r_1,r_2,\ldots,r_k}].$$

Wishart (1949) has given the explicit formulas for several lower cumulants. Some of these are

$$\begin{aligned}
K_{\ldots 21} &= -Np_ip_j(q_i - p_i)\\
K_{.111} &= 2Np_ip_jp_k\\
K_{..31} &= -Np_ip_j(1 - 6p_iq_i)\\
K_{..22} &= -Np_ip_j\{(q_i - p_i)(q_j - p_j) + 2p_ip_j\}\\
K_{\ldots 211} &= 2Np_ip_jp_k(q_i - 2p_i)\\
K_{.1111} &= -6Np_ip_jp_kp_l
\end{aligned}$$

12.3 Properties of Multinomial Distribution

Multinomial Distribution is Reproductive with Respect to N

Let $n_{11}, n_{12}, \ldots, n_{1k}$ have a joint multinomial distribution with parameters $N_1, p_1, p_2, \ldots, p_k$ and $n_{21}, n_{22}, \ldots, n_{2k}$ have also joint multinomial distribution with parameters $N_2, p_1, p_2, \ldots, p_k$. Let us assume that these two sets of variables are mutually independent of each other. Then $n_{11} + n_{21}, n_{12} + n_{22}, \ldots, n_{1k} + n_{2k}$ have also a joint multinomial distribution with parameters $(N_1 + N_2), p_1, p_2, \ldots, p_k$.

We have

$$M_{n_{11},n_{12},\ldots,n_{1k}}(t_1, t_2, \ldots, t_k) = \left(p_1e^{t_1} + p_2e^{t_2} + \ldots + p_ke^{t_k}\right)^{N_1}$$

$$\text{and } M_{n_{21},n_{22},\ldots,n_{2k}}(t_1, t_2, \ldots, t_k) = \left(p_1e^{t_1} + p_2e^{t_2} + \cdots + p_ke^{t_k}\right)^{N_2}.$$

Again

$$\begin{aligned}M_{n_{11}+n_{21},n_{12}+n_{22},\ldots,n_{1k}+n_{2k}}(t_1,t_2,\ldots,t_k)\\ &= M_{n_{11},n_{12},\ldots,n_{1k}}(t_1,t_2,\ldots,t_k)\times M_{n_{21},n_{22},\ldots,n_{2k}}(t_1,t_2,\ldots,t_k)\\ &= \left(p_1e^{t_1}+p_2e^{t_2}+\cdots+p_ke^{t_k}\right)^{N_1}\left(p_1e^{t_1}+p_2e^{t_2}+\cdots+p_ke^{t_k}\right)^{N_2}\\ &= \left(p_1e^{t_1}+p_2e^{t_2}+\cdots+p_ke^{t_k}\right)^{N_1+N_2}\end{aligned}$$

which is the m.g.f. of multinomial distribution with parameters $(N_1+N_2), p_1, p_2, \ldots, p_k$.

❑ **THEOREM 12.1** *If $X_1, X_2, \ldots, X_k$ are independent Poisson variates with parameters $\theta_1, \theta_2, \ldots, \theta_k$ respectively, then the conditional distribution of*

$$X_1 \cap X_2 \cap \cdots \cap X_k | X = X_1 + X_2 + \cdots + X_k \textit{ is multinomial.}$$

Proof. The conditional distribution of $X_1 \cap X_2 \cap \cdots \cap X_k|X$ is

$$\begin{aligned}&P(X_1 = n_1 \cap X_2 = n_2 \cap \cdots \cap X_k = n_k | X = n)\\ &= \frac{P(X_1 = n_1 \cap X_2 = n_2 \cap \cdots \cap X_k = n_k \cap X = n)}{P(X = n)}\\ &= \frac{P(X_1 = n_1 \cap X_2 = n_2 \cap \cdots \cap X_{k-1} = n_{k-1} \cap X_k = n - n_1 - n_2 - \cdots - n_{k-1})}{P(X = n)}\\ &= \frac{P(X_1 = n_1)P(X_2 = n_2)\cdots P(X_{k-1} = n_{k-1})P(X_k = n - n_1 - n_2 - \cdots - n_{k-1})}{P(X = n)}\\ &= \frac{\dfrac{e^{-\lambda_1}\lambda_1^{n_1}}{n_1!}\dfrac{e^{-\lambda_2}\lambda_2^{n_2}}{n_2!}\cdots\dfrac{e^{-\lambda_k}\lambda_k^{n-n_1-n_2-\cdots-n_{k-1}}}{(n-n_1-n_2-\cdots-n_{k-1})!}}{\dfrac{e^{-\lambda}\lambda^n}{n!}}\end{aligned}$$

with

$$\lambda = \sum_{i=1}^{k}\lambda_i,\ n = \sum_{i=1}^{k} n_i\ [\because\ X_1 + X_2 + \cdots + X_k \text{ is also Poisson variate}]$$

$$\begin{aligned}&= \frac{n!}{n_1!n_2!\cdots(n-n_1-n_2-\cdots-n_{k-1})!}\left(\frac{\lambda_1}{\lambda}\right)^{n_1}\left(\frac{\lambda_2}{\lambda}\right)^{n_2}\cdots\left(\frac{\lambda_k}{\lambda}\right)^{n-n_1-\cdots-n_{k-1}}\\ &= \frac{n!}{n_1!n_2!\cdots n_k!}p_1^{n_1}p_2^{n_2}\cdots p_k^{n_k}\end{aligned}$$

with $p_j = \dfrac{\lambda_j}{\lambda}$, $n_k = n - n_1 - n_2 - \cdots - n_{k-1}$. Also, $\sum_{j=1}^{k} p_j = 1$.

Hence the conditional distribution of $X_1 \cap X_2 \cap \cdots \cap X_k|X$ is multinomial.

Mode of the Multinomial Distribution

The mode(s) of the distribution is(are) located near the mean value, i.e.,

$$n_1 \doteq Np_1,\ n_2 \doteq Np_2,\ \ldots,\ n_k = Np_k.$$

The values of $n_1, n_2, \ldots, n_k$ satisfy the relation

$$N_{p_i} - 1 < n_i \le (N + k - 1)p_i; \; i = 1, 2, \ldots, k.$$

If k is small, there may be only one combination of values of $n_1, n_2, \ldots, n_k$ satisfying the relation.

Joint Distribution of Subset of n_j's

Let $N_1, N_2, \ldots, N_s$ be the s subsets of the random variable $n_1, n_2, \ldots, n_k$. Then the joint distribution of $N_1, N_2, \ldots, N_s$ is given by

$$P(N_1, N_2, \ldots, N_s)$$

$$= N! \left[\left(N - \sum_{i-1}^{s} N_i\right)! \prod_{i=1}^{s} N_i!\right]^{-1} \times \left(1 - \sum_{i=1}^{s} p_i\right)^{N - \sum_{i=1}^{s} N_i} \prod_{i=1}^{s} p_i^{N_i}$$

When $s = 1$, we have

$$P(N_1) = \frac{N!}{(N - N_1)! N_1!} (1 - p_1)^{N - N_1} p_1^{N_1}$$

which is a binomial distribution.

Approximations of Multinomial Distribution

The most commonly used approximation of multinomial distribution is chi-squared (χ^2) approximation. As N tends to infinity and $p_1, p_2, \ldots, p_k$ remain constant, the limiting form of $P(n_1, n_2, \ldots, n_k)$ is

$$P(n_1, n_2, \ldots, n_k) \doteq \frac{1}{\sqrt{2\pi N}} \left(\prod_{j=1}^{k} p_j\right)^{-\frac{1}{2}}$$

$$\times \exp\left[-\frac{1}{2}\sum_{j=1}^{k} \frac{(n_j - Np_j)^2}{Np_j} - \frac{1}{2}\sum_{j=1}^{k} \frac{(n_j - Np_j)}{Np_j} + \frac{1}{6}\sum_{j=1}^{k} \frac{(n_j - Np_j)^3}{(Np_j)^2}\right]$$

Neglecting the terms in $N^{-\frac{1}{2}}$, we get

$$P(n_1, n_2, \ldots, n_k) \doteq \frac{1}{\sqrt{2\pi N}} \left(\prod_{j=1}^{k} p_j\right)^{-\frac{1}{2}} \exp\left[-\frac{1}{2}\sum_{j=1}^{k} \frac{(n_j - Np_j)^2}{Np_j}\right]$$

$$\doteq \frac{1}{\sqrt{2\pi N}} \left(\prod_{j=1}^{k} p_j\right)^{-\frac{1}{2}} \exp\left(-\frac{1}{2} X^2\right)$$

where

$$X^2 = \sum \frac{(\text{observed value} - \text{expected value})^2}{\text{expected value}}.$$

The accuracy of this approximation increases as

$$\min(Np_1, Np_2, \ldots, Np_k)$$

increases and decreases with increasing k. Pearson (1900) showed that X^2 is approximately distributed as χ^2 with $(k-1)$ d.f., while

$$E(X^2) = E(\chi^2) = k - 1.$$

But

$$V(X^2) = \left(1 - \frac{1}{N}\right) 2(k-1) + \frac{1}{N}\left(\sum_{j=1}^{k} p_j^2 - k^2\right).$$

12.4 Compound Multinomial Distribution

Let $n_1, n_2, \ldots, n_k$ be the variables which follow multinomial distribution with parameters $N, p_1, p_2, \ldots, p_k$. Let us assume that $p_1, p_2, \ldots, p_k$ follow Dirichlet distribution with parameters $\alpha_1, \alpha_2, \ldots, \alpha_k$. Then the compound multinomial distribution is defined by

$$\begin{aligned} P(n_1, n_2, \ldots, n_k) &= \frac{N!}{\prod_{j=1}^{k} n_j!} E\left[\prod_{j=1}^{k} p_j^{n_j}\right] \\ &= \frac{N!\,\Gamma\left(\sum_{j=1}^{k} \alpha_j\right)}{\Gamma\left(N + \sum_{j=1}^{k} \alpha_j\right)} \prod_{j=1}^{n} \frac{\Gamma(n_j + \alpha_j)}{\Gamma(\alpha_j)} \\ &= \frac{N!}{\left(\sum_{j=1}^{k} \alpha_j\right)^{[N]}} \prod_{j=1}^{k} \frac{\alpha_j^{[n_j]}}{n_j!} \end{aligned} \tag{12.2}$$

with $n_j \geq 0$, $\sum_{j=1}^{k} n_j = N$.

This distribution is called a *Dirichlet* (or *β*-) *compound multinomial distribution.* Ishii and Hayakawa (1960) named this distribution as *multivariate binomial-beta distribution.* The marginal distribution of n_j is a compound binomial-beta distribution, where n_j is a binomial variable with parameters N, p_j. Again, p_j is distributed as beta distribution with parameters

$$\alpha_j \text{ and } \left(\sum_{i=1}^{k} \alpha_i - \alpha_j\right).$$

Thus

$$P(n_j = x) = \frac{\binom{N}{x}}{B\left(\alpha_j, \sum_{i=1}^{k} \alpha_i - \alpha_j\right)} \int_0^1 p_j^{x+\alpha_j-1} (1-p_j)^{N-x+\sum_{i=1}^{k} \alpha_i - \alpha_j - 1} d_{p_j}$$

$$= \binom{N}{x} \frac{\Gamma\left(\sum_{i=1}^{k} \alpha_i\right)}{\Gamma\alpha_j \, \Gamma\left(\sum_{i=1}^{k} \alpha_i - \alpha_j\right)} \frac{\Gamma(x+\alpha_j)\,\Gamma\left(N - x + \sum_{i=1}^{k} \alpha_i - \alpha_j\right)}{\Gamma\left(N + \sum_{i=1}^{k} \alpha_i\right)}$$

$$= \frac{\binom{x+\alpha_j-1}{x}\binom{N-x+\sum_{i=1}^{k} \alpha_i - \alpha_j - 1}{N-x}}{\binom{N+\sum_{i=1}^{k} \alpha_i - 1}{N}}.$$

For integer values of α_j's this is a negative hypergeometric distribution with parameters

$$N + \sum_{i=1}^{k} \alpha_i - 1, \; \alpha_j - 1, N.$$

It is also called *negative multivariate hypergeometric distribution.*

The factorial moments of distribution (12.2) are

$$\mu_{(r_1, r_2, \ldots, r_k)} = E\left[\prod_{j=1}^{k} n_j^{(r_j)}\right]$$

$$= N^{\left(\sum_{j=1}^{k} r_j\right)} \left(\prod_{j=1}^{k} \alpha_j\right)^{[r_j]} \Big/ \left(\sum_{j=1}^{k} \alpha_j\right)^{\left[\sum_{j=1}^{k} r_j\right]}$$

In particular

$$E(n_j) = Np'_j;$$

$$V(n_j) = \left(N + \sum_{i=1}^{k} \alpha_i\right)\left(1 + \sum_{j=1}^{k} \alpha_j\right)^{-1} Np'_j\,(1-p'_j)$$

where

$$p'_j = \alpha_j \left(\sum_{i=1}^{k} \alpha_i\right)^{-1} \quad \text{and} \quad \sum_{i=1}^{k} p'_i = 1.$$

The correlation coefficient of n_j and n_j is

$$-\sqrt{p'_i p'_j\,(1-p'_i)^{-1}\,(1-p'_j)^{-1}}.$$

EXERCISE

1. (a) Multinomial distribution is an extension of binomial distribution. Can you justify it? Find the moments of multinomial distribution. Show that the multinomial distribution can be approximated by chi-square distribution.

(b) In a packet of electronic goods there are 15 mobile phone sets, 5 optical 3D mouses, 5 key boards and 10 pieces of television remote control. A sample of 10 pieces is randomly selected. What is the probability of selection of 3 mobile phone sets, 2 optical mouses, 2 key boards and 3 pieces of remote control?

2. (a) Write down examples of variable which follow multinomial distribution. Show that, if $X_1, X_2, \ldots, X_k$ are k independent Poisson variates, then the conditional distribution of

$$X_1 \cap X_2 \cap \cdots \cap X_k | X = \sum_{i=1}^{k} X_i \text{ is multinomial.}$$

Find the mode of multinomial distribution.

(b) A garments industry produces 40% shirts, 20% jackets, and 40% trousers. From a day's output 10 items are selected at random. Find the probability that, out of 10 items 4 are shirts 3 are jackets and another 3 are trousers.

3. (a) State and prove important properties of multinomial distribution. Show that under certain condition multinomial distribution transforms to binomial distribution. Find mean, variance and covariance of multinomial distribution.

(b) Discuss the chi-square approximation of multinomial distribution.

CHAPTER 13

Logarithmic Series Distribution

13.1 Definition and Applications

The logarithmic series distribution can be defined in terms of the expansion of $-\alpha\log(1-\theta)$. This function $-\alpha\log(1-\theta)$ arises in the experiment of biological sciences. For example, let us consider the experiment of collection of moths by means of light-trap. In this experiment, if the number of species represented by exactly one individual in the sample is n_1, then the number of species represented by two, three, etc., ... individuals are approximately

$$\left(\frac{n_1}{\theta}\right)\frac{\theta^2}{2}, \left(\frac{n_1}{\theta}\right)\frac{\theta^3}{3}, \ldots, \text{ etc.}$$

respectively, where θ is a positive number less than 1. The total number of species is (approximately)

$$\sum_{k=1}^{\infty}\left(\frac{n_1}{\theta}\right)\frac{\theta^k}{k} = -\frac{n_1}{\theta}\log(1-\theta)$$

$$= \frac{n_1}{\theta}\left[\theta + \frac{\theta^2}{2} + \frac{\theta^3}{3} + \cdots\right]$$

which is a similar function to $-\alpha\log(1-\theta)$. In the expansion of this function the k-th term can be written as a function of x which is

$$P(X = k) = \alpha\frac{\theta^k}{k},\ k = 1, 2, \ldots.$$

If

$$\alpha = -\frac{1}{\log(1-\theta)}, \text{ then } \sum_{k=1}^{\infty} P(X = k) = 1.$$

Therefore,

$$P(X = k) = -\frac{1}{\log(1-\theta)}\frac{\theta^k}{k},\ k = 1, 2, \ldots \tag{13.1}$$

is a probability function.

A random variable X has a logarithmic series distribution if its probability function is given by (13.1).

Fisher *et al* (1943) have given a theoretical analysis of the logarithmic series distribution. According to them the number of individuals from a given species, say the i-th, to be caught in a specified trap can be represented by a Poisson variate with parameter λ_i. Let us assume that the λ_i's are the random sample from gamma distribution. Then the expected total number of species represented in a given catch by $k = 1, 2, 3, \ldots$ individuals would be proportional to terms in a negative binomial expansion, truncated by exclusion of the term corresonding to $k = 0$. As the exponent $(-N)$ of the negative binomial tends to zero, then the probability of taking the value k tends to

$$\lim_{N\to 0} \frac{\Gamma(N+k)}{k!\,\Gamma(N+1)} \frac{N}{Q^N - 1}\left(\frac{P}{Q}\right)^k = \alpha\theta^k/k$$

where $\alpha = P/Q$, $k = 1, 2, \ldots$.

This distribution also arises in the study of number of plants observed in a quadrat in quadrat sampling. The growth of population can be represented by a discrete Markov process leading to a negative binomial distribution and the limiting form of the distribution is a logarithmic series distribution. The distribution of number of items of a commodity purchased by a buyer in a specified period of time can also be represented by logarithmic series distribution.

13.2 Moments and Generating Functions of Logarithmic Series Distribution

The moment generating function of this distribution is given by

$$\begin{aligned} M_x(t) &= E\left[e^{tx}\right] = -\frac{1}{\log(1-\theta)} \sum_{x=1}^{\infty} e^{tx} \frac{\theta^x}{x} \\ &= -\frac{1}{\log(1-\theta)} \sum_{x=1}^{\infty} \frac{(\theta e^t)^x}{x} = \frac{\log(1-\theta e^t)}{\log(1-\theta)} \\ &= -\alpha \log\left(1-\theta e^t\right), \text{ with } \alpha = -\frac{1}{\log(1-\theta)}. \end{aligned}$$

Now

$$\begin{aligned} \mu_1' &= \left.\frac{\partial M_x(t)}{\partial t}\right|_{t=0} = \alpha \left.\frac{\theta e^t}{1-\theta e^t}\right|_{t=0} = \frac{\alpha\theta}{1-\theta} \\ \mu_2' &= \left.\frac{\partial^2 M_x(t)}{\partial t^2}\right|_{t=0} = \alpha\theta\left[\frac{e^t}{1-\theta e^t} + \frac{\theta e^{2t}}{(1-\theta e^t)^2}\right]_{t=0} \\ &= \alpha\theta\left[\frac{1}{1-\theta} + \frac{\theta}{(1-\theta)^2}\right] = \alpha\theta\frac{1}{(1-\theta)^2} \\ \mu_2 &= \mu_2' - \mu_1'^2 = \frac{\alpha\theta}{(1-\theta)^2} - \frac{(\alpha\theta)^2}{(1-\theta)^2} = \frac{\alpha\theta(1-\alpha\theta)}{(1-\theta)^2} \end{aligned}$$

$$\mu_3' = \alpha\theta\left[\left\{\frac{e^t}{1-\theta e^t}+\frac{\theta e^{2t}}{(1-\theta e^t)^2}\right\}+\theta\left\{\frac{2e^{2t}}{(1-\theta e^t)^2}+\frac{2\theta e^{3t}}{(1-\theta e^t)^3}\right\}\right]_{t=0}$$

$$= \alpha\theta\left[\frac{1}{(1-\theta)^2}+\frac{2\theta}{(1-\theta)^3}\right]=\frac{\alpha\theta(1+\theta)}{(1-\theta)^3}$$

$$\mu_4' = \alpha\theta\left[\left\{\frac{e^t}{(1-\theta e^t)}+\frac{\theta e^{t}}{(1-\theta e^t)^2}\right\}+\theta\left\{\frac{2e^{2t}}{(1-\theta e^t)^2}+\frac{2\theta e^{3t}}{(1-\theta e^t)^3}\right\}\right.$$

$$\left.+2\theta\left\{\frac{2e^{2t}}{(1-\theta e^t)^2}+\frac{2\theta e^{3t}}{(1-\theta e^t)^3}\right\}+2\theta^2\left\{\frac{3e^{3t}}{(1-\theta e^t)^3}+\frac{3\theta e^{4t}}{(1-\theta e^t)^4}\right\}\right]_{t=0}$$

$$= \alpha\theta\left[\frac{1}{(1-\theta)^2}+\frac{6\theta}{(1-\theta)^3}+\frac{6\theta^2}{(1-\theta)^4}\right]=\frac{\alpha\theta(1+4\theta+\theta^2)}{(1-\theta)^4}$$

$$\mu_3 = \mu_3' - 3\mu_2'\mu_1' + 2\mu_1'^3$$

$$= \alpha\theta\left[\frac{(1+\theta)}{(1-\theta)^3}-\frac{3\alpha\theta}{(1-\theta)^3}+\frac{2(\alpha\theta)^2}{(1-\theta)^3}\right]$$

$$= \frac{\alpha\theta}{(1-\theta)^3}(1+\theta-3\alpha\theta+2\alpha^2\theta^2)$$

$$\mu_4 = \mu_4' - 4\mu_3'\mu_1' + 6\mu_2'\mu_1'^2 - 3\mu_1'^4$$

$$= \alpha\theta\left[\frac{1+4\theta+\theta^2}{(1-\theta)^4}-\frac{4\alpha\theta(1+\theta)}{(1-\theta)^4}+\frac{6(\alpha\theta)^2}{(1-\theta)^4}-\frac{3(\alpha\theta)^3}{(1-\theta)^4}\right]$$

$$= \frac{\alpha\theta}{(1-\theta)^4}\left\{1+4\theta+\theta^2-4\alpha\theta(1+\theta)+6\alpha^2\theta^2-3\alpha^3\theta^3\right\}.$$

The moment ratios are

$$\beta_1 = \frac{1}{\alpha\theta}\frac{\left(1+\theta-3\alpha\theta+2\alpha^2\theta^2\right)^2}{(1-\alpha\theta)^3}$$

$$\text{and } \beta_2 = \frac{1}{\alpha\theta}\frac{\left\{1+4\theta+\theta^2-4\alpha\theta(1+\theta)+6\alpha^2\theta^2-3\alpha^3\theta^3\right\}}{(1-\alpha\theta)^2}.$$

It is observed that if $\theta \to 0$, both β_1 and β_2 tends to zero. This is true even θ tends to 1.

The probability generating function of logarithmic series distribution is given by

$$g(t) = E(t^x) = \alpha\sum_{x=1}^{\infty}\frac{\theta^x}{x}t^x = \alpha\sum_{x=1}^{\infty}\frac{(\theta t)^x}{x} = \frac{\log(1-\theta t)}{\log(1-\theta)}.$$

The r-th factorial moment of this distribution is given by

$$\mu_{(r)} = E\left[x^{(r)}\right] = \alpha\sum_{x=1}^{\infty}\frac{\theta^x x^{(r)}}{x} = \alpha\theta^r\sum_{x=1}^{\infty}\frac{x^{(r)}\theta^{x-r}}{x}$$

$$= \alpha\theta^r\sum_{x=r}^{\infty}(x-1)(x-2)\cdots(x-r+1)\theta^{x-r}$$

$$= \alpha\theta^r\frac{d^{r-1}}{d\theta^{r-1}}\left[\sum_{x=1}^{\infty}\theta^{x-1}\right] = \alpha\theta^r(r-1)!(1-\theta)^{-r}.$$

We have

$$\mu'_{r+1} = \alpha \sum_{x=1}^{\infty} x^{r+1} \frac{\theta^x}{x} \text{ and } \mu'_r = \alpha \sum_{x=1}^{\infty} x^r \frac{\theta^x}{x}$$

$$\frac{d\mu'_r}{d\theta} = \alpha \sum_{x=1}^{\infty} x^{r+1} \frac{\theta^{x-1}}{x} - \frac{\alpha}{\log(1-\theta)} \sum x^r \frac{\theta^x}{x} \frac{1}{1-\theta}$$

$$\theta \frac{d\mu'_r}{d\theta} = \alpha \sum_{x=1}^{\infty} x^{r+1} \frac{\theta^x}{x} - \frac{\alpha\theta}{1-\theta} \mu'_r.$$

$$\therefore \mu'_{r+1} = \theta \frac{d\mu'_r}{d\theta} + \frac{\alpha\theta}{1-\theta} \mu'_r.$$

In a similar way, it can be shown that

$$\mu_{r+1} = \theta \frac{d\mu_r}{d\theta} + r\mu_2\mu_{r-1}.$$

The cumulants of logarithmic series distribution satisfy the recurrence relations

$$K_r = \theta \frac{dK_{r-1}}{d\theta}.$$

The factorial cumulants satisfy the relations

$$K_{(r)} = \theta \frac{dK_{(r-1)}}{d\theta} - (r-1)K_{(r-1)}.$$

13.3 Properties of Logarithmic Series Distribution

Mode of the Distribution

We have

$$P(x) = \alpha\theta^x / x; \; x = 1, 2, \ldots .$$

Thus

$$\frac{P(x+1)}{P(x)} = \frac{x\theta}{x+1} < 1 \; [\because \; \theta < 1] \text{ for all values of } x.$$

From this, we have

$$P(1) > P(2) > P(3) > \cdots$$

i.e., the maximum probability at the initial value of $x = 1$. Hence logarithmic series distribution is unimodal as modal value is at $x = 1$.

For large values of x, we have

$$\frac{P(x+1)}{P(x)} = \frac{\theta^{x+1}}{\theta^x} = \theta$$

which is similar to geometric distribution. We have already discussed that β_1 and β_2 tend to ∞ as θ tends to 0 or as θ tends to 1. Therefore, the distribution has a long positive tail. Also, we have

$$\lim_{\theta\to 0}(\beta_2/\beta_1) = 1$$

$$\text{and } \lim_{\theta\to 1}(\beta_2/\beta_1) = \frac{3}{2}.$$

The cumulative probability of logarithmic series distribution is written as

$$P(X \le x) = \alpha \sum_{j=1}^{x} \frac{\theta^j}{j}.$$

This can be approximated using

$$\sum_{j=1}^{x} \frac{\theta^j}{j} = -\log(1-\theta) - \sum_{j=x+1}^{\infty} \frac{\theta^j}{j}.$$

Here

$$\begin{aligned}\sum_{j=x+1}^{\infty} \frac{\theta^j}{j} &= -\log(1-\theta) - \sum_{j=1}^{x} j^{-1} + \int_0^{1-\theta} \phi^{-1}\{1-(1-\phi)^x\}d\phi \\ &\doteq -\log(1-\theta) - \left(\gamma + \frac{1}{2}x^{-1} - \frac{1}{12}x^{-2} + \log x\right) \\ &\quad + [-E_i\{-x(1-\theta)\} + \gamma + \log\{x(1-\theta)\}] \\ &\doteq -E_i[-x(1-\theta)] - \frac{1}{2}x^{-1} + \frac{1}{12}x^{-2}.\end{aligned}$$

Here

$$E_i(u) = \int_{-\infty}^{u} z^{-1}e^z dz$$

$$\text{and } \gamma = \lim_{x\to\infty}\left\{\sum_{j=1}^{x} j^{-1} - \log x\right\}.$$

If $x(1-\theta)$ is small, then the approximation is given by

$$\sum_{j=x+1}^{\infty} \frac{\theta^j}{j} \doteq -\log[x(1-\theta)] + x(1-\theta) - \gamma. \qquad (13.2)$$

If $x(1-\theta)$ is large, then

$$\begin{aligned}\sum_{j=x+1}^{\infty} \frac{\theta^j}{j} &\doteq \theta^{x+1}(1-\theta)^{-1}[x^{-1} - 1!\{x(x-1)\}^{-1}(1-\theta)^{-1} \\ &\quad + 2!\{x(x-1)(x-2)\}^{-1}(1-\theta)^{-2} - \cdots \\ &\quad + (-1)^r r!\{x(x-1)(x-2)\cdots(x-r)\}^{-1}(1-\theta)^{-r}].\end{aligned}$$

Using (13.2) the approximate median of the distribution is given by

$$\text{Median} \doteq e^{-\gamma}(1-\theta)^{-1/2} = 0.56146(1-\theta)^{-1/2}.$$

13.4 Multivariate Logarithmic Series Distribution

It has already been mentioned in Section 13.1 that logarithmic series distribution can be derived as a limiting form truncated negative binomial distribution where truncation is done at $x = 0$. A similar limiting form of multivariate negative binomial distribution can be derived where the negative binomial distribution is truncated at

$$n_1 = n_2 = \cdots = n_k = 0.$$

From (9.4), we have

$$P_1(n_1, n_2, \ldots, n_k) = \frac{\Gamma\left(N + \sum_{j=1}^{k} n_j\right)}{\prod_{j=1}^{k} n_j! \, \Gamma N \left(1 - Q^{-N}\right)} \prod_{j=1}^{k} (P_i/Q)^{n_j}$$

where $P_1(n_1, n_2, \ldots, n_k)$ is a truncated distribution by omitting $n_1 = n_2 = \cdots = n_k = 0$. As N tends to zero

$$\lim_{N \to 0} \frac{\Gamma\left(N + \sum_{j=1}^{k} n_j\right)}{\Gamma N \left(1 - Q^{-N}\right)} = \lim_{N \to 0} \frac{N(N+1)\cdots\left(N + \sum_{1}^{k} n_j - 1\right)}{\left(1 - Q^{-N}\right)}$$

$$= \left(\sum_{j=1}^{k} n_j - 1\right)! \lim_{N \to 0} N \left(1 - Q^{-N}\right)^{-1} = \frac{\left(\sum_{j=1}^{k} n_j - 1\right)!}{\log Q}.$$

Thus $P_1(n_1, n_2, \ldots, n_k)$ stands as $(N \to 0)$

$$\lim_{N \to 0} P_1(n_1, n_2, \ldots, n_k) = \frac{\left(\sum_{j=1}^{k} n_j - 1\right)!}{\left(\prod_{j=1}^{k} n_j!\right) \log Q} \prod_{j=1}^{k} (P_i/Q)^{n_j}.$$

Let

$$P_i/Q = \theta_i \left(0 < \sum_{i=1}^{k} \theta_i < 1\right) \text{ such that } Q = \left(1 - \sum_{i=1}^{k} \theta_i\right)^{-1}.$$

Then we have

$$P_1(n_1, n_2, \ldots, n_k) = \frac{\left(\sum_{j=1}^{k} n_j - 1\right)!}{\prod_{j=1}^{k} n_j! \left[-\log\left(1 - \sum_{i=1}^{k} \theta_i\right)\right]} \prod_{j=1}^{k} \theta_j^{n_j} \tag{13.3}$$

$$\text{with } n_j \geq 0; \sum_{i=1}^{k} n_i > 0.$$

The distribution (13.3) is called *multivariate logarithmic series distribution* with parameters $\theta_1, \theta_2, \ldots, \theta_k$.

Taking sum on (13.3) and keeping $\sum_{i=1}^{k} n_i = N > 0$ we have

$$\sum P_1(n_1, n_2, \ldots, n_k) = \frac{\left(\sum_{j=1}^{k} n_j - 1\right)!}{-\log\left(1 - \sum_{i=1}^{k} \theta_i\right)} \sum \prod_{j=1}^{k} (\theta_j^{n_j}/n_j!)$$

$$= -\frac{1}{\log\left(1 - \sum_{i=1}^{k} \theta_i\right)} \left(\sum_{j=1}^{k} \theta_j\right)^N \bigg/ N.$$

Thus $\sum_{j=1}^{k} n_j$ has a logarithmic series distribution with parameter $\sum_{j=1}^{k} \theta_j$.

The joint probability generating function is

$$\lim_{N \to 0} \frac{\left(Q - \sum_{i=1}^{k} P_i t_i\right)^{-N} - Q^{-N}}{1 - Q^{-N}} = \frac{-\log\left[1 - \sum_{i=1}^{k} P_i t_i / Q\right]}{\log Q}$$

$$= \frac{\log\left(1 - \sum_{i=1}^{k} \theta_i t_i\right)}{\log\left(1 - \sum_{i=1}^{k} \theta_i\right)}.$$

Putting $t_2 = t_3 = \cdots = t_k = 1$, we get

$$\frac{\log\left(1 - \sum_{i=2}^{k} \theta_i - \theta_1 t_1\right)}{\log\left(1 - \sum_{i=1}^{k} \theta_i\right)}$$

which is the probability generating function of the random variable n_1. Thus the distribution of n_1 is

$$P(n_1 = 0) = \frac{\log\left(1 - \sum_{i=2}^{k} \theta_i\right)}{\log\left(1 - \sum_{i=1}^{k} \theta_i\right)}$$

$$\text{and } P(n_1 = r) = \frac{r^{-1}}{-\log\left(1 - \sum_{i=1}^{k} \theta_i\right)} \left[\frac{\theta_1}{1 - \sum_{i=2}^{k} \theta_i}\right]^r, \quad r \geq 1$$

which is a modified logarithmic series distribution.

The factorial moment of multivariate logarithmic series distribution is given by

$$\mu_{(r_1,r_2,\ldots,r_k)} = E\left[\prod_{j=1}^{k} n_j^{(r_j)}\right] = \left[\frac{\left(\sum_{j=1}^{k} r_j - 1\right)!}{-\log\left(1-\sum_{j=1}^{k}\theta_j\right)} + E\left[\left(\sum_{j=1}^{k} n_j\right)^{\left[\sum_{j=1}^{k} r_j\right]}\right]\right]\prod_{j=1}^{k}\theta_j^{r_j}$$

$$= \frac{\left(\sum_{j=1}^{k} r_j - 1\right)!\prod_{j=1}^{k}\theta_j^{r_j}}{\left[-\log\left(1-\sum_{j=1}^{k}\theta_j\right)\right]\left(1-\sum_{j=1}^{k}\theta_j\right)^{\sum_1^k r_i}} \quad (r_j\text{'s not all zero}).$$

From this, we have

$$E(n_j) = \frac{\theta_j}{-\left(1-\sum_{i=1}^{k}\theta_i\right)\log\left(1-\sum_{i=1}^{k}\theta_i\right)}$$

$$V(n_j) = \frac{\theta_j}{\left(1-\sum_{i=1}^{k}\theta_i\right)\log\left(1-\sum_{i=1}^{k}\theta_i\right)} \times \left[\frac{\theta_j}{1-\sum_{i=1}^{k}\theta_i}\left\{1-\frac{1}{-\log\left(1-\sum_{i=1}^{k}\theta_i\right)}\right\}^{-1}\right]$$

$$\operatorname{cov}(n_i,n_j) = -\frac{\theta_i\theta_j}{\left(1-\sum_{i=1}^{k}\theta_i\right)^2\left[-\log\left(1-\sum_{i=1}^{k}\theta_i\right)\right]}\left[\frac{1}{-\log\left(1-\sum_{i=1}^{k}\theta_i\right)}-1\right].$$

EXERCISE

1. Write down some examples of the variable which follows logarithmic series distribution. Find moment generating function of logarithmic series distribution. Hence or otherwise find the values of β_1 and β_2.
2. Write down some experiments, where experimental results can be expressed by logarithmic series distribution. Show that $k_r = \theta\frac{dk_{r-1}}{d\theta}$, where θ is the parameter of the logarithmic series distribution and k_r is the r-th cumulant of the distribution.
3. Discuss the origin of logarithmic series distribution. For logarithmic series distribution, show that

$$\mu_{r+1} = \theta\frac{d\mu_r}{d\theta} + r\mu_2\mu_{r-1},$$

where θ is the parameter of logarithmic series distribution and μ_r is the r-th central moment of the distribution. Find the mode of logarithmic series distribution.

CHAPTER 14

Some More Discrete Distributions

14.1 Waiting-Time Distributions

The distribution arises in an experiment where units are drawn successively from a finite population. Let us consider that an urn contains N balls, M of which are white and $(N - M)$ are black. Let balls be drawn successively until exactly k white balls are drawn. In such an experiment two cases arise, viz. (a) Hypergeometric case and (b) Binomial case.

14.1.1 Hypergeometric Case

Let the probability of drawing a white ball is p and probability of drawing a black ball is q where $p > 0$, $q > 0$ and $p + q = 1$. We need to determine the probability that exactly x balls will have to be drawn to get k white balls. From the urn white balls can be drawn in $\binom{N}{M}$ ways. Thus the sample points of the experiment are $\binom{N}{M}$.

For each sample point, the value of the random variable x is equal to the number of balls drawn in order to get exactly k white balls. For any value x_1 satisfying both inequalities $0 \leq k \leq x_1 - 1$ and $0 \leq M - k \leq N - x_1$, the number of sample points for which $X = x_1$ is seen to be

$$\binom{x_1 - 1}{k - 1}\binom{N - x_1}{M - k}.$$

If all sample points have equal probabilities, then we have

$$P(X = x_1) = \frac{\binom{x_1-1}{k-1}\binom{N-x_1}{M-k}}{\binom{N}{M}}. \tag{14.1}$$

Thus, we have

$$P(x) = \frac{\binom{x-1}{k-1}\binom{N-x}{M-k}}{\binom{N}{M}}. \tag{14.2}$$

The distribution with (14.2) as probability function is called *hypergeometric waiting-time distribution* since we need to wait up to number of trials x to get k white balls.

Here the sample space of x is the set of integers

$$k, k+1, \ldots, N-M+k, 1 \leq k \leq M.$$

The r-th factorial moment of x with respect to k is given by

$$\begin{aligned} E\left[(x-k)^{(r)}\right] &= \sum_{x=k}^{N-M+k} (x-k)^{(r)} \frac{\binom{x-1}{k-1}\binom{N-x}{M-k}}{\binom{N}{M}} \\ &= (k+r-1)^{[r]} \left[\sum_{x=k+r}^{N-M+k} \binom{x-1}{k+r-1}\binom{N-x}{M-k} \Big/ \binom{N}{M}\right] \\ &= (k+r-1)^{[r]} \binom{N}{M+r} \Big/ \binom{N}{M}. \end{aligned}$$

In particular, when $r = 1$, the mean of x is given by

$$\mu_1' = \frac{k(N+1)}{M+1}.$$

Also

$$\begin{aligned} \mu_2 = & \frac{(N-M)(N-M-1)k(k+1)}{(M+2)(M+1)} \\ & + \frac{k(2k+1)(N+1)}{M+1} - \frac{k^2(N+1)^2}{(M+1)^2} - k(k+1). \end{aligned}$$

14.1.2 Binomial Case

Let us again consider the trial of drawing balls from an urn containing M white and $N-M$ black balls to get exactly k white balls. Assume that trials are performed successively and the outcome of each trial is either a white ball or a black ball, since the trials are independent. The sample space consists of all possible sequences of white balls and black balls which could occur in an indefinitely long sequence of trials. Let x be a random variable indicating the number of trials needed in order to get exactly k white balls, where x can be defined for each sample point.

Let A be an event in the sample space indicating $(k-1)$ white balls in the first $(x-1)$ trials and B be the event of getting a white ball on the x-th trial. Now, if E be the event of making exactly x trials to get exactly k white balls, then $E = A \cap B$. Therefore,

$$P(E) = P(A \cap B) = P(A)P(B/A).$$

But

$$\begin{aligned} P(B/A) &= P(1 \text{ white ball}) = p \\ \text{and } P(A) &= \binom{x-1}{k-1} p^{k-1} q^{x-k}. \\ \therefore\ P(x) = P(E) &= \binom{x-1}{k-1} p^k q^{x-k} \qquad (14.3) \end{aligned}$$

where

$$x = k, k+1, \ldots \; (1 \le k \le M).$$

Here $P(x)$ indicates the probability that one must wait through x independent trials in order to get k white balls. The distribution having probability function (14.3) is called the *binomial waiting-time distribution.* The distribution is also called the negative binomial distribution and the Pascal distribution. Thus the characteristic function of this distribution will be found out similarly as it has been found out in Section 9.3. Here the characteristic function is

$$\begin{aligned}\psi(t) &= \sum_{x=k}^{\infty} \binom{x-1}{k-1} \left(pe^{it}\right)^k \left(qe^{it}\right)^{x-k} \\ &= \left(pe^{it}\right)^k \left(1 - qe^{it}\right)^{-k} = p^k \left(e^{-it} - q\right)^{-k}.\end{aligned}$$

Also we have

$$\mu_1' = \frac{k}{p}, \; \mu_2 = \frac{kq}{p^2}.$$

14.2 Distributions of Runs

A run is a sequence of their sample observations from a population when at least two sets of sample observations from two independent populations are mixed up and arranged in ascending order of magnitude. The theory of runs plays an important role in nonparametric statistical inference. Some of the nonparametric tests involve the distribution of runs.

Runs can be studied from two kinds of observations or from more than two kinds (k kinds). Thus the distribution of runs can be derived from runs of two kinds of elements and from runs of k kinds of elements.

Distribution of Runs from Two Kinds of Elements

Let there be two kinds of elements denoted by S and F, where the number of S's and number of F's are n_1 and n_2 respectively. Let us consider a sample space whose sample points are $\binom{n}{n_1}$ permutations of n_1 S's and n_2 F's, where $n = n_1 + n_2$. The arrangement of S's and F's may be as follows:

$$SSFFFSFFSSSFFFFSSSSFS\cdots.$$

In the arrangement there are runs of S's and F's of different lengths. The length of a run is the number of elements in it. Let r_{1j} be the number of runs of S's of length j and r_{2j} be the number of runs of F's of length j in the above arrangement of S's and F's. Then $r_{11} = 2$, $r_{12} = 1$, $r_{13} = 1$, $r_{14} = 1$, $r_{21} = 1$, $r_{22} = 1$, $r_{23} = 1$, $r_{24} = 1$, where $n_1 = 11$, $n_2 = 10$. The other values of r are zero.

It is observed that

$$\sum_j jr_{1j} = n_1 \text{ and } \sum_j jr_{2j} = n_2.$$

Let

$$r_1 = \sum_j r_{1j} \text{ and } r_2 = \sum_j r_{2j}$$

be the total number of runs of S's and F's respectively. Now, in the sample space we can define the random variables r_{1j} and r_{2j}. We need the probability function of r_{1j} and r_{2j} when all sample points in the sample space are assigned equal probabilities. For a given set of values of r_{1j}, the number of arrangements of r_1 runs of S's is

$$\frac{r_1!}{r_{11}!\, r_{12}! \cdots r_{1n_1}!}.$$

Similarly, the number of arrangements of r_2 runs of F's is

$$\frac{r_2!}{r_{21}!\, r_{22}! \cdots r_{2n_2}!}.$$

Let $f(n_1, n_2)$ be the number of ways of arranging r_1 indistinguishable objects of one kind and r_2 indistinguishable objects of a second kind so that no two objects of the same kind appear together; then we have

$$f(n_1, n_2) = \begin{cases} 0, & |r_1 - r_2| > 1 \\ 1, & |r_1 - r_2| = 1 \\ 2, & r_1 = r_2 \end{cases} \tag{14.4}$$

Thus the total number of ways of getting r_{1j} runs of S's of lengths $j = 1, 2, \ldots, n_1$ and of getting r_{2j} runs of F's of lengths $j = 1, 2, \ldots, n_2$ is

$$\frac{r_1!\, r_2!\, f(n_1, n_2)}{r_{11}!\, r_{12}! \cdots r_{1n_1}!\, r_{21}!\, r_{22}! \cdots r_{2n_2}!} \tag{14.5}$$

There are $\binom{n}{n_1}$ arrangements of n_1 S's and n_2 F's. If equal probabilities are assigned to all these arrangements, then $(n_1 + n_2)$-dimensional random variable

$$r_{ij}\ (i = 1, 2;\ j = 1, 2, \ldots, n_i)$$

will have the probability function

$$P(\{r_{ij}\}) = \frac{r_1!r_2!}{r_{11}!r_{12}! \cdots r_{1n_1}!r_{21}! \cdots r_{2n_2}!} \frac{n_1!n_2!}{n!} f(n_1, n_2).$$

The probability function of the runs of S's that is $P(\{r_{ij}\})$ will be obtained by taking sum of $P(\{r_{ij}\})$ with respect to $r_{21}, r_{22}, \ldots, r_{2n_2}$ such that

$$\sum_j j r_{2j} = n_2 \text{ and } \sum_j r_{2j} = r_2.$$

Thus the probability function of r_{1j} and r_2 is given by

$$P(\{r_{1j}\}, r_2) = \frac{r_1!}{r_{11}! \cdots r_{1n_1}!} \frac{(n_2 - 1)!}{(r_2 - 1)!(n_2 - r_2)!} \frac{n_1!n_2!}{n!} f(n_1, n_2). \tag{14.6}$$

Taking sum of (14.6) with respect to r_2, the probability function of r_{1j} is obtained. At this stage using (14.4) and after some simplification, we have

$$\sum_{r_2=1}^{n_2} \frac{(n_2-1)!}{(r_2-1)!(n_2-r_2)!} f(n_1, n_2) = \frac{(n_2+1)!}{r_1!(n_2-r_1+1)!}$$

$$= \binom{n_2+1}{r_1}.$$

$$\therefore\ P(\{r_{1j}\}) = \frac{r_1!}{r_{11}!\cdots r_{1n_1}!}\binom{n_2+1}{r_1}\Big/\binom{n}{n_1} \qquad (14.7)$$

Similarly, the probability function of $\{r_{2j}\}$ is given by

$$P(\{r_{2j}\}) = \frac{r_2!}{r_{21}!\cdots r_{2n_2}!}\binom{n_1+1}{r_2}\Big/\binom{n}{n_1}.$$

Taking sum on (14.6) with respect to r_{1j} subject to the conditions

$$\sum_j jr_{1j} = n_1 \text{ and } \sum_j r_{1j} = r_1$$

the joint distribution of r_1 and r_2 is obtained. This distribution is given by

$$P(r_1, r_2) = \frac{\binom{n_1-1}{r_1-1}\binom{n_2-1}{r_2-1}}{\binom{n}{n_1}} f(n_1, n_2). \qquad (14.8)$$

Taking sum of (14.8) with respect to r_2, the probability function of r_1 is obtained and is given by

$$P(r_1) = \frac{\binom{n_1-1}{r_1-1}\binom{n_2+1}{r_1}}{\binom{n}{n_1}} \qquad (14.9)$$

which is the probability function of total number of runs of S's. Here $r_1 = 1, 2, \ldots, n_1$.

The s-th factorial moment of this distribution is given by

$$\mu_{[s]} = E\left[r_1^{[s]}\right] = \frac{(n_2+1)^{[s]}}{\binom{n}{n_1}} \sum_{r_1=s}^{n_1} \binom{n_1-1}{r_1-1}\binom{n_2+1-s}{r_1-s}$$

$$= (n_2+1)^{[s]}\binom{n-s}{n_1-s}\Big/\binom{n}{n_1}.$$

In particular, we have

$$\mu_{[1]} = \mu_1' = \frac{n_1(n_2+1)}{n}$$

$$\text{and } \mu_{[2]} = (n_2+1)^{[s]}\frac{n_1(n_1-1)}{n(n-1)}.$$

Also we have

$$\mu_2 = \frac{(n_2+1)^{[2]}(n_1)^{[2]}}{n(n)^{[2]}}.$$

The general factorial moment of (r_1-1) and (r_2-1) is obtained from the distribution (14.8) and is given by

$$E\left[(r_1-1)^{[s_1]}(r_2-1)^{[s_2]}\right] = \frac{(n_1-1)^{[s_1]}(n_2-1)^{[s_2]}}{\binom{n}{n_1}}\binom{n-s_1-s_2}{n_1-s_2}.$$

Sometimes it needs to study the distribution of total runs, say $u = r_1 + r_2$. The probability function of u can be found out from (14.8) by summing $P(r_1, r_2)$ along the line $u = r_1 + r_2$ in the $r_1 r_2$-plane. If u is even, there is only one point at which $P(r_1, r_2)$ has a value different from zero, and if u is odd, there are two such points. The probability function of u is given by

$$P(u) = \begin{cases} \dfrac{2\binom{n_1-1}{\frac{u}{2}-1}\binom{n_2-1}{\frac{u}{2}-1}}{\binom{n}{n_1}}, & \text{if } u \text{ is even} \\[2ex] \dfrac{\binom{n_1-1}{\frac{u}{2}-\frac{1}{2}}\binom{n_2-1}{\frac{u}{2}-\frac{3}{2}} + \binom{n_1-1}{\frac{u}{2}-\frac{3}{2}}\binom{n_2-1}{\frac{u}{2}-\frac{1}{2}}}{\binom{n}{n_1}}, & \text{if } u \text{ is odd.} \end{cases} \tag{14.10}$$

Here

$$E(u) = \frac{2n_1n_2}{n} + 1,\ V(u) = \frac{2n_1n_2(2n_1n_2-1)}{n^2(n-1)}.$$

In the discussion of probability function of runs n_1 and n_2 are assumed fixed. These two quantities may be random variables. Let us consider a given set of independent trials, each trial resulting in an S or F, then n_1 and n_2 would be linearly dependent random variables satisfying the condition $n_1 + n_2 = n$ such that n_1 would have a binomial distribution with parameter n and p, where p is the probability of getting an S in a single trial. In that case the joint probability function of n_1 and n_2 is

$$P(n_1, n_2) = \binom{n}{n_1} p^{n_1} q^{n_2}, \text{ where } n_2 = n - n_1.$$

Now, the joint distribution of r_1, n_1 and n_2 is given by

$$\begin{aligned} P(r_1, n_1, n_2) &= \frac{\binom{n_1-1}{r_1-1}\binom{n_2+1}{r_1}}{\binom{n}{n_1}} P(n_1, n_2) \\ &= \binom{n_1-1}{r_1-1}\binom{n_2+1}{r_1} p^{n_1} q^{n_2}. \end{aligned} \tag{14.11}$$

Summing (14.11) over the range of n_1 and n_2 we shall get the probability function of r_1.

Distribution of Runs From k Kinds of Elements

Let us assume that instead of two kinds of elements S and F there are k kinds of elements $S_1, S_2, \ldots, S_k$, the i-th kind S_i has n_i elements and the total elements are

$n = n_1 + n_2 + \cdots + n_k$. Let r_{ij} be the number of runs of S_i of length j and $r_i = \sum_j r_{ij}$ be the total number of runs of S_i's. Mood (1940) has shown that for the random variable r_{ij} the probability function is

$$P(\{r_{ij}\}) = \frac{n_1!n_2!\cdots n_k!}{n!}\left[\prod_{i=1}^{k}\frac{r_i!}{r_1!\cdots r_{in_i}!}\right] f(r_1, r_2, \ldots, r_k)$$

where $f(r_1, r_2, \ldots, r_k)$ is the number of ways r_1 objects of one kind, r_2 objects of a second kind, and so on, can be arranged so that no two adjacent objects are of the same kind. Now, the probability function $P(\{r_i\})$ is obtained from $P(\{r_{ij}\})$ by a similar way as it has been obtained when $k = 2$. Here $P(\{r_i\})$ is given by

$$P(\{r_i\}) = \frac{n_1!n_2!\cdots n_k!}{n!}\left[\prod_{i=1}^{k}\binom{n_i - 1}{r_i - 1}\right] f(r_1, \ldots, r_k). \tag{14.12}$$

14.3 Distribution Related to Queues

Let us consider the distribution of the total number of customers served before a queue vanishes. Let us assume that there is a single queue and the customers arrive at constant rate λ with random arrival times. Also assume that a customer is served at a constant time β. Let the probability of arrival of a customer during the period $(t, t + (\Delta t))$ be $\lambda(\Delta t) + O(\Delta t)$ and that the probability of arrival of two (or more) customers in this period be $O(\Delta t)$. Let X be the random variable indicating the total number of customers served before the queue vanishes. If there are initially M customers in the queue, then

$$P(X = k) = \frac{M}{(k-M)!}k^{k-M-1}(\lambda\beta)^{k-M}e^{-\lambda\beta k} \tag{14.13}$$

with $k = M, M+1, \ldots$ and $\lambda\beta < 1$.

Here (14.13) represents the probability function of the distribution of total number of customers served before a queue vanishes. The parameters λ, β of the distribution appear in product form $\lambda\beta = \alpha$ (say). Thus

$$P(X = k) = \frac{M}{(k-M)!}k^{k-M-1}\alpha^{k-M}e^{-\alpha k}. \tag{14.14}$$

This distribution is called *Borel-Tanner distribution* with parameters M and α.

The mean and variance of the Borel-Tanner distribution are given, respectively, by

$$\mu_1' = \frac{M}{1-\alpha} \text{ and } \mu_2 = \frac{M\alpha}{(1-\alpha)^3}.$$

The mode of the distribution is approximately between $(X-1)$ and X, where

$$M + 1 \le X \le \frac{2}{3}M^2 + 3M$$

$$\text{and } \alpha e^{1-\alpha} = (X - M)X^{M+\frac{1}{2}}(X-1)^{-(M+\frac{3}{2})}.$$

Haight (1961) has described a distribution analogous to Borel-Tanner. The probability function of it is given by

$$P(X=k)=\frac{M}{k}\binom{2k-M-1}{k-1}\frac{\alpha^{k-M}}{(1+\alpha)^{2k-M}}$$

with $k=r,r+1,\cdots$. The mean and variance of this distribution are given, respectively, by

$$\mu_1'=\frac{M}{1-\alpha},\ \mu_2=\frac{M\alpha(1+\alpha)}{(1-\alpha)^3}.$$

The probability generating function is given by

$$g(t)=(2t)^M\left[1+\alpha+\left\{(1+\alpha)^2-4\alpha t\right\}^{\frac{1}{2}}\right]^{-M}.$$

14.4 Matching Distribution

This distribution arises when we study the random ordering of n numbered subjects. In random ordering the position of random order of some of the subjects may equals the number originally assigned to them. Let us consider that n subjects, each numbered $1,2,\ldots,n$ respectively, are arranged in random order. Let X be the random variable indicating the number of subjects for which position in the random order equals the number originally assigned to them. Clearly, the values of X will be from 0 to n. We need to find the probability of X taking a particular value k (say). This probability is given by

$$P(X=k)=\frac{1}{k!}\left[1-\frac{1}{1!}+\frac{1}{2!}+\cdots+\frac{(-1)^{n-k}}{(n-k)!}\right] \tag{14.15}$$

where $k=0,1,2,\ldots,n$. The distribution of X with probability function (14.15) is called *matching distribution.*

From (14.15) it is observed that $P(X=n-1)=0$ and if $(n-k)$ is large enough $P(X=k)$ can be approximated by

$$P(X=k)\doteq\frac{e^{-1}}{k!} \tag{14.16}$$

which is the probability function of Poisson variate with parameter 1.

The moment generating function of matching distribution can be expressed as follows:

$$M_x(t)=\sum_{x=0}^{n}(-1)^x\frac{(1-e^t)^x}{x!} \tag{14.17}$$

while that of Poisson distribution can be written as

$$M_x(t)=\sum_{x=0}^{\infty}(-1)^x\frac{(1-e^t)^x}{x!}. \tag{14.18}$$

Comparing (14.17) and (14.18) it can be concluded that matching distribution has identical 1st, 2nd, ..., n-th moments as these are for Poisson distribution with parameter 1. Therefore, the mean and variance of matching distribution are $\mu_1' = 1$ and $\mu_2 = 1$, respectively.

An extension of this matching distribution arises when the cards of two identical decks are dealt out. Let us consider that in each deck there are n cards. These cards are of s suits and in each suit there are c cards so that $cs = n$. Then the cards of two decks are dealt out in random order at the same rate. In this experiment a match will occur if the cards dealt from the two decks at the same time are of the same suit. Then the probability function of the number of k matches is given by

$$P(X = k) = \frac{1}{n!}\sum_{j=k}^{n}(-1)^{j-k}\binom{j}{k}(n-j)!H_j \tag{14.19}$$

where H_j is the coefficient of Z^j in the expansion of

$$\left[\sum_{i=0}^{c}\frac{(c!)^2}{i![(c-i)!]^2}Z^i\right]^s.$$

When $c = 1$, the distribution (14.19) transforms to (14.15) and for s large it can be approximated by

$$P(X = k) \doteq \frac{c^k e^{-c}}{k!}\left[1 - \frac{c-1}{2cs}\frac{(c-k)^2 - k}{c}\right]$$

$$\text{and } P(X = k) \doteq \frac{c^k}{k!}\left(1 - \frac{1}{s}\right)^{cs}\left[1 + \frac{1}{2s} + \frac{(c-1)(2c+1-k)k}{2c^2 s}\right].$$

14.5 Discrete Rectangular Distribution

The probability function of this distribution is given by

$$P(X = a + jh) = \frac{1}{M+1};\ j = 0, 1, 2, \ldots, M. \tag{14.20}$$

When $a = 0$ and $h = 1/M$, x takes values

$$0, \frac{1}{M}, \frac{2}{M}, \ldots, 1 \text{ and } P\left(X = \frac{J}{M}\right) = \frac{1}{M+1}. \tag{14.21}$$

Also the probability function can be represented by

$$P\left(X = -\frac{M}{2} + j\right) = \frac{1}{M+1}. \tag{14.22}$$

When $a = 0$ and $h = 1$, we have

$$P(X = j) = \frac{1}{M+1},\ j = 0, 1, 2, \ldots, M. \tag{14.23}$$

The moment generating function of (14.20) is given by

$$\begin{aligned} M_x(t) = E\left[e^{tx}\right] &= \frac{1}{M+1}\sum_{j=0}^{M} e^{t(a+jh)} \\ &= \frac{1}{M+1} e^{ta}\left\{e^{(M+1)th} - 1\right\}\left(e^{th}-1\right)^{-1} \\ &= \frac{1}{M+1} e^{\left\{t\left(a+\frac{1}{2}Mh\right)\right\}} \cdot \frac{\sinh\left\{\frac{1}{2}(M+1)th\right\}}{\sinh\left\{\frac{1}{2}th\right\}}. \end{aligned}$$

The mean and all central moments of odd order of the distribution (14.22) are zero. The central moments of even order are given by

$$\mu_r = \frac{1}{M+1}\sum_{j=0}^{M}\left(j - \frac{1}{2}M\right)^r.$$

In particular

$$\begin{aligned} \mu_2 &= \frac{1}{12}\left\{(M+1)^2 - 1\right\} \\ \text{and } \mu_4 &= \frac{1}{240}\left\{3(M+1)^2 - 7\right\}\left\{(M+1)^3 - 1\right\}. \end{aligned}$$

CHAPTER **15**

Rectangular Distribution

15.1 Definition

Let X be a random variable taking values over an interval

$$\left(M - \frac{W}{2},\ M + \frac{W}{2}\right).$$

Let us assume that the probability density function (p.d.f.) of X is constant over the entire range of X, i.e.,

$$f(x) = \begin{cases} \frac{1}{W}, & M - \frac{W}{2} < x < M + \frac{W}{2} \\ 0, & \text{otherwise.} \end{cases} \tag{15.1}$$

Then X is said to have a continuous rectangular or uniform distribution. Here $M - \frac{W}{2}$ and $M + \frac{W}{2}$ are the two parameters of the distribution.

The graph of this distribution takes the adjoining shape [FIG. 15.1] of a rectangle over the x-axis between the ordinates at

$$x = a = M - \frac{W}{2} \text{ and } x = b = M + \frac{W}{2}.$$

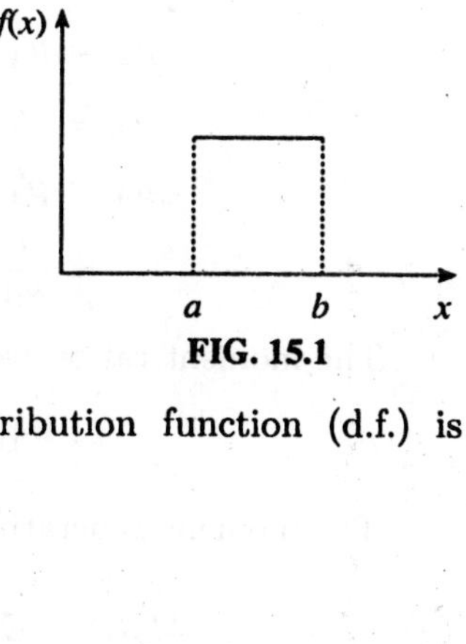

FIG. 15.1

Thus the distribution is called *rectangular distribution.*

For the interval $\left(a = M - \frac{W}{2},\ b = W + \frac{W}{2}\right)$ the distribution function (d.f.) is given by

$$F(x) = \begin{cases} 0, & \text{if } -\infty < x < a \\ \frac{x-a}{b-a}, & \text{if } a \le x \le b \\ 1, & \text{if } b < x < \infty \end{cases}$$

Here $F(x)$ is not continuous at $x = a$ and $x = b$, and hence it is not differentiable at these points. Thus

$$\frac{dF(x)}{dx} = f(x) = \frac{1}{b-a} \neq 0,$$

exists everywhere except at the points $x = a$ and $x = b$. Therefore, the p.d.f. of X is given by (15.1) and the total probability of this function is

$$\int_{M-\frac{W}{2}}^{M+\frac{W}{2}} f(x)dx = \frac{1}{W}\int_{M-\frac{W}{2}}^{M+\frac{W}{2}} dx = \frac{1}{W}\Big[x\Big]_{M-\frac{W}{2}}^{M+\frac{W}{2}} = 1.$$

Here W is the range of the distribution.

15.2 Moments of Rectangular Distribution

If X is a uniform variate taking values over the interval (a, b), then r-th moment about origin is given by

$$\mu'_r = E(x^r) = \frac{1}{b-a}\int_a^b x^r\,dx = \frac{1}{b-a}\left[\frac{x^{r+1}}{r+1}\right]_a^b = \frac{1}{b-a}\,\frac{b^{r+1}-a^{r+1}}{r+1}.$$

In particular

$$\begin{aligned}
\mu'_1 &= \frac{1}{b-a}\,\frac{b^2-a^2}{2} = \frac{a+b}{2}\\
\mu'_2 &= \frac{1}{b-a}\,\frac{b^3-a^3}{3} = \frac{b^2+ab+a^2}{3}\\
\mu'_3 &= \frac{1}{b-a}\,\frac{b^4-a^4}{4} = \frac{(a+b)(a^2+b^2)}{4}\\
\mu'_4 &= \frac{1}{b-a}\,\frac{b^5-a^5}{5} = \frac{1}{5}(a^4+a^3b+a^2b^2+ab^3+b^4)\\
\mu_2 &= \mu'_2-\mu'^2_1 = \frac{a^2+ab+b^2}{3}-\frac{(a+b)^2}{4} = \frac{(b-a)^2}{12}\\
\mu_3 &= \mu'_3-3\mu'_2\mu'_1+2\mu'^3_1 = 0\\
\mu_4 &= \mu'_4-4\mu'_3\mu'_1+6\mu'_2\mu'^2_1-3\mu'^4_1\\
&= \frac{1}{80}[a^4-4a^3b+6a^2b^2-4ab^3+b^4] = \frac{(b-a)^4}{80}.
\end{aligned}$$

The moment ratios are

$$\alpha_3 = \sqrt{\beta_1} = 0 \text{ and } \alpha_4 = \beta_2 = \frac{9}{5} < 3.$$

The moment generating function of rectangular distribution is given by

$$M_x(t) = E\left[e^{tx}\right] = \frac{1}{b-a}\int_a^b e^{tx}dx = \frac{1}{t(b-a)}\left[e^{tb}-e^{ta}\right].$$

15.3 Properties of Rectangular Distribution

The characteristic function of rectangular distribution is given by

$$\psi(t) = E\left[e^{itx}\right] = \frac{1}{b-a}\int_a^b e^{itx}dx = \frac{e^{itb}-e^{ita}}{it(b-a)}.$$

Mean Deviation about Mean

$$\gamma_1 = E|x - \mu_1'| = \frac{1}{b-a}\int_a^b \left| x - \frac{a+b}{2} \right| dx.$$

Let $x - \frac{a+b}{2} = z$, then

$$\gamma_1 = \frac{1}{b-a}\int_{-\frac{(b-a)}{2}}^{\frac{b-a}{2}} |z| dz = \frac{2}{b-a}\int_0^{\frac{b-a}{2}} t dt = \frac{b-a}{4}.$$

If X is a random variable having the rectangular distribution in the interval

$$\left(M - \frac{W}{2},\ M + \frac{W}{2}\right), \text{ then } Y = \frac{X - M + \frac{W}{2}}{W}$$

is a random variable having the rectangular distribution in the interval $(0, 1)$.

Distribution of Sum of n Independent Rectangular Random Variables

Let $X_1, X_2, \ldots, X_n$ be n independent random variables, each having the same rectangular distribution with p.d.f.

$$f(x) = 1,\ 0 < x < 1.$$

The sum of $X_1, X_2, \ldots, X_n$ is denoted by

$$S_n = X_1 + X_2 + \cdots + X_n.$$

Then

$$P(S_n \le x) = \sum_j (-1)^j \binom{n}{j} (x - j)^n / n! \tag{15.2}$$

$[0 \le x \le n]$, where the summation $\sum\limits_j$ is overall $j < x$. When $n = 2$, we have

$$\bar{X}_2 = \frac{1}{2} S_2.$$

The p.d.f. of $\bar{X}_2$ is given by

$$P_{\bar{x}_2}(x) = \begin{cases} (x - a + h)/h^2, & a - h \le x \le a \\ (a + h - x)/h^2, & a \le x \le a + h \end{cases}$$

with $a = h = \frac{1}{2}$. This distribution is a symmetrical triangular distribution.

The standard triangular distribution has p.d.f.

$$P(x) = \begin{cases} 2x/A, & 0 \le x \le A \\ 2(1-x)/(1-A), & A \le x \le 1 \end{cases}$$

If $A = \frac{1}{2}$, the distribution is symmetrical.

The r-th moment about A of this triangular distribution is given by

$$\begin{aligned}\mu_r(x-A) &= E\left[(x-A)^r\right]\\ &= 2(-1)r\int_0^A (A-x)^r\frac{x}{A}dx + 2\int_A^1 (x-A)^r\left(\frac{1-x}{1-A}\right)dx\\ &= \frac{2\left[(-1)^r A^{r+1}+(1-A)^{r+1}\right]}{(r+1)(r+2)}.\end{aligned}$$

Therefore,

$$\begin{aligned}\mu_1' &= A+\frac{1}{3}\left\{-A^2+(1-A)^2\right\} = \frac{1}{3}(1+A)\\ \text{and } \mu_2 &= \mu_2(X-A)-\{\mu_1(X-A)\}^2\\ &= \frac{2[A^3+(1-A)^3]}{3\times 4}-\left[\frac{2[-A^2+(1-A)^2]}{2\times 3}\right]^2\\ &= \frac{1}{6}\{A^3+(1-A)^3\}-\frac{1}{9}(1-2A)^2\\ &= \frac{1}{18}(1-A+A^2).\end{aligned}$$

The mean deviation of triangular distribution is

$$\frac{2}{81A}(1+A)^3 \text{ if } A>\frac{1}{2} \text{ and } \frac{2(2-A)^3}{81(1-A)} \text{ if } A<\frac{1}{2}.$$

For rectangular distribution it has already been observed that $\alpha_3 = 0$, i.e., the distribution is symmetrical. Thus all odd central moments are zero. The r-th even central moment is given by

$$\mu_r = \frac{1}{r+1}\left(\frac{W}{2}\right)^r, \text{ when } M-\frac{W}{2}\le x\le M+\frac{W}{2}.$$

The moments of even order are also given by $\mu_{2n} = \dfrac{a^{2n}}{(2n+1)}$ when

$$f(x) = \frac{1}{2a},\ -a<x<a. \tag{15.3}$$

The m.g.f. of (15.3) is given by

$$\begin{aligned}M_x(t) &= \int_{-a}^{a}\frac{e^{tx}}{2a}dx = \frac{1}{2a}\left|\frac{e^{tx}}{t}\right|_{-a}^{a} = \frac{1}{2at}\left(e^{at}-e^{-at}\right) = \frac{\sinh at}{at}\\ &= \frac{1}{at}\left[at+\frac{(at)^3}{3!}+\frac{(at)^5}{5!}+\cdots\right] = 1+\frac{(at)^2}{3!}+\frac{(at)^4}{5!}+\cdots.\end{aligned}$$

There are no terms with odd powers of t in $M_x(t)$, i.e., all moments of odd order about origin vanish, i.e., μ_{2n+1}' (about origin) $= 0$. In particular

$$\mu_1' \text{ (about origin)} = 0.$$

$$\therefore\ \mu_r' \text{ (about origin)} = \mu_r.$$

Hence

$$\mu_{2n+1} = 0; \; n = 0, 1, 2, \ldots .$$

The moments of even order are given by

$$\mu_{2n} = \text{coefficient of } \frac{t^{2n}}{(2n)!} \text{ in } M_x(t) = \frac{a^{2n}}{(2n+1)}.$$

Distribution of the Difference between Exact Value and Observed Value

Let X be the exact value of a random variable having p.d.f. $P(x)$. Suppose $\hat{X}$ be the observed value which is nearest to X in the set of values $\{\alpha + jh\}$, where j can take any integer value (positive, negative or zero). Let us define $Y = \hat{X} - X$ and assume that Y has a uniform distribution with p.d.f.

$$P(y) = \frac{1}{h}, \; -\frac{1}{2}h \leq y \leq \frac{1}{2}h.$$

Then

$$P\left(\hat{X} = \alpha + jh\right) = \int_{\alpha+\left(j-\frac{1}{2}\right)h}^{\alpha+\left(j+\frac{1}{2}\right)h} P(x)dx.$$

As X and Y are mutually independent, we have

$$K_r(\hat{X}) = K_r(X) + K_r(Y).$$

In particular

$$K_1(X) = K_1(\hat{X}), \; K_2(X) = K_2(\hat{X}) - \frac{1}{12}h^2,$$

$$K_3(X) = K_3(\hat{X}), \; K_4(X) = K_4(\hat{X}) + \frac{1}{120}h^4.$$

These values are used in the derivation of Sheppard's correction to adjust the values of sample moments which are affected by grouping.

EXERCISE

1. (a) What do you mean by rectangular distribution? Mention some experiments, where experimental results can be expressed by discrete and continuous rectangular distribution.

   (b) Write down the properties of rectangular distribution.

   (c) Given

$$f(x) = \begin{cases} \frac{1}{8}, & 0 < x < 8 \\ 0, & \text{otherwise.} \end{cases}$$

   Find (i) $P(X < 3)$, (ii) $P(X > 2)$, (iii) $P(2 < X < 6)$. Calculate mean and variance of X.

2. (a) If X is a rectangular distribution over the range $[a, b]$, then show that the distribution is symmetric and platykurtic. Discuss other properties of rectangular distribution.

   (b) If

$$f(x) = \begin{cases} 1, & 0 < x < 1 \\ 0, & \text{otherwise,} \end{cases}$$

   then find the distribution of $\sum_{i=1}^{n} x_i$, where $x_1, x_2, \ldots, x_n$ are random observations from the distribution. Also, find mean and variance of x.

3. (a) If

$$f(x) = \begin{cases} \frac{1}{2a}, & -a < x < a \\ 0, & \text{otherwise.} \end{cases}$$

   Show that $\mu_{2n} = \frac{a^{2n}}{2n+1}$.

   (b) If

$$f(x_i) = \begin{cases} 1, & 0 < x_i < 1 \\ 0, & \text{otherwise,} \end{cases}$$

   then find the p.d.f. of $Y_i = -2 \ln x_i$ and $\sum_{i=1}^{n} Y_i$.

   (c) The bus leaves after every 10 minutes within 7 : 00 AM – 8 : 00 AM from a bus station. If a person reaches X minutes past 7 : 00 AM, what is the probability that he does not need to wait more than 5 minutes to avail a bus. Assume that X follows uniform distribution over time interval 7 : 00 AM – 7 : 30 AM.

CHAPTER 16

Normal Distribution

16.1 Definition and Remarks

A random variable X is said to be normally distributed if its p.d.f. is given by

$$f(x) = \frac{1}{\sigma\sqrt{2\pi}} e^{-\frac{1}{2}\left(\frac{x-\mu}{\sigma}\right)^2}, -\infty < x < \infty$$

where μ and σ are two parameters of the distribution. Again, let us define a new random variable Z such that $Z = \frac{X-\mu}{\sigma}$. Then

$$f(z) = \frac{1}{\sqrt{2\pi}} e^{-\frac{1}{2}z^2}, -\infty < z < \infty$$

is the p.d.f. of standard normal variate z. This p.d.f. does not depend on the parameters μ and σ.

This distribution was regarded as a convenient approximation of binomial distribution. De Moivre in 1733 obtained this continuous distribution as a limiting case of the binomial distribution and applied it to problems arising in the game of chance. In 1774, Laplace obtained the normal distribution as an approximation to hypergeometric distribution. In 1809, Gauss established techniques based on the normal distribution to study the distribution of errors in Astronomy. He advocated tabulation of the probability integral $[\phi(x)]$.

The normal distribution has a unique position in probability theory, and can be used as an approximation to other distributions. Various efforts were made to establish the normal model as the underlying law ruling all continuous random variables. Thus the name "normal". However, these efforts failed because of false premises. Normal distribution can frequently be applied, with small risk of serious errors, when substantially non-normal distributions correspond more closely to observed values. Thus the normal distribution became widely and uncritically accepted as the basis of much practical statistical work. In most cases the use of normal distribution is based on forms of central limit theorem. By this theorem it can be stated that if $X_1, X_2, \ldots, X_n$ are independent, identically distributed random variables with finite mean and standard deviation, then the distribution of the standardized sum

$$\left(\sum_{i=1}^{n} X_i - nE(X)\right) \Big/ \sqrt{nV(X)}$$

tends to the unit normal distribution as n tends to infinity. The graph of the normal p.d.f. is shown below. The graph is symmetrical with respect to $x = \mu$ with maximum ordinate $1/\sqrt{2\pi}\sigma$ at $x = \mu$. It has inflection points at $x = \mu \pm \sigma$.

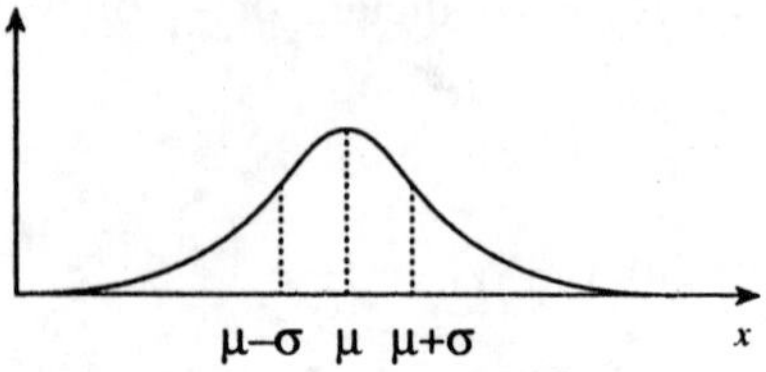

FIG. 16.1 : Graph of normal p.d.f.

16.2 Moments and Generating Functions of Normal Distribution

The moment generating function is given by

$$M_x(t) = E\left[e^{tX}\right] = \frac{1}{\sqrt{2\pi}\sigma}\int_{-\infty}^{\infty} e^{tx} e^{-\frac{1}{2}\left(\frac{x-\mu}{\sigma}\right)^2} dx.$$

Putting $z = \dfrac{x-\mu}{\sigma} \Rightarrow x = \mu + \sigma z$ and $dx = \sigma dz$, we have

$$\begin{aligned} M_x(t) &= \frac{1}{\sqrt{2\pi}}\int_{-\infty}^{\infty} \exp\{t(\mu+\sigma z)\}\exp\left\{-\frac{1}{2}z^2\right\} dz \\ &= e^{\mu t}\frac{1}{\sqrt{2\pi}}\int_{-\infty}^{\infty} \exp\left\{-\frac{1}{2}(z^2 - 2t\sigma z)\right\} dz \\ &= e^{\mu t}\frac{1}{\sqrt{2\pi}}\int_{-\infty}^{\infty} \exp\left[-\frac{1}{2}\{(z-\sigma t)^2 - \sigma^2 t^2\}\right] dz \\ &= e^{\mu t + \frac{1}{2}\sigma^2 t^2}\frac{1}{\sqrt{2\pi}}\int_{-\infty}^{\infty} \exp\left\{-\frac{1}{2}(z-\sigma t)^2\right\} dz \\ &= e^{\mu t + \frac{1}{2}t^2\sigma^2}\frac{1}{\sqrt{2\pi}}\int_{-\infty}^{\infty} \exp\left(-\frac{1}{2}u^2\right) du = e^{\mu t + \frac{1}{2}t^2\sigma^2}\frac{\sqrt{2\pi}}{\sqrt{2\pi}} = e^{\mu t + \frac{1}{2}t^2\sigma^2} \end{aligned}$$

$\therefore$ m.g.f., $M_x(t) = e^{\mu t + \frac{1}{2}t^2\sigma^2}$.

Again, let $Z = \frac{X-\mu}{\sigma} \sim N(0,1)$ where $X \sim N(\mu, \sigma)$. Then

$$M_z(t) = e^{-\mu t/\sigma} M_x(t/\sigma) = \exp(-\mu t/\sigma)\exp\left\{\frac{\mu t}{\sigma} + \frac{t^2}{\sigma^2}\cdot\frac{\sigma^2}{2}\right\} = \exp(t^2/2).$$

The cumulant generating function is given by

$$K_x(t) = \log M_x(t) = \log e^{\mu t + \frac{1}{2}t^2\sigma^2} = \mu t + \frac{1}{2}t^2\sigma^2.$$

In particular

$$\begin{aligned}
\mu_1' &= K_1 = \text{coefficient of } t \text{ in } K_x(t) = \mu \\
\mu_2 &= K_2 = \text{coefficient of } \frac{t^2}{2!} \text{ in } K_x(t) = \sigma^2 \\
\mu_3 &= K_3 = \text{coefficient of } \frac{t^3}{3!} \text{ in } K_x(t) = 0 \\
\mu_4 - 3\mu_2^2 &= K_4 = \text{coefficient of } \frac{t^4}{4!} \text{ in } K_x(t) = 0 \\
\therefore\ \mu_4 &= 3\sigma^4
\end{aligned}$$

Hence $\alpha_3 = \sqrt{\beta_1} = \sqrt{\dfrac{\mu_3^2}{\mu_2^3}} = 0$ and $\alpha_4 = \beta_2 = \dfrac{\mu_4}{\mu_2^2} = 3.$

This means that the distribution is symmetric and mesokurtic.

The moments about origin of normal distribution are obtained from $M_x(t)$ as follows:

We have

$$\begin{aligned}
M_x(t) = &\left[1 + \left(\mu t + \frac{1}{2}t^2\sigma^2\right) + \frac{1}{2!}\left(\mu t + \frac{1}{2}t^2\sigma^2\right)^2\right. \\
&\left. + \frac{1}{3!}\left(\mu t + \frac{1}{2}t^2\sigma^2\right)^3 + \frac{1}{4!}\left(\mu t + \frac{1}{2}t^2\sigma^2\right)^4 + \cdots\right].
\end{aligned}$$

Thus

$$\begin{aligned}
\mu_1' &= \text{coefficient of } t \text{ in } M_x(t) = \mu \\
\mu_2' &= \text{coefficient of } \frac{t^2}{2!} \text{ in } M_x(t) = \mu^2 + \sigma^2 \\
\mu_3' &= \text{coefficient of } \frac{t^3}{3!} \text{ in } M_x(t) = \mu^3 + 3\mu\sigma^2 \\
\mu_4' &= \text{coefficient of } \frac{t^4}{4!} \text{ in } M_x(t) = \mu^4 + 6\mu^2\sigma^2 + 3\sigma^4 \text{ and so on.}
\end{aligned}$$

The central moments of normal distribution can be obtained from m.g.f. about mean. Thus

$$\begin{aligned}
M_{(x-\mu)}(t) &= E\left[e^{t(x-\mu)}\right] = e^{-\mu t}E\left[e^{tx}\right] = e^{-\mu t} \cdot e^{\mu t + \frac{1}{2}t^2\sigma^2} = e^{\frac{1}{2}t^2\sigma^2} \\
&= \left[1 + \frac{t^2\sigma^2}{2} + \frac{(t^2\sigma^2/2)^2}{2!} + \frac{(t^2\sigma^2/2)^3}{3!} + \frac{(t^2\sigma^2/2)^4}{4!} + \cdots\right. \\
&\quad \left. + \frac{(t^2\sigma^2/2)^n}{n!} + \cdots\right]. \qquad (16.1)
\end{aligned}$$

Now the coefficient of $\dfrac{t^r}{r!}$ in (16.1) is the r-th central moment. As there is no term with odd powers of t in (16.1), all moments of odd order about mean are zero. That is

$$\mu_{2n+1} = 0; n = 0, 1, 2, \ldots.$$

Again,

$$\begin{aligned}\mu_{2n} &= \text{coefficient of } \frac{t^{2n}}{(2n)!} \text{ in (16.1)}\\ &= \frac{\sigma^{2n}(2n)!}{n!2^n} = \frac{\sigma^{2n}}{n!2^n}[2n(2n-1)(2n-2)(2n-3)\cdots 5\cdot 4\cdot 3\cdot 2\cdot 1]\\ &= \frac{\sigma^{2n}}{n!2^n}[1\cdot 3\cdot 5\cdots(2n-1][2\cdot 4\cdot 6\cdots(2n-2)2n]\\ &= \frac{\sigma^{2n}}{n!2^n}[1\cdot 3\cdot 5\cdots(2n-1)]2^n[1\cdot 2\cdot 3\cdots n]\\ &= (1\cdot 3\cdot 5\cdots 2n-1)\sigma^{2n}\end{aligned}$$

In particular, $\mu_2 = 1\cdot\sigma^2, \mu_4 = 1\cdot 3\cdot\sigma^4$.

Recurrence Relation for the Moments of Normal Distribution

Even order moments about mean are given by

$$\mu_{2n} = \int_{-\infty}^{\infty}(x-\mu)^{2n}\frac{1}{\sigma\sqrt{2\pi}}e^{-\frac{1}{2}\left(\frac{x-\mu}{\sigma}\right)^2}dx.$$

Putting

$$z = \frac{x-\mu}{\sigma} \Rightarrow x = \mu + \sigma z \text{ and } dx = \sigma dz$$

we have

$$\mu_{2n} = \frac{1}{\sqrt{2\pi}}\int_{-\infty}^{\infty}(\sigma z)^{2n}e^{-\frac{1}{2}z^2}dz = \frac{\sigma^{2n}}{\sqrt{2\pi}}2\int_0^{\infty}z^{2n}e^{-\frac{1}{2}z^2}dz$$

[Since integrand is an even function of z.]

Again, putting

$$\frac{z^2}{2} = t \Rightarrow z = \sqrt{2t} \text{ and } dz = \frac{dt}{\sqrt{2t}}$$

we have

$$\begin{aligned}\mu_{2n} &= \frac{2\sigma^{2n}}{\sqrt{2\pi}}\int_0^{\infty}(2t)^n e^{-t}\frac{dt}{\sqrt{2t}}\\ &= \frac{2^n\sigma^{2n}}{\sqrt{\pi}}\int_0^{\infty}e^{-t}t^{(n+\frac{1}{2})-1}dt = \frac{2^n\sigma^{2n}}{\sqrt{\pi}}\Gamma\left(n+\frac{1}{2}\right)\end{aligned}$$

Similarly, we have

$$\mu_{2n-2} = \frac{2^{n-1}\sigma^{2n-2}}{\sqrt{\pi}}\Gamma\left(n-\frac{1}{2}\right).$$

$$\therefore \frac{\mu_{2n}}{\mu_{2n-2}} = 2\sigma^2\frac{\Gamma\left(n+\frac{1}{2}\right)}{\Gamma\left(n-\frac{1}{2}\right)} = 2\sigma^2\left(n-\frac{1}{2}\right).$$

Hence $\mu_{2n} = \sigma^2(2n-1)\mu_{2n-2}$.

16.3 Fitting of Normal Distribution

Let us consider the following frequency distribution:

Class interval	50–60	60–70	70–80	80–90	90–100	100–110	110–120	120–130
Frequency	8	23	85	175	150	80	20	9

We want to obtain the normal curve which fits the data. Here $N = 550$, $\mu = 89.56$, $\sigma = 12.99$.

Now, the theoretical normal frequencies can be obtained as follows:

Class	Lower class boundary x	$z = \frac{x-\mu}{\sigma}$	$\phi(z) = \frac{1}{\sqrt{2\pi}} \int_{-\infty}^{z} e^{-\frac{1}{2}z^2} dz$	$\Delta\phi(z) = \phi(z+1) - \phi(z)$	Expected frequency $= N\Delta\phi(z)$
Below 50	$-\infty$	$-\infty$	0	0.00114	1
50–60	50	−3.05	0.00114	0.01016	6
60–70	60	−2.28	0.01130	0.05422	30
70–80	70	−1.51	0.06552	0.16413	90
80 90	80	−0.74	0.22965	0.28232	155
90–100	90	0.03	0.51197	0.27617	152
100–110	100	0.80	0.78814	0.15365	84
110–120	110	1.57	0.94179	0.04857	27
120–130	120	2.34	0.99036	0.00870	5
130+	130	3.11	0.99906		

The fitted curve is

$$f(x) = \frac{550}{\sqrt{2\pi} \times 12.99} \exp\left\{-\frac{1}{2}\left(\frac{x - 89.56}{12.99}\right)^2\right\}.$$

The observed frequencies are consistent with the theoretical frequencies as observed

$$\chi^2 = \sum \frac{O_i^2}{E_i} - N = 9.87$$

is less than $\chi^2_{0.05,7} = 14.07$.

16.4 Truncated Normal Distribution

A random variable X is said to follow doubly truncated normal distribution if its p.d.f. is given by

$$\begin{aligned} f(x) &= \frac{1}{\sqrt{2\pi}\sigma} e^{-\frac{1}{2}\left(\frac{x-\mu}{\sigma}\right)^2} \left[\frac{1}{\sqrt{2\pi}\sigma} \int_A^B e^{-\frac{1}{2}\left(\frac{u-\mu}{\sigma}\right)^2} du\right]^{-1} \\ &= \frac{1}{\sigma} z\left(\frac{x-\mu}{\sigma}\right)\left[\psi\left(\frac{B-\mu}{\sigma}\right) - \psi\left(\frac{A-\mu}{\sigma}\right)\right]^{-1} \end{aligned}$$

where $A \leq X \leq B$. Here

$$\frac{1}{\sigma} z \left(\frac{x-\mu}{\sigma} \right) = \frac{1}{\sqrt{2\pi}\sigma} e^{-\frac{1}{2}\left(\frac{x-\mu}{\sigma}\right)^2}$$

$$\psi(u) = \frac{1}{\sqrt{2\pi}} \int_{-\infty}^{u} e^{-\frac{1}{2}x^2} dx$$

A and B are lower and upper truncation points respectively. The degrees of truncation are

$$\psi\left(\frac{A-\mu}{\sigma}\right) \text{ from below and } \left[1 - \psi\left(\frac{B-\mu}{\sigma}\right)\right]$$

from above. If A is replaced by $-\infty$, or B by ∞, the distribution is singly truncated from above or below respectively.

The mean of this truncated normal variable X is given by

$$E(X) = \mu + \frac{z\left(\frac{A-\mu}{\sigma}\right) - z\left(\frac{B-\mu}{\sigma}\right)}{\psi\left(\frac{B-\mu}{\sigma}\right) - \psi\left(\frac{A-\mu}{\sigma}\right)} \sigma.$$

The variance of X is

$$V(X) = \left[1 + \frac{\left(\frac{A-\mu}{\sigma}\right) z\left(\frac{A-\mu}{\sigma}\right) - \left(\frac{B-\mu}{\sigma}\right) z\left(\frac{B-\mu}{\sigma}\right)}{\psi\left(\frac{B-\mu}{\sigma}\right) - \psi\left(\frac{A-\mu}{\sigma}\right)} - \left\{\frac{z\left(\frac{A-\mu}{\sigma}\right) - z\left(\frac{B-\mu}{\sigma}\right)}{\psi\left(\frac{B-\mu}{\sigma}\right) - \psi\left(\frac{A-\mu}{\sigma}\right)}\right\}^2\right] \sigma^2.$$

16.5 Properties of Normal Distribution

Median

Let M be the median of the distribution; then

$$\int_{-\infty}^{M} \frac{1}{\sqrt{2\pi}\sigma} e^{-\frac{1}{2}\left(\frac{x-\mu}{\sigma}\right)^2} dx = \frac{1}{2}$$

$$\text{or,} \quad \frac{1}{\sqrt{2\pi}\sigma} \int_{-\infty}^{\mu} e^{-\frac{1}{2}\left(\frac{x-\mu}{\sigma}\right)^2} dx + \frac{1}{\sqrt{2\pi}\sigma} \int_{\mu}^{M} e^{-\frac{1}{2}\left(\frac{x-\mu}{\sigma}\right)^2} dx = \frac{1}{2}.$$

But

$$\frac{1}{\sqrt{2\pi}\sigma} \int_{-\infty}^{\mu} e^{-\frac{1}{2}\left(\frac{x-\mu}{\sigma}\right)^2} dx = \frac{1}{\sqrt{2\pi}} \int_{0}^{\infty} e^{-\frac{1}{2}z^2} dz = \frac{1}{2}.$$

$$\therefore \quad \frac{1}{2} + \frac{1}{\sqrt{2\pi}\sigma} \int_{\mu}^{M} e^{-\frac{1}{2}\left(\frac{x-\mu}{\sigma}\right)^2} dx = \frac{1}{2}. \tag{16.2}$$

From (16.2) it is clear that

$$\frac{1}{\sqrt{2\pi}\sigma}\int_{\mu}^{M} e^{-\frac{1}{2}\left(\frac{x-\mu}{\sigma}\right)^2} dx = 0.$$

This is possible if $\mu = M$. Hence the median $M = \mu =$ mean of the normal distribution.

Mode

Mode is that value of x for which $f(x)$ is maximum, where

$$f(x) = \frac{1}{\sqrt{2\pi}\sigma} e^{-\frac{1}{2}\left(\frac{x-\mu}{\sigma}\right)^2}, -\infty < x < \infty.$$

The value of mode is available from the solution of $f'(x) = 0$ under the condition that

$$f''(x) < 0.$$

We have

$$\log f(x) = \log \frac{1}{\sqrt{2\pi}\sigma} - \frac{1}{2}\left(\frac{x-\mu}{\sigma}\right)^2.$$

Differentiating with respect to x, we get

$$\frac{1}{f(x)} f'(x) = -\frac{1}{\sigma^2}(x-\mu)$$

$$\text{or,}\quad f'(x) = -\frac{1}{\sigma^2}(x-\mu)f(x) = 0 \Rightarrow -\frac{1}{\sigma^2}(x-\mu) = 0 \; [\because f(x) \neq 0].$$

$$\text{i.e.,}\quad x = \mu.$$

Again,

$$\begin{aligned} f''(x) &= -\frac{1}{\sigma^2}[f(x) + (x-\mu)f'(x)] \\ &= -\frac{1}{\sigma^2}\left[f(x) - \frac{1}{\sigma^2}(x-\mu)^2 f(x)\right] \\ &= -\frac{f(x)}{\sigma^2}\left[1 - \frac{(x-\mu)^2}{\sigma^2}\right]. \end{aligned}$$

At the point $x = \mu$

$$f''(x) = -\frac{1}{\sigma^2}\frac{1}{\sigma\sqrt{2\pi}} < 0.$$

Hence $x = \mu$ is the mode of the normal distribution.

Mean Deviation about Mean (MD)

$$\text{MD} = \int_{-\infty}^{\infty} |x-\mu| \frac{1}{\sqrt{2\pi}\sigma} e^{-\frac{1}{2}\left(\frac{x-\mu}{\sigma}\right)^2} dx.$$

Putting

$$z = \frac{x-\mu}{\sigma} \Rightarrow x = \mu + \sigma z \text{ and } dx = \sigma dz,$$

we have

$$\text{MD} = \frac{\sigma}{\sqrt{2\pi}}\int_{-\infty}^{\infty} |z|e^{-\frac{1}{2}z^2}dz = \frac{\sigma}{\sqrt{2\pi}}2\int_{0}^{\infty} |z|e^{-\frac{1}{2}z^2}dz$$

[since the integrand $|z|e^{-\frac{1}{2}z^2}$ is an even function of z]

$$= \sqrt{\frac{2}{\pi}}\sigma\int_0^{\infty} ze^{-\frac{1}{2}z^2}dz \text{ [since } [0,\infty],\ |z| = z]$$

$$= \sqrt{\frac{2}{\pi}}\sigma\int_0^{\infty} e^{-t}dt \left[\frac{z^2}{2} = t \Rightarrow z = \sqrt{2t} \text{ and } zdz = dt\right]$$

$$= \sqrt{\frac{2}{\pi}}\sigma\left|\frac{e^{-t}}{-1}\right|_0^{\infty} = \sqrt{\frac{2}{\pi}}\sigma \doteqdot \frac{4}{5}\sigma.$$

Area Property

If $X \sim N(\mu, \sigma^2)$, then $P(\mu < X < x_1)$ is given by

$$P(\mu < X < x_1) = \int_{\mu}^{x_1} f(x)dx = \frac{1}{\sqrt{2\pi}\sigma}\int_{\mu}^{x_1} e^{-\frac{1}{2}\left(\frac{x-\mu}{\sigma}\right)^2}dx.$$

Let $z = \dfrac{x-\mu}{\sigma} \Rightarrow x = \mu + \sigma z$ and $dx = \sigma dz$, then we have

$$P(\mu < X < x_1) = P(0 < Z < z_1) = \frac{1}{\sqrt{2\pi}}\int_0^{z_1} e^{-\frac{1}{2}z^2}dz = \int_0^{z_1}\phi(z)dz$$

[since $Z = 0$ when $X = \mu$ and when $X = x_1, z = \frac{x_1-\mu}{\sigma} = z_1$ (say)]

where

$$\phi(z) = \frac{1}{\sqrt{2\pi}}e^{-\frac{1}{2}z^2}$$

is the probability function of standard normal variate. The definite integral $\int_0^{z_1}\phi(z)dz$ is known as normal probability integral and gives the area under standard normal curve between the ordinates at $Z = 0$ and $Z = z_1$.

In particular

$$P(\mu - \sigma < X < \mu + \sigma) = \int_{\mu-\sigma}^{\mu+\sigma}\frac{1}{\sqrt{2\pi}\sigma}e^{-\frac{1}{2}\left(\frac{x-\mu}{\sigma}\right)^2}dx.$$

Putting $z = \dfrac{x-\mu}{\sigma}$, we have

$$P(-1 < Z < 1) = \int_{-1}^{1}\phi(z)dz = 2\int_0^1\phi(z)dz = 2 \times 0.3413 = 0.6826$$

where

$$\int_0^1\phi(z)dz = 0.3413 \text{ is available from table.}$$

Similarly

$$P(\mu - 2\sigma < X < \mu + 2\sigma) = (-2 < Z < 2) = \int_{-2}^{2} \phi(z)dz$$

$$= 2\int_{0}^{2} \phi(z)dz = 2 \times 0.4772 = 0.9544$$

and $$P(\mu - 3\sigma < X < \mu + 3\sigma) = P(-3 < Z < 3) = \int_{-3}^{3} \phi(z)dz$$

$$= 2\int_{0}^{3} \phi(z)dz = 2 \times 0.49865 = 0.9973.$$

The probability that a normal variate X lies outside the limit $\mu \pm 3\sigma$ is given by

$$P(|X - \mu| > 3\sigma) = P(|Z| > 3) = 1 - P(-3 < Z < 3) = 0.0027.$$

Characteristic Function

The characteristic function is given by

$$\psi(t) = E\left[e^{itx}\right] = \frac{1}{\sqrt{2\pi}\sigma}\int_{-\infty}^{\infty} e^{itx}e^{-\frac{1}{2}\left(\frac{x-\mu}{\sigma}\right)^2}dx$$

$$= e^{it\mu - \frac{1}{2}t^2\sigma^2} \quad \text{[on simplification]}.$$

Linear Combination of Normal Variates

Let X_i $(i = 1, 2, \ldots, n)$ be n independent normal variates with mean μ_i and variance σ_i^2 respectively. The m.g.f. of i-th variate is

$$M_{x_i}(t) = \exp\left\{\mu_i t + \frac{1}{2}t^2\sigma_i^2\right\}.$$

Let us consider a linear combination $\sum_{i=1}^{n} a_i X_i$ of X_i with known constants $a_1, a_2, \ldots, a_n$. Then m.g.f. of this linear combination is given by

$$M_{\Sigma a_i x_i}(t) = M_{a_1x_1}(t)M_{a_2x_2}(t)\cdots M_{a_nx_n}(t)$$

[since $x_1, x_2, \ldots, x_n$ are independent].

From the definition of m.g.f. of normal variate, we have

$$M_{a_i x_i}(t) = \exp\left\{a_i\mu_i t + \frac{1}{2}a_i^2\sigma_i^2 t^2\right\}.$$

$$\therefore\ M_{\Sigma a_i x_i}(t) = \exp\left\{(\Sigma a_i\mu_i)t + \frac{1}{2}(\Sigma a_i^2\sigma_i^2)t^2\right\}$$

which is the m.g.f. of a normal variate with mean $\Sigma a_i\mu_i$ and variance $\Sigma a_i^2\sigma_i^2$. Hence

$$\sum_{i=1}^{n} a_i X_i \sim N(\Sigma a_i\mu_i, \Sigma a_i^2\sigma_i^2).$$

In particular, let

$$\sum_{i=1}^{n} a_i X_i = X_1 + X_2 \text{ [when } a_1 = a_2 = 1,\ a_3 = a_4 = \cdots = a_n = 0]$$

then $(X_1 + X_2) \sim N(\mu_1 + \mu_2, \sigma_1^2 + \sigma_2^2)$. But if

$$a_1 = 1, a_2 = -1 \text{ and } a_3 = a_4 = \cdots = a_n = 0,$$

we have

$$\sum_{i=1}^{n} a_i X_i = X_1 - X_2 \text{ and } (X_1 - X_2) \sim N(\mu_1 - \mu_2, \sigma_1^2 + \sigma_2^2).$$

If $a_1 = a_2 = \cdots = a_n = 1$, then

$$\sum^{n} a_i X_i = \sum^{n} X_i \text{ and } (\Sigma X_i) \sim N(\Sigma \mu_i, \Sigma \sigma_i^2)$$

indicates the additive property of normal variates.

Distribution of Sample Mean

Let $x_1, x_2, \ldots, x_n$ be the sample observations drawn independently from normal population with mean μ and variance σ^2. Then the sample mean is $\frac{1}{n}\Sigma x_i$, which is a linear combination with coefficients $a_1 = a_2 = \cdots = a_n = \frac{1}{n}$. Therefore,

$$\frac{1}{n}\sum x_i \sim N\left(\frac{1}{n}\sum^{n} \mu, \frac{1}{n^2}\sum^{n} \sigma^2\right), \text{ i.e., } N\left(\mu, \frac{\sigma^2}{n}\right).$$

Relation among Quartile Deviation (QD), Mean Deviation (MD) and Standard Deviation (SD)

If $X \sim N(\mu, \sigma^2)$, then MD $= \dfrac{4}{5}\sigma$, SD $= \sigma$. Let Q_1 and Q_3 be the first and third quartiles respectively, then $Q_1$and Q_3 are given, respectively, by

$$\int_{-\infty}^{Q_1} \frac{1}{\sqrt{2\pi}\sigma} e^{-\frac{1}{2}\left(\frac{x-\mu}{\sigma}\right)^2} dx = \frac{1}{4}$$

$$\text{and } \int_{Q_3}^{\infty} \frac{1}{\sqrt{2\pi}\sigma} e^{-\frac{1}{2}\left(\frac{x-\mu}{\sigma}\right)^2} dx = \frac{1}{4}.$$

The quartile deviation is given by

$$\text{QD} = \frac{Q_3 - Q_1}{2}.$$

Now,

$$\text{when } X = Q_3,\ Z = \frac{Q_3 - \mu}{\sigma} = z_1 \text{ (say)}$$

$$\text{and when } X = Q_1,\ Z = \frac{Q_1 - \mu}{\sigma} = -z_1 \text{ (say)}.$$

This will be clear from the following figure [FIG. 16.2]:

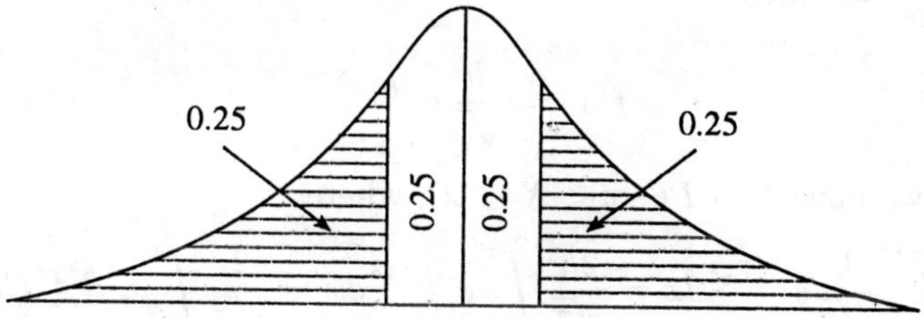

X=Q$_1$ X=μ X=Q$_3$
Z=−z$_1$ X=0 Z=z$_1$

FIG. 16.2

It is obvious, from the figure, that $P(0 < Z < z_1) = 0.25$ which implies $z_1 = 0.67$ [from table]. Again, we have

$$\text{QD} = \frac{Q_3 - Q_1}{2} = \sigma z_1 = 0.67\sigma \doteq \frac{2}{3}\sigma.$$

$$\therefore \text{ QD} : \text{MD} : \text{SD} :: \frac{2}{3}\sigma : \frac{4}{5}\sigma :: \frac{2}{3} : \frac{4}{5} : 1 :: 10 : 12 : 15.$$

Points of Inflexion of Normal Curve

Let

$$f(x) = \frac{1}{\sqrt{2\pi}\sigma} e^{-\frac{1}{2}\left(\frac{x-\mu}{\sigma}\right)^2}, -\infty < x < \infty$$

represents the p.d.f. of normal distribution. Then at the point of inflexion of the normal curve it will be

$$f''(x) = 0 \text{ and } f'''(x) \neq 0,$$

where

$$f''(x) = -\frac{f(x)}{\sigma^2}\left[1 - \left(\frac{x-\mu}{\sigma}\right)^2\right].$$

But

$$f''(x) = 0 \Rightarrow 1 - \frac{(x-\mu)^2}{\sigma^2} = 0 \Rightarrow x = \mu \pm \sigma.$$

It is easy to show that at the points $x = \mu \pm \sigma$, $f'''(x) \neq 0$. Thus the points of inflexion of the normal curve are given by

$$x = \mu \pm \sigma \text{ and } f(x) = \frac{1}{\sqrt{2\pi}\sigma} e^{-\frac{1}{2}}.$$

Error Function

Let X be a random variable with p.d.f.

$$f(x) = \frac{1}{\sqrt{2\pi}\sigma} e^{-\frac{1}{2}x^2/\sigma^2}, -\infty < x < \infty.$$

Putting $h^2 = \frac{1}{2\sigma^2}$, we have

$$f(x) = \frac{h}{\sqrt{\pi}} e^{-h^2x^2}.$$

From this $f(x)$, we have $P = P(-x < X < x)$, where

$$P = \int_{-x}^{x} \frac{h}{\sqrt{\pi}} e^{-h^2x^2} dx = \frac{2h}{\sqrt{\pi}} \int_0^x e^{-h^2x^2} dx = \frac{2}{\sqrt{\pi}} \int_0^x e^{-h^2x^2} (hdx).$$

Using $\psi(y) = \frac{2}{\sqrt{\pi}} \int_0^y e^{-y^2} dy$, P can be rewritten as

$$P = \psi(hx) = \frac{2}{\sqrt{\pi}} \int_0^x e^{-h^2x^2} (hdx).$$

Here the function $\psi(hx)$ is known as the *error function.*

16.6 Compound Normal Distribution

Normal-Normal Distribution

Let $X \sim N(\mu, \sigma^2)$ and $\mu \sim N(\mu_1, \sigma_1^2)$. Then compound normal distribution is given by

$$f(x) = \frac{1}{\sqrt{2\pi}\sigma} \frac{1}{\sqrt{2\pi}\sigma_1} \int_{-\infty}^{\infty} e^{-\frac{1}{2}\left(\frac{x-\mu}{\sigma}\right)^2} e^{-\frac{1}{2}\left(\frac{\mu-\mu_1}{\sigma_1}\right)^2} d\mu.$$

Putting $\frac{\mu - \mu_1}{\sigma_1} = z$, we have

$$f(x) = \frac{1}{\sqrt{2\pi}\sigma} \frac{1}{\sqrt{2\pi}} e^{-\frac{1}{2}\frac{(x-\mu_1)^2}{\sigma^2+\sigma_1^2}} \int_{-\infty}^{\infty} e^{-\frac{1}{2\sigma^2}\left\{Z\sqrt{\sigma_1^2+\sigma^2} - \frac{(x-\mu_1)\sigma_1}{\sqrt{\sigma_1^2+\sigma^2}}\right\}^2} dz.$$

Let

$$\frac{1}{\sigma}\left\{z\sqrt{\sigma_1^2+\sigma^2} - \frac{(x-\mu_1)\sigma_1}{\sqrt{\sigma_1^2+\sigma^2}}\right\} = t$$

then we have

$$f(x) = \frac{1}{\sqrt{2\pi}\sqrt{\sigma_1^2+\sigma^2}} e^{-\frac{1}{2}\frac{(x-\mu_1)^2}{\sigma_1^2+\sigma^2}}, -\infty < x < \infty$$

which is a normal distribution with mean μ_1 and variance $(\sigma_1^2 + \sigma^2)$.

Normal-Gamma Distribution

Let $X \sim N(\mu, \sigma^2)$ and $\frac{1}{\sigma^2}$ is a gamma variate with p.d.f.

$$f\left(\frac{1}{\sigma^2}\right) = \frac{1}{\Gamma\frac{\gamma}{2}(2c)^{\frac{\gamma}{2}}} \left(\frac{1}{\sigma^2}\right)^{\frac{\gamma}{2}-1} e^{-\frac{1}{2c\sigma^2}}, 0 < \frac{1}{\sigma^2} < \infty.$$

Then the compound normal distribution is given by

$$f(x) = \frac{1}{\sqrt{2\pi}\,\Gamma\frac{\gamma}{2}(2c)^{\frac{\gamma}{2}}} \int_0^\infty \left(\frac{1}{\sigma^2}\right)^{\frac{\gamma}{2}-\frac{1}{2}} e^{-\frac{1}{2c\sigma^2}} e^{-\frac{1}{2\sigma^2}(x-\mu)^2} d\left(\frac{1}{\sigma^2}\right)$$

$$= \frac{c^{1/2}}{B\left(\frac{1}{2}, \frac{\gamma}{2}\right)} \left[1 + c(x-\mu)^2\right]^{-\frac{1}{2}(\gamma+1)}$$

$$\left[\frac{1}{\sigma^2}\left\{\frac{1}{2c} + \frac{(x-\mu)^2}{2}\right\} = t \Rightarrow d\left(\frac{1}{\sigma^2}\right) = \frac{dt}{\frac{1}{2c}[1 + c(x-\mu)^2]}\right].$$

16.7 Lognormal Distribution

Let X be a random variable which takes only positive values. Let us define a new random variable $Y = \log X$. If Y is normally distributed, then X is said to have lognormal distribution. More generally, if θ is a number such that $X - \theta$ takes only positive values and if $Z = \log(X - \theta)$ is normally distributed, then the distribution of X is said to be lognormal. The p.d.f. of Y is given by

$$f(y) = \frac{1}{\sqrt{2\pi}\sigma} e^{-\frac{1}{2}\left(\frac{\log x - \mu}{\sigma}\right)^2}, -\infty < y = \log x < \infty$$

where μ and σ are two parameters of the distribution.

The distribution of X can also be defined by the equation

$$U = \gamma + \delta \log(X - \theta)$$

where $U \sim N(0, 1)$ and γ, δ and θ are parameters. In such a case the p.d.f. of X is given by

$$f(x) = \frac{\delta}{(x-\theta)\sqrt{2\pi}} e^{-\frac{1}{2}\{\gamma + \delta \log(x-\theta)\}^2}, \quad x > \theta.$$

If it is assumed that $Z = \log(X - \theta) \sim N(\mu, \sigma^2)$, then $\mu = -\gamma/\delta$ and $\sigma = 1/\delta$. Also, we have

$$U = \frac{1}{\sigma}[\log(X - \theta) - \mu].$$

Then

$$f(x) = \frac{1}{(x-\theta)\sigma\sqrt{2\pi}} e^{-\frac{1}{2}[\{\log(x-\theta)-\mu\}/\sigma^2]}, \quad x > \theta.$$

The graph of the standardized lognormal distribution is shown in the FIG. 16.3 given below:

If θ is known to be zero, then

$$U = \gamma + \delta \log X.$$

Also

$$U = (\log X - \mu)/\sigma.$$

In that case the distribution is called *two-parameter lognormal distribution*, where the parameters are γ, δ or μ, σ. If $\theta \neq 0$, the distribution is called *three-parameter lognormal distribution.*

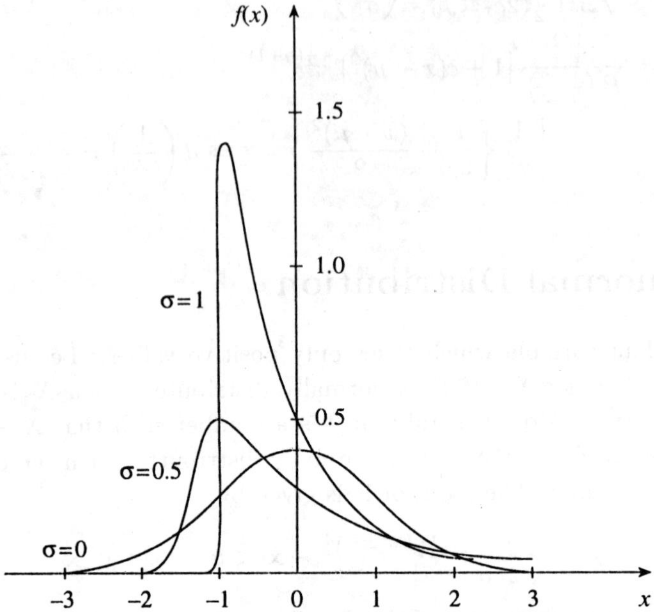

FIG. 16.3 : Graph of standardized lognormal distribution

The two-parameter lognormal distribution arises from the product of n independent positive random variables. Let $X_1, X_2, \ldots, X_n$ be n independent positive random variables, then

$$T_n = \prod_{j=1}^{n} X_j$$

is their product. Also, we have

$$\log T_n = \sum_{j=1}^{n} \log X_j.$$

If the independent random variables, $\log X_j$, are such that the standardized distribution of $\log T_n$ tends to $N(0,1)$ after application of central limit type of results, when n tends to infinity, then the limiting distribution of T_n would be two-parameter lognormal distribution.

Moments of Lognormal Distribution

The r-th moment about origin is given by

$$\begin{aligned}\mu_r' = E(X^r) &= E\left[e^{rY}\right] \quad [\because\ Y = \log X \Rightarrow X = e^Y] \\ &= M_Y(r) = \exp\left\{r\mu + \frac{1}{2}r^2\sigma^2\right\} \quad [\because\ Y \sim N(\mu, \sigma^2)]\end{aligned}$$

In particular,

$$\mu_1' = \exp\left\{\mu + \frac{1}{2}\sigma^2\right\}.$$

If we take $\mu = \log\alpha, \alpha > 0$, i.e., $Y = \log X \sim N(\log\alpha, \sigma^2)$, then

$$\mu_r' = \exp\left\{r\log\alpha + \frac{1}{2}r^2\sigma^2\right\} = \alpha^r \exp\left\{\frac{1}{2}r^2\sigma^2\right\}$$

i.e.,

$$\begin{aligned}\mu_1' &= \alpha e^{\sigma^2/2},\ \mu_2' = \alpha^2 e^{2\sigma^2}\\ \mu_2 &= \mu_2' - \mu_1'^2 = \alpha^2 e^{\sigma^2}\left(e^{\sigma^2} - 1\right).\end{aligned}$$

It can also be written as $\mu_2 = e^{2\mu}e^{\sigma^2}\left(e^{\sigma^2} - 1\right) = W(W-1)e^{2\mu}$ so that $\sigma(x) = e^{\mu}\sqrt{W(W-1)}$, where $W = e^{\sigma^2}$. Also, we have

$$\begin{aligned}\mu_3 &= W^{3/2}(W-1)^2(W+2)e^{3\mu}\\ \mu_4 &= W^2(W-1)^2(W^4+2W^3+3W^2-3)e^{4\mu}.\end{aligned}$$

The moment ratios are

$$\begin{aligned}\alpha_3 &= \sqrt{\beta_1} = (W-1)^{1/2}(W+2)\\ \alpha_4 &= \beta_2 = W^4 + 2W^3 + 3W^2 - 3.\end{aligned}$$

16.8 Importance of Normal Distribution and Lognormal Distribution

Normal Distribution

(i) Normal distribution is almost always used as approximation—either to a theoretical or an unknown distribution. For example, binomial, Poisson, hypergeometric distributions, etc., can be approximated to normal distribution. Many of the sampling distributions, e.g., Student's 't', Snedecor's F, chi-square distribution, etc., tend to normal distribution when sample size is large. The distribution of sample mean, sample variance, etc., tend to normality for large samples. Some skewed distribution can be transformed to normal distribution by simple transformation of variable.

(ii) The sampling distributions, viz., 't', F and χ^2 distributions are derived on the basis of fundamental assumption that the samples have been drawn from normal population. The small sample tests using these three distributions are also performed on the basis of the same assumption.

(iii) When normal distributions are not used, results corresponding to normal theory are useful as standards of comparison. The theory of normal curves can be applied to the graduation of the curves which are not normal.

(iv) Normal theory is applied in statistical quality control in industry for setting control limits.

Lognormal Distribution

(i) The study of lognormal distribution arises in problems of economics, biology, geology and reliability theory. The distribution usefully represented the distribution of size for varied kinds of natural economic units. The distribution of critical dose causing reaction for a number of forms of drug application can be represented with accuracy by a two-parameter lognormal distribution. Lognormal distribution is also applicable to distributions of particle size in naturally occurring aggregates.

(ii) The variables like weight, height, density cannot take negative values and hence the distributions of these variables can better be represented by two-parameter lognormal distribution.

(iii) In a situation where normal distribution is felt to be appropriate, it might be replaced by a suitable lognormal distribution. In that case confidence limits for the coeffcient of variation are easily be constructed.

(iv) The distribution is also applied when certain approximations to the distribution of Fisher's $Z = \frac{1}{2}\log F$ are used. The distribution of Z is much closer to normality than is that of F.

16.9 Bivariate Normal Distribution

The normal distribution is one of the most important frequency functions of one-dimensional random variable used in applied and theoretical statistics. In practice, there are immense use of two or more dimensional random variables in statistics. For two-dimensional random variables bivariate normal distribution is one of the most important frequency functions. In this section we shall investigate the bivariate normal distribution and discuss some of the important properties of it.

Let us consider the function

$$f(x,y) = \frac{1}{2\pi\sigma_1\sigma_2\sqrt{1-\rho^2}}\exp\left[-\frac{1}{2(1-\rho^2)}\left\{\left(\frac{x-\mu_1}{\sigma_1}\right)^2 - \frac{2\rho(x-\mu_1)(y-\mu_2)}{\sigma_1\sigma_2} + \left(\frac{y-\mu_2}{\sigma_2}\right)^2\right\}\right]$$

where

$$-\infty < x < \infty, -\infty < y < \infty, \sigma_1 > 0, \sigma_2 > 0, -1 < \rho < 1.$$

The function $f(x,y)$ is non-negative and it has the properties of a joint p.d.f. of two-dimensional random variables (X,Y). This can be shown as follows:

Let us define $f_1(x)$ by

$$f_1(x) = \int_{-\infty}^{\infty} f(x,y)dy.$$

Now

$$\frac{1}{(1-\rho^2)}\left[\left(\frac{x-\mu_1}{\sigma_1}\right)^2 - 2\rho\frac{(x-\mu_1)(y-\mu_2)}{\sigma_1 \quad \sigma_2} + \left(\frac{y-\mu_2}{\sigma_2}\right)^2\right]$$

can be written as

$$\frac{1}{(1-\rho^2)}\left[\left\{\left(\frac{y-\mu_2}{\sigma_2}\right) - \rho\left(\frac{x-\mu_1}{\sigma_1}\right)\right\}^2 + (1-\rho^2)\left(\frac{x-\mu_1}{\sigma_1}\right)^2\right]$$

$$= \frac{1}{1-\rho^2}\left(\frac{y-b}{\sigma_2}\right)^2 + \left(\frac{x-\mu_1}{\sigma_1}\right)^2, \text{ where } b = \mu_2 + \rho(\sigma_2/\sigma_1)(x-\mu_1).$$

$$\therefore\ f_1(x) = \frac{\exp\left[-(x-\mu_1)^2/2\sigma_1^2\right]}{\sigma_1\sqrt{2\pi}} \int_{-\infty}^{\infty} \frac{\exp\left\{-(y-b)^2/\left[2\sigma_2^2(1-\rho^2)\right]\right\}}{\sigma_2\sqrt{1-\rho^2}\sqrt{2\pi}}dy.$$

The integrand of the integral in $f_1(x)$ may be considered a normal p.d.f. with mean b and variance $\sigma_2^2(1-\rho^2)$. Hence the integral is equal to one.

$$\therefore\ f_1(x) = \frac{1}{\sigma_1\sqrt{2\pi}}e^{-\frac{1}{2\sigma_1^2}(x-\mu_1)^2}, -\infty < x < \infty.$$

Thus

$$\int_{-\infty}^{\infty}\int_{-\infty}^{\infty} f(x,y) = \int_{-\infty}^{\infty} f_1(x)dx = 1.$$

Therefore, the non-negative function $f(x, y)$ is a joint p.d.f. of two continuous random variables X and Y. Here the function $f_1(x)$ is the marginal p.d.f. of X, and $X \sim N(\mu_1, \sigma_1^2)$. Similarly, it can be shown that $Y \sim N(\mu_2, \sigma_2^2)$. The joint p.d.f. of X and Y can be written as

$$\begin{aligned} f(x,y) &= f_1(x)f_2(y/x) \\ &= f_1(x)\frac{1}{\sigma_2\sqrt{1-\rho^2}\sqrt{2\pi}}\exp\left[-\frac{(y-b)^2}{2\sigma_2^2(1-\rho^2)}\right]. \end{aligned}$$

Therefore, the second member in the right hand of $f(x, y)$ is the conditional p.d.f. of Y, given that $X = x$ and this conditional p.d.f. is itself normal with mean

$$b = \mu_2 + \rho(\sigma_2/\sigma_1)(x-\mu_1)$$

and variance $\sigma_2^2(1-\rho^2)$. Thus, the conditional mean of Y, given $X = x$, is linear in x and is given by

$$E(Y/x) = \mu_2 + \rho\frac{\sigma_2}{\sigma_1}(x-\mu_1).$$

As conditional mean of Y, given $X = x$, represents regression line of Y on x, it can be said that ρ is the correlation coefficient of Y and X.

In a similar manner, it can be shown that the conditional distribution of X, given $Y = y$, is

$$N\left[\mu_1 + \rho\frac{\sigma_1}{\sigma_2}(y-\mu_2,\ \sigma_1^2(1-\rho^2)\right].$$

The p.d.f. of X and Y can also be written as

$$f(x,y) = \frac{\sqrt{|\sigma^{ij}|}}{2\pi} e^{-\frac{1}{2}Q(x,y)} \tag{16.3}$$

where $Q = \sigma^{11}(x-\mu_1)^2 + 2\sigma^{12}(x-\mu_1)(y-\mu_2) + \sigma^{22}(y-\mu_2)^2$.

Here the matrix $\left[\sigma^{ij}\right]$, $i,j = 1,2$, with $\sigma^{12} = \sigma^{21}$, is the inverse of the covariance matrix $[\sigma_{ij}]$ of x and y. As $[\sigma_{ij}]$ is positive definite, $\left[\sigma^{ij}\right]$ is also positive definite. Thus Q is a positive definite quadratic form.

We have already proved that

$$\int_{-\infty}^{\infty}\int_{-\infty}^{\infty} f(x,y) = 1. \tag{16.4}$$

This implies that

$$\int_{-\infty}^{\infty}\int_{-\infty}^{\infty} e^{-\frac{1}{2}Q(x,y)}dxdy = \frac{2\pi}{\sqrt{|\sigma^{ij}|}}. \tag{16.5}$$

Means of *X* and *Y*

Differentiating both sides of (16.4) with respect to μ_1 and μ_2, we have

$$\begin{aligned} E\left[\sigma^{11}(x-\mu_1) + \sigma^{12}(y-\mu_2)\right] &= 0 \\ \text{and } E\left[\sigma^{21}(x-\mu_1) + \sigma^{22}(y-\mu_2)\right] &= 0 \end{aligned}$$

which can be written as

$$\begin{aligned} \sigma^{11}E(x-\mu_1) + \sigma^{12}E(y-\mu_2) &= 0 \\ \text{and } \sigma^{21}E(x-\mu_1) + \sigma^{22}E(y-\mu_2) &= 0. \end{aligned}$$

These are the homogeneous linear equations in $E(x-\mu_1)$ and $E(y-\mu_2)$, and since $|\sigma^{ij}| \neq 0$, the unique solutions of these equations are

$$E(x-\mu_1) = 0 \text{ and } E(y-\mu_2) = 0 \Rightarrow E(x) = \mu_1 \text{ and } E(y) = \mu_2.$$

Thus μ_1 and μ_2 are the means of X and Y respectively.

Moment Generating Function

The moment generating function of bivariate normal distribution is given by

$$\begin{aligned} M(t_1,t_2) &= \int_{-\infty}^{\infty}\int_{-\infty}^{\infty} e^{t_1x+t_2y} f(x,y)dxdy \\ &= \int_{-\infty}^{\infty} e^{t_1x} f_1(x) \left[\int_{-\infty}^{\infty} e^{t_2y} f(y/x)dy\right] dx \end{aligned}$$

for all real values of t_1 and t_2. As $f(y/x)$ is normal with mean $\mu_2 + \rho\frac{\sigma_2}{\sigma_1}(x-\mu_1)$ and variance $\sigma_2^2(1-\rho^2)$, then m.g.f. of $Y/X = x$ is

$$\int_{-\infty}^{\infty} e^{t_2y} f(y/x)dy = \exp\left\{t_2\left[\mu_2 + \rho\frac{\sigma_2}{\sigma_1}(x-\mu_1)\right] + \frac{1}{2}t^2\sigma_2^2(1-\rho^2)\right\}.$$

Thus $M(t_1, t_2)$ can be written as

$$M(t_1,t_2) = \exp\left\{t_2\mu_2 - t_2\rho\frac{\sigma_2}{\sigma_1}\mu_1 + \frac{t_2^2\sigma_2^2\left(1-\rho^2\right)}{2}\right\} \times \int_{-\infty}^{\infty} \exp\left[\left(t_1 + t_2\rho\frac{\sigma_2}{\sigma_1}\right)x\right] f_1(x)dx$$

$$= \exp\left\{t_2\mu_2 - t_2\rho\frac{\sigma_2}{\sigma_1}\mu_1 + \frac{t_2^2\sigma_2^2\left(1-\rho^2\right)}{2}\right\} \times \exp\left\{\left(t_1 + t_2\rho\frac{\sigma_2}{\sigma_1}\right)\mu_1 + \frac{\left(t_1+t_2\rho\frac{\sigma_2}{\sigma_1}\right)^2}{2}\sigma_1^2\right\}.$$

$$\left[\because\ f_1(x) \sim N\left(\mu_1, \sigma_1^2\right)\right]$$

$$\therefore\ M(t_1,t_2) = \exp\left\{t_2\mu_2 - t_2\rho\frac{\sigma_2}{\sigma_1}\mu_1 + \frac{t_2^2\sigma_2^2(1-\rho^2)}{2} + \left(t_1 + t_2\rho\frac{\sigma_2}{\sigma_1}\right)\mu_1 + \frac{\left(t_1+t_2\rho\frac{\sigma_2}{\sigma_1}\right)^2}{2}\sigma_1^2\right\}$$

$$= \exp\left\{t_1\mu_1 + t_2\mu_2 + \frac{1}{2}\left(\sigma_1^2t_1^2 + 2\rho\sigma_1\sigma_2t_1t_2 + \sigma_2^2t_2^2\right)\right\}.$$

If we put $\rho = 0$, then

$$M(t_1,t_2) = \exp\left\{t_1\mu_1 + \frac{1}{2}\sigma_1^2t_1^2\right\}\exp\left\{t_2\mu_2 + \frac{1}{2}\sigma_2^2t_2^2\right\}$$
$$= M(t_1,0)M(0,t_2).$$

Thus X and Y are stochastically independent when $\rho = 0$. Again

$$M(t_1,t_2) = M(t_1,0)M(0,t_2) \Rightarrow \exp\left(\rho\sigma_1\sigma_2t_1t_2\right) = 1.$$

This will be true if $\rho = 0$, since each of σ_1 and σ_2 is positive. Thus, it may be concluded that if X and Y have bivariate normal distribution, these will be stochastically independent if and only if $\rho = 0$.

Characteristic Function

The characteristic function of bivariate normal distribution is defined by

$$\psi(t_1,t_2) = \int_{-\infty}^{\infty}\int_{-\infty}^{\infty} e^{it_1x+it_2y} f(x,y)dxdy.$$

Putting $x - \mu_1 = u$ and $y - \mu_2 = v$, we have

$$\psi(t_1,t_2) = \frac{e^{i\mu_1t_1+i\mu_2t_2}}{2\pi\sqrt{|\sigma_{ij}|}}\int_{-\infty}^{\infty}\int_{-\infty}^{\infty} e^{-\frac{1}{2}Q_1(u,v)}dudv$$

where

$$\begin{aligned} Q_1(u,v) &= \sigma^{11}u^2 + \sigma^{22}v^2 + 2\sigma^{12}uv - 2it_1u - 2it_2v \\ &= z_1^2 + z_2^2 + (\sigma_{11}t_1^2 + 2\sigma_{12}t_1t_2 + \sigma_{22}t_2^2) \end{aligned}$$

where

$$\begin{aligned} z_1 &= \sqrt{\sigma^{11}}\left(u + \frac{\sigma^{12}}{\sigma^{11}}v - \frac{it_1}{\sigma^{11}}\right) \\ z_2 &= \sqrt{\frac{|\sigma^{ij}|}{\sigma^{11}}}\left(v + \frac{i\sigma^{12}t_1 - i\sigma^{11}t_2}{|\sigma^{ij}|}\right). \end{aligned}$$

The transformations u and v to z_1 and z_2 provide Jacobian

$$\left|\frac{\partial(u,v)}{\partial(z_1,z_2)}\right| = \frac{1}{\sqrt{|\sigma^{ij}|}}.$$

Thus performing the integration, we have

$$\begin{aligned} \psi(t_1t_2) &= \exp\left[i(\mu_1t_1 + \mu_2t_2) - \frac{1}{2}(\sigma_{11}t_1^2 + \sigma_{22}t_2^2 + 2\sigma_{12}t_1t_2)\right] \\ &= \exp\left[i(\mu_1t_1 + \mu_2t_2) - \frac{1}{2}(\sigma_1^2t_1^2 + \sigma_2^2t_2^2 + 2\rho\sigma_1\sigma_2t_1t_2)\right]. \end{aligned}$$

If $L = c_1x + c_2y$ where c_1 and c_2 are not both zero, the characteristic function of L is given by

$$\psi(c_1t, c_2t) = \exp\left[i(c_1\mu_1 + c_2\mu_2)t - \frac{1}{2}\left(\sigma_{11}c_1^2 + 2\sigma_{12}c_1c_2 + \sigma_{22}c_2^2\right)t^2\right].$$

This indicates that

$$L \sim N\left(\sum_{i=1}^{2} c_i\mu_i, \sum_i^2\sum_j^2 \sigma_{ij}c_1c_2\right).$$

16.10 Multivariate Normal Distribution

The joint p.d.f. of multivariate (p-variate) normal distribution is the extension of the function (16.3) for $p = 2$, where the p.d.f. is given by

$$f(x_1, x_2, \ldots, x_p) = \frac{\sqrt{|\sigma^{ij}|}}{(2\pi)^{p/2}}e^{-\frac{1}{2}Q(x_1,x_2,\ldots,x_p)} \tag{16.6}$$

where

$$Q(x_1, x_2, \ldots, x_p) = \sum_{i,j=1}^{p} \sigma^{ij}(x_i - \mu_i)(x_j - \mu_j),$$

$[\sigma^{ij}]$ being the inverse of the covariance matrix $[\sigma_{ij}]$ and μ_i the means of x_i. Thus the properties of multivariate normal distribution are straight-forward extension of bivariate normal distribution.

The joint p.d.f. of the vector X can also be written as

$$f(x_1, x_2, x_p) = \frac{[|R|]^{1/2}}{(2\pi)^{p/2}} e^{-\frac{1}{2}(X-\boldsymbol{\mu})'R(X-\boldsymbol{\mu})} \tag{16.7}$$

with $-\infty < x_i < \infty$; $i = 1, 2, \ldots, p$. Here

(i) R is the inverse of the variance-covariance matrix of the vector X and is a positive definite matrix whose elements r_{ij} are constants.

(ii) $\boldsymbol{\mu}_i$, which is the i-th element of the vector $\boldsymbol{\mu}$, is a constant.

Since $|R|$ is positive, the quantity $|R|^{1/2}/(2\pi)^{p/2}$ is positive and the exponential function cannot assume negative values, $f(x_1, x_2, \ldots, x_p) > 0$. Now (16.7) will be a p.d.f. if it is possible to show that

$$\int_{-\infty}^{\infty} \int_{-\infty}^{\infty} \cdots \int_{-\infty}^{\infty} f(x_1, x_2, \ldots, x_p) dx_1 dx_2 \ldots dx_p = 1.$$

Let us make the transformation from the $X - \boldsymbol{\mu}$ vector to a vector Z by the relation $X - \boldsymbol{\mu} = Z$, which can be written as

$$(x_i - \boldsymbol{\mu}_i) = z_i; \; i = 1, 2, \ldots, p.$$

Since $dx_i/dz_j = 0$ if $i \neq j$ and $dx_i/dz_i = 1$, the Jacobian of the transformation is unity. Thus we have

$$\begin{aligned} &\int_{-\infty}^{\infty} \int_{-\infty}^{\infty} \cdots \int_{-\infty}^{\infty} \frac{|R|^{1/2}}{(2\pi)^{p/2}} e^{-\frac{1}{2}(X-\boldsymbol{\mu})'R(X-\boldsymbol{\mu})} dx_1 dx_2 \ldots dx_p \\ &= \int_{-\infty}^{\infty} \int_{-\infty}^{\infty} \cdots \int_{-\infty}^{\infty} \frac{|R|^{1/2}}{(2\pi)^{p/2}} e^{-\frac{1}{2}Z'RZ} dz_1 dz_2 \ldots dz_p. \end{aligned}$$

The transformation is one to one, and the entire X space transforms to entire Z space, as indicated by the limits on the integrals on the right-hand side of the equation. Let us make an orthogonal transformation $Z = PT$, where P is an orthogonal matrix and $T = \{t_i\}$ is a vector. We have

$$Z'RZ = T'P'RPT = T'DT$$

as R is positive definite $P'RP = D$, where D is a diagonal matrix with characteristic roots d_i of R. Thus we get

$$\begin{aligned} &\int_{-\infty}^{\infty} \int_{-\infty}^{\infty} \cdots \int_{-\infty}^{\infty} \frac{|R|^{1/2}}{(2\pi)^{p/2}} e^{-\frac{1}{2}Z'RZ} dz_1 dz_2 \ldots dz_p \\ &= \int_{-\infty}^{\infty} \int_{-\infty}^{\infty} \cdots \int_{-\infty}^{\infty} \frac{|R|^{1/2}}{(2\pi)^{p/2}} e^{-\frac{1}{2}T'DT} dt_1 dt_2 \ldots dt_p \\ &= \frac{|R|^{1/2}}{(2\pi)^{p/2}} \int_{-\infty}^{\infty} \int_{-\infty}^{\infty} \cdots \int_{-\infty}^{\infty} e^{-\frac{1}{2}d_1 t_1^2} e^{-\frac{1}{2}d_2 t_2^2} \cdots e^{-\frac{1}{2}d_p t_p^2} dt_1 \cdots dt_p \\ &= \frac{|R|^{1/2}}{(2\pi)^{p/2}} \int_{-\infty}^{\infty} e^{-\frac{1}{2}d_1 t_1^2} dt_1 \int_{-\infty}^{\infty} e^{-\frac{1}{2}d_2 t_2^2} dt_2 \cdots \int_{-\infty}^{\infty} e^{-\frac{1}{2}d_p t_p^2} dt_p. \end{aligned}$$

The range of integration on each t_i is $-\infty < t_i < \infty$, since an orthogonal transformation is simply a rotation. We know

$$\int_{-\infty}^{\infty} e^{-\frac{1}{2}d_i t_i^2} dt_i = \sqrt{\frac{2\pi}{d_i}} \text{ [from univariate normal distribution].}$$

$$\therefore \quad \frac{|R|^{1/2}}{(2\pi)^{p/2}} \int_{-\infty}^{\infty} e^{-\frac{1}{2}d_1 t_1^2} dt_1 \cdots \int_{-\infty}^{\infty} e^{-\frac{1}{2}d_p t_p^2} dt_p = \frac{|R|^{1/2}}{(2\pi)^{p/2}} \frac{(2\pi)^{p/2}}{\prod_{i=1}^{p} (d_i)^{p/2}}.$$

But

$$|D| = \prod_{i=1}^{p} d_i \text{ and } |D|^{1/2} = \prod_{i=1}^{p} d_i^{1/2}.$$

Again

$$|R| = |P'RP| = |D| \Rightarrow |R|^{1/2} = \prod_{i=1}^{p} d_i^{1/2}.$$

Therefore,

$$\frac{|R|^{1/2}}{(2\pi)^{p/2}} \int_{-\infty}^{\infty} e^{-\frac{1}{2}d_1 t_1^2} dt_1 \cdots \int_{-\infty}^{\infty} e^{-\frac{1}{2}d_p t_p^2} dt_p = 1.$$

Thus the function given by (16.7) is a frequency function.

Means of the Random Variables

We have already observed that

$$\int_{-\infty}^{\infty} \int_{-\infty}^{\infty} \cdots \int_{-\infty}^{\infty} \frac{|R|^{1/2}}{(2\pi)^{p/2}} e^{-\frac{1}{2}(X-\boldsymbol{\mu})'R(X-\boldsymbol{\mu})} dx_1 dx_2 \ldots dx_p = 1$$

$$\text{or,} \quad \int_{-\infty}^{\infty} \int_{-\infty}^{\infty} \cdots \int_{-\infty}^{\infty} \frac{[|\sigma^{ij}|]^{1/2}}{(2\pi)^{p/2}} e^{-\frac{1}{2}Q(x_1, x_2, \ldots, x_p)} dx_1 dx_2 \ldots dx_p = 1. \quad (16.8)$$

Differentiating both sides of it with respect to $\boldsymbol{\mu}_i$, we get

$$E\left[\sum_{j}^{p} \sigma^{ij}(x_j - \boldsymbol{\mu}_j)\right] = 0; \; i = 1, 2, \ldots, p$$

$$\text{or,} \sum_{j} \sigma^{ij} E(x_j - \boldsymbol{\mu}_j) = 0. \quad (16.9)$$

Since R is a positive definite matrix, $|R| = |\sigma^{ij}| \neq 0$, the solution of (16.9) is

$$E(x_j - \boldsymbol{\mu}_j) = 0 \Rightarrow E(x_i) = \boldsymbol{\mu}_i; \; i = 1, 2, \ldots, p.$$

Thus $\boldsymbol{\mu}_1, \boldsymbol{\mu}_2, \ldots, \boldsymbol{\mu}_p$ are the means of $x_1, x_2, \ldots, x_p$ respectively.

Covariances of the Random Variables

From (16.6), we have

$$\int_{-\infty}^{\infty} \int_{-\infty}^{\infty} \cdots \int_{-\infty}^{\infty} e^{-\frac{1}{2}Q(x_1, x_2, \ldots, x_p)} dx_1 dx_2 \ldots dx_p = \frac{(2\pi)^{p/2}}{[|\sigma^{ij}|]^{1/2}}.$$

Differentiating both sides of it with respect to σ^{ij} and then multiplying by

$$-(1+\delta_{ij})\frac{[||\sigma^{ij}||]^{1/2}}{(2\pi)^{p/2}}$$

where δ_{ij} is the Kronecker δ, we get

$$E[(x_i-\mu_i)(x_j-\mu_j)]=\sigma_{ij}.$$

This is true for $i=j$ as well as $i\neq j$. Thus covariance of x_i and x_j is σ_{ij}.

Marginal Distribution

Let us find the distribution of x_1 which is given by

$$f_1(x_1)=\int_{-\infty}^{\infty}\int_{-\infty}^{\infty}\cdots\int_{-\infty}^{\infty}\frac{|R|^{1/2}}{(2\pi)^{p/2}}e^{-\frac{1}{2}(X-\mu)'R(X-\mu)}dx_2dx_3\dots dx_p$$

the quadratic form $(X-\mu)'R(X-\mu)$ can be written as

$$(X-\mu)'(X-\mu)=[(X_1-\mu_1),(X_2-M_2)']\begin{bmatrix}R_{11} & R_{12}\\ R_{21} & R_{22}\end{bmatrix}\begin{bmatrix}X_1-\mu_1\\ X_2-M_2\end{bmatrix}$$

where X_2 and M_2 are the vectors containing the last $(p-1)$ elements of X and μ respectively, R_{11} is the element in the first row and column of R. Also, we have

$$\begin{aligned}(X-\mu)'R(X-\mu) &= (x_1-\mu_1)r_{11}(x_1-\mu_1)\\ &\quad+(x_1-\mu_1)R_{12}(x_2-M_2)\\ &\quad+(X_2-M_2)'R_{21}(x_1-\mu_1)\\ &\quad+(X_2-M_2)'R_{22}(X_2-M_2).\end{aligned}$$

We know that R is symmetric and positive definite, and hence $R_{12}=R_{21}'$, R_{22}^{-1} exists and is symmetric, $R_{11}=r_{11}>0$. Thus

$$\begin{aligned}(X-\mu)'R(X-\mu) &= (x_1-\mu_1)(R_{11}-R_{12}R_{22}^{-1}R_{21})(x_1-\mu_1)\\ &\quad+\{[(x_2-M_2)+R_{22}^{-1}R_{21}(x_1-\mu_1)]'\\ &\quad\times R_{22}[(X_2-M_2)+R_{22}^{-1}R_{21}(x_1-\mu_1)]\}.\end{aligned}$$

The first term in the right-hand side of the quadratic form contains only the random variable x_1. Thus we can write

$$f_1(x_1)=\frac{[|R|]^{1/2}}{(2\pi)^{p/2}}Ke^{-\frac{1}{2}(x_1-\mu_1)(R_{11}-R_{12}R_{22}^{-1}R_{21})(x_1-\mu_1)}$$

where

$$K=\int_{-\infty}^{\infty}\int_{-\infty}^{\infty}\cdots\int_{-\infty}^{\infty}\exp\left[-\frac{1}{2}\{X_2[M_2-R_{22}^{-1}R_{21}(x_1-\mu_1)]\}'\right.$$
$$\left.\times R_{22}\{X_2-[M_2-R_{22}^{-1}R_{21}(X_1-\mu_1)]\}dX_2\right.$$

where $dX_2 = dx_2 dx_3 \ldots dx_p$. Here x_1 is not a variable of integration and when x_1 occurs, it is considered as constant. Let $M_2 - R_{22}^{-1}R_{21}(x_1 - \mu_1) = H$, then we have

$$f_1(x_1) = \frac{|R|^{1/2}}{(2\pi)^{p/2}} K e^{-\frac{1}{2}(x_1-\mu_1)(R_{11}-R_{12}R_{22}^{-1}R_{21})(x_1-\mu_1)}$$

where

$$K = \int_{-\infty}^{\infty}\int_{-\infty}^{\infty}\cdots\int_{-\infty}^{\infty} e^{-\frac{1}{2}(X_2-H)'R_{22}(X_2-H)}dX_2.$$

This multiple integral gives

$$K = \frac{(2\pi)^{\frac{1}{2}(p-1)}}{|R_{22}|^{1/2}}.$$

$$\therefore\ f_1(x) = \frac{|R|^{1/2}}{(2\pi)^{p/2}}\frac{(2\pi)^{\frac{1}{2}(p-1)}}{|R_{22}|^{1/2}} e^{-\frac{1}{2}(x_1-\mu_1)(R_{11}-R_{12}R_{22}^{-1}R_{21})(x_1-\mu_1)}.$$

But $|R| = |R_{22}||R_{11} - R_{12}R_{22}^{-1}R_{21}|$. Here $(x_1 - \mu_1)$ and $(R_{11} - R_{12}R_{22}^{-1}R_{21})$ are both scalar quantities and that $(R_{11} - R_{12}R_{22}^{-1}R_{21}) = 1/\sigma_1^2$ (say) is positive definite. Thus

$$f_1(x_1) = \frac{1}{\sigma_1\sqrt{2\pi}} e^{-\frac{1}{2}\left(\frac{x_1-\mu_1}{\sigma_1}\right)^2}, -\infty < x_1 < \infty$$

which is the frequency function of univariate normal distribution. In a similar way the frequency function of any x_i can be found out.

In obtaining the marginal distribution of x_1, we have partitioned the vector X into two components, one containing the scalar quantity x_1 and other is the vector X_2. To find the joint marginal distribution of a subset of vectors, the quadratic form $(X - \mu)'R(X - \mu)$ can be written as

$$(X-\mu)'R(X-\mu) = [(X_1 - M_1), (X_2 - M_2)]' \begin{bmatrix} R_{11} & R_{12} \\ R_{21} & R_{22} \end{bmatrix} \begin{bmatrix} X_1 - M_1 \\ X_2 - M_2 \end{bmatrix}$$

where X_1 contains $s(s < p)$ of the elements of X and $(X_1 - M_1)$ is a vector of order $s \times 1$, R_{11} is a matrix of order $s \times s$. Then the resulting marginal distribution of X_1 is

$$g(X_1) = \frac{e^{-\frac{1}{2}(X_1-M_1)'(R_{11}-R_{12}R_{22}^{-1}R_{21})(X_1-M_1)}}{(2\pi)^{s/2}|R_{11} - R_{12}R_{22}^{-1}R_{21}|^{-\frac{1}{2}}}; -\infty < x_i < \infty, i = 1, 2, \ldots, s.$$

Conditional Distribution

We have already defined

$$f_1(x_1) = \int_{-\infty}^{\infty}\int_{-\infty}^{\infty}\cdots\int_{-\infty}^{\infty} f(x_1, x_2, \ldots, x_p)dx_2dx_3\ldots dx_p.$$

But

$$f_1(x_1) = \frac{f(x_1, x_2, \ldots, x_p)}{f_2(x_2, x_3, \ldots, x_p/x_1)}$$

$$\text{or, } f(x_1, x_2, \ldots, x_p) = f_1(x_1)f_2(x_2, x_3, \ldots, x_p/x_1).$$

But $f(x_1, x_2, \ldots, x_p)$ can be written as

$$f(x_1, x_2, \ldots, x_p) = \frac{|R|^{1/2}}{(2\pi)^{p/2}} e^{-\frac{1}{2}(x_1-\mu_1)(R_{11}-R_{12}R_{22}^{-1}R_{21})(x_1-\mu_1)}$$

$$\times\ e^{-\frac{1}{2}\{X_2-[M_2-R_{22}^{-1}R_{21}(x_1-\mu_1)]\}'R_{22}\{X_2-[M_2-R_{22}^{-1}R_{21}(x_1-\mu_1)]\}}$$

$$= \frac{[|R_{22}||R_{11}-R_{12}R_{22}^{-1}R_{21}|]^{1/2}}{(2\pi)^{1/2}(2\pi)^{\frac{1}{2}(p-1)}} e^{-\frac{1}{2}(x_1-\mu_1)(R_{11}-R_{12}R_{22}^{-1}R_{21})(x_1-\mu_1)}$$

$$\times\ e^{-\frac{1}{2}\{X_2-[M_2-R_{22}^{-1}R_{21}(x_1-\mu_1)]\}'R_{22}\{X_2-[M_2-R_{22}^{-1}R_{21}(x_1-\mu_1)]\}}$$

But

$$f_1(x_1) = \frac{|R_{11}-R_{12}R_{22}^{-1}R_{21}|^{1/2}}{(2\pi)^{1/2}} e^{-\frac{1}{2}(x_1-\mu_1)(R_{11}-R_{12}R_{22}^{-1}R_{21})(x_1-\mu_1)}$$

Therefore,

$$f_2(x_2, x_3, \ldots, x_p/x_1) = \frac{|R_{22}|^{1/2}}{(2\pi)^{\frac{1}{2}(p-1)}}$$

$$\times\ e^{-\frac{1}{2}\{X_2-[M_2-R_{22}^{-1}R_{21}(x_1-\mu_1)]\}'R_{22}\{X_2-[M_2-R_{22}^{-1}R_{21}(x_1-\mu_1)]\}}$$

which is a conditional distribution.

The conditional distribution of one sub-vector given another sub-vector can also be found out in a similar manner. Let the vectors X, μ and variance-covariance matrix V be partitioned as follows:

$$X = \begin{bmatrix} X_1 \\ X_2 \end{bmatrix},\ \mu = \begin{bmatrix} M_1 \\ M_2 \end{bmatrix} \text{ and } V = \begin{bmatrix} V_{11} & V_{12} \\ V_{21} & V_{22} \end{bmatrix} \text{ respectively,}$$

where X_1 is a vector of order $s \times 1$ and V_{11} is a matrix of order $s \times s$. If a set

$$X^* = \begin{bmatrix} X_1 \\ X_2^* \end{bmatrix}$$

the conditional distribution of X_1 given $X_2 = X_2^*$ is

$$h(X_1/X_2^*) = \frac{f(X_1, X_2^*)}{f_2(X_2^*)} = \frac{(1/2\pi)^{p/2}|V|^{\frac{1}{2}} e^{-\frac{1}{2}(X^*-\mu)'V^{-1}(X^*-\mu)}}{(1/2\pi)^{\frac{1}{2}(p-s)}|V_{22}|^{\frac{1}{2}} e^{-\frac{1}{2}(X_2^*-M_2)V_{22}^{-1}(X_2^*-M_2)}}.$$

Replacing X^*, μ and V by their corresponding sub-matrices and on simplification, we get

$$h(X_1/X_2^*) = \frac{e^{-\frac{1}{2}[X_1-M_1-V_{12}V_{22}^{-1}(X_2^*-M_2)][V_{11}-V_{12}V_{22}^{-1}V_{21}]^{-1}[X_1-M_1-V_{12}V_{22}^{-1}(X_2^*-M_2)]}}{(2\pi)^{s/2}\left|V_{11}-V_{12}V_{22}^{-1}V_{21}\right|^{1/2}}$$

where $f(X_1, X_2^*)$ is the joint distribution of the vectors X_1 and X_2 with X_2^* substituted for X_2, and $f_2(X_2^*)$ is the joint marginal distribution of the elements of X_2 substituted at $X_2 = X_2^*$. If the matrix in the exponent of $f(Y)$ is $R = V^{-1}$ and if R can be partitioned corresponding to the partition of V such as $R = \begin{bmatrix} R_{11} & R_{12} \\ R_{21} & R_{22} \end{bmatrix}$, then

$$R_{11} = \left[V_{11} - V_{12}V_{22}^{-1}V_{21}\right]^{-1}$$

where $V_{11} - V_{12}V_{22}^{-1}V_{21}^{-1}$ is positive definite. This conditional distribution is also multivariate normal distribution with mean

$$M_1 + V_{12}V_{22}^{-1}(X_2^* - M_2)$$

and covariance matrix $\left(V_{11} - V_{12}V_{22}^{-1}V_{21}\right)$. This covariance matrix does not depend on X_2^*.

Moment Generating Function

The joint moment generating function of $x_i - \mu_i$, $i = 1, 2, \ldots, p$, is given by

$$M_{X-\mu}(T) = E\left[\exp\left\{\sum_{i=1}^{p} t_i(x_i - \mu_i)\right\}\right] = E\left[e^{(X-\mu)'T}\right]$$

where T is a $p \times 1$ vector with elements t_i.

$$M_{X-\mu}(T) = \frac{|R|^{1/2}}{(2\pi)^{p/2}} \int_{-\infty}^{\infty}\int_{-\infty}^{\infty} \cdots \int_{-\infty}^{\infty} e^{(X-\mu)'T - \frac{1}{2}(X-\mu)'R(X-\mu)} dX \tag{16.10}$$

where $dX = dx_1, dx_2, \ldots, dx_p$. We can write

$$\begin{aligned}
&(X-\mu)'T - \frac{1}{2}(X-\mu)'R(X-\mu) \\
&= -\frac{1}{2}(X - \mu - R^{-1}T)'R(X - \mu - R^{-1}T) + \frac{1}{2}T'R^{-1}T \\
&= -\frac{1}{2}(X - H)'R(X - H) + \frac{1}{2}T'R^{-1}T, \text{ where } H = \mu + R^{-1}T.
\end{aligned}$$

Thus

$$\begin{aligned}
M_{X-\mu}(T) &= e^{\frac{1}{2}T'R^{-1}T} \frac{|R|^{1/2}}{(2\pi)^{p/2}} \int_{-\infty}^{\infty}\int_{-\infty}^{\infty} \cdots \int_{-\infty}^{\infty} e^{-\frac{1}{2}(X-H)'R(X-H)} dx_1 \cdots dx_p \\
&= e^{\frac{1}{2}T'R^{-1}T}.
\end{aligned}$$

If the ij-th element of R^{-1} is written as σ_{ij}, then

$$M_{X-\mu}(T) = \exp\left(\frac{1}{2}\sum_i\sum_j t_i t_j \sigma_{ij}\right).$$

From this moment generating function we can show that the covariance of x_i and $x_j (i \neq j)$ is σ_{ij} and the variance of x_i is the i-th element of R^{-1}. The variance-covariance matrix will be given by

$$\left.\frac{\partial^2 M_{X-\mu}(T)}{\partial t_i \partial t_j}\right|_{T=0}$$

We have

$$\begin{aligned}\frac{\partial^2 M_{X-\mu}(T)}{\partial t_i \partial t_j} &= E[(x_i-\mu_i)(x_j-\mu_j)]\exp\left[\sum_{i=1}^{p} t_i(x_i-\mu_i)\right]\Bigg|_{T=0} \\ &= E[(x_i-\mu_i)(x_j-\mu_j)] \\ &= \begin{cases}\operatorname{cov}(x_i,x_j), & i\neq j \\ V(x_i), & i=j\end{cases}\end{aligned}$$

Here $\frac{\partial^2 M_{X-\mu}(T)}{\partial t_i \partial t_j}$ is evaluated from (16.10). Finally, we can say that

$$\operatorname{cov}(x_i,x_j)=\sigma_{ij}\ (i\neq j).$$

The same result have already been observed in discussing the covariances of the random variables.

Characteristic Function

The characteristic function of p-variate normal distribution is given by

$$\begin{aligned}\psi(T) &= E\left[\exp i\sum_{j=1}^{p} t_j x_j\right] \\ &= \frac{|R|^{\frac{1}{2}}}{(2\pi)^{p/2}}\int_{-\infty}^{\infty}\cdots\int_{-\infty}^{\infty} e^{iX'T-\frac{1}{2}(X-\mu)'R(X-\mu)}dx_1\cdots dx_p.\end{aligned}$$

But

$$\begin{aligned}&iX'T-\frac{1}{2}(X-\mu)'R(X-\mu) \\ &= -\frac{1}{2}(X-\mu-iR^{-1}T)'R(X-\mu-iR^{-1}T)-\frac{1}{2}T'R^{-1}T \\ &= -\frac{1}{2}(X-W)R(X-W)-\frac{1}{2}T'R^{-1}T+i\mu'T\end{aligned}$$

where $W=\mu+iR^{-1}T$. Therefore,

$$\psi(T)=e^{i\mu'T-\frac{1}{2}T'R^{-1}T}\int_{-\infty}^{\infty}\int_{-\infty}^{\infty}\cdots\int_{-\infty}^{\infty}\frac{|R|^{\frac{1}{2}}}{(2\pi)^{p/2}}e^{-\frac{1}{2}(X-W)'R(X-W)}dX$$

where $dX=dx_1,dx_2,\ldots,dx_p$.

$$\therefore\ \psi(T)=e^{i\mu'T-\frac{1}{2}T'R^{-1}T}=\exp\left[i\sum_{j=1}^{p}\mu_j t_j-\frac{1}{2}\sum_{ij=1}^{p}\sigma_{ij}t_i t_j\right].$$

Properties Related to Multivariate Normal Distribution

❐**THEOREM 16.1** *Let X be a $p\times 1$ vector with elements x_i. If X has the multivariate normal distribution with mean vector μ and variance-covariance matrix*

V, then $\sum_{i=1}^{p} a_i x_i$, where a_i are known constants, has the univariate normal distribution with mean $\sum a_i \mu_i$ and variance $\sum_{i=1}^{p}\sum_{j=1}^{p} a_i a_j \sigma_{ij}$, where σ_{ij} is the ij-th element of V.

Proof. Let $Z = \Sigma a_i x_i = X'A$, where $A = (a_i)$. We have

$$\begin{aligned} M_z(t) &= E\left[e^{zt}\right] \\ &= E\left[e^{X'At}\right] \\ &= E\left[e^{(X-\mu)'At+\mu' t}\right] \\ &= e^{\mu' At} E\left[e^{(X-\mu)'At}\right]. \end{aligned}$$

Now, from the definition of $M_{X-\mu}(T)$ of multivariate normal distribution, we can write

$$M_z(t) = e^{\mu' At} e^{\frac{1}{2}t(A'VA)t} = e^{\mu' At + \frac{1}{2}t^2(A'VA)}$$

which is the m.g.f. of univariate normal distribution with mean $\mu' A = \Sigma a_i \mu_i$ and variance $A'VA = \Sigma\Sigma a_i a_j \sigma_{ij}$.

❐ **THEOREM 16.2** *If X is a vector with elements x_i and has the multivariate normal distribution with mean vector μ and variance-covariance matrix V, then x_i are jointly independent if and only if the covariance of x_i and x_j for all $i \neq j$ is zero.*

Proof. If the covariances of x_i and x_j for all $i \neq j$ are zero, the matrix V is diagonal. We have

$$f(x_1, x_2, \ldots, x_p) = \frac{|R|^{\frac{1}{2}}}{(2\pi)^{p/2}} e^{-\frac{1}{2}(X-\mu)' R (X-\mu)}$$

where $R = V^{-1}$. Now, if V is diagonal with elements v_{ii} in the i-th diagonal place, then we have

$$\begin{aligned} f(x_1, x_2, \ldots, x_p) &= \frac{\prod_{i=1}^{p} v_{ii}^{-\frac{1}{2}}}{(2\pi)^{p/2}} e^{-\frac{1}{2}\sum_{i=1}^{p}(x_i-\mu_i)^2 v_{ii}^{-1}} \\ &= \frac{v_{11}^{-\frac{1}{2}}}{(2\pi)^{1/2}} e^{-\frac{1}{2}\left(\frac{X_1-\mu_1}{\sqrt{v_{11}}}\right)^2} \times \frac{v_{22}^{-\frac{1}{2}}}{(2\pi)^{1/2}} e^{-\frac{1}{2}\left(\frac{X_2-\mu_2}{\sqrt{v_{22}}}\right)^2} \times \cdots \\ &\quad \times \frac{v_{pp}^{-1/2}}{(2\pi)^{1/2}} e^{-\frac{1}{2}\left(\frac{X_p-\mu_p}{\sqrt{v_{pp}}}\right)^2} \end{aligned}$$

But, we know that

$$f_i(x_i) = \frac{v_{ii}^{-1/2}}{(2\pi)^{1/2}} e^{-\frac{1}{2}\left(\frac{x_i-\mu_i}{\sqrt{v_{ii}}}\right)^2}.$$

$$\therefore f(x_1, x_2, \ldots, x_p) = f_1(x_1) f_2(x_2) \cdots f_p(x_p).$$

Therefore, if V is diagonal, x_i are jointly independent.

Now we need to show that, if the x_i are jointly independent, V is diagonal. We know that

$$\begin{aligned}\text{cov}(x_i, x_j) &= E\left[(x_i - \mu_i)(x_j - \mu_j)\right] \\ &= v_{ij}, i \neq j \\ &= \int_{-\infty}^{\infty}\int_{-\infty}^{\infty}\cdots\int_{\infty}^{\infty}(x_i - \mu_i)(x_j - \mu_j)f(x_1, x_2, \ldots, x_p)dx_1\cdots dx_p \\ &= \int_{-\infty}^{\infty} f_1(x_1)dx_1 \int_{-\infty}^{\infty} f_2(x_2)dx_2 \cdots \int_{-\infty}^{\infty}(x_i - \mu_i)f_i(x_i)dx_i \cdots \\ &\qquad \int_{-\infty}^{\infty}(x_j - \mu_j)f_j(x_j)dx_j \cdots \int_{-\infty}^{\infty} f_p(x_p)dx_p\end{aligned}$$

[$\because$ x_i and x_j are independent]

But $\int_{-\infty}^{\infty}(x_i - \mu_i)f_i(x_i)dx_i = 0.$

$$\therefore \text{cov}(x_i, x_j) = v_{ij} = 0, \;\; i \neq j.$$

Therefore, if a set of variables are jointly normal, they are independent if and only if they are pairwise uncorrelated.

From this theorem we can say that if $(p \times 1)$ vector X with sub-vectors X_1, $X_2, \ldots, X_q$ $(q \leq p)$ such that

$$X = \begin{bmatrix} X_1 \\ X_2 \\ \vdots \\ X_q \end{bmatrix}$$

has multivariate normal distribution, then the sub-vectors are jointly independent if the joint distribution of X can be expressed as a product of marginal distribution of $X_1, X_2, \ldots, X_q$, i.e., if

$$f(X) = f_1(X_1)f_2(X_2)\cdots f_q(X_q)$$

where $f_i(X_i)$ is the i-th $(i = 1, 2, \ldots, q)$ marginal distribution of i-th sub-vector.

❐ **THEOREM 16.3** *Let $(p \times 1)$ vector X be normally distributed with mean vector μ and variance-covariance matrix V and X be partitioned into q sub-vectors such that*

$$X = \begin{bmatrix} X_1 \\ X_2 \\ \vdots \\ X_q \end{bmatrix}$$

Then a necessary and sufficient condition that the sub-vectors be jointly independent is that all the sub-matrices V_{ij} $(i \neq j)$ be equal to the zero matrix.

•

Proof. Let $E(X_i) = M_i$, then

$$V_{ij} = E[(X_i - M_i)(X_j - M_j)'].$$

If $V_{ij} = 0$, for all $i \neq j$, then $R_{ij} = 0$ $(i \neq j)$, where $R = V^{-1}$. In that case

$$\begin{aligned} f(X) &= \frac{|R|^{1/2}}{(2\pi)^{p/2}} e^{-\frac{1}{2}(X-\mu)'R(X-\mu)} \\ &= \frac{|R|^{1/2}}{(2\pi)^{p/2}} \exp\left[-\frac{1}{2}\sum_{i=1}^{q}(X_i - M_i)'V_{ii}^{-1}(X_i - M_i)\right] \\ &= \prod_{i=1}^{q} K_i e^{-\frac{1}{2}(X_i - M_i)V_{ii}^{-1}(X_i - M_i)} = \prod_{i=1}^{q} f_i(X_i) \end{aligned}$$

where K_i is a constant. Thus

$$f(X) = f_1(X_1)f_2(X_2)\cdots f_q(X_q).$$

This independence is true if $V_{ij} = 0$. Therefore, we need to prove that $V_{ij} = 0$ if independence is true.

Let x_{il} be the l-th element in the sub-vector X_i. Then the lk-th element of the matrix V_{ij} is

$$\begin{aligned} &E[(x_{il} - \mu_{il})(x_{jk} - \mu_{jk})] \\ &= \int_{-\infty}^{\infty}\int_{-\infty}^{\infty}\cdots\int_{-\infty}^{\infty}(x_{il} - \mu_{il})(x_{jk} - \mu_{jk})f_1(X_1)f_2(X_2)\cdots f_q(X_q)dX_1\cdots dX_q \\ &= \int_{-\infty}^{\infty} f_1(X_1)dX_1 \int_{-\infty}^{\infty} f_2(X_2)dX_2 \cdots \int_{-\infty}^{\infty}(x_{il} - \mu_{il})f_i(X_i)dX_i \cdots \\ &\quad \int_{-\infty}^{\infty}(x_{jk} - \mu_{jk})f_j(X_j)dX_j \cdots \int_{-\infty}^{\infty} f_q(X_q)dX_q. \end{aligned}$$

But

$$\int_{-\infty}^{\infty}(x_{il} - \mu_{il})f_i(X_i)dX_i = 0.$$

This integral is zero for all l. Therefore, $V_{ij} = 0$. Here $E(x_{il}) = \mu_{il}$.

❑ **THEOREM 16.4** *If $x_1, x_2, \ldots, x_p$ are jointly normally distributed with quadratic form $Q = (X - \mu)'R(X - \mu)$, where $E(X) = \mu$. The vector of means $E(X) = \mu$ is the solution to the system of equations $\partial Q/\partial X = 0$.*

Proof. We know

$$f(X) = \frac{|R|^{1/2}}{(2\pi)^{p/2}} e^{-\frac{1}{2}(X-\mu)'R(X-\mu)}.$$

The function $f(X)$ will be maximum for a value of X for which

$$Q = 0 = (X - \mu)'R(X - \mu).$$

But R is positive definite, Q can be zero only at the point $X - \mu = 0$, i.e., $X = \mu$. Therefore, μ is the point that maximizes $f(X)$. But $\partial f(X)/\partial X = 0$ gives the point that maximizes $f(X)$. Here the derivatives exist as $f(X)$ is continuous. Again

$$\frac{\partial f(X)}{\partial X} = \frac{|R|^{1/2}}{(2\pi)^{p/2}} e^{-\frac{1}{2}Q} \left(-\frac{1}{2}\right) \frac{\partial Q}{\partial X} = 0.$$

It follows that the vector satisfying

$$\frac{\partial f(X)}{\partial X} = 0$$

is the same as the vector that satisfies $\partial Q/\partial X = 0$. Thus the vector μ is the solution to the system $\partial Q/\partial X = 0$.

Now, we shall present two important results related to multivariate normal distribution. These are:

(i) If the $(p \times 1)$ vector X is distributed normally with mean vector μ and variance-covariance matrix V and if B is a $q \times p$ $(q \leq p)$ matrix of rank q, then the vector $Z = BX$ is distributed as the q-variate normal distribution with mean $B\mu$ and covariance BVB'.

(ii) Let X be a $(p \times 1)$ vector having multivariate normal distribution with mean vector zero and covariance matrix $\sigma^2 I$. Let us define $Z = PX$, where P is an orthogonal matrix, then $Z \sim N(0, \sigma^2 I)$.

16.11 Chief Characteristics of Normal Distribution and Normal Curve

(i) The mean = median = mode = μ of normal distribution.

(ii) The distribution is symmetric and mesokurtic ($\beta_1 = 0, \beta_2 = 3$).

(iii) The normal probability curve takes the following shape:

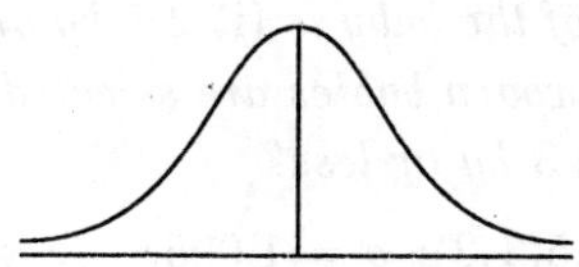

The curve is bell-shaped and is symmetric to $x = -\mu$.

(iv) The points of inflexion of normal distribution is $x = \mu \pm \sigma$. Because

$$f''(x) = -\frac{f(x)}{\sigma^2}\left[1 - \frac{(x-\mu)^2}{\sigma^2}\right] \text{ and } f''(x) = 0 \Rightarrow x = \mu \pm \sigma.$$

Again, $f'''(x) \neq 0$.

(v) If $X \sim N(\mu, \sigma^2)$, then $Z = \frac{X-\mu}{\sigma} \sim N(0,1)$.

(vi) If 95% of the values of a continuous variable X falls in the limit $\bar{x} \pm 2s$, where $\bar{x}$ = mean, s = standard deviation, then X is assumed to be distributed normally.

(vii) If $X \sim N(\mu, \sigma^2)$, then

$$P(\mu - \sigma < X < \mu + \sigma) = 0.6826$$
$$P(\mu - 2\sigma < X < \mu + 2\sigma) = 0.9544$$
$$P(\mu - 3\sigma < X < \mu + 3\sigma) = 0.9973.$$

(viii) For a normal distribution

$$\text{QD} : \text{MD} : \text{SD} :: \frac{2}{3}\sigma : \frac{4}{5}\sigma : \sigma$$
$$\text{or, QD} : \text{MD} : \text{SD} :: 10 : 12 : 15$$

where QD = Quartile deviation, MD = Mean deviation, SD = Standard deviation (σ).

(ix) If the values of normal variate X increases, the maximum probability attains at $x = \mu$ and this probability becomes

$$f(x)_{\max} = \frac{1}{\sqrt{2\pi}}.$$

16.12 Applications of Normal Distribution

Normal distribution plays a vital role in statistical analysis. The quantitative data observed from any random sample is assumed to follow normal distribution if the variable under study is continuous in nature. Even if the data is discrete, it can be transformed by proper transformation of variable to normal variable. Thus normal probability law can be applied to handle any statistical problem related to probability calculation or test of hypothesis when the observations involved are randomly recorded from sampling units.

►**EXAMPLE 16.1** *The birth-weight (in kg) of newborn babies is assumed a random variable X, where $X \sim N(2.5, 1.2)$. A newborn baby is randomly selected. Find the probability that, the weight of the baby is* (i) 2.7 *kg or more,* (ii) 3 *kg or less,* (iii) 2.4 *kg to* 3.2 *kg,* (iv) *If* 10 *such newborn babies are selected at random, what is the probability that* 2 *of them have weights* 3 *kg or less?*

Solution. Given $X \sim N(2.5, 1.2)$; $\sigma = 1.095$.

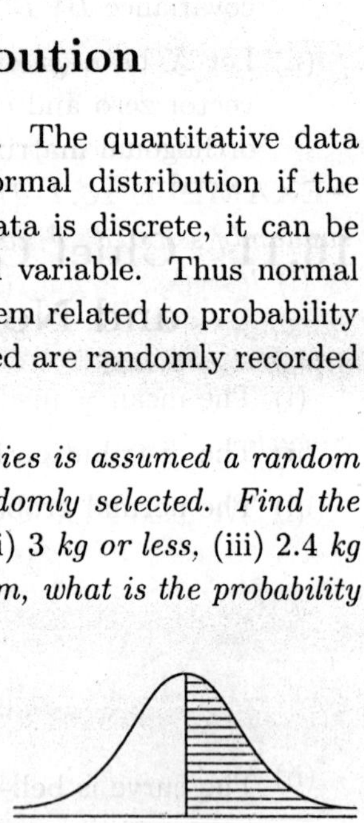
0.18

(i) We need $P(X \geq 2.7)$. We know

$$Z = \frac{X - \mu}{\sigma} = \frac{2.7 - 2.5}{1.095} = 0.18.$$

$$P(X \geq 2.7) = P(Z \geq 0.18) = 1 - P(Z < 0.18)$$
$$= 1 - 0.5714 = 0.4286 \text{ [Using Table 11.1]}$$

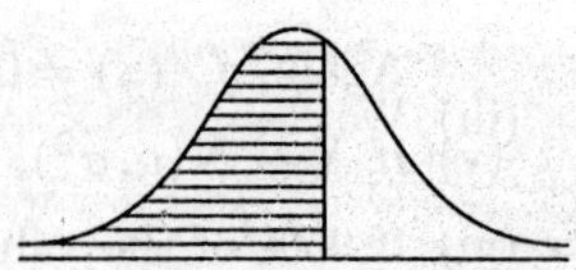
0.46

(ii) We need $P(X \leq 3)$

$$Z = \frac{X - \mu}{\sigma} = \frac{3 - 2.5}{1.095} = 0.46$$
$$P(X \leq 3) = P(Z \leq 0.46) = 0.6772.$$

(iii) We need $P(2.4 < X < 3.2)$

$$Z_1 = \frac{X_1 - \mu}{\sigma} = \frac{2.4 - 2.5}{1.095} = -0.09, \; Z_2 = \frac{X_2 - \mu}{\sigma} = \frac{3.2 - 2.5}{1.095} = 0.64.$$

$$\begin{aligned} P(2.4 < X < 3.2) &= P(-0.09 < Z < 0.64) \\ &= P(Z < 0.64) - P(Z < 0.09) \\ &= 0.7389 - 0.4641 \\ &= 0.2748. \end{aligned}$$

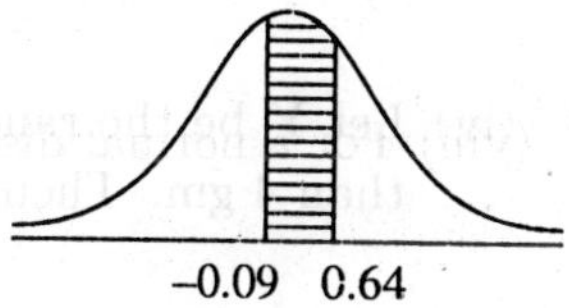

(iv) Let Y be the random variable indicating number of babies having weight less than 3 kg. We have

$$P(X \leq 3) = p = 0.6772, \;\; \therefore \; q = 1 - p = 0.3228.$$

Then

$$P(Y = y) = \binom{10}{y}(0.6772)^y(0.3228)^{10-y}; \; y = 0, 1, 2, \ldots, 10.$$

We need $P(Y = 2)$

$$P(Y = 2) = \binom{10}{2}(0.6772)^2(0.3228)^8 \approx 0.00243.$$

►**EXAMPLE 16.2** *The weight (X gm) of a rose is assumed $N(5, 12.4)$. A rose is randomly selected. Find the probability that the weight of the randomly selected flower is (i) less than 4 gm, (ii) more than 6 gm, (iii) 3 gm to 6 gm. (iv) If 25 roses are randomly selected, what is the probability that at least 3 of them have weights below 4 gm?*

Solution. Given $X \sim N(5, 12.4)$, $\sigma = 3.52$.

(i) We need $P(X < 4)$

$$\begin{aligned} Z &= \frac{X - \mu}{\sigma} = \frac{4 - 5}{3.52} = -0.28 \\ P(X < 4) &= P(Z < -0.28) \\ &= 0.3897. \end{aligned}$$

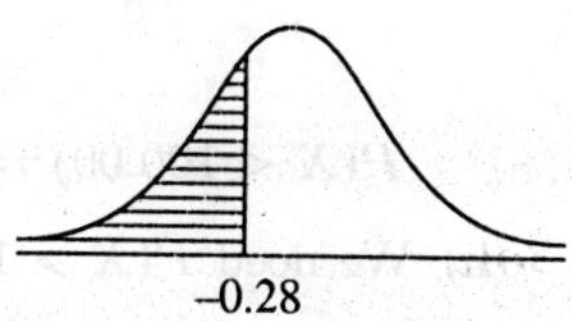

(ii) We need $P(X > 6)$

$$\begin{aligned} Z &= \frac{X - \mu}{\sigma} = \frac{6 - 5}{3.52} = 0.28 \\ P(X > 6) &= P(Z > 0.28) = 1 - P(Z \leq 0.28) \\ &= 1 - 0.6103 = 0.3897. \end{aligned}$$

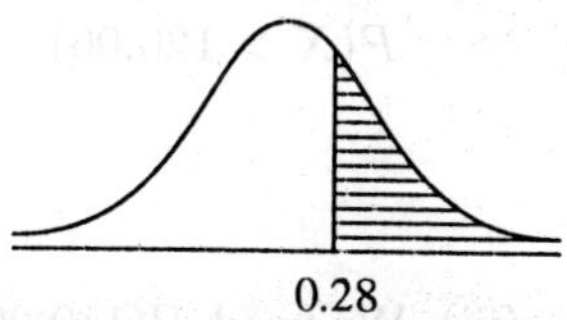

(iii) We need $P(3 < X < 6)$

$$Z_1 = \frac{X_1 - \mu}{\sigma} = \frac{3 - 5}{3.52} = -0.57, \; Z_2 = \frac{X_2 - \mu}{\sigma} = \frac{6 - 5}{3.52} = 0.28$$

$$\begin{aligned} P(3 < X < 6) &= P(-0.57 < Z < 0.28) \\ &= P(Z \leq 0.28) - P(Z \leq -0.57) \\ &= 0.6103 - 0.2843 \\ &= 0.326. \end{aligned}$$

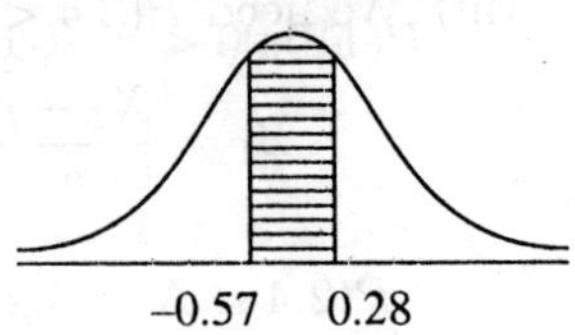

(iv) Let Y be the random variable indicating number of flowers having weights less than 4 gm. Then

$$P(Y = y) = \binom{n}{y} p^y q^{n-y}, \; y = 0; 1, 2, \ldots, n.$$

Given $n = 25, \; p = 0.3897, \; q = 0.6103.$

We need $P(Y \geq 3)$

$$\begin{aligned} P(Y \geq 3) &= 1 - P(Y < 3) = 1 - P(Y = 0) - P(Y = 1) - P(Y = 2) \\ &= 1 - \binom{25}{0}(0.6103)^{25} - \binom{25}{1}(0.3897)(0.6103)^{24} \\ &\quad - \binom{25}{2}(0.3897)^2(0.6103)^{23} \approx 0.999394. \end{aligned}$$

►**EXAMPLE 16.3** *The daily income (X Rs.) of a labourer working in a factory is assumed to be normally distributed with $E(X) = 112.50$ and $V(X) = 1686.45$. A labourer is randomly selected from the factory. Find the probability that his income is* (i) *less than Rs.* 100.00, (ii) *more than Rs.* 120.00, (iii) *between Rs.* 110.00 *to Rs.* 115.00. (iv) *If* 150 *labourers are randomly selected from the factory, find the probability that at best* 2 *of them have income from Rs.* 110.00 *to Rs.* 115.00.

Solution. Given $X \sim N(112.50, 1686.45), \sigma = 41.07.$

(i) We need $P(X < 100.00)$

$$\begin{aligned} Z = \frac{X - \mu}{\sigma} &= \frac{100.00 - 112.50}{41.07} \\ &= -0.30 \end{aligned}$$

$$P(X < 100.00) = P(Z < -0.30) = 0.3821.$$

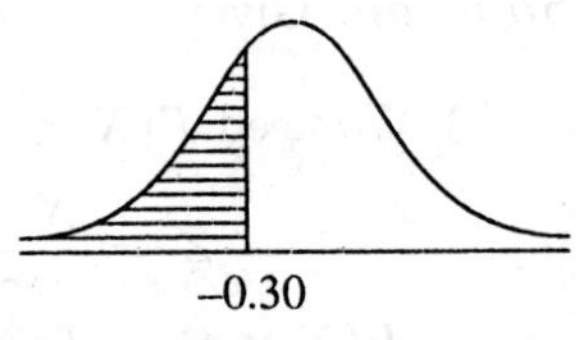

(ii) We need $P(X > 120.00)$

$$\begin{aligned} Z = \frac{X - \mu}{\sigma} &= \frac{120.00 - 112.50}{41.07} = 0.18 \\ P(X > 120.00) &= P(Z > 0.18) \\ &= 1 - P(Z \leq 0.18) \\ &= 1 - 0.5714 = 0.4286. \end{aligned}$$

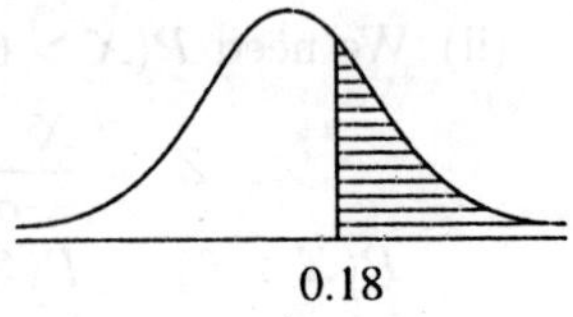

(iii) We need $P(110.00 < X < 115.00)$

$$Z_1 = \frac{X_1 - \mu}{\sigma} = \frac{110.00 - 112.50}{41.07} = -0.06$$

$$Z_2 = \frac{X_2 - \mu}{\sigma} = \frac{115.00 - 112.50}{41.07} = 0.06$$

$$
\begin{aligned}
P(110.00 < X < 115.00) &= P(-0.06 < Z < 0.06) \\
&= P(Z < 0.06) - P(Z < -0.06) \\
&= 0.5239 - 0.4761 \\
&= 0.0478.
\end{aligned}
$$

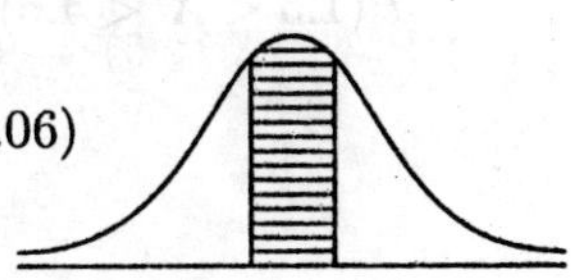

−0.06 0.06

(iv) Let Y be the random variable indicating the number of labourers earning money in the limit Rs. 110.00 to Rs. 115.00. The probability that a labourer earns money in the specified limit is $p = 0.0478$ [as in (iii)]. We have a random sample of $n = 150$ labourers. Since p is small and n is large, Y may be assumed as Poisson variate, where $\lambda = np = 150 \times 0.0478 = 7.17$.

Therefore,

$$P(Y = y) = \frac{e^{-7.17}(7.17)^y}{y!}; \; y = 0, 1, 2, \ldots$$

We need $P(Y \leq 2)$

$$
\begin{aligned}
P(Y \leq 2) &= P(Y = 0) + P(Y = 1) + P(Y = 2) \\
&= e^{-7.17}\left[\frac{(7.17)^0}{0!} + \frac{7.17}{1!} + \frac{(7.17)^2}{2!}\right] \\
&\approx 0.02606.
\end{aligned}
$$

►**EXAMPLE 16.4** *The amount of rain* (X *mm*) *per day during rainy season is assumed to follow* $N(2.6, 34.5)$. *Find the probability that, in a randomly selected day, the amount of rainfall is* (i) 6 *mm or more,* (ii) *less than* 1 *mm,* (iii) 1.5 *mm to* 4.6 *mm.* (iv) *What is the probability that in a randomly selected week at best* 2 *days the rainfall will be between* 1.5 *mm to* 4.6 *mm?*

Solution. Given $X \sim N(2.6, 34.5)$, $\sigma = 5.87$.

(i) We need $P(X > 6)$

$$
\begin{aligned}
Z &= \frac{X - \mu}{\sigma} = \frac{6 - 2.6}{5.87} = 0.58 \\
P(X > 6) &= P(Z > 0.58) = 1 - P(Z < 0.58) \\
&= 1 - 0.719 = 0.281.
\end{aligned}
$$

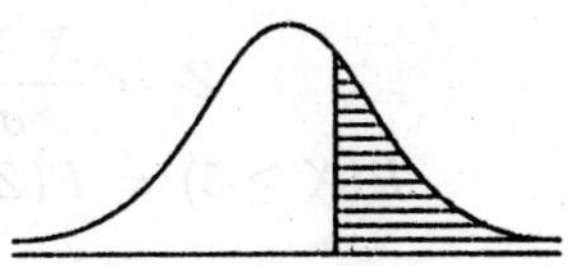

0.58

(ii) We need $P(X < 1)$

$$
\begin{aligned}
Z &= \frac{X - \mu}{\sigma} = \frac{1 - 2.6}{5.87} = -0.27 \\
P(X < 1) &= P(Z < -0.27) \\
&= 0.3936.
\end{aligned}
$$

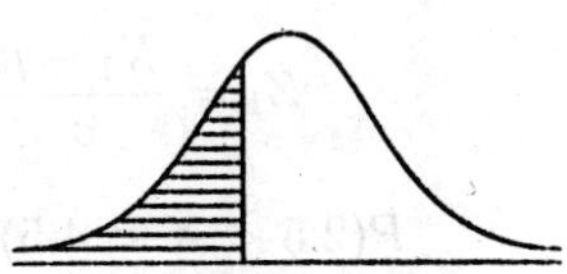

−0.27

(iii) We need $P(1.5 < X < 4.6)$

$$Z_1 = \frac{X_1 - \mu}{\sigma} = \frac{1.5 - 2.6}{5.87} = -0.19, \; Z_2 = \frac{X_2 - \mu}{\sigma} = \frac{4.6 - 2.6}{5.87} = 0.34$$

$$
\begin{aligned}
P(1.5 < X < 4.6) &= P(-0.19 < Z < 0.34) \\
&= P(Z < 0.34) - P(Z < -0.19) \\
&= 0.6331 - 0.4247 \\
&= 0.2084.
\end{aligned}
$$

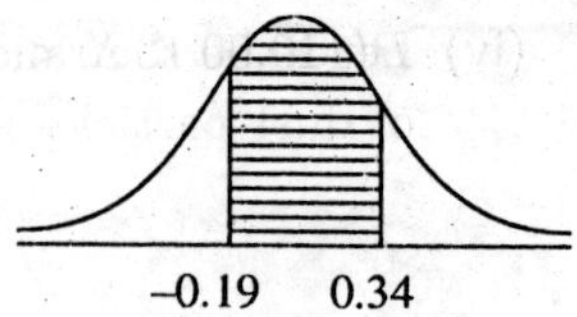

(iv) Let Y be the number of randomly selected days in which the rainfalls range from 1.5 mm to 4.6 mm. The probability of a day having such amount of rainfall is $p = 0.2084$. Therefore, $q = 1 - p = 0.7916$. We have

$$
P(Y = y) = \binom{7}{y}(0.2084)^4(0.7916)^{7-y};\ y = 0, 1, 2, \ldots, 7
$$

$$
\begin{aligned}
\therefore\ P(Y \le 2) &= P(Y = 0) + P(Y = 1) + P(Y = 2) \\
&= \binom{7}{0}(0.7916)^7 + \binom{7}{1}(0.2084)(0.7916)^6 + \binom{7}{2}(0.2084)^2(0.7916)^5 \\
&\approx 0.83722.
\end{aligned}
$$

►**EXAMPLE 16.5** *The amount of dust particles [X microgm/(c.c.)^3] in air in a city is assumed to follow $N(3.2, 2.4)$. Find the probability that, in a randomly selected day, the amount is* (i) *less than or equal to* 2 *microgm/(c.c.)^3,* (ii) 5 *microgm/(c.c.)^3 or more,* (iii) 2.5 *to* 4.5 *microgm/(c.c.)^3.* (iv) *What is the probability that, in a randomly selected week at best* 2 *days the amount of dust particles will be* 2 *microgm/(c.c.)^3 or more?*

Solution. Given $X \sim N(3.2, 2.4)$, $\sigma = 1.55$.

(i) We need $P(X \le 2)$

$$
Z = \frac{X - \mu}{\sigma} = \frac{2 - 3.2}{1.55} = -0.77
$$

$$
\begin{aligned}
P(X \le 2) &= P(Z \le -0.77) \\
&= 0.2206.
\end{aligned}
$$

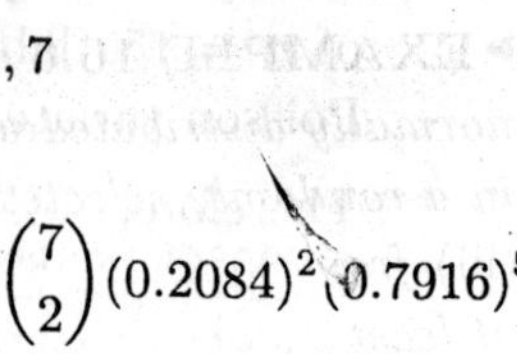

(ii) We need $P(X \ge 5)$

$$
Z = \frac{X - \mu}{\sigma} = \frac{5 - 3.2}{1.55} = 1.16
$$

$$
\begin{aligned}
P(X \ge 5) &= P(Z \ge 1.16) = 1 - P(Z < 1.16) \\
&= 1 - 0.877 = 0.123.
\end{aligned}
$$

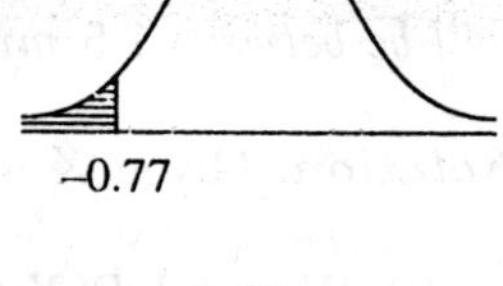

(iii) We need $P(2.5 < X < 4.5)$

$$
Z_1 = \frac{X_1 - \mu}{\sigma} = \frac{2.5 - 3.2}{1.55} = -0.45,\ Z_2 = \frac{X_2 - \mu}{\sigma} = \frac{4.5 - 3.2}{1.55} = 0.84
$$

$$
\begin{aligned}
P(2.5 < X < 4.5) &= P(-0.45 < Z < 0.84) \\
&= P(Z < 0.84) - P(Z < -0.45) \\
&= 0.7995 - 0.3264 \\
&= 0.4731.
\end{aligned}
$$

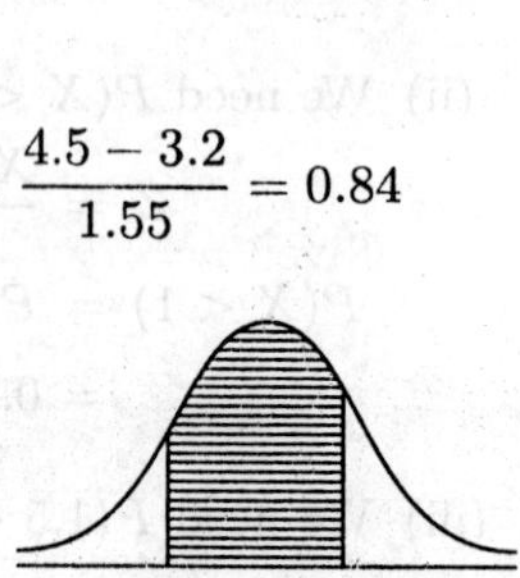

(iv) Let Y be the random variables indicating number of days in which the amount of dust particles is 2.0 microgm/(c.c.)^3 or more. Then

$$P(Y = y) = \binom{n}{y} p^y q^{n-y}; \; y = 0, 1, 2, \ldots, n.$$

Given $n = 7, \; y = 2, \; p = 0.7794, \; q = 0.2206$ [as in (i)].

$$\therefore \; P(Y = 2) = \binom{7}{2}(0.7794)^2(0.2206)^5 \approx 0.00666.$$

►**EXAMPLE 16.6** *The temperature per day during summer* (X°C) *is assumed to be normally distributed with mean* 36.8°C *and variance* 9.7°C^2. *Find the probability that, in a randomly selected day, the temperature is* (i) *less than* 35°C, (ii) *more than* 38°C, (iii) *from* 35°C *to* 38°C. (iv) *What is the probability that in a randomly selected week at least 2 days the temperature will be* 38°C*?* (v) *Find* $P(|X - 36.8| > 4)$.

Solution. Given $X \sim N(36.8, 9.7)$, $\sigma = 3.11$.

(i) We need $P(X < 35)$

$$Z = \frac{X - \mu}{\sigma} = \frac{35 - 36.8}{3.11} = -0.58$$

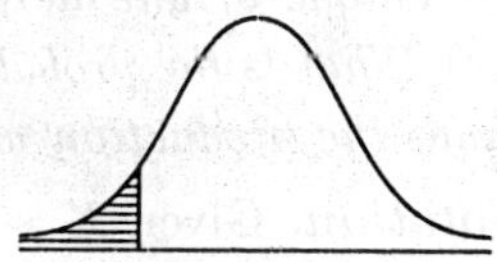

$$P(X < 35) = P(Z < -0.58) = 0.281.$$

(ii) We need $P(X > 38)$

$$Z = \frac{X - \mu}{\sigma} = \frac{38 - 36.8}{3.11} = 0.39$$

$$P(X > 38) = P(Z > 0.39) = 1 - P(Z \leq 0.39) = 1 - 0.6517 = 0.3483.$$

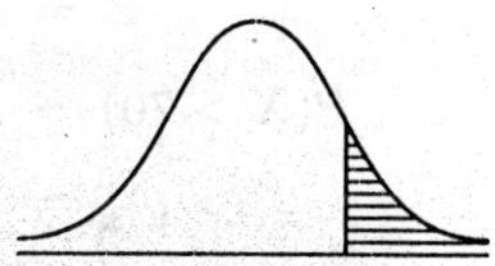

(iii) We need $P(35 < X < 38)$

$$\begin{aligned} P(35 < X < 38) &= P(-0.58 < Z < 0.39) \\ &= P(Z \leq 0.39) - P(Z \leq -0.58) \\ &= 0.6517 - 0.281 = 0.3707. \end{aligned}$$

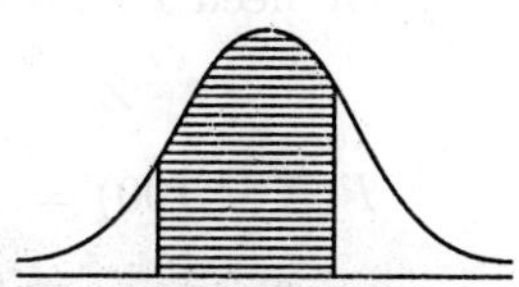

(iv) Let Y be the random variables indicating number of days in which the temperature exceeds 38°C. Then

$$P(Y = y) = \binom{n}{y} p^y q^{n-y}; \; y = 0, 1, 2, \ldots, n.$$

Given $n = 7, \; p = 0.3483$ [as in (ii)], $q = 0.6517$.

$$\begin{aligned} P(Y \geq 2) &= 1 - P(Y < 2) \\ &= 1 - P(Y = 0) - P(Y = 1) \\ &= 1 - \binom{7}{0}(0.6517)^7 - \binom{7}{1}(0.3483)(0.6517)^6 \\ &\approx 0.76329. \end{aligned}$$

(v) We need $P(|X - 36.8|) > 4$

$$P(|X - 36.8| \leq 4) = P(32.8 \leq X \leq 40.8)$$

$$Z_1 = \frac{X_1 - \mu}{\sigma} = \frac{32.8 - 36.8}{3.11} = -1.29, \; Z_2 = \frac{X_2 - \mu}{\sigma} = \frac{40.8 - 36.8}{3.11} = 1.29$$

$$\begin{aligned} P(32.8 < X < 40.8) &= P(-1.29 < Z < 1.29) \\ &= P(Z < 1.29) - P(Z < -1.29) \\ &= 0.9015 - 0.0985 = 0.803 \end{aligned}$$

$$\begin{aligned} \therefore P(|X - 36.8| > 4) &= 1 - P(|X - 36.8| \leq 4) \\ &= 1 - 0.803 = 0.197. \end{aligned}$$

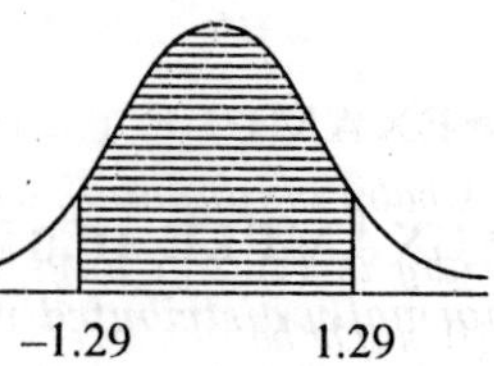

►**EXAMPLE 16.7** *The per acre production of high-yielding variety of rice (X mds/acre) is assumed to be normally distributed with mean* 80 *mds and variance* 196(*mds*)^2. *What is the probability that, in a plot of one acre, there will be production* (i) *more than* 70 *mds,* (ii) *less than* 100 *mds,* (iii) 75 *mds to* 95 *mds? Also, find the percentage of one acre plot of lands in which the stated production will be recorded.* (iv) *What is the probability that in* 5 *plots out of* 10 *randomly selected plots of one acre lands the production will be more than* 70 *mds?*

Solution. Given $X \sim N(80, 196)$, $\sigma = 14$.

(i) We need $P(X > 70)$

$$Z = \frac{X - \mu}{\sigma} = \frac{70 - 80}{14} = -0.71$$

$$\begin{aligned} P(X > 70) &= P(Z > -0.71) = 1 - P(Z \leq 0.71) \\ &= 1 - 0.2389 = 0.7611. \end{aligned}$$

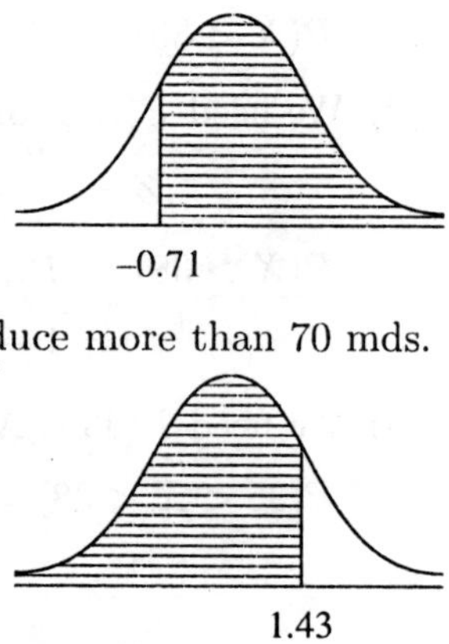

Therefore, in 76.11% of plots of one acre land will produce more than 70 mds.

(ii) We need $P(X < 100)$

$$Z = \frac{X - \mu}{\sigma} = \frac{100 - 80}{14} = 1.43$$

$$\begin{aligned} P(X < 100) &= P(Z < 1.43) \\ &= 0.9236. \end{aligned}$$

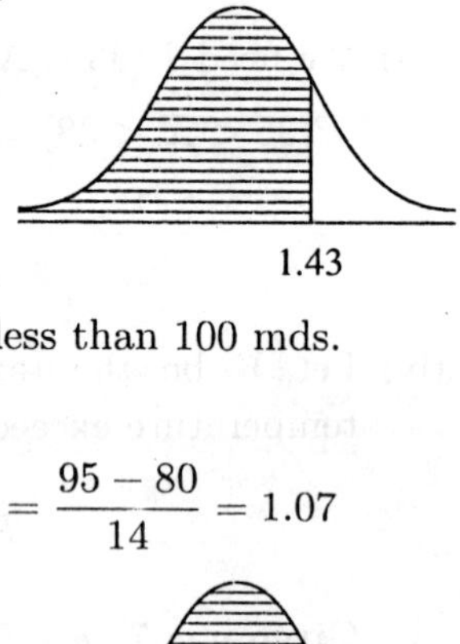

Therefore, in 92.36% of plots of one acre will produce less than 100 mds.

(iii) We need $P(75 < X < 95)$

$$Z_1 = \frac{X_1 - \mu}{\sigma} = \frac{75 - 80}{14} = -0.36, \; Z_2 = \frac{X_2 - \mu}{\sigma} = \frac{95 - 80}{14} = 1.07$$

$$\begin{aligned} P(75 < X < 95) &= P(-0.36 < Z < 1.07) \\ &= P(Z \leq 1.07) - P(Z \leq -0.36) \\ &= 0.8577 - 0.3594 \\ &= 0.4983. \end{aligned}$$

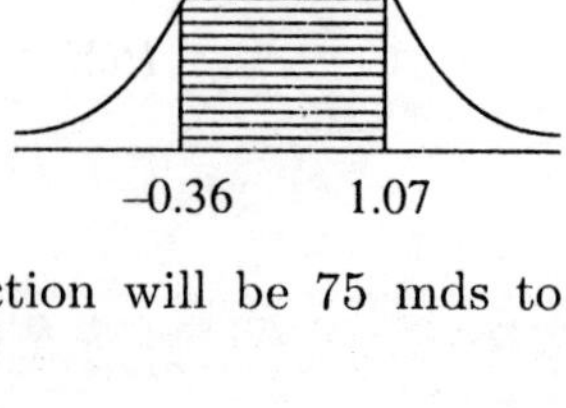

Therefore, in 49.83% of plots of one acre the production will be 75 mds to 95 mds.

(iv) Let Y be the random variable indicating number of plots in which production is more than 70 mds. Then

$$P(Y = y) = \binom{10}{y}(0.7611)^y(0.2369)^{10-y}; \; y = 0, 1, 2, \ldots, 10.$$

$$\therefore \; P(Y = 5) = \binom{10}{5}(0.7611)^5(0.2389)^5 \approx 0.5008.$$

►EXAMPLE 16.8 *The gestation period of babies (X days) is assumed $N(260, 102)$. A baby is randomly selected. Find the probability that the gestation period of the selected baby is* (i) 255 *days or less,* (ii) 268 *days or more,* (iii) 260 *days to* 270 *days.* (iv) *If* 10 *such babies are selected at random, what is the probability that out of* 10 *babies at least* 3 *babies have gestation period* 260–270 *days?*

Solution. Given $X \sim N(260, 102)$, $\sigma = 10.1$.

(i) We need $P(X \leq 255)$

$$Z = \frac{X-\mu}{\sigma} = \frac{255-260}{10.1} = -0.49$$

$$P(X \leq 255) = P(Z \leq -0.49) = 0.3121.$$

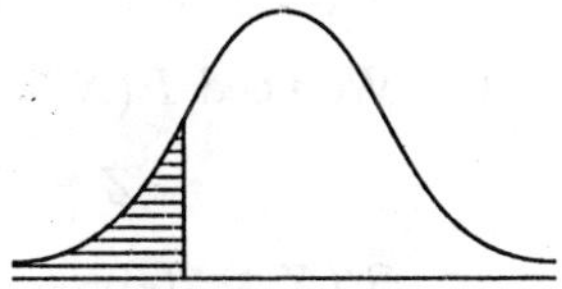

−0.49

(ii) We need $P(X \geq 268)$

$$Z = \frac{X-\mu}{\sigma} = \frac{268-260}{10.1} = 0.79$$

$$P(X \geq 268) = P(Z \geq 0.79) = 1 - P(Z < 0.79)$$

$$= 1 - 0.7852 = 0.2148.$$

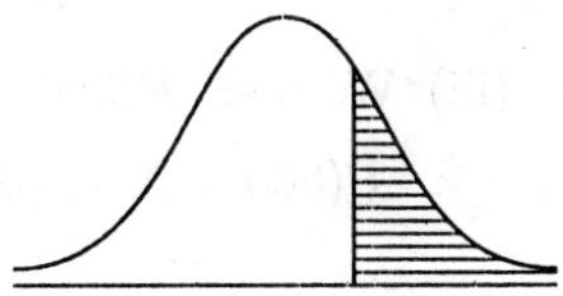

0.79

(iii) We need $P(260 < X < 270)$

$$Z_1 = \frac{X_1-\mu}{\sigma} = \frac{260-260}{10.1} = 0, \; Z_2 = \frac{X_2-\mu}{\sigma} = \frac{270-260}{10.1} = 0.99$$

$$\begin{aligned} P(260 < X < 270) &= P(0 < Z < 0.99) \\ &= P(Z \leq 0.99) - P(0 \leq Z) \\ &= 0.8389 - 0.50 \\ &= 0.3389. \end{aligned}$$

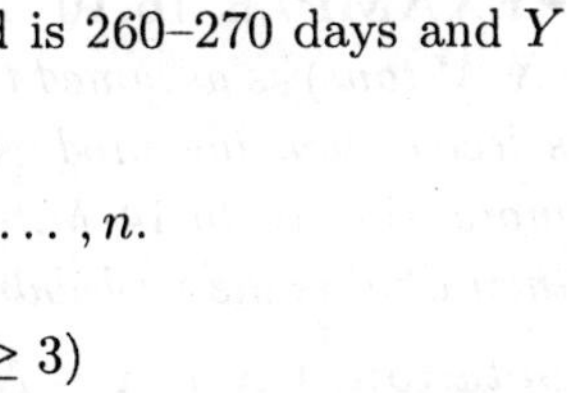

0 0.99

(iv) Let Y be the number of babies whose gestation period is 260–270 days and Y be a random variable. Then

$$P(Y = y) = \binom{n}{y}p^y q^{n-y}; \; y = 0, 1, 2, \ldots, n.$$

Given $n = 10$, $p = 0.3389$, $q = 0.6611$. We need $P(Y \geq 3)$

$$\begin{aligned} P(Y \geq 3) &= 1 - P(Y < 3) = 1 - P(Y = 0) - P(Y = 1) - P(Y - 2) \\ &= 1 - \binom{10}{0}(0.6611)^{10} - \binom{10}{1}0.3389(0.6611)^9 \\ &\quad - \binom{10}{2}(0.3389)^2(0.6611)^8 \approx 0.71373. \end{aligned}$$

▶EXAMPLE 16.9 *The monthly income of foreign currency (X million \$) of a country is assumed to be normally distributed with mean* 600 *million \$ and variance* 10263.5(*million* \$)^2. *Find the probability that, in a randomly selected month, the monthly income is* (i) 500 *million \$ or less,* (ii) 700 *million \$ or more,* (iii) 500–700 *million* \$. (iv) *If* 4 *months are randomly selected, what is the probability that, out of* 4 *months,* 2 *months will earn* 500 *to* 700 *million \$?*

Solution. Given $X \sim N(600, 10263.5)$, $\sigma = 101.31$.

(i) We need $P(X \leq 500)$

$$Z = \frac{X-\mu}{\sigma} = \frac{500-600}{101.31} = -0.99$$

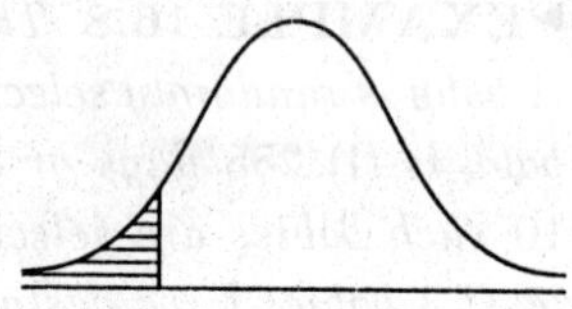

–0.99

$$P(X \leq 500) = P(Z \leq -0.99) = 0.1611.$$

(ii) We need $P(X \geq 700)$

$$Z = \frac{X-\mu}{\sigma} = \frac{700-600}{101.31} = 0.99$$

$$P(X \geq 700) = P(Z \geq 0.99) = 1 - P(Z < 0.99) = 1 - 0.8389 = 0.1611.$$

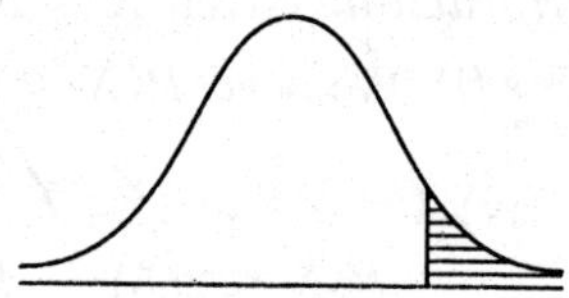

0.99

(iii) We need $P(500 < X < 700)$

$$P(500 < X < 700) = P(-0.99 < Z < 0.99) = P(Z \leq 0.99) - P(Z \leq -0.99) = 0.8389 - 0.1611 = 0.6778.$$

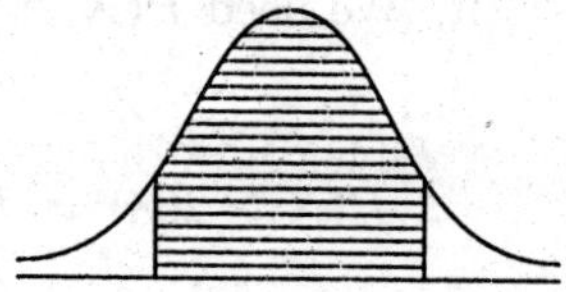

–0.99 0.99

(iv) Let Y be the random variable indicating number of months in which foreign currency earning is 500–700 million \$. Then

$$P(Y = y) = \binom{n}{y} p^y q^{n-y};\ y = 0, 1, 2, \ldots, n.$$

Given $n = 4,\ y = 2,\ p = 0.6778,\ q = 0.3222$. Then

$$P(Y = 2) = \binom{4}{2}(0.6778)^2(0.3222)^2 \approx 0.28616.$$

▶EXAMPLE 16.10 *The production of jute goods per day in a jute industry (X M.tons) is assumed to follow $N(12.5, 22.2)$. Find the probability that, in a randomly selected day, the production of jute goods is* (i) 10 *M.tons or less,* (ii) 15 *M.tons or more,* (iii) 10 *to* 14 *M.tons.* (iv) *If the production data is recorded* 15 *days randomly, then what is the probability that out of* 15 *days* 7 *days the production is* 10 *to* 15 *M.tons?*

Solution. Given $X \sim N(12.5, 22.2)$, $\sigma = 4.71$.

(i) We need $P(X \leq 10)$

$$Z = \frac{X-\mu}{\sigma} = \frac{10-12.5}{4.71} = -0.53$$

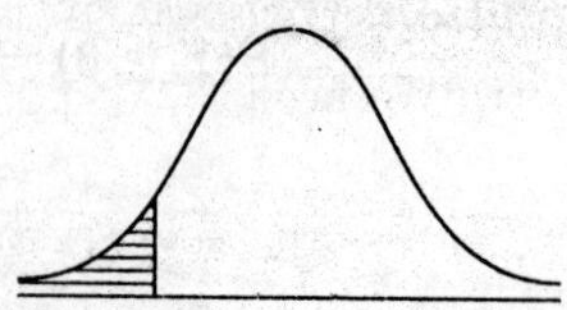

–0.53

$$P(X \leq 10) = P(Z \leq -0.53) = 0.2981.$$

(ii) We need $P(X \geq 15)$

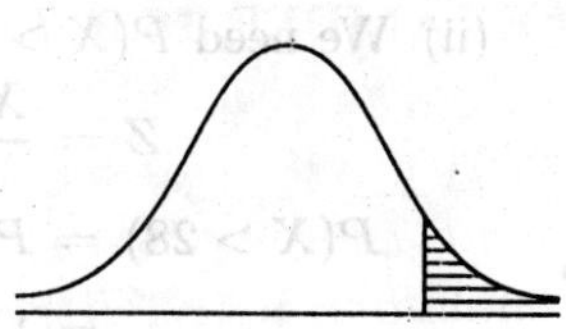

0.53

$$Z = \frac{X-\mu}{\sigma} = \frac{15-12.5}{4.71} = 0.53$$

$$P(X \geq 15) = P(Z \geq 0.53) = 1 - P(Z < 0.53)$$
$$= 1 - 0.7019 = 0.2981.$$

(iii) We need $P(10 < X < 14)$

$$Z_1 = \frac{X_1-\mu}{\sigma} = \frac{10-12.5}{4.71} = -0.53, \; Z_2 = \frac{14-12.5}{4.71} = 0.32$$

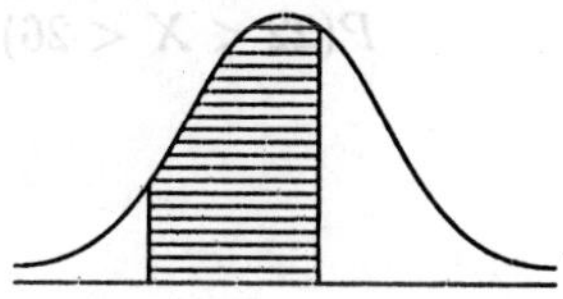

–0.53 0.32

$$\begin{aligned} P(10 < X < 14) &= P(-0.53 < Z < 0.32) \\ &= P(Z \leq 0.32) - P(Z \leq -0.53) \\ &= 0.6255 - 0.2981 \\ &= 0.3274. \end{aligned}$$

(iv) Let Y be the random variable indicating number of days in which production varies from 10 to 15 M.tons. Then

$$P(Y = y) = \binom{n}{y} p^y q^{n-y}.$$

Given $n = 15, \; y = 7,$

$$\begin{aligned} p &= P(10 < X < 15) \\ &= P(-0.53 < Z < 0.53) \\ &= P(Z \leq 0.53) - P(Z \leq -0.53) \\ &= 0.7019 - 0.2981 = 0.4038. \\ \therefore \; q &= 1 - p = 1 - 0.4038 = 0.5962. \end{aligned}$$

We need $P(Y = 7)$

$$P(Y = 7) = \binom{15}{7}(0.4038)^7(0.5962)^8 \approx 0.17982.$$

►**EXAMPLE 16.11** *A garments industry exports different amounts* (X *in* '000 \$) *of garments per month. It is assumed that* $X \sim N(25.5, 48.2)$. *A month is randomly selected. Find the probability that in that month the amount of export is* (i) *less than* 20 *thousand dollars,* (ii) *more than* 28 *thousand dollars,* (iii) 22 *to* 26 *thousand dollars.* (iv) *If* 5 *months are randomly selected, what is the probability that out of* 5 *months at least in* 3 *months the amount of export will be* 22 *to* 26 *thousand dollars?*

Solution. Given $X \sim N(25.5, 48.2), \; \sigma = 6.94.$

(i) We need $P(X < 20)$

$$Z = \frac{X-\mu}{\sigma} = \frac{20-25.5}{6.94}$$
$$= -0.79$$

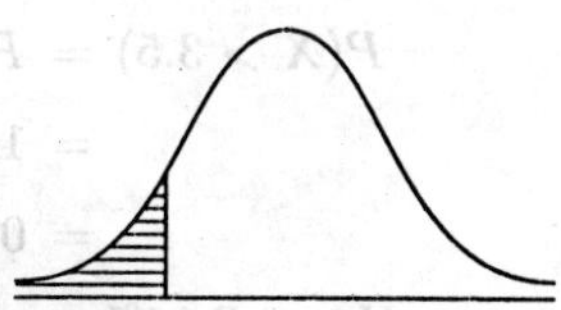

–0.79

$$P(X < 20) = P(Z < -0.79) = 0.2148.$$

(ii) We need $P(X > 28)$

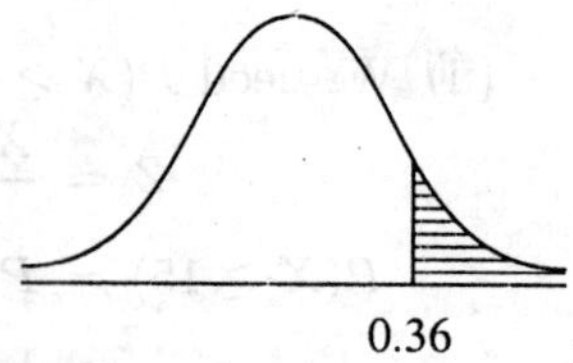

$$Z = \frac{X-\mu}{\sigma} = \frac{28-25.5}{6.94} = 0.36$$

$$P(X > 28) = P(Z > 0.36) = 1 - P(Z \le 0.36)$$

$$= 1 - 0.6406 = 0.3594.$$

(iii) We need $P(22 < X < 26)$

$$Z_1 = \frac{X_1-\mu}{\sigma} = \frac{22-25.5}{6.94} = -0.50,\ Z_2 = \frac{X_2-\mu}{\sigma} = \frac{26-25.5}{6.94} = 0.22$$

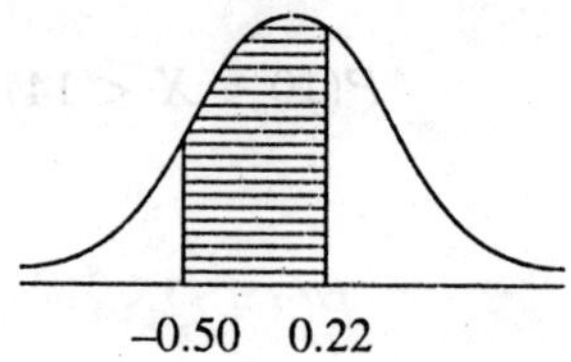

$$\begin{aligned} P(22 < X < 26) &= P(-0.50 < Z < 0.22) \\ &= P(Z \le 0.22) - P(Z \le -0.50) \\ &= 0.5871 - 0.3085 \\ &= 0.2786. \end{aligned}$$

(iv) Let Y be the random variable indicating number of months in which the amount of export is 22 to 26 thousand dollars. Then

$$P(Y = y) = \binom{n}{y} p^y q^{n-y};\ y = 0, 1, 2, \dots, n.$$

Given $n = 5$, $p = 0.2786$, $q = 0.7214$ [as in (iii)].

We need $P(Y \ge 3)$

$$\begin{aligned} P(Y \ge 3) &= 1 - P(Y < 3) \\ &= 1 - P(Y = 0) - P(Y = 1) - P(Y = 2) \\ &= 1 - \binom{5}{0}(0.7214)^5 - \binom{5}{1}(0.2786)(0.7214)^4 \\ &\quad - \binom{5}{2}(0.2786)^2(0.7214)^3 \\ &\approx 0.13595. \end{aligned}$$

►**EXAMPLE 16.12** *The G.P.A.* (X) *of university graduates is assumed* $N(2.85, 0.1225)$. *Find the percentage of students whose G.P.A. is* (i) *more than* 3.5, (ii) *less than* 2.55, (iii) 3.5 *to* 3.9. (iv) *If* 100 *graduates are randomly selected, what is the probability that out of* 100 *graduates at least* 2 *of them have G.P.A. more than* 3.5*?*

Solution. Given $X \sim N(2.85, 0.1225)$, $\sigma = 0.35$.

(i) We need $P(X > 3.5)$

$$Z = \frac{X-\mu}{\sigma} = \frac{3.5-2.85}{0.35} = 1.86$$

$$\begin{aligned} P(X > 3.5) &= P(Z > 1.86) = 1 - P(Z \le 1.86) \\ &= 1 - 0.9686 \\ &= 0.0314. \end{aligned}$$

Hence 3.14% graduates have G.P.A. more than 3.5.

(ii) We need $P(X < 2.55)$

$$Z = \frac{X-\mu}{\sigma} = \frac{2.55-2.85}{0.35} = -0.86$$

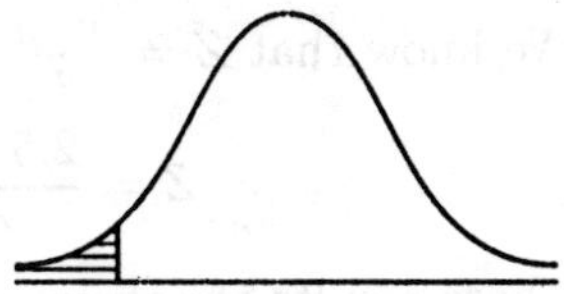

–0.86

$$P(X < 2.55) = P(Z < -0.86) = 0.1949.$$

Hence 19.49% graduates have G.P.A. less than 2.55.

(iii) We need $P(3.5 < X < 3.9)$

$$Z_1 = \frac{X_1-\mu}{\sigma} = \frac{3.5-2.85}{0.35} = 1.86$$

$$Z_2 = \frac{X_2-\mu}{\sigma} = \frac{3.9-2.85}{0.35} = 3.00$$

$$\begin{aligned} P(3.5 < X < 3.9) &= P(1.86 < Z < 3.00) \\ &= P(Z \le 3.00) - P(Z \le 1.86) \\ &= 0.9987 - 0.9685 \\ &= 0.0302. \end{aligned}$$

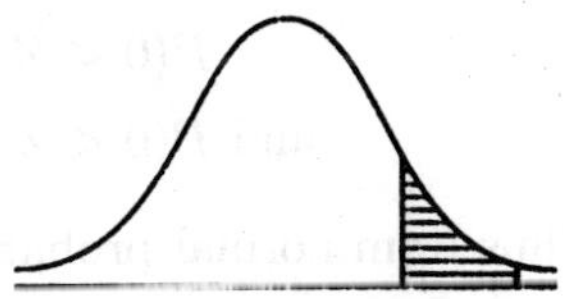

1.86 3.00

Hence 3.02% graduates have G.P.A. 3.5 to 3.9.

(iv) Let Y be the random variable indicating number of graduates whose G.P.A. is 3.5 or more. Then, since the probability of such students is less ($p = 0.0314$) and since $n = 100$ students are selected, Y is assumed to follow Poisson distribution. Therefore

$$P(Y=y) = \frac{e^{-\lambda}\lambda^y}{y!}, \text{ where } \lambda = np = 100 \times 0.0314 = 3.14;\ y = 0, 1, \ldots.$$

We need $P(Y \ge 2)$

$$\begin{aligned} P(Y \ge 2) &= 1 - P(Y < 2) \\ &= 1 - P(Y=0) - P(Y=1) \\ &= 1 - e^{-3.14}\left[\frac{(3.14)^0}{0!} + \frac{3.14}{1!}\right] \approx 0.82081. \end{aligned}$$

►**EXAMPLE 16.13** *If the G.P.A. of any student is not less than 2.5, he is considered graduate. The different grades of graduates are as follows:*

Grade	*A*	*B*	*C*	*D*	*F*
G.P.A	4.00	3.5–3.9	3.0–3.4	2.5–2.9	< 2.5

In an examination 5% *students got A grade and* 2% *students got F grade. What are the percentages of students who get B and C grade?*

Solution. Let X be the random variable indicating G.P.A. of students. Assume that $X \sim N(\mu, \sigma^2)$. Given that

$$P(X < 2.5) = 0.02 \text{ and } P(X \ge 4.0) = 0.05.$$

We know that $Z = \frac{X-\mu}{\sigma}$. Now, if $X = 2.5$ and $X = 4.0$, then

$$Z = \frac{2.5-\mu}{\sigma} = Z_1 \text{ (say) and } Z = \frac{4.0-\mu}{\sigma} = Z_2 \text{ (say).}$$

These results are shown graphically as below:

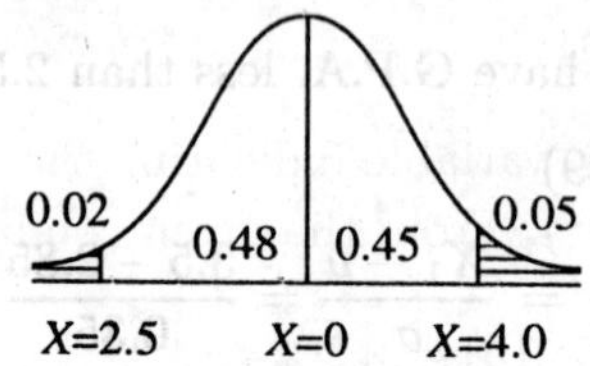

Now

$$P(0 < Z < Z_2) = 0.50 - 0.05 = 0.45$$
$$\text{and } P(0 < Z < Z_1) = P(-Z_1 < Z < 0) = 0.50 - 0.02 = 0.48.$$

Thus from normal probability table, we have $Z_1 = -2.05$, $Z_2 = 1.64$.

$$\therefore\ Z = \frac{2.5-\mu}{\sigma} = -2.05 \text{ and } Z = \frac{4-\mu}{\sigma} = 1.64$$
$$\Rightarrow\ 2.5 = \mu - 2.05\sigma \text{ and } 4 = \mu + 1.64\sigma.$$

Solving these two equations, we have

$$3.69\sigma = 1.5 \text{ or, } \sigma = 0.4065,\ \therefore\ \sigma^2 = 0.1652$$
$$\text{and } \mu = 4 - 1.64 \times 0.04065 = 3.33.$$

Therefore, $X \sim N(3.33, 0.1652)$.

Any student will get B grade if his G.P.A. is $3.5 \leq X \leq 3.9$. Therefore, we need $P(3.5 \leq X \leq 3.9)$. We have

$$Z_1 = \frac{X_1-\mu}{\sigma} = \frac{3.5-3.33}{0.4065} = 0.42$$
$$Z_2 = \frac{X_2-\mu}{\sigma} = \frac{3.9-3.33}{0.4065} = 1.40$$
$$P(3.5 \leq X \leq 3.9) = P(0.42 < Z < 1.40) = P(Z \leq 1.40) - P(Z \leq 0.42)$$
$$= 0.9192 - 0.6628 = 0.2564.$$

Hence 25.64% students get B grade.

Any student will get C grade if $3.0 \leq X \leq 3.4$. We need $P(3.0 \leq X \leq 3.4)$

$$Z_1 = \frac{X_1-\mu}{\sigma} = \frac{3.0-3.33}{0.4065} = -0.81$$
$$Z_2 = \frac{X_2-\mu}{\sigma} = \frac{3.4-3.33}{0.4065} = 0.17$$
$$P(3.0 \leq X \leq 3.4) = P(-0.81 \leq Z \leq 0.17) = P(Z \leq 0.17) - P(Z \leq -0.81)$$
$$= 0.5675 - 0.2090 = 0.3585.$$

Therefore, 35.85% students will get C grade.

►EXAMPLE 16.14 *The owners of garments industry are classified as A, B and C grades according to their 95%, 90% and 85% good products produced in their industries, respectively. The C grades industrialists are not allowed to export their products in foreign countries. The industrialists who has sought permission for foreign export 45% of them are termed as A grade exportes and 10% are C grade producers. Find the percentage of B grade producers.*

Solution. Let X be the random variable indicating the number of defective products. Let us assume that $X \sim N(\mu, \sigma^2)$. From the given condition

$$P(X \leq 5) = 0.45, \;\; P(X \geq 15) = 0.10.$$

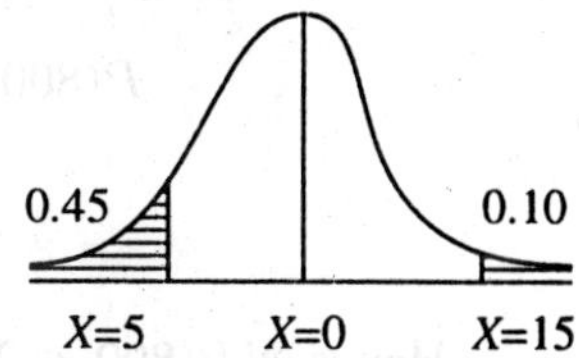

When $X = 5$, $Z = \dfrac{X-\mu}{\sigma} = \dfrac{5-\mu}{\sigma} = -Z_1$ (say)

$X = 15$, $Z = \dfrac{X-\mu}{\sigma} = \dfrac{15-\mu}{\sigma} = Z_2$ (say)

Here

$$P(0 < Z < Z_2) = 0.50 - 0.10 = 0.40 \Rightarrow Z_2 = 1.29$$

$$P(0 < Z < Z_1) = P(-Z_1 < Z < 0) = 0.50 - 0.45 = 0.05 \Rightarrow Z_1 = 1.64$$

$$\therefore \quad \frac{15-\mu}{\sigma} = 1.29 \text{ and } \frac{5-\mu}{\sigma} = -1.64.$$

Solving these two equations, we have $\sigma = 3.41$ and $\mu = 10.60$.

$$\therefore X \sim N(10.60, 11.6281)$$

We need $P(X < 10)$

$$Z = \frac{X-\mu}{\sigma} = \frac{10-10.60}{3.41} = -0.18$$

$$P(X < 10) = P(Z < -0.18) = 0.4286.$$

Hence 42.86% producers are in B grade.

►EXAMPLE 16.15 *Two hundred electric bulbs are given in* 200 *points in a street of a city. It is assumed that the life length (X hours) of any bulb is* $N(900, 4900)$*. How many electric bulbs will be fused* (i) *before* 900 *hours,* (ii) *after* 900 *hours?* (iii) *How many of them will be continued for* 800 *to* 1000 *hours?* (iv) *If* 10 *bulbs are randomly selected, what is the probability that at least* 3 *of them will be continued for* 800 *to* 1000 *hours?*

Solution. Given $X \sim N(900, 4900)$, $\sigma = 70$.

(i) We need $P(X < 900)$

$$Z = \frac{X-\mu}{\sigma} = \frac{900-900}{70} = 0; \; P(X < 900) = P(Z < 0) = 0.50.$$

Therefore, $NP(X < 900) = 200 \times 0.50 = 100$ bulbs will be fused before 900 hours.

(ii) We need $P(X > 900)$

$$Z = \frac{X-\mu}{\sigma} = \frac{900-900}{70} = 0; \; P(Z > 0) = 0.50.$$

Therefore, $NP(Z > 0) = 200 \times 0.50 = 100$ bulbs will be fused after 900 hours.

(iii) We need $P(800 < X < 1000)$

$$\begin{aligned} Z_1 &= \frac{X_1-\mu}{\sigma} = \frac{800-900}{70} = -1.43 \\ Z_2 &= \frac{X_2-\mu}{\sigma} = \frac{1000-900}{70} = 1.43 \\ P(800 < X < 1000) &= P(-1.43 < Z < 1.43) \\ &= P(Z \leq 1.43) - P(Z \leq -1.43) \\ &= 0.9236 - 0.0764 = 0.8472. \end{aligned}$$

Hence $NP(800 < X < 1000) = 200 \times 0.8472 \approx 169$ bulbs may be continued for 800 to 1000 hours.

(iv) Let Y be the number of bulbs which are randomly selected and which continue 800 to 1000 hours. Then

$$P(Y = y) = \binom{n}{y} p^y q^{n-y}; \; y = 0, 1, 2, \ldots, n.$$

Given $n = 10$, $p = 0.8472$, $q = 0.1528$. We need $P(Y \geq 3)$

$$\begin{aligned} P(Y \geq 3) &= 1 - P(Y < 3) = 1 - P(Y = 0) - P(Y = 1) - P(Y = 2) \\ &= 1 - \binom{10}{0}(0.1528)^{10} - \binom{10}{1}(0.8472)(0.1528)^9 \\ &\quad - \binom{10}{2}(0.8472)^2(0.1528)^8 \approx 0.9999901. \end{aligned}$$

►**EXAMPLE 16.16** *The amount of cloth produced by different powerlooms of an area (X m) is assumed normal variable with mean 7000 m and variance 262144 m^2. If 20 powerlooms are randomly selected, then in how many powerlooms will produce (i) 6000 metres or less, (ii) 7500 metres or more, (iii) 6500 to 8000 metres? (iv) What will be the minimum amount of production in 5 large factory? What is the probability of production of 6500 to 8000 metres in at least 3 factories of 20 randomly selected factories?*

Solution. Given $X \sim N(7000, 262144)$, $\sigma = 512$.

(i) We need $P(X \leq 6000)$

$$\begin{aligned} Z &= \frac{X-\mu}{\sigma} = \frac{6000-7000}{512} = -1.95 \\ P(X \leq 6000) &= P(Z \leq -1.95) = 0.0265. \end{aligned}$$

Therefore, out of 20 factories $NP(X \leq 6000) = 20 \times 0.0265 \approx 1$ factory will produce 6000 metres or less.

(ii) We need $P(X \geq 7500)$

$$Z = \frac{X-\mu}{\sigma} = \frac{7500-7000}{512} = 1.95$$

$$P(Z \geq 1.95) = 1 - P(Z < 1.95) = 1 - 0.8365 = 0.1635.$$

Hence $20 \times 0.1635 \approx 3$ factories will produce 7500 metres or more.

(iii) We need $P(6500 < X < 8000)$

$$Z_1 = \frac{X_1-\mu}{\sigma} = \frac{6500-7000}{512} = -0.98$$

$$Z_2 = \frac{X_2-\mu}{\sigma} = \frac{8000-7000}{512} = 1.95$$

$$\begin{aligned} P(6500 < X < 8000) &= P(-0.98 < Z < 1.95) \\ &= P(Z \leq 1.95) - P(Z \leq -0.98) \\ &= 0.9744 - 0.1635 = 0.8109. \end{aligned}$$

Therefore, $20 \times 0.8109 \approx 16$ factories will produce 6500 to 8000 metres.

(iv) Let x_1 be the minimum amount of production in 5 large factories. Then

$$P(X > x_1) = \frac{5}{20} = 0.25.$$

We know that

$$Z = \frac{X_1-\mu}{\sigma} = \frac{x_1-7000}{512} = Z_1 \text{ (say)}.$$

Then

$$P(Z > Z_1) = 0.25 \Rightarrow P(Z < Z_1) = 1 - 0.25 = 0.75 \Rightarrow Z_1 = 0.68.$$

Hence

$$0.68 = \frac{x_1-7000}{512} \text{ or, } x_1 = 7384.16 \text{ m.}$$

(v) Let Y be the random variable indicating number of factories producing 6500 to 8000 metres. Then

$$P(Y = y) = \binom{n}{y} p^y q^{n-y}; \; y = 0, 1, 2, \ldots, n.$$

Given $n = 20$, $p = 0.8109$, $q = 0.1891$. We need $P(Y \geq 3)$

$$\begin{aligned} P(Y \geq 3) &= 1 - P(Y < 3) \\ &= 1 - P(Y=0) - P(Y=1) - P(Y=2) \\ &= 1 - \binom{20}{0}(0.1891)^{20} - \binom{20}{1}(0.1891)(0.8109)^{19} \\ &\quad - \binom{20}{2}(0.1891)^2(0.8109)^{18} \\ &\approx 1.00. \end{aligned}$$

►**EXAMPLE 16.17** *There are 150 students in a course. The students can use any book related to the course in the library. The probability of using any book in the library by any student is 0.25. How many books are needed in the library so that the probability of using any book in any day by any student is 0.75 or more?*

Solution. Let X be the random variable indicating the scope of using a book in the library. Then $X \sim N(np, npq)$, on assumption. Given $n = 150$, $p = 0.25$, $q = 0.75$, then $X \sim N(37.5, 28.125)$. The condition is that $P(X < x) > 0.75$.

Again $Z = \dfrac{X-\mu}{\sigma}$. Also $P(Z < z) > 0.75 \Rightarrow Z = 0.68$.

Hence $0.68 = \dfrac{X-\mu}{\sigma}$ or, $0.68 = \dfrac{X - 37.5}{5.303} \Rightarrow X \approx 41$.

There should be 41 books in the library.

►**EXAMPLE 16.18** *The age distribution of people in an area is assumed normal with mean age 50 years and variance of age is 124.6 years^2. A person is selected randomly from the area. Find the probability that the age of the selected person is* (i) *less than 55 years,* (ii) *more than 60 years,* (iii) *45 years to 65 years.* (iv) *If 10 people are randomly selected from the area, what is the probability that the age of at best two people will be more than 65 years?*

Solution. Let X be the age of the people. Given

$$X \sim N(50, 124.6),\ \sigma = 11.16.$$

(i) We need $P(X < 55)$

$$Z = \frac{X-\mu}{\sigma} = \frac{55-50}{11.16} = 0.45$$

$$P(X < 55) = P(Z < 0.45) = 0.6736.$$

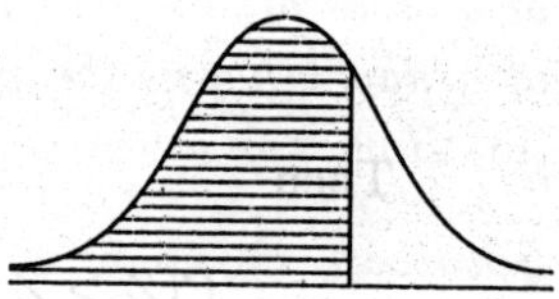

(ii) We need $P(X > 60)$

$$Z = \frac{X-\mu}{\sigma} = \frac{60-50}{11.16} = 0.90$$

$$P(X > 60) = P(Z > 0.90) = 1 - P(Z \le 0.90) = 1 - 0.8159 = 0.1841.$$

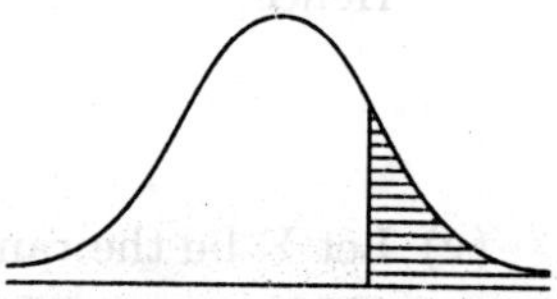

(iii) We need $P(45 < X < 65)$

$$Z_1 = \frac{X_1-\mu}{\sigma} = \frac{45-50}{11.16} = -0.90$$

$$Z_2 = \frac{X_2-\mu}{\sigma} = \frac{65-50}{11.16} = 1.34$$

$$\begin{aligned} P(45 < X < 65) &= P(-0.90 < Z < 1.34) \\ &= P(Z \le 1.34) - P(Z \le -0.90) \\ &= 0.9099 - 0.1841 \\ &= 0.7258. \end{aligned}$$

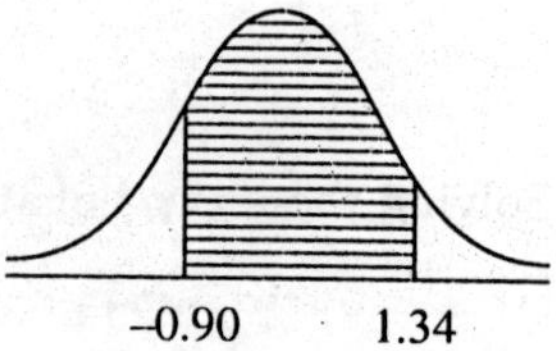

(iv) Let Y be the random variable indicating the number of people of age 65 years or more. Then

$$P(Y = y) = \binom{n}{y} p^y q^{n-y}; \; y = 0, 1, 2, \ldots, n$$

$$p = P(X > 65), \; Z = \frac{X - \mu}{\sigma} = \frac{65 - 50}{11.16} = 1.34$$

$$\therefore p = P(X > 65) = P(Z > 1.34)$$

$$= 1 - P(Z \leq 1.34) = 1 - 0.9099 = 0.0901.$$

$$\therefore q = 1 - p = 0.9099, \; n = 10.$$

We need $P(Y \leq 2)$

$$P(Y \leq 2) = P(Y = 0) + P(Y = 1) + P(Y = 2)$$

$$= \binom{10}{0}(0.9099)^{10} + \binom{10}{1}(0.0901)(0.9099)^9$$

$$+ \binom{10}{2}(0.0901)^2(0.9099)^8 \approx 0.94887.$$

►**EXAMPLE 16.19** *A company produces dresses for babies of different ages. The dresses are numbered from* 1, 2, ... , 15. *It is assumed that the size of a dress* (X) *is normally distributed with mean* μ *and variance* σ^2. *Twenty-five percent of dresses are of size less than* 5 *and* 5% *dresses are of size more than* 10. (i) *What is the percentage of dresses of size* 8 *to* 12? (ii) *If* 5 *dresses are selected randomly from the production of a day, what is the probability that out of* 5 *dresses at least* 2 *are of size* 8 *to* 12? (iii) *How many dresses are of size more than* 12 *if the company produces* 1000 *dresses?*

Solution. Given $X \sim N(\mu, \sigma^2)$, $P(X < 5) = 0.25$ and $P(X > 10) = 0.05$.

Hence $P(X \leq 10) = 0.95$. Again, $P(X < 5) = 0.25 \Rightarrow P(X \geq 5) = 0.75$.

These probabilities are shown graphically below:

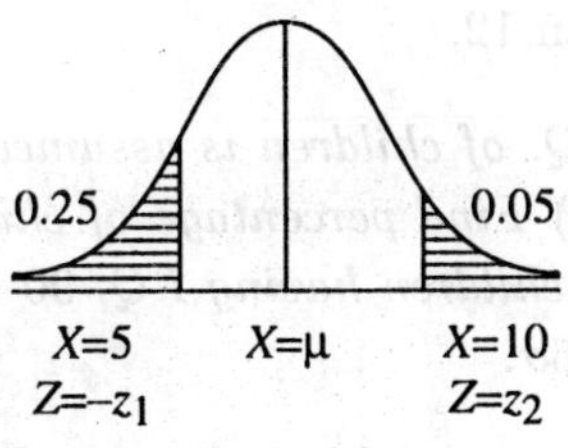

Now

$$P(0 < Z < Z_2) = 0.50 - 0.05 = 0.45$$

$$\text{and } P(-\infty < Z < Z_1) = 0.25 \Rightarrow Z_1 = -0.68.$$

$$\therefore -0.68 = \frac{5 - \mu}{\sigma} \text{ and } 1.64 = \frac{10 - \mu}{\sigma}.$$

Solving these two equations, we have

$$\mu = 6.4655 \text{ and } \sigma = 2.155. \; \therefore \; X \sim N(6.4655, 4.644).$$

(i) We need $(8 < X < 12)$

$$Z_1 = \frac{X_1 - \mu}{\sigma} = \frac{8 - 6.4655}{2.155} = 0.71$$
$$Z_2 = \frac{X_2 - \mu}{\sigma} = \frac{12 - 6.4655}{2.155} = 2.57$$

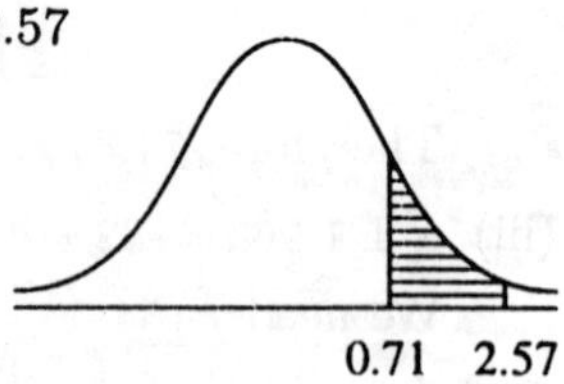

$$\begin{aligned} P(8 < X < 12) &= P(0.71 < Z < 2.57) \\ &= P(Z \le 2.57) - P(Z \le 0.71) \\ &= 0.9949 - 0.7611 = 0.2338. \end{aligned}$$

Hence 23.38% dresses are of size 8 to 12.

(ii) Let Y be the random variable indicating number of dresses of size 8 to 12. Then

$$P(Y = y) = \binom{n}{y} p^y q^{n-y}; \; y = 0, 1, 2, \ldots, n.$$

Given $n = 5$, $p = 0.2338$, $q = 0.7662$. We need $P(Y \ge 2)$

$$\begin{aligned} P(Y \ge 2) &= 1 - P(Y < 2) = 1 - P(Y = 0) - P(Y = 1) \\ &= 1 - \binom{5}{0}(0.7662)^5 - \binom{5}{1}(0.2338)(0.7662)^4 \\ &\approx 0.33305. \end{aligned}$$

(iii) We need $P(X > 12)$

$$\begin{aligned} Z &= \frac{X - \mu}{\sigma} = \frac{12 - 6.4655}{2.155} = 2.57 \\ P(X > 12) &= P(Z > 2.57) = 1 - P(Z \le 2.57) \\ &= 1 - 0.9949 = 0.0051. \end{aligned}$$

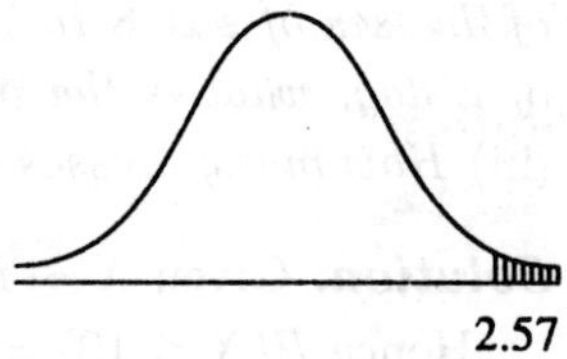

Therefore, out of $N = 1000$ dresses $NP(X > 12) = 1000 \times 0.0051 \approx 5$ dresses will be of size more than 12.

►**EXAMPLE 16.20** *The I.Q. of children is assumed to be normally distributed with mean* 82 *and variance* 196. (i) *Find percentage of children whose I.Q. is* 75 *or more.* (ii) *What is the percentage of children having I.Q.* 90 *or less?* (iii) *Find percentage of children whose I.Q. is* $\mu \pm 1.96\sigma$.

Solution. Let X be the random variable indicating I.Q. of children. Given

$$X \sim N(82, 196), \; \sigma = 14.$$

(i) We need $P(X \ge 75)$

$$\begin{aligned} Z &= \frac{X - \mu}{\sigma} = \frac{75 - 82}{14} = -0.50 \\ P(X \ge 75) &= P(Z \ge -0.50) = 1 - P(Z < -0.50) \\ &= 1 - 0.3085 = 0.6915. \end{aligned}$$

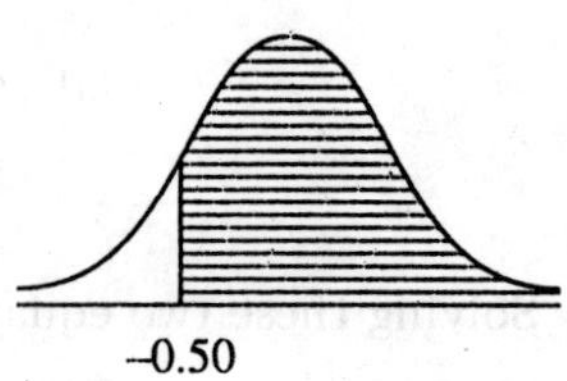

Hence 69.15% children are of I.Q. more than or equal to 75.

(ii) We need $P(X \leq 90)$

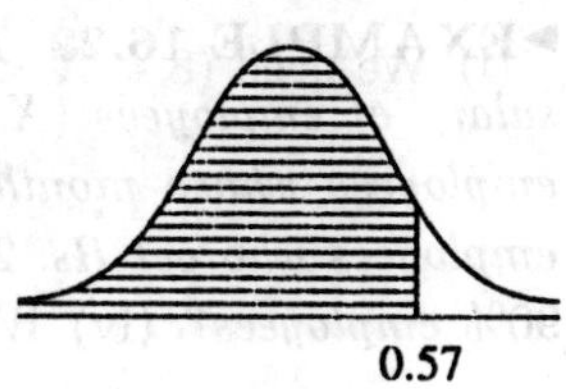

$$Z = \frac{X - \mu}{\sigma} = \frac{90 - 82}{14} = 0.57$$

$$P(X \leq 90) = P(Z \leq 0.57)$$
$$= 0.7157.$$

Therefore, 71.57% children are of I.Q. 90 or less.

(iii) $\mu \pm 1.96\sigma = 82 \pm 1.96 \times 14 = 82 \pm 27.44.$

We need $P(54.56 < X < 109.44)$

$$Z_1 = \frac{X_1 - \mu}{\sigma} = \frac{54.56 - 82}{14} = -1.96, \; Z_2 = \frac{109.44 - 52}{14} = 1.96$$

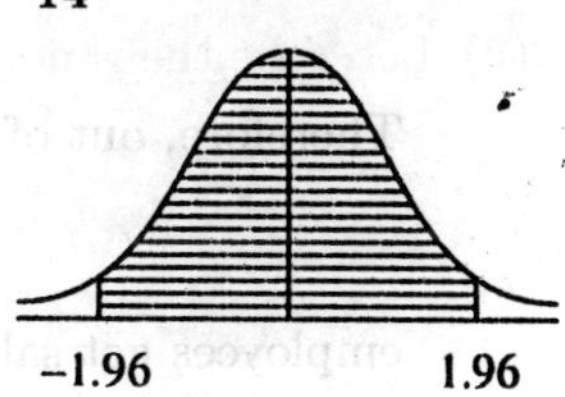

$$P(54.56 < X < 109.44)$$
$$= P(-1.96 < Z < 1.96)$$
$$= P(Z \leq 1.96) - P(Z \leq -1.96)$$
$$= 0.975 - 0.025 = 0.95.$$

Hence 95% children are of I.Q. $\mu \pm 1.96\sigma$.

►**EXAMPLE 16.21** *Ten percent students of a university fail if pass mark is* 70%. *In an examination the average marks and standard deviation of marks are* 55% *and* 8%, *respectively.* (i) *What should be the pass mark if* 95% *students are announced pass?* (ii) *How many students secured marks* 70% *to* 80%*?*

Solution.

(i) Let X be the random variable indicating mark out of 100. Given $X \sim N(55, 64)$. Also

$$P(X < 70) = 0.10 \text{ and } P(X \geq x) = 0.95$$
$$P(Z \geq z) = 0.95 \Rightarrow p(Z < z) = 0.05 \text{ and } Z = -1.64$$

But

$$Z = \frac{X - \mu}{\sigma} \Rightarrow -1.64 = \frac{X - 55}{8}, \; \therefore \; X = 41.88.$$

Hence pass mark should be 41.88%.

(ii) We need $P(70 < X < 80)$

$$Z_1 = \frac{X_1 - \mu}{\sigma} = \frac{70 - 55}{8} = 1.88$$
$$Z_2 = \frac{X_2 - \mu}{\sigma} = \frac{80 - 55}{8} = 3.13$$
$$P(70 < X < 80) = P(1.88 < Z < 3.13)$$
$$= P(Z \leq 3.13) - P(Z \leq 1.88)$$
$$= 0.9991 - 0.9699 = 0.0292.$$

Therefore, 2.92% students get marks 70% to 80%.

►**EXAMPLE 16.22** *The number of employees of a company are* 500. *The monthly salary of employees* (X *Rs.*) *is assumed* $N(4000, 2250000)$. (i) *Find the number of employees whose monthly salary is above Rs.* 8000.00. (ii) *What is the number of employees who get Rs.* 2500.00 *to Rs.* 5500.00? (iii) *What is the maximum salary for* 90% *employees?* (iv) *What is the minimum salary of* 100 *high-salaried employees?*

Solution. Given $X \sim N(4000, 2250000)$, $\sigma = 1500$.

(i) We need $P(X > 8000)$

$$Z = \frac{X-\mu}{\sigma} = \frac{8000-4000}{1500} = 2.67$$
$$P(X > 8000) = P(Z > 2.67) = 1 - P(Z \leq 2.67)$$
$$= 1 - 0.9962 = 0.0038.$$

Therefore, out of $N = 500$ employees

$$N(P > 8000) = 500 \times 0.0038 \approx 2$$

employees get salary more than Rs. 8000.00.

(ii) We need $P(2500 < X < 5500)$

$$Z_1 = \frac{X_1-\mu}{\sigma} = \frac{2500-4000}{1500} = -1$$
$$Z_2 = \frac{X_2-\mu}{\sigma} = \frac{5500-4000}{1500} = 1$$
$$P(2500 < X < 5500) = P(-1 < Z < 1) = P(Z \leq 1) - P(Z \leq -1)$$
$$= 0.8413 - 0.1587 = 0.6826.$$

Hence $500 \times 0.6826 \approx 341$ employees get salary Rs. 2500.00 to Rs. 5500.00.

(iii) $P(X < x) = 0.90 \Rightarrow P(Z < z) = 0.90 \Rightarrow Z = 1.64$

$$Z = \frac{X-\mu}{\sigma} \text{ or, } 1.64 = \frac{X-4000}{1500} \Rightarrow X = 6460.$$

Therefore, the maximum salary of 90% employees is Rs. 6460.00.

(iv) 100 employees are 20% of all employees. Then

$$P(Z > z) = 0.20 \Rightarrow P(Z \leq z) = 0.80 \Rightarrow Z = 0.84.$$

Again,

$$Z = \frac{X-\mu}{\sigma} \text{ or, } 0.84 = \frac{X-4000}{1500} \Rightarrow X = 5260.$$

Therefore, the minimum salary of 100 high-salaried employees is Rs. 5260.00.

►**EXAMPLE 16.23** *A company produces cloth. The average and variance of breadth of cloth are* 1.00 *metre and* 0.0225 *metre*^{2}. *A customer agrees to buy cloth from the company if breadth of cloth of the company is* 1.00 ± 0.02. (i) *What is the probability that the customer cancells the order?* (ii) *Find the percentage of produced cloth which are consistent according to the demand of customer.*

Solution.

(i) Let X be the random variable indicating the amount of breadth of cloth. Given $X \sim N(1.00,\ 0.0225)$, $\sigma = 0.15$. According to the condition of customer, the cloth will be acceptable to the customer if its breadth is 0.98 to 1.02 metre. Hence, we need $P(0.98 < X < 1.02)$. Now,

$$Z_1 = \frac{X_1 - \mu}{\sigma} = \frac{0.98 - 1.00}{0.15} = -0.13$$

$$Z_2 = \frac{X_2 - \mu}{\sigma} = \frac{1.02 - 1.00}{0.15} = 0.13$$

$$\begin{aligned} P(0.98 < X < 1.02) &= P(-0.13 < Z < 0.13) \\ &= P(Z \leq 0.13) - P(Z \leq -0.13) \\ &= 0.5517 - 0.4483 = 0.1034. \end{aligned}$$

Therefore, the probability of producing cloth which is not according to the specification of the customer is

$$1 - P(0.98 < X < 1.02) = 1 - 0.1034 = 0.8966.$$

The customer will cancel the order with probability 0.8966.

(ii) Since $P(0.98 < X < 1.02) = 0.1034$, 10.34% of cloth will be produced according to the demand of customer.

►**EXAMPLE 16.24** *It is decided that the height (X cm) of a candidate to join army should be such that* 75% *youths of the country can apply for the army job. The census report describes that the average height and variance of height of youths of* 18 *to* 30 *years of age are* 153 *cm and* 25 *cm*^{2}*, respectively.* (i) *What should be the minimum height of candidate to join army?* (ii) *What is the percentage of youths of ages* 18 *to* 30 *years whose height is below average height?*

Solution.

(i) Let X be the random variable indicating minimum height to join army. Let us assumed that $X \sim N(\mu, \sigma^2)$, where $\mu = 153, \sigma^2 = 25, \sigma = 5$. Then

$$P(X > x) = 0.75 \Rightarrow P(Z > Z_1) = 0.75 \text{ or, } P(Z \leq Z_1) = 0.25$$

$$\therefore\ Z_1 = -0.67.$$

Now

$$Z_1 = \frac{X - \mu}{\sigma} \text{ or, } -0.67 = \frac{X - 153}{5} \Rightarrow X = 149.65 \text{ cm}.$$

(ii) We need $P(X < 153)$

$$Z = \frac{X - \mu}{\sigma} = \frac{153 - 153}{5} = 0$$

$$P(X < 153) = P(Z < 0) = 0.50.$$

Therefore, 50% youths have height below average height.

▶EXAMPLE 16.25 *The average marks in mathematics, statistics and computer science in an examination are* 85, 70 *and* 86, *respectively. The variances of marks of these* 3 *subjects are, respectively* 25, 48 *and* 36. *A student is randomly selected among the lot of examinees. Find the probability that the total marks of the selected student in all* 3 *subjects is* (i) *less than* 200, (ii) *more than* 250, (iii) 205 *to* 225. (iv) *What is the percentage of students who secured* 80% *marks or more, on an average?*

Solution. Let X_1, X_2 and X_3 be the random variables indicating the marks of mathematics, statistics and computer science, respectively. Given that

$$X_1 \sim N(85, 25), \ X_2 \sim N(70, 48) \text{ and } X_3 \sim N(86, 36).$$

Let $Y = X_1 + X_2 + X_3$, where $Y \sim N(\mu_1 + \mu_2 + \mu_3, \ \sigma_1^2 + \sigma_2^2 + \sigma_3^2)$. Thus

$$Y \sim N(241, 109), \ \sigma = 10.44.$$

(i) We need $P(Y < 200)$

$$Z = \frac{Y - \mu}{\sigma} = \frac{200 - 241}{10.44} = -3.93 \approx -4.00$$
$$P(Y < 200) = P(Z < -4.00) = 0.00003.$$

(ii) We need $P(Y > 250)$

$$Z = \frac{Y - \mu}{\sigma} = \frac{250 - 241}{10.44} = 0.86$$
$$P(Y > 250) = P(Z > 0.86) = 1 - P(Z \leq 0.86)$$
$$= 1 - 0.8051 = 0.1949.$$

(iii) We need $P(205 < Y < 225)$

$$Z_1 = \frac{Y_1 - \mu}{\sigma} = \frac{205 - 241}{10.44} = -3.45$$
$$Z_2 = \frac{Y_2 - \mu}{\sigma} = \frac{225 - 241}{10.44} = -1.53$$
$$P(205 < Y < 225) = P(-3.45 < Z < -1.53)$$
$$= P(Z \leq -1.53) - P(Z \leq -3.45)$$
$$= 0.0630 - 0.0003 = 0.0627.$$

(iv) 80% marks on average equal 240 approximately. We need $P(Y > 240)$

$$Z = \frac{Y - \mu}{\sigma} = \frac{240 - 241}{10.44} = -0.10$$
$$P(Y > 240) = P(Z > -0.10) = 1 - P(Z \leq -0.10)$$
$$= 1 - 0.4602 = 0.5398.$$

Therefore, 53.98% students secured marks 80% or more.

▶EXAMPLE 16.26 *The value of first and last deciles of a normal distribution are* 8 *and* 20, *respectively. If* 10 *observations of this distribution are randomly selected, how many observations of these* 10 *observations will fall in the limit* $\mu \pm 2\sigma$?

Solution. Let X be the normal variable, i.e., $X \sim N(\mu, \sigma^2)$. From the given condition

$$\begin{aligned} P(X < x_1) &= 0.10 \text{ and } P(X > x_2) = 0.10 \\ \Rightarrow P(Z < z_1) &= 0.10 \text{ and } P(Z > z_2) = 0.10 \text{ or, } P(Z \leq z_2) = 0.90, \end{aligned}$$

where $z_1 = -1.28$, $z_2 = 1.29$.

Hence

$$-1.28 = \frac{8 - \mu}{\sigma} \text{ and } 1.29 = \frac{20 - \mu}{\sigma}.$$

Solving these two equations, we have $\mu = 13.98$ and $\sigma = 4.67$, i.e.,

$$X \sim N(13.98, 21.8089).$$

Hence $\mu \pm 2\sigma = 13.98 \pm 2 \times 4.67$. We need

$$P(\mu - 2\sigma < X < \mu + 2\sigma) = P(4.64 < X < 23.32).$$

Now

$$\begin{aligned} Z_1 &= \frac{X_1 - \mu}{\sigma} = \frac{4.64 - 13.98}{4.67} = -2.00 \\ Z_2 &= \frac{X_2 - \mu}{\sigma} = \frac{23.32 - 13.98}{4.67} = 2.00 \\ P(4.64 < X < 23.32) &= P(-2.00 < Z < 2.00) \\ &= P(Z \leq 2.00) - P(Z \leq -2.00) \\ &= 0.9772 - 0.0228 \\ &= 0.9544. \end{aligned}$$

Therefore, out of $N = 10$ observations, $10 \times 0.9544 \simeq 9$ observations or 95.44% observations will fall in the limit $\mu \pm 2\sigma$.

►**EXAMPLE 16.27** *The values of first and third quartiles of a normal distribution are* 12 *and* 25, *respectively.* (i) *How many observations of this distribution will fall within* 12 *to* 25*?* (ii) *How many observations of this distribution will fall within the limit* $\mu \pm 3\sigma$*?*

Solution. Let X be the normal variable, i.e., $X \sim N(\mu, \sigma^2)$. From the given information of quartiles, we can write

$$\begin{aligned} P(X < x_1) &= 0.25 \text{ and } P(X > x_2) = 0.25 \text{ or, } P(X \leq x_2) = 0.75 \\ \Rightarrow P(Z < z_1) &= 0.25 \text{ and } P(Z \leq z_2) = 0.75. \end{aligned}$$

Therefore, $z_1 = -0.67$, $z_2 = 0.67$. Again, we have

$$-0.67 = \frac{x_1 - \mu}{\sigma} \text{ or, } -0.67 = \frac{12 - \mu}{\sigma} \text{ and } 0.67 = \frac{25 - \mu}{\sigma}.$$

Solving these equations, we have $\mu = 18.499$, $\sigma = 9.70$. Therefore,

$$X \sim N(18.499, 94.04).$$

(i) We need $P(12 < X < 25)$

$$Z_1 = \frac{X_1 - \mu}{\sigma} = \frac{12 - 18.499}{9.7} = -0.67$$

$$Z_2 = \frac{X_2 - \mu}{\sigma} = \frac{25 - 18.499}{9.7} = 0.67$$

$$\begin{aligned} P(12 < X < 25) &= P(-0.67 < Z < 0.67) \\ &= P(Z \leq 0.67) - P(Z \leq -0.67) \\ &= 0.7486 - 0.2514 \\ &= 0.4972. \end{aligned}$$

Hence 49.72% observations will fall in the interval 12 to 25.

(ii) $\mu \pm 3\sigma = 18.499 \pm 3 \times 9.70 = 18.499 \pm 29.1$. We need

$$P(\mu - 3\sigma < X < \mu + 3\sigma) = P(-10.601 < X < 47.549)$$

$$Z_1 = \frac{X_1 - \mu}{\sigma} = \frac{-10.601 - 18.499}{9.7} = -3.00$$

$$Z_2 = \frac{X_2 - \mu}{\sigma} = \frac{47.599 - 18.499}{9.7} = 3.00$$

$$\begin{aligned} P(-3.00 < X < 3.00) &= P(Z \leq 3.00) - P(Z \leq -3.00) \\ &= 0.9987 - 0.0013 = 0.9974. \end{aligned}$$

Hence 99.74% observations will fall in the interval $\mu \pm 3\sigma$.

►**EXAMPLE 16.28** *The median and third quartile of a normal distribution are 22 and 28, respectively. If 10 observations are randomly selected from this distribution, what is the probability that 3 observations will be less than 15?*

Solution. Let X be the normal variable, where $X \sim N(\mu, \sigma^2)$. Given

$$\text{Median, } M = 22 \text{ and } Q_3 = 28.$$

Since the distribution is normal,

$$\text{mean} = \text{median} = \text{mode} = \mu = 22.$$

Also, we can write

$$P(X < x_1) = 0.50, \text{ where } x_1 = 22 \text{ (median)}.$$

We can also write

$$P(Z < z_1) = 0.50 \Rightarrow z_1 = 0.00.$$

Again, if x_2 is the Q_3, then

$$P(X > x_2) = 0.25 \text{ or, } P(Z > z_2) = 0.25 \Rightarrow z_2 = 0.67.$$

We can write

$$z_2 = \frac{X - \mu}{\sigma} \text{ or, } 0.67 = \frac{28 - 22}{\sigma} \text{ or, } \sigma = 8.955.$$

Hence $X \sim N(22, 80.196)$. We need $P(X < 15)$

$$Z = \frac{X-\mu}{\sigma} = \frac{15-22}{8.955} = -0.78$$
$$P(X < 15) = P(Z < -0.78) = 0.2177.$$

Let Y be the random variable which indicates value of $X < 15$. Then

$$P(Y = y) = \binom{n}{y} p^y q^{n-y}; \; y = 0, 1, 2, \ldots, n.$$

Given $n = 10$, $p = 0.2177$, $q = 0.7823$. We need $P(Y = 3)$

$$P(Y = 3) = \binom{10}{3}(0.2177)^3(0.7823)^7 \approx 0.222.$$

►**EXAMPLE 16.29** *A company produces iron rod of 2 cm radius. Any customer purchase rods of this company if radius of rods is in the limit* (2 ± 0.02) *cm. The customers usually reject 5% rods of the company due to specification error in the measurement of radius. If a customer wants to purchase rods with radius* (2 ± 0.01) *cm, then what would be the percentage of rejected rods?*

Solution. Let X be the random variable measuring the radius of the rod, where

$$X \sim N(\mu, \sigma^2).$$

From the given condition

$$P(X > x_2) = P(X > 2.02) = 0.05$$
$$\text{and } P(X < x_1) = P(X < 1.98) = 0.05$$
$$\Rightarrow P(Z > z_2) = 0.05 \text{ and } P(Z < z_1) = 0.05.$$

Therefore, $z_1 = -1.64$, $z_2 = 1.64$. Again,

$$z_2 = \frac{X_2 - \mu}{\sigma} \text{ or, } 1.64 = \frac{2.02-\mu}{\sigma}$$
$$\text{and } z_1 = \frac{X_1 - \mu}{\sigma} \text{ or, } -1.64 = \frac{1.98-\mu}{\sigma}.$$

Solving these two equations, we have $\mu = 2.00$, $\sigma = 0.0122$.

$$\therefore X \sim N(2.00, \; 0.00014884)$$

The customer wants to purchase rods if radius is within the limit 2.00 ± 0.01 or, 1.99 to 2.01. We need $P(1.99 < X < 2.01)$

$$Z_1 = \frac{X_1-\mu}{\sigma} = \frac{1.99-2.00}{0.0122} = -0.82$$
$$Z_2 = \frac{X_2-\mu}{\sigma} = \frac{2.01-2.00}{0.0122} = 0.82$$
$$P(1.99 < X < 2.01) = P(-0.82 < Z < 0.82)$$
$$= P(Z \le 0.82) - P(Z \le -0.82)$$
$$= 0.7939 - 0.2061 = 0.5878.$$

The probability that the radius of the rod will be outside the limit 1.99 to 2.01 is

$$1 - P(1.99 < X < 2.01) = 1 - 0.5878 = 0.4122.$$

Therefore, 41.22% rods will be rejected.

►**EXAMPLE 16.30** *The first moment about* 15 *and fourth moment about* 55 *of a normal distribution are* 40 *and* 252, *respectively. Find mean and standard deviation of the distribution.*

Solution. Let X be the normal variable. Given

$$\begin{aligned} E(X-15) &= 40 \Rightarrow E(X) = 40 + 15 = 55 = \mu_1' = \mu \\ E(X-55)^4 &= E[X - E(X)]^4 = \mu_4' = \mu_4 = 252 \ [\because\ E(X) = \mu_1' = \mu]. \end{aligned}$$

We know that $\mu_4 = 3\sigma^4 \Rightarrow 3\sigma^4 = 252$.

$$\therefore \sigma^4 = 84 \text{ or, } \sigma^2 = 9.165.$$

Hence

$$E(X) = \mu = 55 \text{ and } \sigma = \sqrt{9.165} = 3.027.$$

►**EXAMPLE 16.31** *The daily wages of labourers of a factory are as follows:*

Class interval of wages (in Rs.)	75–80	80–85	85–90	90–95	95–100	100+	*Total*
Number of labourers, O_i	20	42	50	175	68	32	$N = 387$

Do you think that the distribution of wage is normal?

Solution. From the given distribution, we have

$$\text{Mean} = \mu = \frac{1}{N}\sum O_i X_i = \frac{35487.5}{387} = 91.70$$

where X_i is the mid-value of i-th class.

$$\begin{aligned} \sigma^2 &= \frac{1}{N}\left[\Sigma O_i X_i^2 - \frac{(\Sigma O_i X_i)^2}{N}\right] \\ &= \frac{1}{387}\left[3268768.75 - \frac{(35487.5)^2}{387}\right] \\ &= 37.73 \\ \therefore \sigma &= 6.14. \end{aligned}$$

If X is the normal variable, we can write $X \sim N(91.70, 37.73)$. We know

$$Z = \frac{X - \mu}{\sigma}.$$

Then

$$P(Z \leq z) = \psi(z) = \int_{-\infty}^{z} f(z).$$

The values of $\psi(z)$ and $\Delta\psi(z) = \psi(z+1) - \psi(z)$ are shown below:

Class interval	Upper boundary of the interval, X	$Z = \frac{X-\mu}{\sigma}$	$\psi(z)$	$\Delta\psi(z) = \psi(z+1) - \psi(z)$	Frequencies Observed O_i	Expected E_i
75–80	79.5	−1.99	0.0233	0.0233	20	≈ 9
80–85	84.5	−1.17	0.1210	0.0977	42	≈ 32
85–90	89.5	−0.36	0.3594	0.2384	50	≈ 92
90–95	94.5	0.45	0.6736	0.3142	175	≈ 122
95–100	99.5	1.27	0.8980	0.2244	68	≈ 87
100+	104.5	2.08	—	0.1020	32	≈ 39
Total	—	—	—	—	387	387

The expected frequencies $[E_i]$ are calculated according to normal probability law. If normal distribution is good for the data, then there will be little discrepancy between O_i and E_i. To verify the suitability of E_i for the data, let us calculate

$$\chi^2 = \sum_{i=1}^{k} \frac{O_i^2}{E_i} - N = 448.47 - 387 = 61.47.$$

But $\chi^2 > \chi^2_{0.05,3} = 7.815$. It indicates E_i's do not conform with O_i's, i.e., the fitted distribution is not suitable for the data. Here two parameters of normal distribution are estimated. Hence d.f. of χ^2 is $(k-2-1) = 6-3 = 3$.

►**EXAMPLE 16.32** *The birth-weights of babies born in a maternity centre are shown below:*

Class interval of birth-weight (X ' kg)	< 2.00	2.00–2.25	2.25–2.50	2.50–2.75	2.75–3.00	3.00–3.25	3.25–3.50
Number of babies, O_i	20	80	250	375	240	110	25

Do you think that the normal distribution is suitable for the above distribution of birth-weight?

Solution. We have

$$\text{Mean} = \mu = \frac{1}{N}\sum_{i=1}^{k} O_i X_i, \text{ where } X_i \text{ is the mid-value of class interval}$$

$$= \frac{2903.75}{1100} = 2.64$$

$$\text{Variance, } \sigma^2 = \frac{1}{N}\left[\Sigma O_i X_i^2 - \frac{(\Sigma O_i X_i)^2}{N}\right] = \frac{1}{1100}\left[7768.4375 - \frac{(2903.75)^2}{1100}\right] = 0.0938$$

$$\therefore \sigma = \sqrt{0.0938} = 0.3063.$$

Therefore, $X \sim N(2.64, 0.0938)$. We know

$$Z = \frac{X-\mu}{\sigma} \text{ and } P(Z \le z) = \int_{-\infty}^{z} f(z)dz = \psi(z).$$

The value of $\psi(z)$ and the expected number of babies are shown below:
Expected number of babies, $E_i = N\Delta\psi(z)$, where $\Delta\psi(z) = \psi(z+1) - \psi(z)$

Class interval	Upper boundary of class interval, X	$Z = \frac{X-\mu}{\sigma}$	$\psi(z)$	$\Delta\psi(z)$	Frequencies Observed O_i	Expected E_i
< 2.00	1.95	−2.25	0.0122	0.0122	20	≈ 13
2.00–2.25	2.20	−1.44	0.0749	0.0627	80	≈ 69
2.25–2.50	2.45	−0.62	0.2676	0.1927	250	≈ 212
2.50–2.75	2.70	0.20	0.5793	0.3117	375	≈ 343
2.75–3.00	2.95	1.01	0.8438	0.2645	240	≈ 294
3.00–3.25	3.20	1.83	0.9664	0.1226	110	≈ 135
3.25–3.50	3.45	2.64	—	0.0336	25	≈ 34
Total	—	—	—	1.00	1100	1100

The E_i's are calculated according to normal probability law. Let us now investigate whether this law is suitable for the data. This can be done by calculating χ^2-statistic, where

$$\chi^2 = \sum \frac{O_i^2}{E_i} - N = 1131.73 - 1100 = 31.73.$$

Since $\chi^2 > \chi^2_{0.05,4} = 9.488$, the data is not suitable for normal probability law.

►EXAMPLE 16.33 *The distribution of life length of electric bulbs produced by a company is given below:*

Life length (X, in days)	10–20	20–30	30–40	40–50	50–60	60–70	70–80	80+	*Total*
Number of bulbs, O_i	30	52	100	150	95	57	20	12	516

Fit normal distribution for the data.

Solution. We have

$$\text{Mean} = \mu = \frac{1}{N}\Sigma O_i X_i = \frac{23450}{516} = 45.45 \text{ days}$$

$$\text{Variance, } \sigma^2 = \frac{1}{N}\left[\Sigma O_i X_i^2 - \frac{(\Sigma O_i X_i)^2}{N}\right] = \frac{1}{516}\left[1192900 - \frac{(23450)^2}{516}\right]$$

$$= 246.5067 \text{ days}^2$$

$$\therefore\ \sigma = \sqrt{246.5067} = 15.70 \text{ days}$$

$\therefore\ X \sim N(45.45,\ 246.5067)$. We know $Z = \frac{X-\mu}{\sigma}$

$$\psi(z) = P(Z \le z) = \int_{-\infty}^{z} f(z)dz,\ \Delta\psi(z) = \psi(z+1) - \psi(z).$$

The values of $\Delta\psi(z)$ and expected number of bulbs, $E_i = N\Delta\psi(z)$ are shown below:

Class interval	Upper boundary of class, X	$Z = \frac{X-\mu}{\sigma}$	$\psi(z)$	$\Delta\psi(z)$	Frequencies	
					Observed O_i	Expected $E_i \approx N\Delta\psi(z)$
10–20	19.5	−1.65	0.0495	0.0495	30	26
20–30	29.5	−1.02	0.1539	0.1044	52	54
30–40	39.5	−0.38	0.3520	0.1981	100	102
40–50	49.5	0.26	0.6026	0.2506	152	129
50–60	59.5	0.89	0.8133	0.2107	95	109
60–70	69.5	1.53	0.9370	0.1237	57	64
70–80	79.5	2.17	0.9850	0.0480	20	25
80+	89.5	2.81	—	0.0150	12	7
Total	—	—	—	1.00	516	516

The observed frequencies are fitted according to normal probability law. The fitting is not good since

$$\chi^2 = \sum \frac{O_i^2}{E_i} - N = 527.28 - 516 = 11.28 > \chi^2_{0.05,5} = 11.07.$$

Here d.f. of χ^2 is $(k-2-1) = 8-3 = 5$.

EXERCISE

1. (a) What is the importance of normal distribution in statistics. Write down some examples of normal vaiable.

   (b) Derive normal p.d.f. as a limiting case of binomial distribution. Write down characteristics of normal distribution.

   (c) The amount of polluted matter in air (microgm/m^3) is assumed to follow $N(5,2)$. Find the probability that in a randomly selected day the amount of polluted matter will (i) exceed 8 microgms/m^3, (ii) be from 4 to 6 microgms/m^3. If polluted matter in air is observed in a randomly selected week, what is the probability that in 3 days it will be 4 to 6 microgms/m^3?

2. (a) Derive normal p.d.f. as a limiting case of Poisson distribution. Write down some examples of normal variable.

   (b) Show that the normal distribution is symmetric and mesokurtic.

   (c) The diameter of iron rod is assumed to follow normal distribution with mean 0.25 cm and standard deviation 0.02 cm. A customer ordered for some rods with specification 0.25 ± 0.0005 cm as diameter. Find the percentage of rods which will be defective according to specification.

3. (a) Write down the chief characteristics of normal distribution. Show that, for normal distribution

$$\mu_{2n} = (1 \cdot 3 \cdot 5 \cdots 2n-1)\sigma^{2n}.$$

(b) If X is a normal variable, then derive the p.d.f. of $A \leq X \leq B$. What do you mean by truncated normal distribution?

(c) The amount of arsenic in water (gms/litre) follows $N(0.2, 0.08)$. A sample of water is collected at random. Find the probability that the sample contains (i) more than 0.3 gms/litre, (ii) less than 0.1 gms/litre and (iii) 0.2 to 0.4 gms/litre arsenic. If 6 samples are investigated, what is the probability that at least two samples contains 0.1 gms/litre of arsenic?

4. (a) Show that under certain conditions binomial distribution transforms to normal distribution. Find moments, median and mode of normal distribution.

(b) If $X_1 \sim N(\mu_1, \sigma^2)$ and $X_2 \sim N(\mu_2, \sigma^2)$, show that $X_1 \pm X_2$ is also normally distributed. If $x_1, x_2, \ldots, x_n$ are random sample observations from normal population, then $\bar{x} = \frac{1}{n}\Sigma x_i \sim N(\mu_1, \sigma^2/n)$.

(c) The mean height of soldiers is 156 cm and variance of height is 9 cm^2. If the distribution of height is normal, then how many soldiers in a regiment of 1000 soldiers have height over 158 cm?

5. (a) Write down the characteristics of normal distribution. Find mean deviation about mean of normal distribution.

(b) What do you mean by compound distribution. If $X \sim N(\mu, \sigma^2)$ and $\mu \sim N(\mu_1, \sigma_1^2)$, then derive the p.d.f. of the compound distribution of X.

(c) Discuss the importance of normal distribution.

6. (a) Distinguish univariate and bivariate normal distribution. Find moment generating function of bivariate normal distribution.

(b) If X_i's $(i = 1, 2, \ldots, n)$ are independent normal variable with mean μ_i and variance σ_i^2, then show that $\Sigma a_i X_i$ is normally distributed, where a_i's are constants.

(c) If $X \sim N(4, 9)$ and $Y \sim N(8, 25)$, then find the mean and variance of $Z = 2X + Y$. Find the percentage of values of Z which will fall within the limit $\bar{Z} \pm 2\sigma$, where σ is the standard deviation of Z.

7. (a) Discuss the importance of normal distribution. Let X be a $p \times 1$ vector having elements x_i. If X has the multivariate normal distribution with mean vector μ and variance-covariance matrix V, then show that $\sum_{i=1}^{p} a_i x_i$ has the univariate normal distribution with mean $\Sigma a_i \mu_i$ and variance $\sum_{j=1}^{p}\sum_{j=1}^{p} a_{ij}\sigma_{ij}$, where a_i's are known constants and σ_{ij} is the element of i-th row and j-th column of V. Also, find the conditional density function of $x_2 x_3 \cdots x_p / x_1$.

(b) If X is a $p \times 1$ vector with elements x_i and has the multivariate normal distribution with mean vector μ and variance-covariance matrix V, then x_i's are jointly independent if and only if $\text{cov}(x_i, x_j) = 0$ for all $i \neq j$.

8. (a) Write down some examples of univariate and multivariate normal distribution. How would you decide that a variable would be normal?

(b) Let $X \sim N(\mu, \sigma^2)$ and $\frac{1}{\sigma^2}$ is a gamma variate with parameter γ, then derive the p.d.f. of the compound distribution of X.

(c) Find median and mode of univariate normal distribution. Also, write down the characteristics of normal distribution.

CHAPTER **17**

Beta Distribution

17.1 Definition

A random variable X has a beta distribution with parameters m and n $(m > 0, n > 0)$ if X has a continuous distribution with p.d.f.

$$f(x) = \frac{1}{B(m,n)} x^{m-1}(1-x)^{n-1}, \ 0 < x < 1. \tag{17.1}$$

This is the standard form of the *beta distribution* with parameters m and n. It is also known as beta distribution of the first kind and written as $\beta_1(m, n)$.

The continuous random variable X which is distributed with probability law

$$f(x) = \frac{1}{B(m,n)} \frac{x^{m-1}}{(1+x)^{m+n}}, \ 0 < x < \infty \tag{17.2}$$

is known as beta variate of the second kind with parameters m and n, and its distribution is called beta distribution of the second kind. It is denoted by $\beta_2(m, n)$ variate.

The family of this beta distribution is composed of all distributions having the p.d.f. of the form

$$f(y) = \frac{1}{B(m,n)} \frac{(y-a)^{m-1}(b-y)^{n-1}}{(b-a)^{m+n-1}} \tag{17.3}$$
$$a \leq y \leq b; \ m > 0 \text{ and } n > 0.$$

The p.d.f. (17.3) transforms to (17.1) if we make the transformation

$$X = (Y - a)/(b - a).$$

Again, if we make the transformation

$$Z = X/(1 - X)$$

we obtain the distribution with p.d.f.

$$f(z) = \frac{1}{B(m,n)} \frac{Z^{m-1}}{(1+Z)^{m+n}}, \ 0 < Z < \infty \tag{17.4}$$

which is the p.d.f. similar to (17.2). The graph of the beta distribution is shown in FIG. 17.1.

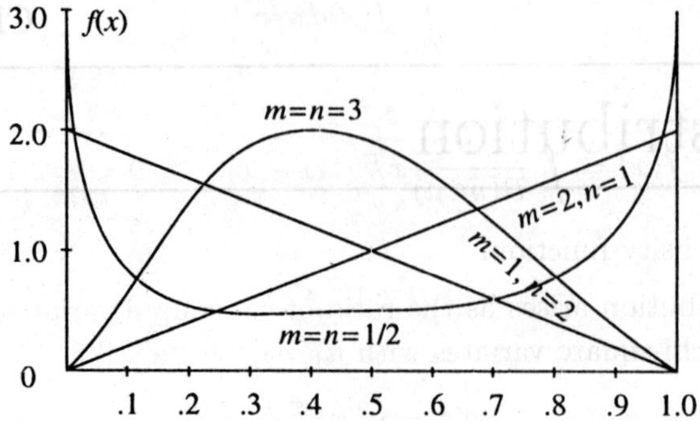

FIG. 17.1 : Graph of beta density functions.

The function (17.1) is non-negative, since $m > 0$ and $n > 0$. Also

$$\int_0^1 x^{m-1}(1-x)^{n-1}dx = B(m,n).$$

This can be shown as follows: From the definition of the gamma function, it can be written as

$$\begin{aligned}\Gamma m \Gamma n &= \int_0^\infty v^{m-1}e^{-v}dv \int_0^\infty u^{n-1}e^{-u}du \\ &= \int_0^\infty \int_0^\infty u^{n-1}v^{m-1}e^{-(u+v)}dudv.\end{aligned}$$

Let

$$z_1 = \frac{u}{u+v} \text{ and } z_2 = u+v.$$

Then $u = z_1 z_2$ and $v = z_2(1-z_1)$. Now

$$J = \frac{\partial uv}{\partial z_1 z_2} = \begin{vmatrix} \frac{\partial u}{\partial z_1} & \frac{\partial v}{\partial z_1} \\ \frac{\partial u}{\partial z_2} & \frac{\partial v}{\partial z_2} \end{vmatrix} = \begin{vmatrix} z_2 & -z_2 \\ z_1 & 1-z_1 \end{vmatrix} = z_2.$$

Here $0 < z_1 < 1$ and $0 < z_2 < \infty$. Therefore

$$\begin{aligned}\Gamma m \Gamma n &= \int_0^1 \int_0^\infty z_1^{n-1}(1-z_1)^{m-1} z_2^{m+n-1} e^{-z_2} dz_1 dz_2 \\ &= \int_0^1 z_1^{n-1}(1-z_1)^{m-1}dz_1 \int_0^\infty e^{-z_2} z_2^{m+n-1} dz \\ &= \Gamma(m+n) \int_0^1 z_1^{n-1}(1-z_1)^{m-1}.\end{aligned}$$

$$\begin{aligned}\therefore \quad \int_0^1 z_1^{n-1}(1-z_1)^{m-1}dz_1 &= B(m,n) \\ &= B(n,m).\end{aligned}$$

From this result we can say that

$$\int_0^1 f(x)dx = 1$$

where

$$f(x) = \int \frac{1}{B(m,n)} x^{m-1}(1-x)^{n-1},\ 0 < x < 1.$$

Thus (17.1) is a density function.

The beta distribution arises as the ratio of chi-square variates. Let $\chi_1^2, \chi_2^2, \dots, \chi_k^2$ be k independent chi-square variates with $n_1, n_2, \dots, n_{n_k}$ d.f.. Then

$$X_1^2 = \chi_1^2/(\chi_1^2 + \chi_2^2)$$

has the standard beta distribution. Here X_1^2 and $(\chi_1^2 + \chi_2^2)$ are mutually independent. If all the k chi-square variates are mutually independent, we can define

$$\begin{aligned} X_1^2 &= \chi_1^2/(\chi_1^2 + \chi_2^2) \\ X_2^2 &= (\chi_1^2 + \chi_2^2)/(\chi_1^2 + \chi_2^2 + \chi_3^2) \\ \dots &\quad \dots \quad \dots \quad \dots \quad \dots \\ X_{k-1}^2 &= (\chi_1^2 + \chi_2^2 + \dots + \chi_{k-1}^2)/(\chi_1^2 + \dots + \chi_k^2) \end{aligned}$$

which are also mutually independent and each of $X_1^2, X_2^2, \dots, X_{k-1}^2$ has the beta distribution.

The beta distribution also arises from the distribution of r-th order statistic when the sample observations are drawn from a rectangular distribution with interval $[0,1]$. Let $y_1, y_2, \dots, y_n$ be independent sample observations from a population with p.d.f.

$$f(y) = 1,\ 0 \le y \le 1.$$

Let Y_r be the r-th $(r = 1, 2, \dots, n)$ order statistic of $y_1, y_2, \dots, y_n$. Then

$$g(Y_r) = \frac{n!}{(r-1)!(n-r)!}[F(y)]^{r-1}[1 - F(y)]^{n-r} f(y).$$

But

$$F(y) = \int_0^y f(y)dy = y.$$

Therefore

$$\begin{aligned} g(Y_r) &= \frac{n!}{(r-1)!(n-r)!} y^{r-1}(1-y)^{n-r} \\ &= \frac{1}{B(r, n-r-1)} y^{r-1}(1-y)^{n-r}, 0 \le Y_r \le 1 \end{aligned}$$

which is the p.d.f. similar to (17.1).

There are many other statistics from which beta distribution arises. The reader may consult Murthy (1960), Bańkövi (1964) and Jöhnk (1964).

17.2 Moments of Beta Distributions

Moments of Beta Distribution of the First Kind

The r-th moment about origin is given by

$$\begin{aligned}\mu'_r = E(x^r) &= \int_0^1 \frac{1}{B(m,n)} x^{m+r-1}(1-x)^{n-1}dx \\ &= \frac{B(m+r,n)}{B(m,n)} \\ &= \frac{\Gamma(m+r)\Gamma(m+n)}{\Gamma m\Gamma(m+n+r)}\end{aligned}$$

In particular,

when $r = 1$,

$$\mu'_1 = \frac{\Gamma(m+1)\Gamma(m+n)}{\Gamma m\Gamma(m+n+1)} = \frac{m}{m+n}$$

when $r = 2$,

$$\mu'_2 = \frac{\Gamma(m+2)\Gamma(m+n)}{\Gamma m\Gamma(m+n+2)} = \frac{m(m+1)}{(m+n+1)(m+n)}$$

when $r = 3$,

$$\mu'_3 = \frac{\Gamma(m+3)\Gamma(m+n)}{\Gamma m\Gamma(m+n+3)} = \frac{m(m+1)(m+2)}{(m+n+2)(m+n+1)(m+n)}$$

when $r = 4$,

$$\mu'_4 = \frac{\Gamma(m+4)\Gamma(m+n)}{\Gamma m\Gamma(m+n+4)} = \frac{m(m+1)(m+2)(m+3)}{(m+n+3)(m+n+2)(m+n+1)(m+n)}.$$

Now,

$$\begin{aligned}\mu_2 &= \mu'_2 - \mu'^2_1 = \frac{m(m+1)}{(m+n+1)(m+n)} - \frac{m^2}{(m+n)^2} \\ &= \frac{mn}{(m+n)^2(m+n+1)} \\ \mu_3 &= \mu'_3 - 3\mu'_2\mu'_1 + 2\mu'^3_1 = \frac{2mn(n-m)}{(m+n+2)(m+n+1)(m+n)^3} \\ \mu_4 &= \mu'_4 - 4\mu'_3\mu'_1 + 6\mu'_2\mu'^2_1 - 3\mu'^4_1 \\ &= \frac{3mn[mn(m+n-6)+2(m+n)^2]}{(m+n)^4(m+n+1)(m+n+2)(m+n+3)}.\end{aligned}$$

The moment ratios are

$$\alpha_3 = \sqrt{\beta_1} = \sqrt{\frac{\mu_3^2}{\mu_2^3}} = \sqrt{\frac{4(n-m)^2(m+n+1)}{(m+n+2)^2 mn}}$$

$$\alpha_4 = \beta_2 = \frac{\mu_4}{\mu_2^2} = \frac{3(m+n+1)[mn(m+n-6)+2(m+n)^2]}{mn(m+n+2)(m+n+3)}.$$

Moments of Beta Distribution of the Second Kind

The r-th moment about origin is given by

$$\begin{aligned}
\mu_r' = E(x^r) &= \frac{1}{B(m,n)}\int_0^\infty \frac{x^{m+r-1}}{(1+x)^{m+n}}dx \\
&= \frac{1}{B(m,n)}\int_0^\infty \frac{x^{m+r-1}}{(1+x)^{m+r+n-r}}dx \\
&= \frac{B(m+r,n-r)}{B(m,n)}\,. \\
\mu_r' &= \frac{\Gamma(m+r)\Gamma(n-r)}{\Gamma m\Gamma n}.
\end{aligned}$$

In particular,

when $r=1$,

$$\mu_1' = \frac{\Gamma(m+1)\Gamma(n-1)}{\Gamma m\Gamma n} = \frac{m}{n-1}$$

when $r=2$,

$$\mu_2' = \frac{\Gamma(m+2)\Gamma(n-2)}{\Gamma m\Gamma n} = \frac{m(m+1)}{(n-1)(n-2)}$$

when $r=3$,

$$\mu_3' = \frac{\Gamma(m+3)\Gamma(n-3)}{\Gamma m\Gamma n} = \frac{m(m+2)(m+1)}{(n-1)(n-2)(n-3)}$$

when $r=4$,

$$\mu_4' = \frac{\Gamma(m+4)\Gamma(n-4)}{\Gamma m\Gamma n} = \frac{m(m+1)(m+2)(m+3)}{(n-1)(n-2)(n-3)(n-4)}$$

$$\mu_2 = \mu_2' - \mu_1'^2 = \frac{m(m+n-1)}{(n-1)^2(n-2)}$$

$$\mu_3 = \mu_3' - 3\mu_2'\mu_1' + 2\mu_1'^3 = \frac{2m(m^2+3mn-3m+n^2-2n+1)}{(n-1)^3(n-2)(n-3)}.$$

17.3 Properties of Beta Distribution

The p.d.f. of family of beta distributions is

$$f(y) = \frac{1}{B(m,n)}\frac{(y-a)^{m-1}(b-y)^{n-1}}{(b-a)^{m+n-1}},\ a \le y \le b.$$

If $n=1$, we get

$$f(y) = \frac{m(y-a)^{m-1}}{(b-a)^m}$$

which is the density of a power function distribution.

Again, from (17.1) for $n=1$, we get

$$f(x) = mx^{m-1},\ 0 < x < 1$$

which is the density of the standard power function distribution.

Cumulative Distribution Function

The cumulative distribution function is given by

$$F(x) = \begin{cases} 0, & x < 0 \\ \dfrac{1}{B(m,n)} \displaystyle\int_0^x x^{m-1}(1-x)^{n-1}dx, & 0 < x < 1 \\ 1, & x > 1 \end{cases}$$

Here

$$I_x(m,n) = \frac{1}{B(m,n)} \int_0^x x^{m-1}(1-x)^{n-1}dx$$

is called the *incomplete beta function ratio.* On the other hand

$$B_x(m,n) = \int_0^x x^{m-1}(1-x)^{n-1}dx$$

is called the *incomplete beta function.*

Harmonic Mean of Beta Distribution

The harmonic mean of beta distribution of the first kind is given by

$$\begin{aligned} \frac{1}{\text{H.M.}} &= \int_0^1 \frac{1}{x} \frac{1}{B(m,n)} x^{m-1}(1-x)^{n-1}dx \\ &= \frac{1}{B(m,n)} \int_0^1 x^{m-1-1}(1-x)^{n-1}dx \\ &= \frac{B(m-1,n)}{B(m,n)} \\ &= \frac{m+n-1}{m-1}. \\ \therefore \text{ H.M.} &= \frac{m-1}{m+n-1}. \end{aligned}$$

The harmonic mean of beta distribution of the second kind is given by

$$\begin{aligned} \frac{1}{\text{H.M.}} &= \frac{1}{B(m,n)} \int_0^\infty \frac{x^{m-2}}{(1+x)^{m+n}}dx \\ &= \frac{1}{B(m,n)} \int_0^\infty \frac{x^{m-1-1}}{(1+x)^{m-1+n+1}}dx \\ &= \frac{B(m-1,n+1)}{B(m,n)} = \frac{n}{m-1}. \\ \therefore \text{ H.M.} &= \frac{m-1}{n}. \end{aligned}$$

The mean deviation is given by

$$\frac{2m^m n^n}{(m+n)^{m+n}} \frac{\Gamma(m+n)}{\Gamma m \Gamma n}.$$

But for large m and n, the mean deviation is approximately given by

$$\sqrt{\frac{2mn}{\pi(m+n)}\frac{1}{m+n}}\left[1+\frac{1}{12(m+n)}-\frac{1}{12m}-\frac{1}{12n}\right].$$

Hence, the ratio of mean deviation to standard deviation is approximately

$$\frac{\text{mean deviation}}{\text{standard deviation}} \doteq \sqrt{\frac{2}{\pi}}\left[1+\frac{7}{12(m+n)}-\frac{1}{12m}-\frac{1}{12n}\right].$$

If $m > 1$ and $n > 1$, then $f(x) \to 0$ as $x \to 0$ or $x \to 1$; if $0 < p < 1$, $f(x) \to \infty$ as $x \to 0$ and if $0 < q < 1$, $f(x) \to \infty$ as $x \to 1$.

If $m > 1$ and $n > 1$, there is single mode of beta distribution of the first kind and mode at $x = (m-1)/(m+n-2)$. However, there is an antimode of $f(x)$ at this value of x when $m < 1$ and $n < 1$, and the distribution is called *U-shaped beta distribution*. Again, if $(m-1)(n-1)$ is not positive, the p.d.f. $f(x)$ does not have a mode for $0 < x < 1$ and the distribution is called *J-shaped beta distribution*. If $p = q$, the distribution is symmetrical at $x = 1/2$.

Thc mean of beta distribution is $m/(m+n)$ which depends on the ratio m/n. If both m and n are increased but the ratio m/n is kept constant, then the variance of the beta distribution decreased. In that case standardized beta distribution tends to the unit normal distribution.

The incomplete beta function ratio can be approximated by

$$I_x(m, n) = h(z)$$

where

$$z = \frac{d}{\left|n-\frac{1}{2}-N(1-x)\right|}\left\{\frac{2}{1+\frac{6}{N}}\left[\left(n-\frac{1}{2}\right)\log\left(\frac{n-\frac{1}{2}}{N(1-x)}\right) + \left(m-\frac{1}{2}\right)\log\left(\frac{m-\frac{1}{2}}{Nx}\right)\right]\right\}^{1/2}.$$

Here

$$N = m+n-1, \text{ the value of } d \text{ is either}$$

$$d = n-\frac{1}{3}-\left(N+\frac{1}{3}\right)(1-x)$$

$$\text{or, } d = n-\frac{1}{3}-\left(N+\frac{1}{3}\right)(1-x)+\frac{x}{5n}-\frac{(1-x)}{m}+\frac{\left(x-\frac{1}{2}\right)}{m+n}.$$

If X is a beta variate of first kind with parameter $(m+1)$ and $(n+1)$, $m > -1$, $n > -1$, then the distribution $U = (1-X)/X$ is a beta variate of the second kind.

We have

$$f(x) = \frac{1}{B(m+1, n+1)}x^m(1-x)^n,\ 0 < x < 1.$$

Let

$$U = (1-X)/X \Rightarrow 1+u = \frac{1}{x}$$

$$\text{or, } x = \frac{1}{(1+u)}, \; dx = -\frac{du}{(1+u)^2}.$$

$$\therefore \; g(u) = \frac{1}{B(m+1,n+1)}\left(\frac{1}{1+u}\right)^m\left(\frac{u}{1+u}\right)^n\frac{1}{(1+u)^2}$$

$$= \frac{1}{B(m+1,n+1)}\frac{u^n}{(1+u)^{m+n+2}}, \; 0 < u < \infty.$$

EXERCISE

1. (a) What is beta variate? Show that ratio of two independent chi-square variate is a beta variate. If $X \sim b(n,p)$ and p is a beta variate, then find the compound distribution of X.

(b) Find mean and variance of beta distribution.

2. (a) What are the different types of beta variate? Find moments of beta variate of second kind.

(b) Let $y_1, y_2, \ldots, y_n$ be random sample observations from uniform distribution with interval $[0,1]$. Then show that the r-th order statistic of y_i's is a beta variate of first kind.

3. (a) If u and v are two independent gamma variates, then show that $z = \frac{u}{u+v}$ is a beta variate. Find mean and variance of z.

(b) Find moments of beta variate of first kind.

CHAPTER **18**

Gamma Distribution

18.1 Definition

If X is a random variable with probability density function

$$f(x) = \frac{1}{\Gamma m \theta^m} e^{-x/\theta} x^{m-1}; 0 < x < \infty \tag{18.1}$$

then X is called *gamma variate* and its distribution is called *gamma distribution.* The general form of the p.d.f. of gamma distribution is

$$f(x) = \frac{1}{\Gamma m \theta^m} e^{-\frac{1}{\theta}(x-\alpha)} (x-\alpha)^{m-1} \tag{18.2}$$

with $m > 0$, $\theta > 0$; $x > \alpha$. This is Pearson's Type III distribution having three parameters m, θ and α. The shape of the curve of this distribution is shown in FIG. 18.1. But if $\alpha = 0$, we have (18.1) and if $\theta = 1$ and $\alpha = 0$, we have

$$f(x) = \frac{1}{\Gamma m} e^{-x} x^{m-1}; 0 < x < \infty \tag{18.3}$$

which is the density of standard gamma variate.

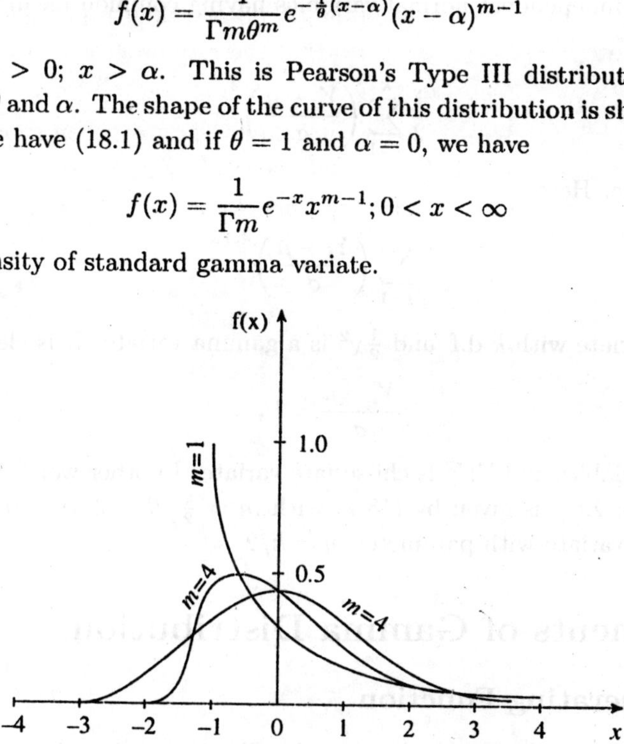

FIG. 18.1 : Graph of gamma density functions.

The gamma distribution appears naturally in the theory associated with normally distributed random variables. For example, let us consider the distribution of

$$Z = \frac{1}{2}\left(\frac{Y-\mu}{\sigma}\right)^2, \text{ where } Y \sim N(\mu, \sigma^2).$$

We have

$$f(y) = \frac{1}{\sqrt{2\pi}\sigma} e^{-\frac{1}{2}\left(\frac{y-\mu}{\sigma}\right)^2}, -\infty < y < \infty.$$

Let $z = \frac{1}{2}\left(\frac{y-\mu}{\sigma}\right)^2 \Rightarrow y = \mu + \sigma\sqrt{2z}$.

$$\therefore\ dy = \frac{\sigma}{\sqrt{2}} z^{-\frac{1}{2}} dz$$

$$\therefore\ g(z) = \begin{cases} \frac{1}{2\Gamma\frac{1}{2}} e^{-z} z^{\frac{1}{2}-1}, & \infty < z < 0 \\ \frac{1}{2\Gamma\frac{1}{2}} e^{-z} z^{\frac{1}{2}-1}, & 0 < z < \infty \end{cases}$$

$$g(z) = \frac{1}{\Gamma\frac{1}{2}} e^{-z} z^{\frac{1}{2}-1}, 0 < z < \infty$$

which is the p.d.f. of standard form of gamma variate with parameter $m = 1/2$. Thus, if $Y_1, Y_2, \ldots, Y_k$ are independent normal variables having common mean μ and common variance σ^2, then

$$\frac{1}{2}\sum_{i=1}^{k}\left(\frac{Y_i-\mu}{\sigma}\right)^2$$

is a gamma variate. Here

$$\sum_{i=1}^{k}\left(\frac{Y_i-\mu}{\sigma}\right)^2$$

is a chi-square variate with k d.f. and $\frac{1}{2}\chi^2$ is a gamma variate. It is clear that

$$\frac{Y_i-\mu}{\sigma} = U_i$$

is unit normal variables and ΣU_i^2 is chi-square variate. In other words we can say that the distribution of ΣU_i^2 is given by (18.2) with $m = \frac{k}{2}$; $\theta = 2$, $\alpha = 0$ and $\frac{1}{2}\Sigma U_i^2$ is a standard gamma variate with parameter $m = k/2$.

18.2 Moments of Gamma Distribution

Moment Generating Function

The moment generating function of gamma distribution is given by

$$M_x(t) = E(e^{tx}) = \frac{1}{\Gamma m \theta^m}\int_0^{\infty} e^{-x\left(\frac{1-\theta t}{\theta}\right)} x^{m-1} dx.$$

Putting $\frac{x(1-\theta t)}{\theta} = z$, we have

$$dx = \frac{\theta dz}{1-\theta t}$$

$$\text{and } M_x(t) = \frac{1}{\Gamma m(1-\theta t)^m}\int_0^\infty e^{-z}z^{m-1}dz = (1-\theta t)^{-m}.$$

Cumulant Generating Function

The cumulant generating function is given by

$$K_x(t) = \log M_x(t) = \log(1-\theta t)^{-m} = m\left(\theta t + \frac{(\theta t)^2}{2} + \frac{(\theta t)^3}{3} + \frac{(\theta t)^4}{4} + \cdots\right).$$

Thus

$$\begin{aligned} K_1 &= \mu_1' = m\theta \\ K_2 &= \mu_2 = m\theta^2 \\ K_3 &= \mu_3 = 2m\theta^3 \\ K_4 &= 6m\theta^4 \text{ and } \mu_4 = K_4 + 3\mu_2^2 = 3m\theta^4(m+2). \end{aligned}$$

Also, we have

$$\begin{aligned} \mu_1' &= \left.\frac{\partial M_x(t)}{\partial t}\right|_{t=0} = m\theta, \\ \mu_2' &= \left.\frac{\partial^2 M_x(t)}{\partial t^2}\right|_{t=0} = m(m+1)\theta^2, \\ \mu_3' &= \left.\frac{\partial^3 M_x(t)}{\partial t^3}\right|_{t=0} = m(m+1)(m+2)\theta^3, \\ \text{and } \mu_4' &= \left.\frac{\partial^4 M_x(t)}{\partial t^4}\right|_{t=0} = m(m+1)(m+2)(m+3)\theta^4. \end{aligned}$$

The moment ratios are

$$\alpha_3 = \sqrt{\beta_1} = \sqrt{\frac{\mu_3^2}{\mu_2^3}} = \sqrt{\frac{4m^2\theta^6}{m^3\theta^6}} = \frac{2}{\sqrt{m}}$$

$$\alpha_4 = \beta_2 = \frac{\mu_4}{\mu_2^2} = \frac{3m\theta^4(m+2)}{m^2\theta^4} = 3 + \frac{6}{m}.$$

Hence gamma distribution is skewed and it is leptokurtic.

If $\theta = 1$, then we have (18.3). In that case $\mu_1' = m$, $\mu_2 = m$, $\mu_3 = 2m$, $\mu_4 = 3m(m+2)$. It is observed that the mean and variance of gamma distribution is same. This phenomenon is similar to Poisson distribution, but Poisson distribution is a discrete one.

The r-th moment about origin of standard gamma distribution is given by

$$\mu_r' = \frac{1}{\Gamma m}\int_0^\infty e^{-x}x^{m+r-1}dx = \frac{\Gamma(m+r)}{\Gamma m}.$$

18.3 Properties of Gamma Distribution

Additive Property

If $X_1, X_2, \ldots, X_k$ are k independent gamma variates with parameters $m_1, m_2, \ldots, m_k$ respectively, then their sum $Y = X_1 + X_2 + \cdots + X_k$ is also a gamma variate with parameter $m_1 + m_2 + \cdots + m_k = M$ (say).

We know that $M_x(t) = (1-t)^{-m}$. Again,

$$\begin{aligned} M_{x_1+x_2+\cdots+x_k}(t) &= M_{x_1}(t)M_{x_2}(t)\cdots M_{x_k}(t) \\ &= (1-t)^{-m_1}(1-t)^{-m_2}\cdots(1-t)^{-m_k} \\ &= (1-t)^{-M} \\ \text{or, } M_y(t) &= (1-t)^{-M} \end{aligned}$$

which is the m.g.f. of a gamma variate with parameter $M = m_1 + m_2 + \cdots + m_k$.

Harmonic Mean

For gamma distribution

$$\begin{aligned} \frac{1}{\text{H.M.}} &= \frac{1}{\Gamma m}\int_0^\infty \frac{1}{x}e^{-x}x^{m-1}dx \\ &= \frac{1}{\Gamma m}\int_0^\infty e^{-x}x^{m-2}dx \\ &= \frac{\Gamma(m-1)}{\Gamma m} \\ &= \frac{1}{m-1}. \\ \therefore\ \text{H.M.} &= (m-1). \end{aligned}$$

The standard gamma distribution has a single mode at $x = m-1$ if $m \geq 1$. If $m < 1$, $f(x)$ tends to infinity as x tends to zero, but $f(x) = 1$ if $x \to 0$ when $m = 1$. If $m = 1$, the distribution becomes standard exponential one.

There are points of inflexion, equidistant from the mode, at

$$x = m - 1 \pm \sqrt{m-1}.$$

The mean deviation of standard gamma distribution is $2m^m e^{-m}/\Gamma m$.

Approximation

The standard gamma distribution tends to unit normal distribution as the value of the parameter m tends to infinity.

We know that $E(x) = V(x) = m$. Let us define standardized variable

$$Z = \frac{X-m}{\sqrt{m}}.$$

The m.g.f. of Z is

$$\begin{aligned}
M_z(t) &= E\left[e^{zt}\right] \\
&= E\left[\frac{x-m}{\sqrt{m}}t\right] = e^{-\sqrt{m}t}E\left[\frac{xt}{\sqrt{m}}\right] \\
&= e^{-\sqrt{m}t}M_x\left(\frac{t}{\sqrt{m}}\right) = e^{-\sqrt{m}t}\left(1-\frac{t}{\sqrt{m}}\right)^{-m}. \\
\therefore\ K_z(t) &= \log M_z(t) = -\sqrt{m}t - m\log\left(1-\frac{t}{\sqrt{m}}\right) \\
&= -\sqrt{m}t + m\left(\frac{t}{\sqrt{m}} + \frac{t^2}{2m} + \frac{t^3}{3m^{3/2}} + \cdots\right) \\
&= \frac{t^2}{2} + \frac{t^3}{3\sqrt{m}} + \cdots \\
\lim_{m\to\infty} K_z(t) &= \frac{t^2}{2} \\
\lim_{m\to\infty} M_z(t) &= e^{t^2/2}
\end{aligned}$$

which is the m.g.f. of standard (unit) normal variate. Hence, gamma variate can be approximated by standard normal variate if $m \to \infty$.

If X has a standard gamma distribution, then the m.g.f. of $\log X$ is

$$M_{\log x}(t) = E\left[e^{t\log x}\right] = E\left[x^t\right] = \frac{\Gamma(m+t)}{\Gamma m}.$$

From this β_1 and β_2 can be found out. For m large

$$\beta_1(\log x) \doteq 1/m \text{ and } \beta_2(\log x) \doteq 3 + \frac{2}{m}.$$

But

$$\beta_1(x) = \frac{4}{m} \text{ and } \beta_2(x) = 3 + \frac{6}{m}.$$

In both the cases $\beta_1 \to 0$ and $\beta_2 \to 3$ if $m \to \infty$. But the distribution of $\log X$ is more nearly normal than the distribution of X.

Cumulative Distribution Function

It is given by

$$F(x) = \frac{1}{\Gamma m}\int_0^x e^{-x}x^{m-1}dx.$$

This is an incomplete gamma function ratio. But

$$\Gamma m F(x) = \int_0^x e^{-x}x^{m-1}dx$$

is sometimes called an *incomplete gamma function*.

Mean of Square Root of a Gamma Variate

Let X be a gamma variate with p.d.f.

$$f(x) = \frac{1}{\Gamma m} e^{-x} x^{m-1}, 0 < x < \infty.$$

Then

$$\begin{aligned} E\left[\sqrt{x}\right] &= \frac{1}{\Gamma m} \int_0^\infty e^{-x} x^{\frac{1}{2}} x^{m-1} dx \\ &= \frac{1}{\Gamma m} \int_0^\infty e^{-x} x^{m+\frac{1}{2}-1} dx \\ &= \frac{\Gamma\left(m+\frac{1}{2}\right)}{\Gamma m}. \end{aligned}$$

❒**THEOREM 18.1** *If X and Y are two independent gamma variates with parameters m and n respectively, then $U = X + Y$ and $V = X/Y$ are independent.*

Proof. We have

$$\begin{aligned} f(x) &= \frac{1}{\Gamma m} e^{-x} x^{m-1}, 0 < x < \infty \\ f(y) &= \frac{1}{\Gamma n} e^{-y} y^{n-1}, 0 < y < \infty \\ g(x, y) &= \frac{1}{\Gamma m \Gamma n} e^{-(x+y)} x^{m-1} y^{n-1}. \end{aligned}$$

Let $u = x + y$, $v = \frac{x}{y}$ i.e., $x = \frac{uv}{1+v}$, $y = \frac{u}{1+v}$, $0 < u < \infty$, $0 < v < \infty$

$$\begin{aligned} J &= \frac{\partial(x, y)}{\partial(u, v)} = \begin{vmatrix} \frac{\partial x}{\partial u} & \frac{\partial y}{\partial u} \\ \frac{\partial x}{\partial v} & \frac{\partial y}{\partial v} \end{vmatrix} = \begin{vmatrix} \frac{v}{1+v} & \frac{1}{1+v} \\ \frac{u}{(1+v)^2} & -\frac{u}{(1+v)^2} \end{vmatrix} = \frac{-uv}{(1+v)^2} \\ |J| &= \frac{uv}{(1+v)^2} \\ g_1(u, v) &= \frac{1}{\Gamma m \Gamma n} e^{-u} \frac{(uv)^{m-1}}{(1+v)^{m-1}} \frac{u^{n-1}}{(1+v)^{n-1}} \frac{uv}{(1+v)^2} \\ &= \frac{1}{\Gamma(m+n) B(m, n)} e^{-u} u^{m+n-1} \frac{v^{m-1}}{(1+v)^{m+n}} \\ &= \frac{1}{\Gamma(m+n)} e^{-u} u^{m+n-1} \frac{1}{B(m, n)} \frac{v^{m-1}}{(1+v)^{m+n}} \\ &= h(u) h_1(v) \\ h(u) &= \int_0^\infty g_1(u, v) dv \\ &= \frac{1}{\Gamma(m+n)} e^{-u} u^{m+n-1} \frac{1}{B(m, n)} \int_0^\infty \frac{v^{m-1}}{(1+v)^{m+n}} dv \\ &= \frac{1}{\Gamma(m+n)} e^{-u} u^{m+n-1}. \end{aligned}$$

Hence U and V are independent, where U is a gamma variate.

❐THEOREM 18.2 *If X and Y are two independent gamma variates with parameters m and n respectively, then $U = X + Y$ and $V = X/(X + Y)$ are distributed independently.*

Proof. We have

$$f(x) = \frac{1}{\Gamma m} e^{-x} x^{m-1},\ 0 < x < \infty$$

$$f(y) = \frac{1}{\Gamma n} e^{-y} y^{n-1},\ 0 < y < \infty$$

$$g(x,y) = \frac{1}{\Gamma m \Gamma n} e^{-(x+y)} x^{m-1} y^{n-1}.$$

Let $u = x + y$ and $v = \frac{x}{x+y}$ i.e., $x = uv$ and $y = u(1-v)$, $0 < u < \infty$; $0 < v < 1$

$$J = \frac{\partial(x,y)}{\partial(u,v)} = \begin{vmatrix} \frac{\partial x}{\partial u} & \frac{\partial y}{\partial u} \\ \frac{\partial x}{\partial v} & \frac{\partial y}{\partial v} \end{vmatrix} = \begin{vmatrix} v & 1-v \\ u & -u \end{vmatrix} = -u$$

$$|J| = u$$

$$g_1(u,v) = \frac{1}{\Gamma m \Gamma n} e^{-u} u^{m-1} v^{m-1} u^{n-1} (1-v)^{n-1} u$$

$$= \frac{1}{\Gamma(m+n) B(m,n)} e^{-u} u^{m+n-1} v^{m-1} (1-v)^{n-1}$$

$$= \frac{1}{\Gamma(m+n)} e^{-u} u^{m+n-1} \frac{1}{B(m,n)} v^{m-1} (1-v)^{n-1}$$

$$h(u) = \int_0^1 g_1(u,v)dv$$

$$= \frac{1}{\Gamma(m+n)} e^{-u} u^{m+n-1} \int_0^1 \frac{1}{B(m,n)} v^{m-1} (1-v)^{n-1} dv$$

$$= \frac{1}{\Gamma(m+n)} e^{-u} u^{m+n-1}$$

which is the p.d.f. of a gamma variate with parameter $m + n$. Therefore,

$$g_1(u,v) = h(u)h_1(v)$$

where $h_1(v)$ is the p.d.f. of a beta variate with parameters m and n. Hence U and V are independent.

►EXAMPLE 18.1 *If X is a gamma variate with parameter β and (m + n), and Y is a beta variate of the first kind with parameters m and n, then find the distribution of U = XY.*

Solution. We have

$$f(x) = \frac{\beta^{m+n}}{\Gamma(m+n)} e^{-\beta x} x^{m+n-1},\ 0 < x < \infty$$

$$f(y) = \frac{1}{B(m,n)} y^{m-1} (1-y)^{n-1},\ 0 < y < 1$$

$$g(x,y) = \frac{\beta^{m+n}}{\Gamma(m+n) B(m,n)} e^{-\beta x} x^{m+n-1} y^{m-1} (1-y)^{n-1}.$$

Let $u = xy$ and $v = y$, then $0 < u < \infty$, $0 < v < 1$
$x = u/v$ and $y = v$

$$J = \begin{vmatrix} \frac{\partial x}{\partial u} & \frac{\partial y}{\partial u} \\ \frac{\partial x}{\partial v} & \frac{\partial y}{\partial v} \end{vmatrix} = \begin{vmatrix} \frac{1}{v} & 0 \\ \frac{-u}{v^2} & 1 \end{vmatrix} = \frac{1}{v};\ |J| = \frac{1}{v}$$

$$g_1(u,v) = \frac{\beta^{m+n}}{\Gamma m \Gamma n} e^{-\frac{\beta u}{v}} \left(\frac{u}{v}\right)^{m+n-1} v^{m-1} \frac{1}{v} (1-v)^{n-1}$$

$$= \frac{\beta^{m+n}}{\Gamma m \Gamma n} e^{-\frac{\beta u}{v}} u^{m+n-1} v^{-n-1} (1-v)^{n-1}$$

$$h(u) = \frac{\beta^{m+n}}{\Gamma m \Gamma n} u^{m+n-1} \int_0^1 e^{-\frac{\beta u}{v}} v^{-n-1} (1-v)^{n-1} dv$$

$$= \frac{\beta^{m+n}}{\Gamma m \Gamma n} u^{m+n-1} \int_0^1 e^{-\frac{\beta u}{v}} \frac{v^{n-1}}{v^{n+1}} \left(\frac{1}{v} - 1\right)^{n-1} dv.$$

Let $\left(\frac{1}{v} - 1\right) = z \Rightarrow v = \frac{1}{1+z}$, $dv = -\frac{dz}{(1+z)^2}$, $\infty \le z \le 0$.

$$h(u) = \frac{\beta^{m+n}}{\Gamma m \Gamma n} u^{m+n-1} \int_0^\infty e^{-\beta u(1+z)} z^{n-1} dz$$

$$= \frac{\beta^{m+n}}{\Gamma m \Gamma n} u^{m+n-1} e^{-\beta u} \int_0^\infty e^{-\beta u z} z^{n-1} dz$$

$$= \frac{\beta^{m+n}}{\Gamma m \Gamma n} u^{m+n-1} e^{-\beta u} \int_0^\infty \frac{(\beta u)^n}{(\beta u)^n} e^{-\beta u z} z^{n-1} dz$$

$$= \frac{\beta^{m+n}}{\Gamma m \Gamma n} u^{m+n-1} e^{-\beta u} \frac{\Gamma n}{(\beta u)^n}$$

$$= \frac{\beta^m}{\Gamma m} e^{-\beta u} u^{m-1},\ 0 < u < \infty.$$

►**EXAMPLE 18.2** *If X is a gamma variate with parameter m, find the distribution of $U = 1/X$.*

Solution. We have

$$f(x) = \frac{1}{\Gamma m} e^{-x} x^{m-1},\ 0 < x < \infty.$$

Let $u = 1/x$ or, $x = 1/u \Rightarrow dx = -du/u^2$.

$$g(u) = \frac{1}{\Gamma m} e^{-\frac{1}{u}} \left(\frac{1}{u}\right)^{m-1} \frac{du}{u^2}$$

$$= \frac{1}{\Gamma m} e^{-\frac{1}{u}} \left(\frac{1}{u}\right)^{m+1},\ 0 < u < \infty.$$

►**EXAMPLE 18.3** *If X and Y are two independent standard normal variates, find the distribution of* (i) X^2, (ii) $X^2 + Y^2$ *and* (iii) X^2/Y^2.

Solution.

(i) Given

$$f(x) = \frac{1}{\sqrt{2\pi}} e^{-\frac{1}{2}x^2}, \ -\infty < x < \infty$$
$$f(y) = \frac{1}{\sqrt{2\pi}} e^{-\frac{1}{2}y^2}, \ -\infty < y < \infty$$
$$M_{x^2}(t) = E\left[e^{tx^2}\right] = \frac{1}{\sqrt{2\pi}} \int_{-\infty}^{\infty} e^{tx^2} e^{-\frac{1}{2}x^2} dx$$
$$= \frac{1}{\sqrt{2\pi}} \int_{-\infty}^{\infty} e^{-\frac{1}{2}x^2(1-2t)} dx.$$

Let

$$\frac{1}{2}x^2(1-2t) = z \text{ or, } x^2 = \frac{2z}{1-2t} \Rightarrow dx = \frac{\sqrt{2}}{\sqrt{1-2t}} \frac{1}{2} z^{-\frac{1}{2}} dz.$$
$$\therefore \ M_{x^2}(t) = \frac{2}{\sqrt{2\pi}} \int_0^{\infty} e^{-z} \frac{\sqrt{2}}{2\sqrt{1-2t}} z^{\frac{1}{2}-1} dz = (1-2t)^{-\frac{1}{2}}$$

which is the m.g.f. of a gamma variate with parameters $\theta = 2$ and $m = 1/2$. Therefore,

$$f(x^2) = \frac{1}{\Gamma\frac{1}{2} 2^{\frac{1}{2}}} e^{-\frac{1}{2}x^2} (x^2)^{\frac{1}{2}-1}, \ 0 < x^2 < \infty.$$

Similarly

$$f(y^2) = \frac{1}{\Gamma\frac{1}{2} 2^{\frac{1}{2}}} e^{-\frac{1}{2}y^2} (y^2)^{\frac{1}{2}-1}, \ 0 < y^2 < \infty.$$

(ii) and (iii) Let

$$u = x^2 + y^2 \text{ and } v = \frac{x^2}{y^2} \Rightarrow x^2 = \frac{uv}{1+v} \text{ and } y^2 = \frac{u}{1+v}; \ |J| = \frac{u}{(1+v)^2}.$$

We have

$$f(x^2, y^2) = \frac{1}{\left(\Gamma\frac{1}{2}\right)^2 2} e^{-\frac{1}{2}(x^2+y^2)} (x^2)^{\frac{1}{2}-1} (y^2)^{\frac{1}{2}-1}$$
$$g(u,v) = \frac{1}{\left(\Gamma\frac{1}{2}\right)^2 2} e^{-u/2} \left(\frac{uv}{1+v}\right)^{\frac{1}{2}-1} \left(\frac{u}{1+v}\right)^{\frac{1}{2}-1} \frac{u}{(1+v)^2}$$
$$= \frac{1}{\Gamma\left(\frac{1}{2}+\frac{1}{2}\right) B\left(\frac{1}{2},\frac{1}{2}\right) 2} e^{-\frac{1}{2}u} u^{1-1} \frac{v^{\frac{1}{2}-1}}{(1+v)}$$
$$g(u,v) = \frac{1}{2} e^{-\frac{1}{2}u} \frac{1}{B\left(\frac{1}{2},\frac{1}{2}\right)} \frac{1}{\sqrt{v}(1+v)}; \ 0 < u < \infty, \ 0 < v < \infty$$
$$h(u) = \int_0^{\infty} \frac{1}{2} e^{-\frac{1}{2}u} \frac{1}{B\left(\frac{1}{2},\frac{1}{2}\right)} \frac{1}{\sqrt{v}(1+v)} du$$
$$= \frac{1}{2} e^{-\frac{1}{2}u} = \frac{1}{\Gamma(1) 2^1} e^{-\frac{1}{2}u} u^{1-1}, \ 0 < u < \infty$$

which is the p.d.f. of a gamma variate with parameters $\theta = 2$ and $m = 1$.

$$h(v) = \frac{1}{B\left(\frac{1}{2}, \frac{1}{2}\right)} \frac{v^{\frac{1}{2}-1}}{(1+v)} \int_0^\infty \frac{1}{2} e^{-\frac{1}{2}u} du = \frac{1}{B\left(\frac{1}{2}, \frac{1}{2}\right)} \frac{v^{\frac{1}{2}-1}}{(1+v)}$$

which is the p.d.f. of a beta variate of the second kind with parameters $m = \frac{1}{2}$ and $n = \frac{1}{2}$.

►**EXAMPLE 18.4** *A random sample of size n is drawn from a gamma distribution with parameters θ and m. Find the distribution of sample mean.*

Solution. We have

$$\begin{aligned} f(x) &= \frac{1}{\Gamma m \theta^m} e^{-x/\theta} x^{m-1}; \; 0 < x < \infty, \; m > 0 \\ \text{and } M_x(t) &= (1 - \theta t)^{-m}. \end{aligned}$$

Also

$$\begin{aligned} M_{\bar{x}}(t) &= M_{(x_1+x_2+\cdots+x_n)/n}(t) = M_{(x_1+x_2+\cdots+x_n)}(t/n) \\ &= M_{x_1}(t/n) M_{x_2}(t/n) \cdots M_{x_n}(t/n) \text{ [since } x_1, x_2, \ldots, x_n \text{ are independent]} \\ &= \left(1 - \frac{\theta t}{n}\right)^{-m} \left(1 - \frac{\theta t}{n}\right)^{-m} \left(1 - \frac{\theta t}{n}\right)^{-m} \cdots \left(1 - \frac{\theta t}{n}\right)^{-m} \\ &= \left[\left(1 - \frac{\theta t}{n}\right)^{-m}\right]^n = \left(1 - \frac{\theta t}{n}\right)^{-mn} = \left(1 - \frac{t}{\frac{n}{\theta}}\right)^{-mn} \end{aligned}$$

which is the m.g.f. of a gamma distribution. Thus

$$f(\bar{x}) = \frac{\left(\frac{n}{\theta}\right)^{mn}}{\Gamma mn} e^{-\frac{n\bar{x}}{\theta}} (\bar{x})^{mn-1}, \; 0 < \bar{x} < \infty$$

18.4 Truncated Gamma Distribution

Let us suppose that some values exceeding a fixed number (which is usually known) N is omitted in studying a standard gamma distribution having p.d.f.

$$f(x) = \frac{1}{\Gamma m} e^{-x} x^{m-1}, \; 0 < x < \infty$$

This is a case of truncation from above. Let us assume a new random variable X_1, where $0 < x_1 \leq N$. Then the p.d.f. of X_1 is given by

$$g(x_1) = \frac{\frac{1}{\Gamma m} e^{-x} x^{m-1}}{\int_0^N \frac{1}{\Gamma m} e^{-x} x^{m-1} dx} = \frac{e^{-x} x^{m-1}}{\int_0^N e^{-x} x^{m-1} dx}$$

The distribution of X_1 is called truncated gamma distribution. The r-th moment above origin of the distribution can be expressed in terms of Pearson's incomplete gamma function, where

$$\mu_r' = m^{(r)} \frac{I(N(m+r)^{-\frac{1}{2}}, m+r-1)}{I(N/\sqrt{m}, m-1)}.$$

18.5 Compound Gamma Distribution

Let X be a gamma variate with p.d.f.

$$f(x) = \frac{1}{\Gamma m \theta^m} e^{-x/\theta} x^{m-1};\ 0 < x < \infty,\ \theta > 0.$$

Let us also assume that $1/\theta$ follows gamma distribution, where

$$h\left(\frac{1}{\theta}\right) = \frac{\lambda^k}{\Gamma k} e^{-\frac{\lambda}{\theta}} \left(\frac{1}{\theta}\right)^{k-1};\ 0 < \frac{1}{\theta} < \infty,\ \lambda > 0.$$

Then the compound distribution of X is given by

$$\begin{aligned} g(x) &= \frac{\lambda^k}{\Gamma m \Gamma k} x^{m-1} \int_0^\infty \frac{1}{\theta^m} e^{-x/\theta} e^{-\frac{\lambda}{\theta}} \left(\frac{1}{\theta}\right)^{k-1} d\left(\frac{1}{\theta}\right) \\ &= \frac{\lambda^k \Gamma(m+k)}{\Gamma m \Gamma k} x^{m-1} (x+\lambda)^{-(m+k)},\ 0 \le x. \end{aligned}$$

Gamma-Gamma Distribution has also been considered by Bhattacharya (1966), where

$$f(x) = \frac{1}{\Gamma m \theta^m} e^{-x/\theta} x^{m-1};\ 0 < x < \infty,\ \theta > 0$$

$$\text{and } h(\theta) = \frac{1}{\Gamma \alpha \theta_1^\alpha} e^{-\theta/\theta_1} \theta^{\alpha-1};\ 0 < \theta < \infty,\ \theta_1 > 0.$$

The compound distribution of X is given by

$$g(x) = \frac{2}{\theta_1 \Gamma m \Gamma \alpha} \left(\frac{x}{\theta_1}\right)^{\frac{1}{2}(m+\alpha)-1} K_{\alpha-m}\left(2\sqrt{x/\theta_1}\right);\ (0 < \theta < \infty)$$

where $K_{\alpha-m}\left(2\sqrt{x/\theta_1}\right)$ is the modified Bessel function of the third kind.

18.6 Generalized Gamma Distribution

Let X be a random variable with p.d.f.

$$f(x) = \frac{1}{\Gamma m} e^{-x} x^{m-1};\ 0 < x < \infty,\ m > 0.$$

Let us take

$$X = \left[\frac{z-\alpha}{\theta}\right]^c,\ c > 0 \tag{18.4}$$

Then

$$h(z) = \frac{c(z-\alpha)^{cm-1}}{\theta^{cm} \Gamma m} \exp\left(-\frac{z-\alpha}{\theta}\right)^c;\ z > 0.$$

If $\alpha = 0$, we have

$$h(z) = \frac{c z^{cm-1}}{\theta^{cm} \Gamma m} \exp\left(-\frac{z}{\theta}\right)^c. \tag{18.5}$$

Stacy (1962) defined this distribution of z as the family of generalized gamma distribution. When $m = 1$, we have

$$h(z) = \frac{cz^{c-1}}{\theta^c} \exp\left(-\frac{z}{\theta}\right)^c \tag{18.6}$$

and is called *Weibull distribution.* Again, if $c = 1$ we have

$$h(z) = \frac{z^{m-1}}{\Gamma m \theta^m} \exp -\frac{z}{\theta}$$

which is the ordinary gamma distribution as has been given by (18.1). If $m = 1/2$, $c = 2$, $\alpha = 0$, we have

$$h(z) = \frac{2}{\theta \Gamma \frac{1}{2}} \exp\left(-\frac{z}{\theta}\right)^2$$

which is the p.d.f. of half-normal distribution.

Since $X = [(z-\alpha)/\theta]^c$ has a standard gamma distribution, we have

$$E[\{(z-\alpha)/\theta\}^r] = E\left[\{(z-\alpha)/\theta\}^{(\frac{r}{c})c}\right] = \Gamma\left(m + \frac{r}{c}\right)/\Gamma m. \tag{18.7}$$

In particular

$$\mu_1' = \Gamma\left(m + \frac{1}{c}\right)/\Gamma m, \ \mu_2' = \Gamma\left(m + \frac{2}{c}\right)/\Gamma m.$$

Other moments can also be deduced from (18.7). If $c = 1$, then

$$\begin{aligned} \mu_1' &= m, \ \mu_2' = m(m+1), \ \mu_3' = m(m+1)(m+2), \\ \mu_4' &= m(m+1)(m+2)(m+3). \end{aligned}$$

EXERCISE

1. (a) What is gamma variate. Show that, if $X \sim N(\mu, \sigma^2)$, then $Y = \frac{1}{2}\left(\frac{X-\mu}{\sigma}\right)^2$ is a gamma variate.
   (b) Derive moment generating function of gamma variate. Hence or otherwise comment on shape characteristics of gamma probability curve.
2. (a) If X and Y are two independent gamma variates, then find the p.d.f. of $U = X+Y, V = \frac{X}{X+Y}$ and $Z = X/Y$.
   (b) Find mean and variance of U and V.
   (c) If X is gamma variate with parameter m, find the p.d.f. of $U = 1/X$.
3. (a) Discuss the properties of gamma distribution.
   (b) If X and Y are two independent standard normal variates, then find the distribution of X^2, (i) $X^2 + Y^2$ and (ii) X^2/Y^2.
   (c) A random sample of n observations is selected from a gamma distribution with parameters θ and m. Find the distribution of sample mean.

CHAPTER **19**

Exponential Distribution

19.1 Definition

Let X be a continuous random variable which takes non-negative values. Let the p.d.f. of X be

$$f(x) = \begin{cases} \theta e^{-\theta x}, & x \geq 0 \\ 0, & \text{otherwise.} \end{cases} \tag{19.1}$$

Then X is said to have an exponential distribution with parameter $\theta > 0$. It is also called *one-parameter exponential distribution.* The more general form of this distribution is

$$f(x) = \theta e^{-\theta(x-a)};\ x > a,\ \theta > 0.$$

When $a = 0$ and $\theta = 1$, we have

$$f(x) = e^{-x};\ x \geq 0$$

which is the p.d.f. of standard exponential distribution. The graph of the distribution is shown below:

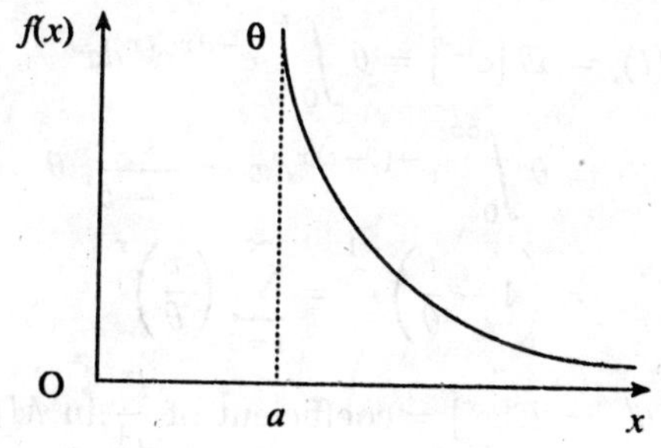

FIG. 19.1 : Graph of exponential density function.

The exponential distribution is applied in a wide variety of statistical analysis. One of the most widely used situations is to describe the events occurring 'at random in time'. For example, suppose that the future 'life time', X, of an individual has the same distribution at any point of its age. That is

$$P[X \leq x_0 + x | X > x_0] = P(X \leq x]$$

for all $x_0 > 0$ and all $x > 0$. If X is a continuous random variable with p.d.f. $f(x)$, then the conditional p.d.f. of X, given that $X \geq x_0$, is

$$f(x)/[1 - F(x_0)]; \; x > x_0 > 0.$$

Since the conditional distribution of the future lifetime $(X - x_0)$ is the same as the unconditional distribution of X, we have, say

$$\frac{f(x_0)}{1 - F(x_0)} = p_0.$$

Now, if $F(x_0) \neq 1, p_0 > 0$ and $F(x)$ satisfies the differential equation

$$\frac{dF(x)}{dx} = p_0[1 - F(x)]$$

where $1 - F(x) \propto e^{-p_0 x}$. If we introduce the condition

$$\lim_{x \to 0} F(x) = 0$$

we have

$$\begin{aligned} 1 - F(x) &= e^{-p_0 x} \\ F(x) &= 1 - e^{-p_0 x} = p_0 \int_0^x e^{-p_0 t} dt. \end{aligned}$$

Thus the p.d.f. of X is of the form (19.1) with $\theta = p_0$.

19.2 Moments and Generating Functions

The m.g.f. of exponential distribution is given by

$$\begin{aligned} M_x(t) &= E\left[e^{tx}\right] = \theta \int_0^\infty e^{-\theta x} e^{tx} dx \\ &= \theta \int_0^\infty e^{-(\theta - t)x} dx = \frac{\theta}{\theta - t}, \; \theta > t \\ &= \left(1 - \frac{t}{\theta}\right)^{-1} = \sum_{r=0}^{\infty} \left(\frac{t}{\theta}\right)^r. \end{aligned}$$

$$\begin{aligned} \therefore \; \mu'_r &= E[x^r] = \text{coefficient of } \frac{t^r}{r!} \text{ in } M_x(t) \\ &= \frac{r!}{\theta^r}; \; r = 1, 2, \ldots \end{aligned}$$

In particular

$$\mu'_1 = \frac{1}{\theta}, \; \mu'_2 = \frac{2}{\theta^2}, \; \mu'_3 = \frac{6}{\theta^3}, \; \mu'_4 = \frac{24}{\theta^4}.$$

Thus

$$\mu_2 = \mu_2' - \mu_1'^2 = \frac{2}{\theta^2} - \frac{1}{\theta^2} = \frac{1}{\theta^2}$$

$$\mu_3 = \mu_3' - 3\mu_2'\mu_1' + 2\mu_1'^3 = \frac{6}{\theta^3} - \frac{6}{\theta^3} + \frac{2}{\theta^3} = \frac{2}{\theta^3}$$

$$\mu_4 = \mu_4' - 4\mu_3'\mu_1' + 6\mu_2'\mu_1'^2 - 3\mu_1'^4 = \frac{24}{\theta^4} - \frac{24}{\theta^4} + \frac{12}{\theta^4} - \frac{3}{\theta^4} = \frac{9}{\theta^4}.$$

The moment ratios are

$$\alpha_3 = \sqrt{\beta_1} = \sqrt{\frac{\mu_3^2}{\mu_2^3}} = 2, \ \alpha_4 = \beta_2 = \frac{\mu_4}{\mu_2^2} = 9.$$

Characteristic Function

The characteristic function is given by

$$\psi(t) = E\left[e^{itx}\right] = \theta \int_0^\infty e^{-\theta x} e^{itx} dx$$

$$= \theta \int_0^\infty e^{-x(\theta - it)} dx$$

$$= \frac{\theta}{\theta - it} = \left(1 - \frac{it}{\theta}\right)^{-1}.$$

The cumulant generating function is given by

$$K_x(t) = \log M_x(t) = -\log\left(1 - \frac{t}{\theta}\right).$$

$$\therefore \ K_r = \frac{(r-1)!}{\theta^r}$$

In particular

$$K_1 = \frac{1}{\theta}, \ K_2 = \frac{1}{\theta^2}, \ K_3 = \frac{2}{\theta^3}, \ K_4 = \frac{6}{\theta^4}.$$

19.3 Properties of Exponential Distribution

If $\theta = 1$, then mean $=$ variance $= 1$ of the standard exponential distribution. Therefore, the standard exponential distribution is standardized one.

Median

Median is given by

$$\theta \int_0^M e^{-\theta x} dx = \frac{1}{2} \text{ or, } -\left[e^{-M\theta} - 1\right] = \frac{1}{2}$$

$$\text{or, } -e^{-M\theta} = -\frac{1}{2} \text{ or, } -M\theta = \log\frac{1}{2} \text{ or, } M = \frac{1}{\theta}\log 2.$$

Mode

The mode of the general form of exponential distribution is at the lowest value, a, of the range of variation.

Mean Deviation

For standardized exponential distribution the mean deviation is given by

$$\text{M.D.} = \int_0^{\infty} |x-1|e^{-x}dx = 2\int_1^{\infty} (x-1)e^{-x}dx = \frac{2}{e}.$$

Therefore

$$\frac{\text{mean deviation}}{\text{standard deviation}} = \frac{2}{e} = 0.736.$$

❐**THEOREM 19.1** *If $X_1, X_2, \ldots, X_n$ are independent random variables and if each of X_i $(i = 1, 2, \ldots, n)$ has an exponential distribution with parameter θ_i, then*

$$Z = \min(X_1, X_2, \ldots, X_n)$$

has exponential distribution with parameter $\sum_{i=1}^{n} \theta_i$.

Proof. We have

$$\begin{aligned} F(z) &= P(Z < z) = 1 - P(Z > z) \\ &= 1 - P[\min(X_1, X_2, \ldots, X_n) > z] \\ &= 1 - P[X_i > z \text{ for all } i = 1, 2, \ldots, n] \\ &= 1 - \prod_{i=1}^{n} P(X_i > z) \\ &= 1 - \prod_{i=1}^{n} [1 - P(X_i < z)] \\ &= 1 - \prod_{i=1}^{n} [1 - F_{X_i}(z)] \end{aligned}$$

where F_{X_i} is the distribution function of X_i.

$$\therefore\ F(z) = 1 - \prod_{i=1}^{n} 1 - (1 - e^{-\theta_i z})$$

$$= \begin{cases} 1 - \exp\left\{\left(-\sum_{i=1}^{n} \theta_i\right) z\right\}, & z > 0 \\ 0, & \text{otherwise.} \end{cases}$$

$$\therefore\ f(z) = \begin{cases} \sum_{i=1}^{n} \theta_i \exp\left\{\left(-\sum_{i=1}^{n} \theta_i\right) z\right\}, & z > 0 \\ 0, & \text{otherwise.} \end{cases}$$

Thus $Z = \min(X_1, X_2, \ldots, X_n)$ is an exponential variate with parameter $\Sigma\theta_i$.

If X_1 and X_2 are independent random variables with absolutely continuous distributions, and $\min(X_1, X_2)$ and $(X_1 - X_2)$ are mutually independent random variables, then both X_1 and X_2 have exponential distributions.

If X_1, X_2 have the same absolutely continuous distribution, then if the distribution of $X_1/(X_1 + X_2)$ is uniform over the range 0 to 1 and is independent of $(X_1 + X_2)$, the common distribution of X_1 and X_2 is exponential. This characteristic of exponential distribution has been shown in Example 26, Chapter 4.

19.4 Related Distributions

Let X be a random variable such that $Y = (X - a)^c$ $(X > a)$ has the exponential distribution with p.d.f.

$$f(y) = \theta e^{-\theta y}, \ y > 0. \tag{19.2}$$

Then X is said to have a Weibull distribution with shape parameter $c (c > -1)$.

If $Y = e^{X}$ has an exponential distribution with p.d.f. of the form (19.2), then X has the distribution of 'extreme value'.

In (19.1) if we assume $\theta = 1/2$, then

$$f(x) = \frac{1}{2} e^{-x/2}; \ x \geq 0$$

which is a chi-square distribution with 2 d.f.. Thus it can be shown that if $X_1, X_2, \ldots, X_n$ are independent random variables each with p.d.f.

$$f(x) = \theta e^{-\theta(x-a)}; \ x > a, \theta > 0$$

then $\bar{X}$ is distributed as $\left(a + \frac{1}{2n\theta}\right) \chi^2$ with $2n$ d.f..

Let $Z_1, Z_2, \ldots, Z_n$ be n independent standard normal variables. Let us define a quadratic form

$$Q = \sum_{i} \sum_{j}^{n} a_{ij} Z_i Z_j$$

where $A = [a_{ij}]$ is a matrix with eigenvalues $\lambda_1, \lambda_2, \ldots, \lambda_n$. Then Q can be written as

$$Q = \sum_{i}^{n} \sum_{j}^{n} a_{ij} Z_i Z_j = \sum_{j=1}^{n} \lambda_j Z_j^{*2}$$

by a linear transformation, where $Z_1^*, Z_2^*, \ldots, Z_n^*$ are independent standard normal variables. If the non-zero λ's are equal in pairs, then

$$\sum_{j=1}^{n} \lambda_j Z_j^{*2}$$

can be expressed as a linear function of independent variables each distributed as χ^2 with 2 d.f., and so exponentially.

Mixture of Exponential Distribution

Let $X_1, X_2, \cdots, X_n$ be n independent random variables each having p.d.f. of the form (19.1). Then the p.d.f. of

$$Y = \sum_{j=1}^{n} C_j X_j \ (C_j \neq C_k)$$

is given by

$$h(y) = \sum_{j=1}^{n} \left(\prod_{j \neq k} (C_j - C_k)^{-1} \right) C_j^{n-2} e^{-y/c_j}; \ (y > 0). \tag{19.3}$$

The distribution (19.3) is the general gamma or the general Erlang distribution. It is a special type of mixture of exponential distributions.

Truncated Exponential Distribution

Let X be a random variable with p.d.f.

$$f(x) = \theta e^{-\theta x}; \ x > 0, \ \theta > 0.$$

Let us assume that the values of X exceeding x_0 be omitted. Then the p.d.f. of truncated exponential distribution is given by

$$f_1(x) = \theta \left\{1 - e^{-\theta x_0}\right\}^{-1} e^{-\theta x}; \ x_0 > x > 0.$$

The mean of this distribution is

$$\frac{1}{\theta} \left[1 - \theta x_0 \left\{e^{\theta x_0} - 1\right\}^{-1}\right].$$

19.5 Double Exponential Distribution (Laplace Distribution)

A random variable X is said to follow standard Laplace distribution if its p.d.f. is given by

$$f(x) = \frac{1}{2} e^{-|x|}, \ -\infty < x < \infty. \tag{19.4}$$

The density function (19.4) has arisen in deriving the distribution of

$$Y = \frac{1}{2}(X_1 - X_2)$$

where X_1 and X_2 are two random variables each having p.d.f.

$$g(x) = \frac{1}{2} e^{-x/2}, \ x > 0.$$

The more general form of this distribution is

$$f(x) = \frac{1}{2\theta} e^{-\frac{1}{\theta}|x-\alpha|}, \ \theta > 0; \ -\infty < x < \infty. \tag{19.5}$$

The distribution (19.5) is called *two-parameter* [θ and α] *Laplace distribution* after the name of Laplace who discovered it in 1774. This is also known as the *first law of Laplace* [Laplace's second law of error is the normal distribution]. If we put $z = \frac{x-\alpha}{\theta}$ in (19.5), we get (19.4). The graph of the distribution is shown below:

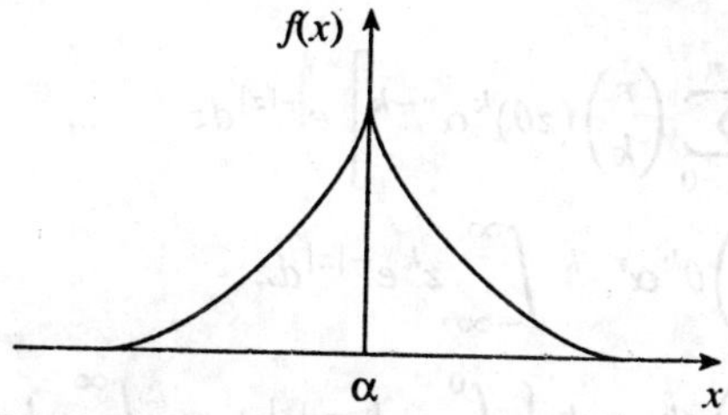

FIG. 19.2 : Graph of laplace density function.

Moments and Generating Functions

The moment generating function of standard Laplace distribution is given by

$$M_x(t) = E\left[e^{tx}\right] = \frac{1}{2}\int_{-\infty}^{\infty} e^{tx} e^{-|x|} dx$$

which on simplification gives

$$M_x(t) = \frac{1}{1-t^2}.$$

The cumulant generating function is $-\log(1-t^2)$. Therefore, the r-th cumulant is given by

$$K_r = 0, \text{ if } r \text{ is odd}$$
$$K_r = 2[(r-1)!], \text{ if } r \text{ is even.}$$

In particular

$$K_1 = \mu_1' = 0, \ K_2 = \mu_2 = 2, \ K_3 = \mu_3 = 0, \ K_4 = 12$$
$$\mu_4 = K_4 + 3\mu_2^2 = 24.$$

The r-th central moment is given by

$$\mu_r = 0, \text{ if } r \text{ is odd}$$
$$\mu_r = r!, \text{ if } r \text{ is even.}$$

The moment ratios are

$$\alpha_3 = \sqrt{\beta_1} = 0 \text{ and } \alpha_4 = \beta_4 = 6.$$

Hence the distribution is symmetrical about $x = 0$ and it is leptokurtic.

Moments about Origin

The r-th moment about origin of the general Laplace distribution is given by

$$\mu_r' = \frac{1}{2\theta}\int_{-\infty}^{\infty} x^r e^{-\frac{1}{\theta}|x-\alpha|}dx.$$

Putting $z = \frac{x-\alpha}{\theta}$, we have

$$\begin{aligned}
\mu_r' &= \frac{1}{2}\int_{-\infty}^{\infty} (z\theta+\alpha)^r e^{-|z|}dz \\
&= \frac{1}{2}\int_{-\infty}^{\infty}\left[\sum_{k=0}^{r}\binom{r}{k}(z\theta)^k\alpha^{r-k}\right]e^{-|z|}dz \\
&= \frac{1}{2}\sum_{k=0}^{r}\binom{r}{k}\theta^k\alpha^{r-k}\int_{-\infty}^{\infty} z^k e^{-|z|}dz \\
&= \frac{1}{2}\sum_{k=0}^{r}\binom{r}{k}\theta^k\alpha^{r-k}\left\{\int_{-\infty}^{0} z^k e^{-|z|}dz + \int_{0}^{\infty} z^k e^{-|z|}dz\right\} \\
&= \frac{1}{2}\sum_{k=0}^{r}\binom{r}{k}\theta^k\alpha^{r-k}\left\{(-1)^k\int_{0}^{\infty} z^k e^{-z}dz + \int_{0}^{\infty} z^k e^{-z}dz\right\} \\
&= \frac{1}{2}\sum_{k=0}^{r}\binom{r}{k}\theta^k\alpha^{r-k}\Gamma(k+1)\{(-1)^k+1\}.
\end{aligned}$$

In particular $\mu_1' = \alpha$, $\mu_2' = \alpha^2 + 2\theta^2$ and hence $\mu_2 = 2\theta^2$.

Characteristic Function

It is given by

$$\begin{aligned}
\psi(t) &= E\left[e^{itx}\right] \\
&= \frac{1}{2}\int_{-\infty}^{\infty} e^{itx}e^{-|x|}dx \\
&= \frac{1}{2}\left[\int_{-\infty}^{\infty}\cos tx e^{-|x|}dx + i\int_{-\infty}^{\infty}\sin tx e^{-|x|}dx\right] \\
&= \frac{1}{2}\cdot 2\int_{0}^{\infty}\cos tx e^{-|x|}dx.
\end{aligned}$$

Since the integrands in the first and second integrals are even and odd functions of x respectively,

$$\begin{aligned}
\therefore\ \psi(t) &= \int_{0}^{\infty}\cos tx e^{-|x|}dx \\
&= 1 - t^2\int_{0}^{\infty} e^{-x}\cos tx\, dx \\
&= 1 - \frac{t^2}{1+t^2} = \frac{1}{1+t^2}.
\end{aligned}$$

Properties of Laplace Distribution

The mean deviation is given by

$$\begin{aligned}\text{M.D.} &= E[|x|] = \frac{1}{2}\int_{-\infty}^{\infty} |x|e^{-|x|}dx \\ &= \frac{1}{2}\left[\int_{-\infty}^{0} |x|e^{-|x|}dx + \int_{0}^{\infty} |x|e^{-|x|}dx\right] \\ &= \frac{1}{2}\left[\int_{0}^{\infty} xe^{-x}dx + \int_{0}^{\infty} xe^{-x}dx\right] \\ &= \int_{0}^{\infty} xe^{-x}dx = 1.\end{aligned}$$

Therefore,

$$\frac{\text{mean deviation}}{\text{standard deviation}} = \frac{1}{\sqrt{2}} = 0.707.$$

Mode

From figure of Laplace density function it is observed that the probability density function of Laplace distribution has a maximum at $x = \alpha$, where there is a cusp.

Relation with Normal Distribution

Let Z_1, Z_2, Z_3 and Z_4 be independent standard normal variables. Let us define the statistic

$$D = \begin{vmatrix} Z_1 & Z_2 \\ Z_3 & Z_4 \end{vmatrix} = Z_1Z_4 - Z_2Z_3.$$

Nyquist *et al* (1954) showed that

$$h(d) = \frac{1}{4}e^{-\frac{|d|}{2}}; \ -\infty < d < \infty$$

which is the general Laplace distribution with $\theta = 2$ and $\alpha = 0$.

Compound Distribution

Let X be a random variable with p.d.f.

$$f(x) = \frac{1}{2\theta}e^{-\frac{1}{\theta}|x-\alpha|}; \ \theta > 0, \ -\infty < x < \infty$$

and α is distributed $N(\mu, \sigma^2)$. Then the compound distribution of X is given by

$$\begin{aligned}h(x) &= \frac{1}{2\theta}e^{\frac{1}{2}\left(\frac{\sigma}{\theta}\right)^2}\left[\phi\left(\frac{x-\mu}{\sigma} - \frac{\sigma}{\theta}\right)\exp\left(-\frac{x-\mu}{\theta}\right)\right. \\ &\left. +\phi\left(-\frac{x-\mu}{\sigma} - \frac{\sigma}{\theta}\right)\exp\left(\frac{x-\mu}{\theta}\right)\right]\end{aligned}$$

where

$$\phi(x) = \frac{1}{\sqrt{2\pi}} \int_{-\infty}^{x} \exp\left(-\frac{1}{2}t^2\right) dt.$$

Here α is considered as random variable.

The compound Laplace distribution can also be obtained assuming θ as random variable. Let the p.d.f. of $1/\theta$ be

$$g(1/\theta) = \frac{1}{\Gamma m \beta^m} e^{-\frac{1}{\beta\theta}} \theta^{-(m-1)};\ 0 < 1/\theta < \infty.$$

Then the p.d.f. of compound Laplace distribution is given by

$$g(x) = \frac{m}{2\beta} \left[1 + |x-\alpha|/\beta\right]^{-(m+1)}.$$

EXERCISE

1. (a) Write down some examples of exponential variable. Why exponential distribution is known as distribution of 'Lack of Memory'?

   (b) Show that exponential distribution is positively skewed and leptokurtic.

   (c) The average time needed to get a serive in a bank is 15 minutes. Find the probability that a man will be served (i) by 15 minutes, (ii) after 20 minutes, (iii) within 15 to 20 minutes. If the man is in queue for half-an-hour, what is the probability that he will be served within 40 minutes?

2. (a) Find moments of the exponential distribution. Hence or otherwise find β_1 and β_2 of the distribution.

   (b) If $X_1, X_2, \ldots, X_n$ are independent exponential variables with parameters $\theta_i (i = 1, 2, \ldots, n)$, then $z = \min(X_1, X_2, \ldots, X_n)$ has exponential distribution with parameter $\sum_{i=1}^{n} \theta_i$.

   (c) The average time needed to get a service in a mobile repairing shop is half-an-hour. Find the probability that a service will be finished (i) within an hour, (ii) after an hour, (iii) within 45 to 70 minutes. If a customer is in queue for an hour, what is the probability that his service will be finished within one anf half-an-hour?

CHAPTER **20**

Weibull Distribution

20.1 Definition

If the random variable

$$Y = \left(\frac{X-\mu}{\lambda}\right)^c$$

has exponential distribution with p.d.f.

$$f(y) = e^{-y},\ y > 0$$

the random variable X has Weibull distribution with parameters $c > 0$, $\lambda > 0$ and μ. The p.d.f. of X is

$$g(x) = \frac{c}{\lambda}\left(\frac{x-\mu}{\lambda}\right)^{c-1} \exp\left[-\left(\frac{x-\mu}{\lambda}\right)^c\right],\ x > \mu. \tag{20.1}$$

The shape of the curve of this distribution is shown in FIG. 20.1.

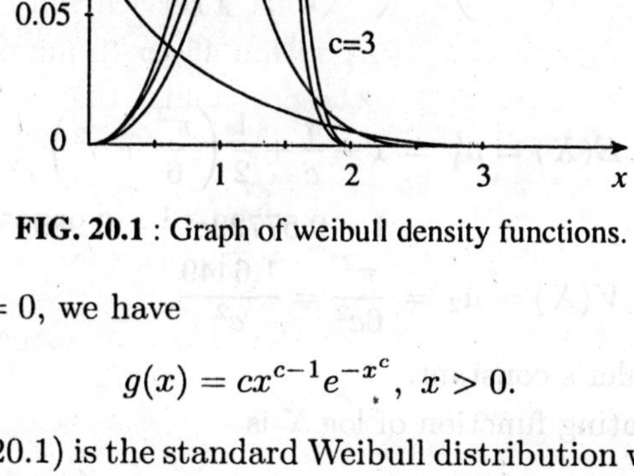

FIG. 20.1 : Graph of weibull density functions.

Putting $\lambda = 1$ and $\mu = 0$, we have

$$g(x) = cx^{c-1}e^{-x^c},\ x > 0. \tag{20.2}$$

The density function (20.1) is the standard Weibull distribution with single parameter c.

The distribution is named after Waloddi Weibull, a Swedish physicist, who used it in 1939 to represent the distribution of the breaking strength of materials, Kao (1958–59) advocated the use of this distribution in reliability and quality control works. The distribution is often suitable where the conditions of 'strict randomness' of the exponential distribution are not satisfied. Thus, this distribution adds flexibility where an exponential distribution might possibly be adequate. However, there is no explicit theoretical reasoning indicating that a Weibull distribution should be used.

The cumulative distribution function of Weibull distribution is

$$F(x) = 1 - \exp\left[-\left\{\frac{x-\mu}{\lambda}\right\}^c\right]. \tag{20.3}$$

This explicit form of cumulative distribution function makes Weibull distribution especially suitable to be used as a tolerance distribution in the analysis of quantal response data. The distribution belongs to the family of generalized gamma distribution (18.3).

20.2 Moments and Generating Function

For standard Weibull distribution we have

$$\begin{aligned} Y &= X^c \; [\because \; \lambda = 1, \; \mu = 0] \\ \Rightarrow X &= Y^{1/c}. \end{aligned}$$

The r-th moment about origin is

$$\begin{aligned} \mu'_r &= E(X^r) = E\left[\left(Y^{1/c}\right)^r\right] = E\left[Y^{r/c}\right] \\ &= \int_0^\infty y^{r/c} e^{-y} dy = \Gamma\left(\frac{r}{c}+1\right). \end{aligned}$$

In particular

$$\begin{aligned} \mu'_1 &= \Gamma\left(\frac{1}{c}+1\right), \; \mu'_2 = \Gamma\left(\frac{2}{c}+1\right), \; \mu'_3 = \Gamma\left(\frac{3}{c}+1\right), \; \mu'_4 = \Gamma\left(\frac{4}{c}+1\right) \\ \mu_2 &= \mu'_2 - \mu'^2_1 = \Gamma\left(\frac{2}{c}+1\right) - \left(\Gamma\left(\frac{1}{c}+1\right)\right)^2. \end{aligned}$$

For large c

$$\begin{aligned} E(X) = \mu'_1 &\doteq 1 - \frac{\gamma}{c} + \frac{1}{2}\left(\frac{\pi^2}{6} + \gamma^2\right)\frac{1}{c^2} \\ &= 1 - 0.57722c^{-1} + 0.98905c^{-2} \\ \text{and } V(X) = \mu_2 &= \frac{\pi^2}{6c^2} = \frac{1.6449}{c^2}. \end{aligned}$$

Here $\gamma = 0.57722$ is Euler's constant.

The moment generating function of $\log X$ is

$$M_{\log x}(t) = E\left[e^{t \log X}\right] = E\left[X^t\right] = \Gamma\left(\frac{t}{c}+1\right).$$

Hence the r-th cumulant of $\log X$ is

$$K_r(\log X) = \left(\frac{1}{c}\right)^r \psi^{(r)}(1)$$

where $\psi(1) = -\gamma$;

$$\begin{aligned}\psi\left(\frac{1}{2}\right) &= -\gamma - 2\log 2 \\ &= -1.96351 \\ \text{and } \psi^{(r)}(\alpha) &= \frac{d^r}{d\alpha^r}\{\psi(\alpha)\}.\end{aligned}$$

In particular

$$\begin{aligned}E[\log X] &= -\frac{\gamma}{c} \\ V(\log X) &= \frac{\pi^2}{6c^2}.\end{aligned}$$

20.3 Properties

For $c > 1$, the p.d.f. tends to zero as x tends to zero, and there is a single mode at

$$x = \lambda[(c-1)/c]^{1/c} + \mu.$$

This value tends to $(\mu + \lambda)$ very rapidly as c tends to infinity. For $0 < x < 1$, the mode is at zero, and $f(x)$ is a decreasing function of x for all $x > \mu$. The median is

$$\lambda[\log 2]^{1/c} + \mu.$$

It has already been observed that the moments vary with the variation of c, so the values of β_1 and β_2. Johnson and Kotz (1970) have given the values of $\sqrt{\beta_1}$ and β_2 for different values of c. For $c = 3.6$ (approx.) $\sqrt{\beta}$ is zero and $\beta_2 = 2.72$. Thus, for values of c in the neighbourhood of 3.6, the Weibull distribution is similar in shape to a normal distribution. For $c > 3.6$ the distribution becomes negatively skewed and platykurtic. For $c < 3.6$ the distribution becomes positively skewed. The value of $\beta_2 = 2.71$ is minimum when $c = 3.35$ (approx.).

Characterization

If $X_1, X_2, \ldots, X_n$ are independent and identically distributed random variables; then $Y = \min(X_1, X_2, \ldots, X_n)$ has a Weibull distribution if and only if the common distribution of the X_i's is a Weibull distribution.

Proof. Suppose $X_1, X_2, \ldots, X_n$ are independent and identically distributed random variables each with Weibull density function

$$f(x) = \frac{c}{\lambda}\left(\frac{x-\mu}{\lambda}\right)^{c-1} \exp\left[-\left(\frac{x-\mu}{\lambda}\right)^c\right], \quad x > \mu.$$

Then

$$\begin{aligned}P(Y > y) &= P[\min(X_1, X_2, \ldots, X_n) > y] \\ &= P\left[\bigcap_{i=1}^{n} X_i > y\right] = \prod_{i=1}^{n} P(X_i > y) \\ &= \{P[X_i > y]\}^n \ [\because X_i\text{'s are independent}].\end{aligned}$$

Now

$$\begin{aligned}P(X_i > y) &= \int_y^\infty \frac{c}{\lambda}\left(\frac{x-\mu}{\lambda}\right)^{c-1} \exp\left[-\left(\frac{x-\mu}{\lambda}\right)^c\right] \\ &= \int_{\left(\frac{y-\mu}{\lambda}\right)^c}^\infty e^{-t} dt \ \left[t = \left(\frac{x-\mu}{\lambda}\right)^c\right] \\ &= \exp\left[-\left(\frac{y-\mu}{\lambda}\right)^c\right].\end{aligned}$$

$$\begin{aligned}\therefore\ P(Y > y) &= \left\{\exp\left[-\left(\frac{y-\mu}{\lambda}\right)^c\right]\right\}^n = \exp\left[-n\left(\frac{y-\mu}{\lambda}\right)^c\right] \\ &= \exp\left[-\left\{\frac{n^{1/c}(y-\mu)}{\lambda}\right\}^c\right].\end{aligned}$$

Hence Y has the same Weibull distribution as each X_i, except that is due to Dubey (1966).

If X has a Weibull distribution with $\mu = c$, then the p.d.f. of $Y = -c\log(X/\lambda)$ is

$$h(y) = e^{-y} e^{-e^{-y}}$$

which is a form of extreme value distribution.

Let $\theta = (1/\lambda)^c$, then (20.1) can be written as

$$g(x) = c\,\theta\, x^{c-1} \exp\left[-\theta x^c\right],\ c > 0.$$

If c is fixed, but λ varies so that the newly defined parameter θ be assumed to have gamma probability density function

$$h_\theta(t) = \frac{1}{\Gamma p}\delta^p t^{p-1} e^{-\delta t};\ \theta > 0;\ p, \delta > 0.$$

Then the p.d.f. of X is

$$\begin{aligned}g(x) &= \frac{\delta^p c x^{c-1}}{\Gamma p}\int_0^\infty t^p \exp\left[-t(x^c+\delta)\right] dt \\ &= p\delta^p c x^{c-1}(x^c+\delta)^{-(p+1)};\ x > 0. \qquad (20.4)\end{aligned}$$

CHAPTER 21

Gumbel Distribution

21.1 Definition and Applications

A random variable X has a Gumbel distribution if its probability density function is of the form

$$f(x) = \frac{1}{\lambda} e^{-\frac{1}{\lambda}(x-\mu)} \exp\left[-e^{-\frac{1}{\lambda}(x-\mu)}\right];\ x > \mu \tag{21.1}$$

where $\lambda > 0$ and $\mu > 0$ are two parameters of the distribution. The standard form of the distribution is obtained by putting $\mu = 0$ and $\lambda = 1$ or equivalently considering the distribution of

$$Y = (X - \mu)/\lambda$$

we have

$$f(y) = \exp(-y - e^{-y});\ y > 0. \tag{21.2}$$

The shape of the curve of this distribution is shown in FIG. 21.1.

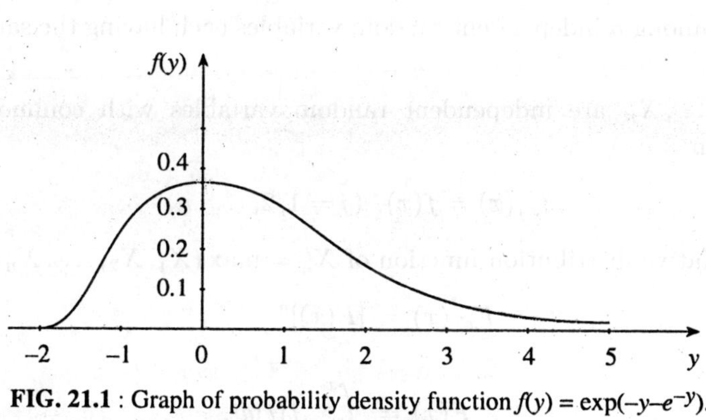

FIG. 21.1 : Graph of probability density function $f(y) = \exp(-y-e^{-y})$.

The distribution (21.1) belongs to the families of

$$\text{Type 1} : \left[P(X \le x) = \exp\left\{-e^{-(x-\mu)/\lambda}\right\}\right]$$

distribution of 'extreme value'. The other two types of families of 'extreme value' distribution are

$$\text{Type 2}: P[X \le x] = \begin{cases} 0, & (x < \mu) \\ \exp\left\{-\left(\dfrac{x-\mu}{\lambda}\right)^{-k}\right\}, & (x \ge \mu) \end{cases}$$

$$\text{Type 3}: P[X \le x] = \begin{cases} \exp\left\{-\left(\dfrac{x-\mu}{\lambda}\right)^{k}\right\}, & (x \le \mu) \\ 1, & (x > \mu) \end{cases}$$

where μ, $\theta > 0$ and $k > 0$ are parameters.

The corresponding distributions of $(-X)$ are also called *extreme value distributions.* The Type 3 distribution of $(-X)$ is a Weibull distribution. The distributions of Type 2 and Type 3 can be transformed to Type 1 distribution by the simple transformations

$$Z = \log(X - \mu); \; Z = -\log(\mu - X)$$

respectively. Due to the use of these simple formulas relating to Z and X, Types 1 and 2 are closely related to the Weibull distribution. Type 1 distribution is sometimes called 'double exponential' distribution.

Extreme value theory seems to have first arisen mainly in connection with the needs of astronomers in utilizing outlying observations. The theoretical developments of the distributions were followed by practical applications using extreme value statistics of distributions of human lifetimes, radioactive emissions, and strength of materials. There are many other fields of applications of extreme value distributions. Some of these are rainfall, flood flows, earthquakes, general meteorological data, aircraft load, corrosion and microorganism survival times.

The distributions can be obtained as limiting distributions (as $n \to \infty$) of the greatest value among n independent random variables each having the same continuous distribution.

If $X_1, X_2, \ldots, X_n$ are independent random variables with common probability density function

$$f_{x_j}(x) = f(x), \; (j = 1, 2, \ldots, n)$$

then the cumulative distribution function of $X'_n = \max(X_1, X_2, \ldots, X_n)$ is

$$F_{x'_n}(x) = [F(x)]^n \tag{21.3}$$

where

$$F(x) = \int_{-\infty}^{x} f(t)dt.$$

As n tends to infinity, it is clear that for any fixed value x

$$\lim_{n\to\infty} F_{x'_n}(x) = \begin{cases} 1, & \text{if } F(x) = 1 \\ 0, & \text{if } F(x) < 1. \end{cases}$$

Let us find the limiting distribution of some sequence of 'reduced' values such as $(a_n X'_n + b_n)$, where a_n, b_n may depend on n, but not on x. Let us denote the limiting cumulative distribution of the 'reduced' greatest value by $G(x)$, where

$$[G(x)]^N = G(a_N x + b_N) \tag{21.4}$$

since the greatest of Nn values $X_1, X_2, \ldots, X_{Nn}$ is also the greatest of the N values, i.e.,

$$\max\left(X_{(j-1)n+1},\ X_{(j-1)n+2},\ \ldots,\ X_{jn}\right);\ j = 1, 2, \ldots, N.$$

Type 1 distribution is obtained by taking $a_N = 1$; Types 2 and 3 by taking $a_N \neq 1$. In this latter case

$$x = a_N x + b_N \text{ if } x = b_N(1 - a_N)^{-1}$$

and from (21.4) it follows that $G\left[b_N(1-a_N)^{-1}\right]$ must equal either 1 or 0. Type 2 corresponds to 1, Type 3 to 0.

Let us now consider the case $a_N = 1$. Then

$$[G(x)]^N = G(x + b_N). \tag{21.5}$$

Also

$$[G(x)]^{NM} = [G(x + b_N)]^M = G(x + b_N + b_M).$$

But from (21.4)

$$[G(x)]^{NM} = G(x + b_{NM}).$$

Also, we have $b_N + b_M = b_{NM}$ whence

$$b_N = \lambda \log N \ (\lambda \text{ a constant}). \tag{21.6}$$

Taking logarithms of (21.5) twice and inserting the value of b_N from (21.6), we have

$$\log N + \log(-\log G(x)) = \log(-\log G(x + \lambda \log N)). \tag{21.7}$$

In other words, when the argument of

$$h(x) = \log(-\log G(x))$$

increases by $\lambda \log N$, $h(x)$ decreases by $\log N$. Hence

$$h(x) = h(0) - x/\lambda. \tag{21.8}$$

Since $h(x)$ increases as x increases, $\lambda > 0$. From (21.8)

$$-\log G(x) = \exp[-(x - \lambda h(0))/\lambda] = \exp[-(x-\mu)/\lambda]$$

where

$$\mu = \lambda \log(-\log G(0)).$$

Hence

$$G(x) = \exp\left[-e^{-(x-\mu)/\lambda}\right]$$

in agreement with Type 1.

Although the distributions are called 'extreme value', it is to be borne in mind that (i) they do not represent distributions of all kinds of 'extreme values' and (ii) they can be used empirically in the same way as other distributions. By replacing X by $-X$, limiting distributions of least values can also be obtained.

21.2 Moments and Properties

The standard form of the distribution $[\mu = 0,\ \lambda = 1]$ is

$$f(y) = \exp\left[-y - e^{-y}\right]$$

where

$$Y = (X - \mu)/\lambda.$$

Let

$$Z = \exp\left[-(X - \mu)/\lambda\right] = e^{-Y}.$$

Then

$$f(z) = e^{-z},\ (z \geq 0)$$

is the density function of exponential distribution. From this we have the moment generating function of Y

$$M_y(t) = E\left[e^{t(X-\mu)/\lambda}\right] = E\left[z^{-t}\right] = \Gamma(1-t);\ t < 1.$$

Replacing t by λt, we have the moment generating function of X

$$M_x(t) = E\left[e^{tX}\right] = e^{t\mu}\Gamma(1 - \lambda t),\ (\theta|t| < 1).$$

The cumulant generating function of X is

$$K_x(t) = \mu t + \log\Gamma(1 - \theta t).$$

Therefore

$$K_1 = \mu_1' = \mu - \lambda\psi(1) = \mu + \gamma\lambda = \mu + 0.57722\lambda$$

In general

$$K_r = (-\lambda)^r \psi^{(r-1)}(1),\ (r \geq 2)$$

In particular

$$K_2 = \mu_2 = \frac{1}{6}\pi^2\lambda^2 = 1.64493\lambda^2.$$

The moment ratios are

$$\alpha_3^2 = \beta_1 = 1.29857; \ \alpha_4 = \beta_2 = 5.4.$$

The distribution of X is positively skewed and leptokurtic.

The distribution is unimodal and the mode is at $X = \mu$. The median $X_{0.5}$ satisfies the equation

$$\exp\left[-\exp\{-(X_{0.5} - \mu)/\lambda\}\right] = 0.5$$

which on simplification gives

$$X_{0.5} = \mu - \lambda \log\log 2 \doteq \mu + 0.36611\lambda.$$

If $Y_1 \leq Y_2 \leq \cdots \leq Y_n$ are the order statistics corresponding to n random variables having the common distribution

$$f(y) = \exp(-y - e^{-y})$$

then the p.d.f. of Y_r is

$$h(y_r) = \frac{n!}{(r-1)!(n-r)!} e^{-y} \times \left[\exp\left(-re^{-y}\right)\right] \sum_{j=0}^{n-r} (-1)^j \binom{n-r}{j} \exp\left(-je^{-y}\right).$$

CHAPTER 22

Cauchy Distribution

22.1 Definition and Origin

A random variable X is said to have a standard Cauchy distribution if its p.d.f. is given by

$$f(x) = \frac{1}{\pi(1+x^2)}, \quad -\infty < x < \infty. \tag{22.1}$$

The general form of this distribution has the p.d.f..

$$f(x) = \frac{1}{\pi\lambda\left[1+\left(\frac{x-\mu}{\lambda}\right)^2\right]}; \ \lambda > 0, \ -\infty < x < \infty. \tag{22.2}$$

Here the parameters μ and λ are location and scale parameters respectively. The density function (22.2) is a special form of Pearson Type VII distribution.

The Cauchy distribution is obtained as the limiting distribution of

$$n^{-1}\sum_{j=1}^{n} X_j \text{ as } n \to \infty$$

where $X_1, X_2, \ldots$ are independent identically distributed random variables with common density function $P_x(x)$ satisfying the conditions

(i) $P_x(0) = 0$;

(ii) $P_x(x)$ continuous at $x = 0$;

(iii) $P_x(x)$ possesses left-hand and right-hand derivatives at $x = 0$.

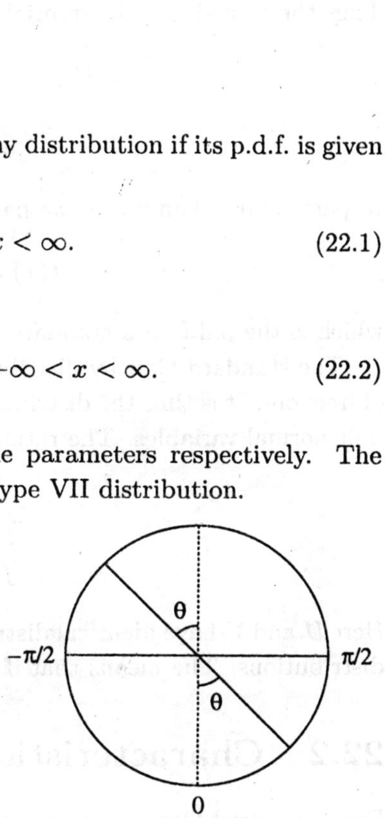

FIG. 22.1

The distribution also arises in describing the distribution of the point of intersection of a fixed straight line. Let us consider a roulette wheel in which the probability of the pointer stopping at any part of the circumference is constant. The situation is represented diagrammatically in FIG. 22.1. We want the probability of θ in the interval $-\pi/2$ to $\pi/2$. The probability for any value of θ in the range $[-\pi/2, \pi/2]$ is constant and consequently θ is a rectangular variate with probability differential given by

$$dP(\theta) = \begin{cases} \frac{1}{\pi}d\theta, & -\pi/2 \leq \theta \leq \pi/2 \\ 0, & \text{otherwise.} \end{cases}$$

Let us now transform to the variable X with the transformation

$$x = r\tan\theta$$

where

$$dx = r\sec^2\theta d\theta; \ -\infty < x < \infty.$$

Thus the probability differential of X is

$$\begin{aligned} dF(x) &= \frac{1}{\pi}\frac{dx}{r\sec^2\theta} = \frac{1}{\pi}\frac{dx}{r\{1+x^2/r^2\}} \\ &= \frac{r}{\pi}\frac{dx}{r^2+x^2}; \ -\infty < x < \infty. \end{aligned}$$

In particular, when $r = 1$, we have

$$f(x) = \frac{1}{\pi}\frac{1}{1+x^2}, \ -\infty < x < \infty$$

which is the p.d.f. of a standard Cauchy variate.

The standard Cauchy distribution is the distribution of central 't' with one degree of freedom. It is thus the distribution of the ratio U/V, where U and V are independent unit normal variables. The ratio U/V has also a Cauchy distribution if

$$\begin{aligned} f(u) &= \frac{\sqrt{2}}{\pi}\left(1+u^4\right)^{-1} \\ f(v) &= \frac{2}{\pi}\left(1+v^4\right)^{-1}. \end{aligned}$$

Here U and V have identical distributions and hence U/V and V/U must have identical distributions. The means that if X has Cauchy distribution so does X.

22.2 Characteristic Function

For a standard Cauchy variate X the characteristic function is given by

$$\psi(t) = \frac{1}{\pi}\int_{-\infty}^{\infty}\frac{e^{itx}}{1+x^2}dx. \tag{22.3}$$

To find the value of $\psi(t)$ let us consider the characteristic function of Laplace distribution

$$f(z) = \frac{1}{2}e^{-|z|}, -\infty < z < \infty$$

where

$$\psi_1(t) = E\left[e^{itz}\right] = \frac{1}{1+t^2}.$$

Since $\psi_1(t)$ is absolutely integrable in $(-\infty, \infty)$, we have by inversion theorem

$$\frac{1}{2}e^{-|z|} = f(z) = \frac{1}{2\pi}\int_{-\infty}^{\infty} e^{-itz}\psi(t)dt = \frac{1}{2\pi}\int_{-\infty}^{\infty} \frac{e^{-itz}}{1+t^2}dt.$$

Thus, we have

$$e^{-|z|} = \frac{1}{\pi}\int_{-\infty}^{\infty} \frac{e^{-itz}}{1+t^2}dt = \frac{1}{\pi}\int_{-\infty}^{\infty} \frac{e^{itz}}{1+t^2}dt \text{ [changing } t \text{ to } -t].$$

On interchanging t and z, we have

$$e^{-|t|} = \frac{1}{\pi}\int_{-\infty}^{\infty} \frac{e^{itz}}{1+z^2}dz. \tag{22.4}$$

Thus from (22.1) and (22.2), we have

$$\psi(t) = e^{-|t|}. \tag{22.5}$$

For general Cauchy distribution with parameters μ and λ, the characteristic function is

$$\psi(t) = e^{i\mu t - \lambda|t|}.$$

This can be shown as follows:

Let

$$Y = \frac{X-\mu}{\lambda} \Rightarrow X = \mu + \lambda Y.$$

Then

$$\begin{aligned}\psi_X(t) &= E\left[e^{itX}\right] = E\left[e^{it(\mu+\lambda Y)}\right] = e^{it\mu}E\left[e^{it\lambda Y}\right]\\ &= e^{it\mu}\psi_Y(t\lambda) = e^{it\mu}e^{-\lambda|t|} = e^{it\mu-\lambda|t|},\ \lambda > 0.\end{aligned}$$

22.3 Properties

Since $\psi'(t)$ in (22.3) does not exist at $t = 0$, the mean of the Cauchy distribution does not exist. The variance of the distribution is given by

$$\mu_2 = \int_{-\infty}^{\infty} (x-\mu)^2 \frac{1}{\pi\lambda\left[1+\left(\frac{x-\mu}{\lambda}\right)^2\right]}dx.$$

But this integral is not convergent and hence μ_2 does not exist. In general, $\mu_r (r \geq 2)$ for Cauchy distribution does not exist.

Already it has been mentioned that μ and λ are the location and scale parameters respectively. If we conventionally agree to assume that the mean of Cauchy distribution exists, then it is located at $x = \mu$. Also, obviously, the probability curve is symmetrical about the point $x = \mu$. Hence the mean, median and mode of this distribution coincide at the point $x = \mu$. The upper and lower quartiles are $\mu \pm \lambda$. The probability density function has points of inflexion at $\mu \pm \lambda/\sqrt{3}$.

❐ **THEOREM 22.1** *If $X_1, X_2, \ldots, X_n$ are independent random variables with probability density functions*

$$P_{x_j}(x) = \frac{1}{\pi\lambda_j}\frac{1}{\left[1+\left(\frac{x-\mu_j}{\lambda_j^2}\right)\right]};\ \lambda_j > 0,\ j = 1, 2, \ldots, n$$

then $S = \sum_{j=1}^{n} X_j$ has a Cauchy probability density function.

Proof. We have

$$\psi_{x_j}(t) = e^{it\mu_j - \lambda_j|t|}.$$

Again

$$\psi_{\sum_{j=1}^{n} X_j}(t) = \prod_{j=1}^{n}\psi_{x_j}(t) = \prod_{j=1}^{n} e^{it\mu_j-\lambda_j|t|} = \exp\left[it\sum_{j=1}^{n}\mu_j - |t|\sum_{j=1}^{n}\lambda_j\right]$$

which is the characteristic function of Cauchy distribution with parameters

$$\sum^{n}\mu_j \text{ and } \sum^{n}\lambda_j.$$

More generally

$$\sum_{j=1}^{n} a_j X_j$$

has a Cauchy distribution with parameters

$$\lambda = \sum_{j=1}^{n}|a_j|\mu_j \text{ and } \mu = \sum_{j=1}^{n} a_j\mu_j.$$

In particular, if

$$\lambda_1 = \lambda_2 = \cdots = \lambda_n = \lambda$$
$$\mu_1 = \mu_2 = \cdots = \mu_n = \mu$$
$$a_1 = a_2 = \cdots = a_n = \frac{1}{n}$$

then

$$\sum_{j=1}^{n} a_j X_j = \frac{1}{n}\Sigma X_j = \bar{X}$$

is the mean of n independent random variables each having Cauchy distribution. Therefore, the mean of Cauchy distribution has also a Cauchy probability density function.

If X_1 and X_2 are independent, and each has a standard Cauchy distribution, then $(a_1X_1 + a_2X_2)$ has a Cauchy distribution and it is distributed as $(|a_1| + |a_2|)X_1$.

There is no standardized form of Cauchy distribution since standardization cannot be made without using (finite) values of mean and standard deviation which do not exist.

22.4 Characterization and Related Distributions

If X_1 and X_2 be two independent random variables with common probability density function

$$f_X(x) = \frac{1}{\sqrt{2\pi}} e^{-\frac{1}{2}x^2}, \ -\infty < x < \infty$$

then the distribution of X_1/X_2 is Cauchy.

From the given condition the joint density function of X_1 and X_2 is

$$f(x_1, x_2) = \frac{1}{2\pi} e^{-\frac{1}{2}(x_1^2+x_2^2)}.$$

Let us make the transformation $u = x_1/x_2$ and $v = x_2$ so that $x_1 = uv$ and $x_2 = v$. The Jacobian of the transformation is $J = v$. Hence the joint p.d.f. of U and V is

$$\begin{aligned} g(u,v) &= \frac{1}{2\pi} \exp\{-(u^2v^2 + v^2)/2\}|J| \\ &= \frac{1}{2\pi} \exp\{-(1+u^2)v^2/2\}|v|, \ -\infty < (u,v) < \infty. \end{aligned}$$

The marginal p.d.f. of U is

$$\begin{aligned} h(u) &= \frac{1}{2\pi} \int_{-\infty}^{\infty} \exp\{-(1+u^2)v^2/2\} v dv \\ &= \frac{2}{2\pi} \int_0^{\infty} \exp\{-(1+u^2)v^2/2\} v dv \\ &= \frac{1}{\pi} \int_0^{\infty} e^{-t} \frac{dt}{(1+u^2)}, \ \left[\frac{v^2}{2}(1+u^2) = t\right] \\ &= \frac{1}{\pi(1+u^2)} \left|-e^{-t}\right|_0^{\infty} \\ &= \frac{1}{\pi(1+u^2)}, \ -\infty < u < \infty \end{aligned}$$

which is the p.d.f. of a standard Cauchy distribution.

If X and Y are independent and have the same distribution, then a necessary and sufficient set of conditions for this to be a Cauchy distribution is (i) the characteristic function $\psi(t) = E\left[e^{itX}\right]$ has a finite non-zero right-hand derivative at $t = 0$, i.e., $\psi_t'(0) = \lambda \neq 0$, and (ii) for any pair of positive real numbers a, b there is a positive real number c (depending on a and b) such that $aX + bY$ has the same distribution of cX(or cY).

Another characterization of Cauchy distribution is that, if $(1 + aX)(a - X)^{-1}$ has the same distribution as X, and $\pi^{-1}\tan^{-1} a$ is not a rational number, then X has a standard Cauchy distribution.

A related characterization is that, if $(\mu^2 + \lambda^2)(2\mu - X)^{-1}$ has the same distribution as X and $\pi^{-1}\tan^{-1}(\mu/\lambda)$ is not a rational number, then X has a Cauchy distribution with parameters μ, λ.

Generalized Cauchy Distribution

A random variable X with p.d.f.

$$f(x) = \frac{k\Gamma h}{2\lambda\Gamma\frac{1}{k}\Gamma\left(h-\frac{1}{k}\right)} \frac{1}{\left[1+\left|\frac{(x-\mu)}{\lambda}\right|^k\right]^h};\ \lambda, k, h > 0$$

has a generalized Cauchy distribution. For $k = 2$ and $\mu = 0$, X is distributed as $\lambda(2h-1)^{-1/2}$ times a central t variable with $(2h-1)$ degrees of freedom, and for $k = 2$, $h = 1$, X is distributed as Cauchy distribution.

CHAPTER **23**

Logistic Distribution

23.1 Definition and Applications

A random variable X is said to have a logistic distribution if its probability density function is given by

$$f(x) = \frac{\exp[-(x-\alpha)/\beta]}{\beta[1+\exp\{-(x-\alpha)/\beta]^2} \tag{23.1}$$

$$= \frac{1}{4\beta}\operatorname{sech}^2\left\{\frac{1}{2}(x-\alpha)/\beta\right\}, \quad -\infty < x < \infty. \tag{23.2}$$

The cumulative distribution function is given by

$$F(x) = [1+\exp\{-(x-\alpha)/\beta\}]^{-1}$$
$$= \frac{1}{2}\left[1+\tanh\left\{\frac{1}{2}(x-\alpha)/\beta\right\}\right].$$

This cumulative logistic distribution can be fitted for fitting the logistic as a growth curve with x representing time. The use of the logistic function as a growth curve can be based on the differential equation

$$\frac{dP}{dt} = \Pr(1+KP) \tag{23.3}$$

where P is the population size at time t and the size $P+\Delta P$ at time $t+\Delta t$. The rate of increase of the population at time t is

$$\frac{dP}{dt} = \lim_{\Delta t \to 0}\frac{\Delta P}{\delta t}.$$

Then the relative growth rate of P is given by

$$\frac{1}{P}\frac{dP}{dt}.$$

When the relative growth rate is changing with time, it is proper to relate these changes to the changes in P. A very plausible assumption for a population that is growing in an area of fixed limits would be that the relative growth rate gradually decreases as 't' and P increases. For this, the decreasing function (23.3) for relative growth rate is assumed. Now, from (23.3), we have

$$\left(\frac{1}{P}+\frac{K}{1-KP}\right)\frac{dP}{dt}=r$$
$$\frac{1}{P}\frac{dP}{dt}+\frac{K}{1-KP}\frac{dP}{dt}=r$$
$$\frac{d\log P}{dt}-\frac{d\log(1-KP)}{dt}=r$$
$$\log P-\log(1-KP)=\int r dt$$
$$\log\frac{P}{1-KP}=rt+C$$
$$\frac{P}{1-KP}=Ae^{rt}$$

where A is a positive constant. Thus, we have

$$P=(1-KP)Ae^{rt}$$
$$P=\frac{Ae^{rt}}{1+AKe^{rt}}=\frac{\frac{1}{K}}{1+\frac{1}{K}\frac{1}{A}e^{-rt}}. \tag{23.4}$$

When $t\to\infty$, $P\to 0$ and $t\to\infty$, $P\to\frac{1}{K}$. Let us take this upper limit by L. Then

$$P=\frac{L}{1+\frac{L}{A}e^{-rt}}. \tag{23.5}$$

Let λ be the value of t for which $P=L/2$. Then, we have

$$\frac{L}{2}=\frac{L}{1+\frac{L}{A}e^{-r\beta}}\Rightarrow A=Le^{-r\beta}.$$

Also,

$$1+\frac{L}{A}e^{-r\beta}=2,\ \therefore\ L=Ae^{r\beta}(2-1).$$

Putting the value of A in (23.5), we get

$$P=\frac{L}{1+e^{r(\beta-t)}}. \tag{23.6}$$

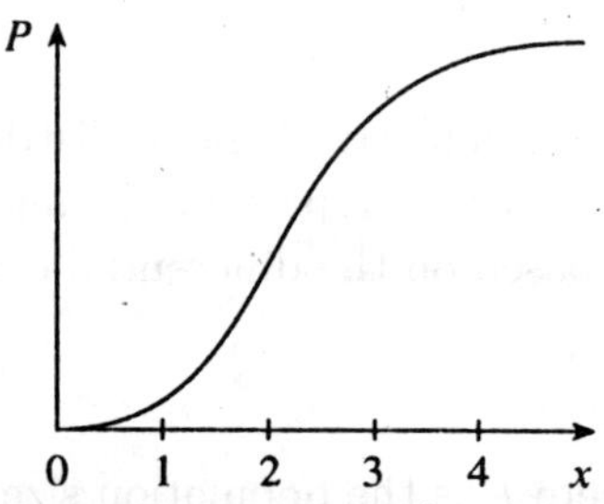

FIG. 23.1 : Graph of logistic distribution.

The shape of the curve of this distribution is shown in FIG. 23.1. This is the form of the logistic curve.

The logistic distribution arises as the limiting distribution of the standardized mid-range of random samples of size n when $n\to\infty$. If $X_1, X_2, \ldots, X_n$ are n independent random variables each having a Type-1 extreme value distribution [Section 23.1], then the limiting distribution (as $n\to\infty$) of the standardized variable corresponding to

$$\sum_{j=1}^{n} j^{-1}X_j$$

is logistic. This result was given by Talacko (1956).

Dubey (1969) has shown that the logistic distribution can be obtained as a mixture of extreme value distributions

$$P[X \leq x|\eta] = \exp[-\eta\beta \exp\{-(x-\alpha)/\beta\}], \; (x > 0; \; \eta, \beta > 0)$$

with η having an exponential distribution with p.d.f.

$$f(y) = \beta e^{-\beta y}; \; y > 0.$$

For then

$$\begin{aligned} P[X \leq x] &= \beta \int_0^{\infty} \exp\left[-\beta y\{1 + \exp(-(X-\alpha)/\beta)\}\right] \\ &= [1 + \exp\{-(x-\alpha)/\beta\}]^{-1}. \end{aligned}$$

The logistic distribution is also used as a substitute for the normal distribution. One of such applications is the representation of the population tolerance distribution. In that case the analysis is carried out in terms of logits instead of probits.

The logit, Y, and the corresponding proportion, P, are connected by the equation

$$P = \left(1 + e^{-Y}\right) \text{ or, } Y = \log[P/(1-P)].$$

If the observed proportions $P_i = d_i/n_i$ of 'deaths' at 'dosages' x_i $(i = 1, 2, \ldots, k)$ are given, the logits $Y_i = \log[P_i/(1-P_i)]$ can be calculated.

Let the model for proportion P be

$$E(P) = \frac{e^{\beta_0 + \beta_1 x}}{1 + e^{\beta_0 + \beta_1 x}} \tag{23.7}$$

so that

$$\log\left[P_i/(1-P_i)\right] = \beta_0 + \beta_1 x.$$

We have

$$V(P_i) = \frac{1}{n_i(1-P_i)P_i}.$$

Let

$$W_i = \frac{1}{V(P_i)} = n_i(1-P_i)P_i.$$

Hence by weighted least square method

$$\hat{\beta}_1 = \frac{\Sigma W_i P_i x_i - \frac{1}{W}\Sigma W_i P_i \Sigma W_i x_i}{\Sigma W_i x_i^2 - \frac{(\Sigma W_i x_i)^2}{W}}, \text{ where } W = \Sigma W_i.$$

Also, we have

$$\hat{\beta}_0 = \bar{P} - \hat{\beta}_1 \bar{x}.$$

Using the value of $\hat{\beta}_0$ and $\hat{\beta}_1$ the model (23.7) can be fitted.

23.2 Moments and Generating Functions

We have

$$f(x) = \frac{\exp\{-(x-\alpha)/\beta\}}{\beta[1+\exp\{-(x-\alpha)/\beta\}^2}.$$

Let us make the transformation $Y = (X-\alpha)/\beta$. This gives the p.d.f. of Y, where

$$f(y) = e^{-y}\left(1+e^{-y}\right)^{-2}. \tag{23.8}$$

The cumulative distribution function of Y is

$$F(y) = \left(1+e^{-y}\right)^{-1}. \tag{23.9}$$

Here (23.8) and (23.9) are standard forms for the logistic distribution.

The m.g.f. of Y with p.d.f. (23.8) is

$$\begin{aligned} M_y(t) &= E\left[e^{tY}\right] \\ &= \int_{-\infty}^{\infty} e^{-(1-t)y}\left(1+e^{-y}\right)^{-2} dy \\ &= \int_0^1 z^{-t}(1-z)^t dz, \quad \left[z = (e^y+1)^{-1}\right] \\ &= B(1-t, 1+t) = \pi t \operatorname{cosec} \pi t. \end{aligned}$$

The moments of Y can be obtained as follows: We have for $r > 0$

$$\begin{aligned} E(|Y|^r) &= 2\int_0^{\infty} y^r e^{-y}\left(1+e^{-y}\right)^2 dy \\ &= 2\int_0^{\infty} y^r \sum_{j=1}^{\infty}(-1)^{j-1} j e^{-jy} dy \\ &= 2\Gamma(r+1)\sum_{j=1}^{\infty}(-1)^{j-1} j^{-r} \text{ (for } r > 0) \\ &= 2\Gamma(r+1)\left(1-2^{-(r-1)}\right)\xi(r) \text{ (for } r > 1) \end{aligned}$$

where

$$\xi(r) = \sum_{j=1}^{\infty} j^{-r}$$

is the Riemann zeta function.

The cumulants are

$$\kappa_r = 6(2^r - 1)B_r \text{ (for } r \text{ even)}$$

where B_r is the r-th Bernoulli number. If r is odd, $\kappa_r = 0$. Therefore,

$$\begin{aligned} \mu_2 &= E(Y^2) = 2\cdot 2(1-2^{-1})\pi^2/6 = \pi^2/3 \\ \mu_4 &= 2\cdot 24(1-2^{-3})\pi^4/90 = 7\pi^4/15. \end{aligned}$$

The moment ratios are

$$\alpha_3 = \sqrt{\beta_1} = 0 \text{ and } \alpha_4 = \beta_2 = 4.2.$$

The characteristic function of Y is

$$\psi(t) = E\left[e^{itY}\right] = \pi t \operatorname{cosech} \pi t.$$

23.3 Properties and Related Distributions

It has already been observed that $\kappa_1 = \mu_1' = 0$. Thus the distribution of Y is symmetrical about $y = 0$.

The mean deviation of Y is

$$2\sum_{j=1}^{\infty}(-1)^{j-1}j^{-1} = 2\log 2.$$

Hence for the logistic distribution

$$\frac{\text{mean deviation}}{\text{standard deviation}} = \frac{2\sqrt{3}\log 2}{\pi} = 0.764.$$

We know $X = \alpha + \beta Y$, where X has the distribution (23.1). Thus

$$\begin{aligned} E(X) &= \alpha + \beta E(Y) = 0 \\ V(X) &= \beta^2 V(Y) = \beta^2\pi^2/3. \end{aligned}$$

The moment ratios of X are same as that of Y.

From the logistic curve (23.6) it is clear that $\dfrac{d^2P}{dt^2}$ is positive if $P < L/2$, it is zero if $P = L/2$ and negative if $P > L/2$. The critical value $L/2$ occurs when $t = \beta$. Thus the curve has a point of inflexion at $t = \beta$ and is concave upwards for $t < \beta$ and convex upwards for $t > \beta$. Again, $\dfrac{dP}{dt} = 0$ for $P = 0$ and $P = L$ which values correspond to $t \to -\infty$ and $t \to \infty$ respectively. Hence the logistic curve has two asymptotes, viz., $P = 0$ and $P = L$.

The Perks distribution [Perks (1932)] has the form

$$f(y) = \sum_{j=0}^{m} a_j e^{-j\theta} y \Big/ \sum_{j=0}^{m_1} b_j e^{-j\theta} y \tag{23.10}$$

and the logistic distribution can be regarded as a member of this large class of distributions, where a_j, b_j, θ are real parameters. If we take $m = 1, m_1 = 2; a_0 = 0$; $b_0 = b_2$ then (23.10) becomes

$$\begin{aligned} f(y) &= \frac{a_1 e^{-y}}{b_0 + b_1 e^{-y} + b_0 e^{-2y}} \qquad (23.11) \\ &= \frac{c_1}{e^y + c_2 + e^{-y}} \text{ (with } c_1 = a_1/b_0; c_2 = b_1/b). \end{aligned}$$

The distribution (23.11) is a symmetrical one. If we put $c_2 = 2$, we get

$$f(y) = \frac{c_1}{2 + e^y + e^{-y}} = \frac{c_1}{e^y[1 + e^{-y}]^2} \tag{23.12}$$

which is the p.d.f. of logistic distribution.

If we put $c_2 = 0$, we get

$$f(y) = \frac{c_1}{e^y + e^{-y}} = \frac{1}{2}c_1 \operatorname{sech} y.$$

The condition

$$\int_{-\infty}^{\infty} f(y)dy = 1$$

requires that $c_1 = 2/\pi$ so that

$$f(y) = \pi^{-1} \operatorname{sech} y. \tag{23.13}$$

The distribution (23.13) is the hyperbolic secant distribution [Talacko (1956)]. If Y has the distribution (23.13), e^Y has half-Cauchy distribution.

Dubey (1966) obtained a generalized logistic distribution which is a mixture of extreme value distributions

$$P[X \le x|\eta] = 1 - \exp[-\eta\beta[\exp\{-(x-\alpha)/\beta\} - 1]], \; (x > \alpha; \; \eta, \beta > 0)$$

with η having a gamma distribution with density function

$$P_\eta(y) = \frac{1}{\Gamma b}\beta'^b y^{b-1} e^{-\beta'_y} \; (y, b, \beta' > 0).$$

The resulting density function of X is

$$g(x) = b\beta'^b e^{(x-\alpha)/\beta} \left[\beta' + \beta e^{(x-\alpha)/\beta}\right]^{-(b-1)}.$$

For $b = 1$ and $\beta' = \beta$, $g(x)$ transforms to the logistic distribution.

CHAPTER **24**

Pareto Distribution

24.1 Definition and Origin

A random variable X is said to have a Pareto distribution if its p.d.f. is given by

$$f(x) = \frac{ak^a}{x^{a+1}};\ a > 0,\ x \geq k > 0. \tag{24.1}$$

The distribution is related to Pareto's law, which was formulated by Italian professor Vilfredo Pareto (1848–1923) for the distribution of income over a population. This law is stated as follows:

$$N = k(x - c)^{-a} = kx^{-a},\ \text{when } c = 0 \tag{24.2}$$

where N is the number of persons having $\geq x$, and k, a are parameters. The original form of the Pareto distribution is represented by

$$P(X \geq x) = \left(\frac{k}{x}\right)^a,\ k > 0,\ a > 0,\ x \geq k \tag{24.3}$$

indicating the probability that the income is equal to or greater than x, where k represents some minimum income. The cumulative distribution function of X corresponding to (24.2) may be written as

$$F(x) = 1 - \left(\frac{k}{x}\right)^a,\ k > 0,\ a > 0,\ x \geq k.$$

The above cumulative distribution is known as *Pareto distribution of the first kind.* The Pareto distributions of the second and third kind have the cumulative distribution functions

$$F(x) = 1 - \frac{k_1}{(x+c)^a} \tag{24.4}$$

$$\text{and } F(x) = 1 - \frac{k_2 e^{-bx}}{(x+c)^a} \tag{24.5}$$

respectively. The density function (24.1) corresponds to the distribution (24.3). The shape of the curve of Pareto law is shown in FIG. 24.1.

It was felt by Pareto that his law was universal and inevitable—regardless of taxation and social and political conditions. But it has been observed that while the fit of the Pareto curve may be rather good at the extremities of the income range, the fit over the whole range is rather poor. On the other hand, the lognormal distribution fits well over a large part of the income range but diverges markedly at the extremities.

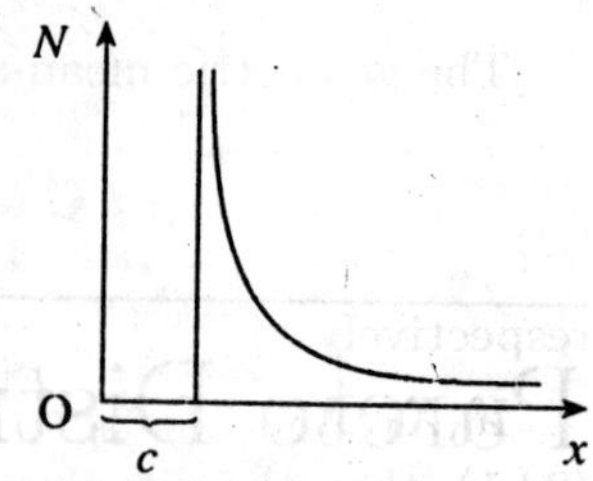

FIG. 24.1 : Pareto's curve.

24.2 Moments and Properties

The r-th moment about origin is given by

$$\mu'_r = E(X^r) = ak^a \int_k^{\infty} x^r x^{-(a+1)} dx$$
$$= ak^a \int_k^{\infty} x^{r-a-1} dx$$
$$= \frac{ak^r}{a-r}, \text{ provided } a > r.$$

In particular

$$\mu'_1 = \frac{ak}{a-1}, \ \mu'_2 = \frac{ak^2}{a-2}, \ \mu'_3 = \frac{ak^3}{a-3}, \ \mu'_4 = \frac{ak^4}{a-4}$$

$$\mu_2 = \mu'_2 - \mu'^2_1 = \frac{ak^2}{a-2} - \frac{(ak)^2}{(a-1)^2} = \frac{ak^2}{(a-1)^2(a-2)}, \text{ if } a > 2$$

$$\mu_3 = \frac{ak^3}{a-3} - \frac{3ak^2}{(a-2)} \frac{ak}{(a-1)} + \frac{2(ak)^3}{(a-1)^3} = \frac{2a(a+1)k^3}{(a-1)^3(a-2)(a-3)}$$

$$\mu_4 = \frac{3a(3a^2+a+2)k^4}{(a-1)^4(a-2)(a-3)(a-4)}, \text{ if } a > 4$$

The moment ratios are

$$\alpha_3 = \sqrt{\beta_1} = \frac{2(a+1)}{a-3}\sqrt{1-\frac{2}{a}}$$

$$\alpha_4 = \beta_2 = \frac{3(a-2)(3a^2+a+2)}{(a-3)(a-4)}$$

The distribution is positively skewed if $a > 3$.

The mean deviation is $\dfrac{2k}{(a-1)}\left(1-\dfrac{1}{a}\right)^{a-1}$, and

$$\frac{\text{mean deviation}}{\text{standard deviation}} = 2\sqrt{\left(1-\frac{2}{a}\right)}\left(1-\frac{1}{a}\right)^{a-1}, \ (a > 2).$$

The mode of the distribution is located at k. The median, $m = k \cdot 2^{1/a}$.

The geometric mean and harmonic mean of the Pareto distribution are

$$\text{G.M.} = k \exp\left(\frac{1}{a}\right) \text{ and H.M.} = k\left(1 + \frac{1}{a}\right)$$

respectively.

If $X_1, X_2, \ldots, X_n$ are independently distributed with common density function (24.1), then the distribution of

$$Y = \sum_{i=1}^{n} \log X_i$$

has the density function

$$h(y) = \frac{a^n}{\Gamma n}(y - n\log k)^{n-1} e^{-a(y-n\log k)}, \quad (y > n\log k).$$

The p.d.f. of $z = e^Y = X_1 X_2 \ldots X_n$ is

$$g(z) = \frac{a^n k^{na}}{\Gamma n}\left[\log\frac{z}{k^n}\right]^{n-1} z^{-(a+1)}, \quad (z > k^n).$$

24.3 Related Distributions

Gumbel (1958) has shown that if T has an exponential distribution with scale parameter θ, then $X = \exp(T) + \epsilon$ has the Pareto distribution having cumulative distribution function

$$F(x) = 1 - (x - \epsilon)^{-a}, \; x \geq 1 + \epsilon$$

with

$$a = 1/\theta, \; a \geq 1$$
$$x > 1 + \epsilon, \; \epsilon \geq 0.$$

Harris (1968) has pointed out that a mixture of exponential distribution, with parameter θ^{-1} having a gamma distribution, and with origin at zero, gives rise to a Pareto distribution. Thus, if

$$P[X \leq x|\theta] = 1 - e^{-x/\theta} \text{ and } \mu = \theta^{-1}$$

has a gamma distribution, then

$$\begin{aligned} P[X \leq x] &= \frac{1}{\Gamma m \beta^m}\int_0^\infty t^{m-1} e^{-t/\beta}\left(1 - e^{-tx}\right) dt \\ &= 1 - \frac{1}{\beta^m \Gamma m}\int_0^\infty t^{m-1} e^{-t\left(x + \frac{1}{\beta}\right)} dt \\ &= 1 - (\beta x + 1)^m \end{aligned}$$

which is the form (24.4).

Ljubo (1965) has described a class of generalized Pareto distributions, the cumulative distribution function of which is given by

$$F(x) = \begin{cases} 1 - \left(\dfrac{k+\alpha}{x+\alpha}\right)^a e^{-\beta(x-k)}, & (x \geq k) \\ 0, & (x < k) \end{cases}$$

with $\beta > 0$.

If X has the probability density function (24.1), then $Y = X^{-1}$ has the density function

$$h_1(y) = ak^a y^{a-1}, \ (0 < y < 1/k).$$

This distribution is a special Pearson Type 1 distribution and is known as *power function distribution.* Its moments are given by

$$\mu'_r = \frac{ak^{-r}}{a+r}.$$

Truncated Pareto Distribution

Goldberg (1967) has found the truncated Pareto distribution with an upper limit (k_1) and lower limit (k), the density function of which is given by

$$g_1(x) = \frac{a}{k}\,[1-(k/k_1)^a]^{-1}\left(\frac{k}{x}\right)^{a+1}, \ (k \leq x \leq k_1).$$

This distribution has been found to fit the distribution of oil-fields, in a specified area, by size.

Champernowne Distribution

This is a distribution to give better representation of income distribution. It has been proposed by Champernowne. Let X be a random variable representing income-power and defined by $X = \log T$, where T is the actual income. Then X has the density function

$$g_2(x) = \frac{n}{\cosh\{\alpha(x-x_0)\} + \lambda} \tag{24.6}$$

where n, α, x_0 and λ are parameters.

The curve of this distribution is symmetric and x_0 is the median value of the income-power. If we let $\log t_0 = x_0$, then the density function of income ($T = e^\lambda$) is given by

$$f(t) = \frac{n}{t\left[\frac{1}{2}\left(\frac{t}{t_0}\right)^{-\alpha} + \lambda + \frac{1}{2}\left(\frac{t}{t_0}\right)^{\alpha}\right]}. \tag{24.7}$$

Here t_0 is the median value of the income.

The proportion of persons with income greater than t can be found out by integrating (24.7). For $-1 < \lambda < 1$, the cumulative distribution function is given by

$$F(t) = 1 - \frac{1}{\theta}\tan^{-1}\left\{\frac{\sin\theta}{\cos\theta + \left(\frac{t}{t_0}\right)^{\alpha}}\right\}, \quad (t > 0) \tag{24.8}$$

where $0 < \theta < \pi$ and $\cos\theta = \lambda$.

For $\lambda = 1$

$$F(t) = 1 - \frac{t_0^{\alpha}}{t_0^{\alpha} + t^{\alpha}}, \quad (t > 0) \tag{24.9}$$

For $\lambda > 1$

$$F(t) = 1 - \frac{1}{2\eta}\log\left\{\frac{t^{\alpha} + e^{\eta}t_0^{\alpha}}{t^{\alpha} + e^{-\eta}t_0^{\alpha}}\right\}, \quad (t > 0) \tag{24.10}$$

where $\cosh\eta = \lambda$.

In all these cumulative distribution functions α is equal to Pareto's constant and t_0 is the median value of income.

The density functions corresponding to (24.8), (24.9) and (24.10) are, respectively

$$f_1(t) = \frac{\alpha\sin\theta}{\theta t\left\{\left(\frac{t}{t_0}\right)^{\alpha} + 2\cos\theta + \left(\frac{t_0}{t}\right)^{\alpha}\right\}},$$

$$f_2(t) = \frac{\alpha}{t\left\{\left(\frac{t}{t_0}\right)^{\alpha/2} + \left(\frac{t_0}{t}\right)^{\alpha/2}\right\}^2}$$

$$\text{and } f_3(t) = \frac{\alpha\sinh\eta}{\eta t\left\{\left(\frac{t}{t_0}\right)^{\alpha} + 2\cosh\eta + \left(\frac{t_0}{t}\right)^{\alpha}\right\}}.$$

Strong Law of Pareto

The Pareto law related to the cumulative distribution function

$$F(x) = \begin{cases} 1 - \left(\frac{k}{x}\right)^{\alpha}, & x \geq k \\ 1, & x < k \end{cases}$$

is called *Strong Law of Pareto*. However, if the form of the cumulative distribution function is represented by

$$1 - F(x) \sim \left(\frac{k}{x}\right)^{\alpha} \quad \text{as } x \to \infty$$

then it is known as *Weak Law of Pareto*. Thus if $\log[1 - F(x)]$ is plotted against $\log x$, then resulting curve would be asymptotic to a straight line with slope equal to $-a$ as $x \to \infty$.

CHAPTER **25**

Inverse Gaussian (Wald) Distribution

25.1 Definition and Origin

A random variable X is said to have inverse Gaussian distribution if its p.d.f. is given by

$$f(x) = \sqrt{\lambda/(2\pi x^3)}\, e^{-\lambda(x-\mu)^2/2\mu^2 x}, \quad (x > 0;\ \lambda > 0;\ \mu > 0) \tag{25.1}$$

This is the standard form of the inverse Gaussian distribution.

The inverse Gaussian distributions have been used to explain the motion of particles influenced by Brownian movement. Let a particle moves along a line with a uniform velocity v. Suppose that the particle is subject to linear Brownian motion which causes it to take a variable amount of time to cover a fixed distance (d). Then the time (X) to cover the distance is a random variable with p.d.f.

$$f_1(x) = \frac{1}{\sqrt{2\pi\beta x^3}} d e^{-(d-vx)^2/2\beta x}, \quad (0 < x) \tag{25.2}$$

where β is the diffusion constant.

Now, if we put $v = d/\mu$ and $\beta = d^2/\lambda$ in (25.2) we get (25.1).

The distribution (25.2) is the distribution of time to cover a fixed distance. But we can consider the random variable distance (D) over which the particle travels in a fixed time (x). The distribution of D is assumed normal with p.d.f.

$$f(d) = \frac{1}{\sqrt{2\pi\beta x}} e^{-(d-vx)/(2\beta x)}. \tag{25.3}$$

Tweedie (1956) examined the cumulant generating functions of (25.2) and (25.3), and observed the inverse relationship between these functions and suggested the name 'Inverse Gaussian' distribution.

Wald (1947) derived this kind of distribution as a limiting form of distribution of sample size in certain sequential probability ratio tests. For example, let $Z_1, Z_2, \ldots,$ be independent random variables each having the same distribution, with finite expected value $E[Z] > 0$ and finite variance $V(Z)$. Let us define a random variable N such that

$$\left[\bigcap_{j=1}^{N-1}\left(\sum_{i=1}^{j} Z_i < K\right)\right] \cap \left(\sum_{i=1}^{N} Z_i \geq K\right) \quad \text{with } K > 0$$

with

$$E[N] = K/E[Z] \text{ and } \lim_{E(N)\to\infty} P[N \leq xE(N)].$$

This limiting probability is given by the integral up to x of the function

$$f_2(x) = \sqrt{\frac{\phi}{2\pi}}\, e^{\phi} x^{-3/2} \exp\left[-\frac{1}{2}\phi\left(x + \frac{1}{x}\right)\right], \ (x > 0) \tag{25.4}$$

with

$$\phi = KE[Z]/V(Z) > 0.$$

Here (25.4) is the standard form of the p.d.f. of Wald distribution.

The most familiar applications of Wald distribution are in sequential analysis. It is also used in various physical applications connected with diffusion process with boundary conditions. It is also used in electronics and radiotechnique. The shape of the curve of this distribution is given below (FIG. 25.1):

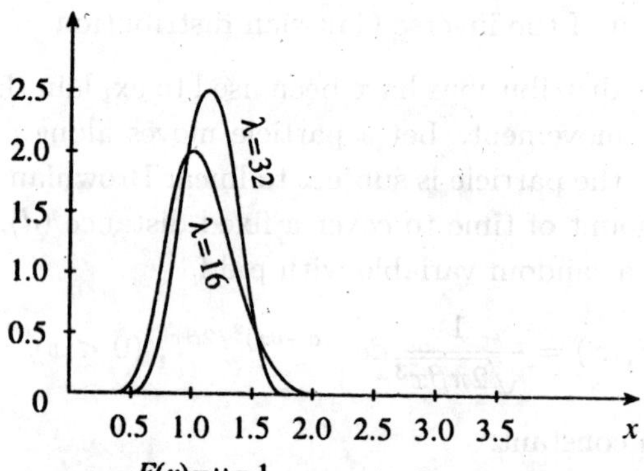

FIG. 25.1 : Graph of Wald density function.

25.2 Moments and Properties

The cumulant generating function of the inverse Gaussian distribution is given by

$$K(t) = \frac{\lambda}{\mu}\left\{1 - \left(\frac{2\mu^2 t}{\lambda}\right)^{1/2}\right\}.$$

In particular

$$K_1 = \mu_1' = \mu,\ K_2 = \mu_2 = \mu^3/\lambda,\ K_3 = \mu_3 = 3\mu^5/\lambda^2$$

$$K_4 = \mu_4 - 3\mu_2^2 = 15\mu^7/\lambda^3 \text{ and hence } \mu_4 = \frac{1}{\lambda^3}\left(15\mu^7 + 3\lambda\mu^6\right).$$

In general, for $r \geq 2$

$$K_r = 1 \cdot 3 \cdot 5 \cdots (2r-3)\mu^{2r-1}\lambda^{1-r}.$$

The moment ratios are

$$\alpha_3 = \sqrt{\beta_1} = 3\sqrt{\frac{\mu}{\lambda}}, \ \alpha_4 = \beta_2 = 3 + \frac{15\mu}{\lambda}.$$

Thus the inverse Gaussian distribution is positive skewed and leptokurtic.

For Wald distribution the r-th cumulant is given by

$$K_r = 1 \cdot 3 \cdot 5 \cdot \cdots (2r-3)\phi^{-(r-1)}$$

whatever be the value of ϕ. In particular

$$\begin{aligned} K_1 &= \mu_1' = 1, \ K_2 = \mu_2 = \phi^{-1}, \ K_3 = \mu_3 = 3\phi^{-2}, \\ K_4 &= 15\phi^{-3}, \ \mu_4 = 15\phi^{-3} + 3\phi^{-2}. \end{aligned}$$

The moment ratios are

$$\alpha_3 = \sqrt{\beta_1} = \frac{3}{\sqrt{\phi}}, \ \alpha_4 = \beta_2 = 3 + \frac{15}{\phi}.$$

Thus the Wald distribution also is positively skewed and leptokurtic.

Reproductive Property of Inverse Gaussian Distribution

Let $X_1, X_2, \ldots, X_n$ be independent random variables each of which distributed according to probability law (25.1) with $\mu = \mu_i$, $\lambda = \lambda_i$ $(i = 1, 2, \ldots, n)$. Then the distribution of

$$\sum_{i=1}^{n} (\mu_i^{-2}\lambda_i X_i)$$

is also of the form (25.1) with

$$\phi = \mu = \sum_{i=1}^{n} \phi_i, \ \lambda = \left(\sum_{i=1}^{n} \phi_i\right)^2$$

where $\phi = \lambda/\mu$ for the inverse Gaussian distribution of form

$$f(x|\mu,\phi) = \left[\mu\phi/2\pi x^3\right]^{1/2} \exp\left\{-\frac{\phi x}{2\mu} + \phi - \frac{\mu\phi}{2x}\right\} \tag{25.5}$$

$$\text{or, } f(x|\phi,\lambda) = \left[\lambda/2\pi x^3\right]^{1/2} \exp\left\{-\frac{\phi^2 x}{2\lambda} + \phi - \frac{\lambda}{2x}\right\}. \tag{25.6}$$

Thus the distribution of the arithmetic mean $\bar{X}$ of n independent random variables with distribution (25.1) has the same form as (25.1) or (25.1) and (25.2) with the same values for μ and α, but with ϕ and λ replaced by $n\phi$ and $n\lambda$ respectively.

As ϕ tends to infinity the standardized distribution tends to a unit normal distribution when μ is fixed.

As μ tends to infinity the distribution of $Y = X^{-1}$ tends to the gamma distribution for fixed λ, the density function of which is given by

$$f(y) = \lambda(2\pi)^{-1/2}y^{-1/2}\exp\left(-\frac{1}{2}\lambda y\right),\ (y > 0).$$

The inverse Gaussian distribution is unimodal and the mode is at

$$X = \mu\left\{\left(1+\frac{9}{4\phi^2}\right)^{1/2} - \frac{3}{2\phi}\right\}.$$

25.3 Related Distributions

Truncated Distribution

If x_1 and x_2 denote the lower and upper truncation points, respectively, of the distribution of X $(0 < x_1 < x_2)$, then the p.d.f. of the doubly truncated inverse Gaussian distribution is given by

$$g(x) = Kx^{-\frac{3}{2}}\exp\left\{-\frac{\lambda x}{2\mu^2} - \frac{\lambda}{2x}\right\},\ x_1 \le x \le x_2$$

where

$$K = \left[\int_{x_1}^{x_2} x^{-\frac{3}{2}}\exp\left\{-\frac{\lambda x}{2\mu^2} - \frac{\lambda}{2x}\right\}dx\right]^{-1}.$$

Random Walk Distribution

If X is a inverse Gaussian variate, then the distribution of $Y = X^{-1}$ is known as *random walk distribution.* Its p.d.f. is given by

$$h(y) = \exp\left\{-\frac{\lambda y}{2} + \frac{\lambda}{\mu} - \frac{\lambda}{2\mu^2 y}\right\}\left[\frac{\lambda}{2\pi y}\right]^{1/2},\ (y > 0)$$

The cumulant generating function of Y is

$$\psi_y(t) = \phi\left(1 - (1 + 2t\lambda^{-1})^{1/2}\right) - \frac{1}{2}\log_e(1 + 2t\lambda^{-1}).$$

The first two cumulants are

$$K_1 = \mu_1' = \frac{1}{\mu} + \frac{1}{\lambda},\ K_2 = \mu_2 = \frac{1}{\lambda\mu} + \frac{2}{\lambda^2}.$$

The positive moments about zero for Y are the same as the negative moments about zero for the inverse Gaussian variate X. These are

$$\mu_1'(y) = \frac{1}{\mu} + \frac{1}{\lambda},\quad \mu_2'(y) = \frac{1}{\mu^2} + \frac{3}{\mu\lambda} + \frac{3}{\lambda^2}$$

$$\mu_3'(y) = \frac{1}{\mu_3} + \frac{6}{\mu^2\lambda} + \frac{15}{\mu\lambda^2} + \frac{15}{\lambda^3}.$$

The mode of the distribution of Y is at $Y = \dfrac{1}{\mu}\left[\left(1+\dfrac{1}{4\phi^2}\right)^{1/2} - \dfrac{1}{2\phi}\right].$

CHAPTER **26**

Chi-square Distribution

26.1 Definition

A random variable X is said to have chi-square distribution if its p.d.f. is given by

$$f(x) = \frac{1}{\Gamma\frac{\gamma}{2}2^{\gamma/2}} e^{-\frac{1}{2}x} x^{\gamma/2-1}, \ 0 < x < \infty.$$

This is nothing but the p.d.f. of a gamma distribution of the form (18.2) with

$$m = \frac{\gamma}{2}; \ \theta = 2; \ \alpha = 0.$$

This particular form of gamma distribution arises from the distribution of

$$\sum_{j=1}^{n} U_j^2, \text{ where } U_1, U_2, \dots, U_n$$

are independent standard normal variables. The distribution is called *chi-square distribution* with γ degrees of freedom. The corresponding random variable is often denoted by χ_γ^2 and the p.d.f. of this random variable is given by

$$f(\chi^2) = \frac{1}{\Gamma\frac{\gamma}{2}2^{\gamma/2}} e^{-\frac{1}{2}\chi^2} (\chi^2)^{\frac{\gamma}{2}-1}, \ 0 < \chi^2 < \infty \tag{26.1}$$

where γ is any positive integer (number).

The chi-square variate can be defined as the square of a standard normal variate. Thus if U is a normal variate with mean μ and variance $\sigma^2[N(\mu, \sigma^2]$, then

$$X = \frac{U-\mu}{\sigma} \sim N(0,1) \text{ and } X^2 = \left(\frac{U-\mu}{\sigma}\right)^2$$

is a chi-square variate with 1 d.f. In general, if $U_1, U_2, \dots, U_n$ are n independent normal variables with mean $\mu_1, \mu_2, \dots, \mu_n$ and variance $\sigma_1^2, \sigma_2^2, \dots, \sigma_n^2$, respectively, then

$$\chi^2 = \sum \left(\frac{U_i - \mu_i}{\sigma_i}\right)^2$$

is a chi-square variate with n d.f. We have already mentioned in Section (18.1) that

$$\frac{1}{2}\sum\left(\frac{U_i-\mu_i}{\sigma_i}\right)^2$$

is a gamma variate with parameter $n/2$. It is observed from (26.1) that chi-square density function does not involve any population parameter.

From (26.1), we have

$$f'(\chi^2)=\left[\frac{n-2-\chi^2}{2\chi^2}\right]f(\chi^2).$$

This $f'(\chi^2)<0$ if $(n-2)\le 0$, since $\chi^2>0$ and $f(\chi^2)$ being p.d.f. is always non-negative. Thus the chi-square probability curve as shown in FIG. 26.1 is monotonically decreasing for 1 and 2 degrees of freedom.

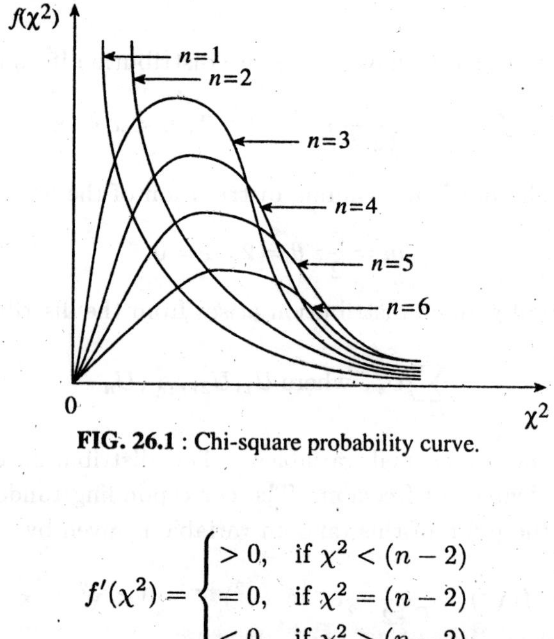

FIG. 26.1 : Chi-square probability curve.

When $n>2$

$$f'(\chi^2)=\begin{cases}>0, & \text{if } \chi^2<(n-2)\\ =0, & \text{if } \chi^2=(n-2)\\ <0, & \text{if } \chi^2>(n-2)\end{cases}$$

From this result it may be concluded that for $n>2$, $f(\chi^2)$ is monotonically increasing if $0<\chi^2<(n-2)$ and monotonically decreasing if $(n-2)<\chi^2<\infty$. The maximum value of $f(\chi^2)$ is attained at $\chi^2=(n-2)$. From the probability curve it is observed that as χ^2 increases, $P(\chi^2)$ decreases rapidly when $n\ge 1$ and finally it tends to zero as χ^2 tends to infinity. For $n>1$, the χ^2-probability curve is positively skewed towards higher values of χ^2. The χ^2-axis is an asymptote to the curve.

26.2 Derivation of the Distribution

The chi-square distribution is a sampling distribution. Its derivation can be done using the techniques discussed in Section (2.6).

Moment Generating Method

Let $u_1, u_2, \ldots, u_n$ be a random sample drawn from $N(\mu, \sigma^2)$. Then by the definition of chi-square variate and standardized normal variate, we can write

$$\chi^2 = \sum Z_i^2 = \sum_{i=1}^{n} \left(\frac{v_i - \mu}{\sigma}\right)^2$$

where Z_i $(i = 1, 2, \ldots, n)$ is standard normal variate. The m.g.f. of Z_i^2 is given by

$$\begin{aligned}
M_{Z_i^2}(t) &= E\left[e^{tZ_i^2}\right] \\
&= \frac{1}{\sqrt{2\pi}} \int_{-\infty}^{\infty} e^{tZ_i^2} e^{-\frac{1}{2}Z_i^2} dZ_i \\
&= \frac{2}{\sqrt{2\pi}} \int_{0}^{\infty} e^{-\frac{1}{2}Z_i^2(1-2t)} dZ_i \\
&= \frac{1}{\sqrt{\pi}} \int_{0}^{\infty} e^{-v_i} v_i^{-\frac{1}{2}} (1-2t)^{-\frac{1}{2}} dv_i \left[v_i = \frac{1}{2} Z_i^2 (1-2t)\right] \\
&= (1-2t)^{-\frac{1}{2}} \frac{1}{\Gamma\frac{1}{2}} \int_{0}^{\infty} e^{-v_i} v^{\frac{1}{2}-1} dv_i \\
&= (1-2t)^{-\frac{1}{2}}.
\end{aligned}$$

Now

$$\begin{aligned}
M_{\Sigma Z_i^2}(t) &= M_{Z_1}(t) M_{Z_2}(t) \cdots M_{Z_n}(t) \text{ [since } Z_i\text{'s are independent]} \\
&= (1-2t)^{-n/2}
\end{aligned}$$

which is the m.g.f. of gamma variate with parameters $m = n/2$ and $\theta = 2$. Therefore, the density function of $\chi^2 = \Sigma Z_i^2$ is

$$f(\chi^2) = \frac{1}{\Gamma\frac{n}{2} 2^{n/2}} e^{-\frac{1}{2}\chi^2} (\chi^2)^{\frac{n}{2}-1}, \; 0 < \chi^2 < \infty.$$

This is the required p.d.f. of χ^2-variate with n d.f.

Transformation of Variable Technique

Let u_1 and u_2 be two random observations from $N(0, 1)$. We need to derive the distribution of $(u_1^2 + u_2^2)$.

The joint p.d.f. of u_1 and u_2 is

$$f(u_1, u_2) = \frac{1}{2\pi} e^{-\frac{1}{2}(u_1^2 + u_2^2)}; \quad -\infty < (u_1, u_2) < \infty.$$

Let $u_1 = r\cos\theta$, $u_2 = r\sin\theta$. Then

$$J = \frac{\partial(x_1, x_2)}{\partial(r, \theta)} = \begin{vmatrix} \dfrac{\partial x_1}{\partial r} & \dfrac{\partial x_2}{\partial r} \\ \dfrac{\partial x_1}{\partial \theta} & \dfrac{\partial x_2}{\partial \theta} \end{vmatrix} = r.$$

We have $u_1^2 + u_2^2 = r^2$; $0 < r < \infty$, $0 < \theta < 2\pi$.

$$\therefore\ f(r, \theta) = \frac{1}{2\pi} e^{-\frac{1}{2}r^2} \cdot r$$

$$g(r) = \frac{1}{2\pi} \int_0^{2\pi} e^{-\frac{1}{2}r^2} r d\theta = \frac{re^{-\frac{1}{2}r^2}}{2\pi} \Big[\theta\Big]_0^{2\pi} = re^{-\frac{1}{2}r^2}$$

$$g(r^2) = \frac{1}{2} e^{-\frac{1}{2}r^2} dr^2,\ 0 < r^2 < \infty$$

$$= \frac{1}{\Gamma\frac{2}{2} 2^{2/2}} e^{-\frac{1}{2}r^2} (r^2)^{\frac{2}{2}-1},\ 0 < r^2 < \infty$$

which is the p.d.f. of chi-square distribution with 2 d.f.

For n random observations $u_1, u_2, \ldots, u_n$, we can use the transformation

$$\begin{aligned} u_1 &= \chi \cos\theta_1 \cos\theta_2 \cdots \cos\theta_{n-1} \\ u_2 &= \chi \cos\theta_1 \cos\theta_2 \cdots \cos\theta_{n-2} \sin\theta_{n-1} \\ u_3 &= \chi \cos\theta_1 \cos\theta_2 \cdots \cos\theta_{n-3} \sin\theta_{n-2} \\ &\cdots \\ x_n &= \chi \sin\theta_1 \end{aligned}$$

Then $u_1^2 + u_2^2 + \cdots + u_n^2 = \chi^2$ and

$$|J| = \chi^{n-1} \cos^{n-2}\theta_1 \cos^{n-3}\theta_2 \cdots \cos\theta_{n-2}$$

where $\chi > 0$; $-\pi < \theta_1 < \pi$ and $-\pi/2 < \theta_i < \pi/2$ for $i = 2, 3, \ldots, (n-1)$.

The joint distribution of $(u_1, u_2, \ldots, u_n)$ is

$$f(u_1, u_2, \ldots, u_n) = \left(\frac{1}{\sqrt{2\pi}}\right)^n \exp\left(-\frac{1}{2}\Sigma u_i^2\right) du_1 du_2 \ldots du_n.$$

Now

$$f(\chi, \theta_1, \theta_2, \ldots, \theta_{n-1}) = \left(\frac{1}{\sqrt{2\pi}}\right)^n \exp\left(-\chi^2/2\right) \chi^{n-1} \cos^{n-2}\theta_1 \cos^{n-3}\theta_2 \cdots \cos\theta_{n-2} d\chi^2 d\theta_1 \cdots d\theta_{n-1}.$$

Integrating over $\theta_1, \theta_2, \ldots, \theta_{n-1}$, we get

$$g(\chi^2) = K \exp(-\chi^2/2)(\chi^2)^{\frac{n}{2}-1},\ 0 < \chi^2 < \infty.$$

The constant K is determined in such a way that

$$\int_0^\infty g(\chi^2) d\chi^2 = 1$$

$$\text{i.e., } K \int_0^\infty \exp(-\chi^2/2)(\chi^2)^{\frac{n}{2}-1} d\chi^2 = 1 \Rightarrow K = \frac{1}{2^{\frac{n}{2}}\Gamma\frac{n}{2}}.$$

$$\therefore\ g(\chi^2) = \frac{1}{2^{\frac{n}{2}}\Gamma\frac{n}{2}} e^{-\frac{1}{2}\chi^2} (\chi^2)^{\frac{n}{2}-1},\ 0 < \chi^2 < \infty$$

which is the p.d.f. of chi-square distribution with n d.f.

The chi-square distribution can also be derived from the sample statistic

$$s^2 = \frac{1}{n}\Sigma(x_i - \bar{x})^2.$$

The derivation is shown by a theorem as follows:

❒ **THEOREM 26.1** *Let $x_1, x_2, \ldots, x_n$ be a random sample from normal population $N(\mu, \sigma^2)$. Then* (i) *$\bar{x}$ and* (ii) *ns^2/σ^2 are independently distributed as normal and chi-square respectively.*

Proof. The joint probability function of $x_1, x_2, \ldots, x_n$ is

$$\begin{aligned} f(x_1, x_2, \ldots, x_n) &= \left(\frac{1}{2\pi\sigma^2}\right)^{n/2} e^{-\frac{1}{2\sigma^2}\sum_{i=1}^{n}(x_i-\mu)^2} \\ &= \left(\frac{1}{2\pi\sigma^2}\right)^{n/2} e^{-\frac{1}{2\sigma^2}\Sigma\{(x_i-\bar{x})+(\bar{x}-\mu)\}^2} \\ &= \left(\frac{1}{2\pi\sigma^2}\right)^{n/2} e^{-\frac{1}{2\sigma^2}\left[\Sigma(x_i-\bar{x})^2+n(\bar{x}-\mu)^2\right]} \end{aligned}$$

Let us perform the orthogonal transformation $Y = AX$ such that

$$\begin{bmatrix} y_1 \\ y_2 \\ \vdots \\ y_n \end{bmatrix} = \begin{bmatrix} 1/\sqrt{n} & 1/\sqrt{n} & \cdots & 1/\sqrt{n} \\ a_{21} & a_{22} & \cdots & a_{2n} \\ a_{31} & a_{32} & \cdots & a_{3n} \\ \vdots & \vdots & \vdots & \vdots \\ a_{n1} & a_{n2} & \cdots & a_{nn} \end{bmatrix} \begin{bmatrix} x_1 \\ x_2 \\ \vdots \\ x_n \end{bmatrix}$$

where

$$\sum_{j=1}^{n} a_{lj}^2 = 1, \; \sum_{j=1}^{n} a_{lj} = 0$$

$$\sum_{j=1}^{n} a_{lj}a_{lj'} = 0, \; \sum_{i=1}^{n-1} a_{lj}^2 = 1 - \frac{1}{n}$$

$$\sum_{i'=1}^{n-1} a_{lj}a_{l'j'} = -\frac{1}{n}; \; l \neq l' = 1, 2, \ldots, n-1.$$

From the transformation, we have

$$y_1 = \frac{1}{\sqrt{n}}\sum_{i=1}^{n} x_i = \sqrt{n}\bar{x}.$$

Again $Y'Y = X'A'AX = X'X$, since $A'A = I$. Thus

$$\sum_{i=1}^{n} y_i^2 = \Sigma x_i^2 = \Sigma(x_i - \bar{x} + \bar{x})^2 = ns^2 + y_1^2.$$

$$\therefore \sum_{i=2}^{n} y_i^2 = ns^2.$$

Thus, we can write

$$\sum_{i=1}^{n}(x_i-\mu)^2=\sum_{i=1}^{n}(x_i-\bar{x})^2+n(\bar{x}-\mu)^2=\sum_{i=2}^{n}y_i^2+n(\bar{x}-\mu)^2.$$

Now,

$$g(y_1,y_2,\ldots,y_n)=\frac{1}{(2\pi\sigma^2)^{n/2}}e^{-\frac{1}{2\sigma^2}\left[\sum\limits_{i=2}^{n}y_i^2+n(\bar{x}-\mu)^2\right]}|J|$$

where $|J|=1$ [$\because$ $Y=AX$ and A is an orthogonal matrix].

$$\therefore\ g(y_1,y_2,\ldots,y_n)=\frac{1}{\sqrt{2\pi}\sigma}e^{-\frac{1}{2\sigma^2}n(\bar{x}-\mu)^2}\frac{1}{(2\pi\sigma^2)^{\frac{n-1}{2}}}e^{-\frac{1}{2\sigma^2}ns^2}$$

$$dG(y_1,y_2,\ldots,y_n).=\frac{1}{\sqrt{2\pi}\sigma}\exp\left[-\frac{1}{\frac{2\sigma^2}{n}}(\bar{x}-\mu)^2\right]$$

$$\times\left(\frac{1}{2\pi\sigma}\right)^{\frac{n-1}{2}}\exp\left[-\frac{1}{2\sigma^2}ns^2\right]dy_1,dy_2,\ldots,dy_n.$$

But $y_1=\sqrt{n}\bar{x}\Rightarrow dy_1=\sqrt{n}d\bar{x}$.

$$\therefore\ dG(y_1,y_2,\ldots,y_n)=\frac{1}{\sqrt{2\pi}\sigma/\sqrt{n}}\exp\left[-\frac{1}{\frac{2\sigma^2}{n}}(\bar{x}-\mu)^2\right]d\bar{x}$$

$$\times\left(\frac{1}{2\pi\sigma}\right)^{\frac{n-1}{2}}\exp\left[-\frac{1}{2\sigma^2}ns^2\right]dy_2,dy_3,\ldots,dy_n.$$

Thus $\bar{x}$ and ns^2/σ^2 are independently distributed, where

$$\bar{x}\sim N\left(\mu,\sigma^2/n\right).$$

Due to orthogonal transformation Y_i $(i=2,3,\ldots,n)$ are independently distributed as $N(0,1)$. Hence

$$\sum_{i=2}^{n}y_i^2\bigg/\sigma^2=\frac{ns^2}{\sigma^2}$$

being the sum of squares of $(n-1)$ independent standard normal variates, is distributed as chi-square with $(n-1)$ d.f. Therefore

$$dP\left(\frac{ns^2}{\sigma^2}\right)=\frac{1}{\Gamma\frac{n-1}{2}}2^{\frac{n-1}{2}}e^{-\frac{1}{2}\left(\frac{ns^2}{\sigma^2}\right)}\left(\frac{ns^2}{\sigma^2}\right)^{\frac{n-1}{2}-1};\ 0<\frac{ns^2}{\sigma^2}<\infty.$$

Thus

$$f\left(s^2\right)=\frac{\left(n/2\sigma^2\right)^{\frac{n-1}{2}}}{\Gamma\frac{n-1}{2}}e^{-\frac{s^2}{2\sigma^2/n}}\left(s^2\right)^{\frac{n-1}{2}-1};\ 0<s^2<\infty$$

which is the p.d.f. of a gamma variate with parameters $(n-1)/2$ and $2\sigma^2/n$.

26.3 Moments and Generating Functions

The m.g.f. of chi-square distribution is

$$\begin{aligned}M_{\chi^2}(t) &= E\left[e^{t\chi^2}\right] = \frac{1}{\Gamma\frac{n}{2}}2^{n/2}\int_0^{\infty} e^{-\frac{1}{2}\chi^2}(\chi^2)^{\frac{n}{2}-1}e^{t\chi^2}\,d\chi^2 \\ &= \frac{1}{\Gamma\frac{n}{2}2^{n/2}}\int_0^{\infty}(\chi^2)^{\frac{n}{2}-1}e^{-\frac{1}{2}\chi^2(1-2t)}d\chi^2.\end{aligned}$$

Let

$$z = \frac{1}{2}\chi^2(1-2t) \Rightarrow d\chi^2 = \frac{2z\,dz}{1-2t},\ 0<z<\infty$$

$$M_{\chi^2}(t) = \frac{1}{(1-2t)^{n/2}\Gamma\frac{n}{2}}\int_0^{\infty} e^{-z}z^{\frac{n}{2}-1}dz = (1-2t)^{-n/2},\ |2t|<1.$$

Using Binomial expansion for negative index, we get

$$\begin{aligned}M_{\chi^2}(t) &= 1+\frac{n}{2}(2t)+\frac{\frac{n}{2}\left(\frac{n}{2}+1\right)}{2!}(2t)^2+\cdots \\ &+\frac{\frac{n}{2}\left(\frac{n}{2}+1\right)\left(\frac{n}{2}+2\right)+\cdots+\left(\frac{n}{2}+r-1\right)}{r!}(2t)^r+\cdots\end{aligned}$$

We know μ_r' is the coefficient of $t^r/r!$ in the expansion of $M_{\chi^2}(t)$. Therefore

$$\begin{aligned}\mu_r' &= 2^r\frac{n}{2}\left(\frac{n}{2}+1\right)\left(\frac{n}{2}+2\right)+\cdots+\left(\frac{n}{2}+r-1\right) \\ &= n(n+2)(n+4)(n+6)+\cdots+\{n+2(r-1)\}.\end{aligned}$$

In particular

$$\begin{aligned}\mu_1' &= n,\ \mu_2' = n(n+2),\ \mu_3' = n(n+2)(n+4), \\ \mu_4' &= n(n+2)(n+4)(n+6) \\ \mu_2 &= \mu_2' - \mu_1'^2 = n(n+2)-n^2 = 2n.\end{aligned}$$

Cumulant Generating Function

The cumulant generating function of chi-square distribution is

$$\begin{aligned}K_{\chi^2}(t) &= \log M_{\chi^2}(t) = -\frac{n}{2}\log(1-2t) \\ &= \frac{n}{2}\left[2t+\frac{(2t)^2}{2}+\frac{(2t)^3}{3}+\frac{(2t)^4}{4}+\cdots\right].\end{aligned}$$

We know

$$K_r = \text{coefficient of } \frac{t^r}{r!} \text{ in } K_{\chi^2}(t) = n2^{r-1}(r-1)!.$$

In particular

$$\begin{aligned}K_1 &= \mu_1' = n,\ K_2 = \mu_2 = 2n,\ K_3 = \mu_3 = 8n, \\ K_4 &= 48n = \mu_4 - 3\mu_2^2,\ \mu_4 = 12n^2+48n.\end{aligned}$$

The moment ratios are

$$\alpha_3 = \sqrt{\beta_1} = \sqrt{\frac{8}{n}}, \quad \alpha_4 = \beta_4 = 3 + \frac{12}{n}$$

Hence, chi-square distribution is positively skewed and lepto kurtic.

Recurrence Relation for the Moments

We know $M_{\chi^2}(t) = (1-2t)^{-n/2}$ and $E(\chi^2) = n$. Then the m.g.f. about mean is

$$M_{(\chi^2-n)}(t) = E\left[e^{t(\chi^2-n)}\right] = e^{-nt}E\left[e^{t\chi^2}\right] = e^{-nt}(1-2t)^{-n/2}.$$

Taking logarithm on both sides, we get

$$\log M_{(\chi^2-n)}(t) = -nt - \frac{n}{2}\log(1-2t).$$

Differentiating with respect to 't', we get

$$\frac{M'_{(\chi^2-n)}(t)}{M_{(\chi^2-n)}(t)} = -n + \frac{n}{2}\frac{2}{(1-2t)} = \frac{2nt}{1-2t}$$

$$(1-2t)M'_{(\chi^2-n)}(t) = 2ntM_{(\chi^2-n)}(t).$$

Differentiating r times with respect to 't' by Leibnitz theorem, we get

$$(1-2t)M^{r+1}_{(\chi^2-n)}(t) + r(-2)M'_{(\chi^2-n)}(t) = 2ntM_{(\chi^2-n)}(t) + 2nrM^{r-1}_{(\chi^2-n)}(t).$$

Putting $t = 0$ and using the relation

$$\mu_r = \left[\frac{d}{dt}M^r_{(\chi^2-n)}(t)\right]_{t=0} = M^r_{(\chi^2-n)}(0)$$

we get

$$\mu_{r+1} - 2r\mu_r = 2nr\mu_{r-1}$$
$$\mu_{r+1} = 2r(\mu_r + n\mu_{r-1}), \ r \geq 1.$$

Thus

$$\mu_2 = 2n\mu_0 = 2n, \ [\because \mu_0 = 1]$$
$$\mu_3 = 4(\mu_2 + n\mu_1) = 8n, \ [\because \mu_1 = 0]$$
$$\mu_4 = 6(\mu_3 + n\mu_2) = 48n + 12n^2.$$

Characteristic Function

The characteristic function is given by

$$\psi(t) = E\left[e^{it\chi^2}\right] = \frac{1}{2^{n/2}\Gamma\frac{n}{2}}\int_0^\infty e^{itx}e^{-\frac{1}{2}\chi^2}(\chi^2)^{\frac{n}{2}-1}\,d\chi^2$$
$$= \frac{1}{2^{n/2}\Gamma\frac{n}{2}}\int_0^\infty e^{-\frac{1}{2}\chi^2(1-2it)}(\chi^2)^{\frac{n}{2}-1}\,d\chi^2 = (1-2it)^{-n/2}.$$

26.4 Properties

Mode

Mode is that value of χ^2 which satisfies the equations

$$f'(\chi^2) = 0 \text{ and } f''(\chi^2) < 0.$$

Here

$$\begin{aligned} f'(\chi^2) &= \frac{d}{d\chi^2}\left[\frac{1}{2^{n/2}\Gamma\frac{n}{2}} e^{-\frac{1}{2}\chi^2}(\chi^2)^{\frac{n}{2}-1}\right] \\ &= \frac{1}{2^{n/2}\Gamma\frac{n}{2}}\left[-\frac{1}{2}e^{-\frac{1}{2}\chi^2}(\chi^2)^{\frac{n}{2}-1} + \left(\frac{n}{2}-1\right)e^{-\frac{1}{2}\chi^2}(\chi^2)^{\frac{n}{2}-1}\right] \\ &= \frac{1}{2^{n/2}\Gamma\frac{n}{2}} e^{-\frac{1}{2}\chi^2}\left[\left(\frac{n}{2}-1\right)(\chi^2)^{-1} - \frac{1}{2}\right](\chi^2)^{\frac{n}{2}-1}. \end{aligned}$$

Thus $f'(\chi^2) = 0$, gives

$$\left(\frac{n}{2}-1\right)\frac{1}{\chi^2} - \frac{1}{2} = 0.$$

Since $e^{-\frac{1}{2}\chi^2}(\chi^2)^{\frac{n}{2}-1} \neq 0$. Therefore,

$$\chi^2 = (n-2).$$

It is easy to show that at the point $\chi^2 = (n-2)$

$$f''(\chi^2) < 0.$$

Hence mode of the distribution is $(n-2)$ for chi-square distribution with n d.f.

Skewness

Karl Pearson's coefficient of skewness is given by

$$\text{skewness} = \frac{\text{mean} - \text{mode}}{\text{standard deviation}} = \frac{n-(n-2)}{\sqrt{2n}} = \sqrt{\frac{2}{n}}.$$

It is seen that skewness is inversely proportional to the d.f. and hence the distribution rapidly tends to symmetry as the d.f. increases.

Mean Deviation

The mean deviation of chi-square distribution is

$$n^{n/2}e^{-n/2} \Big/ \left\{2^{\frac{n}{2}-1}\Gamma\frac{n}{2}\right\}.$$

Approximations

As $n \to \infty$, the standardized chi-square distribution tends to standard normal distribution.

The standardized chi-square distribution is defined by

$$Z = \frac{\chi^2 - n}{\sqrt{2n}}.$$

Now

$$\begin{aligned}
M_z(t) &= E\left[e^{zt}\right] \\
M_{\frac{\chi^2-n}{\sqrt{2n}}}(t) &= E\left[\exp\left(\frac{\chi^2-n}{\sqrt{2n}}\right)t\right] \\
&= e^{-\frac{nt}{\sqrt{2n}}}E\left[e^{\chi^2 t/\sqrt{2n}}\right] = e^{-\frac{nt}{\sqrt{2n}}}M_{\chi^2}\left(t/\sqrt{2n}\right) \\
&= e^{-\frac{nt}{\sqrt{2n}}}\left(1-\frac{2t}{\sqrt{2n}}\right)^{-\frac{n}{2}} = e^{-\frac{t\sqrt{n}}{\sqrt{2}}}\left(1-\frac{t\sqrt{2}}{\sqrt{n}}\right)^{-n/2} \\
K_z(t) &= \log M_z(t) = -\frac{t\sqrt{n}}{2} - \frac{n}{2}\log\left(1-\frac{t\sqrt{2}}{\sqrt{n}}\right) \\
&= -\frac{t\sqrt{n}}{\sqrt{2}} + \frac{n}{2}\left[\frac{t\sqrt{2}}{\sqrt{n}} + \frac{2t^2}{2n} + \frac{t^3 2^{3/2}}{3n^{3/2}} + \cdots\right] = \frac{t^2}{2} + \frac{t^3\sqrt{2}}{3\sqrt{n}} + \cdots \\
\lim_{n\to\infty} K_z(t) &= \frac{t^2}{2}, \text{ i.e., } M_z(t) = e^{t^2/2}
\end{aligned}$$

which is the m.g.f. of standard normal distribution. Therefore, standard chi-square distribution tends to standard normal distribution as n tends to infinity.

►**EXAMPLE 26.1** $\sqrt{2\chi^2} \sim N\left(\sqrt{2n}, 1\right)$, *where n is the d.f. of χ^2 distribution, for large n.*

Solution. We have standardized chi-square Z, where

$$Z = \frac{\chi^2 - n}{\sqrt{2n}}.$$

This $Z \sim N(0,1)$ for large n.

Consider,

$$\begin{aligned}
P\left[\frac{\chi^2-n}{\sqrt{2n}} \le Z\right] &= P\left[\chi^2 \le n + Z\sqrt{2n}\right] \\
&= P\left[\sqrt{2\chi^2} \le \left(2n + Z\sqrt{2n}\right)^{1/2}\right] \\
&= P\left[\sqrt{2\chi^2} \le \sqrt{2n}\left(1+\frac{Z\sqrt{2}}{\sqrt{n}}\right)^{1/2}\right] \\
&= P\left[\sqrt{2\chi^2} \le \sqrt{2n}\left(1+\frac{Z}{\sqrt{2n}} - \frac{Z^2}{4n} + \cdots\right)\right] \\
&= P\left[\sqrt{2\chi^2} \le \sqrt{2n} + Z\right], \text{ for large } n \\
&= P\left[\sqrt{2\chi^2} - \sqrt{2n} \le Z\right], \text{ for large } n.
\end{aligned}$$

Since for large n,

$$Z = \frac{\chi^2 - n}{\sqrt{2n}} \sim N(0,1)$$

we can conclude that $\left(\sqrt{2\chi^2} - \sqrt{2n}\right) \sim N(0,1)$, for large n.

$$\therefore\ \sqrt{2\chi^2} \sim N\left(\sqrt{2n}, 1\right) \text{ for large } n.$$

This result is used to find the value of χ^2 when the d.f., n, is larger than 30. This result does not reflect anything as to how good the approximation is, for moderate values of n. R.A. Fisher has shown that the approximation is improved by taking $\sqrt{(2n-1)}$ instead of $\sqrt{2n}$. Another improvement in approximation is due to Wilson-Hilferty (1931), which is

$$\left\{\left(\chi^2/n\right)^{1/3} - 1 + \frac{2}{9n}\right\}\sqrt{\frac{9n}{2}} \sim N(0,1).$$

An exprcssion for $P[\chi^2 \leq x]$ as a power series in x can be obtained in the following way:

$$\begin{aligned} P\left[\chi^2 \leq x\right] &= \frac{1}{2^{n/2}\Gamma\frac{n}{2}}\int_0^x e^{-\frac{t}{2}} t^{\frac{n}{2}-1} dt \\ &= \frac{1}{2^{n/2}\Gamma\frac{n}{2}}\int_0^x \sum_{j=0}^{\infty}(-1)^j \left(2^j j!\right)^{-1} t^{\frac{n}{2}+j-1} dt \\ &= \frac{2\left(\frac{x}{2}\right)^{n/2}}{\Gamma\frac{n}{2}} \sum_{j=0}^{\infty}(-1)^j \frac{x^j}{(n+2j)2^j j!} \end{aligned} \qquad (1)$$

It converges for all $x > 0$ and, for sufficiently large m, the true value lies between

$$\sum_{j=0}^{m} \text{ and } \sum_{j=0}^{m+1}.$$

The $P[\chi^2 \leq x)$ as given in (1) can also be written in terms of Pearson's incomplete gamma function

$$I_{u,P} = \frac{\left(\frac{u}{2}\sqrt{P+1}\right)^{P+1}}{\Gamma(P+1)} \sum_{j=0}^{\infty}(-1)^j \frac{\left(u\sqrt{P+1}\right)^j}{(P+j+1)2^{j+1}j!}.$$

►**EXAMPLE 26.2** *Show that if n is even,*

$$\begin{aligned} P &= \frac{1}{2^{\frac{n}{2}-1}\Gamma\left(\frac{n}{2}\right)}\int_x^{\infty} \exp\left(-\chi^2/2\right)\chi^{n-1} d\chi \\ &= \exp\left(-\chi^2/2\right)\left[1 + (\chi^2/2) + \chi^4/2\cdot 4 + \cdots + \chi^{n-2}/2\cdot 4\cdots(n-2)\right] \end{aligned}$$

and hence the values of P for a given χ^2 *can be derived from tables of Poisson's exponential limit.*

Solution. Let us consider the incomplete gamma integral

$$I_r = \frac{1}{r!}\int_\beta^\infty e^{-t}t^r\,dt$$

where r is a positive integer. Integrating by parts, we get

$$I_r = \left|-\frac{e^{-t}t^r}{r!}\right|_\beta^\infty + \frac{1}{(r-1)!}\int_\beta^\infty e^{-t}t^{r-1}dt = \frac{e^{-\beta}\beta^r}{r!} + I_{r-1}$$

which is a reduction formula. Repeated application of this gives

$$I_r = \frac{e^{-\beta}\beta^r}{r!} + \frac{e^{-\beta}\beta^{r-1}}{(r-1)!} + \cdots + \frac{e^{-\beta}\beta^2}{2!} + \frac{e^{-\beta}\beta}{1!} + I_0.$$

But

$$I_0 = \int_\beta^\infty e^{-t}dt = \left|-e^{-t}\right|_\beta^\infty = e^{-\beta}.$$

$$\therefore\ I_r = e^{-\beta}\left[1+\beta+\frac{\beta^2}{2!}+\cdots+\frac{\beta^{r-1}}{(r-1)!}+\frac{\beta^r}{r!}\right]$$

Putting

$$\beta = \frac{\chi^2}{2} \text{ and } r = \frac{n-2}{2} = \frac{n}{2}-1$$

(since r is an integer, $n = 2r+2$ must be even), we get

$$\frac{1}{\left(\frac{n}{2}-1\right)!}\int_{\chi^2/2}^\infty e^{-t}t^{\frac{n}{2}-1}dt = \exp\left(-\chi^2/2\right)\left[1+\frac{\chi^2}{2}+\frac{\chi^4}{2\cdot 4}+\cdots+\frac{\chi^{n-1}}{2\cdot4\cdot6\cdots(n-1)}\right] \quad (1)$$

Taking $t = \chi^2/2$ in the integral on the L.H.S., we get

$$\text{L.H.S.} = \frac{1}{\left(\frac{n}{2}-1\right)!}\int_\chi^\infty \exp\left(-\chi^2/2\right)\left(\chi^2/2\right)^{\frac{n}{2}-1} d(\chi^2/2)$$

$$= \frac{1}{2^{\frac{n}{2}-1}\Gamma\frac{n}{2}}\int_\chi^\infty \exp\left(-\chi^2/2\right)\chi^{n-1}d\chi. \quad (2)$$

From (1) and (2) the required result is obtained.

Let the given value of χ^2 be χ_0^2, then

$$P = P\left(\chi^2 > \chi_0^2\right) = \frac{1}{2^{\frac{n}{2}-1}\Gamma\frac{n}{2}}\int_\chi^\infty \exp\left(-\chi^2/2\right)\chi^{n-1}d\chi$$

$$= \exp\left(-\chi_0^2/2\right)\left[1+\frac{\chi_0^2}{2}+\frac{\chi_0^4}{2\cdot4}+\cdots+\frac{\chi_0^{n-1}}{2\cdot4\cdots(n-1)}\right]$$

$$= e^{-\lambda}\left[1+\lambda+\frac{\lambda^2}{2!}+\frac{\lambda^3}{3!}+\cdots+\frac{\lambda^{\left(\frac{n}{2}-1\right)}}{\left(\frac{n}{2}-1\right)!}\right]$$

where $\lambda = \chi_0^2/2$.

The terms on the right-hand side are the successive terms of the Poisson distribution with $\lambda = \chi_0^2/2$. Hence the result.

►**EXAMPLE 26.3** *Normal distribution is a particular case of χ^2-distribution with $n-1$ d.f.*

Solution. When $n=1$, the p.d.f. of χ^2-distribution is

$$f(\chi^2) = \frac{1}{2^{1/2}\Gamma\frac{1}{2}} e^{-\frac{1}{2}\chi^2} (\chi^2)^{\frac{1}{2}-1}, \ 0<\chi^2<\infty$$
$$= \frac{1}{\sqrt{2\pi}} e^{-\frac{1}{2}\chi^2} \frac{1}{\chi}, \ 0<\chi^2<\infty.$$

But $d(\chi^2) = 2\chi d\chi$.

$$\therefore \ f(\chi) = \frac{2}{\sqrt{2\pi}} e^{-\frac{1}{2}\chi^2}, \ 0<\chi<\infty$$
$$= \frac{1}{\sqrt{2\pi}} e^{-\frac{1}{2}\chi^2}, \ -\infty<\chi<\infty.$$

Additive Property

The sum of independent chi-square variates is also chi-square. Let $\chi_1^2, \chi_2^2, \ldots, \chi_k^2$ be the k independent chi-square variates with $n_1, n_2, \ldots, n_k$ d.f. respectively. The sum $\sum_{i=1}^{k} \chi_i^2$ is also a chi-square variate with $n = \sum_{i=1}^{k} n_i$ d.f.

We know, the m.g.f. of chi-square variate is

$$M_{\chi_i^2}(t) = (1-2t)^{-n_i/2}; \ i = 1, 2, \ldots, k.$$

Since $\chi_1^2, \chi_2^2, \ldots, \chi_k^2$ are independent, the m.g.f. of $\sum_{i=1}^{k} \chi_i^2$ can be written as

$$M_{\sum_i \chi_i^2}(t) = M_{\chi_1^2}(t) M_{\chi_2^2}(t) \cdots M_{\chi_k^2}(t)$$
$$= (1-2t)^{-\frac{n_1}{2}} (1-2t)^{-\frac{n_2}{2}} \cdots (1-2t)^{-\frac{n_k}{2}}$$
$$= (1-2t)^{-\frac{1}{2}\sum_{i=1}^{k} n_i}$$
$$= (1-2t)^{-\frac{n}{2}}$$

which is the m.g.f. of a chi-square variate with $n = \Sigma n_i$ d.f. Hence by uniqueness theorem of m.g.f.'s $\sum_{i=1}^{k} \chi_i^2$ is a chi-square variate with $\sum_{i=1}^{k} n_i$ d.f.

26.5 More about Chi-square Distribution

►**EXAMPLE 26.4** *If $X_1, X_2, \ldots, X_n$ are independent random variables having common density function*

$$f(x) = 1, \ 0<x<1.$$

Then if $P = X_1 X_2 \ldots X_n$, show that $-2\log_e P$ has chi-square distribution with $2n$ degrees of freedom.

Solution.

$$-2\log_e P = -2\log_e(X_1X_2\cdots X_n) = \eta_1+\eta_2+\cdots+\eta_n = \sum_{i=1}^{n}\eta_i$$

where $\eta_i = -2\log X_i$. Thus $X_i = e^{-\frac{\eta_i}{2}}$.

The probability function of η_i is given by

$$g(\eta_i) = f(x_i)\left|\frac{dx_i}{d\eta_i}\right| = \left|\exp\left(-\frac{\eta_i}{2}\right)\times\left(-\frac{1}{2}\right)\right|.$$

Since

$$f(x_i) = 1 \text{ and } \frac{dx_i}{d\eta_i} = -\frac{1}{2}e^{-\eta_i/2}$$

$$\therefore\ g(\eta_i) = \frac{1}{2}\exp\left(-\frac{\eta_i}{2}\right) = \frac{1}{2^{2/2}\Gamma\frac{2}{2}}e^{-\frac{\eta_i}{2}}\eta_i^{\frac{2}{2}-1}$$

which is the probability function of a chi-square variate with 2 d.f.

Again, η_i $(i = 1, 2, \ldots, k)$ are independent χ^2-variate with 2 d.f. Hence by additive property of chi-square variate, we have

$$-2\log_e P = \sum_{i=1}^{n}\eta_i$$

is a chi-square variate with $2n$ d.f.

►EXAMPLE 26.5 *The m.g.f. of* $Y = \log\chi^2$ *is*

$$M_Y(t) = 2^t\Gamma\left(\frac{n}{2}+t\right)\Big/\Gamma\frac{n}{2}$$

where χ^2 *is a chi-square variate with n d.f.*

Solution. Let $y = \log\chi^2 \Rightarrow \chi^2 = e^y$ and $d\chi^2 = e^y dy$. We know

$$f(\chi^2) = \frac{1}{2^{\frac{n}{2}}\Gamma\frac{n}{2}}e^{-\frac{1}{2}\chi^2}(\chi^2)^{\frac{n}{2}-1},\ 0<\chi^2<\infty$$

$$g(y) = f(\chi^2)\left|\frac{d\chi^2}{dy}\right|,\ -\infty<y<\infty$$

$$= \frac{1}{2^{\frac{n}{2}}\Gamma\frac{n}{2}}\exp\left[-\frac{1}{2}e^y+\frac{ny}{2}\right]$$

$$M_Y(t) = \frac{1}{2^{\frac{n}{2}}\Gamma\frac{n}{2}}\int_{-\infty}^{\infty}\exp\left[-\frac{1}{2}e^y+\frac{ny}{2}+ty\right]dy$$

$$= \frac{1}{2^{\frac{n}{2}}\Gamma\frac{n}{2}}\int_0^{\infty}e^{-z}(2z)^{\frac{n}{2}+t}\frac{dz}{z},\ [2z = e^y]$$

$$= \frac{2^t}{\Gamma\frac{n}{2}}\int_0^{\infty}e^{-z}z^{\frac{n}{2}+t-1}dz$$

$$= \frac{2^t\Gamma\left(\frac{n}{2}+t\right)}{\Gamma\frac{n}{2}}.$$

►**EXAMPLE 26.6** *If $X_1, X_2, \ldots, X_n$ are normally and independently distributed with mean zero and common variance σ^2, then show that*

$$\left(\sum_{i=1}^{n} X_i^2 - \sum_{i=1}^{p} y_i^2\right) \Big/ \sigma^2$$

is distributed as χ^2 with $(n-p)$ degrees of freedom where

$$y_i = \sum_{j=1}^{n} C_{ij} X_j; \; (i = 1, 2, \ldots, n) \text{ such that } \sum_{j=1}^{n} C_{ij} C_{i'j} = \delta_{ii'}.$$

Here $\delta_{ii'}$ is Kronecker delta.

Solution. Since $\delta_{ii'} = \sum_{j=1}^{n} C_{iji'j}$ = Kronecker delta, we have

$$\sum_{j=1}^{n} C_{ij} C_{i'j} = \begin{cases} 0, & \text{if } i \neq i' \\ 1, & \text{if } i = i' \end{cases}$$

Thus X_i's are transformed to y_i's by a linear orthogonal transformation. Hence by Fisher's lemma y_i $(i = 1, 2, \ldots, n)$ are independent normal variates with mean zero and common variance σ^2.

As the transformation is orthogonal, we have

$$\sum_{i=1}^{n} X_i^2 = \sum_{i=1}^{n} y_i^2.$$

$$\therefore \; \frac{1}{\sigma^2}\left(\sum_{i=1}^{n} X_i^2 - \sum_{i=1}^{p} y_i^2\right) = \frac{1}{\sigma^2}\left(\sum_{i=1}^{n} y_i^2 - \sum_{i=1}^{p} y_i^2\right)$$

$$= \frac{1}{\sigma^2} \sum_{i=p+1}^{n} y_i^2$$

$$= \sum_{i=p+1}^{n} \left(\frac{y_i}{\sigma}\right)^2.$$

But $\sum_{i=p+1}^{n} \left(\frac{y_i}{\sigma}\right)^2$ is the sum of squares of $(n-p)$ independent standard normal variates. Thus its distribution is χ^2 with $(n-p)$ d.f.

►**EXAMPLE 26.7** *If χ_1^2 and χ_2^2 are independently distributed as chi-square variates with n_1 and n_2 degrees of freedom respectively, then show that $U = \chi_1^2 + \chi_2^2$ and $V = \chi_1^2/(\chi_1^2 + \chi_2^2)$ are independentiy distributed as chi-square and beta variate of the first kind respectively.*

Solution. The joint p.d.f. of χ_1^2 and χ_2^2 is

$$f\left(\chi_1^2, \chi_2^2\right) = \frac{1}{2^{\frac{n_1+n_2}{2}} \Gamma\frac{n_1}{2} \Gamma\frac{n_2}{2}} e^{-\frac{1}{2}(\chi_1^2+\chi_2^2)} \left(\chi_1^2\right)^{\frac{n_1}{2}-1} \left(\chi_2^2\right)^{\frac{n_2}{2}-1}; 0 < \left(\chi_1^2, \chi_2^2\right) < \infty.$$

Let us make the transformation

$$u = \chi_1^2 + \chi_2^2 \text{ and } v = \chi_1^2/\left(\chi_1^2 + \chi_2^2\right)$$
$$\Rightarrow \chi_1^2 = uv \text{ and } \chi_2^2 = u(1-v)$$

$$J = \frac{\partial\left(\chi_1^2, \chi_2^2\right)}{\partial(u,v)} = \begin{vmatrix} \frac{\partial \chi_1^2}{\partial u} & \frac{\partial \chi_1^2}{\partial v} \\ \frac{\partial \chi_2^2}{\partial u} & \frac{\partial \chi_2^2}{\partial v} \end{vmatrix} = \begin{vmatrix} v & u \\ 1-v & -u \end{vmatrix} = -u$$

$$\begin{aligned} g(u,v) &= \frac{1}{2^{\frac{n_1+n_2}{2}}\Gamma\frac{n_1}{2}\Gamma\frac{n_2}{2}} e^{-\frac{1}{2}u}(uv)^{\frac{n_1}{2}-1}\{u(1-v)\}^{\frac{n_2}{2}-1}u \\ &= \frac{1}{2^{\frac{n_1+n_2}{2}}\Gamma\left(\frac{n_1+n_2}{2}\right)\beta\left(\frac{n_1}{2},\frac{n_2}{2}\right)} e^{-\frac{1}{2}u}u^{\frac{n_1+n_2}{2}-1}v^{\frac{n_1}{2}-1}(1-v)^{\frac{n_2}{2}-1} \\ &= \frac{1}{2^{\frac{n_1+n_2}{2}}\Gamma\left(\frac{n_1+n_2}{2}\right)} e^{-\frac{1}{2}u}u^{\frac{n_1+n_2}{2}-1} \frac{1}{\beta\left(\frac{n_1}{2},\frac{n_2}{2}\right)} v^{\frac{n_1}{2}-1}(1-v)^{\frac{n_2}{2}-1} \\ &= h_1(u)\, h_2(v). \end{aligned}$$

Here $g(u,v)$ can be expressed as the product of their marginal p.d.f. Hence U and V are independently distributed. Also, it is seen that U is a chi-square variate with (n_1+n_2) d.f. and V is beta variate of the first kind with parameters $\frac{n_1}{2}$ and $\frac{n_2}{2}$.

►**EXAMPLE 26.8** *If χ_1^2 and χ_2^2 are two independent chi-square variates with n_1 and n_2 d.f. respectively, find the distribution of $U = \chi_1^2/\chi_2^2$.*

Solution. The joint p.d.f. of χ_1^2 and χ_2^2 is

$$f\left(\chi_1^2, \chi_2^2\right) = \frac{1}{2^{\frac{n_1+n_2}{2}}\Gamma\frac{n_1}{2}\Gamma\frac{n_2}{2}} e^{-\frac{1}{2}\left(\chi_1^2+\chi_2^2\right)}\left(\chi_1^2\right)^{\frac{n_1}{2}-1}\left(\chi_2^2\right)^{\frac{n_2}{2}-1}; 0 < \left(\chi_1^2, \chi_2^2\right) < \infty.$$

Let us make the transformation

$$u = \chi_1^2/\chi_2^2 \text{ and } v = \chi_2^2$$
$$\Rightarrow \chi_1^2 = uv \text{ and } \chi_2^2 = v$$

$$J = \frac{\partial\left(\chi_1^2, \chi_2^2\right)}{\partial(u,v)} = \begin{vmatrix} \frac{\partial \chi_1^2}{\partial u} & \frac{\partial \chi_1^2}{\partial v} \\ \frac{\partial \chi_2^2}{\partial u} & \frac{\partial \chi_2^2}{\partial v} \end{vmatrix} = \begin{vmatrix} v & u \\ 0 & 1 \end{vmatrix} = v$$

$$g(u,v) = \frac{1}{2^{\frac{n_1+n_2}{2}}\Gamma\left(\frac{n_1+n_2}{2}\right)\beta\left(\frac{n_1}{2},\frac{n_2}{2}\right)} e^{-\frac{1}{2}v(1+u)}(uv)^{\frac{n_1}{2}-1}v^{\frac{n_2}{2}-1}v$$

$$g_1(u) = \frac{1}{\beta\left(\frac{n_1}{2},\frac{n_2}{2}\right)}u^{\frac{n_1}{2}-1}\frac{1}{2^{\frac{n_1+n_2}{2}}\Gamma\left(\frac{n_1+n_2}{2}\right)}\int_0^\infty e^{-\frac{1}{2}v(1+u)}v^{\frac{n_1+n_2}{2}-1}dv.$$

Put

$$v(1+u) = z \Rightarrow v = \frac{z}{1+u} \text{ and } dv = \frac{dz}{1+u} \text{ where } 0 < z < \infty.$$

Now

$$g_1(u) = \frac{1}{\beta\left(\frac{n_1}{2}, \frac{n_2}{2}\right)} \frac{u^{\frac{n_1}{2}-1}}{(1+u)^{\frac{n_1+n_2}{2}}} \frac{1}{2^{\frac{n_1+n_2}{2}}\Gamma\left(\frac{n_1+n_2}{2}\right)} \int_0^{\infty} e^{-\frac{1}{2}z} z^{\frac{n_1+n_2}{2}-1} dz$$

$$= \frac{1}{\beta\left(\frac{n_1}{2}, \frac{n_2}{2}\right)} \frac{u^{\frac{n_1}{2}-1}}{(1+u)^{\frac{n_1+n_2}{2}}}, \quad 0 < u < \infty$$

since

$$\int_0^{\infty} e^{-\frac{1}{2}z} z^{\frac{n_1+n_2}{2}-1} dz = 2^{\frac{n_1+n_2}{2}} \Gamma\left(\frac{n_1+n_2}{2}\right).$$

26.6 Noncentral Chi-square Distribution

If $y_1, y_2, \ldots, y_n$ are independent normal variables with mean $\mu_1, \mu_2, \ldots, \mu_n$, respectively, and with variance equal to unity, then

$$\chi_n'^2 = \sum_{i=1}^{n} y_i^2$$

is distributed as noncentral chi-square with n degrees of freedom and noncentrality parameter $\lambda = \frac{1}{2}\Sigma\mu_i^2$. The frequency function of $\chi_n'^2$ is given by

$$f\left(\chi_n'^2\right) = e^{-\lambda} \sum_{i=0}^{\infty} \frac{\lambda^i \left(\chi_n'^2\right)^{\frac{1}{2}(n+2i)-1} e^{-\frac{1}{2}\chi_n'^2}}{i!\, 2^{\frac{1}{2}(n+2i)} \Gamma\left(\frac{n+2i}{2}\right)}, \quad 0 < \chi_n'^2 < \infty \tag{26.2}$$

Thus the probability density function of noncentral chi-square distribution is the mixture of central chi-square distribution with d.f. $n, n+2, n+4, \cdots$, the corresponding weights being the successive terms of the Poisson distribution with parameter λ.

The distribution of $\chi_n'^2(\lambda)$, λ is the noncentrality parameter, has been derived in quite a number of different ways. Fisher (1928) gave an indirect derivation. Tang (1938) gave dircct derivation. The derivation can be done by a process of induction, first obtaining the distribution of $\chi_1'^2(\lambda)$ and then using the relation

$$\chi_n'^2(\lambda) = \chi_1'^2(\lambda) + \chi_{n-1}^2$$

the noncentral and central χ^2's in the right-hand side being mutually independent. Graybill (1961) used moment generating function technique to derive the distribution. Here Graybill's method of derivation has been discussed.

The m.g.f. of $\chi_n'^2(\lambda) = \sum_{i=1}^{n} y_i^2$ is given by

$$M_{\chi_n'^2}(t) = E\left[e^{\chi_n'^2 t}\right] = E\left[e^{t\Sigma y_i^2}\right]$$

$$= \int_{-\infty}^{\infty}\int_{-\infty}^{\infty}\cdots\int_{-\infty}^{\infty} \frac{1}{(2\pi)^{n/2}} \exp\left[-\frac{1}{2}\Sigma(y_i-\mu_i)^2 + t\Sigma y_i^2\right] dy_1 dy_2 \cdots dy_n$$

Here

$$\begin{aligned}
-\frac{1}{2}\Sigma\left(y_i-\mu_i\right)^2+t\Sigma y_i^2 &= -\frac{1}{2}\left(\Sigma y_i^2-2t\Sigma y_i^2-2\Sigma y_i\mu_i+\Sigma\mu_i^2\right)\\
&= -\frac{1-2t}{2}\left[\Sigma\left(y_i-\frac{\mu_i}{1-2t}\right)^2-\frac{\Sigma\mu_i^2}{(1-2t)^2}+\frac{\Sigma\mu_i^2}{1-2t}\right]\\
&= -\frac{\Sigma\mu_i^2}{2}\left(1-\frac{1}{1-2t}\right)-\frac{1-2t}{2}\sum\left(y_i-\frac{\mu_i}{1-2t}\right)^2.
\end{aligned}$$

Thus

$$\begin{aligned}
M_{\chi_n'^2(\lambda)}(t) &= \exp\left[-\frac{1}{2}\Sigma\mu_i^2\left(1-\frac{1}{1-2t}\right)\right]\int_{-\infty}^{\infty}\int_{-\infty}^{\infty}\\
&\qquad\cdots\int_{-\infty}^{\infty}\exp\left[-\frac{1-2t}{2}\sum\left(y_i-\frac{\mu_i}{1-2t}\right)^2dy_1dy_2\cdots dy_n\right]\\
&= \frac{1}{(1-2t)^{n/2}}\exp\left[-\frac{1}{2}\Sigma\mu_i^2\left(1-\frac{1}{1-2t}\right)\right]\\
&= \frac{\exp\left(-\frac{1}{2}\Sigma\mu_i^2\right)}{(1-2t)^{n/2}}\times\exp\left[\frac{\Sigma\mu_i^2}{2(1-2t)}\right].
\end{aligned}$$

Let $\lambda=\dfrac{1}{2}\Sigma\mu_i^2$. Thus

$$\exp\left[\frac{\Sigma\mu_i^2}{2(1-2t)}\right]=e^{\lambda/(1-2t)}.$$

Now, if we expand $e^{\lambda/(1-2t)}$ into a Taylor series about zero, we get

$$M_{\chi_n'^2(\lambda)}=e^{-\lambda}(1-2t)^{-n/2}\sum_{i=0}^{\infty}\frac{\lambda^i}{i!(1-2t)^i}=e^{-\lambda}\sum_{i=0}^{\infty}\frac{\lambda^i}{i!}(1-2t)^{-(n/2+i)}. \tag{26.3}$$

The p.d.f. (26.2) can be written as

$$f\left(\chi_n'^2\right)=\sum_{i=0}^{\infty}\frac{e^{-\lambda}\lambda^i}{i!}g_{n+2i}\left(\chi_n'^2\right)$$

where $g_{n+2i}\left(\chi_n'^2\right)$ is the central chi-square density function with $(n+2i)$ degrees of freedom. Here $e^{-\lambda}\lambda^i/i!$ is the i-th term of a Poisson distribution. Therefore, the noncentral chi-square is the sum of an infinite number of chi-square frequencies each weighted by a term from the Poisson distribution. We know

$$\int_0^{\infty}g_{n+2i}\left(\chi_n'^2\right)d\left(\chi_n'^2\right)=1.$$

Hence

$$\int_0^{\infty}f\left(\chi_n'^2\right)d\chi_n'^2=\int_0^{\infty}e^{-\lambda}\sum_{i=0}^{\infty}\frac{\lambda^i}{i!}g_{n+2i}\left(\chi_n'^2\right)d\left(\chi_n'^2\right)=e^{-\lambda}\sum_{i=0}^{\infty}\frac{\lambda^i}{i!}=1.$$

Let the m.g.f. of noncentral chi-square distribution be denoted by $m_{\chi_n'^2}(t)$, where

$$\begin{aligned} m_{\chi_n'^2}(t) &= E\left[e^{\chi_n'^2 t}\right] \\ &= \int_0^\infty e^{\chi_n'^2 t}\sum_{i=0}^{\infty}\frac{e^{-\lambda}\lambda^i}{i!} g_{n+2i}\left(\chi_n'^2\right) d\left(\chi_n'^2\right) \\ &= \sum_{i=0}^{\infty}\frac{e^{-\lambda}\lambda^i}{i!}\int_0^\infty e^{\chi_n'^2 t} g_{n+2i}\left(\chi_n'^2\right) d\left(\chi_n'^2\right). \end{aligned}$$

The term in the integral is the m.g.f. of central chi-square distribution with $(n+2i)$ d.f. This m.g.f. is $(1-2t)^{-\frac{1}{2}(n+2i)}$. Therefore

$$\begin{aligned} m_{\chi_n'^2}(t) &= \sum_{i=0}^{\infty}\frac{e^{-\lambda}\lambda^i}{i!}(1-2t)^{-\frac{1}{2}(n+2i)} \qquad (26.4) \\ &= M_{\chi_n'^2}(t). \end{aligned}$$

As the two moment generating functions are same, the distributions of

$$\sum_{i=1}^{n} y_i^2 \text{ and } f\left(\chi_n'^2\right)$$

are same. The two moment generating functions exist in some neighbourhood of zero.

If $\mu_1 = \mu_2 = \cdots = \mu_n = 0$, λ becomes zero. Then noncentral chi-square distribution transforms to the central chi-square distribution.

The cumulative distribution function of $\chi_n'^2(\lambda)$ is

$$\begin{aligned} F\left(\chi_n'^2(\lambda)\right) &= P\left[\chi_n'^2(\lambda) \le x\right] \\ &= e^{-\frac{1}{2}\lambda}\sum_{i=0}^{\infty}\frac{\left(\frac{1}{2}\lambda\right)^i}{i!}\frac{1}{2^{\frac{n}{2}+i}\Gamma\left(\frac{n}{2}+i\right)}\int_0^x e^{-\frac{1}{2}y}y^{\frac{n}{2}+i-1}dy. \end{aligned}$$

The noncentral chi-square distribution arises to study the distribution of sum of squares. Let $X_1, X_2, \ldots, X_n$ be n independent random variables distributed normally with means $\mu_1, \mu_2, \ldots, \mu_n$, respectively and common variance σ^2. Let us define

$$Z_i = \frac{X_i - \mu_i}{\sigma} \Rightarrow X_i = \mu_i + \sigma Z_i.$$

Let the sum of squares of X_i's be

$$S^2 = \sum_{i=1}^{n}(X_i - \bar{X})^2 = \sum_{i=1}^{n}(\mu_i + \sigma Z_i - \bar{\mu} - \sigma\bar{Z})^2$$

where

$$\bar{\mu} = \frac{1}{n}\sum_{i=1}^{n}\mu_i, \; \bar{Z} = \frac{1}{n}\sum_{i=1}^{n}Z_i.$$

Now, applying a transformation from $Z_1, Z_2, \ldots, Z_n$ to $y_1, y_2, \ldots, y_{n-1}, \bar{Z}$ such that

$$\sum_{i=1}^{n} (Z_i - \bar{Z})^2 = \sum_{i=1}^{n-1} y_i^2.$$

Also, we can write

$$\begin{aligned} S^2 &= \sigma^2 \sum_{i=1}^{n} \left(\sigma^{-1}\mu_i + Z_i - \bar{Z} - \sigma^{-1}\bar{\mu}\right)^2 \\ &= \sigma^2 \sum_{i=1}^{n-1} (y_i + \delta_i)^2 \end{aligned}$$

where the δ_i's are linear functions of the μ_i's, and y_i's of Z_i's. Putting $Z_i = 0$ for all i, it follows that $y_i = 0$ for all i and hence

$$\sum_{i=1}^{n-1} \delta_i^2 = \sum_{i=1}^{n} (\mu_i - \bar{\mu})^2 / \sigma^2$$

and so S^2 is distributed as σ^2 times a noncentral chi-square with $(n-1)$ degrees of freedom with noncentrality parameter

$$\sum_{i=1}^{n} (\mu_i - \bar{\mu})^2 / \sigma^2.$$

The noncentral chi-square distribution can be regarded as a *generalized Rayleigh distribution*. It is also called *Rayleigh-Rice* or *Rice distribution*. The distribution was obtained by Fisher in 1928 as a limiting distribution of the multiple correlation coefficient.

26.7 Moments and Generating Functions of Noncentral Chi-square Distribution

If $X \sim N(\mu, 1)$, then the m.g.f. of X^2 is given by

$$\begin{aligned} M_{X^2}(t) &= E\left[e^{tX^2}\right] = \frac{1}{\sqrt{2\pi}} \int_{-\infty}^{\infty} e^{tx^2} e^{-\frac{1}{2}(x-\mu)^2} dx \\ &= \frac{1}{\sqrt{2\pi}} \int_{-\infty}^{\infty} \exp\left[-\left\{\left(\frac{1}{2} - t\right) x^2 - \mu x + \frac{\mu^2}{2}\right\}\right] dx \\ &= \frac{1}{\sqrt{2\pi}} \int_{-\infty}^{\infty} \exp\left[-\left(\frac{1-2t}{2}\right)\left\{x^2 - \frac{2\mu x}{1-2t} + \frac{\mu^2}{1-2t}\right\}\right] dx \\ &= \frac{1}{\sqrt{2\pi}} \int_{-\infty}^{\infty} \exp\left[-\left(\frac{1-2t}{2}\right)\left\{\left(x - \frac{\mu}{1-2t}\right)^2 + \frac{\mu^2}{1-2t} - \frac{\mu^2}{(1-2t)^2}\right\}\right] dx \\ &= \frac{1}{\sqrt{2\pi}} \exp\frac{t\mu^2}{1-2t} \int_{-\infty}^{\infty} \exp\left[-\left(\frac{1-2t}{2}\right)\left(x - \frac{\mu}{1-2t}\right)^2\right] dx. \end{aligned}$$

Let

$$(1-2t)^{1/2}\left\{x-\frac{\mu}{1-2t}\right\} = u \Rightarrow x = \frac{\mu+\mu(1-2t)}{(1-2t)^{1/2}(1-2t)}$$

$$dx = \frac{du}{(1-2t)^{1/2}}$$

$$M_{X^2}(t) = \exp\frac{t\mu^2}{1-2t}\frac{1}{(1-2t)^{1/2}}\frac{1}{\sqrt{2\pi}}\int_{-\infty}^{\infty} e^{-\frac{1}{2}u^2}du$$

$$= (1-2t)^{-1/2}\exp\frac{t\mu^2}{1-2t};\ 1-2t>0 \Rightarrow t<\frac{1}{2}.$$

If $X_1, X_2, \ldots, X_n$ are independent and if X_i is $N(\mu_i, 1)$, then $\Sigma X_i^2 = \chi_n'^2(\lambda)$. Therefore, the m.g.f. of $\chi_n'^2(\lambda)$ is

$$M_{\chi_n'(\lambda)}(t) = M_{\Sigma X_i^2}(t) = M_{X_1}(t)M_{X_2}(t)\cdots M_{X_n}(t)$$

$$= (1-2t)^{-n/2}\exp\left[\frac{t}{1-2t}\sum_{i=1}^{n}\mu_i^2\right].$$

Putting $\lambda = \frac{1}{2}\Sigma\mu_i^2$, we get

$$M_{\chi_n'^2(\lambda)}(t) = (1-2t)^{-n/2}e^{2\lambda t/(1-2t)}$$

where λ is the noncentrality parameter.

$$\therefore\ M_{\chi_n'^2(\lambda)}(t) = (1-2t)^{-n/2}e^{-\lambda}\exp\left(\frac{\lambda}{1-2t}\right)$$

$$= (1-2t)^{-n/2}e^{-\lambda}\sum_{r=0}^{\infty}\left(\frac{\lambda}{1-2t}\right)^r\frac{1}{r!}$$

$$= \sum_{r=0}^{\infty}\frac{e^{-\lambda}\lambda^r}{r!}(1-2t)^{-(\frac{n}{2}+r)};\ t<\frac{1}{2}.$$

If we take $\lambda = \Sigma\mu_i^2$, then

$$M_{\chi_n'^2(\lambda)}(t) = (1-2t)^{-n/2}\exp\frac{\lambda t}{1-2t}.$$

The cumulant generating function is

$$K(t,\lambda) = \log M_{\chi_n'^2(\lambda)}(t) = -\frac{n}{2}\log(1-2t) + \lambda t(1-2t)^{-1}.$$

Hence the r-th cumulant is

$$K_r = 2^{r-1}(r-1)!(n+r\lambda).$$

In particular

$$K_1 = \mu_1' = (n+\lambda),\ K_2 = \mu_2 = 2(n+2\lambda),$$

$$K_3 = \mu_3 = 8(n+3\lambda),\ K_4 = 48(n+4\lambda),$$

$$\mu_4 = K_4 + 3\mu_2^2 = 48(n+4\lambda) + 12(n+2\lambda)^2.$$

The moment ratios are

$$\alpha_3 = \sqrt{\beta_1} = \sqrt{8}(n+3\lambda)/(n+2\lambda)^{3/2}$$
$$\alpha_4 = \beta_2 = 3 + 12(n+4\lambda)/(n+2\lambda)^2$$

If $\lambda = \frac{1}{2}\Sigma\mu_i^2$, then

$$K_r = 2^{r-1}(r-1)!(n+2r\lambda).$$

The characteristic function of $\chi_n'^2(\lambda)$ is

$$(1-2it)^{-n/2}\exp\left[it\lambda(1-2it)^{-1}\right].$$

26.8 Properties of Noncentral Chi-square Distribution

If $Z_1, Z_2, \ldots, Z_n$ are independent noncentral chi-square variates with $n_1, n_2, \ldots, n_k$ degrees of freedom and with $\lambda_1, \lambda_2, \ldots, \lambda_k$ noncentrality parameters, respectively, then $\sum_{i=1}^{k} Z_i$ is also a noncentral chi-square variate with $\sum_{i=1}^{k} n_i = n$ d.f. and noncentrality parameter $\sum_{i=1}^{k} \lambda_i$.

We know

$$M_{\chi_1'^2(\lambda)}(t) = (1-2t)^{-n/2}e^{2\lambda t(1-2t)}.$$

Thus

$$M_{Z_i}(t) = (1-2t)^{-n_i/2}e^{2\lambda_i t(1-2t)}.$$

Again

$$\begin{aligned} M_{\sum_1^k Z_i}(t) &= M_{Z_1}(t)M_{Z_2}(t)\cdots M_{Z_k}(t) \\ &= \prod_{i=1}^{k}(1-2t)^{\frac{-n_i}{2}}e^{2\lambda_i t(1-2t)} \\ &= (1-2t)^{-n/2}e^{2t\Sigma\lambda_i(1-2t)} \end{aligned}$$

which is the m.g.f. of a noncentral chi-square variate with $n = \Sigma n_i$ degrees of freedom and noncentrality parameter $\Sigma\lambda_i$. Therefore, by uniqueness theorem of m.g.f.'s, we can say that

$$\sum_{i=1}^{k} Z_i \sim \chi_n'^2(\Sigma\lambda_i).$$

This property is called the *additive property of noncentral chi-square distribution.*

If Z has a $\chi_n'^2(\lambda)$ distribution and $Z = Z_1 + Z_2 + \cdots + Z_k$, where the Z_i's are independent and identically distributed, then each Z_i has a $\chi_1'^2(\lambda/n)$ distribution.

The standardized noncentral chi-square variate is given by

$$\frac{\chi_n'^2(\lambda) - (n+\lambda)}{[2(n+2\lambda)]^{1/2}}.$$

This standardized variable tends to normality as $n \to \infty$, λ remaining fixed or as $n \to \infty$, n remaining fixed.

From the moment ratios, we have

$$\begin{aligned}\frac{\beta_2 - 3}{\beta_1} &= \frac{3(n+4\lambda)(n+2\lambda)}{2(n+3\lambda)^2} \\ &= \frac{3}{2}\left[1 - \frac{\lambda^2}{(n+3\lambda)^2}\right]\end{aligned}$$

whence

$$\frac{4}{3} \le \frac{\beta_2 - 3}{\beta_1} \le \frac{3}{2}. \tag{26.5}$$

This inequality and the density function (26.2) lead one to expect that a gamma distribution should give a useful approximation of noncentral chi-square distribution. The approximation is done by replacing $\chi_n'^2(\lambda)$ by a multiple of χ^2, $c\chi_i^2$ say, with c and f so chosen that the first two moments of the two variables, $\chi_n'^2(\lambda)$ and $c\chi_i^2$, agree. The approximate values of c and f are

$$c = \frac{n+2\lambda}{n+\lambda}$$

$$f = \frac{(n+\lambda)^2}{(n+2\lambda)} = n + \frac{\lambda^2}{n+2\lambda}.$$

This approximation was suggested by Patnaik (1949).

Using Wilson-Hilferty approximations it can be shown that $\left[\dfrac{\chi_n'^2(\lambda)}{n+\lambda}\right]^{1/3}$ is normal with expected value

$$1 - \frac{2(n+2\lambda)}{9(n+\lambda)} \text{ and variance } \frac{2(n+2\lambda)}{n+\lambda}.$$

Sankaran (1959, 1963) discussed that

(i) $\left\{\chi_n'^2(\lambda) - \frac{1}{2}(n-1)\right\}^{1/2}$ is approximately normal with expected value

$$\left\{1 + \frac{1}{2}(n-1)\right\}^{1/2} \text{ and variance 1.}$$

(ii) $\left\{\left[\chi_n'^2(\lambda) - \frac{1}{3}(n-1)\right]/(n+\lambda)\right\}^{1/2}$ is approximately normal with mean value

$$\left\{1 - \frac{n-1}{3(n+\lambda)}\right\} \text{ and variance } (n+\lambda)^{-1}.$$

(iii) $\left[\dfrac{\chi_n'^2(\lambda)}{n+\lambda}\right]^h$ is approximately normal with expected value

$$1 + h(h-1)\frac{n+2\lambda}{(n+\lambda)^2} - h(h-1)(2-h)(1-3h)\frac{(n+2\lambda)^2}{2(n+\lambda)^4}$$

and variance

$$h^2\frac{2(n+2\lambda)}{(n+\lambda)^2}\left[1-(1-h)(1-3h)\frac{n+2\lambda}{(n+\lambda)^2}\right]$$

where

$$h = 1 - \frac{2}{3}(n+\lambda)(n+3\lambda)(n+2\lambda)^{-2}.$$

26.9 Applications

Central Chi-square

The distribution has a large number of applications in Statistics, specially in testing the hypotheses (i) $H_0 : \sigma^2 = \sigma_0^2$, (ii) $H_0 : \sigma_1^2 = \sigma_2^2 = \cdots = \sigma_k^2$, where $\sigma^2, \sigma_1^2, \ldots, \sigma_k^2$ are the population variances. In social research χ^2-test is widely used to test the independence of attributes. The goodness of fit of any theoretical distribution to the frequency distribution is judged by χ^2-test. The probabilities of test obtained from repeated experiments are combined to give a single test of significance, where combined probabilities follow chi-square distribution. Homogeneity of correlation coefficients obtained from independent samples drawn from same population is also judged using χ^2-test.

Noncentral Chi-square

A useful application of noncentral chi-square distribution is in approximating to the power of χ^2-tests applied to contingency tables. The distribution appears in the calculation of approximate powers of certain nonparametric tests. Noncentral chi-square distribution represents the distribution of sample variance from a normal population with unstable mean. If the data are represented by n independent random variables with identical distributions depending on parameters, then the limiting distribution, under certain conditions, of $-2\log_e$ (likelihood ratio) is a noncentral χ^2-distribution.

EXERCISE

1. (a) Define sampling distribution. What are the important sample statistics? Define chi-square statistic and derive its density function.
   (b) Derive moment generating function of chi-square distribution and hence discuss the properties of chi-square distribution.

2. (a) Define chi-square variate. Let $x \sim N(0, 1)$, show that Σx_i^2 is a chi-square variate, where $x_1, x_2, \ldots, x_n$ are n independent sample observations.
   (b) Let χ_1^2 be a chi-square variate with n_1 d.f. and χ_2^2 be another chi-square variate with n_2 d.f. Assume that χ_1^2 and χ_2^2 are independent. Derive the distribution of

$$U = \chi_1^2 + \chi_2^2 \text{ and } V = \frac{\chi_1^2}{\chi_1^2 + \chi_2^2}.$$

3. (a) Let $x_1, x_2, \ldots, x_n$ be n random sample observations from $N(\mu, \sigma^2)$. Then (i) $\bar{x}$ and (ii) ns^2/σ^2 are independently distributed as normal and chi-square variates, respectively.
   (b) Find moment generating function of chi-square variate and show that chi-square distribution is positively skewed and lepto kurtic.

4. (a) Define chi-square variate. Show that $\sqrt{2\chi^2} \sim N\left(\sqrt{2n}, 1\right)$, where n is the d.f. of χ^2-variate.
   (b) State and prove additive property of chi-square distribution.
   (c) Discuss some applications of chi-square distribution.

5. (a) Define chi-square variate. If $X_1, X_2, \ldots, X_n$ are independet random variables having common density function

$$f(x) = 1, \ 0 < x < 1$$

   then, if $P = X_1 X_2 \cdots X_n$, show that $-2 \ln P$ has χ^2-distribution with $2n$ d.f.
   (b) Show that normal distribution is a particular case of χ^2-distribution with 1 d.f.

6. Define χ^2-variate. If $x \sim N(0, 1)$, then Σx^2 is a χ^2-variate. State and prove some important properties of chi-square distribution. Write down some application of chi-square distribution.

7. (a) If $X_1, X_2, \ldots, X_n$ are normally and independently distributed with mean zero and common variance σ^2, then show that

$$\left(\sum^{n} X_i^2 - \sum^{p} y_i^2\right) \Big/ \sigma^2$$

   is distributed as chi-square with $(n - p)$ d.f., where

$$y_i = \sum_{j=1}^{n} c_{ij} X_j; \ i = 1, 2, \ldots, n$$

   and $\sum_{j=1}^{n} C_{ij} C_{i'j} = \delta_{ii'}$. Here $\delta_{ii'}$ is Kronecker delta.

(b) If χ_1^2 and χ_2^2 are two independent chi-square variates with n_1 and n_2 d.f., respectively, find the distribution of $U = \chi_1^2/\chi_2^2$.

8. Distinguish between central and non-central chi-square variates. Derive the density function of non-central chi-square variate. Under what condition a non-central chi-square distribution becomes central chi-square distribution?

State and prove some important properties of non-central chi-square distribution.

CHAPTER 27

Distribution of Quadratic Forms

27.1 Introduction

Let $Y' = (Y_1, Y_2, \dots, Y_n)$ be a random vector which follows multivariate normal distribution $N(\mu, V)$, where $\mu' = (\mu_1, \mu_2, \dots, \mu_n)$ and V is the variance-covariance matrix. Then $Y'AY = Q$ is the quadratic form associated with a symmetric matrix A. The quadratic form Q can be expressed as

$$Q = \sum_{i}^{n} \sum_{j}^{n} a_{ij} Y_i Y_j.$$

The distribution of Q is known as the distribution of quadratic form. This Q involves in many statistical analysis, specially, to study the distribution of sum of squares when it is calculated on the basis of sample observations drawn from normal population. The analysis of variance is entirely concerned with statistics constructed from quadratic forms in random variables representing original or transformed observations.

The quadratic form is an expression of sum of squares and sum of products. Its distribution is related to central or noncentral chi-square distribution depending on the parameters of multivariate normal distribution from which the sample is drawn. For example, if Y is distributed $N(O, I)$, then $Y'AY$ is distributed as central chi-square distribution. On the other hand, when $Y \sim N(\mu, I)$, then $Y'AY$ is distributed as $\chi'^2(\lambda)$, where $\lambda = \frac{1}{2}\mu' A \mu$ is the noncentrality parameter vector. In this chapter we shall discuss the derivation, properties, application and the conditions under which a quadratic form follows central chi-square distribution.

27.2 Derivation of Distribution

Let us consider that the elements of the vector $Y' = (Y_1, Y_2, \dots, Y_n)$ follow multivariate normal distribution where the elements of mean vector are $\mu_1 = \mu_2 = \cdots = \mu_n = 0$. Then the density function of Y is

$$f(Y) = (2\pi)^{-n/2} |V|^{1/2} \exp\left(-\frac{1}{2} Y' V^{-1} Y\right). \tag{27.1}$$

We know

$$Q = Y'AY = \sum_{i=1}^{n}\sum_{j=1}^{n} a_{ij}Y_iY_j$$

where A is a symmetric matrix with elements a_{ij}. Our objective is to find the distribution of Q, that is to find $P[Q \leq x]$, $-\infty < x < \infty$, where

$$P[Q \leq x] = \int_{Y'AY \leq x} f(Y)dY. \tag{27.2}$$

The term under integration can be expressed in a convenient form. For this, let us consider a transformation

$$y_1 = L^{-1}y$$

where L is a non-singular matrix and is obtained by factorizing V, i.e., $V = LL'$. Since V is positive definite and symmetrical, factorization of V is permissible, where L is a non-singular lower triangular matrix. Since $y_1'y_1 = y'Y^{-1}y$ and $|V| = |L|^2$, $P[Q \leq x]$ can be expressed as

$$P[Q \leq x] = (2\pi)^{-n/2} \int_{y_1'(L'AL)y_1 \leq x} \exp\left(-\frac{1}{2}y_1'y_1\right) dy_1. \tag{27.3}$$

The matrix $L'AL$ is symmetric. Let P be an orthogonal matrix. Then

$$P'(L'AL)P = D$$

where D is the diagonal matrix. The elements of D are the eigenvalues of $L'AL$. Now, we can apply another linear transformation on y_1 such that $W = P'y_1$. Then, we have

$$P[Q \leq x] = (2\pi)^{-n/2} \int_{W'DW \leq x} \exp\left(-\frac{1}{2}W'W\right) dW. \tag{27.4}$$

Thus, the distribution of Q is the same as that of

$$W'DW = \sum_{i=1}^{n} d_{ii}W_i^2$$

where the variables W_i are independent standard normal variables. Here

$$d_{11} \geq d_{22} \geq d_{33} \geq \cdots \geq d_{nn}$$

are the eigenvalues of $L'AL$, or equivalently of VA. Here (27.4) is the cumulative distribution function of Q.

If all the $\boldsymbol{\mu}_i$'s are not equal to, an exactly similar expression for Q can be derived, where

$$Q = \sum_{i=1}^{n} d_{ii}(W_i - w_i)^2$$

where w_i is the same function of the μ_i's as W_i is of the Y''s.

If all the diagonal elements of D are same, i.e., if

$$d_{11} = d_{22} = \cdots = d_{nn} = d \text{ (say)}$$

then $Q = d \sum_{i=1}^{n} W_i^2$. Therefore, Q is distributed as $d\chi_n^2$ with n d.f.

Let $F_n(x; \delta)$ and $F_n(x; \delta; w)$ be the cumulative distribution function of central and noncentral distribution of Q. Then the probability functions are

$$\begin{aligned} f_n(x; \delta) &= \frac{dF_n(x; \delta)}{dx} \\ \text{and } f_n(x; \delta; w) &= \frac{dF_n(x; \delta w)}{dx}, \text{ respectively.} \end{aligned}$$

Also, let us consider that the characteristic functions are $\psi_n(t; \delta)$ and $\psi_n(t; \delta; w)$ in the central and noncentral cases respectively.

Since W_i's are mutually independent standard normal random variables, the characteristic function of $Q = \sum_{j=1}^{n} d_{jj}(W_j - w_j)^2$ is given by

$$\begin{aligned} \psi_n(t; \delta; w) &= E\left[e^{itQ}\right] \\ &= (2\pi)^{-n/2} \int_{-\infty}^{\infty} \int_{-\infty}^{\infty} \cdots \int_{-\infty}^{\infty} \exp\left[-\frac{1}{2}\sum_{j=1}^{n} W_j^2 + it \sum_{j=1}^{n} d_{jj}(W_j - w_j)^2\right] \\ &\qquad \times dW_1 dW_2 \cdots dW_n \\ &= \exp\left(-\frac{1}{2}\sum_{j=1}^{n} w_j^2\right) \exp\left(\frac{1}{2}\sum_{j=1}^{n} \frac{w_j^2}{(1 - 2itd_{jj})}\right) \times \prod_{j=1}^{n} (1 - 2itd_{jj})^{-1/2}. \end{aligned}$$

In central case $w_j = 0$. Therefore,

$$\psi_n(t; \delta) = (1 - 2itd_{jj})^{-n/2}.$$

Using the value of $\psi_n(t; \delta)$ we shall find the general formula for $f_n(x; \delta)$. Let us first consider a special case where the eigenvalues of A are all of even multiplicity. Thus n must be even and in the most general case there are $n/2$ distinct eigenvalues $\beta_1, \beta_2, \ldots, \beta_m$, where $m = n/2$, each of multiplicity two. Therefore, the characteristic function is

$$\psi_n(t, \delta) = \prod_{j=1}^{n} (1 - 2it\beta_j)^{-1}.$$

Assume that A is positive definite so that $\beta_1 > \beta_2 > \cdots > \beta_n > 0$ and all multiplicities are exactly two. Then the partial fraction expansion

$$\psi_n(t; \delta) = \sum_{j=1}^{m} \frac{b_j}{1 - 2it\beta_j} \tag{27.5}$$

where

$$b_j = \beta_j^{m-1} \sum_{\substack{i=1 \\ j\neq 1}}^{m} (\beta_j - \beta_l)^{-1}.$$

Thus we have

$$f_n(x;\delta) = \frac{1}{2}\sum_{j=1}^{m}(b_j/\beta_j)\exp\left(-\frac{1}{2}y/\beta_j\right), \quad x > 0. \tag{27.6}$$

❑ **THEOREM 27.1** *If the random vector Y is distributed $N(O, I)$, a necessary and sufficient condition that the quadratic form $Y'AY$ be distributed as central chi-square with k d.f. is that A be an idempotent matrix of rank k.*

Proof. If A is an idempotent matrix of rank k, then

$$P'AP = \begin{bmatrix} I & O \\ O & O \end{bmatrix}$$

where P is an orthogonal matrix and I is the $k \times k$ unit matrix. Let us define a vector Z of order $n \times 1$ such that $Z = P'Y$. As P is an orthogonal matrix, Z is distributed $N(O, I)$. Thus

$$Y'AY = Z'P'APZ = Z'Z = \sum_{i=1}^{k} Z_i^2$$

which is the sum of squares of independent normal variables with means zero and standard deviation 1. Therefore, $\sum_{i=1}^{k} Z_i^2$ is distributed as central chi-square with k d.f. This completes the sufficiency of the theorem.

To prove necessity, let us assume that $Y'AY$ is distributed as chi-square with k d.f. Since A is symmetric, there exists an orthogonal matrix C such that $C'AC = D$, where D is a diagonal matrix with d_{ii} as i-th diagonal elements. Let $[w_i] = W = C'Y$, then

$$Y'AY = W'C'ACW = W'DW = \sum_{i=1}^{n} W_i^2 d_{ii}.$$

As C is orthogonal, then w_i are independent standard normal variables. Therefore, the moment generating function of

$$u = \sum_{i=1}^{n} W_i^2 d_{ii} \text{ is } \prod_{i=1}^{n}(1 - 2td_{ii})^{-\frac{1}{2}}.$$

Again, the m.g.f. of $Y'AY$ is $(1-2t)^{-k/2}$. As $Y'AY = u$, it is clear that there exists a neighbourhood of zero such that, for all t in that neighbourhood, the quantities of the two moment generating functions are equal. For this to be true, k of the d_{ii} must be equal to unity and the remaining d_{ii} must be equal to zero. But, since d_{ii} are the characteristic roots of A, this implies that A is idempotent matrix of rank k.

In a similar way it can be proved that, if Y is distributed $N(\mu, I)$, then $Y'AY$ is distributed as $\chi_k'^2(\lambda)$, where $\lambda = \frac{1}{2}\mu' A\mu$, if and only if A is an idempotent matrix of rank k. Also, it can be proved that, if Y is distributed $N(\mu, \sigma^2 I)$, then $Y'AY/\sigma^2$ is distributed as $\chi_k'^2(\lambda)$, where $\lambda = \frac{1}{2\sigma^2}\mu' A\mu$, if an only if A is an idempotent matrix of rank k.

❑ **THEOREM 27.2** *If a vector Y is distributed $N(O, V)$, then $Y'BY$ is distributed as central chi-square with k degrees of freedom if and only if BV is an idempotent matrix of rank k.*

Proof. Let C be a matrix such that $C'VC = I$. If we transform Y to Z such that $Z = C'Y$, then Z is distributed $N(O, I)$. We have $Y'BY = Z'C^{-1}BC'^{-1}Z$. But $Z'(C^{-1}BC'^{-1})Z$ is distributed as central chi-square with k d.f. if and only if $C^{-1}BC'^{-1}$ is idempotent of rank k. $C^{-1}BC'^{-1}$ will be idempotent if and only if BV is idempotent. If BV is idempotent, then $BV = BVBV$; and $B = BVB = BC'^{-1}C^{-1}B$, and $C^{-1}BC'^{-1} = C^{-1}BC'^{-1}C^{-1}BC'^{-1}$. Also, if $C^{-1}BC'^{-1}$ is idempotent, then

$$C^{-1}BC'^{-1} = C^{-1}BC'^{-1}C^{-1}BC'^{-1} = C^{-1}BVBC'^{-1}$$

and $BV = BVBV$. Hence BV is idempotent.

❑ **THEOREM 27.3** *If Y is distributed $N(\mu, V)$, then $Y'BY$ is distributed as $\chi_k'^2(\lambda)$, where $\lambda = \frac{1}{2}\mu' B\mu$ and k is the rank of B, if and only if BV is an idempotent matrix.*

Proof. Let C be any matrix such that $C'VC = I$. Then, if $Z = C'Y$, Z is distributed $N(C'\mu, I)$ and $Y'BY = Z'(C^{-1}BC'^{-1}Z)$ is distributed as $\chi_k'^2(\lambda)$ if and only if $C^{-1}BC'^{-1}$ is idempotent, where

$$\lambda = \frac{1}{2}\mu' CC^{-1}BC'^{-1}\mu = \frac{1}{2}\mu' B\mu.$$

Now, we need to show that $C^{-1}BC'^{-1}$ is idempotent if and only if BV is idempotent. But it has been proved in the previous theorem.

Laguerre Series Expansions

We have already mentioned that if

$$d_{11} = d_{22} = \cdots = d_{nn} = d \text{ (say)}$$

then $Q = d\sum_{j=1}^{n} W_i^2$ is distributed as $d\chi_n^2$. If the ratios $d_{11} : d_{22} : \cdots : d_{nn}$ are not far from 1, then $f_n(x; \delta)$ and $F_n(x; \delta)$ can be expanded with the first term corresponding to the distribution of a multiple of a single central χ^2. This can be done by expressing the ratio

$$f_n(x; d_{11}, d_{22}, \ldots, d_{nn})/f_n(x; \beta, \beta, \ldots, \beta)$$

as a series of Laguerre polynomials associated with the denominator, where β is a constant which can be chosen arbitrarily. Here $f_n(x; \beta, \beta, \ldots, \beta) = \beta f_n(x/\beta; \mathbf{1})$, where $\mathbf{1}$ denotes the vector $(1, 1, \ldots, 1)$ and $f_n(x : 1)$ is the probability density function of χ_n^2.

The generalized Laguerre polynomials is given by the equation

$$L_r^{(\gamma)}(x) = \frac{1}{r!} x^{-\gamma} e^{x} \frac{d^r}{dx^r}\left(x^{r+\gamma} e^{-x}\right)$$

$$= \sum_{j=0}^{r} \frac{(-x)^j}{j!(r-j)!} \frac{\Gamma(r+\gamma+1)}{\Gamma(j+\gamma+1)} \text{ for } \gamma > -1 \tag{27.7}$$

where $x^\gamma e^{-x}$ is the weight function.

The polynomials possess the orthogonality properties

$$\int_0^\infty x^\gamma e^{-x} L_r^{(\gamma)}(x) L_{r'}^{(\gamma)}(x)dx = \begin{cases} 0, & (r \neq r') \\ \frac{\Gamma(r+\gamma+1)}{r!}, & (r = r') \end{cases}$$

We want an expansion of $f_n(x;\delta)$ in the form of

$$f_n(x;\delta) = \beta^{-1} f_n(x/\beta; 1) \sum_{j=0}^{\infty} b_j L_j^{(\frac{1}{2}-1)}\left(\frac{1}{2}x/\beta\right). \tag{27.8}$$

We have

$$\psi_n(t;\delta) = \prod_{j=1}^{n}(1-2itd_{jj})^{-\frac{1}{2}} = (1-2it\beta)^{-\frac{n}{2}} \prod_{j=1}^{n}\left[1 - \frac{2it(\beta-d_{jj})}{1-2it\beta}\right]^{-\frac{1}{2}}$$

$$= \sum_{j=0}^{\infty} c_j (1-2it\beta)^{-\frac{1}{2}(n+2j)}(-2it)^j \tag{27.9}$$

where c_j's are obtained from the identity

$$\prod_{j=1}^{n}\left[1-(\beta-\gamma_j)\,s\right]^{-\frac{1}{2}} \equiv \sum_{j=0}^{n} c_j s^j .$$

The expansion (27.9) is valid, and the series is uniform convergent for all t for which

$$\left|\frac{2it(\beta-d_{jj})}{1-2it\beta}\right| < 1$$

for all d_{jj}. This is true for all real t if $\beta > \frac{1}{2}\max d_{jj} = \frac{1}{2}d_{11}$. Now Fourier's inversion formula can be applied to (27.8) which gives

$$f_n(x;\delta) = \frac{1}{2\pi}\int_{-\infty}^{\infty} e^{-itx}\left[\sum_{j=0}^{\infty} c_j(1-2it\beta)^{-\frac{1}{2}(n+2j)}(-2it)^j\right] dt$$

$$= \frac{1}{\beta} f_n\left(\frac{x}{\beta};1\right) \sum_{j=0}^{\infty} c_j \beta^{-j} \frac{j!\,\Gamma\left(\frac{n}{2}\right)}{\Gamma\left(\frac{n}{2}+j\right)} L_j^{(\frac{n}{2}-1)}\left(\frac{1}{2}x/\beta\right) \tag{27.10}$$

The values of c_j's can be obtained from the relation

$$c_r = \frac{\mu_r'(Q;\beta-d_{11},\beta-d_{22},\ldots,\beta-d_{nn})}{2^r\, r!}$$

where we must have $\beta > \frac{1}{2}d_{11}$. In particular

$$\begin{cases} c_0 = 1 \\ c_r = (2r)^{-1}\sum_{j=0}^{r-1} S_{r-j}\, c_j \quad (r \geq 1) \end{cases}$$

where

$$S_r = \sum_{j=1}^{n} (\beta - d_{jj})^r.$$

The value of c_j can be computed in a simple way if

$$k_1(Q;\ \beta - d_{11}, \ldots \beta - d_{nn}) = \mu_1' \ (Q;\ \beta - d_{11}, \ldots, \beta - d_{nn})$$

so that μ_r' equals to μ_r. In order to arrange that this is so, it is necessary to make

$$\sum_{j=1}^{n} (\beta - d_{jj}) = 0 \Rightarrow \beta = \frac{1}{n}\sum_{j=1}^{n} d_{jj}.$$

This simplification can be used only if $\beta > \frac{1}{2}d_{11}$. If $d_{11} = d_{22} = \cdots = d_{nn}$, then $\beta - d_{jj} = 0$ for all j and the series expansion reduces to a single term (of central χ^2 form) which in this case corresponds to the exact distribution. However, if the values of d_{jj} are scattered widely, then choice of $\beta = \frac{1}{n}\Sigma d_{jj}$ may lead to slower convergence than would some other value of β.

The quantities S_r can be evaluated from the relation

$$S_r = t_r(\beta I - VA)^r$$

and $F_n(x;\delta)$ can be obtained by using the relationship

$$(2\beta)^{-1}\int_0^x \left(\frac{1}{2}y/\beta\right)^{\frac{1}{2}n-1} \exp\left(-\frac{1}{2}y/\beta\right) L_j^{\frac{1}{2}n-1}\left(\frac{1}{2}y/\beta\right) dy$$
$$= j^{-1}\left(\frac{1}{2}x/\beta\right)^{\frac{1}{2}n} \exp\left(-\frac{1}{2}x/\beta\right) L_j^{\frac{1}{2}n-1}\left(\frac{1}{2}x/\beta\right), \ (j \geq 1).$$

The expansion is

$$F_n(x;\delta) = \frac{1}{2^{n/2}\Gamma\frac{n}{2}}\int_0^{x/\beta} y^{\frac{1}{2}n-1}e^{-\frac{1}{2}y}dy + \sum_{j=1}^{\infty}\frac{(j-1)!c_j}{\beta^j\Gamma\left(\frac{1}{2}n+j\right)} \times \left(\frac{x}{2\beta}\right)^{\frac{1}{2}n}$$
$$e^{-\frac{1}{2}x/\beta}L_{j-1}^{\frac{1}{2}n}\left(\frac{1}{2}x/\beta\right), \ (y \geq 0) \tag{27.11}$$

The functions (27.10) and (27.11) are uniformly convergent if $y \geq 0$.

Using Laguerre series expansion the density function of noncentral distribution of quadratic form can also be derived. For this, let us consider

$$Q = \sum_{j=1}^{n} d_{jj}(W_j - w_j)^2$$

with some of w_j's not equal to zero. Let us also consider that Q is positive definite, i.e., $d_{11} \geq d_{22} \geq \cdots \geq d_{nn} > 0$.

Now, we need an expansion of the form

$$F_n(x;\delta) = \beta^{-1} f_n\left(\frac{x}{\beta};1\right) \sum_{j=0}^{\infty} b'_j L_j^{(\frac{1}{2}n-1)}\left(\frac{1}{2}x/\beta\right) \tag{27.12}$$

where β is a positive constant. Putting $\theta' = -2it\beta(1-2it\beta)^{-1}$ in

$$\psi_n(t;\delta,w) = \exp\left[\sum_{j=1}^{n}\left(\frac{itd_{jj}w_j^2}{1-2itd_{jj}}\right)\right]\prod_{j=1}^{n}(1-2itd_{jj})^{-\frac{1}{2}} \tag{27.13}$$

we get

$$\psi_n(t;\delta,w) = (1-2it\beta)^{-\frac{1}{2}n} \sum_{j=0}^{\infty} c'_j \beta^{-j} \theta'^j \tag{27.14}$$

provided $|\theta'| < \min_j |1 - d_{jj}/\beta|^{-1}$, where

$$\sum_{j=0}^{\infty} c'_j \beta^{-j} \theta'^j \equiv \exp\left[-\frac{\theta'}{2\beta}\sum_{j=1}^{n}\frac{d_{jj}w_j^2}{1-(1-d_{jj}/\beta)\theta'}\right]$$

$$\times \prod_{j=1}^{n}\{1-(1-d_{jj}/\beta)\theta'\}^{-\frac{1}{2}}. \tag{27.15}$$

In this case the values of c'_j will be

$$c'_0 = 1$$

$$c'_r = (2r)^{-1}\sum_{j=0}^{r-1} S'_{r-j} c'_j \quad (r \geq 1)$$

where

$$S'_r = -\left(\frac{r}{\beta}\right)\sum_{j=1}^{n} d_{jj}w_j^2(1-d_{jj}/\beta)^{r-1} + \sum_{j=1}^{n}(1-d_{jj}/\beta)^r.$$

Inverting (27.14), we get

$$f_n(x;\delta,w) = \beta^{-1} f_n(x/\beta;1)\sum_{j=0}^{\infty} c'_j \frac{j!\,\Gamma\left(\frac{n}{2}\right)}{\Gamma\left(\frac{n}{2}+j\right)} L_j^{(\frac{1}{2}n-j)}\left(\frac{1}{2}x/\beta\right), \ (y \geq 0) \tag{27.16}$$

Thus the cumulative distribution function is

$$F_n(x;\delta,w) = P\left[\chi_n^2 \leq \frac{x}{\beta}\right] + \left(\frac{x}{2\beta}\right)^{\frac{n}{2}} e^{-\frac{1}{2}y/\beta}$$

$$\times \sum_{j=1}^{\infty} c'_j \frac{(j-1)!}{\Gamma\left(\frac{n}{2}+j\right)} L_{j-1}^{(\frac{n}{2})}\left(\frac{1}{2}x/\beta\right). \tag{27.17}$$

27.3 Moments and Generating Function

Let

$$Q = \sum_{j=1}^{n} d_{jj}(W_j - w_j).$$

Then the m.g.f. of Q is given by

$$\begin{aligned} M_Q(t) &= E\left[e^{tQ}\right] \\ &= (2\pi)^{-\frac{n}{2}} \int_{-\infty}^{\infty}\int_{-\infty}^{\infty}\cdots\int_{-\infty}^{\infty} \exp\left[-\frac{1}{2}\Sigma W_j^2 + t\sum_{j=1}^{n} d_{jj}(W_j - w_j)^2\right] \\ &\qquad \times dw_1 dw_2 \cdots dw_n \\ &= \exp\left(-\frac{1}{2}\sum_{j=1}^{n} w_j^2\right)\exp\left(\frac{1}{2}\sum_{j=1}^{n}\frac{w_j^2}{1-2td_{jj}}\right)\prod_{j=1}^{n}(1-2td_{jj})^{-\frac{1}{2}}. \end{aligned}$$

In the central case, each $w_j = 0$, and the m.g.f. is

$$M_Q(t) = \prod_{j=1}^{n}(1-2td_{jj})^{-\frac{1}{2}}.$$

Again, if A is an idempotent matrix of rank k, then

$$M_Q(t) = (1-2t)^{-\frac{n}{2}} \; [\because d_{11} = d_{22} = \cdots = d_{kk} = 1]$$

The cumulant generating function is

$$\begin{aligned} K(Q) &= \log M_Q(t) \\ &= -\frac{1}{2}\sum_{j=1}^{n} w_j^2 + \frac{1}{2}\sum_{j=1}^{n}\frac{w_j^2}{(1-2td_{jj})} - \frac{1}{2}\sum_{j=1}^{n}\log(1-2td_{jj}) \\ &= \sum_{r=1}^{\infty} K_r(Q)\frac{t^r}{r!} \end{aligned}$$

where K_r is the r-th cumulant of Q and is given by

$$K_r = 2^{r-1}(r-1)!\sum_{j=1}^{n}(rw_j^2+1)d_{jj}^r.$$

In the central case

$$K_r = 2^{r-1}(r-1)!\sum_{j=1}^{n} d_{jj}^r.$$

If k of d_{jj}'s are equal to one, then

$$K_r = k2^{r-1}(r-1)!$$

We observe that the distribution of $Y'AY$ is the same as that of $W'DW$, where D

is the diagonal matrix of eigenvalues of VA. Thus

$$\sum_{j=1}^{n} d_{jj}^{r} = \text{tr}(VA)^{r}.$$

Therefore, we have

$$K_r = 2^{r-1}(r-1)!\,\text{tr}(VA)^r.$$

In particular

$$K_1 = \text{tr}(VA) = \mu_1', \ K_2 = \mu_2 = 2\,\text{tr}(VA)^2.$$

If Y is distributed $N(O, \sigma^2 I)$, then $\mu_1' = \sigma^2\,\text{tr}(A)$ since

$$\mu_1' = E(Y'AY) = E\sum_i\sum_j a_{ij}y_iy_j = E\left(\sum_i a_{ii}y_i^2\right) + E\sum\sum_{i\neq j}(y_iy_ja_{ij}).$$

But, if $i \neq j$

$$E(y_iy_j) = E(y_i)E(y_j) = 0.$$

Hence

$$E[Y'AY] = E\sum_i a_{ii}y_i^2 = \sum a_{ii}E(y_i^2) = \sigma^2\sum_i a_{ii} = \sigma^2\,\text{tr}(A).$$

If A is an idempotent matrix of rank k, then $\text{tr}(A) = k$. In that case

$$E(Y'AY) = E(Q) = k\sigma^2.$$

27.4 Independence of Quadratic Forms

❒**THEOREM 27.4** *If Y is distributed as $N(\mu, I)$, the two positive semi-definite quadratic forms $Y'AY$ and $Y'BY$ are independent if and only if $AB = 0$.*

Proof. Let us assume that $AB = 0 \Rightarrow B'A' = 0$. As A and B are symmetric, $AB = BA = 0$; that is to say, A and B commute. As A and B are symmetric, their exists orthogonal transformations P such that $P'AP = D_1$ and $P'BP = D_2$, where D_1 and D_2 are diagonal matrices. If $AB = 0$, it follows that $P'APP'BP = 0$, which gives $D_1D_2 = 0$. Thus, if the i-th diagonal element of D_1 is nonzero, then i-th diagonal of D_2 must be zero. Therefore, we can write

$$D_1 = \begin{bmatrix} D_1^* & 0 & 0 \\ 0 & 0 & 0 \\ 0 & 0 & 0^* \end{bmatrix}, \ D_2 = \begin{bmatrix} 0^* & 0 & 0 \\ 0 & D_2^* & 0 \\ 0 & 0 & 0^* \end{bmatrix}$$

where D_i^* is $n_i \times n_i$ diagonal matrix with rank n_i and 0^* is the $n_3 \times n_3$ null matrix such that $n_1 + n_2 + n_3 = n$. Let $Z = P'Y$, then Z is distributed $N(P'\mu, I)$. Now the quadratic form

$$Y'AY = Z'P'APZ = Z'D_1Z = \sum_{j=1}^{n_1} d^*_{1j} z_j^2$$

where d^*_{1j} is the j-th diagonal element of D^*_1. Similarly, we have

$$Y'BY = Z'P'BPZ = Z'D_2Z = \sum_{j=n_1+1}^{n_1+n_2} d^*_{2j} z_j^2$$

where d^*_{2j} is the j-th diagonal element of D^*_2. Since $Y'AY$ depends only on the first n_1 elements of Z and since $Y'BY$ depends only on the elements of Z from n_1+1 to n_2 and since all the elements of Z are independent, $Y'AY$ and $Y'BY$ are also independent. This completes the sufficiency of the theorem.

Again, let $Y'AY$ and $Y'BY$ be independent. This implies that

$$Y'AY = \sum d^*_{1j} z_j^2 \text{ and } Y'BY = \sum d^*_{2j} z_j^2$$

are independent. These two quantities will be independent if $D_1 D_2 = 0$. But

$$D_1 D_2 = P'APP'BP = P'ABP = 0 \Rightarrow AB = 0.$$

The theorem can be generalized to prove that $Y'B_1Y, Y'B_2Y, \ldots, Y'B_kY$ are jointly independent if and only if $B_iB_j = 0$ for all $i \neq j$.

❑**THEOREM 27.5** *If Y is distributed as $N(\mu, I)$, the set of quadratic forms $Y'A_1Y$, $Y'A_2Y, \ldots, Y'A_kY$, where the rank of A_i is n_i, are jointly independent and $Y'A_iY$ is distributed as $\chi'^2_{n_i}(\lambda_i)$, where $\lambda_i = \frac{1}{2}\mu' A_i \mu$, if any two of the following three conditions are satisfied:*

(i) *Each A_i is idempotent.*

(ii) *$\sum_{i=1}^{k} A_i$ is idempotent.*

(iii) *$A_iA_j = 0$ for all $i \neq j$.*

Proof. From theorem of matrices we know that if $A_1, A_2, \ldots, A_k$ are idempotent matrices and if $\sum_{i=1}^{k} A_i$ is also idempotent, then $A_iA_j = 0$ for all $i \neq j$ or if $A_iA_j = 0$ for all $i \neq j$ and ΣA_i is idempotent, then $A_1, A_2, \ldots, A_k$ are idempotent. Also, we know that if A_i is idempotent of rank n_i, then $Y'A_iY$ is distributed as $\chi'^2_{n_i}(\lambda_i)$, where $\lambda_i = \frac{1}{2}\mu' A_i \mu$. Again, as $A_iA_j = 0$ for all $i \neq j$, $Y'A_1Y, Y'A_2Y, \ldots, Y'A_kY$ are independent.

This theorem leads to a useful theorem which is stated as follows:

If Y is distributed $N(\mu, I)$ and if

$$Y'Y = \sum_{i=1}^{k} Y'A_iY$$

either of the following two conditions is necessary and sufficient for $Y'A_iY$ to be distributed as $\chi'^2_{n_i}(\lambda_i)$, where n_i is the rank of A_i and $\lambda_i = \frac{1}{2}\mu' A_i \mu$, and for the set $Y'A_1Y, Y'A_2Y, \ldots, Y'A_kY$ to be jointly independent:

(i) $A_1, A_2, \ldots, A_k$ are each idempotent matrix.

(ii) $A_iA_j = 0$ for all $i \neq j$.

Another important form of this theorem is as follows:

❏ **THEOREM 27.6** *If Y is distributed as $N(\mu, I)$ and if*

$$Y'Y = \sum_{i=1}^{k} Y'A_iY$$

a necessary and sufficient condition that $Y'A_iY$ be distributed as $\chi'^2_{n_i}(\lambda_i)$, where n_i is the rank of A_i and $\lambda_i = \frac{1}{2}\mu' A_i \mu$, and for the set $Y'A_1Y, Y'A_2Y, \ldots, Y'A_kY$ to be jointly independent, is that the rank of the sum of A_i be equal to the sum of the ranks of the separate A_i.

Proof. As $Y'A_iY$ is quadratic form, A_i is a symmetric matrix $(i = 1, 2, \ldots, k)$. If $A_1, A_2, \ldots, A_k$ are symmetric matrices, then the following conditions are satisfied:

(i) $A_1, A_2, \ldots, A_k$ are each idempotent.

(ii) $B = \sum_{i=1}^{k} A_i$ is idempotent.

(iii) $A_iA_j = 0$ for all $i \neq j$.

Again, if these conditions are satisfied, the rank of B equals the sum of the ranks of A_i. Hence $Y'A_1Y, Y'A_2Y, \ldots, Y'A_kY$ are jointly independent. Also, we know that if the condition (i) or (iii) is satisfied, then $Y'A_iY$ is distributed as $\chi'^2_{n_i}(\lambda_i)$, where $\lambda_i = \frac{1}{2}\mu' A_i \mu$.

❏ **THEOREM 27.7** *If B is a $q \times n$ matrix, A is an $n \times n$ matrix, and Y is distributed as $N(\mu, \sigma^2 I)$, then the linear forms BY are independent of the quadratic form $Y'AY$ if $BA = 0$.*

Proof. Let P be an orthogonal matrix. Then $P'AP = D$, where D is a diagonal matrix. Let us make the transformation $Z = P'Y$. Then Z is distributed as $N(P'\mu, \sigma^2 I)$. Also, consider a matrix C such that $C = BP$ and assume that $D = \begin{bmatrix} D_1 & 0 \\ 0 & 0 \end{bmatrix}$, where D_1 is a diagonal matrix with nonzero elements in the diagonal. Now,

$$BA = 0 = BAP = BPP'AP = CD = \begin{bmatrix} C_{11} & C_{12} \\ C_{21} & C_{22} \end{bmatrix} \begin{bmatrix} D_1 & 0 \\ 0 & 0 \end{bmatrix} = \begin{bmatrix} 0 \\ 0 \end{bmatrix}.$$

This implies that $C_{11}D_1 = 0$ and $C_{21}D_1 = 0$, which imply $C_{11} = 0$ and $C_{21} = 0$. Therefore, C can be written as $C = [0, C_2]$, where $C_2 = \begin{bmatrix} C_{12} \\ C_{22} \end{bmatrix}$. Now

$$BY = BPP'Y = CZ = [0, C_2] \begin{bmatrix} Z_1 \\ Z_2 \end{bmatrix} = C_2 Z_2$$

$$\text{and } Y'AY = Y'PP'APP'Y = Z'DZ = [Z_1' \;\; Z_2'] \begin{bmatrix} D_1 & 0 \\ 0 & 0 \end{bmatrix} \begin{bmatrix} Z_1 \\ Z_2 \end{bmatrix} = Z_1' D_1 Z_1.$$

It is observed that BY depends on the elements of Z in Z_2 and $Y'AY$ depends on the elements of Z in Z_1. But all elements of Z are independent. Hence the elements of Z_1 and Z_2 are independent, which imply that BY and $Y'AY$ are independent.

27.5 Wishart Distribution

The Wishart distribution is related to the matrix-valued quadratic functions of the form XCX', where C is a symmetric matrix and X $(p \times n)$ matrix of sample observations drawn from $N(\mu, \Sigma)$. Among such functions the most important special case is the sample variance-covariance matrix, which is called *Wishart matrix*. Let A be a matrix given by

$$A = (X - \mu E_{1n})(X - \mu E_{1n})'$$

where E_{1n} is a row vector of n elements each of which is unity. The matrix A has $p(p+1)/2$ distinct element $a_{ij}; i, j = 1, 2, \ldots, p; i \leq j$. The distribution of A is the joint distribution of these distinct elements a_{ij} and is known as *Wishart distribution.* The distribution can be considered for the following cases:

(i) The parent population of X is $N(O, I)$ and

$$A = XX' = \sum_{i=1}^{n} X_i X_i'.$$

(ii) The parent population of X is $N(\mu, \Sigma)$ and the matrix A is given by

$$A = (X - \mu E_{1n})(X - \mu E_{1n})'.$$

(iii) The parent population of X is $N(\mu, \Sigma)$ and the matrix A is given by

$$A = (X - \bar{X} E_{1n})(X - \bar{X} E_{1n})' = \sum_{i=1}^{n} (X_i - \bar{X})(X_i - \bar{X})'$$

where

$$\bar{X} = n^{-1} \times E_{n1} = n^{-1} \sum_{i=1}^{n} X_i.$$

The distribution of A is denoted by $A \sim W_p(\Sigma, n)$. If $\Sigma = I_p$ and $\mu = O$, the distribution is said to be in standard form. As the matrix A is the matrix of sum of squares and sum of products, its distribution is a matrix generalization of the univariate chi-square distribution and is given by

$$W_p = K(p, n)|A|^{\frac{n-p-1}{2}} e^{-\frac{1}{2} \operatorname{tr} A}, \ A > 0$$

where

$$1/K(p, n) = 2^{np/2} \prod^{p(p-1)/4} \prod_{i=1}^{p} \Gamma\left(\frac{n+1-i}{2}\right).$$

27.6 Derivation of Wishart Distribution

To derive the distribution, let us use Gram-Schmidt process of orthogonalization on the vectors $X_1, X_2, \ldots, X_p$, such that

$$Y_i = b^{i1} X_1 + b^{i2} X_2 + \cdots + b^{ii} X_i; \; i = 1, 2, \ldots, p.$$

The coefficients $b^{i1}, b^{i2}, \ldots, b^{ii}$ are to be chosen in such a way that Y_i is orthogonal to $Y_1, Y_2, \ldots, Y_i$ and of unit length. Let us write the orthogonal transformation as

$$Y = B^{-1} X$$

where

$$B^{-1} = \begin{bmatrix} b^{11} & 0 & 0 & \cdots & 0 \\ b^{21} & b^{22} & 0 & \cdots & 0 \\ \cdots & \cdots & \cdots & \cdots & \cdots \\ b^{p1} & b^{p2} & \cdots & \cdots & b^{pp} \end{bmatrix}.$$

The matrix B is written as

$$B = \begin{bmatrix} b_{11} & 0 & 0 & \cdots & 0 \\ b_{21} & b_{22} & 0 & \cdots & 0 \\ \cdots & \cdots & \cdots & \cdots & \cdots \\ b_{p1} & b_{p2} & b_{p3} & \cdots & b_{pp} \end{bmatrix}.$$

$$\therefore \; X = BY.$$

Let us define the Wishart matrix A by $A = XX'$. Thus, we have

$$A = XX' = BYY'B' = BB' \; [\because \; YY' = I_p]$$

The matrix B is called the *Bartlett decomposition* of A. We have

$$X_i = b_{i1} Y_1 + b_{i2} Y_2 + \cdots + b_{ii} Y_i.$$

Pre-multiplying by Y_j', we get

$$\begin{aligned} Y_j' X_i &= b_{i1} Y_j' Y_1 + b_{i2} Y_j' Y_2 + \cdots + b_{ii} Y_j' Y_i \\ &= b_{ij} \; [\because \; Y_j' Y_i = 0; Y_j' Y_j = 1] \\ i &= 1, 2, \ldots, i; j = 1, 2, \ldots, i-1 \\ X_i' X_i &= b_{i1}^2 Y_1' Y_1 + b_{i2}^2 Y_2' Y_2 + \cdots + b_{ii}^2 Y_i' Y_i \\ \therefore \; b_{ii}^2 &= X_i' X_i - \sum_{j=1}^{i-1} b_{ij}^2. \end{aligned}$$

Also, we have

$$X_i' X_i = a_{ii} \;\; [\because \; A = XX']$$

Now, from the definition of the population, all the

$$X_{il} \; (i = 1, 2, \ldots, p; \; l = 1, 2, \ldots, n)$$

are $NI(O, I)$. By orthogonal transformation from X_i to $b_{i1}, b_{i2}, \ldots, b_{i\,i-1}$ we have

$$\begin{bmatrix} b_{i1} \\ b_{i2} \\ \vdots \\ b_{i\,i-1} \end{bmatrix} = \begin{bmatrix} Y_1' \\ Y_2' \\ \vdots \\ Y_{i-1}' \end{bmatrix} X_i \quad [\because\ b_{ij} = Y_j' X_i]$$

Here we are getting $(i-1)$ new variables and these transformed variables are orthogonal random variable. Hence $b_{i1}, b_{i2}, \ldots, b_{i\,i-1}$ are $NI(O, 1)$. Therefore,

$$b_{ii}^2 = X_i' X_i - \sum_{j=1}^{i-1} b_{ij}^2 \sim \chi^2 \text{ with } n-(i-1) \text{ d.f.}$$

Therefore, the distribution of B is given by

$$f(B)d(B) = \prod_{i=1}^{p}\prod_{j=1}^{i-1}\left\{\frac{1}{\sqrt{2\pi}} e^{-\frac{1}{2}b_{ij}^2} db_{ij}\right\} \prod_{k=1}^{p}\left\{\chi^2_{n-(k-1)} b_{kk}^2 d\left(b_{kk}^2\right)\right\}$$

$$-\infty < b_{ij} < \infty; 0 < b_{ii} < \infty$$

From this distribution of B we need to find the distribution of A, where

$$A = BB' \Rightarrow |A| = \prod_{i=1}^{p} b_{ii}^2.$$

The Jacobian of the transformation is

$$J(B \to A) = \frac{1}{J(A \to B)} = 2^p \prod_{i=1}^{p} (b_{ii})^{i-p-1}.$$

We have

$$\operatorname{tr} A = \operatorname{tr} BB' = \sum_{i=1}^{p}\sum_{j=1}^{i} b_{ij}^2.$$

Thus the p.d.f. of A becomes

$$W_p(n, I) = K(p, n)|A|^{\frac{1}{2}(n-p-1)} e^{-\frac{1}{2}\operatorname{tr} A}, \quad A > 0$$

where

$$K(p, n) = 2^{\frac{1}{2}np} \pi^{\frac{1}{4}p(p-1)} \prod_{i=1}^{p}\left\{\Gamma\left(\frac{n+1-i}{2}\right)\right\}.$$

Here n is called the *parameter of the Wishart distribution.*

Wishart Distribution in General Case

For general case, the Wishart matrix A is defined by

$$A = (X - \mu E_{1n})(X - \mu E_{1n})'$$

where X is $(p \times n)$ data matrix drawn from $N_p(\mu, \Sigma)$. Let us consider the transformation

$$Y = C^{-1}(X - \mu)$$

where $\Sigma = CC'$, C being the lower triangular matrix. Thus Y is distributed as $N_p(O, I)$ and hence $D = YY'$ follows Wishart distribution. But

$$D = C^{-1}(X - \mu E_{1n})(X - \mu E_{1n})'(C^{-1})' = C^{-1}A(C^{-1})'.$$

Now, we need to transform D to A, where Jacobian of the transformation is given by

$$J(D \to A) = |C^{-1}|^{p+1} = |\Sigma|^{-\frac{1}{2}(p+1)}.$$

Also, we have

$$\begin{aligned} |D| &= |C^{-1}||A||C^{-1}| = |A|/|\Sigma| \\ \text{and } \operatorname{tr} D &= \operatorname{tr} C^{-1}A(C^{-1})' = \operatorname{tr}(CC')^{-1}A = \operatorname{tr} \Sigma^{-1}A. \\ \therefore\ f(A) &= \frac{K(p,n)}{|\Sigma|^{n/2}}|A|^{\frac{1}{2}(n-p-1)}e^{-\frac{1}{2}\operatorname{tr}\Sigma^{-1}A},\ A > 0. \end{aligned}$$

Wishart Distribution when μ is Replaced by $\bar{X}$

Let us assume that the data matrix $X(p \times n)$ is a random sample of size n from $N_p(\mu, \Sigma)$. The likelihood function of the sample observations is

$$L = \frac{1}{(2\pi)^{\frac{1}{2}np}|\Sigma|^{n/2}} e^{-\frac{1}{2}\operatorname{tr}\Sigma^{-1}(X-\mu E_{1n})(X-\mu E_{1n})'},\ -\infty < X < \infty.$$

We need to find the distribution of

$$A = (X - \bar{X}E_{1n})(X - \bar{X}E_{1n})'.$$

The ML estimates of μ and Σ are

$$\hat{\mu} = \bar{X} = n^{-1}XE_{n1} \text{ and } \hat{\Sigma} = S + (\bar{X} - \mu)(\bar{X} - \mu)'$$

where

$$nS = \Sigma(X_i - \bar{X})(X_i - \bar{X})' \text{ or, } S = (X - \bar{X}E_{1n})(X - \bar{X}E_{1n})'.$$

Let us make a transformation $Y = XP$, where P is an orthogonal matrix, such that its last column is $n^{-1/2}E_{n1}$. The Jacobian of the transformation is

$$J(X \to Y) = |P|^p = 1$$

since P is orthogonal. Since the last column of P is $n^{-1/2}E_{n1}$, Y can be partitioned

$$Y = \left[Z : \sqrt{n}\bar{X}\right]$$

where Z is of order $p \times (n-1)$. Also

$$E_{1n}P = \left[0, 0, \cdots, 0, \sqrt{n}\right]$$

as the first $(n-1)$ columns of P are orthogonal to the last one, which is proportional to E_{n1}. Hence

$$\begin{aligned}(X-\mu E_{1n})(X-\mu E_{1n})' &= (X-\mu E_{1n})PP'(X-\mu E_{1n})' \\ &= (Y-\mu E_{1n}P)(Y-\mu E_{1n}P)' \\ &= [Z:\sqrt{n}(\bar{X}-\mu)][Z:\sqrt{n}(\bar{X}-\mu)]' \\ &= ZZ'+n(\bar{X}-\mu)(\bar{X}-\mu)'\end{aligned}$$

where $-\infty < Z < \infty$; $-\infty < \sqrt{n}\bar{X} < \infty$.

Now, the likelihood function can be written as

$$L = \frac{1}{(2\pi)^{\frac{1}{2}np}|\Sigma|^{n/2}} e^{-\frac{1}{2}\operatorname{tr}\Sigma^{-1}[ZZ'+n(\bar{X}-\mu)(\bar{X}-\mu)']}$$

The likelihood function can be factorized which shows that Z and $\sqrt{n}\bar{X}$ are independently distributed. Also, it shows that

$$\sqrt{n}\bar{X} \sim N_p\left(\sqrt{n}\mu, \Sigma\right).$$

The factor containing Z only in the likelihood function shows that it is a random sample of size $(n-1)$ from a p-variate normal population with zero means and variance-covariance matrix Σ. Hence

$$ZZ' \sim W_p(n-1, \Sigma).$$

But, as P is orthogonal

$$ZZ' = YY' - n\bar{X}\bar{X}' = XPP'X' - n\bar{X}\bar{X}' = X\left(I - \frac{1}{n}E_{nn}\right)X = S.$$

Therefore $S \sim W_p(n-1, \Sigma)$.

Distribution of XAX'

Let X be a data matrix of order $p \times n$ and distributed as $N_p(O, \Sigma)$ with density function

$$\frac{1}{(2\pi)^{np/2}|\Sigma|^{n/2}} e^{-\frac{1}{2}\operatorname{tr}\Sigma^{-1}XX'}.$$

Thus the columns of X are independent. Let A be any $n \times n$ idempotent matrix of rank m. Then there exists an orthogonal matrix P such that

$$P'AD = D = \begin{bmatrix} I_m & 0 \\ 0 & 0 \end{bmatrix}.$$

Now, we have

$$\begin{aligned}XAX' &= XPP'APP'X = XPDP'X' \\ &= YDY' \text{ [if } Y = XP] = YY'.\end{aligned}$$

But YY' has Wishart distribution. Hence, XAX' has Wishart distribution with m d.f., i.e., $XAX' \sim W_p(m, \Sigma)$.

Noncentral Wishart Distribution

Let X be a data matrix of order $p \times n$ and distributed as $N_p(M, \Sigma)$ with density function

$$\frac{1}{(2\pi)^{np/2}|\Sigma|^{n/2}} e^{-\frac{1}{2}\operatorname{tr}\Sigma^{-1}(X-M)(X-M)'}$$

Here the columns of X are independently distributed about mean M. In such a case the distribution of $A = XX'$ is called *noncentral Wishart distribution* with noncentrality matrix $\Sigma^{-1}MM'$.

Further, if A is an idempotent matrix of order $n \times n$ and of rank m, then XAX' will have a noncentral Wishart distribution with m d.f., associated matrix Σ, and noncentrality matrix $\Sigma^{-1}MAM'$.

27.7 Characteristic Function and Properties of Wishart Distribution

Let X be a data matrix of order $p \times n$ and distributed as $N(O, \Sigma)$. Let

$$A = [a_{ij}] = XX'$$

be the Wishart matrix. Let ψ be the characteristic function of A. This characteristic function is calculated for the distinct $p(p+1)/2$ elements

$$a_{11}, a_{12}, \ldots, a_{1p}, a_{21}, \ldots, a_{2p}, \ldots, a_{pp}$$

of the matrix A. Therefore,

$$\begin{aligned}
\psi &= \psi(t_{11}, t_{12}, \ldots, t_{1p}, t_{21}, \ldots, t_{2p}, \ldots, t_{pp}) \\
&= E\left[e^{i\left(t_{11}a_{11}+t_{12}a_{12}+\cdots+t_{pp}a_{pp}\right)}\right] \\
&= E\left[\exp\left(\frac{1}{2}i\sum_i^p\sum_j^p(1+\delta_{ij})t_{ij}a_{ij}\right)\right] = E\left[e^{\frac{1}{2}i\operatorname{tr}AK}\right]
\end{aligned}$$

where $t_{ij} = t_{ji}$ for $i > j$, $K_{ij} = (1+\delta_{ij})t_{ij}$, $K = [k_{ij}]$ and δ_{ij} is the Kronecker delta.

$$\begin{aligned}
\therefore\ \psi &= \int_{-\infty}^{\infty}\int_{-\infty}^{\infty}\cdots\int_{-\infty}^{\infty} e^{-\frac{1}{2}\operatorname{tr}AK}\frac{1}{(2\pi)^{np/2}|\Sigma|^{n/2}}e^{-\frac{1}{2}\operatorname{tr}\Sigma^{-1}XX'}dX \\
&= \int_{-\infty}^{\infty}\int_{-\infty}^{\infty}\cdots\int_{-\infty}^{\infty}\frac{1}{(2\pi)^{np/2}|\Sigma|^{n/2}}e^{-\frac{1}{2}\operatorname{tr}(\Sigma^{-1}-iK)XX'}dX \quad [\because\ A = XX']
\end{aligned}$$

We know, if Y is a data matrix of order $p \times m$ and distributed as $N_p(O, G)$, then

$$\begin{aligned}
M\int_{-\infty}^{\infty}\int_{-\infty}^{\infty}\cdots\int_{-\infty}^{\infty} e^{-\frac{1}{2}Y'G'Y}dY &= M\int_{-\infty}^{\infty}\int_{-\infty}^{\infty}\cdots\int_{-\infty}^{\infty} e^{-\frac{1}{2}G^{-1}YY'}dY \\
&= (2\pi)^{mp/2}|G|^{m/2}
\end{aligned}$$

where $M = 1/\left[(2\pi)^{mp/2}|G|^{m/2}\right]$.

Using this result the integral of ψ can be evaluated, where

$$\begin{aligned}\psi &= \frac{1}{(2\pi)^{np/2}|\Sigma|^{n/2}}(2\pi)^{np/2}|\Sigma^{-1} - iK|^{-n/2} \\ &= \frac{|\Sigma^{-1}|^{-n/2}|I - iK\Sigma|^{-n/2}}{|\Sigma^{-1}|^{-n/2}} = \frac{1}{|I - iK\Sigma|^{n/2}}.\end{aligned}$$

Properties

(i) If $A \sim W_p(\Sigma, n)$ and B is a $q \times p$ matrix, then $BAB' \sim W_q(B\Sigma B', n)$.
Proof. Let X be a data matrix of order $p \times n$ and distributed as $N_p(O, \Sigma)$. Then

$$A = XX' \sim W_p(\Sigma, n).$$

Let us make a transformation $Y = BX$. Then $Y \sim N_q(O, B\Sigma B')$. Now, $BAB' = BXX'B' = YY'$. But $YY' \sim W_q(B\Sigma B', n)$. Hence the result.

(ii) If $A \sim W_p(\Sigma, n)$ and b be a $p \times 1$ vector which is independently distributed of A, then $u = b'Ab/b'\Sigma b$ has the χ^2-distribution and is independent of b.

(iii) If $A \sim W_p(\Sigma, n)$, then $|A|/|\Sigma|$ is distributed as the product of p independent chi-square variates with d.f. $n, n-1, \ldots, n-(p-1)$, respectively.

(iv) If $A \sim W_p(\Sigma, n)$, then σ^{pp}/a^{pp} has the chi-square distribution with $n-(p-1)$ d.f., where σ^{pp} and a^{pp} are the last elements of Σ^{-1} and A^{-1}, respectively.

(v) If $A \sim W_p(\Sigma, n)$, then $E(A) = n\Sigma$ and

$$E\left(A^{-1}\right) = (n-p-1)^{-1}\Sigma^{-1}, \text{ if } (n-p-1) > 0.$$

(vi) If X is a data matrix of order $p \times n$ and distributed as $N_p(O, \Sigma)$ and if C is an $(n \times n)$ symmetric matrix, then XCX' has a Wishart distribution if and only if C is idempotent. In such a case $X'CX \sim W_p(\Sigma, r)$, where r is the rank of C.

(vii) If $A_1, A_2, \ldots, A_k$ are k $(p \times p)$ matrices, each of which follows

$$W_p(A_i, n_i, \Sigma) \ (i = 1, 2, \ldots, k)$$

then the matrix $A = \sum_{i=1}^{k} A_i$ has the Wishart distribution $W_p(A, n, \Sigma)$, where

$$n = \sum_{i=1}^{k} n_i.$$

(viii) If the rows of X are independently and identically distributed as $N_p(\mu, \Sigma)$, and if $C_1, C_2, \ldots, C_k$ are symmetric matrices, then $X'C_1X, X'C_2X, \ldots, X'C_kX$ are jointly independent if $C_iC_j = 0$ for all $i \neq j$.

(ix) If $A \sim W_p(\Sigma, n)$ and $B \sim W_p(\Sigma, m)$ are independent, and if $n \geq p$; $m \geq p$, then

$$\phi = |A^{-1}B| = |B|/|A|$$

is proportional to the product of p independent F variables, of which the i-th has degree of freedom $(m-i+1)$ and $(n-i+1)$.

Proof. From property (iii), $|A|$ and $|B|$ are each $|\Sigma|$ times the product of p independent chi-squared variables. Therefore, ϕ is the product of p ratios of independent chi-squared statistics. Thus, the i-th ratio is $(m-i+1)/(n-i+1)$ times an $F_{m-i+1,n-i+1}$ statistic. This is true for all values of i.

(x) If $A \sim W_p(I, m)$ and $B \sim W_p(I, n)$ are independent, $m \geq p$, then

$$\Lambda = |A|/|A+B| = |I + A^{-1}B| \sim \Lambda(p, m, n)$$

has a Wilk's lambda distribution with parameters p, m and n. Again,

$$\Lambda(p, m, n) = \prod_{i=1}^{n} u_i$$

where $u_1, u_2, \ldots, u_n$ are independent variables and

$$u_i \sim B\left(\frac{1}{2}(m+i-p), \frac{1}{2}p\right);\ i = 1, 2, \ldots, n.$$

EXERCISE

1. (a) What do you mean by quadratic form in case of random variables? If $Y \sim N(\mu, I)$, derive the distribution $Y'AY$. What would be the distribution of $Y'AY$ if $\mu = O$?

   (b) If $Y \sim N(\mu, I)$, show that $Y'AY$ and $Y'BY$ are independently distributed.

2. (a) Prove that, for a random vector $Y \sim N(O, I)$ a necessary and sufficient condition that the quadratic form $Y'AY$ be distributed as central chi-square with k d.f. is that A is an idempotent matrix of rank k.

   (b) Prove that, the two positive semi-definite quadratic forms $Y'AY$ and $Y'BY$ are independent if and only if $AB = 0$, where $Y \sim N(\mu, I)$.

3. (a) Define quadratic form in terms of random vector. Derive the distribution of quadratic form.

   (b) Find out the conditions that the k quadratic forms $Y'A_iY (i = 1, 2, \ldots, k)$ are jointly independently distributed as chi-square.

4. (a) Distinguish between Wishart form and quadratic form. If S is sample Wishart matrix, show that $S \sim W_p(n-1, \Sigma)$.

   (b) Let B be a $q \times n$ matrix, A be an $m \times n$ matrix and $Y \sim N(\mu, \sigma^2 I)$. Show that the linear form BY and the quadratic form $Y'AY$ are independent.

CHAPTER **28**

F-Distribution

28.1 Definition and Derivation

If χ_1^2 and χ_2^2 are two independent chi-square variates having n_1 and n_2 degrees of freedom respectively, then the statistic

$$F = \frac{\chi_1^2/n_1}{\chi_2^2/n_2}$$

has the F-distribution with n_1 and n_2 degrees of freedom. Thus, F is the ratio of two independent chi-square variates divided by their respective degrees of freedom. The random variable which has F-distribution is generally denoted by F_{n_1,n_2}. From the definition of F-statistic, we can also define another statistic

$$F_{n_2,n_1}^{-1} = \frac{\chi_2^2/n_2}{\chi_1^2/n_1}.$$

Both the statistics F_{n_1,n_2} and F_{n_2,n_1}^{-1} are the ratios of two independent chi-square variates and hence have the identical distributions.

We have

$$F = \frac{\chi_1^2/n_1}{\chi_2^2/n_2} \text{ or } \frac{n_1 F}{n_2} = \frac{\chi_1^2}{\chi_2^2}.$$

Thus the statistic n_1F/n_2 is the ratio of two independent chi-square variates and hence follows beta distribution of second kind [Section 26.5]. Its density function is given by

$$f\left(\frac{n_1F}{n_2}\right) d\left(\frac{n_1F}{n_2}\right) = \frac{1}{B\left(\frac{n_1}{2}, \frac{n_2}{2}\right)} \frac{\left(\frac{n_1F}{n_2}\right)^{\frac{n_1}{2}-1}}{\left(1+\frac{n_1}{n_2}F\right)^{\frac{n_1+n_2}{2}}} d\left(\frac{n_1}{n_2}F\right), \; 0 < F < \infty.$$

On simplification, we have

$$f(F) = \frac{(n_1/n_2)^{n_1/2}}{B\left(\frac{n_1}{2}, \frac{n_2}{2}\right)} \frac{F^{\frac{n_1}{2}-1}}{\left(1+\frac{n_1}{n_2}F\right)^{\frac{1}{2}(n_1+n_2)}}, \; 0 < F < \infty. \tag{28.1}$$

Derivation of F-Distribution

Let χ_1^2 and χ_2^2 be two independent chi-square variates with n_1 and n_2 degrees of freedom respectively. Thus, we have

$$h\left(\chi_1^2, \chi_2^2\right) = \frac{1}{2^{\frac{n_1+n_2}{2}}\Gamma\frac{n_1}{2}\Gamma\frac{n_2}{2}} e^{-\frac{1}{2}(\chi_1^2+\chi_2^2)} \left(\chi_1^2\right)^{\frac{n_1}{2}-1} \left(\chi_2^2\right)^{\frac{n_2}{2}-1}$$

Let us put

$$F = \frac{\chi_1^2/n_1}{\chi_2^2/n_2} \text{ and } u = \chi_2^2;\ 0 < F < \infty,\ 0 < u < \infty$$

$$\Rightarrow \frac{n_1}{n_2}Fu = \chi_1^2, \chi_2^2 = u.$$

The Jacobian of the transformation is

$$J = \frac{\partial(\chi_1^2, \chi_2^2)}{\partial(F, u)} = \begin{vmatrix} \frac{\partial \chi_1^2}{\partial F} & \frac{\partial \chi_1^2}{\partial u} \\ \frac{\partial \chi_2^2}{\partial F} & \frac{\partial \chi_2^2}{\partial u} \end{vmatrix} = \begin{vmatrix} \frac{n_1}{n_2}u & \frac{n_1}{n_2}F \\ 0 & 1 \end{vmatrix} = \frac{n_1}{n_2}u.$$

Hence

$$g(u, F) = \frac{1}{2^{\frac{1}{2}(n_1+n_2)}\Gamma\frac{n_1+n_2}{2}B\left(\frac{n_1}{2}, \frac{n_2}{2}\right)}$$

$$\times\, e^{-\frac{1}{2}\left(\frac{n_1}{n_2}Fu+u\right)} \left(\frac{n_1}{n_2}Fu\right)^{\frac{n_1}{2}-1} u^{\frac{n_2}{2}-1} \left(\frac{n_1}{n_2}u\right)$$

$$= \frac{(n_1/n_2)^{\frac{n_1}{2}} F^{\frac{n_1}{2}-1}}{2^{\frac{n_1+n_2}{2}}\Gamma\frac{n_1+n_2}{2}B\left(\frac{n_1}{2}, \frac{n_2}{2}\right)} e^{-\frac{1}{2}u\left(1+\frac{n_1}{n_2}F\right)} u^{\frac{n_1+n_2}{2}-1}$$

Let us take

$$\left(1 + \frac{n_1}{n_2}F\right)u = z \Rightarrow u = \frac{z}{\left(1 + \frac{n_1}{n_2}F\right)}$$

$$du = \frac{dz}{\left(1 + \frac{n_1}{n_2}F\right)},\ 0 < z < \infty$$

$$f(F) = \frac{(n_1/n_2)^{\frac{n_1}{2}} F^{\frac{n_1}{2}-1}}{B\left(\frac{n_1}{2}, \frac{n_2}{2}\right)\left(1 + \frac{n_1}{n_2}F\right)^{\frac{n_1+n_2}{2}}}$$

$$\times \int_0^\infty \frac{1}{2^{\frac{n_1+n_2}{2}}\Gamma\frac{n_1+n_2}{2}} e^{-\frac{1}{2}z} z^{\frac{n_1+n_2}{2}-1} dz$$

$$= \frac{(n_1/n_2)^{\frac{n_1}{2}} F^{\frac{n_1}{2}-1}}{B\left(\frac{n_1}{2}, \frac{n_2}{2}\right)\left(1 + \frac{n_1}{n_2}F\right)^{\frac{n_1+n_2}{2}}},\ 0 < F < \infty. \tag{28.2}$$

This F is distributed as variance ratio with n_1 and n_2 d.f. It is called *Snedecor's F-distribution.* Hald (1952) has given the graphs of the probability density function for $n_1 = 10$ and a number of values of n_2. The shape of the graph is as follows:

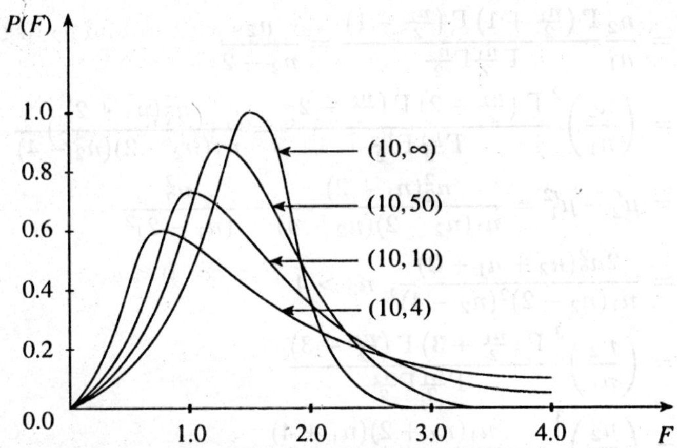

FIG. 28.1 : *F*-Density function

The distribution of F is sampling distribution and does not involve any population parameter. It depends only on the d.f. n_1 and n_2.

28.2 Properties of F-Distribution

The r-th moment about zero of F-distribution is given by

$$\begin{aligned}\mu_r' &= E\left[F^r\right]\\ &= \int_0^\infty F^r f(F)dF\\ &= \frac{(n_1/n_2)^{\frac{n_1}{2}}}{B\left(\frac{n_1}{2},\frac{n_2}{2}\right)}\int_0^\infty F^r \frac{F^{\frac{n_1}{2}-1}}{\left(1+\frac{n_1}{n_2}F\right)^{\frac{n_1+n_2}{2}}}dF.\end{aligned}$$

Let us put

$$\frac{n_1}{n_2}F = u \text{ so that } dF = \frac{n_2}{n_1}du,\ 0 < u < \infty.$$

In that case

$$\begin{aligned}\mu_r' &= \frac{(n_1/n_2)^{\frac{n_1}{2}}}{B\left(\frac{n_1}{2},\frac{n_2}{2}\right)}\frac{n_2}{n_1}\int_0^\infty \frac{(n_2/n_1)^{\frac{n_1}{2}+r-1}}{(1+u)^{\frac{n_1+n_2}{2}}}u^{\frac{n_1}{2}+r-1}du\\ &= \frac{(n_2/n_1)^r}{B\left(\frac{n_1}{2},\frac{n_2}{2}\right)}\int_0^\infty \frac{u^{\frac{n_1}{2}+r-1}}{(1+u)^{\frac{n_1}{2}+r+\frac{n_2}{2}-r}}du\\ &= \left(\frac{n_2}{n_1}\right)^r \frac{1}{B\left(\frac{n_1}{2},\frac{n_2}{2}\right)}B\left(\frac{n_1}{2}+r,\frac{n_2}{2}-r\right),\ \text{provided } r < \frac{n_2}{2}\\ &= \left(\frac{n_2}{n_1}\right)^r \frac{\Gamma\left(\frac{n_1}{2}+r\right)\Gamma\left(\frac{n_2}{2}-r\right)}{\Gamma\frac{n_1}{2}\Gamma\frac{n_2}{2}}.\end{aligned}$$

In particular

$$\begin{aligned}
\mu_1' &= \frac{n_2}{n_1}\frac{\Gamma\left(\frac{n_1}{2}+1\right)\Gamma\left(\frac{n_2}{2}-1\right)}{\Gamma\frac{n_1}{2}\Gamma\frac{n_2}{2}} = \frac{n_2}{n_2-2} \\
\mu_2' &= \left(\frac{n_2}{n_1}\right)^2\frac{\Gamma\left(\frac{n_1}{2}+2\right)\Gamma\left(\frac{n_2}{2}-2\right)}{\Gamma\frac{n_1}{2}\Gamma\frac{n_2}{2}} = \frac{n_2^2(n_1+2)}{n_1(n_2-2)(n_2-4)} \\
\mu_2 &= \mu_2' - \mu_1'^2 = \frac{n_2^2(n_1+2)}{n_1(n_2-2)(n_2-4)} - \frac{n_2^2}{(n_1-2)^2} \\
&= \frac{2n_2^2(n_2+n_1-2)}{n_1(n_2-2)^2(n_2-4)},\ n_2>4 \\
\mu_3' &= \left(\frac{n_2}{n_1}\right)^3\frac{\Gamma\left(\frac{n_1}{2}+3\right)\Gamma\left(\frac{n_2}{2}-3\right)}{\Gamma\frac{n_1}{2}\Gamma\frac{n_2}{2}} \\
&= \left(\frac{n_2}{n_1}\right)^3\cdot\frac{n_1(n_1+2)(n_1+4)}{(n_2-2)(n_2-4)(n_2-6)} \\
&= \frac{n_2^3}{n_1^2}\frac{(n_1+2)(n_1+4)}{(n_2-2)(n_2-4)(n_2-6)} \\
\mu_4' &= \frac{n_2^4}{n_1^3}\frac{(n_1+2)(n_1+4)(n_1+6)}{(n_2-2)(n_2-4)(n_2-6)(n_2-8)}.
\end{aligned}$$

The moment ratios are

$$\alpha_3 = \sqrt{\beta_1} = \sqrt{\frac{8(n_2-4)}{(n_1+n_2-2)n_1}}\cdot\frac{(2n_1+n_2-2)}{(n_2-6)},\ n_2>6$$

$$\alpha_4 = \beta_2 = 3 + \frac{12\{(n_2-2)^2(n_2-4)+n_1(n_1+n_2-2)(5n_2-22)\}}{n_1(n_2-6)(n_2-8)(n_1+n_2-2)}$$

Thus, for $n_2 > 6$ the distribution is positively skewed and it is leptokurtic for $n_2 > 8$.

The characteristic function of F-distribution is confluent hypergeometric function which is given by

$$\psi(t) = E\left[e^{itF}\right] = M\left(\frac{1}{2}n_1; -\frac{1}{2}n_2; \frac{n_2}{n_1}it\right).$$

Mode of F-Distribution

We have

$$f(F) = \frac{(n_1/n_2)^{\frac{n_1}{2}}}{B\left(\frac{n_1}{2},\frac{n_2}{2}\right)}\frac{F^{\frac{n_1}{2}-1}}{\left(1+\frac{n_1}{n_2}F\right)^{\frac{n_1+n_2}{2}}},\ 0<F<\infty$$

$$\log f(F) = C + \left(\frac{n_1}{2}-1\right)\log F - \left(\frac{n_1+n_2}{2}\right)\log\left(1+\frac{n_1}{n_2}F\right)$$

where

$$C = \log\left[\frac{(n_1/n_2)^{n_1/2}}{B\left(\frac{n_1}{2},\frac{n_2}{2}\right)}\right]$$

Mode is that value of F for which $f'(F) = 0$ and $f''(F) < 0$. Here

$$\frac{\partial}{\partial F}[\log f(F)] = \left(\frac{n_1}{2} - 1\right)\frac{1}{F} - \frac{n_1 + n_2}{2}\frac{1}{\left(1 + \frac{n_1}{n_2}F\right)}\frac{n_1}{n_2}$$

$$f'(F) = \frac{\partial}{\partial F}f(F) = 0 \Rightarrow \frac{n_1 - 2}{2F} - \frac{n_1(n_1 + n_2)}{2(n_2 + n_1F)} = 0.$$

$$\therefore\ F = \frac{n_2(n_1 - 2)}{n_1(n_2 + 2)}$$

It is easy to verify that at this point of F, $f''(F) < 0$. Hence, mode of F-distribution is

$$\text{Mode} = \frac{n_2(n_1 - 2)}{n_1(n_2 + 2)} = \frac{n_2}{n_2 + 2}\frac{n_1 - 2}{n_1}.$$

It exists if and only if $n_1 > 2$ and the value of mode is always less than unity. There is a single mode, since $f(F)$ tends to zero as F tends to infinity and $f(F)$ tends to zero if F tends to zero. If $n_1 = 2$, there is a mode at $F = 0$.

The points of inflexion of F-distribution exist for $n_1 = 4$ and are equidistant from mode.

❒ **THEOREM 28.1** *Show that $F_{n_1,n_2} = 1/F_{n_2,n_1}$.*

Proof. We know

$$f(F) = \frac{(n_1/n_2)^{n_1/2}}{B\left(\frac{n_1}{2}, \frac{n_2}{2}\right)}\frac{F^{\frac{n_1}{2}-1}}{\left(1 + \frac{n_1}{n_2}F\right)^{\frac{n_1+n_2}{2}}},\ 0 < F < \infty.$$

Let us put $F = 1/y$ or $y = 1/F \Rightarrow dF = -(1/y^2)dy$.

$$\begin{aligned}\therefore\ g(y) &= \frac{(n_1/n_2)^{\frac{n_1}{2}}}{B\left(\frac{n_1}{2}, \frac{n_2}{2}\right)}\frac{(1/y)^{\frac{n_1}{2}-1}}{\left(1 + \frac{n_1}{n_2y}\right)^{\frac{n_1+n_2}{2}}}\left(-\frac{1}{y^2}\right),\ \infty < y < 0\\ &= \frac{(n_1/n_2)^{\frac{n_1}{2}}}{B\left(\frac{n_1}{2}, \frac{n_2}{2}\right)}\frac{(n_2)^{\frac{n_1+n_2}{2}}y^{\frac{n_1+n_2}{2}-\frac{n_1}{2}+1-2}}{(n_1 + n_2y)^{\frac{n_1+n_2}{2}}},\ 0 < y < \infty\\ &= \frac{(n_2/n_1)^{\frac{n_2}{2}}}{B\left(\frac{n_2}{2}, \frac{n_1}{2}\right)}\frac{y^{\frac{n_2}{2}-1}}{\left(1 + \frac{n_2}{n_1}y\right)^{\frac{n_1+n_2}{2}}},\ 0 < y < \infty.\end{aligned}$$

Thus, $y = 1/F$ is F_{n_2,n_1}-variate.

❒ **THEOREM 28.2** *If X_1 and X_2 are two independent random variables having common density function*

$$f(x) = e^{-x},\ 0 < x < \infty$$

show that $u = x_1/x_2$ has an F-distribution.

Proof. The joint p.d.f. of X_1 and X_2 is

$$f(x_1, x_2) = e^{-(x_1+x_2)}.$$

Let us make the transformation

$$v = x_1 + x_2 \text{ and } u = x_1/x_2$$
$$\Rightarrow x_1 = ux_2 = \frac{uv}{1+u}$$
$$v = x_2(1+u), x_2 = \frac{v}{1+u}$$

The Jacobian of the transformation is

$$J = \frac{\partial(x_1, x_2)}{\partial(u, v)} = \begin{vmatrix} \dfrac{\partial x_1}{\partial u} & \dfrac{\partial x_1}{\partial v} \\ \dfrac{\partial x_2}{\partial u} & \dfrac{\partial x_2}{\partial v} \end{vmatrix}$$

$$= \begin{vmatrix} v\left[\dfrac{1}{1+u} - \dfrac{u}{(1+u)^2}\right] & \dfrac{u}{1+u} \\ -\dfrac{v}{(1+u)^2} & \dfrac{1}{1+u} \end{vmatrix}$$

$$= \frac{v}{(1+u)^2};\ 0 < v < \infty,\ 0 < u < \infty$$

$$\therefore\ g(u, v) = e^{-v}\frac{v}{(1+u)^2}$$

$$h(u) = \frac{1}{(1+u)^2}\int_0^\infty e^{-v} v\, dv = \frac{\Gamma 2}{(1+u)^2},\ 0 < u < \infty$$

$$= \frac{(2/2)^{\frac{2}{2}-1}}{B\left(\frac{2}{2}, \frac{2}{2}\right)} \frac{u^{\frac{2}{2}-1}}{\left(1+\frac{2}{2}u\right)^{\frac{2+2}{2}}},\ 0 < u < \infty$$

which is the density function of $F_{2,2}$.

►**EXAMPLE 28.1** *If X_1 and X_2 are two independent normal variates with means μ_1 and μ_2 and variances σ_1^2 and σ_2^2 respectively, find the distribution of*

$$U = (X_1 - \mu_1)/(X_2 - \mu_2).$$

Solution. Let

$$u = (x_1 - \mu_1)/(x_2 - \mu_2)$$
$$\frac{\sigma_2 u}{\sigma_1} = \left(\frac{x_1 - \mu_1}{\sigma_1}\right) \Big/ \left(\frac{x_2 - \mu_2}{\sigma_2}\right)$$
$$\frac{\sigma_2^2 u^2}{\sigma_1^2} = \left(\frac{x_1 - \mu_1}{\sigma_1}\right)^2 \Big/ \left(\frac{x_2 - \mu_2}{\sigma_2}\right)^2$$
$$= \chi_1^2/\chi_2^2$$

where both χ_1^2 and χ_2^2 are chi-square variates with 1 degree of freedom. Thus, $(\sigma_2^2 u^2/\sigma_1^2)$ is the ratio of two independent chi-square variates and is beta variate of second kind, i.e.,

$$\frac{\sigma_2^2 u^2}{\sigma_1^2} \sim B\left(\frac{1}{2}, \frac{1}{2}\right).$$

Therefore,

$$dF\left(\frac{\sigma_2^2}{\sigma_1^2}u^2\right) = \frac{1}{B\left(\frac{1}{2},\frac{1}{2}\right)}\frac{\left(\frac{\sigma_2^2}{\sigma_1^2}u^2\right)^{\frac{1}{2}-1}}{\left(1+\frac{\sigma_2^2u^2}{\sigma_1^2}\right)^{\frac{1}{2}+\frac{1}{2}}}\, d\left(\frac{\sigma_2^2}{\sigma_1^2}u^2\right),\ 0<u<\infty$$

$$= \frac{1}{B\left(\frac{1}{2},\frac{1}{2}\right)}\frac{\left(\frac{\sigma_2^2}{\sigma_1^2}\right)^{-\frac{1}{2}}\left(\frac{\sigma_2^2}{\sigma_1^2}\right)2u(u^2)^{-\frac{1}{2}}}{1+\frac{\sigma_2^2}{\sigma_1^2}u^2}du = \frac{2}{B\left(\frac{1}{2},\frac{1}{2}\right)}\frac{\frac{\sigma_2}{\sigma_1}du}{\left(1+\frac{\sigma_2^2}{\sigma_1^2}u^2\right)}$$

$$\therefore\ f(u) = \frac{\sigma_2/\sigma_1}{\pi\left(1+\frac{\sigma_2^2}{\sigma_1^2}u^2\right)},\ -\infty<u<\infty$$

$$= \frac{\sigma_1\sigma_2}{\pi(\sigma_1^2+\sigma_2^2u^2)}.$$

If $\sigma_1=\sigma_2=1$, we have $f(u)=\dfrac{1}{\pi(1+u^2)}$, $-\infty<u<\infty$ which is the density function of a standard Cauchy variate.

Relation between F- and χ^2-Distribution

We know

$$f(F) = \frac{(n_1/n_2)^{\frac{n_1}{2}}}{B\left(\frac{n_1}{2},\frac{n_2}{2}\right)}F^{\frac{n_1}{2}-1}\frac{1}{\left(1+\frac{n_1}{n_2}F\right)^{\frac{n_1+n_2}{2}}},\ 0<F<\infty.$$

Let us consider that n_2 tends to infinity; then

$$f(F) = \frac{(n_1/n_2)^{\frac{n_1}{2}}}{\Gamma\frac{n_1}{2}\Gamma\frac{n_2}{2}}\frac{\Gamma\frac{n_1+n_2}{2}F^{\frac{n_1}{2}-1}}{\left(1+\frac{n_1}{n_2}F\right)^{\frac{n_1}{2}+\frac{n_2}{2}}}$$

becomes

$$f(F) = \frac{(n_1)^{\frac{n_1}{2}}}{\Gamma\frac{n_1}{2}2^{\frac{n_1}{2}}}\frac{F^{\frac{n_1}{2}-1}}{\left(1+\frac{n_1}{n_2}F\right)^{\frac{n_1}{2}+\frac{n_2}{2}}}$$

$$\left[\lim_{n_2\to\infty}\frac{\Gamma\frac{n_1+n_2}{2}}{(n_2)^{\frac{n_1}{2}}\Gamma\frac{n_2}{2}} \to \frac{(n_2/2)^{n_1/2}}{(n_2)^{n_1/2}} = \frac{1}{2^{n_1/2}}\ \because\ \frac{\Gamma(n+k)}{\Gamma n}\to n^k \text{ as } n\to\infty\right]$$

Also,

$$\lim_{n_2\to\infty}\left(1+\frac{n_1}{n_2}F\right)^{\frac{n_1+n_2}{2}} = \lim_{n_2\to\infty}\left[\left(1+\frac{n_1}{n_2}F\right)^{n_2}\right]^{1/2}\times\lim_{n_2\to\infty}\left[1+\frac{n_1}{n_2}F\right]^{n_1/2}$$

$$= \exp\left(\frac{n_1F}{2}\right) = \exp\left(\frac{\chi_1^2}{2}\right)$$

$$\left[\because\ \chi_1^2 = n_1F;\ \text{we have } \tfrac{n_1F}{n_2} = \tfrac{\chi_1^2}{\chi_2^2}.\ \text{But as } n_2\to\infty,\ \chi_2^2\to 1\right]$$

We have $\dfrac{\chi_1^2}{n_1} = F \Rightarrow dF = \dfrac{d\chi_1^2}{n_1}$.

Hence in the limit, the p.d.f. of $\chi_1^2 = n_1 F$ becomes

$$\begin{aligned} df(\chi_1^2) &= \frac{(n_1)^{\frac{n_1}{2}} e^{-\frac{1}{2}\chi_1^2}}{2^{\frac{n_1}{2}}\Gamma\frac{n_1}{2}} \left(\frac{\chi_1^2}{n_1}\right)^{\frac{n_1}{2}-1} d\left(\frac{\chi_1^2}{n_1}\right) \\ &= \frac{1}{2^{\frac{n_1}{2}}\Gamma\frac{n_1}{2}} e^{-\frac{1}{2}\chi_1^2} (\chi_1^2)^{\frac{n_1}{2}-1}, \; 0 < \chi_1^2 < \infty. \end{aligned}$$

$\therefore$ As $n_2 \to \infty$, $n_1 F = \chi_1^2$.

We have already mentioned that ratio of two independent chi-square variates is a beta variate of the second kind. Thus

$$f(v) = \frac{1}{B\left(\frac{n_1}{2}, \frac{n_2}{2}\right)} \frac{v^{\frac{n_1}{2}-1}}{(1+v)^{\frac{n_1+n_2}{2}}}, \; 0 < v < \infty \tag{28.3}$$

where $v = \chi_1^2/\chi_2^2$; χ_1^2 and χ_2^2 are two independent chi-square variates with n_1 and n_2 degrees of freedom, respectively. Now, from (28.3) it can be shown that

$$\frac{v}{1+v} = n_1 F(n_2 + n_1 F)^{-1} \tag{28.4}$$

has a standard beta distribution with parameters $\frac{n_1}{2}, \frac{n_2}{2}$.

The density function (28.3) is a form of Pearson Type VI distribution. The most general form of this distribution is

$$f(x) = \frac{\Gamma n_1 (a_2 - a_1)^{n_1 - n_2 - 1}}{\Gamma(n_1 - n_2 - 1)\Gamma(n_2 + 1)} \frac{(x - a_2)^{n_2}}{(x - a_1)^{n_1}} \tag{28.5}$$
$$(n_1 > n_2 > -1; \; x \geq a_2 > a_1)$$

Here n_1, n_2, a_1 and a_2 are four parameters.

The r-th moment of X is

$$\mu_r'(X) = E[(X - a_1)^r] = (a_2 - a_1)^r (n_1 - 1)^{(r)}/(n_1 - n_2 - 2)^{(r)}.$$

Hence

$$\begin{aligned} \mu_1' &= E(X) = a_1 + (n_1 - 1)(a_2 - a_1)/(n_1 - n_2 - 2) \\ &= a_2 + (n_2 - 1)(a_2 - a_1)/(n_1 - n_2 - 2), \; (n_1 - n_2 > 2) \\ \mu_2(X) &= (a_2 - a_1)^2 (n_1 - 1)(n_2 + 1)(n_1 + n_2 - 2)^{-2}(n_1 - n_2 - 3)^{-1}, \; (n_1 - n_2 > 3) \end{aligned}$$

When X has the form of distribution (28.5), then $Y = (X - a_2)/(X - a_1)$ is of standard beta form with parameters $(n_2 + 1)$ and $(n_1 - n_2 - 1)$.

Relation between F- and t-Distribution

The density function of t-distribution [Chapter 29] is given by

$$f(t) = \frac{1}{\sqrt{n} B\left(\frac{1}{2}, \frac{n}{2}\right)} \frac{1}{\left(1 + \frac{t^2}{n}\right)^{\frac{n+1}{2}}}, \; -\infty < t < \infty. \tag{28.6}$$

Also

$$f(F)=\frac{(n_1/n_2)^{\frac{n_1}{2}}}{B\left(\frac{n_1}{2},\frac{n_2}{2}\right)}\frac{F^{\frac{n_1}{2}-1}}{\left(1+\frac{n_1}{n_2}F\right)^{\frac{n_1+n_2}{2}}},\ 0<F<\infty.$$

If we take $n_1=1$ and $n_2=n$, then

$$f(F)=\frac{1}{\sqrt{n}B\left(\frac{1}{2},\frac{n}{2}\right)}\frac{F^{-\frac{1}{2}}}{\left(1+\frac{1}{n}F\right)^{\frac{n+1}{2}}},\ 0<F<\infty.$$

Let $F=t^2\Rightarrow dF=2tdt,\ 0<t^2<\infty.$

$$d(t)=\frac{1}{\sqrt{n}B\left(\frac{1}{2},\frac{n}{2}\right)}\frac{(t^2)^{-\frac{1}{2}}2tdt}{\left(1+\frac{t^2}{n}\right)^{\frac{n+1}{2}}}$$

$$\therefore\ g(t)=\frac{1}{\sqrt{n}B\left(\frac{1}{2},\frac{n}{2}\right)}\frac{1}{\left(1+\frac{t^2}{n}\right)^{\frac{n+1}{2}}},\ -\infty<t<\infty$$

which is similar to (28.6). Therefore, when $n_1=1$, $F=t^2$.

Further, if $n_1=n_2=n$, then $\frac{1}{2}\sqrt{n}\left(F_{n,n}^{\frac{1}{2}}-F_{n,n}^{-\frac{1}{2}}\right)$ has a t-distribution with n degrees of freedom.

F-distribution has also a relationship with binomial distribution, which can be expressed in terms of probability, i.e.,

$$P\left[F_{2(n-r+1),2r}>\frac{1-p}{p}\frac{r}{n-r+1}\right]=P[X\geq r]$$

[r is an integer, $0\leq r\leq n$] where X is a binomial variate with parameters (n,p).

28.3 Fisher's Z-Distribution

We have

$$f(F)=\frac{(n_1/n_2)^{\frac{n_1}{2}}}{B\left(\frac{n_1}{2},\frac{n_2}{2}\right)}\frac{F^{\frac{n_1}{2}-1}}{\left(1+\frac{n_1}{n_2}F\right)^{\frac{n_1+n_2}{2}}},\ 0<F<\infty.$$

Let us make the transformation

$$F=e^{2z}\text{ or }z=\frac{1}{2}\log F\Rightarrow dF=2e^{2z}dz.$$

Then

$$g(z)=\frac{(n_1/n_2)^{\frac{n_1}{2}}}{B\left(\frac{n_1}{2},\frac{n_2}{2}\right)}\frac{\left(e^{2z}\right)^{\frac{n_1}{2}-1}}{\left(1+\frac{n_1}{n_2}e^{2z}\right)^{\frac{n_1+n_2}{2}}}2e^{2z},\ [-\infty<z<\infty]$$

$$=\frac{2\,(n_1/n_2)^{\frac{n_1}{2}}}{B\left(\frac{n_1}{2},\frac{n_2}{2}\right)}\frac{\left(e^{2z}\right)^{\frac{n_1}{2}}}{\left(1+\frac{n_1}{n_2}e^{2z}\right)^{\frac{n_1+n_2}{2}}}.\qquad(28.7)$$

Moment Generating Function of Z-Distribution

The m.g.f. of Z-distribution is given by

$$M_z(t) = E\left[e^{tz}\right] = E\left[e^{\frac{2tz}{2}}\right] = E\left[F^{t/2}\right] \quad \left[\because\ e^{2z} = F \Rightarrow e^{2z\frac{t}{2}} = F^{t/2}\right]$$

Putting $\frac{t}{2} = r$, we have

$$\begin{aligned}
M_z(t) = E[F^r] &= \int_0^\infty F^r f(F)dF \\
&= \frac{(n_1/n_2)^{\frac{n_1}{2}}}{B\left(\frac{n_1}{2}, \frac{n_2}{2}\right)} \int_0^\infty \frac{F^r F^{\frac{n_1}{2}-1}}{\left(1+\frac{n_1}{n_2}F\right)^{\frac{n_1+n_2}{2}}} dF \\
&= \frac{(n_1/n_2)^{\frac{n_1}{2}}}{B\left(\frac{n_1}{2}, \frac{n_2}{2}\right)} \int_0^\infty \frac{F^{r+\frac{n_1}{2}-1}}{\left(1+\frac{n_1}{n_2}F\right)^{r+\frac{n_1}{2}+\frac{n_2}{2}-r}} dF \\
&= \left(\frac{n_2}{n_1}\right)^r \frac{1}{B\left(\frac{n_1}{2}, \frac{n_2}{2}\right)} B\left(\frac{n_1}{2}+r, \frac{n_2}{2}-r\right)
\end{aligned}$$

[see Section 28.2, m.g.f. of F].

Therefore,

$$M_z(t) = \left(\frac{n_2}{n_1}\right)^{t/2} \frac{\Gamma\left(\frac{n_1}{2}+\frac{t}{2}\right)\Gamma\left(\frac{n_2}{2}-\frac{t}{2}\right)}{\Gamma\frac{n_1}{2}\Gamma\frac{n_2}{2}}.$$

The cumulant generating function of Z-distribution is given by

$$\begin{aligned}
K_z(t) &= \log M_z(t) \\
&= \frac{t}{2}\left[\log n_2 - \log n_1\right] + \log\Gamma\left(\frac{n_1}{2}+\frac{t}{2}\right) \\
&\quad + \log\Gamma\left(\frac{n_2}{2}-\frac{t}{2}\right) - \log\Gamma\frac{n_1}{2} - \log\Gamma\frac{n_2}{2}.
\end{aligned}$$

Using Stirling's approximation for $n!$, when n is large, i.e.,

$$\lim_{n\to\infty} \Gamma n+1 = \lim_{n\to\infty} n! \approx \sqrt{2\pi}\, e^{-n} n^{n+\frac{1}{2}}$$

we have

$$\begin{aligned}
K_z(t) &= \frac{t}{2}\left[\log n_2 - \log n_1\right] + \left(\frac{n_1+t-4}{2}+\frac{1}{2}\right) \\
&\quad \times \log\left(\frac{n_1+t+4}{2}\right) - \left(\frac{n_1+t-4}{2}\right) + \log\sqrt{2\pi} \\
&\quad + \left(\frac{n_2-t-4}{2}+\frac{1}{2}\right)\log\left(\frac{n_2-t-4}{2}\right) - \frac{n_2-t-4}{2} \\
&\quad + \log\sqrt{2\pi} - \log\left(\frac{n_1}{2}-1\right)! - \log\left(\frac{n_2}{2}-1\right)! \qquad (28.8)
\end{aligned}$$

In general

$$K_r(z) = 2^{-r}\left[\psi^{(r-1)}\left(\frac{1}{2}n_1\right) + (-1)^r\psi^{(r-1)}\left(\frac{1}{2}n_2\right)\right],\ r \geq 2.$$

An alternative formula is given by

$$K_r(z) = (r-1)!\sum_{j=0}^{\infty}\left[(-1)^r(n_1+2j)^{-r} + (n_2+2j)^{-r}\right].$$

When n_1 and n_2 are even

$$\mu_1' = E(z) = \frac{1}{2}\log\left(\frac{n_2}{n_1}\right) - \Sigma_j^* j^{-1}$$

where Σ_j^* denotes summation from $j = \frac{1}{2}\min(n_1, n_2)$ to $\left[\frac{1}{2}\max(n_1, n_2) - 1\right]$ inclusive and $\Sigma_j^* = 0$ if $n_1 = n_2$.

Also

$$V(z) = 0.822467 - \frac{1}{4}\left(\sum_{j=1}^{\frac{n_1}{2}-1} j^{-2} + \sum_{j=1}^{\frac{1}{2}n_2-1} j^{-2}\right).$$

From (28.8), we have

$$K_1 = \mu_1' = \frac{1}{2}\left(\frac{1}{n_2} - \frac{1}{n_1}\right)$$

$$K_2 = \mu_2 = \frac{1}{2}\left(\frac{1}{n_2} + \frac{1}{n_1} + \frac{1}{n_1^2} + \frac{1}{n_2^2}\right)$$

$$K_3 = \mu_3 = \frac{1}{2}\left[\left(\frac{1}{n_2^2} - \frac{1}{n_1^2}\right) + \left(\frac{1}{n_2^3} + \frac{1}{n_1^3}\right)\right]$$

$$K_4 = \mu_4 - 3\mu_2^2 = \frac{1}{n_1^3} + \frac{1}{n_2^3} + 3\left(\frac{1}{n_1^4} + \frac{1}{n_2^4}\right)$$

Thus β_1 and β_2 can be calculated and it can easily be shown that if n_1 and n_2 tends to infinity $\beta_1 \to 0$ and $\beta_2 \to 3$. Hence Z-distribution can be approximated to normal distribution with expected value $\frac{1}{2}\left(\frac{1}{n_2} - \frac{1}{n_1}\right)$ and variance $\frac{1}{2}\left(\frac{1}{n_1} + \frac{1}{n_2}\right)$.

28.4 Approximations of *F*-Distribution

It has already been mentioned that Z-distribution can be approximated by normal distribution. Thus, $Z = \frac{1}{2}\log F$ has a more nearly normal distribution than does F itself.

We already observed that (28.4) has a standard beta distribution. Thus random variates following F-distributions are easily constructed from beta variates and hence there is a relationship between probability integral of F and incomplete beta function ratio. Thus approximations to incomplete beta function ratio can be applied to the F-distribution.

We know

$$\frac{n_1 F}{n_2} = \frac{\chi_1^2}{\chi_2^2}.$$

Thus F can be approximated using Wilson-Hilferty approximation to the distribution of chi-square, where $F_{n_1,n_2}^{1/3}$ is approximated by the distribution of the ratio of two independent normal variates. The variate $F_{n_1,n_2}^{1/3}$ is approximately distributed as

$$\frac{1 - \frac{2}{9n_1} + U_1\sqrt{\frac{2}{9n_1}}}{1 - \frac{2}{9n_2} + U_2\sqrt{\frac{2}{9n_2}}}$$

where U_1, U_2 are independent standard normal variates. Further approximation of this ratio leads

$$Z = \left[\left(1 - \frac{2}{9n_2}\right) F_{n_1,n_2}^{1/3} - \left(1 - \frac{2}{9n_1}\right)\right]\left(\frac{2}{9n_1} F_{n_1,n_2}^{2/3} + \frac{2}{9n_1}\right)$$

to have a standard normal variate. This approximation is remarkably accurate for $n_2 \geq 10$.

The upper percentage points of $F_{n_1,n_2,\alpha}$ can be improved for $n_2 \leq 10$ by the formula

$$\text{(improved value)} = m \times (\text{calculated value of } Z) + c$$

where m and c depend on n_2 and the percentile, but not on n_1. Ashby (1968) gives values of m and c for $\alpha = 0.95, 0.99$; $n_2 = 1(1)10$. Kelley (1948) recommended to use Z' in place of Z, where $Z' = Z\left(1 + 0.08Z^4 n_2^{-3}\right)$. Aroian (1941) proposed an approximation for large $n_1$and n_2, which is given by

$$t = Z\frac{\sqrt{2(n_1 + n_2 - 1)n_1 n_2}}{n_1 + n_2}$$

where 't' is distributed as t-distribution with $(n_1 + n_2 - 1)$ degrees of freedom.

Another approximation of F is given by

$$\left[\sqrt{1 - \frac{1}{2n_2}} F^{\frac{1}{2}} - \sqrt{1 - \frac{1}{2n_1}}\right]\left[\frac{1}{2n_2} F + \frac{1}{2n_1}\right]^{-\frac{1}{2}}.$$

This statistic is approximately distributed as standard normal variate.

28.5 Noncentral F-Distribution

Definition and Derivation

Let $\chi_1'^2(\lambda_1)$ be noncentral chi-square variate with n_1 degrees of freedom and χ_2^2 is a central chi-square variate with n_2 degree of freedom. Then

$$F' = \frac{\chi'^2(\lambda_1)/n_1}{\chi_2^2/n_2} \tag{28.9}$$

has noncentral **F-distribution with parameters** n_1, n_2 and λ_1. The statistic F' is usually written as $F'_{n_1,n_2}(\lambda_1, 0)$. Here $\chi_1'^2(\lambda_1)$ is independent of χ_2^2 and the statistic F' is known as ***'singly' noncentral*** **F or simply noncentral** F with n_1 and n_2 d.f. The doubly noncentral F is defined by

$$F'_{n_1,n_2}(\lambda_1, \lambda_2) = \frac{\chi_1'^2(\lambda_1)/n_1}{\chi_2'^2(\lambda_2)/n_2} \tag{28.10}$$

where $\chi_1'^2(\lambda_1)$ and $\chi_2'^2(\lambda_2)$ are independently distributed as noncentral chi-square variates with n_1 and n_2 degrees of freedom and noncentrality parameters λ_1 and λ_2 respectively.

The distribution was derived by Tang in 1938, though Patnaik named the distribution in 1949. The derivation of it can be given by a theorem.

❑ **THEOREM 28.3** *Let W be a random variable distributed $\chi_1'^2(p, \lambda)$ and Z be another random variable distributed as χ_2^2 with q degrees of freedom. Then the statistic*

$$u = \frac{Wq}{pZ} \tag{a}$$

is distributed as noncentral F with parameters p, q and λ. The frequency function of u is given by

$$f(u) = \sum_{i=0}^{\infty} \frac{\Gamma\left(\frac{2i+p+q}{2}\right)\left(\frac{p}{q}\right)^{\frac{1}{2}(2i+p)} \lambda^i e^{-\lambda}}{\Gamma\left(\frac{q}{2}\right)\Gamma\left(\frac{2i+p}{2}\right) i!} \frac{u^{\frac{1}{2}(2i+p-2)}}{\left(1+\frac{pu}{q}\right)^{\frac{1}{2}(2i+p+q)}}, \quad 0 \le u < \infty. \tag{b}$$

Proof. The joint distribution of W and Z is given by

$$h(W, Z) = \frac{Z^{\frac{1}{2}(q-2)} e^{-\frac{1}{2}Z}}{2^{\frac{q}{2}}\Gamma\frac{q}{2}} \sum_{i=0}^{\infty} \frac{e^{-\lambda}\lambda^i}{i!} \frac{W^{\frac{1}{2}(2i+p-2)} e^{-\frac{1}{2}W}}{2^{\frac{1}{2}(2i+p)}\Gamma\left(\frac{2i+p}{2}\right)}$$
$$0 \le W < \infty,\ 0 \le Z < \infty.$$

Let us consider the transformation $x = W/Z$ and $v = W \Rightarrow Z = v/x,\ W = v$. The Jacobian of the transformation is

$$J = \frac{\partial(Z, W)}{\partial(x, v)} = \begin{vmatrix} \frac{\partial Z}{\partial x} & \frac{\partial W}{\partial x} \\ \frac{\partial Z}{\partial v} & \frac{\partial W}{\partial v} \end{vmatrix} = \begin{vmatrix} -\frac{v}{x^2} & 0 \\ \frac{1}{x} & 1 \end{vmatrix} = -\frac{v}{x}$$

$$\therefore\ g(x, v) = \frac{e^{-\lambda}}{2^{q/2}\Gamma\frac{q}{2}} \sum_{i=0}^{\infty} \frac{\lambda^i}{i!} \left(\frac{v}{x}\right)^{\frac{1}{2}(q-2)} e^{-\frac{v}{2x}} \frac{v^{\frac{1}{2}(2i+p-2)} e^{-\frac{v}{2}}}{2^{\frac{1}{2}(2i+p)}\Gamma\left(\frac{2i+p}{2}\right)}$$
$$0 \le v < \infty, 0 \le x < \infty.$$

The marginal distribution of x is

$$h(x) = \int_0^\infty g(x, v)dv = \sum_{i=0}^{\infty} C_i x^{-\frac{1}{2}(q+2)} \int_0^\infty v^{\frac{1}{2}(2i+p+q-2)} e^{-\frac{1}{2}v\left(1+\frac{1}{x}\right)} dv$$

where

$$C_i = \frac{e^{-\lambda}\lambda^i}{i!\, 2^{q/2}\Gamma\frac{q}{2}2^{\frac{1}{2}(2i+p)}\Gamma\left(\frac{2i+p}{2}\right)}.$$

Let

$$\frac{v}{2}\left(1+\frac{1}{x}\right) = r \Rightarrow v = \frac{2rx}{1+x}$$

$$\text{and } dv = \frac{2dr}{1+\frac{1}{x}} = \frac{2xdr}{1+x},\ 0 \le r < \infty.$$

$$\therefore\ h(x) = \sum_{i=0}^{\infty} C_i x^{-\frac{1}{2}(q+2)}\left(\frac{2x}{1+x}\right)^{\frac{1}{2}(2i+p+q-2)+1}\int_0^{\infty} e^{-r} r^{\frac{1}{2}(2i+p+q)}$$

$$= \sum_{i=0}^{\infty}\frac{C_i x^{\frac{1}{2}(2i+p-2)}\Gamma\left(\frac{2i+p+q}{2}\right)2^{\frac{1}{2}(2i+p+q)}}{(1+x)^{\frac{1}{2}(2i+p+q)}}. \qquad \text{(a)}$$

Substituting $u = \frac{q}{p}x$ in (a), we get

$$f(u) = \sum_{i=0}^{\infty}\frac{\Gamma\left(\frac{2i+p+q}{2}\right)\left(\frac{p}{q}\right)^{\frac{1}{2}(2i+p)}}{\Gamma\frac{q}{2}\Gamma\left(\frac{2i+p}{2}\right)}\frac{\lambda^i e^{-\lambda}}{i!}\frac{u^{\frac{1}{2}(2i+p-2)}}{\left(1+\frac{p}{q}u\right)^{\frac{1}{2}(2i+p+q)}},\ 0 \le u < \infty. \qquad \text{(b)}$$

Note that, if $\lambda = 0$, the function $f(u)$ is to be evaluated by taking sum only for

$$i = 0.$$

Thus, $f(u)$ becomes

$$f(u) = \frac{\Gamma\left(\frac{p+q}{2}\right)\left(\frac{p}{q}\right)^{p/2}}{\Gamma\frac{q}{2}\Gamma\left(\frac{p}{2}\right)}\frac{u^{\frac{p}{2}-1}}{\left(1+\frac{p}{q}u\right)^{\frac{1}{2}(p+q)}},\ 0 \le u < \infty$$

$$= \frac{(p/q)^{p/2}}{B\left(\frac{p}{2},\frac{q}{2}\right)}\frac{u^{\frac{p}{2}-1}}{\left(1+\frac{p}{q}u\right)^{\frac{1}{2}(p+q)}} \qquad \text{(c)}$$

which is the density function of central F with p and q degrees of freedom.

28.6 Properties of Noncentral F-Distribution

The noncentral F-variate is defined by (28.10), where the numerator and denominator are independent. Therefore, the r-th moment about origin of $F'_{n_1,n_2}(\lambda)$ can be expressed as

$$\mu'_r\left[F'_{n_1,n_2}(\lambda)\right] = \left(\frac{n_2}{n_1}\right)^r \mu'_r\left[\chi_1'^2(\lambda)\right]\mu'_{-r}[\chi_2^2]$$

$$= \left(\frac{n_2}{n_1}\right)^r\frac{\Gamma\left(\frac{n_1}{2}+r\right)\Gamma\left(\frac{n_2}{2}-r\right)}{\Gamma\frac{n_1}{2}\Gamma\frac{n_2}{2}}\sum_{j=0}^{r}\binom{r}{j}\frac{\Gamma\frac{r_1}{2}}{\Gamma\left(\frac{r_1}{2}+j\right)}\left(\frac{1}{2}\lambda n_1\right).$$

In particular

$$\mu_1' = E\left[F'_{n_1,n_2}(\lambda)\right] = \frac{n_2(n_1+\lambda)}{n_1(n_2-2)}, \ (n_2 > 2)$$

$$\mu_2 = \mu_2' - \mu_1'^2 = 2\left(\frac{n_2}{n_1}\right)^2 \frac{(n_1+\lambda)^2 + (n_1+2\lambda)(n_2-2)}{(n_2-2)^2(n_2-4)}, \ (n_2 > 4)$$

$$\mu_3 = \frac{4}{(n_2-4)(n_2-6)} \times \left[4\left(\frac{n_1+\lambda}{n_2-2}\right)^3 + \frac{6(n_1+\lambda)(r_1+2\lambda)}{(n_2-2)^2} + \frac{2(n_1+3\lambda)}{n_2-2}\right], \ (n_2 > 6).$$

The characteristic function of this distribution is given by

$$\psi(t) = \sum_{j=0}^{\infty}\left[\left(\frac{\lambda}{2}\right)^j \Big/ j!\right] e^{-\frac{1}{2}\lambda} M\left(\frac{n_1}{2}+j; -\frac{n_2}{2}; -\frac{n_2}{r_1}it\right).$$

Approximations to $F'_{n_1,n_2}(\lambda)$

Patnaik (1949) suggested an approximation to noncentral F-distribution which can be described as follows:

We know that

$$F'_{n_1,n_2}(\lambda) = \frac{\chi'^2_{n_1}(\lambda)/n_1}{\chi^2_{n_2}/n_2}.$$

But $\chi'^2_{n_1}(\lambda)$ can be approximated by $C\chi^2_n$ with

$$C = (n_1+2\lambda)(n_1+\lambda)^{-1}; \ n = (n_1+\lambda)^2(n_1+2\lambda)^{-1}.$$

Therefore

$$\begin{aligned} F'_{n_1,n_2}(\lambda) &= \frac{C\chi^2_n/n_1}{\chi^2_{n_2}/n_2} = \left(\frac{Cn}{n_1}\right)\frac{\chi^2_n/n}{\chi^2_{n_2}/n_2} = \frac{Cn}{n_1}F_{n,n_2} \\ &= \left[\frac{n_1+2\lambda}{n_1+\lambda}\cdot\frac{(n_1+\lambda)^2}{(n_1+2\lambda)}\right]\frac{1}{n_1}F_{n,n_2} = \frac{n_1+\lambda}{n_1}F_{n,n_2} \\ &= \left(1+\frac{\lambda}{n_1}\right)F_{n,n_2}. \end{aligned}$$

Thus, $P\left\{F'_{n_1,n_2}(\lambda) \geq F_{n_1,n_2}(\alpha)\right\}$, where $\alpha = \int_{F_{n_1,n_2}(\alpha)}^{\infty} h(F_{n_1,n_2})dF_{n_1,n_2}$.

$$\begin{aligned} \therefore \ P\left\{F'_{n_1,n_2}(\lambda) \geq F_{n_1,n_2}(\alpha)\right\} &= P\left\{\frac{n_1+\lambda}{n_1}F_{n,n_2} \geq F_{n_1,n_2}(\alpha)\right\} \\ &= P\left\{F_{n,n_2} \geq \frac{n_1}{n_1+\lambda}F_{n_1,n_2}(\alpha)\right\} \\ &= \alpha \text{ if } \lambda = 0 \\ &= 1 \text{ if } \lambda = \infty. \end{aligned}$$

Again, F_{n,n_2} itself may also be approximated leading to a composite approximation to the distribution of $F'_{n_1,n_2}(\lambda)$. Thus, using Paulson's approximation, Severo and Zelen (1960) suggested that

$$\frac{\left(1-\frac{2}{9n_2}\right)\left(\frac{n_1F'}{n_1+\lambda}\right)^{1/3}-\left[1-\frac{2}{9}(n_1+2\lambda)(n_1+\lambda)^{-2}\right]}{\left[\frac{2}{9}(n_1+2\lambda)(n_1+\lambda)^{-2}+\frac{2}{9n_2}\left(\frac{n_1F'}{n_1+\lambda}\right)^{2/3}\right]^{1/2}}$$

approximately distributed as standard normal variate.

Tiku (1966) fitted the distribution of $F'_{n_1,n_2}(\lambda)$ by $(b+cF_{n',n_2})$, where

$$\begin{aligned} n' &= \frac{1}{2}(n_2-2)\left[\sqrt{\frac{H^2}{H^2-4K^3}}-1\right] \\ c &= \frac{n'}{n_1}(2n'+n_2-2)^{-1}(H/K) \\ b &= -n_2(n_2-2)^{-1}\left(c-1-\lambda n_1^{-1}\right). \end{aligned}$$

Here

$$\begin{aligned} H &= 2(n_1+\lambda)^3+3(n_1+\lambda)(n_1+2\lambda)+(n_2-2)+(n_1+3\lambda)(n_2-2)^2 \\ K &= (n_1+\lambda)^2+(n_2-2)(n_1+2\lambda). \end{aligned}$$

In Section (28.3) it has already been mentioned that $Z = \frac{1}{2}\log F$ can be approximated better than F-distribution itself. By analogy with this result useful approximation can be made by considering the distribution

$$Z'_{n_1,n_2}(\lambda)=\frac{1}{2}\log F'_{n_1,n_2}(\lambda).$$

Since

$$Z'=Z'_{n_1,n_2}(\lambda)=\frac{1}{2}\log\left(\frac{n_2}{n_1}\right)+\frac{1}{2}\log\chi'^2_{n_1}(\lambda)-\frac{1}{2}\log\chi^2_{n_2}$$

the cumulants of Z' are

$$K_1(Z')=\frac{1}{2}\left[\log\left(\frac{n_2}{n_1}\right)+K_1\left(\chi'^2_{n_1}(\lambda)\right)-K_1\left(\chi^2_{n_2}\right)\right].$$

In general $K_r(Z')=2^{-r}\left[K_r\left(\chi'^2_{n_1}(\lambda)\right)+(-1)^rK_r\left(\chi^2_{n_2}\right)\right],\ r\geq 2.$

Distributions Related to Noncentral F-Distribution

Tang (1938) compiled tables that can be used to evaluate

$$\int_0^{F_\alpha} f(F')dF'$$

for certain values of F_α, where F' indicates noncentral F-variate. The tabulated values are given in terms of E^2, where $E^2=n_1F'/(n_2+n_1F')$. The frequency function of E^2

is given by

$$h\left(E^2; n_1, n_2, \lambda\right) = \sum_{i=0}^{\infty} \frac{\Gamma\left(\frac{2i+n_1+n_2}{2}\right)\lambda^i}{\Gamma\frac{n_2}{2}\Gamma\left(\frac{2i+n_1}{2}\right)i!} e^{-\lambda}(E^2)^{(2i+n_1-2)} \times \left(1-E^2\right)^{\frac{1}{2}(n_2-2)}, \; 0 \leq E^2 \leq 1. \quad (28.11)$$

If $\lambda = 0$, $g\left(E^2; n_1, n_2, \lambda = 0\right)$ is the beta distribution. When $\lambda \neq 0$, it is noncentral beta distribution. Now, one may be interested in evaluating the integral

$$\int_0^{E_\alpha^2} g(E^2; n_1 n_2, \lambda) dE^2$$

where the quantity E_α^2 is obtained from the integral

$$\int_{E_\alpha^2}^{1} g(E^2; n_1, n_2, \lambda) dE^2 = \alpha$$

where α is given.

We know that if $\chi_{n_1}^2$ and $\chi_{n_2}^2$ are mutually independent, then $\chi_{n_1}^2 \left(\chi_{n_1}^2 + \chi_{n_2}^2\right)^{-1}$ has a standard beta distribution with parameters $n_1/2$ and $n_2/2$. If $\chi_{n_1}^2$ is replaced by $\chi_{n_1}^{\prime 2}(\lambda_1)$, the resultant distribution is called *noncentral beta distribution* with parameters $n_1/2$, $n_2/2$ and noncentrality parameter λ_1. Here $n_1/2$ and $n_2/2$ are known as *shape parameters*. If both χ^2's are replaced by noncentral χ^2, the resultant distribution is doubly noncentral beta distribution with shape parameters $n_1/2$ and $n_2/2$, and noncentrality parameters λ_1 and λ_2.

Each noncentral χ^2 may be approximated by Patnaik's approximation leading noncentral beta distribution to that of

$$\frac{(n_1 + 2\lambda_1)(n_2 + 2\lambda_2)}{(n_1 + \lambda_1)(n_2 + 1\lambda_2)} \times \begin{bmatrix} \text{beta variable with} \\ \text{parameters } f_1, f_2 \end{bmatrix}$$

where $f_j = (n_j + \lambda_j)^2/(n_j + 2\lambda_j)$; $j = 1, 2$.

We know that $\chi_n^{\prime 2}(\lambda)$ is related to the difference between two independent Poisson variates. By a similar argument it can be shown that, if n_1 is even

$$P\left[F'_{n_1,n_2}(\lambda) < f_0/n_1\right] = P\left[Y - Z \geq \frac{1}{2}n_1\right]$$

where Y and Z are mutually independent, Y has a negative binomial distribution with parameters $\frac{1}{2}n_2$, $f_0 n_2$ and Z has Poisson distribution with parameter $\frac{1}{2}\lambda_1$.

28.7 Applications

Applications of *F*-Distribution

The important application of F-distribution is in the study of distribution of ratios of independent estimators of variance. If

$$\{X_{ij}\} \; (i = 1, 2; j = 1, 2, \ldots, n_i; n_i \geq 2)$$

are sample observations independently drawn from normal populations with mean and variance of X_{ij} as μ_i and σ_i^2 respectively, then

$$S_i^2 = (n_i - 1)^{-1} \sum_{j=1}^{n_i} (X_{ij} - \bar{X}_i)^2 ; \ \bar{X}_i = n_i^{-1} \sum_{j=1}^{n_i} X_{ij}$$

is distributed as $\chi_i^2 \sigma_i^2 (n_i - 1)^{-1}$. The ratio S_1^2/S_2^2 is, therefore, distributed as:

$$\frac{\chi_{n_1-1}^2 \sigma_1^2/(n_1 - 1)}{\chi_{n_2}^2 \sigma_2^2/(n_2 - 1)} \text{ or as } \left(\frac{\sigma_1}{\sigma_2}\right)^2 F_{n_1-1,n_2-1}$$

The statistic S_1^2/S_2^2 (if $S_1^2 > S_2^2$) is used to test the hypothesis $H_0 : \sigma_1^2 = \sigma_2^2$. The hypothesis is rejected if either

$$S_1^2/S_2^2 \leq F_{n_1-1,n_2-1,\alpha_1} \text{ or } S_1^2/S_2^2 \geq F_{n_1-1,n_2-1,1-\alpha_2}$$

with $\alpha_1 + \alpha_2 < 1$. The level of significance of the test is $\alpha = \alpha_1 + \alpha_2$ and the power of the test is given by

$$1 - P\left[F_{n_1-1,n_2-1,\alpha_1} < S_1^2/S_2^2 < F_{n_1-1,n_2-1,1-\alpha_2} | \sigma_1/\sigma_2\right]$$
$$= 1 - P\left[(\sigma_2/\sigma_1)^2 F_{n_1-1,n_2-1,\alpha_1} < F_{n_1-1,n_2-1} < (\sigma_2/\sigma_1)^2 F_{n_1-1,n_2-1,1-\alpha_2}\right].$$

The power is calculated for a specified value of σ_1/σ_2.

The commonest application of F-distribution is in the tests associated with the analysis of variance. Using the analysis of variance technique the equality of several means is tested by F-test statistic when the sample observations

$$\{x_{ij}\} \ (i = 1, 2, \ldots, k; \ j = 1, 2, \ldots, n_i)$$

are drawn from normal population with mean μ_i and variance σ^2. The analysis of variance technique is also used to test the general linear hypothesis about the parameters of a general linear model with normal residuals where the test statistic is the ratio of two independent mean squares and has an F-distribution when null hypothesis is true.

The application of F-distribution in testing the significance of an observed multiple correlation coefficient is well known. Let R be the sample multiple correlation coefficient based on n observations drawn from $(k + 1)$ variate multinormal population, where R measure the amount of correlation of one variate with other k variates. Then the statistic

$$F = \frac{R^2}{1 - R^2} \cdot \frac{n - k - 1}{k}$$

has F-distribution with $(k, n - k - 1)$ degrees of freedom under the null hypothesis (H_0) that the population multiple correlation coefficient is zero.

Under the null hypothesis that the population correlation ratio is zero, the statistic

$$F = \frac{\eta^2}{1 - \eta^2} \cdot \frac{N - h}{h - 1}$$

has an F-distribution with $(h-1)$ and $(N-h)$ degrees of freedom if a sample of size N is drawn from bivariate population, where η is the sample correlation ratio when sample observations are arranged in h arrays. In such a case the statistic

$$F = \frac{\eta^2 - r^2}{1-\eta^2}\frac{N-h}{h-2}$$

also has an F-distribution under the null hypothesis of linearity of regession.

Using the relationships of β and F variables, functions of Wilk's Λ can be expressed in terms of F variable. Thus, the distribution of some of the multivariate test statistics can be approximated by F-distribution. For example,

$$\frac{1-\Lambda(p,m,1)}{\Lambda(p,m,1)} \sim \frac{p}{m-p+1}F_{p,m-p+1}$$

where $\Lambda(p,m,1)$ has Wilk's lambda distribution with parameters p and m.

Applications of Noncentral *F*-Distribution

The noncentral F-distribution is used in the calculation of the power function of tests of general linear hypothesis. Using Tang's (1938) Table the power of the analysis of variance test can be computed. Tang transformed noncental F to E^2 by

$$E^2 = \frac{pF'}{q+pF'}$$

where p and q are the degrees of freedom of F'. Using this transformation, we get

$$\begin{aligned}\alpha &= \int_{E^2_\alpha}^{1} g(E^2; p, q, \lambda = 0)dE^2 \\ &= \int_{F_\alpha}^{\infty} f\left(F'_{p,q}(\lambda = 0)\right) dF'\end{aligned}$$

where $F_\alpha = qE^2_\alpha/(p - pE^2_\alpha)$ and its value for given α can be obtained from Snedecor's table for the F-distribution. For example, let the hypothesis be $H_0 : \lambda = 0$ in the F'-distribution. Let us use a type 1 error size α, where the interval $F_\alpha \leq F' < \infty$ is the critical region. Now, the power of the test $\beta(\lambda)$ is the probability that the observed F' falls in the critical region $F_\alpha \leq F' < \infty$ when $\lambda \neq 0$, and is equal to the integral

$$\beta(\lambda) = \int_{F_\alpha}^{\infty} f\left(F'_{p,q,\lambda}\right) dF'.$$

This $\beta(\lambda)$ can be evaluated transforming F' to E^2 giving

$$\beta(\lambda) = \int_{E^2_\alpha}^{1} g(E^2; p, q, \lambda)dE^2.$$

Also, we have

$$1 - \beta(\lambda) = \int_0^{E^2_\alpha} g(E^2; p, q, \lambda)dE^2.$$

Tang gave values of $\beta(\lambda)$ for various values of p, q, λ and for $\alpha = 0.05$ and 0.01.

Tang evaluated the integral

$$P(\text{II}) = 1 - \beta(\phi) = \int_0^{E_\alpha^2} g(E^2; f_1, f_2, \phi) dE^2$$

for various values of f_1, f_2, ϕ and E_α^2, where

$$p = f_1,\ q = f_2 \text{ and } \phi = \sqrt{\frac{2\lambda}{f_1 + 1}}.$$

More recent application of noncentral F-distribution is in communication theory where it is used for determination of probabilities of errors in transmission over certain types of noise-perturbed channels.

EXERCISE

1. (a) Define F-statistic and derive its distribution.
   (b) Show that the F-distribution is positively skewed and leptokurtic under certain condition on sample size.
2. (a) What do you mean by sampling distribution? If x_1 and x_2 are two independent random observations from a population having p.d.f.

   $$f(x) = e^{-x},\ 0 < x < \infty$$

   then $u = x_1/x_2$ is distributed as variance ratio.
   (b) Let $X_1 \sim N(\mu_1, \sigma_1^2), X_2 \sim N(\mu_2, \sigma_2^2)$. Derive the distribution of

   $$U = (X_1 - \mu_1)/(X_2 - \mu_2).$$

3. (a) Define F-statistic and χ^2-statistic. Show that $\chi^2 = n_1 F$, where F is distributed as variance ratio with n_1 and n_2 d.f.
   (b) What is Fisher's Z-distribution? How it is related to F-distribution?
4. (a) Define sampling distribution and important sample statistics. Show that $F = t^2$, where F is the ratio of two variances and t is the Student's t-statistic.
   (b) Find moments of F-distribution.
5. (a) Distinguish between central F-distribution and non-central F-distribution. Derive non-central F-distribution.
   (b) Mention some applications of F-distribution.
6. Define F-statistic. Show that $F_{n_1,n_2} = 1/F_{n_2,n_1}$. Find mode, mean and variance of F-distribution. Also, find mean and variance of Fisher's Z-distribution.

CHAPTER **29**

t-Distribution

29.1 Introduction and Derivation

Let $X_1, X_2, \ldots, X_n$ be independent random variables, each having the same normal distribution, $N(\mu, \sigma^2)$; then

$$z = \frac{\sqrt{n}\,(\bar{X} - \mu)}{\sigma} \sim N(0,1) \tag{29.1}$$

if n is large. The entire large sample theory is based on the normal test using the statistic (29.1) to test the hypothesis $H_0 : \mu = \mu_0$ (say), provided σ is known. If σ is not known, it is replaced by its unbiased estimator

$$s = \sqrt{(n-1)^{-1}\Sigma\left(X_i - \bar{X}\right)^2}.$$

However, if the sample size n is small, the statistic

$$z_1 = \frac{\sqrt{n}\,(\bar{X} - \mu)}{s} \tag{29.2}$$

is far from normality. The distributions of z and z_1 are not identical. In 1908, W.S. Gosset who wrote under the pen name Student obtained the distribution of the statistic

$$z_2 = (n-1)^{\frac{1}{2}}\left\{\sqrt{n}\,(\bar{X} - \mu)/s\right\}. \tag{29.3}$$

This statistic is distributed as the ratio of standard normal variate, say z, and χ^2_{n-1}, where z and χ^2_{n-1} are mutually independent. In 1925, Fisher defined the statistic

$$t_\gamma = z/\left(\chi^2_\gamma/\gamma\right)^{\frac{1}{2}} \tag{29.4}$$

where $s^2 = (n-1)^{-1}\Sigma(X_i - \bar{X})^2$ is distributed as $\chi^2_\gamma\sigma^2$ with $\gamma = (n-1)$ degrees of freedom and z is a standard normal variate. The statistic t_γ is known as *Student's 't'* and its distribution is called *Student's t-distribution* the p.d.f. of which is given by

$$f(t) = \frac{1}{\sqrt{\gamma}B\left(\frac{1}{2}, \frac{\gamma}{2}\right)} \frac{1}{\left(1 + \frac{t^2}{\gamma}\right)^{\frac{1}{2}(\gamma+1)}}, \quad -\infty < t < \infty. \tag{29.5}$$

This is a special form of Pearson Type VII distribution. The distribution reduces to standard Cauchy distribution when $\gamma = 1$, since

$$f(t) = \frac{1}{\sqrt{1}B\left(\frac{1}{2}, \frac{1}{2}\right)} \frac{1}{(1+t^2)^{\frac{1}{2}(1+1)}} = \frac{1}{\pi(1+t^2)}.$$

Derivation

Student's 't' can also be defined by the statistic

$$t = \frac{\sqrt{n}\left(\bar{X} - \mu\right)}{s}$$

$$\frac{t^2}{n-1} = \frac{n\left(\bar{X}-\mu\right)^2}{(n-1)s^2} = \frac{n\left(\bar{X}-\mu\right)^2/\sigma^2}{(n-1)s^2/\sigma^2}$$

$$= \frac{n\left(\bar{X}-\mu\right)^2/\sigma^2}{nS^2/\sigma^2} \quad \left[S^2 = \frac{1}{n}\Sigma(X_i - \bar{X})^2\right]$$

But $\frac{n(\bar{X}-\mu)^2}{\sigma^2}$ is a chi-square variate with 1 degree of freedom [since $X_1, X_2, \ldots, X_n$ follow $N(\mu, \sigma^2]$ and ns^2/σ^2 is also a chi-square variate with $\gamma = (n-1)$ d.f. We also know that $\bar{X}$ and s^2 [Section 26.2] are independently distributed. Hence $t^2/(n-1)$ is the ratio of two independent chi-square variates where the chi-square in numerator has 1 d.f. and the denominator one has γ d.f. Thus (t^2/γ) is a beta variate of the second kind with parameters $1/2$ and $\gamma/2$, and its density function is given by

$$f(t^2/\gamma)d(t^2/v) = \frac{1}{B\left(\frac{1}{2}, \frac{\gamma}{2}\right)} \frac{(t^2/\gamma)^{\frac{1}{2}-1}}{\left(1+\frac{t^2}{\gamma}\right)^{\frac{1+\gamma}{2}}} d(t^2/\gamma), \; 0 < t^2/\gamma < \infty$$

$$= \frac{2}{\sqrt{\gamma}B\left(\frac{1}{2}, \frac{\gamma}{2}\right)} \frac{dt}{\left(1+\frac{t^2}{\gamma}\right)^{\frac{1+\gamma}{2}}}, \; 0 < t^2 < \infty.$$

$$\therefore \; f(t) = \frac{1}{\sqrt{\gamma}B\left(\frac{1}{2}, \frac{\gamma}{2}\right)} \frac{1}{\left(1+\frac{t^2}{\gamma}\right)^{\frac{1+\gamma}{2}}}, \; -\infty < t < \infty.$$

The distribution can also be derived from Fisher t, where

$$t = \frac{z}{(\chi^2_\gamma/\gamma)^{\frac{1}{2}}}$$

where z and χ^2_γ are independently distributed. The joint density of z and χ^2_γ is given by

$$f\left(z, \chi^2_\gamma\right) = \frac{1}{\sqrt{2\pi}} e^{-\frac{1}{2}z^2} \frac{1}{\Gamma\frac{\gamma}{2}2^{\gamma/2}} e^{-\frac{1}{2}\chi^2} (\chi^2)^{\frac{\gamma}{2}-1};$$

$$-\infty < z < \infty, \; 0 < \chi^2 < \infty$$

Let

$$t = z/\sqrt{\chi^2/\gamma} \text{ and } u = \chi^2 \Rightarrow z = t\sqrt{u/\gamma} \text{ and } \chi^2 = u.$$

The Jacobian of the transformation is

$$J = \frac{\partial(z, \chi^2)}{\partial(t, u)} = \begin{vmatrix} \frac{\partial z}{\partial t} & \frac{\partial z}{\partial u} \\ \frac{\partial \chi^2}{\partial t} & \frac{\partial \chi^2}{\partial u} \end{vmatrix} = \begin{vmatrix} \sqrt{u/\gamma} & \frac{t}{\sqrt{\gamma}}\frac{1}{2}u^{-\frac{1}{2}} \\ 0 & 1 \end{vmatrix}$$

$$= \sqrt{u/\gamma}; \; -\infty < t < \infty, \; 0 < u < \infty.$$

$$\therefore \; g(t, u) = \frac{1}{\sqrt{2\pi}} e^{-\frac{1}{2}\frac{t^2 u}{\gamma}} \frac{1}{\Gamma\frac{\gamma}{2}2^{\gamma/2}} e^{-\frac{1}{2}u} u^{\frac{\gamma}{2}-1} \sqrt{u/\gamma}$$

$$= \frac{1}{\sqrt{\gamma}\Gamma\frac{1}{2}\Gamma\frac{\gamma}{2}2^{\frac{\gamma+1}{2}}} e^{-\frac{1}{2}u(1+t^2/\gamma)} u^{\frac{\gamma-1}{2}}.$$

Let

$$v = \frac{1}{2}u(1 + t^2/\gamma)$$

$$\Rightarrow u = \frac{2v}{(1 + t^2/\gamma)} \text{ and } du = \frac{2dv}{(1 + t^2/\gamma)}, \; 0 < v < \infty$$

$$h(t) = \int_0^\infty g(tv)dv = \frac{1}{\sqrt{\gamma}\Gamma\frac{1}{2}\Gamma\frac{\gamma}{2}2^{\frac{\gamma+1}{2}}} \frac{2^{\frac{\gamma-1}{2}+1}}{\left(1 + \frac{t^2}{\gamma}\right)^{\frac{\gamma+1}{2}+1}} \int_0^\infty e^{-v} v^{\frac{\gamma-1}{2}} dv$$

$$= \frac{1}{\sqrt{\gamma}\Gamma\frac{1}{2}\Gamma\frac{\gamma}{2}} \frac{1}{\left(1 + \frac{t^2}{\gamma}\right)^{\frac{\gamma+1}{2}}} \int_0^\infty e^{-v} v^{\frac{\gamma+1}{2}-1} dv$$

$$= \frac{\Gamma\left(\frac{\gamma+1}{2}\right)}{\sqrt{\gamma}\Gamma\frac{1}{2}\Gamma\frac{\gamma}{2}} \frac{1}{\left(1 + \frac{t^2}{\gamma}\right)^{\frac{\gamma+1}{2}}}$$

$$= \frac{1}{\sqrt{\gamma}B\left(\frac{1}{2}, \frac{\gamma}{2}\right)} \frac{1}{\left(1 + \frac{t^2}{\gamma}\right)^{\frac{\gamma+1}{2}}}, \; \infty < t < \infty.$$

The graph of the probability density function of t-distribution is given in FIG. 29.1.

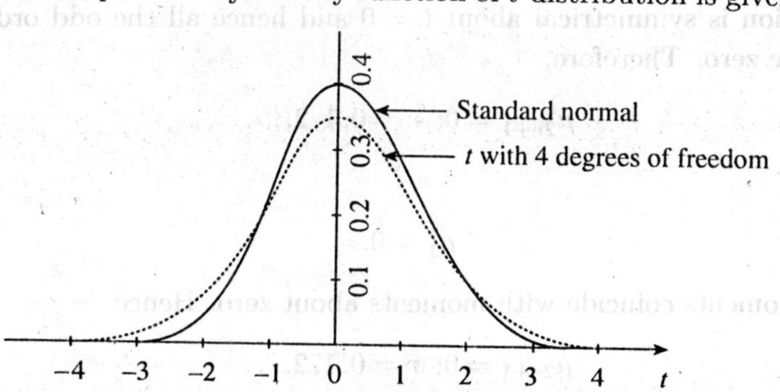

FIG. 29.1 : Graph of t-density function

So far we have discussed the t-statistic based on one sample of size n from $N(\mu, \sigma^2)$. Let there be two samples of sizes n_i drawn independently from two normal populations

$N(\mu_i, \sigma^2)(i = 1, 2)$. Then

$$u = \frac{(\bar{x}_1 - \bar{x}_2) - (\mu_1 - \mu_2)}{\sigma\sqrt{\left(\frac{1}{n_1} + \frac{1}{n_2}\right)}} \sim N(0,1).$$

The pooled sample variance is given by

$$(n_1 + n_2 - 2)s^2 = \sum_{j=1}^{n_1}(x_{1j} - x_1)^2 + \sum_{j=1}^{n_2}(x_{2j} - \bar{x}_2)^2$$

where

$$\bar{x}_i = n_i^{-1}\sum_{j=1}^{n_i} x_{ij}.$$

Thus, $\frac{(n_1+n_2-2)s^2}{\sigma^2}$ is a $\chi^2_{n_1+n_2-2}$ variate. Hence

$$\begin{aligned} t &= \frac{(\bar{x}_1 - \bar{x}_2) - (\mu_1 - \mu_2)}{\sigma\sqrt{\left(\frac{1}{n_1} + \frac{1}{n_2}\right)}} \div \sqrt{\left[\frac{(n_1 + n_2 - 2)s^2}{\sigma^2} \Big/ (n_1 + n_2 - 2)\right]} \\ &= \frac{(\bar{x}_1 - \bar{x}_2) - (\mu_1 - \mu_2)}{s\sqrt{\left(\frac{1}{n_1} + \frac{1}{n_2}\right)}} \qquad (29.6) \end{aligned}$$

is distributed as t with $(n_1 + n_2 - 2)$ degrees of freedom. This result is due to R.A. Fisher. Cacoullos (1965) has shown that

$$\frac{1}{2}\left(\chi_1^2 - \chi_2^2\right)\sqrt{\gamma}\left(\chi_1^2\chi_2^2\right)^{-\frac{1}{2}}$$

has a t_γ-distribution with γ degrees of freedom if χ_1^2 and χ_2^2 are independent chi-square variates, each with γ degrees of freedom.

29.2 Properties of t-Distribution

The t-distribution is symmetrical about $t = 0$ and hence all the odd order moments about origin are zero. Therefore,

$$\mu'_{2r+1} = 0;\ r = 0, 1, 2, \ldots$$

In particular

$$\mu'_1 = 0.$$

Thus central moments coincide with moments about zero. Hence

$$\mu_{2r+1} = 0;\ r = 0, 1, 2, \ldots$$

The moments of even order are given by

$$\mu'_{2r} = E\left[t^{2r}\right] = \frac{2}{\sqrt{\gamma}B\left(\frac{1}{2}, \frac{\gamma}{2}\right)}\int_0^\infty \frac{t^{2r}\,dt}{\left(1 + \frac{t^2}{\gamma}\right)^{\frac{\gamma+1}{2}}}.$$

Let $1+\frac{t^2}{\gamma}=\frac{1}{y} \Rightarrow t^2=\frac{\gamma}{y}-\gamma,\ 2tdt=-\frac{\gamma}{y^2}dy,$

$$
\begin{aligned}
\mu_{2r}' &= \frac{1}{\sqrt{\gamma}B\left(\frac{1}{2},\frac{\gamma}{2}\right)}\int_0^1 \frac{\left(\frac{\gamma}{y}-\gamma\right)^r}{\left(\frac{1}{y}\right)^{\frac{\gamma+1}{2}}}\frac{\gamma}{y^2}\frac{1}{\left[\gamma\left(\frac{1}{y}-1\right)\right]^{1/2}}dy \\
&= \frac{\gamma^r}{B\left(\frac{1}{2},\frac{\gamma}{2}\right)}\int_0^1 y^{\frac{\gamma}{2}-r-1}(1-y)^{r+\frac{1}{2}-1}dy \\
&= \frac{\gamma^r}{B\left(\frac{1}{2},\frac{\gamma}{2}\right)}B\left(\frac{\gamma}{2}-r,r+\frac{1}{2}\right) \\
&= \frac{\gamma^r\Gamma\left(\frac{\gamma}{2}-r\right)\Gamma\left(r+\frac{1}{2}\right)}{\Gamma\frac{1}{2}\Gamma\frac{\gamma}{2}} \\
&= \gamma^r\frac{(2r-1)(2r-3)\cdots 3\cdot 1}{(\gamma-2)(\gamma-4)(\gamma-6)\cdots(\gamma-2r)},\ \frac{\gamma}{2}>r.
\end{aligned}
$$

In particular

$$\mu_2'=\mu_2=\frac{\gamma}{\gamma-2};\ \mu_4=\frac{\gamma^2\cdot 1\times 3}{(\gamma-2)(\gamma-4)},\ \gamma>4.$$

The moment ratios are

$$
\begin{aligned}
\alpha_3 &= \sqrt{\beta_1}=\frac{\mu_3^2}{\mu_2^3}=0 \\
\alpha_4 &= \beta_2=\frac{\mu_4}{\mu_2^2}=\frac{3(\gamma-2)}{\gamma-4}=3+\frac{6}{\gamma-4}.
\end{aligned}
$$

Thus, for large value of γ, β_2 tends to 3. Therefore, t-distribution tends to normal distribution as γ tends to infinity. This can also be shown as follows:

We have

$$f(t)=\frac{1}{\sqrt{\gamma}B\left(\frac{1}{2},\frac{\gamma}{2}\right)}\frac{1}{\left(1+\frac{t^2}{\gamma}\right)^{\frac{\gamma+1}{2}}},\ -\infty<t<\infty.$$

We know

$$
\begin{aligned}
\lim_{\gamma\to\infty}\frac{1}{\sqrt{\gamma}B\left(\frac{1}{2},\frac{\gamma}{2}\right)} &= \lim_{\gamma\to\infty}\frac{1}{\sqrt{\gamma}}\frac{\Gamma(\gamma+1)/2}{\Gamma\frac{1}{2}\Gamma\frac{\gamma}{2}} \\
&= \frac{1}{\sqrt{\gamma}}\frac{1}{\pi}\left(\frac{\gamma}{2}\right)^{1/2} \\
&= \frac{1}{\sqrt{2\pi}}.
\end{aligned}
$$

$$
\begin{aligned}
\therefore\ \lim_{\gamma\to\infty}f(t) &= \lim_{\gamma\to\infty}\frac{1}{\sqrt{\gamma}B\left(\frac{1}{2},\frac{\gamma}{2}\right)}\lim_{\gamma\to\infty}\left[\left(1+\frac{t^2}{\gamma}\right)^{\gamma}\right]^{-\frac{1}{2}}\times\lim_{\gamma\to\infty}\left(1+\frac{t^2}{\gamma}\right)^{-\frac{1}{2}} \\
&= \frac{1}{\sqrt{2\pi}}e^{-\frac{1}{2}t^2}
\end{aligned}
$$

which is the p.d.f. of a standard normal variate.

From the value of β_1 and β_2 it can be said that t-distribution is symmetric and leptokurtic. But it tends to standard normal variate when $\gamma \to \infty$. However, from

$$\mu_2 = \frac{\gamma}{\gamma - 2} \text{ and } \mu_4 = \frac{3\gamma^2}{(\gamma-2)(\gamma-4)}$$

it can be said that, when $\gamma > 2$, $\mu_2 > 1$, i.e., variance of t greater than the variance of standard normal variate. When $\gamma > 4, \beta_2 > 3$, i.e., t-distribution is more flat on the top than the normal curve.

Since $f(-t) = f(t)$, the probability curve is symmetrical about the line $t = 0$. As t increases, $f(t)$ decreases rapidly and tends to zero as $t \to \infty$ so that $t - a$ is an asymptote to the curve. The distribution has a single mode at $t = 0$.

The probability density function of t has points of inflexion

$$t_\gamma = \pm\sqrt{\gamma/(\gamma+2)}.$$

Mean Deviation about Mean

The mean deviation about mean is defined by

$$\begin{aligned}
\text{M.D. } (t) &= E|t - \mu_1'| = E|t|, \quad \because \mu_1' = 0 \\
&= \frac{1}{\sqrt{\gamma}B\left(\frac{1}{2}, \frac{\gamma}{2}\right)} \int_{-\infty}^{\infty} \frac{|t|}{\left(1 + \frac{t^2}{\gamma}\right)^{\frac{\gamma+1}{2}}} dt \\
&= \frac{2}{\sqrt{\gamma}B\left(\frac{1}{2}, \frac{\gamma}{2}\right)} \int_{0}^{\infty} \frac{tdt}{\left(1 + \frac{t^2}{\gamma}\right)^{\frac{\gamma+1}{2}}}.
\end{aligned}$$

Let $t^2/\gamma = y \Rightarrow t^2 = \gamma y$ and $2tdt = \gamma dy$, $0 < y < \infty$.

$$\begin{aligned}
\therefore \text{M.D. } (t) &= \frac{\gamma}{\sqrt{\gamma}B\left(\frac{1}{2}, \frac{\gamma}{2}\right)} \int_{0}^{\infty} \frac{dy}{(1+y)^{\frac{\gamma+1}{2}}} \\
&= \frac{\sqrt{\gamma}}{B\left(\frac{1}{2}, \frac{\gamma}{2}\right)} \int_{0}^{\infty} \frac{y^{1-1}dy}{(1+y)^{\frac{\gamma+1}{2}}} \\
&= \frac{\sqrt{\gamma}}{B\left(\frac{1}{2}, \frac{\gamma}{2}\right)} B\left(1, \frac{\gamma-1}{2}\right) = \frac{\sqrt{\gamma}\,\Gamma\frac{\gamma-1}{2}}{\sqrt{\pi}\,\Gamma\frac{\gamma}{2}}.
\end{aligned}$$

Distribution Function

The distribution function of t can be expressed in terms of incomplete beta function ratio, i.e., if

$$I_x(p, q) = \frac{1}{B(p,q)} \int_0^x t^{p-1}(1-t)^{q-1}dt; \; p > 0, \; q > 0$$

then

$$F(t) = 1 - \frac{1}{2} I_x\left(\frac{\gamma}{2}, \frac{1}{2}\right), \text{ where } x = \left(1 + \frac{t^2}{y}\right)^{-1}.$$

We know

$$F(t) = P(u \le t) = \int_{-\infty}^{t} f(u)du = 1 - \int_{t}^{\infty} f(u)du$$

$$= 1 - \frac{1}{\sqrt{\gamma}B\left(\frac{1}{2},\frac{\gamma}{2}\right)} \int_{t}^{\infty} \frac{1}{\left(1+\frac{u^2}{\gamma}\right)^{\frac{1+\gamma}{2}}} du.$$

Let

$$\frac{1}{z} = 1 + \frac{u^2}{\gamma} \Rightarrow z = \left(1 + \frac{u^2}{\gamma}\right)^{-1}$$

$$du = -\frac{1}{2} z^{-3/2} (1-z)^{-1/2} dz$$

$$\therefore\ F(t) = 1 + \frac{1}{\sqrt{\gamma}B\left(\frac{1}{2},\frac{\gamma}{2}\right)} \int_{\left(1+\frac{t^2}{\gamma}\right)^{-1}}^{0} z^{\frac{\gamma}{2}-1} (1-z)^{-1/2} dz$$

$$= 1 - \frac{1}{\sqrt{\gamma}B\left(\frac{1}{2},\frac{\gamma}{2}\right)} \int_{0}^{\left(1+\frac{t^2}{\gamma}\right)^{-1}} z^{\frac{\gamma}{2}-1} (1-z)^{\frac{1}{2}-1} dz$$

$$= 1 - I_x\left(\frac{\gamma}{2},\frac{1}{2}\right), \text{ where } x = \left(1 + \frac{t^2}{\gamma}\right)^{-1}.$$

29.3 Approximations and Related Distributions

It has already been discussed that t_γ can be regarded as standard normal variate if γ is large. But this approximation is very crude unless γ is at least 30, and unsatisfactory for considerably larger values of γ if extreme tails (say, $|t_\gamma| > 4$) of the distribution are being considered. The statistic $t_\gamma\sqrt{1 - 2\gamma^{-1}}$ may be considered as standard normal variate and this approximation is moderately good if γ is less than 20 or if extreme tails are being considered.

Fisher and Cornish (1960) obtained the approximation of t_γ by

$$t_{\gamma,\alpha} \doteq z_\alpha + \frac{1}{4} z_\alpha \left(z_\alpha^2 + 1\right) \gamma^{-1} + \frac{1}{96} z_\alpha \left(5z_\alpha^4 + 16z_\alpha^2 + 3\right) \gamma^{-2}$$
$$+ \frac{1}{384} z_\alpha \left(3z_\alpha^6 + 19z_\alpha^4 + 17z_\alpha^2 - 15\right) \gamma^{-3}.$$

The statistic

$$\sqrt{2(\gamma+1)C_\gamma} t_\gamma \left[t_\gamma^2 + 2\gamma\right]^{-\frac{1}{2}}$$

has a standard normal distribution. In this approximation $\sqrt{2(\gamma+1)C_\gamma}$ may be replaced by $\sqrt{2\gamma - 1}$ unless γ is small. Here

$$C_\gamma = 1 - \frac{3}{4}(\gamma+1)^{-1} - \frac{7}{32}(\gamma+1)^{-2}.$$

Also $\sinh^{-1}\left(t_\gamma\sqrt{\frac{3}{2}\gamma}\right)$ has a standard normal distribution.

Related Distributions

If x_1 and x_2 are two independent sample observations from $N(\mu, \sigma^2)$, then

$$u = \frac{x_1 + x_2 - 2\mu}{\sqrt{(x_1 - x_2)^2}}$$

has a Student's t-distribution. This can be shown as follows:

As x_1 and x_2 are independent sample observations from $N(\mu, \sigma^2)$

$$\begin{aligned} V(x) &= \left[(x_1 - \bar{x})^2 + (x_2 - \bar{x})^2\right], \text{ where } x = \frac{1}{2}(x_1 + x_2) \\ &= \left(x_1 - \frac{x_1 + x_2}{2}\right)^2 + \left(x_2 - \frac{x_1 + x_2}{2}\right)^2 \\ &= \frac{(x_1 - x_2)^2}{2}. \end{aligned}$$

Again $\frac{x_i - \mu}{\sigma} \sim N(0, 1)$ $(i = 1, 2)$.

$$\therefore \left(\frac{x_1 - \mu}{\sigma} + \frac{x_2 - \mu}{\sigma}\right) \sim N(0, 2), \text{ i.e., } z = \frac{x_1 + x_2 - 2\mu}{\sigma\sqrt{2}} \sim N(0, 1)$$

Therefore $t = \frac{x_1 + x_2 - 2\mu}{s\sqrt{2}}$ has Student's t-distribution, where

$$s^2 = \sum_{i=1}^{2} (x_i - \bar{x})^2, \text{ i.e., } t = \frac{x_1 + x_2 - 2\mu}{\sqrt{\frac{(x_1 - x_2)^2 2}{2}}} = \frac{x_1 + x_2 - 2\mu}{\sqrt{(x_1 - x_2)^2}}$$

has Student's t-distribution.

If $x_{i1}, x_{i2}, \ldots, x_{in_i}$ are independent random variables, each normally distributed with mean μ_i and variance σ_i^2, then for $i = 1, 2, \sqrt{n_i}\,(\bar{x}_i - \mu_i)/s_i$ is distributed as $t_{n_i - 1}$, where

$$s_i^2 = (n_i - 1)^{-1} \sum_{j=1}^{n_i} (x_{ij} - \bar{x}_i)^2.$$

Thompson found that the distribution of $(x_i - \bar{x})/s$ is related to the t-distribution, where x_i is randomly chosen from $x_1, x_2, \ldots, x_n$. If x_i is replaced by x_n and putting $E(x_i) = 0$ and $V(x_i) = 1$, then the distribution is not affected.

We have

$$\begin{aligned} \sum_{i=1}^{n} (x_i - \bar{x})^2 &= x_1^2 + \frac{1}{2}(x_2 - x_1)^2 + \frac{2}{3}\left[x_3 - \frac{1}{2}(x_1 + x_2)\right]^2 + \cdots \\ &\quad + \frac{n(n-1)}{n}\left[x_n - \frac{1}{n-1}(x_1 + x_2 + \cdots + x_{n-1})\right]^2 \\ &= y + \frac{n}{n-1}(x_n - \bar{x})^2 \end{aligned}$$

with y and $(x_n - \bar{x})^2$ mutually independent, and y is distributed as χ^2_{n-2}. Hence

$(x_n - \bar{x})/s$ is **distributed as**

$$\frac{n-1}{\sqrt{n}} z \left(\chi^2_{n-2} + z^2\right)^{-1/2}$$

with z and χ^2_{n-2} mutually independent. If we put $(x_n - \bar{x})/s = t$, then t is distributed as

$$\frac{n-1}{n} \sqrt{n-2}\, t_{n-2} \left[1 + (n-2)t^2_{n-2}\right]^{-1/2}.$$

Weir (1960) has shown that the upper 2.5% points of the statistic

$$\frac{\bar{x}_1 - \bar{x}_2}{\sqrt{\left[\frac{(n_1-1)s_1^2}{(n_1-3)n_1} + \frac{(n_2-1)s_2^2}{(n_2-3)n_2}\right]}}$$

lies between 1.96 and 2, provided $n_1 \geq 6$ and $n_2 \geq 6$. These percentile points are similar to those of t-variate.

Welch (1938) showed that the distribution of $n_1^{-1}s_1^2 + n_2^{-1}s_2^2$ is approximately $C\chi^2_\gamma$, where

$$\begin{aligned} C\gamma &= E\left[n_1^{-1}s_1^2 + n_2^{-1}s_2^2\right] \\ &= n_1^{-1}\sigma_1^2 + n_2^{-1}\sigma_2^2 \\ V\left[n_1^{-1}s_1^2 + n_2^{-1}s_2^2\right] &= 2n_1^{-2}(n_1-1)^{-1}\sigma_1^4 + 2n_2^{-2}(n_2-1)^{-1}\sigma^4 \\ &= 2C^2\gamma \end{aligned}$$

Thus

$$C = \frac{n_1^{-2}(n_1-1)^{-1}\sigma_1^4 + n_2^{-2}(n_2-1)^{-1}\sigma_2^4}{n_1^{-1}\sigma_1^2 + n_2^{-1}\sigma_2^2}$$

$$\text{and } \gamma = \frac{(n_1^{-1}\sigma_1^2 + n_2^{-1}\sigma_2^2)^2}{n_1^{-2}\sigma_1^4(n_1-1) + n_2^{-2}(n_2-1)\sigma_2^4}.$$

Then $(\bar{x}_1 - \bar{x}_2)(n_1^{-1}s_1^2 + n_2^{-1}s_2^2)^{-1/2}$ is approximately distributed as

$$t\gamma = \frac{z\sqrt{n_1^{-1}\sigma_1^2 + n_2^{-1}\sigma_2^2}}{\sqrt{C\gamma}\,(\chi_\gamma/\sqrt{\gamma})}.$$

29.4 Non-central t-Distribution

Let z be a standard normal variate and χ^2_γ be a chi-square variate with γ degrees of freedom. Let us assume that z and χ^2_γ are independent. Then the statistic

$$t' = (z + \delta)/\sqrt{\chi^2_\gamma/\gamma} \tag{29.7}$$

is called a *non-central t* with γ degrees of freedom and its distribution is called *non-central t-distribution* with non-centrality parameter δ. If it is assumed that X is an $N(\mu, 1)$ variate, then $z = x - \mu$. In that case $\delta = \mu$. When X is an $N(\mu, \sigma^2/n)$, $\delta = \frac{\sqrt{n}\mu}{\sigma}$. If $\delta = 0$, the statistic t' reduces to t with γ degrees of freedom.

The non-central t-distribution arises in testing the hypothesis $H_0 : \mu = \mu_0$ using the test statistic

$$t = \frac{\sqrt{n}(\bar{x} - \mu_0)}{s} \tag{29.8}$$

where $\bar{x}$ is the sample mean of n observations drawn independently from normal population $N(\mu, \sigma^2)$ and $s^2 = (n-1)^{-1}\Sigma(x_i - \bar{x})^2$. If $\mu = \mu_0$, then the statistic t is distributed as central t-distribution with $(n-1)$ degrees of freedom. However, if $\mu \neq \mu_0$, the statistic t is distributed as non-central t with non-centrality parameter

$$\delta = \sqrt{n}(\mu - \mu_0)/\sigma.$$

The statistic $t = \sqrt{n}\bar{x}/s$ also distributed as non-central t with $(n-1)$ degrees of freedom and non-centrality parameter $\delta = \sqrt{n}\mu/\sigma$.

Suppose that we want to test the hypothesis $H_0 : \mu_1 = \mu_2$ on the basis of two independent samples drawn from two normal populations $N(\mu_1, \sigma^2)$ and $N(\mu_2, \sigma^2)$, where the sample sizes are n_1 and n_2 respectively. The statistic used for this hypothesis is

$$t = \sqrt{\frac{n_1 n_2}{n_1 + n_2}}(\bar{x}_1 - \bar{x}_2) \Big/ \sqrt{(n_1 + n_2 - 2)^{-1}[(n_1 - 1)s_1^2 + (n_2 - 1)s_2^2]} \tag{29.9}$$

This t is also distributed as central t if $\mu_1 = \mu_2$. However, if $\mu_1 - \mu_2 = \theta$ (say), then t is distributed as non-central t with $(n_1 + n_2 - 2)$ degrees of freedom and non-centrality parameter

$$\delta = \theta\sigma^{-1}\sqrt{\frac{n_1 n_2}{n_1 + n_2}}.$$

Derivation of Non-central t-Distribution

Suppose $X \sim N(\mu, 1)$ and $Y \sim \chi^2_\gamma$ and they are independent. Then the non-central t-statistic is defined by

$$t' = X/\sqrt{Y/\gamma}.$$

The joint density function of X and Y is

$$\begin{aligned} f(x,y) &= \frac{1}{\sqrt{2\pi}} e^{-\frac{1}{2}(x^2-\mu)^2} \frac{1}{2^{\gamma/2}\Gamma\frac{\gamma}{2}} e^{-\frac{1}{2}y} y^{\frac{\gamma}{2}-1}; \\ &\qquad -\infty < x < \infty,\ 0 < y < \infty \\ &= \frac{1}{\sqrt{2\pi}2^{\gamma/2}\Gamma\frac{\gamma}{2}} e^{-\frac{1}{2}(x^2+\mu^2)}\, e^{-\frac{1}{2}y} y^{\frac{\gamma}{2}-1} \sum_{i=0}^{\infty} \frac{(\mu x)^i}{i!}. \end{aligned}$$

Let us consider the transformation

$$t' = \frac{x}{\sqrt{y/\gamma}} \text{ and } z = \sqrt{y}$$

$$\Rightarrow x = zt'/\sqrt{\gamma} \text{ and } y = z^2; \ -\infty < t' < \infty,\ 0 < z < \infty.$$

The Jacobian of the transformation is

$$J = \frac{\partial(x,y)}{\partial(z,t')} = \begin{vmatrix} \frac{\partial x}{\partial t'} & \frac{\partial x}{\partial z} \\ \frac{\partial y}{\partial t'} & \frac{\partial y}{\partial z} \end{vmatrix} = \begin{vmatrix} \frac{z}{\sqrt{\gamma}} & \frac{t'}{\sqrt{\gamma}} \\ 0 & 2z \end{vmatrix} = 2z^2/\sqrt{\gamma}.$$

Now

$$g(t',z) = \frac{e^{-\frac{1}{2}\mu^2}}{\sqrt{\pi}2^{\frac{\gamma-1}{2}}\Gamma\frac{\gamma}{2}\sqrt{\gamma}} e^{-\frac{1}{2}z^2\left(1+\frac{t'^2}{\gamma}\right)} (z^2)^{\frac{\gamma}{2}} \sum_{i=0}^{\infty} \frac{(\mu z t'/\sqrt{\gamma})^i}{i!}.$$

Integrating z over the range 0 to ∞, we get

$$\begin{aligned} h(t') &= \frac{e^{-\frac{1}{2}\mu^2}}{\sqrt{\pi}2^{\frac{\gamma-1}{2}}\Gamma\frac{\gamma}{2}} \sum_{i=0}^{\infty} \frac{(\mu t')^i}{i!\,\gamma^{\frac{1}{2}(i+1)}} \int_0^{\infty} e^{-\frac{1}{2}z^2\left(1+\frac{t'^2}{\gamma}\right)} z^{\gamma+i} dz \\ &= \frac{e^{-\frac{1}{2}\mu^2}}{\sqrt{\pi}2^{\frac{\gamma-1}{2}}\Gamma\frac{\gamma}{2}} \sum_{i=0}^{\infty} \frac{(\mu t')^i}{i!\,\gamma^{\frac{1}{2}(i+1)}} \int_0^{\infty} e^{-\frac{1}{2}z^2\left(1+\frac{t'^2}{\gamma}\right)} (z^2)^{\frac{1}{2}(\gamma+i-1)} z\, dz. \end{aligned}$$

Also, consider the transformation

$$\frac{1}{2}z^2\left(1+\frac{t'^2}{\gamma}\right) = v \Rightarrow z^2 = \frac{2v}{\left(1+\frac{t'^2}{\gamma}\right)} \text{ and } zdz = \frac{dv}{\left(1+\frac{t'^2}{\gamma}\right)},\ 0<v<\infty$$

$$\begin{aligned} g(t',v) &= \frac{e^{-\frac{1}{2}\mu^2}}{\sqrt{\pi}2^{\frac{\gamma-1}{2}}\Gamma\frac{\gamma}{2}} \sum_{i=0}^{\infty} \frac{(\mu t')^i}{i!\,\gamma^{\frac{1}{2}(i+1)}} \int_0^{\infty} \frac{e^{-v}(2v)^{\frac{1}{2}(\gamma+i-1)}}{\left(1+\frac{t'^2}{\gamma}\right)^{\frac{1}{2}(\gamma+i+1)}} dv \\ &= \frac{e^{-\frac{1}{2}\mu^2}}{\sqrt{\pi}\Gamma\frac{\gamma}{2}} \sum_{i=0}^{\infty} \frac{(\mu t')^i 2^{i/2}}{i!\,\gamma^{\frac{1}{2}(i+1)}} \frac{\Gamma\left(\frac{\gamma+i+1}{2}\right)}{\left(1+\frac{t'^2}{\gamma}\right)^{\frac{1}{2}(\gamma+i+1)}} \\ &= \frac{e^{-\frac{1}{2}\mu^2}}{\sqrt{\pi}\Gamma\frac{\gamma}{2}} \sum_{i=0}^{\infty} \frac{2^{i/2}\mu^i\Gamma\left(\frac{\gamma+i+1}{2}\right)}{i!\,\gamma^{\frac{1}{2}(i+1)}} \frac{(t')^i}{\left(1+\frac{t'^2}{\gamma}\right)^{\frac{1}{2}(\gamma+i+1)}}, \\ &\qquad -\infty < t' < \infty. \qquad (29.10) \end{aligned}$$

Note that, if $\mu = 0$, the joint p.d.f. of X and Y becomes

$$f(x,y) = \frac{1}{\sqrt{2\pi}2^{\gamma/2}\Gamma\frac{\gamma}{2}} e^{-\frac{1}{2}(x^2+y)} y^{\frac{\gamma}{2}-1},$$

and due to the transformation, we get

$$\begin{aligned} h(t') &= \frac{1}{\sqrt{\pi}\Gamma\frac{\gamma}{2}} \frac{\Gamma\frac{\gamma+1}{2}}{\gamma^{1/2}} \frac{1}{\left(1+\frac{t'^2}{\gamma}\right)^{\frac{1}{2}(\gamma+1)}} \\ &= \frac{1}{\sqrt{\gamma}B\left(\frac{1}{2},\frac{\gamma}{2}\right)} \frac{1}{\left(1+\frac{t'^2}{\gamma}\right)^{\frac{1}{2}(\gamma+1)}},\ -\infty<t'<\infty \end{aligned}$$

which is the p.d.f. of central t-distribution.

Fisher (1931) gave alternative form of probability density function of t' which can be derived from the definition of t' given in (29.7). The cumulative density function of t' can be written as

$$\begin{aligned} P[t' \leq t] &= P\left[\frac{z+\delta}{\chi_\gamma/\sqrt{\gamma}} \leq t\right] \\ &= P\left[(z+\delta) \leq t\chi_\gamma/\sqrt{\gamma}\right]. \end{aligned}$$

Since χ_γ and z are independent, we can write

$$P[t' \leq t] = \frac{1}{2^{\frac{\gamma}{2}-1}\Gamma\frac{\gamma}{2}} \int_0^\infty \chi^{\gamma-1} e^{-\frac{1}{2}\chi^2} \frac{1}{\sqrt{2\pi}} \times \int_0^{tx/\sqrt{\gamma}} \exp\left[-\frac{1}{2}(z-\delta)^2\right] dz\, d\chi.$$

Differentiating with respect to t gives the probability density function of t', where

$$\begin{aligned} f(t') &= \frac{e^{\frac{1}{2}\delta^2}}{2^{\frac{1}{2}(\gamma-1)}\sqrt{\pi\gamma}\Gamma\frac{\gamma}{2}} \int_0^\infty \chi^\gamma \exp\left[-\frac{1}{2}\left\{\left(1+\frac{t^2}{\gamma}\right)\chi^2 - 2\left(\sqrt{t\gamma}\right)\chi\right\}\right] d\chi \\ &= \frac{\gamma!}{2^{\frac{1}{2}(\gamma-1)}\sqrt{\pi\gamma}\Gamma\frac{\gamma}{2}} \left\{\exp\left[-\frac{\gamma\delta^2}{\gamma+t^2}\right]\right\} \\ &\quad \times \left(\frac{\gamma}{\gamma+t^2}\right)^{\frac{1}{2}(\gamma+1)} Hh_\gamma\left[-\frac{\delta t}{\sqrt{\gamma+t^2}}\right] \end{aligned} \tag{29.11}$$

where

$$Hh_\gamma(x) = (\gamma!)^{-1} \int_0^\infty v^\gamma \exp\left[-\frac{1}{2}(v+x)^2\right] dv.$$

29.5 Properties of Non-central t-Distribution

We have

$$t' = \frac{z+\delta}{\chi_\gamma/\sqrt{\gamma}}.$$

Therefore, the r-th moment about origin can be written as

$$\begin{aligned} \mu_r' &= E\left[t'^r\right] \\ &= \gamma^{\frac{r}{2}} E\left[\chi_\gamma^{-r}\right] E\left[(z+\delta)^r\right] \\ &= \left(\frac{1}{2}\gamma\right)^{\frac{1}{2}r} \frac{\Gamma\left[\frac{1}{2}(\gamma-r)\right]}{\Gamma\frac{\gamma}{2}} \sum_{0\leq i\leq r/2} \binom{r}{2i} \frac{(2i)!}{2^i i!} \delta^{r-2i} \end{aligned}$$

Hogben *et al* (1961) found out the value of μ_r' which is given by

$$\mu_r' = \left(\frac{1}{2}\gamma\right) \frac{1}{2} r \frac{\Gamma\left[\frac{1}{2}(\gamma-r)\right]}{\Gamma\frac{\gamma}{2}} e^{-\frac{1}{2}\delta^2} \left(\frac{d}{d\delta}\right)^r e^{\frac{1}{2}\delta^2}.$$

They also gave the tables of the coefficients in the polynomials in δ which represent the central moments of t'.

According to Merrington and Pearson (1958), the first four moments are

$$\mu_1' = \left(\frac{1}{2}\gamma\right)^{\frac{1}{2}} \frac{\Gamma\left[\frac{1}{2}(\gamma-1)\right]}{\Gamma\frac{\gamma}{2}}\delta$$

$$\mu_2 = \frac{\gamma}{\gamma-2}(1+\delta^2) - \mu_1'^2$$

$$\mu_3 = \mu_1'\left\{\frac{\gamma(2\gamma-3+\delta^2)}{(\gamma-2)(\gamma-3)} - 2\mu_2\right\}$$

$$\mu_4 = \frac{\gamma^2}{(\gamma-2)(\gamma-4)}(3+6\delta^2+\delta^4) - \mu_1'^2\left\{\frac{\gamma[(\gamma+1)\delta^2+3(3\gamma-5)]}{(\gamma-2)(\gamma-3)} - 3\mu_2\right\}.$$

For γ large and for fixed δ

$$\mu_1' \doteq \delta$$

$$\mu_2 \doteq 1 + \frac{1}{2}\frac{\delta^2}{\gamma}$$

$$\mu_3 \doteq \frac{1}{\gamma}\delta\left[3 + \frac{5}{4}\frac{\delta^2}{\gamma}\right].$$

The value of

$$\alpha_3 = \sqrt{\beta_1} \doteq \frac{1}{\gamma}\delta\left(3 - \frac{\delta^2}{\gamma}\right).$$

They also established an approximate relation between β_1 and β_2 which is given by

$$\beta_2 \doteq \frac{1.406(\gamma-3.2)}{\gamma-4}\beta_1 + \frac{3(\gamma-2)}{\gamma-4}.$$

29.6 Approximations and Related Distributions to Non-central t-Distribution

If δ is kept fixed, and γ is increased without limit, the non-central t-distribution tends to normal distribution $N(\delta, 1)$. But if $\gamma(>2)$ is kept fixed and δ is increased without limit, the standardized t' tends to the standardized χ_γ^{-1} distribution. The percentage point t'_α can be approximated by

$$t'_\alpha \doteq \frac{\delta b_\gamma + z_\alpha\sqrt{b_\gamma^2 + (1-b_\gamma^2)(\delta^2 - z_\alpha^2)}}{b_\gamma^2 - z_\alpha^2(1-b_\gamma^2)}$$

where

$$b_\gamma = \gamma^{-\frac{1}{2}}E(\chi_\gamma) = \left(\frac{2}{\gamma}\right)^{\frac{1}{2}}\Gamma\left[\frac{1}{2}(\gamma+1)\right]\Big/\Gamma\frac{\gamma}{2}.$$

For small values of δ and for large values of $\gamma(>20)$, the percentage point $\alpha t'_\alpha$ can be approximated by

$$t'_\alpha \doteq \frac{\gamma}{\gamma-1}\delta b_\gamma + z_\alpha\sqrt{\frac{\gamma}{\gamma-2}(1+\delta^2) - \frac{\gamma^2}{(\gamma-1)^2}\delta^2 b_\gamma^2}.$$

When $\delta = 0$, the approximation becomes

$$t'_\alpha \doteq z_\alpha \sqrt{\frac{\gamma}{\gamma - 2}}.$$

Related Distributions

The non-central t-variate has been defined by

$$t' = \frac{(z + \delta)}{\chi_\gamma / \sqrt{\gamma}}.$$

If χ_γ is replaced by non-central χ'_γ, then

$$t'' = \frac{(z + \delta)}{\chi'_\gamma(\gamma) / \sqrt{\gamma}}.$$

The distribution of t'' is called a *double non-central t-distribution and the distribution* is a mixture of

$$\sqrt{\frac{\gamma}{\gamma + 2i}}\, t'_{\gamma+2i}.$$

The statistic $t'^2/(\gamma + t'^2)$ is distributed as non-central beta distribution with parameters 1 and γ and non-centrality parameter δ^2.

29.7 Applications of t-Distribution

The t-distribution is widely used in constructing tests and confidence intervals relating to sample means, sample correlation coefficient, sample regression coefficient, and partial and multiple correlation coefficients. For example, if $x_1, x_2, \cdots, x_n$ are sample observations drawn from normal population $N(\mu, \sigma^2)$, then to test $H_0 = \mu = \mu_0$, the test statistic is

$$t = \frac{\bar{x} - \mu_0}{s/\sqrt{n}}$$

where

$$\bar{x} = n^{-1} \sum_{i=1}^{n} x_i$$

$$\text{and } s^2 = (n-1)^{-1} \sum_{i=1}^{n} (x_i - \bar{x})^2.$$

This t follows Student's t-distribution.

Let $x_{11}, x_{12}, \cdots, x_{1n_1}$ be a random sample from $N(\mu_1, \sigma^2)$ and $x_{21}, x_{22}, \cdots, x_{2n_2}$ be another random sample from $N(\mu_2, \sigma^2)$. If two samples are independent, then we use the statistic

$$t = \frac{(\bar{x}_1 - \bar{x}_2) - (\mu_1 - \mu_2)}{s\sqrt{\frac{1}{n_1} + \frac{1}{n_2}}}$$

to test the hypothesis $H_0 : \mu_1 = \mu_2$, where

$$\bar{x}_1 = n_1^{-1} \sum_{j=1}^{n_1} x_{1j}$$

$$\bar{x}_2 = n_2^{-1} \sum_{j=1}^{n_2} x_{2j}$$

$$\text{and } s^2 = \frac{(n_1 - 1)s_1^2 + (n_2 - 1)s_2^2}{n_1 + n_2 - 2}$$

with

$$s_i^2 = (n_i - 1)^{-1} \sum_{j=1}^{n_i} (x_{ij} - \bar{x}_i)^2$$

This t has also t-distribution under H_0.

To test the hypothesis $H_0 : \rho = 0$, the statistic used is

$$t = \frac{r\sqrt{n-2}}{\sqrt{1-r^2}}$$

where r is the sample correlation coefficient and ρ is the population correlation coefficient. This t has also Student's t-distribution with $(n - 2)$ degrees of freedom.

The statistic

$$t = (b - \beta) \left[\frac{(n-2)\Sigma(x_i - \bar{x})^2}{\Sigma(y_i - \hat{y})^2} \right]^{\frac{1}{2}}$$

has Student's t-distribution with $(n - 2)$ degrees of freedom under the null hypothesis that the population regression coefficient of y on x is β. Here b is the sample regression coefficient.

In analysis of variance t-tests are used to compare the means of factors in pairs. If the sum of squares of any factor has 1 degree of freedom, the test statistic to test the significance of the effects of factors is F which is identical with t^2.

The non-central t-distribution is used to calculate the power of the t-test.

EXERCISE

1. (a) What do you mean by sampling distribution? Define Student's t-statistic and derive its distribution.

   (b) Show that $F = t^2$, where F is the distribution of variance ratio.

2. (a) Define Student's t-statistic. Show that t-distribution is symmetric and lepto kurtic. Also, show that, under certain condition, t-distribution transforms to normal distribution.

   (b) Let x_1 and x_2 be two independent sample observations from $N(\mu, \sigma^2)$. Show that

   $$u = \frac{x_1 + x_2 - 2\mu}{\sqrt{(x_1 - x_2)^2}}$$

   is distributed as t-distribution. Find mean deviation about mean of t-distribution.

3. (a) Distinguish between central t-distribution and non-central t-distribution. Derive the non-central t-distribution.

   (b) Find moments of central t-distribution.

4. (a) What is t-statistic? Derive central t-distribution and find moments of t-distribution.

   (b) Write down some important applications of t-distribution.

5. (a) Define sampling distribution and t-statistic. Show that $F = t^2$.

   (b) Derive the central t-distribution and show that t-distribution is symmetric and lepto kurtic.

6. (a) Define Student's t-statistic. Show that t-distribution tends to normal distribution under certain condition.

   (b) Distinguish between central t-distribution and non-central t-distribution. Show that non-central t-distribution transforms to central t-distribution.

CHAPTER 30

Distributions of Correlation and Regression Coefficients

30.1 Introduction

Let $(x_{11}, x_{21}), (x_{12}, x_{22}), \ldots, (x_{1n}, x_{2n})$ be a random sample of n pairs of values drawn from a bi-variate normal population whose probability density is given by

$$f(x_1, x_2) = \frac{1}{2\pi\sigma_1\sigma_2\sqrt{1-\rho^2}} \exp\left[-\frac{1}{2(1-\rho^2)}\left\{\left(\frac{x_1-\mu_1}{\sigma_1}\right)^2 - 2\rho\left(\frac{x_1-\mu_1}{\sigma_1}\right)\left(\frac{x_2-\mu_2}{\sigma_2}\right) + \left(\frac{x_2-\mu_2}{\sigma_2}\right)^2\right\}\right] \quad (30.1)$$

$$-\infty < x_1 < \infty, \ -\infty < x_2 < \infty.$$

Let us assume that X_1 and X_2 be linearly related and the linear relationship is represented by

$$x_1 = \alpha + \beta x_2 + \epsilon. \quad (30.2)$$

From the sample observations the parameters α and β can be estimated, where the estimates are

$$a = \bar{x}_1 - b\bar{x}_2 \text{ and } b = \frac{\Sigma(x_{1i} - \bar{x}_1)(x_{2i} - \bar{x}_2)}{\Sigma(x_{2i} - \bar{x}_2)^2}$$

respectively. Also, we can find the sample correlation coefficient which is given by

$$r = \frac{\Sigma(x_{1i} - \bar{x}_1)(x_{2i} - \bar{x}_2)}{\sqrt{[\Sigma(x_{1i} - \bar{x}_1)^2\Sigma(x_{2i} - \bar{x}_2)^2]}} = \frac{\Sigma(x_{1i} - \bar{x}_1)(x_{2i} - \bar{x}_2)}{(n-1)s_1 s_2}$$

where

$$s_1^2 = (n-1)^{-1}\Sigma(x_{1i} - \bar{x}_1)^2 \text{ and } s_2^2 = (n-1)^{-1}\Sigma(x_{2i} - \bar{x}_2)^2.$$

In this chapter the distribution of a, b and r will be discussed. The distribution of r can be considered when (i) $\rho = 0$, and (ii) $\rho \neq 0$, i.e., when X_1 and X_2 are independent and they follow the bi-variate normal distribution which has probability density function (30.1).

30.2 Distribution of r When $\rho = 0$

It is known that the correlation coefficient between X_1 and X_2 is the same as that between standardized variables $(X_1 - \mu_1)/\sigma_1$ and $(X_2 - \mu_2)/\sigma_2$. Therefore, without loss of generality it can be assumed that

$$E(X_1) = E(X_2) = 0 \text{ and } V(X_1) = V(X_2) = 1.$$

Let the variable X_1 be transformed to Y by an orthogonal transformation

$$Y = CX_1$$

where

$$Y_{n\times 1} = (y_1, y_2, \dots, y_n)', X_{1(n\times 1)} = (x_{11}, x_{12}, \dots, x_{1n})'$$

and $C_{n\times n} = (C_{ij})$ is an orthogonal matrix, where

$$C_{11} = C_{12} = \dots = C_{1n} = 1/\sqrt{n}$$

so that

$$y_1 = \frac{1}{\sqrt{n}}(x_{11} + x_{12} + \dots + x_{1n}) = \sqrt{n}\bar{x}_1.$$

Now, by Fisher's lemma $y_1, y_2, \dots, y_n$ are independent normal variates with mean zero and variance 1. Further, due to orthogonal transformation

$$\sum_{i=1}^{n} y_i^2 = \sum_{i=1}^{n} x_{1i}^2 = \sum_{i=1}^{n}(x_{1i} - \bar{x}_1 + \bar{x}_1)^2$$
$$= (n-1)s_1^2 + y_1^2.$$

$$\therefore \sum_{i=2}^{n} y_i^2 = (n-1)s_1^2. \tag{30.3}$$

Again, from the definition of r, it can be written that

$$\sqrt{n-1}\cdot s_1 r = \frac{\sum_{i=1}^{n}(x_{1i} - \bar{x}_1)(x_{2i} - \bar{x}_2)}{\sqrt{n-1}\, s_2} = \frac{\sum_{i=1}^{n}(x_{2i} - \bar{x}_2)x_{1i}}{\sqrt{n-1}\, s_2} = y_2 \text{ (say)} \tag{30.4}$$

Now, under the hypothesis $\rho = 0$, i.e., X_1 and X_2 are independent, we have

$$C_{21} = \frac{(x_{21} - \bar{x}_2)}{s_2\sqrt{n-1}}, C_{22} = \frac{(x_{22} - \bar{x}_2)}{s_2\sqrt{n-1}}, \dots, C_{2n} = \frac{(x_{2n} - \bar{x}_2)}{s_2\sqrt{n-1}}.$$

Since $\dfrac{\Sigma(x_{2i} - \bar{x}_2)^2}{s_2\sqrt{n-1}} = 1$ the condition of orthogonality is satisfied by the coefficient of

y_1 and y_2. Therefore, from (30.3) and (30.4) it follows that

$$\begin{aligned}\sum_{i=3}^{n} y_i^2 &= \sum_{i=2}^{n} y_i^2 - y_2^2 = (n-1)s_1^2 - (n-1)s_1^2 r^2 \\ &= (n-1)s_1^2(1-r^2). \end{aligned} \tag{30.5}$$

Since y_i are independent $N(0,1)$, $(n-1)s_1^2r^2$ is a χ_1^2-variate and $(n-1)s_1^2(1-r^2)$ is a χ_{n-2}^2-variate. Again,

$$\begin{aligned} r^2 &= \frac{(n-1)r^2s_1^2}{(n-1)s_1^2} \\ &= \frac{(n-1)r^2s_1^2}{[r^2+(1-r^2)]s_1^2(n-1)} \\ &= \frac{(n-1)r^2s_1^2}{(n-1)s_1^2r^2+(n-1)s_1^2(1-r^2)} \\ &= \frac{\chi_1^2}{\chi_1^2+\chi_{n-2}^2}. \end{aligned}$$

$\therefore$ r^2 is a $\beta_1\left(\frac{1}{2}, \frac{n-2}{2}\right)$-variate. Hence, the p.d.f. of r^2 is given by

$$f(r^2) = \frac{1}{B\left(\frac{1}{2}, \frac{n-2}{2}\right)} (r^2)^{\frac{1}{2}-1}(1-r^2)^{\frac{n-2}{2}-1}, \; 0 \le r^2 \le 1$$

$$\Rightarrow h(r) = \frac{1}{B\left(\frac{1}{2}, \frac{n-2}{2}\right)} (1-r^2)^{\frac{1}{2}(n-4)}, \; -1 \le r \le 1. \tag{30.6}$$

In Section (17.2) it has been observed that μ_r' for $\beta_1(m,n)$-variate is given by

$$\mu_r' = \frac{\Gamma(m+r)\Gamma(m+n)}{\Gamma m \Gamma(m+n+r)}.$$

For the distribution of r^2, we have $m = 1/2$ and $n = \frac{n-2}{2}$. Therefore, r-th moment about origin of the distribution of r^2 is given by

$$\mu_r'(r^2) = \frac{\Gamma\left(\frac{1}{2}+r\right)\Gamma\left(\frac{1}{2}+\frac{n-2}{2}\right)}{\Gamma\frac{1}{2}\Gamma\left(\frac{1}{2}+\frac{n-2}{2}+r\right)}$$

In particular,

$$\begin{aligned} \mu_1' &= \frac{\Gamma\left(\frac{1}{2}+1\right)}{\Gamma\left(\frac{1}{2}\right)} \frac{\Gamma\left(\frac{1}{2}+\frac{n-2}{2}\right)}{\Gamma\left(\frac{1}{2}+\frac{n-2}{2}+1\right)} = \frac{\frac{1}{2}}{\frac{1}{2}+\frac{n-2}{2}} = \frac{1}{n-1} \\ \mu_2' &= \frac{\frac{1}{2}\left(\frac{1}{2}+1\right)}{\left(\frac{1}{2}+\frac{n-2}{2}+1\right)\left(\frac{1}{2}+\frac{n-2}{2}\right)} = \frac{3}{(n^2-1)} \\ \therefore\ \mu_2 &= \frac{3}{(n^2-1)} - \frac{1}{(n-1)^2} = \frac{2(n-2)}{(n+1)(n-1)^2} \\ \mu_3 &= \frac{8(n-2)(n-3)}{(n+1)^4(n+3)}. \end{aligned}$$

The density function of r is given by (30.4). This density function has been derived when $\rho = 0$. But $E(r) = \rho = 0$, i.e., $\mu_1'(r) = 0$. Again

$$E(r^2) = \frac{1}{n-1}$$

$$\therefore \mu_2(r) = E(r^2) - [E(r)]^2 = \frac{1}{n-1}.$$

Related Distribution

Let us define a statistic

$$v = \frac{\chi_1^2}{\chi_{n-2}^2}$$

which is distributed as $\beta_2\left(\frac{1}{2}, \frac{n-2}{2}\right)$-variate. The statistic v can be expressed in terms of r^2, since

$$v = \frac{(n-1)s_1^2 r^2}{(n-1)s_1^2(1-r^2)} = \frac{r^2}{1-r^2}$$

$\therefore \frac{r^2}{1-r^2}$ is a $\beta_2\left(\frac{1}{2}, \frac{n-2}{2}\right)$-variate.

Also, we have

$$\frac{n-2}{1}v = \frac{(n-2)r^2}{(1-r^2)} \left[\because \frac{n-2}{1}v = \frac{\chi_1^2/1}{\chi_{n-2}^2/n-2}\right].$$

But $\frac{n-2}{1}v$ is an F-variate with 1 and $(n-2)$ degrees of freedom. Thus

$$\frac{r\sqrt{n-2}}{\sqrt{1-r^2}}$$

is a t-variate with $(n-2)$ degrees of freedom $[\because F = t^2]$. Therefore, when $\rho = 0$

$$t = \frac{r\sqrt{n-2}}{\sqrt{1-r^2}}$$

has Student's t-distribution with $(n-2)$ degrees of freedom.

30.3 Distribution of r When $\rho \neq 0$

As before it can be assumed that $E(X_1) = E(X_2) = 0$ and $V(X_1) = V(X_2) = 1$. It has already been observed that $r(1-r^2)^{-1/2}$ has central t-distribution when $\rho = 0$. Again, the conditional distribution of x_{1i} given x_{2i} is normal with mean ρx_{2i} and variance $(1-\rho^2)$. Hence, $r(1-r^2)^{-1/2}$ is distributed as $(n-2)^{-1/2}$ times noncentral t with $(n-2)$ degrees of freedom and noncentrality parameter

$$\sqrt{\sum_{i=1}^{n}(x_{2i}-\bar{x}_2)^2 \frac{\rho}{\sqrt{1-\rho^2}}}.$$

We need the distribution of $r(1-r^2)^{-1/2}$ over the distribution of $x_{21}, x_{22}, \ldots, x_{2n}$. For this, we need to calculate the expected value of the conditional distribution of

$r(1-r^2)^{-1/2}$. The density function of this distribution depends on the values of x_{2i}'s, since its noncentrality parameter involves

$$\sum_{i=1}^{n}(x_{2i}-\bar{x}_2)^2.$$

But the function is distributed as central chi-square with $(n-1)$ degrees of freedom. Let us write $u = r(1-r^2)^{-1/2}$ and $s = \Sigma(x_{2i}-\bar{x}_2)^2$, then the conditional probability density function of u is

$$f(u/s) = \frac{\exp\left[\frac{-\rho^2 s}{2(1-\rho^2)}\right]}{\sqrt{\pi}\Gamma\left(\frac{n}{2}-1\right)}(1+u^2)^{-\frac{1}{2}(n-1)} \times \sum_{i=0}^{\infty}\frac{\Gamma\left(\frac{n-1+i}{2}\right)}{i!}\left(\frac{2\rho^2u^2s}{(1-\rho^2)(1+u^2)}\right)^{\frac{1}{2}i}.$$

Since

$$h(s) = \frac{1}{2^{\frac{1}{2}(n-1)}\Gamma\frac{n-1}{2}}s^{\frac{1}{2}(n-3)}e^{-\frac{1}{2}s},\ s>0$$

and

$$\int_0^{\infty}s^{\frac{1}{2}i}\exp\left[-\frac{1}{2}\rho^2 s\left(1-\rho^2\right)^{-1}\right]h(s)ds = \frac{2^{\frac{1}{2}i}\Gamma\left(\frac{n-1+i}{2}\right)}{\Gamma\left(\frac{n-1}{2}\right)}(1-\rho^2)^{\frac{1}{2}(n+i-1)}$$

$$f(u) = \frac{(1-\rho^2)^{\frac{1}{2}(n-1)}}{\sqrt{\pi}\Gamma\left(\frac{n-1}{2}\right)\Gamma\left(\frac{n}{2}-1\right)}\left(1+u^2\right)^{-\frac{1}{2}(n+1)} \times \sum_{i=0}^{\infty}\frac{(2\rho)^i\left[\Gamma\left(\frac{n-1+i}{2}\right)\right]^2}{i!}\left(\frac{u^2}{1+u^2}\right)^{\frac{1}{2}i}.$$

Finally, making the transformation $u = r(1-r^2)^{-\frac{1}{2}}$, we get

$$h(r) = \frac{(1-\rho^2)^{\frac{1}{2}(n-1)}(1-r^2)^{\frac{1}{2}(n-4)}}{\sqrt{\pi}\Gamma\left(\frac{n-1}{2}\right)\Gamma\left(\frac{n}{2}-1\right)} \times \sum_{i=0}^{\infty}\frac{\left[\Gamma\left(\frac{n-1+i}{2}\right)\right]^2}{i!}(2\rho r)^i,\ -1\le r\le 1. \quad (30.7)$$

Fisher (1915) derived the density function of r when $\rho \neq 0$. The form of this density function is

$$h(r) = \frac{(1-\rho^2)^{\frac{1}{2}(n-1)}(1-r^2)^{\frac{1}{2}(n-4)}}{\pi\Gamma(n-2)}\left(\frac{\partial}{\sin\theta\partial\theta}\right)^{n-2}\frac{\theta}{\sin\theta} \quad (30.8)$$

where

$$\theta = \cos^{-1}(\rho r),\ -1\le r\le 1.$$

Ghosh (1966) obtained the moments of the distribution of r in terms of hypergeometric functions. These are

$$\mu_1' = C_n \rho F\left(\frac{1}{2}, \frac{1}{2}; \frac{1}{2}(n+1); \rho^2\right)$$

$$\mu_2' = 1 - \frac{(n-2)(1-\rho^2)}{n-1} F\left(1, 1; \frac{1}{2}(n+1); \rho^2\right)$$

$$\mu_3' = C_n \left[\rho F\left(\frac{1}{2}, \frac{1}{2}; \frac{1}{2}(n+1); \rho^2\right) - \rho^{-1}(n-1)(n-2)\right]$$
$$\times \left\{F\left(\frac{1}{2}, \frac{1}{2}; \frac{1}{2}(n-1); \rho^2\right) - F\left(\frac{1}{2}, \frac{1}{2}; \frac{1}{2}(n+1); \rho^2\right)\right\}$$

$$\mu_4' = 1 + \frac{(n-2)(n-4)(1-\rho^2)}{2(n-1)} F\left(1, 1; \frac{1}{2}(n+1); \rho^2\right)$$
$$- \frac{n(n-2)(1-\rho^2)}{4\rho^2}\left[F\left(1, 1; \frac{1}{2}(n+1); \rho^2\right) - 1\right]$$

Ghosh also obtained the value of μ_1', μ_2, μ_3 and μ_4 in inverse powers of $m = (n+6)$, where Fisher (1921) showed that z_1 is approximately distributed normally with expected value

$$\xi = \frac{1}{2}\log\frac{1+\rho}{1-\rho}, \text{ where } z_1 = \frac{1}{2}\log\frac{1+r}{1-r}$$

and variance $(n-3)^{-1}$ irrespective of the values of ρ. Since

$$E(z_1) = \xi + \frac{\rho}{2(n-1)}\left\{1 + \frac{5\rho^2}{4(n-1)} + \cdots\right\}$$

$$\text{and } V(z_1) = \frac{1}{n-1}\left\{1 + \frac{4-\rho^2}{2(n-1)} + \frac{22-6\rho^2-3\rho^4}{6(n-1)^2} + \cdots\right\}$$

for large n, $E(z_1) = \xi$ and $V(z_1) = \frac{1}{n-3}$

$$\left[\frac{1}{n-3} = \frac{1}{n-1-2} = \frac{1}{(n-1)\left(1-\frac{2}{n-1}\right)}\right.$$
$$\left. = \frac{1}{n-1}\left\{1 + \frac{2}{n-1} + \frac{4}{(n-1)^2} + \cdots\right\}\right]$$

Also, for the distribution of z_1

$$\beta_1 = \frac{\mu_3^2}{\mu_2^3} = \frac{\rho^6}{(n-1)^3}$$

$$\text{and } \beta_2 = \frac{\mu_4}{\mu_2^2} = 3 + \frac{2}{n-1} + \frac{4+2\rho^2-3\rho^4}{(n-1)^2}.$$

Thus β_1 tends to zero and β_2 tends to 3 if n tends to infinity. Hence distribution of z_1 tends to normal if n is large.

Another kind of approximation can be done as follows: It has already been mentioned that the distribution of $r(1-r^2)^{-1/2}$ for a fixed set of values $x_{21}, x_{22}, \ldots, x_{2n}$ is $(n-2)^{-1/2}$ times noncentral t with $(n-2)$ degrees of freedom and noncentrality parameter

$$\delta = \sqrt{\sum_{i=1}^{n}(x_{2i} - \bar{x}_2)^2}\,\frac{\rho}{\sqrt{1-\rho^2}}.$$

Thus, the distribution of $r(1-r^2)^{-1/2}$ is $(z+\delta)/\chi_{n-2}$ where z is standard normal variate and z and χ_{n-2} are independent. Averaging over the joint distribution of X_2's, it can be shown that the distribution of $r(1-r^2)^{-1/2}$ is

$$\left[z + \chi_{n-1}\rho(1-\rho^2)^{-1/2}\right] \Big/ \chi_{n-2}.$$

For this statistic, it can be written that

$$P[r \le r_1] = P\left[z + \rho(1-\rho^2)^{-1/2}\chi_{n-2} - r_1(1-r_1^2)^{-1/2}\chi_{n-1} \le 0\right].$$

Now, if n is not too small, χ_{n-2} and χ_{n-1} can be approximated by normal variates [$\because$ $\sqrt{2\chi_n^2} - \sqrt{2n-1}$ is approximately standard normal variate]. Then

$$z + \rho(1+\rho^2)^{-1/2}\chi_{n-2} - r_1(1-r_1^2)^{-1/2}\chi_{n-1}$$

is approximately normally distributed with mean value

$$\rho(1-\rho^2)^{-1/2}\left(n-\frac{5}{2}\right)^{1/2} - r_1(1-r_1^2)^{-1/2}\left(n-\frac{3}{2}\right)^{1/2}$$

and with standard deviation

$$\left[1 + \frac{1}{2}\rho^2(1-\rho^2)^{-1} + \frac{1}{2}r_1^2(1-r_1^2)^{-1}\right]^{1/2}.$$

30.4 Distribution of a and b

It has been discussed in Section (30.1) that a and b are the estimates of α and β in the model

$$x_1 = \alpha + \beta x_2 + \epsilon.$$

For this regression we assume that X_2 is fixed, ϵ is $N(0,\sigma^2)$ and X_1 is normally distributed about the regression values with variance σ^2. The fitted regression line from n pairs of sample observations is

$$\hat{x}_1 = a + bx_2$$

where

$$a = \bar{x}_1 - b\bar{x}_2 \text{ and } b = \frac{\sum_{i=1}^{n}(x_{1i}-\bar{x}_1)(x_{2i}-\bar{x}_2)}{\sum_{i=1}^{n}(x_{2i}-\bar{x}_2)^2}.$$

The estimate b can be written as

$$b = \frac{\sum x_{1i}(x_{2i}-\bar{x}_2)}{\sum(x_{2i}-\bar{x}_2)^2} = \frac{\sum_{i=1}^{n}(\alpha+\beta x_{2i}+\epsilon_i)(x_{2i}-\bar{x}_2)}{\sum(x_{2i}-\bar{x}_2)^2}$$

$$= \frac{\beta\sum_{i=1}^{n} x_{2i}(x_{2i}-\bar{x}_2)+\sum_{i=1}^{n}\epsilon_i(x_{2i}-\bar{x}_2)}{\sum(x_{2i}-\bar{x}_2)^2}$$

$$= \beta + \frac{\sum \epsilon_i(x_{2i}-\bar{x}_2)}{\sum(x_{2i}-\bar{x}_2)^2}.$$

Since X_2's are fixed, b is the linear function of normal variate ϵ_i and hence b is also normal with mean value

$$E(b) = \beta + \frac{\sum_{i=1}^{n} E(\epsilon_i)(x_{2i}-\bar{x}_2)}{\sum_{i=1}^{n}(x_{2i}-\bar{x}_2)^2} = \beta.$$

The variance of b is given by

$$V(b) = \frac{\sum_{i=1}^{n}(x_{2i}-\bar{x}_2)^2 V(\epsilon_i)}{\left[\sum_{i=1}^{n}(x_{2i}-\bar{x}_2)^2\right]^2} = \frac{\sigma^2}{\sum_{i=1}^{n}(x_{2i}-\bar{x}_2)^2}.$$

Thus

$$b \sim N\left(\beta, \frac{\sigma^2}{\sum_{i=1}^{n}(x_{2i}-\bar{x}_2)^2}\right).$$

We also have

$$a = \bar{x}_1 - b\bar{x}_2 = \alpha + \beta\bar{x}_2 + \bar{\epsilon} - b\bar{x}_2 = \alpha + \bar{x}_2(\beta - b) + \bar{\epsilon}.$$

Thus a is also a linear function of normal variates with

$$\begin{aligned} E(a) &= \alpha + E\left[\bar{x}_2(\beta-b) + E(\bar{\epsilon})\right] = \alpha \\ \text{and } V(a) &= E[a-\alpha]^2 = E\left[\bar{x}_2(\beta-b)+\bar{\epsilon}\right]^2 = \bar{x}_2^2 V(b) + V(\bar{\epsilon}) \\ &= \frac{\bar{x}_2^2\sigma^2}{\sum_{i=1}^{n}(x_{2i}-\bar{x}_2)^2} + \frac{\sigma^2}{n} = \sigma^2\left[\frac{1}{n} + \frac{\bar{x}_2^2}{\sum_{i=1}^{n}(x_{2i}-\bar{x}_2)^2}\right]. \end{aligned}$$

Therefore

$$a \sim N\left[\alpha, \sigma^2\left(\frac{1}{n} + \frac{\bar{x}_2^2}{\sum_{i=1}^{n}(x_{2i}-\bar{x}_2)^2}\right)\right].$$

Distribution of b When $\rho = 0$

We have

$$b = \frac{\sum_{i=1}^{n}(x_{1i} - \bar{x}_1)(x_{2i} - \bar{x}_2)}{\sum_{i=1}^{n}(x_{2i} - \bar{x}_2)^2}.$$

Also $\hat{x}_1 = a + bx_2 = \bar{x}_1 - b\bar{x}_2 + bx_2 = \bar{x}_1 + b(x_2 - \bar{x}_2)$. The sum of squares of X_1's can be written as

$$\begin{aligned}\sum_{i=1}^{n}(x_{1i} - \bar{x}_1)^2 &= \sum[(x_{1i} - \hat{x}_1) + (\hat{x}_1 - \bar{x}_1)]^2 = \sum_{i=1}^{n}(x_{1i} - \hat{x}_1)^2 + \sum_{i=1}^{n}(\hat{x}_1 - \bar{x}_1)^2 \\ \text{or, } (n-1)s_1^2 &= (n-1)(1-r^2)s_1^2 + (n-1)r^2 s_1^2\end{aligned}$$

where

$$s_1^2 = (n-1)^{-1}\sum_{i=1}^{n}(x_{1i} - \bar{x}_1)^2.$$

We know that $(n-1)\ s_1^2/\sigma_1^2 \sim \chi^2_{n-1}$-variate. Here $\sigma_1^2 = \sigma^2$ if $\rho = 0$. In Section (30.2) it has been observed that $(n-1)(1-r^2)s_1^2$ is a chi-square variate with $(n-2)$ degrees of freedom and $(n-1)r^2 s_1^2$ is also a chi-square variate with 1 degree of freedom. Again

$$\begin{aligned}b &= r\frac{s_1}{s_2} \\ \frac{\sigma_2^2 b^2}{\sigma_1^2} &= r^2\frac{s_1^2/\sigma_1^2}{s_2^2/\sigma_2^2} = \frac{(n-1)r^2 s_1^2/\sigma_1^2}{(n-1)s_2^2/\sigma_2^2} = \frac{\chi_1^2}{\chi^2_{n-1}} = \frac{1}{(n-1)}\frac{\chi_1^2/1}{\chi^2_{n-1}/(n-1)} \\ &= \frac{1}{(n-1)}F_{1,(n-1)}.\end{aligned}$$

Thus

$$b^2\frac{\sigma_2^2}{\sigma_1^2} \text{ is a } \beta_2\left(\frac{1}{2}, \frac{n-1}{2}\right)\text{-variate}$$

$$\text{and } b^2\frac{\sigma_2^2}{\sigma_1^2}(n-1) \text{ is an } F_{1,(n-1)}\text{-variate or } t^2_{(n-1)}\text{-variate } [\because\ F = t^2].$$

Now, the distribution of $t^2_{(n-1)}$ is given by

$$\begin{aligned}h(t^2) &= \frac{1}{\sqrt{(n-1)}B\left(\frac{1}{2}, \frac{n-1}{2}\right)}(t^2)^{-\frac{1}{2}}\left(1 + \frac{t^2}{n-1}\right)^{-\frac{n}{2}}. \\ \therefore\ f(b^2) &= \frac{1}{\sqrt{(n-1)}B\left(\frac{1}{2}, \frac{n-1}{2}\right)}\left\{\frac{b^2\sigma_2^2}{\sigma_1^2}(n-1)\right\}^{-\frac{1}{2}} \times \left(1 + \frac{b^2\sigma_2^2}{\sigma_1^2}\right)^{-\frac{n}{2}}\frac{\sigma_2^2}{\sigma_1^2}(n-1) \\ &= \frac{1}{B\left(\frac{1}{2}, \frac{n-1}{2}\right)}\frac{1}{b}\sigma_1^{n-1}\sigma_2\left(\sigma_1^2 + b^2\sigma_2^2\right)^{-\frac{n}{2}}. \\ \therefore\ f(b) &= \frac{1}{B\left(\frac{1}{2}, \frac{n-1}{2}\right)}\sigma_1^{n-1}\sigma_2\left(\sigma_1^2 + b^2\sigma_2^2\right)^{-\frac{n}{2}},\ -\infty < b < \infty. \qquad (30.9)\end{aligned}$$

Note that there is a relationship between the distribution of b and t. This can be established as follows: We know

$$b^2 = r^2 \frac{s_1^2}{s_2^2} \tag{30.10}$$

where

$$s_1^2 = (n-1)^{-1} \sum_{i=1}^{n} (x_{1i} - \bar{x}_1)^2 \text{ and } s_2^2 = (n-1)^{-1} \sum_{i=1}^{n} (x_{2i} - \bar{x}_2)^2 .$$

Multiplying both sides of (30.10) by

$$\frac{(n-1)s_2^2/\sigma_1^2}{(n-2)s^2/\sigma_1^2}, \text{ where } s^2 = (n-2)^{-1} \sum_{i=1}^{n} (x_{1i} - \hat{x}_1)^2$$

we get

$$\frac{b^2(n-1)s_2^2/\sigma_1^2}{(n-2)s^2/\sigma_1^2} = \frac{r^2(n-1)s_2^2 s_1^2/\sigma_1^2}{(n-2)s^2 s_2^2/\sigma_1^2} = r^2 \frac{(n-1)s_1^2/\sigma_1^2}{(n-2)s^2/\sigma_1^2} = \frac{\chi_1^2}{\chi_{n-2}^2}.$$

Thus

$$\frac{b^2(n-1)s_2^2}{(n-2)s^2}(n-2) \sim F_{1,n-2} = t_{n-2}^2\text{-variate.}$$

$$\therefore\ t = b\sqrt{\frac{(n-1)s_2^2(n-2)}{(n-1)s^2}} = b\left[\frac{\Sigma(x_{2i} - \bar{x}_2)^2(n-2)}{\Sigma(x_{1i} - \hat{x}_1)^2}\right]^{\frac{1}{2}}. \tag{30.11}$$

Thus the statistic (30.11) has t-distribution with $(n-2)$ degrees of freedom, provided $\rho = 0$ [which implies $\beta = 0$]. The statistic (30.11) is used to test the hypothesis

$$H_0 : \beta = 0 \text{ [Section (29.7)]}.$$

30.5 Distribution of Multiple Correlation Coefficient

Let R be the multiple correlation coefficient between a random variable X_k and variables $X_1, X_2, \ldots, X_{k-1}$ in a sample of size n from a k-variate normal population $N_k(\boldsymbol{\mu}, \Sigma)$, where X_k is the dependent variable and $X_1, X_2, \ldots, X_{k-1}$ are independent variables. Let the population multiple correlation coefficient between X_k and $X_1, X_2, \ldots, X_{k-1}$ be P, where

$$P = 1 - \frac{|\Sigma|}{|\Sigma_{k-1}|\sigma_{kk}} \text{ and } R^2 = 1 - \frac{|S|}{|S_{k-1}|s_{kk}}.$$

Here $\Sigma = (\sigma_{ij})$, $S = (s_{ij})$ for all $i, j = 1, 2, \ldots, k$ and

$$s_{ij} = (n-1)^{-1} \sum_{i=1}^{n} (x_{ij} - \bar{x}_i)(x_{ji} - \bar{x}_j).$$

The matrix S and Σ can be partitioned as follows:

$$S = \begin{bmatrix} S_{k-1} & \underline{s} \\ \underline{s}' & s_{kk} \end{bmatrix} \text{ and } \Sigma = \begin{bmatrix} \Sigma_{k-1} & \underline{\sigma} \\ \underline{\sigma}' & \sigma_{kk} \end{bmatrix}$$

where $\underline{\sigma}$ is a $(k-1)$ component vector of elements σ_{ij} and $\underline{s}$ is a $(k-1)$ component vector of elements s_{ij} $(i, j = 1, 2, \ldots, k)$.

Consider a $k \times k$ matrix M such that

$$M = \begin{bmatrix} M_{k-1} & \underline{0} \\ \underline{0} & m_{kk} \end{bmatrix}$$

where

$$M_{k-1} = \begin{bmatrix} \underline{\sigma}'\Sigma_{k-1}^{-1}/P\sqrt{\sigma_{kk}} \\ \underline{m}_2'/(\underline{m}_2'\Sigma_{k-1}\underline{m}_2)^{1/2} \\ \underline{m}_{k-1}'/\left(\underline{m}_{k-1}'\Sigma_{k-1}m_{k-1}\right)^{1/2} \end{bmatrix}, \quad m_{kk} = \frac{1}{\sigma_{kk}^{1/2}}$$

and $\underline{m}_2, \underline{m}_3, \ldots, \underline{m}_{k-1}$ are $(k-2)$ vectors orthogonal to $\underline{\sigma}$. In otherwords

$$\underline{m}_i'\underline{\sigma} = 0 \ (i = 2, 3, \ldots, k-1).$$

Therefore

$$\Sigma^* = \left[\sigma_{ij}^*\right] = M\Sigma M' = \left[\begin{array}{c|c} I_{k-1} & \begin{matrix} P \\ 0 \\ 0 \\ \vdots \\ 0 \end{matrix} \\ \hline P \ 0 \ 0 \ \cdots \ 0 & 1 \end{array}\right] = CC' \text{ (say)}$$

where C is a lower triangular matrix and is defined by

$$C = \left[\begin{array}{c|c} I_{k-1} & \underline{0} \\ \hline P \ 0 \ 0 \ \cdots \ 0 & \sqrt{1-P^2} \end{array}\right].$$

Also consider

$$S^* = \left[s_{ij}^*\right] = MSM' = TT' \text{ (say)}$$

where T is a lower triangular matrix such that

$$A = C^{-1}S^*C'^{-1} = C^{-1}TT'C'^{-1} = BB' \text{ (say)}$$

where $B = [b_{ij}] = C^{-1}T$. Here B is also lower triangular since it is the product of two lower triangular matrices.

We know that $S \sim W_k(\Sigma, n-1)$. Therefore, by the property of Wishart distribution $S^* \sim W_k(\Sigma^*, n-1)$. Also $A \sim W_k(I, n-1)$, where A is in canonical form. As $A = BB'$, B is, therefore, the Bartlett decomposition of the Wishart matrix and hence by decomposition theorem the b_{ij}'s are $N(0,1)$ for $i > j$ and the b_{ii}^2 are χ^2-variates with $\{n-1-(i-1)\}$ degrees of freedom. All the b_{ij}'s and b_{ii} are independently distributed.

The matrix S_{k-1}^* can be written as

$$S_{k-1}^* = M_{k-1}S_{k-1}M_{k-1}', \; s_{kk}^* = m_{kk}^2 s_{kk}$$

where S_{k-1}^* is the matrix of the first $(k-1)$ rows and columns of S^* and s_{kk}^* is the last element of S^*. Also, we can write $|M = m_{kk}|M_{k-1}|$. Hence

$$R^2 = 1 - \frac{|S|}{|S_{k-1}|s_{kk}} = 1 - \frac{|MSM'|}{|M_{k-1}S_{k-1}M_{k-1}'|m_{kk}^2 s_{kk}} = 1 - \frac{|S^*|}{|S_{k-1}^*|s_{kk}^*}.$$

As $S^* = TT'$, $|S^*| = t_{11}^2 t_{22}^2 \cdots\cdots t_{kk}^2$

$$\begin{aligned} |S_{k-1}^*| &= t_{11}^2 t_{22}^2 \cdots\cdots t_{k-1,k-1}^2 \\ s_{kk}^* &= \sum_{i=1}^{k} t_{ki}^2 \end{aligned}$$

where t_{ij} $(i > j)$ are the elements of the matrix T. Hence

$$R^2 = \frac{t_{k1}^2 + t_{k2}^2 + \cdots + t_{k,k-1}^2}{t_{k1}^2 + t_{k2}^2 + \cdots + t_{kk}^2}.$$

But, we have $T = CB$, and hence equating the elements in the last row of T and CB, we have

$$\begin{aligned} t_{k1} &= Pb_{11} + (1-P^2)^{1/2} b_{k1} \\ t_{kj} &= (1-P^2)^{1/2} b_{kj}, \; (j = 2, 3, \ldots, k). \end{aligned}$$

Putting the values of t_{k1} and t_{kj} in R^2, we have

$$\frac{R^2}{1-R^2} = \frac{\left(b_{k1} + Pb_{11}/(1-P^2)^{1/2}\right)^2 + (b_{k2}^2 + b_{k3}^2 + \cdots + b_{k,k-1}^2)}{b_{kk}^2}.$$

We can write $R^2 = u/\left(u + b_{kk}^2\right)$ where

$$u = \left\{b_{k1} + Pb_{11}/(1-P^2)^{1/2}\right\}^2 + \sum_{j=2}^{k-1} b_{kj}^2.$$

We know that $b_{k1}, b_{k2}, \ldots, b_{k,k-1}$ are $N(0,1)$ and hence the conditional distribution of u given b_{11} fixed is noncentral chi-square with $(k-1)$ degrees of freedom and noncentrality parameter

$$\lambda^2 = P^2 b_{11}^2/(1-P^2).$$

Again, b_{kk}^2 is distributed as χ_{n-k}^2. Thus, the joint distribution of u and b_{kk}^2, given b_{11}^2 is given by

$$\chi_{k-1}'^2(\lambda^2)\chi_{n-k}^2 du db_{kk}^2; \; 0 < u < \infty, \; 0 < b_{kk}^2 < \infty.$$

Now, using the transformation

$$R^2 = u/(u + b_{kk}^2) \text{ and } v = u + b_{kk}^2$$

in the joint distribution of $\chi'^2_{k-1}(\lambda^2)$ and χ^2_{n-k} and integrating out u, we get the conditional distribution of R^2, when b_{11}^2 is fixed. The probability density function of this conditional distribution is given by

$$f\left(R^2/b_{11}^2\right) = e^{-\frac{1}{2}\lambda^2} \sum_{i=0}^{\infty} \frac{(\lambda^2/2)^i}{i!} \times \frac{(R^2)^{\frac{1}{2}(k-1)+i-1}(1-R^2)^{\frac{1}{2}(n-k)-1}}{B\left(\frac{k-1}{2}+i, \frac{n-k}{2}\right)} \tag{30.12}$$

$$0 \leq R^2 \leq 1$$

Since $b_{11}^2 \sim \chi^2_{n-1}$, we can write the joint distribution of R^2, given b_{11}^2 is fixed and b_{11}^2 as the product of $f(R^2/b_{11}^2)$ and χ^2_{n-1}. Then substituting $\lambda^2 = P^2 b_{11}^2/(1-P^2)$ and then integrating out b_{11}^2 over the range 0 to ∞, the distribution of R^2 is obtained and it is expressed as

$$(1-P^2)^{\frac{1}{2}(n-1)} B\left(\frac{k-1}{2}, \frac{n-k}{2}\right) F\left(\frac{n-1}{2}, \frac{n-1}{2}; \frac{k-1}{2}; P^2R^2\right) \tag{30.13}$$

where $B\left(\frac{k-1}{2}, \frac{n-k}{2}\right)$ is used for the distribution of R^2 indicating β-variate of first kind,

$$F\left(\frac{n-1}{2}, \frac{n-1}{2}; \frac{k-1}{2}, P^2R^2\right)$$

is hypergeometric function.

If $P = 0$, the distribution of R^2 transforms to $B\left(\frac{k-1}{2}, \frac{n-k}{2}\right)$ and it can be transformed to F-distribution, where

$$F = \frac{n-k}{k-1} \frac{R^2}{1-R^2} \tag{30.14}$$

with $(k-1)$ and $(n-k)$ degrees of freedom.

The distribution of R^2 was originally obtained by Fisher in 1928. He obtained the probability density function of R^2 as

$$f(R^2) = \frac{\Gamma\frac{n}{2}\left(1-P^2\right)^{\frac{1}{2}(n-1)}}{\pi\Gamma\left(\frac{m-1}{2}\right)\Gamma\left(\frac{n-m}{2}\right)} \left(R^2\right)^{\frac{1}{2}(m-1)} \left(1-R^2\right)^{\frac{1}{2}(n-m)-1}$$
$$\times \int_0^{\pi}\int_{-\infty}^{\infty} \frac{\sin^{m-2}\theta}{(\cosh\phi - PR\cos\theta)^{n-1}} d\phi d\theta, \; R^2 > 0.$$

This function can be expressed in terms of hypergeometric function, where

$$f(R^2) = \frac{\left(1-P^2\right)^{\frac{1}{2}(n-1)} \left(R^2\right)^{\frac{m}{2}-1} \left(1-R^2\right)^{\frac{1}{2}(n-m-1)-1}}{B\left[\frac{m}{2}, \frac{n-m-1}{2}\right]}$$
$$\times F\left(\frac{n-1}{2}, \frac{n-1}{2}; \frac{m}{2}, P^2R^2\right), \; R^2 > 0 \tag{30.15}$$

where m is the number of independent variables.

When $P = 0$, the density function transforms to

$$f(R^2) = \frac{1}{B\left(\frac{m}{2}, \frac{n-m-1}{2}\right)} (R^2)^{\frac{m}{2}-1} (1-R^2)^{\frac{1}{2}(n-m-1)-1}, \quad 0 < R^2 < 1$$

which is the p.d.f. of beta variate of the first kind.

Wishart (1931) derived the mean and variance of R^2, where

$$\begin{aligned} E(R^2) &= 1 - \frac{n-m-1}{n-1}(1-P^2)F\left(1, 1; \frac{1}{2}(n+1); P^2\right) \\ &= P^2 + \frac{m}{n-1}(1-P^2) - \frac{2(n-m-1)}{n^2-1}P^2(1-P^2) + O\left(n^{-2}\right) \\ V(R^2) &= \frac{(n-m)^2-1}{n^2-1}\left(1-P^2\right)^2 F\left(2, 2; \frac{1}{2}(n+3); P^2\right) - \left[E(R^2)-1\right]^2 \\ &= \begin{cases} \dfrac{4P^2(1-P^2)^2(n-m-1)^2}{(n^2-1)(n+3)} + O\left(n^{-2}\right), & \text{for } P \neq 0 \\ \dfrac{2m(n-m-1)}{(n-1)^2(n+1)}, & \text{for } P = 0 \end{cases} \end{aligned}$$

Banerjee (1952) derived the r-th moment of R about zero, which is given by

$$\mu_r'(R) = \frac{(1-P^2)^{\frac{1}{2}(n-1)}\Gamma\left(\frac{r+k}{2}\right)}{\Gamma\left(\frac{n-1+r}{2}\right)} D'^r F\left(\frac{1}{2}(n-1+r), \frac{n}{2}; \frac{m}{2}, P^2\right)$$

where D' is the operator $\frac{1}{2}P^3\frac{d}{\partial P}$.

Hodgson (1967) has shown that, the distribution of $R^2(1-R^2)$ is

$$\left[\chi^2_{m-1} + \left\{z + P(1-P^2)^{-\frac{1}{2}}\chi_{n-1}\right\}^2\right] \Big/ \chi^2_{n-m-1}$$

where χ^2's and z are mutually independent. He has also shown that, for a suitable positive number h

$$\frac{(n-m+h-2)^h\left[R^2\left(1-R^2\right)^{-1}\right]^h - \left\{m+2h-2+(n+h-2)P^2\left(1-P^2\right)^{-1}\right\}^h}{\sqrt{2}h\left[(n-m-1)^{2h-1}\left\{R^2\left(1-R^2\right)^{-1}\right\}^{2h} + \left(2+P^2\left(1+P^2\right)^{-1}\right)\left\{m+(n-1)P^2\left(1-P^2\right)^{-1}\right\}^{2h-1}\right]^{\frac{1}{2}}}$$

is approximately distributed as standard normal variate. Again, $\{R^2(1-R^2)^{-1}\}^h$ is approximately normally distributed with mean value

$$(n-m+h-2)^{-1}\left[m+2h-2+(n+h-2)\left\{P^2\left(1-P^2\right)^{-1}\right\}\right]^h$$

and variance

$$2h^2(n-m-1)^{-(2h+1)}\left\{m+(n-1)P^2\left(1-P^2\right)^{-1}\right\}^{2h-1}(2n-m)\left(1-P^2\right)^{-1}.$$

Gajjar (1967) has shown that $\sqrt{n-1}\tanh^{-1}R$ is noncentral χ^2 with m degrees of freedom and noncentrality parameter $(n-1)(\tanh^{-1}P)^2$ when n tends to infinity,

where

$$\tanh^{-1} R = \frac{1}{2} \log \frac{1+R}{1-R}.$$

Khatri (1966) has shown that

$$m^{-1}(n-m-1)(1-P^2)\omega(P)R^2/(1-R^2)$$

is approximately distributed as noncentral F with m, $(n-m-1)$ degrees of freedom and noncentrality parameter

$$\frac{1}{2}P^2\omega(P)\sqrt{(n-1)(n-m-1)}$$

where

$$\omega(P) = \left[m + \left\{n-m-1+\sqrt{(n-1)(n-m-1)}\right\} P^2\right] \times \left[m+(n-m-1)(2-P^2)P^2\right]^{-1}.$$

He has also shown that

$$\frac{(n-m-1)(1-P^2)}{[(n-m-1)P^2+m]} \frac{R^2}{(1-R^2)}$$

is approximately distributed as central F with

$$\left[(n-m-1)P^2+m\right]^2 \left[(n-m-1)P^2(2-P^2)+m\right]^{-1}, \ (n-m-1)$$

degrees of freedom.

30.6 More Distributions Related to Correlation and Regression

Distribution of Partial Correlation Coefficient

Fisher (1924) proved that the sampling distribution of partial correlation coefficient of order k is of the same form as that of total correlation coefficient from bi-variate normal population for sample of size $(n-k)$. Thus, if $\rho_{(k)}$ is the partial correlation coefficient in the population and if $\rho_{(k)} = 0$, then $r^2_{(k)}$ is a beta variate of the first kind with parameter $\frac{1}{2}$ and $\frac{n-k-2}{2}$, where $r^2_{(k)}$ is the sample partial correlation coefficient for sample of size n. Thus

$$t = \frac{r_{(k)}}{\sqrt{1-r^2_{(k)}}}\sqrt{n-k-2}$$

has t-distribution with $(n-k-2)$ degrees of freedom.

If $\rho_{(k)} \neq 0$, one can use Fisher's z-transformation, where

$$z = \frac{1}{2} \log \frac{1+r_{(k)}}{1-r_{(k)}} = \tanh^{-1} r_{(k)}.$$

This z is approximately distributed normally with

$$E(z) = \xi \text{ and } V(z) = \frac{1}{n-k-3} \text{ where } \xi = \tanh^{-1} \rho_{(k)}.$$

Distribution of Sample Regression Line

The sample regression line is

$$E(\hat{y}) = \sum_{i=1}^{k} \hat{\beta}_i x_i$$

where the population regression line is

$$y = \sum_{i=1}^{k} \beta_i x_i + e.$$

Let us assume that $E(e) = 0$ and $E(ee') = \sigma^2 I$, where I is the unit matrix of order $n \times n$. Thus, we have

$$E(y) = \sum_{i=1}^{k} \beta_i x_i.$$

If β_i are known, then for a given set of $x_1, x_2, \ldots, x_k$ the quantity $E(y)$ can be accurately estimated, since

$$E(y) = \sum_{i=1}^{k} \beta_i x_i.$$

However, if the β_i are unknown, the estimate of β_i can be obtained from a set of sample observations of size n. Let the estimate of β_i be $\hat{\beta}_i$ which is obtained by OLS method, such that $E\left(\hat{\beta}_i\right) = \beta_i$ and $V\left(\hat{\beta}_i\right) = \sigma^2 s^{ii}$, where $S = (s_{ij}) = X'X$, s^{ii} is the i-th diagonal element of S^{-1}, X is a matrix of order $(n \times k)$. If we assume that $e \sim NID(0, \sigma^2 I)$, y is also normally distributed. Thus, $E(\hat{y})$ is a random variable and it is also normally distributed. Now, let us define a new random variable

$$u = \sum_{i=1}^{k} \left(\hat{\beta}_i - \beta\right) x_i = X_1' \left(\hat{\beta} - \beta\right)$$

where X_1 is a $(k \times 1)$ vector whose i-th element is x_i. Therefore, u is distributed

$$N(0, \sigma^2 X_1' S^{-1} X_1) \text{ and } \sum_{i=1}^{k} \frac{\left(\hat{\beta}_i - \beta_i\right) x_i}{\sigma \sqrt{X_1' S^{-1} X_1}}$$

is distributed $N(0, 1)$. Also, we know that $W = (n-k)\hat{\sigma}^2/\sigma^2$ is distributed as χ^2_{n-k} and it is independent of u, where

$$\hat{\sigma}^2 = \frac{1}{n-k} \left(Y - X\hat{\beta}\right)' (Y - X\beta).$$

Here Y is an $(n \times 1)$ vector of sample observations of the random variable y. Therefore

$$t = \frac{u}{\hat{\sigma}\sqrt{X_1' S^{-1} X_1}} = \frac{\sum\limits_{i=1}^{k} \left(\hat{\beta}_i - \beta_i\right) x_i}{\hat{\sigma}\sqrt{X_1' S^{-1} X_1}}$$

is distributed as central 't' with $(n-k)$ degrees of freedom.

Distribution of Sample Correlation Ratio When $\eta = 0$

Let X_1 be a random variable which is normally distributed with variance σ_1^2. Let us assume that a random sample of observations of X_1 be drawn and these are classified into k classes, where the number of observations of j-th class is n_j $(j = 1, 2, \ldots, k)$. Also, assume that n pairs of values

$$(x_{1jl}, x_{2j}) \quad \left(l = 1, 2, \ldots, n_j; n = \sum_j n_j\right)$$

can be formed with the values of random variable X_1. Thus, for a particular value x_{2j} there are different values x_{1jl}. Let us assume that these n pairs of values are drawn from a bi-variate population in which X_1 is normally distributed. We can assume that k classes of observations are drawn from k normal populations, where the population mean of j-th class is μ_{1j} and common variance σ_1^2. It is seen that there are n_j observations of the variable X_1 corresponding to the variable X_2. As X_1 and X_2 are correlated values, we can define correlation ratio of X_1 on X_2, where the population correlation ratio, η, is defined by

$$\eta = \frac{\sigma_{\bar{x}_{1j}}^2}{\sigma_1^2}$$

where σ_1^2 is the variance of X_1 and $\sigma_{\bar{x}_{1j}}^2$ is the variance of the weighted means of different classes of X_1. The total sum of squares of all observations x_{1jl} $(j = 1, 2, \ldots, k;$ $l = 1, 2, \ldots, n_j)$ is

$$\sum_{j=1}^{k} \sum_{l=1}^{n_j} (x_{1jl} - \bar{x}_1)^2$$

where $\bar{x}_1$ is the mean of all values of x_1. Let $\bar{x}_{1j}$ be the mean of the values of x_1 corresponding to the value of x_{2j}. Then, we can write

$$\sum_{j=1}^{k} \sum_{l=1}^{n_j} (x_{1jl} - \bar{x}_1)^2 = \sum_{j=1}^{k} \sum_{l=1}^{n_j} (x_{1jl} - \bar{x}_{1j})^2 + \sum_{j=1}^{k} n_j (\bar{x}_{1j} - \bar{x}_1)^2 . \tag{30.16}$$

Now, we can define the sample correlation ratio

$$\hat{\eta}^2 = \frac{\sum_{j=1}^{k} n_j (\bar{x}_{1j} - \bar{x}_1)^2}{\sum_{j=1}^{k}\sum_{l=1}^{n_j} (x_{1jl} - \bar{x}_1)^2} = \frac{\sum_{j=1}^{k} n_j (\bar{x}_{1j} - \bar{x}_1)^2}{(n-1)s_1^2}$$

where

$$s_1^2 = (n-1)^{-1} \sum_{j=1}^{k}\sum_{l=1}^{n_j} (x_{1jl} - \bar{x}_1)^2 .$$

Now, (30.16) can be written as

$$(n-1)s_1^2 = (n-1)s_1^2 (1 - \hat{\eta}^2) + (n-1)s_1^2\hat{\eta}^2.$$

But under the hypothesis of $\eta = 0$, it is clear that there is no association between X_1 and X_2, i.e., the observations x_{1jl} ($j = 1, 2, \ldots, k$; $l = 1, 2, \ldots, n_j$) corresponding to x_{2j} constitute a random sample from X_1. Hence

$$\frac{(\bar{x}_{1j} - \mu_{1j})^2}{\sigma_1^2/n_j} \sim \chi_1^2\text{-variate}$$

$$\frac{\sum_{j=1}^{k} n_j (\bar{x}_{1j} - \bar{x}_1)^2}{\sigma_1^2} = \frac{(n-1)s_1^2\hat{\eta}^2}{\sigma_1^2} \sim \chi_{k-1}^2\text{-variate}$$

$$\text{and } \frac{(n-1)s_1^2}{\sigma_1^2} \sim \chi_n^2\text{-variate.}$$

Thus

$$\frac{(n-1)s_1^2 (1 - \hat{\eta}^2)}{\sigma_1^2} \sim \chi_{n-k}^2\text{-variate.}$$

Therefore, $\hat{\eta}^2$ is a

$$\beta_1\left(\frac{k-1}{2}, \frac{n-k}{2}\right)\text{-variate}$$

$$\text{and } \frac{\hat{\eta}^2}{1-\hat{\eta}^2} \cdot \frac{n-k}{k-1} \text{ is an } F_{k-1,n-k}\text{-variate.}$$

We have

$$E(\hat{\eta}^2) = \frac{(k-1)/2}{\frac{(k-1)}{2} + \frac{(n-k)}{2}} = \frac{k-1}{n-1}$$

$$\text{and } V(\hat{\eta}^2) = \frac{\left(\frac{k-1}{2}\right)\left(\frac{n-k}{2}\right)}{\left(\frac{k-1}{2} + \frac{n-k}{2}\right)^2 \left(\frac{k-1}{2} + \frac{n-k}{2} + 1\right)}$$

$$= \frac{2(k-1)(n-k)}{(n-1)^2(n+1)}.$$

30.7 Applications of Distribution of Correlation Coefficients

It has already been mentioned that [Section 29.7] the statistic $t = r\sqrt{n-2}/\sqrt{1-r^2}$ has Student's t-distribution with $(n-2)$ degrees of freedom when $\rho = 0$. Thus to test the hypothesis $H_0 : \rho = 0$, we use the distribution related to that of r, the correlation coefficient [Section 29.7]. But when $\rho \neq 0$ or when $H_0 : \rho = \rho_0$ (say), we can use the approximate distribution of r as suggested by Fisher (1921). Thus

$$Z = \sqrt{n-3}\,(Z_1 - \xi_0)$$

can be used to test the hypothesis $H_0 : \rho = \rho_0$, where Z is standard normal variate

$$Z_1 = \frac{1}{2}\log\frac{1+r}{1-r},\quad \xi_0 = \frac{1}{2}\log\frac{1+\rho_0}{1-\rho_0}.$$

The hypothesis H_0 is rejected if $|Z| > Z_\alpha$, such that $\rho(Z_\alpha) = 1 - \alpha/2$. The level of significance of this test is α.

Fisher's approximation to the distribution of r can also be used to test the hypothesis $H_0 : \rho_1 = \rho_2$ if two samples of sizes n_1 and n_2, respectively, are drawn from two bi-variate normal populations with correlation coefficients ρ_1 and ρ_2. The test statistic for this hypothesis is

$$Z = \frac{Z_1 - Z_2}{\left[\frac{1}{n_1-3} + \frac{1}{n_2-3}\right]^{\frac{1}{2}}}$$

where $Z_i = \tan^{-1} r_i$ $(i = 1, 2)$ and r_i is the correlation coefficient of sample observations from i-th population. Under $H_0 : \rho_1 = \rho_2$, Z is $N(0, 1)$.

The transformation of r_i to Z_i $(i = 1, 2, \ldots, k)$ is also used to test the hypothesis $H_0 : \rho_1 = \rho_2 = \cdots = \rho_k$, where ρ_i is the population correlation coefficient of i-th population and r_i is the sample correlation coefficient based on n_i observations drawn from i-th bi-variate population. The test statistic for this hypothesis is

$$\chi^2 = \sum_{i=1}^{k} W_i^2 - u_1$$

where $W_i = (n_i - 3)^{1/2} Z_i$,

$$Z_i = \frac{1}{2}\log\frac{1+r_i}{1-r_i} \text{ and } u_1 = \frac{\sum_{i=1}^{k}(n_i-3)^{1/2}W_i}{\left[\sum_{i=1}^{k}(n_i-3)\right]^{\frac{1}{2}}}.$$

This χ^2 has chi-square distribution with $(k-1)$ degrees of freedom. The hypothesis of homogeneity of correlation coefficients will, therefore, be rejected, at a level of significance α, if $\chi^2 > \chi_0^2$, such that $P[\chi^2 \leq \chi_0^2] = 1 - \alpha$.

The application of distribution of multiple correlation coefficient has been discussed in the application of F-distribution. In Section (30.5) it has been shown that under

$H_0: \rho = 0$, $R^2(n-m-1)/(1-R^2)m$ has an F-distribution with $(m, n-m-1)$ d.f. The F-distribution is also used to test $H_0: \eta = 0$.

The transformation of r_i to Z_i is also used to obtain the pooled estimate of ρ when $H_0: \rho_1 = \rho_2 = \cdots = \rho_k$ is true. Under null hypothesis each of $Z_i = \tanh^{-1} r_i$ has variance $(n_i - 3)^{-1}$. Let us consider an arbitrary linear combination of Z_i, such that

$$v = \sum_{i=1}^{k} C_i Z_i.$$

We can choose the values of C_i's to have $E(v) = \tanh^{-1}\rho$ and $V(v)$ is minimum, where

$$V(v) = \sum_{i=1}^{k} [C_i^2/(n_i - 3)].$$

The variance is to be minimized subject to $\Sigma C_i = 1$. This yields

$$C_i = (n_i - 3)/(N - 3k)$$

where $N = \Sigma n_i$. Thus

$$v \doteq \frac{\sum^{k}(n_i - 3)Z_i}{N - 3k}$$

is the better estimator of $\tanh^{-1}\rho$. Hence, the estimator of ρ is $\tanh v$. The minimum variance of v is $1/(N - 3k)$.

EXERCISE

1. Define Student's t-statistic. Let $(x_1, y_1), (x_2, y_2), \cdots, (x_n, y_n)$ be n pairs of values observed from a bi-variate normal distribution. Show that the distribution of b follows Student's t-distribution, where

$$b = \frac{\Sigma(x - \bar{x})(y - \bar{y})}{\Sigma(x - \bar{x})^2}.$$

Find mean and variance of the distribution of b.

2. Let $(x_1, y_1), (x_2, y_2), \ldots, (x_n, y_n)$ be n pairs of values selected from a bi-variate population, where population correlation coefficient is ρ. Derive the distribution of

$$r = \frac{\Sigma(x - \bar{x})(y - \bar{y})}{\sqrt{[\Sigma(x - \bar{x})^2 \Sigma(y - \bar{y})^2]}}$$

if (i) $\rho = 0$ and (ii) $\rho \neq 0$. What is the distribution of $b = \dfrac{\Sigma(x - \bar{x})y}{\Sigma(x - \bar{x})^2}$ when $\rho = 0$?

3. Let $(x_1, y_1), (x_2, y_2), \ldots, (x_n, y_n)$ be n pairs of observations selected from a bi-variate population. Then

$$r = \frac{\Sigma(x - \bar{x})y}{\sqrt{\Sigma(x - \bar{x})^2 \Sigma(y - \bar{y})^2}} \text{ and } z = \frac{1}{2}\ln\frac{1 + r}{1 - r}.$$

Show that z is approximately distributed as normal. Also, show that $(r\sqrt{n-2})/\sqrt{1 - r^2}$ follows Student's t-distributions.

CHAPTER **31**

Distribution of Order Statistics

31.1 Introduction

Let $x_1, x_2, \ldots, x_n$ be a random sample of observations drawn from a continuous population having probability density function $f(x)$ and cumulative density function $F(x)$. The sample observations can be arranged in ascending order of magnitude, such that $x_{(1)} < x_{(2)} < \cdots < x_{(n)}$, where the number $i\ (i = 1, 2, \ldots, n)$ in parenthesis indicate the rank of the observations in the sample. Such an ordered set of new random variables constitutes the order statistics, where $x_{(1)}$ is the smallest observation and is called *first order statistic.* The next higher observation is $x_{(2)}$ which is called *second order statistic.* In general, $x_{(r)}$ is the r-th $(r = 1, 2, \ldots, n)$ order statistic. Clearly, $x_{(r)}$ is a random variable having p.d.f. and c.d.f. Since $x_{(r)}$ is a random variable, its function $R_n = x_{(n)} - x_{(1)}$ is also a random variable. Here R_n is called the *sample range.* If $n = 2m + 1$, then $(m + 1)$-th observation is called *median.* For $n = 2m$

$$\text{Median} = \frac{1}{2}\left[x_{(m)} + x_{(m+1)}\right].$$

In this chapter we shall discuss the distribution of r-th order statistic along with the distributions of range and median. Limiting distribution of r-th order statistic will also be discussed.

31.2 Distribution of r-th Order Statistic

Let $x_1, x_2, \ldots, x_n$ be a random sample of observations from a population having p.d.f. $f(x)$. The r-th order statistic of these sample observations is $x_{(r)}$. We want to derive the distribution of $x_{(r)}$. First, let us consider the distribution of $x_{(1)}$ and $x_{(n)}$, where $r = 1$ and n respectively, and then it will be generalized for $x_{(r)}$.

Clearly, $x_{(n)}$ is a random variable taking values less than or equal to any x_i, i.e.,

$$x_{(n)} \leq x_i\ (i = 1, 2, \ldots, n).$$

Thus, if a number of samples of each of size n are taken, then cumulative relative frequency of occurrence of $x_{(n)} \leq x$ is precisely equal to the cumulative relative frequency of occurrence of $x_{(i)} \leq x$ for all $i = 1, 2, \ldots, n$. Again, all the sample observations are independent and $P\left(x_{(i)} \leq x\right) = F(x)$. It follows that the preceding relative frequency is the product of n identical relative frequencies of the measurement event $x_i \leq x$. Hence the c.d.f. F_n of the random variable $x_{(n)}$ is given by

$$F_n = [F(x)]^n . \tag{31.1}$$

The p.d.f. of the n-th order statistic $x_{(n)}$ is given by

$$f_n\left[x_{(n)}\right] = \frac{dF_n}{dx} = \frac{d\,[F(x)]^n}{dx} = n\,[F(x)]^{n-1}\,f(x). \tag{31.2}$$

Similarly, if $x_{(1)}$ is the first order statistic and if it is larger than a given value x, it follows that all sample values x_i must be larger than x. Thus the cumulative relative frequency of occurrence of $x_{(1)} \geq x$ equals the cumulative relative frequency of $x_i \geq x$, for all $i = 1, 2, \ldots, n$. Thus, we can write

$$[1 - F_1] = [1 - F_n]^n \tag{31.3}$$

where F_1 is the c.d.f. of $x_{(1)}$ and is given by

$$F_1 = 1 - [1 - F_n]^n .$$

Hence the p.d.f. of $x_{(1)}$ is

$$f_1\left[x_{(1)}\right] = \frac{dF_1}{dx} = \frac{d}{dx}\left[1 - \{1 - F_n\}^n\right] = n\,[1 - F_n]^{n-1}\,f(x). \tag{31.4}$$

Now, let us consider the distribution of $x_{(r)}$, the r-th order statistic in a sample. For r-th order statistic there are $(r-1)$ observations which are less than $x_{(r)}$ and $(n-r)$ observations which are greater than $x_{(r)}$. Therefore, the $x_{(r)}$ observations can occur in

$$\frac{n!}{(r-1)!(n-r)!} \text{ ways.}$$

Each combination giving rise to one occurrence of the order statistic $x_{(r)}$ and let us suppose that $x_{(r)}$ falls in the interval x and $x+\Delta x$. It follows that $(r-1)$ sample values fall below x, and $(n-r)$ sample values fall above $x+\Delta x$. For a large number of samples of size n, the relative frequency of $(r-1)$ sample values below x becomes $[F(x)]^{r-1}$, that of one sample value falling in $(x, x+\Delta x)$ approaches $f(x)\Delta x$. The relative frequency of $(n-r)$ sample values falling above $x+\Delta x$ approaches $[1 - F(x+\Delta x)]^{n-r}$. Thus as the sample size n increases, the relative frequency of occurrence of $x_{(r)}$ in $(x, x+\Delta x)$ approaches the limiting value

$$f_r[x_{(r)}]\Delta x = \frac{n!}{(r-1)!(n-r)!}[F(x)]^{r-1} f(x)\Delta x \times [1 - F(x+\Delta x)]^{n-r} . \tag{31.5}$$

Diving both sides by Δx and then letting Δx approach zero, the p.d.f. of $x_{(r)}$ is obtained as follows:

$$f_r[x_{(r)}] = \frac{n!}{(r-1)!(n-r)!}[F(x)]^{r-1}[1 - F(x)]^{n-r} f(x). \tag{31.6}$$

Using similar intuitive argument one can derive the joint distribution of r-th and s-th ($r < s$) order statistics $x_{(r)}$ and $x_{(s)}$. This joint p.d.f. is given by

$$f_{(r,s)}\left[x_{(r)}, x_{(s)}\right] = \frac{n!}{(r-1)!\,(s-r-1)!\,(n-s)!}\left[F[x_{(r)}]\right]^{r-1}$$
$$\left[1-F(x_{(s)})\right]^{n-s}\left[F(x_{(s)})-F(x_{(r)})\right]^{s-r-1} f(x_{(r)})f(x_{(s)}) \qquad (31.7)$$

In particular, when $r = 1$, $s = n$, we have the joint distribution of smallest and largest sample values, where the joint distribution is

$$\begin{aligned} f_{(1,n)}\left[x_{(1)}, x_{(n)}\right] &= \frac{n!}{(n-2)!}\left[F(x_{(n)})-F(x_{(1)})\right]^{n-2} f(x_{(1)})f(x_{(n)}) \\ &= n(n-1)\left[F(x_{(n)})-F(x_{(1)})\right]^{n-2} f(x_{(1)})f(x_{(n)}) \qquad (31.8) \end{aligned}$$

►**EXAMPLE 31.1** *Let $x_1, x_2, \ldots, x_n$ be a random sample drawn from a population having density function*

$$f(x) = 1,\ 0 < x < 1.$$

Find the distribution of $x_{(r)}$. Also, find the mean and variance of this distribution.

Solution. The c.d.f. of the given distribution is

$$F(x) = x.$$

Here $x_{(r)}$ represents the r-th order statistic of the given sample observations. The p.d.f. $x_{(r)}$ is given by

$$f_r\left(x_{(r)}\right) = \frac{n!}{(r-1)!\,(n-r)!}[F(x)]^{r-1}[1-F(x)]^{n-r}f(x).$$

But $F(x) = x$. Thus, we have

$$\begin{aligned} f_r(x_{(r)}) &= \frac{n!}{(r-1)!\,(n-r)!}x_r^{r-1}(1-x_r)^{n-r} \\ &= \frac{1}{B(r, n-r+1)}x_r^{r-1}(1-x_r)^{n-r+1},\ 0 < x_r < 1 \end{aligned}$$

which is the density function of $\beta_1(r, n-r+1)$-variate. Thus

$$E(x_r) = \frac{r}{n+1} \text{ and } V(x_r) = \frac{r(n-r+1)}{(n+1)^2(n+2)}.$$

►**EXAMPLE 31.2** *Let $x_1, x_2, \ldots, x_n$ be a random sample of size n drawn from a population having p.d.f.*

$$f(x) = 2x,\ 0 < x < 1.$$

Find the distribution of (i) $x_{(1)}$, (ii) $x_{(n)}$, (iii) $x_{(r)}$.

Solution. We have

$$F(x) = 2\int_0^x x\,dx = x^2.$$

(i) The p.d.f. of $x_{(1)}$ is given by

$$\begin{aligned} f(x_{(1)}) &= n\,[1-F(x)]^{n-1}\,f(x) \\ &= 2nx_1\left[1-x_1^2\right]^{n-1} \quad [\because\ F(x)=x^2],\ 0<x_{(1)}<1. \end{aligned}$$

(ii) The p.d.f. of $x_{(n)}$ is given by

$$\begin{aligned} f(x_{(n)}) &= n[F(x)]^{n-1}f(x) = 2nx_n(x_n^2)^{n-1} \\ &= 2nx_n^{2n-1},\ 0<x_n<1. \end{aligned}$$

(iii) The p.d.f. of $x_{(r)}$ is given by

$$\begin{aligned} f(x_{(r)}) &= \frac{n!}{(r-1)!\,(n-r)!}[F(x)]^{r-1}[1-F(x)]^{n-r}f(x) \\ &= \frac{2n!}{(r-1)!\,(n-r)!}x_r\cdot x_r^{2r-2}(1-x_r^2)^{n-r} \\ &= \frac{2n!}{(r-1)!\,(n-r)!}(x_r^2)^{r-\frac{1}{2}}(1-x_r^2)^{n-r},\ 0<x_r<1. \end{aligned}$$

►**EXAMPLE 31.3** *Let $x_1, x_2, \ldots, x_n$ be a random sample drawn from logistic distribution with p.d.f.*

$$f(x) = \frac{e^{-x}}{(1+e^{-x})^2} = \frac{1}{4}\operatorname{sech}^2\frac{1}{2}x,\ -\infty<x<\infty.$$

Find the distribution of $x_{(r)}$.

Solution. We have

$$F(x) = \int_{-\infty}^{x}\frac{e^{-x}}{(1+e^{-x})^2}dx = \frac{1}{1+e^{-x}}.$$

Thus the distribution of $x_{(r)}$ is given by

$$\begin{aligned} f(x_{(r)}) &= \frac{n!}{(r-1)!\,(n-r)!}\left[(1+e^{-x})^{-1}\right]^{r-1}\left[\frac{e^{-x}}{1+e^{-x}}\right]^{n-r}\frac{e^{-x}}{(1+e^{-x})^2} \\ &= \frac{n!}{(r-1)!\,(n-r)!}e^{-x(n-r+1)}(1+e^{-x})^{-(n+1)},\ x>0. \end{aligned}$$

The m.g.f. of this distribution is

$$M_{x(r)}(t) = \frac{\Gamma(r+t)\Gamma(n-r-t+1)}{\Gamma r\Gamma(n-r+1)}$$

31.3 Distribution of Range

Let $x_1, x_2, \ldots, x_n$ be a random sample of observations from a population having p.d.f. $f(x)$. The r-th order statistics of these sample observations are

$$x_{(1)} < x_{(2)} < \cdots < x_{(r)} < \cdots < x_{(n)};\ r = 1, 2, \ldots, n.$$

Then the joint distribution of $x_{(1)}$ and $x_{(n)}$ is given by

$$f(x_{(n)}, x_{(1)}) = n(n-1)\left[F(x_{(n)}) - F(x_{(1)})\right]^{n-2} f(x_{(1)})f(x_{(n)}).$$

Let us perform the transformation

$$R = x_{(n)} - x_{(1)} \text{ and } S = x_{(1)} \Rightarrow R + S = x_{(n)}.$$

The Jacobian of the transformation is 1. Here R is the sample range of the observations. The joint p.d.f. of R and S is

$$h(r,s) = n(n-1)[F(r+s) - F(s)]^{n-2} f(s)f(r+s).$$

From this the distribution of R is obtained by integrating out S over the interval $-\infty$ to ∞. Thus

$$g(r) = \int_{-\infty}^{\infty} h(r,s)ds.$$

As an example, let us consider the distribution of range from a sample of n observations drawn from a population having density function

$$f(x) = 1, \; 0 < x < 1.$$

Here $F(x) = x \Rightarrow F(r+s) = r+s$ and $F(s) = s$, where $R = x_{(n)} - x_{(1)}$ and $S = x_{(1)}$. Thus $h(r,s) = n(n-1)r^{n-2};\; 0 < s < 1-r,\; 0 < r < 1.$

$$\begin{aligned} \therefore\; g(r) &= n(n-1)\int_0^{1-r} r^{n-2} ds = n(n-1)r^{n-2}(1-r) \\ &= \frac{1}{B(n-1,2)} r^{(n-1)-1}(1-r). \end{aligned}$$

Gupta and Shah (1965) derived the distribution of range $r = x_{(n)} - x_{(1)}$, where $x_1, x_2, \ldots, x_n$ are the sample observations from logistic distribution. The p.d.f. of r is

$$\begin{aligned} f(r) = &\frac{\sqrt{\pi}\Gamma n}{2\sqrt{2}\Gamma\left(n+\frac{1}{2}\right)}\left[1 + \cosh\frac{1}{2}r\right]^{-(n-\frac{1}{2})} \\ &\times F\left[\frac{1}{2}, \frac{1}{2}; n + \frac{1}{2}; \left(1 - \cosh\frac{1}{2}r\right)\right], \quad r > 0. \end{aligned}$$

Shah (1965) obtained the joint distribution of r and the mid-range $\left[m = \frac{1}{2}\left(x_{(1)} + x_{(n)}\right)\right]$. The joint probability density function is

$$h(r,m) = \frac{n(n-1)\left[\sinh\frac{1}{2}r\right]^{n-2}}{4\left[\cosh m + \sinh\frac{1}{2}r\right]^{n}}, \quad r > 0.$$

►**EXAMPLE 31.4** *A sample of* 4 *observations are drawn from a population having density function*

$$f(x) = 2x, \; 0 < x < 1.$$

The observations are (in ascending order of magnitude)

$$x_{(1)} < x_{(2)} < x_{(3)} < x_{(4)}.$$

Find the distribution of $r = x_{(4)} - x_{(1)}$.

Solution. For the given distribution

$$F(x) = \int_0^x 2x dx = x^2.$$

The joint distribution of $x_{(1)}$ and $x_{(4)}$ is

$$\begin{aligned} h(x_{(1)}, x_{(4)}) &= 4 \times 3 \left[F(x_4) - F(x_1)\right]^2 f(x_1) f(x_4) \\ &= 12 \left[x_{(4)}^2 - x_{(1)}^2\right]^2 \cdot 4x_{(1)} x_{(4)} \\ &= 48 \left[x_{(4)}^2 - x_{(1)}^2\right]^2 x_{(1)} x_{(4)}. \end{aligned}$$

Let $r = x_{(4)} - x_{(1)}$ and $s = x_{(4)}$.

$$\therefore\ x_{(4)} = s \text{ and } x_{(1)} = s - r.$$

The Jacobian of the transformation is $J = 1$. Thus

$$\begin{aligned} g(r, s) &= 48(2sr - s^2)^2 s(s - r);\ r < s < 1,\ 0 < r < 1. \\ \therefore\ g_1(r) &= \int_r^1 g(r, s) ds \\ &= \frac{8r^2}{5}(r^5 - 15r^3 + 50r^2 - 60r + 24),\ 0 < r < 1. \end{aligned}$$

31.4 Limiting Distribution

Let $x_1, x_2, \ldots, x_n$ be a random sample from a population having p.d.f. $f(x)$. Also, consider that $y_1 < y_2 < \cdots < y_n$ are the sample observations in ascending order. Let the distribution function $F_n(y)$ of the random variable y_n depend upon n, a positive integer. If $F(y)$ is a distribution function and if

$$\lim_{n \to \infty} F_n(y) = F(y)$$

for every point y at which $F(y)$ is continuous, then the random variable y_n is said to have a limiting distribution with distribution function $F(y)$.

Let us find the limiting distribution of r-th order statistic. The p.d.f. of r-th order statistic is

$$\begin{aligned} h(y_r) &= \frac{n!}{(r-1)!\,(n-r)!} \left[F(y_r)\right]^{r-1} \left[1 - F(y_r)\right]^{n-r} f(y_r) \\ &= \frac{n!}{(r-1)(n-r)!} \left[F(y)\right]^{r-1} \left[1 - F(y)\right]^{n-r} f(y). \end{aligned}$$

Putting

$$nF(y) = z \Rightarrow [1 - F(y)] = 1 - \frac{z}{n} \text{ and } f(y)dy = \frac{dz}{n}$$

we have

$$h(z_r) = \frac{n!}{(r-1)!\,(n-r)!}\left(\frac{z}{n}\right)^{r-1}\left(1-\frac{z}{n}\right)^{n-r}\frac{1}{n} \tag{31.9}$$

$$= \frac{n!}{n^n(n-z)^r(n-r)!}(n-z)^n\frac{z^{r-1}}{(r-1)!}$$

$$= \frac{(n-r+1)(n-r+2)\cdots n}{(n-z)^r}\left(1-\frac{z}{n}\right)^n\frac{z^{r-1}}{(r-1)!}$$

$$= \frac{n^r\left(1-\frac{r-1}{n}\right)\left(1-\frac{r-2}{n}\right)\cdots 1\left(1-\frac{z}{n}\right)^n}{n^r\left[1-\binom{r}{1}n^{-1}z+\binom{r}{2}n^{-2}z^2-\cdots(-1)^r n^{-r}z^r\right]}\frac{z^{r-1}}{(r-1)!}$$

$$\lim_{n\to\infty} h(z_r) = \lim_{n\to\infty}\frac{\left(1-\frac{r-1}{n}\right)\left(1-\frac{r-2}{n}\right)\cdots 1\left(1-\frac{z}{n}\right)^n z^{r-1}}{\left[1-\binom{r}{1}n^{-1}z+\binom{r}{2}n^{-2}z^2-\cdots(-1)^r n^{-r}z^r\right]}\frac{1}{(r-1)!}$$

$$= \frac{z^{r-1}}{(r-1)!}e^{-z}\quad\left[\because \lim_{n\to\infty}\left(1-\frac{z}{n}\right)^n = e^{-z}\right] \tag{31.10}$$

which is the limiting distribution of the r-th order statistic. In particular, when $r = 1$

$$h(z_1) = e^{-z},\ 0 < z < \infty$$

when $r = n$

$$h(z_n) = \frac{z^{n-1}e^{-z}}{(n-1)!} = \frac{1}{\Gamma n}e^{-z}z^{n-1},\ 0 < z < \infty.$$

Thus $\mu_1(z_n) = n$ and $\mu_2(z_n) = n$.

Limiting Distribution of r-th Largest Sample Value

In a sample of n observations the r-th largest sample is $(n-r+1)$-th order statistic, since the r-th largest sample value is counted from the top starting from y_n, where y_n is the largest value in the sample. Let y_{n-r+1} be the r-th largest sample value. We need the distribution of y_{n-r+1}.

Let us define $z_1 = n[1 - F(y)]$, $0 < z_1 < n$.

The Jacobian of the transformation is $J = \left|\frac{dy}{dz_1}\right| = \frac{1}{n}f(y)$.

We know

$$h(y_{n-r+1}) = \frac{n!}{(n-n+r-1)!\,(n-r+1-1)!}$$
$$\times[F(y)]^{n-r+1-1}[1-F(y)]^{n-n+r-1}\times f(y)$$

$$= \frac{n!}{(r-1)!\,(n-r)!}[F(y)]^{n-r}[1-F(y)]^{r-1}f(y).$$

$$\therefore\ g(z_{n-r+1}) = \frac{n!}{(r-1)!\,(n-r)!}\left[1-\frac{z_1}{n}\right]^{n-r}\left[\frac{z_1}{n}\right]^{r-1}\frac{1}{n} \tag{31.11}$$

which is equivalent to (31.9). Therefore, the limiting distribution of r-th largest value is

$$g(z_{n-r+1}) = \frac{1}{\Gamma r} e^{-z_1} z_1^{r-1}, \; 0 < z_1 < \infty. \tag{31.12}$$

In a similar way the joint distribution of r-th smallest (r-th order statistic) and r-th largest [$s = (n-r+1)$-th order statistic] can be derived from the joint p.d.f. (31.7). For this we need the transformation $z_1 = nF(y)$ and $z_2 = n[1 - F(y)]$. Then the limiting distribution of r-th and s-th order statistics will takes the form

$$h(z_r, z_s) = \frac{1}{(r-1)!} z_1^{r-1} e^{-z_1} \times \frac{1}{(r-1)!} z_2^{r-1} e^{-z_2}; \; 0 \leq z_1, \; z_2 \leq \infty. \tag{31.13}$$

It is implied that $r < s$ so that $r < (n+1)/2$ in (31.13). The r-th smallest sample value is counted from the bottom whereas the r-th largest value is counted from the top.

From (31.13) it is seen that the joint p.d.f. of r-th largest and r-th smallest sample values is the product of the marginal p.d.f.'s (31.10) and (31.12). Thus the limiting distributions of r-th largest and r-th smallest sample values are independent.

31.5 Distribution of Median

Let $x_1, x_2, \ldots, x_n$ be a random sample of n observations drawn from a population having density function $f(x)$. Consider that $y_1 < y_2 < \cdots < y_n$ are the n order statistics of this sample observations. If the sample size n is odd, suppose $n = 2m + 1$, then $(m+1)$-th observation $[y_{m+1}]$ is the median. Thus we need the distribution of y_{m+1}.

The distribution of r-th order statistic is given by

$$h(y_r) = \frac{n!}{(r-1)!\,(n-r)!} [F(y_r)]^{r-1} [1 - F(y_r)]^{n-r} f(y_r).$$

Here $r = m + 1$. Hence, we get

$$h(y_{m+1}) = \frac{n!}{m!\,(n-m-1)!} [F(y_{m+1})]^m [1 - F(y_{m+1})]^{n-m-1} f(y_{m+1}). \tag{31.14}$$

If n is even, say $n = 2m$, then median is the average of m-th and $(m+1)$-th observations. From (31.7), we have

$$h(y_m, y_{m+1}) = \frac{n!}{(m-1)!\,(n-m-1)!} [F(y_m)]^{m-1} [1 - F(y_{m+1})]^{n-m+1} \times f(y_m) f(y_{m+1}).$$

Let

$$\frac{y_m + y_{m+1}}{2} = u \text{ and } y_m = v. \;\therefore\; y_m = v \text{ and } y_{m+1} = 2u - v.$$

The Jacobian of the transformation is $J = 2$. Hence

$$g(u, v) = \frac{n!2}{(m-1)!\,(n-m-1)!} [F(v)]^{m-1} [1 - F(2u-v)]^{n-m-1} f(u) f(2u-v); \; -\infty < u < \infty, \; u < v < \infty. \tag{31.15}$$

The distribution of u is

$$g_1(u) = \int_u^{\infty} g(u,v)dv. \tag{31.16}$$

►**EXAMPLE 31.5** *Suppose $y_1 < y_2 < y_3 < y_4 < y_5 < y_6$ are the order statistics of a sample of 6 observations from a population having p.d.f.*

$$f(x) = 1,\ 0 < x < 1.$$

Find the distribution of median.

Solution. The median is given by $M = (y_3 + y_4)/2$. The joint p.d.f. of y_3 and y_4 is

$$\begin{aligned} h(y_3, y_4) &= \frac{6!}{(3-1)!(4-3-1)!(6-4)!} y_3^{3-1} (y_4 - y_3)^{4-3-1} (1 - y_4)^{6-4} \\ &= 180 y_3^2 (1 - y_4)^2. \end{aligned}$$

Let $(y_3 + y_4)/2 = u$ and $y_4 = v \Rightarrow y_4 = v$ and $y_3 = 2u - v$.
The Jacobian of the transformation is $|J| = 2$.

$$\therefore\ g(u,v) = 360(1-v)^2(2u-v)^2.$$

Thus

$$g_1(u) = 360 \int (1-v)^2(2u-v)^2 dv. \tag{1}$$

The integrand in (1) is nonzero for the intersection of the regions $0 < 2u - v < 1$ and $0 < v < 1$. The region of integration then is the intersection of the three regions

$$u < v,\ \frac{v}{2} < u < \frac{v+1}{2} \text{ and } 0 < v < 1.$$

Thus the limits on the integral in (1) must be $u < v < 2u$ for $0 < u \le 1/2$ and $u < v < 1$ for $\frac{1}{2} < u < 1$. Therefore,

$$g_1(u) = \begin{cases} 12(16u^5 - 25u^4 + 10u^3), & 0 < u < \frac{1}{2} \\ 12(1 - 10u + 40u^2 - 70u^3 + 55u^4 - 16u^5), & \frac{1}{2} < u < 1. \end{cases}$$

Large Sample Approximation of the Distribution of Median

Let $y_1 < y_2 < \cdots < y_n$ be the order statistics of a sample of n observations drawn from a population having density function $f(x)$. Let us consider that $n = 2m + 1$. Then y_{m+1} is the median of the sample observations. We need the distribution of y_{m+1} when n is large.

The distribution of y_{m+1} is given by

$$\begin{aligned} h(y_{m+1}) &= \frac{n!}{m!(n-m-1)!} [F(y_{m+1})]^m [1 - F(y_{m+1})]^{n-m-1} f(y_{m+1}) \\ &= \frac{(2m+1)!}{(m!)^2} \left[\int_{-\infty}^{y_{m+1}} f(x)dx\right]^m \left[\int_{y_{m+1}}^{\infty} f(x)dx\right]^{n-m-1} f(y_{m+1}). \end{aligned}$$

Let M be the population median, so that

$$\int_{-\infty}^{M} f(x)dx = \int_{M}^{\infty} f(x)dx = \frac{1}{2}.$$

Also, let

$$\phi(y_{m+1}) = \int_{M}^{y_{m+1}} f(x)dx$$

Then

$$\int_{-\infty}^{y_{m+1}} f(x)dx = \frac{1}{2} + \phi(y_{m+1}) \tag{31.17}$$

$$\text{and} \quad \int_{y_{m+1}}^{-\infty} f(x)dx = \frac{1}{2} - \phi(y_{m+1}). \tag{31.18}$$

$$\begin{aligned} \therefore\ h(y_{m+1}) &= \frac{(2m+1)!}{(m!)^2}\left[\frac{1}{4} - \phi^2(y_{m+1})\right]^m f(y_{m+1}) \\ &= \frac{(2m+1)!}{(m!)^2}\frac{1}{4^m}\left[1 - 4\phi^2(y_{m+1})\right]^m f(y_{m+1}). \end{aligned} \tag{31.19}$$

It is clear that, $h(y_{m+1})$ will be maximum if $[1 - 4\phi^2(y_{m+1})]^m$ is maximum. Again, $[1 - 4\phi^2(y_{m+1})]^m$ will be maximum if $\phi(y_{m+1})$ is zero. It, therefore, follows that y_{m+1} approaches M. Now, by Taylor's theorem,

$$\begin{aligned} \phi(y_{m+1}) &= \phi(M) + \{\phi(y_{m+1}) - M\}\phi'(M) \\ &= (y_{m+1} - M)f(M) \\ f(y_{m+1}) &= f(M) + (y_{m+1} - M)f'(y_{m+1}) + \cdots \end{aligned}$$

Therefore,

$$\begin{aligned} h(y_{m+1}) &= k[1 - 4(y_{m+1} - M)^2 f^2(M)]^m \\ &\qquad [f(M) + \{y_{m+1} - M\}f'(y_{m+1}) + \cdots] \end{aligned}$$

where k is some constant

$$= k'\{1 - 4(y_{m+1} - M)^2 f^2(M)\}^m \text{ if } (y_{m+1} - M)f'(y_{m+1})$$

is neglected. Here

$$\begin{aligned} k' &= kf(M) \\ &= k'\{1 - 4m(y_{m+1} - M)^2 f^2(M) + \cdots\} \\ &= k'e^{-4m(y_{m+1}-M)^2 f^2(M)} \\ &= k' \exp\left[-\frac{1}{2}(y_{m+1} - M)^2 \Big/ \frac{1}{8mf^2(M)}\right]. \end{aligned}$$

Thus y_{m+1} tends to an $N\left(M, \dfrac{1}{8mf^2(M)}\right)$ for large n.

►**EXAMPLE 31.6** *Let $y_1 < y_2 < \cdots < y_n$ $(n = 2m+1)$ be the order statistics of a random sample of observations from a rectangular population having density function*

$$f(x) = 1, \ 0 < x < 1.$$

Find the distribution of median.

Solution. For the given sample, median $= y_{m+1}$. For $n = 2m+1$

$$h(y_{m+1}) = \frac{n!}{m!\ m!}\left[F(y_{m+1})\right]^m \left[1 - F(y_{m+1})\right]^m f(y_{n+1}).$$

But $F(x) = x$ for the given rectangular distribution. Therefore,

$$\begin{aligned} h(y_{m+1}) &= \frac{(2m+1)!}{m!\ m!} y_{m+1}^m (1 - y_{m+1})^m, \ 0 < y_{m+1} < 1 \\ &= \frac{1}{B(m+1, m+1)} y_{m+1}^m (1 - y_{m+1})^m \end{aligned}$$

which is the p.d.f. of β_1 $(m+1, m+1)$-variate.

►**EXAMPLE 31.7** *Let $y_1 < y_2 < \cdots < y_n$ be the order statistics of sample observations $x_1, x_2, \ldots, x_n$ from a population having density function*

$$f(x) = \frac{1}{2\phi} \exp\left[-|x - \theta|/\phi\right], \ \phi > 0.$$

Find the distribution of r-th order statistic.

Solution. Johnson and Kotz (1970) has given the p.d.f. of r-th order statistic, where the p.d.f. is

$$h(y_r) = \begin{cases} \dfrac{n!}{(r-1)!\ (n-r)!} \dfrac{1}{2\phi} \left\{1 - \dfrac{1}{2} e^{-(\theta - y_r)/\phi}\right\}^{n-r} \\ \quad \times \exp\left[-(r+1)(\theta - y_r)/\phi\right], & y_r \le \theta \\ \dfrac{n!}{(r-1)!\ (n-r)!} \dfrac{1}{2\phi} \left\{1 - \dfrac{1}{2} e^{-(y_r - \theta)/\phi}\right\}^{r-1} \\ \quad \times \exp\left[-(n-r+1)(y_r - \theta)/\phi\right], & y_r \ge \theta. \end{cases}$$

The s-th moment of y_r about θ is given by

$$\begin{aligned} E\left[y_r - \theta\right]^s &= \phi^s \frac{n!\Gamma(s+1)}{(r-1)!\ (n-r)!} \\ &\quad \times \left[(-1)^s \sum_{j=0}^{n-r} (-1)^j \binom{n-r}{j} 2^{-(r+j+1)} (r+j)^{-(s+1)}\right. \\ &\quad \left. + \sum_{j=0}^{r-1} (-1)^j \binom{r-1}{j} 2^{-(n-r+2+j)} (n-r+1+j)^{-(s+1)}\right]. \end{aligned}$$

Johnson and Kotz also considered the distribution of median, where median $= y_{m+1}$

when $n = 2m + 1$. Thus

$$h(y_{m+1}) = \begin{cases} \dfrac{(2m+1)!}{(m!)^2}\dfrac{1}{2\phi}\left\{1 - \dfrac{1}{2}e^{-(\theta - y_{m+1})/\phi}\right\}^m \\ \quad \times \exp\left[-(m+2)(\theta - y_{m+1})/\phi\right], & y_{m+1} \le \theta \\ \dfrac{(2m+1)!}{(m!)^2}\dfrac{1}{2\phi}\left\{1 - \dfrac{1}{2}e^{-(y_{m+1} - \theta)/\phi}\right\}^m \\ \quad \times \exp\left[-(m+1)(y_{m+1} - \theta)/\phi\right], & y_{m+1} \ge \theta. \end{cases}$$

Edwards (1948) derived the cumulative distribution function of range, $r = y_n - y_1$ for $n = 4$ and 5.

For $n = 4$,

$$F(r) = 1 + \frac{15}{8}e^{-r} - 3e^{-2r} + \frac{1}{8}e^{-3r} - \frac{3}{4}re^{-r}(4 + e^{-r}),\ r \ge 0.$$

For $n = 5$,

$$F(r) = 1 + \frac{77}{12}e^{-r} - \frac{57}{8}e^{-2r} - \frac{1}{4}e^{-3r} - \frac{1}{24}e^{-4r} - \frac{5}{4}re^{-r}\left(4 + 3e^{-r}\right),\ r \ge 0.$$

Johnson and Kotz (1970) have also given the p.d.f. of r-th order statistic of sample observations drawn from Cauchy population, Pareto population and Weibull population. They have discussed the properties of the distribution of r-th order statistic.

Let $y_1 < y_2 < \cdots < y_n$ be the order statistics of sample observations drawn from Cauchy population having p.d.f.

$$f(x) = \frac{1}{\pi\lambda}\frac{1}{[1 + \{(x-\theta)/\lambda\}^2]},\quad -\infty < x < \infty.$$

Then the p.d.f. of y_r is given by

$$h(y_r) = \frac{n!}{(r-1)!\,(n-r)!}\left(\frac{1}{2} + \frac{1}{\pi}\tan^{-1}\{(y_r - \theta)/\lambda\}\right)^{r-1} \times \left(\frac{1}{2} - \frac{1}{\pi}\tan^{-1}\{(y_r - \theta)/\lambda\}\right)^{n-r}\frac{1}{\pi\lambda[1 + \{(y_r - \theta)/\lambda\}^2]}.$$

For large n

$$V(y_r) \doteq \frac{1}{n}\lambda^2\pi^2(r/n)(1 - r/n)\operatorname{cosec}^4(\pi r/n).$$

If $y_1 < y_2 < \cdots < y_n$ are the order statistics of sample observations drawn from Pareto distribution, then the p.d.f. of y_r is

$$h(y_r) = \frac{n!}{(r-1)!\,(n-r)!}\left(\frac{a}{y_r}\right)\left(\frac{k}{y_r}\right)^{a(n-r+1)}\left\{1 - \left(\frac{k}{y_r}\right)^a\right\}^{r-1},\ k \le y_r.$$

The s-th moment about zero of this distribution is

$$\mu'_s = \frac{\Gamma(n+1)\Gamma(n-r+1-sa^{-1})k^s}{\Gamma(n+1-sa^{-1})\Gamma(n-r+1)}.$$

The p.d.f. of y_r of the sample observations drawn from Weibull distribution having p.d.f.

$$f(x) = c\alpha^{-1}\{(x-\xi_0)/\alpha\}^{c-1}\exp\left[-\{(x-\xi_0)/\alpha\}^c\right],\ x > \xi_0$$

is given by

$$\begin{aligned} h(y_r) &= \frac{n!}{(r-1)!\,(n-r)!}\{1-\exp[-\{(y_r-\xi_0)/\alpha\}^c]\}^{r-1} \\ &\quad \times \exp\left[-(n-r+1)\{(y_r-\xi_0)/\alpha\}^c\right]\cdot\frac{c\{(y_r-\xi)/\alpha\}^{c-1}}{\alpha} \\ &= \frac{n!\ c\{(y_r-\xi_0)/\alpha\}^{c-1}}{(r-1)!\,(n-r)!\,\alpha}\times\sum_{j=0}^{r-1}\binom{r-1}{j}(-1)^j \\ &\quad \times \exp\left[-(n-r+1-j)\{(y_r-\xi_0)/\alpha\}^c\right],\ y_r > \xi_0. \end{aligned}$$

The s-th moment of y_r about ξ_0 is

$$\begin{aligned} E[(y_r-\xi_0)^s] &= \frac{n!\ \alpha^s}{(r-1)!\,(n-r)!}\Gamma(1+s/c) \\ &\quad \times\sum_{j=0}^{r-1}\binom{r-1}{j}(-1)^j(n-r+1-j)^{-(1+s/c)}. \end{aligned}$$

►**EXAMPLE 31.8** *Let $y_1 < y_2 < \cdots < y_n$ be the order statistics of a random sample drawn from rectangular population having density function*

$$f(x) = \frac{1}{\theta},\ 0 < x < \theta.$$

Find the distribution of the range of sample observations.

Solution. The range is given by $r = y_n - y_1$. The joint p.d.f. of y_1 and y_n is given by

$$h(y_1, y_n) = n(n-1)[F(y_n) - F(y_1)]^{n-2} f(y_1) f(y_n).$$

Here $F(x) = x/\theta$. Therefore,

$$h(y_1, y_n) = n(n-1)\left[\frac{y_n}{\theta} - \frac{y_1}{\theta}\right]^{n-2}\frac{1}{\theta^2} = n(n-1)\frac{1}{\theta^n}\,[y_n - y_1]^{n-2}.$$

Let $r = y_n - y_1$ and $s = y_1$, so that $|J| = 1$, where $0 < s < \theta - r$ and $0 < r < 1$. We have

$$g(r,s) = n(n-1)\frac{1}{\theta^n}r^{n-2}.$$

Thus

$$g_1(r) = n(n-1)\frac{1}{\theta^n}r^{n-2}(\theta - r),\ 0 < r < 1.$$

31.6 Tolerance Limits for Distributions

In Section (2.12) it has mentioned that, if X is a random variable of the continuous type having a p.d.f. $f(x)$ and distribution function $F(x)$, then the random variable $Y = F(x)$ has a uniform distribution with p.d.f.

$$h(y) = 1, \; 0 < y < 1.$$

Now, if $0 < p < 1$, we have

$$P[F(x) \leq p] = \int_0^p dy = p. \tag{31.20}$$

Again, $F(x) = P(X \leq x)$, then $F(x)$ is the fractional part of the probability for the distribution of X which is between $-\infty$ and x, since $P(X = x) = 0$. If $F(x) \leq p$, then no more than $100p$ per cent of the probability for the distribution of X is between $-\infty$ and x. As $Y = F(x)$, $P[F(x) \leq p]$ is the probability that the random variable Y is less than or equal to p is precisely the probability that the random interval $(-\infty, X)$ contains no more than $100p$ per cent of the probability for the distribution.

Now, let us consider that $z_1 < z_2 < \cdots < z_n$ denote the order statistics of a random sample of size n from a distribution of the continuous type with distribution function $F(x)$. We have

$$Y_i = F(z_i); \; i = 1, 2, \ldots, n.$$

The joint distribution of $Y_r = F(z_r)$ and $Y_s = F(z_s)$ is, with $r < s$, given by

$$h(y_r, y_s) = \frac{n!}{(r-1)!\,(s-r-1)!\,(n-s)!} y_r^{r-1}(y_s - y_r)^{s-r-1}(1 - y_s)^{n-s}, \quad 0 < y_r < y_s < 1.$$

From this joint distribution of y_r and y_s we can find the probability

$$\begin{aligned} \gamma &= P[F(z_s) - F(z_r) \geq p] \\ &= P[Y_s - Y_r \geq p] \\ &= \int_0^{1-p} \int_{p+y_r}^{1} h(y_r, y_s) dy_r dy_s. \end{aligned} \tag{31.21}$$

Here the difference $F(z_s) - F(z_r)$ is that fractional part of the probability for the distribution of X that is between z_r and z_s. If p is a positive proper fraction and if

$$F(z_s) - F(z_r) \geq p,$$

then at least $100p$ per cent of the probability for the distribution of X is between z_r and z_s. This probability denoted by γ is for the random interval (z_r, z_s) containing at least $100p$ per cent of the probability for the distribution of X. Here the interval (z_r, z_s) is the *tolerance interval* with tolerance coefficient γ.

To evaluate the probability γ given in (31.21) let us make the transformation

$$Y_s - Y_r = U \text{ and } Y_s = V.$$

Then the joint distribution of U and V is

$$g(u,v) = \frac{n!}{(r-1)!\,(s-r-1)!\,(n-s)!}(v-u)^{r-1}u^{s-r-1}(1-v)^{n-s}, \quad 0<u<v<1$$

$$\text{and } g_1(u) = \frac{n!}{(r-1)!\,(s-r-1)!\,(n-s)!}u^{s-r-1} \times \int_u^1 (v-u)^{r-1}(1-v)^{n-s}dv.$$

Under the integral sign, we make the transformation $(v-u)=t(1-u)$. Then

$$\begin{aligned} g_1(u) &= \frac{n!}{(r-1)!\,(s-r-1)!\,(n-s)!}u^{s-r-1}(1-u)^{r-1} \\ &\quad \times \int_0^1 t^{r-1}[(1-u)-t(1-u)]^{n-s}(1-u)dt \\ &= \frac{n!}{(r-1)!\,(s-r-1)!\,(n-s)!}u^{s-r-1}(1-u)^{n-s+r}B(r,n-s+1) \\ &= \frac{n!}{(s-r-1)!\,(n-s+r)!}u^{s-r-1}(1-u)^{n-s+r}, \ 0<u<1 \end{aligned}$$

which is the p.d.f. of another beta variate. Now, from $g_1(u)$, we get

$$\gamma = P[U \geq p] = \int_p^1 g_1(u)du \tag{31.22}$$

which is the *tolerance coefficient* for any given r and s.

EXERCISE

1. What do you mean by order statistics? Derive the probability density function of r-th order statistic. Let $x_1, x_2, \ldots, x_n$ be n random observations selected from a population having p.d.f.

$$f(x) = \begin{cases} 1, & 0<x<1 \\ 0, & \text{otherwise.} \end{cases}$$

Find the p.d.f. of r-th order statistic of this sample observations. Also, find mean and variance of the distribution.

2. Explain by example the terms order statistics and range. If $x_1, x_2, \ldots, x_n$ are n sample observations, then derive the p.d.f. of the range of these observations. If the sample observations follow the density function

$$f(x) = \begin{cases} 1, & 0<x<1 \\ 0, & \text{otherwise.} \end{cases}$$

Find the p.d.f. of range when $n = 6$.

3. Define order statistic. Derive the probability density function of r-th order statistic. Also, derive the p.d.f. of limiting form of the r-th order statistic. Let $x_1, x_2, \ldots, x_n$ be a random sample of size n from a population having p.d.f.

$$f(x) = \begin{cases} 2x, & 0 < x < 1 \\ 0, & \text{otherwise.} \end{cases}$$

Find the distribution of (i) $x_{(1)}$, (ii) $x_{(n)}$, and (iii) $x_{(r)}$, where $x_{(r)} = r$-th order statistic ($r = 1, r = n, r = r$).

4. Define the terms order statistic, range and median. Derive the p.d.f. of median. Let $y_1 < y_2 < y_3 < y_4 < y_5 < y_6$ be the order statistics of 6 sample observations from a population having p.d.f.

$$f(x) = \begin{cases} 1, & 0 < x < 1 \\ 0, & \text{otherwise.} \end{cases}$$

Find the distribution of median of these sample observations. Also, derive the p.d.f. of y_3. Find mean and variance of both the distributions.

5. What is meant by order statistic? Explain range of a sample observations and derive the distribution of range. Let $y_1 < y_2 < \cdots < y_n$ be the order statistics of n sample observations selected from a population having p.d.f.

$$f(x) = \frac{1}{\theta}, \quad 0 < x < \theta.$$

Find the distribution of (i) range, and (ii) median of the sample observations.

6. Derive the limiting distribution of r-th order statistic and r-th largest observation of a sample. Let $y_1 < y_2 < y_3 < y_4 < y_5$ be the order statistics of 5 sample observations selected from a population having p.d.f.

$$f(x) = \begin{cases} 3x^2, & 0 < x < 1 \\ 0, & \text{otherwise.} \end{cases}$$

Let $z = y_2/y_4$. Derive the p.d.f. of z. Also, find mean and variance of z.

CHAPTER **32**

Point Estimation

32.1 Introduction

The objective of the statistical analysis is to estimate the unknown characteristic, usually known as parameter, and to infer about the parameter. The unknown characteristic of the distribution is not, usually, found out investigating all the population observations, specially if the population is infinite one. For example, let us consider the distribution of production of steel (in metric tons per day) for different years. If the data are considered for 25 years together, the population observations are large in number and one can easily assume that the production data will follow a normal distribution with population mean production per day μ and variance of production σ^2. These two characters μ and σ^2 are unknown and are not, usually, found out using all the population observations. The calculation of μ and σ^2 from all population observations is also not recommended. However, the researcher or the statistician is interested to have an idea about these unknown parameters. As an alternative technique to guess the parameters and to infer about the parameter, the researcher investigate a representative part of the population observations and find a function of sample observations which can provide the information about the parameter.

The function of sample observations is called 'statistic'. If this statistic is used to express a particular value of a parameter, it is called 'estimator' or 'point estimator'. The method of deriving such a function of sample observations to guess the population unknown characteristic is known as 'point estimation'.

Let $x_1, x_2, \ldots, x_n$ be a set of n random sample observations from a population having probability density function $f(x, \theta)$. Here θ, a parameter exists in a real parametric space Ω. The problem is to estimate the parameter θ using the set of sample observations. Let $g(x)$ be a function of sample observations. This function $g(x)$ is to be suggested in such a way that it can provide precise information about θ. A function $g(x)$ will provide precise information if the distribution of similar functions

$$g_1(x), g_2(x), \ldots, g_k(x)$$

are concentrated around the parameter θ. This is possible if the function of sample observations is derived according to some standard principle. The principle of identifying a sample function which will provide precise information about parameter θ is known as 'method of estimation'. The precise information is possible if the 'statistic' possesses some criteria.

32.2 Criterion of a Good Estimator

The important criteria of a good estimator are (a) Unbiasedness, (b) Consistency, (c) Efficiency, and (d) Sufficiency.

Unbiasedness

Let $x_1, x_2, \ldots, x_n$ be a set of random sample observations selected from a population having probability density function (p.d.f.) $f(x, \theta)$. Let $\psi(\theta)$ be a function of θ and $t(x_1, x_2, \ldots, x_n)$ be a function of sample observations. The sample statistic $t(x_1, x_2, \ldots, x_n)$ is an unbiased estimator of $\psi(\theta)$, if

$$E[t(x_1, x_2, \ldots, x_n)] = \psi(\theta).$$

This property of the estimator is known as *unbiasedness of the estimator.*

The unbiasedness property shows that

$$E[t(x_1, x_2, \ldots, x_n)] - \psi(\theta) = 0.$$

Let this property does not exist for a certain function $t_1(x_1, x_2, \ldots, x_n)$ and

$$E[t_1(x_1, x_2, \ldots, x_n)] - \psi(\theta) = \delta$$

where δ is a nonzero element. Then $t_1(x_1, x_2, \ldots, x_n)$ is a biased estimator of $\psi(\theta)$. However, when $\delta = 0$, the statistic becomes unbiased.

►**EXAMPLE 32.1** *Let $x_1, x_2, \ldots, x_n$ be a random sample of size n from a Poisson population with parameter λ. Find unbiased estimators of λ and λ^2. Also, find the variance of the estimator of λ.*

Solution. Given

$$P(X = x) = \frac{e^{-\lambda}\lambda^x}{x!}, \quad x = 0, 1, 2, \ldots.$$

We know that $E(x) = \lambda$ and $E(x^2) = \lambda^2 + \lambda$, $V(x) = \lambda$. Let us define a function of sample statistic

$$t(x_1, x_2, \ldots, x_n) = \frac{1}{n}\Sigma x = \bar{x}.$$

We have

$$E[t(x_1, x_2, \ldots, x_n)] = E(\bar{x}) = \frac{1}{n}\Sigma E(x) = \lambda.$$

Therefore, sample mean $\bar{x}$ is an unbiased estimator of λ. The variance of $\hat{\lambda} = \bar{x}$ is $V(\bar{x}) = \lambda/n$.

Again, let $t_1(x_1, x_2, \ldots, x_n) = \frac{1}{n}\Sigma x^2$. Then

$$E[t_1(x_1, x_2, \ldots, x_n)] = \frac{1}{n}\Sigma E(x^2) = \lambda^2 + \lambda.$$

Now, another function is defined as

$$t_2(x_1, x_2, \ldots, x_n) = \frac{1}{2}\Sigma x^2 - \bar{x}.$$

We have

$$E[t_2(x_1, x_2, \ldots, x_n)] = \frac{1}{n}\Sigma E(x^2) - E(\bar{x}) = \lambda^2.$$

Hence $\left(\frac{1}{n}\Sigma x^2 - \bar{x}\right)$ is an unbiased estimator of λ^2.

►**EXAMPLE 32.2** *Let $x_1, x_2, \ldots, x_n$ be a random sample of size n from a population having p.d.f. $f(x, \theta)$, where*

$$f(x, \theta) = \frac{1}{\theta}e^{-x/\theta};\ 0 < x < \infty,\ 0 < \theta < \infty.$$

Find unbiased estimators of θ and θ^2. Also, find variance of estimator of θ.

Solution. For the given distribution, $E(x) = \theta$ and $V(x) = \theta^2$. We have

$$\bar{x} = \frac{1}{n}\Sigma x \text{ and } E(\bar{x}) = \frac{1}{n}\Sigma E(x) = \theta.$$

Hence sample mean $\bar{x}$ is an unbiased estimator of θ. We know

$$\begin{aligned} V(\bar{x}) &= E(\bar{x}^2) - [E(\bar{x})]^2 \\ E(\bar{x}^2) &= V(\bar{x}) + [E(\bar{x})]^2 = \frac{\theta^2}{n} + \theta^2 \quad \left[\because\ V(\bar{x}) = \frac{1}{n^2}\Sigma V(x) = \frac{\theta^2}{n}\right] \\ &= \theta^2\frac{(n+1)}{n}. \\ \therefore\ E\left[\frac{n\bar{x}^2}{n+1}\right] &= \theta^2. \end{aligned}$$

Hence $\frac{n\bar{x}^2}{n+1}$ is an unbiased estimator of θ^2. The variance of $\hat{\theta} = \bar{x}$ is

$$V(\hat{\theta}) = V(\bar{x}) = \frac{\theta^2}{n}.$$

►**EXAMPLE 32.3** *The random variable X follows probability distribution whose p.d.f. is*

$$f(x) = \begin{cases} \frac{1}{b-a}, & a < x < b \\ 0, & \textit{otherwise}. \end{cases}$$

If the value of 'a' is known, estimate b. Also, find variance of your estimator.

Solution. Let $x_1, x_2, \ldots, x_n$ be random sample observations of size n. We know

$$\begin{aligned} E(x) &= \frac{1}{b-a}\int_a^b x\,dx = \frac{b+a}{2} \\ E(x^2) &= \frac{1}{b-a}\int_a^b x^2\,dx = \frac{b^2+ab+a^2}{3} \\ V(x) &= E(x^2) - [E(x)]^2 = \frac{b^2+ab+a^2}{3} - \frac{(b+a)^2}{4} = \frac{(b-a)^2}{12}. \end{aligned}$$

The sample mean is $\bar{x} = \frac{1}{n}\Sigma x$. We have

$$E(\bar{x}) = \frac{1}{n}\Sigma E(x) = \frac{b+a}{2} \Rightarrow E(2\bar{x}-a) = b.$$

$\therefore$ The unbiased estimator of b is $2\bar{x}-a$, as a is known. The variance of this estimator is

$$\begin{aligned} V(2\bar{x}-a) &= 4V(\bar{x}) = \frac{4}{n^2}\Sigma V(x) \; [\because \; x_i\text{'s are independent}] \\ &= \frac{4}{n}\frac{(b-a)^2}{12} = \frac{(b-a)^2}{3n}. \end{aligned}$$

►**EXAMPLE 32.4** *Let $x_1, x_2, \ldots, x_n$ be a random sample observations from population having p.d.f.*

$$f(x) = \frac{1}{N}, \; x = 1, 2, \ldots, N.$$

Find unbiased estimator of N. Also, find variance of your estimator.

Solution. We have

$$\begin{aligned} E(x) &= \sum_{x=1}^{N} xP(x) = \frac{N+1}{2} \\ E(x^2) &= \frac{1}{N}\sum^{N} x^2 = \frac{(N+1)(2N+1)}{6} \\ V(x) &= \frac{(N+1)(2N+1)}{6} - \frac{(N+1)^2}{4} = \frac{N^2-1}{12}. \end{aligned}$$

We also have

$$E(\bar{x}) = \frac{1}{n}\Sigma E(x) = \frac{N+1}{2} \Rightarrow E(2\bar{x}-1) = N.$$

Hence, the unbiased estimator of N is $(2\bar{x}-1)$. The variance of this estimator is

$$V(2\bar{x}-1) = 4V(\bar{x}) = \frac{4}{n^2}\sum^{n} V(x) = \frac{N^2-1}{3n}.$$

►**EXAMPLE 32.5** *The random sample observations $x_1, x_2, \ldots, x_n$ are assumed to be selected from a population having probability function*

$$P(x=1) = p \text{ and } P(x=0) = 1-p, \; 0 < p < 1.$$

Find unbiased estimators of p and p^2. Also, find the variance of the estimator of p.

Solution. Let $g(x) = \sum^{n} x_i$. The distribution of $g(x)$ is binomial.

Hence $E[g(x)] = np$ and $V[g(x)] = npq$.

Hence $E(\bar{x}) = \frac{1}{n}E[g(x)] = p \Rightarrow \bar{x}$ is an unbiased estimator of p.

The variance of $\hat{p} = \bar{x}$ is

$$V(\hat{p}) = V(\bar{x}) = \frac{1}{n^2}V[g(x)] = \frac{pq}{n}.$$

We have

$$\begin{aligned} s^2 &= \frac{1}{n-1}\left[\Sigma x^2 - \frac{(\Sigma x)^2}{n}\right] \\ g(x) &= \Sigma x = \Sigma x^2 \; [\because\; x_i = 0 \text{ or, } 1]. \\ \therefore\; s^2 &= \frac{1}{n-1}\left[g(x) - \frac{[g(x)]^2}{n}\right] = \frac{1}{n-1}\left[n\bar{x} - n\bar{x}^2\right] = \frac{n\bar{x}}{n-1}[1-\bar{x}]. \end{aligned}$$

We know

$$\begin{aligned} (n-1)s^2 &= \sum^{n}(x-\bar{x})^2 = \sum[(x-\mu)-(\bar{x}-\mu)]^2 \\ &= \Sigma(x-\mu)^2 - n(\bar{x}-\mu)^2 \\ (n-1)E(s^2) &= \Sigma E(x-\mu)^2 - nE(\bar{x}-\mu)^2 = n\sigma^2 - n\frac{\sigma^2}{n} \\ &= (n-1)\sigma^2 \Rightarrow s^2 \text{ is an unbiased estimator of } \sigma^2. \end{aligned}$$

Now

$$\begin{aligned} E(s^2) &= \frac{1}{n-1}E\left[g(x) - \frac{[g(x)]^2}{n}\right] \\ p(1-p) &= \frac{1}{n-1}E\left[g(x) - \frac{[g(x)]^2}{n}\right] \quad [\because\; V(x) = \sigma^2 = p(1-p)] \\ p^2 &= p - \frac{1}{n-1}E\left[g(x) - \frac{[g(x)]^2}{n}\right] \\ &= E\left[\frac{g(x)}{n}\right] - \frac{1}{n-1}E\left[g(x) - \frac{[g(x)]^2}{n}\right] \\ &\Rightarrow \frac{g(x)}{n} - \frac{1}{n-1}\left[g(x) - \frac{[g(x)]^2}{n}\right] \text{ is an unbiased estimator of } p^2. \\ \text{or, } \hat{p}^2 &= \frac{n\bar{x}}{n} - \frac{1}{n-1}\left[n\bar{x} - \frac{n^2\bar{x}^2}{n}\right] = \bar{x} - \frac{n\bar{x}}{n-1}[1-\bar{x}] = \frac{(n\bar{x}-1)\bar{x}}{n-1}. \end{aligned}$$

It is observed that, when $n = 1$, the estimator is not feasible.

►**EXAMPLE 32.6** *The random variable X follows hypergeometric distribution, where*

$$P(x) = \frac{\binom{M}{x}\binom{N-M}{n-x}}{\binom{N}{n}}, \; x = 0, 1, 2, \ldots, \min(n, M).$$

Assume that n and N are known. Find unbiased estimator of M^2.

Solution. We have

$$E[x(x-1)] = \frac{M(M-1)n(n-1)}{N(N-1)} \Rightarrow E\left[\frac{N(N-1)x(x-1)}{n(n-1)}\right] = M^2 - M.$$

Again,

$$E(x) = \frac{nM}{N} \Rightarrow E\left[\frac{Nx}{n}\right] = M.$$

$$\therefore\ E\left[\frac{N(N-1)x(x-1)}{n(n-1)} + \frac{Nx}{n}\right] = M^2$$

$\Rightarrow$ the unbiased estimator of M^2 is $\dfrac{N(N-1)x(x-1)}{n(n-1)} + \dfrac{Nx}{n}$.

►**EXAMPLE 32.7** *Let $x_1, x_2, \ldots, x_n$ be a random sample from binomial population with parameters n and p. If p is known, find unbiased estimator of n and variance of your estimator.*

Solution. Since X follows binomial distribution, we have $E(x) = np$, $V(x) = npq$. Then

$$E(\bar{x}) = \frac{1}{n}\Sigma E(x) = np.$$

Hence, if p is known, the estimator of n is $\bar{x}/p$. The variance of this estimator is

$$V(\bar{x}/p) = \frac{1}{p^2}V(\bar{x}) = \frac{pq}{p^2} = \frac{q}{p}.$$

►**EXAMPLE 32.8** *Let $x_1, x_2, \ldots, x_n$ be a random sample from a population having p.d.f.*

$$P(x) = pq^{x-1},\ x = 1, 2, \ldots;\ 0 < p < 1.$$

Find unbiased estimator of $1/p$. Also, find the variance of your estimator.

Solution. We have

$$E(x) = \sum^{\infty} xpq^{x-1} = \frac{p}{(1-q)^2} = 1/p$$

$$V(x) = \frac{q}{p^2} \Rightarrow E(\bar{x}) = \frac{1}{n}\Sigma E(x) = \frac{1}{p}.$$

Therefore, $\bar{x}$ is an unbiased estimator of $1/p$. The variance of this estimator is

$$V(\bar{x}) = \frac{1}{n^2}\Sigma V(x) = \frac{q}{np^2}.$$

►**EXAMPLE 32.9** *Let $x_1, x_2, \ldots, x_n$ be a random sample from a population having p.d.f.*

$$f(x) = \begin{cases} \frac{2}{\theta^2}(\theta - x), & 0 \le x \le \theta \\ 0, & \textit{otherwise.} \end{cases}$$

Find unbiased estimators of θ and θ^2. Also, find the variance of the estimator of θ.

Solution. We have

$$E(x) = \frac{2}{\theta^2}\int_0^\theta x(\theta - x)dx = \frac{\theta}{3}$$
$$E(x^2) = \frac{2}{\theta^2}\int x^2(\theta - x)dx = \frac{\theta^2}{6}.$$
$$\therefore\ V(x) = E(x^2) - [E(x)]^2 = \frac{\theta^2}{6} - \frac{\theta^2}{3} = \frac{\theta^2}{18}.$$

We also have

$$E(\bar{x}) = \frac{1}{n}\Sigma E(x) = \frac{\theta}{3} \Rightarrow 3\bar{x} \text{ is an unbiased estimator of } \theta.$$
$$V(\hat{\theta}) = 9V(\bar{x}) = \frac{9}{n}\frac{\theta^2}{6} = \frac{3\theta^2}{2n}.$$

Now $V(\bar{x}) = E(\bar{x}^2) - [E(\bar{x})]^2$

$$\frac{\theta^2}{18n} = E(\bar{x}^2) - \frac{\theta^2}{9} \Rightarrow E(\bar{x}^2) = \frac{\theta^2}{18n} + \frac{\theta^2}{9}$$
$$= \frac{\theta^2(1+2n)}{18n} \Rightarrow E\left[\frac{18nx^2}{1+2n}\right] = \theta^2.$$

Therefore, $\dfrac{18n\bar{x}^2}{1+2n}$ is an unbiased estimator of θ^2.

►**EXAMPLE 32.10** *Let $x_1, x_2, \ldots, x_n$ be a random sample from a population having p.d.f.*

$$f(x) = \frac{1}{\theta^n \Gamma n} e^{-x/\theta} x^{n-1};\ 0 < x < \infty,\ \theta > 0.$$

Find unbiased estimators of θ and θ^2. Also, find the variance of the estimator of θ.

Solution. We know

$$E(x) = n\theta,\ V(x) = n\theta^2$$
$$E(\bar{x}) = \frac{1}{n}\Sigma E(x) = n\theta \Rightarrow \hat{\theta} = \frac{\bar{x}}{n}$$
$$V(\hat{\theta}) = \frac{1}{n^2}V(\bar{x}) = \frac{1}{n^4}\Sigma V(x) = \frac{n^2\theta^2}{n^4} = \frac{\theta^2}{n^2}$$
$$V(x) = E(x^2) - E[(x)]^2 \Rightarrow E(x^2) = n\theta^2 + n^2\theta^2 = n\theta^2(n+1)$$
$$E[\Sigma x^2] = \Sigma E(x^2) = n^2\theta^2(n+1) \Rightarrow E\left[\frac{\Sigma x^2}{n^2(n+1)}\right] = \theta^2.$$

Therefore, $\dfrac{\Sigma x^2}{n^2(n+1)}$ is an unbiased estimator of θ^2.

►**EXAMPLE 32.11** *Let $x_1, x_2, \ldots, x_n$ be a random sample of n observations selected from normal population with mean μ and variance σ^2. Find unbiased estimators of μ and σ^2. Also, find the variance of the estimators.*

Solution. We have $E(x) = \mu, V(x) = \sigma^2$. The sample mean is $\bar{x} = \frac{1}{n}\Sigma x$

$$\begin{aligned} E(\bar{x}) &= \frac{1}{n}\Sigma E(x) = \mu, \ \therefore \ \hat{\mu} = \bar{x} \\ V(x) &= E(x^2) - [E(x)]^2 \Rightarrow \sigma^2 + \mu^2 = E(x^2) \\ \therefore \ \frac{1}{n}\Sigma E(x^2) &= \sigma^2 + \mu^2 \text{ or, } E\left[\frac{1}{n}\Sigma x^2 - \mu^2\right] = \sigma^2 \end{aligned}$$

$\therefore \ \hat{\sigma}^2 = \frac{1}{n}\Sigma x^2 - \mu^2$, if μ is known $= \frac{1}{n}\Sigma(x-\mu)^2$.
If μ is not known, we get

$$\begin{aligned} E\sum^{n}(x-\bar{x})^2 &= \sum^{n} E[(x-\mu)-(\bar{x}-\mu)]^2 \\ &= \Sigma E(x-\mu)^2 + nE(\bar{x}-\mu)^2 - 2nE(\bar{x}-\mu) \\ &= n\sigma^2 - n\frac{\sigma^2}{n} = n\sigma^2(n-1). \end{aligned}$$

Therefore, $s = \frac{1}{n}\Sigma(x-\bar{x})^2$ is an unbiased estimator of σ^2 if μ is not known.
We have $\hat{\mu} = \bar{x}$,

$$V(\hat{\mu}) = \frac{1}{n^2}\Sigma V(x) = \frac{n\sigma^2}{n^2} = \frac{\sigma^2}{n}.$$

We know

$$Z = \frac{x-\mu}{\sigma} \sim N(0,1) \Rightarrow Z^2\sigma^2 = (x-\mu)^2$$

is distributed as χ^2 with 1 d.f. Therefore,

$$\begin{aligned} \sigma^2\Sigma Z^2 &= \Sigma(x-\mu)^2 \\ \text{or, } \frac{\sigma^2}{n}\Sigma Z^2 &= \frac{1}{n}\Sigma(x-\mu)^2 \text{ is distributed as } \chi^2 \text{ with } n \text{ d.f.} \\ \therefore \ V\left[\frac{1}{n}\Sigma(x-\mu)^2\right] &= \frac{2n\sigma^4}{n^2} = \frac{2\sigma^4}{n}. \end{aligned}$$

►**EXAMPLE 32.12** *Show that sample r-th moment is unbiased estimator of population r-th moment, if it exists.*

Solution. Let $x_1, x_2, \ldots, x_n$ be a random sample of n observations from a population. The r-th moment of the population is μ'_r and it exists. The r-th sample moment is

$$m'_r = \frac{1}{n}\Sigma x^r.$$

We have $E[m'_r] = \frac{1}{n}\Sigma E(x^r) = \mu'_r \Rightarrow m'_r$ is unbiased estimator of μ'_r.

►**EXAMPLE 32.13** *Let t_1 and t_2 be two estimators of θ. Assume that t_1 and t_2 are independent unbiased estimators of θ. Find unbiased estimators of* (i) θ^2 *and* (ii) $\theta(\theta-1)$.

Solution. Given $E(t_1) = E(t_2) = \theta$. Let $V(t_1) = \sigma_1^2$ and $V(t_2) = \sigma_2^2$.

We have

$$E\left[\frac{1}{2}(t_1+t_2)\right] = \frac{1}{2}(\theta+\theta) = \theta$$

$$\Rightarrow \quad t = \frac{1}{2}(t_1+t_2) \text{ is an unbiased estimator of } \theta.$$

Here

$$\begin{aligned} V(t) &= \frac{1}{4}[V(t_1)+V(t_2)], \ \because \ t_1 \text{ and } t_2 \text{ are independent} \\ &= \frac{1}{4}(\sigma_1^2+\sigma_2^2). \end{aligned}$$

But $V(t) = E(t^2) - [E(t)]^2$

$$[E(t)]^2 = E(t^2) - V(t) = \theta^2 \Rightarrow E\left[\left\{\frac{1}{2}(t_1+t_2)\right\}^2 - \frac{1}{4}\left(\sigma_1^2+\sigma_2^2\right)\right] = \theta^2.$$

$\therefore \ \frac{1}{4}\left[(t_1+t_2)^2 - (\sigma_1^2+\sigma_2^2)\right]$ is unbiased estimator of θ^2.

Therefore, $\frac{1}{4}\left[(t_1+t_2)^2 - (\sigma_1^2+\sigma_2^2)\right] - \frac{1}{2}(t_1+t_2)$ is an unbiased estimator of $\theta(\theta-1)$.

►**EXAMPLE 32.14** *Let t_1 and t_2 be two independent unbiased estimators of θ. Assume that $2V(t_2) = V(t_1)$. If $(k_1t_1 + k_2t_2)$ is the probable minimum variance estimator, find k_1 and k_2.*

Solution. Given $E(t_1) = E(t_2) = \theta$. If $(k_1t_1 + k_2t_2)$ is the unbiased estimator of θ, then $E(k_1t_1 + k_2t_2) = \theta$. This is possible, if

$$E(k_1t_1) + E(k_2t_2) = k_1\theta + k_2\theta = (k_1+k_2)\theta = \theta \Rightarrow k_1 + k_2 = 1.$$

Now, $V[k_1t_1 + k_2t_2]$ will be minimum under the condition $k_1 + k_2 = 1$. We have

$$\begin{aligned} V[k_1t_1+k_2t_2] &= k_1^2V(t_1) + k_2^2V(t_2), \ \because \ t_1 \text{ and } t_2 \text{ are independent} \\ &= V(t_2)[2k_1^2 + k_2^2], \ \because \ 2V(t_2) = V(t_1) \end{aligned}$$

We need to minimize $V[k_1t_1 + k_2t_2]$. For this let us use Lagrange multiplier, where

$$\phi = V(t_2)\left[2k_1^2 + k_2^2\right] - \lambda\left[k_1 + k_2 - 1\right].$$

Now

$$\frac{\partial\phi}{\partial\lambda} = -(k_1+k_2-1) = 0 \tag{1}$$

$$\frac{\partial\phi}{\partial k_1} = V(t_2)4k_1 - \lambda = 0 \tag{2}$$

$$\frac{\partial\phi}{\partial k_2} = V(t_2)2k_2 - \lambda = 0 \tag{3}$$

From (2) and (3), we get

$$\lambda = V(t_2)[2k_1 + k_2] = V(t_2)[k_1 + 1], \quad \because \ k_1 + k_2 = 1.$$

Putting the value of λ in (2), we get

$$V(t_2)4k_1 - V(t_2)(k_1+1) = 0 \Rightarrow k_1 = \frac{1}{3}$$

$$\therefore\ k_2 = 1 - k_1 = 1 - \frac{1}{3} = \frac{2}{3}.$$

►**EXAMPLE 32.15** *Let t is an unbiased estimator of θ. Show that t^2 is not an unbiased estimator of θ^2.*

Solution. Given $E(t) = \theta$

$$\begin{aligned} V(t) &= E(t^2) - [E(t)]^2 = E(t^2) - \theta^2 \\ E(t^2) &= V(t) + \theta^2 \neq \theta^2 \ [\because\ V(t) \neq 0]. \end{aligned}$$

Hence t^2 is not an unbiased estimator of θ^2.

►**EXAMPLE 32.16** *Let $y_1 < y_2 < y_3$ be three ordered sample observations from a population with p.d.f.*

$$f(x,\theta) = \frac{1}{\theta};\ 0 < x < \theta,\ 0 < \theta < \infty.$$

Find unbiased estimator of θ using $y_1 < y_2 < y_3$.

Solution. Since y_1 is the minimum observation in the sample, the distribution of y_1 is

$$f_1(y_1) = n[1 - F(x)]^{n-1} f(x).$$

Here $n = 3$, $F(x) = x/\theta$. Hence

$$\begin{aligned} f_1(y_1) &= 3\left[1 - \frac{y_1}{\theta}\right]^2 \frac{1}{\theta};\ 0 < y_1 < \theta,\ 0 < \theta < \infty \\ E(y_1) &= \int_0^\theta \frac{3y_1}{\theta}\left(1 - \frac{y_1}{\theta}\right)^2 dy_1 = \frac{\theta}{4} \Rightarrow E[4y_1] = \theta. \end{aligned}$$

Therefore, $4y_1$ is an unbiased estimator of θ.

Consistency

Let $x_1, x_2, \ldots, x_n$ be a random sample of n observations from a population having p.d.f. $f(x,\theta)$. Consider that $\hat{\theta}_n$ is any estimator of the parameter θ. The estimator $\hat{\theta}_n$ is said to be consistent estimator of θ if for $\epsilon > 0$ and $\eta > 0$

$$\lim_{n\to\infty} P\left(\left|\hat{\theta}_n - \theta\right| > \epsilon\right) = 0 \text{ or, } P\left(\left|\hat{\theta}_n - \theta\right| < \epsilon\right) \geq 1 - \eta,\ n \geq N$$

where n is any large value of N.

This definition indicates that when sample size n increases $\hat{\theta}_n$ tends to θ and $V(\hat{\theta}_n) \to 0$. This property of the estimator is known as *consistency*

$$\left[n \to \infty, V(\hat{\theta}_n) \to 0\right].$$

The property is a large sample property. It has already been observed that using a random sample of observations from $N(\mu, \sigma^2)$ the sample mean $\bar{x}$ and sample variance

s^2 are unbiased estimators of μ and σ^2, respectively. In estimating these parameters nothing is mentioned about the size of the sample. Even if an estimator is unbiased, there is discrepancy between estimator and parameter. The study of the discrepancy leads us to find interval estimator for any parameter. This is possible by Chebyshev's inequality. This inequality for population mean is

$$P\left[|\bar{x}-\mu| \leq k\frac{\sigma}{\sqrt{n}}\right] = 1 - \frac{1}{k^2}.$$

When $k = 1/\sqrt{\alpha}$, we get $(1-\alpha)100\%$ confidence limit of μ. It is observed that, if n is small, the confidence limit becomes large. The larger the sample size, the smaller the confidence limit. This property of the estimator leads us to have an estimator which will tend to parameter when sample size becomes large.

►**EXAMPLE 32.17** *Let $x_1, x_2, \ldots, x_n$ be a random sample of observations from $N(\mu, \sigma^2)$. Show that the sample mean $\bar{x}$ and sample variance s^2 are consistent estimator of μ and σ^2, respectively.*

Solution. It has already been observed that [Example 32.11]

$$E(\bar{x}) = \mu \text{ and } E(s^2) = \sigma^2$$

where

$$\bar{x} = \frac{1}{n}\Sigma x, \; s^2 = \frac{1}{n}\Sigma(x-\bar{x})^2.$$

The variances of these estimators are

$$V(\bar{x}) = \frac{\sigma^2}{n} \text{ and } V(s^2) = \frac{2\sigma^4}{n}.$$

In both the cases, when $n \to \infty$, variances tend to zero. Hence $\bar{x}$ and s^2 are consistent estimators of μ and σ^2, respectively.

►**EXAMPLE 32.18** *Let $x_1, x_2, \ldots, x_n$ be a random sample of observations from a population with probability function*

$$P(x) = \frac{1}{N}, \; x = 1, 2, \ldots, N.$$

Find consistent estimator of N.

Solution. From Example 32.4, we have

$$E(2\bar{x}-1) = N \text{ and } V(2\bar{x}-1) = \frac{N^2-1}{3n}.$$

Here $(2\bar{x}-1)$ is an unbiased estimator of N. $V(2\bar{x}-1) \to 0$ when $n \to \infty$. Thus $(2\bar{x}-1)$ is also consistent estimator of N.

►**EXAMPLE 32.19** *If $x_1, x_2, \ldots, x_n$ are n observations of a random sample from a population whose probability function is*

$$P(x) = \begin{cases} \frac{1}{b-a}, & a < x < b \\ 0, & \textit{otherwise.} \end{cases}$$

Let 'a' is known. Find consistent estimator of b.

Solution. The unbiased estimator of b, when a is known, is [Example 32.3]

$$\hat{b} = 2\bar{x} - a \text{ and } V(\hat{b}) = \frac{(b-a)^2}{3n}.$$

Since $n \to \infty$, $V(\hat{b}) \to 0, \hat{b} = 2\bar{x} - a$ is consistent estimator of b.

►**EXAMPLE 32.20** *Let $x_1, x_2, \ldots, x_n$ be a random sample of n observations from a population whose probability function is*

$$P(x = 1) = p \text{ and } P(x = 0) = 1 - p, \ 0 < p < 1.$$

Show that the sample mean $\bar{x}$ is a consistent estimator of p.

Solution. It is observed [Example 32.5] that

$$E(\bar{x}) = p \text{ and } V(\bar{x}) = \frac{pq}{n}.$$

This variance tends to zero as $n \to \infty$. Hence $\bar{x}$ is consistent estimator of p.

►**EXAMPLE 32.21** *Let $x_1, x_2, \ldots, x_n$ be a random sample of n observations from a hypergeometric distribution with parameters n, N and M. Assuming n and N known, find consistent estimator of M.*

Solution. We have

$$P(x) = \frac{\binom{M}{x}\binom{N-M}{n-x}}{\binom{N}{n}}, \ x = 0, 1, 2, \cdots, \min(n, M).$$

We have [Example 32.6] $E(x) = \dfrac{nM}{N}$.

Also

$$\begin{aligned}
E(\bar{x}) &= \frac{1}{n}\Sigma E(x) = \frac{nM}{N} \Rightarrow \hat{M} = \frac{N\bar{x}}{n} \\
V(x) &= \frac{nM(N-M)(N-n)}{N^2(N-1)} \\
V(\hat{M}) &= \frac{N^2}{n}\frac{1}{n^2}\Sigma V(x) = \frac{N^2}{n^2}\frac{nM(N-M)(N-n)}{N^2(N-1)} \\
&= \frac{M(N-M)(N-n)}{n(N-1)}.
\end{aligned}$$

Hence as $n \to \infty$, $V(\hat{M}) \to 0$. So $\hat{M}$ is consistent estimator of M.

►**EXAMPLE 32.22** *Let $x_1, x_2, \ldots, x_n$ be a random sample of n observations from binomial population with parameters n and p. Show that the estimator of n, when p is known, is not a consistent estimator.*

Solution. As the sample follows binomial distribution with parameters n and p, we have $E(x) = np$, $V(x) = npq$. Also, we have [Example 32.7]

$$\hat{n} = \bar{x}/p \text{ and } V(\hat{n}) = q/p.$$

But when $n \to \infty$, there is no scope for $V(\hat{n})$ to be zero. Hence $\hat{n} = q/p$ is not consistent estimator of n.

►EXAMPLE 32.23 *Let $x_1, x_2, \ldots, x_n$ be a random sample of n observations from a Poisson population with parameter λ. Find consistent estimator of λ.*

Solution. Example 32.1 shows that

$$E(\bar{x}) = \lambda \text{ and } V(\bar{x}) = \lambda/n.$$

Hence $\bar{x}$ is unbiased estimator of λ and as $n \to \infty$, $V(\bar{x}) \to 0$. Therefore, $\bar{x}$ is consistent estimator of λ.

►EXAMPLE 32.24 *Let $x_1, x_2, \ldots, x_n$ be a random sample of n observations from a population having p.d.f.*

$$f(x) = \frac{1}{\theta}e^{-x/\theta};\ 0 < x < \infty,\ 0 < \theta < \infty.$$

Show that sample mean is consistent estimator of θ.

Solution. Example 32.2 shows that $E(\bar{x}) = \theta$ and $V(\bar{x}) = \theta^2/n$.

Thus, when $n \to \infty$, $V(\bar{x}) \to 0$.

Therefore, $\bar{x}$ is consistent estimator of θ.

►EXAMPLE 32.25 *Let $x_1, x_2, \ldots, x_n$ be a random sample of n observations from Cauchy population. Show that sample mean is not consistent estimator of population parameter.*

Solution. We have

$$f(x) = \frac{1}{\pi}\frac{1}{1+(x-\mu)^2},\quad -\infty < x < \infty.$$

Also

$$f(\bar{x}) = \frac{1}{\pi}\frac{1}{1+(\bar{x}-\mu)^2},\quad -\infty < \bar{x} < \infty.$$

Here μ is the parameter. Now, if $\bar{x}$ is the estimator of μ, $V(\bar{x})$ will not be reduced if n increases [$\because\ V(\bar{x}) = V(x)$]. Therefore, $\bar{x}$ is not a consistent estimator of μ.

►EXAMPLE 32.26 *If $x_1, x_2, \ldots, x_n$ be a random sample of n observations from geometric distribution, then show that the estimator of $1/p$ is consistent, where p is a parameter.*

Solution. We have $P(x) = pq^{x-1};\ x = 1, 2, \ldots;\ 0 < p < 1$.

From Example 32.8, it is observed that

$$E(\bar{x}) = 1/p \text{ and } V(\bar{x}) = \frac{q}{np^2}.$$

As $n \to \infty$, $V(\bar{x}) \to 0$. Hence the estimator of $1/p$ is $\bar{x}$ and it is consistent.

►EXAMPLE 32.27 *Let $x_1, x_2, \ldots, x_n$ be a random sample of n observations from a population having p.d.f.*

$$f(x) = \begin{cases} \frac{2}{\theta^2}(\theta - x), & 0 \le x \le \theta \\ 0, & \text{otherwise.} \end{cases}$$

Show that the estimator of θ is consistent.

Solution. From Example 32.8, we have

$$\hat{\theta} = 3\bar{x} \text{ and } V(\hat{\theta}) = \frac{3\theta^2}{2n}.$$

As $n \to \infty$, $V(\hat{\theta}) \to 0$. Hence $\hat{\theta}$ is consistent estimator of θ.

❒ **THEOREM 32.1** *Let $\hat{\theta}_n$ be an unbiased estimator of θ_n and $V(\hat{\theta}_n) = \sigma_n^2$. Also, assume that as $n \to \infty$, $\theta_n \to \theta$ and $\sigma_n \to 0$. Then $\hat{\theta}_n$ is consistent estimator of θ.*

Proof. By Chebyshev's Lemma, we can write

$$P[|\hat{\theta}_n - \theta_n| < t] > 1 - \frac{\sigma_n^2}{t^2} \; [\because \; E(\hat{\theta}_n) = \theta_n, \; V(\hat{\theta}_n) = \sigma_n^2].$$

Now, if $|\hat{\theta}_n - \theta_n| < t$

$$|\hat{\theta}_n - \theta| = |\hat{\theta}_n - \theta_n + \theta_n - \theta| < |\hat{\theta}_n - \theta_n| + |\theta_n - \theta| < t + |\theta_n - \theta|.$$
$$\therefore \; P[|\hat{\theta}_n - \theta| < t + |\theta_n - \theta|] > P[|\hat{\theta}_n - \theta| < t]$$
$$\text{or, } P[|\hat{\theta}_n - \theta| - |\theta_n - \theta| < t] > P[|\hat{\theta}_n - \theta| < t] = P[|\hat{\theta}_n - \theta_n| < t].$$

But $P[|\hat{\theta}_n - \theta_n| < t] > 1 - \dfrac{\sigma_n^2}{t^2}$.

$$\therefore \; P[|\hat{\theta}_n - \theta| < t + |\hat{\theta}_n - \theta|] \geq 1 - \frac{\sigma_n^2}{t^2}.$$

Now, if $|\theta_n - \theta| = \epsilon$ and $\sigma_n^2/t^2 = \eta$, then

$$P\left[|\hat{\theta}_n - \theta| < t + \epsilon\right] \geq 1 - \eta$$

where ϵ and η are small quantities. Thus $\hat{\theta}_n$ is consistent estimator of θ.

►**EXAMPLE 32.28** *Let $x_1, x_2, \ldots, x_n$ be a random sample from Poisson population with parameter λ. Let us define an estimator T_n of λ, where*

$$T_n = \frac{1}{n+1}\Sigma x_i = \frac{n\bar{x}}{n+1} = \frac{\bar{x}}{1+\frac{1}{n}}.$$

Show that the estimator T_n is consistent estimator of λ but it is not an unbiased estimator.

Solution. We have

$$V(T_n) = \frac{1}{\left(1+\frac{1}{n}\right)^2}V(\bar{x}) = \frac{1}{\left(1+\frac{1}{n}\right)^2}\frac{\lambda}{n}$$

As $n \to \infty$, $V(T_n) \to 0$. Hence T_n is a consistent estimator of λ. Again,

$$E(T_n) = \frac{1}{1+\frac{1}{n}}E(\bar{x}) = \frac{\lambda}{1+\frac{1}{n}} \neq \lambda.$$

Hence T_n is not an unbiased estimator of λ. But, if n is large enough

$$n + 1 \approx n \text{ and } E(T_n) \approx \lambda.$$

It is observed that consistent estimator depends on sample size. But unbiasedness does not depend on sample size. An estimator based on small sample size may be unbiased but it may not be consistent. Thus, it may be concluded that a consistent estimator will be unbiased but an unbiased estimator may or may not be consistent.

Efficiency

We have considered the properties unbiasedness and consistency of estimators. The property consistency indicates that the variance of the estimator becomes minimum if sample size increases. But the property unbiasedness does not depend on large sample size and the variance of the estimator does not become minimum for samples of all sizes. Moreover, two estimators based on same sample size may give two different variances. For example, if $x_1, x_2, \ldots, x_n$ is a random sample of size n from $N(\mu, \sigma^2)$, then the sample mean $\bar{x}$ and the sample median 'm' are unbiased estimators of μ. Here

$$V(\bar{x}) = \frac{\sigma^2}{n} \text{ and } V(m) = \frac{\pi\sigma^2}{2n} = 1.57\frac{\sigma^2}{n}.$$

This implies that $V(\bar{x}) < V(m)$. This minimum variance property of an estimator is known as efficiency of an estimator and this property is an important aspect to be considered to find an estimator.

Efficient Estimator

Let $T_1, T_2, \ldots, T_k$ be k consistent estimators of a parameter θ such that

$$V(T_1) < V(T_2) < \cdots < V(T_k).$$

Here T_1 is an efficient estimator of θ since its variance is lowest among all other consistent estimators.

Relative Efficiency

Let $\hat{\theta}_1$ and $\hat{\theta}_2$ be any two estimators of θ such that $V(\hat{\theta}_1) = v_1 < V(\hat{\theta}_2) = v_2$. Then the relative efficiency of $\hat{\theta}_2$ compared to $\hat{\theta}_1$ is

$$\text{R.E.} = \frac{v_1}{v_2} < 1.$$

This property of the estimator is studied since an estimator having minimum variance is accepted even if it is biased one. Sometimes an unbiased estimator is not available and in that case we need an estimator which has minimum variance.

►**EXAMPLE 32.29** *Let $x_1, x_2, \ldots, x_n$ be a random sample of size n from $N(\mu, \sigma^2)$. Consider two estimators of μ as*

$$T_1 = \frac{1}{n}\sum^{n} x \text{ and } T_2 = \frac{1}{n+1}\sum^{n} x.$$

Find relative efficiency of T_2 compared to T_1.

Solution. We have

$$E(T_1) = E(\bar{x}) = \mu$$
$$E(T_2) = \frac{En\bar{x}}{n+1} = \frac{1}{1+\frac{1}{n}}E(\bar{x}) = \frac{\mu}{1+\frac{1}{n}} \neq \mu.$$

Here T_1 is an unbiased estimator of μ but T_2 is a biased estimator of T_2. Also, we have

$$V(T_1) = V(\bar{x}) = \frac{\sigma^2}{n}$$
$$\text{and } V(T_2) = V\left[\frac{n\bar{x}}{n+1}\right] = \left(\frac{n}{n+1}\right)^2 \frac{\sigma^2}{n}$$
$$= \frac{\sigma^2}{n}\frac{1}{\left(1+\frac{1}{n}\right)^2}.$$

Here $V(T_2) < V(T_1)$. Therefore, T_2 is more efficient than T_1. The relative efficiency of T_2 compared to T_1 is

$$\text{R.E.} = \frac{V(T_1)}{V(T_2)} = \left(1+\frac{1}{n}\right)^2 > 1.$$

Here T_2 is a biased estimator but more efficient. Also, T_2 is a consistent estimator.

Mean Square Error

Let $T = t(x_1, x_2, \ldots, x_n)$ be an estimator of a parametric function $\psi(\theta)$ based on sample observations $x_1, x_2, \ldots, x_n$ selected randomly from a population having p.d.f. $f(x, \theta)$. The mean square error (MSE) of T is

$$\text{MSE } (T) = E[T - \psi(\theta)]^2, \text{ where } E(T) \neq \psi(\theta)$$
$$= E[T - E(T)]^2 + E[\psi(\theta) - E(T)]^2 - 2E[\psi(\theta) - E(T)][T - E(T)]$$
$$= V(T) + E[\psi(\theta) - E(T)]^2, \;\because\; E[T - E(T)] = 0$$
$$= V(T) + (\text{Bias})^2, \text{ where Bias of } T = \psi(\theta) - E(T).$$

It implies that, if T is an unbiased estimator of $\psi(\theta)$, then MSE $(T) = V(T)$, otherwise

$$\text{MSE } (T) > V(T), \text{ i.e., MSE } (T) \geq V(T).$$

The estimator of T_2 in Example 32.29 is biased one. The amount of bias of T_2 is

$$\text{Bias } (T_2) = \mu - E(T_2) = \mu - \frac{n\mu}{n+1} = \frac{\mu}{n+1}.$$

Hence

$$\text{MSE } (T_2) = V(T_2) + [\text{Bias } (T_2)]^2$$
$$= \left(\frac{n}{n+1}\right)^2 \frac{\sigma^2}{n} + \frac{\mu^2}{(n+1)^2}$$
$$= \frac{1}{(n+1)^2}[n\sigma^2 + \mu^2].$$

►**EXAMPLE 32.30** *Let $x_1, x_2, \ldots, x_n$ be a random sample of size n selected from a population having p.d.f.*

$$f(x) = \begin{cases} 1, & \theta < x < \theta + 1 \\ 0, & otherwise. \end{cases}$$

Let $\hat{\theta}_1 = \bar{x} - \frac{1}{2}$ and $\hat{\theta}_2 = \max(x_1, x_2, \ldots, x_n)$. Investigate consistency and efficiency of the estimators.

Solution. We know $E(x) = \int_{\theta}^{\theta+1} x dx = \theta + \frac{1}{2}$.

Also, $E(\bar{x}) = \theta + \frac{1}{2}$

$$E\left(\bar{x} - \frac{1}{2}\right) = \theta \Rightarrow \left(\bar{x} - \frac{1}{2}\right) \text{ is an unbiased estimator of } \theta.$$

$$\begin{aligned} V(x) &= E(x^2) - [E(x)]^2 \\ &= \frac{1}{3}[3\theta^2 + 3\theta + 1] - \left(\theta + \frac{1}{2}\right)^2 = \frac{1}{12} \\ \therefore \; V(\bar{x}) &= \frac{1}{12n} \text{ and } V\left(\bar{x} - \frac{1}{2}\right) = \frac{1}{12n}. \end{aligned}$$

Hence $\hat{\theta}_1$ is a consistent estimator of θ.

We have $\hat{\theta}_2 = \max(x_1, x_2, \ldots, x_n)$. The distribution of $\hat{\theta}_2$ is

$$\begin{aligned} f(\hat{\theta}_2) &= n\left[F_x(\hat{\theta}_2)\right]^{n-1} f_x(\hat{\theta}_2), \text{ where} \\ F(x) &= x - \theta \text{ and } f(\hat{\theta}_2) = n(\hat{\theta}_2 - \theta)^{n-1} \\ E(\hat{\theta}_2) &= \int_{\theta}^{\theta+1} \hat{\theta}_2(\hat{\theta}_2 - \theta)^{n-1} d\hat{\theta}_2 \\ &= \theta + \frac{n}{n+1} \Rightarrow \left[\hat{\theta}_2 - \frac{n}{n+1}\right] \text{ is unbiased estimator of } \theta \\ E(\hat{\theta}_2^2) &= \int_{\theta}^{\theta+1} \hat{\theta}_2^2(\hat{\theta}_2 - \theta)^{n-1} d\hat{\theta}_2 = (\theta+1)^2 - \frac{2\theta}{n+1} - \frac{2}{n+2}. \\ \therefore \; V(\hat{\theta}_2) &= V\left[\hat{\theta}_2 - \frac{n}{n+1}\right] = E(\hat{\theta}_2^2) - [E(\hat{\theta}_2)]^2 \\ &= (\theta+1) - \frac{2\theta}{n+1} - \frac{2}{n+2} - \left(\theta + \frac{n}{n+1}\right)^2 \\ &= \frac{1}{(n+1)^2\left(1 + \frac{2}{n}\right)} \Rightarrow n \to \infty, V(\hat{\theta}_2) \to 0. \end{aligned}$$

Hence $\hat{\theta}_2$ is a consistent estimator of θ. Between these two estimators $\hat{\theta}_1$ will be more efficient estimator if

$$\text{R.E.} = \frac{V(\hat{\theta}_1)}{V(\hat{\theta}_2)} \leq 1 = \frac{1}{12}\left(1 + \frac{1}{n}\right)^2 (n+2) \leq 1.$$

This is true, if $n \leq 7$.

▶EXAMPLE 32.31 *Let $x_1, x_2, \ldots, x_n$ be a random sample of n observations from $N(\theta, \theta)$. Find an unbiased and efficient estimator of θ.*

Solution. Given $E(x) = V(x) = \theta$. The sample mean is $\bar{x} = \frac{1}{n}\Sigma x$ and the sample variance is $s^2 = \frac{1}{n-1}\Sigma(x-\bar{x})^2$.

Now,

$$\begin{aligned} E(\bar{x}) &= \frac{1}{n}\Sigma E(x) = \mu \\ (n-1)s^2 &= \Sigma(x-\bar{x})^2 = \Sigma(x-\theta)^2 - n(\bar{x}-\theta)^2 \\ (n-1)E(s^2) &= \Sigma E(x-\theta)^2 - nE(\bar{x}-\theta)^2 \; [\because \text{mean} = \theta] \\ &= n\theta - n\frac{\theta}{n} = (n-1)\theta \Rightarrow E(s^2) = \theta. \end{aligned}$$

Therefore, both $\bar{x}$ and s^2 are unbiased estimators of θ.

From the given condition

$$V(\bar{x}) = \frac{\theta}{n} \text{ and } V(s^2) = \frac{2\theta^2}{n-1} \quad \left[\because \frac{(n-1)s^2}{\sigma^2} \sim \chi^2_{n-1}\right].$$

It is observed that $V(\bar{x}) < V(s^2)$. Hence $\bar{x}$ is an unbiased and efficient estimator of θ.

▶EXAMPLE 32.32 *Let $x_1, x_2, \ldots, x_n$ be a random sample of n observations from $N(\mu, \sigma^2)$. Let us consider two estimators*

$$\hat{\mu}_1 = \bar{x} \text{ and } \hat{\mu}_2 = \frac{2}{n(n+1)}\sum_{i=1}^{n} ix_i.$$

Compare these two estimators.

Solution. We have already observed that $\hat{\mu}_1 = \frac{1}{n}\Sigma x_i$ is unbiased and consistent estimator of μ. Its variance is $V(\hat{\mu}_1) = \frac{\sigma^2}{n}$. Again

$$E(\hat{\mu}_2) = \frac{2}{n(n+1)}\sum_{i=1}^{n} iE(x_i) = \frac{2}{n(n+1)}\mu\sum_{i=1}^{n} i = \mu.$$

Hence $\hat{\mu}_2$ is an unbiased estimator of μ

$$\begin{aligned} V(\hat{\mu}_2) &= \frac{4}{[n(n+1)]^2}\sum_{i=1}^{n} i^2 V(x_i) \\ &= \frac{4\sigma^2}{[n(n+1)]^2}\sum_{i=1}^{n} i^2 \\ &= \frac{2\sigma^2(2n+1)}{3n(n+1)} \end{aligned}$$

$$\lim_{n\to\infty} V(\hat{\mu}_2) = \lim_{n\to\infty} \frac{2\sigma^2\left(2+\frac{1}{n}\right)}{3n\left(1+\frac{1}{n}\right)} = \lim_{n\to\infty}\frac{4\sigma^2}{3n} = 0.$$

Hence $\hat{\mu}_2$ is a consistent estimator of μ.

Now, the efficiency of $\hat{\mu}_2$ compared to $\hat{\mu}_1$ is given by

$$E = \frac{V(\hat{\mu}_1)}{V(\hat{\mu}_2)} = \frac{\sigma^2}{n}\frac{3n(n+1)}{2\sigma^2(2n+1)} = \frac{3(n+1)}{2(2n+1)} = \frac{3}{2\left(1+\frac{n}{n+1}\right)} < 1.$$

Hence $\hat{\mu}_2$ is less efficient than $\hat{\mu}_1$.

►**EXAMPLE 32.33** *Let $x_1, x_2, \ldots, x_n$ be a random sample of n observations from a population having p.d.f.*

$$f(x) = \begin{cases} \frac{1}{\theta}, & 0 < x < \theta \\ 0, & otherwise. \end{cases}$$

Let $\hat{\theta}_1 = 2\bar{x}$ and $\hat{\theta}_2 = \max(x_1, x_2, \ldots, x_n)$. Investigate unbiasedness, consistency and efficiency of these two estimators.

Solution. We have

$$E(x) = \frac{1}{\theta}\int_0^\theta x dx = \frac{\theta}{2}$$

$$E(\bar{x}) = E\left[\frac{1}{h}\Sigma x\right] = \frac{\theta}{2}$$

$\therefore$ $\hat{\theta}_1 = 2\bar{x}$ is an unbiased estimator of θ.

$$E(x^2) = \frac{1}{\theta}\int_0^\theta x^2 dx = \frac{\theta^2}{3}$$

$$V(x) = E(x^2) - [E(x)]^2 = \frac{\theta^2}{3} - \frac{\theta^2}{4} = \frac{\theta^2}{12}$$

$$V(\hat{\theta}_1) = \frac{4}{n^2}\Sigma V(\bar{x}) = \frac{\theta^2}{3n}$$

As $n \to \infty$, $V(\hat{\theta}_1) \to 0$. Hence $\hat{\theta}_1$ is consistent estimator of θ.

Also, we have

$$f(\hat{\theta}_2) = n\left[F_x(\hat{\theta}_2)\right]^{n-1} f_x(\hat{\theta}_2) = \frac{n\hat{\theta}_2^{n-1}}{\theta^n} \quad \left[\because\ F(x) = \frac{x}{\theta},\ 0 < \hat{\theta}_2 < \theta\right]$$

$$E(\hat{\theta}_2) = \frac{n}{\theta^n}\int_0^\theta \hat{\theta}_2^n d\hat{\theta}_2 = \frac{n}{\theta^n}\frac{\theta^{n+1}}{n+1} = \frac{n}{n+1}\theta.$$

Thus, $\hat{\theta}_2$ is a biased estimator of θ.

$$E(\hat{\theta}_2^2) = \frac{n}{\theta^n}\int_0^\theta \hat{\theta}_2^{n+1} d\hat{\theta}_2 = \frac{n\theta^2}{n+2}$$

$$V(\hat{\theta}_2) = E(\hat{\theta}_2^2) - [E(\hat{\theta}_2)]^2 = \frac{n\theta^2}{n+2} - \frac{n^2\theta^2}{(n+1)^2}$$

$$= \frac{n\theta^2}{(n+2)(n+1)^2} = \frac{\theta^2}{\left(1+\frac{2}{n}\right)n^2\left(1+\frac{1}{n}\right)^2}.$$

As $n \to \infty$, $V(\hat{\theta}_2) \to 0$. Hence $\hat{\theta}_2$ is a consistent estimator of θ.

Now, let us find the efficiency of $\hat{\theta}_2$ compared to $\hat{\theta}_1$. Here

$$E = \frac{V(\hat{\theta}_1)}{V(\hat{\theta}_2)} = \frac{\theta^2}{3n}\frac{(n+2)(n+1)^2}{n\theta^2} = \frac{(n+2)(n+1)^2}{3n^2} > 1.$$

Hence $\hat{\theta}_2$ is more efficient estimator than $\hat{\theta}_1$.

►**EXAMPLE 32.34** *Let $x_1, x_2, \ldots, x_n$ be a random sample of n observations from a population having p.d.f.*

$$f(x) = \begin{cases} \frac{2}{\theta^2}(\theta - x), & 0 \le x \le \theta \\ 0, & \textit{otherwise.} \end{cases}$$

Let $\hat{\theta}_1 = x$ and $\hat{\theta}_2 = 3x$. Compare the efficiency of $\hat{\theta}_1$ compared to $\hat{\theta}_2$. Also, find the MSE of $\hat{\theta}_1$.

Solution. We have

$$E(x) = \frac{2}{\theta^2}\int_0^\theta x(\theta - x)dx = \frac{2}{\theta^2}\left[\frac{\theta^3}{2} - \frac{\theta^3}{3}\right] = \frac{\theta}{3}.$$

Hence $E(\hat{\theta}_1) = E(x) \neq \theta$ and $\hat{\theta}_1$ is a biased estimator of θ. But $E(\hat{\theta}_2) = 3E(x) = \theta$. Hence $\hat{\theta}_2$ is an unbiased estimator of θ.

$$\begin{aligned} E(x^2) &= \frac{2}{\theta^2}\int_0^\theta x^2(\theta - x)dx = \frac{2}{\theta^2}\left[\frac{\theta^4}{3} - \frac{\theta^4}{4}\right] = \frac{\theta^2}{6} \\ V(x) &= E(x^2) - [E(x)]^2 = \frac{\theta^2}{6} - \frac{\theta^2}{9} = \frac{\theta^2}{18} \\ V(\hat{\theta}_1) &= V(x) = \frac{\theta^2}{18} \text{ and } V(\hat{\theta}_2) = 9V(x) = \frac{\theta^2}{2}. \end{aligned}$$

The efficiency of $\hat{\theta}_1$ compared to $\hat{\theta}_2$ is

$$E = \frac{V(\hat{\theta}_2)}{V(\hat{\theta}_1)} = \frac{\theta^2}{2} \times \frac{18}{\theta^2} = 9.$$

Here $\hat{\theta}_1$ is a biased estimator of θ. The amount of bias is

$$\begin{aligned} \text{bias } (\hat{\theta}_1) &= \theta - E(\hat{\theta}_1) = \theta - \frac{\theta}{3} = \frac{2\theta}{3} \\ \text{MSE } (\hat{\theta}_1) &= V(\hat{\theta}_1) + [\text{bias } (\hat{\theta}_1)]^2 = \frac{\theta^2}{18} + \frac{4\theta^2}{9} = \frac{\theta^2}{2}. \end{aligned}$$

Again

$$\text{MSE } (\hat{\theta}_2) = V(\hat{\theta}_2) = \frac{\theta^2}{2} \quad [\because \text{ bias } (\hat{\theta}_2) = 0]$$

Thus, $\hat{\theta}_1$ and $\hat{\theta}_2$ are equally precise. However, $\hat{\theta}_2$ is acceptable since it is unbiased.

►**EXAMPLE 32.35** *Find unbiased estimator of θ^2, where θ is a parameter of general Gamma distribution having p.d.f.*

$$f(x) = \begin{cases} \frac{1}{\theta^m \Gamma m} e^{-x/\theta} x^{m-1}, & 0 < x < \infty;\ \theta > 0 \\ 0, & \textit{otherwise.} \end{cases}$$

Also, find the variance of your estimator. *If $m = 1$, find unbiased estimator of θ. If $\hat{\theta}_1 = \min(x_1, x_2, \ldots, x_n)$ where $x_1, x_2, \ldots, x_n$ be a random sample of n observations from exponential population $(m = 1)$, compare $\hat{\theta}_1$ and an estimator of θ.*

Solution. We have

$$E(x) = \frac{1}{\theta^m \Gamma m}\int_0^\infty e^{-x/\theta}x^m dx = \frac{\Gamma(m+1)}{\theta^m \Gamma m}\theta^{m+1} = m\theta$$

$$E(x^2) = \frac{1}{\theta^m \Gamma m}\int_0^\infty e^{-x/\theta}x^{m+1} dx = \frac{\Gamma(m+2)}{\theta^m \Gamma m}\theta^{m+2} = m(m+1)\theta^2.$$

$\therefore\ E\left[\dfrac{x^2}{m(m+1)}\right] = \theta^2$. Hence $\dfrac{x^2}{m(m+1)}$ is an unbiased estimator of θ^2.

$$V\left[\frac{x^2}{m(m+1)}\right] = \frac{1}{m^2(m+1)^2}V(x^2)$$

$$V(x^2) = E(x^4) - [E(x^2)]^2$$

$$E(x^4) = \frac{1}{\theta^m \Gamma m}\int_0^\infty e^{-x/\theta}x^{m+3} dx$$

$$= \frac{\Gamma(m+4)}{\theta^m \Gamma m}\theta^{m+4}$$

$$= \theta^4 m(m+1)(m+2)(m+3).$$

$$\therefore\ V\left[\frac{x^2}{m(m+1)}\right]^2 = \frac{1}{m^2(m+1)^2}\left[\theta^4 m(m+1)(m+2)(m+3) - m^2(m+1)^2\theta^4\right]$$

$$= \theta^4\left[\frac{(m+2)(m+3)}{m(m+1)} - 1\right] = \frac{(4m+6)\theta^4}{m(m+1)}.$$

Again, if $m = 1$, $E(x) = \theta$. Let

$$\hat{\theta}_2 = \frac{1}{n}\Sigma x = \bar{x},\ E(\hat{\theta}_2) = \frac{1}{n}\Sigma E(x) = \theta.$$

Hence $\hat{\theta}_2 = \bar{x}$ is an unbiased estimator of θ. The variance of this estimator is

$$V(\hat{\theta}_2) = \frac{1}{n^2}\Sigma V(x) = \frac{\theta^2}{n}\ [\text{when } m = 1, V(x) = \theta^2].$$

Again, when $m = 1$

$$f(\hat{\theta}_1) = n\left[1 - F_x(\hat{\theta}_1)\right]^{n-1} f_x(\hat{\theta}_1),\ F(x) = 1 - e^{-x/\theta}.$$

$$\therefore\ f(\hat{\theta}_1) = \frac{n}{\theta}e^{-n\hat{\theta}_1/\theta},\ \hat{\theta}_1 > 0$$

$E(\hat{\theta}_1) = \theta/n \Rightarrow n\hat{\theta}_1$ is an unbiased estimator of θ. But $\hat{\theta}_1$ is a biased estimator of θ.

$$V(\hat{\theta}_1) = \theta^2/n^2$$

Hence $\hat{\theta}_1$ is a consistent estimator of θ. Here efficiency of $\hat{\theta}_1$ compared to $\hat{\theta}_2$ is

$$E = \frac{V(\hat{\theta}_2)}{V(\hat{\theta}_1)} = \frac{\theta^2}{n^2}\frac{n}{\theta^2} = \frac{1}{n} \le 1.$$

►**EXAMPLE 32.36** *Let $x_1, x_2, \ldots, x_n$ be a random sample of n observations from a population having p.d.f.*

$$f(x) = \begin{cases} 2x/\theta^2, & 0 \le x \le \theta \\ 0, & \textit{otherwise.} \end{cases}$$

Find an unbiased estimator of θ. Also, find the variance of your estimator.

Solution. We have

$$E(x) = \frac{2}{\theta^2}\int_0^\theta x^2 dx = \frac{2}{3}\theta.$$

Let $\hat{\theta}_1 = \frac{1}{n}\Sigma x$.

Then $E(\hat{\theta}_1) = \frac{1}{n}\Sigma E(x) = \frac{2}{3}\theta \Rightarrow \hat{\theta}_2 = \frac{3\bar{x}}{2}$ is an unbiased estimator of θ.

$$E(x^2) = \frac{2}{\theta^2}\int_0^\theta x^3 dx = \frac{\theta^2}{2}$$

$$V(x) = E(x^2) - [E(x)]^2 = \frac{\theta^2}{2} - \frac{4\theta^2}{9} = \frac{\theta^2}{18}$$

$$V(\hat{\theta}_2) = \frac{9}{4}V(\bar{x}) = \frac{9}{4}\frac{1}{n^2}\Sigma V(x) = \frac{9}{4n}\frac{\theta^2}{18} = \frac{\theta^2}{8n}.$$

As $n \to \infty$, $V(\hat{\theta}_2) \to 0$. Hence $\hat{\theta}_2$ is also a consistent estimator of θ.

Best Estimator

Let $x_1, x_2, \ldots, x_n$ be a random sample of size n selected from a population having p.d.f. $f(x, \theta)$. Let $T = t(x_1, x_2, \ldots, x_n)$ be an unbiased estimator of a parametric function $\phi(\theta)$. This estimator T is called *best estimator* of $\phi(\theta)$ if its variance is less than or equal to the variance of any other estimator of $\phi(\theta)$.

Best Asymptotically Normal (BAN) Estimator

Let $T_1, T_2, \ldots, T_n$ be the sequence of estimators of a parameter θ. This sequence will be termed as BAN estimator, if the following conditions are satisfied:

(i) If $n \to \infty$, $\sqrt{n}(t_n - \theta) \sim N(0, \sigma^2)$.

(ii) For all $\epsilon > 0$, $\lim_{n\to\infty} P[|t_n - \theta| > \epsilon] = 0$ for all θ.

(iii) Let $\{t_n\}$ be the sequence of another set of consistent estimator such that $\sqrt{n}(t_n - \theta) \sim N(0, \sigma_1^2)$.

(iv) In any open interval σ_1^2 will not be less than σ^2.

For example, let us consider the estimation of μ when $x_1, x_2, \ldots, x_n$ be a random sample of n observations from $N(\mu, \sigma^2)$. Here $\sqrt{n}(\bar{x} - \mu) \sim N(0, \sigma^2)$ if $n \to \infty$. Therefore, $\bar{x}$ is a BAN estimator of μ. However, there may be many BAN estimators of a parameter. For example, let us consider another estimator of μ for $N(\mu, \sigma^2)$. This estimator is

$$t_n = \frac{1}{n+1}\sum_{i=1}^{n} x_i$$

and it is BAN. Under condition (ii) of BAN estimator t_n is weak consistent estimator. Here

$$V(t_n) = V\left[\frac{n\bar{x}}{n+1}\right] = \left(\frac{n}{n+1}\right)^2 \frac{\sigma^2}{n} = \frac{\sigma^2}{n\left(1+\frac{1}{n}\right)^2}.$$

Here as $n \to \infty$, $V(t_n) \to 0$. Hence t_n is consistent estimator but it is not unbiased estimator of μ. Because

$$E(t_n) = \frac{1}{n+1}\sum E(x) = \frac{n\mu}{n+1} \neq \mu.$$

Mean Square Error Consistency

Let $x_1, x_2, \ldots, x_n$ be a random sample of n observations from a population having p.d.f. $f(x, \theta)$. Let

$$T_1 = t_1(x_1), T_2 = t_2(x_1, x_2), \ldots, T_n = t_n(x_1, x_2, \ldots, x_n)$$

be a sequence of n estimators of a parametric function $\phi(\theta)$. This sequence of estimators will be called mean square error consistent estimators, if and only if

$$\lim_{n\to\infty} E\left[T_n - \phi(\theta)\right]^2 = 0$$

where $\theta \in S$ and S is the parametric space. This property of estimators is known as *mean square error consistency*.

From this property of mean square error consistency, it is clear that, when n tends to infinity, the bias and variance of the estimator tend to zero. Because

$$E[T_n - \phi(\theta)]^2 = V(T_n) + [\phi(\theta) - E(T_n)]^2.$$

►**EXAMPLE 32.37** *Let $x_1, x_2, \ldots, x_n$ be a random sample of n observations from a population having p.d.f.*

$$f(x) = \begin{cases} 2(\theta - x)/\theta^2, & 0 \leq x \leq \theta \\ 0, & \textit{otherwise.} \end{cases}$$

Show that $T_n = 3\bar{x}$ is a mean square consistent estimator of θ.

Solution. We have

$$E(x) = \frac{\theta}{3} \quad \text{[Example 32.34]}$$

$$E(\bar{x}) = \frac{1}{n}\Sigma E(x) = \frac{\theta}{3} \Rightarrow T_n = 3\bar{x} \text{ is an unbiased estimator of } \theta.$$

Thus $[\theta - E(T_n)] = 0$. The variance of T_n is

$$\begin{aligned}\text{MSE }(T_n) &= E[T_n - \phi(\theta)]^2 = V(T_n) + [\phi(\theta) - E(T_n)]^2 \\ &= V(T_n)\ [\because\ \{\phi(\theta) - E(T_n)\} = 0].\end{aligned}$$

We have

$$V(x) = \frac{\theta^2}{18},\ V(T_n) = 9V(\bar{x}) = \frac{9\theta^2}{18n} = \frac{\theta^2}{2n}.$$

Thus, $\lim_{n\to\infty} E[T_n - \phi(\theta)]^2 \to 0$. Therefore, $T_n = 3\bar{x}$ is the mean square error consistent estimator of θ.

❑ **THEOREM 32.2** *If an estimator is mean square consistent for any parametric function, then it is simple consistent estimator also, but the converse is not true.*

Proof. Let $T_n = t_n(x_1, x_2, \dots, x_n)$ be the mean square consistent estimator of a parametric function $\phi(\theta)$ based on a set of random sample observations $x_1, x_2, \dots, x_n$ of size n from a population having p.d.f. $f(x, \theta)$. Using Chebyshev's inequality, we have

$$P[|T_n - \phi(\theta)| > k\sigma_T] \leq \frac{1}{k^2}$$

where

$$k = \frac{\epsilon}{\sigma_T} \text{ or, } P[|T_n - \phi(\theta)| > \epsilon] \leq \frac{\sigma_T^2}{\epsilon^2} = \frac{E[T_n - \phi(\theta)]^2}{\epsilon^2}.$$

But

$$\lim_{n\to\infty} E[T_n - \phi(\theta)]^2 \to 0.$$

Hence

$$\lim_{n\to\infty} P[|T_n - \phi(\theta)| > \epsilon] = 0 \text{ or, } \lim_{n\to\infty} P[|T_n - \phi(\theta)| < \epsilon] = 1.$$

Thus, T_n is also a simple consistent estimator of $\phi(\theta)$.

32.3 Minimum Variance Unbiased (MVU) Estimator

In the previous section, it is mentioned that the best estimator is one [depending on the criteria unbiasedness and efficiency] for which the variance of the estimator is minimum. The estimator with minimum variance is also called *MVU estimator.* However, the MVU estimator is not always available. The variance of some estimators cannot be less than a certain bound. In this section, we shall discuss the minimum variance bound (MVB) estimator.

Rao-Cramer Inequality

Under some fairly general conditions the variance of an unbiased estimator cannot be less than a definite value. Such an estimator is called *minimum variance bound (MVB) estimator.* The bound of this variance is known as *Rao-Cramer inequality.*

Let $x_1, x_2, \ldots, x_n$ be a random sample of n observations selected from a population having p.d.f. $f(x, \theta)$, where $\theta \in S$ and S is the parametric space. Let

$$T = t(x_1, x_2, \ldots, x_n)$$

be the unbiased estimator of parametric function $\phi(\theta)$. The estimator T will be the MVB estimator if $f(x, \theta)$ satisfies the following conditions:

(i) The range of x is independent of parameter. For all values of x and θ,

$$\frac{\partial}{\partial \theta} \log f(x, \theta) \text{ will exist.}$$

(ii) $$\frac{\partial}{\partial \theta} \iint \cdots \int \prod_{i=1}^{n} f(x_i, \theta) dx_1 dx_2 \cdots dx_n = \iint \cdots \int \frac{\partial}{\partial \theta} \prod_{i=1}^{n} f(x_i, \theta) dx_1 dx_2 \cdots dx_n.$$

(iii) $$\frac{\partial}{\partial \theta} \iint \cdots \int t(x_1, x_2, \ldots, x_n) \prod_{i=1}^{n} f(x_i, \theta) dx_1 dx_2 \cdots dx_n = \iint \cdots \int t(x_1, x_2, \ldots, x_n) \frac{\partial}{\partial \theta} \prod_{i=1}^{n} f(x_i, \theta) dx_1 dx_2 \cdots dx_n.$$

(iv) $0 < E\left[\left\{\frac{\partial}{\partial \theta} \log f(x, \theta)\right\}^2\right] < \infty$. It implies that, for all values of θ

$$E\left[\frac{\partial}{\partial \theta} \log f(x, \theta)\right]^2 \text{ exists.}$$

The likelihood function for sample observations is

$$L(x_1, x_2, \ldots, x_n) = f(x_1, \theta) f(x_2, \theta) \cdots f(x_n, \theta).$$

Here L is the joint probability density function. We have

$$\iint \cdots \int L(x_1, x_2, \ldots, x_n; \theta) dx_1 dx_2 \cdots dx_n = 1.$$

Differentiating both sides with respect to θ, we get

$$\iint \cdots \int \frac{\partial L}{\partial \theta}(x_1, x_2, \ldots, x_n; \theta) dx_1 dx_2 \cdots dx_n = 0.$$

We have

$$\frac{\partial \log L}{\partial \theta} = \frac{1}{L} \frac{\partial L}{\partial \theta}.$$

Hence

$$E\left[\frac{\partial \log L}{\partial \theta}\right] = E\left[\frac{1}{L} \frac{\partial L}{\partial \theta}\right] = \iint \cdots \int \frac{1}{L} \frac{\partial L}{\partial \theta} L(x_1, x_2, \ldots, x_n; \theta) dx_1 dx_2 \cdots dx_n$$
$$= \iint \cdots \int \frac{\partial L}{\partial \theta} dx_1 dx_2 \cdots dx_n = 0.$$

Again

$$E\left[\frac{\partial \log L}{\partial \theta}\right] = \iint\cdots\int \frac{1}{L}\frac{\partial L}{\partial \theta} L dx_1 dx_2 \cdots dx_n = 0.$$

Differentiating both sides w.r.t. θ, we get

$$\iint\cdots\int \left[\left(\frac{1}{L}\frac{\partial L}{\partial \theta}\right)\frac{\partial L}{\partial \theta} + L\frac{\partial}{\partial \theta}\left(\frac{1}{L}\frac{\partial L}{\partial \theta}\right)\right] dx_1 dx_2 \cdots dx_n = 0$$

$$\text{or, } \iint\cdots\int \left[\left(\frac{1}{L}\frac{\partial L}{\partial \theta}\right)^2 + \frac{\partial^2 \log L}{\partial \theta^2}\right] L dx_1 dx_2 \cdots dx_n = 0$$

$$\text{or, } \iint\cdots\int \left(\frac{1}{L}\frac{\partial L}{\partial \theta}\right)^2 L dx_1 dx_2 \cdots dx_n$$

$$= -\iint\cdots\int \frac{\partial^2 \log L}{\partial \theta^2} L dx_1 dx_2 \cdots dx_n$$

$$\text{or, } E\left(\frac{1}{L}\frac{\partial L}{\partial \theta}\right)^2 = -E\left(\frac{\partial^2 \log L}{\partial \theta^2}\right)$$

$$\text{or, } E\left(\frac{\partial \log L}{\partial \theta}\right)^2 = -E\left(\frac{\partial^2 \log L}{\partial \theta^2}\right).$$

But

$$V\left(\frac{\partial \log L}{\partial \theta}\right) = E\left(\frac{\partial \log L}{\partial \theta}\right)^2 - \left[E\left(\frac{\partial \log L}{\partial \theta}\right)\right]^2$$

$$= -E\left(\frac{\partial^2 \log L}{\partial \theta^2}\right) \quad \left[\because\ E\left(\frac{\partial \log L}{\partial \theta}\right) = 0\right].$$

It has already been assumed that $T = t(x_1 x_2 \cdots x_n)$ be the unbiased estimator of a parametric function $\phi(\theta)$. Then

$$E(T) = \iint\cdots\int T L dx_1 dx_2 \cdots dx_n = \phi(\theta).$$

Differentiating both sides w.r.t θ, we get

$$\iint\cdots\int T\left(\frac{1}{L}\frac{\partial L}{\partial \theta}\right) L dx_1 dx_2 \cdots dx_n = \phi'(\theta) \qquad (32.1)$$

$$\text{or, } \iint\cdots\int T\left(\frac{\partial \log L}{\partial \theta}\right) L dx_1 dx_2 \cdots dx_n = \phi'(\theta)$$

$$\text{or, } \operatorname{cov}\left(T, \frac{\partial \log L}{\partial \theta}\right) = \operatorname{cov}\left(T, \frac{1}{L}\frac{\partial L}{\partial \theta}\right) = \phi'(\theta).$$

We have already observed that

$$\iint\cdots\int \frac{1}{L}\frac{\partial L}{\partial \theta} L\, dx_1 dx_2 \cdots dx_n = 0$$

$$\text{or, } \iint\cdots\int \phi(\theta)\frac{1}{L}\frac{\partial L}{\partial \theta} L dx_1 dx_2 \cdots dx_n = 0. \qquad (32.2)$$

Now, (32.1) – (32.2) gives

$$\iint \cdots \int [T - \phi(\theta)] \frac{1}{L} \frac{\partial L}{\partial \theta} L dx_1 dx_2 \cdots dx_n = \phi'(\theta)$$

$$\text{or, } [\phi'(\theta)]^2 = \left[\iint \cdots \int \{T - \phi(\theta)\} \frac{1}{L} \frac{\partial L}{\partial \theta} L dx_1 dx_2 \cdots dx_n \right]^2$$

Using Cauchy-Schwartz inequality, we can write

$$[\phi'(\theta)]^2 \le \iint \cdots \int [T - \phi(\theta)]^2 L dx_1 dx_2 \cdots dx_n \int \cdots \int \left(\frac{\partial \log L}{\partial \theta} \right)^2 L dx_1 \cdots dx_n$$

$$\text{or, } [\phi'(\theta)]^2 \le V(T) E \left(\frac{\partial \log L}{\partial \theta} \right)^2 \quad \text{or, } V(T) \ge \frac{[\phi'(\theta)]^2}{E \left(\frac{\partial \log L}{\partial \theta} \right)^2}$$

$$\text{or, } V(T) \ge - \frac{(\phi'(\theta))^2}{E \left(\frac{\partial^2 \log L}{\partial \theta^2} \right)}.$$

Here

$$\frac{[\phi'(\theta)]^2}{E \left(\frac{\partial^2 \log L}{\partial \theta^2} \right)}$$

is minimum bound of variance of T. This bound is known as *Cramer-Rao bound* for variance of an unbiased estimator.

Using the probability density function the bound can be written as

$$V(T) \ge \frac{1}{nE \left[\frac{\partial \log f(x, \theta)}{\partial \theta} \right]^2}.$$

If the variance of T is not less than this bound, the estimator is called MVB estimator. For $\phi(\theta) = \theta$, the bound is written as

$$V(T) \ge \frac{1}{E \frac{\partial^2 \log L}{\partial \theta^2}}, \quad \because \ \phi'(\theta) = 1.$$

Also, we can write

$$E[T - \phi(\theta)]^2 \ge \frac{[\phi'(\theta)]^2}{E \left(\frac{\partial \log L}{\partial \theta} \right)^2}.$$

The inequality transforms to equality if $[T - \phi(\theta)]$ is proportional to $\frac{\partial \log L}{\partial \theta}$, i.e.,

$$\frac{\partial \log L}{\partial \theta} \propto [T - \phi(\theta)] \text{ or, } \frac{\partial \log L}{\partial \theta} = A(\theta)[T - \phi(\theta)]$$

where $A(\theta)$ is the function of θ.

Since the sample observations are independent, multiplying both sides by $[T-\phi(\theta)]$, we get

$$E\left(\frac{\partial \log L}{\partial \theta}\right)[T-\phi(\theta)] = A(\theta)E[T-\phi(\theta)]^2$$

$$\text{or, } \phi'(\theta) = A(\theta)V(T) \text{ or, } V(T) = \frac{\phi'(\theta)}{A(\theta)}.$$

Hence T is MVB estimator and its variance is $\dfrac{\phi'(\theta)}{A(\theta)}$.

►**EXAMPLE 32.38** *Let $x_1, x_2, \ldots, x_n$ be a random sample of n observations from $N(\mu, \sigma^2)$. Show that the estimators of μ and σ^2 are MVB estimators.*

Solution. We have

$$L = \left(\frac{1}{2\pi\sigma^2}\right)^{n/2} e^{-\frac{1}{2}\Sigma\left(\frac{x-\mu}{\sigma}\right)^2}$$

$$\log L = -\frac{n}{2}\log 2\pi - \frac{n}{2}\log\sigma^2 - \frac{1}{2\sigma^2}\Sigma(x_i-\mu)^2$$

$$\frac{\partial \log L}{\partial \mu} = \frac{1}{\sigma^2}\Sigma(x_i-\mu) = \frac{n}{\sigma^2}(\bar{x}-\mu) = A(\theta)[T-\phi(\theta)]$$

where

$$A(\theta) = \frac{n}{\sigma^2} \text{ and } T = \bar{x},\ \phi(\theta) = \mu.$$

Thus $\bar{x}$ is MVB estimator of μ and its variance is σ^2/n. Again,

$$\frac{\partial \log L}{\partial \sigma^2} = -\frac{n}{2\sigma^2} + \frac{\Sigma(x_i-\mu)^2}{2\sigma^4} = \frac{n}{2\sigma^4}[s^2-\sigma^2]$$

where

$$s^2 = \frac{1}{n}\Sigma(x_i-\mu)^2.$$

Hence s^2 is MVB estimator of σ^2. The variance of s^2 is $\dfrac{2\sigma^4}{n}$.

►**EXAMPLE 32.39** *Let $x_1, x_2, \ldots, x_n$ be a random sample of n observations from a population whose p.d.f. is*

$$f(x) = \frac{1}{\theta^m \Gamma m} e^{-x/\theta} x^{m-1};\ 0 < x < \infty,\ \theta > 0.$$

Assuming m known, find MVB estimator of θ. Also, find MVB estimator of θ when $m = 1$ and $m = 2$.

Solution. The likelihood function of the sample observations is

$$L = \left(\frac{1}{\theta^m \Gamma m}\right)^n e^{-\frac{n\bar{x}}{\theta}} \prod_{i=1}^{n} x_i^{m-1}$$

$$\log L = -n\log\Gamma m - nm\log\theta - \frac{n\bar{x}}{\theta} + (m-1)\sum^{n}\log x_i$$

$$\frac{\partial \log L}{\partial \theta} = -\frac{nm}{\theta} + \frac{n\bar{x}}{\theta^2} = \frac{mn}{\theta^2}\left[\frac{\bar{x}}{m} - \theta\right].$$

Hence $\bar{x}/m$ is MVB estimator of θ. The variance of this estimator is θ^2/nm.

Assuming $m = 1$, we have

$$L = \frac{1}{\theta^n} e^{-\frac{n\bar{x}}{\theta}} \text{ or, } \log L = -n \log \theta - \frac{n\bar{x}}{\theta}$$

$$\frac{\partial \log L}{\partial \theta} = -\frac{n}{\theta} + \frac{n\bar{x}}{\theta^2} = \frac{n}{\theta^2}[\bar{x} - \theta].$$

Therefore, $\bar{x}$ is MVB estimator of θ and $V(\bar{x}) = \theta^2/n$.

In a similar way, it can be shown that $\bar{x}/2$ is the MVB estimator of θ and its variance is $\theta^2/2n$.

►**EXAMPLE 32.40** *Let $x_1, x_2, \ldots, x_n$ be a random sample of n observations from Poisson population with parameter λ. Find MVB estimator of λ.*

Solution. The likelihood function is

$$L = \frac{e^{-n\lambda} \lambda^{\Sigma x_i}}{\prod^n x_i}$$

$$\log L = -n\lambda + n\bar{x} \log \lambda - \sum^n \log x_i$$

$$\frac{\partial \log L}{\partial \lambda} = -n + \frac{n\bar{x}}{\lambda} = \frac{n}{\lambda}[\bar{x} - \lambda].$$

Hence $\bar{x}$ is MVB estimator of λ and $V(\bar{x}) = \lambda/n$.

►**EXAMPLE 32.41** *Let $x_1, x_2, \ldots, x_n$ be a random sample of n observations from Bernoulli population having parameter p. Find MVB estimator of p. Also, find unbiased estimator p and its efficiency.*

Solution. Given $P(x) = p^x(1-p)^{1-x}$, $x = 0, 1$.

The likelihood function of sample observations is

$$L = \binom{n}{r} p^r (1-p)^{n-r}; \ r = 0, 1, 2, \ldots, n, \ r = \Sigma x_i$$

$$\log L = \log \binom{n}{r} + r \log p + (n-r) \log(1-p)$$

$$\frac{\partial \log L}{\partial p} = \frac{r}{p} - \frac{n-r}{1-p} = \frac{r}{p} + \frac{r}{1-p} - \frac{n}{1-p}$$

$$= \frac{1}{p(1-p)}[r(1-p) + rp - np] = \frac{n}{p(1-p)}\left[\frac{r}{n} - p\right].$$

Hence r/n is MVB estimator of p, where

$$V\left(\frac{r}{n}\right) = \frac{p(1-p)}{n}.$$

Again, let us consider that T is an unbiased estimator of p, where

$$V(T) \geq \frac{1}{nE\left[\frac{\partial \log P(x)}{\partial p}\right]} = \frac{1}{-nE\left[\frac{\partial^2 \log P(x)}{\partial p}\right]}.$$

Here

$$\begin{aligned}
\log P(x) &= x \log p + (1-x)\log(1-p) \\
\frac{\partial \log P(x)}{\partial p} &= \frac{x}{p} - \frac{1-x}{1-p} \\
\frac{\partial^2 \log P(x)}{\partial p^2} &= -\frac{x}{p^2} - \frac{(1-x)}{(1-p)^2} \\
-E\left[\frac{\partial^2 \log P(x)}{\partial p^2}\right] &= \frac{E(x)}{p^2} + \frac{E(1-x)}{(1-p)^2} = \frac{1}{p} + \frac{1}{1-p} = \frac{1}{p(1-p)}.
\end{aligned}$$

Now, if T is an unbiased estimator of p, then

$$V(T) = \frac{p(1-p)}{n}.$$

Hence the efficiency of r/n compared to T is

$$E = \frac{V(T)}{V(r/n)} = \frac{p(1-p)}{n} \cdot \frac{n}{p(1-p)} = 1.$$

Hence r/n is the most efficient estimator of p.

►**EXAMPLE 32.42** *Let $x_1, x_2, \ldots, x_n$ be a random sample of n observations from $N(\mu, \theta)$. If $ns^2/(n-1)$ is an estimator of θ, find its efficiency.*

Solution. Let $T = \dfrac{ns^2}{n-1}$ be the estimator of θ, where

$$V(T) \geq \frac{1}{nE\left[\frac{\partial \log f(x)}{\partial \theta}\right]^2} = \frac{1}{-nE\left[\frac{\partial^2 \log f(x)}{\partial \theta^2}\right]}.$$

We have

$$\begin{aligned}
f(x) &= \frac{1}{\sqrt{2\pi\theta}} e^{-\frac{1}{2\theta}(x-\mu)^2}, \quad -\infty < x < \infty \\
\log f(x) &= -\frac{1}{2}\log\theta - \frac{1}{2}\log 2\pi - \frac{1}{2\theta}(x-\mu)^2 \\
\frac{\partial \log f(x)}{\partial \theta} &= -\frac{1}{2\theta} + \frac{(x-\mu)^2}{2\theta^2} \\
\frac{\partial^2 \log f(x)}{\partial \theta^2} &= \frac{1}{2\theta^2} - \frac{(x-\mu)^2}{\theta^3} \\
-E\left[\frac{\partial^2 \log f(x)}{\partial \theta^2}\right] &= \frac{E(x-\mu)^2}{\theta^3} - \frac{1}{2\theta^2} = \frac{1}{\theta^2} - \frac{1}{2\theta^2} = \frac{1}{2\theta^2}.
\end{aligned}$$

Hence $V(T) = \dfrac{2\theta^2}{n}$. Now, let

$$s^2 = \frac{1}{n}\sum^{n}(x-\bar{x})^2 \Rightarrow ns^2 = \Sigma(x-\mu)^2 - n(\bar{x}-\mu)^2.$$

$$\therefore\ nE(s^2) = \sum^{n} E(x-\mu)^2 - nE(\bar{x}-\mu)^2 = n\theta - n\frac{\theta}{n} = (n-1)\theta.$$

Hence $\dfrac{ns^2}{n-1}$ is an unbiased estimator of θ. Again, $\dfrac{ns^2}{\theta} \sim \chi^2_{n-1}$. Then

$$V\left[\frac{ns^2}{\theta}\right] = 2(n-1) \text{ and } V\left[\frac{ns^2}{(n-1)}\right] = \frac{2(n-1)\theta^2}{(n-1)^2} = \frac{2\theta^2}{n-1}.$$

Therefore, the efficiency of $ns^2/(n-1)$ compared to T is

$$E = \frac{V(T)}{V\left[\frac{ns^2}{n-1}\right]} = \frac{2\theta^2}{n} \cdot \frac{n-1}{2\theta^2} = \frac{n-1}{n} < 1.$$

►**EXAMPLE 32.43** *Let $x_1, x_2, \ldots, x_n$ be a random sample of n observations from a population having p.d.f.*

$$f(x) = \begin{cases} \theta x^{\theta-1}, & 0 < x < 1 \\ 0, & \textit{otherwise.} \end{cases}$$

Find unbiased estimator of θ and efficiency of your estimator.

Solution. We have $\log f(x) = \log\theta + (\theta-1)\log x$

$$\frac{\partial \log f(x)}{\partial\theta} = \frac{1}{\theta} + \log x, \frac{\partial^2 \log f(x)}{\partial\theta^2} = -\frac{1}{\theta^2}, -nE\left[\frac{\partial^2 \log f(x)}{\partial\theta^2}\right] = \frac{n}{\theta^2}.$$

Hence, if T is the MVB estimator of θ, the minimum variance bound of T is θ^2/n.

Let $y_i = -\ln x_i$, $i = 1, 2, \ldots, n$.

$$\therefore\ x_i = e^{-y_i},\ 0 < x_i < 1 \text{ and } 0 < y_i < \infty.$$

The Jacobian of the transformation is $|J| = e^{-y_i}$. Therefore,

$$g(y_i) = \theta e^{-y_i\theta},\ 0 < y_i < \infty.$$

Again, let $z = -\sum_{i=1}^{n} \ln x_i$. Then $f(z) = \dfrac{\theta^n}{\Gamma n} z^{n-1} e^{-z\theta}$, $0 < z < \infty$.

$$\therefore\ E(z) = n/\theta, E(1/z) = \frac{\theta}{n-1}.$$

Hence $\dfrac{n-1}{z}$ is an unbiased estimator of θ, where $n > 2$. Again

$$V\left(\frac{n-1}{z}\right) = \frac{\theta^2}{n-2}.$$

Thus, the efficiency of $\dfrac{n-1}{z}$ compared to T is

$$E = \frac{V(T)}{V\left(\frac{n-1}{z}\right)} = \frac{\theta^2}{n}\frac{n-2}{\theta^2} = \frac{n-2}{n};\ n > 2.$$

It is observed that $E \to 1$ if n increases. Hence $\dfrac{n-1}{z}$ is asymptotically efficient estimator of θ.

►**EXAMPLE 32.44** *Let $x_1, x_2, \ldots, x_n$ be a random sample of n observations from a population having p.d.f.*

$$f(x) = \begin{cases} 1/\theta, & 0 \le x \le \theta \\ 0, & \text{otherwise.} \end{cases}$$

Show that $\left(\frac{n+1}{n}\right) x_{(n)}$ is the most efficient estimator of θ, where

$$x_{(n)} = \max(x_1, x_2, \ldots, x_n).$$

Solution. We have

$$\log f(x) = -\log\theta, \quad \frac{\partial \log f(x)}{\partial \theta} = -\frac{1}{\theta},$$
$$\frac{\partial^2 \log f(x)}{\partial \theta^2} = \frac{1}{\theta^2}, \quad E\left[\frac{\partial \log f(x)}{\partial \theta}\right]^2 = \frac{1}{\theta^2}.$$

Hence MVB is θ^2/n.

We have

$$f[x_{(n)}] = \frac{n}{\theta^n} x_{(n)}^{n-1}, \quad 0 < x_{(n)} < \theta$$
$$E[x_{(n)}] = \frac{n}{n+1}\theta.$$

Hence $\left[\frac{n+1}{n} x_{(n)}\right]$ is an unbiased estimator of θ. The variance of this estimator is

$$V\left[\frac{n+1}{n} x_{(n)}\right] = \left(\frac{n+1}{n}\right)^2 V\left[x_{(n)}\right]$$
$$= \left(\frac{n+1}{n}\right)^2 \frac{n\theta^2}{(n+2)(n+1)^2} = \frac{\theta^2}{n(n+2)}.$$

Therefore, the efficiency of $\left[\frac{n+1}{n} x_{(n)}\right]$ compared to the MVB estimator T is

$$E = \frac{V(T)}{V\left[\frac{n+1}{n} x_{(n)}\right]} = \frac{\theta^2}{n} \frac{n(n+2)}{\theta^2} = n + 2 > 1.$$

Hence $\left[\frac{n+1}{n} x_{(n)}\right]$ is the most efficient estimator of θ. Here the value of T is not found out by Cramer-Rao inequality since the regularity conditions are not satisfied.

Uniformly Minimum Variance Unbiased (UMVU) Estimator

Let a probability distribution be $f(x, \theta)$, where θ contains in the parametric space $S[\theta \in S]$. Consider that the set of unbiased estimators T of θ is U and $V(T)$ exists. The estimator $T_0 \in U$ is uniformly minimum variance unbiased estimator of θ, if

$$E[T_0 - \theta]^2 \le E[T - \theta]^2.$$

In Example 32.42, it is observed that the estimator T of θ has attained minimum variance bound. But $\left[\frac{n+1}{n} x_{(n)}\right]$ is an unbiased estimator with lower variance than the variance of T. Hence $\left[\frac{n+1}{n} x_{(n)}\right]$ is UMVU estimator.

Locally Minimum Variance Unbiased (LMVU) Estimator

Let S be the parametric space of the parameters of a distribution and θ_0 be the parameter in the parametric space $[\theta_0 \in S]$. Consider that the set T of unbiased estimators of θ_0 is $U(\theta_0)$ and $V(T)$ exists. Then $T_0 \in U(\theta_0)$ is called *LMVU etimator*, if for all $T \in U(\theta_0)$

$$E[T_0 - \theta_0]^2 \leq E[T - \theta_0]^2.$$

❒ **THEOREM 32.3** *Let U be the set of estimators T of parameter $\theta \in S$, where S is the parametric space. Consider that $V(T)$ exists and U is non-empty. Also, consider that U_0 is the set of all unbiased estimators η of θ. Then $T_0 \in U$ is UMVU estimator, if and only if*

$$E[\eta T_0] = 0 \text{ for all } \theta \text{ and for all } \eta \in U_0.$$

Proof. Let $T_0 \in U$ is the UMVUE and $E[\eta T_0] \neq 0$ for some θ_0 and for some $\eta_0 \in U_0$. Then for all real values λ, $T_0 + \lambda\eta_0 \in U$. Now, if for $P[\eta_0 = 0] = 1$, $E[\eta_0^2] = 0$, then it must be $E[\eta_0 T_0] = 0$. Consider that $E[\eta_0^2] > 0$ and also consider that

$$\lambda_0 = -E\left[T_0\eta_0\right]/E\left[\eta_0^2\right].$$

Then

$$E[T_0 + \lambda_0\eta_0]^2 = E[T_0^2] - \frac{E^2[\eta_0 T_0]}{E[\eta_0^2]} < E[T_0^2] \tag{a}$$

Now, since $T_0 + \lambda_0\eta_0 \in U$ and $T_0 \in U$, we get from (a) that

$$V\left[T_0 + \eta_0\lambda_0\right] < V(T_0) \tag{b}$$

This is impossible. Hence $E[\eta T_0] = 0$.

Again, let $E[\eta T_0] = 0$ is true for some $T_0 \in U$, where all $\theta \in S$ and $\eta \in U_0$. Let $T \in U_1$, then $T_0 - T \in U_0$ and for every θ, $E[T_0(T_0 - T)] = 0$. Now, by Cauchy-Schwartz inequality, we can write

$$E[T_0^2] = E[TT_0] \leq \{E[T_0^2]\}^{1/2}[E(T^2)]^{1/2}.$$

Now, $E[T_0^2] = 0$; then $P[T_0 = 0] = 1$. Hence $E[\eta T_0] = 0$ is true, otherwise

$$E[T_0^2]^{1/2} \leq [E(T^2)]^{1/2} \text{ or, } V(T_0) \leq V(T).$$

❒ **THEOREM 32.4** *Let the set of unbiased estimators T of $\theta \in S$ be U. Assume that $V(T)$ exists. In such a case there may be at best one UMVUE of θ.*

Proof. If T and T_0 are both UMVUE, then $E[T_0(T - T_0)] = 0$, where T and $T_0 \in U$ and $T - T_0 \in U$. T and T_0 are unbiased estimators of $\theta \in S$. Thus

$$E[T_0^2] = E[TT_0] \text{ or, } \operatorname{cov}[T, T_0] = V(T_0).$$

Since both T_0 and T are considered UMVUE, $V(T) = V(T_0)$. This implies that correlation coefficient of T and T_0 is 1, and hence for $\theta \in S$ and for some a and b

$$P[aT + bT_0 = 0] = 1.$$

Again, since T and T_0 are both unbiased estimators of θ, $P[T = T_0] = 1$ for all θ. This implies that there is only one UMVUE.

❑ **THEOREM 32.5** *Let $\{T_n\}$ be a sequence of UMVUE and T be the estimator of θ such that $V(T)$ exists. Also, consider that for all $\theta \in S$ if $n \to \infty$, then $E[T_n - T]^2 \to 0$. In such a case the estimator T will be UMVUE of θ.*

Proof. Let the set of unbiased estimators of η be U_0 and the estimators are unbiased for θ. Hence for all θ, for all $\eta \in U_0$ and for every $n = 1, 2, \ldots$

$$E[\eta T_0] = 0.$$

Here T is unbiased estimator, since

$$\begin{aligned} |E(T) - \theta| &\le E|T - T_n| \le E^{1/2}[T_n - T]^2. \\ \therefore\ E[\eta T] &= E[\eta T] - E[\eta T_0] = E[\eta(T - T_0)] \\ \text{and}\ \ |E(\eta T)| &\le [E(\eta^2)]^{1/2}[E(T - T_n)^2]^{1/2} \to 0, \text{ when } n \to 0. \end{aligned}$$

This is true for all θ and for all $\eta \in U$. Therefore,

$$E[\eta T] = 0.$$

This is also true for all $\eta \in U_0$ and for all $\theta \in S$. Hence T must be UMVUE.

Definition. If T_1 is the MVU estimator of θ and T_2 is another unbiased estimator of θ, then the correlation coefficient ρ of T_1 and T_2 is $\rho = \sqrt{E}$, where E is the efficiency of T_2 and $E = V(T_1)/V(T_2)$.

❑ **THEOREM 32.6** *If T_1 and T_2 are the MVU estimators of θ, then $T_1 = T_2$ [MVU estimator is unique].*

Proof. Since T_1 and T_2 are unbiased estimators of θ, we can write $E(T_1) = E(T_2) = \theta$. Also, $V(T_1) = V(T_2)$ [$\because$ T_1 and T_2 are MVU estimators]. Let a new estimator be $T_3 = \frac{1}{2}(T_1 + T_2)$, where $E(T_3) = \frac{1}{2}[E(T_1) + E(T_2)] = \theta$. This implies that T_3 is also unbiased estimator of θ. Now

$$\begin{aligned} V(T_3) &= \frac{1}{4}[V(T_1) + V(T_2) + 2\operatorname{cov}(T_1 T_2)] \\ &= \frac{1}{2}[V(T_1) + \operatorname{cov}(T_1, T_2)] = \frac{1}{2}[V(T_1) + \rho_{T_1T_2}V(T_1)] \\ &= \frac{1}{2}[V(T_1)(1 + \rho_{T_1T_2})], \quad \because\ V(T_1) = V(T_2). \end{aligned}$$

Now, since T_1 is the MVU of θ, $V(T_3) \ge V(T_1)$

$$\Rightarrow \frac{1}{2}[V(T_1)(1 + \rho_{T_1T_2})] \ge V(T_1) \text{ or, } \rho_{T_1T_2} \ge 1.$$

But $|\rho_{T_1T_2}| \le 1$ and hence we can assume that $\rho_{T_1T_2} = 1$. But, when $\rho_{T_1T_2} = 1$, there may be a linear relationship between T_1 and T_2. Let this relation be

$$T_1 = \alpha + \beta T_2$$

where α and β are constants and independent of sample observations. Therefore,

$$\begin{aligned} E(T_1) &= \alpha + \beta E(T_2) \Rightarrow \theta = \alpha + \beta\theta \\ \text{and } V(T_1) &= V(\alpha) + \beta^2 V(T_2) = \beta^2 V(T_2). \end{aligned}$$

As $V(T_1) = V(T_2)$, $V(T_1) = \beta^2 V(T_1) \Rightarrow \beta^2 = 1$ or, $\beta = \pm 1$. Since $\rho_{T_1T_2} = 1$, β must be $+1$. But, if $\beta = +1$, $\alpha = 0$. Therefore, $T_1 = T_2$.

❒**THEOREM 32.7** *Let T_1 be the MVU estimator of θ and T_2 be another unbiased estimator of θ. Then the linear combination of T_1 and T_2 will not be MVU estimator of θ.*

Proof. Consider $T_3 = l_1T_1 + l_2T_2$ is the linear combination of T_1 and T_2. This combination becomes unbiased estimator of θ if

$$E(T_3) = l_1E(T_1) + l_2E(T_2) = \theta(l_1 + l_2) = \theta.$$

This is possible if $l_1 + l_2 = 1$. The correlation coefficient of T_1 and T_2 can be written as

$$\rho_{T_1T_2} = \sqrt{E}, \text{ where } E = \frac{V(T_1)}{V(T_2)}.$$

Then

$$\begin{aligned} V(T_3) &= l_1^2 V(T_1) + l_2^2 V(T_2) + 2l_1l_2 \operatorname{cov}(T_1T_2) \\ &= l_1^2 V(T_1) + l_2^2 \frac{V(T_1)}{E} + 2l_1l_2\rho\frac{V(T_1)}{\sqrt{E}}, \quad \because V(T_2) = \frac{V(T_1)}{E} \\ &= V(T_1)\left[l_1^2 + l_2^2 + 2l_1l_2\rho/\sqrt{E}\right]. \end{aligned}$$

$$\begin{aligned} \therefore\ & V(T_3) \geq V(T_1)\left[l_1^2 + l_2^2 + 2l_1l_2\right], \quad \because \rho = \sqrt{E}, \rho^2 = E, \rho^2 \leq 1 \text{ and } \frac{1}{E} \geq 1 \\ & V(T_3) \geq V(T_1), \quad \because l_1 + l_2 = 1. \end{aligned}$$

Hence T_3 cannot be MVU estimator of θ.

►**EXAMPLE 32.45** *Show that the MVUE does not exist for parameter μ of Cauchy population, where*

$$f(x) = \frac{1}{\pi}\frac{1}{1+(x-\mu)^2}, \quad -\infty < x < \infty.$$

Solution. The likelihood function is

$$\begin{aligned} L &= \frac{1}{\pi^n}\prod_{i=1}^{n}\frac{1}{1+(x_i-\mu)^2} \\ \log L &= n\log\pi - \sum_{i=1}^{n}\log\left[1+(x_i-\mu)^2\right] \\ \frac{\partial \log L}{\partial\mu} &= 2\sum_{i=1}^{n}\frac{(x_i-\mu)}{1+(x_i-\mu)^2}. \end{aligned}$$

But $\frac{\partial \log L}{\partial \mu}$ cannot be expressed in the form of $A(\theta)[T - \phi/\theta)]$. Hence MVU estimator does not exist for μ.

►**EXAMPLE 32.46** *Let $x_1, x_2, \ldots, x_n$ be a random sample of n observations from Poisson population with parameter λ. Show that the sample mean $\bar{x}$ and the sample variance s^2 are unbiased estimators of λ but s^2 is not UMVUE.*

Solution. We have

$$P(x) = \frac{e^{-\lambda}\lambda^x}{x!}, \quad x = 0, 1, 2, \ldots$$

We know

$$E(x) = \lambda \text{ and } E(x^2) = \lambda^2 + \lambda, \; V(x) = \lambda$$
$$E(\bar{x}) = \frac{1}{n}\Sigma E(x) = \lambda.$$

This $\bar{x}$ is MVB estimator of λ [Example 32.40] and its variance is $V(\bar{x}) = \lambda/n$. Again,

$$\begin{aligned} s^2 &= \frac{1}{n-1}\Sigma(x_i - \bar{x})^2 = \frac{1}{n-1}\Sigma[(x_i - \lambda) - (\bar{x} - \lambda)]^2 \\ &= \frac{1}{n-1}\left[\Sigma(x_i - \lambda)^2 - n(\bar{x} - \lambda)^2\right] \\ E(s^2) &= \frac{1}{n-1}\left[\Sigma E(x_i - \lambda)^2 - nE(\bar{x} - \lambda)^2\right] \\ &= \frac{1}{n-1}\left[n\lambda - n\frac{\lambda}{n}\right] = \lambda. \end{aligned}$$

Hence s^2 is also an unbiased estimator of λ. Since s^2 is distributed as $\chi^2\lambda$ with $(n-1)$ d.f., we have

$$\frac{(n-1)s^2}{\lambda} \sim \chi^2_{n-1} \Rightarrow s^2 \sim \frac{\lambda}{n-1}\chi^2_{n-1}.$$

$$\therefore \; V(s^2) = \frac{\lambda^2}{(n-1)^2}2(n-1) = \frac{2\lambda^2}{n-1}.$$

Hence s^2 is not an UMVUE of λ.

►**EXAMPLE 32.47** *Let $x_1, x_2, \ldots, x_n$ be a random sample of n observations from Cauchy population with parameter μ. Show that the Cramer-Rao lower bound of variance of an unbiased estimator of μ is $2/n$, where n is the sample size.*

Solution. We have

$$\begin{aligned} f(x) &= \frac{1}{\pi}\frac{1}{1+(x-\mu)^2}, \quad -\infty < x < \infty \\ \log f(x) &= -\log\pi - \log[1 + (x-\mu)^2] \\ \frac{\partial \log f(x)}{\partial\mu} &= \frac{2(x-\mu)}{1+(x-\mu)^2} \\ E\left[\frac{\partial \log f(x)}{\partial\mu}\right]^2 &= E\left[\frac{2(x-\mu)}{1+(x-\mu)^2}\right]^2 = \int_{-\infty}^{\infty} \frac{4(x-\mu)^2}{[1+(x-\mu)^2]^2} f(x)dx \\ &= \frac{1}{\pi}\int_{-\infty}^{\infty} \frac{4(x-\mu)^2}{[1+(x-\mu)^2]^3}dx. \end{aligned}$$

Put $x-\mu=\tan\theta$, then $dx=\sec^2\theta d\theta$.

$$\begin{aligned}
\therefore\ E\left[\frac{\partial}{\partial\mu}\log f(x)\right]^2 &= \frac{2}{\pi}\int_0^{\pi/2}\frac{4\tan^2\theta}{\sec^6\theta}d\theta\\
&= \frac{2}{\pi}\int_0^{\pi/2}\frac{4\sin^2\theta\cos^4\theta}{\cos^2\theta}d\theta\\
&= \frac{2}{\pi}\int_0^{\pi/2}4\sin^2\theta\cos^2\theta d\theta\\
&= \frac{8}{\pi}\int_0^{\pi/2}(\cos^2\theta-\cos^4\theta)d\theta\\
&= \frac{8}{\pi}\left[\frac{1}{2}\cdot\frac{\pi}{2}-\frac{3}{4}\cdot\frac{1}{2}\cdot\frac{\pi}{2}\right]=\frac{8}{\pi}\left[\frac{\pi}{4}-\frac{3\pi}{16}\right]=\frac{1}{2}
\end{aligned}$$

Hence Cramer-Rao lower bound is

$$\frac{1}{nE\left(\frac{\partial\log f(x)}{\partial\mu}\right)^2}=\frac{1}{n\frac{1}{2}}=\frac{2}{n}.$$

Chapman-Robbins-Kiefer (CRK) Inequality

The MVB estimator [Cramer-Rao inequality] is obtained under regularity conditions. But these conditions are not satisfied for certain probability distributions. In such a situation the minimum variance bound of an estimator T of θ in $f(x,\theta)$ is known as *CRK bound.*

❑ **THEOREM 32.8** *Let θ in $f(x,\theta)$ be such that $\theta\in S$, where S is the parametric space and $\phi(\theta)$ is the parametric function of θ. Assume that T is an unbiased estimator of $\phi(\theta)$ for all values of θ and $V(T)$ is finite. If $\theta\neq\psi$, then $f(x,\theta)$ and $f(x,\psi)$ are different. Also, assume that $\psi\in S$ such that $\theta\neq\psi$ and $f(x,\theta)>0$, $f(x,\psi)>0$. Then*

$$V(T)\geq\sup_{\psi\neq\theta}\frac{[\phi(\psi)-\phi(\theta)]^2}{V[f(x,\psi)/f(x,\theta)]}.$$

Proof. Since T is unbiased for ψ, $E(T)=\phi(\psi)$. This is true for all $\psi\in S$. Then for $\psi\neq\theta$

$$\int T\frac{f(x,\psi)-f(x,\theta)}{f(x,\theta)}f(x,\theta)dx=\phi(\psi)-\phi(\theta).$$

Thus, we have

$$\operatorname{cov}\left[T,\frac{f(x,\psi)}{f(x,\theta)}-1\right]=\phi(\psi)-\phi(\theta).$$

Using Cauchy-Schwartz inequality, we get

$$\operatorname{cov}^2\left\{T,\frac{f(x,\psi)}{f(x,\theta)}-1\right\}\leq V(T)V\left[\frac{f(x,\psi)}{f(x,\theta)}-1\right]=V(T)V\left[\frac{f(x,\psi)}{f(x,\theta)}\right].$$

$$\therefore\ V(T)\geq\frac{[\phi(\psi)-\phi(\theta)]^2}{V[f(x,\psi)/f(x,\theta)]}.$$

►**EXAMPLE 32.48** *Show that the variance of the estimator of θ in*

$$f(x,\theta) = \begin{cases} 1/\theta, & 0 < x < \theta \\ 0, & \textit{otherwise.} \end{cases}$$

is less than the MVB.

Solution. The distribution and the parameter do not satisfy the regularity conditions and hence MVB estimator of θ is not available. But, we have

$$E(x) = \frac{1}{\theta}\int_0^\theta x dx = \frac{\theta}{2}$$

and hence $2x$ is an unbiased estimator of θ. The variance of this estimator is

$$V(2x) = \theta^2/3.$$

Here $2x$ is UMVUE. The CRK of the estimator of θ is

$$V(T) \geq \sup_{\psi \neq \theta} \frac{[\phi(\psi) - \phi(\theta)]^2}{V[f(x,\psi)/f(x,\theta)]}$$

where

$$V\left[f(x,\psi)/f(x,\theta)\right] = E\left[\frac{f_\psi(x)}{f_\theta(x)}\right]^2 = 1.$$

Now,

$$E\left[\frac{f(x,\psi)}{f(x,\theta)}\right]^2 = \int_0^\psi \left(\frac{\theta}{\psi}\right)^2 \frac{1}{\theta} dx = \frac{\theta}{\psi}.$$

$$\therefore\ V(T) \geq \sup_{\psi<\theta} \frac{(\psi-\theta)^2}{\frac{\theta}{\psi} - 1}.$$

Because $\psi(\theta - \psi)$ attains maximum value, if $\psi = \theta/2$

$$V(T) = \sup_{\psi<\theta}\{\psi(\theta-\psi)\} = \frac{\theta^2}{4}, \text{ where } T = 2x.$$

$$\therefore\ V(2x) = \frac{\theta^2}{3} > \frac{\theta^2}{4}.$$

Fisher's Information Measure

Let $x_1, x_2, \ldots, x_n$ be a random sample of n observations selected from a population having p.d.f. $f(x,\theta)$. The likelihood function of the sample observations is

$$L = f(x_1,\theta)\ f(x_2,\theta)\ \ldots\ f(x_n,\theta).$$

Assume that L can be differentiated with respect to θ and $f(x,\theta)$ satisfies conditions. The likelihood function also satisfied regularity conditions. Thus, $\dfrac{\partial \log L}{\partial \theta}$ exists. Then Fisher's measure of information is given by

$$I = E\left(\frac{\partial \log L}{\partial \theta}\right)^2 = V\left(\frac{\partial \log L}{\partial \theta}\right).$$

►**EXAMPLE 32.49** *Let $x_1, x_2, \ldots, x_n$ be a random sample of n observations from $N(0, \theta)$. Find Fisher's measure of information in estimating θ.*

Solution. Here

$$f(x, \theta) = \frac{1}{\sqrt{2\pi\theta}} e^{-\frac{1}{2}\frac{x^2}{\theta}}, \quad -\infty < x < \infty$$

$$L = \left(\frac{1}{2\pi\theta}\right)^{n/2} e^{-\frac{1}{2\theta}\Sigma x_i^2}$$

$$\log L = -\frac{n}{2}\log 2\pi - \frac{n}{2}\log\theta - \frac{1}{2\theta}\Sigma x_i^2$$

$$\frac{\partial \log L}{\partial \theta} = -\frac{n}{2\theta} + \frac{1}{2\theta^2}\Sigma x_i^2 = \frac{n}{2\theta^2}\left[\frac{\Sigma x_i^2}{n} - \theta\right].$$

We know

$$V\left[\frac{\partial \log L}{\partial \theta}\right] = E\left(\frac{\partial \log L}{\partial \theta}\right)^2 = -E\left[\frac{\partial^2 \log L}{\partial \theta^2}\right].$$

Here

$$\frac{\partial^2 \log L}{\partial \theta^2} = \frac{n}{2\theta^2} - \frac{\Sigma x_i^2}{\theta^3}$$

$$-E\left[\frac{\partial^2 \log L}{\partial \theta^2}\right] = \frac{E[\Sigma x_i^2]}{\theta^3} - \frac{n}{2\theta^2} = \frac{n\theta}{\theta^3} - \frac{n}{2\theta^2} = \frac{n}{2\theta^2} \quad [\because \Sigma x_i^2 \sim \chi^2\theta \text{ with } n \text{ d.f.}]$$

$$\therefore I = V\left[\frac{\partial \log L}{\partial \theta}\right] = \frac{n}{2\theta^2}.$$

►**EXAMPLE 32.50** *Find the Fisher's measure of information in estimating the parameter λ in*

$$f(x, \lambda) = \frac{1}{\lambda}e^{-\frac{x}{\lambda}}; \; x \geq 0, \; \lambda > 0.$$

Solution. The likelihood function for a random sample of n observations is

$$L = \frac{1}{\lambda^n}e^{-\frac{n\bar{x}}{\lambda}}, \quad \log L = -n\log\lambda - \frac{n\bar{x}}{\lambda}$$

$$\frac{\partial \log L}{\partial \lambda} = -\frac{n}{\lambda} + \frac{n\bar{x}}{\lambda^2}, \quad \frac{\partial^2 \log L}{\partial \lambda^2} = \frac{n}{\lambda^2} - \frac{2n\bar{x}}{\lambda^3}$$

$$-E\left[\frac{\partial^2 \log L}{\partial \lambda^2}\right] = \frac{2n}{\lambda^3}E(\bar{x}) - \frac{n}{\lambda^2} = \frac{2n}{\lambda^3}\lambda - \frac{n}{\lambda^2} = \frac{n}{\lambda^2}.$$

$$\therefore I = V\left[\frac{\partial \log L}{\partial \lambda}\right] = -E\left[\frac{\partial^2 \log L}{\partial \lambda^2}\right] = \frac{n}{\lambda^2}.$$

32.4 Sufficiency

Let $x_1, x_2, \ldots, x_n$ be a random sample of n observations selected from a population whose p.d.f. is $f(x, \theta)$. Let $T = t(x_1, x_2, \ldots, x_n)$ be a function of sample observations which is used to estimate the parametric function $\phi(\theta)$, where $\theta \in S$ (parametric space). If T provides all information of $\phi(\theta)$, then T is called *sufficient statistic.* The estimator

T will provide all information of θ or $\phi(\theta)$ if the joint probability density function of $x_1, x_2, \ldots, x_n$ under the condition that $T = t(x_1, x_2, \ldots, x_n)$ is free of θ, i.e.,

$$P[X_1 = x_1, X_2 = x_2, \ldots, X_n = x_n/T = t] = \frac{P(X_1 = x_1, X_2 = x_2, \ldots, X_n = x_n)}{P(T \neq t)}.$$

Let us explain the condition of sufficiency by an example.

►**EXAMPLE 32.51** *Let $x_1, x_2, \ldots, x_n$ be a random sample of n observations from Poisson population having parameter λ. Let $T = \Sigma x_i = t$ be a statistic. Show that T is a sufficient statistic to estimate λ.*

Solution. The density function of T is

$$P(T = t) = \frac{e^{-n\lambda}\lambda^t}{t!}, \text{ where } t = \Sigma x_i \text{ and } P(x) = \frac{e^{-\lambda}\lambda^x}{x!}.$$

Now

$$\begin{aligned} &P(X_1 = x_1, X_2 = x_2, \ldots, X_n = x_n/T = t) \\ &= \frac{P(X_1 = x_1, X_2 = x_2, \ldots, X_n = x_n)}{P(T = t)} \\ &= \frac{e^{-\lambda}\lambda^{x_1}}{x_1!}\frac{e^{-\lambda}\lambda^{x_2}}{x_2!}\cdots\frac{e^{-\lambda}\lambda^{x_n}}{x_n!}\bigg/\frac{e^{-n\lambda}\lambda^t}{t!} = \frac{t!}{x_1!x_2!\cdots x_n!}. \end{aligned}$$

This conditional distribution of $x_1, x_2, \ldots, x_n$ is free of parameter λ. Therefore, $T = \Sigma x_i = t$ is a sufficient statistic for λ.

In the above example, we have shown that Σx_i is a sufficient statistic. In practice, we need to find out a sufficient statistic from many statistics. This is possible if the joint conditional distribution of different statistics under the condition that a particular statistic has a specified value is free of parameter θ, then the statistic having specified value is a sufficient statistic for parameter θ. Mathematically, this can be expressed as follows:

Let $x_1, x_2, \ldots, x_n$ be a random sample of n observations from a population having p.d.f. $f(x, \theta)$, where $\theta \in S$. The joint p.d.f. of sample observations is

$$L = f(x_1, \theta) f(x_2, \theta) \cdots f(x_n, \theta).$$

Let

$$\begin{aligned} T_1 = t_1(x_1, x_2, \ldots, x_n),\ T_2 = t_2(x_1, x_2, \ldots, x_n), \cdots \\ \cdots, T_n = t_n(x_1, x_2, \ldots, x_n) \end{aligned}$$

be n statistics based on sample observations. Let us assume that the transformation of x to T is one-to-one. Then

$$x_j = t_j(T_1, T_2, \ldots, T_n);\ j = 1, 2, \ldots, n.$$

The Jacobian of the transformation is

$$J = \frac{\partial(x_1, x_2, \ldots, x_n)}{\partial(T_1, T_2, \ldots, T_n)} \neq 0.$$

Then, the joint distribution of $T_1, T_2, \ldots, T_n$ is

$$\begin{aligned} g(T_1, T_2, \ldots, T_n; \theta) &= f\{t_1(T_1, T_2, \ldots, T_n)\} f\{t_2(T_1, T_2, \ldots, T_n)\} \cdots \\ &\quad \cdots f\{t_n(T_1, T_2, \ldots, T_n)\}. \end{aligned}$$

The marginal density function of T_1 is

$$\begin{aligned} g_1(T_1, \theta) &= \int_{-\infty}^{\infty} \cdots \int_{-\infty}^{\infty} g(T_1, T_2, \ldots, T_n, \theta) dT_2 dT_3 \cdots dT_n; \\ &\qquad \text{for continuous random variable} \\ &= \sum_{T_n} \sum_{T_{n-1}} \cdots \sum_{T_2} g(T_1, T_2, \ldots, T_n, \theta); \\ &\qquad \text{for discrete random variable.} \end{aligned}$$

The conditional distribution of $T_2, T_3, \ldots, T_n$ under the condition that $T_1 = t_1$ is

$$h(T_2, T_3, \ldots, T_n / T_1 = t_1) = \frac{g(T_1, T_2, \ldots, T_n, \theta)}{g_1(T_1, \theta)}$$

such that $g_1(T_1, \theta) > 0$.

If this conditional density function is free of θ, then T_1 is a sufficient statistic for θ. But $h(\cdot)$ is not free of θ. This will be free of θ, if

$$g(T_1, T_2, \ldots, T_n, \theta) = g_1(T_1, \theta) h_1(T_2, T_3, \ldots, T_n / t_1).$$

Here $h_1(\cdot)$ is free of θ but $g_1(\cdot)$ is not free of θ. This last equation indicates that, if the joint density function of $x_1, x_2, \ldots, x_n$ can be expressed into two density functions such that one density function is of T and θ, and another one is free of θ, then the statistic T is known as *sufficient statistic.* This criterion of sufficient statistic is known as *Fisher-Neyman criterion of sufficient statistic.*

Fisher-Neyman Lemma

❐ **THEOREM 32.9** *Let $x_1, x_2, \ldots, x_n$ be a random sample of observations from a population having p.d.f. $f(x, \theta)$, $\theta \in S$. Consider a statistic $T_1 = t_1(x_1, x_2, \ldots, x_n)$. The density function of T_1 is $g_1(T_1, \theta)$. This $T_1 = t_1(x_1, x_2, \ldots, x_n)$ is a sufficient statistic for θ, if and only if*

$$\begin{aligned} L &= f(x_1, \theta) f(x_2, \theta), \ldots, f(x_n, \theta) \\ &= g_1[t_1(x_1, x_2, \ldots, x_n), \theta] h(x_1, x_2, \ldots, x_n) \end{aligned}$$

where $h(x_1, x_2, \ldots, x_n)$ does not depend on θ for each value of θ.

Proof. Let

$$L = g_1[t_1(x_1, x_2, \ldots, x_n), \theta] h(x_1, x_2, \ldots, x_n).$$

Consider a one-to-one transformation

$$T_1 = t_1(x_1, x_2, \ldots, x_n), T_2 = t_2(x_1, x_2, \ldots, x_n), \cdots, T_n = t_n(x_1, x_2, \ldots, x_n)$$

such that

$$x_1 = W_1(T_1, T_2, \dots, T_n), x_2 = W_2(T_1, T_2, \dots, T_n), \cdots, x_n = W_n(T_1, T_2, \dots, T_n)$$

are inverse transformation. Let us consider that J is the Jacobian of the transformation, where

$$J = \frac{\partial(x_1, x_2, \dots, x_n)}{\partial(T_1, T_2, \dots, T_n)}.$$

Now, the joint probability density function of $T_1, T_2, \dots, T_n$ is

$$\begin{aligned} g(T_1, T_2, \dots, T_n; \theta) &= f[W_1(T_1, T_2, \dots, T_n), \theta] f[W_2(T_1, T_2, \dots, T_n), \theta] \cdots \\ &\quad \cdots f[W_n(T_1, T_2, \dots, T_n), \theta]\ |J| \\ &= g_1(T_1, \theta) h[W_1(T_1, T_2, \dots, T_n) \cdots W_n(T_1, T_2, \dots, T_n)]\ |J|. \end{aligned}$$

But from the definition of conditional density function, we can write

$$h(T_2, T_3, \dots, T_n/T_1 = t_1; \theta) = \frac{g(T_1, T_2, \dots, T_n; \theta)}{g_1(T_1, \theta)}.$$

Hence

$$h[W_1(T_1, T_2, \dots, T_n), \dots, W_n(T_1, T_2, \dots, T_n)]\ |J| = h(T_2, T_3, \dots, T_n/T_1 = t_1; \theta).$$

Since θ is not included in the transformation, it is also not included in $|J|$. Also, θ is not included in

$$h[W_1(T_1, T_2, \dots, T_n), \dots, W_n(T_1, T_2, \dots, T_n)].$$

This implies that $h(T_2, T_3, \dots, T_n/T_1 = t_1; \theta)$ does not depend on θ. Thus, by definition, T_1 is the sufficient statistic for θ.

Now we need to prove that

$$g(T_1, T_2, \dots, T_n;\ \theta) = g_1(T_1, \theta) h(T_2, T_3, \dots, T_n/T_1)$$

where $h(T_2, T_3, \dots, T_n/T_1)$ does not depend on θ. Let the Jacobian of the transformation of old variables T_1, T_2, $\dots$, T_n to new variables $x_1, x_2, \dots, x_n$ be J_1. Replacing $T_1, T_2, \dots, T_n$ by $t_1(x_1, x_2, \dots, x_n)$, $t_2(x_1, x_2, \dots, x_n), \dots, t_n(x_1, x_2, \dots, x_n)$ and multiplying by $|J_1|$, we get

$$\begin{aligned} & g[t_1(x_1, x_2, \dots, x_n) \dots, t_n(x_1, x_2, \dots, x_n); \theta]\ |J_1| \\ &= g_1[t_1(x_1, x_2, \dots, x_n)] h(T_2, T_3, \dots, T_n/T_1) |J_1|. \end{aligned}$$

But the quantity in left-hand side is the likelihood function

$$L = f(x_1, \theta) f(x_2, \theta) \cdots f(x_n, \theta).$$

Again, $h(T_2, T_3, \dots, T_n/T_1)$ is free of θ and hence it does not depend on θ. Thus

$$h(x_1, x_2, \dots, x_n) = h(T_2, T_3, \dots, T_n/T_1)\ |J_1|$$

does not depend on θ.

►**EXAMPLE 32.52** *Let $x_1, x_2, \ldots, x_n$ be a random sample of n observations from Bernoulli population with parameter $p(0 < p < 1)$. Find sufficient statistic for p.*

Solution. Given $p(x) = p^x(1-p)^{1-x}$, $x = 0, 1$.

Here $L = p^{\Sigma x}(1-p)^{n-\Sigma x} = p^{n\bar{x}}(1-p)^{n-n\bar{x}}$.

Let $T_1 = x_1 + x_2 + \cdots + x_n$, $T_2 = x_2$, $\ldots$, $T_n = x_n$. Then

$$\begin{aligned} g(T_1, T_2, \ldots, T_n) &= p^{T_1}(1-p)^{n-T_1} \\ g_1(T_1, \theta) &= \sum_{T_n}\sum_{T_3}\cdots\sum_{T_n} L(T_1, T_2, \ldots, T_n). \end{aligned}$$

Let $T_1 = 0, P(T_1) = 1 - p$; $T_1 = 1, P(T_1) = p$. Then in n repetition $T_1 = t_1$ success will occur with probability

$$\begin{aligned} g(t_1) &= \binom{n}{t_1} p^{t_1}(1-p)^{n-t_1};\ t_1 = 0, 1, 2, \ldots, n. \\ \therefore\ L &= g(t_1, \theta) \times \frac{1}{\binom{n}{t_1}} \\ &= p^{T_1}(1-p)^{n-T_1}\frac{(x_1 + x_2 + \cdots + x_n)!\,[n - (x_1 + x_2 + \cdots + x_n)]!}{n!} \\ &= p^{\Sigma x_i}(1-p)^{n-\Sigma x_i} h(x_1, x_2, \ldots, x_n) = g(T, \theta)h(x_1, x_2, \ldots, x_n) \end{aligned}$$

where

$$g(T, \theta) = p^{\Sigma x_i}(1-p)^{n-\Sigma x_i} \text{ and } h(x_1, x_2, \ldots, x_n) = \frac{(\Sigma x_i)!\,(n - \Sigma x_i)!}{n!}.$$

Hence $T_1 = \sum_{i=1}^{n} x_i$ is a sufficient statistic for p.

►**EXAMPLE 32.53** *Let $x_1, x_2, \ldots, x_n$ be a random sample of n observations from population having p.d.f.*

$$f(x) = \theta x^{\theta-1};\ 0 < x < 1,\ \theta > 0.$$

Find a sufficient statistic for θ.

Solution. The likelihood function is

$$L = \theta^n \prod_{i=1}^{n} x_i^{\theta-1} = \theta^n \prod_{i=1}^{n} x_i^{\theta} \Big/ \prod_{i=1}^{n} x_i = g(T, \theta)h(x_1, x_2, \ldots, x_n)$$

where

$$g(T, \theta) = \theta^n \prod_{i=1}^{n} x_i^{\theta}, h(x_1, x_2, \ldots, x_n) = \frac{1}{\prod_{i=1}^{n} x_i}.$$

Therefore, $\prod_{i=1}^{n} x_i$ is a sufficient statistic for θ.

►**EXAMPLE 32.54** *Let $x_1, x_2, \ldots, x_n$ be a random sample of n observations from Poisson population having parameter λ. Find a sufficient statistic for λ.*

Solution. We have

$$L = \frac{e^{-n\lambda}\lambda^{\Sigma x_i}}{\prod_{i=1}^{n} x_i!} = g(T,\theta)h(x_1, x_2, \ldots, x_n)$$

where

$$g(T,\theta) = e^{-n\lambda}\lambda^{\Sigma x_i} = e^{-n\lambda}\lambda^{n\bar{x}} \text{ and } h(x_1, x_2, \ldots, x_n) = \frac{1}{\prod_{i=1}^{n} x_i!}.$$

Hence $T = n\bar{x}$ is a sufficient statistic for λ.

►**EXAMPLE 32.55** *Let $x_1, x_2, \ldots, x_n$ be a random sample from $N(0, \theta)$. Find sufficient statistic for θ.*

Solution. The likelihood function is

$$\begin{aligned} L &= \left(\frac{1}{2\pi\theta}\right)^{n/2} e^{-\frac{1}{2\theta}\Sigma x_i^2} \\ &= \frac{1}{(2\pi)^{n/2}(\theta)^{n/2}} e^{-\frac{1}{2\theta}\Sigma x_i^2} = g(T,\theta)h(x_1, x_2, \ldots, x_n) \end{aligned}$$

where

$$g(T,\theta) = \frac{1}{(2\pi)^{n/2}\theta^{n/2}} e^{-\frac{1}{2\theta}\Sigma x_i^2},\ h(x_1, x_2, \ldots, x_n) = 1.$$

Hence Σx_i^2 is a sufficient statistic for θ.

►**EXAMPLE 32.56** *Let $x_1, x_2, \ldots, x_n$ be a random sample of n observations from $N(\mu, \sigma^2)$. Show that $T = (\bar{x}, s^2)$ is the sufficient statistic for parameter vector*

$$\theta = (\mu, \sigma^2).$$

Solution. We have

$$\begin{aligned} L &= \left(\frac{1}{2\pi\sigma}\right)^{n/2} e^{-\frac{1}{2\sigma^2}\Sigma(x_i - \mu)^2} \\ &= \left(\frac{1}{2\pi\sigma}\right)^{n/2} e^{-\frac{1}{2\sigma^2}\Sigma(x_i - \bar{x})^2} e^{n(\bar{x}-\mu)^2/2\sigma^2} \\ &= \left(\frac{1}{2\pi\sigma}\right)^{n/2} e^{-\frac{1}{2\sigma^2}(n-1)s^2} e^{n(\bar{x}-\mu)^2/2\sigma^2} \\ &= g(T,\theta)h(x_1, x_2, \ldots, x_n) \end{aligned}$$

where

$$g(T,\theta) = \left(\frac{1}{2\pi\sigma}\right)^{n/2} e^{-\frac{1}{2\sigma^2}(n-1)s^2} e^{n(\bar{x}-\mu)^2/2\sigma^2};\ h(x_1, x_2, \ldots, x_n) = 1.$$

Hence $T = (\bar{x}, s^2)$ is sufficient statistic for $\theta = (\mu, \sigma^2)$.

►**EXAMPLE 32.57** *Let $x_1, x_2, \ldots, x_n$ be a random sample of n observations from Gamma distribution having parameter θ. Find sufficient statistic for θ.*

Solution. We have

$$f(x) = \frac{1}{\Gamma\theta}e^{-x}x^{\theta-1},\ 0 < x < \infty$$

$$L = \frac{1}{(\Gamma\theta)^n}e^{-n\bar{x}}\prod_{i=1}^{n} x_i^{\theta-1} = \frac{1}{(\Gamma\theta)^n}e^{-n\bar{x}}\frac{\prod_{i=1}^{n} x_i^{\theta}}{\prod_{i=1}^{n} x_i}$$

$$= g(T,\theta)h(x_1, x_2, \ldots, x_n)$$

where

$$g(T,\theta) = \frac{1}{(\Gamma\theta)^n}\prod_{i=1}^{n} x_i^{\theta},\ h(x_1, x_2, \ldots, x_n) = e^{n\bar{x}}/(x_1x_2\cdots x_n).$$

Hence $\prod_{i=1}^{n} x_i$ is a sufficient statistic for θ.

►**EXAMPLE 32.58** *Let $x_1, x_2, \ldots, x_n$ be a random sample of n observations from a Beta distribution having parameter θ. Find sufficient statistic for θ.*

Solution. We have

$$f(x) = \frac{1}{B(\theta,2)}x^{\theta-1}(1-x);\ 0 < x < 1,\ \theta > 0.$$

$$\therefore\ L = \frac{1}{[B(\theta,2)]^n}(x_1x_2\cdots x_n)^{\theta-1}\prod_{i=1}^{n}(1-x_i)$$

$$= \frac{1}{[B(\theta,2)]^n}\frac{(x_1x_2\cdots x_n)^{\theta}}{(x_1x_2\cdots x_n)}\prod_{i=1}^{n}(1-x_i)$$

$$= g(T,\theta)h(x_1, x_2, \ldots, x_n)$$

where

$$g(T,\theta) = \frac{1}{[B(\theta,2)]^n}(x_1x_2\cdots x_n)^{\theta}, h(x_1, x_2, \ldots, x_n) = \frac{\prod_{i=1}^{n}(1-x_i)}{(x_1x_2\cdots x_n)}.$$

Hence $\prod_{i=1}^{n} x_i$ is a sufficient statistic for θ.

►**EXAMPLE 32.59** *Let $x_1, x_2, \ldots, x_n$ be a random sample of n observations from a population having p.d.f.*

$$f(x) = \begin{cases} 1/\theta, & 0 < x < \theta \\ 0, & \text{otherwise.} \end{cases}$$

Find sufficient statistic for θ.

Solution. Let us consider an indicator function $k(a, b)$ such that

$$k(a,b) = \begin{cases} 1, & \text{if } a \le b \\ 0, & \text{if } a > b. \end{cases}$$

Then

$$\begin{aligned} f(x_i) &= \frac{k(0,x_i)k(x_i,0)}{\theta} \\ \text{and } L = \prod_{i=1}^{n} f(x_i) &= \prod_{i=1}^{n} \frac{k(0,x_i)k(x_i,0)}{\theta} \\ &= \frac{1}{\theta^n}\left[k\left(\theta,\ \min_{1\le i\le n} x_i\right)k\left(\max_{1\le i\le n} x_i,\ \theta\right)\right] \\ &= g(T,\theta)h(x_1,x_2,\ldots,x_n). \end{aligned}$$

Hence $T = \max(x_1, x_2, \ldots, x_n)$ is a sufficient statistic for θ.

►**EXAMPLE 32.60** *Let $x_1, x_2, \ldots, x_n$ be a random sample of n observations from a population whose p.d.f. is*

$$f(x) = e^{-(x-\theta)};\ 0 < x < \infty,\ -\infty < \theta < \infty.$$

Find sufficient statistic for θ.

Solution. We have $L = e^{-n\bar{x}}e^{n\theta} \neq g(T,\theta)h(x_1,x_2,\ldots,x_n)$.

That is L cannot be expressed in the form of $g(T,\theta)h(x_1,x_2,\ldots,x_n)$.

Let $y_1 < y_2 < \cdots < y_n$ be the order statistics of sample observations. The p.d.f. of y_1 is

$$g(y_1) = n[1 - F_x(y)]^{n-1} f_x(y),\ 0 < y < \infty.$$

Here

$$\begin{aligned} F(x) &= \int_0^x e^{-(x-\theta)}dx = 1 - e^{-(x-\theta)} \\ \therefore\ g(y_1) &= n\left[e^{-(y_1-\theta)}\right]^{n-1} e^{-(y_1-\theta)} = ne^{-n(y_1-\theta)}. \end{aligned}$$

Hence

$$L = e^{n\theta}e^{-\Sigma x_i} = ne^{-n(y_1-\theta)}\frac{e^{-\Sigma x_i}}{ne^{-ny_1}} = g(T,\theta)h(x_1,x_2,\ldots,x_n)$$

where

$$g(T,\theta) = \frac{ne^{-n(y_1-\theta)}}{ne^{-ny_1}} \text{ and } h(x_1,x_2,\ldots,x_n) = e^{-n\bar{x}}.$$

Hence $T = \min(x_1, x_2, \ldots, x_n)$ is a sufficient statistic for θ.

►**EXAMPLE 32.61** *Let $x_1, x_2, \ldots, x_n$ be a random sample of n observations from a population having p.d.f.*

$$f(x) = \frac{2}{\theta^2}(\theta - x),\ 0 < x < \theta.$$

Find sufficient statistic for θ.

Solution. We have

$$L = \frac{2^n}{\theta^{2n}}\prod_{i=1}^{n}(\theta - x_i).$$

This cannot be expresed in the form of $g(T,\theta)h(x_1,x_2,\dots,x_n)$. Let $y_1 < y_2 < \cdots < y_n$ be the order statistic of sample observations. Then

$$g(y_n) = n\,[F_x(y_n)]^{n-1}\, f_x(y_n)$$

where

$$F(x) = \frac{2}{\theta^2}\int_0^x (\theta - x)dx = \frac{2}{\theta^2}x\left(\theta - \frac{x}{2}\right).$$

$$\therefore\; g(y_n) = 2^n\left[\frac{2}{\theta^2}y_n\left(\theta - \frac{y_n}{2}\right)\right]^{n-1}\frac{2}{\theta^2}(\theta - y_n),\; 0 < y_n < \theta$$

$$= n\left(\frac{2}{\theta^2}\right)^n y_n^{n-1}\left(\theta - \frac{y_n}{2}\right)^{n-1}(\theta - y_n).$$

$$\therefore\; L = \frac{2^n}{\theta^{2n}}\prod_{i=1}^{n}(\theta - x_i) = \frac{2^n}{\theta^{2n}}(\theta - x_1)(\theta - x_2)\cdots(\theta - x_n)$$

$$= \frac{2^n}{\theta^{2n}}\frac{nx^{n-1}(\theta - \frac{x}{2})^{n-1}}{nx^{n-1}(\theta - \frac{x}{2})^{n-1}}\prod_{i=1}^{n-1}(\theta - x_i)(\theta - x_n)$$

$$= g(T,\theta)h(x_1,x_2,\dots,x_n)$$

where

$$g(T,\theta) = \frac{2^n}{\theta^{2n}}nx^{n-1}\left(\theta - \frac{x}{2}\right)^{n-1}(\theta - x)\prod_{i=1}^{n}(\theta - x_i)\Big/\left(\theta - \frac{x}{2}\right)^{n-1}$$

$$h(x_1,x_2,\dots,x_n) = \frac{1}{nx^{n-1}}.$$

Hence $\max(x_1,x_2,\dots,x_n)$ is a sufficient statistic for θ.

❑ **THEOREM 32.10** *Let $x_1,x_2,\dots,x_n$ be a random sample of n observations from a population whose p.d.f. is $f(x,\theta)$. Let $T_1 = t_1(x_1,x_2,\dots,x_n)$ be a sample statistic. This statistic will be sufficient statistic for θ, if and only if the likelihood function can be expressed as a product of two functions k_1 and k_2, where*

$$k_1 = [t_1(x_1,x_2,\dots,x_n),\theta] \text{ and } k_2 = k_2(x_1,x_2,\dots,x_n).$$

This should be true for all values of $T_1 = t_1(x_1,x_2,\dots,x_n)$ and k_2 free of θ.

Proof. The likelihood function is

$$L = f(x_1,\theta),\; f(x_2,\theta),\dots,\; f(x_n,\theta).$$

From Fisher-Lehman lemma, we can write

$$g(T_1,T_2,\dots,T_n;\theta) = k_1(T_1,\theta)k_2[W_1,W_2,\dots,W_n]\,|J|$$

where $W_i = W_i(T_1,T_2,\dots,T_n)$; $i = 1,2,\dots,n$. From the above equation, we have

$$g_1(T_1,\theta) = \int_{-\infty}^{\infty}\cdots\int_{-\infty}^{\infty} g(T_1,T_2,\dots,T_n;\theta)\, dT_2dT_3\cdots dT_n$$

$$= k_1(T_1,\theta)\int_{-\infty}^{\infty}\cdots\int_{-\infty}^{\infty}|J|\; k_2(W_1,W_2,\dots,W_n)\; dT_2\cdots dT_n.$$

Here the function k_2 does not depend on θ for each value of $T_1 = t_1(x_1, x_2, \ldots, x_n)$. $|J|$ is related to transformation of variables, hence it is also free of θ. Therefore,

$$\int_{-\infty}^{\infty} \cdots \int_{-\infty}^{\infty} |J| \; k_2(W_1, W_2, \ldots, W_n) dT_2 \cdots dT_n$$

is the function of only T_1. Let this function be $u(T_1)$. Now, if $u(T_1) = 0$, $g_1(T_1, \theta) = 0$ and if $u(T_1) > 0$, then

$$k_1[t_1(x_1, x_2, \ldots, x_n), \theta] = \frac{g_1[t_1(x_1, x_2, \ldots, x_n), \theta]}{h[t_1(x_1, x_2, \ldots, x_n)]}.$$

Then

$$L = g_1[t_1(x_1, x_2, \ldots, x_n), \theta] \frac{k_2(x_1, x_2, \ldots, x_n)}{h[t_1(x_1, x_2, \ldots, x_n)]}.$$

Since k_2 does not depend on h or θ, T_1 is the sufficient statistic for θ. Therefore, if L is factorized, k_1 will be the p.d.f. of T_1. Here k_1 is identified as g_1.

Sufficiency and Unbiasedness

Let $x_1, x_2, \ldots, x_n$ be a random sample of n observations selected from a population whose p.d.f. is $f(x, \theta)$. Consider that

$$T_1 = t_1(x_1, x_2, \ldots, x_n) \text{ and } T_2 = t_2(x_1, x_2, \ldots, x_n)$$

are two unbiased estimators of θ and $T = t(x_1, x_2, \ldots, x_n)$ is a sufficient statistic of θ. Assume that T_1 is a function of $T[T_1 = f(T)]$. Since T is sufficient statistic for θ and T_1 is a function of T, as an estimator of θ the statistic T_1 is more meaningful than T_2 and $V(T_1) \leq V(T_2)$. If this property of variance does not exist, there is no need to study the characteristic sufficiency. This means that, if any statistic 'h' is unbiased for θ, then we need to study whether 'h' can be improved using the sufficient statistic of θ.

Since T is the sufficient statistic for θ, the conditional distribution of $(x_1, x_2, \ldots, x_n)$ under the condition that $T = t$ will be independent of θ. This is achieved from

$$E[h(x_1, x_2, \ldots, x_n)/T = t] = T_1$$

where T_1 itself is an estimator of θ. Again,

$$\begin{aligned} E[T_1] &= E[h(x_1, x_2, \ldots, x_n)/T = t] \\ &= E[h(x_1, x_2, \ldots, x_n)] = \theta, \quad \because \; h \text{ is unbiased estimator of } \theta. \end{aligned}$$

Therefore, T_1 is an unbiased estimator of θ and this estimator is sufficient statistic for θ.

Again, from the definition of conditional variance, we can write

$$V(h) = V(T_1) + E[V(h/T_1)] > V(T_1).$$

This is true since $V(h/T_1) > 0$ and $E[V(h/T_1)] > 0$. Thus it may be mentioned that, if T is the best sufficient statistic for θ, then T_1 is the best estimator.

►EXAMPLE 32.62 *Let $x_1, x_2, \ldots, x_n$ be a random sample of n observations from a population whose p.d.f. is*

$$f(x) = \frac{1}{\theta^m \Gamma m} e^{-x/\theta} x^{m-1}; \; 0 < x < \infty, \; \theta > 0.$$

When m is known the estimator of θ is a function of sufficient statistic where the sufficient statistic is MVB estimator.

Solution. From Example 32.39, we can write

$$L = \left(\frac{1}{\theta^m \Gamma m}\right)^n e^{-n\bar{x}/\theta} \prod_{i=1}^{n} x_i^{m-1} = g(T/\theta) h(x_1, x_2, \ldots, x_n)$$

where

$$g(T/\theta) = \left(\frac{1}{\theta^m \Gamma m}\right)^n e^{-n\bar{x}/\theta}, h(x_1, x_2, \ldots, x_n) = \prod_{i=1}^{n} x_i^{m-1}.$$

Hence $\bar{x}$ is a sufficient statistic for θ. Again,

$$\frac{\partial \log L}{\partial \theta} = \frac{mn}{\theta^2}\left[\frac{\bar{x}}{m} - \theta\right].$$

Here the estimator of θ is $\bar{x}/m$ and this estimator is MVB estimator and it is a function of sufficient statistic $\bar{x}$.

►EXAMPLE 32.63 *Let $x_1, x_2, \ldots, x_n$ be a random sample of n observations from a population whose p.d.f. is*

$$f(x) = \frac{1}{\theta} e^{-x/\theta}, \; 0 < x < \infty.$$

Show that the estimator of θ^2 is a function of sufficient statistic and the sufficient statistic is MVB estimator.

Solution. We have

$$L = \frac{1}{\theta^n} e^{-\frac{1}{\theta} n\bar{x}} = g(T/\theta) h(x_1, x_2, \ldots, x_n)$$

where

$$g(T/\theta) = \frac{1}{\theta^n} e^{-\frac{1}{\theta} n\bar{x}}, \; h(x_1, x_2, \ldots, x_n) = 1.$$

Therefore, $\bar{x}$ is the sufficient statistic for θ. Again,

$$\frac{\partial \log L}{\partial \theta} = \frac{n}{\theta^2}(\bar{x} - \theta).$$

Hence $\bar{x}$ is the MVB estimator of θ. Now, $E(\bar{x}^2) = V(\bar{x}) + [E(\bar{x})]^2$, where $V(\bar{x}) = \theta^2/n$. Also, $E(\bar{x}) = \theta$. Thus

$$E(\bar{x}^2) = \frac{\theta^2}{n} + \theta^2 = \frac{n+1}{n}\theta^2.$$

Hence $\dfrac{n\bar{x}^2}{n+1}$ is an unbiased estimator of θ^2 and this estimator is a function of sufficient statistic.

Sufficiency and Minimum Variance

The property sufficiency and minimum variance of an estimator is explained by Rao-Blackwell theorem.

❒**THEOREM 32.11 [Rao-Blackwell Theorem]** *Let T be a sufficient statistic for a parameter θ in $f(x,\theta)$. Here T and θ may be vectors. Consider another estimator T_1 of θ. If $g(\theta)$ is any function of θ, then*

$$V(T_1) = E[T_1 - g(\theta)]^2 \geq E[h(T) - g(\theta)]^2$$

where $h(T) = E(T_1/T = t)$ is free of θ and $h(T)$ is unbiased for $g(\theta)$, if T_1 is unbiased.

Proof. It is mentioned that T is sufficient statistic for θ and hence $E[T_1/T = t]$ is independent of θ and $h(t)$ is independent of θ. Therefore,

$$E[T_1] = E[E(T_1/T)] = E[h(T)].$$

Again

$$\begin{aligned} E[h(T)\{T_1 - h(T)\}] &= E[h(T)\{E(T_1 - h(T)/T)\}] \\ &= 0, \quad \because \; E(T_1/T = t) = h(t). \end{aligned}$$

Now

$$\begin{aligned} V(T_1) = E[T_1 - g(\theta)]^2 &= E[T_1 - h(T) + h(T) - g(\theta)]^2 \\ &= E[T_1 - h(T)]^2 + E[h(T) - g(\theta)]^2 \\ &\geq E[h(T) - g(\theta)]^2 \Rightarrow V(T_1) \geq V[h(T)]. \end{aligned}$$

The sign of equality holds good if $T_1 = h(T)$. Thus, it can be said that, the minimum variance unbiased estimator of $g(\theta)$ is a function of sufficient statistic. In other word, it can be said that, the MVB estimator is available if there exists sufficient statistic.

We know that, if sufficient statistic exists, the likelihood function is factorized as follows:

$$L = g(T/\theta)h(x_1, x_2, \ldots, x_n)$$

where T is a sufficient statistic for θ. Here

$$\frac{\partial \log L}{\partial \theta} = \frac{\partial \log g(T/\theta)}{\partial \theta}, \quad \because \; h(x_1, x_2, \ldots, x_n) \text{ is free of } \theta.$$

Again, MVB exists if $\dfrac{\partial \log L}{\partial \theta}$ can be written as

$$\frac{\partial \log L}{\partial \theta} = A(\theta)[T - \phi(\theta)].$$

►**EXAMPLE 32.64** *Let $x_1, x_2, \ldots, x_n$ be a random sample of n observations from a population whose p.d.f. is*

$$f(x) = \frac{1}{\theta} e^{-x/\theta}.$$

Show that the estimator of θ is sufficient and MVB.

Solution. Here

$$L = \frac{1}{\theta^n} e^{-\frac{n\bar{x}}{\theta}} = g(T/\theta)\ h(x_1, x_2, \dots, x_n)$$

where

$$g(T/\theta) = \frac{1}{\theta^n} e^{-\frac{n\bar{x}}{\theta}},\ h(x_1, x_2, \dots, x_n) = 1.$$

Hence $\bar{x}$ is sufficient statistic for θ. Again,

$$\frac{\partial \log L}{\partial \theta} = \frac{n}{\theta^2}[\bar{x} - \theta].$$

Thus $\bar{x}$ is MVB estimator with $V(\bar{x}) = \theta^2/n$.

Complete Sufficiency

Let $T(x_1, x_2, \dots, x_n)$ be a sufficient statistic for θ and $f_i(T)$; $i = 1, 2, \dots, k$ be the i-th function of T. If $E[f_i(T)] \neq 0$ for $f_i(T) \neq 0$, then T is called *complete sufficient statistic* and this property of estimator is known as *complete sufficiency.* It is to be noted that, if complete sufficient statistic exists, then each function of it is UMVUE of expected value of that function.

Complete Probability Function

Let $f(x, \theta)$, $\theta \in S$ be a family of probability density function. This family is called *complete family of probability distribution* if for all $\theta \in S$

$$E[h(x)] = 0.$$

It implies that $P[h(x) = 0] = 1$.

The statistic $T(x_1, x_2, \dots, x_n)$ is known as *complete sufficient statistic* if the distribution of T is complete probability distribution.

►**EXAMPLE 32.65** *Let $x_1, x_2, \dots, x_n$ be a random sample of n observations from a population whose p.d.f. is*

$$f(x, \theta) = \frac{1}{\theta};\ 0 < x < \theta,\ 0 < \theta < \infty.$$

Find complete sufficient statistic for θ.

Solution. The p.d.f. of family of X will be complete if $E[h(x)] = 0$, when $\theta > 0$. We need to show that

$$E[h(x)] = \int_0^\theta \frac{1}{\theta} h(x)dx = 0.$$

This is possible if for all values of x, $h(x) = 0$. Let $h(x)$ is continuous function of X. Then for $\theta > 0$,

$$\int_0^\theta h(x)dx = 0.$$

Differentiating both sides with respect to θ, we get

$$h(\theta) = 0,\ \theta > 0 \Rightarrow h(x) = 0,\ x > 0.$$

Hence all p.d.f. of family of X are complete.

Let $y_1, y_2, \ldots, y_n$ be the ordered statistics of $x_1, x_2, \ldots, x_n$. Then

$$f(y_n, \theta) = \frac{n}{\theta^n} y_n^{n-1},\ 0 < y_n < \theta.$$

Here y_n is complete sufficient statistic. Because

$$E[T(y_n)] = \int_0^\theta T(y_n) \cdot \frac{n}{\theta^n} y_n^{n-1} dy_n.$$

Now, for all $\theta, E[T(y_n)] = 0$, if and only if

$$\frac{n}{\theta^n}\int_0^\theta T(y_n)y_n^{n-1}dy_n = 0 \text{ or, } \int_0^\theta T(y_n)y_n^{n-1}dy_n = 0,\ \theta > 0.$$

Differentiating both sides with respect to θ, we get

$$T(y_n)\theta^{n-1} = 0.$$

But $\theta^{n-1} \neq 0$. Hence $T(y_n) = 0, \theta > 0$ and y_n is the complete sufficient statistic for θ.

Exponential Class of Probability Density Functions

Let $f(x, \theta)$; $\theta \in S$ be the family of p.d.f., where the parametric space is

$$S = \{0, \gamma < \theta < \delta\}$$

and γ and δ are known values. If this $f(x, \theta)$ can be expressed as

$$f(x, \theta) = \begin{cases} \exp[P(\theta)k(x) + \delta(x) + q(\theta)], & a < x < b \\ 0, & \text{otherwise} \end{cases}$$

then $f(x, \theta)$ is a member of exponential class of probability density functions. This member is a regular member of exponential class of family, if

(i) for $\gamma < \theta < \delta$ none of a or b does not depend on θ,

(ii) for $\gamma < \theta < \delta$, $P(\theta)$ is nontrivial continuous function of θ,

(iii) none of $k'(x)$ is zero and $S(x)$ is the continuous function of x within the limit $a < x < b$.

Under the conditions $f(x, \theta)$ is the member of continuous type of regular exponential class of family. This member becomes discrete type, if

(i) $\{x : x = a_1, a_2, \ldots\}$ does not depend on θ,

(ii) for $\gamma < \theta < \delta$, $P(\theta)$ is the nontrivial continuous function of θ,

(iii) $k(x)$ is the nontrivial function of x, where the set of x is $\{x : x = a_1, a_2, \ldots\}$.

►**EXAMPLE 32.66** *Show that $N(\mu, \sigma^2)$ is a member of regular exponential family of distribution. Also, show that $T = \Sigma x$ is sufficient statistic for μ.*

Solution. We have

$$\begin{aligned} f(x) &= \frac{1}{\sigma\sqrt{2\pi}} e^{-\frac{1}{2}\left(\frac{x-\mu}{\sigma}\right)^2}, \ -\infty < x < \infty \\ &= \exp\left[-\left(\frac{x-\mu}{\sigma}\right)^2 - \ln\sqrt{2\pi} - \ln\sigma\right] \\ &= \exp\left[\frac{\mu x}{\sigma^2} - \frac{x^2}{2\sigma^2} - \ln\sqrt{2\pi\sigma^2} - \frac{\mu^2}{2\sigma^2}\right]. \end{aligned}$$

Let σ^2 be a positive constant. Here

$$\begin{aligned} P(\theta) &= \frac{\mu}{\sigma^2}, \ k(x) = x, \\ s(x) &= -\frac{x^2}{2\sigma^2} - \ln\sqrt{2\pi\sigma^2}, \ q(\theta) = -\frac{\mu^2}{2\sigma^2}. \end{aligned}$$

Let $x_1, x_2, \ldots, x_n$ be a random sample of $N(\mu, \sigma^2)$. Then

$$\begin{aligned} L &= \left(\frac{1}{2\pi\sigma^2}\right)^{n/2} e^{-\frac{1}{2\sigma^2}\Sigma(x-\mu)^2} \\ &= \exp\left[\frac{\mu\Sigma x}{\sigma^2} - \frac{\Sigma x^2}{2\sigma^2} - \ln(2\pi\sigma^2)^{n/2} - \frac{n\mu^2}{2\sigma^2}\right] \\ &= \exp\left[P(\theta)\sum^{n} k(x) + \sum^{n} s(x) + nq(\theta)\right] \\ &= \exp\left[P(\theta)\Sigma k(x) + nq(\theta)\right]\exp[\Sigma s(x)]. \end{aligned}$$

Here L is factorized into two factors, where one factor is a function of $T = \Sigma k(x) = \Sigma x$ and parameter $\theta = \mu$ [when σ^2 is known]. Hence $T = \Sigma x$ is sufficient statistic for μ.

Here $g(T, \theta) = R(y)\exp[P(\theta)y + nq(\theta)]$.

Definition. Let $x_1, x_2, \ldots, x_n$ be a random sample of n observations from a population whose p.d.f. is $f(x, \theta)$, $\theta \in S$, $\gamma < \theta < \delta$. The density function $f(x, \theta)$ is a regular member of exponential class of p.d.f. Then $T = \Sigma k(x)$ is a sufficient statistic for θ and $g(t, \theta)$, $\gamma < \theta < \delta$ is the member of p.d.f. of T. This T is the complete sufficient statistic for θ.

In the above example $T = \Sigma k(x) = n\bar{x}$ is the complete sufficient statistic for μ when σ^2 is known. Here $E(T) = \theta$ and $h(T) = T/n = \bar{x}$ is the continuous function of T, where T is unbiased estimator of μ. Again, since T is sufficient statistic, it is MVUE. Thus $\bar{x}$ is the unique best estimator of μ.

►**EXAMPLE 32.67** *Let $x_1, x_2, \ldots, x_n$ be a random sample of n observations from Poisson population having parameter θ. Find complete sufficient statistic for θ.*

Solution. We know

$$P(x) = \frac{e^{-\theta}\theta^x}{x!}, \ x = 0, 1, 2, \ldots; \ 0 < \theta < \infty.$$

Here

$$\begin{aligned} P(x) &= \exp[x \ln \theta - \ln(x!) - \theta] \\ L &= \exp[\Sigma x \ln \theta - \Sigma \ln(x!) - n\lambda] \\ &= \exp[\Sigma x \ln \theta - n\lambda] \exp[-\Sigma \ln(x!) \end{aligned}$$

where $T = \Sigma k(x) = \Sigma x$ is complete sufficient statistic for θ. Again,

$$E(T) = n\lambda \text{ and } h(T) = T/n = \bar{x}$$

is the sufficient statistic for θ and hence it is MVUE. Thus $\bar{x}$ is the unique best estimator of θ.

►**EXAMPLE 32.68** *Let $x_1, x_2, \ldots, x_n$ be a random sample of n observations from a population having p.d.f.*

$$f(x, \theta) = \frac{1}{6\theta^4} x^3 e^{-x/\theta}; \; 0 < x < \infty, \; 0 < \theta < \infty.$$

Find complete sufficient statistic for θ.

Solution. Here

$$\begin{aligned} L &= \frac{1}{(6\theta^4)^n} \prod_{i=1}^{n} x_i^3 e^{-n\bar{x}/\theta} \\ &= \exp\left[-\frac{\Sigma x}{\theta} + \Sigma \ln x_i^3 - n \ln(6\theta^4)\right] \\ &= \exp\left[-\frac{\Sigma x}{\theta} - n \ln(6\theta^4)\right] \exp[\Sigma \ln x_i^3]. \end{aligned}$$

Thus $T = \Sigma k(x)$ is the complete sufficient statistic for θ. Here $T = \Sigma x = n\bar{x}$. Now, $h(T) = T/n = \bar{x}$ is the sufficient statistic for θ. As $\bar{x}$ is sufficient statistic, it is MV estimator. Again,

$$E(x) = 4\theta \text{ or, } E(x/4) = \theta \text{ or, } E\left(\frac{n\bar{x}}{4}\right) = \theta.$$

Hence $T/4n$ is the unbiased estimator and it is sufficient statistic and so it is the unique best estimator of θ.

k-Parameter Exponential Class of Probability Density Function

Let $x_1, x_2, \ldots, x_n$ be a random sample of n observations from a population whose p.d.f. is $f(x; \theta_1, \theta_2, \ldots, \theta_k)$. This density function is called k-parameter exponential class of probability density function if it is expressed as follows:

$$f(x; \theta_1, \theta_2, \ldots, \theta_k) = \exp\left[\sum^{k} p_j(\theta_1, \theta_2, \ldots, \theta_k) k_j(x) + S(x) + q(\theta_1, \theta_2, \ldots, \theta_k)\right]$$

where $a < x < b$. This probability density function is called *regular member* of k-parameter exponential class of family, if

(i) a and b do not depend on any one or all of $\theta_1, \theta_2, \dots, \theta_k$,

(ii) $P_j(\theta_1, \theta_2, \dots, \theta_k)$ is the continuous function of θ_j, $\gamma_j < \theta_j < \delta_j$ and the function is nontrivial,

(iii) $k'_j(x)$; $j = 1, 2, \dots, k$ is continuous function for $a < x < b$ and no function is linearly homogeneous of others,

(iv) for $a < x < b$ the function $S(x)$ is the continuous function of x.

The above conditions are valid for continuous exponential class of family of p.d.f. For discrete exponential class of family of probability function the conditions are

(i) The set $\{x : x = a_1, a_2, \dots\}$ does not depend on any parameter,

(ii) $P_j(\theta_1, \theta_2, \dots, \theta_k)$; $j = 1, 2, \dots, k$ is nontrivial and independent continuous function of θ_j; $\gamma_j < \theta_j < \delta_j$,

(iii) $k_j(x); j = 1, 2, \dots, n$ is the nontrivial function of set of $x\{x : x = a_1, a_2, \dots\}$ and no one is the function of other.

For example, let us consider the $N(\mu, \sigma^2)$ where

$$\begin{aligned} f(x; \mu, \sigma^2) &= \frac{1}{\sigma\sqrt{2\pi}} \exp\left[-\frac{1}{2\sigma^2}(x-\mu)^2\right], \ -\infty < x < \infty \\ &= \exp\left[\frac{\mu x}{\sigma^2} - \frac{x^2}{2\sigma^2} - \frac{\mu^2}{2\sigma^2} - \ln\sqrt{2\pi\sigma^2}\right] \end{aligned}$$

where

$$\begin{aligned} k_1(x) &= x^2;\ k_2(x) = x;\ j = 1, 2 \\ P_1(\theta_1, \theta_2) &\doteq P(\mu, \sigma^2) = -\frac{1}{2\sigma^2},\ P_2(\theta_1, \theta_2) = P_2(\mu, \sigma^2) = \mu/\sigma^2 \\ q(\mu, \sigma^2) &= -\frac{\mu^2}{2\sigma^2} - \ln\sqrt{2\pi\sigma^2};\ S(x) = 0. \end{aligned}$$

Joint Sufficient Statistic

Let $x_1, x_2, \dots, x_n$ be a random sample of n observations from a population whose p.d.f. is $f(x; \theta_1, \theta_2)$; $(\theta_1, \theta_2) \in S$, where S is parametric space. Let us consider n statistics

$$T_1 = t_1(x_1, x_2, \dots, x_n),\ T_2 = t_2(x_1, x_2, \dots, x_n), \cdots$$
$$\cdots,\ T_n = t_n(x_1, x_2, \dots, x_n).$$

Assume that the transformation of x to T is one-to-one and Jacobian of the transformation is available. Now, if $g(T_1, T_2, \dots, T_n)$ is the joint density function of $T_1, T_2, \dots, T_n$ and $g_{12}(T_1, T_2)$ is the joint density function of T_1 and T_2, then T_1 and T_2 are called *joint sufficient statistics* of θ_1 and θ_2, if and only if the conditional distribution of $T_3, T_4, \dots, T_n$ does not depend on θ_1 or θ_2 or on both, i.e.,

$$h(T_3, T_4, \dots, T_n/T_1 = t_1, T_2 = t_2) = \frac{g(T_1, T_2, \dots, T_n;\ \theta_1, \theta_2)}{g_{12}(T_1, T_2;\ \theta_1, \theta_2)}$$

or, $g(T_1, T_2, \dots, T_n;\ \theta_1, \theta_2) = h(T_3, T_4, \dots, T_n/T_1 = t_1, T_2 = t_2) g_{12}(T_1, T_2;\ \theta_1, \theta_2).$

Thus, using Fisher-Neyman criterion, we can write

$$L = g_{12}[t_1(x_1, x_2, \ldots, x_n)t_2(x_1, x_2, \ldots, x_n); \theta_1, \theta_2]h(x_1, x_2, \ldots, x_n).$$

If L can take the above form, T_1 and T_2 are called *joint sufficient statistic* of θ_1 and θ_2.

As an example, we can mention the case of Example 32.56, where $T(\bar{x}, s^2)$ is the joint sufficient statistic for $\{\mu, \sigma^2\}$.

►**EXAMPLE 32.69** *Let $x_1, x_2, \ldots, x_n$ be a random sample of n observations from a population whose p.d.f. is*

$$f(x; \theta_1, \theta_2) = \begin{cases} 1/2\theta_2; & \theta_1 - \theta_2 < x < \theta_1 + \theta_2, \ -\infty < \theta_1 < \infty \\ & \qquad\qquad\qquad\qquad\qquad -\infty < \theta_2 < \infty \\ 0, & \text{otherwise.} \end{cases}$$

Find joint sufficient statistic for θ_1 and θ_2.

Solution. We have

$$L = \left(\frac{1}{2\theta_2}\right)^n.$$

Here the regularity conditions do not valid and for that reason the sufficient statistic is to be found out from the distribution of order statistic. Let $y_1 < y_2 < \cdots < y_n$ be the order statistics of $x_1, x_2, \ldots, x_n$. Then

$$h_{1n}(y_1, y_n; \theta_1, \theta_2) = \frac{n(n-1)}{(2\theta_2)^n}(y_n - y_1)^{n-2}; \ \theta_1 - \theta_2 < y_1 < y_2 < \theta_1 + \theta_2.$$

Thus L can be written as

$$\begin{aligned} L &= \frac{n(n-1)}{(2\theta_2)^n}(y_n - y_1)^{n-2}\frac{1}{n(n-1)(y_n - y_1)^{n-2}} \\ &= g_{12}(T_1, T_2; \theta_1, \theta_2)h(x_1, x_2, \ldots, x_n) \end{aligned}$$

where

$$\begin{aligned} g_{12}(T_1, T_2; \theta_1, \theta_2) &= \frac{n(n-1)}{(2\theta_2)^n}(y_n - y_1)^{n-2} \\ \text{and } h(x_1, x_2, \ldots, x_n) &= 1/\left[n(n-1)(y_n - y_1)^{n-2}\right]. \end{aligned}$$

Hence y_1 and y_n are the joint sufficient statistics of θ_1 and θ_2.

Joint Complete Sufficient Statistics

Let $x_1, x_2, \ldots, x_n$ be a random sample of n observations from a population whose density function is $f(x; \theta_1, \theta_2, \ldots, \theta_k)$, where $(\theta_1, \theta_2, \ldots, \theta_k) \in S$. This density function belongs to k-parameter exponential family of distribution if

$$f(x; \theta_1, \theta_2, \ldots, \theta_k) = \exp\left[\sum^k P_j(\theta_1, \theta_2, \ldots, \theta_k)K_j(x) + S(x) + q(\theta_1, \ldots, \theta_k)\right].$$

The joint p.d.f. of $x_1, x_2, \ldots, x_n$ becomes

$$L(x_1, x_2, \ldots, x_n)$$
$$= \exp\left[\sum_j^k P_j(\theta_1, \theta_2, \ldots, \theta_k) \sum_i^n K_i(x_i) + \sum_i^n S(x_i) + nq(\theta_1, \theta_2, \ldots, \theta_k)\right]$$
$$= \exp\left[\sum_j P_j(\theta_1, \theta_2, \ldots, \theta_k) \sum_i^n K_i(x_i)\right] \exp\left[\sum_i^n S(x_i)\right].$$

Therefore,

$$T_1 = \sum_{i=1}^n K_1(x_i),\ T_2 = \sum_{i=1}^n K_2(x_i),\ \ldots,\ T_k = \sum_{i=1}^n K_k(x_i)$$

are the joint complete sufficient statistics of $\theta_1, \theta_2, \ldots, \theta_k$.

As an example, we can mention the sufficient statistics of Example 32.56 $[T(\bar{x}, s^2)]$ since the distribution of that example is a 2-parameter exponential class of probability distribution and $T(\bar{x}, s^2)$ is joint sufficient statistics of μ and σ^2. Here

$$Z_1 = \frac{T_1}{n} = \bar{x},\ Z_2 = \frac{1}{n-1}\sum(x_i - \bar{x})^2.$$

The transformation of Z_1 and Z_2 from T_1 and T_2 is one-to-one and hence Z_1 and Z_2 is joint complete sufficient statistic of μ and σ^2. Again, $E(Z_1) = \mu$ and $E(Z_2) = \sigma^2$. Thus Z_1 and Z_2 are unbiased estimators of μ and σ^2, respectively. Since Z_1 and Z_2 are sufficient statistics, they are MVU also.

►**EXAMPLE 32.70** *Let $x_1, x_2, \ldots, x_n$ be a random sample of n observations from a population whose p.d.f. is*

$$f(x; \alpha, \beta) = \frac{1}{\beta^\alpha \Gamma\alpha} e^{-x/\beta} x^{\alpha-1},\ 0 < x < \infty;\ \alpha > 0,\ \beta > 0.$$

Find joint complete sufficient statistic of α and β.

Solution. The likelihood function is

$$L = \left[\frac{1}{\beta^\alpha \Gamma\alpha}\right]^n e^{-n\bar{x}/\beta} \prod_{i=1}^n x_i^{\alpha-1}$$
$$= \exp\left[-\frac{\Sigma x}{\beta} + \alpha\sum_{i=1}^n \ln x_i - \sum_i^n \ln x_i - n\ln\Gamma\alpha - n\ln\beta^\alpha\right].$$

Hence $T_1 = \Sigma x$, $T_2 = \Sigma \ln x_i$ are the joint complete sufficient statistics of α and β since the distribution is 2-parameter exponential family of distribution.

For the given distribution if β is known, then

$$E(x/\beta) = \alpha \Rightarrow E(\bar{x}/\beta) = \alpha.$$

Therefore, $T_1/n\beta$ is unbiased estimator of α. Again, this estimator is a function of sufficient statistic and hence it is MVUE. Hence for known β, $\bar{x}/\beta$ is the best estimator.

Minimal Sufficient Statistics

The set of joint complete sufficient statistics is called *minimal sufficient statistics*, if and only if the set is the function of set of other joint sufficient statistics. Thus for k-parameter exponential class of distribution

$$T_1 = \sum_i^n K_1(x_i),\ T_2 = \sum_i^n K_2(x_i),\ \ldots,\ T_k = \sum_i^n K_k(x_i)$$

are the joint sufficient statistics of $\theta_1, \theta_2, \ldots, \theta_k$ and this set of statistics is minimal sufficient statistics.

As an example, we can mention the joint sufficient statistics of μ and σ^2 in $N(\mu, \sigma^2)$, where $T(\bar{x}, s^2)$ is the joint sufficient statistics of μ and σ^2. This set of sufficient statistics is also minimal sufficient statistics.

Lehmann-Scheffe Theorem

❒ **THEOREM 32.12** *Let $x_1, x_2, \ldots, x_n$ be a random sample of n observations from a population whose p.d.f. is $f(x, \theta)$, $\theta \in S$. Let $T = t(x_1, x_2, \ldots, x_n)$ be the complete sufficient statistics for θ and $T_1 = f(T)$ is the function of T. This function is unbiased estimator of $\phi(\theta)$. Then T_1 is the MVUE of $\phi(\theta)$.*

Proof. Let T' be the unbiased estimator of $\phi(\theta)$ and T' is also the function of $T[T' = f(T)]$. Then $E[T_1 - T'] = 0$. Here $(T_1 - T')$ is the function of T. Since T is complete sufficient statistic $P[f(T) = f(T_1)] = 1$ and this is true for all $\theta \in S$. Hence there exists an unbiased estimator of $\phi(\theta)$. Let this unbiased estimator be 't'. Since $E[t/T] = \phi(\theta)$, T_1 must be $F(t/T)$. Again, since T is the complete sufficient statistic $V(T) \leq V(t)$. This is true for all $\theta \in S$.

Unique Best Statistic

Let $x_1, x_2, \ldots, x_n$ be a random sample of n observations from a population having p.d.f. $f(x, \theta)$, $\theta \in S$. Consider that $T = t(x_1, x_2, \ldots, x_n)$ is the sufficient statistic for θ and $g(T, \theta)$ is the p.d.f. of complete sufficient statistic. Now, if there exists any continuous function of T which is unbiased for θ, then that function is the unique best unbiased estimator of θ.

►**EXAMPLE 32.71** *Let $x_1, x_2, \ldots, x_n$ be a random sample of n observations from Poisson population having parameter θ. Find UMVUE of $e^{-\theta}$.*

Solution. We know

$$P(x) = \frac{e^{-\theta}\theta^x}{x!},\ x = 0, 1, 2, \ldots .$$

Let $T = \Sigma x$ is the sufficient statistic of θ. Also, let $\psi(T_1)$ is the estimator of $e^{-\theta}$. Now, we can find the value of $\psi(T_1)$ from $E[T_1/T = t]$. To find this expected value, let us define a dummy variable $u(x_1)$ such that $u(x_1) = 1$ if $x_1 \neq 0$ and $u(x_1) = 0$, if $x_1 = 0$.

Then

$$\begin{aligned}
E[u(x_1)] &= 1 \cdot P(x_1 = 0) + 0 \cdot P(x_1 \neq 0) = P(x_1 = 0) = e^{-\theta} \\
\psi(T) &= E[u(x_1)/T = t] \\
&= 1P(x_1 = 0/\Sigma x_i = t) + 0P(x_1 \neq 0/\Sigma x_i = t) \\
&= P(x_1 = 0/\Sigma x_i = t) \\
&= \frac{P[x_1 = 0, \Sigma x_i = t]}{P[\Sigma x_i = t]} = \frac{P[x_1 = 0]\Sigma\left[P \sum_{i=2}^{n} x_i = t\right]}{P[\Sigma x_i = t]} \\
&= \frac{e^{-1}e^{-(n-1)\theta}[(n-1)\theta]^t/t!}{e^{-n\theta}(n\theta)^t/t!} = \left(\frac{n-1}{n}\right)^t, \ n > 1.
\end{aligned}$$

Therefore, $\left(\frac{n-1}{n}\right)^{\Sigma x_i}$ is the UMVUE of $e^{-\theta}$.

►**EXAMPLE 32.72** *Let $x_1, x_2, \ldots, x_n$ be a random sample of n observations from a population whose p.d.f. is*

$$f(x) = \theta e^{-\theta x}, \ 0 < x < \infty.$$

Find UMVUE of (i) $1/\theta$, (ii) θ *and* (iii) $e^{-k\theta}$.

Solution.

(i) The likelihood function is $L = \exp[-\theta\Sigma x_i + n \ln \theta]$.
Therefore, $T = \Sigma x_i$ is the complete sufficient statistic for θ. Again,

$$E(x) = \theta \int_0^\infty x e^{-\theta x} dx = \frac{1}{\theta} \Rightarrow E(\bar{x}) = \frac{1}{\theta}.$$

Hence $\bar{x}$ is an unbiased estimator of $1/\theta$. Since $T = \Sigma x_i$ is the complete sufficient statistic, $\bar{x} = T/n$ is the function of complete sufficient statistic and hence $\bar{x}$ is UMVUE. Then $V(\bar{x}) = \frac{1}{n\theta^2}$.

(ii) Let the estimator of θ be $C/\Sigma x_i$, where C is a constant which depends on n.

$$E\left[\frac{C}{\Sigma x_i}\right] = CE\left[\frac{1}{T}\right] = C\int_0^\infty \frac{1}{T} f(T) dT$$

But the distribution of T is general gamma distribution. Hence

$$\begin{aligned}
E\left[\frac{C}{\Sigma x_i}\right] &= \frac{C}{\Gamma n}\int_0^\infty \frac{1}{T}\theta^n T^{n-1} C^{-\theta T} dT \\
&= \frac{C}{\Gamma n}\int_0^\infty \theta^n T^{n-2} e^{-\theta T} dT = \frac{\theta C}{\Gamma n}\int_0^\infty u^{n-2} e^{-u} du \\
&= \frac{\theta C}{\Gamma n}\Gamma(n-1) = \frac{\theta C}{n-1}, \text{ where } n > 1 \\
E\left[\frac{n-1}{\Sigma x_i}\right] &= \theta \Rightarrow (n-1)/\Sigma x_i \text{ is UMVUE of } \theta \\
V\left[\frac{n-1}{\Sigma x_i}\right] &= \frac{\theta^2}{n-2}, \ n > 2.
\end{aligned}$$

(iii) Let $u(x_1)$ be a dummy variable such that $u(x_1) = 1$, if $x_1 \geq k$ and $u(x_1) = 0$, if $x_1 < k$. Then

$$E[u(x_1)] = 1 \cdot P(x_1 \geq k) + 0 \cdot P(x < k) = P(x_1 \geq k) = e^{-k\theta}.$$

Then $E[u(x_1)/T = t]$ is the UMVUE of $e^{-k\theta}$, where $T = \sum^{n} x_i$

$$\begin{aligned} E[u(x_1)/T = t] &= 1 \cdot P[x_1 \geq k/T = t] + 0 \cdot P[x < k/T = t] \\ &= P[x_1 \geq T = t]. \end{aligned}$$

The conditional distribution of x_1 under the condition $T = t$ is

$$\begin{aligned} fx_1(x_1/T = t)\Delta x_1 &= \frac{fx_1T(x_1t_1)\Delta x_1 \Delta t}{f(t)\Delta t} \\ &= \frac{P[x_1 < X_1 < x_1 + \Delta x_1 \cdot t < T < t + \Delta t]}{\frac{1}{\Gamma n}\theta e^{-\theta t}(\theta t)^{n-1}\Delta t} \\ &= \frac{P[x_1 < X_1 < x_1 + \Delta x_1 \cdot t - x_1 < \Sigma x_i < (t + x_1) + \Delta t]}{\frac{1}{\Gamma n}e^{-\theta t}(\theta t)^{n-1}\theta \Delta t} \\ &= \frac{\theta e^{-\theta x}\frac{1}{\Gamma n-1}e^{-\theta(t-x_1)[\theta(t-x_1)]^{n-2}}\theta \Delta t \Delta x_1}{\frac{1}{\Gamma n}e^{-\theta t}(\theta t)^{n-1}\theta \Delta t} \\ &= \frac{(n-1)(t-x_1)^{n-2}\Delta x_1}{t^{n-1}}, \quad x_1 < t. \end{aligned}$$

Now,

$$\begin{aligned} E[u(x_1)/T = t] &= P[x_1 \geq k/T = t] \\ &= \int_k^t \frac{(n-1)(t-x_1)^{n-2}dx_1}{t^{n-1}} \\ &= \frac{n-1}{t^{n-1}}\int_k^t (t-x_1)^{n-2}dx_1. \end{aligned}$$

Let $(t - x_1) = y \Rightarrow dx_1 = -dy,\ t - k < y < 0.$

$$\begin{aligned} \therefore\ E[u(x_1)/T = t] &= \frac{n-1}{t^{n-1}}\int_0^{t-k} y^{n-2}dy \\ &= \frac{n-1}{t^{n-1}}\frac{(t-k)^{n-1}}{n-1} \\ &= \left(1 - \frac{k}{t}\right)^{n-1}. \end{aligned}$$

Therefore, $\left(1 - \dfrac{k}{\Sigma x_i}\right)^{n-1}$ is UMVUE of $e^{-k\theta}$.

►**EXAMPLE 32.73** *Let $x_1, x_2, \ldots, x_n$ be a random sample of n observations from Bernoulli population with parameter p. Find unique best estimator of $p(1-p)/n$.*

Solution. From Example 32.41, it is observed that $\frac{r}{n} = \bar{x}$ is the MVB estimator of p. This $\bar{x}$ is unique best estimator and its variance is $p(1-p)/n$. Here $T = \Sigma x_i$ is sufficient statistic [Example 32.52] and hence $\bar{x} = T/n$ is a function of sufficient statistic.

Consider an estimator $\frac{T}{n}\left(1-\frac{T}{n}\right)/n$. Then

$$\begin{aligned}
E\left[\frac{T}{n}\left(1-\frac{T}{n}\right)/n\right] &= \frac{1}{n}E\left[\frac{T}{n}\left(1-\frac{T}{n}\right)\right] = \frac{1}{n^2}E(T) - \frac{1}{n^3}E(T^2) \\
&= \frac{np}{n^2} - \frac{1}{n^3}[np(1-p)+n^2p^2], \quad \because \; V(T) = npq \\
&= \frac{p}{n} - \frac{p(1-p)}{n^2} - \frac{p^2}{n} = \frac{p(1-p)}{n} - \frac{p(1-p)}{n^2} \\
&= \frac{p(1-p)}{n}\left[1-\frac{1}{n}\right] = \frac{n-1}{n^2}p(1-p).
\end{aligned}$$

$\frac{n}{n-1}\left[\frac{T}{n}\left(1-\frac{T}{n}\right)/n\right]$ is an unbiased estimator of $\frac{p(1-p)}{n}$. Here

$$\frac{n}{n-1}[\bar{x}(1-\bar{x})/n] = \frac{\bar{x}(1-\bar{x})}{n-1}$$

is an unbiased estimator of $\frac{p(1-p)}{n}$. This estimator is unique best estimator.

►**EXAMPLE 32.74** *Let $x_1, x_2, \ldots, x_n$ be a random sample of n observations from $N(0,\theta)$. Find unique best estimator of θ.*

Solution. We have $L = \exp\left[-\frac{\Sigma x_i^2}{\theta} - \ln(2\pi\theta)^{n/2}\right]$.

Therefore, $T = \Sigma x_i^2$ is a complete sufficient statistic for θ. Since $x_i \sim N(0,\theta)$, $x_i^2 \sim \chi^2\theta$ with 1 d.f. and $\Sigma x_i^2 \sim \chi^2\theta$ with n d.f. Hence

$$\begin{aligned}
E(\Sigma x_i^2) &= n\theta \text{ and } V[\Sigma x_i^2] = 2n\theta^2 \\
V[\Sigma x_i^2] &= E(\Sigma x_i^2)^2 - [E(\Sigma x_i^2)]^2 \\
E[(\Sigma x_i^2)^2] &= V[\Sigma x_i^2] + [E(\Sigma x_i^2)]^2 = 2n\theta^2 + n^2\theta^2 = n\theta^2(n+2) \\
\text{or, } E\frac{(\Sigma x_i^2)^2}{n(n+2)} &= \theta^2.
\end{aligned}$$

Hence $\frac{(\Sigma x_i^2)^2}{n(n+2)}$ is the unique best estimator of θ^2.

Location Parameter

Consider that $f(x,\theta);\, \theta \in S$ is a member of probability density function. The parameter θ in $f(x,\theta)$ is called *location parameter* if and only if $f(x,\theta) = h(x-\theta)$, where $h(x-\theta)$ is a function of $(x-\theta)$. In other words, θ is called *location parameter* if the p.d.f. of x is $f(x,\theta)$ and if the distribution of $(x-\theta)$ does not depend on θ. For example, let us consider the p.d.f. of x, where x follows $N(\mu,\sigma^2)$. Here μ is called *location parameter* since

$$z = \frac{x-\mu}{\sigma} \sim N(0,1) \text{ and } f(z) = \frac{1}{\sqrt{2\pi}}e^{-\frac{1}{2}z^2}.$$

The distribution of $z = \frac{x-\mu}{\sigma}$, where σ is assumed to be a constant, is independent of μ.

Scale Parameter

Consider that $f(x,\theta)$, $\theta \in S$ is a member of probability density function. The parameter θ in $f(x,\theta)$ is called *scale parameter*, if and only if

$$f(x,\theta) = \frac{1}{\theta} h\left(\frac{x}{\theta}\right)$$

where $h\left(\frac{x}{\theta}\right)$ is the function of $\frac{x}{\theta}$. This implies that θ will be scale parameter if the distribution of $\frac{x}{\theta}$ is independent of θ. Thus, when $x \sim N(\mu, \sigma^2)$ and $z = \dfrac{x-\mu}{\sigma} \sim N(0,1)$, then σ is called *scale parameter* since $h(z)$ is free of σ.

Estimator of Location Parameter

Let $x_1, x_2, \dots, x_n$ be a random sample of n observations from a population having p.d.f. $f(x,\theta)$, $\theta \in S$ and θ is the location parameter. Consider that $T = t(x_1, x_2, \dots, x_n)$ is an estimator of θ, where

$$T = \frac{\displaystyle\int \theta L(x,\theta) d\theta}{\displaystyle\int L(x,\theta) d\theta}$$

where $L(x,\theta)$ is the likelihood function of sample observations. The estimator T is known as Pitman estimator and its mean square error is uniquely minimum.

►**EXAMPLE 32.75** *Let $x_1, x_2, \dots, x_n$ be a random sample of n observations from a population whose p.d.f. is*

$$f(x) = 1,\ \theta - \frac{1}{2} < x < \theta + \frac{1}{2}.$$

Find Pitman estimator of θ.

Solution. Let $z = x - \theta \Rightarrow dx = dz, -\frac{1}{2} < z < \frac{1}{2}$. Then $f(z) = 1$.

Let $T = t(x_1, x_2, \dots, x_n)$ be an estimator of θ. Then

$$T = \frac{\displaystyle\int \theta L(x,\theta) d\theta}{\displaystyle\int L(x,\theta) d\theta} = \frac{\displaystyle\int \theta d\theta}{\displaystyle\int d\theta}, \quad \because\ L(x) = 1.$$

As $\theta - \frac{1}{2} < x < \theta + \frac{1}{2}$, we can write $x - \frac{1}{2} < \theta < x + \frac{1}{2}$. Again, $\theta - \frac{1}{2} < y_1 < y_2 \cdots < y_n < \theta + \frac{1}{2}$, where $y_1 < y_2 < \cdots < y_n$ are the order statistics of sample observations. Therefore, $y_n - \frac{1}{2} < \theta < y_1 + \frac{1}{2}$. Then

$$T = \frac{\displaystyle\int_{y_n - \frac{1}{2}}^{y_1 + \frac{1}{2}} \theta d\theta}{\displaystyle\int_{y_n - \frac{1}{2}}^{y_1 + \frac{1}{2}} d\theta} = \frac{1}{2} \frac{\left(y_1 + \frac{1}{2}\right)^2 - \left(y_n - \frac{1}{2}\right)^2}{\left(y_1 + \frac{1}{2}\right) - \left(y_n - \frac{1}{2}\right)}$$

$$= \frac{1}{2}\left[\left(y_1 + \frac{1}{2}\right) + \left(y_n - \frac{1}{2}\right)\right] = \frac{y_1 + y_n}{2}.$$

Location Invariant

Let $x_1, x_2, \ldots, x_n$ be a random sample of n observations from a population whose p.d.f. is $f(x, \theta)$; $\theta \in S$, where θ is a location parameter. Let $T = t(x_1, x_2, \ldots, x_n)$ be an estimator of θ. Now, if the sample observations $x_1, x_2, \ldots, x_n$ are changed by $x_1 + C, x_2 + C, \ldots, x_n + C$, where C is a constant, then the estimator of θ may be $T = t(x_1 + C, x_2 + C, \ldots, x_n + C)$. Now, if $T_1 = t(x_1, x_2, \ldots, x_n) + C$ for all x and θ, then T will be location parameter.

Let

$$T = \frac{1}{n}\sum^{n} x_i \text{ and } T_1 = \frac{\Sigma(x_i + C)}{n}.$$

Here $T_1 = \dfrac{\Sigma x_i}{n} + C = T + C$. Therefore, $T = \bar{x}$ is location invariant.

Sufficiency and Pitman Estimator

❑ **THEOREM 32.13** *The Pitman estimator for location parameter is sufficient statistic.*

Proof. Let

$$T_1 = t_1(x_1, x_2, \ldots, x_n),\ T_2 = t_2(x_1, x_2, \ldots, x_n) \ldots$$
$$\ldots,\ T_k = t_k(x_1, x_2, \ldots, x_n)$$

be a set of sufficient statistics. Then

$$L(x_1, x_2, \ldots, x_n/\theta) = g(T_1, T_2, \ldots, T_k, \theta)h(x_1, x_2, \ldots, x_n).$$

Now, if $T = t(x_1, x_2, \ldots, x_n)$ is the estimator of location parameter, then

$$\begin{aligned} T &= \frac{\displaystyle\int \theta L(x_1, x_2, \ldots, x_n;\ \theta)d\theta}{\displaystyle\int L(x_1, x_2, \ldots, x_n)d\theta} \\ &= \frac{\displaystyle\int \theta g(T_1, T_2, \ldots, T_k; \theta)h(x_1, x_2, \ldots, x_n)d\theta}{\displaystyle\int g(T_1, T_2, \ldots, T_k; \theta)h(x_1, x_2, \ldots, x_n)d\theta} \\ &= \frac{\displaystyle\int \theta g(T_1, T_2, \ldots, T_k; \theta)d\theta}{\displaystyle\int g(T_1, T_2, \ldots, T_k; \theta)d\theta}. \end{aligned}$$

Here T is a function of sufficient statistic.

Estimation of Scale Parameter

Let $x_1, x_2, \ldots, x_n$ be a random sample of n observations from a population whose p.d.f. is $f(x, \theta)$; $\theta \in S$, where $\theta > 0$ is a scale parameter and $x_i > 0$. If $T = t(x_1, x_2, \ldots, x_n)$

is the estimator of scale parameter θ, then T should be

$$T = \frac{\displaystyle\int_0^\infty \frac{1}{\theta^2} L(x_1, x_2, \ldots, x_n; \theta) d\theta}{\displaystyle\int_0^\infty \frac{1}{\theta^3} L(x_1, x_2, \ldots, x_n; \theta) d\theta}.$$

This estimator T is also a function of sufficient statistic.

If the estimator $T = t(x_1, x_2, \ldots, x_n)$ is expressed as

$$T_1 = t_1(Cx_1, Cx_2, \ldots, Cx_n) = Ct_1(x_1, x_2, \ldots, x_n) = Ct(x_1, x_2, \ldots, x_n)$$

where $C > 0$, then T is called *scale invariant.* For example, let

$$T = \frac{1}{n}\Sigma x_i = \bar{x}, T_1 = \frac{\Sigma C x_i}{n} = C\bar{x} = CT$$

then T is scale invariant.

►EXAMPLE 32.76 *Let $x_1, x_2, \ldots, x_n$ be a random sample of observations from a population having p.d.f.*

$$f(x) = \frac{1}{\lambda} e^{-x/\lambda}; \ 0 < x < \infty, \ \lambda > 0.$$

Find Pitman estimator of λ.

Solution. Let $z = x/\lambda$, then $x = \lambda z$ and $dx = \lambda dz$; $0 < z < \infty$. We can write

$$f(z) = e^{-z}, \ 0 < z < \infty.$$

Hence λ is scale parameter. Again, let $T = t(x_1, x_2, \ldots, x_n)$ is the Pitman estimator of λ, where

$$T = \frac{\displaystyle\int_0^\infty \frac{1}{\lambda^2} L(x_1, x_2, \ldots, x_n; \lambda) d\lambda}{\displaystyle\int_0^\infty \frac{1}{\lambda^3} L(x_1, x_2, \ldots, x_n; \lambda) d\lambda} = \frac{\displaystyle\int_0^\infty \frac{1}{\lambda^{n+2}} \exp\left[-\frac{1}{\lambda}\Sigma x_i\right] d\lambda}{\displaystyle\int_0^\infty \frac{1}{\lambda^{n+3}} \exp\left[-\frac{1}{\lambda}\Sigma x_i\right] d\lambda}.$$

Consider that $\dfrac{\Sigma x_i}{\lambda} = \theta, \ \lambda = \dfrac{\Sigma x_i}{\theta}, \ d\lambda = -\dfrac{\Sigma x_i}{\theta^2} d\theta.$

$$\therefore \ T = \frac{\displaystyle\int_0^\infty \left(\frac{\theta}{\Sigma x_i}\right)^{n+2} e^{-\theta} \frac{\Sigma x_i}{\theta^2} d\theta}{\displaystyle\int_0^\infty \left(\frac{\theta}{\Sigma x_i}\right)^{n+3} e^{-\theta} \frac{\Sigma x_i}{\theta^2} d\theta} = \Sigma x_i \frac{\displaystyle\int_0^\infty \theta^n e^{-\theta} d\theta}{\displaystyle\int_0^\infty \theta^{n+1} e^{-\theta} d\theta}$$

$$= \Sigma x_i \frac{\Gamma(n+1)}{\Gamma(n+2)} = \frac{\Sigma x_i}{n+1}.$$

This T is the function of sufficient statistic. Again, we know that UMVUE [Example 32.39] of λ is $\bar{x}$, where $V(\bar{x}) = \lambda^2/n$. Here T is Pitman estimator, where

$$V(T) = \frac{\Sigma V(x)}{(n+1)^2} = \frac{n\lambda^2}{(n+1)^2} = \frac{\lambda}{n\left(1 + \frac{1}{n}\right)^2} < V(\bar{x}).$$

EXERCISE

1. (a) Define parameter, estimator, estimate and statistic with examples.
   (b) What are the different criteria of a good estimator?
   (c) Find MVB estimator of λ, where λ is the parameter of a Poisson population.

2. (a) When would you say that an estimator is good? Show, by an example, that a consistent estimator need not be unbiased.
   (b) Describe sufficiency of a statistic. Show that the estimator of μ in $N(\mu, \sigma^2)$ is sufficient statistic.
   (c) What do you mean by complete sufficiency and joint complete sufficiency? Find joint complete sufficient statistics for μ and σ^2 in $N(\mu, \sigma^2)$.

3. (a) Describe unbiased estimator and consistent estimator.
   (b) If $x_1, x_2, \ldots, x_n$ is a random sample of observations from $N(\mu, \sigma^2)$ and

$$T = \frac{1}{n} \sum_{i=1}^{n} |x_i - \mu|$$

   then show that $T\sqrt{\pi/2}$ is an unbiased estimator of σ.
   (c) Find unbiased estimator of 'a' if b is known in

$$f(x) = \frac{1}{b-a}, \quad a < x < b.$$

   Also, find the variance of your estimator.
   (d) If $\bar{x}_1, \bar{x}_2, \ldots, \bar{x}_k$ are the sample means based on samples of $n_1, n_2, \ldots, n_k$, respectively observations and if

$$t = \frac{n_1\bar{x}_1 + n_2\bar{x}_2 + \cdots + n_k\bar{x}_k}{k}$$

   is an estimator of μ. Find value of k so that 't' is unbiased for μ.

4. (a) Explain the criterion of a good estimator.
   (b) For a geometric distribution having p.d.f.

$$f(x) = \theta(1-\theta)^{x-1}, \; x = 1, 2, \ldots \; ; \; 0 < \theta < 1.$$

   Find unbiased estimator of (i) $1/\theta$ and (ii) $\theta(1-\theta)$.
   (c) For the hypergeometric distribution having probability function

$$P(x) = \frac{\binom{M}{x}\binom{N-M}{n-x}}{\binom{N}{n}}.$$

   Assuming n and N known, find unbiased estimator of M^2.

5. (a) Explain consistency, efficiency, unbiasedness and sufficiency of an estimator.
   (b) If t_1 and t_2 be two independent and unbiased estimators of θ, find unbiased estimators of (i) θ^2 and (ii) $\theta(\theta - 1)$.
   (c) Let $\hat{\theta}_n$ be an unbiased estimator of θ_n and $V(\hat{\theta}_n) = \sigma_n^2$. Assume that as $n \to \infty$, $\theta_n \to \theta$, $\sigma_n \to 0$. Then $\hat{\theta}$ is a consistent estimator of θ.

6. (a) Explain with examples the best estimator and best asymptotically normal estimator.

(b) If an estimator is mean square consistent for any parametric function, then it is simple consistent estimator of also, but the converse is not true. Prove it.

(c) Let $x_1, x_2, \ldots, x_n$ be a random sample of n observations from a population having p.d.f.

$$f(x) = \begin{cases} 1, & \theta < x < \theta + 1 \\ 0, & \text{otherwise.} \end{cases}$$

Let $\hat{\theta}_1 = \bar{x} - \frac{1}{2}$ and $\hat{\theta}_2 = \max(x_1, x_2, \ldots, x_n)$. Investigate the consistency and efficiency of the estimators.

7. (a) Explain minimum variance unbiased estimator with examples. If 't' is an unbiased estimator of θ, find Cramer-Rao bound of t.

(b) Let $x_1, x_2, \ldots, x_n$ be a random sample of n observations from a population having p.d.f.

$$f(x) = \begin{cases} \frac{2}{\theta^2}(\theta - x), & 0 \le x \le \theta \\ 0, & \text{otherwise.} \end{cases}$$

Let $\hat{\theta}_1 = x$ and $\hat{\theta}_2 = 3x$. Find relative efficiency of $\hat{\theta}_1$ compared to $\hat{\theta}_2$.

(c) Given

$$f(x) = \frac{1}{\theta^m \Gamma m} e^{-x/\theta} x^{m-1};\ 0 < x < \infty,\ \theta > 0.$$

Assuming m known, find MVB estimator of θ. What would be the MVB estimator if $m = 2$?

8. (a) Explain uniformly minimum variance estimator and locally minimum variance unbiased estimator.

(b) If T_1 and T_2 are the MVU estimators of θ, then $T_1 = T_2$. Prove it.

(c) If λ is the parameter of a Poisson population, then sample variance s^2 is an UMVUE of λ. Prove it.

(d) Let $x_1, x_2, \ldots, x_n$ be a random sample of n observations from Cauchy population with parameter μ. Show that the Cramer-Rao lower bound of variance of an unbiased estimator of μ is $2/n$.

9. (a) Explain sufficiency and joint sufficiency of estimators. Show that, if t is a consistent estimator of θ, then e^t is a consistent estimator of e^θ.

(b) State and prove Fisher-Neyman lemma related to sufficient statistic.

(c) Let $x_1, x_2, \ldots, x_n$ be a random sample of n observations from a population with p.d.f.

$$f(x) = \theta x^{\theta - 1},\ 0 < x < 1.$$

Show that $T = x_1, x_2, \ldots, x_n$ is a sufficient statistic for θ. Also, show that the estimator of θ^2 is a function of sufficient statistic.

10. (a) Define efficient estimator. State and prove Rao-Blackwell theorem related to sufficient statistic and minimum variance estimator.

(b) What do you mean by complete sufficiency? Find complete sufficient statistic for θ in

$$f(x,\theta) = \frac{1}{\theta};\ 0 < x < \theta,\ 0 < \theta < \infty.$$

(c) If θ is the parameter of a Poisson distribution, find unique best estimator of θ.

11. (a) Define joint sufficient statistics. Find joint sufficient statistics for μ and σ^2 in $N(\mu, \sigma^2)$.

(b) Given

$$f(x) = \frac{1}{\beta^\alpha \Gamma\alpha} e^{-x/\beta} x^{\alpha-1};\ 0 < x < \infty,\ \alpha > 0,\ \beta > 0.$$

Find joint complete sufficient statistics for α and β.

(c) What is meant by best unique estimator?

12. (a) Establish a necessary and sufficient condition for an unbiased estimator to be MVU estimator.

(b) Explain the concept of best unbiased estimator. Obtain the Cramer-Rao lower bound for the variance of the unbiased estimator of a parametric function.

(c) Let $x_1, x_2, \ldots, x_n$ be a random sample of n observations from a population with p.d.f.

$$f(x) = \theta e^{-\theta x},\ 0 < x < \infty.$$

Find UMVUE of (i) $1/\theta$, (ii) θ.

13. (a) Define sufficient statistic. State and prove the factorization theorem on sufficiency.

(b) If T_1 and T_2 are both efficient estimators with variance v and if

$$T = \frac{1}{2}(T_1 + T_2)$$

show that the variance of T is $\frac{v}{2}(1+\rho)$, where ρ is the correlation coefficient of T_1 and T_2. Also, show that when $\rho = 1$, T is efficient estimator.

(c) Let $x_1, x_2, \ldots, x_n$ be a random sample of n observations from $N(0, \theta)$. Find unique best estimator of θ.

14. (a) What are location and scale parameters? Show that Pitman estimator for location parameter is sufficient statistic.

(b) Define sufficiency, efficiency and consistency. State and prove Fisher-Neyman lemma related to sufficient statistic.

(c) Let $x_1, x_2, \ldots, x_n$ be a random sample of n observations from a population with p.d.f.

$$f(x) = \frac{1}{B(\theta, 2)} x^{\theta-1}(1-x);\ 0 < x < 1,\ \theta > 0.$$

Find sufficient statistic for θ.

15. (a) Define minimum variance estimator. Show that the variance of the estimator θ in

$$f(x) = \frac{1}{\theta},\ 0 < x < \theta$$

is less than the MVB.

(b) If T_1 be the MVU estimator and T_2 be another unbiased estimator of θ, then the linear combination of T_1 and T_2 will not be MVU estimator of θ. Prove.

(c) Let the set of unbiased estimators T of $\theta \in S$ be U. Assume that $V(T)$ exists. In such a case there may be at best one UMVUE of θ. Prove.

16. (a) Explain the criteria of a good estimator. Let $x_1, x_2, \ldots, x_n$ be a random sample of n observations from Bernoulli population with parameter p. Find unbiased estimator of p and its efficiency.

(b) What do you mean by mean square error consistency? Let $x_1, x_2, \ldots, x_n$ be a random sample of n observations from a population with p.d.f.

$$f(x) = \begin{cases} 2x/\theta^2, & 0 \le x \le \theta \\ 0, & \text{otherwise.} \end{cases}$$

Find an unbiased and consistent estimator of θ.

(c) Find an unbiased estimator of parameter θ in

$$f(x) = \frac{1}{\theta^m \Gamma m} e^{-x/\theta} x^{m-1}; \; 0 < x < \infty, \; \theta > 0.$$

Also, find the variance of your estimator.

CHAPTER **33**

Method of Estimation

33.1 Introduction

In the previous chapter we have discussed the criteria of a good estimator. Also, we have found out unbiased, consistent, efficient, MVB and sufficient estimator of different types of parameters. We have found the unique best estimator of parameter. However, there are some other methods to derive estimators. The methods are (i) Maximum Likelihood (ML) Method, (ii) Method of Least Squares (LS), (iii) Method of Moment, (iv) Method of Minimum Chi-square, and (v) Bayes Method of Estimation. In the present chapter we shall discuss these methods of estimation with examples.

33.2 Maximum Likelihood (ML) Method

Let $x_1, x_2, \ldots, x_n$ be a random sample of n observations from a population with p.d.f. $f(x, \theta)$. The joint probability density function of these sample observations is

$$\begin{aligned} L(x_1, x_2, \ldots, x_n; \theta) &= f(x_1, \theta) f(x_2, \theta) \ldots f(x_n, \theta) \\ &= \prod_{i=1}^{n} f(x_i, \theta). \end{aligned}$$

Here the joint density function is also known as *likelihood function* and is usually denoted by L. Let $T = t(x_1, x_2, \ldots, x_n)$ be an estimator of θ. The ML estimator states that, if θ in L is replaced by T, then the value of L will be maximum and in such a case T is called *maximum likelihood estimator.*

Since logarithm of any function is monotone, the estimator which maximize L also maximizes $\ln L$. Hence to find ML estimator we can maximizes $\ln L$. The estimator which maximizes $\ln L$ is found out by Maxima and Minima Principle. Let $\hat{\theta} = T$ be the estimator of θ. Then the value of $\hat{\theta}$ is to be found out solving the equation

$$\frac{\partial \log L}{\partial \theta} = 0.$$

The function $\ln L$ will be maximum for $\hat{\theta}$ if $\dfrac{\partial^2 \log L}{\partial \theta^2} < 0$ when θ in $\dfrac{\partial^2 \log L}{\partial \theta^2}$ is replaced by $\hat{\theta}$. However, the differentiation of $\ln L$ is possible if some regularity conditions are satisfied by the probability density function. If regularity conditions are not satisfied, we cannot solve the equation

$$\frac{\partial \log L}{\partial \theta} = 0.$$

In such a situation the maximum likelihood estimator is the order statistic which maximizes $f(x, \theta)$ or L.

►**EXAMPLE 33.1** *Let $x_1, x_2, \ldots, x_n$ be a random sample of n observations from $N(\mu, \theta^2)$. Find ML estimators of μ and σ^2.*

Solution. We have

$$f(x) = \frac{1}{\sigma\sqrt{2\pi}} \exp\left[-\frac{1}{2}(x-\mu)^2\right], \quad -\infty < x < \infty.$$

Then

$$L = \left(\frac{1}{2\pi\sigma^2}\right)^{n/2} \exp\left[-\frac{1}{2\sigma^2}\sum_{i=1}^{n}(x_i-\mu)^2\right]$$

$$\log L = -\frac{n}{2}\log 2\pi - \frac{n}{2}\log\sigma^2 - \frac{1}{2\sigma^2}\Sigma(x_i-\mu)^2.$$

Now,

$$\frac{\partial \log L}{\partial \mu} = \frac{2}{2\sigma^2}\Sigma(x_i-\mu) = 0 \Rightarrow \hat{\mu} = \bar{x}.$$

Again,

$$\frac{\partial \log L}{\partial \sigma^2} = -\frac{n}{2}\frac{1}{\sigma^2} + \frac{\Sigma(x_i-\mu)^2}{2\sigma^4} = 0$$

$$\text{or, } \frac{n}{2\sigma^4}\left[\frac{1}{n}\Sigma(x_i-\mu)^2 - \sigma^2\right] = 0 \Rightarrow \hat{\sigma}^2 = \frac{1}{n}\Sigma(x_i-\mu)^2$$

where μ is assumed to be known.

►**EXAMPLE 33.2** *Let $x_1, x_2, \ldots, x_n$ be a random sample of n observations from Bernoulli population with parameter p. Find ML estimators of* (i) p, (ii) p^2.

Solution.

(i) We have $P(x) = p^x(1-p)^{1-x};\ x = 0, 1.$

Then

$$L = p^{\Sigma x}(1-p)^{n-\Sigma x} = p^{n\bar{x}}(1-p)^{n-n\bar{x}}$$

$$\partial \log L = n\bar{x}\log p + (n - n\bar{x})\log(1-p)$$

$$\frac{\partial \log L}{\partial p} = \frac{n\bar{x}}{p} - \frac{n(1-\bar{x})}{1-p} = 0$$

$$\text{or, } \frac{\bar{x}}{p} = \frac{1-\bar{x}}{1-p} \Rightarrow \hat{p} = \bar{x}.$$

(ii) Let $p^2 = \theta \Rightarrow p = \sqrt{\theta}$.

Then

$$L = \left(\sqrt{\theta}\right)^{n\bar{x}} \left(1 - \sqrt{\theta}\right)^{n - n\bar{x}}$$

$$\log L = \frac{n\bar{x}}{2} \log \theta + (n - n\bar{x}) \log \left(1 - \sqrt{\theta}\right)$$

$$\frac{\partial \log L}{\partial \theta} = \frac{n\bar{x}}{2\theta} - \frac{n - n\bar{x}}{1 - \sqrt{\theta}} \frac{1}{2\sqrt{\theta}} = 0$$

$$\text{or, } \frac{\bar{x}}{\sqrt{\theta}} = \frac{1 - \bar{x}}{1 - \sqrt{\theta}} \Rightarrow \sqrt{\theta} = \bar{x}$$

$$\text{or, } \theta = \bar{x}^2 \Rightarrow \hat{\theta} = \hat{p}^2 = \left(\frac{\Sigma x}{n}\right)^2.$$

The estimate of p can also be obtained alternatively, where L can be written as

$$L = \binom{n}{r} p^r (1 - p)^{n-r}; \quad r = \Sigma x_i, \quad i = 0, 1, 2, \ldots, n$$

$$\log L = \log \binom{n}{r} + r \log p + (n - r) \log(1 - p)$$

$$\frac{\partial \log L}{\partial p} = \frac{r}{p} - \frac{n - r}{1 - p} = 0 \Rightarrow \hat{p} = \frac{r}{n} = \frac{\Sigma x}{n} = \bar{x}.$$

►**EXAMPLE 33.3** *Let $x_1, x_2, \ldots, x_n$ be a random sample of n observations from Poisson population with parameter θ. Find ML estimator of θ.*

Solution. Given

$$P(x) = \frac{e^{-\theta} \theta^x}{x!}, \quad x = 0, 1, 2, \ldots$$

$$L = \frac{e^{-n\theta} \theta^{\Sigma x_i}}{\prod_{i=1}^{n} x_i!}, \quad \log L = -n\theta + n\bar{x} \log \theta - \log \prod_{i=1}^{n} x_i$$

$$\frac{\partial \log L}{\partial \theta} = -n + \frac{n\bar{x}}{\theta} = 0 \text{ or, } \frac{\bar{x}}{\theta} = 1 \Rightarrow \hat{\theta} = \bar{x}.$$

►**EXAMPLE 33.4** *Let $x_1, x_2, \ldots, x_n$ be a random sample of n observations from a population with p.d.f.*

$$f(x) = \theta x^{\theta - 1}; \; 0 < x < 1, \; 0 < \theta < \infty.$$

Find ML estimator of θ.

Solution.

$$L = \theta^n \prod_{i=1}^{n} x_i^{\theta - 1}, \quad \log L = n \log \theta + (\theta - 1) \sum_{i=1}^{n} \log x_i$$

$$\frac{\partial \log L}{\partial \theta} = \frac{n}{\theta} + \sum_{i=1}^{n} \log x_i = 0 \Rightarrow \hat{\theta} = -\frac{n}{\Sigma \log x_i}.$$

►**EXAMPLE 33.5** *Let $x_1, x_2, \ldots, x_n$ be a random sample of n observations from a population with p.d.f.*

$$f(x) = \frac{1}{\theta}, \quad 0 < x < \theta.$$

Find ML estimator of θ.

Solution. $L = \dfrac{1}{\theta^n}$. Here regularity conditions are not satisfied and hence

$$\frac{\partial \log L}{\partial \theta} = 0$$

will not give $\hat{\theta}$. We know that $0 < y_1 < y_2 < \cdots < y_n < \theta$, where y_i's are order statistics of $x_1, x_2, \ldots, x_n$. The likelihood function will be maximum if θ takes minimum value. The minimum value of θ is y_n. Hence the estimator of θ is $\hat{\theta} = y_n$ = the largest order statistic.

►**EXAMPLE 33.6** *Let $x_1, x_2, \ldots, x_n$ be a random sample of n observations from a population having p.d.f.*

$$f(x) = \frac{1}{2}e^{-|x-\theta|}; \quad -\infty < x < \infty, \ -\infty < \theta < \infty.$$

Find ML estimator of θ.

Solution. We have

$$L = \frac{1}{2}e^{-\Sigma|x-\theta|}.$$

Here L will be maximum if $\Sigma|x - \theta|$ is minimum. This is possible if θ is the median. Hence sample median is the estimator of θ.

►**EXAMPLE 33.7** *Let $x_1, x_2, \ldots, x_n$ be a random sample of n observations from a population having p.d.f.*

$$f(x) = e^{-(x-\theta)}; \quad \theta \le x \le \infty, \ -\infty < \theta < \infty.$$

Find ML estimator of θ.

Solution. We have

$$L = e^{-n(\bar{x}-\theta)}, \quad \log L = -n\bar{x} + n\theta, \quad \frac{\partial \log L}{\partial \theta} = n.$$

Hence $\dfrac{\partial \log L}{\partial \theta} = 0$ does not give the estimator of θ. [Here the distribution of X does not follow regularity conditions.]

The likelihood function will be maximum if θ takes maximum value in terms of x. From the range of x's, we have

$$\theta \le y_1 < y_2 < \cdots < y_n$$

where $y_1, y_2, \ldots, y_n$ are the order statistics of sample observations. The maximum value of θ is y_1 which is the first order statistic of $x_1, x_2, \ldots, x_n$. Hence, $\hat{\theta} = y_1$.

►**EXAMPLE 33.8** *Let $x_1, x_2, \ldots, x_n$ be a random sample of n observations from a population whose p.d.f. is*

$$f(x) = \frac{1}{\Gamma m \theta^m} x^{m-1} e^{-x/\theta}; \; m > 0, \; 0 < x < \infty.$$

Assuming (i) *m known,* (ii) *$m = 1$,* (iii) *$m = 2$, find ML estimator of θ.*

Solution.

(i) We have

$$L = \frac{1}{(\Gamma m)^n \theta^{mn}} e^{-\frac{n\bar{x}}{\theta}} \prod_{i=1}^{n} x_i^{m-1}$$

$$\log L = -n \log \Gamma m - mn \log \theta - \frac{n\bar{x}}{\theta} + (m-1) \sum_{i=1}^{n} \log x_i$$

$$\frac{\partial \log L}{\partial \theta} = \frac{-mn}{\theta} + \frac{n\bar{x}}{\theta^2} = \frac{nm}{\theta^2} \left[\frac{\bar{x}}{m} - \theta\right] = 0$$

$$\therefore \; \hat{\theta} = \frac{\bar{x}}{m}, \text{ where } m \text{ is known.}$$

(ii) When $m = 1$, $f(x) = \dfrac{1}{\theta} e^{-x/\theta}$

$$L = \frac{1}{\theta^n} e^{-\frac{n\bar{x}}{\theta}}, \; \log L = -n \log \theta - \frac{n\bar{x}}{\theta}$$

$$\frac{\partial \log L}{\partial \theta} = -\frac{n}{\theta} + \frac{n\bar{x}}{\theta^2} = \frac{n}{\theta^2} [\bar{x} - \theta] = 0 \Rightarrow \hat{\theta} = \bar{x}.$$

(iii) When $m = 2$, $f(x) = \dfrac{1}{\theta^2} x e^{-x/\theta}$

$$L = \frac{1}{\theta^{2n}} e^{-\frac{n\bar{x}}{\theta}} \prod_{i=1}^{n} x_i$$

$$\log L = -2n \log \theta - \frac{n\bar{x}}{\theta} + \sum_{i=1}^{n} \log x_i$$

$$\frac{\partial \log L}{\partial \theta} = -\frac{2n}{\theta} + \frac{n\bar{x}}{\theta^2} = \frac{2n}{\theta^2} \left[\frac{\bar{x}}{2} - \theta\right] = 0 \Rightarrow \hat{\theta} = \frac{\bar{x}}{2}.$$

►**EXAMPLE 33.9** *Let $x_1, x_2, \ldots, x_n$ be a random sample of n observations from a population having p.d.f. $f(x) = \beta e^{-\beta(x-\alpha)}$; $\alpha \leq x \leq \infty$, $\beta > 0$. Find ML estimators of α and β.*

Solution. The likelihood function is

$$L = \beta^n e^{-\beta \Sigma (x-\alpha)}, \; \log L = n \log \beta - n\beta(\bar{x} - \alpha), \; \frac{\partial \log L}{\partial \alpha} = +n\beta.$$

Hence $\dfrac{\partial \log L}{\partial \alpha} = 0$ does not give ML estimator of α. Let $y_1 < y_2 < \cdots < y_n$ be the order statistics of the sample observations. Then, we have $\alpha \leq y_1 < y_2 < \cdots < y_n < \infty$. The likelihood function will be maximum if α takes maximum value in terms of x. The maximum value of α is y_1, the first order statistic. Hence, $\hat{\alpha} = y_1$. Again,

$$\frac{\partial \log L}{\partial \beta} = \frac{n}{\beta} - n(\bar{x} - \alpha) = 0 \Rightarrow \bar{x} - \alpha = \frac{1}{\beta} \text{ or, } \hat{\beta} = \frac{1}{\bar{x} - \alpha}.$$

As $\hat{\alpha} = y_1$, then $\hat{\beta} = \dfrac{1}{\bar{x} - y_1}$.

►**EXAMPLE 33.10** *Let $x_1, x_2, \ldots, x_n$ be a random sample of n observations from a population having p.d.f.*

$$f(x) = \frac{1}{\Gamma\theta}\left(\frac{\theta}{\alpha}\right)^{\theta} e^{-\frac{\theta}{\alpha}x} x^{\theta-1};\ 0 \leq x \leq \infty,\ \theta > 0.$$

Find ML estimators of α and θ.

Solution. The likelihood function is

$$L = \frac{1}{(\Gamma\theta)^n}\left(\frac{\theta}{\alpha}\right)^{n\theta} e^{-\frac{n\theta\bar{x}}{\alpha}} \prod_{i=1}^{n} x_i^{\theta-1}$$

$$\log L = n\log\Gamma\theta + n\theta\log\frac{\theta}{\alpha} - \frac{n\theta\bar{x}}{\alpha} + (\theta-1)\sum_{i=1}^{n}\log x_i$$

$$= n\log\Gamma\theta + n\theta[\log\theta - \log\alpha] - \frac{n\theta\bar{x}}{\alpha} + (\theta-1)\sum_{i=1}^{n}\log x_i$$

$$\frac{\partial \log L}{\partial \alpha} = -\frac{n\theta}{\alpha} + \frac{n\theta\bar{x}}{\alpha^2} = 0 \text{ or, } \frac{n\theta}{\alpha^2}[\bar{x} - \alpha] = 0 \Rightarrow \hat{\alpha} = \bar{x}$$

For large n,

$$\frac{\partial \log\Gamma\theta}{\partial\theta} \approx \log\theta - \frac{1}{2\theta}.$$

Then

$$\frac{\partial \log L}{\partial \theta} = -n\left[\log\theta - \frac{1}{2\theta}\right] - n\log\alpha + \frac{n\theta}{\theta} + n\log\theta - \frac{n\bar{x}}{\alpha} + \sum_{i=1}^{n}\log x_i$$

$$= \frac{n}{2\theta} - n\log\alpha + n - \frac{n\bar{x}}{\alpha} + n\log G$$

where G is the geometric mean. Now, $\dfrac{\partial \log L}{\partial \theta} = 0$ gives

$$\frac{1}{2\theta} - \log\alpha + 1 - \frac{\bar{x}}{\alpha} + \log G = 0.$$

Again, $\alpha = \bar{x}$. Therefore, we get

$$\frac{1}{2\theta} = \log\bar{x} - \log G \text{ or, } \frac{1}{2\theta} = \log\frac{\bar{x}}{G} \text{ or, } 2\theta = \frac{1}{\log\frac{\bar{x}}{G}}.$$

$$\therefore\ \hat{\theta} = \frac{1}{2\log\bar{x}/G}.$$

►**EXAMPLE 33.11** *Let $x_1, x_2, \ldots, x_n$ be a random sample of n observations from a population with p.d.f.*

$$f(x) = \frac{1}{\theta_2} \exp\left[-\frac{1}{\theta_2}(x-\theta_1)\right]; \; \theta_1 \leq x < \infty, \; -\infty < \theta_1 < \infty,$$
$$0 < \theta_2 < \infty.$$

Find ML estimators of θ_1 and θ_2.

Solution. We have

$$L = \frac{1}{\theta_2^n} \exp\left[-\frac{1}{\theta_2}\Sigma(x-\theta_1)\right].$$

The regularity conditions are not satisfied. Hence the estimate of θ_1 is to be obtained from order statistic of sample observations. We have $y_1 < y_2 < \cdots < y_n$ as the order statistics of sample observations. Hence

$$\theta_1 \leq y_1 < y_2 < \cdots < \infty.$$

As θ_2 is positive, L will be maximum, if $\Sigma(x-\theta_1)$ is minimum. $\Sigma(x-\theta_1)$ will be minimum, if θ_1 becomes maximum. The maximum value of θ_1 is y_1. Therefore, $\hat{\theta}_1 = y_1$. Again,

$$\log L = -n \log \theta_2 - \frac{1}{\theta_2}\Sigma(x-\theta_1)$$
$$\frac{\partial \log L}{\partial \theta_2} = -\frac{n}{\theta_2} + \frac{\Sigma(x-\theta_1)}{\theta_2^2} = 0 \Rightarrow \hat{\theta}_2 = \bar{x} - \hat{\theta}_1 = \bar{x} - y_1.$$

►**EXAMPLE 33.12** *Let $x_1, x_2, \ldots, x_n$ be a random sample of n observations from a population with p.d.f.*

$$f(x) = \frac{1}{b-a}, \quad a < x < b.$$

Find ML estimators of a and b.

Solution. We have

$$L = \frac{1}{(b-a)^n}, \quad \log L = -n \log(b-a).$$

Thus, neither $\dfrac{\partial \log L}{\partial a} = 0$ nor $\dfrac{\partial \log L}{\partial b} = 0$ will give the estimates of a and b. Here, we get

$$a < y_1 < y_2 < \cdots < y_n < b$$

where $y_1 < y_2 < \cdots < y_n$ are the order statistics of sample observations. The likelihood function will be maximum, if $(b-a)$ is minimum. Here, $(b-a)$ will be minimum, if b is minimum and a is maximum. The maximum value of a is y_1 and the minimum value of b is y_n.

$$\therefore \; \hat{a} = y_1 \text{ and } \hat{b} = y_n.$$

►**EXAMPLE 33.13** *Let $x_1, x_2, \ldots, x_n$ be a random sample of n observations from population having p.d.f.*

$$P(X = x_1) = \frac{1-\theta}{2}, \; P(X = x_2) = \frac{1}{2}, \; P(X = x_3) = \frac{\theta}{2}, \; 0 \le \theta \le 1.$$

Find ML estimator of θ.

Solution. Let the number of observations in i-th sample be n_i $(i = 1, 2, 3)$ such that $n_1 + n_2 + n_3 = n$. The likelihood function is

$$L = \left(\frac{1-\theta}{2}\right)^{n_1} + \left(\frac{1}{2}\right)^{n_2} + \left(\frac{\theta}{2}\right)^{n_3}$$

$$\log L = n_1 \log \frac{1-\theta}{2} + n_2 \log \frac{1}{2} + n_3 \log \frac{\theta}{2}$$

$$\frac{\partial \log L}{\partial \theta} = \frac{-n_1}{\frac{1-\theta}{2}} + n_3 \frac{1}{\frac{\theta}{2}} = 0 \Rightarrow \frac{-n_1}{1-\theta} + \frac{n_3}{\theta} = 0$$

$$\text{or, } (1-\theta)n_3 - n_1\theta = 0 \text{ or, } \theta = \frac{n_3}{n_1 + n_3} \text{ if } n_1 > 0, n_3 > 0.$$

If $n_1 = n_3 = 0$, then $\hat{\theta}$ will take value in the interval $[0, 1]$. Again, if $n_3 = 0, \hat{\theta} = 0$.

►**EXAMPLE 33.14** *Let $x_1, x_2, \ldots, x_n$ be a random sample of n observations from a population with probability function*

$$P(x) = \frac{1}{N}, \; x = 1, 2, \ldots, N.$$

Find ML estimator of N.

Solution. The likelihood function is

$$L = \frac{1}{N^n}.$$

Here regularity conditions are not satisfied and hence, $\frac{\partial \log L}{\partial N} = 0$ does not give the estimator of N. However, let $y_1 < y_2 < \cdots < y_n$ be the order statistics of sample observations. Then $1 \le y_n \le N$. Here likelihood function will be maximum if N takes minimum value. The minimum value of N is y_n. Hence, the ML estimator of N is $\hat{N} = y_n$.

►**EXAMPLE 33.15** *For the hypergeometric distribution of the type*

$$P(x) = \frac{\binom{M}{n}\binom{N-M}{n-x}}{\binom{N}{n}}, \quad \max(0, n - N + M) \le x \le \min(N, M).$$

Find ML estimator of N.

Solution. Let the ML estimator of N be $\hat{N} = \hat{n}(X)$. Consider

$$R = \frac{P(X = x)}{P(X = x-1)} = \frac{N-n}{N} \frac{N-M}{N-M-n+x}.$$

Now, for which value of N, $R > 1$, the value of $P(X = x)$ will increase with the increase in the value of N. Again, for which value of N, $R < 1$, $P(X = x)$ will decrease with the increase in the value of N. Thus

$$R > 1, \text{ if and only if, } N < \frac{nM}{x};$$
$$R < 1, \text{ if and only if, } N > \frac{nM}{x}.$$

Therefore, $P(X = x)$ will be maximum, if $N = \dfrac{nM}{x}$.

Hence, $\hat{N} = \dfrac{nM}{[x]}$ where $[x]$ is the largest observation of x.

►**EXAMPLE 33.16** *Find the ML estimators of parameters in bivariate normal distribution.*

Solution. If X and Y follow bivariate normal distribution, then

$$f(x,y) = \frac{1}{2\pi\sigma_1\sigma_2\sqrt{1-\rho^2}} \exp\left[-\frac{1}{2(1-\rho^2)}\left\{\left(\frac{x-\mu_1}{\sigma_1}\right)^2 - 2\rho\frac{(x-\mu_1)(y-\mu_2)}{\sigma_1\sigma_2} + \left(\frac{y-\mu_2}{\sigma_2}\right)^2\right\}\right].$$

The likelihood function is

$$L = \frac{1}{(2\pi)^n(\sigma_1^2\sigma_2^2)^{n/2}(1-\rho^2)^{n/2}} \exp\left[-\frac{1}{2(1-\rho^2)}\left\{\sum\left(\frac{x-\mu_1}{\sigma_1}\right)^2 - 2\rho\frac{\Sigma(x-\mu_1)(y-\mu_2)}{\sigma_1\sigma_2} + \sum\left(\frac{y-\mu_2}{\sigma_2}\right)^2\right\}\right]$$

$$\log L = -n\log 2\pi - \frac{n}{2}\log\sigma_1^2 - \frac{n}{2}\log\sigma_2^2 - \frac{n}{2}\log(1-\rho^2) - \frac{1}{2(1-\rho^2)} \left[\sum\left(\frac{x-\mu_1}{\sigma_1}\right)^2 - 2\rho\frac{\Sigma(x-\mu_1)(y-\mu_2)}{\sigma_1\sigma_2} + \sum\left(\frac{y-\mu_2}{\sigma_2}\right)^2\right]$$

$$\frac{\partial \log L}{\partial \mu_1} = \frac{n}{2(1-\rho^2)}\left[\frac{2(\bar{x}-\mu_1)}{\sigma_1^2} - \frac{2\rho(\bar{y}-\mu_2)}{\sigma_1\sigma_2}\right] = 0 \qquad (1)$$

$$\frac{\partial \log L}{\partial \mu_2} = \frac{n}{2(1-\rho^2)}\left[\frac{2(\bar{y}-\mu_2)}{\sigma_2^2} - \frac{2\rho(\bar{x}-\mu_1)}{\sigma_1\sigma_2}\right] = 0 \qquad (2)$$

$$\frac{\partial \log L}{\partial \sigma_1^2} = -\frac{1}{2\sigma_1^2(1-\rho^2)}\left\{n(1-\rho^2) - \frac{\Sigma(x-\mu_1)^2}{\sigma_1^2} + \frac{\rho\Sigma(x-\mu_1)(y-\mu_2)}{\sigma_1\sigma_2}\right\} = 0 \qquad (3)$$

$$\frac{\partial \log L}{\partial \sigma_2^2} = -\frac{1}{2\sigma_2^2(1-\rho^2)}\left\{n(1-\rho^2) - \frac{\Sigma(y-\mu_2)^2}{\sigma_1^2} + \frac{\rho\Sigma(x-\mu_1)(y-\mu_2)}{\sigma_1\sigma_2}\right\} \qquad (4)$$

$$\frac{\partial \log L}{\partial \rho} = \frac{1}{1-\rho^2}\left[n\rho - \frac{1}{1-\rho^2}\left\{\frac{\rho\Sigma(x-\mu_1)^2}{\sigma_1^2} + \frac{\rho\Sigma(y-\mu_2)^2}{\sigma_2^2} - (1+\rho^2)\frac{\Sigma(x-\mu_1)(y-\mu_2)}{\sigma_1\sigma_2}\right\}\right] = 0. \qquad (5)$$

Now, from (3) and (4), we get

$$n(1-\rho^2) = \frac{\Sigma(x-\mu_1)^2}{\sigma_1^2} - \rho\frac{\Sigma(x-\mu_1)(y-\mu_2)}{\sigma_1\sigma_2} \tag{6}$$

$$n(1-\rho^2) = \frac{\Sigma(y-\mu_2)^2}{\sigma_2^2} - \rho\Sigma\frac{(x-\mu_1)(y-\mu_2)}{\sigma_1\sigma_2}. \tag{7}$$

From (5), we get

$$n(1-\rho^2) = \frac{1-\rho^2}{\rho}\frac{\Sigma(x-\mu_1)(y-\mu_2)}{\sigma_1\sigma_2}. \tag{8}$$

$$\therefore\ \hat{\rho} = \frac{\Sigma(x-\mu_1)(y-\mu_2)}{n\sigma_1\sigma_2}.$$

Solving other equations, we get

$$\hat{\mu}_1 = \bar{x},\ \hat{\mu}_2 = \bar{y},\ \hat{\sigma}_1^2 = \frac{1}{n}\Sigma(x-\mu_1)^2,\ \hat{\sigma}_2^2 = \frac{1}{n}\Sigma(y-\mu_2)^2.$$

33.3 Properties of ML Estimators

Let $f(x,\theta)$ be a member of family of probability density function, where $\theta \in S$. Assume that $f(x,\theta)$ is the p.d.f. of continuous random variable and for all values of x and $\theta \in S$

(i) $\dfrac{\partial \log f(x,\theta)}{\partial\theta}$, (ii) $\dfrac{\partial^2 \log f(x,\theta)}{\partial\theta^2}$ and (iii) $\dfrac{\partial^3 \log f(x,\theta)}{\partial\theta^3}$ exist.

(iv) For all $\theta \in S$, $\displaystyle\int_{-\infty}^{\infty} \frac{\partial^2 \log f(x,\theta)}{\partial\theta^2} = 0.$

(v) For all $\theta \in S$, $\displaystyle -\infty < \int_{-\infty}^{\infty} \frac{\partial^2 \log f(x,\theta)}{\partial\theta^2} f(x) < 0.$

(vi) Let $H(x)$ be a function of x. Then for all $\theta \in S$,

$$\left|\frac{\partial^3 \log f(x,\theta)}{\partial\theta^3}\right| < H(x) \text{ and } \int_{-\infty}^{\infty} H(x)f(x,\theta)dx = M(\theta) < \infty.$$

(vii) Let $g(\theta)$ be a function of θ such that for all $\theta \in S$, the function $g(\theta)$ can be differentiated twice. Also, $g(\theta)$ is positive. Then

$$\left|\frac{\partial^2}{\partial\theta^2}\left[g(\theta)\frac{\partial \log f(x,\theta)}{\partial\theta}\right]\right| < H(x) \text{ and } \int_{-\infty}^{\infty} H(x)f(x,\theta)dx < \infty.$$

Under the above assumptions the ML etimators possess some characteristics. These are discussed below:

❒ THEOREM 33.1 *Under assumption* (i), (ii), (iii), (iv), (v) *and* (vi), *if* $n \to \infty$,

$$\frac{\partial \log L}{\partial\theta} = 0$$

has a consistent solution with probability 1. *That is, the ML estimator is consistent.*

Proof. We have seen that the ML estimator of λ in Poisson distribution is given by

$$\frac{\partial \log L}{\partial \lambda} = \frac{n\bar{x}}{\theta} - n = 0 \text{ [Example 33.3]}$$

$$\text{or, } \frac{n}{\theta}[\bar{x} - \theta] = 0, \text{ where } \hat{\theta} = \bar{x}.$$

But $\bar{x}$ is an MVB estimator with $V(\bar{x}) = \dfrac{\theta}{n}$ [Example 32.40]. If $n \to \infty$, $V(\bar{x}) \to 0$ and hence, $\bar{x}$ is a consistent estimator of θ.

❒ **THEOREM 33.2** *Under assumptions* (i) *to* (vi) *mentioned above, if the solution of ML estimator's equation is* $\hat{\theta}_n$, *then it is asymptotically normal variable. However, in some cases the assumption* (vi) *is not valid. In those cases, under assumptions* (i), (iv), (v) *and* (vii) *Kulldorf* (1957) *has shown that* $\hat{\theta}_n \to N(0,1)$.

Proof. As an example of 'asymptotically normality' of ML estimator, let us consider the ML estimator of θ in $N(0, \theta)$. Here the likelihood function of $x_1, x_2, \ldots, x_n$ sample observations is

$$L = \left(\frac{1}{\sqrt{2\pi\theta}}\right)^n e^{-\frac{1}{2}\Sigma x^2/\theta}, \quad -\infty < x < \infty$$

$$\log L = -\frac{n}{2}\log\pi - \frac{n}{2}\log\theta - \frac{\Sigma x^2}{2\theta}$$

$$\frac{\partial \log L}{\partial \theta} = -\frac{n}{2\theta} + \frac{\Sigma x^2}{2\theta^2} = 0.$$

$$\therefore\ \hat{\theta} = \frac{1}{n}\Sigma x_i^2.$$

Here, $E(x^2) = \theta, V(x^2) = 2\theta^2$

$$\frac{\partial \log f(x,\theta)}{\partial \theta} = -\frac{1}{2\theta} + \frac{x^2}{2\theta^2}$$

$$\frac{\partial^2 \log f(x,\theta)}{\partial \theta^2} = \frac{1}{2\theta^2} - \frac{x^2}{2\theta^3} \quad E\left[\frac{\partial^2 \log f(x,\theta)}{\partial \theta^2}\right] = \frac{1}{2\theta^2}.$$

It is to be noted that $\hat{\theta}_n \to \theta$ as $n \to \infty$.

$$\therefore\ \sqrt{n}(\hat{\theta}_n - \theta) = \sqrt{2\theta}\frac{\Sigma x_i^2 - n\theta}{\sqrt{2n\theta}} \to N(0, 2\theta^2) \text{ as } n \to \infty.$$

But $\left|\dfrac{\partial^3 \log f(x,\theta)}{\partial \theta^3}\right| \not< H(x)$ since

$$\frac{\partial^3 \log f(x,\theta)}{\partial \theta^3} = -\frac{1}{\theta^3} + \frac{3x^2}{\theta^4} \to \infty \text{ as } \theta \to 0.$$

Therefore, in such a case Kulldorf's (1957) theorem is applicable.

❒ **THEOREM 33.3** *Let the estimator of* θ *in* $f(x, \theta)$, $\theta \in S$ *be* T, *where* T *is a sufficient statistic. If the ML estimator of* θ *exists, then it is the function of sufficient statistic.*

Proof. If T is a sufficient statistic of θ in $f(x, \theta)$, then

$$\begin{aligned} L &= g(T/\theta)h(x_1, x_2, \ldots, x_n) \\ \log L &= \log g(T/\theta) + \log h(x_1, x_2, \ldots, x_n) \\ \frac{\partial \log L}{\partial \theta} &= \frac{\partial \log g(T/\theta)}{\partial \theta} = 0 \Rightarrow \psi(T) = 0 \end{aligned}$$

where

$$\psi(T) = \frac{\partial \log g(T/\theta)}{\partial \theta}.$$

Hence, if the solution of $\frac{\partial \log L}{\partial \theta} = 0$ exists, the solution is the function of sufficient statistic.

As an example, we can mention the case of estimator of θ in

$$f(x, \theta) = \frac{1}{\theta} e^{-x/\theta}, \; 0 < x < \infty.$$

Here

$$\begin{aligned} \frac{\partial \log L}{\partial \theta} &= -\frac{n}{\theta} + \frac{n\bar{x}}{\theta^2} = \frac{n}{\theta^2}[\bar{x} - \theta] \\ \frac{\partial \log L}{\partial \theta} &= 0 \Rightarrow \hat{\theta} = \bar{x} \end{aligned}$$

Again,

$$\log L = \frac{1}{\theta^n} e^{-n\bar{x}/\theta} = g(T/\theta)h(x_1, x_2, \ldots, x_n)$$

where

$$g(T/\theta) = \frac{1}{\theta^n} e^{-n\bar{x}/\theta}, \; h(x_1, x_2, \ldots, x_n) = 1.$$

Hence $\bar{x}$ is the sufficient statistic for θ. Thus, ML estimator is a function of sufficient statistic. Also, this estimator $\bar{x}$ is consistent since $V(\bar{x}) = \frac{\theta^2}{n} \to 0$ as $n \to \infty$.

❑ **THEOREM 33.4** *Let $x_1, x_2, \ldots, x_n$ be a random sample of n observations from a population with p.d.f. $f(x, \theta)$, $\theta \in S$. Consider that the MVB estimator of θ is available. Then there is unique solution of ML equation.*

Proof. The likelihood function is

$$L = f(x_1, \theta) f(x_2, \theta) \cdots f(x_n, \theta).$$

If MVB estimator exists, then

$$\frac{\partial \log L}{\partial \theta} = A(\theta)[t - \phi(\theta)]$$

where t is an estimator of $\phi(\theta)$ and this 't' is an MVB estimator. Again, ML equation to estimate $\phi(\theta)$ is

$$\frac{\partial \log L}{\partial \theta} = 0 \Rightarrow A(\theta)[t - \phi(\theta)] = 0.$$

Hence unique value of t is available. Here $A(\theta)$ cannot be zero since 't' is MVB and $V(t) = 1/A(\theta)$.

We know, MVB exists if

$$V(t) \geq -\frac{[\phi'(\theta)]^2}{E\left[\dfrac{\partial^2 \log L}{\partial \theta^2}\right]} \text{ and hence, } E\left[\frac{\partial^2 \log L}{\partial \theta^2}\right] \neq 0.$$

In the example of estimation of θ in

$$f(x, \theta) = \frac{1}{\theta} e^{-x/\theta}, \ 0 < x < \infty$$

we have

$$\frac{\partial \log L}{\partial \theta} = \frac{n}{\theta^2}[\bar{x} - \theta]$$

where $\bar{x}$ is an MVB estimator of θ. Also, we have $\frac{n}{\theta^2}[\bar{x} - \theta] = 0 \Rightarrow \hat{\theta} = \bar{x}$. This solution of θ is unique. Since ML estimator is MVB, we can say that ML estimator is more efficient. However, all ML estimators may not be efficient. As an example, we can mention the ML estimator of θ in

$$f(x) = \frac{1}{\theta}, \ 0 < x < \theta$$

where $\hat{\theta} = y_n$, the highest observations in the sample. This $\hat{\theta}$ is not an MVB estimator.

❐ **THEOREM 33.5** *Let $f(x, \theta)$, $\theta \in S$ be a member of family of p.d.f. and L is the likelihood function of $f(x, \theta)$. Consider $S \leq R_k$, $k \geq 1$, where R_k is the range space of X. Further, consider that h is a function such that there is a mapping from S to Λ and the interval of Λ is $R_p (1 \leq p \leq k)$. In such a case, if $\hat{\theta}$ is the ML estimator of θ, then $h(\hat{\theta})$ is the ML estimator of $h(\theta)$.*

Proof. The theorem can be proved by an example. Let us consider the probability function $P(x = 1) = p$, $P(x = 0) = 1 - p$, where $\hat{p} = \bar{x}$, $0 \leq p \leq 1$. Now, let $h(p) = p(1 - p)$, $\Lambda = [0, 1/4]$. Here $h(p)$ is not one-to-one function. The ML estimator of $h(p) = p(1 - p)$ is $\bar{x}(1 - \bar{x})$. This property of ML estimator is known as invariance property. According to this property, if $\hat{\sigma}^2 = \frac{1}{n}\Sigma(x - \mu)^2$ is ML estimator of σ^2 in $N(\mu, \sigma^2)$, then $\hat{\sigma} = \sqrt{\frac{1}{n}\Sigma(x - \mu)^2}$ and $\log \sigma^2$ is estimated by $\log\left[\frac{1}{n}\Sigma(x - \mu)^2\right]$, where μ is known.

33.4 Method of Least Squares

Let a sample of size n be randomly selected from a population whose probability density function may be known or may be unknown. Assume that each sample unit provides information on variables $Y, X_1, X_2, \ldots, X_k$, where the variable Y is assumed to be dependent on other variables $X_1, X_2, \ldots, X_k$. The variables $X_1, X_2, \ldots, X_k$ are known as independent or explanatory variables. Here $(k + 1)$ variables are studied from each sample point and hence the distribution of variables is in the discipline of multivariate distribution. Since the variables are observed from each sample point, these are assumed to be related either linearly or non-linearly.

Let us consider that there exists a linear relationship of the variables, where Y depends on X's. The linear model for the variable is assumed to be

$$y = \beta_0 + \beta_1 x_1 + \beta_2 x_2 + \cdots + \beta_k x_k + \epsilon \tag{33.1}$$

where β_i's $(i = 0, 1, 2, \ldots, k)$ are the regression parameters. The parameters $\beta_1, \beta_2, \ldots, \beta_k$ are the partial regression coefficients of Y on X's. Here ϵ is a random component measuring the impacts of other possible variables which are not included in the model.

In matrix notation, the model (33.1) is written as

$$Y = X\beta + U \tag{33.2}$$

where

$$Y = \begin{bmatrix} y_1 \\ y_2 \\ \vdots \\ y_n \end{bmatrix}_{n\times 1}, \quad X = \begin{bmatrix} 1 & x_{11} & x_{21} & \cdots & x_{k1} \\ 1 & x_{12} & x_{22} & \cdots & x_{k2} \\ \vdots & \vdots & \vdots & \vdots & \vdots \\ 1 & x_{1n} & x_{2n} & \cdots & x_{kn} \end{bmatrix}_{n\times(k+1)}$$

$$\beta = \begin{bmatrix} \beta_0 \\ \beta_1 \\ \beta_2 \\ \vdots \\ \beta_k \end{bmatrix}_{(k+1)\times 1}, \quad U = \begin{bmatrix} \epsilon_1 \\ \epsilon_2 \\ \vdots \\ \epsilon_n \end{bmatrix}_{n\times 1}$$

The problem is to estimate the parameter vector β. The estimation is done in such a way that the sum of squares due to error is minimum, where error sum of squares is given as follows:

We have

$$\hat{U} = Y - X\hat{\beta}$$

$$\phi = \hat{U}'\hat{U} = (Y - X\hat{\beta})'(Y - X\hat{\beta}) = Y'Y - 2X'X\hat{\beta}' + \hat{\beta}'X'X\hat{\beta}.$$

We need to find the value of $\hat{\beta}$ so that ϕ becomes minimum. This is possible, if

$$\frac{\partial \phi}{\partial \hat{\beta}} = 0. \tag{33.3}$$

Solving this equation, we shall get the value of $\hat{\beta}$, where $\hat{\beta}$ is called estimator by Method of Least Squares.

Assumptions for Estimation

(i) $E(U) = 0$, $E(UU') = \sigma^2 I_n$.

(ii) $r(X) = k + 1$; this assumption indicates that the explanatory variables are uncorrelated.

(iii) The X_i's are measured without error.

(iv) The X_i $(i = 1, 2, \ldots, k)$ variables are non-random.

(v) X and U are independent.

Under the above assumptions, the solutions of equations shown in (33.3) give

$$\hat{\beta} = (X'X)^{-1}X'Y.$$

Here $\hat{\beta}$ is the least squares estimator of β.

Now, let us consider the estimation of regression parameters on the basis of one dependent and one independent variable. The model assumed is

$$y = A + Bx + \epsilon \tag{33.4}$$

$$\text{or, } \epsilon = y - A - Bx$$

$$\Sigma\epsilon^2 = \Sigma(y - A - Bx)^2 = \text{sum of squares of error.}$$

If $\epsilon \sim N(0, \sigma^2)$, then

$$f(\epsilon) = \frac{1}{\sqrt{2\pi\sigma^2}} e^{-\frac{1}{2\sigma^2}(y - A - Bx)^2}.$$

The likelihood function is

$$L = \left(\frac{1}{2\pi\sigma^2}\right)^{n/2} \exp\left[-\frac{1}{2\sigma^2}\Sigma(y - A - Bx)^2\right].$$

According to ML method of estimation we need to find the value of A and B in such a way that L becomes maximum. This is possible if $\Sigma(y - A - Bx)^2$ is minimum. This latter condition of finding the value of A and B is the condition of method of least squares. Thus the estimators of the parameters in linear model are equivalent to ML estimators, if the error term is normally distributed with mean zero and common variance σ^2.

Now, let us consider the estimation of A and B in model (33.4). Let the estimated error according to model (33.4) be

$$e = y - a - bx.$$

Then sum of squares of error is

$$\phi = \Sigma e^2 = \Sigma(y - a - bx)^2.$$

The method of least squares states that the values of 'a' and 'b' are to be found out solving the equations

$$\frac{\partial\phi}{\partial a} = 0 \text{ and } \frac{\partial\phi}{\partial b} = 0.$$

Here

$$\frac{\partial\phi}{\partial a} = 0 \Rightarrow -2\Sigma(y - a - bx) = 0 \text{ or, } a = \bar{y} - b\bar{x}.$$

Again,

$$\frac{\partial \phi}{\partial b} = 0 \Rightarrow -2\Sigma(y - a - bx)x = 0 \text{ or, } \Sigma(y - \bar{y} + b\bar{x} - bx)x = 0$$

$$\text{or, } \Sigma\{(y - \bar{y}) - b(x - \bar{x})\}x = 0 \text{ or, } \Sigma(y - \bar{y})x = b\Sigma(x - \bar{x})x$$

$$\text{or, } \Sigma(y - \bar{y})(x - \bar{x}) = b\Sigma(x - \bar{x})^2, \because \bar{x}\Sigma(y - \bar{y}) = 0, \bar{x}\Sigma(x - \bar{x}) = 0$$

$$\therefore b = \frac{\Sigma(y - \bar{y})(x - \bar{x})}{\Sigma(x - \bar{x})^2} = \frac{SP(xy)}{SS(x)} = \frac{\frac{1}{n}\Sigma(y - \bar{y})(x - \bar{x})}{\frac{1}{n}\Sigma(x - \bar{x})^2} = \frac{\text{Cov}(x, y)}{V(x)}.$$

►**EXAMPLE 33.17** *The following data represent the number of earning members* (x) *and total monthly income* (y) *(in 000 Rs.) of 15 randomly selected families in an area:*

x	1	1	1	1	1	1	1	2	2	2	3	3	3	4	2
y	20.0	15.5	12.0	8.6	6.0	8.0	6.0	33.5	40.6	50.0	65.6	20.0	38.0	100.0	35.0

Assuming a linear relationship of y on x, estimate the parameters by method of least squares.

Solution. Let us assume that x and y are linearly related, where the linear function is

$$y = A + Bx + \epsilon.$$

The problem is to estimate A and B, where

$$b = \frac{SP(xy)}{SS(x)} \text{ and } a = \bar{y} - b\bar{x}.$$

Here

$$SP(xy) = \Sigma xy - \frac{\Sigma x \Sigma y}{n} = 1165.1 - \frac{28 \times 458.8}{15} = 308.67$$

$$SS(x) = \Sigma x^2 - \frac{(\Sigma x)^2}{n} = 66 - \frac{(28)^2}{15} = 13.73.$$

$$\therefore b = \frac{SP(xy)}{SS(x)} = \frac{308.67}{13.73} = 22.48,$$

$$a = \bar{y} - b\bar{x} = 30.59 - 22.48 \times 1.87 = -11.45.$$

►**EXAMPLE 33.18** *The following data represent the body weight* (y *kg*)*, body length* (x_1 *cm*) *and body breadth* (x_2 *cm*) *of 12 randomly selected sea fish:*

x_1	12	20	14	25	18	16	10	18	18	20	16	12
x_2	4	7	6	12	10	8	4	8	9	10	7	4
y	0.5	0.8	0.7	2.0	1.2	0.9	0.4	0.9	1.4	1.5	0.8	0.6

Assume the linear model of y, x_1 *and* x_2 *as*

$$y = \beta_0 + \beta_1 x_1 + \beta_2 x_2 + \epsilon.$$

Estimate the parameters β_0, β_1 *and* β_2 *by method of least squares.*

Solution. From the model, we can write

$$\begin{aligned} \bar{y} &= \beta_0 + \beta_1\bar{x}_1 + \beta_2\bar{x}_2 + \bar{\epsilon} \\ \text{or, } y - \bar{y} &= \beta_1(x_1 - \bar{x}_1) + \beta_2(x_2 - \bar{x}_2) + (\epsilon - \bar{\epsilon}) \\ Y &= \beta_1 X_1 + \beta_2 X_2 + U \end{aligned}$$

where $Y = y - \bar{y}, X_1 = x_1 - \bar{x}_1, X_2 = x_2 - \bar{x}_2, U = \epsilon - \bar{\epsilon}$. Then

$$\begin{bmatrix} \hat{\beta}_1 \\ \hat{\beta}_2 \end{bmatrix} = \begin{bmatrix} \Sigma X_1^2 & \Sigma X_1 X_2 \\ \Sigma X_1 X_2 & \Sigma X_2^2 \end{bmatrix}^{-1} = \begin{bmatrix} \Sigma X_1 Y \\ \Sigma X_2 Y \end{bmatrix}$$

$$\hat{\beta}_0 = \bar{y} - \hat{\beta}_1\bar{x}_1 - \hat{\beta}_2\bar{x}_2.$$

Here

$$\begin{aligned} \Sigma X_1^2 &= \Sigma(x_1 - \bar{x}_1)^2 = \Sigma x_1^2 - \frac{(\Sigma x_1)^2}{n} = 3494 - \frac{(199)^2}{12} = 192.92 \\ \Sigma X_2^2 &= \Sigma(x_2 - \bar{x}_2)^2 = \Sigma x_2^2 - \frac{(\Sigma x_2)^2}{n} = 735 - \frac{(89)^2}{12} = 74.92 \\ \Sigma X_1 Y &= \Sigma(x_1 - \bar{x}_1)(y - \bar{y}) = \Sigma x_1 y - \frac{\Sigma x_1 \Sigma y}{n} = 213.25 - \frac{199 \times 11.7}{12} = 19.175 \\ \Sigma X_2 Y &= \Sigma(x_2 - \bar{x}_2)(y - \bar{y}) = \Sigma x_2 y - \frac{\Sigma x_2 \Sigma y}{n} = 99.4 - \frac{89 \times 11.7}{12} = 12.625 \\ \Sigma X_1 X_2 &= \Sigma(x_1 - \bar{x}_1)(x_2 - \bar{x}_2) = \Sigma x_1 x_2 - \frac{\Sigma x_1 \Sigma x_2}{n} = 1586 - \frac{199 \times 89}{12} = 110.083 \end{aligned}$$

$$\begin{aligned} \therefore \begin{bmatrix} \hat{\beta}_1 \\ \hat{\beta}_2 \end{bmatrix} &= \begin{bmatrix} 192.92 & 110.083 \\ 110.083 & 74.92 \end{bmatrix}^{-1} \begin{bmatrix} 19.175 \\ 12.625 \end{bmatrix} \\ &= \begin{bmatrix} 0.032081 & -0.047139 \\ -0.047139 & 0.082610 \end{bmatrix} \begin{bmatrix} 19.175 \\ 12.625 \end{bmatrix} = \begin{bmatrix} 0.02002 \\ 0.13906 \end{bmatrix} \\ \bar{y} &= 0.975, \ \bar{x}_1 = 16.583, \ \bar{x}_2 = 7.417 \\ \therefore \ \hat{\beta}_0 &= \bar{y} - \hat{\beta}_1\bar{x}_1 - \hat{\beta}_2\bar{x}_2 \\ &= 0.975 - 0.02002 \times 16.583 - 0.13906 \times 7.417 \\ &= -0.3884 \\ \therefore \ \hat{\beta}_0 &= -0.3884, \ \hat{\beta}_1 = 0.02002, \ \hat{\beta}_2 = 0.13906. \end{aligned}$$

33.5 Properties of Least Squares Estimator

1. The least squares estimator is unbiased and its variance is given by

$$V(\hat{\beta}) = \sigma^2(X'X)^{-1}.$$

Proof. The linear model assumed is

$$Y = X\beta + U, \text{ where } E(U) = 0, \ E(UU') = \sigma^2 I_n.$$

The least squares estimator of β is

$$\begin{aligned}\hat{\beta} &= (X'X)^{-1}X'Y = (X'X)^{-1}X'(X\beta+U)\\ &= (X'X)^{-1}(X'X)\beta + (X'X)^{-1}X'U\\ &= \beta + (X'X)^{-1}X'U \text{ or, } \hat{\beta}-\beta = (X'X)^{-1}X'U\\ E(\hat{\beta}-\beta) &= (X'X)^{-1}X'E(U) = 0\end{aligned}$$

$\therefore$ $\hat{\beta}$ is unbiased.

$$\begin{aligned}V(\hat{\beta}) &= E(\hat{\beta}-\beta)(\hat{\beta}-\beta)' = E[(X'X)^{-1}X'UU'X(X'X)^{-1}]\\ &= (X'X)^{-1}X'E[(UU')X(X'X)^{-1}]\\ &= \sigma^2(X'X)^{-1}X'X(X'X)^{-1} = \sigma^2(X'X)^{-1}.\end{aligned}$$

2. The least squares estimator is linear function of observations, unbiased and has minimum variance. This property is expressed by a theorem.

❐**THEOREM 33.6** *The least squares estimator is BLUE (best, linear, unbiased estimator).*

Proof. The assumed model is

$$Y = X\beta + U, \text{ where } E(U) = 0,\ E(UU') = \sigma^2 I_n.$$

The least squares estimator is

$$\hat{\beta} = (X'X)^{-1}X'Y.$$

Again, $E(\hat{\beta}) = \beta$, Property (1).

Here $\hat{\beta}$ is a linear function of observation vector Y and it is unbiased. We need to show that $V(\hat{\beta})$ is minimum.

Let $C'\beta$ is a function of parameter vector β, where the elements in C are known and C is a row vector. Since $\hat{\beta}$ is unbiased for β, $C'\hat{\beta}$ is unbiased for $C'\beta$. The variance of $C'\hat{\beta}$ is

$$V(C'\hat{\beta}) = \sigma^2 C'(X'X)^{-1}C.$$

If this variance is not minimum, let us assume another estimator of $C'\beta$, where the estimator is $b = a'Y$. Here the elements in a are assumed to be known. This estimator is also a linear function of observations in Y vector. This estimator will be unbiased if

$$\begin{aligned}E(b) &= a'E(Y) = Ea'(X\beta+U) = E[a'X\beta + a'U]\\ &= a'X\beta + a'E(U) = a'X\beta, \quad \because\ E(U)=0\\ &= C'\beta, \text{ if } a'X = C'.\end{aligned}$$

Now,

$$\begin{aligned}V(b) &= E[b-E(b)][b-E(b)]' = E[b-a'X\beta][b-a'X\beta]'\\ &= E[a'(X\beta+U) - a'X\beta][a'(X\beta+U)-a'X\beta]'\\ &= E[a'UU'a] = a'E(UU')a = \sigma^2 a'a, \quad \because\ E(UU') = \sigma^2 I_n.\end{aligned}$$

Here $V(b)$ will be minimum if $a'a$ is minimum under the condition that $a'X = C'$. to minimise $a'a$, let $\psi = a'a - 2\lambda'(X'a - C)$, where λ is a Lagrange multiplier vector.

$$\frac{\partial \psi}{\partial \lambda} = 2(X'a - C) = 0 \qquad \text{(a)}$$

$$\frac{\partial \psi}{\partial a} = 2a - 2X\lambda = 0 \qquad \text{(b)}$$

Pre-multiplying (b) by X', we get $X'a = X'X\lambda$. Again, from (a), we get $X'a = C$. Therefore,

$$C = X'X\lambda \text{ or, } \lambda = (X'X)^{-1}C.$$

Putting the value of λ in (b), we get

$$a - X(X'X)^{-1}C = 0 \Rightarrow a = X(X'X)^{-1}C.$$

Hence $b = a'Y = C'(X'X)^{-1}X'Y = C'\hat{\beta}$. Thus the variance of $C'\hat{\beta}$ is minimum.

3. If X and Y are related by a non-linear function, the least squares estimator can also be found out after transformation of the non-linear function into a linear function. For example, let

$$y = \alpha X^{\beta} \epsilon.$$

Then

$$\begin{aligned} \log y &= \log \alpha + \beta \log X + \log \epsilon \\ Y &= A + BX + e, \text{ where } Y = \log y, A = \log \alpha, B = \beta. \end{aligned}$$

Here $\hat{A} = \bar{Y} - \hat{B}\bar{X}$, $\hat{B} = \hat{\beta} = \dfrac{\Sigma(X - \bar{X})(Y - \bar{Y})}{\Sigma(X - \bar{X})^2}$ and $\hat{\alpha} = \text{Antilog } \hat{A}$.

Estimation of Variance of $\hat{\beta}$

We have $V(\hat{\beta}) = \sigma^2(X'X)^{-1}$. Its estimator is $v(\hat{\beta}) = s^2(X'X)^{-1}$, where s^2 is an estimator of σ^2. We need to find the value of s^2.

The model is $Y = X\beta + U$, $E(U) = 0$, $E(UU') = \sigma^2 I_n$.

Now

$$\begin{aligned} \hat{U} &= Y - X\hat{\beta} = Y - X(X'X)^{-1}X'Y \\ &= X\beta + U - X(X'X)^{-1}X'(X\beta + U) \\ &= U - X(X'X)^{-1}X'U = [I_n - X(X'X)^{-1}X']U \\ \hat{U}'\hat{U} &= U'[I_n - X(X'X)^{-1}X'][I_n - X(X'X)^{-1}X']U. \end{aligned}$$

But $[I_n - X(X'X)^{-1}X']$ is an idempotent and symmetric matrix. Then

$$\begin{aligned} \hat{U}'\hat{U} &= U'[I_n - X(X'X)^{-1}X']U \\ E[\hat{U}'\hat{U}] &= E\{U'[I_n - X(X'X)^{-1}X']U\} \\ &= E\,\text{tr}\{U'[I_n - X(X'X)X']U\} \end{aligned}$$

$\because U'[I_n - X(X'X)^{-1}X']U$ is a scalar quantity.

$$\begin{aligned}
E[\hat{U}'\hat{U}] &= E\,\text{tr}(UU')[I_n - X(X'X)^{-1}X'] \\
&= \sigma^2\,\text{tr}[I_n - X(X'X)^{-1}X'] \\
&= \sigma^2[\text{tr}I_n - \text{tr}X(X'X)^{-1}X'] \\
&= \sigma^2[n - \text{tr}(X'X)(X'X)^{-1}] \\
&= \sigma^2[n-k-1], \quad \because \text{rank } (X) = k+1.
\end{aligned}$$

$$\therefore\ s^2 = \hat{\sigma}^2 = \frac{\hat{U}'\hat{U}}{n-k-1}.$$

Here

$$\begin{aligned}
\hat{U}'\hat{U} &= (Y - X\hat{\beta})'(Y - X\hat{\beta}) \\
&= Y'Y - Y'X\hat{\beta} - \hat{\beta}X'Y + \hat{\beta}'X'X\hat{\beta} \\
&= Y'Y - \hat{\beta}'X'Y - \hat{\beta}'X'Y + \hat{\beta}'X'X(X'X)^{-1}X'Y \\
&= Y'Y - \hat{\beta}'X'Y \\
&= \text{Total sum of squares} - \text{Sum of squares estimates.} \\
\therefore\ s^2 &= MS\ (\text{Error}).
\end{aligned}$$

As an example, we can estimate $V(\hat{\beta})$ using the data of Example 33.18, where

$$\begin{aligned}
SS(\text{Total}) &= \Sigma y^2 - \frac{(\Sigma y)^2}{n} = 13.81 - \frac{(11.7)^2}{12} = 2.4025 \\
SS(\text{Regression}) &= \hat{\beta}_1 SP(x_1 y) + \hat{\beta}_2 SP(x_2 y) \\
&= 0.02002 \times 19.175 + 0.13906 \times 12.625 = 2.151., \\
\therefore\ SS(\text{Error}) &= SS(\text{Total}) - SS(\text{Regression}) \\
&= 2.4025 - 2.151 = 0.2515 \\
s^2 &= \frac{SS(\text{Error})}{n-k-1} = \frac{0.2515}{12-2-1} = 0.02794 \\
V(\hat{\beta}) &= s^2(X'X)^{-1} \\
(X'X)^{-1} &= \begin{bmatrix} 0.032081 & -0.047139 \\ -0.047139 & 0.082610 \end{bmatrix} \text{ in deviation form.} \\
\therefore\ V(\hat{\beta}_1) &= 0.02794 \times 0.032081 = 0.000896 \\
V(\hat{\beta}_2) &= 0.02794 \times 0.08261 = 0.00231 \\
\text{cov}(\hat{\beta}_1, \hat{\beta}_2) &= 0.02794 \times (-0.047139) = -0.00132.
\end{aligned}$$

The variance of $\hat{\beta}_0$ is obtained from $v(\hat{\beta}) = [X'X]^{-1}s^2$, where

$$(X'X)^{-1} = \begin{bmatrix} 12 & 199 & 89 \\ 199 & 3494 & 1586 \\ 89 & 1586 & 735 \end{bmatrix}^{-1} = \begin{bmatrix} 1.82263 & -0.17678 & 0.16077 \\ -0.17678 & 0.03109 & -0.04569 \\ 0.16077 & -0.04569 & 0.08049 \end{bmatrix}.$$

$$\therefore\ V(\hat{\beta}_0) = 1.82263 \times 0.02794 = 0.05092.$$

33.6 Method of Moment

Let $x_1, x_2, \ldots, x_n$ be a random sample of n observations from a population having p.d.f. $f(x; \theta_1, \theta_2, \ldots, \theta_k)$, where the parameters $\theta_1, \theta_2, \ldots, \theta_k$ are in the parametric space S. Let the r-th moment about zero of this distribution is μ'_r, where $\mu'_r = E(X^r)$ and $\mu'_r = h(\theta_1, \theta_2, \ldots, \theta_k)$. Here $h(\theta_1, \theta_2, \ldots, \theta_k)$ is a function of parameters $\theta_1, \theta_2, \ldots, \theta_k$. The sample moment about zero for this population is

$$m'_r = \frac{1}{n} \sum_{i=1}^{n} x_i^r.$$

Since $\mu'_r = h(\theta_1, \theta_2, \ldots, \theta_k)$, we can write

$$\theta_r = g(\mu'_1, \mu'_2, \ldots, \mu'_r), \text{ when } r = k. \tag{33.4}$$

The method of moment states that m'_r is to be replaced by $\hat{\theta}_r$. Thus the estimator is $\hat{\theta}_r = m'_r$.

As an example, let us consider the estimation of θ in Poisson distribution, where

$$P(x) = \frac{e^{-\theta}\theta^x}{x!}, \quad x = 0, 1, 2, \ldots.$$

Here

$$E(x) = \mu'_1 = e^{-\theta} \sum_{x=0}^{\infty} \frac{x\theta^x}{x!} = \theta.$$

Again, if $x_1, x_2, \ldots, x_n$ be a random sample of n observations, then the sample mean is

$$m'_1 = \frac{1}{n}\Sigma x_i = \bar{x}.$$

Method of moment states that m'_1 is to be replaced by $\hat{\theta}$. Thus

$$\hat{\theta} = m'_1 = \bar{x}.$$

►**EXAMPLE 33.19** *Let $x_1, x_2, \ldots, x_n$ be a random sample of n observations from a population whose p.d.f. is*

$$f(x) = \frac{1}{b-a}, \quad a < x < b.$$

Find the estimator of a and b by method of moment.

Solution. We have

$$\mu'_1 = E(x) = \int_a^b \frac{x}{b-a} dx = \frac{a+b}{2}$$

$$E(x^2) = \frac{1}{b-a} \int_a^b x^2 dx = \frac{b^3 - a^3}{3(b-a)}$$

$$\mu_2 = E(x^2) - [E(x)]^2 = \frac{b^2 + ab + a^2}{3} - \left(\frac{a+b}{2}\right)^2 = \frac{(b-a)^2}{12}$$

$$= \mu'_2 - \mu'^2_1.$$

Solving the equations given for μ_1' and μ_2, we get

$$ab = \left(\mu_1' - \sqrt{3\mu_2}\right)\left(\mu_1' + \sqrt{3\mu_2}\right).$$

Thus, we can write $a = \mu_1' - \sqrt{3\mu_2}$, $b = \mu_1' + \sqrt{3\mu_2}$. Now, the first and second moments about zero from sample are

$$m_1' = \bar{x}_1,\ m_2 = \frac{1}{n}\Sigma(x - \bar{x})^2.$$

Replacing μ_1' and μ_2 by m_1' and m_2, respectively, we get

$$\hat{a} = \bar{x}_1 - \sqrt{\frac{3}{n}\Sigma(x - \bar{x})^2} \text{ and } \hat{b} = \bar{x} + \sqrt{\frac{3}{n}\Sigma(x - \bar{x})^2}.$$

►**EXAMPLE 33.20** *Let $x_1, x_2, \ldots, x_m$ be a random sample of m observations from binomial population with parameters n and p. Find the estimators of n and p by method of moment.*

Solution. Given

$$P(x) = \binom{n}{x} p^x q^{n-x},\quad x = 0, 1, 2, \ldots, n.$$

We know, $\mu_1' = E(x) = np$, $\mu_2 = \mu_2' - \mu_1'^2 = V(x) = npq$. Solving these two equations, we have

$$p = \frac{\mu_1'}{n} \text{ and } n = \frac{\mu_1'^2}{\mu_1' + \mu_1'^2 - \mu_2'}.$$

From sample, we have

$$m_1' = \frac{1}{m}\sum_{i=1}^{m} x_i = \bar{x},\ m_2' = \frac{1}{m}\sum_{i=1}^{m} x_i^2.$$

Therefore,

$$\hat{p} = \frac{\bar{x}}{n},\ \hat{n} = \frac{\bar{x}^2}{\bar{x} + \bar{x}^2 - m_2'}.$$

►**EXAMPLE 33.21** *Let $x_1, x_2, \ldots, x_n$ be a random sample of n observations from a population having p.d.f.*

$$f(x) = \frac{1}{\theta},\ 0 \le x \le \theta.$$

Find the estimator of θ by method of moment.

Solution. We have

$$\mu_1' = E(x) = \frac{1}{\theta}\int_0^{\theta} x\,dx = \frac{\theta}{2} \Rightarrow \theta = 2\mu_1'.$$

The sample mean is

$$m_1' = \frac{1}{n}\Sigma x = \bar{x};\ \therefore\ \hat{\theta} = 2\bar{x}.$$

►**EXAMPLE 33.22** *Let $x_1, x_2, \ldots, x_n$ be a random sample of n observations from a population having p.d.f.*

$$f(x) = \theta e^{-\theta x},\ 0 \le x \le \infty.$$

Find the estimators of θ, θ^2 and θ^3 by method of moment.

Solution. The moment generating function of the distribution is

$$M_x(t) = E\left[e^{tX}\right] = \theta \int_0^\infty e^{tx} e^{-\theta x} dx = \theta \int_0^\infty e^{-x(\theta-t)} dx$$

$$= \frac{\theta}{\theta - t} = \left(1 - \frac{t}{\theta}\right)^{-1} = 1 + \frac{t}{\theta} + \frac{t^2}{\theta^2} + \frac{t^3}{\theta^3} + \cdots$$

$$\therefore\ \mu_1' = \frac{1}{\theta},\ \mu_2' = \frac{2}{\theta^2},\ \mu_3' = \frac{6}{\theta^3} \text{ or, } \theta = \frac{1}{\mu_1'},\ \theta^2 = \frac{2}{\mu_2'},\ \theta^3 = \frac{6}{\mu_3'}.$$

The sample moments are

$$m_1' = \frac{1}{n}\Sigma x = \bar{x},\ m_2' = \frac{1}{n}\Sigma x^2,\ m_3' = \frac{1}{n}\Sigma x^3.$$

$$\therefore\ \hat{\theta} = \left|\frac{1}{m_1'}\right| = \frac{1}{\bar{x}},\ \hat{\theta^2} = \frac{2}{m_2'},\ \hat{\theta^3} = \frac{6}{m_3'}.$$

►**EXAMPLE 33.23** *Let $x_1, x_2, \ldots, x_n$ be a random sample of n observations from a population having p.d.f.*

$$f(x) = e^{-(x-\theta)};\ -\infty < \theta < \infty,\ 0 < x < \infty.$$

Find estimator of θ and e^θ by method of moment.

Solution. We have

$$M_x(t) = E\left[e^{tX}\right] = \int_0^\infty e^{tx} e^{-(x-\theta)} dx = e^\theta \int_0^\infty e^{-x(1-t)} dx$$

$$= \frac{e^\theta}{1-t} = e^\theta[1 + t + t^2 + t^3 + \cdots].$$

Here $\mu_1' = e^\theta \Rightarrow \theta = \ln \mu_1'$.

The sample moment is $m_1' = \frac{1}{n}\Sigma x_i = \bar{x}$.

$$\therefore\ \hat{\theta} = \ln m_1' = \ln \bar{x} \text{ and } \left(\widehat{e^\theta}\right) = m_1' = \bar{x}.$$

►**EXAMPLE 33.24** *Let $x_1, x_2, \ldots, x_n$ be a random sample of n observations from a population having p.d.f.*

$$f(x) = \frac{\beta^\alpha}{\Gamma\alpha} x^{\alpha-1} e^{-\beta x};\ 0 \le x \le \infty,\ \alpha > 0,\ \beta > 0.$$

Find the estimator of α and β by method of moment.

Solution. The r-th general moment about zero is

$$\mu_r' = E(x^r) = \frac{\beta^\alpha}{\Gamma\alpha} \int_0^\infty e^{-\beta x} x^{\alpha+r-1} dx.$$

Let $\beta x = z \Rightarrow dx = dz/\beta$ and $x = z/\beta$. Then

$$\mu_r' = \frac{\beta^\alpha}{\Gamma\alpha}\int_0^\infty e^{-z}z^{\alpha+r-1}\frac{dz}{\beta\beta^{\alpha+r-1}} = \frac{\Gamma(\alpha+r)}{\beta^r\Gamma\alpha}.$$

$$\therefore\ \mu_1' = \frac{\Gamma(\alpha+1)}{\beta\Gamma\alpha} = \frac{\alpha}{\beta}$$

$$\mu_2' = \frac{\Gamma(\alpha+2)}{\beta^2\Gamma\alpha} = \frac{\alpha(\alpha+1)}{\beta^2}.$$

Solving these two equations, we get

$$\frac{\mu_2'}{\mu_1'^2} = \frac{\alpha(\alpha+1)}{\beta^2}\frac{\beta^2}{\alpha^2} = 1+\frac{1}{\alpha}.$$

$$\therefore\ \alpha = \frac{\mu_1'^2}{\mu_2'-\mu_1'^2} \text{ and } \beta = \frac{\alpha}{\mu_1'} = \frac{\mu_1'^2}{(\mu_2'-\mu_1'^2)\mu_1'} = \frac{\mu_1'}{\mu_2'-\mu_1'^2}.$$

The sample moments are

$$m_1' = \frac{1}{n}\Sigma x = \bar{x},\ m_2' = \frac{1}{n}\Sigma x^2.$$

$$\therefore\ \hat{\alpha} = \frac{\bar{x}^2}{m_2'-m_1'^2},\ \hat{\beta} = \frac{\bar{x}}{m_2'-m_1'^2}.$$

►**EXAMPLE 33.25** *Let $x_1, x_2, \ldots, x_n$ be a random sample of n observations from first kind of beta distribution with parameter α and β. Find the estimators of α and β by method of moment.*

Solution. Given

$$f(x) = \frac{1}{B(\alpha,\beta)}x^{\alpha-1}(1-x)^{\beta-1};\ (\alpha,\beta)>0,\ 0<x<1$$

$$\mu_r' = \frac{1}{B(\alpha,\beta)}\int_0^1 x^{\alpha+r-1}(1-x)^{\beta-1}dx = \frac{B(\alpha+r,\beta)}{B(\alpha,\beta)}.$$

$$\therefore\ \mu_1' = \frac{B(\alpha+1,\beta)}{B(\alpha,\beta)} = \frac{\alpha}{\alpha+\beta}$$

$$\mu_2' = \frac{B(\alpha+2,\beta)}{B(\alpha,\beta)} = \frac{\alpha(\alpha+1)}{(\alpha+\beta)(\alpha+\beta+1)}.$$

Now

$$\frac{\mu_2'}{\mu_1'} = \frac{\alpha+1}{\alpha+\beta+1}$$

$$\alpha\mu_2'+\beta\mu_2'+\mu_2' = \alpha\mu_1'+\mu_1'$$

$$\alpha(\mu_2'-\mu_1') = \mu_1'-\mu_2'-\beta\mu_2' = \mu_1'-\mu_2'+\frac{\alpha(\mu_1'-1)\mu_2'}{\mu_1'},$$

$$\because\ \alpha\mu_1'+\beta\mu_1' = \alpha \text{ or, } \beta = -\frac{\alpha(\mu_1'-1)}{\mu_1'}$$

$$\therefore\ \alpha = \frac{\mu_1'-\mu_2'}{(\mu_2'-\mu_1')-\dfrac{(\mu_1'-1)\mu_2'}{\mu_1'}}.$$

The sample moments are

$$m_1' = \frac{1}{n}\Sigma x = \bar{x},\ m_2' = \frac{1}{n}\Sigma x^2.$$

Therefore,

$$\hat{\alpha} = \frac{\bar{x} - m_2'}{(m_2' - \bar{x}) - \dfrac{(\bar{x}-1)m_2'}{\bar{x}}} \text{ and } \hat{\beta} = -\frac{\hat{\alpha}(\bar{x}-1)}{\bar{x}}.$$

►**EXAMPLE 33.26** *Let $x_1, x_2, \ldots, x_n$ be a random sample of n observations from $N(\mu, \sigma^2)$. Find estimators of μ and σ^2 by method of moments.*

Solution. We have

$$f(x) = \frac{1}{\sigma\sqrt{2\pi}} \exp\left[-\frac{1}{2\sigma^2}(x-\mu)^2\right],\ -\infty < x < \infty$$

$$M_x(t) = e^{\mu t + \frac{1}{2}t^2\sigma^2} = 1 + \left(\mu t + \frac{1}{2}t^2\sigma^2\right) + \frac{1}{2}\left(\mu t + \frac{1}{2}t^2\sigma^2\right)^2 + \cdots$$

$$\therefore\ \mu_1' = \mu,\ \mu_2' = \sigma^2 + \mu^2;\ \therefore\ \sigma^2 = \mu_2' - \mu_1'^2.$$

The sample moments are $m_1' = \frac{1}{n}\Sigma x = \bar{x}, m_2' = \frac{1}{n}\Sigma x^2.$

$$\therefore\ \hat{\mu}_1' = \bar{x},\ \hat{\sigma}^2 = m_2' - m_1'^2.$$

Some Remarks about Estimators by Method of Moment

The sample observations $x_1, x_2, \ldots, x_n$ from $f(x, \theta)$ are randomly selected and x_i^r $(i = 1, 2, \ldots, n)$ independently distributed. Hence, if $E(x_i^r)$ exists, then by weak law of large numbers

$$m_r' = \frac{1}{n}\Sigma x_i^r \to \mu_r' = E(x_i^r)$$

with probability p. Therefore:

(i) Sample moments are consistent estimators of population moments. The estimators by method of moment are consistent and unbiased.

(ii) The estimator by method of moment follows normal distribution asymptotically. But the estimator is not always efficient.

(iii) The estimator by method of moment is less efficient compared to ML estimator. However, if the probability density function is a member of exponential family of distributions, then ML estimator and estimator by method of moment are equally efficient.

33.7 Method of Minimum Chi-square

Let $x_1, x_2, \ldots, x_n$ be a random sample of n observations from a population whose p.d.f. is $f(x, \theta)$, where the range of x is $-\infty < x < \infty$. Consider that the range space of

x can be divided into k parts and the different parts are $R_1, R_2, \ldots, R_k$. Thus the sample observations must be included in either of these k parts. Let the probability of inclusion of any sample observation in i-th part be $P_i(\theta)$ $[i = 1, 2, \ldots, k]$ such that

$$P_i(\theta) = P[x : x \in R_i] = \int_{R_i} f(x)dx$$

$$\text{and } \sum_{i=1}^{k} P_i(\theta) = 1.$$

Let the number of observations of x in i-th part be n_i such that

$$\sum_{i=1}^{k} n_i = n.$$

Since the probability of x to be included in i-th part is $P_i(\theta)$, the expected number of x to be included in i-th part is $nP_i(\theta)$. Thus, using observed n_i and expected n_i the χ^2-statistic is

$$\chi^2 = \sum_{i=1}^{k} \frac{[n_i - nP_i(\theta)]^2}{n_i P_i(\theta)}.$$

The value of θ is to be estimated in such a way that the value of χ^2 becomes minimum. The minimum value of χ^2 is zero. This is possible, if $n_i = nP_i(\theta)$.

►**EXAMPLE 33.27** *Let $x_1, x_2, \ldots, x_n$ be a random sample of n observations from Bernoulli population with parameter θ. Find the estimator of θ by method of minimum chi-square.*

Solution. Given $P(x) = \theta^x(1-\theta)^{1-x}$, $x = 0, 1$.

Thus $P(x = 0) = 1 - \theta$, $P(x = 1) = \theta$. Let there be n_0 elements of x, the value of which is zero and number of observations of x having value 1 be n_1. Here $n_0 = n - n_1$. The range space is divided into 2 parts. Here

$$\chi^2 = \sum_{i=0}^{1} \frac{[n_i - nP_i(\theta)]^2}{nP_i(\theta)}$$

$$= \frac{[n_0 - n_1 - n(1-\theta)]^2}{n(1-\theta)} + \frac{(n_1 - n\theta)^2}{n\theta} = \frac{(n_1 - n\theta)^2}{n} \frac{1}{\theta(1-\theta)}.$$

Here χ^2 becomes minimum, if $\theta = n_1/n$.
Therefore, $\hat{\theta} = n_1/n$.

33.8 Bayes Method of Estimation

Let $x_1, x_2, \ldots, x_n$ be a random sample of n observations from a population having p.d.f. $f(x, \theta)$, $\theta \in S$ and S is the parametric space. The parameter θ may be a random variable. For example, in Bernoulli population the probability of $x = 0$ is p and probability of $x = 1$ is $1 - p$, where $0 < p < 1$. Here p is the parameter in Bernoulli population but it varies within the range 0 to 1 from experiment to experiment. The sample observations of different experiments may be considered as the sample of a large population and p observed in different experiments may be considered as random variable. The distribution of p is uniform within the range 0 to 1 and this distribution is known as 'a priori' distribution. Let the p.d.f. of p be $f_1(p)$. Therefore, the distribution of random variable X depends on the 'a priori' distribution of parameter p or θ and this distribution is known as *conditional distribution.*

Since the sample observations are $x_1, x_2, \ldots, x_n$, their joint distribution is

$$g(x_1, x_2, \ldots, x_n/\theta) = f(x_1, \theta)f(x_2, \theta) \cdots f(x_n, \theta).$$

Again, the joint distribution of sample observations and θ is

$$k(x_1, x_2, \ldots, x_n; \theta) = g(x_1, x_2, \ldots, x_n/\theta)f_1(\theta).$$

Hence the marginal distribution of $x_1, x_2, \ldots, x_n$ is

$$h(x_1, x_2, \ldots, x_n) = \int g(x_1, x_2, \ldots, x_n/\theta)f_1(\theta)d\theta.$$

The conditional distribution of θ is

$$f(\theta/x_1, x_2, \ldots, x_n) = \frac{k(x_1, x_2, \ldots, x_n; \theta)}{h(x_1, x_2, \ldots, x_n)}.$$

Now, the estimator θ is found out using the conditional probability $f(\cdot)$, where this probability is based on 'Bayes Probability Theory' and hence the estimator of θ is known as *Bayes estimator.*

Let $T = t(x_1, x_2, \ldots, x_n)$ be a sample statistic used as an estimator of θ [probably, sufficient statistic]. The particular value of T is assumed to be $W(t)$ and T is the point estimator of θ. The function $W(t)$ helps in deciding the point estimator of θ and hence it is called *decision function* or *decision rule.* The value of decision function is known as *decision.*

The decision may be right or may be wrong. To study the precision of decision one needs to study the difference between θ and $W(t)$. Consider that different values of $W(t)$ are observed. Let $L[\theta, W(t)]$ be a positive quantity observed from the differences of θ and $W(t)$. This L is called 'Loss' function and $E[L] = R(\theta, W)$ is known as 'Risk' function, where

$$R(\theta, W) = E\{L[\theta, W(t)]\} = \int_{-\infty}^{\infty} L[\theta, W(t)]g(t; \theta)dt.$$

If T is the continuous variable, the decision function is to be selected in such a way that for all $\theta \in S$, the value $R(\theta, W)$ becomes minimum. In practice, this is almost impossible since for a particular value of θ, $R(\theta, W)$ may be minimum but it may not be minimum for other values of θ. Therefore, the value of $T[\hat{\theta}]$ is to be selected in such a way that $V(\hat{\theta})$ becomes minimum, where $V(\hat{\theta}) = E[\hat{\theta} - \theta]^2$. This variance is to be calculated using the conditional distribution of θ, where the conditional distribution is based on Bayes probability law. Thus the estimator $\hat{\theta}$ is known as Bayes estimator.

►**EXAMPLE 33.28** *Let $x_1, x_2, \dots, x_n$ be a random sample of n observations from Bernoulli population with parameter θ. Find Bayes estimator of θ.*

Solution. We have

$$P(x) = \theta^x (1-\theta)^{1-x};\ x = 0, 1;\ 0 < \theta < 1.$$

The prior probability of θ is

$$f_1(\theta) = 1,\ 0 \le \theta \le 1.$$

The joint density function of $x_1, x_2, \dots, x_n$ and θ is

$$k(x_1, x_2, \dots, x_n; \theta) = g(x_1, x_2, \dots, x_n/\theta) f_1(\theta) = \theta^{\Sigma x}(1-\theta)^{n-\Sigma x}$$

where the conditional density function of $x_1, x_2, \dots, x_n$ given θ is

$$g(x_1, x_2, \dots, x_n/\theta) = f(x_1, \theta) f(x_2, \theta) \cdots f(x_n, \theta) = \theta^{\Sigma x}(1-\theta)^{n-\Sigma x}.$$

Therefore, the marginal density function of $x_1, x_2, \dots, x_n$ is

$$h(x_1, x_2, \dots, x_n) = \int_0^1 \theta^{\Sigma x}(1-\theta)^{n-\Sigma x} d\theta = B[\Sigma x + 1, n - \Sigma x + 1].$$

The conditional density function of θ given $x_1, x_2, \dots, x_n$ is

$$f(\theta/x_1, x_2, \dots, x_n) = \frac{k(x_1, x_2, \dots, x_n; \theta)}{h(x_1, x_2, \dots, x_n)} = \frac{\theta^{\Sigma x}(1-\theta)^{n-\Sigma x}}{B[\Sigma x + 1, n - \Sigma x + 1]}.$$

We have

$$\begin{aligned}
V(\hat{\theta}) &= E\left[\hat{\theta} - \theta\right]^2 = E\left[\hat{\theta}^2 - 2\hat{\theta}\theta + \theta^2\right] \\
&= \int_0^1 \left(\hat{\theta}^2 - 2\hat{\theta}\theta + \theta^2\right) f(\theta/x_1, x_2, \dots, x_n) d\theta \\
&= \hat{\theta}^2 \int_0^1 \frac{\theta^{\Sigma x}(1-\theta)^{n-\Sigma x}}{B[\Sigma x + 1, n - \Sigma x + 1])} d\theta - 2\hat{\theta} \int \frac{\theta^{\Sigma x+1}(1-\theta)^{n-\Sigma x}}{B[\Sigma x + 1, n - \Sigma x + 1]} d\theta \\
&\quad + \int_0^1 \frac{\theta^{\Sigma x+2}(1-\theta)^{n-\Sigma x}}{B[\Sigma x + 1, n - \Sigma x + 1]} d\theta \\
&= \hat{\theta}^2 - 2\hat{\theta} \frac{B[\Sigma x + 2, n - \Sigma x + 1]}{B[\Sigma x + 1, n - \Sigma x + 1]} + \frac{B[\Sigma x + 3, n - \Sigma x + 1]}{B[\Sigma x + 1, n - \Sigma x + 1]} \\
&= \hat{\theta}^2 + \frac{(\Sigma x + 2)(\Sigma x + 1)}{(n+3)(n+2)} - 2\hat{\theta} \frac{\Sigma x + 1}{n+2}.
\end{aligned}$$

Now

$$\frac{V(\hat{\theta})}{\partial \hat{\theta}} = 0 \Rightarrow 2\hat{\theta} - \frac{2(\Sigma x + 1)}{n+2} = 0.\ \therefore\ \hat{\theta} = \frac{(\Sigma x + 1)}{n+2} = \frac{n\bar{x} + 1}{n+2}.$$

❒ **THEOREM 33.7** *The quadratic loss function in estimating θ in $f(x, \theta)$; $\theta \in S$ is $L(\theta, W(t)] = (\theta - W)^2$. If θ is estimated with respect to this loss function, then the Bayes estimator is $W(t) = E[\theta/X = x]$.*

Proof. If $f_1(\theta)$ is the 'a priori' p.d.f. of θ for continuous random variable, then

$$\begin{aligned} R(\theta, W) &= \int h(x)\{f[\theta - W(t)]^2 f(\theta/x)d\theta\}dx \\ &= \int f_1(\theta)[\{f(W(t) - \theta)^2 f(x/\theta)]d\theta \end{aligned}$$

where $h(x)$ is the marginal distribution of X, f is the conditional distribution. The Bayes rule states that θ is to be estimated in such a way that $R(\theta, W)$ becomes minimum. The minimum $R(\theta, W)$ indicates that

$$\int [\theta - W(t)]^2 f(\theta/X = x)d\theta$$

should be minimum. This quantity becomes minimum, if $W(t) = E[\theta/X]$. Here

$$W(t) = \int \theta f(\theta/x)d\theta$$

is the Bayes estimator.

Let us estimate θ^2 of Bernoulli population with parameter θ using the loss function stated above. According to the loss function theorem, we have

$$\begin{aligned} W(t) &= E(\theta^2/X) \\ &= \int_0^1 \frac{\theta^2\theta^{\Sigma x}(1-\theta)^{n-\Sigma x}}{B[\Sigma x + 1, n - \Sigma x + 1]}d\theta = \int_0^1 \frac{\theta^{\Sigma x+2}(1-\theta)^{n-\Sigma x}}{B[\Sigma x + 1, n - \Sigma x + 1]}d\theta \\ &= \frac{B[\Sigma x + 3, n - \Sigma x + 1]}{B[\Sigma x + 1, n - \Sigma x + 1]} = \frac{[\Sigma x + 1][\Sigma x + 2]}{(n+2)(n+3)}. \\ \therefore\ \hat{\theta}^2 &= \frac{(\Sigma x + 1)(\Sigma x + 2)}{(n+2)(n+3)}. \end{aligned}$$

Definition. Let $f(x, \theta)$ be a member of p.d.f., where $\theta \in S$ and W^* be the Bayes estimator of θ. If for this estimator the risk function $R(\theta, W^*)$ is constant on S, then W^* is called *minimax estimator.*

►**EXAMPLE 33.29** *Let $x_1, x_2, \ldots, x_n$ be a random sample of n observations from a population having p.d.f.*

$$f(x, \theta) = \theta e^{-\theta x},\ x > 0.$$

The 'a priori' distribution of θ is

$$f_1(\theta) = \frac{1}{\theta},\ 0 < \theta < \infty.$$

Find Bayes estimator of θ.

Solution. We have

$$\begin{aligned} g(x_1, x_2, \ldots, x_n/\theta) &= \theta^n e^{-\theta\Sigma x} \\ k(x_1, x_2, \ldots, x_n/\theta) &= g(x_1, x_2, \ldots, x_n/\theta)f_1(\theta) = \theta^{n-1}e^{-\theta\Sigma x}. \\ \therefore\ h(x_1, x_2, \ldots, x_n) &= \int_0^\infty \theta^{n-1}e^{-\theta\Sigma x}d\theta = \frac{\Gamma n}{(\Sigma x)^n}. \end{aligned}$$

Hence

$$f(\theta/x_1,x_2,\ldots,x_n) = \frac{k(x_1,x_2,\ldots,x_n;\theta)}{h(x_1,x_2,\ldots,x_n)} = \frac{(\Sigma x)^n}{\Gamma n}\theta^{n-1}e^{-\theta\Sigma x}.$$

$$\therefore\; E[\theta/X] = \int_0^\infty \theta f(\theta/x_1,x_2,\ldots,x_n)d\theta$$

$$= \frac{(\Sigma x)^n}{\Gamma n}\int_0^\infty \theta e^{-\theta\Sigma x}d\theta = \frac{(\Sigma x)^n}{\Gamma n}\frac{\Gamma(n+1)}{(\Sigma x)^{n+1}} = \frac{n}{\Sigma x}.$$

$$\therefore\; \hat{\theta} = \frac{n}{\Sigma x} = \frac{1}{\bar{x}}.$$

►EXAMPLE 33.30 *Let $x_1, x_2, \ldots, x_n$ be a random sample of n observations from a Poisson population with parameter θ such that* (i) $f_1(\theta) = e^{-\theta}$, $\theta > 0$ *and* (ii) $f_1(\theta) = \frac{1}{\alpha}e^{-\theta/\alpha}$, $\theta > 0$. *Using quadratic loss function find Bayes estimator of θ.*

Solution.

(i) Given $P(x) = \dfrac{e^{-\theta}\theta^x}{x!}$, $x = 0, 1, 2, \ldots$. Then

$$g(x_1,x_2,\ldots,x_n/\theta) = \frac{e^{-n\theta}\theta^{\Sigma x}}{\prod_{i=1}^n x_i!}$$

$$k(x_1,x_2,\ldots,x_n;\theta) = \frac{e^{-(n+1)\theta}\theta^{\Sigma x}}{\prod_{i=1}^n x_i!}$$

$$h(x_1,x_2,\ldots,x_n) = \int_0^\infty \frac{e^{-(n+1)\theta}\theta^{\Sigma x}}{\prod x_i!}d\theta.$$

Let $(n+1)\theta = z$, $d\theta = \dfrac{dz}{n+1}$, $\theta = \dfrac{z}{(n+1)}$. Then

$$h(x_1,x_2,\ldots,x_n) = \int_0^\infty \frac{e^{-z}z^{\Sigma x}}{(n+1)^{\Sigma x+1}\prod x_i!}dz = \frac{\Gamma(\Sigma x+1)}{(n+1)^{\Sigma x+1}\prod_{i=1}^n x_i!}.$$

Hence

$$f(\theta/x_1,x_2,\ldots,x_n) = \frac{k(x_1,x_2,\ldots,x_n;\theta)}{h(x_1,x_2,\ldots,x_n)} = \frac{e^{-(n+1)\theta}\theta^{\Sigma x}(n+1)^{\Sigma x+1}}{\Gamma(\Sigma x+1)}$$

$$E[\theta/X] = \int_0^\infty \theta f(\theta/x_1,x_2,\ldots,x_n)d\theta$$

$$= \frac{(n+1)^{\Sigma x+1}}{\Gamma(\Sigma x+1)}\int_0^\infty \theta^{\Sigma x+1}e^{-(n+1)\theta}d\theta$$

$$= \frac{(n+1)^{\Sigma x+1}}{\Gamma(\Sigma x+1)}\int_0^\infty e^{-z}z^{\Sigma x+1}\frac{dz}{(n+1)^{\Sigma x+2}}$$

$$= \frac{1}{(\Sigma x)!}\frac{\Gamma(\Sigma x+2)}{(n+1)} = \frac{\Sigma x+1}{n+1} = \frac{n\bar{x}+1}{n+1}.$$

$$\therefore\; \hat{\theta} = \frac{n\bar{x}+1}{n+1}.$$

(ii) We have

$$k(x_1, x_2, \dots, x_n) = \frac{\theta^{\Sigma x} e^{-n\theta}}{\prod x_i!} \frac{1}{\alpha} e^{-\theta/\alpha}$$

$$\text{and } h(x_1, x_2, \dots, x_n) = \frac{1}{\alpha \prod x_i!} \int_0^\infty e^{-\theta\left(n+\frac{1}{\alpha}\right)} \theta^{\Sigma x} d\theta$$

$$= \frac{\Gamma(\Sigma x + 1)}{\left(n + \frac{1}{\alpha}\right)^{\Sigma x + 1}} \frac{1}{\theta \prod_{i=1}^{n} x_i!}.$$

$$\therefore\ f(\theta/x_1, x_2, \dots, x_n) = \frac{\left(n + \frac{1}{\alpha}\right)^{\Sigma x + 1} \theta^{\Sigma x} e^{-\theta\left(n+\frac{1}{\alpha}\right)}}{\Gamma(\Sigma x + 1)}.$$

$$E[\theta/X] = \int_0^\infty \theta f(\theta/x_1, x_2, \dots, x_n)$$

$$= \int_0^\infty \frac{\left(n + \frac{1}{\alpha}\right)^{\Sigma x + 1} \theta^{\Sigma x + 1} e^{-\theta\left(n+\frac{1}{\alpha}\right)}}{\Gamma(\Sigma x + 1)} d\theta = \frac{(\Sigma x + 1)\alpha}{(n\alpha + 1)}.$$

$$\therefore\ \hat{\theta} = \frac{(\Sigma x + 1)^\alpha}{n\alpha + 1}.$$

►EXAMPLE 33.31 *Let X be a binomial random variable with parameters n and p, where*

$$f_1(p) = 1,\ 0 < p < 1.$$

Find Bayes estimator of p.

Solution. Here

$$g(x/p) = \binom{n}{x} p^x q^{n-x},\ x = 0, 1, 2, \dots, n$$

$$k(x, p) = g(x/p) f_1(p) = \binom{n}{x} p^x (1-p)^{n-x}$$

$$h(x) = \int_0^1 k(x, p) dp = \int_0^1 \binom{n}{x} p^x (1-p)^{n-x} dp$$

$$= \binom{n}{x} B[x+1, n-x+1]$$

$$f(p/x_1, x_2, \dots, x_n) = \frac{k(x, p)}{h(x)} = \frac{\binom{n}{x} p^x (1-p)^{n-x}}{\binom{n}{x} B[x+1, n-x+1]}$$

$$E(p/X) = \int_0^1 p f(p/x_1, x_2, \dots, x_n) dp$$

$$= \int_0^1 \frac{1}{B[x+1, n-x+1)]} p^{x+1} (1-p)^{n-x} dp$$

$$= \frac{B[x+2, n-x+1]}{B[x+1, n-x+1]} = \frac{x+1}{n+2}.$$

$$\therefore\ \hat{p} = \frac{x+1}{n+2}.$$

EXERCISE

1. (a) Distinguish between estimator and statistic with examples. What are the criteria of a good estimator?

   (b) State and explain the principle of maximum likelihood estimation of population parameter.

   (c) Find the maximum likelihood estimator of θ in

$$f(x,\theta) = \begin{cases} (1+\theta)x^{\theta}; & \theta > 0,\ 0 < x < \theta \\ 0, & \text{otherwise.} \end{cases}$$

2. (a) What is meant by estimation? Explain in short the different methods of estimation.

   (b) Show that if an ML estimator exists, it is a function of MVB estimator and a function of sufficient statistic.

   (c) Find the ML estimator of $\dfrac{\alpha}{\alpha+2}$ in

$$f(x) = \frac{1}{B(\alpha,2)}(1-x)x^{\alpha-1};\ 0 \le x \le 1,\ \alpha > 0.$$

3. (a) Distinguish between method of moment and ML method as a method of estimation.

   (b) State some properties of ML estimator and show that these properties exist in practice.

   (c) Find the ML estimator and estimator by method of moment of θ in

$$f(x,\theta) = \frac{1}{\theta},\ 0 \le x \le \theta.$$

4. (a) Explain the method of least squares. Show that least squares estimator is BLUE.

   (b) The data on family income (x_1, in 000 taka), family member (x_2) and family expenditure (y, in 000 taka) of 10 randomly selected families are given below:

x_1	5.5	10.0	14.6	15.0	4.5	8.5	20.0	2.5	9.5	10.0
x_2	4	6	5	4	6	3	5	4	3	4
y	4.2	8.5	7.5	14.5	4.5	6.2	15.0	2.5	6.0	8.0

   Estimate the impact of income and family member on family expenditure by method of least squares. Also, estimate variance of your estimators.

5. (a) Describe clearly the criterion of a good estimator. Show that, if a sufficient statistic exists, the ML estimator is a function of sufficient statistic.

   (b) For the bivariate normal distribution $N(\mu_1, \mu_2, \sigma_1^2, \sigma_2^2, \rho)$ find the ML estimators of the parameters.

   (c) Is efficient estimator unique?

6. (a) Define with examples estimator, statistic, parameter. Explain different methods of estimation.

(b) Find the estimator of θ in Poisson population with parameters θ by method of moment.

(c) The sample values from a population with p.d.f.

$$f(x) = (1+\theta)x^{\theta};\ 0 < x < 1,\ \theta > 0$$

are given below:

x : 0.85, 0.65, 0.70, 0.88, 0.64, 0.42, 0.56, 0.72, 0.34, 0.40.

Find estimate of θ by method of moment.

7. (a) Explain the concepts of efficiency and sufficiency in statistical estimation.

(b) Show that ML estimator is consistent.

(c) Find estimator of θ_1 and θ_2 in

$$f(x) = \frac{1}{\theta_2 - \theta_1},\ \theta_1 < x < \theta_2$$

by method of moment and by method of maximum likelihood.

8. (a) Explain method of least squares and method of moment.

(b) State and prove the properties of ML estimator.

(c) Find Bayes estimator λ in Poisson distribution, if

$$f_1(\lambda) = 1,\ 0 < \lambda < 1.$$

9. (a) Distinguish between method of least squares and ML method of estimation. Show that LS estimator is BLUE.

(b) The temperature (x_1, °celsius), humidity (x_2, %) and rainfall (x_3, mm) during 10 randomly selected days during rainy season are given below:

x_1	34.2	30.6	32.6	34.5	32.5	30.6	34.2	31.6	32.4	33.2
x_2	90	85	92	93	86	92	97	88	91	88
x_3	0.2	4.5	0.5	4.5	2.8	6.5	0.2	8.6	5.0	6.4

Estimate the impacts of temperature and rainfall on humidity by method of least squares. Also, estimate the standard error of your estimators.

10. (a) Define consistent estimator. Show that ML estimator is consistent.

(b) Explain method of moment in estimating population parameter. Describe some of its characters.

(c) Find the estimator of α and β in

$$f(x) = \frac{1}{B(\alpha,\beta)} x^{\alpha-1}(1-x)^{\beta-1};\ (\alpha,\beta) > 0,\ 0 < x < 1$$

by method of moment.

11. (a) Explain method of least squares. When least squares estimator is equivalent to ML estimator? How would you estimate the variance of least squares estimator?

(b) Find ML estimator of θ in

$$f(x) = \theta x^{\theta-1};\ 0 < x < 1,\ 0 < \theta < \infty.$$

Also, find the estimator by method of moment.

(c) State the factorisation theorem on sufficiency. Obtain a sufficient statistic for θ in

$$f(x) = \frac{1}{\theta},\ 0 < x < \theta.$$

12. (a) Define estimator, parameter with examples. Explain the criterion of a good estimator.

(b) Show that ML estimator is MVB if it exists. Write down the characteristics of estimator by method of moment.

(c) Find the estimator of α and β in

$$f(x) = \beta e^{-\beta(x-\alpha)};\ 0 < x < \infty,\ \beta > 0$$

by ML method.

13. (a) Explain necessary and sufficient condition for an unbiased estimator to be MVU estimator.

(b) Find ML estimator of μ and σ^2 in $N(\mu, \sigma^2)$.

(c) State the properties of ML estimator.

14. (a) Distinguish between ML method and method of least squares as method of estimation. Show that least squares estimator is BLUE.

(b) Find ML estimator of p, where p is the parameter in binomial population.

(c) Find ML estimator of θ_1 and θ_2 in

$$f(x) = \frac{1}{\theta_2}\exp\left[-\frac{1}{\theta_2}(x-\theta_1)\right];\ \theta_1 < x < \infty,$$
$$-\infty < \theta_1 < \infty,\ 0 < \theta_2 < \infty.$$

15. (a) Discuss the necessity of estimation of population parameter. What are the different methods of estimation. Which method would you recommend to have a good estimator?

(b) Distinguish between method of moment and ML method of estimation.

(c) Find ML estimator of θ in

$$f(x) = 1,\ \theta - \frac{1}{2} < x < \theta + \frac{1}{2}.$$

(d) Find estimator of θ in

$$f(x) = \theta^x(1-\theta)^{1-x};\ x = 0, 1;\ 0 < \theta < 1$$

by method of moment.

CHAPTER **34**

Test of Hypothesis

34.1 Introduction

In the previous chapter also it is mentioned that the objective of the statistical analysis is to estimate the unknown population characteristic (parameter) and to infer about that parameter. The inference is drawn on the basis of probability distribution of the sample estimator. The process of making inference about parameter using sample statistic (estimator) is known as *test of hypothesis* or *test of statistical hypothesis.*

The statistical hypothesis is basically two types, viz., (a) Parametric hypothesis and (b) Non-parametric hypothesis. The unknown population characteristic is called *parameter* and any specification about one or more parameters is termed as *parametric hypothesis.* This is done when the distribution of population observations is specified or known. For example, the distribution of measurement errors in measuring the ages of a randomly selected group of people can be assumed normally distributed with population mean error μ and population variance error σ^2. Here μ and σ^2 are termed as parameters. Any specification about μ or σ^2 which is under verification is known as *hypothesis.* Again, the distribution of family planning adopter couples, unless it is assumed, is usually not specified and hence its parameter is also not mentioned. In such a case, one may verify the characteristics related to population or populations but not the parameter of the population. Any specification of population or sample of the population except the parameter of the population is known as *non-parametric hypothesis.*

Whatever be the hypothesis, parametric or non-parametric, the decision regarding population characteristic is based on sample statistic. The sample statistic follows some probability laws. The probability calculated for any particular value of the statistic helps the researcher in deciding the rejection or acceptance of the hypothesis. The whole procedure of finding out a sample statistic and finally rejecting or accepting the hypothesis are subject matters of test of hypothesis. In this chapter we shall discuss parametric hypothesis.

34.2 Terms Related to Test of Hypothesis

STATISTICAL TEST. The statistical procedure used in finding out a sample statistic using random sample observations and which helps in deciding about the possible rejection or acceptance of a statistical hypothesis is known as *statistical test.* For example, let $x_1, x_2, \ldots, x_n$ be the n birth-weights of babies born in a hospital during a year. The problem is to verify whether the average birth-weight is 3.00 kg or not. If it is assumed that the birth-weights $X \sim NID(\mu, \sigma^2)$, then a sample statistic related to mean birth-weight can be found out as

$$z = \frac{\bar{x} - \mu}{\sigma/\sqrt{n}} \sim N(0, 1).$$

This sample statistic follows normal distribution and for a specific value of μ and for known value of σ the value of z can be calculated and its probability can also be found out. The probability will help us in deciding about the specified value μ. The whole process of decision is a subject matter of statistical test.

HYPOTHESIS. Any specification regarding a parameter or a population which is under investigation is known as *hypothesis.* The hypothesis is of two types, viz., (a) Null hypothesis and (b) Alternative hypothesis.

NULL HYPOTHESIS. The hypothesis which is used for possible rejection under verification on the basis of sample statistic is known as *null hypothesis.* It is usually denoted by H_0. Thus, if $X \sim N(\mu, \sigma^2)$ and if we want to verify whether μ has a specific value μ_0 or not, we specify the null hypothesis $H_0 : \mu = \mu_0$.

ALTERNATIVE HYPOTHESIS. Any statement against the null hypothesis is known as *alternative hypothesis.* Thus, when $H_0 : \mu = \mu_0$ is under investigation, it is investigated against an alternative statement $H_A : \mu \neq \mu_0$ or $H_A : \mu > \mu_0$ or $H_A : \mu < \mu_0$. The alternative hypothesis is denoted by H_A.

TEST STATISTIC. The function of sample observations which is used to verify the null hypothesis is known as *test statistic.* Thus, if the phenomenon under investigation is to test the null hypothesis $H_0 : P = P_0$ (a known value), where P is the proportion of adopter couples in a society, then the sample statistic is

$$z = \frac{p - P_0}{\sqrt{\frac{P_0 Q_0}{n}}} \sim N(0, 1).$$

Here n is the sample size of couples under investigation, p is the sample proportion of adopter couples and $Q_0 = 1 - P_0$. The statistic 'z' is test statistic.

CRITICAL REGION OR REJECTION REGION. The area constituted with the values of sample statistic due to which the null hypothesis is rejected even when it is true is called *critical region.* It is denoted by W.

ACCEPTANCE REGION. The area constituted with the values of sample statistic due to which the null hypothesis is accepted when it is true is called *acceptance region.*

TYPE-I AND TYPE-II ERROR. The error that creeps in taking decision to reject the null hypothesis when null hypothesis is true is called *type-I error.* The probability of this error is denoted by α. The error that creeps in taking decision to accept the null hypothesis even when it is not true is known as *type-II error.* The probability of this error is denoted by β.

LEVEL OF SIGNIFICANCE. The error that creeps in rejecting the null hypothesis when it is true is known as type-I error. The probability of this error α is also known as *level of significance.* Mathematically, we write

$$P(X \in W/H_0) = \alpha.$$

Here X denotes the sample statistic and W is the critical region.

p-VALUE. The minimum value of α for which the null hypothesis is rejected is known as *p-value.* If the p-value is less than or equal to α, the null hypothesis is rejected, otherwise it is accepted. If $g(x)$ is a function of sample statistic related to a test, then

$$\int_X^{\infty} g(x)dx = p \text{ or, } \int_{-\infty}^{-X} g(x)dx = p$$

where X is a particular value of X.

ONE-SIDED TEST. If the critical region is considered at one end of the probability distribution of test statistic, the test is called *one-sided test.* It depends on alternative hypothesis. For example, let $H_A : \mu > \mu_0$, then

$$\int_{z_0}^{-\infty} f(z)dz = p$$

where z_0 is a particular value of function of sample mean. The test related to $H_0 : \mu = \mu_0$ against $H_A : \mu > \mu_0$ is known as one-sided test. Again, if $H_A : \mu < \mu_0$, then

$$\int_{-\infty}^{-z_0} f(z)dz = p.$$

In such a case also the test is one-sided test. The former test is right-sided test and the latter test is left-sided test.

TWO-SIDED TEST. If in a statistical test, the critical regions are considered in left as well as right side of the distribution of sample statistic, then the test is called *two-sided test.* The alternative hypothesis is $H_A : \mu \neq \mu_0$. In such a case, if $p \leq \alpha/2$, the null hypothesis is rejected, otherwise it is accepted.

POWER OF TEST. The probability of rejecting the null hypothesis when alternative hypothesis is really true. It is denoted by $1-\beta$. This is the probability of right decision regarding the null hypothesis. Mathematically,

$$\text{Power} = 1 - \beta = P(X \in W/H_A).$$

►EXAMPLE 34.1 *A company supplies computer spare parts in packets. Usually, 2% packets of a day's out are found defective. In a day company produces 200 packets. Do you expect that in a randomly selected day, the defective packets exceed 2%?*

Solution. Usually, the average defective packets per day is $\lambda = 0.02 \times 200 = 4$. We need to test $H_0 : \lambda = 4$ against $H_A : \lambda > 4$. The null hypothesis will be rejected if the number of defective packets exceeds 4. Then

$$\begin{aligned}\alpha &= \text{Probability of rejecting } H_0 \text{ when } H_0 : \lambda = 4\\ &= \sum_{x=5} \frac{e^{-\lambda}\lambda^x}{x!}\\ &= 1 - \sum_{n=0}^{4} \frac{e^{-4}4^x}{x!}\\ &= 0.3643\\ \beta &= \text{Probability of accepting } H_0 \text{ when } \lambda > 4\\ &= \sum_{x=0}^{3} \frac{e^{-4}4^x}{x!}\\ &= 0.4335.\end{aligned}$$

Then, the power of test is $1 - \beta = 1 - 0.4335 = 0.5665$.

SIMPLE HYPOTHESIS. If in investigating the significance of a hypothesis all parameters of the distribution or the distribution itself are specified, then the hypothesis is known as *simple hypothesis.* For example, let $X \sim N(\mu, \sigma^2)$. The object is to test the significance of $H_0 : \mu = \mu_0$ against $H_A : \mu \neq \mu_0$. Here, one can assume that σ^2 is known. In such a case the hypothesis is called *simple hypothesis.* Again, one can assume that σ^2 is unknown. Since both the parameters of the distribution are not specified, the hypothesis is not simple and the hypothesis is known as *composite hypothesis.*

COMPOSITE HYPOTHESIS. If by a hypothesis all the parameters of a distribution or distribution itself are not specified, the hypothesis is called *composite hypothesis.*

PARAMETRIC HYPOTHESIS. Any hypothesis related to parameter(s) of a distribution is known as *parametric hypothesis.* For example, ten coins are tossed and we need to test the significance of $H_0 : p = 1/2$, i.e., the coins are unbiased. The problem is to investigate the number of heads (H). If the coins are unbiased, $P(H) = P(T) = \frac{1}{2} = p$. This is a case of binomial distribution. There are two parameters of binomial distribution. These are n and p. Both parameters are specified. Hence the hypothesis is a parametric and simple hypothesis.

NON-PARAMETRIC HYPOTHESIS. The hypothesis which does not specify anything about the parameter(s) of a distribution is called *non-parametric hypothesis.* For example, a group of students are selected for some social activities. One may be interested to verify the hypothesis that the selection is random. Here nothing is mentioned about the parameter. So the hypothesis is non-parametric hypothesis.

34.3 Best Critical Region (BCR) and Most Powerful (MP) Test

In performing any test there may be chances of increasing Type-I or Type-II errors. It is not possible to reduce two types of errors simultaneously. Hence, the usual technique is to reduce the Type-II error. This is done in such a way that the critical region is found out keeping Type-I error fixed. Since α is the probability of Type-I error, the critical region of size α is to be found out in such a way that the power of the test is not less than the power of a test of same size of critical region. Such a critical region with maximum power is called *best critical region* (*BCR*). Mathematically, BCR can be defined as follows:

Let $H_0 : \theta = \theta_0$ and $H_A : \theta = \theta_1$, where θ is a parameter in the parametric space. The critical region W is known as BCR, if

(i) $P[X \in W/H_0] = \alpha$,

(ii) $P[X \in W/H_1] \geq P[X \in W_h/H_1]$,

where W_h is any other critical region such that

$$P[X \in W_h/H_0] = \alpha.$$

MOST POWERFUL (MP) TEST. The test which is performed on the basis of BCR is called *most powerful test.*

METHOD OF FINDING BCR. For simple hypothesis against a simple alternative hypothesis, the BCR of size α is found out using the following theorem:

❏**THEOREM 34.1 [Neyman-Pearson Theorem]** : *Let $x_1, x_2, \ldots, x_n$ be a random sample of size n selected from a population having p.d.f. $f(x, \theta)$. Let the simple hypothesis be $H_0 : 0 = \theta_0$, against a simple alternative hypothesis $H_A : \theta = \theta_1$. The likelihood functions under H_0 and H_A are $L(x/H_0)$ and $L(x/H_A)$, respectively. Now, for any positive constant k, if*

$$\frac{L(x/H_0)}{L(x/H_A)} \leq k \in C \text{ and } \frac{L(x/H_0)}{L(x/H_A)} \geq kC \tag{a}$$

then C is the BCR of size α.

Proof. Since the size of C is α, we can write

$$\alpha = \int_C L(x/H_0)dx.$$

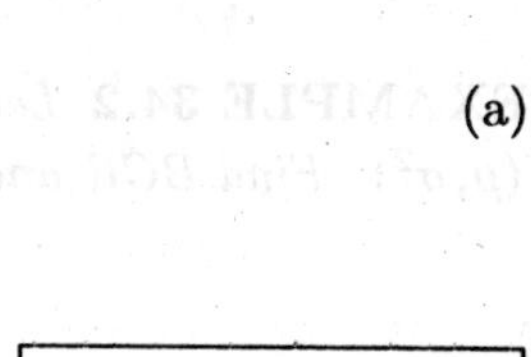

FIG. 34.1

Let D be the another critical region of size α. Here C and D are the sub-space in an n-dimensional space, where C and D can be shown as in FIG. 34.1.

Then

$$\int_C L(x/H_0)dx = \int_D L(x/H_0)dx = \alpha. \tag{b}$$

Here $dx = dx_1 dx_2 \cdots dx_n$. Now,

$$C = (C \cap \bar{D}) \cup (C \cap D) \text{ and } D = (\bar{C} \cap D) \cup (C \cap D).$$

So, we can write

$$\int_{C\cap\bar{D}} L(x/H_0)dx = \int_{\bar{C}\cap D} L(x/H_0)dx.$$

The theorem states that

$$\frac{L(x/H_0)}{L(x/H_A)} \le k \in C \Rightarrow L(x/H_A) \ge \frac{L(x/H_0)}{k}.$$

Hence

$$\begin{aligned}\int_C L(x/H_A)dx &= \int_{C\cap D} L(x/H_A)dx + \int_{C\cap\bar{D}} L(x/H_0)dx \\ &\ge \int_{C\cap D} L(x/H_A)dx + \int_{\bar{C}\cap D} \frac{L(x/H_0)}{k}dx \\ \Rightarrow \int_C L(x/H_A)dx &\ge \int_{C\cap D} L(x/H_A)dx + \frac{1}{k}\int_{\bar{C}\cap D} L(x/H_0)dx. \qquad \text{(c)}\end{aligned}$$

Again, $L(x/H_A) \le \dfrac{L(x/H_0)}{k}$ when $k \in C$

$$\int_{\bar{C}\cap D} \frac{L(x/H_0)}{k}dx \ge \int_{\bar{C}\cap D} L(x/H_A)dx. \qquad \text{(d)}$$

Now, from equations (c) and (d), we can write

$$\int_C L(x/H_A)dx \ge \int_{C\cap D} L(x/H_A)dx + \int_{\bar{C}\cap D} L(x/H_A)dx = \int_D L(x/H_A)dx.$$

$$\therefore \int_C L(x/H_A)dx \ge \int_D L(x/H_A)dx.$$

Therefore, the power of $C \ge$ power of D. Thus, C is the BCR.

►**EXAMPLE 34.2** *Let $x_1, x_2, \ldots, x_n$ be a random sample of n observations from $N(\mu, \sigma^2)$. Find BCR and power of the test in testing the significance of*

$$H_0 : \mu = \mu_0 \text{ against } H_A : \mu = \mu_1.$$

Assume that σ^2 is known.

Solution. The likelihood function under H_0 is

$$L(x/H_0) = L(x/\mu_0) = \left(\frac{1}{\sigma\sqrt{2\pi}}\right)^n e^{-\frac{1}{2}\sum\left(\frac{x-\mu_0}{\sigma}\right)^2}.$$

The same under alternative hypothesis is

$$L(x/H_A) = L(x/\mu_1) = \left(\frac{1}{\sigma\sqrt{2\pi}}\right)^n e^{-\frac{1}{2}\left(\frac{x-\mu_1}{\sigma}\right)^2}.$$

Let us now apply Neyman-Pearson lemma. We have

$$\frac{L(x/H_0)}{L(x/H_A)} \le k \text{ or, } \frac{\exp\left[-\frac{1}{2\sigma^2}\Sigma(x-\bar{x}+\bar{x}-\mu_0)^2\right]}{\exp\left[-\frac{1}{2\sigma^2}\Sigma(x-\bar{x}+\bar{x}-\mu_1)^2\right]} \le k$$

$$\text{or, } \frac{\exp\left[-\frac{n}{2\sigma^2}\{s^2+(\bar{x}-\mu_0)^2\}\right]}{\exp\left[-\frac{n}{2\sigma^2}\{s^2+(\bar{x}-\mu_1)^2\}\right]} \le k$$

$$\text{or, } \exp\left[-\frac{n}{2\sigma^2}\{2\bar{x}(\mu_1-\mu_0)+(\mu_0^2-\mu_1^2)\}\right] \le k.$$

Taking log on both sides, we get

$$\bar{x}(\mu_1-\mu_0) \ge \frac{\mu_1^2-\mu_0^2}{2} - \frac{\sigma^2}{n}\log k.$$

Now, (i) If $\mu_1 > \mu_0$, then the BCR is

$$\bar{x} \ge \frac{\mu_1+\mu_0}{2} - \frac{\sigma^2}{n}\frac{\log k}{\mu_1-\mu_0} \Rightarrow \bar{x} \ge k_1 \text{ (say);}$$

(ii) If $\mu_1 < \mu_0$, then the BCR is

$$\bar{x} \le \frac{\mu_1+\mu_0}{2} + \frac{\sigma^2}{n}\frac{\log k}{\mu_0-\mu_1} \Rightarrow \bar{x} \le k_2 \text{ (say).}$$

The values of k_1 and k_2 are to be found out in such a way that under H_0 $P(\bar{x} \ge k_1) = \alpha$ and $P(\bar{x} \le k_2) = \alpha$.

Under H_0 the distribution of $\bar{x}$ is $N(\mu_0, \sigma^2/n)$. Then, we can write

$$\frac{\sqrt{n}}{\sigma\sqrt{2\pi}}\int_{-\infty}^{k_2} e^{-\frac{n}{2\sigma^2}(\bar{x}-\mu_0)^2}\,d\bar{x} = \alpha \tag{1}$$

$$\text{and } \frac{\sqrt{n}}{\sigma\sqrt{2\pi}}\int_{k_1}^{\infty} e^{-\frac{n}{2\sigma^2}(\bar{x}-\mu_0)^2}\,d\bar{x} = \alpha. \tag{2}$$

Let $z = \dfrac{\sqrt{n}(\bar{x}-\mu_0)}{\sigma}$, then (1) and (2) become

$$\frac{1}{\sqrt{2\pi}}\int_{-\infty}^{\sqrt{n}(k_2-\mu_0)/\sigma} e^{-\frac{1}{2}z^2}\,dz = \alpha \tag{3}$$

$$\text{and } \frac{1}{\sqrt{2\pi}}\int_{\sqrt{n}(k_1-\mu_0)/\sigma}^{\infty} e^{-\frac{1}{2}z^2}\,dz = \alpha. \tag{4}$$

Now, for any given value of α, the value of k_α is found out, where

$$\frac{1}{\sqrt{2\pi}}\int_{k_\alpha}^{\infty} e^{-z^2/2}dz = \alpha.$$

Thus, from (3) and (4), we get $k_2 = \mu_0 - \dfrac{\sigma}{\sqrt{n}}k_\alpha$ and $k_1 = \mu_0 + \dfrac{\sigma}{\sqrt{n}}k_\alpha$.

Let $\alpha = 0.05$, then $\dfrac{k_1-\mu_0}{\sigma/\sqrt{n}} = -1.645$.

Therefore, if $\bar{x} \ge \mu_0 + \dfrac{\sigma}{\sqrt{n}}k_\alpha$, H_0 is rejected.

Power of the Test

CASE 1. $\mu_1 > \mu_0$
The test function

$$= \int_{k_1}^{\infty} f(\bar{x}/H_A)d\bar{x} = 1 - \int_{-\infty}^{\frac{k_1-\mu_1}{\sigma/\sqrt{n}}} \frac{1}{\sqrt{2\pi}} e^{-\frac{1}{2}z^2} dz$$

$$= 1 - \int_{-\infty}^{\frac{1}{\sigma}\left(\mu_0 + \frac{\sigma}{\sqrt{n}}k_\alpha - \mu_1\right)\sqrt{n}} f(z)dz = 1 - F(m)$$

where

$$m = \frac{1}{\sigma}\left(\mu_0 + \frac{\sigma}{\sqrt{n}}k_\alpha - \mu_1\right)\sqrt{n}.$$

CASE 2. $\mu_1 < \mu_0$.
The test function

$$= \int_{-\infty}^{\frac{(k_2-\mu_1)\sqrt{n}}{\sigma}} \frac{1}{\sqrt{2\pi}} e^{-\frac{1}{2}z^2} dz$$

$$= \int_{-\infty}^{\left(\mu_0 - \frac{\sigma}{\sqrt{n}}k_\alpha - \mu_1\right)\frac{\sqrt{n}}{\sigma}} f(z)dz = F(m_1)$$

where

$$m_1 = \frac{\sqrt{n}}{\sigma}\left(\mu_0 - \frac{\sigma}{\sqrt{n}}k_\alpha - \mu_1\right).$$

34.4 Uniformly Most Powerful (UMP) Test

In finding BCR we considered simple null hypothesis against simple alternative hypothesis. In practice, the null hypothesis may be simple but not the alternative hypothesis. The most powerful critical region arises in such a situation where null hypothesis is simple but alternative hypothesis is composite. To test the significance of H_0 against composite alternative hypothesis H_A the critical region W of size α is uniformly most powerful critical region if in testing simple null hypothesis against every simple alternative hypothesis H_A the critical region W of size α is BCR. The test based on uniformly most powerful (UMP) critical region to test simple null hypothesis H_0 against composite alternative hypothesis H_A is called *UMP test* of size α. For example, let us consider the BCR in testing $H_0 : \mu = \mu_0$ against $H_A : \mu = \mu_1$ in case of $N(\mu, \sigma^2)$ when σ^2 is known. It is observed in Example 34.2 that, to test H_0 against $H_A : \mu_1 < \mu_0$ the BCR is $\bar{x} \leq k_\alpha$. This BCR is true for any value of μ_1 less than μ_0. Therefore, $\bar{x} \leq k_1$ is UMP critical region. The test for $H_0 : \mu = \mu_0$ against $H_A : \mu_1 < \mu_0$ is UMP test. In a similar way if $\mu_1 > \mu_0$ the BCR is $\bar{x} \geq k_2$. Here also the UMP test is available. However, in testing $H_0 : \mu = \mu_0$ against $H_A : \mu \neq \mu_0$ no UMP test is available. Because in such a situation no uniform BCR is available. Therefore, UMP critical region is not available always.

►**EXAMPLE 34.3** *The daily production of electric bulb in an industry is assumed to follow* $N(\mu, 100)$. *The data are given from randomly selected* 10 *days, where the production of bulbs (in* 1000 *pieces)* 5.6, 8.7, 10.4, 9.6, 7.8, 6.6, 8.8, 9.2, 8.7, 8, 0. *Find BCR in testing* $H_0 : \mu = \mu_0 = 8.5$ *against* $H_A : \mu = \mu_1 = 9$. *Perform the test at* $\alpha = 0.05$ *level.*

Solution. We have $\bar{x} = 8.34, n = 10$. Here $H_0 = \mu = \mu_0 = 8.5$ and $H_A : \mu = \mu_1 = 9$; $\mu_1 > \mu_0$. The BCR is

$$\bar{x} \geq \frac{\mu_1 + \mu_0}{2} - \frac{\sigma^2}{n} \frac{\log k}{\mu_1 - \mu_0} = k_1 \text{ (say)}$$

where

$$k_1 = \mu_0 + \frac{\sigma}{\sqrt{n}} k_\alpha.$$

But, when $\alpha = 0.05, k_\alpha = 1.645$. Hence the BCR is $\bar{x} \geq \mu_0 + \frac{\sigma}{\sqrt{n}} k_\alpha$. We have

$$\mu_0 + \frac{\sigma}{\sqrt{n}} k_\alpha = 8.5 + \frac{10}{3.16}(+1.645) = 13.71.$$

Since $\bar{x} < 13.71$, H_0 is accepted.

Monotone Likelihood Ratio

Let $f(x, \theta)$ be a family of distribution. Consider that for the parameter θ there are two values θ_1 and $\theta_2 (\theta_1 < \theta_2)$ and for these two values the two likelihood functions are $L(x/\theta_1)$ and $L(x/\theta_2)$, respectively. Then the monotone likelihood ratio of $f(x, \theta)$ is available if for $T(x)$ the ratio

$$\frac{L(x/\theta_1)}{L(x/\theta_2)}$$

is not a decreasing function. Here $T(x)$ is a sample statistic.

❒**THEOREM 34.2** *Let* $f(x, \theta)$ *be a p.d.f. of a random variable* X*, where* θ *is the real parameter. Let* $T(x)$ *be a sample statistic of sample observations* $x_1, x_2, \ldots, x_n$. *Now, if for any* $T(x)$ *the MLR of* $f(x, \theta)$ *is available, then in testing* $H_0 \leq \theta_0$ *against* $H_A : \theta > \theta_0$ *the UMP test is available.*

Proof. The null hypothesis is $H_0 : \theta = \theta_0$ and the alternative hypothesis is $H_A : \theta = \theta_1 (> \theta_0)$. By Neyman-Pearson lemma, we can write

$$\frac{L(x/H_A)}{L(x/H_0)} \geq k. \tag{a}$$

According to the definition of MLR, we can say that the

$$\text{MLR} = \frac{L(x/\theta_2)}{L(x/\theta_1)}$$

is not decreasing through $T(x)$. Since θ_1 and $\theta_2 (\theta_1 < \theta_2)$ are two values of θ, the BCR shown in (a) can be written as

$$T(x) \geq k_1 \in W. \tag{b}$$

Let the level of significance and power of this test be α and $P(\theta)$, respectively. Then, we can write $P(\theta_0) = \alpha$ and for $H_0 : \theta = \theta_1$ and $\theta = \theta_2(> \theta_1)$ the BCR is

$$\frac{L(x/\theta_2)}{L(x/\theta_1)} \geq k_2.$$

This BCR can also be determined using $T(x)$. Let $T(x) \geq k_3$. Now, if it is considered that $k_3 = k_1$, the critical region so obtained is exactly equal to W and this critical region is the best critical region for level of significance $\alpha_1 = P(\theta_1)$ to test the hypothesis $\theta = \theta_1$ and $\theta = \theta_2$. Hence the test is most powerful (MP) test and for any θ_1 and $\theta_2(> \theta_1)$ we have $P(\theta_2) \geq P(\theta_1)$. Hence for $P(\theta) < 1$ the power $P(\theta)$ is exactly increasing function.

The equation (b) can be used as critical region to test the null hypothesis H_0 : $\theta < \theta_0$. Here the level of significance is less than or equal to α. In that case for any alternative hypothesis $\theta = \theta_1(> \theta_0)$ the power of test will be maximum. Therefore, for $H_0 : \theta \leq \theta_0$ and for $H_A : \theta > \theta_0$ the critical region shown in (b) is UMP.

►**EXAMPLE 34.4** *Let $x_1, x_2, \ldots, x_n$ be a random sample of n observations from $N(\mu, 1)$. Do you think that the mean is less than or equal to a specific value?*

Solution. Let the specific value is μ_0. We need to test the significance of hypothesis $H_0 : \mu \leq \mu_0$ against $H_A : \mu > \mu_0$. The likelihood function of the sample observations is

$$L = \left(\frac{1}{2\pi}\right)^n e^{-\frac{1}{2}\sum\limits^{n}(x_i - \mu)^2}.$$

Now,

$$\frac{L(x/\mu_A)}{L(x/\mu_0)} = \exp\left[n\bar{x}(\mu_A - \mu_0) - \frac{n}{2}(\mu_A^2 - \mu_0^2)\right].$$

This function is a monotone increasing function of $\bar{x}$. Hence Theorem 34.2 can be applied to test $H_0 : \mu \leq \mu_0$ against $H_A : \mu > \mu_0$ and for this test UMP critical region can be found out. Here the UMP critical regions are $\bar{x} \geq k_1$ and $\bar{x} \leq k_2$, where

$$k_1 = \mu_0 + \frac{1}{\sqrt{n}}k_\alpha, \; k_2 = \mu_0 - \frac{1}{\sqrt{n}}k_\alpha.$$

The value of k_α is found out from normal probability table such that

$$\frac{1}{\sqrt{2\pi}}\int_{k_\alpha}^{\infty} e^{-z^2/2}dz = \alpha.$$

The above example relates with one-sided test which is based on Theorem 34.2. Let us now discuss the problem for two-sided alternatives.

❑**THEOREM 34.3** *Let the likelihood function of sample observations $x_1, x_2, \ldots, x_n$ be $L(x/\theta)$. If the following regularity conditions are satisfied, then the UMP critical region is not attained to test the significance of $H_0 : \theta = \theta_0$ against $H_A : \theta \neq \theta_0$:*

(i) *$L(x/\theta)$ is continuous for X.*

(ii) *$\dfrac{\partial L(x/\theta)}{\partial\theta}$ exists and is continuous.*

(iii) $\dfrac{\partial L(x/\theta)}{\partial \theta}$ *is not zero for any value of* θ.

(iv) $\dfrac{\partial}{\partial \theta}\displaystyle\int_W L(x/\theta)dx = \int_W \dfrac{\partial}{\partial \theta} L(x/\theta)d\theta$, *where W is critical region.*

Under regularity condition (i), *we find*

$$\frac{L'(x/\theta)}{L(x/\theta)} = \left[\frac{\partial \log L(x/\theta)}{\partial \theta}\right]_{\theta=\theta_0} = \textit{constant.} \tag{a}$$

This condition is essential for UMP test. The condition (a) *attains if the regularity condition* (iv) *is valid.*

►**EXAMPLE 34.5** *Let* $x_1, x_2, \ldots, x_n$ *be random sample of n observations from a population having p.d.f.*

$$f(x) = \begin{cases} e^{-(x-\theta)}, & \theta \le x \le \infty \\ 0, & \textit{otherwise.} \end{cases}$$

Find UMP critical region to test $H_0 : \theta = \theta_0$ *against* $H_A : \theta \neq \theta_0$.

Solution. The likelihood function of sample observations is

$$L(x/\theta) = \exp[-n(\bar{x} - \theta)].$$

Under (a), we get

$$\left.\frac{\partial \log L}{\partial \theta}\right]_{\theta=\theta_0} = n \text{ (constant).}$$

Hence BCR can be found out. The lower limit of X depends on θ. Hence the sufficient statistic of θ is lowest value of X, say x_1. Then $P(x_1 < \theta/H_A) = 0$. So we have

$$\frac{L(x/\theta_0)}{L(x/\theta_1)} = \begin{cases} \infty, & \text{if } x_1 < \theta_1 \\ e^{n(\theta_0-\theta_1)}, & \text{otherwise.} \end{cases}$$

By Neyman-Pearson lemma, we get

$$\frac{L(x/\theta_0)}{L(x/\theta_1)} \le k \Rightarrow e^{n(\theta_0-\theta_1)} \le k. \tag{1}$$

This (1) is the BCR, where $P\left[e^{n(\theta_0-\theta_1)} \le k/H_0\right] = \alpha$. The left-hand side of (1) does not depend on θ. Hence for level of significance α and for $x_1 \ge \theta_1$, every critical region is available by (1). The power of each critical region is same. Therefore, for $\theta_1 > \theta_0$ or $\theta_1 < \theta_0$

$$\frac{L(x/\theta_0)}{L(x/\theta_1)} = \begin{cases} \infty, & \text{if } \theta_0 \le x_1 < \theta_1 \\ e^{n(\theta_0-\theta_1)} < 1, & \text{if } x_1 \ge \theta_1 > \theta_0 \\ e^{n(\theta_0-\theta_1)} > 1, & \text{if } x_1 \ge \theta_0 > \theta_1 \\ 0, & \text{if } \theta_1 \le x_1 < \theta_0. \end{cases}$$

Hence the BCR is $x_1 < \theta_0$ and $x_1 > \theta_0 + C_\alpha$, where $P[x_1 > (\theta_0 + C_\alpha)/H_0] = \alpha$.

34.5 Sufficient Statistic and UMP Test

Definition. Let $x_1, x_2, \ldots, x_n$ be a random sample of n observations from a population having p.d.f. $f(x, \theta)$. The likelihood function of these sample observations can be written as

$$L(x/\theta) = \exp[A(\theta)t(x) + \beta(\theta) + C(x)]. \tag{34.1}$$

Here $t(x)$ is the sufficient statistic for θ and $A(\theta)$ is the monotone function of θ. Then for $H_0 : \theta = \theta_0$ and $H_A : \theta = \theta_1 > \theta_0$ there exists UMP test and it is given by

$$\phi(x) = \begin{cases} 1, & t(x) > C \\ \phi_0, & t(x) = C \\ 0, & t(x) < C \end{cases} \text{ if } A(\theta) \text{ is monotone increasing function of } \theta$$

$$\text{and } \phi(x) = \begin{cases} 1, & t(x) < C \\ \phi_0, & t(x) = C \\ 0, & t(x) > C. \end{cases} \text{ if } A(\theta) \text{ is monotone decreasing function of } \theta$$

Again, to test $H_0 : \theta = \theta_0$ and $H_A : \theta_1 < \theta_0$, UMP test is given by

$$\phi(x) = \begin{cases} 1, & t(x) < C \\ \phi_0, & t(x) = C \\ 0, & t(x) > C \end{cases} \text{ if } A(\theta) \text{ is monotone increasing function}$$

$$\text{and } \phi(x) = \begin{cases} 1, & t(x) > C \\ \phi_0, & t(x) = C \\ 0, & t(x) < C. \end{cases} \text{ if } A(\theta) \text{ is decreasing function}$$

Here $\phi(x)$ is the critical function or test function.

►**EXAMPLE 34.6** *Let $x_1, x_2, \ldots, x_n$ be a random sample of n observations from Poisson population with parameter λ. Test the hypothesis*

$$H_0 : \lambda = \lambda_0 \text{ against } H_A : \lambda = \lambda_1 > \lambda_0.$$

Solution. The likelihood function is

$$L(x/\theta) = \frac{e^{-n\lambda}\lambda^{n\bar{x}}}{\prod_{i=1}^{n} x_i!} = \exp\left[\Sigma x_i \log\lambda - n\lambda - \Sigma \log x_i!\right].$$

Here $t(x) = \Sigma x_i$ is the sufficient statistic of λ and $A(\theta) = \log\lambda$ and it is increasing function. Hence using definition, the null hypothesis will be rejected, if

$$t(x) > C, \text{ where } \int_C^\infty f(t)dt = \alpha,\ P[t(x) > C/H_0] = \alpha.$$

Hence $\sum_{i=1}^{n} x_i > C$ is the UMP critical region.

►**EXAMPLE 34.7** *Let $x_1, x_2, \ldots, x_n$ be a random sample of n observations selected from a population having p.d.f.*

$$f(x,\theta) = \theta e^{-\theta x};\ x > 0, \theta > 0.$$

Test $H_0 : \theta = \theta_0$ against $H_A : \theta = \theta_1 > \theta_0$.

Solution. The likelihood function is

$$L(x/\theta) = \theta^n e^{-\theta \sum^n x_i}$$

$$L(x/\theta) = \exp\left[n \log\theta - \theta \sum^n x_i\right]$$

where $t(x) = \Sigma x_i$ is the sufficient statistic for θ and $A(\theta) = -\theta$ is decreasing function of θ. Therefore, for $H_0 : \theta = \theta_0$ against $H_A : \theta = \theta_1 > \theta_0$ the UMP critical region is

$$\sum^n x_i < C \text{ such that } P[\Sigma x_i < C/H_0] = \alpha.$$

It is mentioned that for the distribution except those mentioned in definition if UMP test is attained, the test may or may not be based on sufficient statistic.

►**EXAMPLE 34.8** *Let $x_1, x_2, \ldots, x_n$ be a random sample of n observations from a population having p.d.f.*

$$f(x;\theta,\sigma) = \frac{1}{\sigma} e^{-\left(\frac{x-\theta}{\sigma}\right)};\ 0 < x < \infty,\ 0 < \sigma.$$

Find UMP critical region to test $H_0 : \theta = \theta_0$, $\sigma = \sigma_0$ against $H_A : \theta = \theta_1 < \theta_0$, $\sigma = \sigma_1 < \sigma_0$.

Solution. By Neyman-Pearson lemma, we can write

$$\frac{L(x/H_0)}{L(x/H_A)} \le k$$

$$\text{or, } \left(\frac{\sigma_1}{\sigma_0}\right)^n \exp\left[\frac{\bar{x}-\theta_1}{\sigma_1} - \frac{\bar{x}-\theta_0}{\sigma_0}\right] n \le k$$

$$\text{or, } n\log\frac{\sigma_1}{\sigma_0} + n\left(\frac{\bar{x}-\theta_1}{\sigma_1} - \frac{\bar{x}-\theta_0}{\sigma_0}\right) \le \log k$$

$$\text{or, } n\log\frac{\sigma_1}{\sigma_0} + n\bar{x}\left(\frac{1}{\sigma_1} - \frac{1}{\sigma_0}\right) + n\left(\frac{\theta_0}{\sigma_0} - \frac{\theta_1}{\sigma_1}\right) \le \log k$$

$$\Rightarrow \bar{x}\left(\frac{1}{\sigma_1} - \frac{1}{\sigma_0}\right) \le k_1 \text{ (say)}$$

$$\Rightarrow \bar{x} \le k_2 \text{ (say)}$$

Here UMP test is performed on the basis of $\bar{x}$. However, $\bar{x}$ is not sufficient statistic for θ and σ.

►**EXAMPLE 34.9** *Let $x_1, x_2, \ldots, x_n$ be a random sample of n observations from a population having p.d.f.*

$$f(x,\theta) = \frac{1+\theta}{(x+\theta)^2},\ 1 \le x \le \infty.$$

Show that UMP test is not possible to test the hypothesis

$$H_0 : \theta = \theta_0 \text{ against } H_A : \theta = \theta_1 > \theta_0.$$

Solution. By Neyman-Pearson lemma, we can write $\dfrac{L(x/\theta_0)}{L(x/\theta_1)} \le k$. Thus

$$\frac{(1+\theta_0)^n \prod_{i=1}^{n} \frac{1}{(x_i+\theta_0)^2}}{(1+\theta_1)^n \prod_{i=1}^{n} \frac{1}{(x_i+\theta_1)^2}} \le k$$

$$\text{or, } \left(\frac{1+\theta_0}{1+\theta_1}\right)^n \prod_{i=1}^{n} \left(\frac{x_i+\theta_1}{x_i+\theta_0}\right)^2 \le k$$

$$\text{or, } 2\sum_{i=1}^{n} \log\left(\frac{x_i+\theta_1}{x_i+\theta_0}\right) \ge \log k + n\log\left(\frac{1+\theta_0}{1+\theta_1}\right).$$

Here the left-hand side cannot be expressed separately in terms of sample observations and the result depends on null hypothesis and alternative hypothesis. Hence BCR is not available and UMP test cannot be done.

34.6 Unbiased Test

Definition. Let α be the level of significance of test ϕ and W be the critical region of the test. Then, for one-sided or two-sided alternative

$$\begin{aligned} P[X \in W/H_0] &= E[\phi(x)]_{\theta=\theta_0} = \alpha \\ \text{and } P[X \in W/H_A] &= E[\phi(x)]_{\theta=\theta_1} = 1-\beta. \end{aligned}$$

Now, for all possible alternative hypothesis, if $P[X \in W/H_A] \ge \alpha$, then W is called the unbiased critical region and the related test is known as unbiased test.

❑ **THEOREM 34.4** *The level of significance of test will never exceed the power of the test in testing simple null hypothesis against simple alternative hypothesis, i.e.,* $(1-\beta) \ge \alpha$.

Proof. Let W be the BCR at α level of significance in testing simple null hypothesis against simple alternative hypothesis. Then

$$\int_W L(x/H_0)dx = \alpha.$$

Now, by Neyman-Pearson theorem, we can write

$$\frac{L(x/H_0)}{L(x/H_A)} \le k \in W$$

$$\text{or, } kL(x/H_A) \ge L(x/H_0) \text{ or, } k\int_W L(x/H_A)dx \ge \int_W L(x/H_0)dx.$$

$$\therefore\ k(1-\beta) \ge \alpha. \tag{a}$$

Again, we can write [by Neyman-Pearson theorem]

$$\frac{L(x/H_0)}{L(x/H_A)} \geq k \in S - W$$

$$\text{or, } kL(x/H_A) \leq L(x/H_0) \text{ or, } k\int_{S-W} L(x/H_A)dx \leq \int_{S-W} L(x/H_0)dx.$$

$$\therefore\ k\beta \leq (1-\alpha). \tag{b}$$

Now, from (a) and (b), we get

$$k(1-\beta)(1-\alpha) \geq k\alpha\beta \text{ or, } (1-\beta)(1-\alpha) \geq \alpha\beta \Rightarrow (1-\beta) \geq \alpha.$$

Uniformly Most Powerful Unbiased (UMPU) Test

Let ϕ be a test to test the null hypothesis

$$H_0 : \theta = \theta_0 \text{ against } H_A : \theta = \theta_1 \neq \theta_0,\ \theta_1 \in S.$$

Also, consider that W be the critical region for this test at level of significance α. Then according to the definition of unbiased test, we have

(i) $E[\phi(x)]_{\theta=\theta_0} = P[X \in W/\theta_0] = \alpha$

(ii) $E[\phi(x)]_{\theta=\theta_1} \geq E[\phi(x)]_{\theta=\theta_0}\ \forall\ \theta_1 \in S.$ (34.2)

Let ϕ' be another unbiased test which satisfies the condition (i) and (ii) above and

$$E[\phi(x)]_{\theta=\theta_1} \geq E[\phi'(x)]_{\theta=\theta_1}\ \forall\ \theta_1 \in S.$$

Then ϕ is the UMPU test at level of significance α.

►**EXAMPLE 34.10** *Let $x_1, x_2, \ldots, x_n$ be a random sample of n observations from $N(\mu, \sigma^2)$, where σ^2 is known. Test the hypothesis*

$$H_0 : \mu = \mu_0 \text{ against } H_A : \mu = \mu_1 \neq \mu_0.$$

Solution. Using the result of Example 34.2, we can write for $\mu_1 > \mu_0$, the BCR is $\bar{x} \geq k_1$ and for $\mu_1 < \mu_0$, the BCR is $\bar{x} \leq k_2$, where

$$k_1 = \mu_0 + \frac{\sigma}{\sqrt{n}} z_\alpha,\ k_2 = \mu_0 - \frac{\sigma}{\sqrt{n}} z_\alpha.$$

Here $P(z \leq z_{\alpha_2}) = \alpha_2$, $P(z \geq z_{\alpha_1}) = \alpha_1$, and $\alpha_1 + \alpha_2 = \alpha$. If $\alpha_1 = \alpha_2$, the test is unbiased. We need to prove it.

Now, let $\mu_1 > \mu_0$, then the test function is

$$\phi(x) = \int_{k_1}^{\infty} f(\bar{x}/H_A)d\bar{x} = 1 - F\left(\frac{k_1 - \mu_1}{\sigma/\sqrt{n}}\right).$$

Since $k_1 = \mu_0 + \frac{\sigma}{\sqrt{n}} z_{\alpha_1}$ we can write

$$\phi(x) = F\left(\frac{\mu_1 - k_1}{\sigma/\sqrt{n}}\right) = F\left(\frac{\mu_1 - \mu_0}{\sigma/\sqrt{n}} - z_{\alpha_1}\right).$$

Again, when $\mu_1 < \mu_0$, the test function is

$$\phi(x) = 1 - F\left(\frac{\mu_1 - \mu_0}{\sigma/\sqrt{n}} + z_{\alpha_2}\right), \text{ where } k_2 = \mu_0 - z_{\alpha_2}\frac{\sigma}{\sqrt{n}}.$$

Therefore, the test function becomes

$$\phi = F\left(\frac{\mu_1 - \mu_0}{\sigma/\sqrt{n}} - z_{\alpha_1}\right) + 1 - F\left(\frac{\mu_1 - \mu_0}{\sigma/\sqrt{n}} + z_{\alpha_2}\right)$$

where ϕ is the function of $\mu_1 - \mu_0$.

$$\therefore\ \phi = \frac{1}{\sqrt{2\pi}}\int_{-\infty}^{\frac{\mu_1-\mu_0}{\sigma/\sqrt{n}} - z_{\alpha_1}} e^{-\frac{1}{2}\bar{x}^2}\, d\bar{x} + 1 - \frac{1}{\sqrt{2\pi}}\int_{-\infty}^{\frac{\mu_1-\mu_0}{\sigma/\sqrt{n}} + z_{\alpha_2}} e^{-\frac{1}{2}\bar{x}^2}\, d\bar{x}.$$

Now, differentiating ϕ with respect to $\mu_1 - \mu_0$, we get

$$\frac{\partial\phi}{\partial(\mu_1 - \mu_0)} = \phi' = \frac{1}{\sqrt{2\pi}}\left[e^{-\frac{1}{2}\left\{\frac{\mu_1-\mu_0}{\sigma/\sqrt{n}} - z_{\alpha_1}\right\}^2}\frac{\sqrt{n}}{\sigma}\right] - \frac{1}{\sqrt{2\pi}}\left[e^{-\frac{1}{2}\left\{\frac{\mu_1-\mu_0}{\sigma/\sqrt{n}} + z_{\alpha_2}\right\}^2}\right]$$

$$= \left[\frac{n}{2\pi\sigma^2}\right]^{1/2}\exp\left[-\frac{1}{2}\left\{\frac{\mu_1 - \mu_0}{\sigma/\sqrt{n}} - z_{\alpha_1}\right\}^2 - \frac{1}{2}\left\{\frac{\mu_1 - \mu_0}{\sigma/\sqrt{n}} + z_{\alpha_2}\right\}^2\right].$$

Putting $\phi' = 0$, we get

$$\left(\frac{\mu_1 - \mu_0}{\sigma/\sqrt{n}} - z_{\alpha_1}\right)^2 = \left(\frac{\mu_1 - \mu_0}{\sigma/\sqrt{n}} + z_{\alpha_2}\right)^2 \text{ or, } -z_{\alpha_1} = z_{\alpha_2} \text{ or, } \alpha_1 = \alpha_2.$$

Since $\alpha_1 = \alpha_2$, the power curve will be minimum at $\mu_1 = \mu_0$; otherwise, it will be minimum at $\mu_1 \neq \mu_0$. When H_0 is not true, the probability of rejecting H_0 is less. This means that when H_0 is true, the probability of accepting it becomes more. The critical value is shown in the following graph (FIG. 34.2):

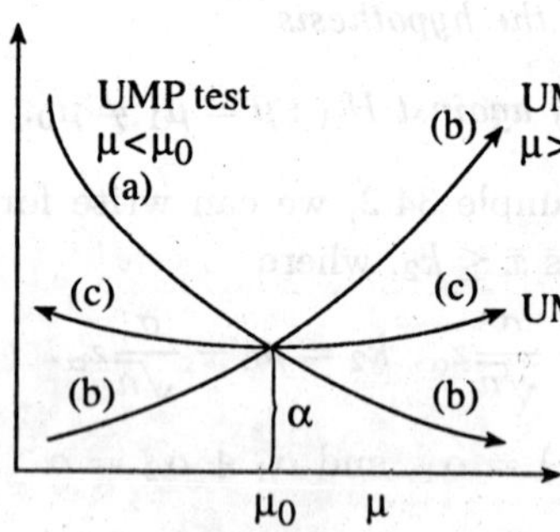

FIG. 34.2

Here the power curves (a) and (b) are for one-sided test, and these tests are not unbiased. The power curve (c) is the most powerful among all the unbiased tests.

Though (b) is not for unbiased test, still we should use it for $H_0 : \mu = \mu_0$ against $H_A : \mu = \mu_1 > \mu_0$. Again, for $H_0 : \mu = \mu_0$ against $H_A : \mu = \mu_1 < \mu_0$ (a) should be used. For all alternative hypotheses $\mu_1 \neq \mu_0$ there is risk in accepting H_0 when H_A is true. However, the risk is reduced if power curve (c) is used.

34.7 Test of Composite Hypothesis

We have already mentioned that all the parameters of a distribution may or may not be specified by any hypothesis. For example, let us consider a p.d.f. $f(x;\theta,\delta)$, where $\theta = (\theta_1, \theta_2, \ldots, \theta_r)$ and $\delta = (\delta_1, \delta_2, \ldots, \delta_s)$. The problem is to test the significance of the null hypothesis

$$H_0 : \theta = \theta_0 \text{ against } H_A : \theta = \theta_1.$$

Here nothing is mentioned or specified about the parameter δ. We have $(r+s)$ parameters but r of them are specified by the hypothesis. Hence it is a composite hypothesis and δ is the nuisance parameter. The problem is to find the critical region W to test the null hypothesis at α level of significance. The value of W is to be found out so that

$$\int_W L(x/H_0)dx = \alpha \text{ and } \int_W L(x/H_A)dx \text{ is maximum}$$

where H_A is the simple alternative hypothesis. Here α is the function of unspecified δ. The problem is to find W so that α is free of δ.

Similar Region Test

Let the test function be ϕ to test the hypothesis $H_0 : \theta = \theta_0$ in such a way that for all values of nuisance parameter δ

$$E[\phi(x)/\theta_0] = \alpha.$$

Such a test of H_0 at α level of significance is known as similar region test. The problem of test of composite hypothesis is to find similar critical region and is to select a region from similar regions similar to sample space so that for a simple alternative hypothesis the critical region becomes most powerful. However, the similar region is not always available. The region is available only if for every nuisance parameter or for joint nuisance parameters the sufficient statistic is available.

Similar Region Test and Sufficient Statistic

The phenomenon is explained by a theorem.

❐ **THEOREM 34.5** *Let $x_1, x_2, \ldots, x_n$ be a random sample of n observations from a population having p.d.f. $f(x,\theta)$. The null hypothesis and alternative hypothesis is for θ. Let W be the critical region of size α and under null hypothesis and alternative hypothesis T is the sufficient statistic for θ. Then for H_0 and H_A the critical region can be formed, where the region is a function of sufficient statistic and it will be similar to level of significance and power of test.*

Proof. Let $\phi(x)$ be a function, where

$$\phi(x) = \begin{cases} 1, & \text{if } x \in W \\ 0, & \text{otherwise.} \end{cases}$$

Then

$$E[\phi(x)] = \int \phi(x)L(x/\theta)dx = \alpha.$$

Using Neyman factorization lemma, we can factorize the likelihood function of sample observations $x_1, x_2, \ldots, x_n$ such that

$$\begin{aligned} L(x/\theta) &= g(T/\theta)h(x/T). \\ \therefore\ E[\phi(x)] &= \int \phi(x)g(T/\theta)h(x/T)dx \\ &= E_T[E\{\phi(x)\}/T]. \end{aligned} \quad \text{(a)}$$

Here T is the sufficient statistic for θ, $E[\phi(x)/T]$ is the function of T and this function does not depend on θ. Hence, when H_0 and H_A are both true, then from (a) it is clear that a critical region similar to sample space is available which is based on T. The level of significance and power are equal to those of W.

Completeness of Statistics

Let $f(x; \theta_1, \theta_2, \ldots, \theta_k)$ be a family of distribution. This family will be called complete if for any function $\phi(x)$

$$E[\phi(x)] = \int \phi(x)f(x; \theta_1\theta_2, \ldots, \theta_k)dx = 0 \quad (34.3)$$

where $\phi(x)$ does not depend on θ.

Neyman Structure

Let $x_1, x_2, \ldots, x_n$ be a random sample of n observations from a population having p.d.f. $f(x; \theta, \delta)$, where $\theta = (\theta_1, \theta_2, \ldots, \theta_r)$ and $\delta = (\delta_1, \delta_2, \ldots, \delta_s)$. Let $T = (t_1, t_2, \ldots, t_n)$ be a set of complete sufficient statistic of δ when $H_0 : \theta = \theta_0$. Then for the size of the test α each similar region satisfies the following condition:

$$E[\phi(x; T, \theta_0)] = \alpha \ \ \forall\ T.$$

If a test satisfies the above condition, then we can say that for given T there exists Neyman structure.

Now, according to Neyman structure for all similar regions W

$$E[\phi(x; T, \theta_0] = \alpha \text{ and } E[\phi(x); T, \theta_1] = 1 - \beta$$

where $H_0 : \theta = \theta_0$ and $H_A : \theta = \theta_1$. Now, if T is constant for nuisance parameter δ, the conditional distribution $f(x/T)$ does not depend on δ. Hence the composite hypothesis H_0 transforms to simple H_0 along with nuisance parameter δ. Under H_A also the statistic T is sufficient statistic. Hence, keeping T constant, the composite H_A can be transformed to simple H_A. Therefore, the most powerful similar region (MPSR) transforms to BCR for testing simple null hypothesis H_0 against simple alternative hypothesis H_A.

However, if T is not sufficient statistic of δ under H_0, the alternative hypothesis H_A against H_0 can be considered the class of simple alternatives of H_A. Now, under Neyman-Pearson lemma the region can be found out

(i) $\int_{W/T} \phi(x) f(x/\theta_0, T) dx = \alpha$ and

(ii) $\int_{W/T} \phi(x) f(x/\theta_1, T) dx \geq \int_{W/T} \phi'(x) f(x/\theta_1, T) dx.$

Here $\phi'(x)$ is any other test function which satisfies condition (i) above and $\theta_1 \in S$. The BCR so obtained is called MPSR. In a similar way, for all classes of alternatives if the test is most powerful, then the critical region so obtained is known as UMPSR.

►**EXAMPLE 34.11** *Let $x_1, x_2, \ldots, x_n$ be a random sample of n observations from $N(\mu, \theta)$. Find BCR to test the null hypothesis*

$$H_0 : \mu = \mu_0 \text{ against } H_A : \mu = \mu_1 > \mu_0.$$

Solution. Here nothing is mentioned about the variance term θ. Hence the null hypothesis is composite. The likelihood function under H_0 is

$$L(x/H_0) = \left(\frac{1}{2\pi\theta}\right)^{\frac{n}{2}} \exp\left[-\frac{1}{2\theta}\sum^{n}(x_i - \mu_0)^2\right].$$

Here

$$T = \sum_{i=1}^{n}(x_i - \mu_0)^2$$

is the sufficient statistic for θ. But it is not sufficient statistic under H_A. However, the statistic 'T' is complete. This can be verified as follows:

The density function can be expressed as an exponential family of distribution, where

$$f(x, \delta) = \exp[\delta x + W(x) + a(\delta)], \ -\infty \leq x \leq \infty. \tag{1}$$

Let $\phi(x)$ is the function of x which does not depend on δ. Then, by (1) we can write

$$E[\phi(x)] = \int_{-\infty}^{\infty} \phi(x) \exp[\delta x + W(x) + a(\delta)] dx = 0$$

$$\text{or, } \exp[a(\delta)] \int_{-\infty}^{\infty} \{\phi(x) \exp[W(x)]\} e^{\delta x} dx = 0. \tag{2}$$

Here (2) is the two-sided Laplace transformation. This means that

$$e^{\delta x} L\left(\phi(x) e^{W(x)}\right) = 0 \text{ or, } \phi(x) e^{W(x)} = 0.$$

Hence $\phi(x) = 0$ identical, because $e^{W(x)} \neq 0$. Hence $f(x, \delta)$ is a complete family of distribution.

Here T/θ follows chi-square distribution with n d.f. Then

$$f\left(\frac{T}{\theta}\right) = \frac{1}{2^{n/2}\Gamma\frac{n}{2}}e^{-T/2\theta}\left(\frac{T}{\theta}\right)^{\frac{n}{2}-1}, \; 0 < \frac{T}{\theta} < \infty$$

$$= \frac{1}{2^{n/2}\Gamma\frac{n}{2}}e^{-T/2\theta}(T)^{\frac{n}{2}-1}\frac{1}{(\theta)^{\frac{n}{2}-1}} \text{ or, } f\left(\frac{T}{\theta}\right) \propto (T)^{\frac{n}{2}-1}e^{-T/2\theta}$$

$$= \exp\left[-\frac{T}{2\theta} + \frac{n-2}{2}\ln T - n\ln T\right].$$

This last function is a special form of (1) and hence 'T' is the complete statistic. Therefore, $H_A : \mu = \mu_1$, $\theta = \theta_1$ can be considered as alternative hypothesis.

Now, we have

$$L(x/T, \mu_0) = \frac{\left(\frac{1}{2\pi\theta}\right)^{n/2}\exp\left[-\frac{1}{2\theta}\sum^{n}(x_i-\mu_0)^2\right]}{2^{n/2}\Gamma\frac{n}{2}e^{-T/2\theta}\left(\frac{T}{\theta}\right)^{\frac{n}{2}-1}}$$

$$\text{and } L(x/\mu_1, \theta_1) = \left(\frac{1}{\sqrt{2\pi\theta_1}}\right)^n \exp\left[-\frac{1}{2\theta_1}\sum^{n}(x_i-\mu_1)^2\right].$$

By Neyman-Pearson theorem, we can write

$$\frac{L(x/H_0)}{L(x/H_A)} \le k \text{ or, } \frac{\Gamma\frac{n}{2}\left(\sqrt{2\theta_1}\right)^n}{T^{\frac{n}{2}-1}2\theta_1}\exp\left[-\sum^{n}(x_i-\mu_1)^2\right] \le k$$

$$\text{or, } \sum_{i=1}^{n}(x_i-\mu_1)^2 \le k_1 \text{ (say)}$$

where k_1 is the function of T and $\sqrt{\theta_1}$.

We can write

$$\sum_{i=1}^{n}(x_i-\mu_0)^2 + n(\mu_1-\mu_0)^2 + 2n(\mu_0-\mu_1)(\bar{x}-\mu_0) \le k_1$$

$$\Rightarrow (\mu_0-\mu_1)(\bar{x}-\mu_0) \le k_2 \text{ (say)}.$$

Now, if $\mu_1 > \mu_0$, the critical region will be

$$(\bar{x}-\mu_0) \ge k_2, \text{ when } x \in W/T.$$

If $\mu_1 < \mu_0$, the critical region will be

$$(\bar{x}-\mu_0) \le k_1, \text{ when } x \in W/T.$$

We can write

$$T = \sum^{n}(x_i-\mu_0)^2 = \sum^{n}(x_i-\bar{x})^2 + n(\bar{x}-\mu_0)^2.$$

This implies that, if the value of $\bar{x}$ increases, $\sum^{n}(x_i-\bar{x})^2$ decreases.

Then

$$\sum^{n}(x_i - \mu_0)^2 = \sum^{n}(x_i - \bar{x})^2 \left[1 + \frac{n(\bar{x} - \mu_0)^2}{\sum\limits_{i=1}^{n}(x_i - \bar{x})^2}\right]$$

$$= \sum_{i=1}^{n}(x_i - \bar{x})^2 \left[1 + \frac{T^2}{n-1}\right].$$

Hence, for $H_0 : \mu = \mu_0$ against $H_A : \mu = \mu_1 < \mu_0$ the BCR is the smallest value of

$$\frac{n(\bar{x} - \mu_0)^2}{\Sigma(x_i - \bar{x})^2} = \frac{T^2}{n-1}$$

and for $H_0 : \mu = \mu_0$ against $H_A : \mu = \mu_1 > \mu_0$ the BCR is the largest value of

$$\frac{n(\bar{x} - \mu_0)^2}{\Sigma(x_i - \bar{x})^2} = \frac{T^2}{n-1}$$

where T is the t-statistic and under H_0 the 't' distribution does not depend on θ. Hence without any restriction we can find UMPSR for $H_A : \mu = \mu_1 < \mu_0, H_A : \mu = \mu_1 > \mu_0$ using t-statistic. Therefore, for $H_0 : \mu = \mu_0$ against $H_A : \mu = \mu_1 > \mu_0$ the BCR is $t \geq C$, where $P[t \geq C/H_0] = \alpha$ and for $H_A : \mu = \mu_1 < \mu_0$ the BCR is $t \leq C'$, where $P[t \leq C'] = \alpha$.

►**EXAMPLE 34.12** *A random sample of* 14 *days is selected to estimate the number of road accidents per year and the number of accidents* (x) *of the selected days are recorded:*

$$x : 7, 10, 12, 18, 15, 15, 17, 16, 10, 12, 8, 18, 11, 5.$$

Find UMP and BCR to test the hypothesis

$$H_0 : \mu = \mu_0 = 10 \text{ against } H_A : \mu = \mu_1 < 10$$

assuming that the sample is selected from $N(\mu, \sigma^2)$.

Solution. We have $n = 14$, $\bar{x} = 12.43$, $s^2 = 17.49$. Let us perform the test at $\alpha = 0.05$ level of significance. Assume that $\sigma^2 = 20$ (a known value). Then for $H_A : \mu_1 < \mu_0$, the BCR is

$$\bar{x} \leq k_2 \text{ (say), where } k_2 = \mu_0 - \frac{\sigma}{\sqrt{n}}k_\alpha.$$

Since the sample is assumed to be drawn from normal population $k_\alpha = 1.645$; therefore,

$$k_2 = 10 - \sqrt{\frac{20}{14}}(1.645) = 8.03.$$

Since $\bar{x} > k_2$, H_0 is not rejected. This test is UMP. Let $\mu_1 = 9$, then the test function is

$$\phi(x) = \int_{-\infty}^{(k_2-9)\frac{\sqrt{n}}{\sigma}} f(z)dz.$$

Here

$$\frac{k_2 - 9}{\sqrt{\frac{\sigma^2}{n}}} = \frac{8.03 - 9}{\sqrt{\frac{20}{14}}} = -0.81, \ \phi(x) = 0.209.$$

Hence the power of the test is $[1 - \phi(x)] = 0.79$.

Let us assume that σ^2 is unknown. Then the BCR (UMPSR) is $t \leq C'$, where

$$P[t \leq C'/H_0] = \alpha = 0.05.$$

Here

$$\frac{t^2}{n-1} = \frac{n(\bar{x} - \mu_0)^2}{\Sigma(x_i - \bar{x})^2} = \frac{14(12.43 - 10)^2}{227.37} = 0.36$$

$$\therefore \ t = 2.17.$$

Here $C' = 1.771$. Since $t > C'$, H_0 is not rejected.

34.8 Likelihood Ratio Test (LR Test)

Let $x_1, x_2, \ldots, x_n$ be a random sample of n observations from a population whose p.d.f. is $f(x; \theta_1, \theta_2, \ldots, \theta_k)$, where $\theta_j \in S$ and S is the parametric space ($j = 1, 2, \ldots, k$). Here $f(x_i; \theta_1, \theta_2, \ldots, \theta_k)$; $i = 1, 2, \ldots, n$ is a family of distribution. The problem is to test the hypothesis

$$H_0 : (\theta_1, \theta_2, \ldots, \theta_k) \in s$$

$$\text{against } H_A : (\theta_1, \theta_2, \ldots, \theta_k) \in \bar{s}, \ \bar{s} = S - s$$

where s is the sub-space in S. The likelihood function of sample observations is

$$L(x) = \prod_{i=1}^{n} f(x_i; \theta_1, \theta_2, \ldots, \theta_k). \tag{34.4}$$

We can write two likelihood functions based on s and S, respectively as follows:

$$L(s) = \prod_{i=1}^{n} f(x_i; \theta_1, \theta_2, \ldots, \theta_k); (\theta_1, \theta_2, \ldots, \theta_k) \in s \tag{34.5}$$

$$L(S) = \prod_{i=1}^{n} f(x_i; \theta_1, \theta_2, \ldots, \theta_k); (\theta_1, \theta_2, \ldots, \theta_k) \in S. \tag{34.6}$$

Using maximum likelihood method of estimation, we can write

$$\frac{\partial L}{\partial \theta_j} = 0; \ j = 1, 2, \ldots, k.$$

Putting the values of estimates in (34.5) and (34.6), the two likelihood functions are written as $L(\hat{s})$ and $L(\hat{S})$. The ratio of these two likelihood functions is

$$\lambda(x_1, x_2, \ldots, x_n) = \lambda = \frac{L(\hat{s})}{L(\hat{S})}. \tag{34.7}$$

The function λ is free of parameter. Hence $\lambda \geq 0$. Again,

$$s \in S \Rightarrow L(\hat{s}) \in L(\hat{S}) \Rightarrow \lambda \leq 1.$$

Therefore, $0 \leq \lambda \leq 1$.

The critical region to test H_0 against H_A is $0 \leq \lambda \leq \lambda_0$, where λ_0 is any number (< 1) and it can be found out for any value of α such that

$$P[\lambda \leq \lambda_0/H_0] = \alpha \text{ or, } \int_0^{\lambda_0} g(\lambda/H_0) = \alpha$$

where $g(\lambda)$ is the p.d.f. of λ.

The test which is based on λ shown above is called ***likelihood ratio test.***

If the p.d.f. of λ is not known, λ_0 cannot be found out. However, any function of λ may be known and the distribution of that function is also known. In such a case the value of λ_0 is obtained using the following theorem:

❐ **THEOREM 34.6** *Let* $\lambda = \dfrac{L(\hat{s})}{L(\hat{S})}$ *is the ratio of likelihood functions and* $U = \phi(\lambda)$ *is a monotone function [increasing or decreasing] of* λ*. Any test based on U is equivalent to likclihood ratio test and the critical region for U is*

$$\phi(0) < U < \phi(\lambda_0), \text{ if } U \text{ is monotone increasing function of } \lambda$$
$$\text{and } \phi(0) > U > \phi(\lambda_0), \text{ if } U \text{ is monotone decreasing function of } \lambda.$$

Proof. The critical region of likelihood ratio is $0 < \lambda < \lambda_0$, where

$$P[\lambda < \lambda_0/H_0] = \int_0^{\lambda_0} g(\lambda/H_0)d\lambda = \alpha.$$

Consider the monotone increasing function of λ as $U = \phi(\lambda)$. The level of significance α is given by

$$\alpha = \int_0^{\lambda_0} g(x/H_0)dx = \int_{\phi(0)}^{\phi(\lambda_0)} h(u/H_0)du$$

where $h(u/H_0)$ is the p.d.f. of U under null hypothesis. Then the critical region $s_0 = \{0 < \lambda < \lambda_0\}$ will be transformed to

$$s_0 = \{\phi(0) < U < \phi(\lambda_0)\}.$$

Again, if U is a monotone decreasing function of λ, then $s_0 = \{\phi(\lambda_0) < U < \phi(0)\}$.

Characteristics of LR Test

(i) Under simple null hypothesis the distribution of λ remains unchanged. However, the distribution may not be same if the null hypothesis is composite. In such a case the distribution of λ is to be ascertained for different parameters and λ_0 is to be sought out in such a way that for all parameters in S

$$\int_0^{\lambda_0} g(\lambda/H_0)d\lambda \leq \infty.$$

(ii) If sample size is large, the test is performed using the following theorem:

❑ **THEOREM 34.7** *Let $x_1, x_2, \ldots, x_n$ be a random sample of n observations from $f(x; \theta_1, \theta_2, \ldots, \theta_k)$, where S is the k-dimensional parametric space. Assume that the null hypothesis is composite, i.e.,*

$$H_0 : \theta_1 = \theta_1', \theta_2 = \theta_2', \ldots, \theta_r = \theta_r' \ (r < k).$$

Here $k - r$ are nuisance parameters. Then under H_0, $-2 \ln \lambda$ is asymptotically distributed as χ^2 with r d.f.

(iii) If there is no nuisance parameter in the distribution of λ, then the LR test transforms to similar region test.

(iv) λ is always the function of sufficient statistic.

34.9 Likelihood Ratio Test for Population Mean of Normal Population

Let $x_1, x_2, \ldots, x_n$ be a random sample of n observations from $N(\mu, \sigma^2)$, where μ and σ^2 are unknown. The problem is to test the significance of

$$H_0 : \mu = \mu_0 \text{ against } H_A : \mu \neq \mu_0.$$

The parametric space is

$$S = \{\mu, \sigma^2\}; -\infty < \mu < \infty, \ 0 \leq \sigma^2 \leq \infty.$$

The parametric sub-space is

$$s = \{\mu, \sigma^2; \ \mu = \mu_0, \ 0 \leq \sigma^2 \leq \infty\}.$$

The likelihood function of the sample observations is

$$L(x_1, x_2, \ldots, x_n) = \frac{1}{(2\pi\sigma^2)^{n/2}} \exp\left[-\frac{1}{2\sigma^2}(x - \mu)^2\right].$$

The maximum likelihood estimators of μ and σ^2 are, respectively

$$\frac{\partial \log L}{\partial \mu} = 0 \Rightarrow \hat{\mu} = \bar{x}$$

$$\frac{\partial \log L}{\partial \sigma^2} = 0 \Rightarrow \hat{\sigma}^2 = s^2 = \frac{1}{n}\Sigma(x_i - \bar{x})^2.$$

Therefore,

$$L(\hat{S}) = \left(\frac{1}{2\pi s^2}\right)^{n/2} \exp\left[-\frac{1}{2s^2}\Sigma(x - \bar{x})^2\right].$$

Under H_0 the maximum likelihood estimator of σ^2 is

$$\hat{\sigma}_0^2 = \frac{1}{n}\sum^{n}(x - \mu_0)^2 = s_0^2 \text{ (say)} = s^2 + (\bar{x} - \mu_0)^2.$$

Therefore,

$$L(\hat{s}) = \left(\frac{1}{2\pi s_0^2}\right)^{n/2} \exp\left[-\frac{1}{2s_0^2}\Sigma(x-\mu_0)^2\right].$$

Now, the likelihood ratio criterion is

$$\lambda = \frac{L(\hat{s})}{L(\hat{S})} = \frac{\left(\frac{1}{2\pi s_0^2}\right)^{n/2} \exp\left[-\frac{1}{2s_0^2}\Sigma(x-\mu_0)^2\right]}{\left(\frac{1}{2\pi s^2}\right)^{n/2} \exp\left[-\frac{1}{2s^2}\Sigma(x-\bar{x})^2\right]}$$

$$= \left(\frac{s^2}{s_0^2}\right)^{n/2} = \left[\frac{s^2}{s^2+(\bar{x}-\mu_0)^2}\right]^{n/2} = \left[\frac{1}{1+\frac{(\bar{x}-\mu_0)^2}{s^2}}\right]^{n/2}.$$

But under $H_0 : \mu = \mu_0$, $\dfrac{\sqrt{n}(\bar{x}-\mu_0)}{s_1}$ is distributed as Student's 't' with $(n-1)$ d.f., where

$$s_1^2 = \frac{1}{n-1}\Sigma(x_i-\bar{x})^2 = \frac{ns^2}{n-1}.$$

$$\therefore \quad \frac{\sqrt{n}(\bar{x}-\mu_0)}{s_1} = t \text{ or, } \frac{t}{\sqrt{n-1}} = \frac{\bar{x}-\mu_0}{s}.$$

Then

$$\lambda = \left[\frac{1}{1+\frac{t^2}{n-1}}\right]^{n/2} = \phi(t^2) \text{ (say)}.$$

Hence instead of λ we can use t^2 as test statistic and the critical region becomes

$$W_0 = \{0 < \lambda < \lambda_0\}$$

$$W_0 = \{\phi(t^2) < \lambda_0\}$$

$$W_0 = \left[\frac{1}{1+\frac{t^2}{n-1}}\right]^{n/2} < \lambda_0 \Rightarrow \left[\frac{1}{1+\frac{t^2}{n-1}}\right] < \lambda_0^{2/n} \Rightarrow \left(1+\frac{t^2}{n-1}\right) > \lambda_0^{n/2}$$

$$\text{or, } t^2 > \left(\lambda_0^{n/2}-1\right)/(n-1) \Rightarrow t^2 > k^2 \text{ (say)}.$$

Therefore, the critical region is

$$|t| = \left|\frac{\sqrt{n}(\bar{x}-\mu_0)}{s_1}\right| > k$$

where

$$P[|t| > k/H_0] = \alpha.$$

Since the distribution of 't' is Student's 't', the value of k is $100\frac{\alpha}{2}\%$ upper value of 't' with $(n-1)$ d.f. Thus

$$k \sim t_{\frac{\alpha}{2},n-1} \text{ and } P\left[|t| > t_{\frac{\alpha}{2},n-1}\right] = \frac{\alpha}{2}.$$

The null hypothesis $H_0 : \mu = \mu_0$ is rejected if

$$\left|\frac{\sqrt{n}(\bar{x} - \mu_0)}{s_1}\right| > t_{\frac{\alpha}{2}, n-1}.$$

►**EXAMPLE 34.13** *The fasting blood sugar level of randomly selected* 60 *patients are given below:*

145, 160, 135, 130, 165, 140, 130, 95, 100, 92, 90, 110, 100, 98, 105, 100, 85, 120, 95, 125, 95, 95, 90, 100, 95, 90, 90, 100, 100, 95, 135, 165, 200, 150, 165, 175, 168, 170, 165, 170, 180, 200, 250, 260, 150, 165, 165, 170, 170, 145, 160, 170, 180, 200, 250, 160 145, 155, 210, 250.

Assuming normal distribution of the distribution of fasting blood sugar, perform LR test to test the null hypothesis $H_0 : \mu = \mu_0 = 175$ *against* $H_A : \mu \neq \mu_0$.

Solution. We have

$$n = 60, \bar{x} = 168.85, s_1^2 = 945.4856.$$

Here

$$\lambda = \left[\frac{1}{1 + \frac{(\bar{x} - \mu_0)^2}{s^2}}\right]^{n/2} = \left[\frac{1}{1 + \frac{(168.85 - 175)^2}{961.5111}}\right]^{60/2} = 0.31.$$

Since $n = 60$ is assumed to be large, we can use $-2\ln\lambda$, where $-2\ln\lambda$ is χ^2 with 1 d.f. We have $\chi^2 = 2.34$. H_0 is accepted since $\chi^2_{0.05,1} = 3.84$. Again,

$$|t| = \frac{\sqrt{n}(\bar{x} - \mu_0)}{s_1} = \frac{\sqrt{60}(168.85 - 175)}{\sqrt{945.4856}} = 1.55.$$

Since $|t| < t_{0.025,59} = 2.001$, H_0 is accepted.

34.10 Likelihood Ratio Test for Equality of Means of Two Normal Populations

Let $x_{11}, x_{12}, \ldots, x_{1n_1}$ and $x_{21}, x_{22}, \ldots, x_{2n_2}$ be two random samples of sizes n_1 and n_2, respectively from $N(\mu_1, \sigma_1^2)$ and $N(\mu_2, \sigma_2^2)$. The parametric space is

$$S = \{\mu_1, \mu_2; \sigma_1^2, \sigma_2^2\}.$$

The problem is to test the significance of null hypothesis $H_0 : \mu_1 = \mu_2 = \mu$ against $H_A : \mu_1 \neq \mu_2$.

CASE 1. $\sigma_1^2 = \sigma_2^2 = \sigma^2$ (say)

Here,

$$S : \{\mu_1, \mu_2, \sigma^2\}; \ -\infty < \mu_i < \infty, \ \sigma^2 > 0; \ i = 1, 2$$
$$s : \{\mu, \sigma^2\}; \ -\infty < \mu < \infty, \ \sigma^2 > 0.$$

Then

$$L(S) = \left(\frac{1}{2\pi\sigma^2}\right)^{\frac{n_1+n_2}{2}} \exp\left[-\frac{1}{2\sigma^2}\{\Sigma(x_{1i}-\mu_1)^2 + \Sigma(x_{2i}-\mu_2)^2\}\right].$$

The ML estimators of μ_1, μ_2 and σ^2 are

$$\begin{aligned}
\hat{\mu}_1 &= \bar{x}_1,\ \hat{\mu}_2 = \bar{x}_2,\\
\hat{\sigma}^2 &= \frac{1}{n_1+n_2}\left[\Sigma(x_{1i}-\bar{x}_1)^2 + \Sigma(x_{2i}-\bar{x}_2)^2\right]\\
&= \frac{1}{n_1+n_2}[n_1 s_1^2 + n_2 s_2^2].\\
\therefore\ L(\hat{S}) &= \left[\frac{n_1+n_2}{2\pi(n_1 s_1^2 + n_2 s_2^2)}\right]^{\frac{n_1+n_2}{2}} e^{-\frac{1}{2}(n_1+n_2)}
\end{aligned}$$

$$\left[\because \exp\left[-\frac{1}{2\hat{\sigma}^2}\{\Sigma(x_{1i}-\bar{x}_1)^2 + \Sigma(x_{2i}-\bar{x}_2)^2\}\right] = \exp\left[-\frac{(n_1+n_2)\hat{\sigma}^2}{2\hat{\sigma}^2}\right]\right]$$

Again,

$$L(s) = \left(\frac{1}{2\pi\sigma^2}\right)^{\frac{n_1+n_2}{2}} \exp\left[-\frac{1}{2\sigma^2}\{\Sigma(x_{1i}-\mu)^2 + \Sigma(x_{2i}-\mu)^2\}\right].$$

The ML estimators of μ and σ^2 are, respectively

$$\begin{aligned}
\hat{\mu} &= \frac{n_1\bar{x}_1 + n_2\bar{x}_2}{n_1+n_2}\\
\text{and } \hat{\sigma}^2 &= \frac{1}{n_1+n_2}\left[\Sigma(x_{1i}-\hat{\mu})^2 + \Sigma(x_{2i}-\hat{\mu})^2\right].
\end{aligned}$$

But

$$\Sigma(x_{1i}-\hat{\mu})^2 = \Sigma(x_{1i}-\bar{x}_1)^2 + n_1(\bar{x}_1-\hat{\mu})^2 = n_1 s_1^2 + n_1 n_2^2\frac{(\bar{x}_1-\bar{x}_2)^2}{(n_1+n_2)^2}.$$

Similarly,

$$\begin{aligned}
\Sigma(x_{2i}-\hat{\mu})^2 &= n_2 s_2^2 + n_1^2 n_2\frac{(\bar{x}_1-\bar{x}_2)^2}{(n_1+n_2)^2}.\\
\therefore\ \hat{\sigma}^2 &= \frac{1}{n_1+n_2}\left[n_1 s_1^2 + n_2 s_2^2 + n_1 n_2\frac{(\bar{x}_1-\bar{x}_2)^2}{n_1+n_2}\right].\\
\therefore\ L(\hat{s}) &= \left[\frac{n_1+n_2}{2\pi\left\{n_1 s_1^2 + n_2 s_2^2 + \frac{n_1 n_2(\bar{x}_1-\bar{x}_2)^2}{n_1+n_2}\right\}}\right]^{\frac{n_1+n_2}{2}} e^{-\frac{1}{2}(n_1+n_2)}.
\end{aligned}$$

Therefore,

$$\lambda = \frac{L(\hat{s})}{L(\hat{S})} = \left[\frac{n_1 s_1^2 + n_2 s_2^2}{n_1 s_1^2 + n_2 s_2^2 + \frac{n_1 n_2(\bar{x}_1-\bar{x}_2)^2}{n_1+n_2}}\right]^{\frac{n_1+n_2}{2}} = \left[\frac{1}{1 + \frac{n_1 n_2(\bar{x}_1-\bar{x}_2)^2}{(n_1+n_2)(n_1 s_1^2 + n_2 s_2^2)}}\right]^{\frac{n_1+n_2}{2}}.$$

Under H_0, we have

$$t = \frac{\bar{x}_1 - \bar{x}_2}{\sqrt{s^2\left(\frac{1}{n_1} + \frac{1}{n_2}\right)}} \sim t_{n_1+n_2-2}$$

where

$$s^2 = \frac{1}{n_1 + n_2 - 2}[n_1 s_1^2 + n_2 s_2^2].$$

Now

$$t^2 = \frac{n_1 n_2 (\bar{x}_1 - \bar{x}_2)^2}{s^2(n_1 + n_2)}.$$

Hence

$$\lambda = \left[\frac{1}{1 + \frac{t^2}{n_1+n_2-2}}\right]^{\frac{n_1+n_2}{2}} = \left[1 + \frac{t^2}{n_1 + n_2 - 2}\right]^{-\frac{1}{2}(n_1+n_2)}.$$

Here 't' can also be used as test statistic instead of λ. Then, the critical region $W_0 = \{0 < \lambda < \lambda_0\}$ transforms to

$$t^2 \geq \left[\frac{1}{\lambda_0^{\frac{2}{n_1+n_2}}} - 1\right](n_1 + n_2 - 2) = a^2 \text{ (say)} \Rightarrow |t| \geq a.$$

Therefore, the critical region W_0 is $|t| \geq a$ such that

$$P[|t| \geq a/H_0] = \alpha.$$

The d.f. of 't' is $(n_1 + n_2 - 2)$, where

$$|t| = \frac{|x_1 - \bar{x}_2|}{\sqrt{s^2\left(\frac{1}{n_1} + \frac{1}{n_2}\right)}}.$$

If $|t| \geq t_{\frac{\alpha}{2}}, n_1 + n_2 - 2$, H_0 is rejected.

The test discussed above is unbiased one. Since under H_0 the distribution of 't' does not depend on σ^2, the critical region W_0 is similar region.

CASE 2. $\sigma_1^2 \neq \sigma_2^2$

Here,

$$S : \{\mu_1, \mu_2, \sigma_1^2, \sigma_2^2;\ -\infty < \mu_i < \infty;\ \sigma_i^2 > 0;\ i = 1, 2\}$$
$$\text{and } s : \{\mu, \sigma_1^2, \sigma_2^2;\ -\infty < \mu < \infty;\ \sigma_i^2 > 0;\ i = 1, 2\}.$$

$$\therefore\ L(S) = \left(\frac{1}{2\pi\sigma_1^2}\right)^{\frac{n_1}{2}} \left(\frac{1}{2\pi\sigma_2^2}\right)^{\frac{n_2}{2}} \times \exp\left[-\frac{1}{2\sigma_1^2}\Sigma(x_{1i} - \mu_1)^2\right] \exp\left[-\frac{1}{2\sigma_2^2}\Sigma(x_{2i} - \mu_2)^2\right].$$

The ML estimators of μ_1, μ_2, σ_1^2 and σ_2^2 are, respectively

$$\hat{\mu}_1 = \bar{x}_1,\ \hat{\mu}_2 = \bar{x}_2,$$
$$\hat{\sigma}_1^2 = \frac{1}{n_1}\Sigma(x_{1i} - \bar{x}_1)^2 = s_1^2,\ s_2^2 = \hat{\sigma}_2^2 = \frac{1}{n_2}\Sigma(x_{2i} - \bar{x}_2)^2.$$
$$\therefore\ L(\hat{S}) = \left(\frac{1}{2\pi s_1^2}\right)^{\frac{n_1}{2}} \left(\frac{1}{2\pi s_2^2}\right)^{\frac{n_2}{2}} e^{-\frac{1}{2}(n_1 + n_2)}$$
$$L(\hat{s}) = \left(\frac{1}{2\pi\sigma_1^2}\right)^{\frac{n_1}{2}} \left(\frac{1}{2\pi\sigma_2^2}\right)^{\frac{n_2}{2}} \exp\left[-\frac{1}{2\sigma_1^2}\Sigma(x_{1i} - \mu)^2\right] \times \exp\left[-\frac{1}{2\sigma_2^2}\Sigma(x_{2i} - \mu)^2\right].$$

The ML estimators of σ_1^2, σ_2^2 and μ are

$$\hat{\sigma}_1^2 = \frac{1}{n_1}\Sigma(x_{1i} - \mu)^2,\ \hat{\sigma}_2^2 = \frac{1}{n_2}\Sigma(x_{2i} - \mu)^2$$
$$\frac{\partial \log L(\hat{s})}{\partial \mu} = 0 \Rightarrow \frac{\Sigma(x_{1i} - \mu)}{\sigma_1^2} + \frac{\Sigma(x_{2i} - \mu)}{\sigma_2^2} = 0$$
$$\text{or, } \frac{n_1\Sigma(x_{1i} - \mu)}{\Sigma(x_{1i} - \mu)^2} + \frac{n_2\Sigma(x_{2i} - \mu)}{\Sigma(x_{2i} - \mu)^2} = 0 \text{ or, } \frac{n_1^2(\bar{x}_1 - \mu)}{\Sigma(x_{1i} - \mu)^2} + \frac{n_2^2(\bar{x}_2 - \mu)}{\Sigma(x_{2i} - \mu)^2} = 0.$$

It is difficult to find the value of $\hat{\mu}$ from the above equation. Since the ratio of variances exists, the distribution of λ is not easily available. Hence the critical region $W_0 : \{0 < \lambda < \lambda_0\}$ cannot be found out. However, for definite values of sample sizes mathematical method may be applied to solve the above equation and in that case critical region may be found out.

►**EXAMPLE 34.14** *The following data represent the time (in hours) taken to develop a computer program by two groups of computer scientists:*

Group 1 : 2.4, 1.5, 2.0, 1.9, 2.3, 2.9, 2.8, 2.6, 3.5, 4.5, 3.7, 5.5, 2.8, 6.5, 5.6

Group 2 : 2.6, 1.5, 1.3, 2.4, 2.9, 3.0, 3.1, 3.9, 2.9, 3.9, 4.8

Assuming normality of data, perform LR test to test the significance of

$$H_0 : \mu_1 = \mu_2 \textit{ against } H_A : \mu_1 \neq \mu_2.$$

Solution. Let $x_{1i} \sim N(\mu_1,\ \sigma_1^2)$, $x_{2i} \sim N(\mu_2,\ \sigma_2^2)$, where x_{1i} and x_{2i} are the observations of Group 1 and Group 2 respectively. Given $n_1 = 15$, $n_2 = 11$, we have $\bar{x}_1 = 3.37$, $\bar{x}_2 = 2.94$, $s_1^2 = 2.27$, $s_2^2 = 1.05$.

$$\therefore\ s^2 = \frac{(n_1 - 1)s_1^2 + (n_2 - 1)s_2^2}{n_1 + n_2 - 2} = 1.76.$$

Let us assume that $\sigma_1^2 = \sigma_2^2 = \sigma^2$. Now, according to LR test criterion, the test statistic is

$$t = \frac{\bar{x}_1 - \bar{x}_2}{\sqrt{s^2\left(\frac{1}{n_1} + \frac{1}{n_2}\right)}} = 0.81.$$

Here $|t| < t_{0.025,24} = 2.064$, so H_0 is accepted.

34.11 Test of Equality of Means of Several Normal Populations

Let x_{ij} $(i = 1, 2, \ldots, k;\ j = 1, 2, \ldots, n_i)$ be the sample observations of k independent samples selected from i-th normal population $[N(\mu_i, \sigma_i^2)]$. Assume that

$$\sigma_1^2 = \sigma_2^2 = \cdots = \sigma_k^2 = \sigma^2 \text{ (unknown, say).}$$

The problem is to test the significance of $H_0 : \mu_1 = \mu_2 = \cdots = \mu_k$ against H_A : at least one of the equality does not hold good. Here

$$S : \{(\mu_1, \mu_2, \ldots, \mu_k), \sigma^2\};\ -\infty < \mu_i < \infty,\ \sigma^2 > 0$$
$$\text{and } s : \{\mu, \sigma^2\},\ -\infty < \mu < \infty;\ \sigma^2 > 0.$$

Now

$$L(S) = \left(\frac{1}{2\pi\sigma^2}\right)^{\frac{n}{2}} \exp\left[-\frac{1}{2\sigma^2}\sum_i^k \sum_j^n (x_{ij} - \mu_i)^2\right],\ n = \sum_{i=1}^k n_i.$$

The ML estimators of μ_i are $\hat{\mu}_i = \bar{x}_i$ $(i = 1, 2, \ldots, k)$. The ML estimator of σ^2 is

$$\hat{\sigma}^2 = \frac{1}{n}\sum_i^k \sum_{j=1}^{n_i} (x_{ij} - \bar{x}_i)^2 = \frac{S_W}{n}$$

$$L(s) = \left(\frac{1}{2\pi\sigma^2}\right)^{\frac{n}{2}} \exp\left[-\frac{1}{2\sigma^2}\sum\sum(x_{ij} - \mu)^2\right].$$

The ML estimators of μ and σ^2 are

$$\hat{\mu} = \bar{x},\ \hat{\sigma}^2 = \frac{1}{n}\sum\sum(x_{ij} - \bar{x})^2 = \frac{S_T}{n},\ S_T = \text{Total sum of squares.}$$

Then

$$L(\hat{S}) = \left(\frac{n}{2\pi S_W}\right)^{\frac{n}{2}} e^{-\frac{n}{2}} \text{ and } L(\hat{s}) = \left(\frac{n}{2\pi S_T}\right)^{\frac{n}{2}} e^{-\frac{n}{2}}.$$

$$\therefore\ \lambda = \frac{L(\hat{s})}{L(\hat{S})} = \left(\frac{S_W}{S_T}\right)^{\frac{n}{2}}.$$

But

$$S_T = \sum\sum(x_{ij} - \bar{x})^2 = \sum\sum(x_{ij} - \bar{x}_i)^2 + \sum\sum(\bar{x}_i - \bar{x})^2 = S_W + S_B.$$

$$\therefore\ \lambda = \left(\frac{S_W}{S_W + S_B}\right)^{\frac{n}{2}} = \left(\frac{1}{1 + \frac{S_B}{S_W}}\right)^{\frac{n}{2}}.$$

We know that under H_0

$$F = \frac{S_B/(k-1)}{S_W/(n-k)} \sim F_{(k-1),(n-k)}.$$

Then

$$\frac{(k-1)F}{n-k} = \frac{S_B}{S_W}, \quad \therefore \ \lambda = \left[1 + \frac{(k-1)F}{n-k}\right]^{n/2}$$

where λ is the monotone function of F and hence instead of λ, F can be used as test statistic. Then the critical region becomes

$$\left[1 + \frac{(k-1)F}{(n-k)}\right]^{-\frac{n}{2}} > \lambda_0 \Rightarrow F > \left(\lambda_0^{\frac{n}{2}} - 1\right)\frac{n-k}{k-1} = a \text{ (say)}$$

where $P[F \geq a/H_0] = \alpha$.

In practice, if $F \geq F_{\alpha;(k-1),(n-k)}$, H_0 is rejected, where $F_{\alpha;(k-1),(n-k)}$ is the upper $100\alpha\%$ value of F distribution with $(k-1)$ and $(n-k)$ d.f.

34.12 Test for the Variance of a Normal Population

Let $x_1, x_2, \ldots, x_n$ be a random sample of n observations from $N(\mu, \sigma^2)$. We need to test the significance of

$$H_0 : \sigma^2 = \sigma_0^2 \text{ (a known value) against } H_A : \sigma^2 \neq \sigma_0^2.$$

The parametric space is

$$S : \{\mu, \sigma^2;\ -\infty < \mu < \infty;\ \sigma^2 > 0\}$$
$$s : \{\mu, \sigma_0^2;\ -\infty < \mu < \infty,\ \sigma_0^2 > 0\}.$$

The ML estimator of μ is $\hat{\mu} = \bar{x}$ and $\hat{\sigma}^2 = s^2$. Then the likelihood function

$$L(\hat{S}) = \left(\frac{1}{2\pi s^2}\right)^{\frac{n}{2}} \exp\left[-\frac{1}{2s^2}\Sigma(x_i - \bar{x})^2\right] = \left(\frac{1}{2\pi s^2}\right)^{\frac{n}{2}} e^{-\frac{n}{2}}$$

$$L(\hat{s}) = \left(\frac{1}{2\pi\sigma_0^2}\right)^{\frac{n}{2}} \exp\left[-\frac{1}{2\sigma_0^2}\Sigma(x_i - \bar{x})^2\right] = \left(\frac{1}{2\pi\sigma_0^2}\right)^{\frac{n}{2}} \exp\left[-\frac{ns^2}{2\sigma_0^2}\right].$$

$$\therefore \ \lambda = \frac{L(\hat{s})}{L(\hat{S})} = \left(\frac{s^2}{\sigma_0^2}\right)^{\frac{n}{2}} \exp\left[-\frac{1}{2}\left(\frac{ns^2}{\sigma_0^2} - n\right)\right].$$

Under H_0, the statistic $\chi^2 = \dfrac{ns^2}{\sigma^2}$ is distributed as chi-square with $(n-1)$ d.f. Therefore,

$$\lambda = \left[\frac{\chi^2}{n}\right]^{n/2} \exp\left[-\frac{1}{2}(\chi^2 - n)\right].$$

Here λ is a monotonic function of χ^2. Hence the test can be performed using χ^2-statistic. The critical region is $0 < \lambda < \lambda_0$ is equivalent to

$$\left[\frac{\chi^2}{n}\right]^{n/2} \exp\left[-\frac{1}{2}(\chi^2 - n)\right] < \lambda_0.$$

Since χ^2 has chi-square distribution with $(n-1)$ d.f., the critical region is determined by a pair of intervals $0 < \chi^2 < \chi_1^2$ and $\chi_2^2 < \chi^2 < \infty$, where χ_1^2 and χ_2^2 are to be determined

$$P(\chi_0^2 > \chi_1^2) = \frac{\alpha}{2} \text{ and } P(\chi^2 > \chi_2^2) = 1 - \frac{\alpha}{2}.$$

In other words, $\chi_1^2 = \chi^2_{\frac{\alpha}{2},n-1}$ and $\chi_2^2 = \chi^2_{(1-\frac{\alpha}{2}),(n-1)}$ where χ_2^2 is the upper $100\alpha\%$ value of chi-square distribution. Since $H_A : \sigma^2 \neq \sigma_0^2$, the test is a two-tailed test. If the alternative hypothesis $H_A : \sigma^2 < \sigma_0^2$, the critical region is $\chi^2 < \chi_\alpha^2$; if $H_A : \sigma^2 > \sigma_0^2$, the critical region is $\chi^2 > \chi_\alpha^2$.

►**EXAMPLE 34.15** *The following information are related to the ages (in years) at marriage of* 10 *randomly selected women in a society. Do you think that the sample observations follow* $N(27, 2)$? $[n = 10, \bar{x} = 27.1, s^2 = 6.1]$

Solution. Here $n = 10$, $\bar{x} = 27.1$, $s^2 = 6.1$.

We need to test $H_0 : \sigma^2 = 2 = \sigma_0^2$ against (i) $H_A : \sigma^2 > 2$, (ii) $H_A : \sigma^2 < 2$ and (iii) $H_A : \sigma^2 \neq 2$.

The test statistic is

$$\chi^2 = \frac{(n-1)s^2}{\sigma_0^2} = \frac{54.9}{2} = 27.45.$$

(i) Since $\chi^2 > \chi^2_{0.05,9} = 16.92$, H_0 is rejected in favour of H_A (i).

(ii) Since $\chi^2 > \chi^2_{0.95,9} = 3.32$, H_0 is rejected in favour of H_A (ii).

(iii) $\chi^2_{0.025,9} = 19.0$ and $\chi^2_{0.975} = 2.70$. The calculated χ^2 does not fall within the limit 2.70 to 19.9; hence the null hypothesis is rejected.

Also, we need to test $H_0 : \mu = 27$ against $H_A : \mu \neq 27$. Here σ^2 is not known. Therefore, the test statistic is

$$t = \frac{\bar{x} - \mu_0}{s/\sqrt{n}} = \frac{27.1 - 27.0}{\sqrt{6.1/10}} = 0.13 \; (\mu_0 = 27).$$

Since $|t| < t_{0.025,9} = 2.262$, H_0 is accepted. Here mean of the distribution is 27 but variance is not 2.

►**EXAMPLE 34.16** *Five new varieties of rice were cultivated in different homogeneous plots in an agricultural research station. The production data* (x_{ij} *kg*) *of different varieties of rice are shown below:*

Rice variety	*Production* (x_{ij} *kg/plot*)	*Total* $R_i = \Sigma x_{ij}$	*No. of plots* n_i
1	10.2, 10.5, 11.2, 10.8, 10.2, 10.0	62.9	6
2	12.4, 10.8, 13.6, 12.8, 13.6	63.2	5
3	9.5, 9.8, 9.4, 9.6, 9.7	48.0	5
4	13.2, 13.4, 12.6, 12.8	52.0	4
5	14.2, 14.0, 11.6, 14.6, 14.5	68.9	5
Total		$G = 295$	$n = 25$

Assuming the normality of x_{ij} $[x_{ij} \sim N(\mu_i, \sigma_i^2)]$, test the significance of

$$H_0 : \mu_1 = \mu_2 = \mu_3 = \mu_4 = \mu_5$$

against H_A : at least one of the equality does not hold good.

Solution. We have $n = \Sigma n_i = 25,\ G = 295,\ k = 5$

$$S_T = \sum\sum x_{ij}^2 - \text{CT},\ \text{CT} = \frac{G^2}{n} = 3481 = 3557.48 - 3481 = 76.48$$

$$S_B = \Sigma\frac{R_i^2}{n_i} - \text{CT} = 63.49,\ S_W = S_T - S_B = 12.99$$

$$\therefore\ F = \frac{S_B/(k-1)}{S_W/(n-k)} = \frac{63.49/4}{12.99/20} = 24.43.$$

Since $F > F_{0.05;4,20} = 2.87$, H_0 is rejected.

34.13 Test of Equality of Variances of Two Normal Populations

Let x_{1j} $(j = 1, 2, \ldots, n_1)$ and x_{2j} $(j = 1, 2, \ldots, n_2)$ be the n_1 and n_2 random sample observations from $N(\mu_1, \sigma_1^2)$ and $N(\mu_2, \sigma_2^2)$, respectively. The parameters in the parametric space are $S : \{\mu_1, \mu_2, \sigma_1^2, \sigma_2^2\}$. The problem is to test the significance of null hypothesis $H_0 : \sigma_1^2 = \sigma_2^2$ against $H_A : \sigma_1^2 \neq \sigma_2^2$. Here

$$-\infty < \mu_i < \infty;\ \sigma_i^2 > 0;\ i = 1, 2.$$

The likelihood function is

$$L(S) = \left(\frac{1}{2\pi\sigma_1^2}\right)^{\frac{n_1}{2}} \left(\frac{1}{2\pi\sigma_2^2}\right)^{\frac{n_2}{2}} \times \exp\left[-\frac{1}{2\sigma_1^2}\Sigma(x_{1j} - \mu_1)^2\right] \exp\left[-\frac{1}{2\sigma_2^2}\Sigma(x_{2j} - \mu_2)^2\right].$$

Again, $s : \{\mu_1, \mu_2, \sigma^2\}$. The ML estimators of μ_i and σ_i^2 are

$$\hat{\mu}_i = \bar{x}_i,\ \sigma_i^2 = \frac{1}{n_i}\sum_j^{n_i}(x_{ij} - \bar{x}_i)^2 = s_i^2;\ i = 1, 2.$$

Therefore,

$$L(\hat{S}) = \left(\frac{1}{2\pi s_1^2}\right)^{\frac{n_1}{2}} \left(\frac{1}{2\pi s_2^2}\right)^{\frac{n_2}{2}} e^{-\frac{1}{2}(n_1 + n_2)}.$$

Under null hypothesis the ML estimators of μ_i and σ^2 are

$$\hat{\mu}_i = \bar{x}_i;\ i = 1, 2$$

$$\hat{\sigma}^2 = \frac{1}{n_1 + n_2}\left[\Sigma(x_{1j} - \bar{x}_1)^2 + \Sigma(x_{2j} - \bar{x}_2)^2\right] = \frac{1}{n_1 + n_2}\left[n_1 s_1^2 + n_2 s_2^2\right].$$

Therefore,

$$L(\hat{s}) = \left[\frac{n_1 + n_2}{2\pi(n_1 s_1^2 + n_2 s_2^2)}\right]^{\frac{n_1+n_2}{2}} e^{-\frac{1}{2}(n_1 + n_2)}.$$

Now, the λ criterion is

$$\lambda = \frac{L(\hat{s})}{L(\hat{S})} = \frac{(n_1 + n_2)^{\frac{n_1+n_2}{2}}}{(n_1 s_1^2 + n_2 s_2^2)^{\frac{n_1+n_2}{2}}} (s_1^2)^{\frac{n_1}{2}} (s_2^2)^{\frac{n_2}{2}}$$

$$= \frac{(n_1 + n_2)^{\frac{1}{2}(n_1 + n_2)} (n_1 s_1^2)^{\frac{n_1}{2}} (n_2 s_2^2)^{\frac{n_2}{2}}}{n_1^{\frac{n_1}{2}} n_2^{\frac{n_2}{2}} (n_1 s_1^2 + n_2 s_2^2)^{\frac{1}{2}(n_1 + n_2)}}$$

$$= \frac{(n_1 + n_2)^{\frac{1}{2}(n_1 + n_2)} (n_1 s_1^2)^{\frac{n_1}{2}} (n_2 s_2^2)^{\frac{n_2}{2}}}{n_1^{\frac{n_1}{2}} n_2^{\frac{n_2}{2}} (n_2 s_2^2)^{\frac{1}{2}(n_1 + n_2)} \left(1 + \frac{n_1 s_1^2}{n_2 s_2^2}\right)^{\frac{1}{2}(n_1 + n_2)}}$$

$$= \frac{(n_1 + n_2)^{\frac{1}{2}(n_1 + n_2)} (n_1 s_1^2 / n_2 s_2^2)^{\frac{n_1}{2}}}{n_1^{\frac{n_1}{2}} n_2^{\frac{n_2}{2}} \left(1 + \frac{n_1 s_1^2}{n_2 s_2^2}\right)^{\frac{1}{2}(n_1+n_2)}}.$$

We know that

$$F = \frac{s_{11}^2}{s_{21}^2} = \frac{\frac{1}{(n_1 - 1)}\Sigma(x_{1j} - \bar{x}_1)^2}{\frac{1}{(n_2 - 1)}\Sigma(x_{2j} - \bar{x}_2)^2}$$

where s_{11}^2 and s_{21}^2 are unbiased estimators of variances. Thus

$$F = \frac{(n_2 - 1) n_1 s_1^2}{(n_1 - 1) n_2 s_2^2}, \quad \because \; s_i^2 = \frac{1}{n_i} \sum_j^{n_i} (x_{ij} - \bar{x}_i)^2.$$

This F is distributed as variance ratio with $(n_1 - 1)$ and $(n_2 - 1)$ d.f. Hence

$$\frac{n_1 - 1}{n_2 - 1} F = \frac{n_1 s_1^2}{n_2 s_2^2}.$$

Now, λ can be written as

$$\lambda = \frac{(n_1 + n_2)^{\frac{1}{2}(n_1 + n_2)}}{n_1^{\frac{n_1}{2}} n_2^{\frac{n_2}{2}}} \left[\frac{\left(\frac{n_1 - 1}{n_2 - 1} F\right)^{\frac{n_1}{2}}}{\left(1 + \frac{n_1 - 1}{n_2 - 1} F\right)^{\frac{1}{2}(n_1 + n_2)}}\right].$$

Since λ is the function of F, we can use F-statistic to test the null hypothesis. Then the critical region W_0 will be $\{F \le F_1 \text{ and } F \ge F_2\}$, where F_1 and F_2 are

$$P[F \le F_1/H_0] = \frac{\alpha}{2} \text{ and } P[F \ge F_2/H_0] = \frac{\alpha}{2}$$

where $F_1 = F_{\frac{\alpha}{2};(n_1-1),(n_2-1)}$ and $F_2 = F_{\left(1-\frac{\alpha}{2}\right);(n_1-1),(n_2-1)}$.

Here $F_{\alpha;(n_1-1),(n_2-1)}$ is the lower $100\alpha\%$ value of F distribution.

►**EXAMPLE 34.17** *The following data represent the number of ever-born children of 20 couples. Ten couples adopted any one of the family planning method:*

Adopted family planning	*No. of ever-born children*	*Total*
Yes	x_{1j} : 1, 2, 2, 2, 1, 3, 2, 1, 1, 3	18
No	x_{2j} : 3, 3, 4, 4, 6, 4, 5, 6, 4, 7	46

Assuming $x_{1j} \sim N(\mu_1, \sigma_1^2)$ and $x_{2j} \sim N(\mu_2, \sigma_2^2)$, test the significance of $H_0 : \sigma_1^2 = \sigma_2^2$ against $H_A : \sigma_1^2 \ne \sigma_2^2$.

Solution. Here $n_1 = n_2 = 10$, $\bar{x}_1 = 1.8$, $\bar{x}_2 = 4.6$. The unbiased estimates of variances are

$$s_1^2 = \frac{1}{n_1 - 1}\Sigma(x_{1j} - \bar{x}_1)^2 = 0.62,$$

$$s_2^2 = \frac{1}{n_2 - 1}\Sigma(x_{2j} - \bar{x}_2)^2 = 1.82.$$

$$\therefore\ F = \frac{s_1^2}{s_2^2} = \frac{0.62}{1.82} = 0.34.$$

This F has $(9, 9)$ d.f. and $F_{0.05;9,9} = 3.18$. Since $F < F_{0.05;9,9}$, H_0 is accepted.

34.14 Test of Equality of Variances of Several Normal Populations

Let x_{ij} $(i = 1, 2, \ldots, k;\ j = 1, 2, \ldots, n_i)$ be the independent random sample of n_i observations selected from $N(\mu_i, \sigma_i^2)$. The problem is to test the significance of the null hypothesis $H_0 : \sigma_1^2 = \sigma_2^2 = \cdots = \sigma_k^2$ against H_A : at least one of the equality does not hold good. The parametric space is

$$S : \{\mu_1, \mu_2, \ldots, \mu_k; \sigma_1^2, \sigma_2^2, \ldots, \sigma_k^2\}; -\infty < \mu_i < \infty, \sigma_i^2 > 0.$$

Under H_0, $s : \{\mu_1, \mu_2, \ldots, \mu_k;\ \sigma^2\}$, $i = 1, 2, \ldots, k; \sigma^2 > 0$.

The ML estimators of μ_i and σ_i^2 are

$$\hat{\mu}_i = \bar{x}_i, \hat{\sigma}_i^2 = S_i^2 = \frac{1}{n_i}\sum_j (x_{ij} - \bar{x}_i)^2;\ i = 1, 2, \ldots, k.$$

$$\therefore\ L(\hat{S}) = \prod_{i=1}^{k}\left(\frac{1}{2\pi S_i^2}\right)^{\frac{n_i}{2}} e^{-\frac{1}{2}n_i} = e^{-\frac{1}{2}n}\prod_{i=1}^{n}\left(\frac{1}{2\pi S_i^2}\right)^{\frac{n_i}{2}}, n = \sum^{k} n_i.$$

We have

$$L(s) = \left(\frac{1}{2\pi\sigma^2}\right)^{\frac{n}{2}} \exp\left[-\frac{1}{2\sigma^2}\sum\sum(x_{ij}-\mu_i)^2\right].$$

Under H_0,

$$\hat{\mu}_i = \bar{x}_i,\ \hat{\sigma}^2 = \frac{1}{n}\sum^{k}\sum^{n_i}(x_{ij}-\bar{x}_i)^2 = \frac{1}{n}\sum_{i=1}^{k} n_i S_i^2 = S^2 \text{ (say)}.$$

$$\therefore\ L(\hat{s}) = \left(\frac{n}{2\pi\Sigma n_i S_i^2}\right)^{\frac{n}{2}} e^{-\frac{1}{2}n}.$$

Hence the λ criterion is

$$\lambda = \frac{L(\hat{s})}{L(\hat{S})} = \frac{\dfrac{n}{2\pi\Sigma n_i S_i^2}}{\prod_{i=1}^{k}\left(\dfrac{1}{2\pi S_i^2}\right)^{\frac{n_i}{2}}} = \frac{n^{\frac{1}{2}n}\prod_{i=1}^{k}(S_i^2)^{\frac{1}{2}n_i}}{(\Sigma n_i S_i^2)^{\frac{n}{2}}} = \frac{\prod_{i=1}^{k}(S_i^2)^{\frac{n_i}{2}}}{\left(\dfrac{1}{n}\Sigma n_i S_i^2\right)^{\frac{n}{2}}} = \frac{\prod_{i=1}^{k}(S_i^2)^{\frac{n_i}{2}}}{(S^2)^{\frac{n}{2}}}$$

$$= \prod_{i=1}^{k}\left(\frac{S_i^2}{S^2}\right)^{\frac{n_i}{2}}.$$

This function is a complex one and it is difficult to find the distribution of λ. However, if n_i are large, $-2\ln\lambda$ is asymptotically distributed as χ^2 with $(k-1)$ d.f., where

$$\chi^2 = -2\ln\prod_{i=1}^{k}\left(\frac{S_i^2}{S^2}\right)^{\frac{n_i}{2}} = -2\left[\sum^{k}\frac{n_i}{2}\ln S_i^2 - \frac{n}{2}\ln S^2\right].$$

Bartlett (1937) found out nearer value of χ^2, where $\chi^2 = \dfrac{-2\ln\lambda_1}{1+\frac{1}{3(k-1)}\left[\Sigma\frac{1}{n_i}-\frac{1}{n}\right]}$, where λ_1 is obtained from λ. This is done replacing n_i by (n_i-1) in calculating S_i^2. Under null hypothesis the Bartlett's χ^2-statistic stands

$$\chi^2 = \frac{-\Sigma(n_i-1)\ln\frac{S_i^2}{S^2}}{1+\frac{1}{3(k-1)}\left[\Sigma\frac{1}{n_i-1}-\frac{1}{n-k}\right]} = \frac{(n-k)\ln s^2 - \Sigma(n_i-1)\ln s_i^2}{1+\frac{1}{3(k-1)}\left[\Sigma\frac{1}{n_i-1}-\frac{1}{n-k}\right]}.$$

This χ^2 has $(k-1)$ d.f. The critical region is $W_0 : \left\{\chi^2 \geq \chi^2_{(1-\alpha),(k-1)}\right\}$, where $\chi^2_{(1-\alpha),(k-1)}$ is the $100\alpha\%$ value of $(k-1)$ d.f.

►**EXAMPLE 34.18** *The fasting blood sugar (FBS; x_{ij}) of 4 groups of patients selected randomly from 4 hospitals are given below:*

Group	*Fasting blood sugar, x_{ij}*
1	x_{1j} : 180, 150, 98, 120, 100, 125, 160, 175, 170, 180, 125
2	x_{2j} : 200, 100, 120, 95, 95, 98, 90, 160, 180, 96, 86, 125, 150
3	x_{3j} : 250, 160, 180, 190, 130, 120, 110, 105, 118
4	x_{4j} : 120, 125, 160, 170, 118, 110, 115, 118, 122, 140, 150

Assuming $x_{ij} \sim N(\mu_i, \sigma_i^2)$; $i = 1, 2, 3, 4,$; $j = 1, 2, \ldots, n_i$, *test the significance of*

$$H_0 : \sigma_1^2 = \sigma_2^2 = \sigma_3^2 = \sigma_4^2$$

against H_A : *at least one of the equality does not hold good.*

Solution. We have

$$n_1 = 11, n_2 = 13, n_3 = 9, n_4 = 11, n = 44.$$

$$s_i^2 = \frac{1}{n_i - 1}\Sigma(x_{ij} - \bar{x}_i)^2, \ i = 1, 2, 3, 4; \ k = 4$$

$$s_1^2 = 987.09, \ s_2^2 = 1433.06, \ s_3^2 = 2316.28, \ s_4^2 = 407.25$$

$$s^2 = \frac{1}{n-k}\Sigma(n_i - 1)s_i^2 = 1273.60, \ (n-k)\ln s^2 - \Sigma(n_i - 1)\ln s_i^2 = 0.60,$$

$$1 + \frac{1}{3(k-1)}\left[\sum \frac{1}{n_i - 1} - \frac{1}{n-k}\right] = 1.042.$$

$$\therefore \ \chi^2 = 0.576 < \chi^2_{0.05,3} = 7.81; \ H_0 \text{ is accepted.}$$

34.15 Combination of Test

Let x_{ij} $(i = 1, 2, \ldots, k; \ j = 1, 2, \ldots, n_i)$ be the sample observations of k samples randomly selected from a population $f(x, \theta)$, where the i-th sample has n_i observations. The problem is to test the significance of $H_0 : \theta = \theta_0$ against $H_A : \theta \neq \theta_0$. The hypothesis can be tested by deriving k test statistics using k sample observations. These k tests can be combined for final conclusion regarding the hypothesis.

Let $p_1, p_2, \ldots, p_k$ be the p-values obtained from k tests using k sample observations separately, where $0 < p_i < 1$. These k p-values can be combined to get a single p-value for the hypothesis. The combination of test using k p-values is done by a theorem discussed below:

❑ **THEOREM 34.8** *Let the variables* $X_1, X_2, \ldots, X_k$ *are independently distributed with probability density function* $f(x) = 1, \ 0 < x < 1$. *Then if* $P = X_1 X_2 \ldots X_k$, $-2\ln P$ *has a* χ^2*-distribution with* $2k$ *d.f.*

Proof. We have

$$-2\ln P = -2\ln(X_1, X_2, \ldots, X_n) = \psi_1 + \psi_2 + \cdots + \psi_k = \sum_{i=1}^{k} \psi_i$$

where $\psi_i = -2\ln X_i \Rightarrow X_i = \exp(\psi_{i/2})$.

The probability function of ψ_i is given by

$$g(\psi_i) = f(x_i)\left|\frac{dx_i}{d\psi_i}\right| = 1\left|\exp\left(-\frac{\psi_i}{2}\right) \times \left(-\frac{1}{2}\right)\right| = \frac{1}{2}\exp\left(-\frac{\psi_i}{2}\right).$$

Thus $g(\psi_i)$ is distributed as χ^2 with 2 d.f. Again, each of ψ_i are independently distributed as χ^2 with 2 d.f. Therefore, $\Sigma\psi_i$ is distributed as χ^2 with $2k$ d.f. Hence $-2\ln P$ is distributed as χ^2 with $2k$ d.f.

In practice the p-values of k test statistics can be found out and they can be combined, where the combined p-values are $P = p_1 p_2 \cdots p_k$, where p_i is the p-value of i-th test. Now $-2\ln P \sim \chi^2$ with $2k$ d.f. If $-2\ln P \geq \chi^2_{\alpha,2k}$, H_0 is rejected. This later conclusion is made combining all the k tests.

►**EXAMPLE 34.19** *Five random samples are selected from a population by* 5 *investigators to study the mean number of ever-born children. Each investigator has selected* 10 *couples. The mean number of ever-born children per couple are observed as*

$$\bar{x}_1 = 2.2,\ \bar{x}_2 = 3.7,\ \bar{x}_3 = 2.5,\ \bar{x}_4 = 2.0,\ \bar{x}_5 = 4.2.$$

The variances of ever-born children from different groups of couples are observed as

$$s_1^2 = 0.0503,\ s_2^2 = 2.7361,\ s_3^2 = 5.0586,\ s_4^2 = 1.2526,\ s_5^2 = 14.4052.$$

The objective is to test the significance of null hypothesis $H_0 : \mu = 2.0$ *against* $H_A : \mu \neq 2.0$. *Do the sample observations provide information in favour of null hypothesis?*

Solution. Each investigator perform separate t-test to test the null hypothesis

$$H_0 : \mu = 2.0.$$

The test statistics are

$$t_1 = \frac{\bar{x}_1 - \mu_0}{s_1/\sqrt{n}} = \frac{2.2 - 2.0}{\sqrt{\frac{0.0503}{10}}} = 2.81,\ P[t \geq 2.81] = 0.01 = p_1$$

$$t_2 = \frac{\bar{x}_2 - \mu_0}{s_2/\sqrt{n}} = \frac{3.7 - 2.0}{\sqrt{\frac{2.7361}{10}}} = 3.25,\ P[t \geq 3.25] = 0.005 = p_2$$

$$t_3 = \frac{\bar{x}_3 - \mu_0}{s_3/\sqrt{n}} = \frac{2.5 - 2.0}{\sqrt{\frac{5.0586}{10}}} = 0.70,\ P[t \geq 0.70] = 0.25 = p_2$$

$$t_4 = \frac{\bar{x}_4 - \mu_0}{s_4/\sqrt{n}} = \frac{2.0 - 2.0}{\sqrt{\frac{1.2526}{10}}} = 0.00,\ P[t \geq 0.00] = 0.05 = p_3$$

$$t_5 = \frac{\bar{x}_5 - \mu_0}{s_5/\sqrt{n}} = \frac{4.2 - 2.0}{\sqrt{\frac{14.4052}{10}}} = 1.83,\ P[t \geq 1.83] = 0.05 = p_4$$

$-2\ln P = p_1 \cdot p_2 \cdot p_3 \cdot p_4 \cdot p_5 = 29.96.$
Here $-2\ln P \sim \chi^2$ with $2k$ d.f.; $k = 5$.
Since $-2\ln P > \chi^2_{0.05,10} = 18.3$, H_0 is rejected.
The conclusion is drawn using all the 5 samples of observations.

34.16 Sequential Probability Ratio Test (SPRT)

The test of hypothesis and the process of finding BCR depend on sample size n, where n is regarded as fixed. The test is performed under the condition that the second kind of error β is minimized keeping the first kind of error α fixed. However, in some tests the sample size n is regarded as random variable, and α and β are assumed fixed. Such a test is usually termed as SPRT which was proposed by A. Wald.

Let $x_1, x_2, \ldots, x_n$ be a random sample of n observations selected from a population having p.d.f. $f(x, \theta)$. The problem is to test the hypothesis $H_0 : \theta = \theta_0$ against an alternative hypothesis $H_A : \theta = \theta_1$. The likelihood functions of sample observations of size n under H_0 and H_A are, respectively

$$L_0 = \prod_{i=1}^{n} f(x_i, \theta_0) \text{ and } L_A = \prod_{i=1}^{n} f(x_i, \theta_1).$$

The ratio of likelihood functions is

$$\lambda_n = \frac{L_A}{L_0} = \frac{\prod f(x_i, \theta_1)}{\prod f(x_i, \theta_0)} = \prod_{i=1}^{n} \frac{f(x_i, \theta_1)}{f(x_i, \theta_0)}.$$

The SPRT to test the significance of H_0 against H_A is defined as follows:

At each stage of experiment the likelihood ratio λ_n $(m = 1, 2, \ldots, n)$ is computed.

(i) If $\lambda_m \geq A$, we terminate the process with the rejection of H_0.

(ii) If $\lambda_m \leq B$, we terminate the process with the acceptance of H_0.

(iii) If $B < \lambda_m < A$, we continue sampling by taking n additional observations.

Here A and B $(B < A)$ are constants which are determined by the relation

$$A = \frac{1-\beta}{\alpha}, \; B = \frac{\beta}{1-\alpha}$$

where α and β are the probabilities of type I and type II errors, respectively.

From the likelihood ratio we can write

$$\log \lambda_m = \sum_{i=1}^{n} \log \frac{f(x_i, \theta_1)}{f(x_i, \theta_0)} = \sum_{i=1}^{n} z_i \text{ where } z_i = \log \frac{f(x_i, \theta_1)}{f(x_i, \theta_0)}.$$

Now, in terms of z_i the SPRT is defined as follows:

(i) If $\Sigma z_i \leq \log A$, reject H_0.

(ii) If $\Sigma z_i \leq \log B$, reject H_A.

(iii) If $\log B < \Sigma z_i < \log A$, continue sampling by taking an additional observation. The process of taking additional observation is continued until

$$\log B < \Sigma z_i < \log A$$

breaks down at either end. It has been proved that SPRT eventually terminates with probability one.

Operating Characteristic (OC) Function of SPRT

The OC function in SPRT for testing $H_0 : \theta = \theta_0$ against $H_A : \theta = \theta_1$ is defined by

$$L(\theta) = \text{Probability of accepting } H_0 : \theta = \theta_0 \text{ when } \theta \text{ is the true value of the parameter.}$$

The power function is defined by

$$P(\theta) = \text{Probability of rejecting } H_0 \text{ when } \theta \text{ is the true value.}$$

Thus, we have $L(\theta) = 1 - P(\theta)$.

The OC function of SPRT for testing $H_0 : \theta = \theta_0$ against alternative hypothesis $H_A : \theta = \theta_1$, where θ is the parameter of $f(x, \theta)$ is given by

$$L(\theta) = \frac{A^{\psi(\theta)} - 1}{A^{\psi(\theta)} - B^{\psi(\theta)}} \tag{34.8}$$

where for each value of θ, the value of $\psi(\theta) \neq 0$ is to be determined so that

$$E\left[\frac{f(x, \theta_1)}{f(x, \theta_0)}\right]^{\psi(\theta)} = 1. \tag{34.9}$$

The constants A and B have already been defined.

Average Sample Number (ASN)

The sample size n in sequential testing is a random variable which can be determined in terms of the true density function $f(x, \theta)$. The average sample number function for SPRT for testing $H_0 : \theta = \theta_0$ against $H_A : \theta = \theta_1$, where θ is a parameter in $f(x, \theta)$ is given by

$$E(n) = \frac{L(\theta) \log B + [1 - L(\theta)] \log A}{E(z)}$$

where

$$z = \log\left[\frac{f(x, \theta_1)}{f(x, \theta_0)}\right], \quad A = \frac{1-\beta}{\alpha}, \quad B = \frac{\beta}{1-\alpha}.$$

►**EXAMPLE 34.20** *Let $x_1, x_2, \ldots, x_m$ be a random sample of m observations from $N(\mu, \sigma^2)$. Deduce SPRT statistic to test the significance of $H_0 : \mu = \mu_0$ against $H_A : \mu = \mu_1$. Also, find OC function and ASN function.*

Solution. We know

$$f(x) = \frac{1}{\sigma\sqrt{2\pi}} e^{-\frac{1}{2}\left(\frac{x-\mu}{\sigma}\right)^2}, \quad -\infty < x < \infty.$$

Assume that σ^2 is known. Then

$$\frac{f(x_i, \mu_1)}{f(x_i, \mu_0)} = \frac{\exp\left[-\frac{1}{2}\left(\frac{x_i-\mu_1}{\sigma}\right)^2\right]}{\exp\left[-\frac{1}{2}\left(\frac{x_i-\mu_0}{\sigma}\right)^2\right]} = \exp\left[-\frac{1}{2\sigma^2}\left\{(x_i-\mu_1)^2 - (x_i-\mu_0)^2\right\}\right]$$

$$= \exp\left[-\frac{1}{2\sigma^2}\left\{(\mu_0-\mu_1)(2x_i-\mu_0-\mu_1)\right\}\right].$$

$$\therefore\ z_i = \log\frac{f(x_i, \theta_1)}{f(x_i, \theta_0)} = \log\frac{f(x_i, \mu_1)}{f(x_i, \mu_0)} = \frac{\mu_1-\mu_0}{\sigma^2}\left[x_i - \frac{\mu_0+\mu_1}{2}\right]$$

$$\therefore\ \log\lambda_m = \sum_{i=1}^{m} z_i = \frac{\mu_1-\mu_0}{\sigma^2}\left[\Sigma x_i - \frac{m(\mu_0+\mu_1)}{2}\right].$$

Hence the SPRT for $H_0 : \mu = \mu_0$ against $H_A : \mu = \mu_1$ is given by

(i) Reject H_0 if

$$\frac{\mu_1 - \mu_0}{\sigma^2}\left[\Sigma x_i - \frac{m(\mu_0 + \mu_1)}{2}\right] \geq \log\left(\frac{1-\beta}{\alpha}\right)$$

$$\text{or, } \sum_{i=1}^{m} x_i \geq \frac{\sigma^2}{\mu_1 - \mu_0}\log\left(\frac{1-\beta}{\alpha}\right) + \frac{m(\mu_0 + \mu_1)}{2}; \ \mu_1 > \mu_0$$

(ii) Accept H_0 if

$$\frac{\mu_1 - \mu_0}{\sigma^2}\left[\Sigma x_i - \frac{m(\mu_0 + \mu_1)}{2}\right] \leq \log\left(\frac{\beta}{1-\alpha}\right)$$

$$\text{or, } \sum_{i=1}^{m} x_i \leq \frac{\sigma^2}{\mu_1 - \mu_0}\log\left(\frac{\beta}{1-\alpha}\right) + \frac{m(\mu_0 + \mu_1)}{2}$$

(iii) Continue taking additional observations as long as

$$\log\left(\frac{\beta}{1-\alpha}\right) < \frac{\mu_1 - \mu_0}{\sigma^2}\left[\Sigma x_i - \frac{m(\mu_0 + \mu_1)}{2}\right] < \log\left(\frac{1-\beta}{\alpha}\right)$$

$$\text{or, } \frac{\sigma^2}{\mu_1 - \mu_0}\log\left(\frac{1-\beta}{\alpha}\right) + \frac{m(\mu_0 + \mu_1)}{2} < \Sigma x_i < \frac{\sigma^2}{\mu_1 - \mu_0}\log\left(\frac{\beta}{1-\alpha}\right) + \frac{m(\mu_1 + \mu_0)}{2}.$$

OC Function

Let us find $\psi = \psi(\theta) \neq 0$. Using the equation (34.9), we have

$$\int_{-\infty}^{\infty}\left[\frac{f(x, \mu_1)}{f(x, \mu_0)}\right]^{\psi} f(x, \mu)dx = 1$$

$$\Rightarrow \frac{1}{\sigma\sqrt{2\pi}}\int_{-\infty}^{\infty}\exp\left[-\frac{1}{2}\left(\frac{x-\mu}{\sigma}\right)^2\right]\left[\exp\left\{-\frac{1}{2\sigma^2}(\mu_1 - \mu_0)(-2x + \mu_0 + \mu_1)\right\}\right]^{\psi} dx = 1$$

$$\text{or, } \frac{1}{\sigma\sqrt{2\pi}}\int_{-\infty}^{\infty}\exp\left[-\frac{1}{2\sigma^2}\{x^2 - 2x(\mu_1 - \mu_0)h + \theta) + \mu^2 + (\mu_1^2 - \mu_0^2)\psi\}\right] dx = 1. \quad (34.10)$$

Let $\lambda = (\mu_1 - \mu_0)\psi + \mu$, $\lambda^2 = (\mu_1^2 - \mu_0^2)\psi + \mu$. Then LHS becomes

$$\frac{1}{\sigma\sqrt{2\pi}}\int_{-\infty}^{\infty}\exp\left[-\frac{1}{2\sigma^2}(x-\lambda)^2\right]dx$$

and this integration gives 1. Hence λ and λ^2 are the solutions of the (34.10). Thus $\psi = \psi(\theta)$ is the solution of (34.10) and is given by

$$(\mu_1^2 - \mu_0^2)\psi + \mu^2 = [(\mu_1 - \mu_0)\psi + \mu]^2$$

$$\text{or, } (\mu_1^2 - \mu_0^2)\psi = (\mu_1 - \mu_0)^2\psi^2 + 2\mu(\mu_1 - \mu_0)\psi.$$

Since $\psi = \psi(\theta) \neq 0$ and $\mu_1 \neq \mu_0$, on dividing throughout by $(\mu_1 - \mu_0)$, we get

$$(\mu_1 + \mu_0) = (\mu_1 - \mu_0)\psi + 2\theta \Rightarrow \psi(\theta) = \frac{\mu_1 + \mu_0 - 2\mu}{\mu_1 - \mu_0}.$$

Substituting for $\psi(\theta)$ in (34.9) we get the required term for OC function.

ASN Function

We have

$$z = \log \frac{f(x,\mu_1)}{f(x,\mu_0)} = \frac{\mu_1 - \mu_0}{\sigma^2}\left[x - \frac{\mu_0 + \mu_1}{2}\right].$$

$$\begin{aligned}\therefore\ E(z) &= \frac{\mu_1 - \mu_0}{2\sigma^2}[2E(x) - \mu_0 - \mu_1]\\ &= \frac{\mu_1 - \mu_0}{2\sigma^2}[2\mu - \mu_0 - \mu_1].\end{aligned}$$

Substituting this in (34.10), we get the required ASN function.

34.17 Application of Different Tests

34.17.1 Tests Related to Means

►**EXAMPLE 34.21** *The rainfalls (in mm) in different randomly selected days in a rainy season are given below:*

Rainfall, x_i (in mm) : 4.5, 8.6, 10.2, 4.2, 1.5, 15.6, 24.7, 30.8, 10.6.

If it is assumed that (i) $x_i \sim N(\mu, \sigma^2)$, (ii) $x_i \sim N(\mu, 16)$, *then test the hypothesis*

$$H_0 : \mu = 15.0 \textit{ against } H_A : \mu \neq 15.$$

Solution. We have $n = 9$, $\bar{x} = 12.31$, $s^2 = 96.06$.

(i) Since σ^2 is unknown and n is small (< 30), the test statistic is

$$t = \frac{\bar{x} - \mu}{s/\sqrt{n}} = \frac{12.31 - 15.0}{9.80/3} = -0.82.$$

Since $|t| < t_{0.025,8} = 2.306$, H_0 is accepted.

(ii) Since $\sigma^2 = 16$ is known, the test statistic is

$$z = \frac{\bar{x} - \mu}{\sigma/\sqrt{n}} = \frac{12.31 - 15.0}{4/3} = -2.02.$$

Since $|z| > 1.96$, H_0 is rejected.

►**EXAMPLE 34.22** *The birth-weights (in kg) of randomly selected babies selected from one urban hospital and from one rural hospital are given below:*

Hospital	*Birth-weight of baby*
Urban, x_{1j}	2.8, 3.0, 3.4, 2.6, 2.7, 3.2, 3.0, 3.1, 2.9, 2.9, 3.0
Rural, x_{2j}	2.4, 2.2, 2.4, 2.6, 2.7, 3.0, 3.1, 2.2, 2.6, 2.7, 2.8, 2.5

Assume that

(i) $x_{1j} \sim N(\mu_1, \sigma_1^2)$; $x_{2j} \sim N(\mu_2, \sigma_2^2)$

(ii) $x_{1j} \sim N(\mu_1, 0.36)$; $x_{2j} \sim N(\mu_2, 0.80)$.

Test the significance $H_0 : \mu_1 = \mu_2$ *against* $H_A : \mu_1 \neq \mu_2$.

Solution.

(i) Given

$$n_1 = 11,\ n_2 = 12,\ \bar{x}_1 = 2.96,\ \bar{x}_2 = 2.60,\ s_1^2 = 0.0505,\ s_2^2 = 0.08.$$

Since σ_1^2 and σ_2^2 are unknown and $n_1 < 30$, $n_2 < 30$, let us assume that $\sigma_1^2 = \sigma_2^2$. Then the test statistic is

$$t = \frac{\bar{x}_1 - \bar{x}_2}{\sqrt{s^2\left(\frac{1}{n_1} + \frac{1}{n_2}\right)}}, \text{ where } s^2 = \frac{(n_1-1)s_1^2 + (n_2-1)s_2^2}{n_1+n_2-2} = 0.066$$

$$= \frac{2.96 - 2.60}{\sqrt{0.066\left(\frac{1}{11} + \frac{1}{12}\right)}} = 3.36.$$

Since $|t| > t_{0.05;21} = 2.069$, H_0 is rejected.

(ii) Since σ_1^2 and σ_2^2 are known, the test statistic to be used is

$$z = \frac{\bar{x}_1 - \bar{x}_2}{\sqrt{\frac{\sigma_1^2}{n_1} + \frac{\sigma_2^2}{n_2}}} = \frac{2.96 - 2.60}{\sqrt{\frac{0.30}{11} + \frac{0.80}{12}}} = 1.14.$$

Since $|z| < 1.96$, H_0 is accepted.

►**EXAMPLE 34.23** *The following data are the random sample of observations recorded from an industry producing juice packets per hour in different days:*

x_i : 50, 55, 62, 67, 45, 68, 70, 62, 73, 64, 75, 55, 50, 68, 64, 60, 66, 60, 56, 59, 60, 60, 63, 67, 66, 68, 70, 65, 54, 55, 70, 66, 67.

Under the assumption $x_i \sim N(\mu, \sigma^2)$, test the significance of

$$H_0 : \mu = 60 \text{ against } H_A : \mu \neq \mu_0.$$

Solution. We have

$$n = 33,\ \bar{x} = 60.51,\ s^2 = \frac{1}{n-1}\Sigma(x_i - \bar{x})^2 = 291.23.$$

Since n is large (> 30), the test statistic is

$$z = \frac{\bar{x} - \mu_0}{s/\sqrt{n}} = \frac{60.51 - 60.0}{2.9707} = 0.17.$$

Since $|z| < 1.96$, H_0 is accepted.

►**EXAMPLE 34.24** *The marks of two groups of students in a Computer Science Course. (marks out of 100) are given below:*

Group–1, x_{1j} : 60, 88, 72, 75, 90, 85, 66, 40, 42, 36, 55, 64, 77, 80, 86, 92, 85, 87, 83, 70, 60, 65, 76, 77, 80, 84, 86, 87, 94, 82, 70

Group–2, x_{2j} : 40, 38, 66, 86, 94, 97, 98, 66, 76, 78, 80, 84, 85, 88, 50, 55, 58, 60, 66, 74, 70, 70, 70, 70, 66, 68, 66, 60, 60, 64, 66, 72, 80, 82

Do you think that the performances of two groups of students are similar?

Solution. Assume that

$$x_{1j} \sim N(\mu_1,\ \sigma_1^2),\ x_{2j} \sim N(\mu_2,\ \sigma_2^2).$$

We need to test the significance of

$$H_0 : \mu_1 = \mu_2 \text{ against } H_A : \mu_1 \neq \mu_2.$$

We have $n_1 = 31,\ n_2 = 34,\ \bar{x}_1 = 74.0,\ \bar{x}_2 = 70.68,$

$$s_1^2 = \frac{1}{n_1 - 1}\sum(x_{1j} - \bar{x}_1)^2 = 234.73,$$

$$s_2^2 = \frac{1}{n_1 - 1}\sum(x_{2j} - \bar{x}_2)^2 = 202.04.$$

Here σ_1^2 and σ_2^2 are not known but the sample sizes are large $[n_1 > 30, n_2 > 30]$. We can apply z-test, where

$$z = \frac{\bar{x}_1 - \bar{x}_2}{\sqrt{\frac{s_1^2}{n_1} + \frac{s_2^2}{n_2}}} = \frac{74.0 - 70.68}{\sqrt{\frac{234.73}{31} + \frac{202.04}{34}}} = 0.90.$$

Since $|z| < 1.96$, H_0 is accepted. The performances of two groups of students are similar.

34.17.2 Test of Equality of Two Correlated Means

Let x_{1j} $(j = 1, 2, \ldots, n)$ be n observations recorded from a randomly selected group of sample points on an occasion. The same group of sample points are also investigated on a second occasion and the values of the same variable are recorded. The recorded values are x_{2j}. Since x_{1j} and x_{2j} are observed from the same sample, these are assumed to be correlated. It is assumed that $x_{1j} \sim N(\mu_1, \sigma^2)$ and $x_{2j} \sim N(\mu_2, \sigma^2)$. The problem is to test the significance of $H_0 : \mu_1 = \mu_2$ against $H_A : \mu_1 \neq \mu_2$.

The test statistic is

$$t = \frac{\bar{d}}{\sqrt{s_d^2/n}}, \text{ where } d_j = x_{1j} - x_{2j};\ j = 1, 2, \ldots, n$$

$$\bar{d} = \tfrac{1}{2}\Sigma d_j$$

$$s_d^2 = \frac{1}{n-1}\left[\Sigma d_j^2 - \frac{(\Sigma d_j)^2}{n}\right]$$

This 't' is called paired 't' and it is distributed as Student's 't' with $(n-1)$ d.f.

►**EXAMPLE 34.25** *To study the impact of supracide on heart rate of a variety of slug, the slugs are kept in supracide for a week. The heart rate of slugs before (x_{1j}, heart-beat/minute) and after (x_{2j}, heart-beat/minute) keeping the slugs in supracide are given below:*

x_{1j}	10	8	12	9	9	7	8	9	10
x_{2j}	7	6	8	7	7	6	7	8	8

Do you think that supracide is creating problem to the slugs?

Solution. We have $d_j = x_{1j} - x_{2j}$: 3, 2, 4, 2, 2, 1, 1, 1, 2; $n = 9$, $\bar{d} = 2$, $s_d^2 = 1$.

$$\therefore\ t = \frac{\bar{d}}{\sqrt{s_d^2/n}} = \frac{2}{\sqrt{1/9}} = 6.00.$$

Since $|t| > t_{0.025,8} = 2.262$, H_0 is rejected. The heart-beats are reduced due to supracide.

34.17.3 Test of Proportion

Let there be N units in a population and A of them possess a characteristic. Then the proportion of units possessing the characteristic is $P = A/N$. The problem is to test the significance of $H_0 : P = P_0$ against $H_A : P \neq P_0$.

Consider that n units from this population are randomly selected and on investigation it is found that 'a' of them possess the characteristic under study. The sample proportion of units possessing the characteristic is $p = a/n$. Then the test statistic to test the hypothesis is

$$z = \frac{p - P_0}{\sqrt{\frac{P_0 Q_0}{n}}} \sim N(0, 1), \text{ where } Q_0 = 1 - P_0.$$

If $|z| \geq 1.96$, H_0 is rejected at 5% level of significance.

►**EXAMPLE 34.26** *In an area* 70% *people are assumed to be literate. Fifty people of the area are randomly selected and found that* 15 *people are literate. Do the data conform with the assumption?*

Solution. Let P be the proportion of literate people in the study area. We need to test $H_0 : P = P_0 = 0.70$ against $H_A : P \neq P_0$. From sample, we have

$$p = \frac{a}{n} = \frac{15}{50} = 0.30.$$

The test statistic is

$$z = \frac{p - P_0}{\sqrt{\frac{P_0 Q_0}{n}}} = \frac{0.30 - 0.70}{\sqrt{\frac{0.70 \times 0.30}{50}}} = -6.17.$$

Since $|z| > 1.96$, H_0 is rejected at 5% level of significance.

34.17.4 Test of Equality of Two Proportions

Let there be N_1 units in a population and A_1 of them possess a certain characteristic. The proportion of population units possessing the characteristic is $P_1 = A_1/N_1$. Let another population be of size N_2 and A_2 of them possess the same characteristic under study. The proportion of units possessing the characteristic in the second population is $P_2 = A_2/N_2$. The problem is to test the significance of null hypothesis $H_0 : P_1 = P_2$ against $H_A : P_1 \neq P_2$.

Consider that n_1 units from first population and n_2 units from second population are randomly selected. Out of n_1 units a_1 units possess the characteristic and out of

n_2 units a_2 units possess the characteristic under study. The sample proportions of units possessing the characteristic are

$$p_1 = \frac{a_1}{n_1} \text{ and } p_2 = \frac{a_2}{n_2}.$$

The test statistic for the hypothesis is

$$z = \frac{p_1 - p_2}{\sqrt{PQ\left(\frac{1}{n_1} + \frac{1}{n_2}\right)}} \sim N(0,1)$$

where

$$P = \frac{a_1 + a_2}{n_1 + n_2} \text{ and } Q = 1 - P.$$

If $|z| \geq 1.96$, H_0 is rejected at 5% level of significance.

►**EXAMPLE 34.27** *Out of 100 students in a course 15 students got A grade and out of 75 students in another course 8 students got A. Do you think that the proportions of students of getting grade A in two courses are same?*

Solution. Let

$$P_1 = \text{proportion of students getting grade } A \text{ in Course-1},$$
$$P_2 = \text{proportion of students getting grade } A \text{ in Course-2}.$$

We need to test the hypothesis $H_0 : P_1 = P_2$ against $H_A : P_1 \neq P_2$. The sample proportions are

$$p_1 = \frac{15}{100} = 0.15, \; p_2 = \frac{8}{75} = 0.11$$
$$p = \frac{a_1 + a_2}{n_1 + n_2} = \frac{15 + 8}{100 + 75} = 0.13, \; q = 1 - P = 0.87.$$

Therefore, the test statistic is

$$z = \frac{p_1 - p_2}{\sqrt{PQ\left(\frac{1}{n_1} + \frac{1}{n_2}\right)}} = \frac{0.15 - 0.11}{\sqrt{0.13 \times 0.87\left(\frac{1}{100} + \frac{1}{75}\right)}} = 0.78.$$

Since $|z| < 1.96$, H_0 is accepted.

►**EXAMPLE 34.28** *A group of people were vaccinated against Hepatitis B virus. After sometime 100 of them were randomly selected and found that 25 of selected people were affected by Hepatitis B. Another group of 200 people were randomly selected who were not vaccinated against Hepatitis B. Among them 80 were found affected by Hepatitis B. Do you think that the vaccination is effective against Hepatitis B?*

Solution. Here $n_1 = 100$, $a_1 = 25$, $n_2 = 200$, $a_2 = 80$

$$p_1 = \frac{a_1}{n_1} = 0.25, \; p_2 = \frac{a_2}{n_2} = \frac{80}{200} = 0.40$$
$$p = \frac{a_1 + a_2}{n_1 + n_2} = \frac{25 + 80}{100 + 200} = 0.35, \; q = 0.65.$$

Let P_1 = proportion of affected people who were vaccinated,
P_2 = proportion of affected people who were not vaccinated.

We need to test $H_0 : P_1 = P_2$ against $H_A : P_1 \neq P_2$. The test statistic is

$$z = \frac{p_1 - p_2}{\sqrt{pq\left(\frac{1}{n_1} + \frac{1}{n_2}\right)}} = \frac{0.25 - 0.40}{\sqrt{0.35 \times 0.65\left(\frac{1}{100} + \frac{1}{200}\right)}} = -2.57.$$

Since $|z| > 1.96$, H_0 is rejected. Therefore, vaccination is effective against Hepatitis B since $p_1 < p_2$.

To test the significance of $H_0 : P_1 = P_2$ against $H_A : P_1 \neq P_2$, where P_1 and P_2 are proportions of population units possessing a particular characteristic, the test statistic is

$$z = \frac{p_1 - p_2}{\sqrt{pq\left(\frac{1}{n_1} + \frac{1}{n_2}\right)}}.$$

Here under $H_0 : P_1 = P_2 = P$ (say), where estimate of P is

$$\hat{P} = \frac{a_1 + a_2}{n_1 + n_2} \text{ and } \hat{Q} = 1 - \hat{P}.$$

This estimate is unbiased, since

$$\begin{aligned} E(\hat{p}) &= \left[\frac{a_1 + a_2}{n_1 + n_2}\right] = E\left[\frac{n_1p_1 + n_2p_2}{n_1 + n_2}\right] = \frac{1}{n_1 + n_2}[n_1E(p_1) + n_2E(p_2)] \\ &= \frac{1}{n_1 + n_2}[n_1P_1 + n_2P_2] = P \; [\because \text{ under } H_0 : P_1 = P_2 = P]. \end{aligned}$$

In two sample cases, we may be interested to test the significance of the difference between p_1 and p, where $p = \dfrac{n_1p_1 + n_2p_2}{n_1 + n_2}$ is a pooled estimate of population proportion using both the samples. Here $V(p_1 - p) = V(p_1) + V(p) - 2\,\text{cov}(p_1, p)$.

Now,

$$\begin{aligned} \text{cov}(p_1, p) &= E[p_1 - E(p_1)][p - E(p)] \\ &= E\left[\{p_1 - E(p_1)\}\frac{1}{n_1 + n_2}\{(n_1p_1 + n_2p_2) - E(n_1p_1 + n_2p_2)\}\right] \\ &= \frac{1}{n_1 + n_2}E[\{p_1 - E(p_1)\}\{n_1(p_1 - E(p_1)) + n_2(p_2 - E(p_2))\}] \\ &= \frac{1}{n_1 + n_2}[n_1E\{p_1 - E(p_1)\}^2 + n_2E\{(p_1 - E(p_1))(p_2 - E(p_2))\}] \\ &= \frac{1}{n_1 + n_2}[n_1V(p_1) + n_2\,\text{cov}(p_1, p_2)] \\ &= \frac{1}{n_1 + n_2}n_1V(p_1), \because \text{cov}(p_1, p_2) = 0 \\ &= \frac{n_1}{n_1 + n_2}\frac{pq}{n_1} = \frac{pq}{n_1 + n_2}. \end{aligned}$$

Thus, the test statistic to test the significance of $H_0 : p_1 = p$ against $H_A : p_1 \neq p$ is

$$z = \frac{p_1 - p}{\sqrt{V(p_1 - p)}} \sim N(0,1).$$

Here

$$\begin{aligned} V(p_1 - p) &= V(p_1) + V(p) - 2\operatorname{cov}(p_1, p) \\ &= \frac{pq}{n_1} + \frac{pq}{n_1 + n_2} - \frac{2pq}{n_1 + n_2} = pq\left[\frac{n_2}{n_1(n_1 + n_2)}\right]. \end{aligned}$$

Thus

$$z = \frac{p_1 - p}{\sqrt{\dfrac{n_2}{n_1 + n_2}\dfrac{pq}{n_1}}} \sim N(0,1).$$

Sometimes sample proportions are not found out but on the basis of samples the equality of two population proportions is to be verified. For example, let us consider that the proportion of literate people in a state is 0.68 and that in another state is 0.42. If from former state N_1 observations ($N_1 = 1000$) and from latter state $N_2 = 500$ people are selected, would we expect that among the selected two groups of people the proportions will be true? Here clearly $P_1 \neq P_2$. The difference in proportions is hidden in the sample. The objective is to test the null hypothesis $H_0 : P_1 = P_2$ against $H_A : P_1 \neq P_2$.

The test statistic is

$$z = \frac{P_1 - P_2}{\sqrt{\dfrac{P_1 Q_1}{n_1} + \dfrac{P_2 Q_2}{n_2}}} \sim N(0,1).$$

In our given example $P_1 = 0.68, P_2 = 0.42, n_1 = 1000, n_2 = 500$. Hence

$$z = \frac{0.68 - 0.42}{\sqrt{\dfrac{0.68 \times 0.32}{1000} + \dfrac{0.42 \times 0.58}{500}}} = 9.79.$$

Since $|z| > 1.96$, H_0 is rejected at 5% level of significance. The difference in proportions of literate people may be observed also in samples.

►**EXAMPLE 34.29** *In a random sample of* 200 *couples of childbearing ages, it is found that* 120 *have adopted family planning method. In another random sample of* 300 *married couples of childbearing ages in rural area, it is found that* 150 *have adopted family planning method. Do you think that the proportion of adopter couples in rural area is significantly lower than the proportion of adopter couples in two samples together?*

Solution. Let p_1 = proportion of adopter couples in rural area. The estimate of proportion of adopter couples in two samples together is p, where

$$p = \frac{n_1 p_1 + n_2 p_2}{n_1 + n_2}$$

where

$$p_1 = \frac{a_1}{n_1} = \frac{150}{300} = 0.5,\ p_2 = \frac{a_2}{n_2} = \frac{120}{200} = 0.6.$$

We need to test the hypothesis

$$H_0 : p_1 = p \text{ against } H_A : p_1 \neq p.$$

The test statistic is

$$z = \frac{p_1 - p}{\sqrt{\frac{n_2}{n_1 + n_2}\frac{pq}{n_1}}} \sim N(0,1).$$

Here

$$p = \frac{n_1 p_1 + n_2 p_2}{n_1 + n_2} = \frac{300 \times 0.5 + 200 \times 0.6}{300 + 200} = 0.54,\ q = 0.46.$$

$$\therefore\ z = \frac{0.50 - 0.54}{\sqrt{\frac{200}{300+200}\frac{0.54 \times 0.46}{300}}} = -2.20.$$

Since $|z| > 1.96$, H_0 is rejected, i.e., the proportion of rural adopter couples is lower than the overall proportion of adopter couples.

►**EXAMPLE 34.30** *In a month there are 125 births in hospital A, of which 55 are affected by jaundice, while in two consecutive months the proportion of jaundice-affected babies among total of 260 babies is 0.35. Is there any significant difference in the proportions of jaundice-affected babies in two months?*

Solution. We have

$$n_1 = 125,\ n_1 + n_2 = 260,\ n_2 = 260 - 125 = 135$$

$$p_1 = \frac{a_1}{n_1} = \frac{55}{125} = 0.44.$$

Given

$$p = \frac{n_1 p_1 + n_2 p_2}{n_1 + n_2} = 0.35 \text{ or, } 125 \times 0.44 + 135 p_2 = 0.35 \times 260 \text{ or, } p_2 = 0.27.$$

Therefore, $q = 0.65$. We need to test the significance of

$$H_0 : P_1 = P_2 \text{ against } H_A : P_1 \neq P_2.$$

The test statistic is

$$z = \frac{p_1 - p_2}{\sqrt{PQ\left(\frac{1}{n_1} + \frac{1}{n_2}\right)}} \sim N(0,1)$$ where P and Q are p and q respectively under H_0

$$= \frac{0.44 - 0.27}{\sqrt{0.35 \times 0.65\left(\frac{1}{125} + \frac{1}{135}\right)}} = 2.87.$$

Since $|z| > 1.96$, H_0 is rejected.

34.17.5 Test of Equality of Several Proportions

Let k samples be selected from k populations, where i-th sample has O_i $(i = 1, 2, \dots, k)$ observations. Total number of sample observations are

$$N = \sum_{i=1}^{k} O_i.$$

The proportion of sample observations of i-th sample is

$$P_i = \frac{O_i}{N}, i = 1, 2, \dots, k.$$

The problem is to test the significance of null hypothesis $H_0 : P_1 = P_2 = \dots = P_k$ against H_A : at least one of the equality does not hold good.

Under H_0 the expected number of observations (frequencies) is

$$E_i = \frac{N}{k}, \text{ where } \sum_{i=1}^{k} E_i = N = \sum_{i=1}^{k} O_i.$$

If H_0 is true, there should be insignificant deviation between O_i and E_i. Hence the difference $(O_i - E_i)$, $i = 1, 2, \dots, k$ can be used for test statistic. Such a test statistic is

$$\chi^2 = \sum_{i=1}^{k} \frac{(O_i - E_i)^2}{E_i} = \sum_{i=1}^{k} \frac{O_i^2}{E_i} - N.$$

This χ^2 is distributed as chi-square with $(k - 1)$ d.f., provided no $E_i < 5$. If any $E_i < 5$, it is suggested that the corresponding E_i should be added to the nearest class. However, Koehler and Larntz (1980) have shown that, if any E_i is not less than 0.25, the χ^2-statistic provides satisfactory result. Now, any calculated χ^2 is greater than or equal to $\chi^2_{0.05,k-1}$; H_0 is rejected.

►**EXAMPLE 34.31** *A sample of* 200 *tribal people is selected and the people are classified according to different sub-tribes. The number of people in sub-tribes are shown below:*

	People in sub-tribes					*Total*
Number	*1*	*2*	*3*	*4*	*5*	
of people	20	45	18	36	81	200

Do you think that the proportions of people in different sub-tribes are homogeneous?

Solution. Let P_i $(i = 1, 2, 3, 4, 5)$ be the proportion of i-th sub-tribe. We need to test

$$H_0 : P_1 = P_2 = P_3 = P_4 = P_5 \text{ against } H_A : P_1 \neq P_2 \neq P_3 \neq P_4 \neq P_5.$$

Under H_0, $E_i = \frac{N}{k} = \frac{200}{5} = 40$. The test statistic is

$$\chi^2 = \sum_{i=1}^{N} \frac{O_i^2}{E_i} - N = \frac{1}{40}(20^2 + 45^2 + 18^2 + 36^2 + 81^2) - 200 = 65.15.$$

Since $\chi^2 > \chi^2_{0.05,4} = 9.49$, H_0 is rejected.

►**EXAMPLE 34.32** *A student of Botany has collected* 60 *leaves from a forest and classified these into* 4 *classes. The number of leaves of different classes are shown below:*

	Class of leaves				*Total*
Number	*1*	*2*	*3*	*4*	
of leaves	27	4	8	21	60

Do you think that the collected leaves are in the ratio

$$1:2:3:4::4:1:2:3?$$

Solution. We need to test the hypothesis $H_0 : P_1 : P_2 : P_3 : P_4$ against H_A : proportions do not follow the given ratio. Under H_0, the expected frequencies E_i are

$$E_1 = \frac{4\times N}{10} = 24,\ E_2 = \frac{1\times N}{10} = 6$$
$$E_3 = \frac{2\times N}{10} = 12,\ E_4 = \frac{3\times N}{10} = 18.$$

Therefore, $\chi^2 = \sum \frac{O_i^2}{E_i} - N = 2.875.$

Since $\chi^2 < \chi^2_{0.05,3} = 7.81$, H_0 is accepted.

34.17.6 Test of Independence

Let there be a sample of n observations, where sample observations are classified into r classes and observations of a particular class are again be classified into c classes according to two characters A and B respectively. Assume that O_{ij} is the number of observation of j-th level of B corresponding to i-th level of A $(i = 1, 2, \ldots, r;\ j = 1, 2, \ldots, c)$ such that

$$n = \sum_i^r \sum_j^c O_{ij}.$$

The O_{ij} observations can be represented in an $r \times c$ table as follows:

$r \times c$ contingency table

A \ B	B_1	B_2	$\cdots$	B_j	$\cdots$	B_c	*Total* $O_{i\cdot}$
A_1	O_{11}	O_{12}	$\cdots$	O_{1j}	$\cdots$	O_{1c}	$O_{1\cdot}$
A_2	O_{21}	O_{22}	$\cdots$	O_{2j}	$\cdots$	O_{2c}	$O_{2\cdot}$
$\vdots$	$\vdots$	$\vdots$	$\vdots$	$\vdots$	$\vdots$	$\vdots$	$\vdots$
A_i	O_{i1}	O_{i2}	$\cdots$	O_{ij}	$\cdots$	O_{ic}	$O_{i\cdot}$
$\vdots$	$\vdots$	$\vdots$	$\vdots$	$\vdots$	$\vdots$	$\vdots$	$\vdots$
A_r	O_{r1}	O_{r2}	$\cdots$	O_{rj}	$\cdots$	O_{rc}	$O_{r\cdot}$
Total $O_{\cdot j}$	$O_{\cdot 1}$	$O_{\cdot 2}$	$\cdots$	$O_{\cdot j}$	$\cdots$	$O_{\cdot c}$	n

The presentation of total investigated observations into a table containing r rows and c columns as shown above is called $(r \times c)$ *contingency table.* For example, let us consider the classification of 100 diabetic patients according to their heart problem as follows:

Level of diabetes, A	*Level of heart problem, B*		*Total*
	Not serious	*Serious*	
Serious	$10 = a$	$50 = b$	$60 = a + b$
Not serious	$28 = c$	$12 = d$	$40 = c + d$
Total	$38 = a + c$	$62 = b + d$	$100 = n$

One of the problems of analysis of such data is to test the independence of two characters A and B, i.e., to check whether serious heart problem is associated with serious diabetes level. The null hypothesis to be tested is

$$H_0 : \text{the two attributes } A \text{ and } B \text{ are independent}$$

$$\text{against } H_A : \text{the two attributes are not independent.}$$

Under H_0,

$$\frac{a}{a+c} = \frac{b}{b+d} = \frac{a+b}{n} = \frac{O_{1\cdot}}{n}$$

$$\frac{c}{a+c} = \frac{d}{b+d} = \frac{c+d}{n} = \frac{O_{2\cdot}}{n}$$

where $O_{11} = a, O_{12} = b, O_{21} = c$ and $O_{22} = d$ are called *cell frequencies*; $O_{i\cdot}$ and $O_{\cdot j}$ are called *marginal total frequencies* and n is the *grand total frequency*; O_{ij} are called *observed frequency* of j-th level of B corresponding to i-th level of A. The frequencies under H_0 are known as *expected frequencies* and is denoted by E_{ij}, where

$$E_{ij} = \frac{O_{i\cdot} \times O_{\cdot j}}{n}; \; i = 1, 2, \ldots, r; \; j = 1, 2, \ldots, c.$$

For a (2×2) contingency table E_{ij}'s are

$$E_{11} = \frac{(a+c)(a+b)}{n}, \; E_{12} = \frac{(b+d)(a+b)}{n},$$

$$E_{21} = \frac{(a+c)(c+d)}{n} \text{ and } E_{22} = \frac{(b+d)(c+d)}{n}.$$

If the null hypothesis is true, the deviation of O_{ij} and E_{ij} may be used as a criterion of test statistic since $(O_{ij} - E_{ij}) = 0$ indicates that the hypothesis is perfectly true.

Assumption for Test

(i) The attributes are independent.

(ii) Marginal total frequencies and grand total frequencies are fixed.

The second assumption indicates that $(r - 1)(c - 1)$ of E_{ij} can be calculated independently using

$$E_{ij} = \frac{O_{i\cdot} \times O_{\cdot j}}{n}.$$

This means that, in any row $(c-1)$ values of E_{ij} can be calculated by the formula. The last one is

$$O_{i.} - \sum_{j=1}^{c-1} E_{ij};\ i = 1, 2, \ldots, r.$$

Similar is the case in columns. For j-th column the expected frequency of last column is

$$O_{.j} - \sum_{i=1}^{r-1} E_{ij}.$$

The test statistic is

$$\chi^2 = \sum_{i=1}^{r}\sum_{j=1}^{c} \frac{(O_{ij} - E_{ij})^2}{E_{ij}} = \sum_{i}^{r}\sum_{j}^{c} \frac{O_{ij}^2}{E_{ij}} - n.$$

This χ^2 is distributed as chi-square with $(r-1)(c-1)$ d.f., provided no $E_{ij} < 5$.

In practice any one or more E_{ij} values can be less than 5. In that case the expected frequencies which are less than 5 are added to the nearest cell so that all $E_{ij} > 5$. The d.f. of χ^2 in such a case is $(r_1 - 1)(c_1 - 1)$, where r_1 and c_1 are reduced number of rows and columns, respectively.

For 2×2 contingency table, the χ^2-statistic takes the form

$$\chi^2 = \frac{n(ad - bc)^2}{(a+b)(a+c)(c+d)(b+d)}$$

and this χ^2 has 1 d.f.

The χ^2 distribution is a continuous distribution but is used for the case of discrete frequencies. Hence the test is approximate one. However, the approximation is good enough, if $n \geq 40$. For $n < 40$ and if any $E_{ij} < 5$ in a 2×2 contingency table, Yate proposed a correction for better approximation of chi-square distribution. The χ^2-statistic under Yate's correction for (2×2) contingency table is

$$\chi^2 = \frac{n\left[|ad - bc| - \frac{n}{2}\right]^2}{(a+b)(a+c)(c+d)(b+d)}.$$

34.17.7 Pirie and Hamdan Test

A modified test have been proposed by Pirie and Hamdan (1972) in case of (2×2) contingency table when total frequency is smaller. According to them

$$\chi^2 = \frac{n\left[|ad - bc| - \frac{n}{4}\right]^2}{(a+b)(a+c)(c+d)(b+d)}.$$

The assumption is that the marginal total frequencies are fixed. If the marginal totals are random, then the related χ^2-statistic is

$$\chi^2 = \frac{n\left[|ad - bc| - \frac{1}{2}\right]^2}{(a+b)(a+c)(c+d)(b+d)}.$$

The above test statistic provides suitable approximation of chi-square distribution when the sample size is smaller ($n < 40$). When $n \leq 20$, Fisher's exact test is used to test the independence of two attributes. However, Conover (1999) has proposed a modified test which can be used instead of Fisher's exact test. The modified test statistic is

$$z = \frac{a - \dfrac{(a+b)(a+c)}{n}}{\left[\dfrac{(a+b)(a+c) - (c+d)(b+d)}{n^2(n-1)}\right]^{\frac{1}{2}}} \sim N(0,1).$$

Therefore, if $|z| \geq 1.96$, H_0 of independence of two attributes is rejected at 5% level of significance.

►**EXAMPLE 34.33** *A randomly selected group of 30 students have been classified according to their results of School Final Examinaton and Higher Secondary Examination. The students in different groups are shown below:*

Result of School Final Examination (in grade)	*Result of Higher Secondary Examination (in grade)*		*Total*
	A	*Not A*	
A	5	7	12
Not A	3	15	18
Total	8	22	30

Are the results of two examinations significantly associated?

Solution. Here $n = 30 < 40$, $E_{11} = 3.2 < 5$. Therefore, one should not use usual χ^2-statistic. The χ^2-statistic with Yate's correction becomes

$$\chi^2 = \frac{n\left[|ad - bc| - \frac{n}{2}\right]^2}{(a+b)(a+c)(c+d)(b+d)} = \frac{30[|75-21| - 15]^2}{12 \times 8 \times 18 \times 22} = 1.20.$$

Since $\chi^2 < \chi^2_{0.05,1} = 3.84$, H_0 is accepted. Here

$$H_0 : \text{the two results are independent}$$
$$\text{against } H_A : \text{the two results are associated.}$$

According to Pirie and Hamdan test, the test statistic is

$$\chi^2 = \frac{n\left[|ad - bc| - \frac{n}{4}\right]^2}{(a+b)(a+c)(c+d)(b+d)} = \frac{30[|75-21| - 7.5]^2}{12 \times 8 \times 18 \times 22} = 1.71.$$

This test also is in favour of accepting the null hypothesis, since

$$\chi^2 < \chi^2_{0\cdot 05,1} = 3.84.$$

►**EXAMPLE 34.34** *A group of 15 newborn babies are classified according to gestation period and presence of Hepatitis B virus in their blood. The number of babies in* *each class are shown below:*

Gestation Period (in days)	*Presence of Hepatitis B virus*		*Total*
	Yes	*No*	
270	1	4	5
< 270	3	7	10
Total	4	11	15

Do you think that the babies who born before normal gestation period are more affected by Hepatitis B?

Solution. Here $n = 15 < 20$. We should use Fisher's exact test to test the presence of independence of two attributes. However, instead of using Fisher's exact test, we can use the test statistic proposed by Conover (1999), where the test statistic is

$$z = \frac{a - \frac{(a+b)(a+c)}{n}}{\left[\frac{(a+b)(a+c)(b+d)(c+d)}{n^2(n-1)}\right]^{\frac{1}{2}}} \sim N(0,1) = \frac{1 - \frac{5\times 4}{15}}{\left[\frac{5\times 4\times 10\times 11}{(15)^2(15-1)}\right]^{\frac{1}{2}}} = -0.40.$$

Since $|z| < 1.96$, H_0 is accepted. Here the null hypothesis is

H_0 : the presence of Hepatitis B virus does not depend on gestation period

H_A : Hepatitis B virus and gestation period are associated.

34.17.8 Mantel-Haenszel Test

This test is also used to test the independence of two attributes but the independence of attributes is observed from several samples. Consider that there are k samples, where each set of sample observations are classified into (2×2) contingency table. The sample observations of l-th sample are arranged into (2×2) contingency table as follows:

First sample				*Second sample*				*l-th sample*			
$A \backslash B$	B	*Not B*	*Total*	$A \backslash B$	B	*Not B*	*Total*	$A \backslash B$	B	*Not B*	*Total*
A	a_1	b_1	a_1+b_1	A	a_2	b_2	a_2+b_2	A	a_l	b_l	a_l+b_l
Not A	c_1	d_1	c_1+d_1	*Not A*	c_2	d_2	c_2+d_2	*Not A*	c_l	d_l	c_l+d_l
Total	a_1+c_1	b_1+d_1	n_1	*Total*	a_2+c_2	b_2+d_2	n_2	*Total*	a_l+c_l	b_l+d_l	n_l

$l = 1, 2, \ldots, k.$

Let P_{ijl} be the probability of inclusion of any sample observation in i-th row and j-th column of l-th sample ($i = 1,2;\ j = 1,2;\ l = 1,2,\ldots,k$). We need to test the significance of

$$H_0 : P_{11l} = P_{21l} \text{ against } H_A : P_{11l} > P_{21l} \text{ or, } P_{11l} < P_{21l}$$

$$\text{or, } H_0 : P_{11l} \geq P_{21l} \text{ against } H_A : P_{11l} \leq P_{21l}$$

$$\text{or, } H_0 : P_{11l} \leq P_{21l} \text{ against } H_A : P_{11l} \geq P_{21l}.$$

The test statistic for the hypothesis is

$$z = \frac{\sum_{l=1}^{k} a_l - \sum_{l=1}^{k} \frac{(a_l+b_l)(a_l+c_l)}{n_l}}{\left[\sum_{l=1}^{k} \frac{(a_l+b_l)(a_l+c_l)(b_l+d_l)(c_l+d_l)}{n_l^2(n_l-1)}\right]^{\frac{1}{2}}} \sim N(0,1).$$

However, for lower-tailed test 0.5 is to be added with denominator and for upper-tailed test 0.5 is to be subtracted from denominator.

The assumption for the test is that the marginal and grand totals are fixed. However, if the marginal totals are random, then the test statistic is

$$z = \frac{\sum a_l - \sum_{l=1}^{k} \frac{(a_l+b_l)(a_l+c_l)}{n_l}}{\left[\sum \frac{1}{n_l^3}(a_l+b_l)(a_l+c_l)(b_l+d_l)(c_l+d_l)\right]^{\frac{1}{2}}} \sim N(0,1).$$

►**EXAMPLE 34.35** *A farmer has purchased seeds of maize of* 3 *varieties and as a test case cultivated these seeds in experimental plots. The seeds are collected from* 2 *selling centres. After sowing seeds the farmer has observed the amount of germinated seeds. The information of germinated seeds of* 3 *varieties of two selling centres are shown below:*

Variety-1			
Selling centre	*Amount of seeds*		*Total*
	Germinated	*Not germinated*	
1	7	3	10
2	9	1	10
Total	16	4	20

Variety-2			
Selling centre	*Amount of seeds*		*Total*
	Germinated	*Not germinated*	
1	6	2	8
2	6	1	7
Total	12	3	15

Variety-3			
Selling centre	*Amount of seeds*		*Total*
	Germinated	*Not germinated*	
1	5	7	12
2	6	2	8
Total	11	9	20

Do you think that the rates of germination of all varieties collected from all centres are same?

Solution. Let P_{i1l} be the probability of germination of seeds and P_{i2l} be the probability of not germination of seeds of l-th variety collected from first centre. We need to test $H_0 : P_{11l} = P_{21l}$ against $H_A : P_{11l} > P_{21l}$.

Assumption: The marginal total number of seeds are fixed. The test statistic is

$$z = \frac{\sum_{i=1}^{k} a_l - \sum \frac{(a_l+b_l)(a_l+c_l)}{n_l}}{\left[\sum_{i=1}^{k} \frac{(a_l+b_l)(a_l+c_l)(b_l+d_l)(c_l+d_l)}{n_l^2(n_l-1)}\right]^{\frac{1}{2}}}$$

Here $k = 3$,

$$z = \frac{(7+6+5) - \left\{\frac{16\times10}{20} + \frac{8\times12}{15} + \frac{12\times11}{20}\right\}}{\left[\frac{10\times16\times10\times4}{(20)^2(20-1)} + \frac{8\times12\times7\times3}{(15)^2(15-1)} + \frac{12\times11\times8\times9}{(20)^2(20-1)}\right]^{\frac{1}{2}}}$$

$$= \frac{18-21}{[0.8421+0.64+1.2505]^{1/2}} = \frac{-3}{1.6531} = -1.81.$$

Since the test is upper-tailed, the test statistic should be

$$z = \frac{-3}{1.6531+0.5} = -1.39.$$

Since $|z| < 1.64$, H_0 is accepted at 5% level of significance.

►**EXAMPLE 34.36** *A randomly selected group of patients are investigated and classified according to their smoking habit and heart problem. The number of patients in each class is shown below:*

Smoking habit	*Heart problem*		*Total*
	Yes	*No*	
Yes	55	15	70
No	10	25	35
Total	65	40	105

Do you think that smoking affects heart?

Solution. We need to test the significance of

H_0 : heart problem is independent of smoking

against H_A : heart problem arises due to smoking.

The test statistic is

$$\chi^2 = \frac{n[ab-bc]^2}{(a+b)(a+c)(b+d)(c+d)} = \frac{105(55\times25-10\times15)^2}{70\times65\times35\times40} = 24.73.$$

Since $\chi^2 > \chi^2_{0.05,1} = 3.84$, H_0 is rejected. Smoking affects the heart.

►**EXAMPLE 34.37** *A randomly selected group of couples of childbearing ages are classified according to family planning adoption behaviour and wife's education. The number of females in each class is shown below:*

Female's education	*Family planning adoption*		*Total*
	Yes	*No*	
Illiterate	4	6	10
Educated	8	12	20
Total	12	18	30

Do you think that female's education influences the family planning adoption?

Solution. The hypothesis to be tested is

H_0 : family planning adoption is independent of female's education

against H_A : educated females are more prone to adopt family planning.

Here $n < 40$, hence we can use Yate's correction in calculating χ^2-statistic. The test statistic is

$$\chi^2 = \frac{n\left[|ad - bc| - \frac{n}{2}\right]^2}{(a+b)(a+c)(b+d)(c+d)} = \frac{30[|48-48|-15]^2}{10 \times 12 \times 18 \times 20} = 0.16.$$

Since $\chi^2 < \chi^2_{0.05,1} = 3.84$, H_0 is accepted.

►**EXAMPLE 34.38** *A researcher in the field of Botany has collected some leaves from a forest on 4 occasions. The collected leaves are classified into 2 classes according to the species A and B. Are the proportions of species A in each occasion same?*

Table of O_{ij}

	Occasion				
Species	*1*	*2*	*3*	*4*	*Total*
A	5	12	8	10	35
B	30	36	22	45	133
Total	35	48	30	55	168

Solution. Let P_i be the proportion of species A collected in i-th occasion. We need to test the significance of

$$H_0 : P_1 = P_2 = P_3 = P_4$$

against H_A : at least one of the equality does not hold good.

Under H_0, the expected frequencies are

Table of E_{ij}

	Occasion				
Species	*1*	*2*	*3*	*4*	*Total*
A	7.3	10	6.2	11.5	35
B	27.7	38	23.8	43.5	133
Total	35	48	30	55	168

$$\therefore \chi^2 = \sum_{i=1}^{2}\sum_{j=1}^{4} \frac{O_{ij}^2}{E_{ij}} - n = 170.33 - 168 = 2.33.$$

Since $\chi^2 < \chi^2_{0.05,3} = 7.81$, H_0 is accepted.

►**EXAMPLE 34.39** *In a nursery an experiment is conducted to increase the production of rose using* 5 *different doses of nitrogen. Seventy plants are under investigation and on these plants urea fertilizer has been applied. The number of plants which have given increased number of flower are recorded and given below:*

Table of O_{ij}

Increased	*Levels of nitrogen*					
number of flower	*1*	*2*	*3*	*4*	*5*	*Total*
Yes	2	5	12	10	11	40
No	7	8	8	3	4	30
Total	9	13	20	13	15	70

Do you think that due to changes in levels of urea proportion of plants having increased flower are also increased?

Solution. Let P_i = proportion of plants having increased flower. We need to test the significance of

$$H_0 : P_1 = P_2 = P_3 = P_4 = P_5$$

against H_A : at least one of the equality does not hold good.

The expected number of plants under H_0 are shown below:

Table of E_{ij}

Increased	*Levels of nitrogen*					
number of flower	*1*	*2*	*3*	*4*	*5*	*Total*
Yes	5.14	7.43	11.43	7.43	8.57	40
No	3.86	5.57	8.57	5.57	6.43	30
Total	9	13	20	13	15	70

The test statistic is

$$\chi^2 = \sum_i^r \sum_j^c \frac{O_{ij}^2}{E_{ij}} - n$$

$$= \frac{7^2}{12.57} + \frac{12^2}{11.43} + \frac{(7.43)^2}{10} + \frac{11^2}{8.57} + \frac{15^2}{9.43} + \frac{8^2}{8.57} + \frac{3^2}{5.57} + \frac{4^2}{6.43} - 70$$

$$= 71.57 - 70 = 1.57.$$

Here $E_{21} = 3.86 < 5$, hence it is added to $E_{22} = 5.57$. Now, the number of columns reduced to 4.

Since $\chi^2 < \chi^2_{0.05,3} = 7.81$, H_0 is accepted.

34.17.9 G-Test

Let N observations under investigation be divided into k classes. The proportion of observation of i-th class is P_i, where

$$P_i = \frac{A_i}{N};\ i = 1, 2, \ldots, k$$

and A_i = number of observations of i-th class.

The problem is to test the significance of null hypothesis

$$H_0 : P_1 = P_2 = \cdots = P_k$$

against H_A : at least one of the equality does not hold good.

To test this hypothesis we can use G-test instead of χ^2, where

$$G = 2\sum_{i=1}^{k} O_i \ln \frac{O_i}{E_i}.$$

This G is distributed as chi-square with $(k-1)$ d.f.

►EXAMPLE 34.40 *A group of butterflies were trapped and then classified according to species. The number of butterflies in different species are shown below:*

	No. of species of butterfly (O_{ij})				
No. of	*1*	*2*	*3*	*4*	*Total*
butterflies	40	28	24	36	128

Are the proportions of butterflies of different species similar?

Solution. Let P_i = proportion of i-th species ($i = 1, 2, 3, 4$).

$$H_0 : P_1 = P_2 = P_3 = P_4$$

against H_A : at least one of the equality does not hold good.

Under H_0, $E_i = \frac{N}{k} = \frac{128}{4} = 32$. The test statistic is

$$G = 2\sum_{i=1}^{k} O_i \ln \frac{O_i}{E_i} = 2[8.9257 + 3.7389 - 6.9044 + 4.2402] = 5.0452.$$

Since $G < \chi^2_{0.05,3} = 7.81$, H_0 is accepted.

34.17.10 Coefficient of Association

We have discussed the test of independence of two attributes, where each attribute is divided into several classes. The test statistic used is

$$\chi^2 = \sum_i^r \sum_j^c \frac{O_{ij}^2}{E_{ij}} - n$$

where O_{ij} and E_{ij} are the observed and expected frequencies of j-th class of attribute corresponding to i-th class of attribute A; $n = \Sigma\Sigma O_{ij} = \Sigma\Sigma E_{ij}$. If the hypothesis of independence of attributes are rejected, one needs to measure the amount of association of two attributes. The important measure of association is known as 'Coefficient of Mean Square Contingency' which was proposed by Karl Pearson. This coefficient for $(r \times c)$ contingency table is

$$C = \sqrt{\frac{\chi^2}{\chi^2 + n}}.$$

The value of C is naturally less than 1. The maximum value of C is

$$C = \sqrt{\frac{r-1}{r}}.$$

For 2×2 contingency table Yule proposed one coefficient of association which is known as Yule's coefficient of colligation. It is given by

$$Q = \frac{ad - bc}{ab + bc}.$$

This Q ranges -1 to 1. If the attributes A and B are positively associated, the value of Q will be positive, and Q will be negative, if A and B are negatively associated.

►**EXAMPLE 34.41** *Fifty students of a school are randomly selected and investigated to collect information regarding their academic performance and calorie consumption per day. The students are classified according to these two attributes as follows:*

Number of students in different classes $[O_{ij}]$

Academic result	*A, Daily calorie consumption*		Total
B	*Sufficient*	*Not sufficient*	
Good	9	3	12
Not good	19	19	38
Total	28	22	50

Do you think that the calorie consumption influences the academic result? Find the amount of association of calorie consumption and academic result.

Solution. H_0 : The two attributes A and B are independent.

H_A : The two attributes are not independent.

$$\chi^2 = \frac{n(ad-bc)^2}{(a+b)(a+c)(c+d)(b+d)} = \frac{50(19\times 9 - 19\times 3)^2}{12\times 28\times 38\times 22} = 2.31.$$

Since $\chi^2 < \chi^2_{0.05,1} = 3.84$, H_0 is accepted. The amount of association is

$$C = \sqrt{\frac{\chi^2}{\chi^2+n}} = 0.21.$$

Yule's coefficient of colligation is

$$Q = \frac{ad-bc}{ad+bc} = 0.5.$$

34.17.11 Test of Significance of Regression Coefficient

Let (x_i, y_i); $i = 1, 2, \ldots, n$ be n pairs of values observed from n units selected at random from a population, where the variable Y is known as dependent variable and X is the independent or explanatory variable. It is assumed that X and Y are linearly related, and the model for this relation is

$$y = A + Bx + \epsilon$$

where B is the regression coefficient of Y on X, A is the value of y in absence of X and ϵ is a random component. It is assumed that $\epsilon \sim NID(0, \sigma^2)$. The objective of the analysis is to study the significance of the relationship of X and Y. If Y does not depend on X, B should be zero. Therefore, the hypothesis to be tested is

$$H_0 : B = 0 \text{ against } H_A : B \neq 0.$$

The fitted regression line is $y = a + bx$, where b is the sample regression coefficient and is measured by

$$b = \frac{\Sigma(x-\bar{x})(y-\bar{y})}{\Sigma(x-\bar{x})^2} = \frac{SP(xy)}{SS(x)};\ a = \bar{y} - b\bar{x}.$$

The test statistic for the hypothesis is

$$t = \frac{b-B}{\sqrt{\frac{s^2}{SS(x)}}},\ \text{where } s^2 = \frac{1}{n-2}[SS(y) - bSP(xy)].$$

This 't' follows Student's 't' distribution with $(n-2)$ d.f. Hence, if $|t| \geq t_{\frac{\alpha}{2}, n-2}$, H_0 is rejected at α% level of significance.

The test statistic to test the significance of $H_0 : A = 0$ against $H_A : A \neq 0$ is

$$t = \frac{a - A}{\sqrt{s^2\left[\frac{1}{n} + \frac{\bar{x}^2}{SS(x)}\right]}}.$$

This 't' is also distributed as Student's 't' with $(n-2)$ d.f.

►**EXAMPLE 34.42** *The following data represent the temperature ($x°$, celsius) and humidity (y, %) of several randomly selected days:*

$$x : 25, 25, 25, 25, 25, 26, 26, 26, 26, 26, 27, 27, 29, 30, 30$$
$$y : 70, 75, 80, 75, 77, 70, 72, 80, 78, 80, 78, 78, 82, 80, 80$$

$$x : 27, 27, 27, 28, 28, 28, 28, 28, 29, 29, 29, 29, 30, 30, 30$$
$$y : 77, 80, 80, 80, 80, 75, 82, 80, 78, 78, 75, 82, 79, 80, 79.$$

Test the significance of the regression. Do you think that the overall regression is also significant?

Solution. Here $n = 30$, $\Sigma y = 2340$, $\Sigma x = 825$, $\Sigma x^2 = 22775$, $\Sigma y^2 = 182816$.

$$SS(x) = \Sigma x^2 - \frac{(\Sigma x)^2}{n} = 87.5,\ SS(y) = \Sigma y^2 - \frac{(\Sigma y)^2}{n} = 296$$

$$SP(xy) = \Sigma xy - \frac{\Sigma x \Sigma y}{n} = 77,\ b = \frac{SP(xy)}{SS(x)} = \frac{77}{87.5} = 0.88$$

$$a = \bar{y} - b\bar{x} = 78 - 0.88 \times 27.5 = 53.8.$$

The hypothesis to be verified is $H_0 : B = 0$ against $H_A : B \neq 0$.

$$t = \frac{b - B}{\sqrt{\frac{s^2}{SS(x)}}},\ \text{where } B = 0 \text{ under } H_0$$

$$s^2 = \frac{1}{n-2}[SS(y) - bSP(xy)] = \frac{1}{28}[296 - 0.88 \times 77] = 8.15.$$

$$\therefore t = \frac{0.88}{\sqrt{\frac{8.15}{87.5}}} = 2.88.$$

Since $|t| > t_{0.025,28} = 2.048$, H_0 is rejected. Humidity significantly increases with the increase in temperature.

We also need to test the significance of $H_0 : A = 0$ against $H_A : A \neq 0$.

$$t = \frac{a - A}{\sqrt{s^2\left[\frac{1}{n} + \frac{\bar{x}^2}{SS(x)}\right]}} = \frac{53.8}{\sqrt{8.15\left[\frac{1}{30} + \frac{(27.5)^2}{87.5}\right]}}, \quad \because A = 0 \text{ under } H_0$$

$$= 6.40.$$

Since $|t| > t_{0.025,28} = 2.048$, H_0 is rejected.

Here all the regression parameters are found significant. Therefore, the overall regression is also significant.

Equality of Two Regression Coefficients

Let (x_{ij}, y_{ij}) be a set of n_i observations $(i = 1, 2)$ recorded from n_i units selected from i-th population by a random process. It is assumed that the variable Y depends on variable X. The linear relationship of Y and X is formulated by $y_1 = A_1 + B_1x_1 + \epsilon_1$ and $y_2 = A_2 + B_2x_2 + \epsilon_2$.

We need to test the hypothesis $H_0 : B_1 = B_2$ against $H_A : B_1 \neq B_2$.

The test statistic is

$$t = \frac{b_1 - b_2}{\sqrt{s^2\left[\frac{1}{SS(x_1)} + \frac{1}{SS(x_2)}\right]}}.$$

This 't' follows Student's 't' distribution with $2(n-2)$ d.f. Here

$$b_1 = \frac{SP(x_1y_1)}{SS(x_1)}, \; b_2 = \frac{SP(x_2y_2)}{SS(x_2)}$$

$$s^2 = \frac{1}{2}(s_1^2 + s_2^2), \; s_1^2 = \frac{1}{n_1 - 2}[SS(y_1) - b_1SP(x_1y_1)]$$

$$s_2^2 = \frac{1}{n_2 - 2}[SS(y_2) - b_2SP(x_2y_2)]$$

$$SS(x_1) = \sum_j x_{1j}^2 - \frac{(\Sigma x_{1j})^2}{n_1}, \; SS(x_2) = \sum_j x_{2j}^2 - \frac{(\Sigma x_{2j})^2}{n_2}$$

$$SS(y_1) = \sum y_{1j}^2 - \frac{(\Sigma y_{1j})^2}{n_1}, \; SS(y_2) = \sum y_{2j}^2 - \frac{(\Sigma y_{2j})^2}{n_2}$$

$$SP(x_1y_1) = \sum x_{1j}y_{1j} - \frac{\Sigma x_{1j}\Sigma y_{1j}}{n_1}, \; SP(x_2y_2) = \sum x_{2j}y_{2j} - \frac{\Sigma x_{2j}\Sigma y_{2j}}{n_2}.$$

►**EXAMPLE 34.43** *In collecting the information on newborn babies from two separate hospitals, the following information are recorded:*

Hospital-1 :	*Gestation period,* x_{1j} :	270	260	250	265	266	262
	Birth-weight, y_{1j} *(in lb)* :	7.0	6.5	6.2	7.0	6.1	5.8
Hospital-2 :	*Gestation period,* x_{2j} :	263	265	270	255	260	260
	Birth-weight, y_{2j} *(in lb)* :	6.3	7.0	7.1	5.5	6.5	6.8

Do you think that the influence of gestation period on birth-weight of babies in two hospitals are similar?

Solution. $H_0 : B_1 = B_2$ against $H_A : B_1 \neq B_2$. Here

$$SS(x_1) = 236.83, \; SS(y_1) = 1.21, \; SP(x_1y_1) = 7.57$$

$$b_1 = \frac{SP(x_1y_1)}{SS(x_1)} = 0.032, \; n = 6$$

$$s_1^2 = \frac{1}{n-2}[SS(y_1) - b_1SP(x_1y_1)] = 0.242$$

$$SS(x_2) = 130.83, SS(y_2) = 1.73, SP(x_2y_2) = 12.47$$

$$b_2 = \frac{SP(x_2y_2)}{SS(x_2)} = 0.095$$

$$s_2^2 = \frac{1}{n-2}[SS(y_2) - b_2SP(x_2y_2)] = 0.136$$

$$s^2 = \frac{1}{2}[s_1^2 + s_2^2] = \frac{1}{2}[0.242 + 0.136] = 0.189$$

$$t = \frac{b_1 - b_2}{\sqrt{s^2\left[\frac{1}{SS(x_1)} + \frac{1}{SS(x_2)}\right]}} = \frac{0.032 - 0.095}{\sqrt{0.189\left[\frac{1}{236.83} + \frac{1}{130.83}\right]}}$$

$$= -1.34.$$

Since $|t| < t_{0.025,8} = 2.306$, H_0 is accepted.

34.17.12 Test of Significance of Correlation Coefficient

Let $(x_1, y_1), (x_2, y_2), \ldots, (x_n, y_n)$ be n pairs of values observed from n randomly selected sample points from a bivariate normal population. The sample correlation coefficient of the variables X and Y is given by

$$r = \frac{\Sigma(x - \bar{x})(y - \bar{y})}{\sqrt{\Sigma(x - \bar{x})^2\Sigma(y - \bar{y})^2}} = \frac{SP(xy)}{\sqrt{SS(x)SS(y)}}.$$

Let the population correlation coefficient of X and Y be ρ. If $X$and Y is significantly correlated, $\rho \neq 0$. Therefore, we need to test the significance of

$$H_0 : \rho = 0 \text{ against } H_A : \rho \neq 0.$$

The test statistic under null hypothesis is

$$t = \frac{r\sqrt{n-2}}{\sqrt{1-r^2}}.$$

This 't' is distributed as Student's 't' with $(n - 2)$ d.f.

►**EXAMPLE 34.44** *The following data represent the body-breadth (x, in cm) and body-weight (y, in gm) of 14 randomly selected sea fishes:*

x	0.5	0.6	0.8	0.4	0.5	0.7	1.0	1.0	0.6	0.7	1.5	0.5	0.5	0.6
y	10	15	25	12	15	14	25	28	18	20	40	18	15	20

Do you think that the body-weight and body-breadth of fishes are significantly correlated?

Solution. We have $n = 14$

$$SS(x) = \Sigma x^2 - \frac{(\Sigma x)^2}{n} = 1.1093$$
$$SS(y) = \Sigma y^2 - \frac{(\Sigma y)^2}{n} = 795.2143$$
$$SP(xy) = \Sigma xy - \frac{\Sigma x \Sigma y}{n} = 27.9357, \ \Sigma x = 9.9, \ \Sigma y = 275$$
$$\Sigma x^2 = 8.11, \ \Sigma y^2 = 6197, \ \Sigma xy = 222.4$$
$$\therefore \ r = \frac{SP(xy)}{\sqrt{SS(x)SS(y)}} = \frac{27.9357}{\sqrt{1.1093 \times 795.2143}} = 0.94$$
$$H_0 : \rho = 0 \text{ against } H_A : \rho \neq 0.$$

The test statistic is

$$t = \frac{r\sqrt{n-2}}{\sqrt{1-r^2}} = \frac{0.94\sqrt{14-2}}{\sqrt{1-(0.94)^2}} = 9.54.$$

Since $|t| > t_{0.025,12} = 2.179$, H_0 is rejected. The body-weight is significantly correlated with body-breadth.

34.17.13 Test Regarding Specific Value of Correlation Coefficient

Let $(x_1, y_1), (x_2, y_2), \ldots, (x_n, y_n)$ be n pairs of values observed from n randomly selected sample points from a bivariate normal population. It is assumed that the variables X and Y are correlated and the correlation coefficient has a specific value ρ_0. The problem is to verify whether this assumption conforms with the data. That is, we need to test the significance of

$$H_0 : \rho = \rho_0 \text{ against } H_A : \rho \neq \rho_0.$$

The sample correlation coefficient is

$$r = \frac{\Sigma(x-\bar{x})(y-\bar{y})}{\sqrt{\Sigma(x-\bar{x})^2\Sigma(y-\bar{y})^2}} = \frac{SP(xy)}{\sqrt{SS(x)SS(y)}}.$$

The test statistic to test the significance of H_0 is based on r, where r is transformed to z, which is known as Fisher's z-transformation, where

$$z = \frac{1}{2}\ln\frac{1+r}{1-r}.$$

In a similar way, the correlation coefficient under H_0 is also transformed to m, where

$$m = \frac{1}{2}\ln\frac{1+\rho_0}{1-\rho_0}.$$

Then, the test statistic using z and m is

$$d = (z-m)\sqrt{n-3} \sim N(0,1).$$

►EXAMPLE 34.45 *The following data represent the study hour (x) and combined grade point average (y) of 12 randomly selected students of a university:*

x	12	15	14	12	18	16	14	12	13	15	16	15
y	2.0	3.5	3.8	3.2	3.8	2.4	2.5	1.6	2.0	3.4	3.8	2.0

Do you think that the correlation coefficient of study hour and the combined grade point average of students is 0.5?

Solution. We have $n = 12$, $\Sigma x = 172$, $\Sigma x^2 = 2504$, $\Sigma y = 34$, $\Sigma y^2 = 103.94$, $\Sigma xy = 496.9$.

$$SS(x) = 38.67, \; SS(y) = 7.61, \; SP(xy) = 9.57$$

$$r = \frac{SP(xy)}{\sqrt{SS(x)SS(y)}} = \frac{9.57}{\sqrt{38.67 \times 7.61}} = 0.56$$

$$H_0 : \rho = \rho_0 = 0.5, \; H_A : \rho \neq \rho_0 = 0.5$$

$$z = \frac{1}{2} \ln \frac{1+r}{1-r} = \frac{1}{2} \ln \frac{1.56}{0.44} = 0.57$$

$$m = \frac{1}{2} \ln \frac{1+\rho_0}{1-\rho_0} = \frac{1}{2} \ln \frac{1.50}{0.5} = 0.55.$$

The test statistic is

$$d = (z - m)\sqrt{n-3} = (0.57 - 0.55)\sqrt{12-3} = 0.06.$$

Since $|d| < 1.96$, H_0 is accepted.

34.17.14 Test of Equality of Two Correlation Coefficients

Let $(x_{11}, y_{11}), (x_{12}, y_{12}), \ldots, (x_{1n_1}, y_{1n_1})$ be n_1 pairs of values observed from n_1 sample points selected randomly from a bivariate normal population for which the population correlation coefficient of X_1 and Y_1 be ρ_1. The sample correlation coefficient is r_1. Again, let us consider that $(x_{21}, y_{21}), (x_{22}, y_{22}), \ldots, (x_{2n_2}, y_{2n_2})$ be another n_2 pairs of values observed from randomly selected sample points from another bivariate normal population for which the population correlation coefficient is ρ_2. The sample correlation coefficient is r_2. The problem is to test the significance of $H_0 : \rho_1 = \rho_2$ against $H_A : \rho_1 \neq \rho_2$.

The test statistic is

$$d = \frac{z_1 - z_2}{\sqrt{\left[\frac{1}{n_1-3} + \frac{1}{n_2-3}\right]}} \sim N(0,1)$$

where

$$z_1 = \frac{1}{2} \ln \frac{1+r_1}{1-r_1}, \; Z_2 = \frac{1}{2} \ln \frac{1+r_2}{1-r_2}.$$

►EXAMPLE 34.46 *The correlation coefficient of temperature and humidity of 30 randomly selected days during summer is recorded as $r_1 = 0.72$. The same is recorded from 25 randomly selected days during winter as $r_2 = 0.58$. Do you think that the* *correlation coefficients of temperature and humidity during summer and winter are similar?*

Solution. We have $n_1 = 30$, $r_1 = 0.72$; $n_2 = 25$, $r_2 = 0.58$. Let ρ_i $(i = 1, 2)$ be the population correlation coefficient of temperature and humidity in i-th season. We need to test the significance of

$$H_0 : \rho_1 = \rho_2 \text{ against } H_A : \rho_1 \neq \rho_2.$$

We also have

$$z_1 = \frac{1}{2}\ln\frac{1+r_1}{1-r_1} = \frac{1}{2}\ln\frac{1+0.72}{1-0.72} = 0.91$$
$$z_2 = \frac{1}{2}\ln\frac{1+r_2}{1-r_2} = \frac{1}{2}\ln\frac{1+0.58}{0.42} = 0.66.$$

The test statistic is

$$d = \frac{z_1 - z_2}{\sqrt{\left[\frac{1}{n_1-3} + \frac{1}{n_2-3}\right]}} = \frac{0.91 - 0.66}{\sqrt{\left[\frac{1}{27} + \frac{1}{22}\right]}} = 0.87.$$

Since $|d| < 1.96$, H_0 is accepted.

34.17.15 Test of Equality of Several Correlation Coefficients

Let n_i $(i = 1, 2, \ldots, k)$ be the randomly selected sample size from i-th bivariate normal population whose population correlation coefficient of two variables X and Y is ρ_i. Assume that the samples are independently selected from the populations. The sample correlation coefficient of X and Y from i-th sample is found out as r_i. The problem is to test the significance of the null hypothesis

$$H_0 : \rho_1 = \rho_2 = \cdots = \rho_k = \rho$$
against H_A : at least one of the equality sign does not hold good.

The test statistic for the null hypothesis is

$$\chi^2 = \sum_{i=1}^{k}(n_i - 3)z_i^2 - \frac{[\Sigma(n_i - 3)z_i]^2}{\Sigma(n_i - 3)}.$$

This χ^2 is distributed as chi-square distribution with $(k-1)$ d.f.

►**EXAMPLE 34.47** *From 5 different hospitals 5 random samples of patients are independently selected to study the correlation coefficient of blood pressure and age of patients. The sample sizes (n_i) and the sample correlation coefficients (r_i) are recorded as follows:*

$$(n_1 = 15,\ r_1 = 0.78);\ (n_2 = 16,\ r_2 = 0.58),\ (n_3 = 18,\ r_3 = 0.45);$$
$$(n_4 = 25,\ r_4 = 0.60);\ (n_5 = 25,\ r_5 = 0.69).$$

Do you think that the correlation coefficients of age and blood pressure for all patients in different hospitals are homogeneous?

Solution. Let ρ_i $(i = 1, 2, \ldots, k)$ be the population correlation coefficient of age and blood pressure of patients of i-th hospital. We need to test the significance of

$$H_0 : \rho_1 = \rho_2 = \cdots = \rho_5$$

against H_A : at least one of the equality sign does not hold good.

Here

$$z_i = \frac{1}{2} \ln \frac{1 + r_i}{1 - r_i}, \quad i = 1, 2, 3, 4, 5.$$
$$z_1 = 0.92, \ z_2 = 0.66, \ z_3 = 0.48, \ z_4 = 0.69, \ z_5 = 0.85.$$

The test statistic is

$$\chi^2 = \Sigma(n_i - 3)z_i^2 - \frac{(\Sigma(n_i - 3)z_i)^2}{\Sigma(n_i - 3)}$$
$$= 12 \times (0.92)^2 + 13 \times (0.66)^2 + 15 \times (0.48)^2 + 22 \times (0.60)^2 + 22 \times (0.85)^2 - \frac{(12 \times 0.92 + 13 \times 0.66 + 15 \times 0.48 + 22 \times 0.60 + 22 \times 0.85)^2}{12 + 13 + 15 + 22 + 22}$$
$$= 43.0906 - \frac{3448.0384}{84} = 2.04.$$

Since $\chi^2 < \chi^2_{0.05,4} = 9.488$, H_0 is accepted.

34.17.16 Test of Significance of Multiple Correlation Coefficient

A randomly selected sample of n observations from multivariate distribution provides information of k variables $X_1, X_2, \ldots, X_k$, where the variable X_1 depends on the remaining $(k - 1)$ variables. The sample multiple correlation coefficient from this sample is $R_{1.23\cdots k}$, where

$$R^2_{1.23\cdots k} = \frac{SS \text{ (Regression)}}{SS(x_1)}$$

where

$$SS(x_1) = \Sigma(x_1 - \bar{x}_1)^2, SS \text{ (Regression)} = \sum_{i=2}^{k} \hat{\beta}_i SP(x_1 x_i).$$

Here

$$\hat{\beta} = (\hat{\beta}_0 \hat{\beta}_2 \cdots \hat{\beta}_k)' = (X'X)^{-1} X'X_1$$

$$X = \begin{bmatrix} 1 & x_{21} & x_{31} & \cdots & x_{k1} \\ 1 & x_{22} & x_{32} & \cdots & x_{k2} \\ \cdots & \cdots & \cdots & \cdots & \cdots \\ 1 & x_{2n} & x_{3n} & \cdots & x_{kn} \end{bmatrix}, \quad X_1 = \begin{bmatrix} x_{11} \\ x_{12} \\ \vdots \\ x_{1n} \end{bmatrix}$$

The problem of analysis is to test the significance of $H_0 : \rho = 0$ against $H_A : \rho > 0$, where ρ is the population correlation coefficient X_1 with other X_i's $(i = 2, 3, \ldots, k)$.

The test statistic is

$$F = \frac{R^2}{1-R^2}\frac{n-k}{k-1}$$

This 'F' is distributed as variance ratio with $(k-1)$ and $(n-k)$ d.f.

▶**EXAMPLE 34.48** *A group of 25 students are selected from a university. For each student the variables study hour (x_2), academic result (x_1) and father's income (x_3) are recorded. From the recorded information the correlation coefficient of x_1 and x_2, x_1 and x_3, and x_2 and x_3 are observed as $r_{12} = 0.65, r_{13} = 0.75, r_{23} = 0.52$, respectively. Do you think that the variable x_1 is significantly correlated simultaneously with x_2 and x_3?*

Solution. Given $r_{12} = 0.65$, $r_{13} = 0.75$ and $r_{23} = 0.52$, $k = 3$. The coefficient of multiple correlation of x_1 with x_2 and x_3 is given by

$$\begin{aligned} R^2_{1.23} &= \frac{r^2_{12} + r^2_{13} - 2r_{12}r_{13}r_{23}}{1 - r^2_{23}} \\ &= \frac{(0.65)^2 + (0.75)^2 - 2 \times 0.65 \times 0.75 \times 0.52}{1 - (0.52)^2} = 0.6552. \end{aligned}$$

$$\therefore\ R = \sqrt{R^2_{1.23}} = 0.8094.$$

Let the population multiple correlation coefficient be ρ. We need to test

$$H_0 : \rho = 0 \text{ against } H_A : \rho > 0.$$

The test statistic is

$$F = \frac{R^2}{1-R^2}\frac{n-k}{k-1} = \frac{0.6552}{1-0.6552}\frac{22}{2} = 20.90.$$

Since $F > F_{0.05;2,22} = 2.93$, H_0 is rejected. Academic result significantly correlated with study hour and father's income.

EXERCISE

1. (a) Define test of significance, null hypothesis, critical region, level of significance, type-I and type-II errors with examples.
   (b) State and prove Neyman-Pearson lemma for testing a simple hypothesis against a simple alternative hypothesis.
   (c) Let $x_1, x_2, \ldots, x_n$ be a random sample of n observations from Poisson population with parameter θ. Obtain most powerful critical region of size α for testing the null hypothesis $H_0 : \theta = \theta_0$ against $H_A : \theta = \theta_1 > \theta_0$.
2. (a) What do you mean by test of hypothesis. Explain different terms related to test of hypothesis.
   (b) Explain MP test and UMP test. Prove that, if $T(x)$ is sample statistic through which monotone likelihood ratio of $f(x, \theta)$ is available, then there exists UMP test for testing $H_0 : \theta \leq \theta_0$ against $H_A : \theta \geq \theta_0$.
   (c) Find BCR to test $H_0 : \mu = \mu_0$ against $H_A : \mu \neq \mu_0$ for $N(\mu, \sigma^2)$.

3. (a) Distinguish MP test and UMP test. Define the terms composite hypothesis and unbiased test. Explain with suitable theoretical basis how you find the most powerful test for testing a simple hypothesis against a simple alternative hypothesis.

(b) Let $x_1, x_2, \ldots, x_n$ be a random sample of n observations from a population having p.d.f.

$$f(x) = \begin{cases} e^{-(x-\theta)}, & 0 \le x \le \infty \\ 0, & \text{otherwise.} \end{cases}$$

Find UMP critical region for testing $H_0 : \theta = \theta_0$ against $H_A : \theta = \theta_1 \neq \theta_0$.

4. (a) Expain the terms type-1 and type-2 errors and level of significance with examples. Also, describe the importance of these in Neyman-Pearson testing theory.

(b) Let x_1, x_2 be a random sample of size 2 from a population having p.d.f.

$$f(x, \theta) = \frac{1}{\theta} e^{-x/\theta}; \; 0 < x < \infty, \; \theta > 0.$$

Find a critical region for testing $H_0 : \theta = 2$ against $H_A : \theta = 1$. Also, find the power function of the test.

5. (a) Explain the concept of a critical region and power of a test in statistical inference.

(b) Suggest suitable tests of significance, at any given level α, for testing the following hypothesis:

(i) $H_0 : \sigma^2 = \sigma_0^2$ using sample of n observations from $N(\mu, \sigma^2)$;

(ii) $H_0 : \mu_1 = \mu_2$ using two independent random samples of sizes n_1 and n_2 from $N(\mu_1, \sigma_1^2), N(\mu_2, \sigma_2^2)$ assuming $\sigma_1^2 = \sigma_2^2$.

(c) The following data represent the production of potato (kg/plot) recorded from two farms:

Farm-1, x_{1j} : 18.5, 20.6, 17.2, 19.8, 21.7, 14.6, 19.2

Farm-2, x_{2j} : 20.0, 18.6, 21.4, 20.5, 22.6, 27.4, 18.6, 19.7, 20.8

Do you think that the average production of potato in two farms are similar?

6. (a) What is meant by test of significance? What is likelihood ratio test? Derive the LR test statistic for testing mean of normal population.

(b) Distinguish z-test and t-test. Explain the assumptions for t-test.

A random sample of birth-weights of newborn babies are observed for 10 babies and are given below:

Birth-weight (x, in kg) : 2.8, 2.5, 3.0, 3.4, 2.6, 2.2, 2.4, 2.8, 3.0, 3.0.

Is the mean birth-weight 3.0? Write down the assumptions for your test.

7. (a) Derive Bartlett's test for testing the equality of k population variances for k normal populations.

(b) Let $x_1, x_2, \ldots, x_n$ be a random sample of n observations from a population with p.d.f.

$$f(x) = \theta x^{\theta - 1}; \; 0 < x < 1, \; \theta > 1.$$

Find BCR for testing $H_0 : \theta = 1$ against $H_A : \theta = 2$.

8. (a) Define likelihood ratio test. Under what circumstances would you recommend this test?

(b) Show that the likelihood ratio test for testing the equality of variances of two normal populations is the usual F-test.

(c) Given two random sample of observations:

x_{1j} : 12, 14, 13, 18, 16, 20, 11, 13, 15, 14, 19, 20

x_{2j} : 8, 12, 15, 18, 24, 16, 7, 15, 12, 11, 13, 24, 20, 19

Do you think that the two samples are selected from two populations having same variance?

9. (a) Define MP test, UMP test, unbiased test. State and prove the theorem related to similar region test and sufficient statistic.

(b) Let $x_1, x_2, \ldots, x_n$ be a random sample of n observations from a population having p.d.f.

$$f(x) = \frac{1}{\sigma} \exp\left[-\left(\frac{x-\mu}{\sigma}\right)\right]; \; 0 < x < \infty, \; \sigma > 0.$$

Find BCR for testing $H_0 : \mu = \mu_0$ against $H_A : \mu = \mu_1 < \mu_0$.

(c) In a village there are 500 people 60% of which are literate. In an another village the size of population is 800 and 40% are literate. Are the proportion of literate persons in two villages same?

10. (a) Define unbiased test. Prove that the power of the test cannot be less than the level of significance for testing simple null hypothesis against simple alternative hypothesis.

(b) Let $x_1, x_2, \ldots, x_n$ be a random sample of n observations from a population having p.d.f.

$$f(x) = \theta e^{-\theta x}; \; x > 0, \; \theta > 0.$$

Find BCR for testing $H_0 : \theta = \theta_0$ against $H_A : \theta = \theta_1 > \theta_0$.

(c) The following data represent the FBS of a group of patients on two occasions, one after treatment (x_2) and one before treatment (x_1):

x_1 : 250, 200, 160, 180, 260, 200, 270, 185, 190, 220, 160

x_2 : 180, 200, 140, 175, 200, 210, 280, 160, 140, 200, 100

Is the treatment effective in reducing FBS?

11. (a) Define test of hypothesis, critical region, null hypothesis, level of significance with examples.

(b) State and prove Neyman-Pearson lemma related to test of simple hypothesis against a simple alternative hypothesis.

(c) What is meant by test of independence? The following data represent the distribution of couples according to family planning adoption and female's working condition:

Female's working condition	*Family planning adoption*		*Total*
	Yes	No	
House-wife	25	105	130
Work outside	50	20	70
Total	75	125	200

Are family planning adoption associated with working condition?

12. (a) Explain the terms:

(i) Type-I and Type-II errors, (ii) BCR,

(iii) Power function of a test, (iv) Level of significance,

(v) Simple and composite hypothesis.

(b) Derive the LR test statistic for testing the equality of k means of normal populations.

(c) What is meant by goodness of fit?

A random sample of 100 sea fishes are collected from a market and these are classified according to species:

Species	1	2	3	4	5	*Total*
No. of fishes	28	32	16	12	12	100

Are the proportions of fishes of different species same?

13. (a) Explain simple hypothesis and composite hypothesis. Explain the use of Neyman-Pearson lemma in obtaining the BCR by an illustration.

(b) What is the need of Yate's correction in test of indepence? Write down the assumption in deriving χ^2-test statistic in case of $(r \times c)$ contingency table.

(c) From five independent random samples of sizes $n_1 = 10$, $n_2 = 15$, $n_3 = 8$, $n_4 = 12$, $n_5 = 20$, the correlation coefficients are observed as $r_1 = 0.68$, $r_2 = 0.56$, $r_3 = 0.75$, $r_4 = 0.82$, $r_5 = 0.64$. Do you think that the samples have been selected from one population?

14. (a) Define test of significance. Explain its necessity in statistics. Explain the method for testing the equality of two proportions.

(b) What is contingency table? Explain the method of test of independence of contingency table.

(c) A group of students of a university have been classified according to the academic performance measured by grade obtained in an examination and by discipline. The number of students of different groups are shown below:

Grade	*Discipline*			*Total*
	Computer science	*Computer engineering*	*Electrical engineering*	
A	10	12	15	37
B	18	10	8	36
C	7	18	22	47
D	4	8	7	19
Total	39	48	52	139

Are grades associated with discipline?

CHAPTER **35**

Interval Estimation

35.1 Introduction

Let $x_1, x_2, \ldots, x_n$ be a random sample of n observations selected from a population having p.d.f. $f(x, \theta)$, $\theta \in S$, where S is the parametric space. The parameter θ or its any function $\phi(\theta)$ is unknown. The problem is to estimate the parameter or any parametric function. Let $T = t(x_1, x_2, \ldots, x_n)$ be any function of sample observations and $g(T; \theta)$ is the distribution of T. In Chapter 33, it is mentioned that T, the function of sample statistic is a point estimator of θ. The criterion of this estimator has also been studied. The estimator is unique best estimator, if T is unbiased and if it has minimum variance compared to the variance of any other estimator and also if the statistic T is a sufficient statistic for θ. .

The estimator T may be the best one to estimate θ. However, there is discrepancy between T and θ. Let this discrepancy is $T - \theta = \epsilon$, where ϵ is a very small quantity. Therefore, $P[T - \theta < \epsilon]$ can be used as a measure precision of the estimator T. The estimator T becomes more precise, if the error $T - \theta$ does not exceed a pre-determined value of ϵ, i.e., $P[T - \theta < \epsilon]$ should be more. The lower bound of ϵ can be ascertained in such a way that

$$P[|T - \theta| < \epsilon] > 1 - \alpha, \text{ for all } \theta \in S$$

where α is a small positive quantity. This probability can be expressed as

$$P[T - \epsilon \leq \theta \leq T + \epsilon] \geq 1 - \alpha, \text{ for all } \theta \in S.$$

This probability indicates that θ will fall within a limit $T \pm \epsilon$ with probability $1 - \alpha$. Here $T \pm \epsilon$ is the interval based on sample observations, where θ falls with probability $(1 - \alpha)$. The estimation of such an interval of $T \pm \epsilon$ is known as *interval estimation.*

It is observed that in interval estimation the basic point is to decide a statistic T and then decision is needed about its suitability as an estimator of θ.

Since it is assumed that T follows the distribution $g(T/\theta)$, for any calculated value of T, the probability of error $|T - \theta|$ can be found out, where the probability, for small quantity α, is

$$P[T - \epsilon \leq \theta \leq T + \epsilon] > 1 - \alpha.$$

Here $[T - \epsilon, T + \epsilon]$ are the confidence interval, $T - \epsilon$, $T + \epsilon$ are the confidence limits and $(1 - \alpha)$ is the confidence coefficient. In practice, $T - \epsilon$ and $T + \epsilon$ are the functions of sample observations. Let

$$T - \epsilon = l(x_1, x_2, \dots, x_n) \text{ and } T + \epsilon = u(x_1, x_2, \dots, x_n)$$

where $l(\cdot)$ is the lower limit of confidence interval and $u(\cdot)$ is the upper limit of confidence interval. The confidence interval $[l, u]$ is correct up to $100(1 - \alpha)\%$.

In practice, T is used as an estimator of θ and all probable intervals with probability $(1 - \alpha)$ are found out. Out of all probable intervals the smallest one is accepted as an interval estimator of θ. If a smallest interval is available, it is called 'Tight'. If the distribution of T is symmetric about θ, the confidence interval using T becomes smallest, if the interval is constructed with equal tails. That is, ϵ is to be determined in such a way that $P[|T - \theta| \leq \epsilon] = 1 - \alpha$. This is possible, if

$$P[T - \theta \geq \epsilon] = \frac{\alpha}{2} \text{ and } P[T - \theta \leq -\epsilon] = \frac{\alpha}{2}.$$

The confidence interval what has been discussed above is two-sided. In practice, one-sided confidence interval can also be constructed. This interval is given by

$$P[T_1 < \phi(\theta)] = 1 - \alpha \text{ and } P[T_2 > \phi(\theta)] = 1 - \alpha.$$

Here $T_1 = t_1\ (x_1, x_2, \dots, x_n)$ and $T_2 = t_2\ (x_1, x_2, \dots, x_n)$ are two functions of sample statistics. The function T_1 is called *one-sided upper confidence interval of* $\phi(\theta)$.

35.2 Method of Construction of Confidence Interval

There are two methods of finding confidence interval. These are (i) Pivotal method and (ii) Statistical method. Let us discuss the Pivotal method first.

35.2.1 Pivotal Method to Find Confidence Interval

Let $x_1, x_2, \dots, x_n$ be a random sample of n observations selected from a population having p.d.f. $f(x, \theta)$. Let $Q = q(x_1, x_2, \dots, x_n; \theta)$ be the function of sample observations and parameter. Now, if the distribution of Q is free of θ, Q is called 'Pivotal' quantity. For example, if the sample observations are selected from $N(\mu, \sigma^2)$, then

$$Q = \frac{\bar{x} - \mu}{\sigma/\sqrt{n}} \sim N(0, 1), \text{ if } \sigma^2 \text{ is known.}$$

Also $Q = \dfrac{\bar{x} - \mu}{s/\sqrt{n}} \sim N(0, 1)$ if n is large, where s^2 is an unbiased estimator of σ^2. Here the distribution of Q is free of parameter μ. Therefore, both the Q is 'Pivotal' quantity. Even, if the sample size is small, the distribution of $Q = \dfrac{\bar{x} - \mu}{s/\sqrt{n}}$ is free of parameter μ.

Since Q is the pivotal quantity, we need to find two values of Q, say q_1 and q_2, so that for $0 < 1-\alpha < 1$, $P[q_1 < Q < q_2] = 1-\alpha$. Here for all possible samples $q_1 < Q < q_2$, if and only if $T_1(x_1, x_2, \ldots, x_n) < \phi(\theta) < T_2(x_1, x_2, \ldots, x_n)$. Then $[T_1, T_2]$ is called $100(1-\alpha)\%$ confidence interval of $\phi(\theta)$. It is noted that, the inequality $q_1 < Q < q_2$ can be inverted to the inequality

$$T_1(x_1, x_2, \ldots, x_n) < \phi(\theta) < T_2(x_1, x_2, \ldots, x_n).$$

For example, for sample from $N(\mu, \sigma^2)$, $q_1 < \dfrac{\bar{x}-\mu}{\sigma/\sqrt{n}} < q_2$ can be inverted into

$$\bar{x} - q_1\frac{\sigma}{\sqrt{n}} < \mu < \bar{x} + q_2\frac{\sigma}{\sqrt{n}}.$$

Therefore, $\left[\bar{x} - q_1\dfrac{\sigma}{\sqrt{n}} < \mu < \bar{x} + q_2\dfrac{\sigma}{\sqrt{n}}\right]$ is the $100(1-\alpha)\%$ confidence interval for μ. However, the pivotal quantity $q_1 < Q < q_2$ is available but it cannot be inverted and hence the confidence interval of $\phi(\theta)$ is not found out by this method.

It is to be noted that q_1 and q_2 are independent of parameter θ since the distribution of Q is free of θ. Again, for a fixed $(1-\alpha)$ there may different pairs of q_1 and q_2, and hence there may be different pairs of T_1 and T_2. Therefore, the values of T_1 and T_2 are to be so chosen so that the distance $T_2 - T_1$ becomes minimum. Here $T_2 - T_1$ is called the distance of confidence interval. If the distance is a random variable, then the values of q_1 and q_2 are to be selected in such a way that the average distance becomes minimum. If

$$Q = \frac{\bar{x}-\mu}{s/\sqrt{n}}$$

is the pivotal quantity, the distance of confidence interval becomes $(q_2 - q_1)s/\sqrt{n}$ and this distance is a random variable. The distance should be minimum, and q_1 and q_2 could be minimised under the condition

$$\int_{q_1}^{q_2} f(t)dt = 1-\alpha. \tag{35.1}$$

Here $f(t)$ is density function of Student's 't' with $(n-1)$ d.f. Differentiating both sides of (35.1) with respect to q_1, we get

$$f(q_2)\frac{dq_2}{dq_1} - f(q_1) = 0.$$

Here q_2 is the function of q_1. Now, to minimise $L = \dfrac{s}{\sqrt{n}}(q_2 - q_1)$ we need to put $\dfrac{\partial L}{\partial q_1} = 0$. Therefore,

$$\frac{\partial L}{\partial q_1} = \frac{s}{\sqrt{n}}\left(\frac{dq_2}{dq_1} - 1\right) = 0.$$

But

$$\frac{s}{\sqrt{n}}\left(\frac{dq_2}{dq_1} - 1\right) = \frac{s}{\sqrt{n}}\left(\frac{f(q_1)}{f(q_2)} - 1\right) = 0.$$

This is possible if $f(q_1) = f(q_2)$. Thus L will be minimum if $q_1 = q_2$. But, if $q_1 \neq q_2$

$$\int_{q_1}^{q_2} f(t)dt \neq 1 - \alpha.$$

Hence L will be minimum, if $q_1 = -q_2$.

►**EXAMPLE 35.1** *Let $x_1, x_2, \ldots, x_n$ be a random sample of n observations from $N(\mu, \sigma^2)$. Find $100(1-\alpha)\%$ confidence interval for μ and σ^2.*

Solution. **Confidence interval for μ.** Let us assume that σ^2 is known. Then

$$Q = \frac{\bar{x} - \mu}{\sigma/\sqrt{n}} = z \sim N(0, 1)$$

is a pivotal quantity. Therefore, the $100(1-\alpha)\%$ confidence interval for μ is

$$\bar{x} - q_1 \frac{\sigma}{\sqrt{n}} < \mu < \bar{x} + q_2 \sigma/\sqrt{n}.$$

Here q_1 and q_2 are to be replaced by $z_1 = z_{\frac{\alpha}{2}}$ and $z_2 = z_{\frac{1-\alpha}{2}}$. Therefore,

$$\bar{x} - z_{\frac{\alpha}{2}} \frac{\sigma}{\sqrt{n}} < \mu < \bar{x} + z_{1-\frac{\alpha}{2}} \frac{\sigma}{\sqrt{n}}$$

is the required confidence interval for μ. Here, if $\alpha = 0.05$, then for 95% confidence interval $z_{\frac{\alpha}{2}} = 1.96 = z_{1-\frac{\alpha}{2}}$.

When σ^2 is not known. If σ^2 is not known, it is replaced by

$$s^2 = \frac{1}{n-1} \Sigma(x_i - \bar{x})^2.$$

In such a case the pivotal quantity $Q = \dfrac{\bar{x} - \mu}{s/\sqrt{n}}$ follows Student's 't' distribution with $(n-1)$ d.f. Therefore, the $100(1-\alpha)\%$ confidence interval for μ is

$$\bar{x} - q_1 \frac{s}{\sqrt{n}} < \mu < \bar{x} + q_2 \frac{s}{\sqrt{n}}$$

where q_1 and q_2 are to be found out in such a way that

$$\int_{q_1}^{q_2} f(t)dt = 1 - \alpha.$$

Confidence interval for σ^2. The estimate of σ^2 is

$$s^2 = \frac{1}{n-1} \sum^{n} (x_i - \bar{x})^2.$$

Therefore, the pivotal quantity is

$$Q = \frac{(n-1)s^2}{\sigma^2} \sim \chi^2_{n-1}.$$

The distribution of Q is free of σ^2. Hence the necessary $100(1-\alpha)\%$ confidence interval is obtained from the equation

$$P[q_1 < Q < q_2] = 1 - \alpha.$$

Here

$$q_1 \leq \frac{(n-1)s^2}{\sigma^2} \leq q_2 \text{ or, } \frac{(n-1)s^2}{q_2} \leq \sigma^2 \leq \frac{(n-1)s^2}{q_1}.$$

Since Q is distributed as chi-square, we can write

$$P[Q \geq \chi^2_\alpha] = \int_{\chi^2_\alpha}^{\infty} f(Q)dQ = \alpha.$$

Then

$$P\left[\chi^2_{1-\frac{\alpha}{2}} < Q < \chi^2_{\frac{\alpha}{2}}\right] = 1-\alpha$$

$$P\left[\chi^2_{1-\frac{\alpha}{2}} < \frac{(n-1)s^2}{\sigma^2} < \chi^2_{\frac{\alpha}{2}}\right] = 1-\alpha$$

$$P\left[\frac{(n-1)s^2}{\chi^2_{\frac{\alpha}{2}}} \leq \sigma^2 \leq \frac{(n-1)s^2}{\chi^2_{1-\frac{\alpha}{2}}}\right] = 1-\alpha.$$

Hence $\left[\frac{(n-1)s^2}{\chi^2_{\frac{\alpha}{2}}}, \frac{(n-1)s^2}{\chi^2_{1-\frac{\alpha}{2}}}\right]$ is the $100(1-\alpha)\%$ confidence interval of σ^2. Here $\chi^2_{\frac{\alpha}{2}}$ and $\chi^2_{1-\frac{\alpha}{2}}$ are the tabulated value of χ^2 with $(n-1)$ d.f.

►**EXAMPLE 35.2** *The following observations are the birth-weights (x, kg) of* 11 *randomly selected newborn babies. Assuming normality of data, find* 95% *confidence interval for μ and σ^2. To find confidence interval for μ assume* (i) $\sigma^2 = 1.5$ *and* (ii) σ^2 *unknown:*

$$x: 2.8, 2.4, 3.2, 3.4, 2.9, 2.7, 2.8, 2.4, 2.5, 2.7, 2.8.$$

Solution. We have

$$\bar{x} = \frac{30.6}{11} = 2.78$$

$$s^2 = \frac{1}{n-1}\left[\Sigma x^2 - \frac{(\Sigma x)^2}{n}\right] = \frac{1}{10}\left[86.08 - \frac{(30.6)^2}{11}\right] = 0.0956$$

95% confidence interval for μ when $\sigma^2 = 1.5$ is

$$\bar{x} - z_{\frac{\alpha}{2}}\sqrt{\frac{\sigma^2}{n}} < \mu < \bar{x} + z_{1-\frac{\alpha}{2}}\sqrt{\frac{\sigma^2}{n}}$$

$$\text{or, } 2.78 - 1.96\sqrt{\frac{1.5}{11}} < \mu < 2.78 + 1.96\sqrt{\frac{1.5}{n}},$$

$$2.06 < \mu < 3.50.$$

Therefore, 95% confidence interval for μ is $[2.06, 3.50]$. However, if σ^2 is unknown, the confidence interval is given by

$$\bar{x} - t_{\frac{\alpha}{2}}\frac{s}{\sqrt{n}} < \mu < \bar{x} + t_{\frac{\alpha}{2}}\frac{s}{\sqrt{n}}$$

$$\text{or, } 2.78 - 2.228\sqrt{\frac{0.0956}{11}} < \mu < 2.78 + 2.228\sqrt{\frac{0.0956}{11}},$$

$$2.57 < \mu < 2.99.$$

Hence the confidence interval is [2.57, 2.99]. Again, 95% confidence interval for σ^2 is

$$\frac{(n-1)s^2}{\chi^2_{\frac{\alpha}{2}}} < \sigma^2 < \frac{(n-1)s^2}{\chi^2_{1-\frac{\alpha}{2}}}.$$

Here χ^2 has 10 d.f. Therefore,

$$\frac{(11-1)0.0956}{20.5} < \sigma^2 < \frac{(11-1)0.0956}{3.247},$$

$$0.047 < \sigma^2 < 0.294.$$

Hence 95% confidence interval for σ^2 is [0.047, 0.294].

►**EXAMPLE 35.3** *Let $x_1, x_2, \ldots, x_n$ be a random sample of n observations from population having p.d.f.*

$$f(x) = \frac{1}{\theta}, \quad 0 < x < \theta.$$

Find $100(1-\alpha)\%$ confidence interval for θ.

Solution. The distribution of n-th order statistic is

$$g\left(x_{(n)}\right) = \frac{nx^{n-1}}{\theta^n}, \quad 0 < x < \theta.$$

Here $x_{(n)}$ is the n-th order statistic and it is the estimator of θ. Therefore, the pivotal quantity is $Q = x_{(n)}/\theta$. But the distribution of θ is not free of θ and hence let us make a transformation $z = x/\theta$. Therefore, we have

$$dz = \frac{dx}{\theta}, 0 < z < 1 \text{ and } h(z) = nz^{n-1}, 0 < z < 1.$$

Here we need to find q_1 and q_2 such that $P[q_1 < Q < q_2] = 1 - \alpha$ where $Q = x/\theta$. Let z_α is the value of z, where

$$P[z_\alpha < z < 1] = \alpha \Rightarrow n\int_{z_\alpha}^{1} z^{n-1}dz = \alpha$$

$$\text{or, } n\left.\frac{z^n}{n}\right|_{z_\alpha}^{1} = \alpha \Rightarrow 1 - z_\alpha^n = \alpha, \text{ where } z_\alpha = (1-\alpha)^{\frac{1}{n}}.$$

Hence, $P\left[(1-\alpha)^{1/n} < z < 1\right] = \alpha$ and we get

$$(1-\alpha)^{1/n} < \frac{x}{\theta} < 1 \text{ or, } 1 < \frac{\theta}{x} < 1/(1-\alpha)^{1/n} \text{ or, } x < \theta < x/(1-\alpha)^{1/n}$$

where x is the n-th order statistic of sample observations. Therefore, $100(1-\alpha)\%$ confidence interval of θ is $\left[x, x/(1-\alpha)^{1/n}\right]$.

Pivotal Quantity Using Distribution Function

Let $x_1, x_2, \ldots, x_n$ be a random sample of n observations from a population having p.d.f. $f(x,\theta)$. Let the distribution function of the p.d.f. is $F(x,\theta)$, where $F(x,\theta)$ is continuous over the values of x. Now, by probability integral transformation, we get

$$g\{F(x_i,\theta)\} = 1,\ 0 < F(x_i,\theta) < 1.$$

The distribution of $-\log F(x_i,\theta)$ is e^{-u}, $0 < u < \infty$. Hence

$$P[-\log F(x_i,\theta) \geq u] = P[\log F(x_i,\theta) \leq -u]$$
$$\Rightarrow P[F(x_i,\theta) \leq e^{-u}] = e^{-u},\ 0 < u < \infty.$$

It is to be noted that $-\Sigma \log F(x_i,\theta)$ is distributed as gamma distribution. Therefore, to find confidence interval of θ, we need to use

$$P\left[-\log q_2 < -\sum_{i=1}^{n} \log F(x_i,\theta) < -\log q_1\right]$$
$$= \int_{-\log q_2}^{-\log q_1} \frac{1}{\Gamma n} z^{n-1} e^{-z} dz$$
$$= P\left[q_1 < \prod_{i=1}^{n} F(x_i,\theta) < q_2\right],\ 0 < q_1 < q_2 < 1.$$

Therefore, $\prod_{i=1}^{n} F(x_i,\theta)$ or, $-\sum_{i=1}^{n} \log F(x_i,\theta)$ is the pivotal quantity. However, to find confidence interval for θ, $\prod_{i=1}^{n} F(x_i,\theta)$ must be monotonic function of θ.

35.2.2 Confidence Interval Using Large Sample

Let $x_1, x_2, \ldots, x_n$ be a random sample of n observations from a population having p.d.f. $f(x,\theta), \theta \in S$. It is assumed that n is large. The likelihood function of the sample observations is

$$L = f(x_1,\theta) f(x_2,\theta) \cdots f(x_n,\theta).$$

It is assumed that the regularity conditions are valid for the distribution and hence the ML estimator of θ is obtained solving the equation

$$\frac{\partial \log L}{\partial \theta} = 0.$$

Again, it is also known that the ML estimator is asymptotically distributed as normal and

$$E\left[\frac{\partial \log L}{\partial \theta}\right] = 0,\ V\left[\frac{\partial \log L}{\partial \theta}\right] = -E\left[\frac{\partial^2 \log L}{\partial \theta^2}\right].$$

Therefore, if the sample size is large

$$z = \frac{\frac{\partial \log L}{\partial \theta}}{\sqrt{V\left(\frac{\partial \log L}{\partial \theta}\right)}} \sim N(0,1).$$

Therefore, the $100(1-\alpha)\%$ confidence interval of θ is obtained from the following equation:

$$P\left[|z| \le z_{\frac{\alpha}{2}}\right] = 1-\alpha \text{ or, } \frac{1}{\sqrt{2\pi}}\int_{-z_{\alpha/2}}^{z_{\alpha/2}} e^{-\frac{1}{2}z^2}dz = 1-\alpha.$$

Here

$$Q = \frac{\partial \log L}{\partial \theta} \text{ and } P[q_1 < Q < q_2] = 1-\alpha.$$

The values of q_1 and q_2 are obtained from Normal Probability Table.

►**EXAMPLE 35.4** *Let $x_1, x_2, \ldots, x_n$ be a random sample of n observations from a population with p.d.f.*

$$f(x) = \frac{1}{\theta}e^{-x/\theta};\ 0 < x < \infty,\ \theta > 0.$$

Find $100(1-\alpha)\%$ *confidence interval* (i) *using* $F(x)$ *and* (ii) *assuming* n *large.*

Solution.

(i) The distribution function of X is

$$F(x) = \frac{1}{\theta}\int_0^x e^{-x/\theta}dx = 1 - e^{-x/\theta}.$$

Hence $1 - F(x) = e^{-x/\theta}$. Let

$$z = -\log[1-F(x)] = -\log e^{-x/\theta} = \frac{x}{\theta}.$$

Here z is a gamma variate with parameter $m = 1$. We can write

$$g(z) = e^{-z} = \frac{1}{\Gamma 1}e^{-z}z^{1-1},\ 0 < z < \infty.$$

Hence $u = \Sigma z = \dfrac{n\bar{x}}{\theta}$ is also a gamma variate with parameter n and

$$h(u) = \frac{1}{\Gamma n}e^{-u}u^{n-1},\ 0 < u < \infty.$$

Thus $n\bar{x}/\theta$ can be used as pivotal quantity. Therefore, $100(1-\alpha)\%$ confidence interval for θ is found out from the equation

$$P[q_1 < Q < q_2] = 1-\alpha \text{ or, } P\left[q_1 < \frac{n\bar{x}}{\theta} < q_2\right] = 1-\alpha$$

$$\text{or, } P\left[\frac{1}{q_2} < \frac{\theta}{n\bar{x}} < \frac{1}{q_1}\right] = 1-\alpha \text{ or, } P\left[\frac{n\bar{x}}{q_2} < \theta < \frac{n\bar{x}}{q_1}\right] = 1-\alpha.$$

Here the values of q_1 and q_2 are obtained from Incomplete Gamma Table, where

$$\int_{q_1}^{q_2} h(u)du = 1-\alpha.$$

Therefore, $100(1-\alpha)\%$ confidence interval for θ is $\left[\dfrac{n\bar{x}}{q_2},\ \dfrac{n\bar{x}}{q_1}\right]$.

(ii) $L = \frac{1}{\theta^n} e^{-n\bar{x}/\theta}, \ \log L = -n \log \theta - \frac{n\bar{x}}{\theta}$

$$\frac{\partial \log L}{\partial \theta} = -\frac{n}{\theta} + \frac{n\bar{x}}{\theta^2} = \frac{n}{\theta^2}[\bar{x} - \theta], \ \frac{\partial^2 \log L}{\partial \theta^2} = \frac{n}{\theta^2} - \frac{2n\bar{x}}{\theta^3}.$$

$$\therefore V\left(\frac{\partial \log L}{\partial \theta}\right) = -E\left[\frac{\partial^2 \log L}{\partial \theta^2}\right] = -\frac{n}{\theta^2} + \frac{2nE(\bar{x})}{\theta^3} = \frac{n}{\theta^2}.$$

Now, we can write

$$z = \frac{\partial \log L / \partial \theta}{\sqrt{V\left(\frac{\partial \log L}{\partial \theta}\right)}} = \frac{\frac{n}{\theta^2}(\bar{x} - \theta)}{\sqrt{n/\theta^2}} \sim N(0, 1).$$

Therefore, $100(1-\alpha)\%$ confidence interval for θ is found out from the equation

$$P\left[-z_{\frac{\alpha}{2}} < z < z_{\frac{\alpha}{2}}\right] = 1 - \alpha$$

$$\text{or, } P\left[-z_{\frac{\alpha}{2}} < \frac{\sqrt{n}(\bar{x} - \theta)}{\theta} < z_{\frac{\alpha}{2}}\right] = 1 - \alpha$$

$$\text{or, } P\left[-z_{\frac{\alpha}{2}} + \sqrt{n} < \frac{\sqrt{n}\bar{x}}{\theta} < z_{\frac{\alpha}{2}} + \sqrt{n}\right] = 1 - \alpha$$

$$\text{or, } P\left[\frac{1}{z_{\frac{\alpha}{2}} + \sqrt{n}} < \frac{\theta}{\sqrt{n}\bar{x}} < \frac{1}{-z_{\frac{\alpha}{2}} + \sqrt{n}}\right] = 1 - \alpha$$

$$\text{or, } P\left[\frac{\sqrt{n}\bar{x}}{\sqrt{n} + z_{\frac{\alpha}{2}}} < \theta < \frac{\sqrt{n}\bar{x}}{\sqrt{n} - z_{\frac{\alpha}{2}}}\right] = 1 - \alpha$$

$$\text{or, } P\left[\frac{\bar{x}}{1 + \frac{1}{\sqrt{n}} z_{\frac{\alpha}{2}}} < \theta < \frac{\bar{x}}{1 - \frac{1}{\sqrt{n}} z_{\frac{\alpha}{2}}}\right] = 1 - \alpha.$$

Here $z_{\frac{\alpha}{2}}$ is found out from normal probability table. Therefore, $100(1-\alpha)\%$ confidence interval for θ is $\left[\frac{\bar{x}}{1 + \frac{1}{\sqrt{n}} z_{\frac{\alpha}{2}}}, \frac{\bar{x}}{1 - \frac{1}{\sqrt{n}} z_{\frac{\alpha}{2}}}\right]$.

►**EXAMPLE 35.5** *Find* $100(1-\alpha)\%$ *confidence interval of* p *in* $b(n,p)$ *using large sample.*

Solution. We have

$$P(x) = \binom{n}{x} p^x q^{n-x}, \ x = 0, 1, 2, \ldots, n$$

$$L = \binom{n}{x} p^x q^{n-x}, \ \log L = \log \binom{n}{x} + x \log p + (n - x) \log(1 - p)$$

$$\frac{\partial \log L}{\partial p} = \frac{x}{p} - \frac{n-x}{1-p} = \frac{n}{p(1-p)}\left[\frac{x}{n} - p\right], \ \frac{\partial^2 \log L}{\partial p^2} = -\frac{x}{p^2} - \frac{n-x}{(1-p)^2}$$

$$V\left[\frac{\partial \log L}{\partial p}\right] = -E\left[\frac{\partial^2 \log L}{\partial p^2}\right] = \frac{E(x)}{p^2} - \frac{E(n-x)}{1-p^2} = \frac{n}{p(1-p)}.$$

$$\therefore \ z = \frac{\frac{n}{p(1-p)}\left(\frac{x}{n} - p\right)}{\sqrt{\frac{n}{p(1-p)}}} \sim N(0,1) = \frac{\sqrt{n}}{\sqrt{p(1-p)}}\left(\frac{x}{n} - p\right) \sim N(0,1).$$

Hence $100(1-\alpha)\%$ confidence interval for p is found out from the equation

$$P\left[-z_{\frac{\alpha}{2}} < z < z_{\frac{\alpha}{2}}\right] = 1-\alpha$$

$$\text{or, } P\left[-z_{\frac{\alpha}{2}} < \frac{\sqrt{n}}{\sqrt{p(1-p)}}\left(\frac{x}{n}-p\right) < z_{\frac{\alpha}{2}}\right] = 1-\alpha$$

$$\text{or, } P\left[|z| \le z_{\frac{\alpha}{2}}\right] = 1-\alpha, \text{ where } z = \frac{x-np}{\sqrt{np(1-p)}}.$$

$$\therefore \frac{(x-np)^2}{np(1-p)} = z^2_{\frac{\alpha}{2}} \text{ or, } (x-np)^2 = np(1-p)z^2_{\frac{\alpha}{2}}.$$

Here the equation is a quadratic equation in p. Solving this equation, we have

$$p = \frac{\left(2nx + nz^2_{\frac{\alpha}{2}}\right) \pm \sqrt{\left[2nx + nz^2_{\frac{\alpha}{2}} - 4\left(n^2 + nz^2_{\frac{\alpha}{2}}\right)x^2\right]}}{2\left(n^2 + nz^2_{\frac{\alpha}{2}}\right)}.$$

►**EXAMPLE 35.6** *Let $x_1, x_2, \ldots, x_n$ be a random sample of n observations from a population with p.d.f.*

$$f(x) = \theta x^{\theta-1}, \; 0 < x < 1.$$

Find $100(1-\alpha)\%$ confidence interval for θ.

Solution. We have

$$F(x,\theta) = x^\theta, \; 0 < x < 1.$$

$$\therefore P\left[q_1 < \prod_{i=1}^{n} F(x_i,\theta) < q_2\right] = 1-\alpha$$

$$\text{or, } P\left[q_1 < \prod_{i=1}^{n} x_i^\theta < q_2\right] = 1-\alpha$$

$$\text{or, } P\left[\log q_1 < \theta \prod_{i=1}^{n} \log x_i < \log q_2\right] = 1-\alpha$$

$$\text{or, } P\left[-\log q_2 < -\theta \prod_{i=1}^{n} \log x_i < -\log q_1\right] = 1-\alpha$$

$$\text{or, } P\left[\frac{\log q_2}{\log \prod_{i=1}^{n} x_i} < \theta < \frac{\log q_1}{\log \prod_{i=1}^{n} x_i}\right] = 1-\alpha.$$

Hence $100(1-\alpha)\%$ confidence interval of θ is

$$\left[\frac{\log q_2}{\log \prod_{i=1}^{n} x_i}, \frac{\log q_1}{\log \prod_{i=1}^{n} x_i}\right].$$

►**EXAMPLE 35.7** *Let $x_1, x_2, \ldots, x_n$ be a random sample of n observations from a population having p.d.f.*

$$f(x) = \frac{1}{\theta}e^{-x/\theta};\ 0 < x < \infty,\ \theta > 0.$$

(i) *Find $100(1-\alpha)\%$ confidence interval for θ.*

(ii) *What would be the confidence interval if n is large?*

Solution.

(i) We have

$$F(x) = \frac{1}{\theta}\int_0^x e^{-t/\theta}dt = 1 - e^{-x/\theta};\ \therefore\ 1 - F(x) = e^{-x/\theta}.$$

Let $z = -\log[1 - F(x)] = -\log e^{-x/\theta} = \dfrac{x}{\theta}$.
Here z is a gamma variate with parameter $m = 1$, i.e.,

$$g(z) = e^{-z} - \frac{1}{\Gamma 1}e^{-z}z^{1-1},\ 0 < z < \infty.$$

Hence $u = \Sigma z = \dfrac{n\bar{x}}{\theta}$ is a gamma variate with parameter n. Thus

$$h(u) = \frac{1}{\Gamma n}e^{-u}u^{n-1},\ 0 < u < \infty.$$

Now, we can use $n\bar{x}/\theta$ as the pivotal quantity. Hence $100(1-\alpha)\%$ confidence limit is obtained from the solution of the equation

$$P[q_1 < Q < q_2] = 1 - \alpha \text{ or, } P\left[q_1 < \frac{n\bar{x}}{\theta} < q_2\right] = 1 - \alpha$$

$$\text{or, } P\left[\frac{1}{q_2} < \frac{\theta}{n\bar{x}} < \frac{1}{q_1}\right] = 1 - \alpha \text{ or, } P\left[\frac{n\bar{x}}{q_2} < \theta < \frac{n\bar{x}}{q_1}\right] = 1 - \alpha.$$

Here q_1 and q_2 are to be found out from Incomplete Gamma Table, where

$$\int_{q_1}^{q_2} h(u)du = 1 - \alpha.$$

Therefore, $100(1-\alpha)\%$ confidence interval is $\left[\dfrac{n\bar{x}}{q_2}, \dfrac{n\bar{x}}{q_1}\right]$.

(ii) We have

$$L = \frac{1}{\theta^n}e^{-n\bar{x}/\theta},\ \log L = -n\log\theta - \frac{n\bar{x}}{\theta}$$

$$\frac{\partial \log L}{\partial \theta} = \frac{-n}{\theta} + \frac{n\bar{x}}{\theta^2} = \frac{n}{\theta^2}[\bar{x} - \theta],\quad \frac{\partial^2 \log L}{\partial \theta^2} = \frac{n}{\theta^2} - \frac{2n\bar{x}}{\theta^3}$$

$$V\left[\frac{\partial \log L}{\partial \theta}\right] = -E\left[\frac{\partial^2 \log L}{\partial \theta^2}\right] = -\frac{n}{\theta^2} + \frac{2nE(\bar{x})}{\theta^3} = \frac{n}{\theta^2}.$$

$$\therefore\ z = \frac{\frac{n}{\theta^2}(\bar{x} - \theta)}{\sqrt{n/\theta^2}} \sim N(0, 1).$$

Hence $100(1-\alpha)\%$ confidence interval for θ is to be found out from the equation

$$P\left[-z_{\alpha/2} < z < z_{\alpha/2}\right] = 1-\alpha$$

$$\text{or,}\quad P\left[-z_{\alpha/2} < \frac{\frac{n}{\theta^2}(\bar{x}-\theta)}{\sqrt{\frac{n}{\theta^2}}} < z_{\alpha/2}\right] = 1-\alpha$$

$$\text{or,}\quad P\left[-z_{\alpha/2}+\sqrt{n} < \frac{\sqrt{n}\bar{x}}{\theta} < z_{\alpha/2}+\sqrt{n}\right] = 1-\alpha$$

$$\text{or,}\quad P\left[\frac{1}{z_{\alpha/2}+\sqrt{n}} < \frac{\theta}{\sqrt{n}\bar{x}} < \frac{1}{-z_{\alpha/2}+\sqrt{n}}\right] = 1-\alpha$$

$$\text{or,}\quad P\left[\frac{\sqrt{n}\bar{x}}{z_{\alpha/2}+\sqrt{n}} < \theta < \frac{\sqrt{n}\bar{x}}{\sqrt{n}-z_{\alpha/2}}\right] = 1-\alpha.$$

Hence $100(1-\alpha)\%$ confidence interval for θ is $\left[\dfrac{\bar{x}}{1+\frac{1}{\sqrt{n}}z_{\frac{\alpha}{2}}}, \dfrac{\bar{x}}{1-\frac{1}{\sqrt{n}}z_{\frac{\alpha}{2}}}\right]$.

►**EXAMPLE 35.8** *Let $x_1, x_2, \ldots, x_n$ be a random sample of n observations from a population having p.d.f.*

$$f(x) = \frac{1}{\Gamma\lambda}\left(\frac{\lambda}{\theta}\right)^{\lambda} e^{-\lambda/\theta^x} x^{\lambda-1};\ 0 < x < \infty,\ \lambda > 0.$$

Let the ML estimator of θ be $\hat{\theta} = \bar{x}$. Then assuming n large, find $100(1-\alpha)\%$ confidence interval of θ.

Solution. We have

$$\frac{\partial \log L}{\partial\lambda} = \frac{1}{2\lambda} + 1 - \log\theta - \frac{\bar{x}}{\theta} + \log G, \text{ where } \log G = \frac{1}{n}\sum_{i=1}^{n}\log x_i$$

$$\frac{\partial \log L}{\partial\lambda} = \frac{1}{2\lambda} + \log\frac{G}{\theta} + 1 - \frac{\bar{x}}{\theta}$$

$$\frac{\partial^2 \log L}{\partial\lambda^2} = -\frac{1}{2\lambda^2},\quad V\left(\frac{\partial \log L}{\partial\lambda}\right) = -E\left[\frac{\partial^2 \log L}{\partial\lambda^2}\right] = \frac{1}{2\lambda^2}.$$

$$\therefore\ z = \frac{\frac{1}{2\lambda} + \log\frac{G}{\theta} + 1 - \bar{x}}{1/\sqrt{2\lambda^2}} \sim N(0,1)$$

$$= \frac{1-2\lambda u}{\sqrt{2}},\ \text{where } u = -\left[\log\frac{G}{\theta} + 1 - \frac{\bar{x}}{\theta}\right].$$

Hence $100(1-\alpha)\%$ confidence interval for λ is obtained from the equation

$$P\left[-z_{\frac{\alpha}{2}} < z < z_{\frac{\alpha}{2}}\right] = 1-\alpha$$

$$\text{or,}\quad P\left[-z_{\frac{\alpha}{2}} < \frac{1-2\lambda u}{\sqrt{2}} < z_{\frac{\alpha}{2}}\right] = 1-\alpha$$

$$\text{or,}\quad P\left[1-\sqrt{2}z_{\frac{\alpha}{2}} < 2\lambda u < 1+\sqrt{2}z_{\frac{\alpha}{2}}\right] = 1-\alpha.$$

Therefore, $\left[1-\sqrt{2}z_{\frac{\alpha}{2}}, 1+\sqrt{2}z_{\frac{\alpha}{2}}\right]$ is the required $100(1-\alpha)\%$ confidence interval of λ.
Here, if $\hat{\theta} = \bar{x}$, then $u = \log\dfrac{\bar{x}}{G}$.

35.2.3 Statistical Method of Confidence Interval

Let $x_1, x_2, \ldots, x_n$ be a random sample of n observations from a population having p.d.f. $f(x, \theta)$, $\theta \in S$. Let θ_0 be the real value of θ and $T = t(x_1, x_2, \ldots, x_n)$ be a sample statistic. This T may be sufficient statistic. It may also be ML estimator and it is found out in such a way that $g(T, \theta)$ exists. Assume that the p.d.f. of T is $g(T, \theta)$ and it is continuous. Now, define two functions of θ, say $h_1(\theta)$ and $h_2(\theta)$, such that

$$\int_{-\infty}^{h_1(\theta)} g(T, \theta)dt = p_1 \text{ and } \int_{h_2(\theta)}^{\infty} g(T, \theta)dt = p_2$$

where $0 < p_1, 0 < p_2$ and $p_1 + p_2 = 1$. Graphically, this probability can be shown as follows in FIG. 35.1:

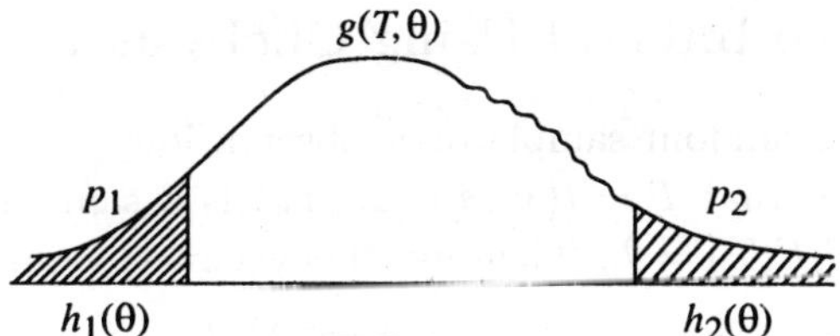

FIG. 35.1

Here $h_1(\theta) < h_2(\theta)$. Let us assume that $h_1(\theta)$ and $h_2(\theta)$ are the monotone increasing functions of θ. The graphs of $h_1(\theta)$ and $h_2(\theta)$ against θ will take the following shape (in FIG. 35.2). Let t_0 be the value of T, $v_1 = v_1(t_0), v_2 = v_2(t_0)$, where v_1 and v_2 are functions of t_0.

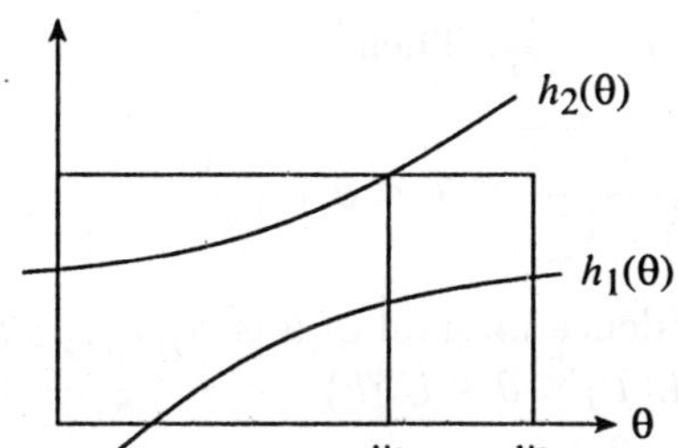

FIG. 35.2

Here the values of p_1 and p_2 are to be determined in such a way that the confidence limits have shortest distance. The method is applicable ever $h_1(\cdot)$ and $h_2(\cdot)$ are not monotonic functions. Here (v_1, v_2) is the $(1 - p_1 - p_2)100\%$ confidence interval for θ_0.

►**EXAMPLE 35.9** *Let $x_1, x_2, \ldots, x_n$ be a random sample of n observations from Bernoulli population having parameter θ. Find $100(1-\alpha)\%$ confidence interval for θ.*

Solution. We know

$$P(x = 1) = \theta, \; P(x = 0) = 1 - \theta; \; x = 0, 1.$$

Let $T = \Sigma x$ be the sufficient statistic for θ; then

$$g(T, \theta) = \binom{n}{t} \theta^t (1 - \theta)^{n-t}, \; t = 0, 1, 2, \ldots, n.$$

The values of v_1 and v_2 are found out in such a way that

$$p_1 = \sum_{t=0}^{t_0} \binom{n}{t} \theta^t (1-\theta)^{n-t} \text{ and } p_2 = \sum_{t=t_0}^{n} \binom{n}{t} \theta^t (1-\theta)^{n-t}.$$

Solving the above two equations, we shall get v_1 and v_2.

Let $p_1 = 0.0510$, $p_2 = 0.0026$. Then to determine $(1 - p_1 - p_2)100\% = 94.64$ confidence interval we have to use 'Binomial Probability Table'. Now, if n and t_0 are known, the value θ can be found out. Let $n = 10$, $t_0 = 2$. Then for $p_1 = 0.0547$ and $p_2 = 0.0115$, the values of θ are 0.50 and 0.05. Hence 93.38% confidence interval for θ is $[0.05, 0.50]$.

35.2.4 Confidence Interval Using Chebyshev's Inequality

Let $x_1, x_2, \ldots, x_n$ be a random sample of n observations from a population having $f(x,\theta), \theta \in S$. Consider that $T = t(x_1, x_2, \ldots, x_n)$ is a sample statistic and it is an unbiased estimator of $\theta[E(T) = \theta]$. Then by Chebyshev's inequality, we can write

$$P[|T-\theta| \leq \epsilon] \geq 1 - \frac{V(T)}{\epsilon^2}.$$

Let $\epsilon = k\sqrt{V(T)}$; then

$$P[|T-\theta| \leq k\sqrt{V(T)}] \geq 1 - \frac{1}{k^2}.$$

Again, let $1 - \alpha = 1 - \frac{1}{k^2} \Rightarrow k = \frac{1}{\sqrt{\alpha}}$. Then

$$P\left[\theta - \sqrt{\frac{V(T)}{\alpha}} \leq T \leq \theta + \sqrt{\frac{V(T)}{\alpha}}\right] \geq 1 - \alpha.$$

Hence the $100(1-\alpha)\%$ confidence interval of θ is $[L(T), U(T)]$, where $L(T)$ and $U(T)$ are the functions of T, i.e., $L(T) \leq \theta \leq U(T)$.

►**EXAMPLE 35.10** *Let $x_1, x_2, \ldots, x_n$ be a random sample of n observations from a population having p.d.f.*

$$f(x) = \begin{cases} \frac{1}{\theta} e^{-x/\theta}, & x \geq 0 \\ 0, & \textit{otherwise.} \end{cases}$$

Find $100(1-\alpha)\%$ confidence interval using one sample observation.

Solution. Let $u = \frac{x}{\theta}$, $du = \frac{dx}{\theta}$, where $u \geq 0$. The distribution of u is

$$g(u) = e^{-u}, \quad u \geq 0.$$

Here $g(u)$ is free of θ. Therefore, x/θ can be used as a pivotal quantity. Thus q_1 and q_2 are to be found out in such a way that

$$P[q_1 < Q < q_2] = 1 - \alpha \text{ or, } P\left[q_1 < \frac{x}{\theta} < q_2\right] = 1 - \alpha.$$

Here

$$\int_0^{q_1} g(u)du = \frac{\alpha}{2} \text{ or, } \int_0^{q_1} e^{-u}du = \frac{\alpha}{2} \Rightarrow q_1 = -\log\left(1-\frac{\alpha}{2}\right).$$

Again,

$$\int_q^{\infty} e^{-u}du = \frac{\alpha}{2} \Rightarrow q_2 = -\log\frac{\alpha}{2}.$$

Hence

$$P\left[-\log\left(1-\frac{\alpha}{2}\right) < \frac{x}{\theta} < -\log\frac{\alpha}{2}\right] = 1-\alpha$$

$$\text{or, } P\left[\frac{1}{-\log\frac{\alpha}{2}} < \frac{\theta}{x} < \frac{1}{-\log\left(1-\frac{\alpha}{2}\right)}\right] = 1-\alpha$$

$$\text{or, } P\left[\frac{x}{-\log\frac{\alpha}{2}} < \theta < \frac{x}{-\log\left(1-\frac{\alpha}{2}\right)}\right] = 1-\alpha.$$

Therefore, $100(1-\alpha)\%$ confidence interval of θ is $\left[\frac{x}{-\log\frac{\alpha}{2}}, \frac{x}{-\log\left(1-\frac{\alpha}{2}\right)}\right]$.

►**EXAMPLE 35.11** *Let $x_1, x_2, \ldots, x_n$ be a random sample of n observations from a population with p.d.f.*

$$f(x) = \frac{1}{\theta}, \; 0 < x < \theta.$$

Find $100(1-\alpha)\%$ confidence interval for θ using Chebyshev's inequality.

Solution. For $f(x) = 1/\theta$, $E(x) = \theta/2$, $V(x) = \theta^2/12$. Hence $T = 2x$ is the unbiased estimate of θ, where $V(T) = \theta^2/3$. Therefore,

$$\theta - \frac{\theta}{\sqrt{3\alpha}} \le 2x \le \theta + \frac{\theta}{\sqrt{3\alpha}} \text{ or, } \frac{2x}{1+\frac{1}{\sqrt{3\alpha}}} \le \theta \le \frac{2x}{1-\frac{1}{\sqrt{3\alpha}}}$$

$\left[\frac{2x}{1+\frac{1}{\sqrt{3\alpha}}}, \frac{2x}{1-\frac{1}{\sqrt{3\alpha}}}\right]$ is the $100(1-\alpha)\%$ confidence interval for θ.

It is to be noted that, if $0 < \alpha < 1/3$, then $1 - \frac{1}{\sqrt{3\alpha}} < 0$ and $1 + \frac{1}{\sqrt{3\alpha}} > 2$. In such a case the confidence interval by Chebyshev's inequality is not meaningful since the interval becomes $[x, \infty]$. The length of the interval becomes large.

35.3 Uniformly Most Accurate (UMA) Confidence Interval

Let $x_1, x_2, \ldots, x_n$ be a random sample of n observations from a population with p.d.f., $f(x, \theta)$, $\theta \in S$. Let $T = t(x_1, x_2, \ldots, x_n)$ be a statistic to estimate $\psi(\theta)$. Now, if $P[T \le \psi(\theta)] = 1-\alpha$, then T is called *lower-tailed* $100(1-\alpha)\%$ *confidence interval* of $\psi(\theta)$. In such a situation, for all $\theta' < \theta$, if $P[T_1 \le \psi(\theta')]$ becomes minimum, then T_1

is called *lower-tailed UMA confidence interval.* In a similar situation, if $P[T_2 \geq \psi(\theta')]$ becomes minimum, then T_2 is called *upper-tailed UMA confidence interval.*

Definition. Let $\{T = t(x_1, x_2, \ldots, x_n)\}$ be a set of $100(1-\alpha)\%$ confidence interval of θ in $f(x, \theta)$. This set belongs to the member of UMA confidence interval, if for all T and T_1

$$P[T \in \theta] \leq P[T_1 \leq \theta].$$

►**EXAMPLE 35.12** *Let $x_1, x_2, \ldots, x_n$ be a random sample of n observations selected from a population with p.d.f.*

$$f(x) = \frac{2x}{\theta}e^{-x^2/\theta}, \ x > 0.$$

Find $100(1-\alpha)\%$ confidence interval for θ.

Solution. We have

$$L = \left(\frac{2}{\theta}\right)^n \prod_{i=1}^{n} x_i e^{-\Sigma x_i^2/\theta} = \exp\left[-\frac{\Sigma x^2}{\theta} + \Sigma \log x_i + n \log\frac{2}{\theta}\right].$$

Therefore, Σx^2 is the sufficient statistic for θ. Now, let

$$\begin{aligned} u_i &= x_i^2 \Rightarrow 2x dx = du, \ u > 0. \\ \therefore \ f(u) &= \frac{1}{\theta}e^{-u/\theta}, \ u > 0. \end{aligned}$$

Again, let

$$w = \frac{2u}{\theta} \Rightarrow du = \frac{\theta}{2}dw, \ w > 0.$$

Then

$$g(w) = \frac{1}{2}e^{-w/2} = \frac{1}{2^{2}2\Gamma 2}e^{-\frac{1}{2}w}w^{22-1}, \ w > 0.$$

Here w is the chi-square variate with 2 d.f. Since $x_1, x_2, \ldots, x_n$ are independent observations, $u_1, u_2, \ldots, u_n$ and $w_1, w_2, \ldots, w_n$ are also independent. Therefore, $\sum_{i=1}^{n} w_i$ is also chi-square variate with $2n$ d.f. The distribution is free of θ. But

$$\sum_{i=1}^{n} w_i = \frac{2}{\theta}\Sigma u_i = \frac{2}{\theta}\Sigma x_i^2.$$

Therefore, $\frac{2}{\theta}\Sigma x_i^2$ can be used as pivotal quantity and hence $100(1-\alpha)\%$ confidence interval for θ is found out from the equation

$$\begin{aligned} &P[q_1 < Q < q_2] = 1 - \alpha \\ \text{or,} \quad &P\left[q_1 < \frac{2}{\theta}\Sigma x_i^2 < q_2\right] = 1 - \alpha \\ \text{or,} \quad &P\left[\frac{1}{q_2} < \frac{\theta}{2\Sigma x_i^2} < \frac{1}{q_1}\right] = 1 - \alpha \\ \text{or,} \quad &P\left[\frac{2\Sigma x_i^2}{q_2} < \theta < \frac{2\Sigma x_i^2}{q_1}\right] = 1 - \alpha. \end{aligned}$$

Since $\frac{2}{\theta}\Sigma x_i^2 \sim \chi^2_{2n}$, q_1 and q_2 are to be found out from tabulated value of χ^2 with $2n$ d.f., where

$$q_1 \sim \chi^2_{2n,1-\alpha/2} \text{ and } q_2 \sim \chi^2_{2n,\alpha/2}.$$

Therefore, $100(1-\alpha)\%$ confidence interval of θ is

$$\left[\frac{2\Sigma x_i^2}{\chi^2_{2n,\alpha/2}}, \frac{2\Sigma x_i^2}{\chi^2_{2n,1-\alpha/2}}\right].$$

35.4 Unbiased Confidence Interval

Let $x_1, x_2, \ldots, x_n$ be a random sample of n observations from a population with p.d.f. $f(x,\theta), \theta \in S$. Let $T = t(x_1, x_2, \ldots, x_n)$ be any statistic for θ and Q be the pivotal quantity to find confidence interval for θ, where Q is the function of T. Using this pivotal quantity, if it is possible to find

$$P[l(x_1, x_2, \ldots, x_n) < \theta < u(x_1, x_2, \ldots, x_n)] = 1-\alpha$$

then $[l(x_1, x_2, \ldots, x_n), u(x_1, x_2, \ldots, x_n)]$ is the $100(1-\alpha)\%$ confidence interval for θ. The limits $l(\cdot)$ and $u(\cdot)$ become unbiased, if for $\theta = \theta'$

$$P(\theta, \theta') = P[l(x_1, x_2, \ldots, x_n) < \theta' < u(x_1, x_2, \ldots, x_n) = 1-\alpha \tag{35.2}$$

and for $\theta \neq \theta'$

$$P(\theta, \theta') = P[l(x_1, x_2, \ldots, x_n) < \theta' < u(x_1, x_2, \ldots, x_n)] < 1-\alpha. \tag{35.3}$$

If $P(\theta, \theta')$ depends on $\gamma = \gamma(\theta, \theta')$, then

$$P(\gamma) = \begin{cases} 1-\alpha, & \text{if } \theta' = \theta \\ < 1-\alpha, & \text{if } \theta' \neq \theta. \end{cases}$$

It is to be noted that $P(\gamma)$ becomes maximum, if $\theta = \theta'$.

►**EXAMPLE 35.13** *Let $x_1, x_2, \ldots, x_n$ be a random sample of n observations from $N(\mu, \sigma^2)$. Find $100(1-\alpha)\%$ unbiased confidence interval for σ^2.*

Solution. The unbiased estimator of σ^2 is

$$s^2 = \frac{1}{n-1}\Sigma(x-\bar{x})^2.$$

The pivotal quantity is $Q = \frac{(n-1)s^2}{\sigma^2}$, where Q is distributed as chi-square with $(n-1)$ d.f. Therefore,

$$P\left[q_1 < \frac{(n-1)s^2}{\sigma^2} < q_2\right] = 1-\alpha$$

$$\text{or,} \quad P\left[\frac{(n-1)s^2}{q_2} < \sigma^2 < \frac{(n-1)s^2}{q_1}\right] = 1-\alpha.$$

Then

$$\begin{aligned} P(\sigma^2, \sigma'^2) &= P\left[\frac{(n-1)s^2}{q_2} < \sigma^2 < \frac{(n-1)s^2}{q_1}\right] \\ &= P\left[\frac{(n-1)s^2}{\sigma^2 q_2} < \frac{\sigma'^2}{\sigma^2} < \frac{(n-1)s^2}{\sigma^2 q_1}\right] \\ &= P\left[\frac{Q}{q_2} < \gamma < \frac{Q}{q_1}\right], \ \gamma = \frac{\sigma'^2}{\sigma^2}. \\ \therefore \ P(\gamma) &= P[q_1\gamma < Q < q_2\gamma]. \end{aligned}$$

Now, using (35.2) and (35.3) in section 35.4, we get $P(1) = 1 - \alpha$ and $P(\gamma) < 1 - \alpha$. Hence q_1 and q_2 are to be determined in such a way that

$$P(1) = 1 - \alpha \tag{1}$$

$$\text{and} \quad \left.\frac{\partial P(\gamma)}{\partial \gamma}\right|_{\gamma=1} = q_2 f_{n-1}(q_2) - q_1 f_{n-1}(q_1) = 0. \tag{2}$$

Here f_{n-1} is the d.f. of Q. Now, if q_1 and q_2 are available, the unbiased confidence interval of σ^2 is

$$\left[\frac{(n-1)s^2}{q_2}, \frac{(n-1)s^2}{q_1}\right].$$

This confidence interval is uniformly most accurate unbiased (UMAU) confidence interval.

►**EXAMPLE 35.14** *Let $x_1, x_2, \ldots, x_n$ be a random sample of n observations selected from a population with p.d.f.*

$$f(x) = \frac{1}{\theta} e^{-x/\theta}, \ x > 0.$$

Find $100(1-\alpha)\%$ *confidence interval for θ using $2\Sigma x/\theta$ as pivotal quantity.*

Solution. From Example 35.4, we get

$$\begin{aligned} f(z) &= \frac{1}{\Gamma n} e^{-z} z^{n-1}, \ z > 0, \text{ where} \\ u &= x/\theta \text{ and } f(u) = e^{-u}, \ u > 0 \\ \text{and } z &= \sum^{n} u_i = \Sigma x_i/\theta. \end{aligned}$$

Therefore, if $2\Sigma x/\theta$ is used as pivotal quantity, then $100(1-\alpha)\%$ confidence interval for θ is found out from the equation

$$\begin{aligned} &P[q_1 < Q < q_2] = 1 - \alpha \\ \text{or,} \ &P\left[q_1 < \frac{2\Sigma x}{\theta} < q_2\right] = 1 - \alpha \\ \text{or,} \ &P\left[\frac{2\Sigma x}{q_2} < \theta < \frac{2\Sigma x}{q_1}\right] = 1 - \alpha. \end{aligned}$$

Then

$$\begin{aligned} P(\theta,\theta') &= P\left[\frac{2\Sigma x}{q_2} < \theta' < \frac{2\Sigma x}{q_1}\right] = 1-\alpha \\ &= P\left[\frac{2\Sigma x}{\theta q_2} < \frac{\theta'}{\theta} < \frac{2\Sigma x}{\theta q_1}\right] = 1-\alpha \\ &= P\left[\frac{Q}{q_2} < \gamma < \frac{Q}{q_1}\right] = 1-\alpha. \\ \therefore\ P(\gamma) &= P[q_1\gamma < Q < q_2\gamma]. \end{aligned}$$

Hence $P(1) = 1-\alpha$ and $P(\gamma) < 1-\alpha$.

Here $Q = \dfrac{2\Sigma x}{\theta}$ follows χ^2-distribution with $(n-1)$ d.f. Therefore, q_1 and q_2 are to be found out in such a way that

$$P(1) = 1-\alpha \tag{1}$$

$$\text{and}\quad \left.\frac{dP(\gamma)}{d\gamma}\right|_{\gamma=1} = q_2 f_n(q_2) - q_1 f_n(q_1) = 0. \tag{2}$$

Here f_n is the d.f. of χ^2-variate. Now, solving (1) and (2) the values of q_1 and q_2 are found out. Therefore, $100(1-\alpha)\%$ confidence interval of θ is

$$\left[\frac{2\Sigma x}{q_2}, \frac{2\Sigma x}{q_1}\right].$$

35.5 Confidence Interval for the Difference of Two Means

Let $x_1, x_2, \ldots, x_{n_1}$ be a random sample of n_1 observations from $N(\mu_1, \sigma_1^2)$ and $y_1, y_2, \ldots, y_{n_2}$ be another random sample of n_2 observations from $N(\mu_2, \sigma_2^2)$. Using these two sample observations we need to find $100(1-\alpha)\%$ confidence interval for $\mu_1 - \mu_2$.

Let us assume that $\sigma_1^2 = \sigma_2^2 = \sigma^2$ (say). The unbiased estimate of common population variance σ^2 is obtained as follows:

The sample variances of first and second samples are, respectively

$$s_1^2 = \frac{1}{n_1-1}\sum_{i=1}^{n_1}(x_i-\bar{x})^2 \text{ and } s_2^2 = \frac{1}{n_2-1}\sum_{j=1}^{n_2}(y_j-\bar{y})^2, \text{ where}$$

$$\bar{x} = \frac{1}{n_1}\Sigma x_i \text{ and } \bar{y} = \frac{1}{n_2}\Sigma y_j.$$

We know

$$E(\bar{x}) = \mu_1, E(\bar{y}) = \mu_2 \Rightarrow E(\bar{x}-\bar{y}) = \mu_1 - \mu_2.$$

Therefore,

$$z = \frac{(\bar{x}-\bar{y}) - E(\bar{x}-\bar{y})}{\sqrt{V(\bar{x}-\bar{y})}} \sim N(0,1).$$

Here

$$\begin{aligned} V(\bar{x}-\bar{y}) &= V(\bar{x})+V(\bar{y}), \because \text{ the samples are independent} \\ &= \frac{\sigma_1^2}{n_1}+\frac{\sigma_2^2}{n_2}=\sigma^2\left(\frac{1}{n_1}+\frac{1}{n_2}\right), \quad \because \sigma_1^2=\sigma_2^2=\sigma^2. \\ \therefore z &= \frac{(\bar{x}-\bar{y})-(\mu_1-\mu_2)}{\sqrt{\sigma^2\left(\frac{1}{n_1}+\frac{1}{n_2}\right)}} \sim N(0,1). \end{aligned}$$

Let

$$\begin{aligned} \chi^2 &= \left[\sum_i (x_i-\bar{x})^2+\Sigma(y_j-\bar{y})^2\right]/\sigma^2 \\ &= \frac{\Sigma(x_i-\bar{x})^2}{\sigma^2}+\frac{\Sigma(y_j-\bar{y})^2}{\sigma^2}=\frac{(n_1-1)s_1^2}{\sigma^2}+\frac{(n_2-1)s_2^2}{\sigma^2}. \end{aligned}$$

But $\frac{(n_1-1)s_1^2}{\sigma^2}$ and $\frac{(n_2-1)s_2^2}{\sigma^2}$ are distributed independently as chi-square with (n_1-1) and (n_2-1) d.f., respectively. Therefore, χ^2 defined above is distributed as chi-square with $(n_1-1)+(n_2-1)=(n_1+n_2-2)$ d.f.

Here

$$s^2=\frac{(n_1-1)s_1^2+(n_2-1)s_2^2}{n_1+n_2-2}$$

is an unbiased estimate of σ^2, since

$$\begin{aligned} E(s^2) &= \frac{1}{(n_1+n_2-2)}E[(n_1-1)s_1^2+(n_2-1)s_2^2] \\ &= \frac{1}{(n_1+n_2-2)}[(n_1-1)\sigma^2+(n_2-1)\sigma^2]=\sigma^2. \end{aligned}$$

Hence Fisher's 't' is defined by

$$t=\frac{z}{\sqrt{\frac{\chi^2}{n_1+n_2-2}}}=\frac{(\bar{x}-\bar{y})-(\mu_1-\mu_2)}{\sqrt{\sigma^2\left(\frac{1}{n_1}+\frac{1}{n_2}\right)}}\Bigg/\sqrt{\frac{\chi^2}{n_1+n_2-2}}=\frac{(\bar{x}-\bar{y})-(\mu_1-\mu_2)}{s\sqrt{\left(\frac{1}{n_1}+\frac{1}{n_2}\right)}}.$$

This 't' has (n_1+n_2-2) d.f. and it is free of the parameters $\mu_1-\mu_2$. Hence this statistic can be used as pivotal quantity.

Now, the $100(1-\alpha)\%$ confidence interval can be found out from the equation

$$P[q_1<Q<q_2]=1-\alpha$$

$$\text{or, } P\left[q_1<\frac{(\bar{x}-\bar{y})-(\mu_1-\mu_2)}{s\sqrt{\left(\frac{1}{n_1}+\frac{1}{n_2}\right)}}<q_2\right]=1-\alpha$$

$$\text{or, } P\left[q_1 s\sqrt{\left(\frac{1}{n_1}+\frac{1}{n_2}\right)}<(\bar{x}-\bar{y})-(\mu_1-\mu_2)<q_2 s\sqrt{\left(\frac{1}{n_1}+\frac{1}{n_2}\right)}\right]=1-\alpha.$$

Here q_1 and q_2 are obtained from

$$\int_{q_1}^{q_2} f(t)dt = 1-\alpha = \int_{-t_{\frac{\alpha}{2}}}^{t_{\frac{\alpha}{2}}} f(t)dt = 1-\alpha.$$

Therefore,

$$P\left[-t_{\frac{\alpha}{2}} s\sqrt{\left(\frac{1}{n_1}+\frac{1}{n_2}\right)} < (\bar{x}-\bar{y})-(\mu_1-\mu_2) < t_{\frac{\alpha}{2}} s\sqrt{\left(\frac{1}{n_1}+\frac{1}{n_2}\right)}\right] = 1-\alpha$$

$$P\left[\bar{x}-\bar{y}-t_{\frac{\alpha}{2}} s\sqrt{\left(\frac{1}{n_1}+\frac{1}{n_2}\right)} < \mu_1-\mu_2 < \bar{x}-\bar{y}+t_{\frac{\alpha}{2}} s\sqrt{\left(\frac{1}{n_1}+\frac{1}{n_2}\right)}\right] = 1-\alpha.$$

Hence the $100(1-\alpha)\%$ confidence interval for $\mu_1-\mu_2$ is

$$\left[\bar{x}-\bar{y}-t_{\frac{\alpha}{2}} s\sqrt{\left(\frac{1}{n_1}+\frac{1}{n_2}\right)},\ \bar{x}-\bar{y}+t_{\frac{\alpha}{2}} s\sqrt{\left(\frac{1}{n_1}+\frac{1}{n_2}\right)}\right].$$

But, if $(x_1, y_1), (x_2, y_2), \ldots, (x_n, y_n)$ are pairs of random sample observations of size n from a bivariate normal distribution, then $100(1-\alpha)\%$ confidence interval for $\mu_1-\mu_2$ is

$$\left[\bar{D}-t_{\frac{\alpha}{2}}\sqrt{\frac{\Sigma(D-\bar{D})^2}{n(n-1)}},\ \bar{D}+t_{\frac{\alpha}{2}}\sqrt{\frac{\Sigma(D-\bar{D})^2}{n(n-1)}}\right]$$

where

$$D_i = x_i - y_i;\ i=1,2,\ldots,n$$

$$\text{and } \bar{D} = \frac{1}{n}\sum^{n} D_i.$$

►**EXAMPLE 35.15** *The following data represent the body-weights (in gm) of two different types of insects:*

Insect-1, x_i : 2.4, 1.5, 2.0, 1.9, 2.3, 2.9, 2.8, 2.6, 3.5, 4.5, 3.7, 5.5, 2.8, 6.5, 5.6

Insect-2, y_j : 2.6, 1.5, 1.3, 2.4, 2.9, 3.0, 3.1, 3.9, 2.9, 3.9, 4.8

It is assumed that

$$x_i \sim N(\mu_1,\ \sigma_1^2),\ y_j \sim N(\mu_2,\ \sigma_2^2),\ \sigma_1^2=\sigma_2^2=\sigma^2.$$

Find 95% *confidence interval for* $\mu_1-\mu_2$.

Solution. Given $n_1 = 15$, $n_2 = 11$. We have

$$\bar{x} = 3.37,\ \bar{y}=2.94,\ s_1^2 = 2.27,\ s_2^2 = 1.05$$

$$s^2 = \frac{(n_1-1)s_1+(n_2-1)s_2^2}{n_1+n_2-2} = \frac{1.4\times 2.27+10\times 1.05}{15+11-2} = 1.76.$$

Since s^2 has $(n_1+n_2-2) = 24$ d.f., $t_{0.025,24} = 2.064$. Therefore, 95% confidence interval is obtained from

$$P\left[\bar{x}-\bar{y}-t_{0.025}\sqrt{s^2\left(\frac{1}{n_1}+\frac{1}{n_2}\right)}<\mu_1-\mu_2<\bar{x}-\bar{y}+t_{0.025}\sqrt{s^2\left(\frac{1}{n_1}+\frac{1}{n_2}\right)}\right]=0.95$$

$$\text{or, } P\left[0.43-2.064\sqrt{1.76\left(\frac{1}{15}+\frac{1}{11}\right)}<\mu_1-\mu_2<0.43+2.064\sqrt{1.76\left(\frac{1}{15}+\frac{1}{11}\right)}\right]=0.95$$

$$\text{or, } P[0.43-0.57<\mu_1-\mu_2<0.43+0.57]=0.95.$$

Therefore, 95% confidence interval of $\mu_1-\mu_2$ is $[-0.14, 1.00]$.

►**EXAMPLE 35.16** *Let $x_1, x_2, \ldots, x_n$ be a random sample of n observations from a Poisson population with parameter θ. Find $100(1-\alpha)\%$ confidence interval for θ.*

Solution. We have

$$\begin{aligned}
P(x) &= \frac{e^{-\theta}\theta^x}{x!}, \quad x=0,1,2,\ldots \\
L &= \frac{e^{-n\theta}\theta^{n\bar{x}}}{\prod_{i=1}^{n} x_i} \\
\log L &= -n\theta+n\bar{x}\log\theta-\log\prod_{i=1}^{n} x_i \\
\frac{\partial \log L}{\partial\theta} &= -n+\frac{n\bar{x}}{\theta}=\frac{n}{\theta}[\bar{x}-\theta] \\
\frac{\partial^2 \log L}{\partial\theta^2} &= -\frac{n\bar{x}}{\theta^2} \\
V\left[\frac{\partial \log L}{\partial\theta}\right] &= -E\left[\frac{\partial^2 \log L}{\partial\theta^2}\right]=\frac{nE(\bar{x})}{\theta^2}=\frac{n}{\theta}. \\
\therefore\ z &= \frac{\frac{n}{\theta}(\bar{x}-\theta)}{\sqrt{n/\theta}}=\frac{\sqrt{n}(\bar{x}-\theta)}{\sqrt{\theta}}\sim N(0,1).
\end{aligned}$$

Hence $100(1-\alpha)\%$ confidence interval is obtained from the following equation:

$$\begin{aligned}
&P\left[-z_{\frac{\alpha}{2}}<z<z_{\frac{\alpha}{2}}\right]=1-\alpha \\
\text{or, } &P\left[-z_{\frac{\alpha}{2}}<\frac{\sqrt{n}(\bar{x}-\theta)}{\sqrt{\theta}}<z_{\frac{\alpha}{2}}\right]=1-\alpha \\
\text{or, } &P\left[-z_{\frac{\alpha}{2}}\sqrt{\frac{\theta}{n}}<\bar{x}-\theta<z_{\frac{\alpha}{2}}\sqrt{\frac{\theta}{n}}\right]=1-\alpha \\
\text{or, } &P\left[\bar{x}-z_{\frac{\alpha}{2}}\sqrt{\frac{\theta}{n}}<\theta<\bar{x}+z_{\frac{\alpha}{2}}\sqrt{\frac{\theta}{n}}\right]=1-\alpha.
\end{aligned}$$

Thus, if θ is known, then $100(1-\alpha)\%$ confidence interval for θ is

$$\left[\bar{x}-z_{\frac{\alpha}{2}}\sqrt{\frac{\theta}{n}},\ \bar{x}+z_{\frac{\alpha}{2}}\sqrt{\frac{\theta}{n}}\right].$$

But, if θ is not known, the confidence interval is obtained as follows:

$$P\left[\left|\frac{\sqrt{n}(\bar{x}-\theta)}{\sqrt{\theta}}\right| \leq z_{\frac{\alpha}{2}}\right] = 1-\alpha \Rightarrow \left|\frac{\sqrt{n}(\bar{x}-\theta)}{\sqrt{\theta}}\right| = z_{\frac{\alpha}{2}}$$

$$\frac{n(\bar{x}-\theta)^2}{\theta} = z_{\frac{\alpha}{2}}^2 \text{ or, } \theta^2 - 2\bar{x}\theta + \bar{x}^2 - \frac{\theta}{n}z_{\frac{\alpha}{2}}^2 = 0$$

$$\text{or, } \theta = \frac{1}{2}\left[2\bar{x} + \frac{z_{\frac{\alpha}{2}}^2}{n} \pm \sqrt{\left(2\bar{x}+\frac{z_{\alpha/2}^2}{n}\right)^2 - 4\bar{x}^2}\right].$$

35.6 Confidence Interval for the Difference of Two Proportions

Let $x_1, x_2, \ldots, x_n$ be a random sample of n observations from a binomial population with parameters n and P_1. Consider another sample observations $y_1, y_2, \ldots, y_m$ from a binomial population with parameters m and P_2. The x_i and y_j ($i = 1, 2, \ldots, n$; $j = 1, 2, \ldots, m$) values measure the qualitative characters of sampling units. Let A_1 of the first population units possess the characteristic under study, where N_1 be the population size. Out of N_2 population units of the second population A_2 possess the characteristic. Then

$$P_1 = \frac{A_1}{N_1} \text{ and } P_2 = \frac{A_2}{N_2}.$$

The problem is to test the significance of null hypothesis

$$H_0 : P_1 = P_2 = P \text{ against } H_A : P_1 \neq P_2.$$

If a_1 is the number of units in the sample possessing the characteristic in first sample and a_2 is the number of units in the second sample possessing the characteristic under study, then

$$p_1 = \frac{a_1}{n} \text{ and } p_2 = \frac{a_2}{m}, \text{ where } E(p_1) = P_1, E(p_2) = P_2.$$

Under H_0, $E(p_1 - p_2) = P$. Since the samples are independent

$$V(p_1 - p_2) = \frac{P_1Q_1}{n} + \frac{P_2Q_2}{m}$$

and under H_0,

$$V(p_1 - p_2) = PQ\left(\frac{1}{n} + \frac{1}{m}\right), \text{ where } Q = 1 - P.$$

Therefore, the test statistic for the hypothesis is

$$z = \frac{p_1 - p_2 - (P_1 - P_2)}{\sqrt{PQ\left(\frac{1}{n}+\frac{1}{m}\right)}} \sim N(0, 1).$$

Therefore, the z can be used as a pivotal quantity.

Now, the $100(1-\alpha)\%$ confidence interval can be found out from the equation

$$P[q_1 < Q < q_2] = 1-\alpha$$

$$\text{or, } P\left[q_1 < \frac{(p_1-p_2)-(P_1-P_2)}{\sqrt{PQ\left(\frac{1}{n}+\frac{1}{m}\right)}} < q_2\right] = 1-\alpha$$

$$\text{or, } P\left[q_1\sqrt{PQ\left(\frac{1}{n}+\frac{1}{m}\right)} < (p_1-p_2)-(P_1-P_2) < q_2\sqrt{PQ\left(\frac{1}{n}+\frac{1}{m}\right)}\right] = 1-\alpha.$$

Here q_1 and q_2 are found out from

$$\int_{q_1}^{q_2} f(z) = 1-\alpha \text{ or, } \int_{-z_{\frac{\alpha}{2}}}^{z_{\frac{\alpha}{2}}} f(z)dz = 1-\alpha.$$

Therefore,

$$P\left[-z_{\frac{\alpha}{2}}\sqrt{PQ\left(\frac{1}{n}+\frac{1}{m}\right)} < (p_1-p_2)-(P_1-P_2) < z_{\frac{\alpha}{2}}\sqrt{PQ\left(\frac{1}{n}+\frac{1}{m}\right)}\right] = 1-\alpha$$

$$\text{or, } P\left[(p_1-p_2)-z_{\frac{\alpha}{2}}\sqrt{PQ\left(\frac{1}{n}+\frac{1}{m}\right)} < P_1-P_2 < (p_1-p_2)+z_{\frac{\alpha}{2}}\sqrt{PQ\left(\frac{1}{n}+\frac{1}{m}\right)}\right] = 1-\alpha.$$

Hence $100(1-\alpha)\%$ confidence interval for P_1-P_2 is

$$\left[(p_1-p_2)-z_{\frac{\alpha}{2}}\sqrt{PQ\left(\frac{1}{n}+\frac{1}{m}\right)},\ (p_1-p_2)+z_{\frac{\alpha}{2}}\sqrt{PQ\left(\frac{1}{n}+\frac{1}{m}\right)}\right].$$

If $P_1 \neq P_2$,

$$V(p_1-p_2) = \frac{P_1Q_1}{n}+\frac{P_2Q_2}{m}.$$

Then the confidence interval becomes

$$\left[(p_1-p_2)-z_{\frac{\alpha}{2}}\sqrt{\frac{P_1Q_1}{n}+\frac{P_2Q_2}{m}},\ (p_1-p_2)+z_{\frac{\alpha}{2}}\sqrt{\frac{P_1Q_1}{n}+\frac{P_2Q_2}{m}}\right].$$

►**EXAMPLE 35.17** *Ten students got grade A in a course. There were 50 students in this course. In another course there were 40 students and 5 of them got grade A. Find 95% confidence interval for the difference in proportions of students who got A grade.*

Solution. We have $n = 50,\ m = 40,\ a_1 = 10,\ a_2 = 5$

$$p_1 = \frac{a_1}{n} = \frac{10}{50} = 0.2,\ p_2 = \frac{a_2}{m} = \frac{5}{40} = 0.125.$$

If it is assumed that the population proportions are same, then the estimate of proportion for both the population is

$$p = \frac{a_1+a_2}{n+m} = \frac{10+5}{50+40} = 0.17,\ q = 1-p = 0.83.$$

Therefore, 95% confidence interval for $P_1 - P_2$ is found out from the equation

$$P\left[p_1 - p_2 - z_{\frac{\alpha}{2}}\sqrt{pq\left(\frac{1}{n}+\frac{1}{m}\right)} < P_1 - P_2 < p_1 - p_2 + z_{\frac{\alpha}{2}}\sqrt{pq\left(\frac{1}{n}+\frac{1}{m}\right)}\right] = 1-\alpha$$

$$\text{or, } \left[p_1 - p_2 - 1.96\sqrt{pq\left(\frac{1}{n}+\frac{1}{m}\right)} < P_1 - P_2 < p_1 - p_2 + 1.96\sqrt{pq\left(\frac{1}{n}+\frac{1}{m}\right)}\right] = 0.95.$$

Therefore, the confidence interval is

$$\left[(0.2-0.125) - 1.96\sqrt{0.17\times 0.83\left(\frac{1}{50}+\frac{1}{40}\right)}, (0.2-0.125)+1.96\sqrt{0.17\times 0.83\left(\frac{1}{50}+\frac{1}{40}\right)}\right]$$

or, $[0.075 - 0.080,\ 0.075 + 0.080]$

or, $[-0.005,\ 0.155]$.

EXERCISE

1. Distinguish between point estimation and interval estimation. Discuss the pivotal method of finding confidence interval of θ in $f(x, \theta)$.
   If $x \sim N(\mu, \sigma^2)$, find $100(1-\alpha)\%$ confidence interval for μ and σ^2 by pivotal method.
2. Explain confidence interval of θ in $f(x,\theta)$. Also, discuss the statistical method of finding confidence interval of θ.
   Let $x_1, x_2, \ldots, x_n$ be random sample observations from a population having p.d.f.
   $$f(x) = \begin{cases} \frac{1}{\theta}, & 0 < x < \theta \\ 0, & \text{otherwise.} \end{cases}$$
   Find $100(1-\alpha)\%$ confidence interval for θ.
   The following are the time required to send e.mails (x, minutes):
   $$x\ :\ 1.2,\ 2.5,\ 0.6,\ 2.8,\ 3.0,\ 3.2,\ 3.4,\ 1.5,\ 1.6$$
   Find 95% confidence interval for population mean and variance.
3. Let $x_1, x_2, \ldots, x_n$ be sample observations selected from a population having p.d.f. $f(x,\theta)$. Find $100(1-\alpha)\%$ confidence interval for θ if (i) n is small, and (ii) n is large.
   If $f(x,\theta) = \begin{cases} \theta x^{\theta-1}, & 0 < x < 1 \\ 0, & \text{otherwise.} \end{cases}$
   Find $100(1-\alpha)\%$ confidence interval for θ.
4. Define point estimation and interval estimation. Discuss the method of finding confidence interval using Chebyshev's inequality.
   Let $x_1, x_2, \ldots, x_n$ be random sample observations from a p.d.f.
   $$f(x) = \frac{2x}{\theta}e^{-x^2/\theta},\ x > 0.$$
   Find $100(1-\alpha)\%$ confidence interval for θ.

5. Define uniformly most accurate confidence interval and unbiased confidence interval. Let $x_{11}, x_{12}, \ldots, x_{1n_1}$ be sample observations from $N(\mu_1, \sigma_1^2)$ and $x_{21}, x_{22}, \ldots, x_{2n_2}$ be another sample observations from $N(\mu_2, \sigma_2^2)$. Find $100(1-\alpha)\%$ confidence interval for $\mu_1 - \mu_2$. Also, find unbiased confidence interval for σ_1^2.

6. Explain point estimation, interval estimation and unbiased confidence interval. If $x_2, x_2, \ldots, x_n$ are sample observations from Poisson population having parameter λ, find $100(1-\alpha)\%$ confidence interval for λ when n is large.

   The times (in minute) required in repairing mobile phone observed from two repairing shops are given below:

   Shop-1 (x_{1j}) : 10.5, 15.8, 30.6, 30.0, 25.8, 25.0

   Shop-2 (x_{2j}) : 18.7, 20.0, 25.0, 30.6, 35.0, 33.8, 40.0, 45.0

   Find 95% confidence interval for the difference of the population means.

7. Define interval estimation. What are the different methods of finding confidence interval for the population parameter. Discuss the method of finding confidence interval for the difference of two population proportions.

   Two students A and B are asked to develop a computer programs for solving a mathematical problem. A becomes successful in doing the work in 6 cases out of 10. The corresponding figures for B are 18 out of 25. Find 95% confidence interval for the difference in the proportions of success of two students.

CHAPTER **36**

Nonparametric Methods

36.1 Introduction

One of the important aspects of statistical analysis is to infer about population parameter. The inference is drawn on the basis of sample statistic and using the sample statistic we usually perform some statistical tests. The basic assumption of parametric statistical test is that the sample observations are randomly and independently selected from normal population. But if the variable under study is qualitative in character and if the sample size is small, the normality assumption does not remain valid and in that case the distribution of sample statistic is not known. Therefore, parametric method of analysis, specially parametric test, is no longer valid.

The distribution of some variable may not be known or it is not suitable to assume the normality of data. Since the distribution is not known or is not specified, the parameters of the distribution are also not specified. In such case also parametric method of analysis, specially parametric test, is not possible. The problems are obviated using nonparametric method of analysis. One of the important aspects of nonparametric method is the nonparametric test. In this chapter, we shall discuss some nonparametric tests. The important nonparametric tests are (i) Sign test, (ii) Median test, (iii) Wilcoxon signed-rank test, (iv) Run test, (v) Mann-Whitney U-test, (vi) Test for normality, (vii) Test for equality of variances, (viii) Test of independence, (ix) Test for equality of several means, etc. In this chapter, we shall discuss some of the tests with examples.

Since the basic requirement of parametric test is the normality of data at hand, let us first discuss the 'Test for Normality' of data. If any data is suitable for normality assumption, we can use parametric method of analysis with that data. The test of normality of data is based on one sample of observations. The other one sample nonparametric tests are for

(a) testing the equality of non-central values of sample and population,

(b) testing the equality of observed and expected frequencies (test of goodness-of-fit, which has been discussed in Chapter 33),

(c) testing the equality of observed and expected proportions,

(d) testing for the specific form of the population distribution, and

(e) testing the randomness of data.

The tests based on one sample are (i) Binomial test, (ii) Sign test, (iii) Wilcoxon signed-rank test, etc.

Before discussing the different nonparametric tests, let us discuss the merits and demerits of nonparametric tests.

36.2 Advantages and Disadvantages of Nonparametric Methods

Advantages

(i) Nonparametric methods are very simple and easy to apply. The sampling theory of data for nonparametric analysis is not complicated.

(ii) No assumption is needed about the parent distribution of data for analysis.

(iii) Nonparametric methods are very suitable to analyse the data which are measured in nominal scale or by ranks.

(iv) The data on social aspects, educational statistics, data on psychometry usually do not follow normal distribution. Hence nonparametric method is suitable to handle such types of data.

Disadvantages

(i) Parametric tests are more powerful than nonparametric tests. Therefore, nonparametric tests can only be used when data were measured in nominal scale or ordinal scale.

(ii) Nonparametric methods deal with test of hypothesis only. No nonparametric method of estimation is available.

The nonparametric methods are based on order statistics. However, certain assumptions are needed for nonparametric methods. These are (i) sample observations are independent, (ii) the variable under study is continuous, (iii) the probability density function is continuous, and (iv) lower order moments exist.

36.3 Test for Normality

The normal distribution plays an important role in statistical analysis. All parametric tests are based on normal distribution. The normality of data is attained if the sample size is large. This is due to central limit theorem. But in the field of research in Life Sciences, the usual analysis is performed with small number of sample observations. In such a situation the study of normality of data is very important. However, non-normality does not invalidate the analytical procedure, specially in performing test of hypothesis, since nonparametric test is available for different types of hypothesis. In the present section we shall discuss the test of normality of data.

Let $x_1, x_2, \ldots, x_n$ be a random sample of n observations from a population having c.d.f. $F(x)$. We need to test the significance of

H_0 : $F(x)$ is normally distributed with unknown mean and variance

against H_A : $F(x)$ is not the c.d.f. of normal distribution.

Shapiro and Wilk (1968) have discussed the test procedure of the above hypothesis. Their proposed test statistic is

$$W = \frac{1}{D}\left[\sum_{i=1}^{k} a_i \left(X^{(n-i+1)} - X^{(i)}\right)\right], \quad k \doteq n/2, \text{ where}$$

$$D = \sum_{i=1}^{n}(x_i - \bar{x})^2, X^{(i)} = i\text{-th order statistic of } x_1, x_2, \ldots, x_n$$

$$a_i = \text{coefficient of Shapiro-Wilk test related to } i\text{-th order statistic.}$$

The quantities of W and the values of a_i or different 'i' and n are provided by Shapiro and Wilk [Table 36.1]. If $W < W_p(p = 0.05$ [Table 36.2], H_0 is rejected.

Conover (1990) has proposed more accurate p-values of W-test. The Shapiro-Wilk test is applicable for $n < 50$. For $n > 50$, Shapiro and Francia (1972) have proposed similar test. Moreover, the test of normality has been discussed by Lillifors (1967), Bowman and Shenton (1975), Pearson, D'Agostino and Bowman (1977).

►**EXAMPLE 36.1** *The following data represent the ages (in years) of* 18 *randomly selected students in a primary school:*

Ages (x, *in years*) : 9, 12, 11, 10, 12, 9, 7, 8, 8, 7, 10, 10, 8, 7, 9, 10, 11, 12.

Do you think that the observations follow normal distribution?

Solution. Let $F(x)$ be the c.d.f. of normal distribution. We need to test the significance of H_0 : $F(x)$ is the c.d.f. of normal distribution having unknown mean and variance, against H_A : $F(x)$ is not the c.d.f. of normal distribution.

The observations in ascending order are:

7, 7, 7, 8, 8, 8, 9, 9, 9, 10, 10, 10, 10, 11, 11, 12, 12, 12; $n = 18$.

The calculation of $[X^{(n-i+1)} - X^{(i)}]$ along with a_i are shown below:

i	a_i	$X^{(n-i+1)} - X^{(i)}$	i	a_i	$X^{(n-i+1)} - X^{(i)}$
1	0.4886	5	6	0.1197	2
2	0.3253	5	7	0.0837	1
3	0.2553	5	8	0.0496	1
4	0.2027	3	9	0.0163	1
5	0.1587	3			

$$D = \sum (x_i - \bar{x})^2 = 50.44$$

$$W = \frac{1}{D}\left[\sum_{i=1}^{9} a_i \left(X^{(n-i+1)} - X^{(i)}\right)\right] = \frac{6.8165}{50.44} = 0.0013 < W_p = 0.897.$$

Hence H_0 is rejected. The data does not follow the normal distribution.

Table 36.1 : Coefficient of Shapiro-Wilk Test

i \ n	2	3	4	5	6	7	8	9	10
1	0.7071	0.7071	0.6872	0.6646	0.6431	0.6233	0.6052	0.5988	0.5739
2	—	0.0000	0.1667	0.2413	0.2846	0.3031	0.3164	0.3244	0.3291
3	—	—	—	0.0000	0.0875	0.1401	0.1743	0.1976	0.2141
4	—	—	—	—	—	0.0000	0.0561	0.0947	0.1224
5	—	—	—	—	—	—	—	0.0000	0.0399
i \ n	**11**	**12**	**13**	**14**	**15**	**16**	**17**	**18**	**19**
1	0.5601	0.5475	0.5359	0.5251	0.5150	0.5056	0.4968	0.4886	0.4808
2	0.3315	0.3325	0.3325	0.3318	0.3306	0.3290	0.3273	0.3253	0.3232
3	0.2260	0.2347	0.2412	0.2460	0.2495	0.2521	0.2540	0.2553	0.2561
4	0.1429	0.1586	0.1707	0.1802	0.1878	0.1939	0.1988	0.2027	0.2059
5	0.0695	0.0922	0.1099	0.1240	0.1353	0.1447	0.1524	0.1587	0.1641
6	0.0000	0.0303	0.0539	0.0727	0.0880	0.1005	0.1109	0.1197	0.1271
7	—	—	0.0000	0.0240	0.0433	0.0593	0.0725	0.0837	0.0932
8	—	—	—	—	0.0000	0.0196	0.0359	0.0496	0.0612
9	—	—	—	—	—	—	0.0000	0.0163	0.0303
10	—	—	—	—	—	—	—	—	0.0000
i \ n	**20**	**21**	**22**	**23**	**24**	**25**	**26**	**27**	**28**
1	0.4734	0.4643	0.4590	0.4542	0.4493	0.4450	0.4407	0.4366	0.4328
2	0.3211	0.3185	0.3156	0.3126	0.3098	0.3069	0.3043	0.3018	0.2992
3	0.2565	0.2578	0.2571	0.2563	0.2554	0.2543	0.2533	0.2522	0.2510
4	0.2085	0.2119	0.2131	0.2139	0.2145	0.2148	0.2151	0.2152	0.2151
5	0.1686	0.1736	0.1764	0.1787	0.1807	0.1822	0.1836	0.1848	0.1857
6	0.1334	0.1399	0.1443	0.1480	0.1512	0.1539	0.1563	0.1584	0.1601

i \ n	20	21	22	23	24	25	26	27	28
7	0.1013	0.1092	0.1150	0.1201	0.1245	0.1283	0.1316	0.1346	0.1372
8	0.0711	0.0804	0.0878	0.0941	0.0997	0.1046	0.1089	0.1128	0.1162
9	0.0422	0.0530	0.0618	0.0696	0.0764	0.0823	0.0876	0.0923	0.0965
10	0.0140	0.0263	0.0368	0.0459	0.0539	0.0610	0.0672	0.0728	0.0778
11	—	0.0000	0.0122	0.0228	0.0321	0.0403	0.0476	0.0540	0.0598
12	—	—	—	0.0000	0.0107	0.0200	0.0284	0.0358	0.0424
13	—	—	—	—	—	0.0000	0.0094	0.0178	0.0253
14	—	—	—	—	—	—	—	0.0000	0.0084
15	—	—	—	—	—	—	—	—	—

i \ n	29	30	31	32	33	34	35	36	37
1	0.4291	0.4254	0.4220	0.4188	0.4156	0.4127	0.4096	0.4068	0.4040
2	0.2968	0.2944	0.2921	0.2898	0.2876	0.2854	0.2834	0.2813	0.2794
3	0.2499	0.2487	0.2475	0.2462	0.2451	0.2439	0.2427	0.2415	0.2403
4	0.2150	0.2148	0.2145	0.2141	0.2137	0.2132	0.2127	0.2121	0.2116
5	0.1864	0.1870	0.1874	0.1878	0.1880	0.1882	0.1883	0.1883	0.1883
6	0.1616	0.1630	0.1641	0.1651	0.1660	0.1667	0.1673	0.1678	0.1683
7	0.1395	0.1415	0.1433	0.1449	0.1463	0.1475	0.1487	0.1496	0.1505
8	0.1192	0.1219	0.1243	0.1265	0.1284	0.1301	0.1317	0.1331	0.1344
9	0.1002	0.1036	0.1066	0.1093	0.1118	0.1140	0.1160	0.1179	0.1196
10	0.0822	0.0862	0.0899	0.0931	0.0961	0.0988	0.1013	0.1036	0.1056
11	0.0650	0.0697	0.0739	0.0777	0.0812	0.0844	0.0873	0.0900	0.0924
12	0.0483	0.0537	0.0585	0.0629	0.0669	0.0706	0.0739	0.0770	0.0798
13	0.0320	0.0381	0.0435	0.0485	0.0530	0.0572	0.0610	0.0645	0.0677
14	0.0159	0.0227	0.0289	0.0344	0.0395	0.0441	0.0484	0.0523	0.0559
15	0.0000	0.0076	0.0144	0.0206	0.0262	0.0314	0.0361	0.0404	0.0444
16	—	—	0.0000	0.0068	0.0131	0.0187	0.0239	0.0287	0.0331
17	—	—	—	—	0.0000	0.0062	0.0119	0.0172	0.0220
18	—	—	—	—	—	—	0.0000	0.0057	0.0110
19	—	—	—	—	—	—	—	—	0.0000
20	—	—	—	—	—	—	—	—	—

i \ n	38	39	40	41	42	43	44	45
1	0.4015	0.3989	0.3964	0.3940	0.3917	0.3894	0.3872	0.3850
2	0.2774	0.2755	0.2737	0.2719	0.2701	0.2684	0.2667	0.2651
3	0.2391	0.2380	0.2368	0.2357	0.2345	0.2334	0.2323	0.2313
4	0.2110	0.2104	0.2098	0.2091	0.2085	0.2078	0.2072	0.2065
5	0.1881	0.1880	0.1878	0.1876	0.1874	0.1871	0.1868	0.1865
6	0.1686	0.1689	0.1691	0.1693	0.1694	0.1695	0.1695	0.1695
7	0.1513	0.1520	0.1526	0.1531	0.1535	0.1539	0.1542	0.1545
8	0.1356	0.1366	0.1376	0.1384	0.1392	0.1398	0.1405	0.1410
9	0.1211	0.1225	0.1237	0.1249	0.1259	0.1269	0.1278	0.1286
10	0.1075	0.1092	0.1108	0.1123	0.1136	0.1149	0.1160	0.1170
11	0.0947	0.0967	0.0986	0.1004	0.1020	0.1035	0.1049	0.1062
12	0.0824	0.0848	0.0870	0.0891	0.0909	0.0927	0.0943	0.0959
13	0.0706	0.0733	0.0759	0.0782	0.0804	0.0824	0.0842	0.0860
14	0.0592	0.0622	0.0651	0.0677	0.0701	0.0724	0.0745	0.0765
15	0.0481	0.0515	0.0546	0.0575	0.0602	0.0628	0.0651	0.0673
16	0.0372	0.0409	0.0444	0.0476	0.0506	0.0534	0.0560	0.0584
17	0.0264	0.0305	0.0343	0.0379	0.0411	0.0442	0.0471	0.0497
18	0.0158	0.0203	0.0244	0.0283	0.0318	0.0352	0.0383	0.0412
19	0.0053	0.0101	0.0146	0.0188	0.0227	0.0263	0.0296	0.0328
20	—	0.0000	0.0049	0.0094	0.0136	0.0175	0.0211	0.0245
21	—	—	—	0.0000	0.0045	0.0087	0.0126	0.0163
22	—	—	—	—	—	0.0000	0.0042	0.0081
23	—	—	—	—	—	—	—	0.0000
24	—	—	—	—	—	—	—	—
25	—	—	—	—	—	—	—	—

i \ n	46	47	48	49	50
1	0.3830	0.3808	0.3789	0.3770	0.3751
2	0.2635	0.2620	0.2604	0.2589	0.2574
3	0.2302	0.2291	0.2281	0.2271	0.2260
4	0.2058	0.2052	0.2045	0.2038	0.2032
5	0.1862	0.1859	0.1855	0.1851	0.1847
6	0.1695	0.1695	0.1693	0.1692	0.1691
7	0.1548	0.1550	0.1551	0.1553	0.1554
8	0.1415	0.1420	0.1423	0.1427	0.1430
9	0.1293	0.1300	0.1306	0.1312	0.1317
10	0.1180	0.1189	0.1197	0.1205	0.1212
11	0.1073	0.1085	0.1095	0.1105	0.1113
12	0.0972	0.0986	0.0998	0.1010	0.1020
13	0.0876	0.0892	0.0906	0.0919	0.0932
14	0.0783	0.0801	0.0817	0.0832	0.0846
15	0.0694	0.0713	0.0731	0.0748	0.0764
16	0.0607	0.0628	0.0648	0.0667	0.0685
17	0.0522	0.0546	0.0568	0.0588	0.0608
18	0.0439	0.0465	0.0489	0.0511	0.0532
19	0.0357	0.0385	0.0411	0.0436	0.0459
20	0.0277	0.0307	0.0335	0.0361	0.0386
21	0.0197	0.0229	0.0259	0.0288	0.0314
22	0.0118	0.0153	0.0185	0.0215	0.0244
23	0.0039	0.0076	0.0111	0.0143	0.0174
24	—	0.0000	0.0037	0.0071	0.0104
25	—	—	—	0.0000	0.0035

Table 36.2 : Quantiles $[W_p]$ of Shapiro-Wilk Test Statistic

n \ α	0.01	0.05	n \ α	0.01	0.05
1	—	—	26	0.891	0.920
2	—	—	27	0.894	0.923
3	0.753	0.767	28	0.896	0.924
4	0.687	0.748	29	0.898	0.926
5	0.686	0.762	30	0.900	0.927
6	0.713	0.788	31	0.902	0.929
7	0.730	0.803	32	0.904	0.930
8	0.749	0.818	33	0.906	0.931
9	0.764	0.829	34	0.908	0.933
10	0.781	0.842	35	0.910	0.934
11	0.792	0.850	36	0.912	0.935
12	0.805	0.859	37	0.914	0.936
13	0.814	0.866	38	0.916	0.938
14	0.825	0.874	39	0.917	0.939
15	0.835	0.881	40	0.919	0.940
16	0.844	0.887	41	0.920	0.941
17	0.851	0.892	42	0.922	0.942
18	0.858	0.897	43	0.923	0.943
19	0.863	0.901	44	0.924	0.944
20	0.868	0.905	45	0.926	0.945
21	0.873	0.908	46	0.927	0.945
22	0.878	0.911	47	0.928	0.946
23	0.881	0.914	48	0.929	0.947
24	0.884	0.916	49	0.929	0.947
25	0.888	0.918	50	0.930	0.947

36.4 Test for Randomness

Let there be 8 balls in a box, 4 of which are white and another four are black. Assume that the balls are drawn from the box one by one. Then the arrangement of balls may be

(i) $WWWWBBBB$, (ii) $WBWBWBWB$, (iii) $WWBWBBWB, \ldots$.

The question arises, whether the above arrangements of drawn balls indicate random selection of ball from the box. The first arrangement indicates that all white balls are selected first one by one and then the black balls. Here 'W' is used to denote white ball and 'B' is used to denote black ball. It is difficult to claim that the ball drawn in the first case is random. Therefore, we need to decide about the randomness of data.

It is observed that in the first arrangement there is a sequence of W followed by another sequence of B. Here any sequence of a particular 'letter' or particular 'sign' is known as 'Run'. In the first case the number of run is 2 and this is the arrangement showing minimum run. Hence depending on number of runs, we can decide about the randomness of data.

36.4.1 Run Test

The objective of run test is to test the significance of

$$H_0 : \text{the number of runs is random,}$$
$$\text{against } H_A : \text{the number of runs is not random.}$$

Now, let us discuss the run test. There are three types of tests based on run. These are (i) test based on total runs, (ii) test based on length of longest run, and (iii) test based on runs up and down. Let us now discuss the run test based on total runs.

RUN TEST BASED ON TOTAL RUNS. Let there be n elements in a sample, the sample observations can be divided into two parts. In one part there are n_1 observations and in another part there are n_2 observations such that $n = n_1 + n_2$. Consider that the observations are arranged according to ascending order of magnitude or according to the steps of inclusion of the observations in the sample. Let there be r_1 runs of first kind of observations and r_2 runs of second kind of observations when these are arranged. Here total runs are $R = r_1 + r_2$. Let r_0 be a particular value of R. Then the distribution of R is

$$f(R = r_0) = \begin{cases} \dfrac{2\dbinom{n_1-1}{\frac{r}{2}-1}\dbinom{n_2-1}{\frac{r}{2}-1}}{\dbinom{n_1+n_2}{n_1}}, & \text{if } r_0 \text{ is even number} \\[2ex] \dfrac{\dbinom{n_1-1}{\frac{r-1}{2}}\dbinom{n_2-1}{\frac{r-3}{2}}+\dbinom{n_1-1}{\frac{r-3}{2}}\dbinom{n_2-1}{\frac{r-1}{2}}}{\dbinom{n_1+n_2}{n_1}}, & \text{if } r_0 \text{ is odd number} \end{cases}$$

$$r_0 = 2, 3, 4, \ldots, (n_1 + n_2).$$

Using the above probability distribution of R the decision is made regarding the critical values of r_0. Let us explain the method of decision by an example.

►**EXAMPLE 36.2** *The following data represent the marks of* 15 *randomly selected students in a Statistics course:*

$$54, 64, 83, 66, 52, 86, 80, 45, 60, 65, 68, 55, 50, 38, 75.$$

Do you think that the majority students are good in results?

Solution. The median performance is $Me = 64$. The majority students may be considered as good if their results exceeds median. In that case the value of r_0 will be smaller if the students are not selected randomly. Hence the null hypothesis and alternative hypothesis will be

$$H_0 : \text{the students are randomly selected,}$$
$$\text{against } H_A : \text{the students are not randomly selected.}$$

The number of observations less than or equal to median is $n_1 = 8$ and hence $n_2 = 7$ (number of observations exceeding median). The observations are now rewritten by +ve and −ve signs according to the value exceeding median and less than or equal to median, respectively. Thus, we have

$$- - + + - + + - - + + - - - +.$$

Here $r = 8$.

Now, the distribution of $r_0 = 8$ is shown below:

r	2	3	4	5	6	7	8
$f(R = r_0)$	0.00031	0.00202	0.01305	0.03589	0.0979	0.14685	0.21756

It is noted that $\sum_{i=2}^{8} f(R = r_i) = 0.36668 > 0.05$. Again, $\sum_{i=2}^{7} f(R = r_i) = 0.14912 > 0.05$. Therefore, if $\alpha = 0.05$ is used as level of significance, r_0 does not fall in the critical region. Hence there is no evidence against null hypothesis.

It is known that, if maximum students get good marks, then there is chance of smaller number of runs. Let these small runs be $r \leq 4$. Here the exact level of significance of rejecting the null hypothesis when $r \leq 4$ is

$$\sum_{i=2}^{4} f(R = r_i) = 0.01533.$$

►**EXAMPLE 36.3** *Ten male and 10 female candidates for a job are called for interview according to the following sequence:*

$$M\,M\,F\,F\,F\,M\,F\,M\,M\,M\,M\,F\,F\,M\,F\,M\,M\,F\,F\,F$$

Do you think that the candidates are called to appear at the board of interview randomly?

Solution. Here $n_1 = n_2 = 10$, $r_0 = 10$. The probabilities for different values of r_0 are shown below:

r_0	2	3	4	5	6	7
$f(R = r_0)$	0.00001	0.00097	0.00088	0.00351	0.01403	0.03274
r_0	15	16	17	18	19	20
$f(R = r_0)$	0.03274	0.01403	0.00351	0.00088	0.000097	0.00001

$$H_0 : \text{the candidates are called randomly,}$$
$$\text{against } H_A : \text{the candidates are not called randomly.}$$

According to alternative hypothesis the total number of runs may be too less or too large. Hence, the critical region for the test may be in the lower-tail or in the upper-tail. It is observed that

$$f(r_0 \leq 6) + f(r_0 \geq 16) = 0.037054 < 0.05.$$

Hence the critical region will be $r \leq 6$ and $r \geq 16$. But $r = 10$, which is beyond the critical region and hence null hypothesis is accepted, i.e., the candidates are called randomly.

The run test can also be performed using the distribution of r_1 or r_2, where

$$f(r_1) = \frac{\binom{n_1 - 1}{r_1 - 1}\binom{n_2 + 1}{r_1}}{\binom{n_1 + n_2}{n_1}}, \quad r_1 = 1, 2, \dots, n_1$$

$$\text{and } f(r_2) = \frac{\binom{n_1 - 1}{r_2 - 1}\binom{n_1 + 1}{r_2}}{\binom{n_1 + n_2}{n_2}}, \quad r_2 = 1, 2, \dots, n_2.$$

►**EXAMPLE 36.4** *To study the effectiveness of a newly discovered medicine the medicine has been fed to rats one by one. The effectiveness of medicine is measured by +ve sign and −ve sign, if the medicine is effective and not effective, respectively. The sequence of successes and failures are shown below:*

$$+ + + - - + - - + + + + - - - - + + + + +$$

Do you think that the medicine is effective?

Solution.

$$H_0 : \text{the impact of medicine is random,}$$
$$H_A : \text{the medicine is effective [more +ve signs are expected].}$$

Here $n_1 = 13$, $n_2 = 8$, $r_1 = 4$

$$f(r_1 = 4) = \frac{\binom{12}{3}\binom{9}{4}}{\binom{21}{13}} = 0.1362.$$

Since more +ve signs are expected, the runs of +ve signs should be minimum. However, it is observed that $f(r_1 \leq 4) > 0.05$. Hence, observed $r_1 = 4$ does not fall in the critical region, and there is no evidence against null hypothesis.

Method of Test for Large Sample

Let us consider that the sample size n is large such that $n_1 > 20$, $n_2 > 20$, where $n = n_1 + n_2$. The critical values for such n_1 and n_2 are not available. Also, it is time-consuming in calculating the probabilities for different values of r_0. It is observed that

$$E(r_0) = 1 + \frac{2n_1n_2}{n_1+n_2}, \quad V(r_0) = \frac{2n_1n_2(2n_1n_2 - n_1 - n_2)}{(n_1+n_2)^2(n_1+n_2-1)}.$$

Hence the distribution of R can be approximated by normal distribution, where normal test statistic is

$$z = \frac{r_0 - E(r_0)}{\sqrt{V(r_0)}} \sim N(0,1).$$

If $|z| \geq 1.96$, H_0 in favour of two-sided alternative is rejected at 5% level of significance. Again, if $|z| \geq 1.64$, H_0 is rejected at 5% level of significance in favour of too less or too many runs [one-sided alternative].

The values of $E(r_0)$ and $V(r_0)$ are based on the distribution of total runs R. However, there is another distribution of r_1 and according to the distribution of r_1, we have

$$E(r_1) = \frac{n_1(n_2+1)}{n_1+n_2}, \quad V(r_1) = \frac{n_1(n_2+1)(n_1-1)}{(n_1+n_2)^2} \frac{n_2}{n_1+n_2-1}.$$

A modified z-statistic is given as follows:

$$z = \frac{r_0 \pm 0.5 - 1 - \frac{2n_1n_2}{n}}{\left[\frac{2n_1n_2}{n^2(n-1)}(2n_1n_2 - n)\right]^{1/2}} \sim N(0,1).$$

Here $n = n_1 + n_2$. The adjustment factor $+0.5$ is used, if $r_0 < 1 + \frac{2n_1n_2}{n}$ and -0.5 is used, if $r_0 > 1 + \frac{2n_1n_2}{n}$. For one-sided alternative hypothesis (too small number of runs) p-value is in the lower-tail and for too large number of runs p-value is the upper-tailed probability.

►**EXAMPLE 36.5** *In a zoological experiment a researcher has collected one kind of slug by placing quadrat. The number of slugs per quadrat are shown below:*

5, 8, 12, 4, 3, 7, 9, 6, 12, 18, 10, 4, 1, 7, 10, 11, 5, 7, 6, 4, 4, 3, 3, 8, 2, 2, 9, 9, 5, 6, 6, 1, 1, 2, 4, 10, 10, 9, 7, 3, 13, 10, 4, 1, 7.

Is the distribution of number of slugs random?

Solution. The median of the distribution is $M = 6$, $n = 45$, $n_1 = 24$ (number of observations less than or equal to median), $n_2 = 21$, $r_0 = 20$. Therefore,

$$E(r_0) = 1 + \frac{2n_1n_2}{n_1+n_2} = 1 + \frac{2 \times 24 \times 21}{24+21} = 23.4$$

$$V(r_0) = \frac{2n_1n_2(2n_1n_2 - n_1 - n_2)}{(n_1+n_2)^2(n_1+n_2-1)}$$

$$= \frac{2 \times 24 \times 21(2 \times 24 \times 21 - 24 - 21)}{(24+21)^2(24+21-1)} = 10.89.$$

$$\therefore \; z = \frac{r_0 - E(r_0)}{\sqrt{V(r_0)}} = \frac{20 - 23.4}{\sqrt{10.89}} = -1.03.$$

Now, if

H_0 : **the distribution of num**ber of slugs is random,
H_A : the distribution is not random.

H_0 is accepted, since $|z| < 1.96$.

36.4.2 Test of Randomness Using χ^2-Statistic

We have studied 'Run' test as a method of test of randomness, where randomly selec... observations are classified into two classes based on 'median' or any specific value in the data set. The random observations are, nowadays, generated using computer program and such generated data are assumed pseudorandom numbers. Therefore, we need to test the randomness of such generated data.

The digits included in numbers are from 0 to 9 and hence in any random number a digit can occur 10% time. This means that the digits in any random number are uniformly distributed. Therefore, to test the randomness of the computer-generated random numbers, we need to test the significance about the uniform distribution of the digits. The null hypothesis for that is

H_0 : the numbers are random,
against H_A : some digits are included in large numbers than others (numbers are not random).

The test statistic for this is χ^2, where χ^2 is used to test the goodness-of-fit test. The χ^2-statistic is

$$\chi^2 = \sum_{i=1}^{k} \frac{(O_i - E_i)^2}{E_i} = \sum_{i=1}^{k} \frac{O_i^2}{E_i} - n.$$

The proof of this statistic is provided by a theorem.

❐ **THEOREM 36.1** *For a sample of size n, if the number of favourable points to an event 'i' is O_i and its expected number is $E_i = np_i$ ($i = 1, 2, \ldots, k$; $n = \Sigma O_i$, where n is large), where p_i is the probability of occurrence of the event 'i', then*

$$\chi^2 = \sum_{i=1}^{k} \frac{(O_i - E_i)^2}{E_i} = \sum_{i=1}^{k} \frac{O_i^2}{E_i} - n.$$

Proof. Let us consider that the total number of observations are divided into k classes and the probability of inclusion of any observation in class 'i' be p_i ($i = 1, 2, \ldots, k$). Then the probability of inclusion of O_i observations in i-th class is

$$P = \frac{n!}{O_1!\, O_2! \cdots O_k!} p_1^{O_1} p_2^{O_2} \cdots p_k^{O_k} \text{ such that } \sum_{i=1}^{k} p_i = 1,\ n = \sum_{i=1}^{k} O_i.$$

If n is large enough and if O_i ($i = 1, 2, \ldots, k$) is not small, then by Stirling's approximation, we can write

$$\lim_{n\to\infty} n! \approx \sqrt{2\pi} e^{-n} n^{n+\frac{1}{2}}.$$

Then

$$P \approx \frac{\sqrt{2\pi}e^{-n}n^{n+\frac{1}{2}}p_1^{O_1}p_2^{O_2}\cdots p_k^{O_k}}{\left(\sqrt{2\pi}\right)^k e^{-(O_1+O_2+\cdots+O_k)}O_1^{O_1+\frac{1}{2}}O_2^{O_2+\frac{1}{2}}\cdots O_k^{O_k+\frac{1}{2}}}$$

$$= \frac{e^{-n}n^{n+\frac{1}{2}}\left(\frac{np_1}{O_1}\right)^{O_1+\frac{1}{2}}\left(\frac{np_2}{O_2}\right)^{O_2+\frac{1}{2}}\cdots\left(\frac{np_k}{O_k}\right)^{O_k+\frac{1}{2}}}{\left(\sqrt{2\pi}\right)^{k-1} e^{-n}n^{(O_1+O_2+\cdots+O_k)}(p_1p_2\cdots p_k)^{\frac{1}{2}}}$$

$$= C\prod_{i=1}^{k}\left(\frac{np_i}{O_i}\right)^{O_i+\frac{1}{2}}, \text{ where}$$

$$C = \frac{1}{(2\pi)^{\frac{k-1}{2}}n^{\frac{k-1}{2}}(p_1p_2\cdots p_k)^{\frac{1}{2}}}$$

is a term which is free of O_i.

Now we have

$$\log P = \log C + \sum_{i=1}^{k}\left(O_i + \frac{1}{2}\right)\log\left(\frac{np_i}{O_i}\right)$$

$$\Rightarrow \log\frac{P}{C} = \sum_{i=1}^{k}\left(O_i + \frac{1}{2}\right)\log\left(\frac{np_i}{O_i}\right) = \sum_{i=1}^{k}\left(O_i + \frac{1}{2}\right)\log\left(\frac{E_i}{O_i}\right),$$

$$\because E_i = np_i = \text{number of expected observations in } i\text{-th class},$$

i.e., $E(O_i) = E_i = np_i$ $(i = 1, 2, \ldots, k)$.

Let

$$U_i = \frac{O_i - E_i}{\sqrt{E_i}} \text{ or, } O_i - E_i = U_i\sqrt{E_i} \text{ or, } O_i = E_i + U_i\sqrt{E_i}.$$

Now, putting the value of O_i in $\log\frac{P}{C}$, we get

$$\log\frac{P}{C} \approx \sum_{i=1}^{k}\left(E_i + U_i\sqrt{E_i} + \frac{1}{2}\right)\log\left(\frac{E_i}{E_i + U_i\sqrt{E_i}}\right)$$

$$= \sum_{i=1}^{k}\left(E_i + U_i\sqrt{E_i} + \frac{1}{2}\right)\log\left[\frac{1}{1 + U_i/\sqrt{E_i}}\right]$$

$$= -\sum_{i=1}^{k}\left(E_i + U_i\sqrt{E_i} + \frac{1}{2}\right)\log[1 + U_i/\sqrt{E_i}].$$

If it is considered that U_i is smaller than E_i, then the expansion of

$$\log\left\{1 + U_i/\sqrt{E_i}\right\}$$

in ascending powers of $U_i/\sqrt{E_i}$ is valid.

$$\therefore \log\frac{P}{C} \approx -\sum_{i=1}^{k}\left(E_i + U_i\sqrt{E_i} + \frac{1}{2}\right)\left[\frac{U_i}{\sqrt{E_i}} - \frac{1}{2}\frac{U_i^2}{E_i} + O\left(\frac{1}{E_i^{3/2}}\right)\right].$$

Neglecting higher powers of E_i, we get

$$\log \frac{P}{C} \approx -\sum U_i\sqrt{E_i} - \frac{1}{2}U_i^2 + U_i^2 + O\left(\frac{1}{E_i^{3/2}}\right).$$

[Since n is large, $E_i = np_i$ is expected large and hence higher powers of E_i can be neglected.]

Moreover,

$$\sum_{i=1}^{k} U_i/\sqrt{E_i} = \sum_{i=1}^{k}(O_i - E_i) = \sum_{i=1}^{k} O_i - \sum_{i=1}^{k} E_i$$

$$= \Sigma O_i - n\Sigma p_i = n - n = 0 \; [\because \Sigma O_i = n, \Sigma p_i = 1].$$

$$\therefore \log\frac{P}{C} \approx -\left[\sum_{i=1}^{k} U_i\sqrt{E_i} + \frac{1}{2}\sum_{i=1}^{k} U_i^2 + O\left(\frac{1}{E_i^{1/2}}\right)\right] \approx -\frac{1}{2}\Sigma U_i^2$$

$$\Rightarrow P \approx C\exp\left(-\frac{1}{2}\sum_{i=1}^{k} U_i^2\right).$$

Hence U_i is distributed as independent normal variates. Therefore,

$$\chi^2 = \sum_{i=1}^{k} U_i^2 = \sum_{i=1}^{k}\frac{(O_i - E_i)^2}{E_i} = \sum_{i=1}^{k}\frac{O_i^2}{E_i} - n$$

is distributed as χ^2 with $(k-1)$ d.f. The linear restriction for this distribution is

$$\sum_{i=1}^{k} U_i/\sqrt{E_i} = \sum(O_i - E_i) = 0 \Rightarrow \sum_i O_i = \sum_i E_i = n.$$

It is to be noted that

(i) The χ^2-test is a test for large sample of size n and it is an approximate test.

(ii) The sample observations are to be selected independently.

(iii) The class frequencies are such that

$$\sum_{i=1}^{k} O_i = \sum_{i=1}^{k} E_i = n.$$

(iv) The value of n should be more than 40 or 50.

(v) No class frequency should be less than 5. If any $E_i < 5$, it is to be added with the nearest class frequency. However, Koehler and Larntz (1980) have mentioned that, if $n \geq 10, k \geq 3, n^2/k \geq 10$, then χ^2-test will provide better result.

(vi) No assumption is made about the parent distribution of sample observations and hence the χ^2-statistic is considered as a nonparametric test.

(vii) If in any case any class frequency is added to the nearest frequency of a class, then total number of classes are reduced to k_1 (say) and the d.f. of χ^2 becomes $(k-1)$.

(viii) For uniform distribution $E_i = np_i$ and $p_i = \frac{n}{k}$. But, if it is needed to estimate p_i, it may need to estimate the 'c' number of parameters of the distribution. For example, if it is assumed that a given set of observations follow Poisson distribution, then to find p_i, the parameter of Poisson population is to be estimated. In such a case, if 'c' numbers of parameters are estimated, then the d.f. of χ^2 will be $(k - c - 1)$.

►**EXAMPLE 36.6** *To conduct a social survey a researcher has selected* 50 *households from* 500 *households. The selected households bearing numbers are as follows:*

001	060	120	180	240	300	360	420	480
010	070	130	190	250	310	370	430	490
020	080	140	200	260	320	380	440	
030	090	150	210	270	330	390	450	
040	100	160	220	280	340	400	460	
050	110	170	230	290	350	410	470	

Is the sample random?

Solution. Here $n = 50$. There are $3 \times 50 = 150$ digits. The frequency distribution of digits is

Digit	0	1	2	3	4	5	6	7	8	9	*Total*
Frequency, O_i	64	16	15	15	15	5	5	5	5	5	150
Expected frequency, E_i	15	15	15	15	15	15	15	15	15	15	150

$$H_0 : \text{the sample is random,}$$
$$H_A : \text{the sample is not random.}$$

Here there are 10 digits. Each digit has the chance of inclusion in a number is $p_i = 0.1$ and hence $E_i = np_i = 15$. The χ^2-statistic is

$$\chi^2 = \sum_{i=1}^{k} \frac{O_i^2}{E_i} - n = 343.47 - 150 = 193.47.$$

Since $\chi^2 > \chi^2_{0.05,9} = 16.919$, H_0 is rejected. The sample is not random.

36.5 Test for Equal Variances

Let $x_1, x_2, \ldots, x_n$ be a random sample of n observations from a population having c.d.f. $F(x)$. Again, $y_1, y_2, \ldots, y_m$ be another set of m sample observations from a population having c.d.f. $F(y)$. It is assumed that

$$E(X) = \mu_x \text{ and } E(Y) = \mu_y.$$

Here μ_x may be equal to μ_y or $\mu_x \neq \mu_y$.

The problem is to test the significance of

$H_0 : V(X) = V(Y)$ [X and Y are distributed similarly except $\mu_x \neq \mu_y$]
against H_A : (i) $V(X) < V(Y)$ [One-sided lower-tailed test]
(ii) $V(X) > V(Y)$ [Upper one-tailed test]
(iii) $V(X) \neq V(Y)$ [Two-tailed test].

Method of Test

(i) Calculate $u_i = |x_i - \mu_x|$ $(i = 1, 2, \ldots, n)$. If μ_x is unknown, its estimator $\bar{x}$ is to be used. Similarly, $v_j = |y_j - \mu_y|$ $(j = 1, 2, \ldots, m)$ are to be calculated.

(ii) The u_i and v_j values are assigned rank to the combined $(n + m)$ observations.

(iii) Assuming no tie among u_i and v_j observations, the test statistic is

$$T = \sum_{i=1}^{n}[R(u_i)]^2, \text{ where } R(u_i) = \text{rank of } u_i.$$

If tie is observed, then the test statistic is

$$T_1 = \frac{T - n\bar{R}^2}{\left[\frac{nm}{N(N-1)} \sum_{i=1}^{n+m} R_i^4 - \frac{nm}{N-1}(\bar{R}^2)^2\right]^{1/2}}$$

where

$$\bar{R}^2 = \frac{1}{N}\left[\sum_{i=1}^{n}[R(u_i)]^2 + \sum_{j=1}^{m}[R(v_j)]^2\right], \quad N = n + m$$

$$\sum_{i=1}^{n+m} R_i^4 = \sum_{i=1}^{n}[R(u_i)]^4 - \sum_{j=1}^{m}[R(v_j)]^4.$$

Iman (1976) has calculated the percentiles of T or T_1. The null hypothesis is rejected in favour of H_A (iii), if T or $T_1 > (1 - \frac{\alpha}{2})$ percentile or T or $T_1 < \frac{\alpha}{2}$ percentile. H_0 in favour of H_A (i) is rejected, if T or $T_1 < \alpha$ percentile, and H_0 in favour of H_A (ii) is rejected, if T or $T_1 > (1 - \alpha)$ percentile.

For large sample $(n > 10, m > 10)$, T or T_1 is to be compared with W_p, where

$$W_p = \frac{n(N+1)(2N+1)}{6} + Z_p\left[\frac{nm(N+1)(2N+1)(8N+11)}{180}\right]^{1/2}$$

when Z_p is the percentile of standard normal variate. The statistic

$$z_0 = \frac{T - n(N+1)(2N+1)/6}{[mn(N+1)(2N+1)(8N+11)/180]^{1/2}} \sim N(0, 1).$$

The p-value for lower-tailed test is $P(z \leq z_0)$ and that is for upper-tailed test is

$$P(z \geq z_0) = 1 - P(z \leq z_0).$$

The p-value for two-tailed test is $2P(z \leq z_0)$. At 5% level of significance H_0 is rejected in favour of two-sided alternative, if $|z| > 1.96$.

►**EXAMPLE 36.7** *The following data represent the body-breadth (in cm) of two different types of sea fishes:*

Type of fish	*Body-breadth (in cm)*
F_1, x	5.14, 4.78, 3.26, 5.63, 4.95, 5.93, 4.32; $n = 7$
F_2, y	4.16, 6.34, 2.15, 5.95, 4.66, 5.73, 6.08; $m = 7$

Are the distribution of body-breadth of two types of fishes similar with respect to variances?

Solution.

H_0 : the distributions of body-breadth of two types of fishes are similar [except $\mu_x \neq \mu_y$],

against H_A : $V(X) \neq V(Y)$.

The estimates of $\mu_x = \bar{x} = 4.86$ and $\mu_y = \bar{y} = 4.98$. Then

$$u_i = |x_i - \bar{x}| \;:\; 0.28,\ 0.08,\ 1.60,\ 0.77,\ 0.09,\ 1.07,\ 0.54$$
$$v_j = |y_i - \bar{y}| \;:\; 0.82,\ 1.36,\ 2.83,\ 0.97,\ 0.32,\ 0.75,\ 1.10$$

Combined u_i and v_j : 0.08, 0.09, 0.28, 0.32, 0.54, 0.75, 0.77, 0.82, 0.97 in ascending order

Rank :	1	2	3	4	5	6	7	8	9
Variable :	u	u	u	v	u	v	u	v	v
Combined u_i and v_j :	1.07	1.10	1.36	1.60	2.83				
Rank :	10	11	12	13	14				
Variable :	u	v	v	u	v				

$$\therefore T = \sum_{i=1}^{7} [R(u_i)]^2 = 357.$$

We have $T_{0.025} = 274, T_{0.975} = 741$. Since $T_{0.025} < T < T_{0.975}$, H_0 is accepted.

►**EXAMPLE 36.8** *In an agricultural experiment two levels of nitrogen as urea are applied to produce maize. The data of maize production (kg/plot) are shown below:*

Levels of urea	*Amount of maize (kg/plot)*
N_1, x	10.5, 6.3, 11.4, 9.6, 12.3, 14.6, 8.7, 7.5; $n = 8$
N_2, y	12.46, 13.6, 10.7, 8.3, 14.5, 15.6, 9.4; $m = 7$

Are the distributions of maize production using N_1 and N_2 similar in respect of variance?

Solution.

H_0 : the distributions of X and Y are similar [$\mu_x \neq \mu_y$],

against H_A : $V(X) < V(Y)$.

$$N = n + m = 8 + 7 = 15, \ \bar{x} = 10.11, \ \bar{y} = 12.07$$

$$u_i = |x_i - \bar{x}| \ : \ 0.39, 3.81, 1.29, 0.51, 2.19, 4.49, 1.41, 2.61$$

$$v_j = |y_i - \bar{y}| \ : \ 0.39, 1.53, 1.37, 3.77, 2.43, 3.53, 2.67$$

Combined u_i and v_j in ascending order	$R(u_i)$	$R(v_j)$	$[R(u_i)]^2$	$[R(v_j)]^2$	$[R(u_i)]^4$	$[R(v_j)]^4$
0.39	1.5	—	2.25	—	5.0625	—
0.39	—	1.5	—	2.25	—	5.0625
0.51	3	—	9	—	81	—
1.29	4	—	16	—	256	—
1.37	—	5	—	25	—	625
1.41	6	—	36	—	1296	—
1.53	—	7	—	49	—	2401
2.19	8	—	64	—	4096	—
2.43	—	9	—	81	—	6561
2.61	10	—	100	—	10000	—
2.67	—	11	—	121	—	14641
3.53	—	12	—	144	—	20736
3.77	—	13	—	169	—	28561
3.81	14	—	196	—	38416	—
4.49	15	—	225	—	50625	—

$$\Sigma[R(u_i)]^4 = 104775.0625$$

$$\Sigma[R(v_j)]^4 = 73530.0625$$

$$\bar{R}^2 = \frac{1}{N}\left[\Sigma[R(u_i)]^2 + \Sigma[R(v_j)]^2\right]$$

$$= \frac{1}{15}[648.25 + 591.25] = 82.63$$

$$\Sigma R_i^4 = \Sigma[R(u_i)]^4 + \Sigma[R(v_j)]^4 = 178305.125$$

$$T = \Sigma[R(u_i)]^2 = 648.25$$

$$T_1 = \frac{T - n\bar{R}^2}{\left[\frac{nm}{N(N-1)}\Sigma R_i^4 - \frac{nm}{N-1}(\bar{R}^2)\right]^{1/2}}$$

$$= \frac{648.25 - 8 \times 82.63}{\left[\frac{8\times 7}{15\times 14}178305.125 - \frac{8\times 7}{15-1}(82.63)^2\right]^{1/2}} = \frac{-12.73}{142.26} = -0.09.$$

Since $T_1 < T_{0.05} = 426$, H_0 is rejected, i.e., the variances of two distributions are not same.

Table 36.3 : Quantiles of test statistics for testing equality of variances.

n	p	m :	3	4	5	6	7	8	9	10
3	0.025		14	14	21	26	29	30	35	41
	0.05		21	21	26	30	38	42	49	54
	0.975		77	110	138	170	213	257	308	362
	0.95		70	101	129	161	197	238	285	333
4	0.025		30	39	50	54	63	71	78	90
	0.05		39	50	57	66	78	90	102	114
	0.975		126	165	206	255	311	374	439	510
	0.95		119	154	197	246	294	350	413	476
5	0.025		66	79	88	100	114	130	145	162
	0.05		75	88	103	120	135	155	175	195
	0.975		183	235	297	363	433	508	592	680
	0.95		178	228	282	342	410	479	558	639
6	0.025		115	130	143	164	184	208	231	255
	0.05		124	139	164	187	211	239	268	299
	0.975		259	331	406	486	574	670	771	880
	0.95		255	319	386	463	545	634	730	831
7	0.025		172	195	217	245	274	305	338	372
	0.05		188	212	240	274	308	344	384	425
	0.975		356	443	536	635	741	856	979	1108
	0.95		347	428	515	608	707	814	929	1051
8	0.025		249	276	311	345	384	425	468	513
	0.05		268	300	340	381	426	473	524	576
	0.975		476	579	689	807	935	1071	1215	1368
	0.95		464	560	664	776	896	1023	1159	1303
9	0.025		342	380	423	469	517	570	624	682
	0.05		365	406	457	510	567	626	689	755
	0.975		615	741	870	1009	1158	1317	1485	1662
	0.95		601	717	840	972	1112	1261	1420	1587
10	0.025		457	505	560	616	677	741	808	879
	0.05		486	539	601	665	734	806	883	963
	0.975		778	925	1078	1241	1413	1596	1788	1991
	0.95		765	901	1045	1197	1360	1533	1715	1907

►**EXAMPLE 36.9** *A group of couples from a rural area have been randomly selected and their number of ever-born children (x) are recorded. Another group of couples from an urban area have been selected to note down their number of ever-born children (y). The information are given below. Are the two distributions of number of ever-born children similar in respect of variance?*

$$x \ : \ 2, 4, 5, 3, 2, 2, 6, 4, 3, 4, 3, 6, 8$$
$$y \ : \ 3, 2, 4, 3, 3, 3, 2, 4, 4, 1, 3, 4$$

Solution. $H_0 : V(x) = V(y)$, against $V(x) \neq V(y)$.
Given $n = 13, \ m = 12, \ N = n + m = 25, \ \bar{x} = 4, \ \bar{y} = 3$

$$u_i = |x_i - \bar{x}| \ : \ 2, 0, 3, 1, 2, 2, 2, 0, 1, 0, 1, 2, 4$$
$$v_j = |y_j - \bar{y}| \ : \ 0, 1, 1, 0, 0, 0, 1, 1, 1, 2, 0, 1$$

Combined u_i and v_j in ascending order	$R(u_i)$	$R(v_j)$	$[R(u_i)]^2$	$[R(u_i)]^4$	$[R(v_j)]^2$	$[R(v_j)]^4$
0	4.5	—	20.25	410.0625	—	—
0	4.5	—	20.25	410.0625	—	—
0	4.5	—	20.25	410.0625	—	—
0	—	4.5	—	—	20.25	410.0625
0	—	4.5	—	—	20.25	410.0625
0	—	4.5	—	—	20.25	410.0625
0	—	4.5	—	—	20.25	410.0625
0	—	4.5	—	—	20.25	410.0625
1	13	—	169	28561	—	—
1	13	—	169	28561	—	—
1	13	—	169	28561	—	—
1	—	13	—	—	169	28561
1	—	13	—	—	169	28561
1	—	13	—	—	169	28561
1	—	13	—	—	169	28561
1	—	13	—	—	169	28561
1	—	13	—	—	169	28561
2	20.5	—	420.25	176610.0625	—	—
2	20.5	—	420.25	176610.0625	—	—
2	20.5	—	420.25	176610.0625	—	—
2	20.5	—	420.25	176610.0625	—	—
2	20.5	—	420.25	176610.0625	—	—
2	—	20.5	—	—	420.25	176610.0625
3	24	—	576	331776	—	—
4	25	—	625	390625	—	—
Total	—	—	3870	1692364.5	1535.5	350026.375

$$\bar{R}^2 = \frac{1}{N}\left[\Sigma[R(u_i)]^2 + \Sigma[R(v_j)]^2\right] = \frac{1}{25}[3870 + 1535.5] = 216.22$$

$$\Sigma R_i^4 = \Sigma[R(u_i)]^4 + \Sigma[R(v_j)]^4 = 2042390.875$$

$$T = \Sigma[R(u_i)]^2 = 3870$$

$$\therefore\ T_1 = \frac{T - n\bar{R}^2}{\left[\frac{nm}{N(N-1)}\Sigma R_i^4 - \frac{nm}{N-1}(\bar{R}^2)^2\right]^{1/2}}$$

$$= \frac{3870 - 13 \times 216.22}{\left[\frac{13\times 12}{25\times 24}2042390.875 - \frac{13\times 12}{24}(216.22)^2\right]^{1/2}}$$

$$= \frac{1059.14}{476.5916} = 2.22.$$

Now,

$$z = \frac{T_1 - n(N+1)(2N+1)/6}{[nm(N+1)(2N+1)(8N+11)/180]^{1/2}} \sim N(0,1)$$

$$= \frac{2.22 - 13 \times 26 \times 51/6}{[13 \times 12 \times 26 \times 51 \times 211/180]^{1/2}}$$

$$= \frac{-2870.78}{492.4238} = -5.83.$$

Since $|z| > 1.96$, H_0 is rejected. The two distributions of number of ever-born children are not similar in respect of variance.

36.5.1 Test of Equality of Several Variances

Let $x_{i1}, x_{i2}, \ldots, x_{in_i}$ be random sample of n_i observations from i-th $(i = 1, 2, \ldots, k)$ population, where $E(x_i) = \mu_i$. Let the c.d.f. of x_i be $F_i(x_i)$. We need to test the significance of H_0

$$H_0 : F_i(x_i)\text{'s are similar with respect to variances,}$$
$$\text{against } H_A : V(x_1) \neq V(x_2) \neq \cdots \neq V(x_k).$$

Let $u_{ij} = |x_{ij} - \mu_i|; \quad j = 1, 2, \ldots, n_i$
$\quad i = 1, 2, \ldots, k$

Here total number of u_{ij}'s is $n = \sum_{i=1}^{k} n_i$. Let $R(u_{ij})$ be the rank of the combined u_{ij}'s; then

$$R_i = \sum_{j=1}^{n_i} R(u_{ij}).$$

The test statistic to test the hypothesis is

$$T_2 = \frac{1}{D^2}\left[\sum_{i=1}^{k} \frac{S_i^2}{n_i} - N(\bar{S})^2\right]$$

where

$$S_i = \text{sum of squares of } R_{ij} = \sum_{j=1}^{n_i} R_{ij}^2$$

$$\bar{S} = \frac{1}{N}\sum_{i=1}^{k} S_i$$

$$D^2 = \frac{1}{N-1}\left[\sum_{i=1}^{N}[R(u_{ij})]^4 - N(\bar{S})^2\right].$$

The statistic T_2 is approximately distributed as chi-square with $(k-1)$ d.f. If $T_2 \geq \chi^2_{\alpha,k-1}$, H_0 is rejected.

The statistic D^2 is calculated assuming that some u_{ij}'s are equal. However, if there is no tie in u_{ij}'s, then

$$D^2 = N(N+1)(2N+1)(8N+11)/180 \text{ and } \bar{S}^2 = (N+1)(2N+1)/6.$$

If H_0 is rejected, the equality of variances in pairs can be compared. The critical values to compare variances in pairs is given by

$$\text{CD} = t_{\frac{\alpha}{2}}\left[\left\{\frac{D^2(N-1-T_2)}{N-k}\right\}^{1/2}\left\{\frac{1}{n_i}-\frac{1}{n_l}\right\}^{1/2}\right].$$

If $\left|\frac{S_i}{n_i} - \frac{S_l}{n_l}\right| > \text{CD}$, H_0 is rejected; $i \neq l = 1, 2, \ldots, k$.

►**EXAMPLE 36.10** *Some illiterate, primary-educated and higher-educated ever-married women are randomly and independently selected to estimate their number of ever-born children. The collected information of number of ever-born children are given below:*

Female's education	*Number of ever-born children*
Illiterate, x_{1j}	7, 5, 6, 4, 3, 2, 8, 7, 6, 5
Primary, x_{2j}	6, 7, 4, 4, 3, 5, 6, 4, 5, 5
Higher, x_{3j}	2, 3, 2, 4, 2, 3, 2, 1, 2, 1

Are the distributions of number of ever-born children similar in respect of variance?

Solution.

H_0 : except mean the distribution of number of ever-born children are similar,

H_A : at least variances of two groups of children are not same.

Given

$$n_1 = n_2 = n_3 = 10, \; N = n_1 + n_2 + n_3 = 30.$$

The estimates of population means are

$$\hat{\mu}_1 = \bar{x}_1 = 5.3, \hat{\mu}_2 = \bar{x}_2 = 4.9, \; \hat{\mu}_3 = \bar{x}_3 = 2.2.$$

$u_{1j} = \lvert x_{1j} - \bar{x}_1 \rvert$	$u_{2j} = \lvert x_{2j} - \bar{x}_2 \rvert$	$u_{3j} = \lvert x_{3j} - \bar{x}_3 \rvert$	*Rank of u_{ij}*		
			R_{1j}	R_{2j}	R_{3j}
1.7	1.1	0.2	23.5	18.5	6
0.3	2.1	0.8	9.5	27	13.5
0.7	0.9	0.2	11.5	16	6
1.3	0.9	1.8	22	16	25
2.3	1.9	0.2	28	26	6
3.3	0.1	0.8	30	2	13.5
2.7	1.1	0.2	29	18.5	6
1.7	0.9	1.2	23.5	16	20.5
0.7	0.1	0.2	11.5	2	6
0.3	0.1	1.2	9.5	2	20.5

$$S_1 = \Sigma R_{1j}^2 = 4558.5,\ S_2^2 = \Sigma R_{2j}^2 = 2869.5,$$

$$S_3^2 = \Sigma R_{3j}^2 = 2010.0,\ \bar{S} = \frac{1}{N}\Sigma S_i = 314.6$$

$$D^2 = \frac{1}{N-1}\left[\sum^{N}[R(u_{ij})]^4 - N(\bar{S})^2\right]$$

$$= \frac{1}{30-1}\left[5263521.75 - 30(314.6)^2\right] = 79114.72$$

$$T_2 = \frac{1}{D^2}\left[\sum_{i=1}^{k}\frac{S_i^2}{n_i} - N(\bar{S})^2\right]$$

$$= \frac{1}{79114.72}\left[3305405.25 - 30(314.6)^2\right] = 4.25.$$

Since $T_2 < \chi^2_{0.05,2} = 5.99$, H_0 is accepted. The three distributions are similar in respect of variance.

36.6 Tests Related to Classification of Observations

In discussing the test of randomness, we have mentioned that the sample observations are divided into two classes with respect to median or any specific value. The classification of sample observations may be done with respect to many other factors. For example, let us consider the classification of a group of couples of childbearing ages into adopter and non-adopter of family planning, the number of insects died or not died due to kept them in a pesticide, the number of diabetic patients classified into level of FBS after a treatment, etc.

Let there be n sample observations. Consider that any object in the sample can be divided into two classes and let p be the probability of inclusion of an object into a particular class. Then $q = 1 - p$ is the probability of not inclusion of the objects in that particular class. Let x objects out of n objects are included in a particular class.

Then the probability of x objects to be included in the particular class is

$$P(x) = \binom{n}{x} p^x q^{n-x}, \quad x = 0, 1, 2, \ldots, n.$$

This $P(x)$ is a binomial probability function and using this function we can test the significance of

$$H_0 : p = p_0 \text{ (a specified value)},$$
$$\text{against } H_A : p \neq p_0 \text{ or } H_A : p > p_0 \text{ or } H_A : p < p_0.$$

The test of such hypothesis is known as 'Binomial' test.

The classification of sample observations can also be made with respect to medians. Let M be the median of a set of sample observations. Usually, half of sample observations are less than the median and another half is greater than the median. In that case probability of an observation to become less than or more than median value is $p = 1/2$. In such a case, we can test the hypothesis

$$H_0 : p = p_0 = \frac{1}{2},$$
$$\text{against } H_A : p > \frac{1}{2} \text{ or, } H_A : p < \frac{1}{2} \text{ or, } H_A : p \neq \frac{1}{2}.$$

Let x be the number of observations which are either less than or more than median. Then

$$P(x) = \binom{n}{x} \left(\frac{1}{2}\right)^n, \quad x = 0, 1, 2, \ldots, n$$

is the probability of inclusion of x observations in either less than median class or in the above median class. The inclusion of an observation is decided by assigning a +ve, if it exceeds median and by a −ve sign, if it does not exceed median. Let x be the number of signs (which is minimum) for a particular class of observations. Since signs are used for classification, the test is known as 'Sign' test. We shall now discuss 'Binomial' and 'Sign' test in the following sub-sections.

36.6.1 Binomial Test

Let there be n random sample observations, x of which belongs to a class and $(n - x)$ belong to another class. Assume that p be the probability of inclusion of a sample observation to a particular class. Then the probability of inclusion of x objects in that particular class is (x is the number of observations a class which is minimum)

$$P(x) = \binom{n}{x} p^x q^{n-x}, \quad x = 0, 1, 2, \ldots, n.$$

The objective of the analysis is to test the significance of

$$H_0 : p = p_0 \text{ (a known value)},$$
$$\text{against } H_A : \text{(i) } p \neq p_0, \text{ (ii) } H_A : p > p_0, \text{ (iii) } H_A : p < p_0.$$

Under H_0 the test statistic is

$$P(x) = \sum_{i}^{x} \binom{n}{i} p_0^i q_0^{n-i}, \quad i = 0, 1, 2, \ldots, x.$$

The probability $P(x)$ can be calculated using the formula. However, the probabilities for different values of n and p are given in 'Binomial Table' [Conover (1999)].

Mode of Decision

(i) **The null hypothesis is rejected in favour of**

$$H_A : p \neq p_0 \text{ if } 2P(x) \leq 0.05.$$

(ii) **The null hypothesis is rejected in favour of**

$$H_A : p > p_0 \text{ or } H_A : p < p_0 \text{ if } P(x) \leq 0.05.$$

In each case the level of significance is $\alpha = 0.05$.

▶**EXAMPLE 36.11** ***A company discovered a medicine for the patients of cancer and claimed that 70% cancer patients would survive at least 5 years if they would use the medicine. A researcher randomly selected 10 such patients and investigated their history. He found that out of 10 such patients 4 died before 5-th year of use of medicine. Do you think that the claim of the company is acceptable?***

Solution. **Given $n = 10$, $p_0 = 0.70$, $q_0 = 0.30$, $x = 4$.**

$$P(x) = \sum_{i=0}^{4} \binom{n}{i} (0.7)^i (0.3)^{n-i}$$

$$= \sum_{i=0}^{4} \binom{10}{i} (0.7)^i (0.3)^{10-i} = 0.0473.$$

According to the company's claims $p \geq 0.70$. Therefore,

$$H_0 : p \geq 0.70, \text{ against } H_A : p < 0.70.$$

Since $P(x) < 0.05$, H_0 is rejected in favour of $p_0 < 0.70$.

▶**EXAMPLE 36.12** ***Twelve diabetic patients are randomly selected from a hospital record and observed that 5 of them suffered from heart problem also. Are the diabetic patients equally probable in suffering from heart disease?***

Solution. **Given $n = 12$, $x = 5$.**

Let p_1 = probability of a patient suffered from heart problem,

p_2 = probability of a patient not suffered from heart problem.

$H_0 : p_1 = p_2$, against $H_0 : p_1 \neq p_2$.

$$\therefore\ P(x) = \sum_{i=0}^{5} \binom{n}{i} \left(\frac{1}{2}\right)^{12} = 0.387.$$

Since $2P(x) = 0.774 > 0.05$, H_0 is accepted.

Binomial Test for Large Sample

If sample observations n is large enough, one can use z-test to test the hypothesis of $H_0 : p = p_0$. Here

$$z = \frac{x - np}{\sqrt{npq}} \sim N(0, 1).$$

We have $M_x(t) = (q + pe^t)^n$. But when n is large

$$\lim_{n\to\infty} M_z(t) = \exp\left[\frac{t^2}{2}\right].$$

This $M_z(t)$ is moment generating function of standard normal variate.

In practice, we can use

$$z = \frac{(x \pm 0.5) - np_0}{\sqrt{np_0q_0}}$$

where p_0 is the probability of inclusion of an observation into a particular class under H_0. Here 0.5 is used as a continuity correction since z is a normal variate which is replaced for a discrete variable x. Here $+0.5$ is used when $x < np_0$ and -0.5 is used when $x > np_0$.

The confidence limit is $x_{\text{CL}} = np_0 - z_{\frac{\alpha}{2}}\sqrt{np_0q_0}$ and $x_{\text{CU}} = np_0 + z_{\frac{\alpha}{2}}\sqrt{np_0q_0}$. Therefore, if $x \le x_{\text{CL}}$ or $x \ge x_{\text{CU}}$, H_0 is rejected. Hence for lower-tailed test the p-value is

$$p_1 = P\left[z \le \frac{x + 0.5 - np_0}{\sqrt{np_0q_0}}\right] \text{ and}$$

for upper-tailed test the p-value is

$$p_2 = P\left[z \ge \frac{x - 0.5 - np_0}{\sqrt{np_0q_0}}\right].$$

For two-tailed test p-value will be either $2p_1$ or $2p_2$.

►**EXAMPLE 36.13** *Six per cent students out of* 50 *students got grade A in a course. Can it be expected that* 10% *of the same group of students will get grade A in a similar course?*

Solution. It is expected that p is the proportion of students who will get A grade in a similar course. Then

$$H_0 : p = p_0 = 0.10, \text{ against } H_A : p \ne p_0.$$

Here $n = 50$, $x = 3(6\%)$, $E(x) = np_0 = 5$, $V(x) = np_0q_0 = 4.5$, $x < np_0$. Therefore,

$$z = \frac{x + 0.5 - np_0}{\sqrt{np_0q_0}} = \frac{3.5 - 5}{\sqrt{4.5}} = -0.71.$$

Since $|z| < 1.96$, there is no evidence against H_0. The p-value of the z-test is

$$P = 2P(z \le -0.71) = 0.4778.$$

36.6.2 One-Sample Sign Test

Let there be n observations in a random sample. Let us assume that the median of the population from which the sample is drawn is M and it is known. Now, if the sample observation is X, then

$$P(X \leq M) = P(X \geq M) = \frac{1}{2}.$$

Therefore, we need to test the significance of

$$H_0 : p = p_0 = \frac{1}{2},$$

$$\text{against (i) } H_A : p \neq \frac{1}{2}, \text{ (ii) } H_A : p > \frac{1}{2}, \text{ (iii) } H_A : p < \frac{1}{2}.$$

The test is based on the sign of $D = X - M$. The sign is $-$ve, if $X > M$ and it will be +ve, if $X < M$. Therefore, the sample observations are divided into two classes, where p is the probability of inclusion of any sample observations into a particular class. Let the number of observations in that particular class be x. In practice, x is the number of +ve or $-$ve signs which is minimum. Then the test statistic is

$$P(x) = \sum_{i=0}^{x} \binom{n}{i} \left(\frac{1}{2}\right)^n.$$

The test is similar to binomial test and hence conclusion is made in a similar way as it is done in case of binomial test.

►**EXAMPLE 36.14** *The following data represent the ages of some randomly selected high blood pressure patient:*

Ages (in years)
25, 28, 35, 42, 41, 26, 32, 37, 50, 62, 40, 41

Do you think that high blood pressure is observed among the lower-aged group of people? It is assumed that the median age of lower-aged group of people is $M = 30$ years.

Solution. The null hypothesis to be tested is

$$H_0 : M = 30, \text{ against } H_A : M < 30.$$

According to null hypothesis, it is expected that $D = X - M$ will give +ve sign in maximum cases. Here

$$D = X - M : - - + + + - + + + + + + \quad (n = 12)$$

Therefore, $x = 3$

$$P(x) = \sum_{i=1}^{3} \binom{12}{i} \left(\frac{1}{2}\right)^{12} = 0.073.$$

[Under H_0, $P(X \leq M) = P(X \geq M) = \frac{1}{2} = p$]. Here $P(x) > 0.05$ and there is no evidence against H_0.

►**EXAMPLE 36.15** *The amounts of rainfall in a city in different days during a rainy season are as follows:*

Amount of rainfall (in mm)
5.5, 1.6, 2.7, 10.8, 8.7, 3.4, 5.6, 8.2, 9.6, 4.2, 1.5, 1.7, 2.0, 3.1, 2.5, 1.9, 4.5, 3.6, 1.8, 9.3, 10.4, 5.3, 6.0, 6.1, 4.2, 3.3, 9.7.

Is the median of the amount of rainfall 4?

Solution. The null hypothesis to be tested is

$$H_0 : M = 4\left(P(X \le M) = P(X \ge M) = \frac{1}{2}\right),$$

against $H_A : M \neq 4$.

Here

$$D = X - M : + - - + + - + + + + - - - - - - + - - + + + + + + - +$$

Here $n = 27,\ x = 12$ (minimum total sign)

$$E(x) = np = 13.5,\ V(x) = npq = 6.75,\ x < np.$$

Therefore,

$$z = \frac{x + 0.5 - np}{\sqrt{npq}} = \frac{12.5 - 13.5}{\sqrt{6.75}} = -0.38.$$

Since $|z| < 1.96$, H_0 is accepted.

36.7 Two-Sample Sign Test

Let $x_1, x_2, \dots, x_n$ be a random sample of n observations from a population whose c.d.f., is $F(x)$. Consider that $y_1, y_2, \dots, y_n$ are another set of n random sample of observations from a population whose c.d.f. is $F(y)$. In practice, the y_i $(i = 1, 2, \dots, n)$ values may be the values of the variable X recorded after the application of a treatment. If the treatment has no significant impact, it is expected that $F(x) \equiv F(y)$. Here the values of X and Y are correlated. The objective of the analysis is to study the impact of the treatment. Hence the null hypothesis to be tested is

$$H_0 : F(x) = F(y), \text{ against } H_A : F(x) \neq F(y).$$

If the treatment has no effect, $F(x) = F(y)$. In that case

$$P(X < Y) = P(X > Y) = \frac{1}{2}.$$

Therefore, the signs of $D = X - Y$ can be used for the test statistic. The signs are divided into two classes and hence the binomial probability law is used for the test statistic.

Under H_0, half of signs are expected to be +ve and another half is expected to be −ve. But, in practice, the signs with minimum number is used to calculate the probability. Let this number be x. Then

$$P(x) = \sum_{i=0}^{x} \binom{n}{i} \left(\frac{1}{2}\right)^n.$$

Here, if $D_i = X_i - Y_i = 0$, then this observation is dropped from the analysis and in that case the number of observations becomes $n_1 < n$, where $n_1 = (n -$ number of observations which gives $X_i - Y_i = 0)$.

The value of x, however, is decided on the basis of alternative hypothesis. For example, let $H_A : P(X < Y) > \frac{1}{2}$. This alternative hypothesis implies that there should be maximum number of −ve signs in $D_i = X_i - Y_i$. Hence the number of positive signs will be the value of x. The test in such a case is known as lower-tailed test. Again, if $H_A : P(X > Y) > \frac{1}{2}$, more +ve signs in $D_i = X_i - Y_i$ are expected. Hence the value of x is minimum number of −ve signs. The test in such a case is upper-tailed test. For two-tailed test alternative hypothesis is

$$H_A : P(X > Y) = P(X < Y) \neq \frac{1}{2}.$$

Since the test statistic is binomial probability function, the conclusion is drawn similarly as it is done in case of binomial test. The assumption for the test is that the data are measured in ordinal scale. Moreover, it is assumed that $D_i = X_i - Y_i$ follow bivariate symmetric distribution.

►**EXAMPLE 36.16** *The body-condition score of cows before the start of lactation period (x) and after the end of lactation period (y) are recorded from randomly selected group of cows and given below:*

$$x : 4.5, 5.0, 4.8, 4.0, 4.7, 4.9, 5.0, 5.0, 5.0$$
$$y : 4.0, 4.5, 5.0, 4.5, 4.2, 4.5, 4.5, 4.5, 4.5$$

Do you think that the body-condition score is reduced after the end of lactation period?

Solution. Here $n = 9$.

$$D_i = X_i - Y_i : + + - - + + + + +$$

If there is no impact on body-condition score during lactation period, we can expact that $P(X < Y) = P(X > Y) = \frac{1}{2}$. If body-condition score is decreased after lactation period, we can expect that $P(X > Y) > \frac{1}{2}$. Therefore, the null and alternative hypothesis should be

$$H_0 : P(X < Y) = P(X > Y) = \frac{1}{2}$$

$$\text{against } H_A : P(X > Y) > \frac{1}{2},$$

According to alternative hypothesis maximum +ve signs are expected in

$$D_i = X_i - Y_i.$$

Therefore, x = minimum signs (−ve) and $x = 2$. Hence

$$P(x) = \sum_{i=0}^{2} \binom{n}{i} \left(\frac{1}{2}\right)^n = \sum_{i=0}^{2} \binom{9}{i} \left(\frac{1}{2}\right)^9 = 0.090.$$

Since $P(x) > 0.05$, there is no evidence against null hypothesis. The body-condition score is not reduced after the end of lactation period.

►**EXAMPLE 36.17** *Ten diabetic patients have tried to reduce their blood sugar level by diet control. The difference in blood sugar levels of those patients are observed by sign of $D = X - Y$, where X = FBS before diet control and Y = FBS after diet control. The D_i values are given below:*

$$D_i = X_i - Y_i : + - - + + + - + - -$$

Do you think that diet control is effective in reducing fasting blood sugar (FBS) level?

Solution. If diet control is effective in reducing blood sugar level, then $P(+\text{ve}) \geq P(-\text{ve})$. Therefore, the null and alternative hypothesis will be

$$H_0 : P(+\text{ve}) = P(-\text{ve}) = \frac{1}{2},$$

$$\text{against } H_A : P(+\text{ve}) < P(-\text{ve}).$$

Here $n = 10$, $x = 5$. Therefore,

$$P(x) = \sum_{i=0}^{5} \binom{n}{i} \left(\frac{1}{2}\right)^n = \sum_{i=0}^{5} \binom{10}{i} \left(\frac{1}{2}\right)^{10} = 0.623.$$

Since $P(x) > 0.05$, there is no evidence against null hypothesis.

►**EXAMPLE 36.18** *A group of 30 people have started physical exercise to reduce their body-weight. Their body-weights (in kg) are recorded before (x) and after (y) the physical exercise. The data are given below:*

$$x : 56, 58, 62, 57, 56, 60, 64, 66, 67, 62, 64, 67, 68, 55, 70$$
$$y : 55, 58, 60, 56, 56, 58, 62, 60, 63, 60, 63, 65, 66, 58, 70$$

$$x : 62, 61, 64, 60, 58, 57, 64, 52, 56, 57, 58, 60, 62, 64, 67$$
$$y : 63, 62, 62, 61, 55, 56, 63, 54, 57, 55, 60, 58, 61, 63, 66$$

Is there any change in body-weights due to physical exercise?

Solution. If there is no change in body-weights due to physical exercise,

$$P(X > Y) = P(X < Y) = \frac{1}{2}.$$

Hence

$$H_0 : P(X < Y) = P(X > Y) = \frac{1}{2},$$

$$\text{against } H_A : P(X < Y) \neq P(X > Y) \neq \frac{1}{2}.$$

Here

$$D_i = X_i - Y_i \; : \; +0++0-++++++++-0-- \\ +-++++--+-++++$$

$x = 8$ (minimum number of signs).

$$E(x) = np = 13.5, \; V(x) = npq = 6.75, \; x < np.$$

$$\therefore \; z = \frac{x + 0.5 - np}{\sqrt{npq}} = \frac{8.5 - 13.5}{\sqrt{6.75}} = -1.92.$$

Since $|z| < 1.96$, so H_0 is accepted. The p-value for the test is

$$P = 2P(z \leq -1.92) = 0.0548.$$

36.8 Similarity and Dissimilarity between Sign Test and Binomial Test

Similarity

Binomial probability law is used in both tests.

Dissimilarity

(i) The null hypothesis for binomial test is that any observation is included in any particular class, where total observations are classified into two classes. The null hypothesis for sign test is that the sample observations are selected from a population having a specified median.

(ii) The probability of inclusion of any observation to a particular class is p, where $0 < p < 1$. But in sign test any observation will be less than or more than the median and this will happen with probability $p = 1/2$.

36.9 Wilcoxon Signed Rank Test

This test is similar to two-sample sign test except that $D = X - Y$ is measured by magnitude and direction instead of direction only. For example, let us consider that $X : 6, 8, 12$ and $Y : 7, 9, 11$. Then $D : X - Y : -1, -1, 1$. Again, consider another set of values $X : 15, 18, 24$ and $Y : 16, 20, 17$. Then $D = X - Y : -1, -2, +7$. Here in both sets of values the signs of differences are similar but the magnitudes of the differences are not similar. In the former case, $D = X - Y : -1, -1, 1$ and in the latter case, $D = X - Y : -1, -2, 7$. Since the magnitudes of the differences are not same in both cases, it is better to consider the values of D in terms of both magnitude and direction. Then the values of D are classified into two classes, where positive values of D are included in one group and negative values are included in another group. Any value of $D = 0$ is deleted from the analysis. The test in which the magnitude and direction of any D_i $(i = 1, 2, \ldots, n)$ is considered is known as *Wilcoxon signed rank test*.

The objective of the test is same as it is for sign test. Hence the null and alternative hypotheses are

$$H_0 : P(X < Y) = P(X > Y) = \frac{1}{2},$$

$$\text{against (i) } H_A : P(X < Y) \neq P(X > Y) \neq \frac{1}{2},$$

$$\text{(ii) } H_A : P(X > Y) > \frac{1}{2},$$

$$\text{(iii) } H_A : P(X < Y) < \frac{1}{2}.$$

In practice, $D_i = X_i - Y_i$ $(i = 1, 2, \ldots, n)$ is calculated and then rank of $|D_i|$ is found out. The smallest $|D_i|$ is given rank 1 and the largest D_i is given rank n on the assumption that there is no tie in the values of X and Y $(D_i \neq 0)$. The ranks of D_i are then identified by their sign. If two or more D_i values are same, average rank is calculated for those D_i values. Now, let T^+ be the total of ranks of +ve D_i values and T^- be the total of ranks of −ve D_i values $(i = 1, 2, \ldots, n)$. Wilcoxon (1949) has found out the critical values of T-statistic for different values of level of significance of α.

The T-value (either T^+ or T^-) which is smaller is used as a test statistic. Then, if any T-value is less than the tabulated value of T, H_0 is rejected. The T-value used as test statistic depends on alternative hypothesis. For example, let

$$H_0 : P(X < Y) = P(X > Y) = \frac{1}{2}$$

$$\text{and } H_A : P(X < Y) > \frac{1}{2}.$$

In such a case we expect maximum −ve D_i. Hence T^+ is to be used as test statistic. If

$$H_A : P(X > Y) > \frac{1}{2}$$

we expect maximum +ve D_i and hence T^- is to be used as test statistic.

Method of Conclusion

Let T_0 be the value of T^+ or T^-. Then, if

(i) $2P(T^+ \geq T_0)$ or $2P(T^- \geq T_0) \leq 0.05$, H_0 is rejected in favour of two-sided alternative;

(ii) $P(T^+ \geq T_0)$ or $P(T^- \geq T_0) \leq 0.05$, H_0 is rejected in favour of one-sided alternative.

It is observed that the Wilcoxon signed rank test depends on the ranks of n values. The ranks are $D_i \neq 0$ and $D_i \neq D_{i'}$ $(i = 1, 2, \ldots, n); 1, 2, \ldots, n$. Some of the ranks will be with −ve sign and some are with +ve sign. Here $D_i = X_i - Y_i$. If $H_0 : M = M_0$, where M is the median of the population from which sample is selected. Then $D_i = X_i - M_0$, where D_i is the value of observation taking i-th position when the data are in array.

Let $r(\cdot)$ be the rank of value of any observation. Then

$$T^+ = \sum_{i=1}^{n} Z_i r(|D_i|),\ T^- = \sum_{i=1}^{n}(1 - Z_i)r(|D_i|)$$

$$T^+ - T^- = 2\sum_{i=1}^{n} Z_i r(|D_i|) - \frac{n(n+1)}{2}, \text{ where}$$

$$Z_i = \begin{cases} 1, & \text{if } D_i > 0 \\ 0, & \text{if } D_i < 0. \end{cases}$$

Without loss of generality let us consider that $|D_1|, |D_2|, \ldots, |D_n|$ are order statistics. Therefore, $r(|D_i|)$ can be replaced by 'i'. Thus, we can write $Z_{(i)}$ instead of Z_i. Then

$$Z_{(i)} = \begin{cases} 1, & \text{if } |D| = |X - Y| \text{ or } D = |X - M_0| \text{ has rank positive} \\ 0, & \text{if rank of } |D| = |X - Y| \text{ or } D = |X - M_0| \text{ is negative.} \end{cases}$$

Then, we can write

$$T^+ = \sum_{i=1}^{n} iZ_{(i)}.$$

Here $Z_{(i)}$ is the independent Bernoulli variate. But the distribution of the variable is not identical. We know that for Bernoulli variate

$$E[Z_{(i)}] = \theta_i,\ V[Z_{(i)}] = \theta_i(1 - \theta_i),\ \text{cov}[Z_{(i)}, Z_{(j)}] = 0$$

where

$$\theta_i = P[Z_{(i)} = 1] = P[r(|D_j|)] = i \cap D_j > 0.$$

This is true for some values of $j = 1, 2, \ldots, n$

$$\begin{aligned} \theta_i &= P[i\text{-th order statistic of the positive differences} \\ &\qquad \text{among } |D_1|, |D_2|, \ldots, |D_n|] \\ &= \int_0^\infty \frac{n!}{(i-1)!(n-i)!}\left[P_{|D|}(u)\right]^{i-1}\left[1 - F_{|D|}(u)\right]^{n-i} f_D(u)du \\ &= n\binom{n-1}{i-1}\int_0^\infty \left[F_D(u) - F_D(-u)\right]^{i-1}\left[1 - F_D(u) + F_D(-u)\right]^{n-i} f_D(u)du. \end{aligned}$$

If X is symmetric about median M_0, then the value of θ under null hypothesis exists. That is $D = X - M_0$ is symmetric about zero. Hence

$$F_D(0) = \frac{1}{2},\ F_D - (u) = 1 - F_D(u).$$

Now, putting the value of F_D in θ_i, we get

$$\begin{aligned} \theta_i &= n\binom{n-1}{i-1}\int_0^\infty \left[2F_D(u) - 1\right]^{i-1}\left[2 - 2F_D(u)\right]^{n-i} f_D(u)du \\ &= \frac{n}{2}\binom{n-1}{i-1}\int_0^1 v^{i-1}(1-v)^{n-i}dv = \frac{n}{2}\binom{n-1}{i-1}B(i, n-i+1) = \frac{1}{2}. \end{aligned}$$

Therefore,

$$E(T^+) = \sum_{i=1}^{n} i\theta_i, \; V(T^+) = \sum_{i=1}^{n} i^2(1-\theta_i)\theta_i$$

$$\text{and } E(T^+/H_0) = \frac{n(n+1)}{4}, \; V(T^+/H_0) = \frac{n(n+1)(2n+1)}{24}.$$

If n is large,

$$z = \frac{T^+ - E(T^+)}{\sqrt{V(T^+)}} \sim N(0,1) \text{ and } z = \frac{T^- - E(T^-)}{\sqrt{V(T^-)}} \sim N(0,1).$$

In practice the value of T^+ and T^- are almost equal. Therefore, if either T^+ or T^- is very small, H_0 cannot be considered as true. For two-tailed test the value of T^+ or T^{-1}, which is smaller, can be compared with the critical values of T. Wilcoxon (1949) has presented the values of T-statistic for different values of α. These values are presented in the following table:

Table 36.4 : Critical values of T-statistic for Wilcoxon signed rank test

Sample size	*Level of significance for one-tailed test*		
	0.025	0.01	0.005
	Level of significance for two-tailed test		
n	0.05	0.02	0.01
6	0	—	—
7	2	0	—
8	4	2	0
9	6	3	2
10	8	5	3
11	11	7	5
12	14	10	7
13	17	13	10
14	21	16	13
15	25	20	16
16	30	24	20
17	35	28	23
18	40	33	28
19	46	38	32
20	52	43	38
21	59	49	43
22	66	56	49
23	73	62	55
24	81	69	61
25	89	77	68

If the value of T^+ or T^- (which is applicable) is less than or equal to tabulated T, H_0 is rejected.

►**EXAMPLE 36.19** *In a hospital some children are under investigation to check up their body-temperature (°F). The children are given Crocin when doctor has observed that the temperature of the children rises. The temperatures of the children before (X) and after (Y) use of medicine are given below:*

$$X : 99.2,\ 100.4,\ 101.6,\ 99.4,\ 102.7,\ 100.0,\ 102.5$$
$$Y : 100.0,\ 100.5,\ 100.0,\ 98.0,\ 99.0,\ 98.0,\ 98.0$$

Is the medicine effective in reducing temperature?

Solution. If the medicine is effective in reducing temperature, then $P(X > Y) \geq \frac{1}{2}$. Hence, we need to test the significance of null hypothesis

$$H_0 : P(X > Y) \geq \frac{1}{2}, \text{ against } H_A : P(X > Y) < \frac{1}{2}.$$

According to H_A the maximum sign of $(X - Y)$ should be −ve. Hence T^+ should be the test statistic. Here

$$D = X - Y : -0.8,\ -0.1,\ 1.6,\ 1.4,\ 3.7,\ 2,\ 4.5$$
$$\text{Rank } (d_i) : -2,\ -1,\ 4,\ 3,\ 6,\ 5,\ 7$$
$$T^+ = 25$$

For $n = 7$ and $\alpha = 0.025$ (one-tailed test) tabulate $T = 2 < T^+$. Hence there is no evidence against null hypothesis. The medicine is effective in reducing temperature.

►**EXAMPLE 36.20** *In a dairy farm an experiment is conducted to study the impact of a dry concentrate in increasing milk production of cows. The milk production of cows before using the concentrate (X, kg/day) and after using the concentrate (Y, kg/day) are given below:*

$$X : 18.5,\ 17.6,\ 16.2,\ 20.6,\ 22.6,\ 18.4,\ 25.0,\ 19.0,\ 20.0$$
$$Y : 20.2,\ 21.4,\ 18.4,\ 18.2,\ 20.7,\ 19.6,\ 24.6,\ 18.7,\ 21.0$$

Is the dry concentrate good to increase milk production?

Solution. The concentrate may be good or not. Hence

$$H_0 : P(X < Y) = P(X > Y) = \frac{1}{2},$$
$$H_A : P(X < Y) \neq P(X > Y) \neq \frac{1}{2}.$$

Here

$$D = X - Y : -1.7,\ -3.8,\ -2.2,\ 2.4,\ 1.9,\ -1.2,\ 0.4,\ 0.3,\ -1.0$$
$$\text{Rank } (d_i) : -5,\ -9,\ -7,\ 8,\ 6,\ -4,\ 2,\ 1,\ -3$$

Hence, $T^+ = 17$, $T^- = 28$, $n = 9$, $\alpha = 0.05$, $T = 6$. Since $T^- > T$, there is no evidence against null hypothesis. The dry concentrate is good to increase the milk production.

►**EXAMPLE 36.21** *The following data represent the marks of a Statistics course of some randomly selected students:*

60, 66, 80, 86, 75, 70, 82, 80, 62, 64, 40, 45, 46, 55, 50.

Is the sample selected from a population having median 65*?*

Solution.

$$H_0 : M = 65,$$
$$H_A : M \neq 65.$$
$$D = X - M : -5, 1, 15, 21, 10, 5, 17, 15, -3, -1, -25, -20, -19, -10, -15$$
$$\text{Rank } (d_i) : -4.5, 1.5, 9, 14, 6.5, 4.5, 11, 9, -3, -1.5, 15, -13, -12, -6.5, -9$$
$$T^+ = 70.5,\ T^- = 49.5,\ n = 15,\ \alpha = 0.05,\ T = 25.$$

Since $T^{-1} > T$, H_0 is accepted. The population median of marks may be taken as 65.

►**EXAMPLE 36.22** *In a school* 30 *students are randomly selected for special coaching so that the results (in %) of the students in their final H.S. Examination are increased. The results in the test examination (X in %) and the results after final examination (Y in %) are given below:*

X : 50, 68, 70, 72, 65, 77, 80, 86, 80, 62, 60, 66, 70, 75, 77,
Y : 55, 65, 70, 75, 60, 75, 85, 85, 84, 66, 68, 70, 68, 76, 75,
X : 58, 67, 68, 70, 72, 74, 66, 65, 66, 70, 80, 84, 83, 70, 71
Y : 60, 70, 65, 75, 78, 73, 70, 71, 74, 77, 82, 86, 85, 77, 80.

Do you think that the coaching is effective for the improved examination result?

Solution. If coaching is effective $P(X < Y) \geq \frac{1}{2}$. Hence

$$H_0 : P(X < Y) \geq \frac{1}{2},$$
$$H_A : P(X < Y) < \frac{1}{2}.$$

Here maximum sign of $(X - Y)$, according to H_A, is expected –ve. Hence T^+ is to be used as test statistic.

$D = (X - Y)$	−5	3	0	−3	5	2	−5	1	−4	−4
Rank (d_i)	−20.5	12.5	—	−12.5	20.5	7	−20.5	2	−16.5	−16.5

$D = (X - Y)$	−8	−4	2	−1	−2	−2	−3	3	−5	−6
Rank (d_i)	−27.5	−16.5	7	−2	−7	−7	−12.5	12.5	−20.5	−23.5

$D = (X - Y)$	1	−4	−6	−8	−7	−2	−2	−2	−7	−9
Rank (d_i)	2	−16.5	−23.5	−27.5	−25.5	−7	−7	−7	−25.5	−29

$n = 29$ ($\because$ $d_3 = 0$, it is deleted from the analysis).

$T^+ = 63.5$. Here n is large. So, we use z-test statistic,

$$z = \frac{T^+ - E(T^+)}{\sqrt{V(T^+)}} = \frac{63.5 - 435}{\sqrt{2138.75}} = -8.03$$

$$E(T^+) = \frac{n(n+1)}{2} = 435, \; V(T^+) = \frac{n(n+1)(2n+1)}{24} = 2138.75.$$

Since $|z| > 1.64$, H_0 is rejected. The level of significance is $\alpha = 0.05$.

The Wilcoxon signed rank test is also known as median test since D_i is calculated by $D_i = X_i - M$, where M is median.

36.10 Cox and Stuart Test

The objective of the test is to verify whether there is any trend in the sample observations. If any kind of trend is present in the data, the data may be non-random. Hence the test regarding the presence of trend can also be used as a test of randomness of data.

Let $x_1, x_2, \ldots, x_n$ be a random sample of n observations from a population having unknown distribution. The sample observations can be divided into two parts. Let x_c be the value which lies in the middle of the sample observations, where $C = \frac{n_1}{2}$ (if n is even number), $C = \frac{1}{2}(n+1)$, if n is odd number. The sample observations are divided into two groups with respect to x_c. In the first group, the observations are below x_c and in the second group the observations are above x_c. Then, pairs of values with i-th observation of first group and i-th observation of second group can be formed. Thus, the pairs are

$$(x_1, x_{1+c}), (x_2, x_{2+c}), \ldots, (x_{n_1-c}, x_{n_1}).$$

It is assumed that there is no tie in observations within a pair. If tie is observed, the pair of observations is deleted from the analysis. Let us consider that, finally there be n_1 pairs of values. If any $x_i < x_{i+c}$, the pair is denoted by $-$ve sign, if $x_i > x_{i+c}$, the pair is denoted by $+$ve sign.

If there is no trend in the data (data is random) $P(+\text{ve}) = P(-\text{ve}) = \frac{1}{2}$. Therefore, the null hypothesis to be tested is

$$H_0 : \text{the data are random,}$$
$$\text{against } H_A : \text{the data are non-random.}$$

Since the pairs of values are divided into two classes, the number of pairs in any class follows binomial distribution with the probability that any pair belongs to a class is $\frac{1}{2}$. Hence the probability that in a class there will be x pairs is given by

$$P(x) = \sum_{i=0}^{x} \binom{n_1}{i} \left(\frac{1}{2}\right)^{n_1}.$$

Here x is the number of pairs with $+$ve or $-$ve sign (which is smaller). Hence, if $2P(x) < 0.05$, H_0 is rejected in favour of non-randomness of data.

If $H_0 : P(+\text{ve}) \geq \frac{1}{2}$, against $H_A : P(+\text{ve}) < \frac{1}{2}$, the first value in the pair will be smaller in maximum cases. Hence x will be the number of $-$ve signs and the test is one-tailed. In a similar way, the one-tailed right-sided test

$$\left[H_0 : P(+\text{ve}) \leq \frac{1}{2},\ H_A : P(+\text{ve}) > \frac{1}{2}\right]$$

can also be performed.

►**EXAMPLE 36.23** *The temperature of a city during month of May are recorded as follows:*

Temperature (in °C)

30.5, 30, 31.2, 31.5, 30.6, 32.4, 32.5, 32.5, 33.0, 34.2, 36.0, 38.5, 39.4, 39.8, 38.5, 37.6, 36.2, 35.0, 35.2, 37.4, 36.2, 35.8, 37.2, 38.0, 39.5, 40.2, 39.6, 38.2, 39.0, 39.2, 41.0.

Is there any upward trend in the amount of temperature?

Solution. Here $n = 31$, $C = \frac{n+1}{2} = 16$ ($\because$ n is odd). Therefore, deleting the temperaturc of 16th day we need to group the observations in pairs, where the pairs will be as follows [here $x_c = 37.6$]:

$$(30.5, 36.2),\ (30, 35),\ (31.2, 35.2),\ (31.5, 37.4),\ (30.6, 36.2),$$
$$(32.4, 35.8),\ (32.5, 37.2),\ (32.5, 38),\ (33.0, 39.5),\ (34.2, 40.2),$$
$$(36.0, 39.6),\ (38.5, 38.2),\ (39.4, 39.0),\ (39.8, 39.2),\ (38.5, 41.0).$$

The null hypothesis to be tested is

H_0 : there is no upward trend in the amount of temperature,

against H_A : there is upward trend in the amount of temperature.

According to H_A the first observation in a pair should be smaller $[x_i < x_{i+c}]$. Therefore, we can expect smaller number of $(x_i > x_{i+c})$. Any pair is assigned $-$ve sign, if $x_i > x_{i+c}$. Hence x = number of negative signs assigned to pairs. Here $n_1 = 15, x = 3$ and the test statistic is

$$P(x) = \sum_{i=0}^{x} \binom{n_1}{i} \left(\frac{1}{2}\right)^{n_1} = \sum_{i=0}^{3} \binom{15}{i} \left(\frac{1}{2}\right)^{15} = 0.0176 < 0.05.$$

Hence H_0 is rejected. Here for level of significance of $\alpha = 0.05$ the critical value of x should be less than 4. The observed value of $x = 3 < 4$. Therefore, null hypothesis is rejected. There is a upward trend in the data of temperature.

Since binomial probability law is used for the test of presence of trend in a set of data, we can use z-test, if n is large, where

$$z = \frac{x - np}{\sqrt{npq}} \sim N(0, 1).$$

Here $p = q = \frac{1}{2}$.

36.11 Kolmogorov-Smirnov Test

Let $x_1, x_2, \ldots, x_m$ be a random sample of m observations from a population having distribution function $F(x)$. Consider another sample of n observations $y_1, y_2, \ldots, y_n$ which are selected from a population having distribution function $F(y)$. It is assumed that both $F(x)$ and $F(y)$ are continuous functions. The problem is to test the significance of null hypothesis

$$H_0 : F_x(x) = F_y(x) \; \forall \; x,$$
$$\text{against } H_A : F_x(x) \neq F_y(x).$$

The order statistics of x_i's and y_j's are respectively

$$x_{(1)} < x_{(2)} < \cdots < x_{(m)}; \; y_{(1)} < y_{(2)} < \cdots < y_{(n)}.$$

Let $S_m(x)$ and $T_n(x)$ be the ratio of X and Y values which do not exceed x when both x_i's and y_j's are mixed according to their order statistics. For example, let

$$X : 40, 45, 55, 45, 30, 50$$
$$Y : 60, 50, 50, 45, 50, 55$$

The order statistics of X's are

$$x_{(1)} = 30 < x_{(2)} = 40 < x_{(3)} = 45 = x_{(4)} < x_{(5)} = 50 < x_{(6)} = 55.$$

The order statistics of Y's are

$$y_{(1)} = 45 < y_{(2)} = 50 = y_{(3)} = y_{(4)} < y_{(5)} = 55 < y_{(6)} = 60.$$

Now, the mixed series according to order statistics of X's and Y's is

$$x_{(1)} = 30 < x_{(2)} = 40 = x_{(3)} = 45 = x_{(4)} = y_{(1)} < y_{(2)} = 50 = y_{(3)} = y_{(4)}$$
$$= x_{(5)} < x_{(6)} = 55 = y_{(5)} < y_{(6)} = 60.$$

Here for different values of X and Y the $S_m(x)$ and $T_n(x)$ are shown below:

X :	30	40	45	50	55	60
$S_m(x)$:	1/6	2/6	4/6	5/6	1	1
$T_n(x)$:	0	0	1/6	4/6	5/6	1

The distribution of $S_m(x)$ and $T_n(x)$ are as follows:

$$S_m(x) = \begin{cases} 0, & \text{if } x < x_{(1)} \\ k/m, & \text{if } x_{(k)} \leq x \leq x_{(k+1)}, \; k = 1, 2, \ldots, m-1 \\ 1, & \text{if } x \geq x_{(m)} \end{cases}$$

$$\text{and } T_n(x) = \begin{cases} 0, & \text{if } x < y_{(1)} \\ k/n, & \text{if } y_{(k)} \leq x \leq y_{(k+1)}, \; k = 1, 2, \ldots, n-1 \\ 1, & \text{if } x \geq y_{(n)} \end{cases}$$

Then the test statistic for the hypothesis is

$$D_{m,n} = \sup_x |S_m(x) - T_n(x)|.$$

The null hypothesis is rejected at $100\alpha\%$ level of significance, if

$$D_{m,n} \geq D_{m,n;\alpha}, \text{ where under } H_0$$
$$P[D_{m,n} \geq D_{m,n;\alpha}] = \alpha.$$

If the distribution is continuous, the sampling distribution of D can be found out. Gibbons (1976) has proposed the exact p-value for the test.

The alternative hypothesis may be related to one-sided test, where

$$H_0 : F_y(x) \leq F_x(x), \; \forall \; x \text{ and } F_y(x) < F_x(x), \text{ for some } x.$$

Then the test statistic is

$$D^+_{m,n} = \sup_x [S_m(x) - T_n(x)]$$

and the critical region is

$$D^+_{m,n} \geq D^+_{m,n;\alpha}.$$

Again, if

$$H_A : F_y(x) \geq F_x(x), \; \forall \; x \text{ and } F_y(x) > F_x(x), \text{ for some } x$$

then the test statistic is

$$D^-_{m,n} = \sup[S_m(x) - T_n(x)].$$

The critical region is

$$D^-_{m,n} \geq D^-_{m,n;\alpha}.$$

Rohatgi (1976) has presented the values of

$$D^+_{m,n;\alpha} \text{ and } D^-_{m,n;\alpha}.$$

In our given example, we have

x :	30	40	45	50	55	60
$\lvert S_m(x) - T_n(x)\rvert$:	1/6	2/6	3/6	1/6	1/6	0

$$\therefore \sup_x |S_m(x) - T_n(x)| = \frac{3}{6}.$$

Here $m = n = 6$. The value of

$$D_{m,n:\alpha} = D_{6,6;0.05} = 4/6 > D_{6,6}.$$

Hence there is no evidence against the null hypothesis, i.e., the two samples are selected from the same population.

36.12 Mann-Whitney U-Test

Let $x_1, x_2, \ldots, x_m$ be a random sample of m observations from a population having distribution function $F(x)$. Consider another sample of n observations $y_1, y_2, \ldots, y_n$ from a population having distribution function $F(y)$. The problem is to test the significance of null hypothesis

$$\begin{aligned} H_0 &: F_x(x) = F_y(x), \\ \text{against (i) } H_A &: F_x(x) \geq F_y(x), \\ \text{(ii) } H_A &: F_y(x) \leq F_x(x) \text{ and} \\ \text{(iii) } H_A &: F_x(x) \neq F_y(x). \end{aligned}$$

According to the procedure of U-test, we need to test the hypothesis

$$H_0 : P(X < Y) = \frac{1}{2} \text{ against } H_A : P(X < Y) \neq \frac{1}{2} \text{ (two-tailed)}$$

$$\text{or, } H_0 : P(X < Y) \leq \frac{1}{2} \text{ against } H_A : P(X < Y) > \frac{1}{2} \text{ (one-tailed)}$$

$$\text{or, } H_0 : P(X < Y) \geq \frac{1}{2} \text{ against } H_A : P(X < Y) < \frac{1}{2} \text{ (one-tailed).}$$

The reason is that, if $F_x(x) = F_y(x)$, then

$$P(X < Y) = P(X > Y) = \frac{1}{2}.$$

The test statistic depends on the number of X before Y when m values of X and n values of Y are mixed and arranged in ascending order. For example, let us consider a set of x values

$$x : 2,\ 3,\ 5,\ 8,\ 7,\ 10$$

and a set of y values

$$y : 4,\ 6,\ 9,\ 12.$$

Then the x and y values in ascending order are

$$x_1 < x_2 < y_1 < x_3 < y_2 < x_4 < x_5 < y_3 < x_6 < y_4.$$

Here number of y before x's is

$$0 + 0 + 1 + 2 + 2 + 3 = 8 = U.$$

The U value is also calculated by the formula

$$U = S - \frac{m(m+1)}{2}$$

where S is the sum of rank of x's. In our given example,

$$\begin{aligned} S &= 1 + 2 + 4 + 6 + 7 + 9 = 29. \\ \therefore\ U &= 29 - \frac{6 \times 7}{2} = 8. \end{aligned}$$

Alternatively,

$$U = \sum_{i=1}^{m}\sum_{j=1}^{n} D_{ij},$$

where

$$D_{ij} = \begin{cases} 1, & \text{if } y_j < x_i, \text{ for all } i = 1, 2, \ldots, m \\ 0, & \text{if } y_j > x_i, \ j = 1, 2, \ldots, n. \end{cases}$$

The 100α% values of U are shown in Statistics books as quantiles of Mann-Whitney U-statistic. Now, the null hypothesis will be rejected in favour of two-sided alternative, if $U < W_{\frac{\alpha}{2}}$ or $U > nm - W_{\frac{\alpha}{2}}$, where $W_{\frac{\alpha}{2}}$ is $100\frac{\alpha}{2}$% quantiles of U.

The value of $U = 0$, if all values of Y are greater than all values of X. Again, $U = mn$, if all values of Y are less than all values of X. Thus, $0 \le U \le mn$. The critical values are proposed on the basis of values of U.

Decision Rule

H_0	H_A	H_0 is rejected, if
$F_x(x) = F_y(x)$	$F_x(x) \ge F_y(x)$	$U' \le C_\alpha$
$F_x(x) = F_y(x)$	$F_x(x) \le F_y(x)$	$U \le C_\alpha$
$F_x(x) = F_y(x)$	$F_x(x) \ne F_y(x)$	$U \le C_{\frac{\alpha}{2}}$ or, $U' \le C_{\frac{\alpha}{2}}$

Here

$$U' = \sum_{i=1}^{m}\sum_{j=1}^{n} (1 - D_{ij}).$$

The probability of $U = u_i$ is also calculated by the formula

$$P[U = u_i] = \frac{\text{Number of } u_i \text{ up to } i\text{-th } u_i}{\binom{m+n}{m}}.$$

►**EXAMPLE 36.24** *The following data represent the monthly salary (in* **000** *Rs.) of two groups of employees in a factory:*

Monthly salary (in 000 Rs.)
Group-1, X : 5.65, 8.62, 10.36, 14.58, 7.32, 11.45
Group-2, Y : 4.82, 9.75, 15.60, 10.45, 18.60, 5.62

Are the employees of two groups getting similar salary?

Solution. Here $m = n = 6$. Let $F(x)$ and $F(y)$ be the distribution of $X$and Y, respectively. We need to test

$$H_0 : F_x(x) = F_y(x), \text{ for all } x,$$

$$\text{against } H_A : F_x(x) \ne F_y(x), \text{ for some } x.$$

Here the mixed values of X and Y in ascending order are as follows:

$$4.82 < 5.62 < 5.65 < 7.32 < 8.62 < 9.75 < 10.36 < 10.45$$
$$< 11.45 < 14.58 < 15.60 < 18.60$$
$$y_1 < y_2 < x_1 < x_2 < x_3 < y_3 < x_4 < y_4 < x_5 < x_6 < y_5 < y_6$$
$$\therefore\ S = 3+4+5+7+9+10 = 38.$$
$$\therefore\ U = S - \frac{m(m+1)}{2} = 38 - \frac{6\times 7}{2} = 17.$$

Now, for $m = n = 6$, $W_{0.025} = 27$, $nm - W_{\frac{\alpha}{2}} = 6\times 6 - 27 = 9$. Here $U < W_{0.025}$ and $U > nm - W_{\frac{\alpha}{2}}$. Therefore, H_0 is rejected. The two groups of employees are not getting similar salary.

The procedure discussed above is useful if both n and m are small. For large values of n and m, we need to apply z-test, where

$$z = \frac{U - \frac{mn}{2}}{\sqrt{\frac{mn(m+n+1)}{12}}} \sim N(0,1).$$

Here

$$E(U) = \frac{mn}{2},\ V(U) = \frac{mn(m+n+1)}{12}.$$

However, as U is an integer z is adjusted by adding 0.5 with U. Thus

$$z = \frac{(U+0.5) - \frac{mn}{2}}{\sqrt{\frac{mn(m+n+1)}{12}}} \sim N(0,1).$$

36.13 Wilcoxon Test

Let $x_1, x_2, \ldots, x_m$ be a random sample of m observations from a population whose distribution function is $F_x(x)$. Again, let us consider that $y_1, y_2, \ldots, y_n$ are another random sample observations of size n selected from a population having distribution function $F_y(x)$. We need to verify whether these two populations are same or not in respect of measure of central tendency. The null hypothesis to be tested is

$$H_0 : F_x(x) = F_y(x), \text{ for all } x,$$
$$\text{against } H_A : F_y(x) = F_x(x-\theta), \text{ for some } x \text{ and for some } \theta \neq 0.$$

Here H_A is called 'Location' alternative.

If $\theta < 0$, the median of X variable will be greater than the median of Y variable and the ranks of X values will be bigger than the ranks of Y values. In such circumstance, Wilcoxon (1945) has proposed that $H_A : \theta < 0$ is accepted, if the sum of the ranks of values of X is larger or $H_A : \theta > 0$ is acceptable, if the sum of the ranks of X values is very small. The alternative hypothesis $H_A : \theta \neq 0$ is acceptable, if the sum of the ranks of X values is either very big or very small. Thus the test statistic is a function of ranks and the test statistic is written as

$$W_N = \sum_{i=1}^{N} iZ_i, \; N = m + n$$

where

$$Z_i = \begin{cases} 1, & \text{if the } i\text{-th value is of } X \text{ variable when two samples are combined and arranged in array} \\ 0, & \text{otherwise.} \end{cases}$$

If there is no tie, then under H_0

$$E[W_N] = \frac{m(N+1)}{2} \text{ and } V(W_N) = \frac{mn(N+1)}{12}.$$

The mean and variance of W_N are calculated as follows:

$$W_N = \sum_{i=1}^{N} iZ_i; \text{ where } Z_i \text{ is indicator variable}$$

$$E[W_N/H_0] = \sum_{i=1}^{N} iE(Z_i/H_0) \text{ and } V(W_N/H_0) = \Sigma i^2 V(Z_i/H_0).$$

Since Z_i is indicator variable

$$f_i(Z_i) = \begin{cases} \frac{m}{N}, & \text{if } Z_i = 1;\ i = 1, 2, \ldots, N \\ \frac{n}{N}, & \text{if } Z_i = 0 \\ 0, & \text{otherwise.} \end{cases}$$

$$\therefore\ E(Z_i) = \frac{m}{N},\ V(Z_i) = \frac{m}{N} - \frac{m^2}{N^2} = \frac{mn}{N^2}$$

$$\therefore\ E[W_N/H_0] = \frac{m}{N}\sum_{i=1}^{N} i = \frac{m(N+1)}{2} \text{ and } V[W_N/H_0] = \frac{mn}{N^2}\Sigma i^2 = \frac{mn(N+1)}{12}.$$

The maximum and minimum value of W_N are, respectively

$$\sum_{i=N-m+1}^{N} i = \frac{m(2N-m+1)}{2} \text{ and } \sum_{i=1}^{m} i = \frac{m(m+1)}{2}.$$

Using the values of W_N the decision is made as follows:

Null hypothesis H_0	*Alternative hypothesis H_A*	*H_0 is rejected, if*
$\theta = 0$	$\theta > 0$	$W_N \geq W(\alpha, m, n)$
$\theta = 0$	$\theta < 0$	$W_N \leq [m(N+1) - W(\alpha, m, n)]$
$\theta = 0$	$\theta \neq 0$	$W_N \geq W(\alpha_2, m, n)$ or, $W_N \leq [m(N+1) - W(\alpha_1, m, n)]$

where $\alpha = \alpha_1 + \alpha_2$ and $W(\alpha, m, n)$ is a value such that $P[W_N \geq W(\alpha, m, n)] = \alpha$. The null distribution of W_N can be found out from the lowest and highest value of W_N. As an example, we can consider the two samples of sizes $m = 3$, $n = 4$. Therefore, $N = m + n = 7$. These seven values can be arranged in array in $\binom{7}{4} = 35$ ways. However, for such an example, $6 \leq W_N \leq 18$. Therefore, we do not need to show the $\binom{7}{4} = 35$ arrangements. The null distribution of W_N is shown below:

W_N	*Rank of X values*	*Frequency*
6	1, 2, 3	1
7	1, 2, 4	1
8	1, 3, 4; 1, 2, 5	2
9	2, 3, 4; 1, 3, 5; 1, 2, 6	3
10	1, 2, 7; 1, 3, 6; 1, 4, 5; 2, 3, 5	4
11	2, 3, 6; 2, 4, 5; 1, 4, 6; 1, 3, 7	4
12	1, 5, 6; 1, 4, 7; 2, 4, 6; 2, 3, 7; 3, 4, 5	5
13	1, 5, 7; 2, 4, 7; 2, 5, 6; 3, 4, 6	4
14	1, 6, 7; 2, 5, 7; 3, 4, 7; 3, 5, 6	4
15	2, 6, 7; 3, 5, 7; 4, 5, 6	3
16	3, 6, 7; 4, 5, 7	2
17	4, 6, 7	1
18	5, 6, 7	1

It is observed that W_N is symmetric about mean $\bar{W} = \frac{m(N+1)}{2} = 12$. Since W_N is symmetric, if half of its values are found out, the remaining half can be also be found out. However, for large N the distribution of W_N can be found out by a recurrence relation similar to Mann-Whitney U-test. Let $r_{m,n}(k)$ be the number of arrangements of m values of X and n values of Y. Then

$$r_{m,n}(k) = r_{m-1,n}(k - N) + r_{m,n-1}(k) \text{ and } f(k) = \frac{r_{m,n}(k)}{\binom{m+n}{m}}.$$

Gibbons (1976) and Wilcoxon (1947) have presented the critical values of W_N for different values of m and n.

For large sample, the test statistic is

$$z = \frac{W_N - \frac{m(N+1)}{2}}{\sqrt{\left[\frac{mn(N+1)}{12}\right]}} \sim N(0, 1).$$

Since W_N is integer, the corrected z is calculated adding or subtracting 0.5 with W_N. Thus

$$z = \frac{(W_N \pm 0.5) - \frac{m(N+1)}{2}}{\sqrt{\left[\frac{mn(N+1)}{12}\right]}}.$$

If $N \geq 12$, z tends to normal variate.

►**EXAMPLE 36.25** *The ages of 8 randomly selected employees of a garments industry-A and the ages of another group of 8 employees of industry-B are given below:*

$$A,\ X\ :\ 19\ \ 55\ \ 50\ \ 40\ \ 51\ \ 53\ \ 54\ \ 56$$
$$B,\ Y\ :\ 31\ \ 18\ \ 41\ \ 17\ \ 24\ \ 15\ \ 52\ \ 37$$

Do the two ages differ in respect of central values?

Solution. We need to test the significance of

$$H_0\ :\ F_x(x) = F_y(x),$$
$$\text{against } H_A\ :\ F_y(x) = F_x(x-\theta),\ \theta \neq 0.$$

Both the ages are in array as follows:

Ages:	15	17	18	19	24	31	37	40	41	50	51	52	53	54	55	56
Rank:	1	2	3	4	5	6	7	8	9	10	11	12	13	14	15	16
Variable:	Y	Y	Y	X	Y	Y	Y	X	Y	X	X	Y	X	X	X	X

Here

$$W_N = \sum_{i=1}^{N} iZ_i = 4+8+10+11+13+14+15+16 = 91.$$

Here $N = 16$, $m = n = 8$. From table, we have $W_N > W(0.007, 8, 8) = 91$. Since the test is two-sided $\alpha = \alpha_1 + \alpha_2 = 0.007 + 0.007 = 0.014 < 0.05$, H_0 is rejected. The ages of two groups of employees differ in respect of central value.

36.14 Median Test

Let $x_1, x_2, \ldots, x_{n_1}$ be a random sample of n_1 observations from a population having distribution function $F_x(x)$. Let $y_1, y_2, \ldots, y_{n_2}$ be another random samples of size n_2 from another population having distribution function $F_y(x)$. The problem is to test the significance of null hypothesis that both the populations have same median.

$$H_0\ :\ F_y(x) = F_x(x),$$
$$\text{against } H_A\ :\ F_y(x) = F_x(x-\theta),\ \theta \neq 0.$$

This alternative hypothesis is also a location alternative and hence similar to the hypothesis considered in Wilcoxon test. However, there is difference in the test procedure.

Let us consider that the median of the combined sample observations is M. Also, consider that m_1 out of n_1 observations exceed median M and out of n_2 observations m_2 exceed median M. Then the joint distribution of m_1 and m_2 observations is hypergeometric distribution with probability function

$$P(m_1, m_2) = \frac{\binom{n_1}{m_1}\binom{n_2}{m_2}}{\binom{n_1+n_2}{m_1+m_2}}.$$

If $m_1 < n_1/2$, then the critical region corresponding to the size of level of significance α is given by $m_1 < m_1'$ where m_1' is computed from the equation

$$\sum_{m_1=1}^{m_1'} P(m_1, m_2) = \alpha.$$

The distribution of m_1 under H_0 is also hypergeometric with mean and variance, respectively

$$\begin{aligned} E(m_1) &= \frac{n_1}{2}, \text{ if } N = n_1 + n_2 \text{ is even} \\ &= \frac{n_1(N-1)}{2N}, \text{ if } N \text{ is odd;} \\ V(m_1) &= \frac{n_1 n_2}{4(N-1)}, \text{ if } N \text{ is even} \\ &= \frac{n_1 n_2 (N+1)}{4N^2}, \text{ if } N \text{ is odd.} \end{aligned}$$

Therefore, for large samples, we can use

$$z = \frac{m_1 - E(m_1)}{\sqrt{V(m_1)}} \sim N(0, 1).$$

In practice, the test of equality of two populations in respect of median can be performed using χ^2-test, where the sample observations are classified into two classes, one contains the observations which exceed median and another class contains the observations which are less than or equal to the median of combined sample observations. This classification is done for both the sample observations. Thus, we have a 2×2 contingency table of sample observations. If the two sample are selected from the populations having same median, the proportions of observations exceeding median and less than or equal to median should be same. That is, if P_{ij} be the proportion of observations exceeding median of i-th sample, then

$$P_{11} = \frac{m_1}{n_1} \text{ and } P_{21} = \frac{m_2}{n_2} \ (j = 1, 2)$$

$$P_{12} = \frac{n_1 - m_1}{n_1} \text{ and } P_{22} = \frac{n_2 - m_2}{n_2}.$$

We expect $P_{11} = P_{12}$ or $P_{11} = P_{21}$ or $P_{21} = P_{22}$. Therefore, the test of equality of common median for two populations transforms to

$$H_0 : P_{11} = P_{21}, \text{ against } H_A : P_{11} \neq P_{21}.$$

Hence the hypothesis can be tested by χ^2-test statistic, where

$$\chi^2 = \sum_{i=1}^{2} \sum_{j=1}^{2} \frac{(O_{ij} - E_{ij})^2}{E_{ij}} = \sum \sum \frac{O_{ij}^2}{E_{ij}} = N.$$

Here $m_1 = O_{11}, m_2 = O_{12}, O_{21} = n_1 - m_1, O_{22} = n_2 - m_2, N = n_1 + n_2$. This χ^2 is distributed as chi-square with 1 d.f.

The details of χ^2-test has been discussed in Chapter 34.

►**EXAMPLE 36.26** *The amount of rainfalls during rainy seasons in different randomly selected days in two different cities are given below:*

Rainfall (in mm/day)
City-A, X : 4.5, 0.1, 0.4, 16.2, 4.2, 5.6, 4.3, 7.9, 11.2, 17.6, 6.4
City-B, Y : 1.2, 2.4, 3.6, 9.7, 16.2, 10.4, 0.2, 0.3, 12.4, 13.6, 8.7

Are the rainfalls in two cities similar in respect of median rainfall?

Solution. Let $F_x(x)$ be the distribution function of X and $F_y(x)$ be the distribution function of Y. We need to test

$$H_0 : F_x(x) = F_y(x), \text{ against } H_A : F_x(x) \neq F_y(x)$$
$$\text{or, } H_0 : F_y(x) = F_x(y), \text{ against } H_A : F_y(x) = F_x(x - \theta),\ \theta \neq 0.$$

The median of the combined sample observations is $M = 6$. The sample observations are classified as follows:

2 × 2 *Contingency table*

No. of observations	X	Y	*Total*
$> M$	$5 = a$	$5 = b$	10
$\leq M$	$6 = c$	$6 = d$	12
Total	11	11	22

Since $N < 40$, we can use

$$\chi^2 = \frac{N\left[|ad - bc| - \frac{N}{4}\right]^2}{(a+b)(a+c)(c+d)(b+d)} \text{ [Conover (1999)]}$$
$$= \frac{22\left[|30 - 30| - \frac{22}{4}\right]^2}{10 \times 11 \times 12 \times 11} = 0.046.$$

Since $\chi^2 < \chi^2_{0.05,1} = 3.84$, H_0 is accepted. The two samples are selected from a population with same median.

Median Test for Equality *k* Independent Populations

Let x_{jl} be the l-th sample observations selected from j-th population ($j = 1, 2, \ldots, k$; $l = 1, 2, \ldots, n_j$) having distribution function $F(x)$. We need to test the hypothesis that

$$H_0 : F_1(x) = F_2(x) = \cdots = F_k(x),$$

against H_A : at least one of the equality does not hold good in respect of median of the distribution.

Let us consider that M be the median of the combined sample observations and O_{ij} be the number of observations of j-th sample in i-th class, where ($i = 1, 2$; $i = 1$, a class assigned for those observations which are greater than median of combined sample observations M; $i = 2$, otherwise). Then we have a $(2 \times k)$ contingency table of observations, where the table is as follows:

$2 \times k$ Contingency table

No. of observations	*Samples*							*Total*
	1	*2*	$\cdots$	*j*	$\cdots$	*k*		
$> M$	O_{11}	O_{12}	$\cdots$	O_{1j}	$\cdots$	O_{1k}		$O_{1\cdot}$
$\leq M$	O_{21}	O_{22}	$\cdots$	O_{2j}	$\cdots$	O_{2k}		$O_{2\cdot}$
Total	$O_{\cdot 1}$	$O_{\cdot 2}$	$\cdots$	$O_{\cdot j}$	$\cdots$	$O_{\cdot k}$		n

Then the test statistic for the hypothesis is

$$\chi^2 = \sum_{i=1}^{2}\sum_{j=1}^{k}\frac{(O_{ij} - E_{ij})^2}{E_{ij}} = \sum_{i=1}^{2}\sum_{j=1}^{k}\frac{O_{ij}^2}{E_{ij}} - n.$$

This χ^2 is distributed as chi-square distribution with $(k-1)$ d.f., if no $E_{ij} < 5$ $[E_{ij} < 0.25, n \geq 10, k \geq 3, n^2/k \geq 10$; Koehler and Larntz (1980)]. Here

$$E_{ij} = \frac{O_{i\cdot} \times O_{\cdot j}}{n}.$$

►**EXAMPLE 36.27** *Four independent samples of primary school children are selected from four primary schools. The IQ of the children are measured and given below:*

IQ of children												
Group-1, X_1 :	62,	72,	73,	73,	75,	62,	70,	66,	64,	50,	55,	80
Group-2, X_2 :	73,	74,	75,	66,	60,	66,	67,	68,	70,	70,	70,	72
Group-3, X_3 :	56,	55,	58,	69,	72,	56,	78,	79,	74,	73,	68,	69
Group-4, X_4 :	50,	52,	80,	85,	60,	61,	58,	59,	72,	66,	63,	63

Are the 4 groups of children similar in respect of median?

Solution.

$H_0 : F_1(x) = F_2(x) = F_3(x) = F_4(x)$,

H_A : at least one of the equality does not hold good in respect of median.

We have $n_1 = n_2 = n_3 = n_4 = 12$. Median of combined sample observations is $M = 68$. The sample observations can be arranged in $(2 \times k)$ contingency table as follows:

$(2 \times k)$ *Contingency table for* (O_{ij})

No. of observations	*Samples*				*Total*
	1	*2*	*3*	*4*	
$> M$	6	7	7	3	23
$\leq M$	6	5	5	9	25
Total	12	12	12	12	48

All $E_{1j} = 5.75$ and all $E_{2j} = 6.25$. Therefore,

$$\chi^2 = \sum\sum \frac{O_{ij}^2}{E_{ij}} - n = 50.89 - 48 = 2.89.$$

Since $\chi^2 < \chi^2_{0.05,3} = 7.81$, H_0 is accepted. All the children of all primary scholls are similar in median IQ level.

36.15 Confidence Interval for Median

Let $x_1, x_2, \ldots, x_n$ be a random sample of n observations selected from a population having distribution function $F_x(x)$. Assume that the median of the distribution is M. Let $x_{(r)}$ and $x_{(s)}$ be the r-th and s-th order statistics of the sample observations, where $r < s$. Now, $x_{(r)} < M$, if minimum r observations are less than median. Again, $M < x_{(s)}$, if at best $(s-1)$ observations are less than median. Then

$$x_{(r)} < M < x_{(s)}.$$

Therefore,

$$P\left[x_{(r)} < M < x_{(s)}\right] = P[r \le S_+ \le S-1] \text{ or, } P[r \le S_- \le S-1].$$

Here

$$\begin{aligned} S_+ &= \text{number of observations giving +ve sign for } x_i - M_0 \\ \text{and } S_- &= \text{number of observations showing −ve sign for } x_i - M_0, \end{aligned}$$

where M_0 is the assumed median.

The problem is to find $100(1-\alpha)\%$ confidence interval for M. If we are interested to test the significance of

$$H_0 : M = M_0, \text{ against } H_A : M \neq M_0$$

then let us consider that the lower-tailed critical value on the basis of n sample observations be $k'_{\alpha/2}$. In such a case the lower-tailed value of confidence interval will be those values which will be in $\left(k'_{\alpha/2}+1\right)$-th position. Therefore, $100(1-\alpha)\%$ confidence interval for median will be given by

$$x_{\left(k'_{\alpha/2}+1\right)} < M < x_{k_{\alpha/2}} \text{ or, } x_{\left(k'_{\alpha/2}+1\right)} \le M \le x_{\left(n-k'_{\alpha/2}\right)}.$$

For $n \le 20$, $k'_{\alpha/2}$ is available from any Binomial Probability Table. However, if $n > 20$, the critical value is obtained from normal probability table. In such a situation, if $z_{\alpha/2}$ is the value of normal variate, 0.5 is to be added with the normal variate value to avoid the problem of discontinuity in the order statistics. Thus, we have

$$\begin{aligned} z_{\frac{\alpha}{2}} &= \frac{k_{\alpha/2} - 0.5 - 0.5n}{0.5\sqrt{n}} \Rightarrow k_{\alpha/2} = 0.5n + 0.5 + 0.5z_{\alpha/2}\sqrt{n} \\ \text{and } -z_{\frac{\alpha}{2}} &= \frac{k'_{\alpha/2} + 0.5 - 0.5n}{0.5\sqrt{n}} \Rightarrow k'_{\alpha/2} = 0.5n - 0.5 - 0.5z_{\alpha/2}\sqrt{n}. \end{aligned}$$

►**EXAMPLE 36.28** *The following data represent the daily production of electricity (X in MW) in a thermal project:*

$$X \ : \ 200, 180, 175, 210, 205, 190, 185, 170, 215, 200, 185.$$

In normal condition the project produces 200 MW as a median production level. Do the data conform with the median production? Find confidence interval for the median production.

Solution. We need to test the significance of

$$H_0 : M = M_0 = 200, \text{ against } H_A : M \neq M_0.$$

We have

$$D_i = X_i - M_0 \ : \ 0, -20, -25, 10, 5, -10, -15, -30, 15, 0, -15.$$

Here $S_- = 6$, $S_+ = 3$ (two observations are deleted since these two provide $X_i - M = 0$). The probability is

$$2P(x) = 2\sum_{i=1}^{x}\binom{n}{i}\left(\frac{1}{2}\right)^n = 2\sum_{i=0}^{3}\binom{9}{i}\left(\frac{1}{2}\right)^9 = 2 \times 0.2539 = 0.5078 > 0.05.$$

Hence, there is no evidence against null hypothesis.

Here for $S_+ \leq 3$, the exact probability level is 0.5078. Therefore, $k_{\alpha/2} = 3$ and at $(1 - 0.5078) = 0.4922$ or 49.22% level the confidence interval is

$$X_{(4)} \leq M \leq X_{(6)} \text{ or, } 185 \leq M \leq 200.$$

36.16 Nonparametric Test for Two-Way Classification

The sign test is used to test the significance of difference of two treatments when treatments are applied on the same experimental units. The experimental results generate two sets of sample observations which are related. We say that the sample observations are correlated. For example, let us consider the case of studying the impact of a medicine which is used for diabetic patients. The blood sugar levels of patients can be studied before and after the use of medicine. Thus, we shall get two correlated samples of blood sugar levels. The objective is to test the hypothesis $H_0 : F(x) = F(y)$.

The treatment (medicine) can also be given to the patients at different stages or on different occasions. The treatments are applied on the same experimental unit repeatedly. This will generate several correlated samples.

Let x_{ij} be the observation recorded on j-th occasion from i-th experimental unit ($i = 1, 2, \ldots, n$; $j = 1, 2, \ldots, k$). It is assumed that there are k treatments which are applied on n experimental units. The objective of the experiment is to test the significance in the difference of treatments. Let us consider that x_{ij} observation has a distribution the distribution function of which is $F_j(x)$ $(j = 1, 2, \ldots, k)$. If treatments are similar or if there is no significant difference among treatments, then

$$F_1(x) = F_2(x) = \cdots = F_k(x).$$

It is assumed that $x_{1j}, x_{2j}, \ldots, x_{nj}$ observations are mutually independent. Therefore, under the assumption and experimental conditions mentioned above, we need to test the null hypothesis

$$H_0 : F_1(x) = F_2(x) = \cdots = F_k(x),$$

against H_A : at least one of the equality does not hold good.

The procedures to test the significance of the above null hypothesis are (i) Friedman's Test, (ii) Quade Test, and (iii) Page Test. Let us first discuss the Friedman's Test.

36.16.1 Friedman's Test

Friedman's tet is based on ranks of k observations recorded for k treatments on one occasion and on all ranks for n occasions. Let R_{ij} be the rank of j-th observation for j-th treatment recorded on i-th occasion, where $R_{ij} = 1$, if x_{ij} is smallest one and $R_{ij} = k$, if x_{ij} is largest. Hence, if there is no tie in the observations of i-th occasion, then the ranks of data of i-th occasion will be $1, 2, \ldots, k$. Let

$$R_i = \sum_{j=1}^{k} R_{ij} = k(k+1)/2.$$

Also, we can write $R_j = \sum_{i=1}^{n} R_{ij}$. Here the values of R_i are fixed since k treatments are used in the experiment. However, the number of replications (blocks) for treatments may not be same and hence R_j values ($j = 1, 2, \ldots, k$) may not be fixed. If treatments are replicated equally, the expected values of R_j are expected to be equal and the expected values should be equal to average value of R_j. The average value of R_j is $k(k+1)/2$, since $\Sigma R_j = nk(k+1)/2$.

If the average of $R_j = k(k+1)/2$, the mean deviation of R_j values about average R_j values will be zero. Here the mean square deviations of treatments is

$$S = \sum_{j=1}^{k} \left[R_j - \frac{k(k+1)}{2}\right]^2 = \sum R_j^2 - \frac{(\Sigma R_j)^2}{k}$$

$$= \sum_{j=1}^{k} \left[\sum_{i=1}^{n} \left(R_{ij} - \frac{k+1}{2}\right)\right]^2.$$

The value of S will be minimum if there is no difference among treatments. Hence the value of S can be used to test the significance of null hypothesis.

The ranks of i-th occasion (block) are assigned completely randomly and hence the observations of i-th block are arrangement of first k natural numbers (if there is no tie). Hence in $(n \times k)$ arrangements of ranks there are $(k!)^n$ sets of observations and each one is equally likely. Hence the probability distribution of S is

$$f(s) = \frac{u_s}{(k!)^n}, \text{ where } u_s \text{ is the value of } S.$$

Kendall (1962) has determined the distribution of s for n small and $k = 3, 4, 5$.
Let $\frac{k+1}{2} = \mu$, then

$$S = \sum_{i=1}^{n}\sum_{j=1}^{k}(R_{ij} - \mu)^2 + 2\sum_{i=1}^{n}\sum\sum_{l \le j \le p \le k}\sum(R_{ij} - \mu)(R_{pj} - \mu)$$

$$= n\sum_{i=1}^{n}(i - \mu)^2 + 2U = \frac{nk(k^2-1)}{12} + 2U.$$

The moments of S, specially the mean and variance of S, are obtained from the distribution of U. This is shown below:

The value of S depends on rank. Let us now describe the Karl Pearson's rank correlation. Let $(x_1, y_1), (x_2, y_2), \ldots, (x_k, y_k)$ be k pairs of values. The rank correlation coefficient of these k pairs of values is

$$R = \frac{\sum_{j=1}^{k}(x_j - \bar{x})(y_j - \bar{y})}{\left[\sum_{j=1}^{k}(x_j - \bar{x})^2 \sum_{j=1}^{k}(y_j - \bar{y})^2\right]^{1/2}}.$$

Let R_j = the rank of x_j and S_j = the rank of y_j. Then for the values of $R_j = r_j$ and $S_j = s_j$, we have for $(x_j, y_j), (r_1, s_1), (r_2, s_2), \ldots, (r_k, s_k)$. But

$$\sum_{j=1}^{k} r_j = \sum_{j=1}^{k} s_j = \frac{k(k+1)}{2}, \ \bar{r} = \bar{s} = \frac{k+1}{2}$$

$$\sum_{j=1}^{k}(r_j - \bar{r})^2 = \sum_{j=1}^{k}(s_j - \bar{s})^2 = \sum_{j=1}^{k}\left(j - \frac{k+1}{2}\right)^2$$

$$= \frac{k(k^2-1)}{2}.$$

Putting these values in R above, we get

$$R = \frac{12\sum_{j=1}^{k}(R_j - \bar{R})(S_j - \bar{S})}{k(k^2-1)} = \frac{12\left[\sum_{j=1}^{k}R_jS_j - \frac{1}{4}k(k+1)^2\right]}{k(k^2-1)}$$

$$= \frac{12\sum_{j=1}^{k}R_jS_j}{k(k^2-1)} - \frac{3(k+1)}{k-1}.$$

It is observed that the distribution of S discussed above is the distribution of U. Again, U and R are correlated [$\because$ U is the covariance of ranks]. Hence, using the distribution of R, we can find the mean and variance of U. Since r_j's are the values of

k sample observations,

$$f(r_j) = \frac{1}{k},\ r_j = 1, 2, \ldots, k$$
$$f(r_j s_j) = \frac{1}{k(k-1)};\ r_j,\ s_j = 1, 2, \ldots, k\ (r_j \neq s_j).$$

Therefore,

$$E(r_j) = \frac{k+1}{2},\ V(r_j) = \frac{k^2-1}{12}$$
$$\begin{aligned}\text{cov}(r_j, s_j) &= \frac{1}{k(k-1)}\left[\sum_{j=1}^{k}\sum_{p=1}^{k} pj - \frac{1}{k^2}\left(\sum_{j=1}^{k} j\right)^2\right] \\ &= \frac{1}{k^2(k-1)}\left[k\,(\Sigma j)^2 - k\Sigma j^2 - (k-1)(\Sigma j)^2\right] \\ &= \frac{-1}{k^2(k-1)}\left[\frac{k^2(k+1)(2k+1)}{6} - \frac{k^2(k+1)^2}{4}\right] \\ &= -\frac{k+1}{12}.\end{aligned}$$

Hence

$$E(R_{ij}) = \frac{k+1}{2},\ V(R_{ij}) = \frac{k^2-1}{12},\ \ \text{cov}(R_{ij}, R_{iq}) = -\frac{k+1}{12}.$$

Again, since the observations of each block are independent $\text{cov}(R_{ij}, R_{pq}) = 0$.

$$\therefore E(U) = k\binom{n}{2}\ \text{cov}(R_{ij}, R_{pq}) = 0$$

$V(U) = E(U^2)$, where

$$\begin{aligned}U^2 &= \sum_{i=1}^{k}\ \sum_{l\leq j\leq p\leq n} (R_{ij}-\mu)^2(R_{pj}-\mu)^2 \\ &\quad + 2\sum\sum_{1\leq i\leq q\leq k}\ \sum\sum_{1\leq i\leq p\leq n}\ \sum\sum_{1\leq r\leq j\leq n} (R_{ij}-\mu)(R_{pj}-\mu)(R_{sq}-\mu)(R_{rq}-\mu).\end{aligned}$$

But, if $i \neq p$, R_{ij} and R_{pq} are independent. Hence

$$\begin{aligned}E[U^2] &= \sum_{j=1}^{k}\ \sum\sum_{1\leq i\leq p\leq n} V(R_{ij})V(R_{pj}) + 2\sum\sum_{1\leq i\leq q\leq k} \text{cov}(R_{ij}, R_{iq})\,\text{cov}(R_{pj}R_{pq}) \\ &= k\binom{n}{2}\frac{(k^2-1)^2}{144} + 2\binom{k}{2}\binom{n}{2}\frac{(k+1)^2}{144} \\ &= k^2\binom{n}{2}(k+1)^2\frac{k-1}{144}.\end{aligned}$$

$$\therefore\ E(S) = \frac{nk(k^2-1)}{12},\ V(S) = \frac{nk^2(n-1)(k-1)(k+1)^2}{72}.$$

Now, we can define a linear function of S, where

$$\chi^2 = \frac{12S}{nk(k+1)} = \frac{12\sum_{j=1}^{k} R_j^2}{nk(k+1)} - 3n(k+1).$$

$$\therefore\ E(\chi^2) = k-1,\ V(\chi^2) = \frac{1}{n}2(n-1)(k-1) \approx 2(k-1).$$

The first two moments of the above χ^2 are approximately equal to the first two moments of chi-square distribution. Hence the χ^2-statistic discussed above is distributed as chi-square distribution with $(k-1)$ d.f. This is known as Friedman's χ^2-statistic.

We have nk observations and nk ranks for the experimental results. The total sum of squares of these nk ranks is

$$s_t = \sum_{i=1}^{n}\sum_{j=1}^{k}(R_{ij} - \bar{r})^2 = n\sum_{i=1}^{n}\left(i - \frac{k+1}{2}\right)^2 = nk\frac{(k^2-1)}{12}.$$

Therefore, Friedman's χ^2-statistic can be written as

$$\chi^2 = \frac{(k-1)S}{s_t}.$$

Though s_t is fixed, we can partition this total sum of squares according to analysis of variance technique as follows:

$$\bar{r} = \sum_{i=1}^{n}\sum_{j=1}^{k}\frac{R_{ij}}{nk} = \frac{k+1}{2}$$

$$\bar{R}_j = \frac{R_j}{n} = \frac{1}{n}\sum_{i=1}^{n} R_{ij}$$

$$\bar{R}_i = \frac{1}{k}\sum_{j=1}^{k} R_{ij}$$

$$\begin{aligned}
s_t &= \sum_{i=1}^{n}\sum_{j=1}^{k}(R_{ij} - \bar{r})^2 = \Sigma\Sigma(R_{ij} - \bar{R}_i + \bar{R}_i - \bar{r})^2 \\
&= \Sigma\Sigma(R_{ij} - \bar{R}_i)^2 + n\Sigma(\bar{R}_i - \bar{r})^2 + 2\Sigma(\bar{R}_i - \bar{r})\Sigma(R_{ij} - \bar{R}_i) \\
&= \Sigma\Sigma(R_{ij} - \bar{R}_i)^2 + \frac{1}{n}\sum_{j=1}^{k}\{\bar{R}_j - [n(k+1)/2]\}^2 \\
&= \Sigma\Sigma(R_{ij} - \bar{R}_i)^2 + \frac{s}{n} = nk(k^2-1)/12.
\end{aligned}$$

Hence the test statistic χ^2 can be transformed to

$$F = \frac{(n-1)\chi^2}{n(k-1) - \chi^2}.$$

This F has $(k-1)$ and $(n-1)(k-1)$ d.f.

To derive the χ^2-statistic it is assumed that x_{ij} values are all different for all $i = 1, 2, \ldots, n$. However, there may be ties in the observations of a block. Then the χ^2-statistic is to be adjusted. The adjusted χ^2-statistic is

$$\chi_a^2 = \frac{1}{A - C}\left[(k-1)\sum_{j=1}^{k}\left\{R_j - \frac{n(k+1)}{2}\right\}^2\right]$$

where

$$A = \sum_{i=1}^{n}\sum_{j=1}^{k} R_{ij}^2, \; C = nk(k+1)^2/4.$$

Iman and Devenport (1980) have shown that the adjusted χ_a^2-statistic is more approximate to the chi-square distribution. Conover (1999) has proposed F-statistic using χ_a^2-statistic, where

$$F = \frac{(n-1)\chi_a^2}{n(k-1) - \chi^2}.$$

The above test procedure is used to test the significance in the difference of treatment effects in two-way analysis of variance. Therefore, if the null hypothesis of equality of treatment effects are rejected, the treatments in pairs are needed to be compared. The difference between j-th and l-th treatment is significant, if

$$|R_j - R_l| > t_{\frac{\alpha}{2}}\left[\frac{2nA - \Sigma R_j^2}{(n-1)(k-1)}\right]^{\frac{1}{2}}$$

where R_j and R_l are the total of ranks of j-th and l-th treatment; $t_{\frac{\alpha}{2}}$ is the tabulated value of t at $100\alpha\%$ level of significance with $(n-1)(k-1)$ d.f.

►**EXAMPLE 36.29** *Five patients went to a diagnostic centre to check their fasting blood sugar level. But after the test results they were not satisfied with the result. For confirmatory decision they went to another* 3 *centres and arranged to measure the fasting blood sugar levels. The results are shown below:*

Patients	*Fasting blood sugar levels in different diagnostic centres*			
	1	*2*	*3*	*4*
1	180	200	190	180
2	220	220	200	220
3	118	140	140	120
4	150	150	160	150
5	130	130	150	130

Are the measurements of fasting blood sugar level in different centres similar?

Solution.

H_0 : there is no difference in the measurements in different centres,

H_A : the centres are not similar in measuring the fasting blood sugar level.

Here $n = 5, k = 4$. The observations are given ranks as follows:

Patients	Centres 1	2	3	4	Total R_i
1	1.5	4	3	1.5	10
2	3	3	1	3	10
3	1	3.5	3.5	2	10
4	2	2	4	2	10
5	2	2	4	2	10
Total R_j	9.5	14.5	15.5	10.5	

$$A = \Sigma\Sigma R_{ij}^2 = 143$$

$$C = \frac{nk(k+1)^2}{4} = 125$$

Since ties are observed, we have to use adjusted χ^2-statistic, where

$$\chi_a^2 = \frac{1}{A-C}\left[(k-1)\sum_{j=1}^{4}\left\{R_j - \frac{n(k+1)}{2}\right\}^2\right] = \frac{1}{143-125}[(4-1)\times 26] = 4.33$$

$$F = \frac{(n-1)\chi_a^2}{n(k-1)-\chi_a^2} = \frac{(5-1)4.33}{5(4-1)-4.33} = 1.62.$$

Since $F < F_{0.05;3,12} = 3.49$, H_0 is accepted. The measurements in different centres are homogeneous.

36.16.2 Quade Test

This test is also used to test the homogeneity of k treatment effects, where treatments are applied in different blocks of k plots each. The method of analysis is similar as it is described for Friedman's test. However, Quade (1972) proposed F-statistic in a different form, where

$$F = \frac{(n-1)B}{A-B}.$$

Here

$$A = \sum_{i=1}^{n}\sum_{j=1}^{k} S_{ij}^2, \quad B = \frac{1}{n}\sum_{j=1}^{k} S_j^2,$$

$$S_{ij} = Q_i\left[R_{ij} - \frac{k+1}{2}\right], \quad S_j = \sum_{i=1}^{n} S_{ij},$$

$$Q_i = \text{Rank of range of } i\text{-th block}$$

$$= \text{Rank } (\max x_{ij} - \min x_{ij});\ i = 1, 2, \ldots, n.$$

This F is distributed as variance ratio with $(k-1)$ and $(n-1)(k-1)$ d.f.

The pairwise comparison is also made, if the treatment effects are found significantly different. The j-th and l-th treatment will be significantly different, if

$$|S_j - S_l| \geq t_{\frac{\alpha}{2}}\left[\frac{2n(A-B)}{(n-1)(k-1)}\right]^{1/2}, \quad j \neq l = 1, 2, \ldots, k.$$

This test is suitable even if there are ties in the observations of a block.

The test can be applied to the data of Example 36.29. Here $Q_1 = 3$, $Q_2 = 3$, $Q_3 = 5$, $Q_4 = 1$, $Q_5 = 3$. S_{ij} observations are as follows:

$[S_{ij}]$

Patients	*Centres*			
	1	*2*	*3*	*4*
1	−3	4.5	1.5	−3
2	1.5	1.5	−4.5	1.5
3	−7.5	5	5	−2.5
4	−0.5	−0.5	1.5	−0.5
5	−1.5	1.5	4.5	−1.5
Total S_j	−11.0	9.0	8.0	−6.0

$$A = \Sigma\Sigma S_{ij}^2 = 210.0$$

$$B = \frac{1}{n}\Sigma S_j^2 = 60.4$$

$$\therefore\ F = \frac{(n-1)B}{A-B} = \frac{(5-1)60.4}{210.0-60.4} = 1.61.$$

Here also the null hypothesis is accepted.

36.16.3 Page Test

This test is also used for the same purpose as Friedman's test. Friedman's test statistic is adjusted, if there are ties in observations of a block. Page (1963) has proposed alternative test procedure for the same purpose, where the test statistic is

$$T = \sum_{j=1}^{k} jR_j = R_1 + 2R_2 + \cdots + kR_k$$

where R_j's are arranged in ascending order of magnitude. Thus

$R_1 =$ the lowest value of rank of treatment totals found in Friedman's test,

$R_k =$ the highest value of R_j's observed in Friedman's test.

In other words $R_{(1)} < R_{(2)} < \cdots < R_{(k)}$, where $R_{(j)}$ is the j-th order statistic of R_j's. Here $R_j = \sum_{i=1}^{n} R_{ij}$ and R_{ij} is the rank of j-th observation in i-th block.

Page (1963) has found out the quantiles of T-statistic. However, for n large, normal test can be performed, where

$$z = \frac{T - nk(k+1)^2/4}{[n(k^3-k)^2/144(k-1)]^{1/2}} \sim N(0,1).$$

The null hypothesis for this test is

$$H_0\ :\ \mu_1 = \mu_2 = \cdots = \mu_k,$$
$$\text{against } H_A\ :\ \mu_1 \le \mu_2 \le \mu_3 \le \cdots \le \mu_k$$

where μ_j is the mean of j-th treatment (j-th population). The alternative hypothesis is one-sided and hence if $|z| < 1.64$, H_0 is rejected at 5% level of significance.

36.16.4 Cochran's Test

This test is also used to test the homogeneity of treatment effects in two-way classification. But the test does not depend on the ranks of x_{ij} values, where x_{ij} is the observation of j-th treatment recorded in i-th block or on i-th occasion. The test is used to study the effectiveness of application of treatments. Thus, there is one treatment the result of which is known from some previous knowledge. Therefore, the other treatments are applied to study the superiority of other treatments over the existing one. For example, let us consider that patients of different ages are under investigation to reduce their blood pressure level. The patients of each age or age group may constitute a block. In each block there may be many plots or units to apply the treatment. Here the treatment is the different doses of a medicine or different medicines. The blood pressure level will either be reduced or not reduced due to the use of medicine. If the medicine is effective, the blood pressure level will be reduced, otherwise it will not be reduced.

According to the above discussion, it is clear that the x_{ij} observations can be measured by an indicator variable, where

$$\begin{aligned} x_{ij} &= 1, \text{ if the treatment is effective,} \\ &= 0, \text{ otherwise; } i = 1, 2, \ldots, n; \ j = 1, 2, \ldots, k. \end{aligned}$$

Using this indicator variable values Cochran has proposed a test statistic to test the homogeneity of treatment effects. According to him, the test statistic is

$$\chi^2 = \frac{(k-1)[k\Sigma C_j^2 - (\Sigma C_j)^2]}{k\Sigma R_i - \Sigma R_i^2}$$

where

$$C_j = \sum_{i=1}^{n} x_{ij}, R_i = \sum_{j=1}^{k} x_{ij}.$$

This χ^2 is distributed as chi-square distribution with $(k-1)$ d.f.

►**EXAMPLE 36.30** *The heart-rate per minute of same slugs kept in different pesticides are shown below:*

	Pesticides				
Slugs	*Glyphosate*	*Sumicidin*	*Supracide*	*Pomex*	*Control*
1	7	9	9	9	11
2	8	9	8	10	13
3	7	9	9	9	14
4	8	9	8	9	12
5	11	10	8	11	11
6	12	10	9	10	12

Are the varieties of pesticide effective in reducing heart-rate of slugs?

Solution. Here $k = 4$, $n = 6$. The measurements of x_{ij} values [$i = 1, 2, 3, 4, 5, 6$; $j = 1, 2, 3, 4$] are shown below:

Slugs	*Pesticides*				*Total, R_i*
	Glyphosate	*Sumicidin*	*Supracide*	*Pomex*	
1	1	1	1	1	4
2	1	1	1	1	4
3	1	1	1	1	4
4	1	1	1	1	4
5	0	1	1	0	2
6	0	1	1	1	3
Total, C_j	4	6	6	5	21

Here, x_{ij} = 1, if the heart-rate is less than that under control

= 0, otherwise.

H_0 : all pesticides are similar in reducing heart-rate,

H_A : pesticides are not effective in reducing heart-rate.

$$\chi^2 = \frac{(k-1)[k\Sigma C_j^2 - (\Sigma C_j)^2]}{k\Sigma R_i - \Sigma R_i^2} = \frac{(4-1)[4 \times 113 - (21)^2]}{4 \times 21 - 77} = 4.71.$$

Since $\chi^2 < \chi^2_{0.05,3} = 7.81$, H_0 is accepted.

36.16.5 Durbin Test

This test is also used to test the significance of the difference of treatment effects in two-way classification except that the number of treatments per block are not same. Usually, it is used to test the homogeneity of treatment effects used for BIB design with parameter b, v, r, k and λ.

Let x_{ij} be the result of j-th treatment in i-th block ($i = 1, 2, \ldots, b; j = 1, 2, \ldots, v$). However, in i-th block there are k plots and k treatments are randomly allocated to the plots. The observations of i-th block are assigned ranks. Let R_{ij} = rank of j-th treatment in i-th block. Since there are k treatments in i-th block, the values of R_{ij} in i-th block will be from 1 to k, where 1 is assigned to lowest observation of x_{ij} and k is assigned to highest observation of x_{ij} for any $i = 1, 2, \ldots, b$. The test statistic for the hypothesis of homogeneity of treatment effects is

$$D = \frac{12(v-1)}{vr(k-1)(k+1)} \sum_{j=1}^{v} \left(R_j - \frac{r(k+1)}{2}\right)^2.$$

This D is distributed as chi-square with $(v-1)$ d.f. It is assumed that there is no tie in the observations of i-th block. Here $R_j = \sum_{i=1}^{b} R_{ij}$.

In practice, there may be ties in the observations of i-th block. In such a case the

D-statistic is to be adjusted. The adjusted D-statistic is

$$D_1 = \frac{(v-1)[\Sigma R_j^2 - r_C]}{A - C}, \text{ where } A = \sum_{i=1}^{b}\sum_{j=1}^{v} R_{ij}^2,\ C = \frac{bk(k+1)^2}{4}.$$

This D_1 is more nearer to chi-square variate with $(v-1)$ d.f.

►**EXAMPLE 36.31** *An experiment with 7 treatments are conducted to study the variability of 7 varieties of maize through a BIB design of parameter b, v, r, k and λ. The arrangement of treatments and the production of maize (x_{ij} kg/plot) are given below:*

Arrangement of treatments and production of maize (kg/plot)

Blocks	Treatments		
1	1–14.5	2–18.0	3–17.6
2	1–15.0	4–10.2	5–9.6
3	1–14.8	6–16.4	7–12.5
4	2–18.5	4–11.0	6–16.7
5	2–18.2	5–10.4	7–13.1
6	3–17.4	4–11.5	7–13.0
7	3–18.4	5–11.9	6–16.8

Analyse the data and comment on treatment effects.

Solution. We have $b = v = 7$, $r = k = 3$, $\lambda = 1$. The observations by ranks are shown below:

Blocks	Treatments (R_{ij})		
1	1–1	2–2	3–3
2	1–3	4–2	5–1
3	1–2	6–3	7–1
4	2–3	4–1	6–2
5	2–3	5–1	7–2
6	3–3	4–1	7–2
7	3–3	5–1	6–2

$$R_j = \sum_{i=1}^{b} R_{ij}$$

$R_1 = 6,\ R_2 = 8,\ R_3 = 9,\ R_4 = 4,$

$R_5 = 3,\ R_6 = 7,\ R_7 = 5.$

Since there is no tie in the observations of a block, we have to use the test statistic

$$D = \frac{12(v-1)}{vr(k-1)(k+1)}\sum_{j=1}^{v}\left[R_j - \frac{r(k+1)}{2}\right]^2 = \frac{12(7-1)}{7\times 3(3-1)(3+1)} \times 28 = 12.$$

Since $D < \chi^2_{0.05,6} = 12.6$, H_0 is accepted, where H_0 : the treatments are homogeneous.

The test statistic D and D_1 are approximately distributed as χ^2 with $(v-1)$ d.f. Conover (1999) has proposed F-statistic for the same hypothesis, where

$$F = \frac{D_1/v-1}{[b(k-1) - D_1]/(bk - b - v + 1)}.$$

The d.f. of F is $(v-1)$ and $(bk - b - v + 1)$ and it follows variance ratio distribution.

36.17 Kruskall-Wallis Test

This test is also known as Kruskall-Wallis one-way analysis of variance test. The test is used to test the equality of several treatment effects, where treatments are under investigation through completely randomized design.

Let x_{ij} be the j-th sample observations selected from i-th population $[i = 1, 2, \ldots, k;\ j = 1, 2, \ldots, n_i]$. It is assumed that the median of i-th population is M_i. The problem is to test the significance of null hypothesis

$$H_0 : M_1 = M_2 = \cdots = M_k,$$

against H_A : at least one of the equality does not hold good.

The test is based on ranks of x_{ij} of $n = \sum^{k} n_i$ observations. Let R_{ij} be the ranks of j-th observation of i-th treatment. Then the total of i-th treatment (i-th sample) and mean of i-th treatment are, respectively

$$R_i = \sum_{j=1}^{n_i} R_{ij} \text{ and } \bar{R}_i = \frac{1}{n_i}\sum_{j=1}^{n_i} R_{ij} = \frac{R_i}{n_i}.$$

The grand mean of n sample observations is

$$\bar{R} = \frac{1}{n}\sum_{i=1}^{k} R_i = \frac{n+1}{2}.$$

The test statistic is based on R_i values, where the test statistic is

$$H = \frac{12}{n(n+1)}\sum_{i=1}^{k} n_i(R_i - \bar{R})^2 = \frac{12}{n(n+1)}\sum_{i=1}^{k}\frac{R_i^2}{n_i} - 3(n+1)$$

where H is distributed as chi-square distribution with $(k-1)$ d.f.

Null Distribution of H

Under H_0, in each case, n_1 ranks of first sample, n_2 ranks of second sample, $\ldots, n_k$ ranks of k-th sample are equally likely $\left[n!\Big/\prod_{i=1}^{k} n_i\right]$. Let $k = 3, n_1 = n_2 = n_3 = 2$. Then

$$H = \frac{12}{6\times 7}\sum_{i=1}^{3}\frac{R_i^2}{2} - 21 = \frac{1}{7}B - 21, \text{ where } B = \sum_{i=1}^{3} R_i^2.$$

For the above-mentioned 3 samples the equally likely arrangements are

$$[6!/(2!\ 2!\ 2!)] = 90.$$

Let us show some of the arrangements of ranks for 3 samples. From the arrangement of ranks, B values can also be calculated. Here B values as well as corresponding H

values are shown below for 15 arrangements of ranks.

Arrangements		Ranks of Treatments			B	H
		1	*2*	*3*		
1		1	3	5	179	4.57
		2	4	6		
	R_i	3	7	11		
2		1	3	4	173	3.71
		2	5	6		
	R_i	3	8	10		
3		1	3	4	171	3.43
		2	6	5		
	R_i	3	9	9		
4		1	2	5	173	3.71
		3	4	6		
	R_i	4	6	11		
5		1	2	4	165	2.25
		3	5	6		
	R_i	4	7	10		
6		1	2	4	161	2.00
		3	6	5		
	R_i	4	8	9		
7		1	2	3	155	1.14
		4	5	6		
	R_i	5	7	9		
8		1	2	5	171	3.43
		4	3	6		
	R_i	5	5	11		
9		1	2	3	153	0.86
		4	6	5		
	R_i	5	8	8		
10		1	2	4	161	2.00
		5	3	6		
	R_i	6	5	10		
11		1	2	3	153	0.86
		5	4	6		
	R_i	6	6	9		
12		1	2	4	155	1.14
		6	3	5		
	R_i	7	5	9		
13		1	2	3	149	0.29
		6	4	5		
	R_i	7	6	8		
14		1	2	4	147	0.00
		6	5	3		
	R_i	7	7	7		
15		1	2	3	149	0.29
		5	6	4		
	R_i	6	8	7		

The H values of 90 arrangements can be found out in a similar way. There will be 6 repetitions of the above 15 distributions and H values of each 6 repetitions will be same. Under H_0

$$P(H = 4.57) = \frac{6}{90},\ P(H = 3.71) = \frac{12}{90},\ P(H = 3.43) = \frac{12}{90},$$
$$P(H = 2.57) = \frac{6}{90},\ P(H = 2.00) = \frac{12}{90},\ P(H = 1.14) = \frac{12}{90},$$
$$P(H = 0.86) = \frac{12}{90},\ P(H = 0.29) = \frac{12}{90},\ P(H = 0.00) = \frac{6}{90}.$$

Hence for any particular value h of H the probability is given by

$$f_n(h) = t(h)\prod_{i=1}^{k}\frac{n_i}{n!},\ \text{where } t(h) \text{ is the number of } h.$$

Distribution of Large Sample

It is assumed that the n_i values of i-th sample are randomly selected. It is also assumed that the n_i values are randomly selected observations from population of finite size n and selection is made without replacement. The mean and variance of the population are, respectively

$$\mu = \sum_{i=1}^{n}\frac{1}{n} = \frac{n+1}{2} \text{ and } \sigma^2 = \sum_{i=1}^{n}\frac{1}{n}\left[i - \frac{n+1}{2}\right]^2 = \frac{n^2-1}{12}.$$

The average ranks of i-th sample is $\bar{R}_i = R_i/n_i$. For sample mean of any sample from finite population, we have

$$E(\bar{R}_i) = \mu,\ V(\bar{R}_i) = \frac{n-n_i}{n}\frac{S^2}{n_i},\ \text{where}$$

$$S^2 = \frac{1}{n-1}\sum_{i=1}^{k}\sum_{j=1}^{n_i}(R_{ij} - \bar{R})^2 = \frac{(n-n_i)\sigma^2}{n_i(N-1)}.$$

Therefore, replacing μ and σ^2, we get

$$E(\bar{R}_i) = \frac{n+1}{2},\ V(\bar{R}_i) = \frac{(n+1)(n-n_i)}{12n_i}.$$

If n_i becomes large, we can define

$$z_i = \frac{\bar{R}_i - \frac{n+1}{2}}{\sqrt{\left[\frac{(n+1)(n-n_i)}{12n_i}\right]}} \sim N(0,1).$$

Hence z_i^2 is distributed as chi-square with 1 d.f. But all z_i are not clearly independent since

$$\sum_{i=1}^{k} n_i\bar{R}_i = \sum_{i=1}^{n} i = \frac{n(n+1)}{2}.$$

However, if n_i is not small enough

$$H = \sum_{i=1}^{k} n_i(\bar{R}_i - \bar{R})^2 = \sum_{i=1}^{k} \frac{n - n_i}{n} z_i^2$$

is distributed as χ^2 with $(k-1)$ d.f.

The distribution is considered assuming no tie in the observations. However, in practice, there may be ties in observations of two or more samples. Then adjusted H statistic is given by

$$H_1 = \frac{H}{1 - g\sum_{i=1}^{k} T_i/n(n^2-1)}.$$

Here g = number of tied observations, $T_i = t_i(t_i^2 - 1)$, t_i = number of i-th tied observations. If there is no tie, $g = n, t_i = 1, T_i = 0$ for all $i = 1, 2, \ldots, n$.

Multiple Comparison

If null hypothesis of homogeneity of k medians is rejected, we can compare $k(k-2)/2$ pairs of medians. The comparison can be performed using Mann-Whitney U-test. However, we can use Dunn's (1964) test, where equality of two medians (two treatments) is rejected if

$$|\bar{R}_i - \bar{R}_{i'}| \geq z\sqrt{\frac{n(n+1)}{12}\left(\frac{1}{n_i} + \frac{1}{n_{i'}}\right)};\; i \neq i' = 1, 2, \ldots, k.$$

The critical values of z for $k = 2, 3, 4, 5$ are presented below:

Table 36.5 : Critical values of Z used for multiple comparison related to Kruskall-Wallis test

k \ α	*Critical values at 100α% level of significance*					
	0.30	0.25	0.20	0.15	0.10	0.05
2	1.036	1.150	1.282	1.440	1.645	1.960
3	1.645	1.732	1.834	1.960	2.128	2.394
4	1.960	2.037	2.128	2.241	2.394	2.638
5	2.170	2.241	2.326	2.432	2.576	2.807

If all treatments have equal number of observations $[n_1 = n_2 = \cdots = n_k]$, the equality of two treatments' median is rejected when

$$|\bar{R}_i - \bar{R}_{i'}| \geq Z\sqrt{\frac{k(n+1)}{6}}.$$

►**EXAMPLE 36.32** *Five varieties of maize are cultivated to study the productivity of varieties. Each variety is cultivated in 6 plots, where maize varieties are randomly* *allocated to different plots. The amounts of production (in kg/plot) of each variety are shown below:*

Production of maize variety (kg per 10′ × 10′ plot)

Variety	*Replications*					
	1	*2*	*3*	*4*	*5*	*6*
1	11.2	11.4	10.8	11.5	11.8	11.6
2	13.6	13.4	12.6	13.9	13.7	14.0
3	10.6	10.5	9.7	10.2	10.3	10.0
4	14.2	14.4	15.0	15.2	15.3	14.9
5	16.0	16.2	16.5	15.8	15.5	16.3

Are the maize variety different?

Solution. We have $k = 5$, $n = 30$. The ranks of production data are shown below:

Variety	*Replications*						*Total* R_i	*Mean* $\bar{R}_i$
	1	*2*	*3*	*4*	*5*	*6*		
1	8	9	7	10	12	11	57	9.5
2	15	14	13	17	16	18	93	15.5
3	6	5	1	3	4	2	21	3.5
4	19	20	22	23	24	21	129	21.5
5	27	28	30	26	25	29	165	27.5

Therefore,

$$H = \frac{12}{n(n+1)} \sum_{i=1}^{k} \frac{R_i^2}{n_i} - 3(n+1) = \frac{12}{30 \times 31} 9367.5 - 3 \times 31 = 27.87.$$

Since $H > \chi^2_{0.05,4} = 9.49$, H_0 is rejected. Here

$$H_0 : \text{all maize varieties are similar.}$$

Here $n_1 = n_2 = n_3 = n_4 = n_5 = 6$. The critical value at 5% level of significance to compare pairs of maize variety is

$$Z\sqrt{\frac{k(n+1)}{6}} = 2.807\sqrt{\frac{5 \times 31}{6}} = 14.59.$$

It is observed that Variety-5 significantly differs from Variety-1 and Variety-3. Variety-4 differs from Variety-2.

36.18 Confidence Interval for Quantile

Let $x_1, x_2, \ldots, x_n$ be a random sample of n observations selected from a population having distribution function $F(x)$. It is assumed that the sample observations are independent and identically distributed. Consider that the sample observations are arranged in array, where the r-th and s-th order statistics are shown below:

$$x_{(1)} \leq x_{(2)} \leq \cdots \leq x_{(r)} \leq \cdots \leq x_{(s)} \leq \cdots \leq x_{(n)}.$$

Here $1 \leq r \leq s \leq n$. Let P be the number between zero and one. We need to find the confidence interval for P-th quantile of the distribution.

The confidence interval of P-th quantile will be the value of r-th and s-th order statistic. Since $0 \leq P \leq 1$, we have to use the values of x for the order statistics, where x is the number of success out of n Bernoulli trials. Let us consider that we need $100(1-\alpha)\%$ confidence interval for P-th quantile. As an example, let us consider the case of confidence interval of first quartile, where $P = 0.25$. Since $100(1-\alpha)\%$ confidence interval is needed, we have to find the value of x against n for which the probability is $\alpha_1 = \alpha/2$. Let us find a value of x for which α_1 is left to $\alpha/2$. Let this value of x is x_1. Hence r is taken $x_1 + 1$. Again, we need to find another value of x for which the probability is $\alpha_2 = 1 - \alpha/2$. Consider that such a value is x_2. This value of x is $s - 1$. Hence $x_2 + 1$ is taken as s. Therefore, the confidence interval for first quartile will be found out from the equation

$$P[x_{(r)} \leq x_{0.25} \leq x_{(s)}] = 1 - \alpha_1 - \alpha_2.$$

Therefore, the $100(1-\alpha)\%$ confidence interval for first quartile is $[x_{(r)}, x_{(s)}]$. Here

$$\alpha = \alpha_1 + \alpha_2.$$

►**EXAMPLE 36.33** *The milk production (kg/day) of* 20 *randomly selected cows are given below:*

16.5, 12.3, 18.6, 28.7, 22.4, 25.4, 17.2, 15.6, 30.4, 18.7,
20.0, 22.4, 18.5, 19.6, 24.7, 30.8, 29.0, 26.3, 21.4, 20.5.

Find confidence interval for 4*-th decile of the data.*

Solution. The observations in array are

$$12.3 < 15.6 < 16.5 < 17.2 < 18.5 < 18.6 < 18.7 < 19.6 < 20.0 < 20.5$$
$$< 21.4 < 22.4 = 22.4 < 24.7 < 25.4 < 26.4 < 28.7 < 29.0 < 30.4 < 30.8.$$

Since we need confidence interval for 4-th decile, $P = 0.4$. Here $n = 20$. Let us try to find confidence interval with confidence coefficient approximately 0.95. Thus, we have for $x = 3$, $[P(X \leq 3) = 0.016]$. Hence $r = 4$. Here instead of $\alpha/2 = 0.025$, we are taking $\alpha_1 = 0.016$. Again, for $x = 11$, $[P(X \leq 11 = 0.9435]$. Therefore, $s = 12$. Here instead of $(1 - \alpha/2)$, we are taking $\alpha_2 = 1 - 0.9435 = 0.0565$. Since $\alpha_1 = 0.016$ for $x \leq 3$, the required value of x is far left to $r = 3$. So we take $r = 3 + 1 = 4$. Again, the probability closest to 0.975 is 0.9435. This probability is less than required probability 0.975. Hence s can be taken 12. Therefore,

$$P[x_{(r)} \leq x_{0.40} \leq x_{(s)} = 1 - 0.016 - 0.0535 = 0.9305.$$

Here $r = 4$ and $s = 12$. Therefore, 93.05% confidence interval for 4-th decile is $[x_{(4)}, x_{(12)}]$ or $[17.2, 22.4]$.

Confidence Interval for Large Value of n

For large value of n the binomial variable transforms to normal variable. Hence r and s values are found out by

$$r = np + z_{\alpha/2}\sqrt{np(1-p)}$$
$$\text{and } s = np + W_{1-\alpha/2}\sqrt{np(1-p)}.$$

Then the confidence interval is found out from the equation

$$P[x_{(r)} \le x_p \le x_{(s)}] = 1 - \alpha_1 - \alpha_2$$

where $\alpha_1 = \alpha/2, \alpha_2 = 1 - \alpha/2$. Here z is the normal variate. For 95% confidence interval $z_{\alpha/2} = -1.96$ and $z_{1-\alpha/2} = 1.96$.

Using the data of Example 36.33, we have

$$\begin{aligned} r &= 20 \times 0.4 - 1.96\sqrt{20 \times 0.4 \times 0.6} \\ &= 8 - 4.29 = 3.71 \approx 4 \\ s &= 20 \times 0.4 + 1.96\sqrt{20 \times 0.4 \times 0.6} \\ &= 8 + 4.29 = 12.29 \approx 12 \end{aligned}$$

Hence 95% confidence interval for 4-th decile is

$$[X_{(4)}, x_{(12)}] \text{ or, } [17.2, 22.4].$$

36.19 Tolerance Limits

Confidence interval is found out for any parameter of the distribution. For this we use sample observations. In certain situation we need to find the interval in which a certain portion of the sample observations or population observations will fall. That is, we need a limit which will give an interval within which at least a proportion q of the population lies with probability $1 - \alpha$. Such an interval gives the limits which are known as *tolerance limits.*

Let $x_1, x_2, \ldots, x_n$ be a random sample of n observations from a population. We want to know how large n should be so that we can be 95% sure that at least 90% of the population lies between $x_{(1)}$ and $x_{(n)}$, where $x_{(1)}$ and $x_{(n)}$ are the smallest and largest observations in the sample. The generalized problem is to know how large n should be so that at least a proportion q of the population is between $x_{(r)}$ and $x_{(n+1-m)}$ with probability $1 - \alpha$ or more. The numbers q, r, m and $1 - \alpha$ are known beforehand, and only n needs to be determined.

The tolerance limit what already has been mentioned is two-sided tolerance limit. We can also decide n so that at least a proportion q of the population is greater than $x_{(r)}$ with probability $1 - \alpha$ or at least a proportion q of the population is less than $x_{(n+1-m)}$ with probability $1 - \alpha$. These latter limits are known as one-sided tolerance limits.

The tolerance limit discussed above is given by

$$n \approx \frac{1}{4}\chi^2_{1-\alpha}\frac{1+q}{1-q} + \frac{1}{2}(r+m-1)$$

where $\chi^2_{1-\alpha}$ is the $(1-\alpha)$ quantile of a chi-square distribution with $2(r+m)$ d.f. This n gives a probability limit

$$P\left[x_{(r)} \leq \text{ at least a proportion } q \text{ of the population} \leq x_{(n+1-m)}\right] \geq 1-\alpha.$$

For one-sided tolerance limit let either r or m equal zero, where $x_{(0)}$ and $x_{(n+1)}$ are considered to be $-\infty$ to ∞ respectively, and proceed as described.

►**EXAMPLE 36.34** *Find the sample size so that 90% of the observations fall within the sample range with 95% confidence.*

Solution. Ninety percent of the sample observations will fall within $x_{(1)}$ to $x_{(n)}$. Here $r=1, m=1$. The sample size is

$$\begin{aligned} n &\approx \frac{1}{4}\chi^2_{0.05,4}\frac{1+q}{1-q} + \frac{1}{2}(r+m-1); \; q = 0.90 \\ &= \frac{1}{4} \times 9.49 \times \frac{1+0.90}{1-0.90} + \frac{1}{2} \\ &= 45.57 \approx 46. \end{aligned}$$

►**EXAMPLE 36.35** *What must be the sample size to be 95% sure that at least 95% of the population equals or exceeds $x_{(10)}$?*

Solution. Here $r=10,\; m=0,\; q=0.95$. Hence

$$\begin{aligned} n &\approx \frac{1}{4}\chi^2_{0.05,20}\frac{1+q}{1-q} + \frac{1}{2}(r+m-1) \\ &= \frac{1}{4} \times 31.4 \times \frac{1+0.95}{1-0.95} + \frac{1}{2}(10+0-1) \\ &= 310.65 \approx 311. \end{aligned}$$

►**EXAMPLE 36.36** *What must be the sample size to be 90% sure that at least 80% of the population will be less than or equal to $x_{(15)}$?*

Solution. Here $r=0,\; m=15,\; q=0.8$. We have

$$\begin{aligned} n &\approx \frac{1}{4}\chi^2_{0.05,30}\frac{1+q}{1-q} + \frac{1}{2}(r+m-1) \\ &= \frac{1}{4} \times 40.3 \times \frac{1+0.8}{1-0.8} + \frac{1}{2}(0+15-1) \\ &= 97.67 \approx 98. \end{aligned}$$

EXERCISE

1. (a) Distinguish between parametric test and nonparametric test. What is the advantage of nonparametric test?

   (b) Derive sign test, stating clearly the assumptions made.

   (c) The following data represent the production of jute goods (in M. ton) per day in two different industries:

   Industry-1, X : 4.2, 4.6, 4.0, 4.8, 5.0, 4.6, 4.4, 4.2, 4.0
   Industry-2, Y : 4.7, 6.2, 5.8, 5.2, 5.1, 3.8, 6.4, 5.7, 6.4, 6.7

   Are the productions in two industries similar? Use Mann-Whitney U-test. Write down the assumptions needed for the test.

   (d) Are the two samples selected from two populations with equal variances?

   (e) Do you think that the sample observations follow normal distribution?

2. (a) Explain what is meant by nonparametric methods. Derive the run test, stating clearly the assumptions made.

   (b) Ten coins are tossed and the number of heads (H) and tails (T) are found as follows:

   $$H\ T\ H\ T\ T\ T\ H\ H\ H\ H$$

   Do you think that the coin-tossing experiment is done randomly?

   (c) From 100 students 20 students are selected for an inquiry. The students bearing serial number in the frame are given below:

   51, 16, 85, 19, 95, 86, 18, 83, 91, 66, 57,
   48, 76, 87, 98, 33, 44, 56, 68, 70.

   Are these numbers selected randomly?

3. (a) What is the difference between confidence limits and tolerance limits? Set up a nonparametric procedure for finding the confidence interval for a median of a distribution.

   (b) Find sample size so that 80% of the observations fall within the range of observation with 90% confidence.

   (c) Develop the χ^2-test to test the randomness of data.

4. (a) Discuss the relative merits and demerits of parametric and nonparametric tests.

   (b) What is the difference between binomial test and sign test? Develop binomial test to classify a group of sample observations.

5. (a) What is meant by confidence interval? Find $100(1-\alpha)\%$ confidence interval for the parameter in Poisson population.

   (b) Explain different methods of deriving confidence interval for population parameter.

   (c) Find $100(1-\alpha)\%$ confidence interval for the parameter θ in

   $$f(x,\theta) = \begin{cases} \theta e^{-\theta x}, & 0 \le x \le \infty \\ 0, & \text{otherwise} \end{cases}$$

   from a sample of large size.

6. (a) Explain uniformly most accurate confidence interval and unbiased confidence interval.
   (b) Find confidence interval for μ and σ^2 in $N(\mu, \sigma^2)$.

7. (a) What is the utility of median test? Describe similarity and dissimilarity between sign test and median test.
   (b) Derive the test statistic for testing the equality of treatment effects when treatments are allocated randomly in different blocks.
   (c) The following data represent the life length (in hours) of electric bulbs produced by 4 different companies:

Company	*Life length (in hours) of bulbs*
A	240, 265, 242, 350, 442, 228, 550
B	348, 470, 382, 375, 390, 468, 395
C	400, 410, 412, 448, 450, 441, 470
D	225, 162, 392, 285, 200, 400, 278

   Are the electric bulbs produced by different companies similar?

8. (a) Distinguish between parametric and nonparametric tests. What is the utility of nonparametric test?
   (b) What is the difference between Wilcoxon test and Wilcoxon signed rank test? Derive the test statistic to test the significance of median of a population.
   (c) A researcher in the field of Botany has collected number of leaves from a forest. The leaves are classified in different species as follows:

Species	1	2	3	4
Number of leaves	50	168	273	42

   Do you think that the leaves are in the ratio 1 : 3 : 4 : 1? Explain the assumptions needed for your test.

9. (a) Describe the median test when there are k independent samples.
   (b) What is the difference between Friedman's test and Durbin test? Derive the test statistic to test the equality of several treatment means.
   (c) The systolic blood pressure (mm of Hg) of several groups of patients are recorded as follows:

Group-1	80, 82, 95, 132, 112, 96, 160, 210, 96
Group-2	110, 100, 132, 120, 140, 145, 162
Group-3	96, 100, 108, 98, 146, 148, 150, 130, 135
Group-4	88, 97, 105, 105, 108, 109, 115, 118

   Are the blood pressure levels of all groups similar?

10. (a) What is the need of nonparametric test? Write down the advantages and disadvantages of nonparametric tests.
    (b) What are the different nonparametric tests for classification of sample observations? Derive the test statistic for any one of the test.

(c) The age (in years) of some primary school children are as follows:

12, 8, 6, 10, 11, 14, 8, 5, 5, 6, 7, 7, 10, 9.

Is the median of the age is 8 years?

11. (a) What is the difference between sign test and Wilcoxon signed rank test? Develop the Wilcoxon signed rank test statistic to test the hypothesis of equality of two populations.

(b) Write down the conditions for better approximation of χ^2-statistic to chi-square distribution.

(c) A group of babies were fed a new protein to increase body-weight. The body-weights are measured before giving protein (X, kg) and after giving protein (Y, kg):

X : 8.6, 6.8, 6.4, 7.3, 8.5, 6.6, 7.5, 8.2, 7.4, 10.2

Y : 10.2, 6.4, 6.2, 7.0, 9.5, 6.0, 7.0, 8.8, 9.2, 11.5

Is the protein effective in increasing body-weight?

(d) A group of male and female travellers are standing in a line to collect tickets as follows:

$M\ M\ F\ F\ F\ F\ M\ F\ F\ M\ M\ M\ F\ M\ F\ M\ M\ F\ F\ F\ M\ M\ M\ M\ M\ F\ F\ M$

Do you think that the travellers are standing randomly?

References

[1] Anscombe, F.J. (1948) The transformation of Poisson, binomial and negative binomial data, Biometrika, 35, 246-254.

[2] Aroian, L.A. (1941) A study of R.A. Fisher's Z-distribution and the related F-distribution, Ann. Math. Stat., 12, 429-448.

[3] Ashby, T. (1968) A modification to Paulson's approximation to the variance ratio distribution, The Computer Journal, 11, 209-210.

[4] Banerjee, D.P. (1952) On the moments of the multiple correlation coefficient in samples from normal populations, Jour. Indian Soc. Agril. Stats., 4, 88-90.

[5] Bánkóvi, G. (1964) A note on the generation of beta distribution and gamma distributed random variables, Publications of the Mathematical Institute, Hungarian Academy of Sciences, Ser. A, 9, 555-562.

[6] Bardwell, G.E. and Crow, E.L. (1964) A two-parameter family of hyper-Poisson distributions. Jour, Amer, Stat. Assoc., 59, 133-141.

[7] Barton, D.E. (1957) The modality of Neyman's contagious distribution of Type A, Trabajos de Estadistica, 8, 13-22.

[8] Bhattacharya, S.K. (1966) A modified Bessel function model in life testing, Metrika, 11, 133-144.

[9] Bury, K.V. (1974) Statistical Models in Applied Science, John Wiley and Sons, New York.

[10] Cacoullos, T. (1965) A relation between t- and F-distributions, Jour. Amer. Stat. Assoc., 60, 528-531.

[11] Carver, H.C. (1919) On the graduation of frequency distributions, Proc. Casualty Act, Soc. Am., 6, 52-72.

[12] Castro, G. de (1952) Note on differences of Burnoulli and Poisson variables, Portuguliae Mathematica, 11, 173-175.

[13] Champernowne, D.G. (1952) The graudation of income distributions, Econometrica, 20, 591-615.

[14] Conover, J. (1999) Practical nonparametric statistics, John Wiley and Sons, Inc., New York.

[15] Cramer, H. (1937) Random variables and probability distributions, Cambridge tracts in mathematics, No. 36, Cambridge University Press.

[16] Cramer, H. (1946) Mathematical methods of statistics, Princeton University Press, Princeton.

[17] Davies, O.L. (1933) On asymptotic formulae for the hypergeometric series, Part I, Biometrica, 25, 295-322; Part II, 26, 59-107.

[18] Dubey, S.D. (1966) Characterization theorems for several distributions and their applications, Jour. Industrial Math., 16, 1-22.

[19] Dubey, S.D. (1969) A new derivation of the logistic distribution, Naval Research Logistics Quarterly, 16, 37-40.

[20] Durbin, J. (1951) Incomplete blocks in ranking experiments, British Jour. Psychology, 4, 85-90.

[21] Edwards, L. (1948) The use of normal significance limits when the parent populaton is of Laplace form, Jour. of the Inst. of Act, Student's Soc., 8, 87-99.

[22] Fisher, R.A. (1915) Frequency distribution of the values of the correlation coefficient in samples from an indefinitely large population, Biometrika, 10, 507-521.

[23] Fisher, R.A. (1921) On the probable error of a coefficient of correlation deduced from a small sample, Metron, 1, 3-32.

[24] Fisher, R.A. (1925) Applications of Student's distribution, Metron, 5, 90-104.

[25] Fisher, R.A. (1928) The general sampling distribution of the multiple correlation coefficient, Proc. of the Royal Soc. of London, 121 A, 551-562.

[26] Fisher, R.A. (1931) Introduction of Table of H functions (In Airey, J.R. (1931) Table of H functions, Table XV, in British Association Mathematical Tables, 1, British Association), pp. xxvi-xxxv.

[27] Fisher, R.A., Corbet, A.S. and Williams, C.B. (1943) The relation between the number of species and the number of individuals in a random sample of an animal population, Jour. Animal Ecology, 12, 42-57.

[28] Fisher, R.A. and Cornish E.A. (1960) The percentile points of distributions having known cumulants, Technometrics, 2, 209-225.

[29] Freeman, M.F. and Tukey, J.W. (1950) Transformations related to the angular and square root, Ann. Math. Stat., 21, 607-611.

[30] Friedman, M. (1937) The use of ranks to avoid the assumptions of normality implicit in the analysis of variance, Jour. Amer. Stat. Assoc., 32, 675-701.

[31] Gajjar, A.V. (1967) Limiting distributions of certain transformations of multiple correlation coefficient, Metron, 26, 189-193.

[32] Ghosh, B.K. (1966) Asymptotic expansions for the moments of the distribution of correlation coefficient, Biometrika, 53, 258-262.

[33] Gibbons, J.D. (1971) Nonparametric Statistical Inference, McGraw-Hill Book Company, New York.

[34] Gibbons, J.D. (1971) Nonparametric statistical inference, McGraw-Hill Book Company, New York.

[35] Goldberg, G. (1967) The Pareto law and the pyramid distribution, Publication 505, Houston, Texas: Shell Development.

[36] Grab, E.L. and Savage, J.R. (1954) Tables of the expected values of $1/x$ for positive Burnoulli and Poisson variables, Jour. Amer. Stats. Assoc., 58, 468-473.

[37] Graybill, F.A. (1961) An introduction to Linear Models, Vol. 1, McGraw-Hill Book, Inc., New York.

[38] Greenwood, M. and Yule, G.U. (1920) An inquiry into the nature of frequency distributions of multiple happenings with particular reference to the occurrence of multiple attacks of disease or repeated accidents, Jour. Royal Stat. Soc. A, 83, 255-279.

[39] Guldberg, A. (1931) On discontinuous frequency-functions and statistical series, Skandinavisk Aktuarietidskrift, 14, 167-187.

[40] Guldbert, S. (1935) Recurrences formulae for the semi-invariants of some discontinuous frequency functions of n variables, Skandinavisk Aktuarietidskrift, 18, 270-278.

[41] Gumbel, E.J. (1958) Statistics of Extremes, Columbia University Press, New York.

[42] Gupta, S.S. and Shah, B.K. (1965) Exact moments and percentage points of the order statistics and the distribution of the range from the logistic distribution, Ann. Math. Stat., 36, 907-920.

[43] Gurland, J. (1963) A method of estimation for some generalized Poisson distributions, Proc. of the International Symposium on Discrete Distributions, Montreal, 141-158.

[44] Haight, F.A. (1961) A distribution analogous to the Borel-Tanner, Biometrika, 48, 167-173.

[45] Hald, H. (1952) Statistical Tables and Formulas, John Wiley and Sons, New York.

[46] Hald, A. (1960) The compound hypergeometric distribution and a system of single sampling inspection plans based on prior distributions and costs, Technometrics, 2, 275-340.

[47] Hald, A. (1968) The mixed binomial distribution and the posterior distribution of p for a continuous prior distribution. Jour. Royal Stat. Soc., Ser. B, 30, 359-367.

[48] Harkness, W.L. (1965) Properties of the extended hypergeometric distribution, Ann. Math. Stat., 36, 938-945.

[49] Harris, C.M. (1968) The Pareto distribution as a queue service discipline, Operations Research, 16, 307-313.

[50] Hodgson, V. (1967) On the sampling distribution of the multiple correlation coefficient (Presented at the IMS meetings in Washington, D.C).

[51] Hogg, R.V. and Craig, A.T. (1970) Introduction to Mathematical Statistics, 3rd Ed., Macmillan Publishing Co., Inc., New York.

[52] Hogben, D., Pinkham, R.S. and Wilk, M.B. (1961) The moments of the noncentral t-distribution, Biometrika, 48, 465-468.

[53] Holgate, P. (1964) Estimation for the bivariate Poisson distribution, Biometrika, 51, 241-245.

[54] Irwing, J.O. (1937) The frequency distribution of the difference between two independent variates following the same Poisson distribution, Jour. Royal Stat. Soc., Ser. A, 100, 415-416.

[55] Jöhnk, M.D. (1964) Erzeugung von beta-verteilten and gamma-verteilten Zufallszahlen, Metrika, 8, 5-15.

[56] Johnson, N.L. and Kotz, S. (1969) Discrete Distributions, John Wiley and Sons, New York.

[57] Johnson, N.L. and Kotz, S. (1970) Continuous Univariate Distributions-1, John Wiley and Sons, New York.

[58] Johnson, N.L. and Kotz, S. (1970) Continuous Univariate Distributions-2, John Wiley and Sons, New York.

[59] Johnson, N.L. and Kotz, S. (1972) Distributions in statistics: Continuous Multivariate distributions, John Wiley and Sons, New York.

[60] Kao, J.H.K. (1958) Computer methods for estimating Weibull parameters in reliability studies, Transactions of IRE-Reliability and Quality Control, 13, 15-22.

[61] Kao, J.H.K. (1959) A graphical estimation of mixed Weibull parameters in life testing electron tubes, Technometrics, 1, 389-407.

[62] Katti, S.K. and Rao, A.V. (1965) The log-zero-Poisson Distribution, Florida State University, Statistics Report, M 106.

[63] Katz, L. (1948) Frequency functions defined by the Pearson difference equation (Abstract), Ann. Math. Stat, 19, 120.

[64] Katz, L. (1963) Unified treatment of a broad class of discrete probability distributions, Proceedings of the International Symposium on Discrete Distributions, Montreal, 175-182.

[65] Kelley, T.L. (1948) The Kelley Statistical Tables (Revised) Cambridge, Mass, Harvard University Press.

[66] Khatri, C.G. (1959) On certain properties of power-series distributions, Biometrika, 46, 486-490.

[67] Khatri, C.G. (1966) A note on a large sample distribution of a transformed multiple correlation coefficient, Ann. of the Inst. Stat. Math., Tokyo, 18, 375-380.

[68] Koehler, K.J. and Larntz, K. (1980) An empirical investigation of goodness-of-fit statistics for sparse multinomials, Jour. Amer. Stat. Assoc., 75, 336-344.

[69] Lexis, W. (1875) Einleitung in der Theorie der Bevolkering-Statistik, Strasburg: Teubner.

[70] Lilliefors, H.W. (1967) On the Kolmogorov-Smirnov test for normality with mean and variance unknown, Jour. Amer. Stat. Assoc., 62, 399-402.

[71] Ljubo, M. (1965) Curves and concentration indices for certain generalized Pareto distributions, Statistical Review, 15, 257-260.

[72] Makabe, H. and Morimura, H. (1955) A normal approximation to the Poisson distribution, Reports on Statistical Application Research, JUSE, 4, 37-46.

[73] Merrington, Maxine and Pearson, E.S. (1958) An approximation to the distribution of noncentral t, Biometrika, 45, 484-491.

[74] Mood, A.M. (1940) The distribution theory of runs, Ann. Math. Stat., 17, 367-392.

[75] Murthy, V.K. (1960) On the distribution of the sum of circular serial correlation coefficients and the effect of non-normality on its distribution, (Abstract), Ann. Math. Stat., 31, 239-240.

[76] Noack, A. (1950) A class of random variables with discrete distributions, Ann. Math. Stat., 21, 127-132.

[77] Nyquist, H., Rice, S.O. and Riordan, J. (1954) The distribution of random determinants, Quarterly of Applied Mathematics, 42, 97-104.

[78] Ord, J.K. (1967) On a system of discrete distributions, Biometrika, 54, 649-656.

[79] Page, E.B. (1963) Ordered hypothesis for multiple treatments: A significance test for linear ranks, Jour. Amer. Stat. Assoc., 67, 215-216.

[80] Patnaik, P.B. (1949) The noncental χ^2- and F distributions and their applications, Biometrika, 36, 202-232.

[81] Pearson, E.S., D' Agostino, R.B. and Bowman, K.O. (1977) Tests for departures from normality: Comparison of powers, Biometrika, 64, 231-246.

[82] Pearson, K. (1900) On the criterion that a given system of deviations from the probable in the case of a correlated system of variables is such that it can reasonably be supposed to have arisen from random sampling, Philosophical Magazine, 5th Series, 50, 157-175.

[83] Pearson, K. (1906) On the curves which are most suitable for describing the frequency of random samples of a population, Biometrika, 5, 172-175.

[84] Perks, W.F. (1932) On some experiments in the graduation of mortality statistics. Jour. Inst. Actuaries, 58, 12-57.

[85] Pirie, W.R. and Hamdan, M.A. (1972) Some revised continuity corrections for discrete distributions, Biometrics, 28, 693-701.

[86] Quade, D. (1972) Analysing randomized blocks by weighted rankings. Technical Report SW |8|72. Stichting Mathematisch Centrum, Amsterdam (5.8)

[87] Romanovsky, V. (1923) Note on the moments of the binomial $(q+p)^N$ about its mean, Biometrika, 15, 410-412.

[88] Sankaran, M. (1963) Approximations to the noncentral chi-square distribution, Biometrika, 50, 199-204.

[89] Severo, N. and Zelen, M. (1960) Normal approximation to the chi-square and noncentral F probability functions, Biometrika, 47, 411-416.

[90] Shah, B.K. (1965) Distribution of mid-range and semi-range from logistic population, Jour. Indian Stat. Assoc., 3, 185-188.

[91] Shapiro, S.S. and Wilk, M.B (1965) An analysis of variance test for normality (complete samples), Biometrika, 52, 591-611.

[92] Shenton, L.R. (1949) On the efficiency of the method of moments and Neyman's type A distribution, Biometrika, 36, 450-454.

[93] Shenton, L.R. and Bowman, K.O. (1967) Remarks on large sample estimators for some discrete distributions, Technometrics, 9, 587-598.

[94] Shumway, R. and Gurland, J. (1960) Fitting the Poisson-binomial distribution, Biometrics, 16, 522-533.

[95] Skellam, J.G. (1946) The frequency distribution of the difference between two Poisson variables belonging to different populations, Jour. Royal Stat. Soc. A, 109, 296.

[96] Slakter, M.J. (1973) Large values for the number of groups with the Pearson chi-square goodness-of-fit test, Biometrika, 60, 420-421.

[97] Stacy, E.W. (1962) A generalization of the gamma distribution, Ann. Math. Stat, 33, 1187-1192.

[98] Student (1908) On the probable error of the mean, Biometrika, 6, 1-125.

[99] Talacko, J. (1956) Perk's distributions and their role in the theory of Wiener's stochastic variables, Trabajos de Estadistica, 7, 159-174.

[100] Tang, P.C. (1938) The power function of the analysis of variance tests with tables and illustrations of their use, Statistical Research Memoirs, 2, 126-149.

[101] Tiku, M.L. (1966) A note on approximating to the central F distribution, Biometrika, 53, 606-610.

[102] Tweedie, M.C.K. (1947) Functions of statistical variate with given means, with special reference to Laplace distribution, Proc. Camb. Phil. Soc., 43, 41-49.

[103] Tweedie, M.C.K. (1956) Statistical properties of inverse Gaussion distributions, 1, Ann. Math. Stat., 28, 362-377.

[104] Wald, A. (1947) Sequential analysis, John Wiley and Sons, New York.

[105] Wallenius, K.T. (1963) Biased Sampling: The noncentral Hypergeometric Probability Distribution, PhD Thesis, Stanford University.

[106] Weir, J.B. de V. (1960) The significance of the difference between two means when the population variances may be unequal, Nature, London, 187, 438.

[107] Welch, B.L. (1938) The significance of the difference between two means when the population variances are unequal, Biometrika, 29, 350-362.

[108] Wilson, E.B. and Hilferty, M.M. (1931) The distribution of chi-square, Proceedings of the National Academy of Sciences, Washington, 17, 684-688.

[109] Wishart, J. (1931) The mean and second moment coefficient of the multiple correlation coefficient in samples from a normal population, Biometrika, 22, 353-361.

[110] Wishart, J. (1949) Cumulants of multivariate multinomial distributions, Biometrika, 36, 47-58.

Index